International Conference on Energy, Environment and Material Systems (EEMS 2017)

E3S Web of Conferences Volume 19 (2017)

Polanica-Zdroj, Poland
13 - 15 September 2017

Editors:

Malgorzata Wzorek
Grzegorz Krolczyk
Anna Krol

ISBN: 978-1-5108-5009-5

Printed from e-media with permission by:

Curran Associates, Inc.
57 Morehouse Lane
Red Hook, NY 12571

proceedings.com

Some format issues inherent in the e-media version may also appear in this print version.

Printed by Curran Associates, Inc. (2017)

For additional information, please contact EDP Sciences – Web of Conferences at the address below.

EDP Sciences – Web of Conferences
17, Avenue du Hoggar
Parc d'Activité de Courtabœuf
BP 112
F-91944 Les Ulis Cedex A
France

Phone: +33 (0) 1 69 18 75 75
Fax: +33 (0) 1 69 28 84 91

contact-edps@webofconferences.org

Additional copies of this publication are available from:

Curran Associates, Inc.
57 Morehouse Lane
Red Hook, NY 12571 USA
Phone: 845-758-0400
Fax: 845-758-2633
Email: curran@proceedings.com
Web: www.proceedings.com

TABLE OF CONTENTS

EEMS

International Conference Energy, Environment and Material Systems

13–15th September 2017
Polanica-Zdrój

OPOLE UNIVERSITY OF TECHNOLOGY

FACULTY OF MECHANICAL ENGINEERING

CEKIT
Centre of Education, Competences, Innovation and Technology

International Conference Energy, Environment and Material Systems

13–15th September 2017
Polanica-Zdrój, Poland

PATRONAGE SPONSORS & PARTNERS

HONORARY PATRONAGE

Ministry of Science and Higher Education
Republic of Poland

prof. Marek Tukiendorf
Rector of Opole University of Technology

SPONSORS

PGE GiEK S.A.
Oddział Elektrownia Opole

Wik Opole

eco Energetyka Cieplna Opolszczyzny SA

eco Jelenia Góra

Warta cementownia

PARTNERS

The Association of TOP 500 Innovators

POLARIS

OBERON 3D metrologia

Stowarzyszenie Producentów Energii z Odpadów

ITA

optotom

Park Naukowo-Technologiczny w Opolu Sp. z o.o.

TEAM

NETZSCH

MEDIA PARTNERS

INSTAL

TERAZ ŚRODOWISKO .pl

MAGAZYN BIOMASA Ogólnopolski miesięcznik klasy biznes

TEHNIČKI VJESNIK TECHNICAL GAZETTE

kierunekenergetyka.pl

REO.PL Rynek Energii Odnawialnej

NOWOCZESNE TECHNOLOGIE

BIZNES ALERT

Radio DOXA

65 lat RADIO OPOLE

SCIENTIFIC COMMITTEE:

- Prof. Janusz POSPOLITA, Opole University of Technology, Poland – chairman,
- Prof. Jerzy BUZEK, Chairman of Industry, Research and Energy ITRE Committee in EU Parliament,
- Prof. Tomasz BOCZAR, Opole University of Technology, Poland,
- Prof. Andrea CAPODAGLIO, University of Pavia, Italy,
- Prof. Somnath CHATTOPADHYAYA, Indian School of Mines, India,
- Ass. Prof. Agostina CHIAVOLA, Sapienza University of Rome, Italy,
- Prof. Robert ĆEP, VSB Ostrava, Czech Republic,
- Prof. Lidia DĄBEK, Kielce University of Technology, Poland,
- Ass. Prof. Bolesław DOBROWOLSKI, Opole University of Technology, Poland,
- Prof. Neven DUIĆ, Univeristy of Zagreb, Croatia,
- Doc. Radomír GOŇO, VSB Ostrava, Czech Republic,
- Prof. Yashar JAVADI, IAU-Semnan Branch, Iran,
- Prof. Orest KOCHAN, Lviv Polytechnic National University, Ukraine,
- Prof. Dražan KOZAK, University of Osijek, Croatia,
- Prof. Zdravko KRAVANJA, University of Maribor, Slovenia,
- Ass. Prof. Anna KRÓL, Opole University of Technology, Poland,
- Ass. Prof. Grzegorz KRÓLCZYK, Opole University of Technology, Poland,
- Prof. Stanisław LEGUTKO, Poznan University of Technology, Poland,
- Prof. Zhixiong LI, University of New South Wales, Sydney, Australia,
- Prof. Tadeusz ŁAGODA, Opole University of Technology, Poland,
- Prof. Natasa MARKOVSKA, Macedonian Academy of Sciences and Arts, Macedonia,
- Prof. Thomas G. MATHIA, L.T.D.S. Ecole Centrale de Lyon, France,
- Prof. Tomasz MRÓZ, Poznan University of Technology, Poland,
- Prof. Vincenzo NADDEO, Univeristy of Salerno, Italy,
- Prof. Wojciech NOWAK, AGH University of Science and Technology, Poland,
- Prof. Pero RAOS, University of Osijek, Croatia,
- Prof. Alessandro RUGGIERO, University of Salerno, Italy,
- Prof. Joachim SCHENK, Leipzig University of Applied Sciences, Germany,
- Prof. Aleksandar SEDMAK, University of Belgrade, Serbia,
- Prof. Biljana SKRBIC, University of Novi Sad, Serbia,
- Prof. Janusz SKOREK, Silesian University of Technology, Poland,
- Ass. Prof. Norbert SZMOLKE, Opole University of Technology, Poland,
- Prof. Leon TRONIEWSKI, Opole University of Technology, Poland,
- Prof. Roman ULBRICH, Opole University of Technology, Poland,
- Prof. Michał WIECZOROWSKI, Poznan University of Technology, Poland,
- Prof. Aneta WIERZBICKA, Lund University, Sweden,
- Prof. Stanisław WITCZAK, Opole University of Technology, Poland,
- Ass. Prof. Małgorzata WZOREK, Opole University of Technology, Poland,
- Doc. Ersel YILMAZ, Adnan Menderes University, Turkey,
- Prof. Ireneusz ZBICIŃSKI, Lodz University of Technology, Poland,
- Prof. Andrzej ZIĘBIK, Silesian University of Technology, Poland,
- Ass. Prof. Izabela ZIMOCH, Silesian University of Technology, Poland.

ORGANIZING COMMITTEE:

- Ass. Prof. Anna KRÓL – chairman,
- Ass. Prof. Małgorzata WZOREK – vice-chairman,
- Ass. Prof. Paweł FRĄCZ,
- Ass. Prof. Grzegorz KRÓLCZYK,
- PhD. Eng. Stanisław ANWEILER,
- PhD. Eng. Iwona KŁOSOK-BAZAN,
- PhD. Eng. Mariusz TAŃCZUK,
- MSc. Adriana SOKULSKA,
- MSc. Anna ZABAWCZUK,
- MSc. Kamila MIZERNA,
- Regina OBROCKA.

The growing demand for goods and services is inevitably linked to production of energy and materials. In order to follow sustainable development principles, currently applied manufacturing processes must be innovative and based on scientific principles.

The International Conference on Energy, Environment and Material Systems EEMS 2017 was organized to provide a forum for researchers and practitioners from around the World who are aware that industry and production processes are changing energy technologies and material systems will be rapidly and creatively evolving in the future.

EEMS 2017 Conference was held on 13-15 September 2017 in Polanica Zdrój, Poland, and organized by the Faculty of Mechanical Engineering at Opole University of Technology and Centre of Education, Competence, Innovations and Technology (CEKiT). The conference captured various practical approaches towards the progress in energy, environment and production technologies contributing to the increase of competitiveness of industry with regard to the sustainable development, preservation of the environment and natural resources. The conference aimed at creating synergies between 'practice and research' increasing the potential and commercial viability of research and development in the field of advanced technologies, materials in energy and environment industries. The conference was attended by delegates from different countries - Croatia, France, Germany, Ukraine, Spain, Turkey and Poland. Participants submitted above 130 abstracts and more than 100 papers were presented in oral and poster sections.

Together with the attendees and reviewers we were able create a very successful symposium. We express our appreciation to all contributors for the papers published in the conference proceedings and the reviewers for their efforts and hard work to ensure the high scientific quality of the conference. We do believe that the conference allowed for a valuable exchange of views between participants.

Proceedings Editors

edp sciences

E3S Web of Conferences

Statement of Peer review

In submitting conference proceedings to E3S Web of Conferences, I certify to the Publisher that I adhere to the **Policy on Publishing Integrity** of the journal in order to safeguard good scientific practice in publishing.

1. All articles have been subjected to peer review administered by the proceedings editors.
2. Reviews have been conducted by expert referees, who have been requested to provide unbiased and constructive comments aimed, whenever possible, at improving the work.
3. Proceedings editors have taken all reasonable steps to ensure the quality of the materials they publish and their decision to accept or reject a paper for publication has been based only on the merits of the work and the relevance to the journal.

Title, date and place of the conference

International Conference Energy, Environment and Material Systems (EEMS 2017)
13th-15th September 2017, Polanica Zdrój, Poland

Proceedings editor(s):

Ass. Prof. Małgorzata Wzorek
Ass. Prof. Grzegorz Królczyk
Ass. Prof. Anna Król

Date and editor's signature

1/09/2017

17, avenue du Hoggar - PA de Courtabœuf – BP 112 - 91944 Les Ulis Cedex A (France)
Tél. : 33 (0)1 69 18 75 75 - Fax : 33(0)1 69 07 45 17 – www.edpsciences.org

Comparison of low frequency signals emitted by wind turbines of two different generator types

Tomasz Malec[1], *Tomasz* Boczar[2,*], *Daria* Wotzka[2], and *Paweł* Frącz[2]

[1] KFB Polska Spółka z o.o., ul. Mydlana 7, 51-502 Wrocław, Poland
[2] Opole University of Technology, ul. Prószkowska 76, 45-758 Opole, Poland

Abstract. Paper presents results of comparative analysis of infrasound noise generated by wind turbines of two types: asynchronous type REPOWER MM92 with power equal to 2 MW and synchronous type Vensys 62 with power equal to 1.2 MW. Frequency spectra of sound pressure levels generated during operation by both turbines for exemplary chosen wind speed values are depicted. Within the shown spectra the resonant frequencies have been indicated, for which sound pressure variations over time are shown. Based on the achieved frequency spectra it was stated that in general the asynchronous type turbine produces lower pressure levels, which are less stable over time, and indicates higher pressure values around the resonant frequencies as compared to the synchronous type turbine. Also it was stated that the asynchronous type turbine is more influenced by the wind conditions and generates higher pressure values by higher wind speeds then the synchronous type turbine. The main contribution of this paper lies in indication that the type of wind turbine generator has significant impact on the level of infrasound noise emitted to the environment.

1 Introduction

The problem of formation, propagation, level and possible impact of infrasound, generated by artificial sources on the natural environment and especially on living organisms, is still valid and important [1]. This is particularly true for wind power engineering, which infrastructure is dynamically developing in recent years in the world. The possibility of measurement and understanding the phenomena related to generation and propagation of low-frequency acoustic signals can contribute to determine the impact of sound pressure levels, emitted by wind turbines with set parameters, on the natural environment. The results obtained due to experimental tests and measurements on real units can be referenced to the maximum allowable infrasound noise levels, which are defined in the standards related to the working environment (Regulation of the Minister of Labor and Social Policy 2002 on the maximum allowable concentrations and intensities of harmful factors at work). Infrasound noise generated by modern high-power wind turbines is a problem frequently raised for opinions on their impact on the environment. There exist a large number of publications relating to the measurement, analysis and computer modeling of infrasound emitted by wind turbines. To mention a few, but far not all: [2-12]. Experimental setups for measuring of infrasound are described among others in [13-14], where authors describe a digital data acquisition system that can be used to record signals from infrasound events. Methods for technical condition monitoring and diagnosis are presented e.g. in [15-18]. If interested in other studies regarding generation of noise and vibration caused by electric power devices the reader is referenced to e.g. [20-21].

In contrast, none of the publicly available work applies to the analysis of the impact of the type of generator on the results obtained. The main cognitive purpose of the research works performed by our team is to investigate the phenomenon of formation and propagation of low frequency acoustic waves emitted due to wind turbine operation. The performed research focuses on determining the impact of e.g.: construction solutions, power capacity, atmospheric conditions and wind speed on the values of basic parameters characterizing the emitted low-frequency noise and spatial distributions of the generated sound pressure. In previous studies vibrations of wind turbines of various power capacities as well as sound pressure levels of the emitted signals in the audible range have been measured. Preliminary measurements and analyses results of infrasound and vibrations generated by different kinds of turbines have been presented among others in [22-26].

The main objective of this paper was to present results of a comparative analysis aimed at determination of the infrasonic noise spectrum for different types of generators and at indication whether differences in the noise spectra are important and that should be considered when choosing the type of generator turbine. There exist a number of different types of turbines. A detailed characteristic of various configurations and design techniques is given among others in [27]. In this paper we examine two different constructions of wind turbines: type asynchronous and type synchronous, which differ

* Corresponding author: t.boczar@po.opole.pl

according to the applied power generator and method of transferring the wind energy to the generator circuit. In the asynchronous unit, wind energy is transferred to the generator via transmission system, adjusting the rotor speed to the nominal speed of the generator. Synchronous units are not equipped with a transmission system. The whole energy generated by the rotor is directly transmitted to the generator, which has a large number of pole pairs, between which switching occurs depending on the number of rotations.

2 Methodology and characteristics of the turbine under study

For both types of constructions the low frequency measurements were performed basing on regulations of the PN-EN 61400-11 standard. The standard allows determination of the acoustic power level of wind turbine for different wind speeds (in practice, from 5 to 10 m/s) in a reproducible manner. Such measurements enable for determination of sound power level in the audible frequency range (from 50 Hz to 10 000 Hz) including tonality and directionality (optional). The standard precisely defines all requirements for the measuring equipment, measurement procedures, analysis methods of measurement results and the uncertainty analysis of the registered data. The standard also permits extension of the research on the infrasound range, but it does not define exactly how to measure it. The present study focuses on the measurement of the sound pressure level. This allows for identification of the range of potential impacts of infrasound at the measuring point location. Simultaneously to the measurement of sound level, the wind speed and direction, as well as other basic meteorological parameters were recorded in accordance with the standard (PN-EN 61400-11:2002 and 2004). The measurement was carried out through a continuous multi-hour measurement using sampling frequency equal to 51.2 kHz (full registration with listening capability). Then, within the analysis of the recorded signal, the FFT analysis in the band up to 200 Hz, with a resolution of 1 Hz (calculations made in LabShop B & K) was performed. The readings of measured values were performed with a second sampling using the multibufor tool.

The subject of acoustic background was also considered and the disturbing events were observed and registered during the measurements. All identified interferences were excluded from the data set subjected for analyzes. It is to mention that for measurements a turbine located at the periphery of the farm was selected, taking into account the wind direction on a given measurement day (avoiding the presence of other turbines on the measuring axis of the turbine-microphone). In addition, both turbines are located away from national roads and motorways. In order to avoid the risk of interference the measurements were performed in accordance to the requirements of EN 61400-11 standard, thus in the vicinity of the infrasound source. This allows achieving the highest accuracy in the „in situ" assessment of typical wind farms under normal operation.

The objects under study were two wind turbines: the first with asynchronous and the second with synchronous generator of rated power equal to 2.0 MW and 1.2 MW respectively. The asynchronous turbine is part of a newly commissioned wind farm, consisting of 15 similar turbines with a total electrical power equal to 30MW. This is a REPOWER wind turbine, type MM92, of electrical power equal to 2MW. Diameter of the turbine blades equals 92 m and the tower height is 80 m. The considered turbine is an outside standing turbine, thus the impact of infrasound noise, emitted by the other turbines, is minimized. The synchronous type turbine is a five years old Vensys device, type 62, with rated power equal to 1.2 MW. Diameter of the turbine blades is 62m and the tower height is 69m. This turbine does not belong to a larger farm; however it is located in a neighborhood of other turbines. Nevertheless, also in this case the impact of low frequency noise, emitted by other turbines, is minimized.

The acoustic pressure levels were measured using a set from Brüel & Kjær. This set consisted of a microphone, type 4190, designed for highly accurate measurements in free field, which was connected to LAN-XI 3050-A-060, a PULSE instrument. The registered frequencies were in the range from 1.2 Hz to 20 kHz, for which the transducer response is linear within ±3 dB. A double cover was used in accordance with PN-EN 61400-11:2002 and 2004 standard, For the noise data analysis, the frequency range was narrowed to infra and low frequencies, up to 100Hz. The applied Brüel & Kjær LAN-XI instrument is equipped with specialized software, PULSE LabShop, which allows for pre-processing of the recorded data. For post-processing of the measured signals the MATLAB environment was applied.

Parallel to sound pressure level also the meteorological conditions were measured using a DAVIS weather station –the Vantage Pro2 Instrument. This device is equipped with a unit of integrated sensors, which record data that is further transmitted to a PC computer wirelessly: temperature was measured within the range from -40 °C to +65 °C, with accuracy equal to +/- 0.1 °C, atmospheric pressure was measured in the range (810-1080) hPa with an accuracy of +/-1 hPa, humidity was measured in the range from 0 to 100%, with an accuracy of 3%,air flow velocity was measured in the range (0-68) m/s, with an accuracy of +/- 0.2 m/s. A summary over meteorological parameters, occurring during measurements is presented in Table 1.

During tests the wind speed values fluctuated significantly in the range from 1.3 m/s to 9.8 m/s, which allowed for performing analysis within a relatively wide range of wind speed impact on the infrasound level generated by the examined turbines. Other meteorological parameters and turbine operating conditions remained at a constant level and did not change significantly during the measurements. This article does not include results of uncertainty measurements.

 DOI: 10.1051/e3sconf/20171901001

Table 1. Summary over the meteorological parameters, occurring during measurements (AT - asynchronous turbine, ST - synchronous turbine).

		Maximal value		Minimal value		Mean value		Standard deviation	
Parameter	**Unit**	AT	ST	AT	ST	AT	ST	AT	ST
Temperature	°C	24.8	5.9	17.1	5.1	21.2	5.6	2.3	0.3
Humidity	%	71	79	44	75	56.5	77.5	7.5	1.0
Dew Point	°C	13.2	2.3	11.2	1.6	12.0	2.0	0.4	0.2
Pressure	hPa	1001.6	1021.9	1000.2	1021.2	1000.8	1021.5	0.4	0.1
Wind speed	m/s	7.6	9.8	1.3	3.6	4.8	6.2	1.4	1.2
Precipitation	mm	0	0	0	0	0	0	0	0

The measurements of low frequency noise generated by the considered turbines were performed with accordance to standard, in a single reference point at a distance equal to 100 m and 131 m, respectively. The location of individual components of the measurement system in relation to the test turbine is presented in Figs. 1 and 2.

Fig. 1. Location of the infrasound noise level measuring point at the asynchronous type turbine.

3 Results and discussion

The scope of works, results of which are presented below, includes analyses of frequency spectra of sound pressure levels generated during operation by both considered turbine types. Regarded data apply to three characteristic values of wind speed: 5.4m/s, 6.3 m/s and 7.6 m/s.

Fig. 2. Location of the infrasound noise level measuring point at the synchronous type turbine.

These values exemplify in a representative manner the range of significant changes in recorded signals. The lowest selected wind speed of 5.4m/s regards the impact on the turbine immediately after its start-up. The middle value of 6.3m/s affects the noise spectrum generated during turbine operation in the center of its energy efficiency. The highest measured wind speed value equal to 7.6m/s implies infrasound noise emission at the maximum turbine energy efficiency, per day.

In Figs. 3 and 4 frequency spectra of sound pressure levels generated by operating of synchronous and asynchronous type wind turbines, for the selected wind speed values: 5.4 m/s, 6.3 m/s and 7.6 m/s, are presented. In the case of asynchronous type turbine, an increase in wind speed value results in visible growth of the sound pressure level of the generated noise; this applies to the whole range of the considered frequencies.

The largest values of measured signals were observed in the infrasound band from 1 Hz to 10 Hz. Several characteristic harmonic components can also be distinguished. The most significant is to be found in the range from 20 to 31 Hz. Within this range we observe narrower picks for higher wind speeds and wider picks for lower wind speeds. Also, the pick location depends on the wind speed value and is placed for higher wind speeds by higher frequencies.

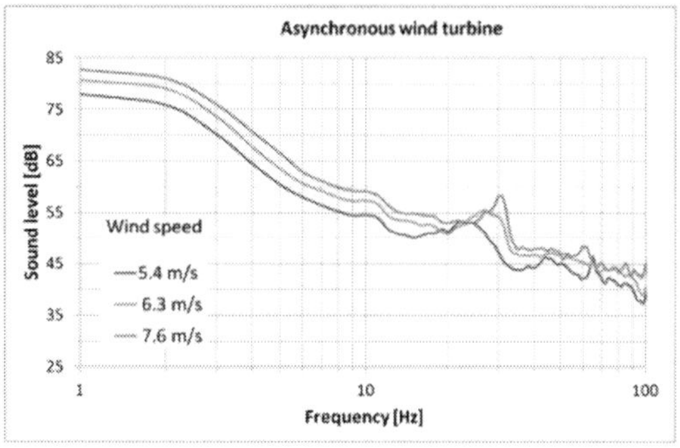

Fig. 3. Frequency spectra of sound pressure levels generated by operating asynchronous type wind turbine, for chosen wind speed values: 5.4, 6.3 and 7.6 m/s.

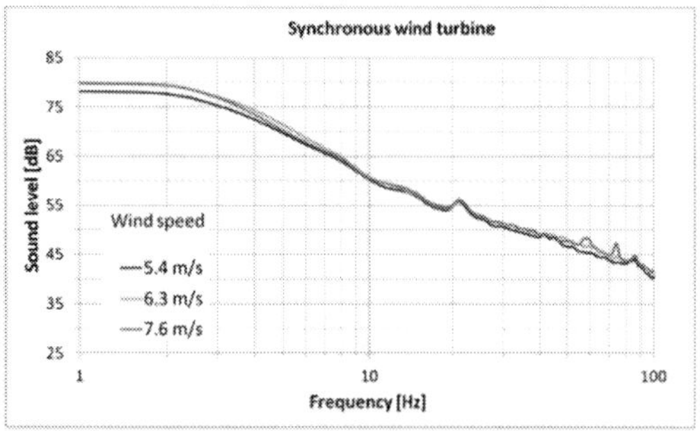

Fig. 4. Frequency spectra of sound pressure levels generated by operating synchronous type wind turbine, for chosen wind speed values: 5.4, 6.3 and 7.6 m/s.

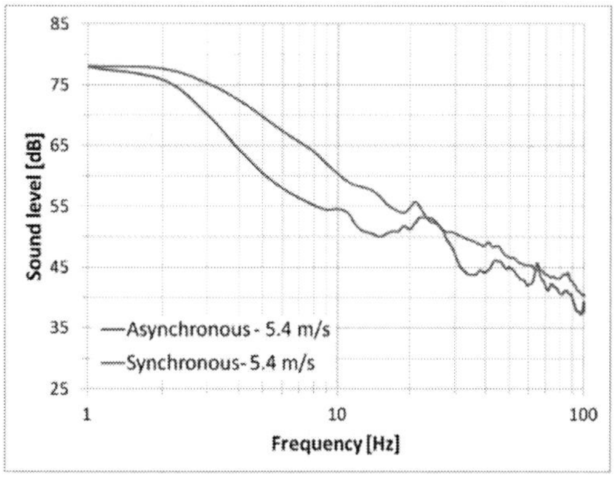

Fig. 5. Frequency spectra of sound pressure levels generated during normal operation of both turbines at wind speed value equal to 5.4 m/s; Pearson correlation coefficient $r = 0.94$.

In the case of synchronous type turbine the wind speed value does not have so significant impact as it was for the other turbine type. In the frequency range up to 2.5 Hz there is no difference in the measured pressure levels for wind speeds above 6.3 m/s. Also the observed picks do not depict much higher sound pressure levels and achieve only several decibels more than it arises from the trend. All the harmonic components are found in the same frequency range independently of the wind speed.

In order to compare frequency spectra of the two considered turbine types with dependence to the wind speed, in Figures 5 and 6 appropriate characteristics are shown. For each dataset pair, the Pearson correlation coefficient r was calculated. This indicator allows for qualitative assessment of correlations between the recorded signals.

Comparison of presented in Fig. 7 sound level spectra of both turbine types, calculated for signals generated at wind speed equal to 5.4 m/s, indicates major acoustic impact of the synchronous type turbine, values of which for the considered frequency range are higher than levels emitted by the asynchronous type turbine.

The calculated Pearson correlation coefficient between both signals amounts $r = 0.94$ and indicates their strong linear correlation. Similar observations were done for data gathered at wind speed values equal to 6.3 m/s (see Fig. 6) and 7.6 m/s (see Fig. 7). Some significant differences are remarked in the frequency range up to about 2 Hz, where by median wind speeds, the measured sound levels remain similar for both considered turbine types.

For the asynchronous type turbine, by wind speeds equal to 6.3 m/s, the measured sound levels are higher only within the harmonic frequencies range. The correlation of both signals is also strong and amounts $r = 0.94$.

During operation by highest wind speed values we observe that up to about 2.4 Hz the asynchronous type turbine generates higher sound levels then the synchronous type turbine.

In the range from 2.4 Hz to 45 Hz, with except of the resonant frequencies, the synchronous type turbine generates higher sound levels. Above the frequency value equal to 45 Hz the depicted values remain very similar. The correlation of both signals is also strong and amounts $r = 0.95$.

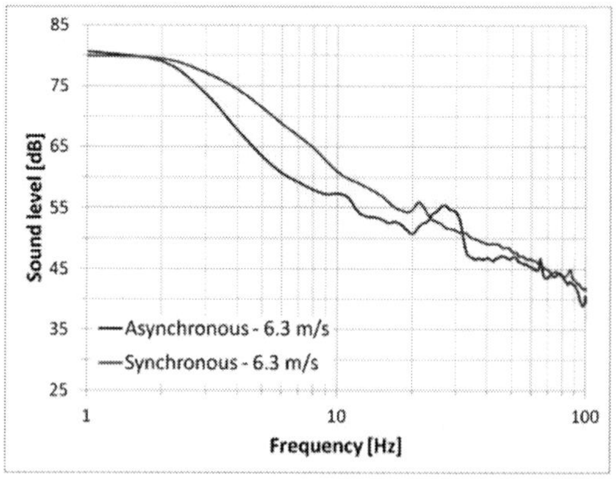

Fig. 6. Frequency spectra of sound pressure levels generated during normal operation of both turbines at wind speed value equal to 6.3 m/s; Pearson correlation coefficient $r = 0.94$.

As already mentioned in the achieved spectra some resonant frequencies have been indicated. Indication of local maxima aims to demonstrate the tonal noise of wind turbines.

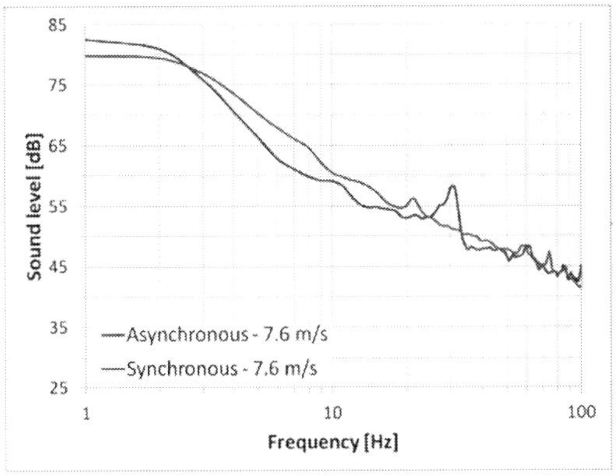

Fig. 7. Frequency spectra of sound pressure levels generated during normal operation of both turbines at wind speed value equal to 7.6 m/s; Pearson correlation coefficient $r = 0.95$.

These frequencies are related to the frequencies of the turbine itself and are associated with rotational speeds of the rotor, generator, and in the case of an asynchronous turbine also the transmission. In order to determine the exact location of emission of measured tonal component it is necessary to conduct more detailed studies associated with continuous monitoring of operating parameters of the constituent elements of the turbine. For asynchronous type turbine the following values were set: 22Hz for wind speed equal to 5.4 m/s, 27 Hz for wind speed equal to 6.3 m/s and 31Hz for wind speed equal to 7.6 m/s. In the case of synchronous turbine the frequency equal to 21 Hz was set for all wind speed values. For these values sound pressure level variations over time were estimated for both considered turbine types separately for each resonance and wind speed. These dependencies are shown in Figs. 7 to 13. For all the datasets the arithmetic mean and standard deviation values were calculated and are summarized in Table 2.

Table 2. Parameters of the infrasound signal gathered by chosen wind speed values and resonant frequencies: average sound pressure level and standard deviation.

Turbine type	Resonant frequency f[Hz]	Wind speed v[m/s]	Mean $L_{Average}$ [dB]	Standard deviation s [dB]
asynchronous turbine	31	7.6	55.1	5.2
	27	6.3	55.5	3.0
	22	5.4	53.2	5.6
synchronous turbine	21	7.6	56.3	1.9
	21	6.3	55.9	1.8
	21	5.4	55.8	1.8

On the basis of the calculated standard deviation values the range of variation of the measured signals can be implied. It was stated that turbine design type has an impact on the dynamics of sound level changes in time. The asynchronous type turbine indicates relatively high fluctuations in the value of the measured sound pressure levels; the standard deviation value varies from 3.0 dB to 5.6 dB for different wind speeds. In contrast, the synchronous type turbine generates sound, which indicate standard deviations around 1.8 dB regardless the wind speed value.

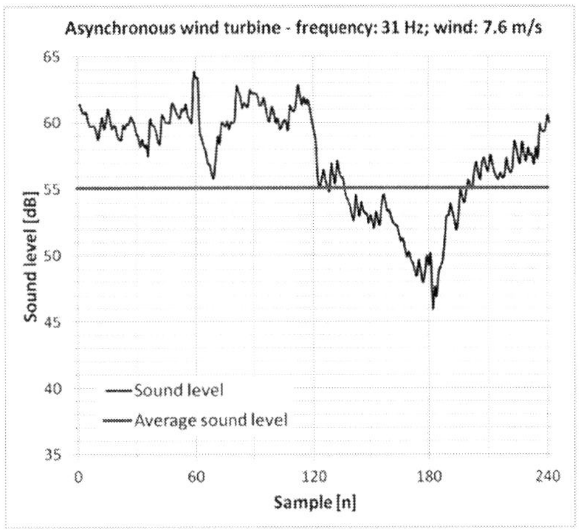

Fig. 8. Variations of sound pressure levels generated by asynchronous type turbine. Depicted data considers resonant frequency value equal to f=31Hz and wind speed value equal to 7.6m/s.

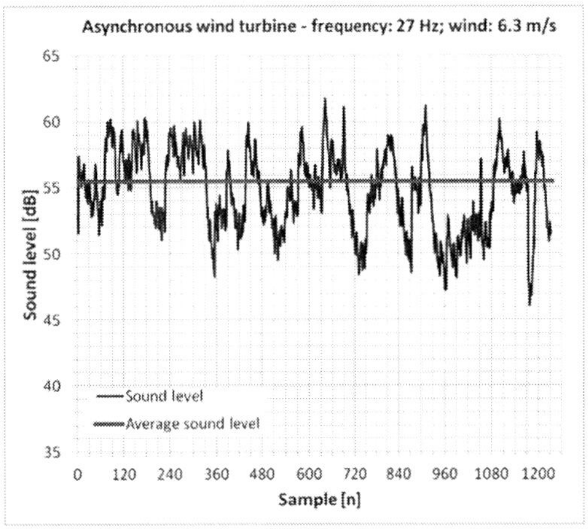

Fig. 9. Variations of sound pressure levels generated by asynchronous type turbine. Depicted data considers resonant frequency value equal to f=27Hz and wind speed value equal to 6.3 m/s.

When comparing the presented spectra of both turbines (see Figs. 2 and 6), it is to observe that for asynchronous turbine the dynamic operating range of the measured signal is greater. In the spectrum of asynchronous turbine, there are characteristic harmonic components with increased level of sound and also ranges that indicate relatively low levels, as if part of the components was damped. Instead, in the spectrum of synchronous turbine, the dynamic of changes in the measured signal is much lower and any dominant component is relatively difficult to distinguish. Comparing these two spectra, it can also be stated that asynchronous turbine operation is characterized by

DOI: 10.1051/e3sconf/20171901001

lower sound levels in the infrasound band, as if they were effectively reduced or dumped (at the design stage). The design of asynchronous wind turbine is more modern than the design of synchronous turbine, what may give some explanation for the observed differences. In both cases, the values of measured sound pressure levels are comparable, wherein more recent turbine design has distinctly better acoustic performance with much lower noise levels.

Correlation between individual spectra is very strong, Pearson correlation coefficient $r > 0.9$.

Stability analysis of noise level generated as a function of time (see Figs. 8-13).

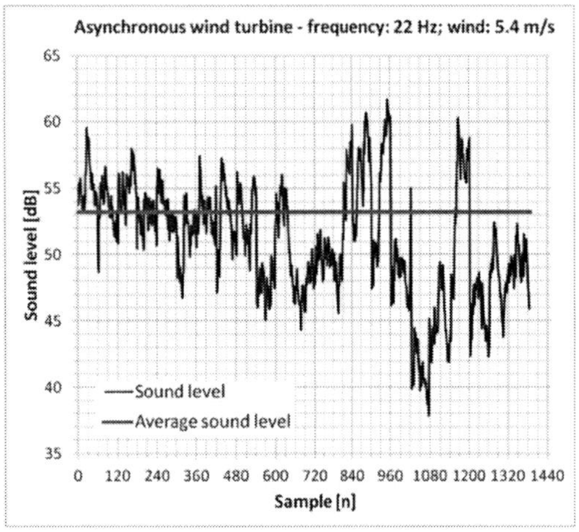

Fig. 10. Variations of sound pressure levels generated by asynchronous type turbine. Depicted data considers resonant frequency value equal to f=22Hz and wind speed value equal to 5.4 m/s.

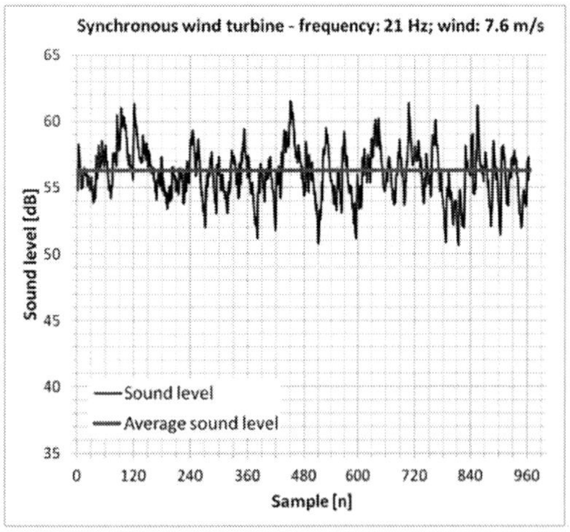

Fig. 11. Variations of sound pressure levels generated by synchronous type turbine. Depicted data considers resonant frequency value equal to f=21Hz and wind speed value equal to 7.6m/s.

It is performed for selected resonant frequencies, revealed that although all signals have a similar mean value, the range of changes, expressed as standard deviation of the sample, depends on the design type of electrical power generation system.

Presented figures, which illustrate time changes of the measured values of sound pressure level, clearly indicate that fluctuation range of infrasound noise can go up to approx. 10dB, which corresponds to changes in the perceived noise level as half shift.

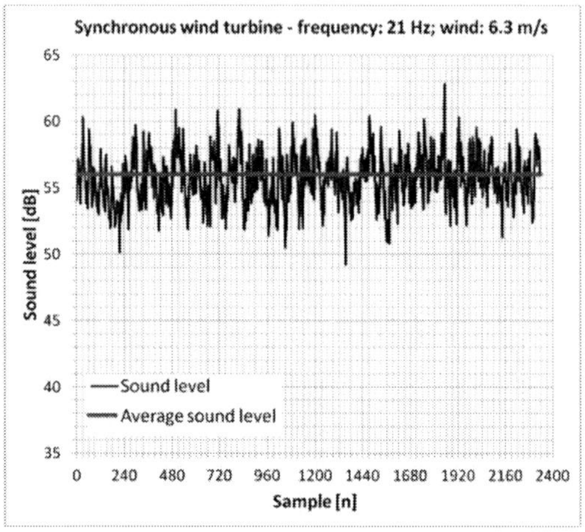

Fig. 12. Variations of sound pressure levels generated by synchronous type turbine. Depicted data considers resonant frequency value equal to f=21Hz and wind speed value equal to 6.3 m/s.

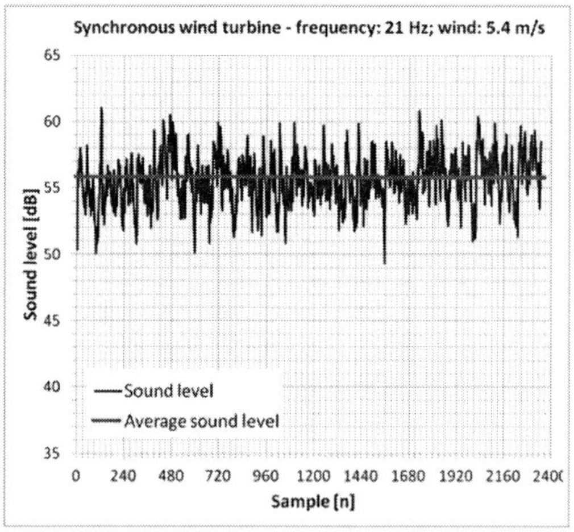

Fig. 13. Variations of sound pressure levels generated by asynchronous type turbine. Depicted data considers resonant frequency value equal to f=21Hz and wind speed value equal to 5.4 m/s.

4 Conclusions

Based on the presented results, it can be stated that operation of wind turbines, regardless of the generator type, causes generation of noise containing infrasound components. However, the registered levels are far below the threshold values, above which they would be harmful to humans [3].

Further it was stated that both analyzed turbines indicate relatively linear decrease of the measured sound pressure levels within increasing frequencies. In both cases several harmonic components can be distinguished, for which the sound pressure level is higher than it arises from the trend.

In general it can be stated that the asynchronous type turbine produces lower pressure levels, which are less stable over time, and indicates higher pressure values around the resonant frequencies as compared to the synchronous type turbine. Also it can be stated that the asynchronous type turbine is more influenced by the wind conditions and generates higher pressure values by higher wind speeds then the synchronous type turbine.

Presented results depict a part of research studies related to development and creation of numerical wind turbine model that takes into account the main sources of generation of infrasound, i.e. the mechanical noise coming from the gondola and the aerodynamic noise generated due to operation of rotating blades. In particular, the model will take into account the wind turbine constructional and technical parameters related to type, geometry and size of the turbine, the generated power and the principle of its operation.

Acknowledgment

The work was co-financed from funds of the National Science Centre (NCS) as part of the OPUS programme, project no.: 2015/17/B/ST8/03371.

References

1. J.H. Schmidt, M. Klokker, *Health Effects Related to Wind Turbine Noise Exposure: A Systematic Review*. Tong S., ed. PLoS ONE, 9: dx.doi.org/10.1371/journal.pone.0114183 (2014)
2. G. Bellhose, *Low frequency noise and infrasound from wind turbines generators: a literature review Bel Acoustic Consulting*, (New Zeland, 2004)
3. D.W.Colby, R. Dobie, G. Leventhall, D.M. Lipscomb, R.J. McCunney, M.T. Seilo, B. Sondergaard, *Wind turbine sound and health effects. An Expert Panel Review* (2009)
4. D.C. Fernandez Valencia, *Infrasound and its effects on humans, Spatial Audio DESC 9137, technical report,* University of Sydney (2007)
5. K.L. Hansen, G. Hessler, C.H. Hansen, B. Zajamsek, *Prediction of infrasound and low frequency noise propagation for modern wind turbines, a proposed supplement to ISO9613-2, 6th International Conference on Wind Turbine Noise*, Glasgow (2015)
6. J. Jakobsen. J. of Low Freq. Noise, Vibr. Act. Contr. **24**, 145-155 (2005)
7. M. Kłaciński, T. Wszołek, Arch. Acous. **39**, 3-10 (2014)
8. C.Y.H. Qibai, H. Shi, J. Low Freq. Noise, Vibr. and Act. Contr. **23**, 71-76 (2004)
9. E. Pedersen, J. Bouma, R. Bakker, F. Van Den Berg, *Proceedings of the 7th European Conference on Noise Control, EURO-NOISE, Acoustics*, Paris, France, 4049–4054 (2008)
10. A. J. Swerdlow, Health Protect. Agency (2010)
11. H. Tachibana, *Proceedings of Inter-noise: 43rd International Congress on Noise Control Engineering,* Melbourne, Australia, 1-10 (2014)
12. S. Wegner, R. Bareiss, G. Guidati, *Wind Turbine Noise*, Springer, Berlin (1996)
13. J. Chilo, T.A. Lindblad, *Proc. of 4th IEEE Workshop on Intelligent Data Acquisition and Advanced Computing Systems: Technology and Applications*, 35-37 Dortmund (2007)
14. M. Golec, Z. Golec, C. Cempel, Diag.'l **37**,115-120 (2006)
15. Z. Hameed, Y.S. Hong, Y.M. Cho, S.H. Ahn, C.K. Song, Renewable Sust Energy Rev. **13**, 1-39 (2009)
16. F. Toja-Silva, A. Colmenar-Santos, M. Castro-Gil, Renewable Sustain. Energy Rev. **24**, 364-378 (2013)
17. G.P. Van den Berg, *Noise notes*, J. Low Freq. Noise, Vibr. Act. Contr. **24**, 1-23 (2005)
18. B. Yu, Y. Li, X. Wu, Z. Yang, *Proceedings of Power Electronics and Machines in Wind Applications,*Lincoln, NE, 1-7 (2009)
19. S. Borucki, A. Cichoń, Przegląd Elektrotechniczny **86**, 45-47 (2010)
20. A. Cichoń, S. Borucki, T. Boczar, Arch. Acous. **36**, 823-830 (2011)
21. H. Majchrzak, A. Cichoń, S. Borucki, Arch. Acous. **42**, 29–35 (2017)
22. T. Boczar, T. Malec, D. Wotzka, Acta Phys. Pol. A **122**, 850-853 (2012)
23. R. Pierzga, T. Boczar, D. Wotzka, D. Zmarzły, Acta Phys. Pol. A **124**, 542-545 (2013)
24. R. Pierzga, T. Boczar, D. Wotzka, Acta Phys. Pol. A **128**, 294-298 (2015)
25. M. Szmechta, Acta Phys Pol A **124**, 574-577 (2013)
26. D. Wotzka, T. Boczar, T. Malec, R. Pierzga, Acta Phy. Pol. A **124**, 595-597 (2013)
27. M. Mahmood, A. Bhutta, N. Hayat, A.U. Ali, R.Sh. Jamil, Z. Hussain, Renewable Sust. Energy Rev. **16**, 1926-1939 (2012)

Statistical analysis of the electric energy production from photovoltaic conversion using mobile and fixed constructions

Artur Bugała[1,*], *Karol* Bednarek[1], *Leszek* Kasprzyk[1], and *Andrzej* Tomczewski[1]

[1]Institute of Electrical Engineering and Electronics, Poznan University of Technology, Piotrowo 3a Street, 60-965 Poznań, Poland

Abstract. The paper presents the most representative – from the three-year measurement time period – characteristics of daily and monthly electricity production from a photovoltaic conversion using modules installed in a fixed and 2-axis tracking construction. Results are presented for selected summer, autumn, spring and winter days. Analyzed measuring stand is located on the roof of the Faculty of Electrical Engineering Poznan University of Technology building. The basic parameters of the statistical analysis like mean value, standard deviation, skewness, kurtosis, median, range, or coefficient of variation were used. It was found that the asymmetry factor can be useful in the analysis of the daily electricity production from a photovoltaic conversion. In order to determine the repeatability of monthly electricity production, occurring between the summer, and summer and winter months, a non-parametric Mann-Whitney U test was used as a statistical solution. In order to analyze the repeatability of daily peak hours, describing the largest value of the hourly electricity production, a non-parametric Kruskal-Wallis test was applied as an extension of the Mann-Whitney U test. Based on the analysis of the electric energy distribution from a prepared monitoring system it was found that traditional forecasting methods of the electricity production from a photovoltaic conversion, like multiple regression models, should not be the preferred methods of the analysis.

1 Introduction

Human beings obtain energy in various ways to subsequently use it to meet their specific functional purposes. The most operationally convenient form of energy is electric energy as it can be transported, stored, and processed. Currently a lot of attention is paid to all the solutions connected with using energy from ecological sources, such as photovoltaic systems, wind turbines, biomass, tidal energy, etc. [1-15].

During the processes of design and execution of power supply systems from renewable sources of energy various types of needs and requirements have to be considered. As the energy is generated and supplied to the recipients – that is currents and voltages occur in the elements of the system – the occurrence and impact of electromagnetic fields should be taken into account. This is connected both with the correctness of operation of all the elements of the system (associated with generation, processing and transport of energy) as well as the impact on the environment (ecological reasons), but also rationalization of energy management. It is important to consider these impacts on various levels: both high current systems (connected with energy generation and supply) and microstructures (photovoltaic conversion of energy or operation of electronic elements) [16-27].

All human activity is associated with optimization behaviours and considerations. People try to achieve their goals in a way most beneficial for them. This is associated with a great variety of optimization criteria, methods of creating criterial functions and the used methods, which depend on a given task [4, 5, 22, 28-36].

The study discusses optimum energy yield achieved from the photovoltaic conversion for the PV modules installed in a fixed and two-axis tracking construction. Due to strongly stochastic character of obtaining energy (sunlight depending on seasons of the year, weather conditions, etc.), the results of the conducted tests were subject to statistical processing, which constitutes the main part of this article [4, 5, 37-43].

2 Parameters of Statistical Analysis

The study presents the exemplary courses of daily variation of electrical energy production with the use of photovoltaic modules in the fixed and two-axis tracking construction for the selected day of the summer, autumn, spring, and winter month. On the basis of the analysis of daily distribution of values of the registered electrical energy, it is possible to indicate the days with almost symmetrical shape of the course of this value in relation to the afternoon hours and the days characterized by strong irregularities. The skewness coefficient described by the following relation was used as a measure of asymmetry [43]:

$$A_s = \frac{\bar{x} - M_o}{s} \quad (1)$$

where:

$\bar{x}$ – average value,
M_o – median,
s – standard deviation.

* Corresponding author: artur.bugala@put.poznan.pl

The higher value of the skewness coefficient, the bigger asymmetry in relation to the average value. The asymmetry factor equalling zero indicates the symmetry of the variable's distribution, the positive value indicates right side asymmetry, while the negative value left side asymmetry.

The coefficient of variation *v* is the quotient of variation of a given value around the average value from the population (standard deviation of the population) to the designated average value. It is assumed that if the determined coefficient of variation does not exceed 10%, the features indicate statistically insignificant variation [42]. Big values of standard deviation in relation to the average value may lead to the limitation of the quality of the forecasting model.

Kurtosis is a measure of the flatness or peakedness of the obtained distribution in relation to the normal distribution. Its value equals zero for the theoretical distribution.

Repeatability of the monthly electrical energy production from photovoltaic conversion was defined with the use of the non-parametric Mann-Whitney U test. The measure of central tendency of the test is median. Zero and alternative hypothesis may be described as follows:

- H_0: distribution of average ranks of observations in the analysed groups does not differ significantly – the samples come from one population (the medians of the tested variable in both groups do not significantly differ between one another);
- H_1: distribution of average ranks of observations in the analysed groups differs significantly – the samples come from different populations (the medians of the tested variable differ between one another).

Depending on the population's size, it is possible to calculate the value of the test statistics and on its basis the probability value p, which is compared with the significance level α from the distribution tables, using the following relation [43]:

- for small size of the population:

$$U = n_1 n_2 + \frac{n_1(n_1+1)}{2} - R_1 \tag{2}$$

where:
n_1, n_2 – respective sizes of the first and second population;
R_1, R_2 – sums of ranks of the elements from the first and second population;

- for big size of the population:

$$Z = \frac{U' - \frac{n_1 n_2}{2}}{\sqrt{\frac{n_1 n_2 (n_1+n_2+1)}{12} - \frac{n_1 n_2 \sum (t^3 - t)}{12(n_1+n_2)(n_1+n_2-1)}}} \tag{3}$$

$$U' = n_1 n_2 + \frac{n_2(n_2+1)}{2} - R_2 \tag{4}$$

where:
t – number of cases included in the tied rank.

The choice of the right hypothesis can be presented with the following relation:

$$p \le \alpha \Rightarrow H_1 \cup p > \alpha \Rightarrow H_0 \tag{5}$$

Repeatability of the daily peak hours in which the highest value of hourly electrical energy production from photovoltaic conversion is observed was described with the use of the non-parametric Kruskal-Wallis test. This test is used to verify the hypothesis about lack of significance of the differences between the medians of the tested variable in a few populations ($k > 2$). It should be assumed that distributions of the measured value are close to each other.

The zero and alternative hypotheses can be recorded as follows:

- H_0: distribution of average ranks of observations in the analysed k groups does not differ significantly – the samples come from one population (the medians of the tested variable in all groups do not significantly differ between one another);
- H_1: distribution of average ranks of observations in the analysed k groups differs significantly (the medians of the tested variable differ between one another in individual groups).

The value of test statistics H can be determined using the following relation [41]:

$$H = \left(\frac{12}{N(N+1)} \sum_{j=1}^{k} \left(\frac{(\sum_{i=1}^{n_j} R_{i,j})^2}{n_j}\right)\right) - 3(N+1) \tag{6}$$

$$N = \sum_{j=1}^{k} n_j \tag{7}$$

where:
n_j – size of the populations for (j=1,2,…,k),
$R_{i,j}$ – ranks assigned to the variable value for (i = 1,2,…,n_j), (j = 1,2,…,k),
t – number of cases included in the tied rank.

The probability value p determined on the basis of the test statistics H should be compared with the assumed significance level α.

The choice of the appropriate hypothesis can be presented with the following relation:

$$p \le \alpha \Rightarrow H_1 \cup p > \alpha \Rightarrow H_0 \tag{8}$$

3 Results of Statistical Analysis

The selected results of statistical analysis of daily distribution of electrical energy production for the recommended days of the summer, autumn, spring, and winter months, including determination of the average value, standard deviation, skewness and coefficient of variation for the photovoltaic module installed in the fixed (us) and two-axis tracking construction (un) were presented in Table 1. The method of determination of the recommended days of the year with their calculation for the typical meteorological year of Poznan was presented in [38].

Table 1. Results of statistical analysis of distribution of the daily electrical energy production for the selected days of the months: July, December, April, and January.

parameter	12.07.14		03.07.14		26.07.14	
	un	us	un	us	un	us
average/ day	84.63	49.92	74.29	45.75	70.67	44.58
average/ sunny hours	135.40	85.57	111.4	78.43	113.07	76.43
standard deviation/day	76.91	62.19	69.06	57.93	69.08	54.54
standard deviation/ sunny hours	50.89	59.82	54.90	56.48	53.30	51.64
skewness/ day	0.00	0.84	0.09	0.93	0.19	0.79
coefficient of variation/ day	0.91	1.25	0.93	1.27	0.98	1.22
coefficient of variation/ sunny hours	0.38	0.70	0.49	0.72	0.47	0.68
	18.12.14		11.12.14		29.12.14	
average/ day	0.79	1.25	9.29	6.83	6.83	4.79
average/ sunny hours	4.75	6.00	37.17	27.33	23.43	16.43
standard deviation/day	1.98	2.70	22.71	15.21	14.07	9.70
standard deviation/ sunny hours	2.17	2.53	32.05	19.10	17.04	11.45
skewness/ day	2.72	2.28	2.93	2.70	2.33	2.23
coefficient of variation/ day	2.50	2.16	2.44	2.23	2.06	2.02
coefficient of variation/ sunny hours	0.46	0.42	0.86	0.70	0.73	0.70
	11.04.14		25.04.14		02.04.14	
average/ day	37.29	23.29	55.42	40.21	55.54	41.58
average/ sunny hours	63.93	46.58	95.00	74.23	111.08	83.17
standard deviation/day	49.01	30.95	69.14	55.76	69.60	59.00
standard deviation/ sunny hours	49.13	28.81	66.59	56.69	59.31	59.19
skewness/ day	1.34	0.99	0.74	1.04	0.85	1.17
coefficient of variation/ day	1.31	1.33	1.25	1.39	1.25	1.42
coefficient of variation/ sunny hours	0.77	0.62	0.70	0.76	0.53	0.71

Table 1 (cont.)

	18.01.15		08.01.15		15.01.15	
average/ day	1.29	1.71	1.42	1.75	11.71	8.46
average/ sunny hours	6.20	8.20	8.50	7.00	46.83	29.00
standard deviation/day	3.21	3.52	3.58	3.92	22.58	17.06
standard deviation/ sunny hours	4.35	2.48	4.09	4.97	19.84	20.05
skewness/ day	3.06	1.94	2.88	2.59	1.71	2.12
coefficient of variation/ day	2.48	2.06	2.53	2.24	1.93	2.02
coefficient of variation/ sunny hours	0.70	0.30	0.48	0.71	0.42	0.69

In the analysed case the value A_s for the groups of days of the summer month (03.07.2014, 12.07.2014, 26.07.2014) is close to zero, indicating almost symmetrical distribution. For winter months, the increase in the asymmetry factor both for the value of electrical energy produced by the photovoltaic module in the fixed and two-axis tracking construction can be observed. The highest values of this parameter, exceeding 2, were observed for the winter months (08.01.2015, 15.01.2015, 18.01.2015) with strongly random distribution of the surface density of the solar power radiation. The symmetry of distribution of the daily production of electrical energy determines its bigger predictability, which makes it easier to make a decision on the method of installation of photovoltaic modules in the photovoltaic system and to determine the possibility of covering the daily energy demand of the powered objects during morning to evening hours.

In order to determine the repeatability of the courses of the monthly electrical energy production from photovoltaic conversion occurring between summer months and summer and winter months, a non-parametric Mann-Whitney U test was used, which consisted in comparing two independent groups. In this case the normal distribution of quantitative variables and the equal size of the groups is not required. It is difficult to meet the last condition in the analysed cases due to various numbers of days in individual months of the year. Also, the used non-parametric method is relevant for populations of small size, where variables are measured in a quantitative, ordinal or dichotomous scale. The starting point was formulating test hypotheses: zero H_0 and alternative H_1:

- H_0: distribution of average ranks of observations in the analysed groups (months of the year) does not differ significantly;
- H_1: there are significant differences in the distributions of variables in both groups (months of the year).

The observations with equal value in the created ordered series were assigned ranks – also tied ranks. The detailed results of the Mann-Whitney U test obtained in *Statistica* software for June, July, August, December, and January are presented in Table 2.

DOI: 10.1051/e3sconf/20171901002

Table 2. Results of Mann-Whitney U test for June, July, August, December, January 2014 and 2015.

period	month: July 2014, December 2014								
variable	sum of rang A	sum of rang B	U	Z	p	Z with correction	p	N valid A	N valid B
energy	1239.00	714.00	218.00	3.688586	0.000226	3.702927	0.000213	31	31
period	month: June 2014, January 2015								
variable	sum of rang A	sum of rang B	U	Z	p	Z with correction	p	N valid A	N valid B
energy	1367.00	524.00	28.00	6.297056	0.000000	6.298888	0.000000	30	31
period	month: June 2014, July 2014								
variable	sum of rang A	sum of rang B	U	Z	p	Z with correction	p	N valid A	N valid B
energy	950.00	941.00	454.00	-0.151476	0.879601	-0.151494	0.879586	31	30
period	month: June 2014, August 2014								
variable	sum of rang A	sum of rang B	U	Z	p	Z with correction	p	N valid A	N valid B
energy	1008.50	882.50	386.50	1.125247	0.260485	1.125322	0.260454	30	31
period	month: December 2014, January 2015								
variable	sum of rang A	sum of rang B	U	Z	p	Z with correction	p	N valid A	N valid B
energy	860.50	1092.50	364.50	-1.626080	0.103935	-1.654200	0.098087	31	31

The determined value of test probability p indicates that while comparing the summer and winter months (June – January, July – December), the assumed significance level α has bigger values. Due to that there are grounds to reject the zero hypothesis and to assume the alternative hypothesis according to which there are significant differences in the monthly distributions of electrical energy production from photovoltaic conversion for both periods. This can also be confirmed in the results of monthly distribution of electrical energy production for the installed fixed and tracking photovoltaic modules for the indicated time periods. Figure 1 and Figure 2 present the distribution of electrical energy obtained from photovoltaic modules installed in a fixed and 2 – axis tracking configuration.

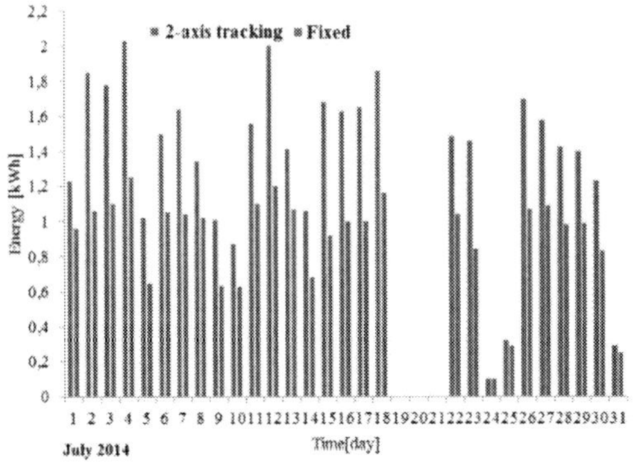

Fig.1. Monthly production of electrical energy by the fixed and tracking photovoltaic modules, July 2014.

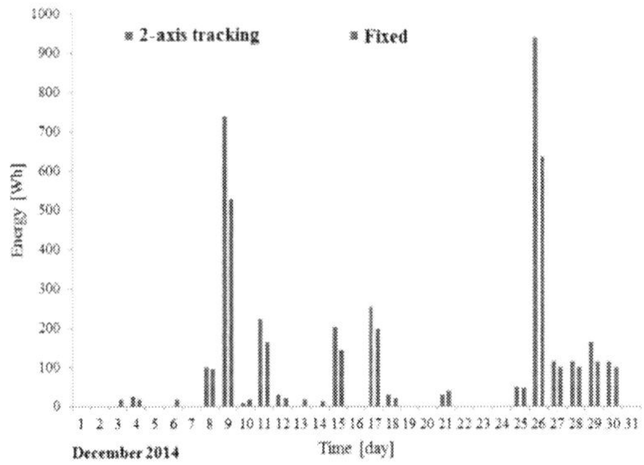

Fig.2. Monthly production of electrical energy by the fixed and tracking photovoltaic modules, December 2014.

In the case of summer months (June – July, June – August) no grounds to reject the zero hypothesis were identified. The monthly distributions of electrical energy do not significantly differ between one another. Also, no statistically significant differences were demonstrated for the two winter months: December 2014 and January 2015.

In order to analyse the repeatability of the daily peak hours in which the highest value of hourly electrical energy production is observed, respective 60-minutes periods of the day were indicated for each day of the analysed month: May, June, July and August 2014. Comparison of four independent groups was conducted using the non-parametric Kruskal-Wallis test which is the extension of the Mann-Whitney test. Both tests do not require to meet many assumptions characteristic for parametric tests. The statistical significance of the Kruskal-Wallis test indicates

* Corresponding author: artur.bugala@put.poznan.pl

differences between the tested groups. The starting point was formulation of test hypotheses: zero H_0 and alternative H_1:

- H_0: distribution of average ranks in the analysed groups (months of the year) does not differ significantly;
- H_1: there are significant differences between the groups in relation to the repeatability of peak hours.

The value of parameter $p = 0.3279$ determined as a part of *Statistica* software is higher than the assumed significance level $\alpha = 0{,}05$. Hence, there are no grounds to reject the zero hypothesis assuming lack of significant differences in the periods of peak hours for the analysed months of the year. Therefore the occurring differences are not statistically significant. The additionally conducted Mann-Whitney U test for two groups confirmed the occurrence of significant differences between the peak hours (average ranks of observation) for the summer month (June 2014) and winter month (January 2015) with the assumed significance level $\alpha = 0.05$. The results of the tests are presented in Table 3.

Table 3. Results of the Mann-Whitney U tests confirming the significance of differences between the peak hours.

period	month: June 2014, January 2015			
sum of rang A	sum of rang F	p	N valid A	N valid F
354.00	207.00	0.030572	24	9

4 Conclusions

The used non-parametric methods are particularly relevant for small populations. Both the Mann-Whitney U test and non-parametric Kruskall-Wallis test demonstrated that there are significant differences between the values of the registered daily electrical energy and instantaneous power from the photovoltaic conversion for summer and winter, summer and autumn and summer and spring months. However, no statistically significant differences between the variable measured for months of the same season of the year were identified with the assumed significance level.

The analysis of variation and repeatability of monthly electrical energy production for almost the whole measurement year was also completed to include determination of, among other things, monthly average value, standard deviation, median, range and coefficient of variation.

Both the quantitative and the qualitative comparative analysis of the monthly electrical energy production from photovoltaic conversion indicates lower variation of courses for the summer months. Basic parameter in the presented analysis is the coefficient of variation in the meaning of ratio of standard deviation to arithmetic mean. The coefficient of variation value changes within the range from 0.36 – 0.51 for the months from June to August 2014. This is almost fivefold lower value compared to the results obtained for December the same year and over threefold lower compared to January the following year. The value of coefficient of variation exceeding 0.60 indicates that a significant percentage of the average value is its standard deviation. Hence, the distribution of the measured value is not homogeneous and consequently the arithmetic mean should not be the main statistic measure. Using only the value of the standard deviation in the comparative analysis may turn out to be insufficient due to considerable differences in the average value of the monthly production of electrical energy for the extreme months of the measurement year.

The determined median value with considerably higher values for the summer months indicates that half of all the observations is found respectively below or over its value in the ordered series. The range has comparable values for the months from the same period (season of the year). However, this is not the measure resistant to the outliers which can disturb the variation of the measured value, at the same time not resulting in the change of the range's value.

The conducted statistical analysis may be useful for evaluation of energy potential of central Poland represented by the city of Poznan, with regard to planning new investments in the sector of renewable energy resources, mainly using technologies based on conversion of solar radiation energy to electrical energy or heat.

References

1. K. Abdelhamid, R. Ibtiouen, Solar Energy, **103**, 171-182 (2014)
2. M. Bortolini, M. Gamberi, A. Graziani, Energy Conv. Manag. **86**, 81-92 (2014)
3. K. Branker, M. J.M. Pathak, J.M. Pearce, Renew. Sust. Energ. Rev. **15**, 4470-448 (2011)
4. A. Bugała, G. Frydrychowicz-Jastrzębska, D. Janczak, J. Dach, P. Boniecki, *International Multidisciplinary Scientific Geoconference* (SGEM, 337-344, 2016)
5. A. Bugała, G. Frydrychowicz-Jastrzębska, D. Janczak, W. Czekała, J. Dach, M. Zaborowicz, *International Conference on Energy and Environment*, (ICEE, 721-727, 2015)
6. D. Burzyński, L. Kasprzyk, E3S Web of Conferences, **14**, 01041 (2017)
7. N. Faghih, J. Zarinchang, Eksploat. Niezawodn., **14**, 234-248 (2012)
8. D. Głuchy, D. Kurz, G. Trzmiel, Prz. Elektrotechniczny, **90/4**, 78-80 (2014)
9. D. Janczak, J. Dach, A. Bugała, P. Sołowiej, M. Neugebauer, *International Multidisciplinary Scientific Geoconference* (SGEM, 243-250, 2016)
10. A. Kaabeche, M. Belhamel, R. Ibtiouen, Solar Energy, **85**, 2407-2420 (2011)
11. M. Kolhe, IEEE Trans. Energy Convers. **24**, 511-519 (2009)

12. M. Maleki, F. Pourfayaz, Solar Energy, **115**, 471-483 (2015)
13. M. B. Shadmand, M. Pasupuleti, R. Balog, IEEE Energy Conversion Congress and Exposition, 3626-3632, (2011)
14. K. Skowronek, G. Trzmiel, Prz. Elektrotechniczny, **83/11**, 108-110, (2007)
15. J. Szymenderski; D. Typanska, *Computational Problems of Electrical Engineering* (IEEE Xplore, 2016)
16. K. Bednarek, R. Nawrowski, A. Tomczewski, Electromagnetic Fields in Electrical Engineering, Studies in Applied Electromagnetics and Mechanics, vol. **22**, IOS Press, Amsterdam, Berlin, Oxford, Tokyo, Washington, 363-368 (2002)
17. K. Bednarek, R. Nawrowski, A. Tomczewski, Prz. Elektrotechniczny, **84/1**, 62-64 (2008)
18. K. Budnik, W. Machczynski, J. Szymenderski, Archives of Electrical Engineering, **64**(2), 227 (2015)
19. K. Ciosk, Prz. Elektrotechniczny, **88/12B**, 47-49 (2012)
20. A. Dobrzycki, S. Mikulski, W. Opydo, *International Conference on High Voltage Engineering and Application* (ICHVE, 1-4, 2014)
21. G. Dombek, Z. Nadolny, Eksploat. Niezawodn., 18, 413-417 (2016)
22. L. Kasprzyk, A. Tomczewski, K. Bednarek, Prz. Elektrotechniczny, **87/12b**, 82-85 (2011)
23. M. Lebioda, J. Rymaszewski, E. Korzeniewska, *International Conference on Microtechnology and Thermal Problems in Electronics* (MicroTherm, 2014)
24. L. Nowak, J. Mikolajewicz, W. Pietrowski, Prz. Elektrotechniczny, **85/6**, 63-66 (2009)
25. W. Opydo, A. Dobrzycki, Electrical Engineering, **94/1**, 37-48 (2012)
26. R. Pawlak, E. Korzeniewska, Z. Stempien, *Selected Issues of Electrical Engineering and Electronics* (WZEE, 2016)
27. Ł. Putz, M. Kurzawa, *Computational Problems of Electrical Engineering* (IEEE Xplore, 2016)
28. K. Bednarek, Prz. Elektrotechniczny, **84/12**, 138-141 (2008)
29. K. Bednarek, J. Jajczyk, Prz. Elektrotechniczny, **85/12**, 29-32 (2009)
30. J. Jajczyk, L. Kasprzyk, *International Conference on Heat Transfer, Thermal Engineering and Environment* (WSEAS, 226-231, 2008)
31. J. Jajczyk, *Computational Problems of Electrical Engineering* (IEEE Xplore, 1-3, 2016)
32. L., Kasprzyk, Prz. Elektrotechniczny, **88/7B**, 131-133 (2012)
33. Ł. Knypiński, L. Nowak, C. Jędryczka, COMPEL, **34/3**, 882-892(2015)
34. Ł. Knypiński, L. Nowak, A. Demenko, COMPEL, **34/2**, 448-455 (2015)
35. A. Lewicki, A. Bugała, H. Piekarska-Boniecka, A. Lewicka, P. Boniecki, M. Cieślik, *International Conference on Chemical and Biological Sciences* (ICCBS 2016)
36. R. Nawrowski, K. Bednarek, A. Tomczewski, *International Conference on Parallel Computing in Electrical Engineering* (PARELEC, 218-222, 2000)
37. A. Bugała, G. Frydrychowicz-Jastrzębska, Z. Zbytek, J. Dach, D. Janczak, *International Conference on Frontiers of Sensors Technologies* (ICFST, 2016), *MATEC Web of Conferences*, **59** (2016)
38. G. Frydrychowicz-Jastrzębska, A. Bugała, Energies **8**/2, 1025-1041 (2015)
39. D. Głuchy, D. Kurz, G. Trzmiel, Prz. Elektrotechniczny, **89/6**, 281-283 (2013)
40. G.S. Maddala, *Ekonometria*, (PWN, Warszawa 2014)
41. M. Sobczyk, *Statystyka* (PWN, Warszawa, 2016)
42. A. Zeliaś, *Metody statystyczne*, (Polskie Wydawnictwo Ekonomiczne, Warszawa, 2000)
43. http://www.statsoft.pl [access 20.01.2017]

 DOI: 10.1051/e3sconf/20171901003

Determination of the number of Vertical Axis Wind Turbine blades based on power spectrum

Waldemar Fedak[1], *Stanisław* Anweiler[1,*], *Wojciech* Gancarski[1], and *Roman* Ulbrich[1]

[1]Opole University of Technology, Department of Mechanical Engineering, 5 Mikołajczyka st., 45-758 Opole, Poland

Abstract. Technology of wind exploitation has been applied widely all over the world and has already reached the level in which manufacturers want to maximize the yield with the minimum investment outlays. The main objective of this paper is the determination of the optimal number of blades in the Cup-Bladed Vertical Axis Wind Turbine. Optimizing the size of the Vertical Axis Wind Turbine allows the reduction of costs. The maximum power of the rotor is selected as the performance target. The optimum number of Vertical Axis Wind Turbine blades evaluation is based on analysis of a single blade simulation and its superposition for the whole rotor. The simulation of working blade was done in MatLab environment. Power spectrum graphs were prepared and compared throughout superposition of individual blades in the Vertical Axis Wind Turbine rotor. The major result of this research is the Vertical Axis Wind Turbine power characteristic. On the basis of the analysis of the power spectra, optimum number of the blades was specified for the analysed rotor. Power spectrum analysis of wind turbine enabled the specification of the optimal number of blades, and can be used regarding investment outlays and power output of the Vertical Axis Wind Turbine.

1 Introduction

European Union directives, as well as increasing environmental awareness, together with decreasing energy resources of the planet, are forcing us to look for non-conventional and renewable energy sources at the same time. As of today, the use of these resources for economic reasons is less profitable than obtaining energy from fossil fuels, so renewables must be subsidized or protected by the government [1]. However, this does not remove the search for more efficient solutions. Economic growth, energy demand, energy security and sustainable development require application of renewable energy sources, especially in urban areas [2].

Wind energy has become an important source of electricity production due to its abundance in nature and low environmental pollution. Advanced techniques have been developed to achieve as much energy as possible with the energy conversion equipment – wind turbines [3]. At present, there are two main fields of harvesting wind energy. High altitude and low altitude winds exploitation. Although high altitude winds are considered to be the most promising renewable energy source in the future [4], low altitude wind technologies are well settled and developed. There are two categories of modern low altitude wind turbines, namely horizontal axis wind turbines (HAWT) and vertical axis wind turbines (VAWT) , which are used mainly for electricity generation and pumping water. The main advantage of VAWT is a single moving part (the rotor) where no yaw mechanisms are required, thus simplifying the design configurations significantly [5].

VAWT can be divided according to the way of wind energy conversion to the rotor.
1) Active VAWT – the use of aerodynamic effect (lift) – Darrieus and H-rotors.
2) Reactive VAWT – the use of thrusting force (drag) – Savonius and differently called cup /bucket /drum /carousel /scoop bladed-rotors.

Recently a hybrid approach to VAWT is becoming more noticeable [6], where active rotors are combined with reactive ones. Smaller, reactive VAWT are mounted on the larger active VAWT's shaft. Reactive VAWT is used as starter for the active VAWT. Hybrid approach to electrical generation is also observed in developing hybrid renewable energy plants with solar and wind power [7]. Sustainable development requires also high economic performance analysis of the renewable investments (for instance with the use of discounted method [8]), so comparison between HAWT and VAWT economic performance is also available [9].

In general, the electrical generation industry is currently in a period of considerable flux caused by (i) obsoletion of central generation, (ii) a desire to decarbonise network generation and (iii) a need to diversify fuel sources to enhance energy security. The effect of these drivers will clearly effect change in network scale infrastructure but they may also offer an opportunity for distributed power generation. There is a growing body of opinion that policy should be developed

* Corresponding author: s.anweiler@po.opole.pl

to promote the widespread adoption of highly distributed, micro-scale generation in the built environment in general and in the domestic sector in particular. This form of distributed generation is widely seen as a beneficial development in terms of both energy security and decarbonisation of the delivery of electricity to the built environment. Governments are seeking to enshrine the adoption of micro generation technologies through the building regulations, setting a target for net zero carbon new build housing. Micro wind is one of the technologies represented in this sector that could develop into a market-mature technology applicable for widespread adoption. Previously the majority of micro wind turbines were installed for educational reasons (at schools or environment centres) and very few domestic systems existed, especially in urban areas. Installations were dominated by Horizontal Axis Wind Turbines in the 2.5–20 kW range. However more recently the small wind turbine industry has started to focus on developing the smaller (<1.5 kW) building integrated turbine market. In particular there has been a push to develop rooftop installations that are suitable for domestic properties in the urban environment. These technologies are generating positive interest from both public and private sectors, with their availability extending to Do It Yourself stores [10]. The type of turbine to be used in a particular project depends mostly on the local circumstances, such as housing density, altitude and the wind velocity, direction and their variability. In the area of high building density there is most demand for electricity but also it is extremely difficult to determine the wind direction. On the basis of authors observations it seems that the cup-bladed rotors are the least affected by the direction of wind vector. That is why the authors took a closer look to reactive VAWT with the use of thrusting force cup-bladed carousel. Although these types of VAWT are known for long time, there is not much research put on gathering high quality experimental data [11]. There are some trials but in a very simple manner [12]. Literature analysis reveals the gap in the research field of the reactive vertical axis wind turbines, especially for cup bladed rotors and their mathematical models. This is the motivation and background of the authors research.

2 Basic problem

The main problem in the design and maximization the performance of a wind turbine, beside the shape and size of the rotor and the blades, is to determine the number of the blades mounted on the rotor. Each cup-bladed VAWT has specific, optimal number of blades, depending on various factors, such as size and materials used. Number of blades can vary, the bigger amount of the blades, the more they will obstruct one another during operation – slipstream. This will reduce the performance of VAWT. For the analysis, a set of blades in the shape of a hollow semi-cylinder, completed with semi-circular domes have been used. The study involved a series of rotors fitted with 2 to 7 blades. Fig. 1 presents the blade and sample rotor consisting of 5 blades. VAWT with only 2 blades will not work at all, but it is taken into account as a reference.

Detailed literature review shows lack of experimental data according drag force research in the field of drag turbines, especially physical measurements of forces on the blades. All gathered data refer to numerical research for Savonius type rotors. The carousel type rotors are not represented in the literature. To start mechanical tests one must first make some theoretical assumptions and verify them. These assumptions are the main problem and the purpose of the research.

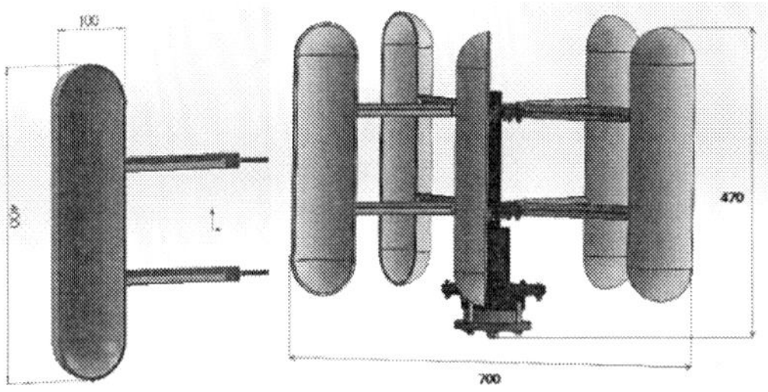

Fig. 1. CAD model of the cup-blade and the 5-bladed carousel rotor.

3 Methods

Cup-bladed carousel turbine is operating on the same principle as the Savonius turbine and is one of the simplest wind turbines. Aerodynamically, it is a drag-type device, consisting of certain number of airfoils blades scoops. Because of the curvature, the blades experience less drag when moving against the wind than when moving with the wind. The differential drag causes the turbine to spin. Because they are drag-type devices, these turbines extract much less of the wind's power than other similarly-sized lift-type turbines. Cup-bladed VAWT rotor rotates under the torque difference caused by different resistance (drag) that make the concave and the convex side of the blade. In order to make theoretical calculations certain rules and simplifications have been set. Due to the fact that the movement of air masses is not homogeneous and is usually turbulent, it was decided to perform the initial calculations for laminar flow. The simplified calculation, when considering the impact of the air stream on the blade does not take into account what is happening around the blades. Only impact of the air stream on the cup-blade. For the computing purposes the width of the air stream is comparable to the width of the blade. The blade movement speed is zero. Another assumption is that the energy of the wind stream is completely transformed into kinetic energy of the blade [13]. Schematic view of the hydrodynamic thrust on the obstacle from the concave face is shown in Fig. 2.

To calculate the thrust of the rotary turbine blade on a single concave side facing the thrust, as shown in Fig. 2, equation (1) was used, characterized by the hydrodynamic pressure [13]. Formerly used for Savonius rotor [14], next for straight blade [12].

$$P = A \cdot \rho \cdot v^3 \cdot (1 + \cos\delta) \qquad (1)$$

 DOI: 10.1051/e3sconf/20171901003

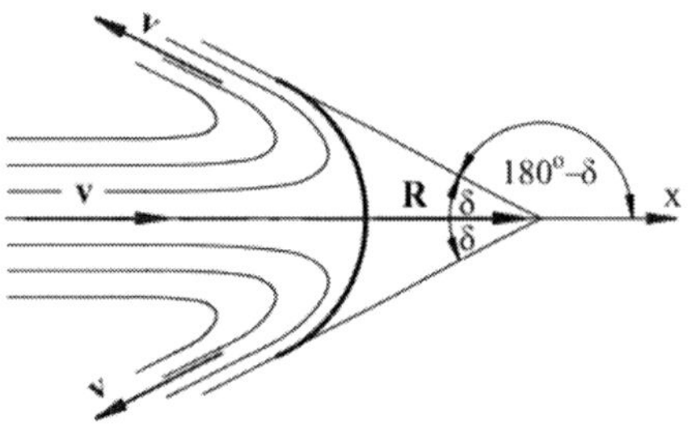

Fig. 2. Scheme of hydrodynamic thrust on the concave side of an obstacle.

The same equation is used in [15] to assess such values as rotor blade configuration, aspect ratio, overlap ratio, number of buckets to finalize the design of Savonius rotor. In contrast to the convex side of the blade facing the thrust, the correct is the relationship (2).

$$P = A \cdot \rho \cdot v^3 \cdot (1 - \cos\delta) \qquad (2)$$

In order to calculate the power spectrum computing environment MatLab was used. Using the above equations (1) and (2) simulation of power that produces a single blade and the whole rotor was made. For a system comprising a rotor consisting of a given number of blades a schematic calculation algorithm is presented in Fig. 3.

Fig. 3. Schematic representation of MatLab computing algorithm for Power Spectrum.

When determining the relationship between the size and shape of the blade and the force of thrust in order to simplify, it is assumed that the blade is fixed as in equation (1) and (2). In this case, the whole energy of moving air is absorbed by the blade. Energy formed on the blade as a whole is converted to stresses in the structure. Betz limit said that the energy of the movement of air masses can be converted to work only with 59.2% efficiency [18], which is true for horizontal axis devices and vertical airfoils [12]. However, in this case, this energy is entirely converted into a structure stress.

The determination of the previously defined values with the assumption that the blade is in motion should be done according to equations (3) when moving against the wind and (4) when moving with the wind.

$$P_{z=\rho A v^2 (v-u)(1-\cos\delta)} \qquad (3)$$

$$P_{n=\rho A v^2 (v+u)(1-\cos\delta)} \qquad (4)$$

Calculations are performed for rotors with number of i= 2 to 7 blades and for the wind speed of v= 3 to 14 m/s. A similar method was used for analysis of this issue in the modelling of the Savonius turbine in [14]. Later Muller et al. used it for flat plate shaped wind blade [12].

The most recent approach to VAWT related with the number of blades was made by [16] and further research to improve the drag wind turbine performance was done by [17].

For the purpose of the experiment, turbine power coefficient *kmax* was defined. It is the ratio of the maximum power of the turbine *Nmax* for the maximum power of a single blade *nmax* (5).

$$k^i_{\max} = \frac{N^i_{\max}}{n^i_{\max}} \qquad (5)$$

The power turbine model expressed by a *kmax* coefficient depends mainly on three factors – the number of blades in a turbine (i), wind speed (v) and aspect ratio (δ). For the study, a constant aspect ratio adopted for the specific type of blade at the δ = 0.9 while wind speed and the number of blades have been edited in order to select the optimum value of the number of blades in a specific turbine.

4 Results and discussion

With the help of MatLab environment and the collected data the power spectrum graphs for selected cases of turbines have been prepared (Table 1).

Table 1. Power spectrum graph and $k_{\max}$ coefficient for selected rotors depending on number of blades (v = 3.0 m/s).

Number of blades i	Power spectrum (for details see Fig. 4-7)	k_{max}
2	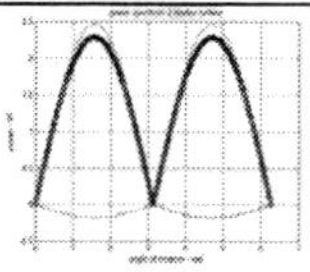	0.92
3	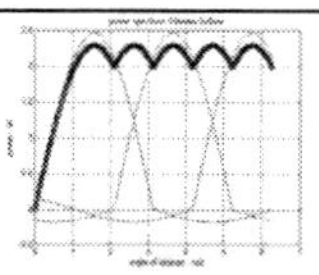	0.93
4	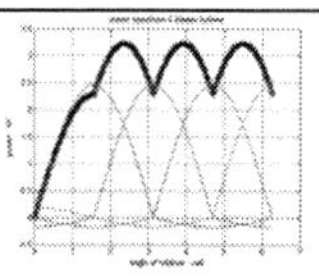	1.30
5	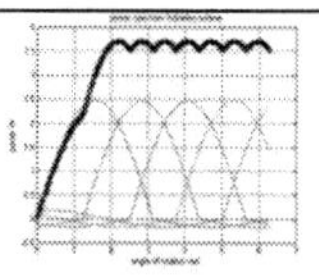	1.50
7	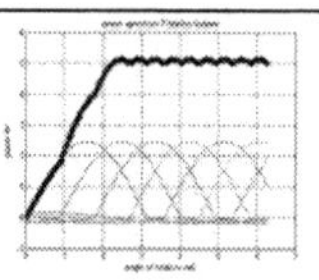	2.08

 DOI: 10.1051/e3sconf/20171901003

The kmax ratio for given VAWT have been calculated. Calculations were performed for the four variants of the turbine rotor, consisted of 2, 3, 4, 5, 7 blades. Summary of cases are presented in Table. 1. The bold curve of the graphs in Table 1 represents the total power generated by the rotor. Thin coloured curves represent the power generated by a single blade. These graphs are thumbnails, full scale graphs will be presented and discussed in detail further in the paper. Presented in Table 1, the *kmax* ratio shows how much more power is the rotor generating compared to a single blade. The 2-bladed and 3-bladed rotor generates less power than a single blade. The 4-bladed rotor generates slightly more power than a single blade, but power spectrum shows high pulsation. There is almost no pulsation for 5-bladed and 7-bladed rotor and their power is clearly higher than a single blade, but the problem appears with the thin lines in the graph. Power spectra for the single blade are getting closer to each other, which means that aerodynamic shadow and slipstream effects are getting more significant. Table 2 shows the value of *kmax* ratio for the 5-bladed turbine depending on the wind speed.

Table 2. Power spectrum and k_{max} coefficient for 5-bladed rotor, depending on wind velocity.

Wind velocity v (m/s)	Power Spectrum (for details see Fig. 4-7)	k_{max}
3	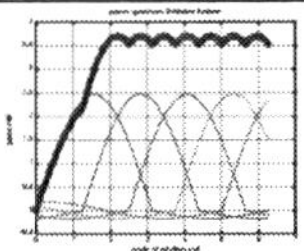	1.50
4	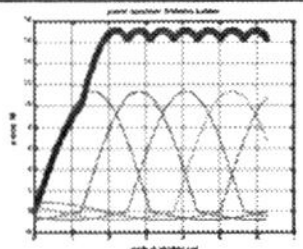	1.52
8	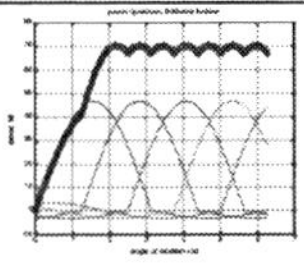	1.49
11	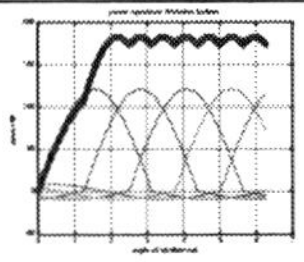	1.50
14	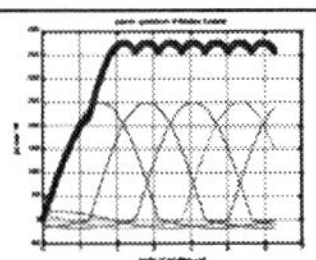	1.50

The bold curve in Table 2 represents the total power of the rotor during the operation. Thin curves represent the power of a single blade. Again these graphs are thumbnails, full scale graphs will be presented and discussed in detail further in the paper. Presented in Table 2, the *kmax* ratio along with rising wind speed shows stability, thus stable operation of the rotor. As a result of experiments and calculations, based on the power spectrum plots for selected turbines, coefficient *kmax* can vary and in the case of this research is in the range from 0.92 to 2.12. The results are shown in Table 3.

Table 3. The k_{max} coefficient depending on the number of blades (wind velocity v = 8.0 m/s).

Number of blades i	**3**	**4**	**5**	**7**
Single blade power n^i_{max} (W)	47	47	47	47
Rotor power N^i_{max} (W)	44	62	70	100
k^i_{max}	0.93	1.32	1.49	2.12

The following figures (Figs. 4-7) show the specific results of the calculations in the form of the power spectrum graphs with a detailed discussion.

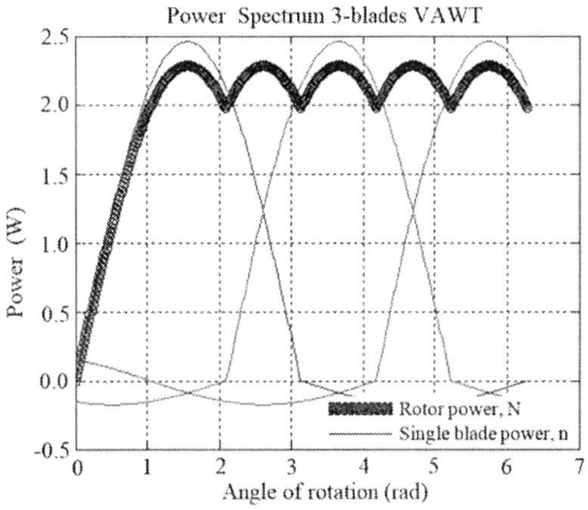

Fig. 4. The power spectrum graph for the 3-bladed VAWT.

The 3-bladed turbine, which power spectrum graph is shown in Fig. 4, produces less power than a single blade. The power coefficient *kmax* is less than 1 (see Table 1). This is due to the thrust equilibrium when the turbine loses power and tends to stop. Thanks to the power of inertia and the difference in the coefficient of drag, turbine has the ability to generate constant power. Although theoretic power of the 3-bladed rotor is smaller than a single blade, thus such turbine is inefficient.

The 4-bladed turbine with power spectrum graph shown in Fig. 5 produces slightly more power than a single blade. The power coefficient *kmax* = 1.3 (see Table 1). However during the rotation at certain points the power of the rotor is smaller than the power of a single blade, which is still undesirable. As in the previous case, during the rotation of the rotor the torque has a high ripple – power spectrum shows high pulsation.

 DOI: 10.1051/e3sconf/20171901003

The 5-bladed turbine has a power spectrum presented in Fig. 6. It produces 1.5 times more power than a single blade. The power coefficient *kmax* = 1.5 (see Table 1).The power spectrum is satisfactory with little pulsation – operation of such rotor should be stable. However in this case more attention should be put on the thin lines, representing power spectra for single blades. They are getting closer to each other. This is the effect of the aerodynamic shadow.

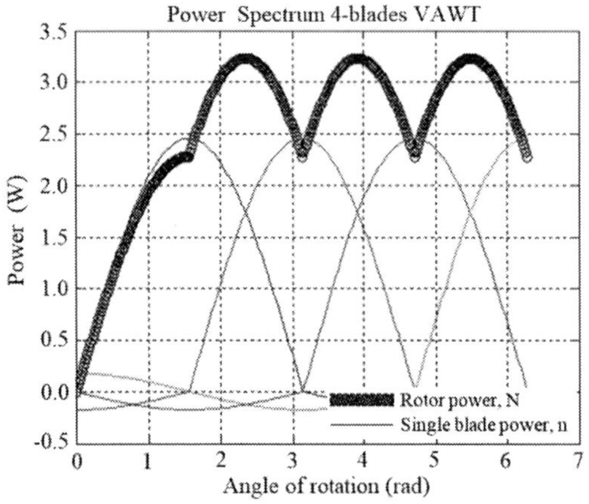

Fig. 5. The power spectrum graph for the 4-bladed VAWT.

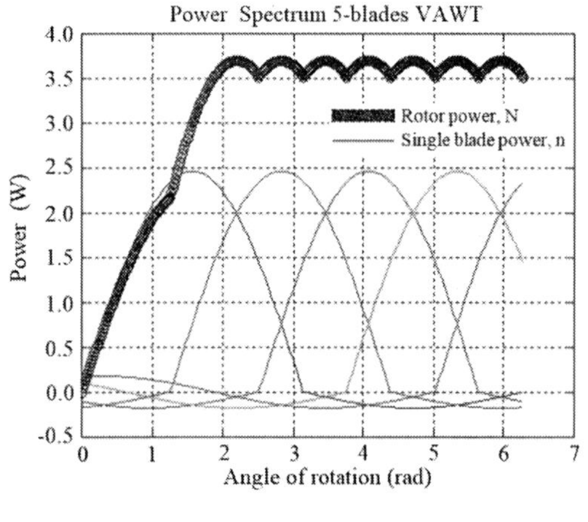

Fig. 6. The power spectrum graph for the 5-bladed VAWT.

Power spectrum graph of the 7-bladed turbine is shown in Fig. 7. This rotor has a coefficient *kmax* = 2.0. The power spectrum is substantially above the power of a single blade. There is almost no pulsation, which does not cause instabilities of torque and provides a stable rotational speed of the turbine. Unfortunately it is clear that along with the increasing number of blades increases aerodynamic shadow effect.

It is efficient to apply the turbine with 5 or 7 blades. It would still follow the plot of the power spectrum of the 9-bladed turbine, but as we saw that with the 7-bladed turbine the power graphs of individual blades overlap which generates slipstream stopping the rotation. This is true for relatively small turbines (up to 10 KW). There is little information on how bigger rotors will operate.

Considering the application of the Betz limit one should think about how this limitation affects the movement of the VAWT blades. Does the speed of the blade may be less than the minimum air speed behind the blade, which is about 1/3 of the wind speed? In the case of carousel wind turbines limitations resulting from the Betz limit are not applicable, because, as is evident from the observations: case 1) the unloaded, free rotating turbine – the blades move at the same speed as the wind and do not generate power; case 2) the fully stopped turbine – the blades do not move and all the energy goes for the generation of stresses in the structure. Between these two cases, there is a whole series of cases in which the turbine is moving at any speed less than the wind speed, and depending on the load, with less speed than resulting from Betz limit. Therefore, the wind speed behind the blades of carousel turbines does not matter because it's all about the thrust generated on the blade on both sides. Further studies are currently being conducted to help establish all these relations.

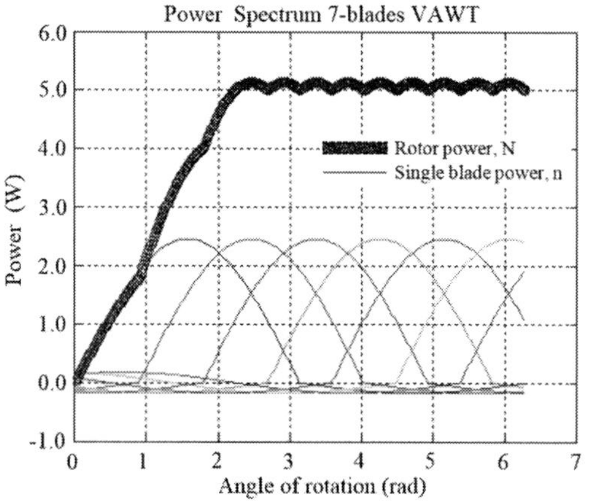

Fig. 7. The power spectrum graph for the 7-bladed VAWT.

5 Conclusions

The object of the study was to determine the optimum number of blades to be mounted on Vertical Axis Wind Turbine. Studies were made on the basis of thrust on the stationary blade. For this research authors resigned to examine complex phenomena occurring around the blades, i.e. aerodynamic shadow, turbulence around the blades and other aerodynamic forces. Assumed simplified (sinusoidal) change of thrust, and wind energy is completely transferred to the blade. This allowed to avoid Betz's limit [14]. These assumptions are valid for a steady state – stationary blade of the turbine set concave face perpendicular to the wind direction. Simplifications helped to determine if the parameter k_{max} can be assigned for the evaluation. As it has been shown in this study, turbine power depends on the number of blades. With the increasing number of blades mounted on the rotor, the torque also increases, and thus the rotation of turbine is smoother. Unfortunately along with the increasing number of blades, slipstream effect increases too. Selection of the number of blades to the turbine has a direct impact on the efficiency and economy. Too small number of blades will cause inappropriate efficiency of the turbine and therefore inadequate power. Too many blades cause, among others, the increase of the weight and the production costs.

By determining the k_{max} ratio there is possibility for a simple and rapid initial profitability characterization of bucket type wind turbines on specific weather conditions. The k_{max} ratio is specific to each single turbine and may be affected basically by wind speed variability and the number of blades. If the k_{max} ratio for the given turbine is constant over the entire range of wind velocity, that is to say that the turbine is designed properly and operates with a stable speed.

On the basis of the k_{max} ratio analysis one can select the number of blades for the turbine, which will be best fitted to the local nature of the wind. This is very important in the urban areas with high density of various buildings. Research shows that the coefficient k_{max} may take a value in the range of 0.9 to 10. For the k_{max} ratio of 1.0 and less, does not pay to construct the turbine at all. The value of 10 is practically unattainable, because the turbine would have a huge size, which is economically viable. For small turbines with power 1 – 10 kW the k_{max} coefficient can be in the range of 1.5 – 3.0. However, for k_{max} greater than 2.0 one should pay particular attention to whether the slipstream effect will not reduce the power of the VAWT. Carousel drag-type VAWT are still regarded as less useful due to their low efficiency. However, due to its advantages mainly low noise, as only can be used in built-up areas. Advances in materials technology and electronics preordained that this type turbines have become interesting again. They are useful also for powering small autonomous installations where it is uneconomical to install grid. Reactive VAWT, thanks to its small size and weight have little inertia, therefore are able to capture wind energy in its gusts. Since a large proportion of wind energy is contained in the gusts, the possibility to use them is a huge advantage over the HAWT.

Nomenclature

A	–	surface area of the blade, m^2
N	–	maximum power of the turbine, W
P	–	thrust force, kg/m^2
i	–	number of blades,
k	–	turbine power coefficient,
n	–	maximum power of a single blade, W
u	–	blade's linear velocity, m/s
v	–	wind speed, m/s
δ	–	the blade form factor,
ρ	–	air density, kg/m^3

References

1. T. Boczar, *Wykorzystanie Energii Wiatru* (WPAK, Gliwice, 2010)
2. W.T. Chong, A. Fazlizan, S.C. Poh, K.C. Pan, W.P. Hew, F.B. Hsiao, Appl. Energy **112**, 601-609 (2013)
3. H. Lu , P. Zeng, L. Lei, Y. Yang, Y. Xu, L. Qian , **88**, pp 535-544 (2014)
4. N. Duić, M. Ban, L. Perković, P. Silva, N. Kranjčević, Appl. Energy **101**, 151-160 (2013)
5. M. Islam, D.S.K. Ting, A. Fartaj, Renewable Sust. Energy Rev. **12**, 1087-1109 (2008)
6. S. Bhuyan, A. Biswas, Energy Convers. Manage. **87**, 859-867 (2014)
7. E. Kabalci, Energy Convers. Manage. **72**, 51–59 (2013)
8. M. Wzorek, M. Tańczuk, Waste Manage. Res. **33**, 8 704-714 (2015)
9. J.D. Bishop, G.A. Amaratunga, Energ Convers. Manage. **49**, 6, 1652-1661 (2008)
10. A.D. Peacock, D. Jenkins, M. Ahadzi, A. Berry, S. Turan, Energ. Buildings **40**, 7, 1324-1333 (2008)
11. Q. Li, T. Maeda, Y. Kamada, J. Murata, K. Furukawa, M. Yamamoto, Energy **90**, 784-795 (2015)
12. G. Muller, M.F. Jentsch, E. Stoddart, Renewable Energy **34**, 1407-1412 (2009)
13. K. Jeżowiecka-Kabsch, H. Szewczyk, *Mechanika Płynów* (OWPW, Wrocław, 2001)
14. J.-L. Menet, Renewable Energy **29**, 1843-1862 (2004)
15. I. Al-Bahadly, Energy Sustainable Develop. **13**, 3, 159-165 (2009)
16. F. Wenehenubun, A. Saputra, H. Sutanto, Energy Procedia **68**, 297-304 (2015)
17. A. R. El-Baz, K. Youssef, M. H. Mohamed, Renewable Energy **86**, 89-98 (2016)
18. O. Ozgener, Energ. Convers. Manage. **47**, 1326–1337 (2006)

DOI: 10.1051/e3sconf/20171901004

Analysis of wind energy generation possibilities with various rotor types at disadvantageous wind condition zones

Andrzej Bieniek[1,*]

[1]Opole University of Technology, Faculty of Mechanical Engineering, 45-271 Opole, Poland

Abstract. The paper describe possibilities of energy generation using various rotor types but especially with multi-blade wind engine operates in the areas with unfavourable wind condition. The paper presents also wind energy conversion estimation results presented based on proposed solution of multi-blade wind turbine of outer diameter of 4 m. Based on the wind distribution histogram from the disadvantage wind condition zones (city of Basel) and taking into account design and estimated operating indexes of the considered wind engine rotor an annual energy generation was estimated. Also theoretical energy generation using various types of wind turbines operates at disadvantage wind conditions zones were estimated and compared. The conducted analysis shows that introduction of multi-blade wind rotor instead of the most popular 3- blades or vertical axis rotors results of about 5% better energy generation. Simultaneously there are energy production also at very disadvantages wind condition at wind speed lower then 4 $m \cdot s^{-1}$. Based on considered construction of multi-blade wind engine the rise of rotor mounting height from 10 to 30 m results with more then 300 % better results in terms of electric energy generation.

1 Introduction

World demand for electricity is growing year by year. As a result, various solutions are being sought to meet these growing needs. According to some sources [1], the increase in electricity demand between 2000 and 2008 increased by 22.2% and continues to grow. Much of the demand for electricity is covered by power plants that use fossil fuels (coal, gas, oil) (Fig. 1). Total share of fossil fuels in world energy production in 2011 amounted to almost 68%. The share of renewable energy in this case amounts to about 4.8%, but as shown by statistics, it is growing rapidly in recent years [2].

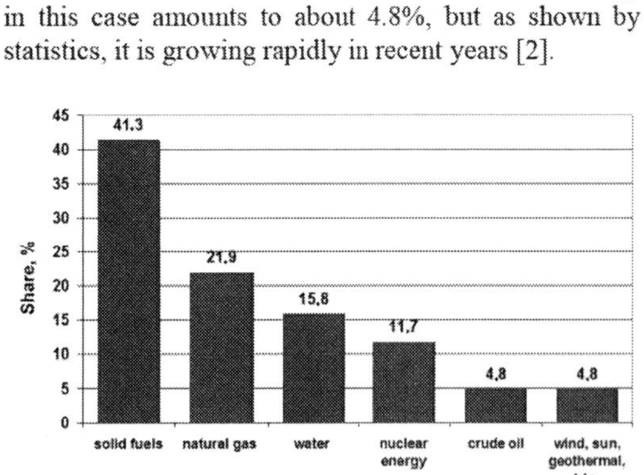

Fig. 1. Share of fuels in global energy production in 2011 [2].

Particularly there is high dynamics of energy growth produced by the use of wind energy [3]. In Europe in the recent years there is a rapid development of wind energy observed (Fig. 2). There is growing importance to produce energy mainly for own use with the use of small wind power plants. As example in one of the EU-countries in the strategic document of the National Action Plan for the Promotion of Renewable Energy Sources - small wind energy (ie. for power below 100 kW) was separated for the first time as a separate category and at the same time it was stated that in Poland by 2020, 550 MW power in this sector. This represents approximately one hundred thousand installations.

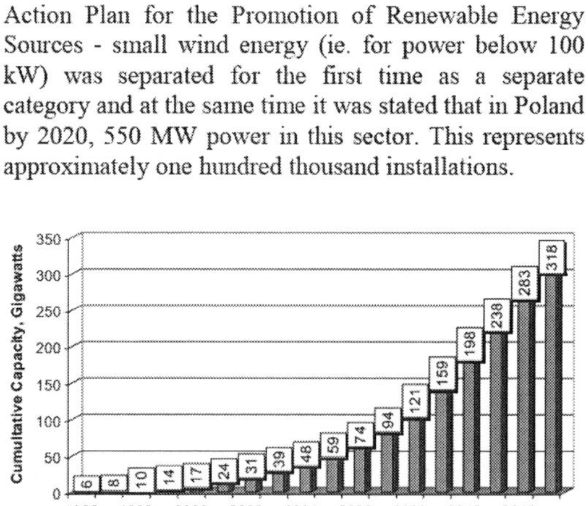

Fig. 2. Wind energy production in years from 1996 to 2013 [3].

As expected, many of these installations will be small home installations with a power of the order of 3-5 kW installed on buildings or masts up to 30 m high. Statistical analysis show that a small wind turbines with a total power capacity of 7 MW was installed in Poland by the end of 2010, of which over 5.2 MW were small wind turbines producing energy for their own needs. As shown it is only about 1.5% of power capacity planed to started up to 2020 year [4]. The most sold wind turbines

* Corresponding author: a.bieniek@po.opole.pl

 DOI: 10.1051/e3sconf/20171901004

in Poland are the smallest pico- and micro-turbines from the power range up to 5 kW. Most of these turbines operate in autonomous systems with about 53% with batteries or 27% of domestic heating water. The most economical version of "surplus" storage is the connection to the power grid, howewer few investors decides to do so because of technical difficulties, and possibly due to the most unfavorable legislation in this regard.

2 Europe's wind energy potential

Determination of wind energy resources requires a multi-year wind speed measurement. As the statistical data of many world agencies show, it is difficult to forecast wind speeds in the so at short and also on the long term future due to their high variability. On the basis of years of research and wind measurements there are wind maps developed by the European Environment Agency (EEA). It can be stated that Europe is a continent in which only favorable wind conditions exist in a relatively restricted area [5]. The relatively favorable windiness conditions can be considered in areas with an average wind speed of more than 6 $m \cdot s^{-1}$ (Fig. 3) measured at 80 m above ground level or 120 m above water level.

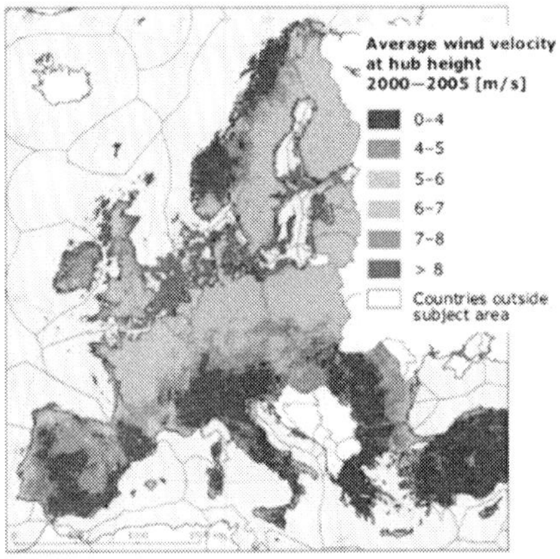

Fig. 3. Wind field data after correction (by hub height of 80 m onshore/ 120m offshore) [5].

Similar is also the distribution of the zones in which wind turbines placed at a height of 80/120 m above the surface can reach its maximum energy generation (Fig. 4). By analyzing the EEA, it can be seen that the most advantageous area is the northern part of Europe along the coast located, where the annual number of hours for maximum capacity reach more than 2000 hours. There are also significantly less favorable zones such as the areas of the northern part of the Scandinavian peninsula and southern Europe, where the full potential of a wind power plant is achievable for significantly less than 1000 hours a year. On the other hand, the assessment of the use of wind energy in a particular location, especially for small wind turbines installed at relatively low altitudes, should be carried out on a case-by-case basis. It is difficult to assess and predict the wind speed profile at a given area due to the lack of detailed data, especially long-term archival wind speed data.

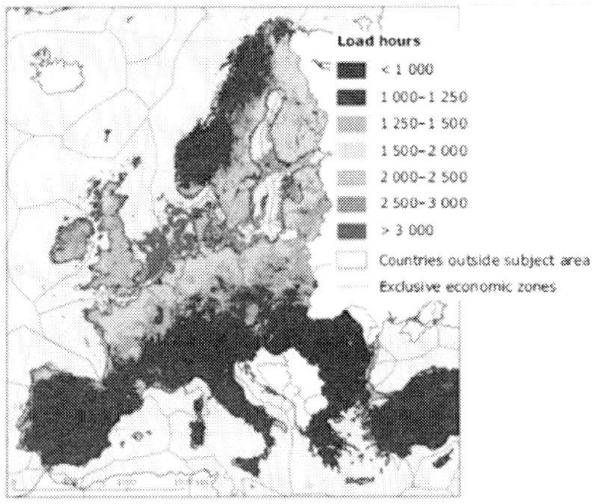

Fig. 4. Distribution of full load hours in Europe (by hub height of 80 m onshore/ 120m offshore) [5].

In this case, the analysis of the pre-selection from several locations can use the data available to the nearest location for which the data is available. Then is possible using empirical relations to estimate the energy production capacity of a particular location.

3 Short- and long-term windiness characteristic

An important issue for cost-effectiveness of the investment wind turbine is a presence of the favorable windiness conditions. The average value of wind speed that determines the electric energy generation. Significant is both short-term and long-term changes in wind speed. The profile of these changes is equally important. In the case of short-term changes (turbulence and gusts), the limits (maximum and minimum) of wind velocity may vary widely. In the perspective of several days, there are also significant differences in wind speed mainly observed.

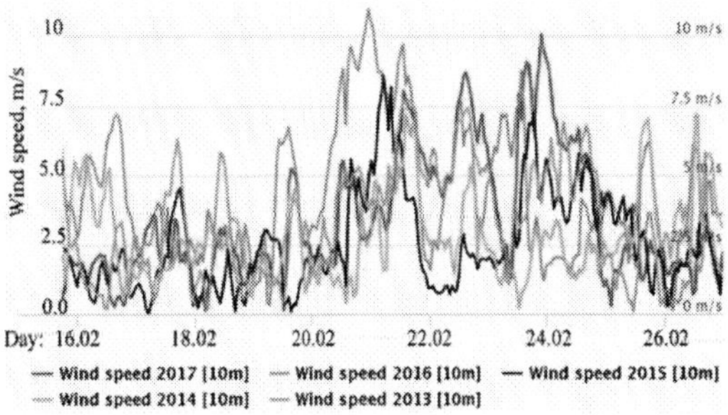

Fig. 5. Wind speed fluctuations in from 16 to 26 February over the past 5 years in Basel [6].

Especially in areas with less favorable wind properties, there may be longer periods when the wind speed falls below the electricity energy production limit (Fig. 5). These case can be remain even several days or longer when the wind speed falls below 4 m·s^{-1}. For most of the wind turbine designs, this is a critical value below which, it is not possible to start a wind turbine rotor and energy generation.

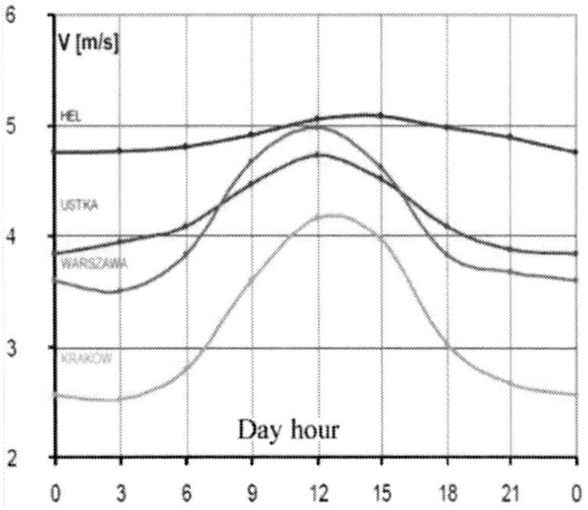

Fig. 6. Variability of wind speed in daily terms [7].

The direct use of wind energy by an individual user should also be take into account the daily wind speed distribution. In this case mainly the highest wind speeds are recorded in the afternoon hours (Fig. 6).

4 Efficiency of wind energy conversion

A very important feature in assessing the use of wind power is the characteristic of the efficiency of the wind energy conversion achieved by individual wind turbine rotor designs. The greatest possible theoretical efficiency of wind energy conversion by a wind turbine is described by the Betz curve, whose maximum is reached to 0.59. In turn [9,10], the total energy of the air stream (wind) can be saved by the formula:

$$P_{wind} = 0.5\rho A V^3 \tag{1}$$

were:

ρ – air density, (1.16 kg· m^{-3})

A – stream thrust area, m^2

V – wind speed, m·s^{-1}

The wind energy conversion efficiency depends to a large extent on turbine rotor design, differences in the characteristics of this important feature are significant. Fig. 7 shows efficiency of wind energy conversion curves ξ and the torque curves ψ (at the bottom of the figure) for the various types of rotors. Rotor type II, IV and V achieve the highest efficiency of 0.35 – 0.37 (Fig. 7) [8]. Achieving the maximum efficiency however, requires the rotor to operate optimally at the speed rating Z of turbine, that determines the ratio of the peripheral speed of the outer rotor to the wind speed:

$$Z = \frac{\omega R}{V} \tag{2}$$

were:

ω - rotor angular speed, rad·s^{-1}

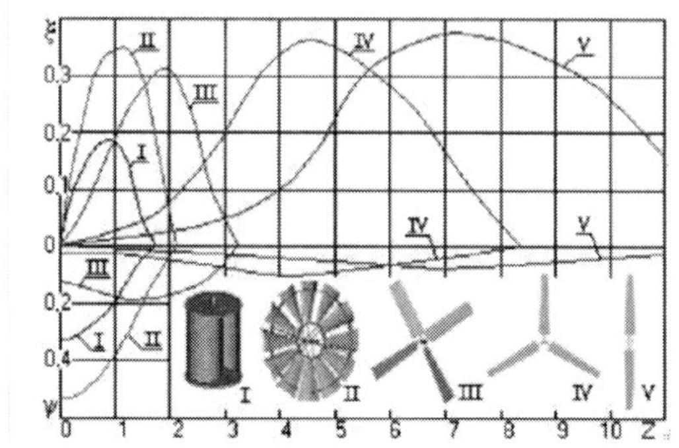

Fig. 7. Key parameters of various types of wind rotors depending of speed rating Z [8].

Knowing the course of the wind energy conversion curve, it is possible to record the useful power generated in a wind turbine using the following formula:

$$P_{turb} = 0.5\rho V^3 \pi R^2 \xi \tag{3}$$

were:

R – rotor radius, m

ξ – wind energy conversion efficiency, -

The efficiency of wind energy conversion is dependent on many factors (including flow patterns, rotor type and structure), and its estimation is possible also in the case of wind speed measurement before and after the rotor:

$$\xi = \frac{1}{2}\left(1-\left(\frac{V_2}{V_1}\right)^2\right)\left(1+\frac{V_2}{V_1}\right) \tag{4}$$

were:

V_1 – upstream wind speed, m·s^{-1}

V_2 – downstream wind speed, m·s^{-1}

The assessment of the various design of rotors should be also pay attention to the torque curve ψ, from which the course depends, among others the minimum wind speed at which rotor rotation and power generation are possible. Definitely the most advantageous course of the discussed curve is related to type II impeller, i.e. multi-blade rotor. At present, in the case of low power wind turbines, three-blade rotors (type IV) are most commonly used. They are characterized by a favorable maximum wind energy conversion value but require a minimum wind speed of about 4 m·s^{-1} for their start. Experimental studies carried out by the author for the model of multiblade rotor with 24 blades rotor of of 0.2 m outer diameter showed that the starting velocity for such a construction is approx. 1.5 m·s^{-1}

5 Eco-energy generation potential with small multi-blade wind turbines

As shown by statistics compiled by some meteorological measuring stations in the area involved unfavorable wind wind speeds exceeding 4 m·s^{-1} has a small share (Fig. 8). At the same time, the wind speed is characterized by high day variability (Fig. 6), which may additionally cause downtime in the power plant due to too low wind speed. The measured mean wind speed variability for the city of Kraków lying in the disadvantageous windy zone is in the range of 2.5 - 4.2 m·s^{-1}, whereas the daily wind share of speeds higher than 4 m·s^{-1} does not exceed 20% [7]. However, in the coastal areas in the very favorable windy zones, the daily share of winds of more than 4 m·s^{-1} can reach 100% (Baltic coast). As can be seen for many frequently used wind turbine constructions, wind speeds occurring in areas with poor wind conditions are insufficient for their operation.

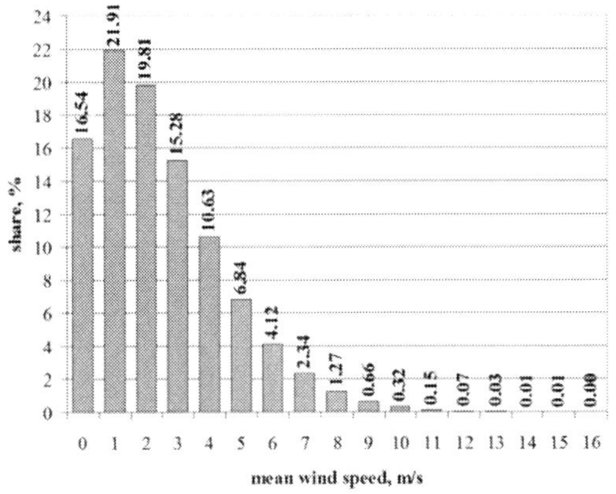

Fig. 8. Wind speed histogram for the city of Basel from 2010 to 2015 [6].

Taking into account the distribution of average wind speeds in zones with unfavorable winds conditions, the possibility of producing electricity for different types of rotors was compared. For the comparison was selected 3 types of rotor construction: a very commonly used in areas with favorable wind parameters rotor type IV, rotor type II (multilayer windmill / wind engine) and vertical rotor type I [8]. Rotors with vertical rotation axis usually have a significantly lower maximum wind energy conversion efficiency (ξ = 0.18). The advantage of type I and II rotors is they start and operation possibility at a relatively low wind speed of about 1.5 m·s^{-1}. Considering the higher efficiency of the type II rotor vs. rotor type I and the low startup wind speed (about 1.5 m·s^{-1}), further analysis was based on a small wind turbine equipped with a multi-blade rotor. The advantage of this solution is that nominal power values are also high at wind speeds of more than 10 m·s^{-1}, similarly for triple blades rotor of type IV. Further analysis was carried out for a multi-blade turbine consisting of 24 blades forming a rotor with an outer diameter of 4 m This rotor constructions was modeled for the purpose of this analysis (Fig. 9).

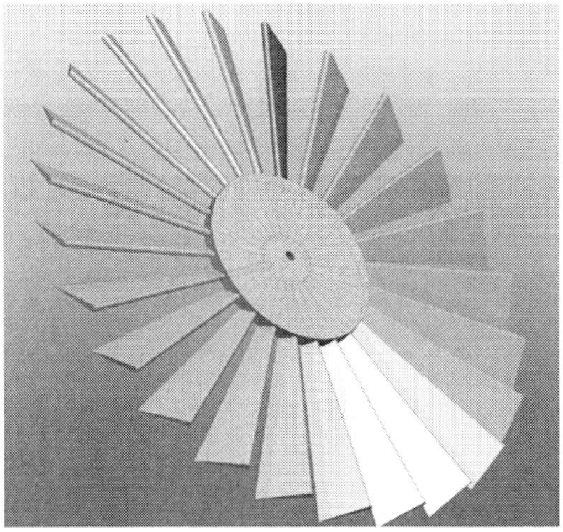

Fig. 9. 3D model of 24 blade wind engine rotor.

Based on the theoretical data [4,9,11], an estimation of the energy efficiency characteristic of the rotor model was performed (Fig. 10).

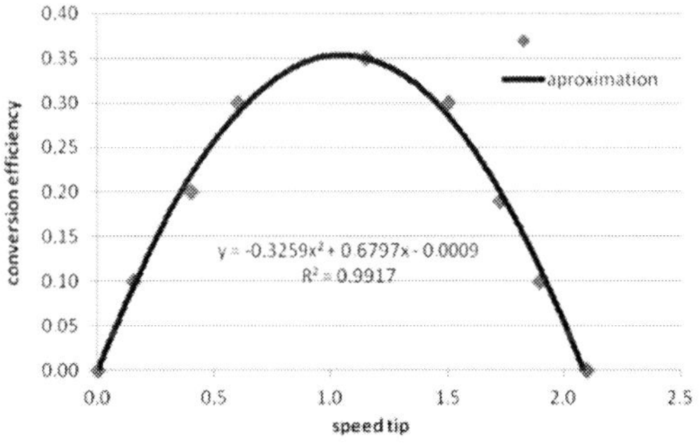

Fig. 10. Theoretical multi-blade rotor energy conversion efficiency.

Theoretical maximum efficiency of a model rotor is used (ξ = 0.35) for Z = 1.04. In a large range of speed rating Z from 0.9 to 1.2, the efficiency of the turbine is almost constant (the deviation from the maximum value is not more than 2% relative to the maximum value). This facilitates obtaining the optimum turbine performance over a wide range of rotational speed of the turbine.

On this basis, the maximum power of the turbine equipped with the rotor of the above design was estimated (Fig. 11).

In order to achieve the maximum efficiency of wind energy conversion, the variables for speed rating Z_tip1=0.9 and Z_tip2=1.2 (as values for almost constant turbine efficiency) were defined as the minimum and maximum rotational speed of the rotor. For that range of speed rating (2) the wind engine with multi-blade rotor achieves maximum conversion efficiency (Fig. 12).

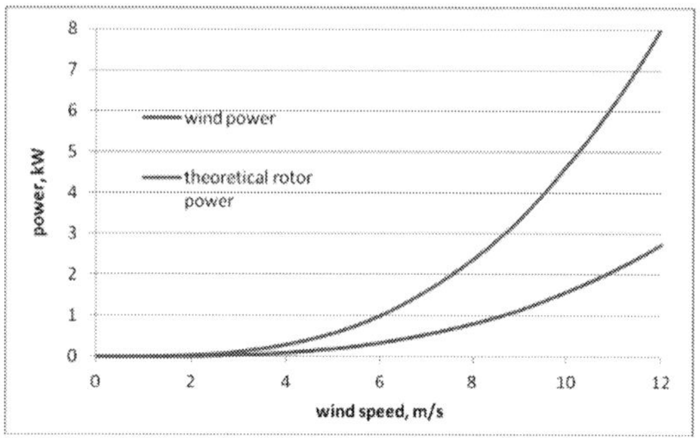

Fig. 11. Theoretical power of wind and modelled multi-blade rotor.

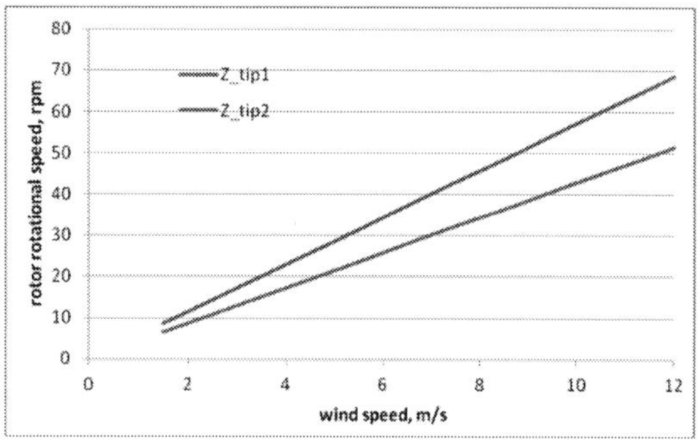

Fig. 12. Optimal rotor rotational speed vs. wind speed.

Based of the wind distribution histogram for the city of Basel (Fig. 8) and considering the construction of the modeled rotor, and in particular its estimated parameters (Fig. 10-12), the annual production was calculated with the details of its components for each wind speed, expressed in absolute terms (kWh). The value of the electricity produced was compiled with results based on the theoretical properties of other types of rotors [8] - Fig. 13.

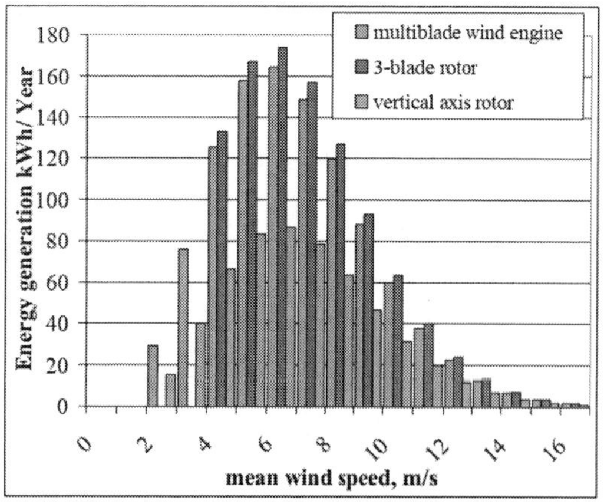

Fig. 13. Components of electricity production with various rotor types for the city of Basel located at a height of 10m.

The analysis of Fig. 8 and 13 shows that, despite the fact that the share of wind speed from 1 to 2 $m \cdot s^{-1}$ is about 40%, the greatest amount of energy is produced at a wind speed of 5-6 $m \cdot s^{-1}$. Such applications should be taken into account in optimizing the operation of the turbine for particular wind speed ranges. Based on the distribution of wind velocity and theoretical data on the efficiency of wind energy conversion to mechanical energy, taking into account the efficiency of current generators, the possibility of producing electricity was estimated using various types of wind turbines: multi-blade, 3-blade and vertical rotor.

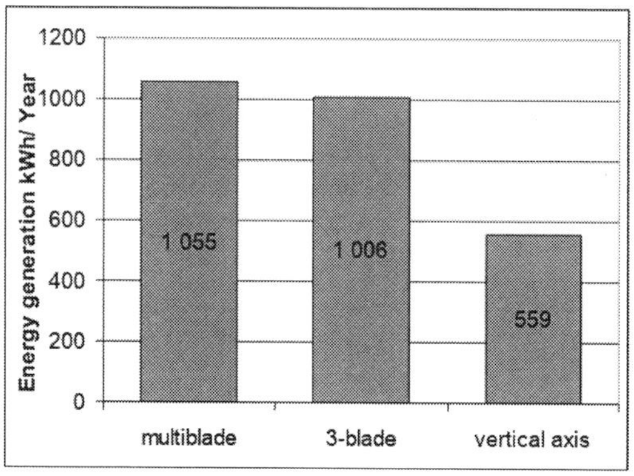

Fig. 14. Comparison of annual energy production by various rotor types.

The results are shown in Fig. 14. The use of a multi-blade rotor for less favorable wind conditions gives about 5% better results in terms of energy generation and is far more effective than a vertical axis rotor.

Considering the legal status that allow use of wind turbines mounted at a height of up to 30 m with less number of permits required by such installations, an analysis of the impact of height turbine placement on the estimated value of annual energy production was carried out. Using formula [9, 10, 12]:

$$V_h = V_0 \left(\frac{h}{h_0} \right)^{\alpha} \quad (5)$$

were:

V_h- wind speed at height h, $m \cdot s^{-1}$
V_o- wind speed at measured height, $m \cdot s^{-1}$
h - reference height, m
h_o- measurement height, m
α - terrain factor,

the wind velocity was estimated. Based of the average wind speed for Basel of V= 3.337 $m \cdot s^{-1}$ (using Weibull formula) at the reference level of 10 m and also accepting terrain factor for the: undeveloped area α = 0.143, built-up area α = 0.27 and the urban α = 0.35 the average wind speed for different elevations above the ground are estimated (Fig. 15) [12].

Based on the wind speed histogram for the city of Basel (Fig. 8), an estimation of wind speed was made at altitudes of 5 - 30 m and the annual total energy generation was compared with the modeled multi-blade wind turbine (Fig. 16).

DOI: 10.1051/e3sconf/20171901004

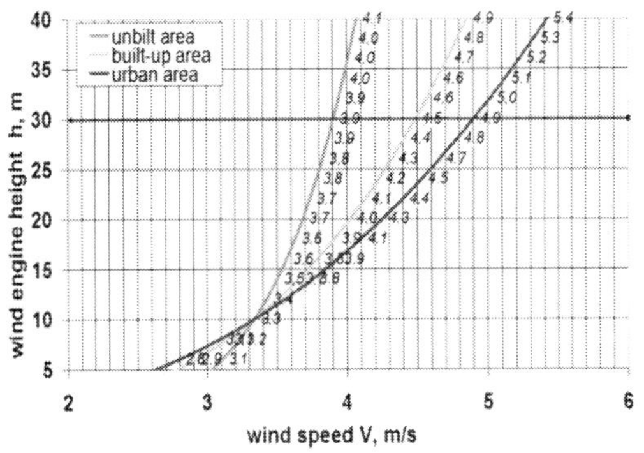

Fig. 15. Wind speed at different areas types by various rotor mounting height.

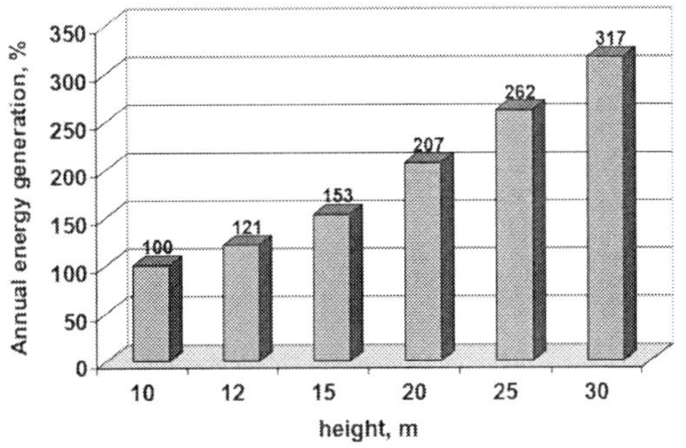

Fig. 16. Comparison of annual energy production by various height of rotor.

As the height of the rotor is increased, the amount of generated energy is increasing rapidly due to the more favorable wind parameters (speed). In this case rotor height increase from 10 to 30 m causes increase the energy generation of more then 300%.

6 Conclusions

Europe is a continent with relatively more areas with disadvantages windinies condition. At this zones share of wind speed higher then 4 m·s^{-1} does not exceed about 20 %. As shown by preliminary analyzes carried out in this studies in areas with less favorable wind conditions, it is appropriate to consider low power wind turbines with multi-blade rotor due to the startup with advantageous value of wind speed and high torque at low wind speeds. The large share of wind speed (e.g. in Basel province) below the starting speed of many popular three-blades rotors makes these turbines less profitable under such wind conditions. In turn, due to the low rotational speed of the multi-blade turbine rotor, the use of innovative low speed rotary power generators based on neodymium magnets technology should be considered. Introduction in these areas multi-blade wind engines are more profitable then other types of rotors. That results with about 5 % higher energy generation then by popular 3-blades turbines and almost 90% more energy generation then vertical axes type. Simultaneously multi-blade rotors ensure more stable energy generation – even at small wind speed lower then 4 m·s^{-1}. At the same time, besides choosing the optimum location for the turbine assembly, it is of utmost importance for the energy production effect the height of rotor location. Conducted in this paper analysis shows possibilities to increase energy generation for example of more then 300 % at raising of rotor mounting height from 10 to 30 m.

References

1. IEA, *Energy Balances of Non-OECD Countries*, (Eenergiläget and Siffror, 2011)
2. IEA, *Keyworld energy statistics* (IEA, 2013)
3. GWEC, *Global Wind Report Annual Market Update* (2014)
4. IEO, *Rynek małych elektrowni wiatrowych w Polsce* (EC BREC, 2010)
5. EEA technical report 6/2009, *Europea's onshore and offshore wind energy potential* (EEA, Copenhagen, 2009)
6. www.meteoblue.com (11.09.2017)
7. M. Brzezioski, J. Choszowski, J. Homoncik, K. Pierzga, K. Ząber, Z. Ząber, *Małe przydomowe elektrownie wiatrowe* (ZZ Dr Ząber, 2012)
8. K. Nalepa, W. Miąskowski, P. Pietkiewicz, J. Piechocki, P. Bogacz, *Poradnik małej energetyki wiatrowej* (WFOŚiGW, Olsztyn, 2011)
9. M. Sathyajith, G. S. Philip, *Advances in wind energy conversion technology* (Springer, 2011)
10. Stiebler M., *Wind energy systems for electric power generation* (Springer, 2008)
11. P. L. Fraenkel, *Water lifting devices*, Wi. Pow. FAO **43**, (1986)
12. E. Głodek, *Możliwości wykorzystania energii wiatru do wytwarzania energii elektrycznej w województwie opolskim*, Wyd. Inst. Śląsk, Opole (2008)

 DOI: 10.1051/e3sconf/20171901005

Investigation of irregularity of flow distribution across the tube bundle in shell-and-tube heat exchanger with segmental baffles using PIV method

Grzegorz Ligus[1,*], and *Szymon* Kołodziej[2]

[1]Opole University of Technology, Department of Environmental Engineering, 5 Mikolajczyka St., Opole, Poland
[2]Opole University of Technology, Department of Vehicles, 5 Mikolajczyka St., Opole, Poland

Abstract. This paper presents the results of an experimental study into flow through a tube bundle applied in shell-and-tube heat exchanger with segmental baffles built on an industrial scale. The reported research applied Digital Particle Image Velocimetry (PIV) technique. The study offered the determination of the velocity vector fields, which were subsequently used for identification and measurement of irregularity of flow distribution across the tube bundle. Differences in the intensity of local flow velocities in shell side were identified by velocity profiles analysis. Furthermore correlation between the map of local velocity and its location in particular tube rows was described. As a result, the three main regions of flow velocity variation throughout cross-section of the shell side were identified.

1 Introduction

Among the equipment applied for heat exchange, shell-and-tube heat exchangers hold an established position in the industry. They are successfully applied in the processes involving heating, cooling, heat recovery, boiling, distillation and condensation. The wide range of the application of this type of exchangers comes as a consequence of several important advantages of such equipment: high value of the heat transfer coefficient in relation to the space that is occupied by them accompanied by their low weight, as well as established manufacturing and exploitation technology combined with considerable flexibility in terms of the operating parameters of this type of apparatus [1]. The shell side region is occupied by various type of baffles, which are applied with the objective of increasing the efficiency of the heat transfer in the shell-and-tube heat exchangers. Such baffles are based on a number of design improvements, which play the role of increasing the turbulence of the medium that flows around the tube bundle. As a consequence, heat exchange coefficient increases as well [2,3]. The basic types of baffles including segmental, grid, disk, helicoidal and others differ in terms of the hydrodynamic characteristics, practical application and the cost of production. Among these types of baffles, the considerable applicability coupled with the high increase of the heat exchange coefficient can be achieved due to the application of segmental baffles [1]. The only problems associated with their application are associated with the considerable pressure drop resulting from the disturbance generated by the flow through the shell side region [4,5]. This directly affects the increase of the exploitation cost of the heat exchangers. In addition, the existence of multiple rows of tubes and baffles in the shell side leads to a further additional interference in the velocity of the liquid in the shell side [6,7]. The differences in the intensity of local flow velocities consequently leads to the occurrence of various heat transfer coefficients for particular tube rows. In the consideration of the above, it is important to relate the map of local velocity in the shell side to the particular tube rows. The use of the Digital Particle Image Velocimetry (PIV) method forms one of the measurement methods with a potential application in the identification and measurement of the flow velocity in the shell side. PIV applies non-invasive, optical measurement techniques useful in the experimental research of aero- and hydrodynamic parameters [8]. An adequately implemented measurement conducted by use of this method is considered to be able to provide highly measurable results, even in applications with a complex geometry [9,10,11,12]. This statement definitely finds an application in the case of shell side of a heat exchanger discussed in this paper.

2 Measurement setup and methodology

The use of Digital Particle Image Velocimetry (PIV) involved the necessity to develop and build a heat exchanger with a maximum possible transparency of the shell and tube bundle [8,13]. This task was achieved by preparing a model of a heat exchanger produced from polymethyl methacrylate (PMMA). The tested model of the heat exchanger was built from a shell with an internal diameter of 240 mm and a length of 2000 mm and

* Corresponding author: g.ligus@po.opole.pl

comprised a tube bundle with the internal diameter of 20 mm. The testing applied a rectangular tube layout with the pitch ratio of 0.625. Ten segmental baffles were installed along the tube bundle, whose baffle cut was equal to 25%. To eliminate the gas phase from the system, the top section of the uppermost baffle comprised an air vent. The above model of a heat exchanger forms the principal part of the experimental set-up (Fig. 1). The arrangement of the tube bundle along with the markings used in the investigation is shown in Fig. 2.

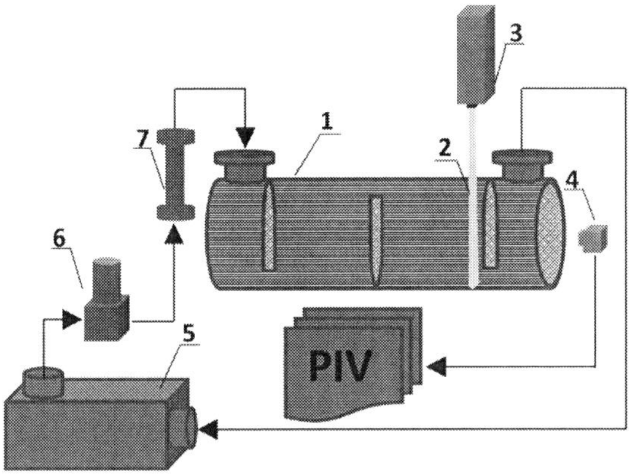

Fig. 1. Experimental set-up with cross section of the heat exchanger: 1. Model of heat exchanger, 2. Laser sheet, 3. Laser, 4. Camera CCD, 5. Water tank, 6. Pump, 7. Flowmeter.

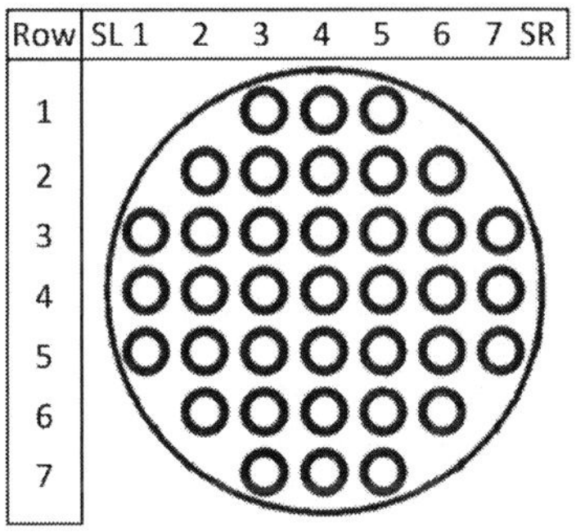

Fig. 2. Arrangement of the tube bundle: 1, 2, 3, 4, 5, 6, 7 – numbers of the tubes in a row; SL – left side of the shell, SR – right side of the shell.

Water stored in a tank with the capacity of 200 dm^3 played the role of the circulating medium. The circulation of the medium through the heat exchanger was possible due to the use of a multistage water pump, which was controlled by a signal transmitted from the electromagnetic flowmeter. The applied method also imposed the need to use seeding particles in the circulating medium. The presented study applied 3 g PSP polyamide with the particle diameter of 20 μm and a density of 1.03g/cm^3. These particles are characterized by a considerable value of the refractive index, as it is equal to 1.5. Due to the above characteristics, in many works [13,14,15], PSP particles are recommended for applications in systems including water phase flow in particular for the case of the complex stream trajectory. In order to obtain a constant water homogenization with seeding particles, an agitator was installed in the tank. The studies into hydrodynamic parameters of the heat exchanger model were conducted for three flow rates of the medium equal to: 5.0 m^3/h, 7.5 m^3/h and 10 m^3/h. A CCD EO-4M Dantec Dynamics FlowSence camera was located in the optical track of the set-up, which provided the registration of the shell side region of the heat exchanger in the transverse direction across the tube bundle before the final baffle. The imaging space occupied the entire cross-sectional field of the heat exchanger. The registration was performed at a resolution of 2048 x 2048 pixels in the double frame mode with a frequency of 10 Hz. The selection of the time between pulses was made on the basis of initial tests and it was set at the level of 500 μs for all flow conditions. Each of the measurement series comprised 250 double frame images. The area that was subjected for observation purposes was illuminated by a laser light (Nd:YAG Dantec Dynamics DualPower TR) by application of the laser sheet technique. The laser generated a coherent light beam with a length of 532nm and pulse energy of 1200mJ. During the research, the light sheet crossed the shell tube at distance of 10 mm before the final baffle. The calculations with regard to the velocity vector fields performed by application of the PIV method were conducted in the rectangular regions of interest, each of which involved the space between the adjacent tube rows that was included to the greatest possible extent. The velocity profiles were determined in the symmetry axes of the measurement regions of interest. The analysis applying PIV was conducted for each of the 250 pairs of the registered double frame images. Fig. 3 contains an overview of the PIV methodology applied during the calculations and analysis. These calculations were performed by application of the Dantec Dynamics Dynamic Studio 4.10. The determination of the velocity vector fields applied Adaptive Correlation algorithm, and it involved the measurement of the displacement between two patterns of pixel sets in selected image section, which were found to be representative for the flow images. In turn, they were represented by two-dimensional grey-level distributions. On the basis of the correlations of two subsequent flow images with the specific pixel patterns and by application of Fast Fourier Transforms, displacement vectors of images with a given grey-level distribution were calculated. This data resulted in development of a vector velocity field. The Fast Fourier Transform forms a calculation algorithm of the Discreet Fourier Transform (DFT) expressed by the formula:

$$F(i,k) = \beta_O \sum_{m=0}^{M-1}\sum_{n=0}^{N-1} O(m,n)\exp\left(\frac{-j2\pi mi}{M}\right)\exp\left(\frac{-j2\pi nk}{N}\right) \quad (1)$$

for m = 0,1,..., M-1; n = 0,1,..., N-1

where:

$$O(m,n) = \beta_F \sum_{i=0}^{M-1}\sum_{k=0}^{N-1} F(i,k)_O \exp\left(\frac{j2\pi mi}{M}\right)\exp\left(\frac{j2\pi nk}{N}\right) \quad (2)$$

 DOI: 10.1051/e3sconf/20171901005

for i = 0,1,..., M-1; k = 0,1,..., N-1

$$\beta_O \cdot \beta_F = \frac{1}{M \cdot N} \tag{3}$$

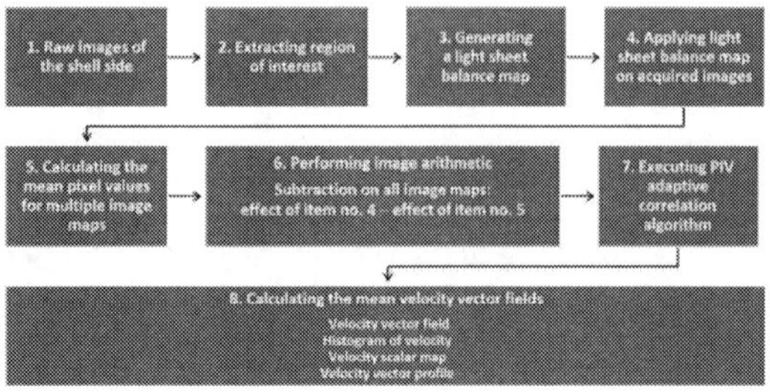

Fig. 3. Overview of the PIV methodology.

Besides the knowledge of the intervals separating the registration of both images, the determination of the vector velocity profile by application of the PIV method involved the need to get to know an arbitrary characteristic dimension of the registered image with the aim of calibrating the linear scale of the phenomenon. A familiar dimension representing the internal diameter of the internal shell of the heat exchanger was applied for this purpose.

3 Results and discussion

The imaging of shell side across the tube bundle in a real heat exchanger performed by the application of the proposed optical track was associated with the occurrence of the phenomenon of perspective in the tube bundle in the section from the plane of the laser sheet to the tube sheet of the heat exchanger. The effect of the phenomenon of perspective is presented in Fig. 4. The same effect was evident in the work [16].

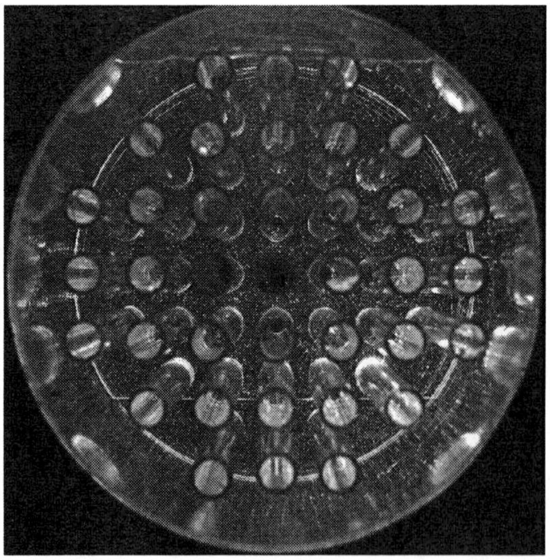

Fig. 4. Effect of the perspective in the shell-and-tube heat exchanger during across tube bundle imaging.

Perspective forms a considerable impediment in the assessment of the hydrodynamics of particular streams and imposes the limitation of the regions of interest. The perspective coefficient calculated as the ratio of the actual external tube diameter in the laser sheet plane to the diameter of the pipe end in the tube sheet captured by the image was determined to be equal to 0.84. This corresponds to the loss of 16% of the region between the tube rows in the focal point of and the adjacent tube rows and 100% for the successive rows. The visualization of velocity vector field under such conditions is shown in Fig. 5. No visible continuity of seeding particles movement is an significant barrier in the description of flow hydrodynamics using PIV method.

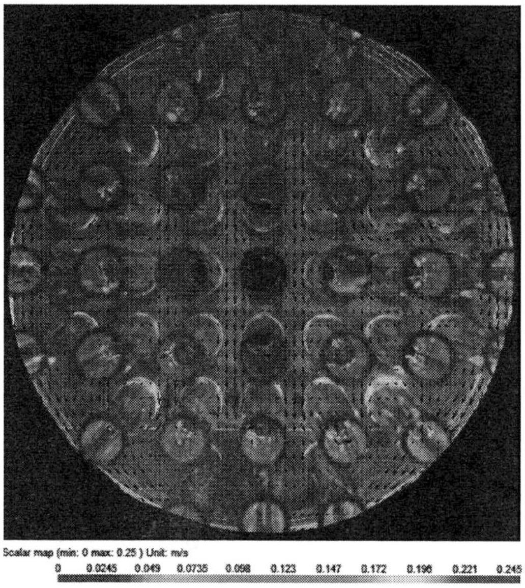

Fig. 5. Visualization of velocity vector field throughout the whole cross-section of the shell side.

As a consequence, the measurements are limited to the analysis in the tube rows that are situated in the focal point. A similar solution was used at work [16]. In the consideration of the fact that the applied measurement methodology involves the analysis of the velocity profiles in the symmetry axes of the zones located between the tube bundles and velocity vector field determined in undistorted region of interest, the loss of the zone applied to measurements did not affect the results of the analysis since calculation algorithms are not limited by the phenomenon of perspective.

The assessment of irregularity of fluid distribution during its flow across a tube bundle began by determination of the velocity distribution symmetry. This goal was achieved by determination of the velocity profile across the tube bundle in the plane across all tube rows above the 4th row counting from the top (Fig. 6).

Velocity profiles analysis allowed to identify common areas of irregularity liquid distribution in shell side. This irregularity was manifested by the variation of

 DOI: 10.1051/e3sconf/20171901005

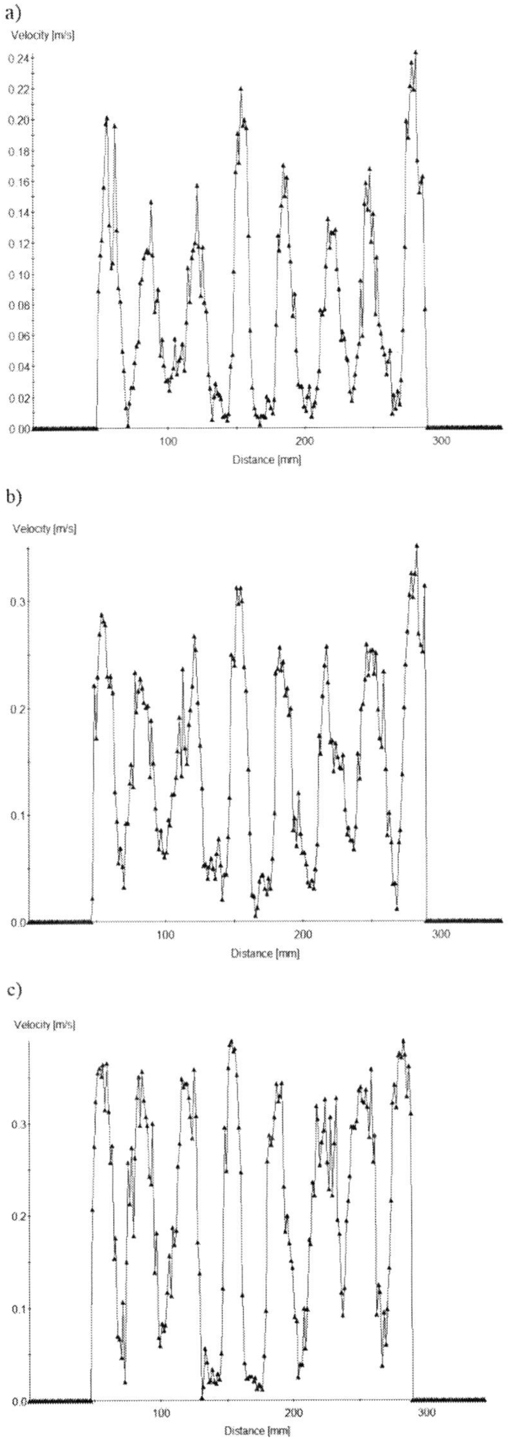

Fig. 6. Velocity profiles above 4th tube row: a) 5 m³/h, b) 7.5 m³/h, c) 10 m³/h.

liquid flow velocity in the analogous region of the tube bundle along the horizontal axis above 4th tubes row. Increased maximum local velocity accrue to external left and right regions of shell side from SL to 1 and from 7 to SR according tube arrangement shown in the Fig. 2. Another one high velocity region was located in the center of shell side (3-4 and 4-5 regions). In the above regions of flow velocity variation, maximum value of the local velocity, depending on flow rate, was higher of 32%, 22% and 8%, respectively. However, the symmetrical nature of velocity variations disappears with increasing flow rate (Fig. 7). For a flow rate equal to 10 m³/h, the difference in velocity variation in particular row of tubes does not exceed 10%. Regions characterized by increased local velocity are also symmetrical and uniformly situated on both sides of the shell side cross-section. The symmetry axis divides the heat exchanger into left and right-hand sizes, which are similar in terms of the hydrodynamic parameters. The testing measurements and calculations can by performed in limited area to a half of the cross-section of a heat exchanger. In certain conditions, this operation can be simplified in this manner without the impact on the quality of the results. In connection with this, the further part of the study focused on the assessment of the irregularity of fluid distribution only from the 4th to the 7th tube rows (i.e. on the left side of the transverse section of the tube bundle in the heat exchanger).

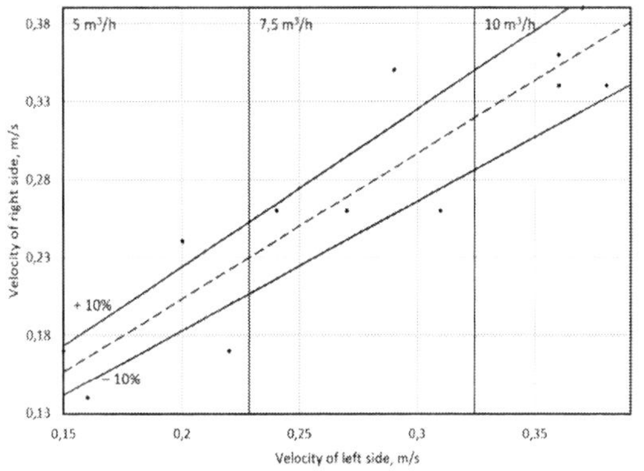

Fig. 7. Influence of flow intensity on the symmetry of liquid flow in the shell side.

The vector velocity fields for the above four cases of location in shell side are given in Table 1 and 2. On the basis of this analysis, it was possible to confirm irregularity of flow distribution by determine the trend representing the mean value of flow velocity. This procedure was performed using the histogram function (Fig. 8-10). The space between the 7th tube row and shell is characterized by the higher value of mean flow velocity regardless of the flow rate through the heat exchanger. In this region mean value of velocity equal to: 0.19 m/s, 0.29 m/s and 0.37 m/s in depending on increasing flow rate used in the experiment. This is due to the fact that liquid flowing around the shell of the heat exchanger does not encounter obstacles in the form of tube rows and consequently, there is no increased local pressure drop in this region and flow assumes considerable value of the mean velocity.

 DOI: 10.1051/e3sconf/20171901005

Table 1. Velocity vector field in particular region of interest between rows of tubes.

Region of interest (location in the shell side)					
4 – 5			5 – 6		
Flow rate, m^3/h					
5	7.5	10	5	7.5	10

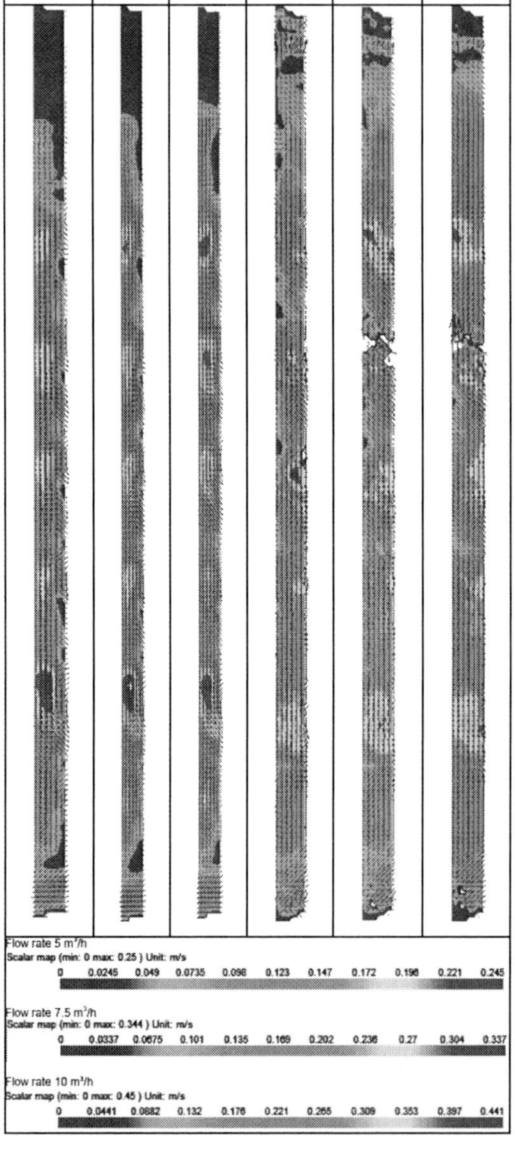

Table 2. Velocity vector field in particular region of interest between rows of tubes or rows of tubes and shell.

Region of interest (location in the shell side)					
6 – 7			7 – SR		
Flow rate, m^3/h					
5	7.5	10	5	7.5	10

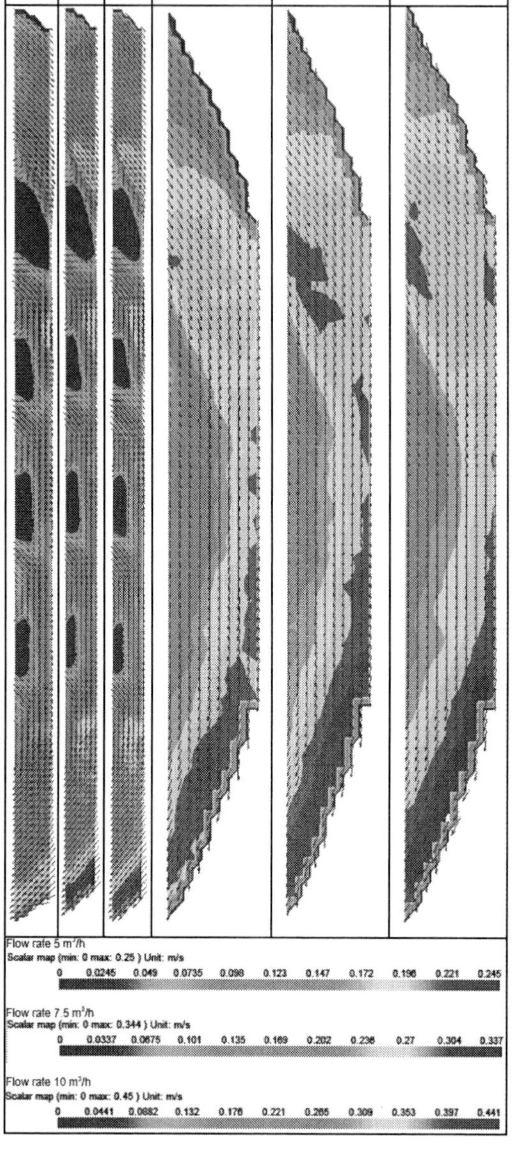

High mean value of flow velocity was also observed in the central part of the shell side (region 4-5). The mean values of flow velocity for similar flow rates were 37%, 41% and 37% lower than the external region. In this case mean value of flow velocity is strongly determined by perpendicular flow to the tube bundle. The areas between these regions can be considered as a transient flow with a characteristic decrease of mean velocity value. On one hand, we have to do with an increased local pressure drop due to the existence of adjacent tube bundles, on the other, the liquid in this region has a lower initial velocity since the flow direction of this stream is not predominantly directed perpendicular to the tube. This demonstrates the existence of regions with various hydrodynamic characteristics. As a result, the study in this paper identified the existence of three irregularity regions. Two external regions between the last rows of tubes and the shell, the central region and the transition region located between the two. Irregularity distribution of liquid flow in this region can negatively affect the performance of the heat exchanger, especially at low flow rate. This applies both to the effect on the heat transfer coefficient and to the susceptibility to settling contamination in the tube bundle.

a)

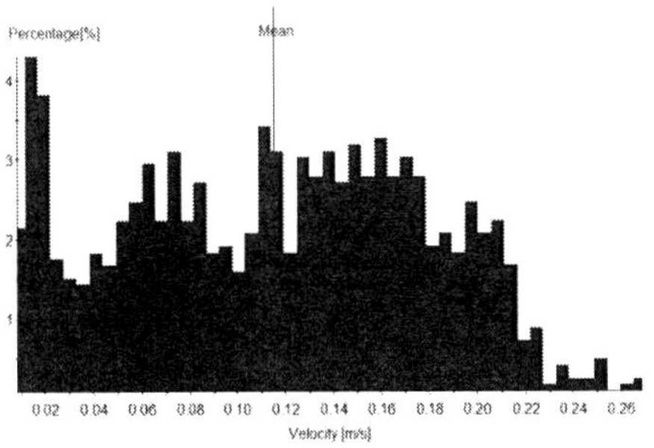

b)

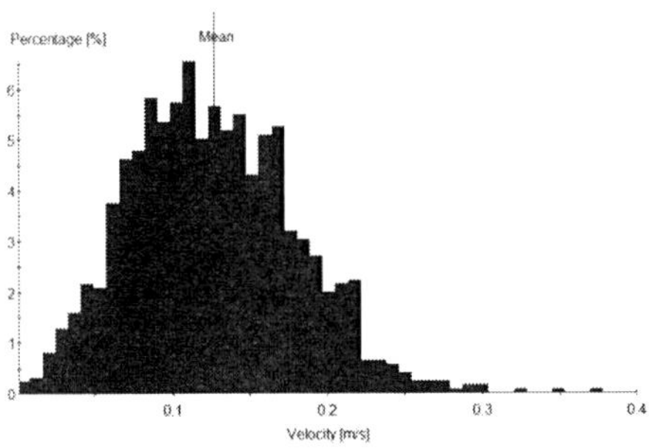

c)

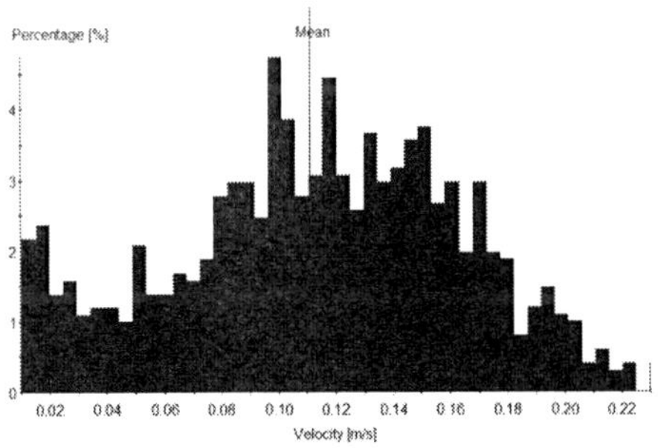

d)

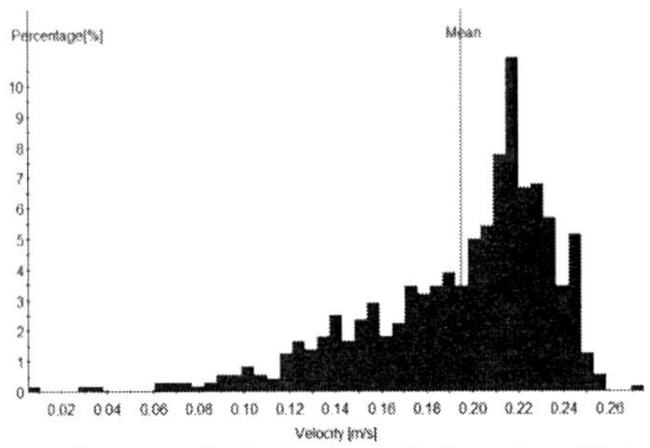

Fig. 8. Histogram of velocity for particular region of interest (flow rate 5 m^3/h): a) 4–5, b) 5–6, c) 6–7, d) 7–SR.

a)

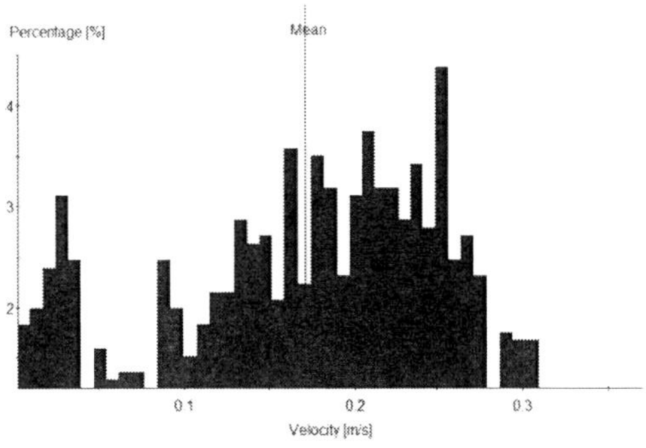

b)

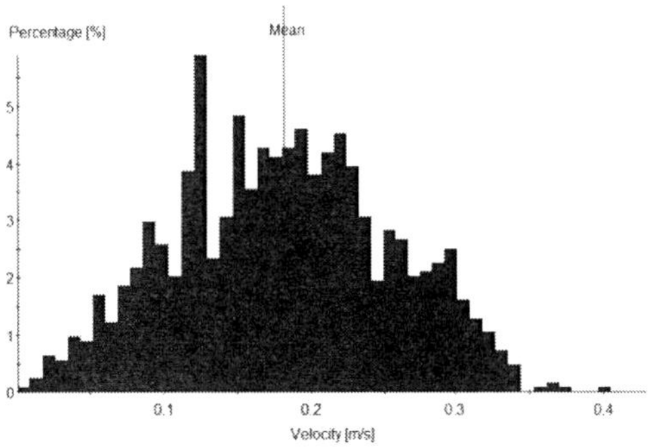

c)

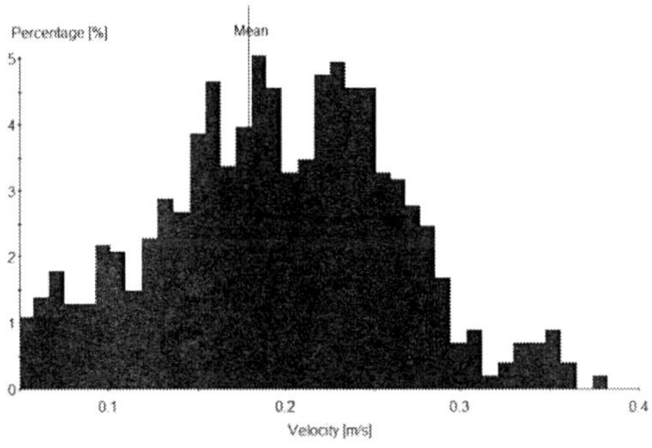

d)

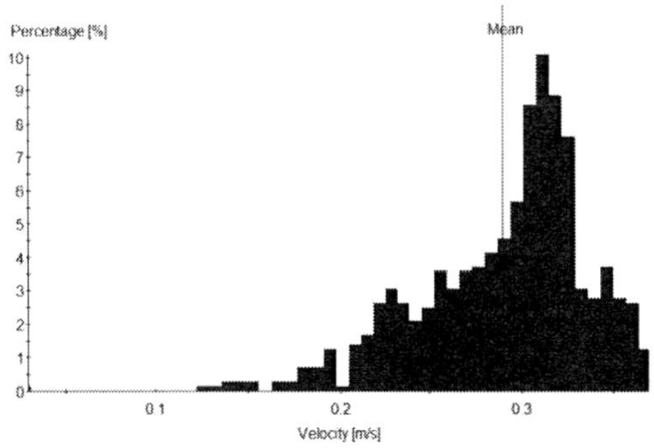

Fig. 9. Histogram of velocity for particular region of interest (flow rate 7.5 m^3/h): a) 4–5, b) 5–6, c) 6–7, d) 7–SR.

DOI: 10.1051/e3sconf/20171901005

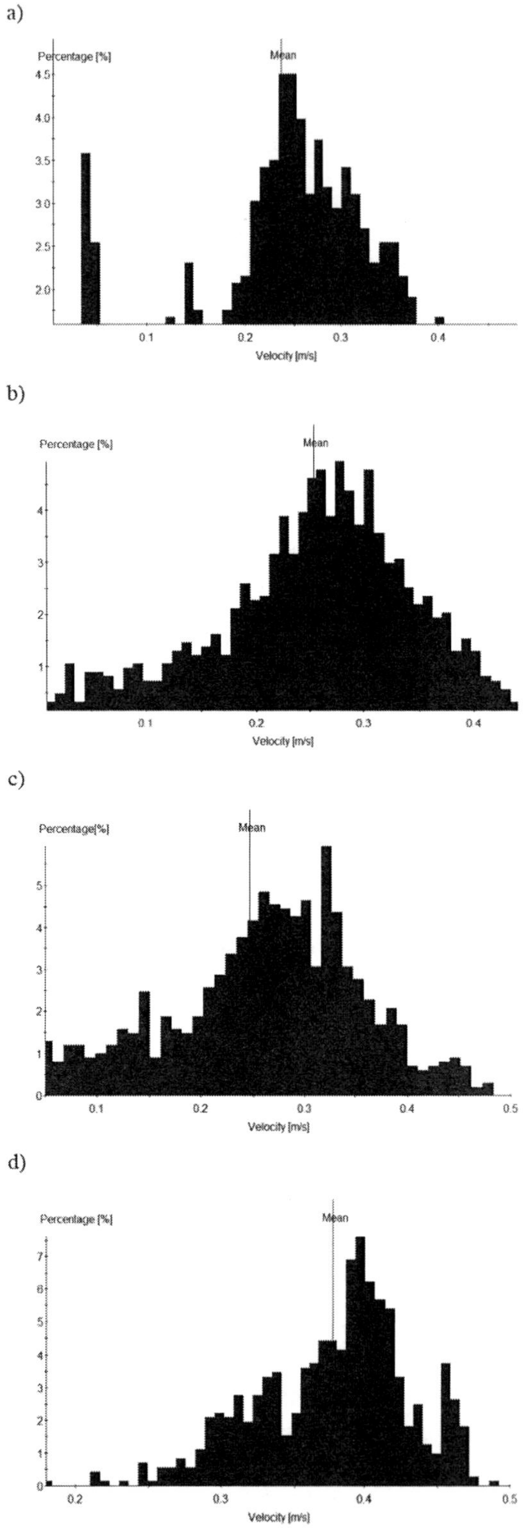

Fig. 10. Histogram of velocity for particular region of interest (flow rate 10 m^3/h):
a) 4–5, b) 5–6, c) 6–7, d) 7–SR.

4 Conclusions

The study of the hydrodynamics of the liquid flow through a tube bundle of a heat exchanger with segmental baffles confirmed the effectiveness of the application of Digital Particle Image Velocimetry (PIV) technique applied for the diagnostics of this type of apparatus. However imaging of the shell side across the tube bundle performed by proposed optical track was associated with the occurrence of the phenomenon of perspective. This phenomenon was a substantial barrier in full imaging of the liquid flow in the shell side. In order to carry out a complete assessment of the velocity vector fields across tube bundle, the investigation should be conducted on a simplified model of heat exchanger with a tube bundle ending directly behind the laser sheet. The results of the study confirmed the existence of a symmetrical flow across the entire cross-section of the tube bundle. The symmetry axis divides the heat exchanger into left and right-hand sizes, which are similar in terms of velocity vector fields parameters. The testing performed into flow that is transverse to the tube bundle orientation can be therefore limited to a half of the cross-section of a heat exchanger. In certain conditions, the procedures involving data processing and analysis can be simplified in this manner without the impact on the quality of the results. The analysis of the vector velocity fields indicated that the distribution of the mean velocities of the liquid flow varies in the particular tube rows. This demonstrates the existence of irregularity in hydrodynamic characteristics. As a result, the investigation in this paper identified the existence of three main regions of the cross-section of shell side. Two symmetrical external regions between the last rows of tubes and the shell, the central region and the transition region located between the two. Differences in local flow velocities in these regions affect the performance of the heat exchanger. The problems discussed in this paper should be confirmed throughout a more extensive study program accounting for other geometries of tube bundle and various flow rates and the relationships above with the heat transfer coefficient.

References

1. Standards of the Tubular Exchanger Manufacturer Association, (TEMA, 9th Edition, New York, 2007)
2. S. Kakacs, L. Hongtan, A. Pramuanjaroenkij, *Heat Exchangers, Selection, Rating, and Thermal Design,* (CRC Press, 2012)
3. E.A.D. Saunders, *Features relating to thermal design in Heat Exchanger Design Handbook*, (Hemisphere, Washington, 1983)
4. A.F. Jozaei, A. Baheri, M.K., Hafshejani, A. Arad, WASJ **18**, 7 (2012)
5. B. Parikshit, K.R. Spandana, V. Krishna, T.R. Seetharam, K.N. Seetharamu, IJHMT **84** (2015)
6. J. Guziałowska, G. Ligus, R. Ulbrich, Proceedings of 5th International Conference on Transport

phenomena in Multiphase Systems (Bialystok, 2008)

7. T.H. Chang, Ch. Lee, H. Lee, K.S. Lee, J. Therm. Sci. **24**, 4 (2015)
8. R.J. Adrian, J. Westerweel, *Particle Image Velocimetry* (Cambridge University Press, New York, 2011)
9. C. Iwaki C., K.H. Cheong, H. Monji, G. Matsui, Exp. Fluids **37** (2004)
10. M. Karaś, D. Zając, R. Ulbrich, Inż. Ap. Chem., **52**, 5 (2013)
11. S.S. Paul, M.F. Tachie, S.J. Ormiston, IJHFF, **28**, 2 (2007)
12. Ch. Bin, G. Liejin, J. Therm. Sci. **9**, 3 (2000)
13. M. Yataghene, F Fayolle, J. Legrand, ATE **31** (2011)
14. G.L. Morrison, I. Budihardjo, M. Behnia, Solar Energy **78** (2005)
15. N. Liu, W. Wang, J. Han, M. Zhang, P. Angeli, Ch. Wu, J. Gong, CES **152** (2016)
16. Sunghyuk T.K. Hyoung, W.R. Bo, J.S. Hyung, ANE, **98** (2016)

DOI: 10.1051/e3sconf/20171901006

Characterization and integration of oxidation catalysts at small-scale biomass combustion furnaces

Mirjam Matthes[1,*], *Ingo* Hartmann[1], *Joachim* Schenk[2], and *Dirk* Enke[3]

[1]Deutsches Biomasseforschungszentrum gGmbH, Department Thermo-chemical conversion, 04347 Leipzig, Germany
[2]Hochschule für Technik, Wirtschaft und Kultur Leipzig, Faculty of Mechanical and Energy Engineering, 04277 Leipzig, Germany
[3]Universität Leipzig, Faculty of Chemistry and Mineralogy, 04103 Leipzig, Germany

Abstract. Small-scale biomass combustion is a major part in heat supply from renewable resources. Drawbacks to the environmental background are the pollutant emissions, which are formed as a result of maloperation, suboptimal furnace construction or the biomass fuel composition. The named primary factors can be influenced by several measures, but the achievable emission results are limited. To provide real clean combustion technology with nearly zero pollutant emissions, secondary emission reduction measures are necessary. One of these measures is the application of catalytic flue gas cleaning as integrated or downstream solution. Catalysis is already a state of the art element in many processes and following this, some studies reveal already its potential to reduce CO, VOC as well as particle emissions in small-scale biomass combustion systems. However, a wide application of catalysts in wood combustion units didn't take place so far, because the challenging process conditions demand a proper integration and highly stable and active catalytic materials. For the achievement of well-functioning combustion systems with catalyst solutions a procedure for application-oriented characterization is presented. Initial investigations with commercially available catalysts have shown that the gas hourly space velocity and the oxygen content have the most significant influence on the conversion rate of carbon monoxide and nitrogen oxide. Two samples with different active phases have been compared, one with solely metal oxides and one with metal oxides and noble metals. The one with noble metals showed as expected a higher activity, but also a higher stability.

1 Introduction

The generation of energy from renewable resources is increasingly in demand to reduce the accumulation of the greenhouse gas carbon dioxide in the atmosphere and accordingly stop global warming. Bioenergy – energy produced out of biomass – provides already 14 % of total energy consumption [1]. The highest share in this sector results from heat production for buildings with traditional and also modern facilities. Biomass combustion in small-scale systems represents here a considerable part. Small combustion furnaces for domestic heating ranging from simple fireplaces in developing countries to modern stoves and pellet boilers in areas like Europe and the United States of America exist all over the world. Dependent on a plurality of parameters like furnace construction, fuel composition and combustion control, a large variety of pollutants is emitted in the flue gas of small-scale biomass combustion systems [2]. Incomplete combustion and unfavourable conditions may lead to significant carbon monoxide (CO), volatile organic compound (VOC), soot, polycyclic aromatic hydrocarbon (PAH) or even dioxin and furan emissions. Beside from organic compounds, other emissions like nitrogen oxides, sulphur oxides or hydrogen chloride can be emitted in relevant concentrations, especially using alternative, agricultural fuels. The emission of the mentioned pollutants reduces the positive environmental effect of the renewable energy generation. In conclusion of this, the legislative regulations for the operation of small-scale furnaces have been tightened during the last years [3, 4, 5]. Therefore, emission reduction measures, primary as well as secondary ones, have to be applied to evolve a truly environmentally friendly energy generation.

2 Catalysis in small-scale biomass combustion

2.1 State of the art in small-scale systems

The application of catalysts is already established in several industrial processes. We can for example no longer imagine automobiles without a catalyst. The ambition to use also catalysts for emission reduction in small-scale biomass combustion systems increased in the last years with the interest to use renewable energy sources and in parallel growing environmental

* Corresponding author: Mirjam.Matthes@dbfz.de

awareness. The main goal by now is to reduce incomplete combusted or formed carbonaceous compounds by integrated or downstream positioned catalysts. Though, first efforts are also have been made to reduce nitrogen oxide emissions by selective catalytic reduction [6].

Suitable catalyst materials for total oxidation of carbon compounds are noble metals as well as mixed metal oxides. In general, noble metals dispersed on alumina supports are common and commercially available, because of their pronounced high activity. Some catalyst solutions for biomass combustion systems are already offered at the market. Basically, there are two ways for the application of catalysts at stoves and boilers. Some manufacturers developed already firing systems with an integrated catalyst [7]. This is favourably, because the facility can be adjusted to the modified pressure and fluid flow conditions with catalyst. By this means comparatively low pollutant emission can be achieved due to optimal construction and catalytic enhanced reaction processes. In the framework of a joint research project for example the constructive optimization and the development of a special down-draft wood log stove accompanied by the development and integration of catalytic devices has been carried out [8, 9]. The other business branch comprises retrofitting of already operating systems. Here are also several solutions at the market by now [10, 11, 12]. However, there is only rare documentation of the long-term stability and no comprehensive analysation of the behaviour from the used catalysts during the manifold conditions occurring in stoves and boilers.

The major part of described research in literature deals with the effect of catalysts at single-room heaters [9, 13-20]. Only some results have been presented from boilers with catalysts [21-23]. The batch-combustion, carried out in furnaces using wood logs as fuel, results in strongly fluctuating conditions. The flue gas composition as well as the temperature varies in the course of one burning cycle. The conditions in pellet boilers are more stable during regular operation, but here also differing states like start and burnout phases or part load operation occur. There the highest pollutant concentrations are emitted and the catalytic effectivity is most important. Most of the described samples consisted of platinum and palladium on ceramic or metallic support materials often with an γ-Al_2O_3-washcoat. In contrast to that, Bensaid et al. [16] and Carnö et al. [17] analyzed lanthanum-chromium-perovskites and copper-manganese-oxides. In general, noble metals show a significant higher oxidation activity, though metal oxides can be beneficial with respect to their stability under some conditions. The direct comparison of a platinum with a copper-manganese-oxides containing catalyst showed a considerably higher activity for carbon monoxide and volatile organic compounds oxidation of the noble metal version [17].

2.2 Challenges and chances for emission reduction with state of the art catalysts

In the first place, investigations on the reduction of carbon monoxide and volatile organic compounds by oxidation catalysts are pursued and described. Even though the catalytic oxidation of carbon monoxide can already take place at room temperature [24], the so far at biomass combustion systems applied catalysts show a considerable reduction at temperatures above 250-300 °C [14, 20, 23]. Beside a preferably low reaction temperature the catalysts in small-scale stoves and boilers also have to feature a wide range and frequent changes in temperature. If the catalyst is applied downstream of a stove, temperatures between 250 and 500 °C can be present [14, 16, 20]. When the catalyst position is inside the furnace, the temperature can rise up to 750-850 °C [9, 17, 22], dependent on the distance to the flame. Since the highest pollutant emissions are generated during start and burnout phases, when the temperatures are low, catalyst heating systems are sometimes applied to gain higher pollutant reduction [14, 21-23]. Literature data show that catalytic conversion rates vary in a wide range and can be hardly compared because of the multitude of differences in the process conditions and mostly not precisely described catalyst probes. In general, the reduction rates for carbon monoxide are clearly higher than for VOC. The often applied noble metal catalysts based on platinum and palladium provide a high activity for CO oxidation. With a good dimensioned and active catalyst reduction rates over 90 % can be realized [19, 20, 21]. Nevertheless, very high pollutant concentrations [13, 18], catalyst bypassing due to leakages [15] or insufficient process temperatures [23] can restrict the performance of the applied catalysts. The reduction rates for VOC are often lower than the rates for CO or at most equal. The compound methane, which has usually the highest share in VOC flue gas concentration from biomass combustion systems [14, 25], is most difficult to oxidize. In this context, a high conversion rate for all VOC except methane [17] and also a fast deactivation of catalysts solely for the oxidation of methane [19] has been observed.

Beside the reduction of gaseous pollutants by catalysis also particulate matter (PM) can be removed from the flue gas. Different mechanisms have to be taken into account. Soot or carbonaceous particle precursors can be catalytically oxidized and accordingly the particle amount in the flue gas is reduced. Particles, organic as well as inorganic ones, can furthermore be deposited at the catalyst surface. As a result the PM concentration is reduced, but also the catalytic active surface is covered and accordingly a deactivation takes place. This is especially critical when alternative non-wood fuels with a higher particle formation are combusted [23]. Nevertheless, several described research results with single-room heaters have shown that a reduction of 30 - 50 % PM is possible [13, 19, 20].

An important issue for the use of catalysts in practical operation is their long-term stability. Only a few research investigations on this topic are published by

DOI: 10.1051/e3sconf/20171901006

now [9, 19, 20, 23]. It has been observed that the activity and stability of the applied catalysts can vary dependent on the existing boundary conditions. Furthermore, maintenance of the catalyst during long-term operation is necessary. Because of the dependency of the catalytic process from various conditions, the aimed pollutant reduction is often not achieved in practice or only for a limited operation time. Also, during regular operation of small-scale combustion systems temperature and flue gas composition vary to a certain extend. A detailed characterization of catalyst activity and a proper integration is necessary to ensure the intended effectivity and long-term stability.

3 Application-related characterization and integration of catalysts

3.1 Essential boundary conditions

The following main boundary conditions have to be mentioned in conjunction with catalysis at small-scale biomass combustion systems: gas composition, temperature, pressure and fluid flow conditions. The integration of a catalyst, especially close to the combustion chamber, may influence the initial conditions and can as a result change the combustion process. In Fig. 1 the sequence of combustion and catalysis with its boundary conditions and interdependencies is presented as schematic view.

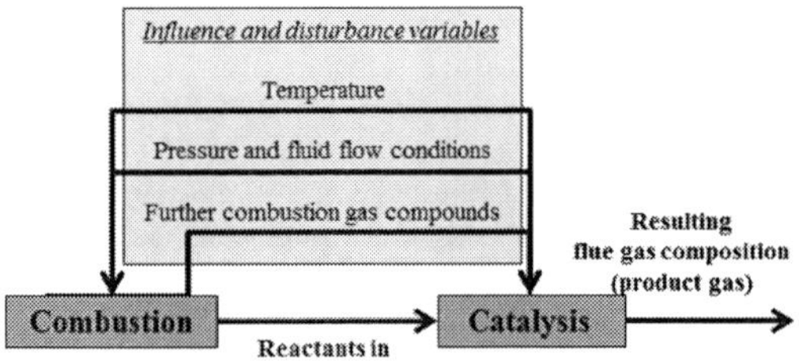

Fig. 1. Catalytic process in biomass combustion systems with boundary conditions and interdependencies.

In general, dependent on the position in the combustion systems the temperature ranges from maximum values between 800 and 1200 °C to values between 100 and 350 °C at the outlet of the furnace. Either by natural or induced draft the combustion takes usually place having a negative pressure in the combustion chamber, which has a level of 15 - 30 Pa.

The pressure drop of a used catalyst has to be very low to prevent an unfavorable or even security-relevant effect on the combustion process. Systems operating with a chimney fan have a higher tolerance than those working with natural draft.

Beside the static pressure, the catalyst changes also the fluid flow conditions in the furnace. This may induce a positive or negative effect on the occurring reactions.

The main flue gas components resulting of an almost total combustion of biomass are nitrogen, carbon dioxide, water and oxygen. Because of partial incomplete combustion smaller or larger quantities of carbon monoxide, volatile up to semi-volatile organic compounds as well as soot are formed. Apart from these components several other elements are present in the fuel, in which the concentration varies according to the fuel type.

A relevant influence on the flue gas composition have the following elements: N, K, P, Ca, Mg, S, Cl. In dependence of several process parameters, as for example temperature and air supply, the flue gas components SO_2, NO_X, PCDD/F, HCl and ash occur.

3.2 Characterization procedure

A comprehensive catalyst utilization in different single-room heaters as well as central heating furnaces can only be achieved by an universal catalyst characterization for the application. So far, the integration of catalysts in combustion systems was based largely on trial and error. Several optimization loops are necessary, for example when the catalyst usage conditions are not met in all operation phases or changes in the construction of the furnace are necessary to maintain good combustion conditions with the additional pressure drop of an integrated catalyst.

A detailed characterization of the catalyst usage conditions as well as an analysis of the catalyst monolith induced process changes can minimize further following expenses for the integration in special stoves and boilers. In conclusion of these considerations, a procedure for the characterization of promising catalysts for small-scale biomass combustion systems has been developed to examine stepwise the effectivity and also the operation range. The general procedure is shown in Fig. 2.

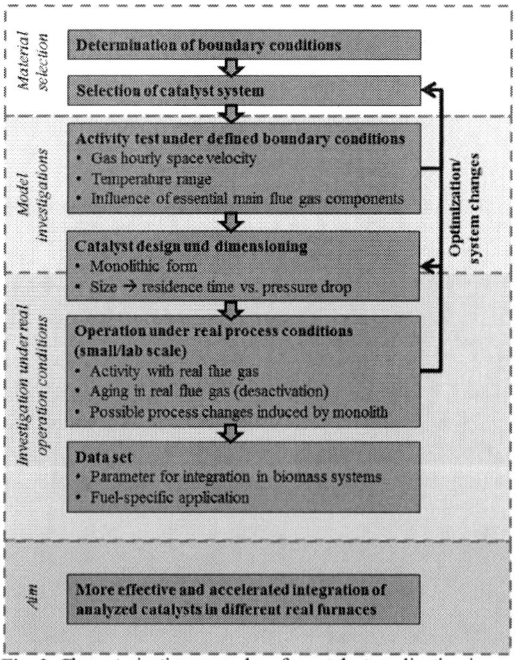

Fig. 2. Characterization procedure for catalyst application in small-scale biomass combustion systems.

The procedure is realized with two laboratory facilities. In the first one the activity of the catalyst probes for selected pollutants is analyzed with a model gas mixture containing basic flue gas components. The facility allows an analysis of the operation range of the catalyst probe and if this range covers the operation phases of biomass combustion systems. In Fig. 3 the facility is shown as a scheme. The main flue gas components for biomass combustion on a quantity basis are nitrogen, carbon dioxide, oxygen and water. These four compounds are the basis of the model gas. All other elements have a clearly lower concentration and their amount strongly depends on the conditions of the combustion process. In general, incomplete oxidized carbon compounds as CO and VOC, nitrogen oxides and particulate matter are the most relevant compounds in the lower concentration range.

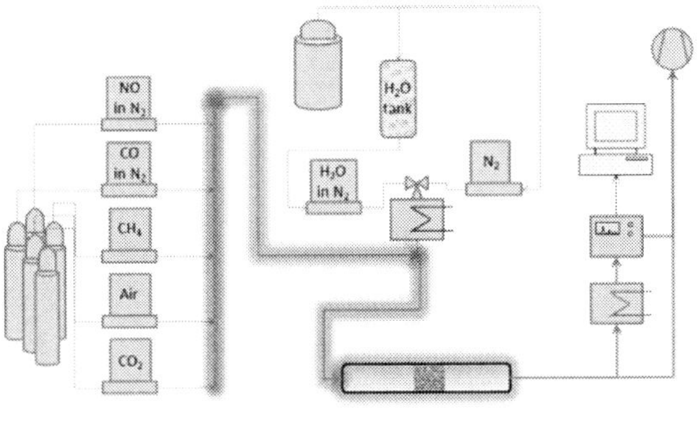

Fig. 3. Scheme of laboratory facility for characterization with model gas.

Design of Experiments (DoE) methods are used to determine the catalyst operation range considering the concentration range of the flue gas components. The gas concentrations are analyzed with an infrared and a paramagnetic oxygen analyzer (type EL 3020, ABB). The influence of particulate matter is though not examined on the model gas level by now.

The real flue gas conditions and at the same time the influence of PM on the catalyst activity are analyzed with a second laboratory facility. The combustion of comminuted biomass is realized on a very small scale with a furnace having a nominal heat output of 1 - 2 kW. A scheme of the facility is shown in Fig. 4. The construction according to the downdraught combustion principle with mass flow controllers for air supply and a dosage system for the supply of comminuted biomass allows the defined adjustment of different combustion states.

The characterization of the catalyst performance and the influences of the monolith on the combustion process are carried out by comparative operation of the facility with and without catalyst. The flue gas composition as well as temperature and pressure at different positions are measured continuously.

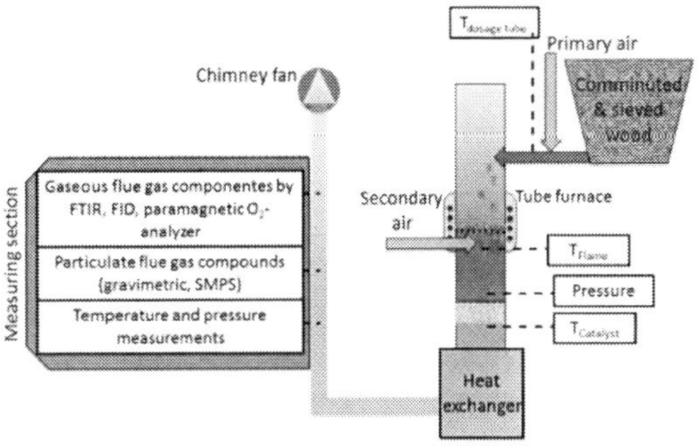

Fig. 4. Scheme of laboratory facility for characterization under real combustion conditions (T – temperature; FTIR - Fourier transform infrared spectroscope; FID – Flame ionization detector; SMPS – Scanning mobility particle sizer).

3.3. Initial findings by model gas investigations

Some commercially available catalyst probes have been tested to check their applicability for the use in biomass combustion furnaces. The following results have been determined with two samples having both an α-alumina foam as support and a high-temperature stable γ-alumina washcoat. One probe has a solely metal oxide active phase. The active phase of the other one is out of metal oxides and noble metals. The precise composition is not disclosed by the manufacturer. The probes have been tested in a model gas based on a potential flue gas composition of combustion furnaces. The following three steps have been performed. At first, the probes have been heated twice to 700 °C with corresponding cooling phases. Afterwards they have been heated to 500 °C for five hours, while the flue gas concentration has been varied according to determine influences of several gas components on the conversion rate by DoE-methods. In the final third step the probes have been heated again twice to 700 °C. As model gas for the start and final test the following composition has been set: 15 Vol.-% CO_2, 8 Vol.-% H_2O, 5.5 Vol.-% O_2, 1700 ppm CO, 175 ppm CH_4 and 100 ppm NO. CO and CH_4 have been chosen as representative for the gaseous incomplete oxidized pollutants in the flue gas. No catalytic conversion has been observed for methane using the described catalyst samples. The CO conversion differed according to the used probe and also dependent on the period of use. The temperature-conversion-curves are shown in Fig. 5 for both catalysts during the start and final step of the model gas tests.

The sample with noble metals showed as expected a clearly higher conversion rate and also a lower light-off temperature. Furthermore, a conclusion about the stability of the probes has been visible.

The probe with noble metals remained stable with certain variations during the model gas characterization.

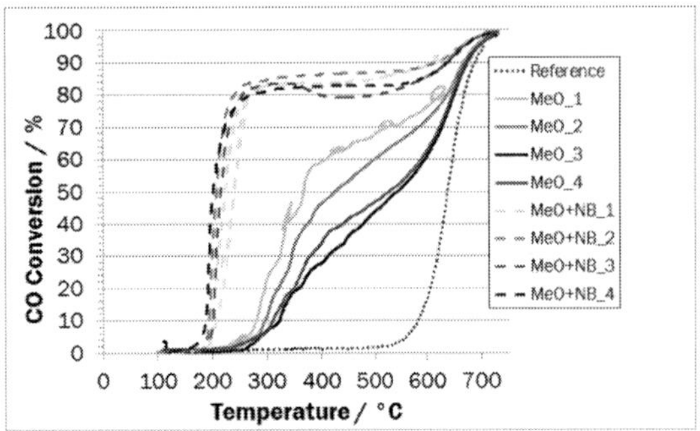

Fig. 5. Carbon monoxide conversion of two commercially available catalyst probes – one with a metal oxide (MeO) active phase and one with a phase composed of metal oxides and noble metals (MeO+NB).

The activity of the probe with the solely metal oxide active phase decreased considerably during the first and second step of the model gas experiments. In the third step (3rd and 4th heating run, see Fig. 5) the sample has been stabilized in all probability, because the conversion rate of the fourth run was slightly higher than the third one.

The second step of the model gas characterization has been carried out by variation of the parameters shown in Table 1 using two levels. The levels have been defined based on conditions occurring during real operation. The temperature and the CO_2 concentration have been kept constant at 500 °C and 12.0 vol.-% respectively.

Table 1. Parameter and levels for step two of model gas characterization

Parameter/Level	1	2
CO/CH_4 [ppm]	300/160	1000/175
O_2 vol.-%	3.0	14.0
NO [ppm]	100	300
H_2O [vol.-%]	5.0	10.0
GHSV [h^{-1}]	50,000	100,000

For evaluation of the data the effects of each parameter has been determined with the following equation:

$$Effect = \bar{y}_{level\ 1} - \bar{y}_{level\ 2} \qquad (1)$$

The average value of the results with level 1 of one parameter is subtracted with the average value of level 2. The calculated conversion rate has been chosen for evaluation of the effects. The analysis of the reference data showed that a standard deviation of 2 - 3 % exists for the concentration of the gas components CO, CH_4 and NO. Therefore, small conversion changes can`t be detected.

Table 2. Average conversion rates for both CO/CH_4 levels and effects (change of conversion) for O_2 and GHSV changes.

Values in %		CO/CH_4 level 1	CO/CH_4 level 2	Effect O_2 ↓	Effect GHSV↓
MeO	CO	40.1	40.1	1.4	15.6
	NO	13.6	12.6	-9.1	2.6
MeO +NB	CO	69.0	71.8	3.5	14.9
	NO	9.5	6.2	-7.0	1.7

In Table 2 some data for the clearest influences are given. The results revealed that the change in gas hourly space velocity (GHSV) induced the most significant change of the conversion rate in the analysed parameter range. The CO conversion was 14.9 % and 15.6 % higher with the lower GHSV. Furthermore, the change of the O_2 concentration from 14.0 to 3.0 vol.-% induced a decrease in NO conversion. The CO conversion increased in contrast. Previous measurements with an FTIR analyser showed that NO is converted to NO_2. The reduced conversion rate of NO with lower oxygen content can be influenced by two effects. A lower availability of oxygen can result in a lower conversion rate. Furthermore, NO_2 formed by the oxidation of NO can act as oxidizing agent for example for the conversion of CO. Oxidation with NO_2 and accordingly back reaction to NO increases potentially with decreasing O_2 concentration. Beside those general tendencies for both catalyst materials also a difference between the catalyst with and without noble metals was visible. The catalyst with noble metals showed an increased conversion rate with a higher concentration of CO/CH_4. In contrast to this, the NO conversion rate decreased. For the solely metal oxide catalyst only a small decrease of NO conversion appeared. The effects for H_2O and NO changes have been negligible as well all effects on CH_4 conversion.

Experiments with the same catalyst systems used during model gas analysis are currently carried out in the laboratory facility shown in Fig. 4 under real gas conditions. The evaluation of the data in comparison with the model gas analysis aims to reveal the effect of further flue gas components on stability, especially the influence of particulate matter.

4 Conclusions

Catalysts are potentially a useful device in small-scale biomass combustion systems to reduce the pollutant emissions to a lower level. They are able to reduce a wide variety of pollutants for example soot, hydrocarbons and carbon monoxide. Some research results show already that a reduction, especially of carbon monoxide, is possible also over several weeks of operation. Nevertheless, the demands on activity and

 DOI: 10.1051/e3sconf/20171901006

stability for pollutant reduction are high. An appropriate analysis and characterization of suitable catalyst materials is therefore necessary to promote their effective integration and accordingly their acceptance and utilisation. A promotion of use-oriented catalyst analysis in the case of small-scale biomass combustion is intended by the presented characterization methods. A deactivation of catalysts has been observed in different cases. Investigations described in literature with real combustion systems revealed already the negative effect of particles deposited on the catalyst surface. In the described model gas experiments no particles have been present. Therefore, the deactivation of the metal oxide sample without noble metals can be assigned to thermal effects or the influence of a gaseous component. No visual apparent loss of material has been observed.

The described research work was amongst others supported by funds of the Federal Ministry of Food and Agriculture (BMEL) based on a decision of the Parliament of the Federal Republic of Germany via the Federal Office for Agriculture and Food (BLE) under the innovation support programme.

References

1. REN21, 2016, *Renewables 2016 Global Status Report* (Paris: REN21 Secretariat), ISBN 978-3-9818107-0-7
2. S. Ozgen, S. Caserini, S. Galante, M. Giugliano, E. Angelino, A. Marongiu, F. Hugony, G. Migliavacca, C. Morreale, Atmos. Environ. **94** (2014)
3. J. Villneuve, J.H. Palacios, P. Savoie, S. Godbout, Bioresour. Technol. **111** (2012)
4. United States Environmental Protection Agency, FACT SHEET: Summary of Requirements for Woodstoves and Pellet Stoves, https://www.epa.gov/residential-wood-heaters/fact-sheet-summary-requirements-woodstoves-and-pellet-stoves (accessed online April 2017)
5. European Environmental Bureau (EEB), Air & domestic heating, *Gasping for Air - Twelve Factsheets on Air Pollution in the EU*, http://www.eeb.org/index.cfm/library/?firstpublications=1&month=0&Air=1&year=0 (accessed online April 2017)
6. M. Matthes, M. König, I. Hartmann, Kombinierte Minderung von Staub und Stickoxiden an Biomassefeuerungen, Proceedings of *8. Abscheider-Fachgespräch „Partikelabscheider in Biomass-efeuerungen"* (Straubing, 2017)
7. United States Environmental Protection Agency, List of EPA Certified Wood Stoves, https://www.epa.gov/compliance/list-epa-certified-wood-stoves (accessed online April 2017)
8. DBFZ Deutsches Biomasseforschungszentrum gemeinnützige GmbH, Novel low-emission stove (DBU-NEKO) – press release, https://www.dbfz.de/en/focus-areas/reference-projects/novel-low-emission-stove-dbu-neko.html (Access online April 2017)
9. R. Bindig, S. Butt, I. Hartmann, D. Dvoracek, W.-D. Einicke, D. Enke, B. Specht, F. Werner, *DBFZ Report Nr. 27*, Neuartiger emissionsarmer Kaminofen (DBU-NEKO) (Leipzig, 2016) ISBN: 978-3-946629-05-4
10. Firecat, http://www.firecatcombustors.com/ (Accessed online April 2017)
11. DR PLEY ENGINEERING & CONSULTING (BUCHAREST) SRL, ChimCat® OXI, http://dr-pley.com/en/imprint/
12. Blue Fire GmbH, The catalyst for stoves and small heating systems, http://blue-fire.org/en/
13. A. Hukkanen, T. Kaivosoja, O. Sippula, K. Nuutinen, J. Jokiniemi, J. Tissari, Atmos. Environ. **50** (2012)
14. F. Ozil, V. Tschamber, F. Haas, G. Trouvé, Fuel Process. Technol. **90** (2009)
15. R. Mack, H. Hartmann, Performance of catalytic and non-catalytic foam ceramic elements in log wood stoves, Proceedings of *24th Eur. Biomass Conf. and Exhibition* (Amsterdam, 2016)
16. S. Bensaid, F.A. Deorsola, D. Fino, N. Russo, Catal. Today **197** (2012)
17. J. Carnö, M. Berg, S. Järås, Fuel **75**, 959-965 (1996)
18. T. Kaivosoja, A. Virén, J. Tissari, J. Ruuskanen, J. Tarhanen, O. Sippula, J. Jokiniemi, Chemosphere **88** (2012)
19. R. Bindig, S. Butt, I. Hartmann, M. Matthes, C. Thiel, Catalysts **2** (2012)
20. M. Matthes, I. Hartmann, Chem. Eng. Technol. **40**, 2 (2017)
21. Y. Roy, M. Lefsrud, V. Orsat, F. Filion, J. Bouchard, Q. Nguyen, L.-M. Dion, A. Glover, E. Madadian, C. P. Lee, Biomass Bioenerg. **66** (2014)
22. M. Ferrandon, M. Berg, E. Björnbom, Catal. Today **53** (1999)
23. M. Matthes, I. Hartmann, A. Groll, U. Riebel, Biomass Conv. Bioref. **6** (2016)
24. M.-G. Jeong, I. H. Kim, S. W. Han, D. H. Kim, Y. D. Kim, J Mol. Catal. A-Chem. **141** (2016)
25. M. Olsson, *Residential biomass combustion – emissions of organic compounds to air from wood pellets and other new alternatives*, Thesis for the degree of doctor at Chalmers Uni. Technol. (Göteborg, 2006) ISSN 0346-718X

 DOI: 10.1051/e3sconf/20171901007

Projected configuration of a coal-fired district heating source on the basis of comparative technical-economical optimization analysis

Mariusz Tańczuk[1,2*], *Wojciech* Radziewicz[2,3], *Eligiusz* Olszewski[2], and *Janusz* Skorek[4]

[1]Opole University of Technology, Department of Environmental Engineering, ul. St. Mikołajczyka 5, 45-271 Opole, Poland
[2]Energetyka Cieplna Opolszczyzny SA, ul. Harcerska 15, 45-118 Opole, Poland
[3]Opole University of Technology, Department of Power Plant, Diagnostics and Computer Engineering, ul. Prószkowska 76, 45-758 Opole, Poland
[4]Silesian University of Technology, Institute of Thermal Technology, ul. Konarskiego 22, 44-100 Gliwice, Poland

Abstract. District heating technologies should be efficient, effective and environmentally friendly. The majority of the communal heating systems in Poland produce district hot water in coal-fired boilers. A large number of them are considerably worn out, low-efficient in the summer time and will not comply with forthcoming regulations. One of the possible solution for such plants is repowering with new CHP systems or new boilers fuelled with fuels alternative to coal. Optimisation analysis of the target configuration of municipal heat generating plant is analysed in the paper. The work concerns repowering the existing conventional heat generating plant according to eight different scenarios of the plant configuration meeting technical and environmental requirements forecasted for the year of 2035. The maximum demand for heat of the system supplied by the plant is 185 MW. Taking into account different technical configurations on one side, and different energy and fuel prices on the other side, the comparative cost-benefits analysis of the assumed scenarios has been made. The basic economical index NPV (net present value) has been derived for each analysed scenario and the results have been compared and discussed. It was also claimed that the scenario with CHP based on ICE engines is optimal.

1 Introduction

Energy and climate strategy of the European Union (EU) bases on three urgently important actions: improving energy efficiency of the processes, applying renewable resources in energy generation and cutting down carbon dioxide (CO_2) emissions. Many countries have adopted energy policies to achieve such goals as increasing share of renewable resources in final energy generation and applying more efficient technologies based on cogeneration of heat and power (CHP) or trigeneration (CCHP). Despite the current changes in the world energy sector as well as the upcoming strategies for the period 2020-2050 [1], inefficient energy systems, often based on solid fuels, mainly coal, still continue to dominate in the structure of the electricity and heat production systems in certain countries of the world, including countries in Europe [2].

Previous and current actions taken to move the energy sector towards environmental friendly and economical efficient shape has focused on both electricity and heat. Centralized generation and distribution of heat via efficient district heating (DH) systems plays an important role in increasing efficiency of energy systems as a whole and seems to be one of a key components in achieving strategic EU goals.

Improving the district heating systems has become a tendency in Europe since at least 20 years. Nowadays, the projects such as Heat Road Map Europe or STRATEGO (Enhanced Heating and Cooling Plans in EU) set out directions in the field of DH systems modernization and development. [3]. One of the important proposals for future sustainable cities is, undoubtedly "Fourth generation district heating" (4DH) focusing on deep integration of district heating in such cities. According to [4] future DH systems will be characterized by a wide use of combined heat and power together with the utilisation of heat from waste-to-energy and various industrial surplus heat sources as well as the inclusion of renewables [5] and advanced monitoring systems [6]. Moreover, the expected combination of district heating and cooling (DHC) systems can help with applying aforementioned solutions [7]. DHC systems are already found economically feasible, as justified in [8].

District heating system across Europe are differentiated by the technology, temperature of water and delivery extent as well as by the resources used for heat (and electricity) generation. Compared to other European countries, particularly these located in western and northern part of Europe, the coal is a dominant energy source in Central and East Europe DH

*Corresponding author: m.tanczuk@po.opole.pl

applications. According to [9] district heating systems supply with heat around 29% of residential buildings in that region. An interesting and worth attention case is Poland with 41% share of DH systems [10] and in the same time with coal as a main source of input energy of heat generating plants. The hard coal remains dominant resource for decades, however its share in heat production is expected to go down as stated in long-term forecasts available in [11].

Vast number of DH system around the world, particularly supplied with coal, require or will require smaller or bigger modifications (i.e. retrofitting) to meet new strict environmental and efficiency standards as well as to have their feasibility improved. In [2] authors claim that due to the continuous restrictions in the dust and sulfur dioxide (SO_2) emissions, the heating industry in Poland faces the necessity of spending whacking expenses of new dedusting and desulfurization installations. The CO_2 limitations are also cost-creative and will require modernization of DH facilities not only in Poland. Eventually, another reason for reconfiguration of the plants and starting necessary investment actions is technical wear of the existing boilers, low-efficient production of heat for network hot water during the off-peak periods and far from optimal configuration of the existing DH sources.

One of the most common DH solutions for managing the current and upcoming conditions is to apply repowering plants with the CHP, usually gas-fired systems. As stated in [12], cogeneration of heat and power technology, compared to separate technologies, always offers the possibility of increasing the efficiency of the energy production and results in a decrease of primary energy use. Energy efficient DH systems also require usage of renewable resources as a measure for giving the positive impact for DH energy efficiency [13]. Regardless of the applied technology of DH plants, energy efficiency of the process should always be high. Optimum efficiency level should, however, result from economic factors. A DH system with an optimally selected and configured source is one, which not only leads to the savings in the use of the primary energy but also has to offer economic benefit.

The further part of this paper presents issues connected with the thermal and economic efficiency of a coal-fired municipal heat generating plant as a source of DH system. The projected configuration has been assumed and the optimization analysis of the plant modernization by installing new units or repowering existing ones, including CHP systems is the subject of the paper.

The aim of presented work is to find optimal DH source configuration as a retrofitting solution for coal-fired municipal heat generating plant.

2 System description

The investigation concerns repowering the existing conventional heat generating plant according to eight different scenarios of the plant configuration meeting technical and environmental requirements forecasted for the year of 2035. The existing plant is a heat source for high temperature municipal district heating system with the maximum demand for heat reaching 185 MW. The hot water of nominal pressure p_n = 1.6 MPa and nominal (maximum) supply temperature $t_{s,n}$ =150 °C is a heat carrier in the system.

The plant consists of the following heat generating units:

- one pulverized hard coal-fired water boiler PCB (labelled as WP120) with the nominal heat capacity 125 MW,
- two hard coal stoker-fired water boilers SCB (labelled as WR25 no. 1 and. 2) with the nominal heat capacity 33 MW each,
- high-efficient cogeneration unit based on a gas turbine unit GTU fired with natural gas of a nominal electric capacity 7.4 MW and integrated with a heat recovery boiler HRB of a nominal heat capacity 14.7 MW,
- high efficient cogeneration unit based on a steam turbine STU supplied with hard coal stoker-fired steam boiler SSCB (labelled as OR50) of a steam capacity 50 t/h.

The STU cogeneration unit is integrated with a steam-water heat exchanger and reaches the nominal thermal capacity 29 MW and nominal electric capacity 10.7 MW. In the heat-only operation mode, thermal capacity increases up to 40 MW. Apart from the described units, there is also additional production facility installed in the plant:

- one pulverized hard coal-fired water boiler PCB (labelled as WR40) – of 40 MW of thermal capacity remaining permanently disconnected from the plant infrastructure,
- one gas boiler of 25 MW of thermal capacity installed as a peak-load unit.

The schematic diagram of the existing plant is shown in Fig. 1. The total installed heat capacity of the analyzed plant reaches almost 260 MW (without 40 MW of disconnected boiler). The total electricity power output equals to 18.1 MW.

Table 1 presents main performance data of the plant supplying district heating system under investigation.

Table 1. Performance data of the DH system – current state.

Parameter		Value
Annual heat production	Q_{DH}	1 582 500 GJ
Annual electricity production	E_{el}	106 000MWh
Annual heat sale	S_{DH}	1 369 700 GJ
Delivery heat losses	Q_L	212 800 GJ
Annual heat production in cogeneration	Q_{CHP}	935 500 GJ
Demand for chemical energy	E_{ch}	2 298 900 GJ
Demand for chemical energy for electricity production	$E_{ch,el}$	455 000 GJ
Demand for chemical energy for heat production	$E_{ch,h}$	1 843 900 GJ

 DOI: 10.1051/e3sconf/20171901007

The load duration curve of the district heating system supplied by the plant is presented in the Figure 2. As it can be seen, the plant remains under operation during the whole year time supplying city DH system with heat in order to provide central heating of residential buildings (in the heating season) and domestic hot water (in the season and off-season). The difference between maximum demand in the winter time that equals 185 MW and rather constant demand during off-season (around 15 MW) is significant and worth noting. A much lower demand for heat by DH in the summer time often results in reduced boiler load and, consequently, can lead to low efficiency of the plant. Such inconvenience occurs with non-optimized plants with oversized units or not enough number of units. One of the solution for drawback of off-season efficiency decrease is installing properly sized CHP unit operated at the base load during the possible longest time a year.

The considered system has already been retrofitted by building two cogeneration units: gaseous unit GTU with gas turbine fed with natural gas and the unit based on a steam turbine STU supplied by steam generated in a stoker-fired boiler fuelled with hard coal. As it can be seen in the table 1, the share of heat produced in cogeneration process reaches almost 60%. Installing CHP facilities was the first stage of repowering and thus the plant requires further modification to meet new requirements and up-coming regulations. Accordingly, different scenarios of district heating plant configuration meeting technical and environmental requirements has been assumed and presented in the further part of the paper.

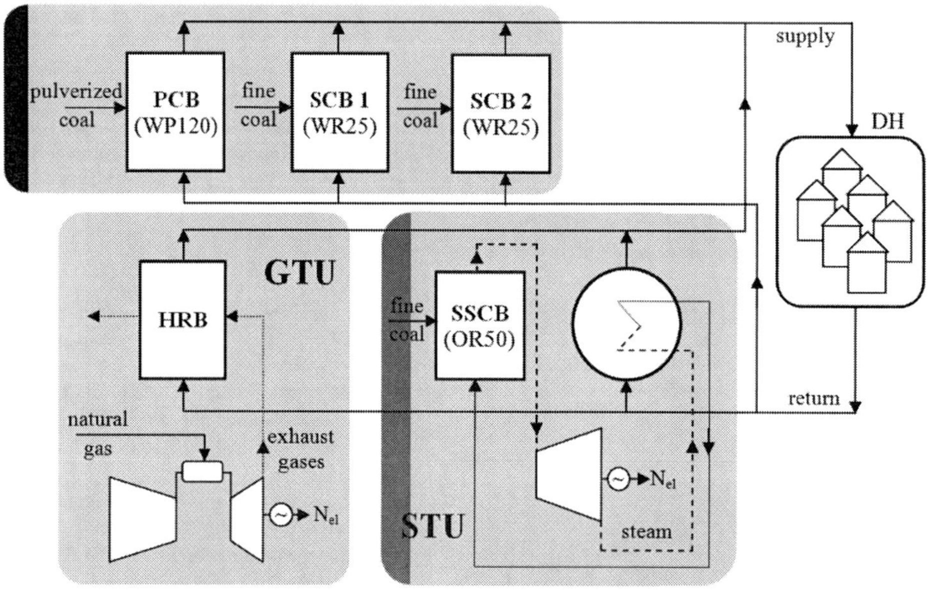

Fig. 1. Schematic diagram of the current configuration of analyzed DH plant.

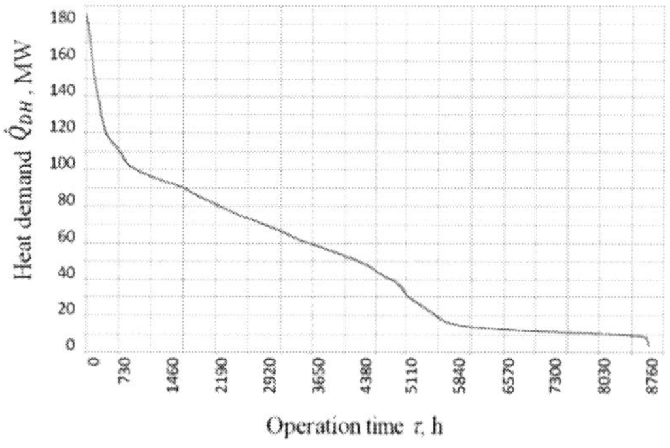

Fig. 2. The load duration curve of the analyzed district heating system.

3 Methods and input data

The research goal is to investigate technical and economic feasibility of DH plant configured according to assumed scenarios. For the analysed type of modernization of heat source of DH, the analysis consists of thermodynamic and economic evaluation linked by a direct correlation. The economic indexes calculated within the analysis are a measure made of a number of parameters. The majority of them is the function of the thermodynamic quantities resulting from the technical characteristics of equipment as well as configuration and operation of the thermal plant.

3.1 Thermodynamic evaluation

The basis for the thermodynamic evaluation was the heat load duration curve (Fig. 2) given as an input data for the analysis. The energy balance calculations were conducted for each eight investigated configuration scenarios where assumed units of a given size and given efficiency were composed in heat generating DH plant. Analysis has been carried out with use of hourly optimization method. Similar approach of hour by hour simulation was applied in the research dedicated to the polish municipal DH case in [2] as well as in the case study of city Zagreb presented in [14].

Table 2 presents main design parameters of the units used for configuring the plant according to the following scenarios:

- scenario **S1** – based on the current configuration of the plant (Fig. 1) with no retrofitting actions, taking into account only the necessary maintenance and modernization work due to technical condition of units and environmental restrictions (necessity of construction of flue gas desulphurisation and denitrification),
- scenario **S2** – existing configuration of the plant extended by the additional cogeneration unit with biomass steam boiler BB and extraction steam turbine (Fig. 3),
- scenario **S7** - based on S6 configuration but with ICE unites operated at the base load,
- scenario **S8** - based on S6 configuration with GTU unit removed.

On the basis of assumed technical configurations of the plant and in accordance with design parameters of applied units as well as with given load duration curve, hour by hour simulation of plant annual operation has been conducted for each of the scenario. In the result, the annual heat and electricity production can be derived with use of the following formulas:

$$Q_{u_i} = \int_{\tau=0}^{\tau=8760} (\dot{Q}_{u_i})\, d\tau \tag{1}$$

where Q_{u_i} is the annaul heat production by the particular unit i, $\dot{Q}_{u_i}$ is the current heat capacity of the unit I of 5 ICE-based gaseous units (Fig. 4), with steam turbine unit STU operated at the base load, provided by the simulation procedure, τ – time, in hours

$$N_{el,u_i} = \int_{\tau=0}^{\tau=8760} (\dot{N}_{el,u_i})\, d\tau \tag{2}$$

Table 2. Main design parameters of the units used for plant configuration.

No.	Unit	Max./min. thermal output, MW	Nominal electric output, MW	Nominal efficiency, %	Remarks
1	Pulverized coal-fired water boiler PCB	120/55	-	91.5	existing and operated
2	Coal stoker-fired water boilers SCB 1	33/10	-	84.3	existing and operated
3	Coal stoker-fired water boilers SCB 2	33/10	-	84.3	existing and operated
4	Coal stoker-fired steam boiler SSCB	40/20	-	85.0	existing and operated
5	Back-presure steam turbine STU	-	10.7	29.0	existing and operated
6	STU turbine heat exchanger	29/7	-	-	existing and operated
7	Coal stoker-fired water boilers SCB 3	40/10	-	85.0	currently disconnected
8	Gas turbine unit GTU with HRB boiler	13.5/13.5	7.0	82.0	existing and operated
9	Gas boiler	25/5	-	92.0	existing, peak-load
10	Biomass stoker-fired steam boiler BB	13.7/7.6	-	82	planned
11	Extraction-condensing steam turbine	-	6.7	31.0	planned
12	Pulverized coal-fired water boiler PCB 2	81/32	-	91.5	planned
13	Gaseous ICE-based CHP – 5 units	5 x 3.5/0.7	5 x 4.5	83.5	planned

- scenario **S3** - based on S2 configuration with GTU unit removed and with steam turbine unit STU operated at the base load,
- scenario **S4** - existing configuration of the plant with PCB replaced by currently disconnected repowered SCB3, peak-load gas boiler in use.
- scenario **S5** - existing configuration of the plant with PCB replaced by repowered PCB2,
- scenario **S6** - existing configuration of the plant with additional CHP block installed on the basis of 5 ICE-based gaseous units (Fig. 4), with steam turbine unit STU operated at the base load,

where: N_{el,u_i} is the annaul electricity production of the particular unit i, $\dot{N}_{el,u_i}$ is the current electric power output of the unit i provided by the simulation procedure.

Within the calculation procedure the annual demand for the particular fuels was derived on the basis of nominal efficiency of operated units (Table 2) corrected in line with load characteristics of units. The low heating values were assumed 23 MJ/kg for hard coal, 36 MJ/m^3 for natural gas and 11 MJ/kg for wooden biomass. The annual carbon dioxide emissions were calculated on the basis of derived fuel quantities and assumed emission factors. The thermodynamic evaluation was the basis for further economic analysis.

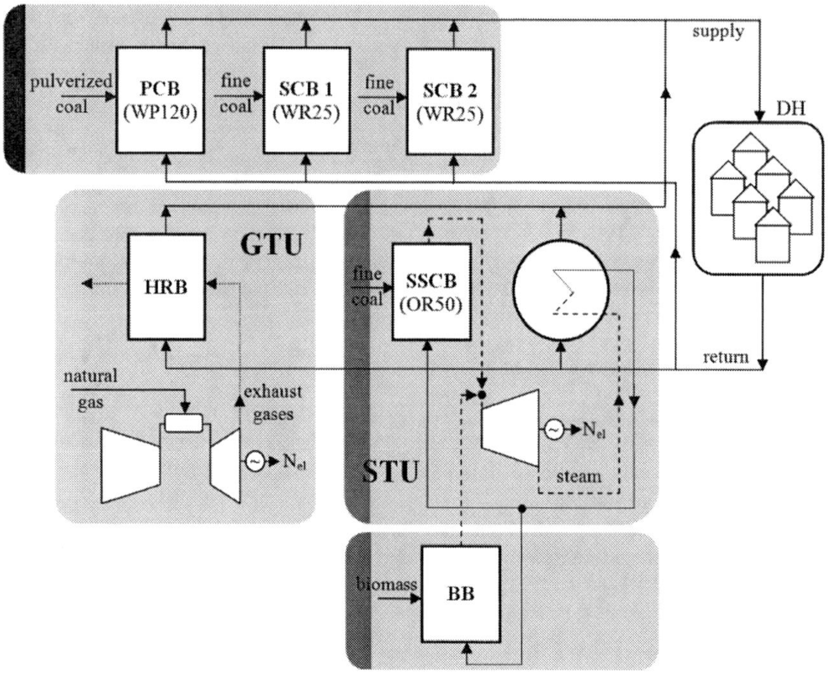

Fig. 3. Schematic diagram of DH plant configuration in scenario **S2**.

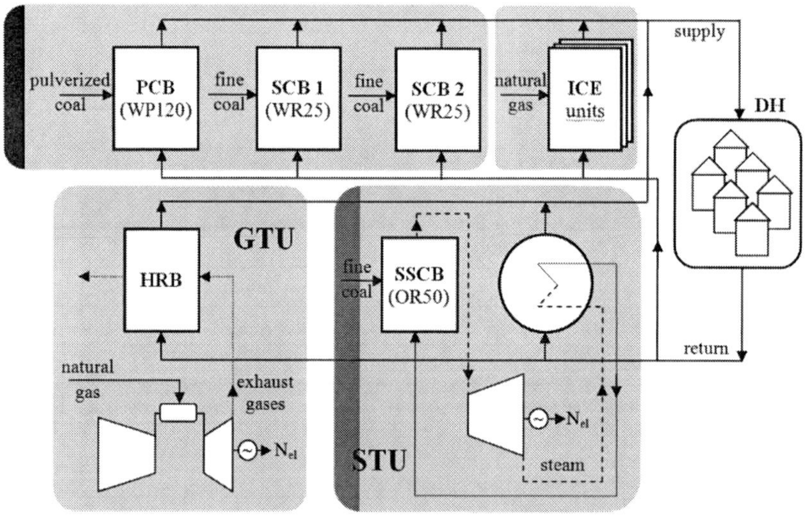

Fig. 4. Schematic diagram of DH plant configuration in scenario **S6**.

3.2 Economic evaluation

The goal of the optimisation procedure is to find the optimal configuration of the DH plant regarding the economic feasibility. The economic evaluation bases in this case on discounted method, with Net Present Value (NPV) as the result of the calculations. The chosen method is widely used in cost-effective designs [15], lifetime optimisation [16] and in comparative techno-economic analysis [17]). To determine the NPV index, cash flow parameter *CF* is used [18], in accordance with the following formula:

$$NPV(t = N) = -J_0 + \sum_{t=1}^{N} \frac{CF_t}{(1+r_d)^t} \qquad (3)$$

where t is the time of operation in years, $N$is assumed lifetime of the project expressed in years, J_0 is the total investment outlays in particular scenario, CF_t – annual

DOI: 10.1051/e3sconf/20171901007

cash flow calculated for each scenario (in EUR), r_d – rate of discount.

Annual cash flows were derived with use of the formula (4):

$$CF_t = C_\tau + S_\tau + T_\tau + L \quad (4)$$

whereC_t is the annual cost of operation, including fuel costs, maintenance costs, overhauls costs, environmental costs and cost of buying CO_2 allowances (European Union Allowances – EUA), S_t is the annual sale income (revenues from the sales of heat into the district heating system and revenues due to selling electricity to the grid), T_t is the yearly income tax, L is the liquidation value of the project.

Apart from NPV index, the applied method enables to calculate also Net Present Value Ratio (NPVR) indicator which is expressed by the equation:

$$NPVR = NPV/J_0 \quad (5)$$

3.2.1 Assumptions for economic analysis

To calculate the value of the annual cash flow CF it was necessary to adopt economic assumptions and additional data. The most important of them are:

- investment outlays for each scenario, specified in Table 3,
- overhauls costs for each scenario, specified in Table 3 as total overhauls costs (in 18 years of operations),
- the sales price of heat p_h = 9.65 EUR/GJ,
- the sales price of electricity s_{el} = 35.29 EUR/MWh [19]
- the price of cogeneration tags (yellow tags) s_{YT} = 25.88 EUR/MWh [19],
- the price of gaseous cogeneration tags (yellow tags) s_{YT} = 25.88 EUR/MWh [19],
- the price of solid fuel cogeneration tags (red tags) s_{RT} = 2.35 EUR/MWh [19],
- the price of renewable tags (green tags) p_{GT} = 23.53 EUR/MWh [19],
- the price of hard coal p_c = 2.81 EUR/GJ,
- the price of biomass p_b = 4.94 EUR/GJ,
- the price of natural gas p_{ng} = 5.61 EUR/GJ [19],
- the price of carbon dioxide CO_2 allowances p_{EUA} = 7 EUR/t CO_2.

The further, additional assumptions have been also made, as follows:

- currency exchange rates 4.25 PLN/€,
- operation period of the system: N = 18 years,
- priority in the production of electricity for the internal load of the heat generating plant is taken into consideration, excess quantities are sold,
- quantitative data regarding the production and sales in the subsequent years of the operation of the system are derived from the calculations performed in the thermodynamic evaluation,
- discount rate was adopted at the level of r_d=7.5%.

Table 3. Investment outlays J_0 and cost of overhauls C_R for analyzed scenarios.

Scenario	Outlays, mln EUR	Entire (18 years) overhauls, mln EUR
S1	12.07	26.39
S2	6.73	23.25
S3	6.73	21.51
S4	3.48	23.82
S5	37.95	27.40
S6	2.53	41.70
S7	2.53	36.09
S8	2.53	35.41

4 Results and discussion

As a result of the research the hourly simulations for each considered DH plant scenario were conducted. The result operating configurations have been presented on the DH heat load durations curves (Fig. 5). On the basis of simulation the quantitative data needed for further economic analysis were obtained.

Table 5 summarizes the annual demand for fuel in the analyzed configuration cases and the annual quantity of heat and electricity production in each DH plant scenario. It also contains CO_2 emissions calculated during the evaluation. It is worth noting that much less CO_2 emissions occur in case of biomass scenarios (S2 and S3). The positive effect of cogeneration should be emphasized for ICE modules scenarios (S6 to S8): in these cases the share of electricity production is significantly larger compared to other configuration alternatives.

Table 4 contains results of economic evaluation. Results have also been presented graphically in the Figure 6.

Table 4. Results of economic analysis conducted for the analyzed scenarios.

Scenario	NPV, mln EUR	NPVR
S1	33.65	2.79
S2	28.71	4.27
S3	30.59	4.55
S4	34.12	9.80
S5	21.88	0.58
S6	37.64	14.88
S7	34.59	13.67
S8	33.64	13.30

For easier results comparison, NPV cells in Table 5 have been marked with colors, where red indicates low values (worse feasibility) and green stands for highest ones (best feasibility).

DOI: 10.1051/e3sconf/20171901007

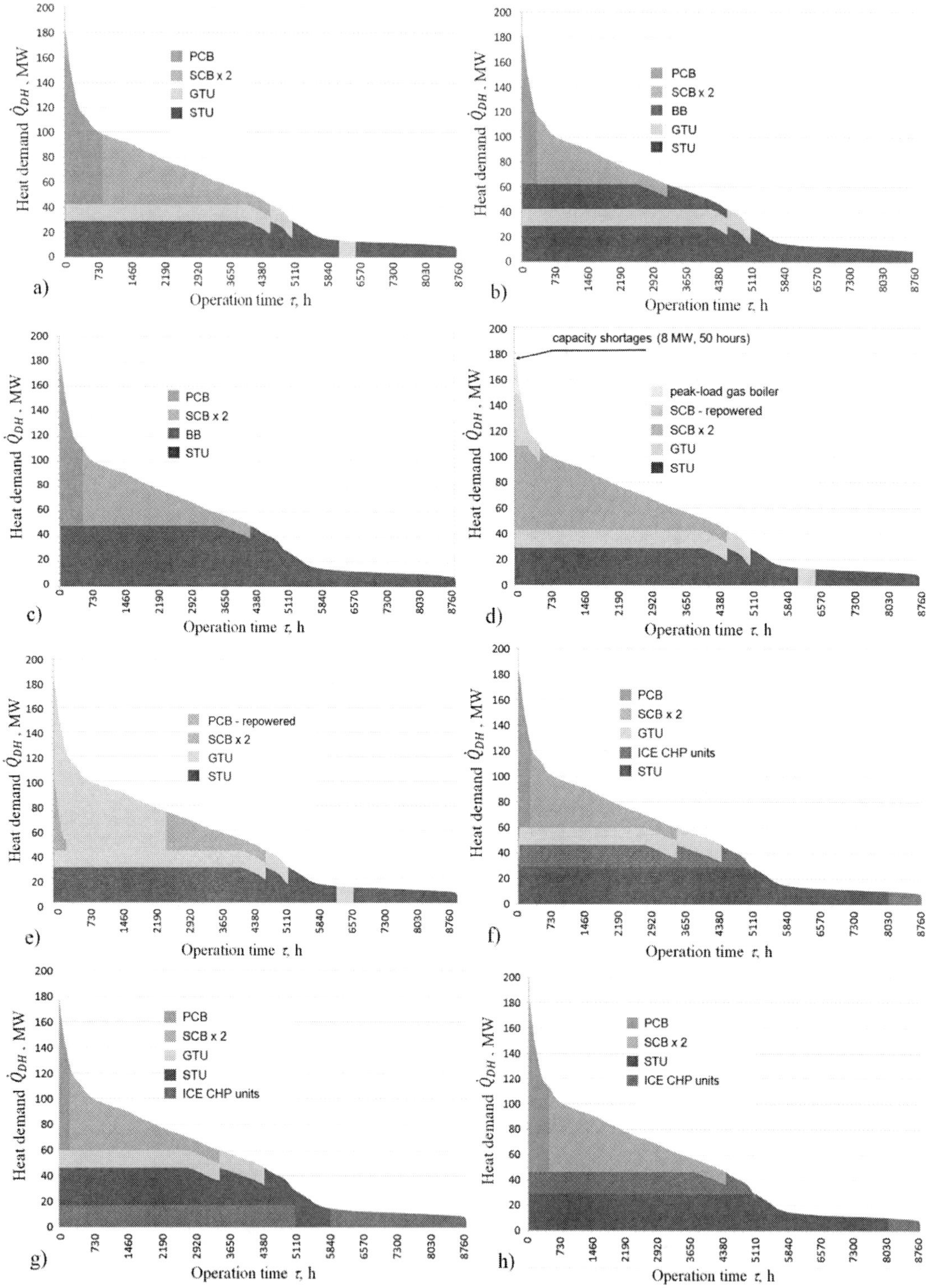

Fig. 5. Results of the simulation of the annual operation of DH plant units conducted for the assumed configuration scenarios: a) S1, b) S2, c) S3, d) S4, e) S5, f) S6, g) S7 and h) S8.

Table 5. Quantitative annual operation data of each analyzed plant configuration scenario.

Scenario	Heat generation, GJ	Electricity generation, MWh	Demand for coal, Mg	Demand for natural gas, MWh	Demand for biomass, Mg	CO_2 emission, Mg
S1	1 571 317	106 037	78 642	149 397	0	199 102,5
S2	1 571 317	102 750	59 135	140 140	44 704	157 302,0
S3	1 571 317	68 502	69 385	0	50 333	151 527,4
S4	1 570 249	106 037	79 106	152 162	0	200 622,2
S5	1 571 317	106 037	77 719	149 397	0	197 087,7
S6	1 571 317	200 316	66 569	362 528	0	211 775,0
S7	1 571 317	238 109	56 498	484 061	0	212 038,0
S8	1 571 317	177 642	76 153	260 957	0	214 102,1

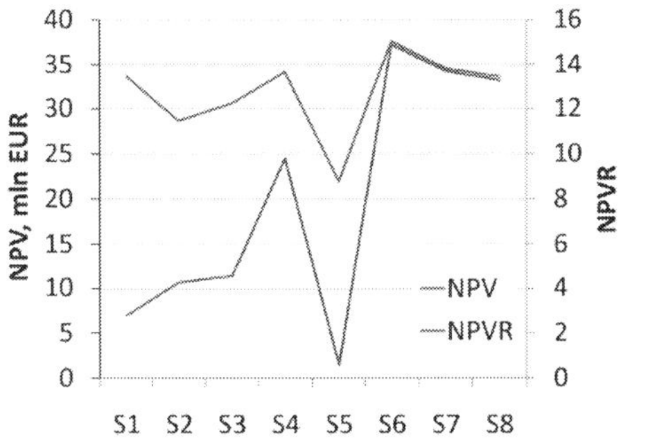

Fig. 6. NPV and NPVR indexes derived for analyzed scenarios.

As it can be concluded both from the Table 5 and Figure 6, the highest feasibility, regarding both criteria, NPV and NPVR, occurs for scenarios where CHP units based on natural gas-fired ICE modules are implemented. NPV values in these cases stands between 13 and 15 mln EUR, while for the weakest scenario (S5) hardly exceeds 9 mln EUR. Results gained for S8 case shows that cutting off gas turbine unit from ICE-based scenario lowers profitability of the project: NPV goes down by 1 mln EUR noticed in scenario S7 or even by 4 mln EUR of scenario S6.

It is important to notice here that NPV and NPV-to-investment ratio trends don't coincide in case of scenario S1, where relatively high investment expenses are involved compared to expected profit.

4.1 Sensitivity analysis

In order to evaluate the relation between project value NPV and the key economic parameters, a sensitivity analysis has been performed for analyzed scenarios (Table 6). The procedure of economic evaluation was subsequently repeated for the alternative values of key economic parameters: natural gas and coal prices, electricity selling price, yellow tags price and EUA prices.

The variations of the prices, except for EUA, were assumed as ±10%. In case of EUA two price alternatives were taken into account:

- constant price 7 EUR/Mg of emitted CO_2,

Table 6. Results of sensitivity analysis: NPV in mln EUR.

Scenario	EUA allowanceprices, EUR/Mg CO_2		Natural gasprices		Hard coalprices		Electricityprices		Yellow tags prices		
	7	Variable from 6 to 23	-10%	10%	-10%	10%	-10%	10%	-100%	-10%	10%
S1	31,8	21,8	34,0	29,3	31,8	31,8	28,9	34,4	23,6	30,9	32,4
S2	27,1	19,1	29,6	24,9	27,1	27,1	24,7	29,8	19,3	26,4	28,0
S3	28,9	21,3	30,0	27,8	28,9	28,9	26,7	31,1	24,2	28,4	29,3
S4	32,2	22,2	34,7	29,8	32,2	32,2	29,6	34,9	24,2	31,3	32,9
S5	20,7	10,9	23,1	18,4	20,7	20,7	18,0	23,3	12,7	20,0	21,6
S6	35,6	28,4	40,9	30,4	35,6	35,6	31,1	40,2	11,1	33,1	38,0
S7	34,7	27,3	38,9	30,4	34,7	34,7	30,4	38,7	13,6	32,4	36,7
S8	31,8	23,6	35,1	28,4	31,8	31,8	28,2	35,3	15,6	30,2	33,6

- variable price in the analyzed time range from 6 EUR in 2018 to 23 EUR in 2035, linearly.

Additionally, case with no support system for cogeneration (yellow certificate prices equal 0) is also analyzed.

Again, for easier results comparison, NPV cells in Table 6 have been marked with colors, where red indicates low feasibility compared to green standing for high one. As the presented values show, scenarios S1 based on current plant configuration and scenarios with ICE cogeneration units are the most sensitive ones. Increase of natural gas prices by 10% or decrease of electricity prices by 10% turns ICE-based scenarios less profitable then scenario S1. Termination of cogeneration support system (no yellow tags and no alternative for that system) makes the configuration with plant retrofitted by ICE units unbeneficial.

5 Final conclusions

The presented study is an example of how the proper configuration of the plant supplying district heating system may positively affect the feasibility of the plant operation. Moreover, the analysis shows how to respond to the expected changes in regulations and environmental requirements to preserve the best possible economic results of the plant.

The calculations conducted for the basic economic assumptions showed higher profitability of the cases with natural gas-fired ICE cogeneration modules installed in the plant. However, the sensitivity analysis shows possibility of changing the results of economic analysis so the ICE cases turn into worse configuration option and the current, coal-based facility stays most profitable. It can be also concluded from the analysis that economic results of operating retrofitting municipal district heating plant results from the CO_2 allowances prices. Having that in mind, the biomass-fuelled units may give the positive input for the project feasibility.

Accordingly, the following final conclusions can be derived from the undertaken analysis:

- projecting the configuration of the district heating plant requires good prediction on electricity and fuels' prices,
- configuration of the system should always be preceded by thermal and economic optimization,
- retrofitting coal-based DH plant by the cogeneration units fuelled with natural gas may be the solution for on-going changes in environmental and energy efficiency regulations,
- implementation of the biomass district heating facilities in coal-based DH systems can improve feasibility under expected legislative conditions,
- however, if the biomass is expected to get the better of natural gas oriented configurations it requires appropriate mechanisms.

Acknowledgement

The authors would like to thank the ECO SA – Heat Engineering Company engineers for the cooperation and for the access to operational data of heat generating district heating plant. The authors also gratefully acknowledge the possibility of using the design data of analyzed unites to the ECO SA company.

References

1. European Commission, "A roadmap for moving to a competitive low carbon economy in 2050," Brussels, 2011.
2. M. Tańczuk, J. Skorek, P. Bargiel, Energy Convers. Manage. **149** (2017)
3. Y. Dou et al. Resour. Conserv. Recycl., (2016), http://dx.doi.org/10.1016/ j.resconrec.2016.03.006. In Press
4. H. Lund, S. Werner, R. Wiltshire, S. Svendsen, J. E. Thorsen, F. Hvelplund and B. V. Mathiesen, Energy, **68** (2014)
5. N. Duić N, M.A. Rosen, Energy Convers. Manage. **87** (2014)
6. S. Anweiler, D. Piwowarski, J.Clean.Prod. **155** (2017).
7. A. Lake, B. Rezaie, S. Beyerlein, Renewable Sustainable Energy Rev. **67** (2017)
8. M. Ameri, Z. Besharati, Energy Build. **110** (2016)
9. Intelligent Energy Europe Programme. Energy Efficiency Trends and Policies in the EU an analysis based on the ODYSSEE and MURE databases (2015)
10. Euroheat & Power, http://euroheat.org, 2017 [accessed 15 May 2017]
11. District Heating and Cooling Country by Country. 2015 Survey. Euroheat & Power.
12. M. Badami, M. Mura, P. Campanile, F. Anzioso, Energy **33** (2008)
13. M. Elci, A. Oliva, S. Herkel, K. Klein, A. Ripka, Energy Procedia **70** (2015)
14. M. Pavičević, T. Novosel, T. Pukšec, N. Duić, Energy (2017) https://doi.org/10.1016/j.energy 2017.06.105 (In Press)
15. A. Vlysidis, M. Binns, C. Webb, C.A. Theodoropoulos, Energy Convers. Manage. **103** (2015)
16. A. Nemet, J.J. Klemes, Z. Kravanja, Energy **57** (2013)
17. M. Wzorek, M. Tańczuk, Waste Manage. Res. **33**, 7 (2015)
18. J. Kalina, J Energy Convers. Manage. **51** (2010)
19. Energy Regulatory Office, http://www.ure.gov.pl/en/ 2017 [accessed 15 May 2017].

Co-firing coal and biomass blends and their influence on the post-combustion CO_2 capture installation

Angelika Więckol-Ryk[1,*], and *Adam* Smoliński[2]

[1]Central Mining Institute, Department of Risk Assessment in Industry, Katowice, Poland
[2]Central Mining Institute, Department of Energy Saving and Air Protection, Katowice, Poland

Abstract. Co-firing of biomass with coal for energy production is a well-known technology and plays an important role in the electricity sector. The post-combustion capture integrated with biomass-fired power plants (Bio-CCS) seems to be a new alternative for reducing greenhouse gas emissions.
This study refers to the best known and advanced technology for post-combustion CO_2 capture (PCC) based on a chemical absorption in monoethanolamine (MEA). The co-firing of hard coal with four types of biomass was investigated using a laboratory fixed bed reactor system. The comparison of gaseous products emitted from the combustion of coal and different biomass blends were determined using gas chromatography.
Research proved that co-firing of biomass in fossil fuel power plants is beneficial for PCC process. It may also reduce the corrosion of CO_2 capture installation. The oxygen concentration in the flue gases from hard coal combustion was comparable with the respective value for a fuel blend of biomass content of 20% w/w. It was also noted that an increase in biomass content in a sample from 20 to 40 % w/w increased the concentration of oxygen in the flue gas streams. However, this concentration should not have a significant impact on the rate of amine oxidative degradation.

1 Introduction

Combustion of biomass and coal for power production helps to reduce greenhouse gas (GHG) emissions from the existing coal-fired power plants. Carbon dioxide, methane, nitrous oxide and perfluorocarbons are the main anthropogenic sources of GHG. For the electricity sector, co-firing of biomass with fossil fuels has a positive impact on CO_2 reduction since it is counted as neutral. Biomass has quite different chemical composition in comparison with hard coal, that is why burning it with coal blends also reduces NO_x and SO_x levels. Co-firing has a positive influence on the environment and economy depending on the chemical composition of the biomass used [1]. It minimizes the amount of waste, soil and water pollution and may also reduce fuel costs.

Recent studies [2, 3] have shown that integrated CO_2 capture and storage (CCS) with biomass-fired power plant is also considered. Enforcing CCS in combination with biomass co-firing power plants provides a further possibility of reducing CO_2 emissions for the electricity [4]. Nowadays, there is an on-going discussion about positive and negative aspects of incorporating CCS into biomass-fired power plants. One of the reasons behind current studies is to find the answer to the following question: "Would the incorporated CCS with biomass fired power plant be an attractive option to further minimalize the cost of CO_2 and generate an additional revenue as CO_2 credit?" [3].

In 2014, greenhouse gas emissions were reduced by 22.9 % compared with the ones in 1990 in 28 Member States of the European Union (EU-28). The overall highest GHG emissions of the EU-28 were in Germany (21.9%) and the lowest in Malta 0.08%. Poland was 24th in terms of the amount of GHG emissions (8.64%) and it overtook Italy (9.69%), France (10.76%), the United Kingdom (12.6%) and aforementioned Germany. The greenhouse gas emissions in Poland in 2014 were reduced by 19.3% when compared with 1990 [5]. After the recent ratification, the EU target is to reduce GHG levels by at least 20% until 2020 and 40% until 2030. The proposed legislation includes tightening the EU Emissions Trading System (ETS) for the period after 2020.

At present, CO_2 fees and ETS credit to cover its CO_2 emissions allows EU-28 to stabilize energy price. From 2008 to 2014, the cost of unit CO_2 emission increased by 21% and stayed at the same level for the last two years [6]. However, calculation of CO_2 emissions and Green Certificates shows that the price may increase significantly in the future. One of the theoretical price scenarios of the U.S. Environmental Protection Agency's (EPA) set Low, Mid, and High case forecasts for CO_2 prices. The CO_2 price per ton according to forecast will increase from \$15 to \$25 in 2022, from \$21 to \$43 in 2030 and from \$36 to \$110 in 2050. In order to avoid paying such high fees, it seems to be effective to carry

* Corresponding author: awieckol@gig.eu

DOI: 10.1051/e3sconf/20171901008

out further research on CCS incorporated into biofuels. The IEA confirms that the cost of reducing GHG without CCS would be over 70% [2]. The combination of CO_2 capture with sustainable biomass conversion (Bio-CCS) seems to be the only large scale and the most promising technology that can remove CO_2 from the atmosphere. It has already been raised in the discussion over the EU Energy Roadmap 2050.

It was presented that currently carbon capture and storage (CCS) is at a phase of demonstration but in the future it could be applied to various energy plants, especially incorporating co-firing or co-gasification of sustainable biomass feedstock, or 100% biomass energy plants, biofuel production facilities or bio refineries.

The aim of this study is to analyse and describe the co-firing process of biomass with coal and to present its influence on capturing CO_2 from the flue gases by chemical absorption.

2 Biomass combustion with CO_2 capture

Combustion of biomass with fossil fuels has a good influence on the environment and the economy of power generation. Carbon dioxide is emitted into the atmosphere during biomass fuels burning, yet CO_2 is also absorbed during its growth thus the combustion of biomass is considered to be CO_2-neutral.

The main components of biomass are carbon, oxygen and hydrogen but they also contain nitrogen, sulphur and small amounts of chlorides. Biomass fuels may differ from each other and the composition of flue gas depends on the type of biomass during combustion. The main component of ash is Ca, K, Si, Mg, Mn, Al, Fe, P, Na and Zn. Due to the fact that biomass fuels have little concentration of sulphur in comparison to hard coal they emit lower amounts of SO_2 when they are fired and co-fired as blends. Co-firing of biomass with coal allows us also to pay less carbon tax due to lower CO_2 emission [4].

The IEA Greenhouse Gas R&D Programme (IEAGHG) [3] estimated the potential advantage of biomass fired power plant with post-combustion capture of CO_2. The study focused on techno-economic evaluation of different cases of biomass combustion and biomass co-firing power plants with and without CO_2 capture. The economic analysis of electric power production was based on the following cost: total investment, coal and biomass, Green Certificates and the ETS. The analysis showed that in case the ETS value rises according to the forecasts from 2030 [7], the existing biomass-fired or co-fired power plant with CO_2 capture will be comparable to power plants without CO_2 capture.

The use of biomass for energy production and conversion to other products plays an important role in Bio-CCS technology nowadays. Bio-CCS is used in many different facilities from the power, industrial and fuel sectors. It delivers power and heat; helps the atmospheric carbon dioxide recycling and does not contribute to negative emissions technologies (NETs). Moreover, biomass allows thermo-chemical and biochemical production of biofuels such us biomethane, bioethane or lignocelulosis.

The IEAGHG studies [8] estimated a technical and economic potential of Bio-CCS about 10 $GtCO_2$/yr in 2050. The following Bio-CCS routes included: biomass integrated gasification combined cycle with CCS (BIGCC-CCS), circulating fluidised bed combustion with CCS (CFB-CCS), integrated gasification combined cycle with CCS (IGCC-CCS), and pulverised coal fired power plant with CCS (PC-CCS). Post-combustion capture with chemical absorption is very extended technology for removing CO_2 from industrial flue gases. The most popular absorbent used in this process is aqueous solution of monoethanolamine (MEA). This technology is well-known and advanced but requires high regeneration energy. The MEA-based CO_2 capture presents also many problems such as equipment corrosion and amine degradation [9].

Oxygen and impurities contained in flue gas: acid gases (SO_x, NO_x, H_2S), heavy metals, dust particulates [10] as well as hydrocarbons [11] can react with the solvent and lead to many degradation products. Oxidative degradation is the dominant type of amine degradation in the post-combustion capture process (PCC) based on MEA [12]. It mainly takes place in the absorber section where oxygen concentration is more than $5\%_{vol}$ [13].

The most common amine degradation products are ammonia, formaldehyde and carboxylic acids such as acetic or oxalic one [14, 15]. All of them reduce the pH of amine solution and lead to equipment corrosion or technological problems like amine losses or foaming [16]. Oxygen is also well-known source of amine oxidative degradation products [17, 18, 19]. The reactions of MEA and an organic or inorganic acids lead to formation the Heat Stable Amine Salts (HSAS) [20, 21]. According to [22] even a small amount of HSS may lead to amine degradation so the concentration of them in MEA solution should not be more than 500 ppm. To avoid amine degradation and losing expensive solvent the concentration of SO_x in the flue gases should not exceed 10 ppm [23].

The study of flue gas composition during biomass combustion and co-firing has not been widely published yet, but may show the advantages and disadvantages of the Bio-CCS and its influence on CO_2 post-combustion capture system.

Although no post-combustion capture units have been installed yet at full scale, large demonstration projects based on amine scrubbing are being prepared. Combining biomass with CCS technologies (Bio-CCS), gasification for power production and conversion to biofuels are the promising technology routes in the power and transport sectors up to 2030 and 2050.

3 Material and methods

3.1 Material

One Polish hard coal (samples nos. C) provided by Ziemowit coal mine and the following biomass samples:

Helianthus Tuberosus, *Salix Viminalis*, olive residues (*Olea europaea.)*, and *Sida Hermaphrodita* – denoted as samples nos. CB1, CB2, CB3 and CB4, respectively were used in the study presented. A sample of 10g of hard coal or its blends were prepared in accordance with the PN-G-04506:1996 standard. Prior to the analysis, the fuels were grounded using mill with a particle size between 1 and 3 mm and dried in the oven at 105°C for 24h in order to remove moisture.

The physico-chemical parameters included: total moisture, ash, volatiles and elemental composition such as total sulphur, carbon, hydrogen, nitrogen and oxygen were determined by the accredited Laboratory of the Department of Solid Fuels Quality Assessment of the Central Mining Institute. The results were shown in Table 1. The elemental composition of coal and biomass was the same but biomass contained more oxygen, total moisture, volatiles, hydrogen and nitrogen compared to coal samples. The amount of biomass total sulfur was almost four times lower than a coal sample.

Table 1. Physical and chemical characteristics of the studied fuels.

Parameter	Fuel sample				
	C	B1	B2	B3	B4
Moisture content, W^a, %wt	11.49	8.37	4.74	7.54	8.76
Ash content, A^a, % wt	5.42	5.93	1.51	4.11	2.63
Volatiles content, V^a, % wt	32.31	66.81	73.16	77.50	71.47
Total sulphur content, S_t^a, % wt	1.32	0.26	0.05	0.05	0.04
Carbon content, C_t^a, % wt	64.90	46.16	52.19	49.93	47.18
Hydrogen content, H_t^a, % wt	3.63	5.55	6.22	6.21	5.68
Nitrogen content, N^a, % wt	0.96	1.71	<LD	1.59	<LD
Oxygen content, O^a, % wt	13.24	32.28	35.29	42.04	35.73

C-hard coal, Biomass: B1-*Helianthus Tuberosus*, B2-*Salix Viminalis*, B3-*Olea europaea*, B4-*Sida Hermaphrodita*; ^{a}-*analytical state*, <LD-below the limit of detection.

A 10 g sample of hard coal and its biomass blends was placed at the bottom of the reactor at a ratio of 10, 20, 30 and 40 % w/w for B1 and 20 % w/w for B2, B3 and B4 respectively.

3.2 Methods

The tests were conducted in the laboratory scale fixed bed reactor set-up of the Laboratory of Advanced Energy Technologies, the Department of Energy Saving and Air Protection of the Central Mining Institute [24, 25] (Fig. 1).

The compositions of dry and clean samples of synthesis gas were analysed automatically via the two-channel gas chromatograph, Agilent 3000A included in two thermal conductivity detectors (TDC). The PLOT U column with TCD1A detector was applied for separation of CO_2, C_2H_6 and H_2S. The MS5A PLOT column with TCD2A detector was applied for separation of H_2, O_2, N_2, CH_4 and CO. Helium and argon were used as the carrier gases in two columns, respectively.

To protect the sample against an entrainment by the gaseous media and for the better temperature distribution, coal and coal blends samples was placed at the bottom of the reactor between quartz wool.

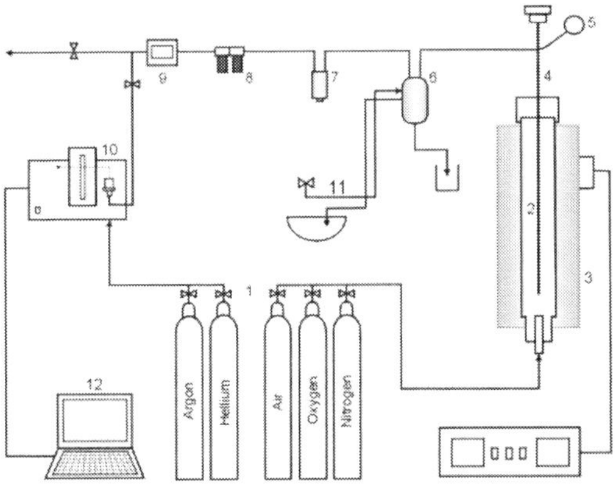

Fig.1. The laboratory-scale fixed bed reactor installation for coal combustion and gasification: 1-gases inlets, 2-fixed bed reactor, 3-resistance furnace, 4-thermocouple, 5-manometer, 6- water trap,7-solid particles filter, 8-gas dehumidifilter, 9-mass flow meter, 10-gas chromatograph, 11- gas cooling water system, 12-computer.

The reactor was heated up at a rate of 1.33°C/s to the temperature of 900°C in the air atmosphere and with flow $(90\pm6)\times10^{-3}$ cm s^{-1}.

The amount of air required for the complete combustion of the selected carbon sample and its biomass blends was calculated stoichiometrically.

The oxygen concentration is closely related to the flow rate of air supplied to the coal boiler and the excess air fuel ratio (AFR). The value of the excess AFR depends primarily on the type of fuel combusted, the combustion process as well as the design of the boiler. Changing the excess AFR results in the amount of gas products combustion. Insufficient amount of air causes the formation of carbon monoxide and hydrocarbon, whereas excess of air leads to nitrogen oxides emission. For the total combustion, the tested samples of the AFR ratio was within the range 1.3-1.4.

Before the gas composition was analysed with the GC, the outlet gas was cooled and dried from liquid products in the water trap. The amount of gaseous products mixture (O_2, H_2, CO, CO_2, CH_4, C_2H_6, H_2S) was measured with the mass flow meter. Time-interval length resulting from the settings of the chromatographic method was 240s. The number of time intervals during the experiment for each sample was 16.

The volume values of the individual gas components of in the co-firing process of coal with biomass V_c was calculated from the equation (1):

DOI: 10.1051/e3sconf/20171901008

$$V_c = \sum_{i=1}^{n} \frac{C_{ci}F_i}{100} t_i \qquad (1)$$

where:

V_c – value of i_{th} gas component obtained during the experiment, m^3

C_{ci} – volume concentration of c_{th} gas component in i_{th} time interval, % vol.

F_i – gas flow rate in i_{th} time interval, m^3s^{-1}

t_i – length of i_{th} time interval, s

n – number of time intervals during the experiment.

4 RESULTS AND DISCUSSION

4.1 Influence of biomass type on gaseous product in co-firing process

The temperature inside the reactor measured for each tested fuel sample was comparable and was within the range of 70-930°C. The first flue-gas samples were determined 240 seconds after the start of the combustion process and contained significant amounts of airborne oxygen. Therefore, for a comparative analysis of flue-gas composition, a time interval of 720-2880s was adopted, corresponding to the combustion temperature of about 250-850°C. The temperature increase during the experiment for each examined sample was comparable as shown in Fig. 2.

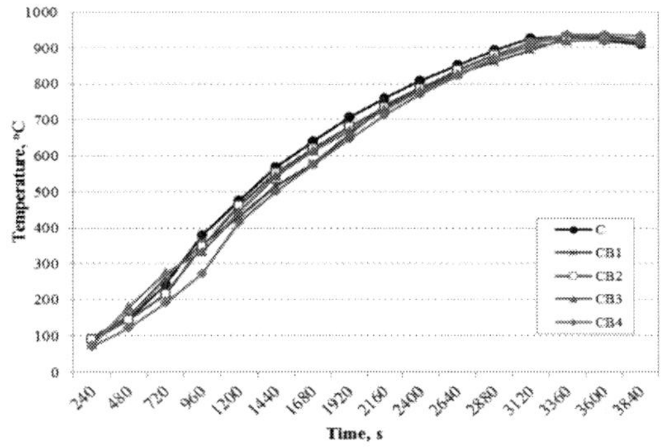

Fig.2. Temperature increase inside the reactor during the experiment for the tested fuel samples.

The main components of the gases released during the co-firing of coal and biomass blends were CO_2, CO, O_2 and N_2. There was also less C_2H_6 and H_2, H_2S and other components, i.e. hydrocarbons, sulphur and nitrogen oxides. The measured maximum concentration of CO_2 for the pure carbon sample during the tested time interval was up to 16.8% vol., and it was close to the maximum concentration of CO_2 emitted during combustion of coal blends with biomass CB1, CB2, CB3 and CB4, that is 15.4% vol., 17.0% vol., 18.2% vol. and 16.8% vol., respectively. The change in CO_2 concentration during the experiment is shown in Fig. 3.
In the CO_2 capture process using post-combustion method, the oxygen content in the flue gas has the greatest impact on the MEA degradation process.

In the CO_2 capture process using post-combustion method, the oxygen content in the flue gas has the greatest impact on the MEA degradation process.

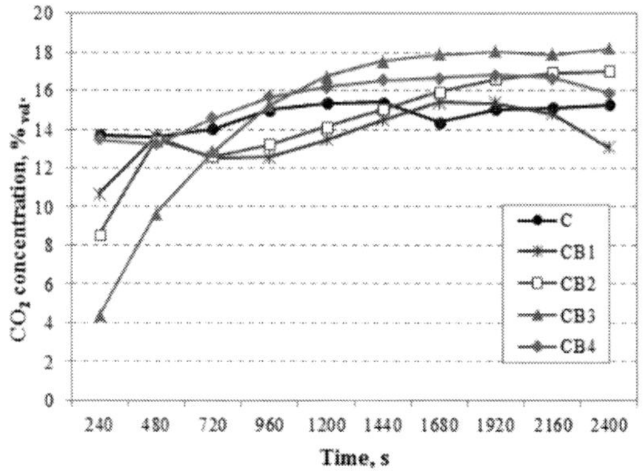

Fig. 3. Change in carbon dioxide concentration over time for the tested fuel samples.

The analysis of oxygen concentration during the experiment showed its slight increase during the combustion of coal and biomass blends in comparison with the coal sample. Most of the oxygen in the range of 250-850°C was obtained for BC3 mixture. It was related to the chemical composition of biomass as well as its ash and moisture contents. As shown in Fig. 4, all biomass blends burned more slowly, and the loss of oxygen was slower.

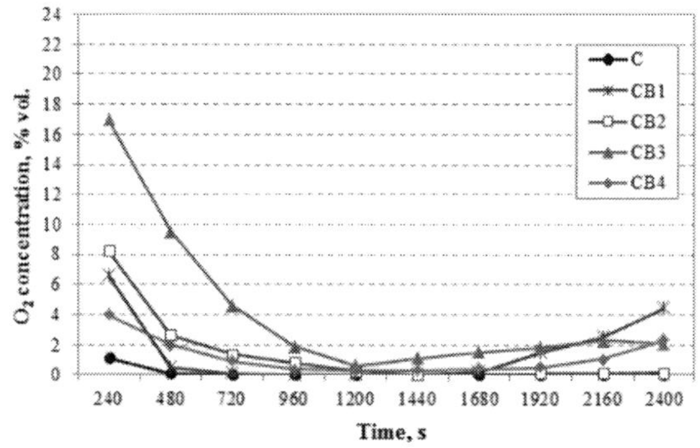

Fig. 4. Change in oxygen concentration over time for the tested fuel samples.

Taking into account the fact that the degradation of the MEA solution in the post-combustion process starts only with the oxygen content in the flue gases above 5% vol., it can be stated that the combustion of the coal-biomass blends will not affect significantly the increase in the rate of amine oxidative degradation of amine solution.
The mean concentration of gaseous products, excluding nitrogen, is shown in Fig. 5, and the volume of all gaseous products and the total volume of gas blends V_c calculated from the equation (1) are shown in Table 2.

The analysis showed that methane content in the combusted coal sample was 1.9% vol, and the presence of this component in the gases after combustion of coal and biomass blends was within the range of 1.8-2.5%.

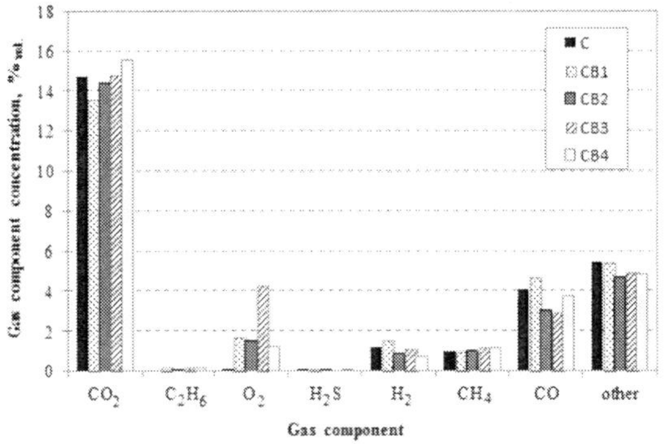

Fig. 5. Average concentration of gas components for the tested coal and biomass blends.

The content of C_2H_6 in all combusted samples was close to zero.

In the combustion processes of solid fuels, its sulphur content is converted into hydrogen sulphide and sulphur oxides. The technology of CO_2 capture process using amine method is very sensitive to the sulphur content in flue gases and the permissible SO_x level should not exceed 10ppm. Because of the low share of sulphur in the chemical composition of biomass compared to the pure carbon sample, the concentration of H_2S in biomass blends flue gas (0.03-0.04% vol.) was three times lower than in the sample of the combusted coal (0.13% vol.). Based on this analysis, it can be stated that the biomass content in the tested fuel blends will also have an impact on the reduction of sulphur oxides whose presence above 10ppm is one of the main causes of MEA degradation.

Table 2. Volume of individual gas components and total volume of gas for the tested coal and biomass blends.

Gas component	Volume, 10^{-3} m^3				
	C	CB1	CB2	CB3	CB4
CO_2	4.99	5.22	4.85	5.41	5.78
C_2H_6	0.01	0.01	0.01	0.02	0.01
O_2	0.04	0.56	0.49	1.60	0.46
N_2	25.15	28.45	25.88	27.11	28.04
H_2S	0.03	0.00	0.00	0.00	0.00
H_2	0.42	0.36	0.24	0.2	0.18
CH_4	0.14	0.11	0.13	0.17	0.18
CO	1.37	1.16	0.95	0.33	0.29
other	1.86	2.05	1.60	1.78	2.24
V_c, m^3	34.00	37.88	34.20	36.60	37.16

Due to the fact that H_2 and CO do not affect amine degradation process, they were not analysed more extensively. It was noted, however, that the presence of biomass in fuel blends reduced their content in flue gases.

For most biomass and coal blends, the increase in the total volume of flue gas was observed. The relative increase in total volume of gas for CB1, CB3 and CB4 samples in relation to pure carbon sample C was 11%, 8% and 9%, respectively. For the sample CB2, the total volume of gas was comparable.

4.2 Influence of biomass concentration on gaseous product in co-firing process

The research results of co-firing the selected coal and biomass type (*Helianthus Tuberosus*) with 20, 30 and 40% w/w are presented in Fig. 6. For a better comparison of the obtained results, only flue gas samples were analysed within the temperature range of 350-850°C when the amount of gaseous products was the highest.

The differences in concentrations of individual gas components were observed regarding the amount of combusted biomass. The average concentration and volume of carbon dioxide for all fuel blends were comparable. It was noted, however, that the share of other gaseous products increased or decreased depending on the biomass concentration. The biggest upward trend has been visible in the case of oxygen. The mean concentration of this component in the analysed time interval was 1.61% vol. for the sample with a biomass content of 20% w/w and 3.10 and 4.98% vol. for the remaining samples respectively. With the increase in the amount of biomass in the sample, the amount of hydrocarbons in the flue gas decreased. The average concentration of CH_4 was 0.96, 0.32 and 0.00% vol. and the concentration of C_2H_6 was 0.10, 0.03 and 0.00% vol., respectively. The downward concentration trend was also observed for H_2S (0.04, 0.02, 0.00% vol.), H_2 (1.50, 0.46, 0.02% vol.), CO (4.70, 1.11, 0.15% vol.) and the remaining gases (5.38, 4.77, 2.25% vol.).

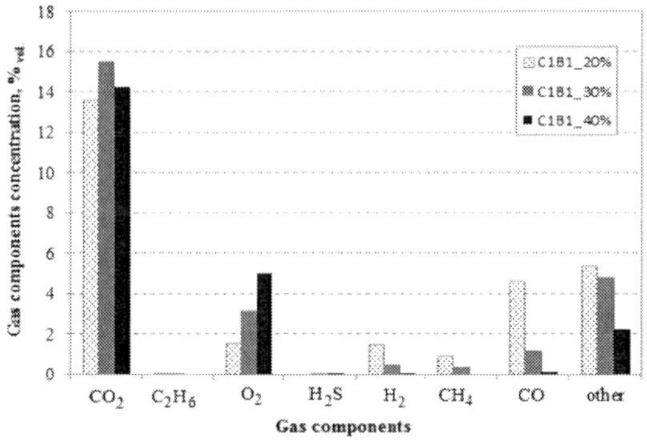

Fig. 6. Average concentration of gas components for the analysed coal and biomass blends.

The volume of the individual gas components and the total volume are shown in Table 3.

As shown in the Table 3, the increase in the percentage share of biomass in the fuel resulted in a decrease in the total volume of the separated gas. Compared to the CB1_20% sample, the volume decrease was 11% for CB1_30% and 18% for CB1_40%.

Table 3. Volume of individual gas components and total volume of gas for the tested coal and biomass blends.

Gas component	Volume, 10^{-3} m^3		
	CB1_20%	CB1_30%	CB1_40%
CO_2	5.22	5.21	4.39
C_2H_6	0.01	0.01	0.00
O_2	0.56	1.04	1.52
N_2	28.45	25.06	24,17
H_2S	0.00	0.00	0.00
H_2	0.36	0.15	0.01
CH_4	0.11	0.11	0.00
CO	1.16	0.37	0.05
other	2.05	1.59	0.07
V_c	37.88	33.60	30.84

5 Conclusions

The conducted studies showed no increased impact of amine solvent degradation on CO_2 capture process using post-combustion method.

During co-firing of coal with biomass energy crops in the amount up to 20% w/w, there is a similar percentage of gaseous products as in the case of pure coal combustion. The oxygen content which contributes to amine degradation is comparable to the various types of biomass and safe for the process to be conducted properly. With an increase in the share of biomass in fuel mixtures, the average oxygen concentration and its volume in the total gas increases proportionately to the amount of biomass. Such flue gas blends subjected to carbon dioxide capture process should be controlled for the risk of oxidative degradation.

The advantage of using an increased amount of biomass is the low emission of sulphur-containing gases as well as the reduced amount of hydrocarbons. All these compounds also have an effect on amine degradation, that is why their reduced content in flue gases has a beneficial effect on the CO_2 capture. The conducted studies also showed that the increase in the share of biomass to coal in the range of 30-40% w/w results in an increase in the total volume of flue gas by 11-18% compared to blends containing 20% w/w of biomass, with similar CO_2 emission.

The results of this research indicate the beneficial effect of the post-combustion method integrated with a biomass-fired power plant.

Acknowledgments

This work was supported by the Ministry of Science and Higher Education. Poland. under Grant no. 11100117.

References

1. A. Demirbas, Progress Energy Combust. Sci. **30**, 219 (2004)
2. EBTP-ZEP, Biomass with CO_2 Capture and Storage (Bio-CCS). The way forward for Europe (2012)
3. R. Domenichini, F. Gasparini, P. Cotone, S. Santos, Energy Procedia **4,** 1851 (2011)
4. Z. Khorshidi, M. T. Ho, and D. E. Wiley, Energy Procedia **37**, 6035 (2013)
5. ZEP, The costs of CO_2 capture, transport and storage: Post-demonstration CCS in the EU, European Technology Platform for Zero Emission Fossil Fuel Power Plants, 2011
6. Regulation of the Minister of Environmental Protection, Monitor Polski **718** (2016) [in Polish]
7. P. Luckow, E. A. Stanton, S. Fields, W. Ong, B. Biewald, S. Jackson, J. Fisher, *Spring 2016 National Carbon Dioxide Price Forecast. Synapse Energy Economics*, 2016
8. J. Kemper, Int. J. Greenhouse Gas Control **40**, 401 (2015)
9. A. Krzemień, A. Więckol-Ryk, A. Smoliński, A. Koteras, L. Więcław-Solny, J. Loss Prev. Process Ind. **43**, 189 (2016)
10. D. Adams, *Flue gas treatment for CO_2 capture.* IEA Clean Coal Centre, CCC/169 (2010)
11. F. S. Manning, R. E. Thompson, *Oilfield Processing of Petroleum: Natural gas* (PellWell Publishing Company Tusla, Oklahoma, 1991)
12. C. Susan, T. Rochelle *First National Conference on Carbon Sequestration* Austin, Texas, 2001
13. J. Kittel, E. Fleury, B. Vuillemin, S. Gonzalez, F. Ropital, R. Oltra, Mater. Corros. **63**, 3, 223 (2012)
14. G. S. Goff, G.T. Rochelle, Ind. Eng. Chem. Res. **43**, 6400 (2004)
15. S. B. Fredriksen, K-J. Jens. Energy Procedia **37**, 1770 (2013)
16. N. Verma, A. Verama, Fuel Proces. Technol. 90, 483-489 (2009)
17. H. Lepaumiera, E.F. da Silva, A. Einbu, A. Grimstvedt, J. N. Knudsen, K. Zahlsen, H. F. Svendsen, Energy Procedia **4,** 1652 (2011)
18. C. Gouedard, D. Picq, F. Launay, P. L. Carrette, Int. J. Greenh. Gas Control **10**, 244 (2012)
19. S. J. Vevelstada, A. Grimstvedtb, J. Elnana, E. F. da Silva., H. F. Svendsena, Int. J. Greenhouse Gas Control **18**, 88 (2013)
20. A. L. Kohl, R. B. Nielsen, *Gas purification - Fifth Edition* (Gulf Publishing Company, Houston, Texas, 1997)
21. T. Supap, R. Idem, and P. Tontiwachwuthikul, Energy Procedia **4**, 591 (2011)
22. P. C. Rooney, T. R. Bacon, M. S. DuPart, Hydrocarb. Process. **75**, 3, 95 (1996)
23. R. M. Davidson, *Post-combustion carbon capture from coal fired plants – solvent scrubbing* (IEA Clean Coal Centre, CCC/125, 2007)
24. A. Smoliński, N. Howaniec, K. Stańczyk, Renew. Energy **36**, 1836 (2011)
25. A. Smoliński, Energy Convers. Manag. **52**, 37 (2011)

Technical and environmental performance of 10 kW understocker boiler during combustion of biomass and conventional fuels

Robert Junga[1,*], *Małgorzata* Wzorek[2], and *Mirosława* Kaszubska[2]

[1] Department of Thermal Engineering and Industrial Facilities, Faculty of Mechanical Engineering, Opole University of Technology, 5 Mikolajczyka St., 45-271 Opole, Poland
[2] Department of Process Engineering, Faculty of Mechanical Engineering, Opole University of Technology, 5 Mikolajczyka St., 45-271 Opole, Poland

Abstract. This paper treats about the impact fuels from biomass wastes and coal combustion on a small boiler operation and the emission of pollutants in this process. Tests were performed in laboratory conditions on a water boiler with retort furnace and the capacity of 10 kW. Fuels from sewage sludge and agriculture wastes (PBZ fuel) and a blend of coal with laying hens mature (CLHM) were taken into account. The results in emission changes of NOx, CO_2, CO and SO_2 and operating parameters of the tested boiler during combustion were investigated. The obtained results were compared with corresponding results of flame coal (GFC). Combustion of the PBZ fuel turned out to be a stable process in the tested boiler but the thermal output has decreased in about 30% compared to coal combustion, while CO and NO_x emission has increased. Similar effect was observed when 15% of the poultry litter was added to the coal. In this case thermal output has also decreased (in about 20%) and increase of CO and NOx emission was observed. As a conclusion, it can be stated that more effective control system with an adaptive air regulation and a modified heat exchanger could be useful in order to achieve the nominal power of the tested boiler.

1 Introduction

In Poland, coal is still the primary heat source, both in the industrial and private sectors. Sustained need for this kind of energy source results from economic and social factors, mainly the relation between the competitive price and quality in terms of energetic potential, domestic availability of relatively shallows deposits and well-developed mining infrastructure. The long-term history of coal consumption in Poland results in most of the operated small scale technical facilities being adopted to its combustion.

Despite of the aforementioned consideration, in recent years there has been an evident change in the approach to energy management and sources. More attention is paid to the use of renewable energy as the source, which may prevent from the environmental damage resulting from the use of fossil fuels, in particular the coal. The main source of renewable energy is the biomass, which represents 80% of the renewable energy on the Polish market [1]. In energy processing the most popular are forest residues and agricultural products specially grown for this purpose (corn, sugar beets, canola, oats, etc.) [2,3]. However, the use of unclassified assortments of cereals or grains as well as livestock residues and sewage sludge produced in wastewater treatment, may become an interesting thermochemical conversion target [4,5].

Biomass renewable energy production trends mentioned above, forces national boiler manufacturers to offer a large range of equipment for domestic use designed only for biomass utilization or multi-fuel boilers with the possibility of burning coal and biomass separately or in co-combustion manner. This multi-fuel thermal conversion is also used in existing poorly controlled stoves and boilers. Data on the biomass thermal treatment in this kind of devices, regarding their energetic sustainability and environmental impact, are still insufficient.

Although the research on woody and agricultural biomass combustion and co-firing in small scale units has been widely reported in literature. For example authors presented testes of combustion agriculture wastes such as: sunflower husk, sugar can, nut shells [6] and coffee ground [7] in small scale boilers. Additional in literature [8] can be found information about modeling gaseous emissions during combustion of agriculture wastes in small scale boilers.

There are only few publications on the scope of livestock waste and sewage sludge. Allouis et al. [9] investigated the co-combustion of pelletized poultry manure waste with olive pomace in a 50 kW burner installed in a water-cooled fire-place. Whereas, Krugly et al. [10] discussed burning of the sewage sludge pellets in a commercial 13 kW pellet boiler.

Therefore, the main aim of this research is to investigate the feasibility of burning or co-burning of the biomass, less commonly processed in household heat production.

* Corresponding author: r.junga@po.opole.pl

Emphasis is put on identification of the main design and operational parameters and their improvements as well as the environmental impact they may have. The combustion tests were performed in a laboratory conditions, on a water boiler with retort furnace with a capacity of 10 kW designed for coal or biomass burning.

2 Materials and methods

2.1 Fuel characterization and thermogravimetric experiments

The combustion tests were performed with the gas-flame coal (GFC) as a reference fuel, blends of sewage sludge, agriculture waste (named PBZ), and blends of coal with laying hens manure (CLHM).

The gas-flame coal was provided by Polish mine KWK Bobrek-Piekary. The PBZ fuel was made of the communal sewage sludge, a product from a mechanical and biological wastewater treatment plant located in Opole, south part of Poland, and the unclassified oats grain from farm located in Krapkowice. The laying hens manure was collected from the poultry farm located in Wlostowa, Opole district.

The fuels were sampled and analyzed in accordance with the relevant EU standard methods for the biomass samples, and national standards for coal (Table 1). Fuel parameters were determined in Laboratory for Instrumental Analysis in Environmental Engineering and Energetics located at the Opole Technical University.

Table 1. Standards for fuels analysis.

Parameter	Coal	Biomass
Moisture	PN-80/G-04511	PN-EN ISO 18134
Volatile matter	PN-G-04516	PN-EN ISO 18123
Fixed carbon	calculated by difference	calculated by difference
Ash	PN-ISO 117 (at 550 °C)	PN-EN ISO 18122
Carbon	PN-EN ISO 16948	PN-G-04571
Hydrogen	PN-EN ISO 16948	PN-G-04571
Sulphur	PN-EN ISO 16994	PN-G-04584
Nitrogen	PN-EN ISO 16948	PN-G-04571
Oxygen	calculated by difference	calculated by difference
Higher heating value	PN-ISO 1928	PN-EN 14918
Lower heating value	calculated	calculated

The thermogravimetric TG measurements were performed on the NETZSCH STA F3 Jupiter device. The fuel amount of 20 ± 0.1 mg was heated in Al_2O_3 crucible from 293 K to 1273 K at 10 K/min heating rate. The protective and purge gas was high purity air, and its flow rates were 20 and 50 mL/min, respectively. All the tests were carried out three times to ensure the reproducibility of the results.

2.2 Combustion testes in a small scale boiler

The research was performed on commercially available ERK water understocker furnace boiler of 10 kW nominal capacity, designed for burning coal or biomass. The unit was installed at the test stand located in Combustion Technology Laboratory being a part of The Department of Thermal Engineering and Industrial Facilities at Opole University of Technology (50.652 °N, 17.902 °E). The schematic layout of the experimental setup is presented in Fig. 1.

The setup, in accordance with the EN 303-5 standard, included several connected functional units allowing to perform boiler test under specific conditions. The equipment used for testing incorporated the following: investigated boiler (1), piping system, water storage tank (2), variable speed pump (3), three-way control value (4) with microprocessor controller, heat exchanger (5) with radiators dissipating the heat outside, and measurement devices.

Paired PT100 resistive sensors were used to measure water outlet (T_o) and return (T_r) temperature, whereas Apator Powogaz vane-wheel meter was used to measure water flow (F). Based on the herein mentioned indicators boiler's power was determined with the heat meter (6). Gaseous compounds concentration (O_2, CO, CO_2, NO_x and SO_2) in the flue gas was continuously measured with the Madur portable emission monitoring system including the conditioner unit PGD-100 (7) and the Photon gas analyser (8) calibrated with appropriate gas. The temperature between the furnace and the deflector was measured with a portable DelthaOHM HD2107.2 temperature recorder coupled with N-type thermocouple. Table 2 shows the specification of the equipment used during the test. Boiler's efficiency was calculated with the indirect method presented in [11].

The flame-gas coal GFC delivered in separate bags was scattered, mixed and left for several days in the laboratory to unify moisture content. The pre-drayed coal was mixed with the poultry manure in 1:0.176 proportions not being formed to obtain the CLHM fuel. Whereas the PBZ was formed from sewage sludge and unclassified oats grain in the 1:0.25 proportion [12].

Granulation of the obtained biomass fuels met the requirements of the fuels for the tested boiler.

The burner, heat exchanger and fuel tank were cleaned before each test to unify the initial conditions.

The electronically controlled pump provided a steady water flow, whereas the three-way valve with controller ensured stability of the return water temperature standing as 55°C. When the fire-up phase was completed, the fuel feed rate and combustion air flow were changed until the power was as nearest to its nominal value as possible and the emission reached the lowest level.

 DOI: 10.1051/e3sconf/20171901009

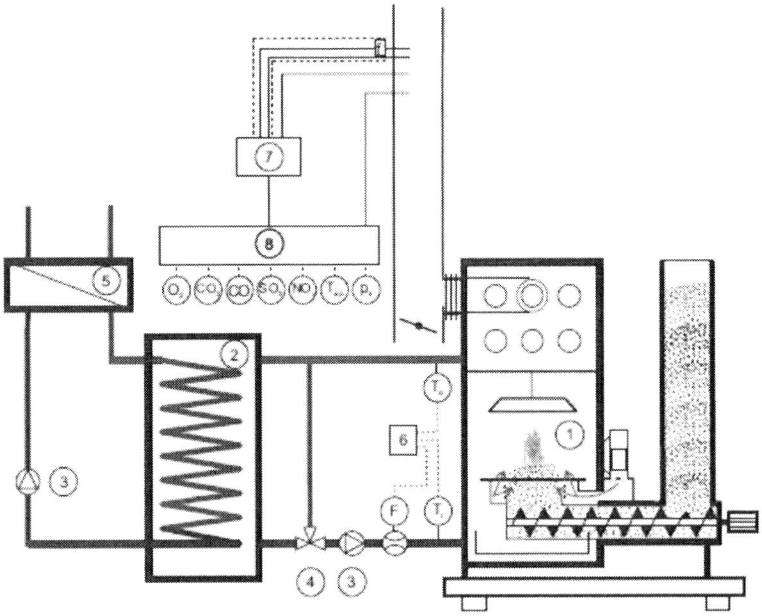

Fig. 1. Scheme of the experimental setup.

Table 2. Specification of the monitoring equipment.

Parameter	Method	Unit	Range	Accuracy	Resolution
O_2	Electrochemical sensor	%	0 ÷ 21	+/- 0.01 abs. or 5 rel.	0.01
CO_2	NDIR detector	%	0 ÷ 25	+/- 0.1 abs. or 4 rel.	0.01
CO	NDIR detector	ppm	0 ÷ 20,000	+/- 4 abs. or 4 rel.	1
NO	NDIR detector	ppm	0 ÷ 5,000	+/- 5 abs. or 5 rel.	1
NO_2	NDIR detector	ppm	0 ÷ 1,000	+/- 5 abs. or 5 rel.	1
SO_2	NDIR detector	ppm	0 ÷ 5,000	+/- 5 abs. or 5 rel.	1
Flu gas temperature	K-type thermocouple	°C	-10 ÷ 1,000	+/- 2	1
Combustion chamber temp.	N-type thermocouple	°C	0 ÷ 1,250	+/- 1.5	1
Water temperature	PT100 resistive sensor	°C	0 ÷ 150	+/- 0.4	0.1
Water flow rate	Vane-wheel	m^3/h	0.24 ÷ 6.0	3*	0.005

* – for hot water, abs. – absolute, rel. – relative.

3 Results and discussion

3.1 Fuel characteristics

The tests were conducted on three different fuels with various ranges of physicochemical parameters as shown in Table 3. As a main approach in this study it was assumed that the reference fuel would be the bituminous coal, as recommended for this type of boiler. According to the manufacturer's guidelines, the device is also suitable for biomass combustion, therefore capabilities of co-burning of coal with the addition of waste biomass, as well as combustion of raw biomass wastes have been tested. Regarding the coal and poultry manure blend, the biomass content was at the level of 15%, to emphasize some mechanisms occurring during the co-combustion process. This share is somewhat more than 10%, i.e. the amount more practically recommended for co-burning [13].

DOI: 10.1051/e3sconf/20171901009

Generally, it is well known that fuel parameters deeply affect combustion process. 10% moisture content was at the similar level in all the tested fuels therefore it should not have a leading role in the process. Whereas the remaining physicochemical parameters of the fuels varied significantly. The lowest, i.e. 6.9% ash content was observed in burned coal, addition of poultry manure to the GFC increased the amount of the ash to about 10%, whereas PBZ was the highest ash fuel with about 20% of mineral matter. The inorganic matter affects the combustion profiles but it may also cause problems with the slagging and effective removal [14]. The carbon content is related to the calorific value [15], and is the determinant of energy concentration. Both parameters were the highest for coal, while addition of waste biomass led to a noticeable decrease in the C and LHV value. The content of other macro elements in the fuel results in a formation of toxic compounds in the combustion process. Particular attention was paid to the presence of nitrogen and sulfur in fuels as they are closely related to the mechanisms of NO_x and SO_2 formation in furnace [14,16] and their emission to the atmosphere, especially in terms of legal requirements.

Among the tested fuels, the GFC contained the lowest amount of N (i.e. 1.50%), while the PBZ contained much more N than the other fuels (3.89%). Regarding the S content, tendencies of these parameters were reversed.

Application of temperature-controlled combustion along with the thermogravimetry on a microscale provided valuable hints on how to increase combustion efficiency as well as how to improve boilers' design and maintenance. Fig. 2 shows that profile of weight loss rate during coal combustion differs significantly from the one of a biomass as a result of different proportions between the volatile matter and fixed carbon of those fuels. In case of coal combustion one of the stages observed (Table 4. Stage II), indicates that released volatile matter passes fluently into the char combustion. However, co-combustion with a certain amount of poultry litter takes place in two stages (Table 4, Stage I and II), which is appropriate for the biomass. The DTG profile for PBZ is typical for the waste biomass combustion with a distinct separate peak for calcium carbonate decomposition (Table 4, Stage III) [17].

The ignition temperature T_i of CLHM is lower than the one of GFC, due the biomass presence in the blend

Table 3. Fuels parameters (air dried basis).

Parameter	Unit	GFC	LHM	PBZ
Moisture	wt. %	7.93	12.18	10.5
Volatile matter	wt. %	31.47	58.85	41.35
Fixed carbon	wt. %	53.7	3.22	28.28
Ash	wt. %	6.9	25.75	19.87
Carbon	wt. %	69.46	33.72	34.10
Hydrogen	wt. %	4.67	5.19	4.23
Sulphur	wt. %	0.53	0.39	0.11
Nitrogen	wt. %	1.50	3.68	3.89
Oxygen	wt. %	9.01	19.09	27.23
Higher heating value	MJ/kg	27.650	13.772	16.128
Lower heating value	MJ/kg	26.458	12.356	14.949

wt % - weight percent

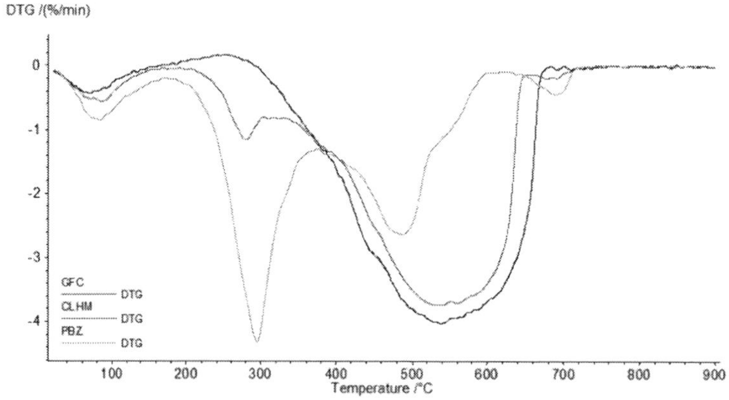

Fig. 2. Derivative thermogravimetric (DTG) curves of tested fuels.

Table 4. Parameters of combustion.

Sample	T_i	T_b	t_b	Stage I		Stage II		Stage III	
				DTG I	T I	DTG II	T II	DTG III	T III
	°C	°C	min	%/min	°C	%/min	°C	%/min	°C
GFC	404	685	28	-	-	-4.00	536.2	-	-
CLHM	249	656	41	-1.16	279.1	-3.72	536.3	-0.19	685.2
PZB	254	627	36	-4.30	295.5	-2.64	486.1	-044	688.3

T_i – ignition temperature,
T_b – final temperature of char combustion,
t_b – total combustion time.

and its early devolatilization. Therefore, addition of the biomass could be in favor of coal ignition initiation, but in practice in the operated units, the early release of volatile matter has led to uncontrolled combustion at low temperatures [13].

It must also be taken into consideration also, that the maximal loss rates DTG_{max} for coal (Stage II, Table 4) occured in higher temperatures, while for the biomass it took place at the devolatilization stage (Stage I, Table 4). High reactivity of the biomass, results in faster and more intensive burning of this material, than in case of the coal, and as a consequence leads to uneven fuel bed burning. In the operation of large industrial boilers, separate fuel delivery systems are used to offset the effect. Regarding the automatic small scale boilers, the fuel rate is controlled by the working period of the feeder. Those inappropriate interval adjustments create difficulties during conventional operation, especially in case of combustion blends such as the CLHM. Increased unburned organic matter in the bottom ash or "crater" combustion are examples of wrong fuel feeding settings.

3.2 Boiler technology

The first objective of the boiler test was to observe the settings of the feeder operating parameters to achieve thermal output with fixed fuel as closest to the nominal value as possible. In case of the coal as the reference fuel the on/off working periods were 3/27 s, which corresponds to fuel mass flow of 1.50 kg/s, with boiler performance being very close to the nominal value (Table 5). However, addition of 15% of the livestock waste biomass, having considerably different bulk density, changed the fuel mass flow and consequently energy input to the boiler. As a result, during CLHM tests the boiler was performing at the 71% load. The mass flow of the PBZ was changed significantly, but it only reached the 51% load. Attempts to increase the amount of fuel caused problems described in more detail in section 3.1, i.e. large loss of ignition matter in the bottom ash for CLHM, and unstable fire bed for PBZ, which led to unexpected shutdown of the boiler.

The flue gas temperature during coal combustion and coal with poultry manure co-firing, increased significantly to 200 °C, resulting in a high physical energy heat loss, at the level of 17% for the GFC, and 20% in case of the CLHM. This indicates, the need to change the design of the heat exchanger to increase heat transfer.

Furthermore, the tested boiler is of a compact size and a horizontal heat exchanger at the top, where considerable quantity of the fly ash is deposited, reducing the heat transfer. Both combustion and co-firing attempts occurred with no ash lumping on the burner and no slagging and the bottom ash was quite fine. However, the combustion of the PBZ with large amount of the ash caused its excessive accumulation in the ash box and limited boiler's operation time.

Table 5. Selected operating parameters for each fuel in the best conditions obtained.

Parameter	Unit	GFC	CLHM	PBZ
Feed mass flow	kg/h	1.50	1.34	1.69
On/off working period	s	3/27	3/24	6/22
Excess air ratio*, λ	-	2.00 (0.11)	2.49 (0.24)	2.72 (0.32)
Combustion chamber temp.*	°C	802 (43)	770 (46)	703 (59)
Flue-gas temperature*	°C	233 (3)	212 (7)	161 (2)
Inlet water temperature*	°C	59 (1)	59 (1)	58 (1)
Outlet water temperature*	°C	67 (1)	65 (1)	63 (1)
Thermal output*	kW_{th}	9.0 (0.7)	7.1 (0.6)	5.9 (0.4)

* average values with standard deviations

 DOI: 10.1051/e3sconf/20171901009

3.3 Gaseous emissions and boiler's efficiency

Arrangement of the combustion process to achieve appropriate emission level (it mainly concerns the CO and NO_x), when increasing to the maximum boiler's, is based on certain principles. The main techniques are: sufficient air excess delivery and high combustion temperature reasurance, appropriate mixing between the fuel and the air, but also proper air distribution, which are often emphasized during biomass combustion in low power boilers [14-16,18].

The tested boiler, as low-budget device has been equipment only with primary air distribution system. The most beneficial excess air ratio was obtained manually with the use of a single fan, unfortunately it was impossible to apply more advanced air control system. As it is shown in Table 4 and Table 6 the obtained average λ ratios corresponding to minimum average values of CO reached 2.00, 2.49 and 2.79 during combustion of the GFC, CLHM and the PBZ respectively. Regarding coal combustion the average CO emission (591 mg/Nm^3 at 10% O_2 in dry gas) has met the requirements of EN 303-5 standard for automatic Class 4 boiler for fossil fuel. The CO emission for the CLHM, although twice as higher as the one for coal, has also met those requirements. PBZ combustion gave visibly higher CO values, at the level of around 1,400 mg/Nm^3 at 10% O_2 d.g., yet still much lower as reported by Krugly [10] for pellet sewage sludge combustion 13,698 ± 6,762 mg/Nm^3.

Regarding the operation more valuable information is shown in Fig. 4. The CO emission during GFC combustion has changed significantly in relatively narrow range of excess air ratio, whereas during the co-firing and raw biomass combustion CO fluctuations were more intense in a wider range of λ values. There may be few explanations of this phenomenon: (a) the combustion chamber (i.e. size, shape and wall material) was designed for coal, addition of highly volatile and less energetics biomass changed the burning profile and temperature significantly; (b) lack of secondary air distribution system, prevented good mixing of flue gases with air and effective oxidation; (c) control system based on discrete on-off regulation, without possibilities to adjust the amount of excess combustion air or fuel.

The NO_x emissions from automatically fired biomass boilers are subject to legal restrictions only in Austria, while in the Netherlands and Great Brittan the NO_x have only been limited for boiler's subsidies. The Ecodesign regulation for domestic boilers (in force in the EU from 2020), will limit the NO_x emission to 200 mg/m^3 and 350 mg/m^3 for biomass and the fossil fuel boilers respectively. As it is shown in Table 6, the analyzed boiler does not currently meet those requirements.

As stated in Fig. 5 the highest NO_x emission were observed for the PBZ, the N-richest fuel. As expected, the Fuel-NO_x mechanism was the main source in NO_x formation, which confirms the combustion temperatures below 800 °C in each test. Also the NO_x/Fuel-N conversion rates (CR) were 5.7%, 4.7% and 2.1% for the GFC, CLHM and the PBZ respectively. The CR decreasing tendency with lowering nitrogen bound in the fuel is often reported [6,8] and could be explained based on the Fuel-NO_x formation mechanism. Confirmation of the dominant role of this mechanism during the tests, allows to introduce certain technologies or fuels which may limit the said mechanism, i.e. air staging or flue gas recirculation.

The SO_2 emission is a result of fuel sulfur oxidation [15]. The SO_2 levels obtained during PBZ combustion were visibly lower than those of the other fuels, due to the S content in the fuel. The fuel-S to SO_2 conversion rates were relatively low 52%, 47% and 61% for combustion of the GFC, CLHM and the PBZ respectively, therefore part of the fuel sulfur remains as solid compound in the ash. This effect can be explained by the low temperature regime in the combustion chamber and in the bottom ash. Similar results were also obtained during combustion of fuel based on sewage sludge and coal slime [19].

Table 6. Flue gas composition* and loss of ignition matter in the bottom ash

Parameter	Unit	GFC	CLHM	PBZ
O_2	vol. % d.g.	10.95 (0.52)	12.93 (0.74)	13.4 (0.87)
CO_2	vol. % d.g.	9.41 (0.50)	7.61 (0.69)	7.78 (0.77)
CO	mg/Nm^3 (10% O_2 d.g.)	591 (190)	945 (271)	1139 (339)
NO_x	mg/Nm^3 (10% O_2 d.g.)	388 (20)	433 (32)	796 (65)
SO_2	mg/Nm^3 (10% O_2 d.g.)	772 (30)	733 (46)	371 (12)
Solid unburned matter	kg/kg solid residue	0.0571	0.1032	0.2014

* - average values with standard deviations
d.g. – dry gas

 DOI: 10.1051/e3sconf/20171901009

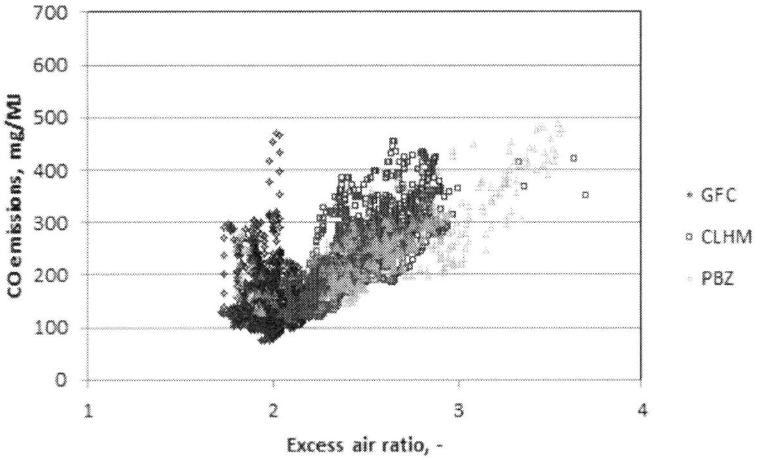

Fig. 4. CO emissions vs excess air ratio.

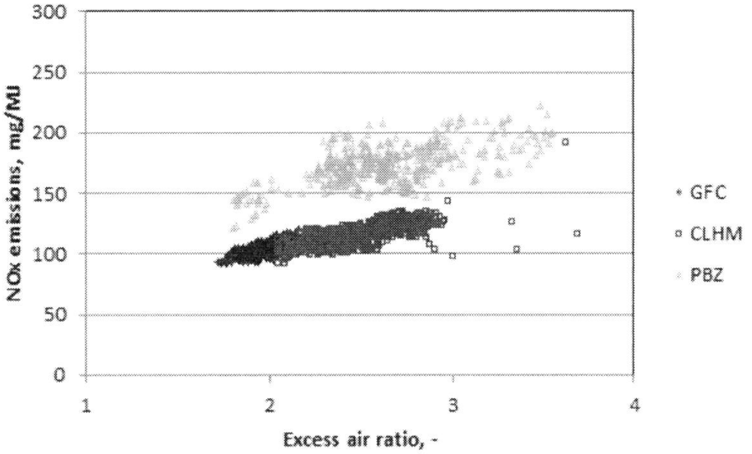

Fig. 5. NO_x emissions vs excess air ratio.

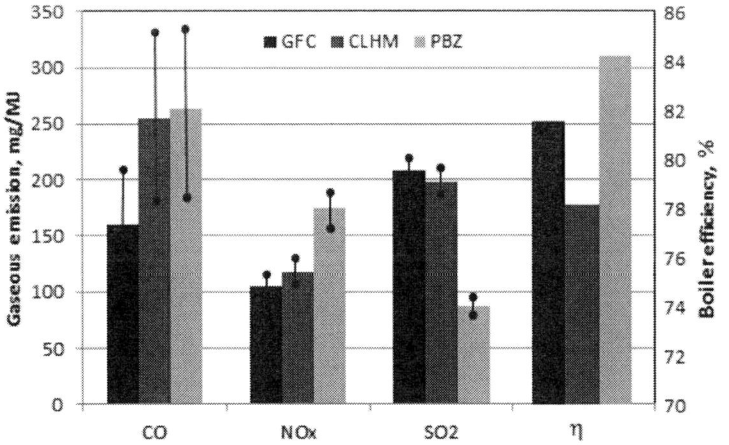

Fig. 6. Specific gaseous emissions (average values with standard deviations) and thermal efficiency.

As presented in Fig. 6 boiler's efficiency was 81.5% for the GFC, 78.1% for the CLHM, and 84.2% during the PBZ combustion. It must be clearly stated that among the researched fuel's combustion only the coal could fulfill manufacturer's guidelines in terms of nominal power.

5 Conclusions

Based on the obtained results and preceding discussion, the main conclusions can be formulated:

- the physicochemical properties of the prepared fuels differ from one another, which is reflected in various combustion profiles affecting the combustion process in the investigated boiler,
- the tested multi-fuel boiler achieved the nominal power only during coal combustion, where the CO emission and efficiency met the requirements for Class 4 Boiler according to the EN 303-5:2012 standard,
- more effective control system, with adaptive air regulation and modified heat exchanger with automatic cleaning, should improve classification performance,
- the addition of 15% poultry litter to the coal imposed changes in the fuel and air feeding, causing a decrease in thermal output (with about 20%) as well as increase in CO and NO_x emissions. Improvement of these parameters requires changes in the combustion chamber and more advanced air distribution system,
- blend of sewage sludge with oats grain in the form of granules could provide a stable combustion in the tested boiler, yet the thermal output has decreased with about 30% compared to the one obtained from coal burning, but also CO and NO_x emissions increased significantly. The PBZ combustion in the subject unit requires changes not only in the combustion chamber itself (mainly related to the increase of the combustion temperature and better air flow to mix flue gas) and its operation but also system for automatic bottom ash removal.

The presented results cover certain range of technical and environmental issues related to combustion and co-combustion of the biomass waste with coal in a small scale boiler, but it must be clearly stated, that full adaptation of this type of fuels requires analysis of the pollution emission on a broader scale, e.g. HCl, HF or NH_3.

References

1. Central Statistical Office, Energy, Warsaw 2015
2. J. Werther, M. Saenger, E.U. Hartge, T. Ogada, Z. Siagi, Prog. Energy Combust. Sci. **26**, 1, (2000)
3. S.G. Sahu, N. Chakraboty, P. Sarkar, Renewable Sustainable Energy Rev. **39**, (2014)
4. H. Haberl, T. Beringer, C. Sribas, C. Bhattachaeya K. Heinz, Curr. Opin. Environ. Sustainability, **2**, 5-6 (2010)
5. M. Wzorek, Waste Manage. Res. **32**, 5 (2014), doi.org/10.1177/0734242X14529618
6. E. Cardozo, C. Erlich, L. Alejo, T.H. Fransson, Fuel, **114** (2014)
7. L. Limousy, M. Jeguirim, P. Dutournie, N. Kraie, M. Lajili, R. Said, Fuel **107** (2013)
8. S. Fournel, B. Marcos, S. Godbout, M. Heitz, Boresourse Technol. **179** (2015)
9. C. Allouis, M. Gallo, R. Nigro, R. Pagliara, XXXV Meeting of the Italian Section of the Combustion Institute (2012)
10. E. Krugly, D. Martuzevicius, E. Puida, K. Buinevicius, I. Stasiulaitiene, I. Radziuniene, A. Minikauskas, L. Kliucininkas, Energy Fuels, **28**, 8, 5057–5066 (2014)
11. R. Junga, P. Chudy, J. Pospolita, Measurement **97**, 186–194 (2017)
12. M. Wzorek, Valorization of sewage sludge and agricultural waste into fuel, 3rd International Conference on Engineering for Waste and Biomass Valorisation, Beijing (2010)
13. M. Otero, C. Díez, L.F. Calvo, A.I. García, A Morán, Biomass Bioenergy **22** (2002) 319 – 329 (2002)
14. L. Carvalho, E. Wopienka, Ch. Pointner, J. Lundgren, V. K. Verma, W. Haslinger, Ch. Schmidl, Appl. Energy **104** 286 – 296 (2013)
15. P. Brassard, J.H. Palacios, S. Godbout, D. Bussières, R. Lagacé, J.P. Larouche, F.Pelletier, Bioresource Technol. **155**, 300 – 306 (2014)
16. M. Díaz-Ramírez, F. Sebastián, J. Royo, A. Rezeau, Appl. Energy **115**, 360–373 (2014)
17. R. Junga, W. Knauer, P. Niemiec, M. Tańczuk, Renewable Energy **111**, 245 – 255 (2017)
18. J. Lundgren, E. Pettersson, Bioresource Technol. **100**, 3121–3126 (2009)
19. M. Wzorek, M Kozioł, W. Ścierski, J. Air Waste Manage. Assoc. **60** (2012)

Sorption characteristic of coal as regards of gas mixtures emitted in the process of the self-heating of coal

Karolina Wojtacha-Rychter[1,*], and *Adam* Smoliński[2]

[1]Central Mining Institute, Department of Mining Aerology, Pl. Gwarków 1, 40-166 Katowice, Poland
[2]Central Mining Institute, Department of Energy Saving and Air Protection, Pl. Gwarków 1, 40-166 Katowice, Poland

Abstract. One of the most challenging tasks in the coal mining sector is the detection of endogenous fire risks. Under field conditions, the distance between the points where samples for the analyses are collected and the actual place where coal self-heating takes place may be quite remote. Coal is a natural sorbent with a diverse character of pore structures which are surrounded by fractures and cleavage planes constituting ideal spaces for the flow and adsorption of gases. The gases (methane, ethane, ethylene, propane, propylene, acetylene, carbon dioxide, carbon monoxide, hydrogen) released from the source of fire migrate through the seam and may be subject to adsorption, or they may cause the desorption of gases accumulated in coal. Therefore, the values of reference sample concentrations may be overstated or understated, respectively. The objective of this experimental study was to investigate the adsorption phenomena accompanying the flow of a multi-component gas mixture through a coal bed which may occur *in situ*. The research was conducted by means of a method based on a series of calorimetric/chromatographic measurements taken to determine the amount of gases released during coal heating at various temperatures under laboratory conditions. Based on the results obtained in the course of the experiments, it was concluded that the amount of gas adsorbed in the seam depends on the type of coal and the gas. Within the multi-component gas mixture, hydrocarbons demonstrated the largest sorption capacity, especially as concerns propylene.

1 Introduction

Endogenic fire has been one of the most serious types of hazard affecting workers safety, economic loss and environment pollution in coal mining. This phenomenon is a consequence of coal self-heating caused by the accumulation of heat released from the contact of hard coal with oxygen from the surrounding air. A number of techniques based on optical microscopy and thermal studies have been investigated to determine the susceptibility of coal to spontaneous combustion [1-3]; however, in order to evaluate the state of spontaneous combustion, mine operators apply fire indices [4-6]. The concept behind this method is based on the following: 1) the monitoring of mine air composition in selected places underground, 2) the assumption that the profile of the emission of gaseous oxidation products is characteristic of the temperature of the coal, and 3) determining the fire indices as calculated on the basis of the concentrations of gases in mine air. Typical composition of the gas mixture emitted during the coal self-heating process is as follows: carbon monoxide, carbon dioxide, hydrogen, methane, propane and ethane, propylene, ethylene and acetylene. As the temperature of the heating medium increases, the concentration of the above gases in the mine atmosphere increases. The relations between coal temperature, the value of fire indices and gas composition are established during the oxidation of a reference coal sample at different temperature ranges under laboratory conditions. Nevertheless, the results of the test carried out in laboratory condition cannot corresponded with the results obtained under field conditions because gases released from the source of fire migrate through the seam and may be subjected to adsorption or desorption on coal. The distance between the point where samples for the analyses are collected and the actual place where coal self-heating takes place may be quite remote. Since coal represents a fractured porous medium with a diverse character of the pore structure, the pores themselves are surrounded by zones of fractures and cleavage planes which enable gas permeability and adsorption. Gas sorption capacity depends primarily on the presence of micro-pores whereas transport pores i.e. meso-pores and macro-pores, as well as fractures are the decisive factors in gas flow as well as the ability of coal to permeability [7]. The structure of the pores varies with carbon content: the macro-pores dominate in lower rank coals while high rank coals contain mainly micro-pores [8]. Additionally, the geometries of gases molecules strongly affect the selectivity of gas transport in coal. Molecules with the small kinetic diameter can permeate more

Corresponding author: kwojtacha@gig.eu

DOI: 10.1051/e3sconf/20171901010

successfully [9]. Coal behaves as a binary system due to the presence of hydrophobic and hydrophilic areas on the surface of coal grains. Hydrophobic properties of coal surface are determined by aliphatic/aromatic hydrocarbons fragments of coal structure and fragments containing sulfur atoms (SH groups, thioethers) [10]. Hydrophilic properties of coal surface correlate to the occurrence of numerous active functional groups: methyl, hydroxyl, carboxyl and carbonyl. These groups give the apolar basic mass of hard coal a certain degree of polarity [11]. The ratio of hydrophilic and hydrophobic areas varies depending on the degree of metamorphism. As the carbon content decreases, the oxygen content increases; simultaneously, the number of the hydrophilic areas increases. Consequently, the greatest amount of sorbed gases is observed in the case of coals which are characterized by a low degree of metamorphism. The sorption phenomenon accompanying the flow of gases through a coal bed may affect the values of gases concentrations in coal sample collected under field condition which may be overstated or understated. Research on gas sorption process on coal has been carried out mostly on pure-single gases or gas mixtures within the context of methane recovery. Under real conditions, gases emitted from the source of coal self-heating are a mixture of products characteristic of different temperatures of coal [12]; hence, it is justified to conduct research on the influence of a multi-component gas mixture flow through a coal bed on selected gases concentration in mine air.

2 Materials

For the purpose of the study, three coal samples were collected from the Upper Silesia Coal Basin. The fresh coals used throughout the study were crushed and sieved to grains of size ranging from 0.5 to 0.7 mm and from 1.0 to 2.0 mm. Finer coal fractions were used for the heating of coal in furnace but larger fractions was used as coal bed. Afterwards, the selected fraction of sample was divided to obtain a representative volume by means of the quartering method. Samples were coded with the consecutive numbers: sample 1, sample 2, sample 3. The proximate and ultimate analyses have been carried out on selected samples according to the following Polish standards: PN-ISO 1171:2002 and PN-G-04560:1998 (ash content); PN-G-04516:1998, PN-G-04560:1998 and PN ISO-562:2000 (volatile matter content); PN-G-0,4511:1980 (total moisture content); PN-G-04571:1998 (carbon, hydrogen and nitrogen content), PN-G-04584:2001 as well as PN-ISO 334:1997 (sulfur content), and PN-ISO 7404-3:2001 (maceral groups content and mineral substance content). The results of the analyses of samples in analytical state are presented in Table 1.

3 Measurement methodology

The apparatus stand used for the research is presented in Fig. 1. The system consists of two integrated main devices: a reaction chamber (furnace) for heating and oxidizing the coal samples at different temperatures and a quartz glass tube with the dimensions of 0.025 m x 1m for examining the sorption of gas mixture. A series of experiments were conducted according to the following procedure. Samples of the grain size from 1.0 to 2.0 mm and weight of 0.4 kg were placed in an air-tight container. The container with a given sample was located in the centre of the reaction chamber. A thermocouple was inserted into the container with the coal sample for monitoring the temperature of coal in the reactor. The furnace was heated up to the desired final temperature. When the predetermined temperature was reached, a stream of synthetic air comprising of O_2 – 20.5% vol. and N_2 – 78.5% vol. under constant pressure of $1\cdot10^5$ Pa was passed through the container; the airflow rate through the container was $5.56\cdot10^{-7}m^3/s$. As a result of the coal combustion, gaseous products were released and streamed through the glass tube filled with coal of the grain size ranging from 0.5 to 0.7 mm and the weight of 0.2 kg. At the inlet and outlet of the tube, gas samples were collected in 1.5-L Tedlar bags. The first gas collection temperature was 50^0C. The cycle of the experiment for one sample involved coal heating at the temperatures of 50^0C, 100^0C, 150^0C and 200^0C, respectively. The choice of the grain size of coal and the airflow rate were determined by the technical limitations of the apparatus.

Table 1. Characteristics of the samples.

Parameter	Unit	Sample 1	Sample 2	Sample 3
Carbon	% w/w	74.66	81.01	64.18
Hydrogen	% w/w	4.16	4.26	4.13
Oxygen	% w/w	9.02	6.83	12.78
Sulfur	% w/w	0.32	0.57	0.85
Ash	% w/w	7.44	5.06	7.52
Moisture	% w/w	3.02	1.34	9.69
Volatiles	% w/w	29.76	28.33	30.55
Mineral matter	% vol.	1	0	3
Vitrinite	% vol.	45	26	70
Liptinite	% vol.	11	16	8
Inertinite	% vol.	44	58	22
Vitrinite reflectance	% vol.	0.76	0.88	0.52

Gas chromatography was used to determine the concentrations of each of the components of the mixture of gases released during the process of coal heating. For the purpose of obtaining optimal detector selectivity and sensitivity, suitable intensities of the carrier gas stream flow along with the temperatures of the detector and column were chosen. High purity argon was used as a carrier gas for hydrogen but in the case of carbon dioxide, carbon monoxide, ethane, ethylene, propane, propylene and acetylene pure helium was applied. The flame ionization detector (FDI) was applied to determine ethane, ethylene, propane, propylene and acetylene whereas for determining carbon dioxide, carbon monoxide and hydrogen, the thermal conductivity

 DOI: 10.1051/e3sconf/20171901010

detector (TCD) was used. The compounds were burnt in a hydrogen/synthetic air flame. The application of special concentrator columns enabled to determine trace levels of ethane, ethylene, propane, propylene, acetylene within the scope of low-temperature oxidation.

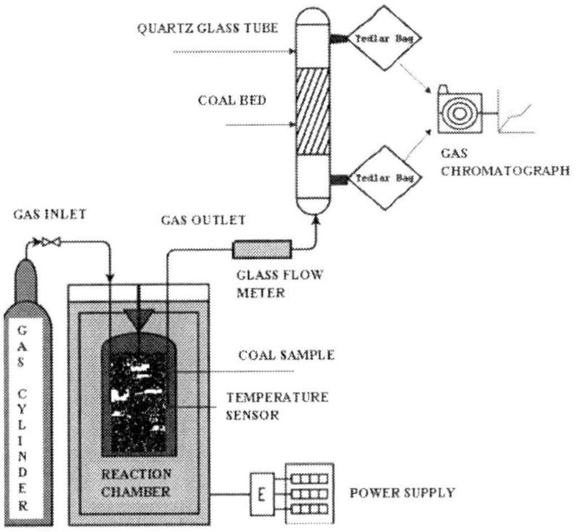

Fig. 1. Schematic of the set test stand.

4 Results and discussion

Experimental data summarized in Figs. 2, 3 and 4 were presented as the percentage of gas sorbed in the coal bed for a given temperature. Based on the results, two characteristic regularities were observed, namely: coal sorption capacity was the highest as regards propylene and acetylene and the lowest in the case of carbon monoxide and hydrogen.

Carbon and oxygen content in coal are the most significant factors determining the behavior of coal with reference to sorbed hydrogen [13]. The percentage of hydrogen molecules in sorbed gas increases along the increase of carbon percentage and the decrease of oxygen amount. It was observed that the highest amount of sorbed hydrogen was gained for coal containing from 70 to 80% of carbon [14]. This behavior reflects the fact that the hydrophobic character of coal surface constitutes a factor which favors the sorbtion of hydrogen molecules. Furthermore, it has been found that sorption of carbon monoxide is closely related to the number of accessible active sites on coal surface [15]. The number of polar active sites centers in coal is proportional to oxygen content and inversely proportional to its carbon content.

The adsorption of saturated hydrocarbons is lower than the sorption of unsaturated hydrocarbons despite the fact that the concentration of propane and ethane was 2-3 times higher in the inlet gas stream. The sorption mechanism of non-polar molecules of propane and ethane is associated with a dispersion interaction [13, 16, 17]. The varied reactivity of propane and ethane can be related to the energy value of the C-H bond. Table 2 presents the typical bond energies for a C-H bond [18]. Propane which contains one secondary carbon is more reactive than ethane which has two primary C-H bonds. The high energy value of the primary C-H bond of 420 kJ/mol reflects the fact that ethane reacts with more difficulty with polar fragments of the coal structure.

Table 2. The value of bond energy for C-H bond.

	Primary	Secondary	Tertiary	Allylic	Vinylic
Energy [kJ/mol]	420	401	390	361	445

The sorption of unsaturated hydrocarbons: propylene, ethylene and acetylene has been previously discussed in the literature. Multiple bonds between the two atoms of carbon make gaseous substance more chemically active and undergo reactions with polar fragments of the coal structure [19, 20]. A double bond (C=C) in propylene and ethylene has one strong sigma and one weak π bonds; a triple bond (C≡C) in acetylene has one sigma bond with two remaining being π bonds. Additionally, a small size of the kinetic diameter of these molecules enables the gases to permeate to the micro-pores which are inaccessible for gases characterizing of larger kinetic diameters. However, the results of the experiment demonstrate that sorption preference is not always directly correlated with the dimension of the molecule. Acetylene is sorbed in a smaller amount than propylene despite the small kinetic diameter of 0.33 nm (kinetic diameter for propylene is 0.45 nm). Alkenes contain structural fragments characteristic of the alkanes. These groups may react in the same conditions as the alkanes. Hence, the varied reactivity of propylene and ethylene can also be linked to the value of the energy of the C-H bond. Ethylene presents vinylic C-H bond, while propylene has both vinylic and allylic C-H bonds [18]. The energy of allylic bonds is 361 kJ/mol, the energy of vinylic C-H bonds are higher than 400 kJ/mol. It may be therefore assumed that propylene is more reactive with polar fragments of coal than ethylene. To sum up, the difference in the reactivity between the unsaturated hydrocarbons and saturated hydrocarbons could be explained by the reactivity of the carbon-carbon bond and the value of the energy of carbon-hydrogen bond. Alkenes are relatively stable compounds, but are more reactive than alkanes either because of the reactivity of the carbon–carbon π-bond or the presence of allylic C-H bond.

In the literature, the sorption of carbon dioxide was mostly considered in the aspect of enhanced coal bed methane recovery. According to literature [21-23], the CO_2 affinity for the coal structure is twice as much as the methane, and methane is readily desorbed from the coal and replaced by carbon dioxide. The phenomenon of preferential carbon dioxide adsorption on natural coals is connected with a linear molecule shape, small size of the kinetic diameter of molecules (0.330 nm), large adsorption affinity (energy), high quadrupole moment and low value of the activation energy [24, 25].

DOI: 10.1051/e3sconf/20171901010

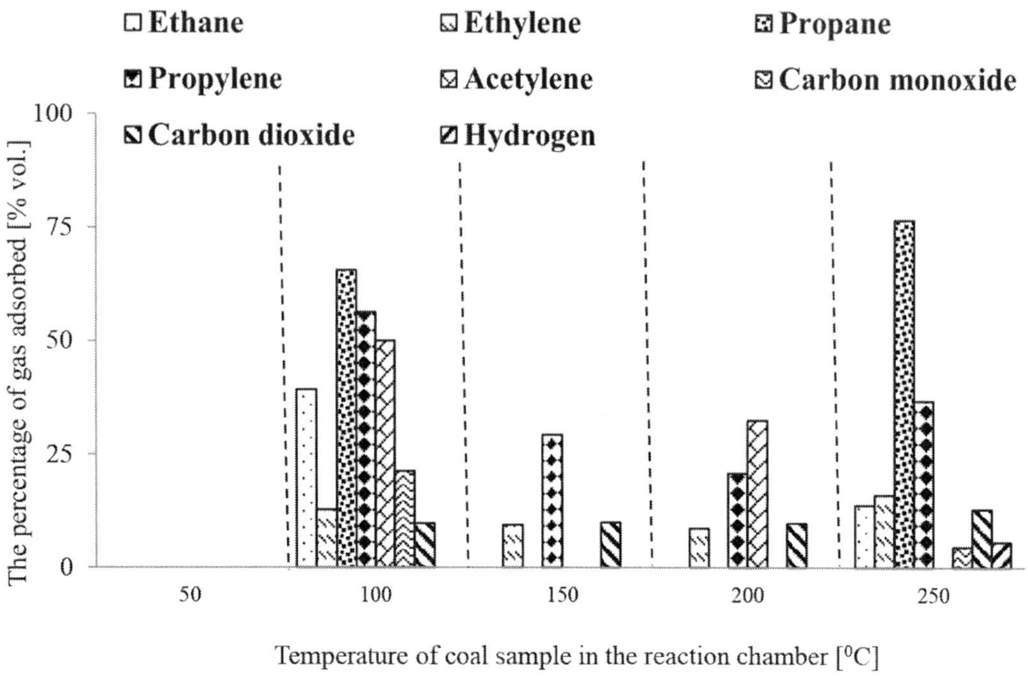

Fig. 2. The percentage of gas adsorbed in the coal sample 1.

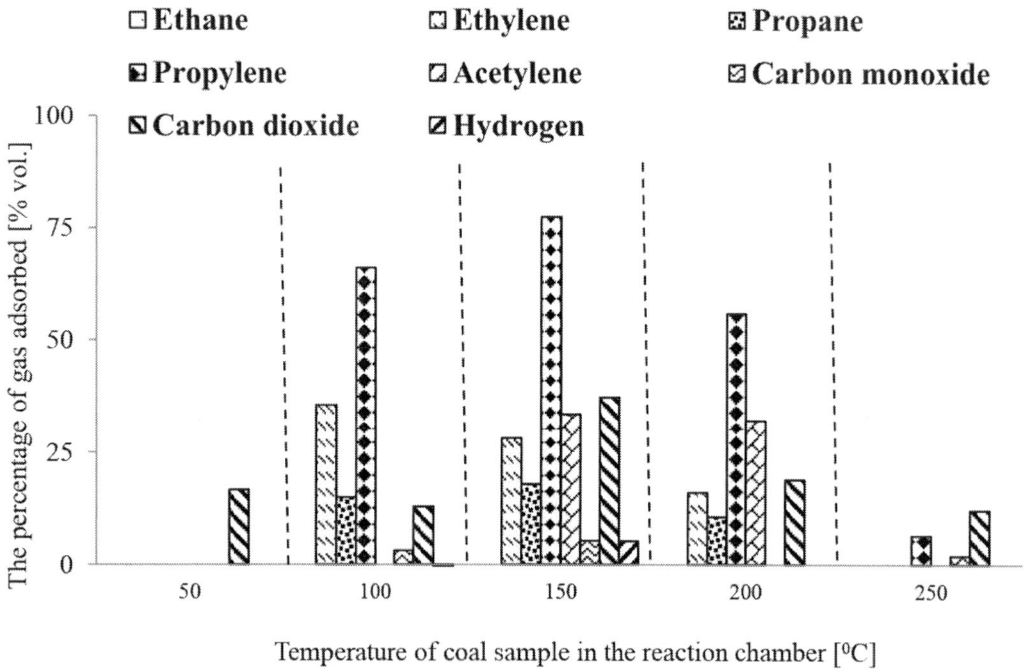

Fig. 3. The percentage of gas adsorbed in the coal sample 2.

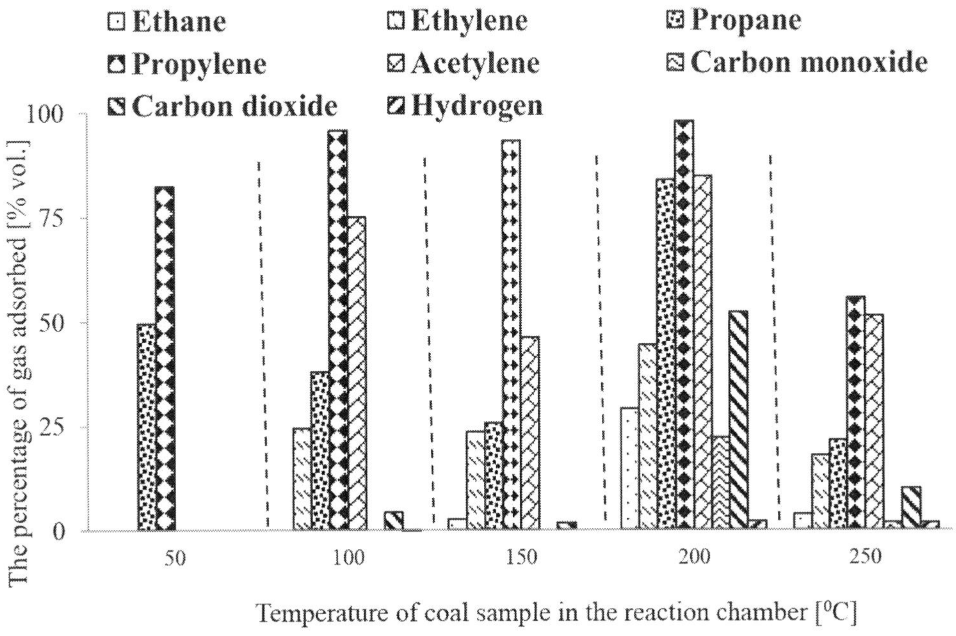

Fig. 4. The percentage of gas adsorbed in the coal sample 3.

Based on the analysis illustrated in Table 1, each sample represented different coal rank group. Sample 3 has the highest carbon content, followed by sample 1 and sample 2. The lowest degree of metamorphism of sample 3 corresponds to the highest oxygen content. It is generally known that coal with low carbon content and high oxygen content has the highest sorption capacity. Sample 3 sorbed acetylene and propylene at all temperatures. Interestingly, despite the high content of mineral matter, low sorption capacity was not observed for sample 3. Most probably, the composition of mineral substance had a positive influence on coal sorption capacity. Similar observations were made by Deng et al. [26] who concluded that mineral composition may influence the volume of pores consequently reducing or enhancing the gas sorption capacity. Additionally, they found that clay, the most commonly occurring inorganic constituent of coals, has strong positive correlation with surface area and pore volumes. The petrographic analysis of coals indicated that sample 3 has the highest content of vitrinite and the lowest content of liptinite and inertinite. Virtinite group is less porous than inertinite but has high sorption capacity due to its high micropore fraction [27]. This relation may correspond to specific surface area and sorption capacity of sample 3. It was also reported in the literature that vitrinite contains more aromatic components than inertinite and liptinite groups which contain more aliphatic groups [28-30]. The hydrophobic properties differ for each group: aromatic groups are less hydrophobic than aliphatic groups [8]. The sorption capacity of samples 1 and 2 was lower probably as result of the increase in the degree of condensation of aromatic rings in internal vitrinite structure with the increase in coal rank and the decrease in micropores [29].

5 Conclusions

The study on the influence of a multi-component gas mixture flow through a coal bed allowed for the characterization of selective sorption of gases in relation to different coal rank groups. On the basis of the research experiments, it can be concluded that the sorption capacity of coal with respect to multi-component gas mixtures as a sorbate depends on the interaction between gases molecules and coal structure. In the course of the experiment, propylene was sorbed in the largest amount. This could be explained by both the occurrence of the reactivity of the double π-bond (C-C) and the allylic bond (C-H). Carbon monoxide and hydrogen constitute gases which were sorbed in the smallest amounts. The sorption capacity of the three coal samples tested changed depending on the coal rank. Sample 3 with the lowest degree of metamorphism and high content of oxygen as well as vitrinite and mineral matter reveals great sorption capacity. The high content of oxygen indicates that the surface of sample 3 may be strongly hydrophilic due to some oxygen-functional groups present on the coal surface. Test results confirm a positive correlation between mineral matter content and sorption capacity of coal. In conclusion, fire indices calculated on the basis of the concentrations of gases with high sorption ability can be a source of errors due to the sorption taking place, especially if the coal displays high sorption capacity.

Acknowledgements
This work was supported by the Ministry of Science and Higher education, Poland [grant number 11305117]

References

1. A.C. Smith, C.O. Lazzara, *Spontaneous Combustion Studied of U.S. Coals* (US Bureau of Mines Report of Investigations RI 9079, 1987)
2. B. Beamish, R. Beamish, *Experience with using a moist coal adiabatic oven testing method for spontaneous combustion assessment* (Proceeding in 11th Underground Coal Operators Conference, 2011)
3. N.K. Mohalik, E. Lester, I.S. Lowndes, Int. J. Min. Reclam. Env. (2016)
4. A. Adamus, P. Prokop, H. Šancer, *Binary Indicators of Thermal Oxidation of OKR Coal* (In Proceedings of the Int. Conf. of the "10th International Bureau of Mining Thermophysics, 2005)
5. J. Xie, S. Xue, W. Cheng, G. Wang, Int. J. Coal. Geol. **85** (2011)
6. A.K. Singh, R.V.K. Singh, M.P. Singh, H. Chandra, N.K. Shukla, Int. J. Coal. Geol. **69**, 3 (2007)
7. B. Nie, X. Liu, L. Yang , J. Meng, X. Li, Fuel **158** (2015)
8. J.G. Speight, *Handbook of Coal Analysis. Second Edition* (Wiley, 2015)
9. X. Cui, R.M. Bustin, G. Dipple, Fuel **83**, 3 (2004)
10. A. Marzec, Gospod. Surowcami Min. **12**, 1 (1996)
11. P. Baran, G.S. Jodłowski, A. Krzyżanowski, K. Zarębska K, GGE **40**, 3 (2014)
12. J. Cygankiewicz, *Proceedings of the 4th schoolof mine ventilation* (Section of MineVentilation Committee on Mining Polish Academy of Science, 2006)
13. A. Dudzińska, J. Earth Sci. **28** (2017)
14. J. Cygankiewicz, A. Dudzińska, M. Żyła, Adsorption **18**, 3 (2012)
15. J. Cygankiewicz, A. Dudzińska, M. Żyła, Gospod. Surowcami Min. **25**, 1 (2009)
16. A. Dudzińska, M. Żyła, J. Cygankiewicz, Arch. Min. Sci. **58**, 3 (2013)
17. J. Cygankiewicz, M. Żyła, A. Dudzińska, Karbo 3 (2012)
18. D. Duprez, F. Cavani, *Handbook of Advanced Methods and Processes in Oxidation Catalysis: From Laboratory to Industry* (Imperial College Press, 2014)
19. A. Krzyżanowski, K. Zarębska, Gospod. Surowcami Min. **23** (2007)
20. A. Dudzińska, Przegląd Górniczy **72**, 2 (2016)
21. A. Bush, Y. Gensterblum, B.M. Krooss, Int. J. Coal. Geol. **55**, 2-4 (2003)
22. A. Bush, B.M. Krooss, Y. Gensterblum, F.van Bergen, H.J.M. Pagnier, J. Geochem. Explor. **78-79** (2003)
23. C.F. Rodrigues, M.J. Lemos de Sousa, Int. J. Coal. Geol. **48**, 3-4 (2002)
24. A. Dudzińska, J. Earth. Sci. **28** (2017)
25. G. Ceglarska-Stefańka, K. Zarębska, Int. J. Coal. Geol. **62** (2005)
26. Ch. Deng, D. Tang, S. Liu, H. Xu, S. Tao, Int. J. Coal. Geol. **25** (2015)
27. L.A. Harris, Ch.S. Yust, Fuel **55**, 3 (1976)
28. M. Mastalerz, J.C. Hower, D.N. Taulbee, Geol. Acta **11** (2013)
29. A. Komorek A, Arch. Min. Sci. **61**, 4 (2016)
30. A. Ślączka, A. Wasilczyk, Górnictwo i Geologia **7**, 3 (2012)

Comparative analysis of heat pump and biomass boiler for small detached house heating

Tomasz Olkowski[1,*], *Seweryn* Lipiński[1], and *Aneta* Olędzka[1]

[1]Department of Electric and Power Engineering, Electronics and Automation, Faculty of Technical Sciences, University of Warmia and Mazury in Olsztyn, 11 Oczapowskiego St., 10-736 Olsztyn, Poland

Abstract. The purpose of the work is to answer the question - which of the two selected heat sources is more economically beneficial for small detached house: heat pump or biomass boiler fuelled with wood-pellets? The comparative analysis of these sources was carried out to discuss the issue. First, cost of both, equipment and operation of selected heat systems were analysed. Additionally, CO_2 emission levels associated with these heat systems were determined. The comparative analysis of the costs of both considered heat systems showed that equipment cost of heat pump system is considerably bigger than the cost of biomass boiler system. The comparison of annual operation costs showed that heat pump operation cost is slightly lower than operation cost of biomass boiler. The analysis of above results shows that lower operation cost of heat pump in comparison with biomass boiler cost lets qualify heat pump as more economically justified only after 38 years of work. For both analysed devices, CO_2 emission levels were determined. The considerations take into account the fact that heat pump consumes electricity. It is mostly generated through combustion of coal in Poland. The results show that in Poland biomass boiler can be described as not only more economically justified system but also as considerably more ecological.

1 Introduction

The analysis of the Central Statistical Office (CSO) data [1] with regard to the number of residences delivered for use over the period from 1991 to 2016 shows the growing trend of private construction, i.e. detached houses (including single-family houses) in relation to the total construction. An average from the analysed 26 years is approx. 58 thousand detached houses of private construction per year, i.e. 49% of total construction.

These buildings need individual heat sources to in order to heat space, as well as domestic hot water (d.h.w.). In currently built Polish detached houses electricity and heat consumption is significantly lower than in the older ones [2], but still households' energy consumption represents a significant part (33%) of energy balance in Poland [3, 4]. Space heating has the largest share in the structure of energy consumption by Polish households (approx. 69%) [2]. It suggests that space heating represents a heavy financial burden for households' owners. That is why it is important to use the least costly heat sources, but of course considering energy efficiency and environmental requirements.

It is difficult to select heating devices from those being used in detached houses, which meet all the above-mentioned conditions. However, it can be assumed that heating devices are more ecological when they use renewable energy sources (RES) [5], and when they are energy-efficient [6, 7].

High efficiency of a heating equipment results in lower operational costs, but advanced technical solutions are usually needed to produce such devices. For that reason, the energy-efficient equipment has often a higher price [8] and that is why the selection of a heating system should always be preceded by a thorough economic analysis.

In this paper, two kinds of heating systems were considered. These are system with a heat pump (HP), as well as system with a wood pellet firing boiler (biomass boiler). Both are RES-based and seem the most ecological and energy-efficient [9, 10]. The main purpose of the work is to answer the following question: which of the two selected heat sources is more economically beneficial for small detached house? The comparative analysis of these sources was carried out to discuss the issue. First, cost of both, equipment and operation of selected heat systems were analysed. Additionally, CO_2 emission levels associated with these heat systems were determined. It should be underlined that this analysis concerns only Polish conditions, what can have a significant impact on the obtained results.

Similar analyses for other countries can be found e.g. in following papers: [8,11-13].

2 Material and methods

The analysis was carried out for a non-existing detached house. It was designed for the sole purpose of this study.

* Corresponding author: tomasz.olkowski@uwm.edu.pl

 DOI: 10.1051/e3sconf/20171901011

The „Audytor OZC 6.6 Pro Edu" and „Audytor C.O. 4.0 Edu" software were used during design.

The house is supposed to be located in north-eastern part of Poland in the town of Kętrzyn, i.e. in the fourth climatic zone with reference external design temperature equal to -22°C [14]. The house has one habitation floor over the ground level and non-habitable (unheated) attic space. The house is to be built from cellular concrete blocks. External and internal wall thickness amount to 24 and 12 cm, respectively. The ceiling is to be made out of oak wood, which is a very good insulator. The whole building is to be placed on the foundation footings and benches directly on the ground. The total habitation area of the house amounts to 141.45 m^2 (Table 1); the number of home residents is five persons. It was assumed that the presented house does not have access to the natural gas network but it is connected to the electricity grid, water supply and sewage system.

Table 1. The area and the heating demand of rooms in the analysed detached house.

	Area [m^2]	Heating demand [W]
Hall	9.9	540
Boiler-room	6.66	0
Kitchen	16.92	989
Bathroom	11.7	701
Room I	18.68	898
Room II	15.75	772
Living room	61.85	2431
TOTAL	141.45	6331

Table 1 shows the design heat load for each room in the house. The design heat load of boiler-room is equal to zero, because boiler-rooms are designed as unheated areas and so the total heated area equals 134.8 m^2.

The analysis of the data from Table 1 shows that rooms heating energy demand (Φ_{co}) amounts to 6.3 kW. Domestic hot water heating energy demand is also needed to be known (Φ_{cwu}) in order to calculate the total energy demand of the house. The daily domestic hot water energy demand can be calculated using formula (1), taking into account 20% heat loss caused by water flow in the pipelines:

$$\Phi_{cwu} = Q_{cwu} / 24\text{ h} + 20\%, \quad (1)$$

where Q_{cwu} is daily d.h.w. heat demand.

Daily domestic hot water heat demand can be calculated using following formula:

$$Q_{cwu} = c\,\rho\,V\,(t_{cw} - t_{zw}), \quad (2)$$

where c is specific heat of water (4,19 kJ/(kg K)), ρ is water density (1000 kg/m^3), V is daily water demand of water to be heated (0,5 m^3), t_{cw} is hot water temperature (55°C) and t_{zw} is cold water temperature (10°C).

A daily demand of water to be heated (V) was assumed based on [15]. According to it, daily use of a water in a dwelling house connected to sewage system equals 100 dm^3 per one home resident. Using above data, Φ_{cwu} = 1.32 kW was calculated and so the total design heat load of the house (Φ_C) is possible to be obtained using the formula:

$$\Phi_C = \Phi_{co} + \Phi_{cwu}. \quad (3)$$

Based on the above data, the total design heat load of the house equals 7.6 kW. It was assumed that one of the two selected heat sources, i.e. heat pump or biomass boiler fuelled with wood-pellets can be exploited in the considered house.

Energy balance of the house is shown in Fig. 1. It includes heat losses and gains. The results shown in Fig. 1 show that supposed solar heat gains and heat gains from household devices represent an important share of the house energy balance. It will allow to reduce the planned heating system cost.

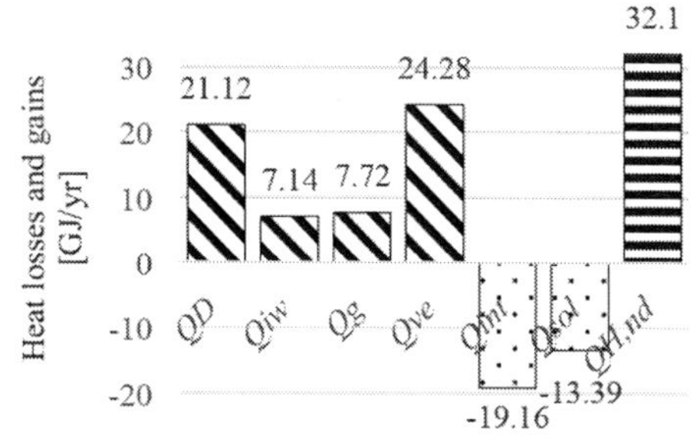

Fig. 1. Energy balance of the house [GJ/yr]: Q_D – heat losses through external walls, Q_{iw} – heat losses through internal walls, Q_g – heat losses to the ground, Q_{ve} – natural ventilation heat losses, Q_{int} – heat gains from household devices, Q_{sol} – gains from solar heat, $Q_{H,nd}$ – annual heat needs of the house with the use of heat gains factor $\eta_{H,gn}$ = 0.87.

For each of the considered heat sources an economic analysis was made. The purpose of this analysis is identification and comparison of cost of equipment and operation of heating devices and systems that make up a given heat source. In this case, the investment costs include the purchase costs of equipment, but do not include manual labour costs. Operating costs include the purchase of fuel and electricity and the cost of equipment service. Based on the economic analysis, a simple payback period of investment in the considered heat sources was calculated.

In addition, the ecological analysis of the heat sources was carried out. The CO_2 emissions from fuels combustion and electricity consumption were analysed. Under Polish conditions, electricity is mainly produced by hard coal combustion, which results in significant CO_2 emissions.

3 Identification of heat sources for comparative analysis

3.1 Heat pump

First of the selected systems is heat pump with a 9 kW maximum heating power output.

DOI: 10.1051/e3sconf/20171901011

Table 2 shows other parameters of the chosen heat pump.

Table 2. Parameters of the selected heat pump.

Parameter	Value
Heating power	9 kW
Power consumption	1.7 kW
Cooling power	7.3 kW
COP	5.2
Supply temperature of heating installation (at heat pump outlet)	35°C
Brine temperature (at heat pump inlet)	5°C

Such power of heat pump results from the possible interruptions in the electricity supply during periods of peak load in the electrical grid [16]; in order to ensure proper work, an oversize device is required. A time of interruption in the electricity supply was assumed as equal to 2 h, with the oversizing factor equal to 1.1 and so the heat pump with 8.4 kW minimum heating power output is required.

As a lower heat source, a flat ground heat exchanger was selected. It is immersed in the ground of aquifers gravel and sand, with a unit heating power output of 32 W/m^2. This is one of the best types of soil for flat heat exchangers, because the more wet the soil, the higher the heat transfer coefficient, and so the more favourable heat transfer conditions [17].

The view of the habitation floor with layout of heating surfaces and shore zones is shown in Fig. 2.

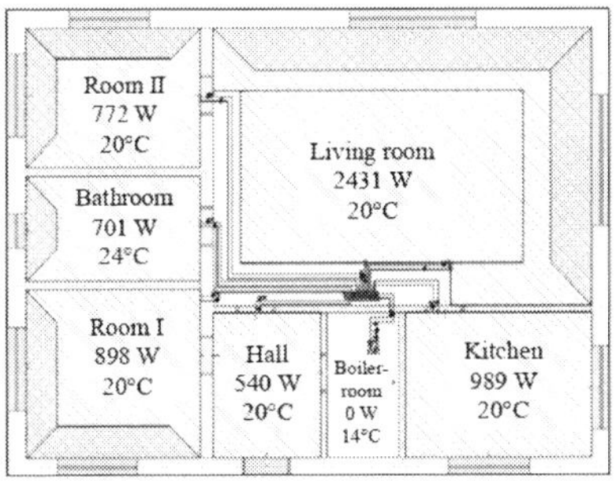

Fig. 2. The view of the habitation floor with layout of heating surfaces and shore zones.

An underfloor heating works as a higher heat source for the heat pump. Such configuration will ensure proper system operation, i.e. protection of the floor from overheating as well as work of the heating device with the highest possible efficiency. An installation of underfloor heating with a 5433 W heating power was designed.

The reason for the lack of shore zone in the kitchen (Fig. 2) is the relatively high heat gain from household goods - identified in the energy balance of the house (Fig. 1). In the hall, no shore zone was designed as its heat demand would be covered by heat gains from neighbouring rooms. The living room was divided into two heating surfaces with heating loops of 87 and 119 m, to avoid exceeding the allowable heating loop length equal to 120 m.

The equipment and operation costs have been identified for such heating system. The individual components of the equipment cost are given in Table 3.

Table 3. The equipment cost of heat pump including installations of lower and higher heat sources.

Specification	Cost [USD]
Heat pump with built-in domestic hot water tank	6255.33
Lower heat source – ground heat exchanger installation	
Lower heat source equipment (collector pipes – 285 m, manifold – 3 circuits)	439.39
Glandless circulation pump, safety group, diaphragm expansion vessel	368.00
Glycol – approx. 300 dm^3	1030.40
Total (lower heat source)	1837.79
Higher heat source – heating flat installation	
Pipes of heating loops – 563 m	580.21
Insulation	414.61
Clips, edge band, manifold – 8 circuits, circulation pump, ball valve (9 pcs)	538.01
Automatic balancing valve (8 pcs)	785.06
Plasticizer, adhesive tape, couplings, expansion joint, expansion tape, peschel	123.16
Flush-mounted cabinet	62.31
Expansion tank closed for rooms heating – 8 dm^3	17.66
Total (higher heat source)	2521.02
Total equipment cost	**10614.16**

Equipment cost may be reduced by subsidies. The subvention can be obtained from the National Fund for Environmental Protection and Water Management under the „Prosument" program [18]. The amount of subsidy is up to 15% of eligible costs, which is approx. 1,821.60 USD in this case. Thus, if the subsidy is taken into account, the equipment cost for the heat pump installation is equal to 8,792.54 USD. This amount will be taken into account during calculation of the payback period of investment in heat pump installation.

During calculation of the operation cost, it was assumed (based on [19]), that the working time of the heat pump for Kętrzyn is 2,426 h/yr. It allows calculating the annual electricity consumption by the heat pump as 4,124.2 kWh/yr. Results of the operation cost calculation are given in Table 4.

Table 4. The operation cost for heat pump heating system.

Specification	Cost [USD/yr]
Annual cost of electricity consumption by heat pump	618.63
Annual cost of electricity consumption by circulation pumps of the lower and higher heat sources	15.00
Service of the heat pump equipment	122.67
Total operation cost	**756.30**

DOI: 10.1051/e3sconf/20171901011

The electricity price of 0.15 USD/kWh was assumed for a single-zone tariff for households (G11). The electricity consumption of each of the used circulation pumps (one in the lower and one in the higher installation) is approx. 50 kWh/yr. Service (including the heat exchanger cleaning) is approx. 122.67 USD/yr. Although equipment cost can be a serious financial burden for an investor, total operation cost (Table 4) of 756.30 USD/yr is not too high in Polish conditions.

3.2 Biomass boiler

A biomass boiler is an alternative for the heat pump. The correct choice of the boiler depends mainly on the design heat load of the building. Other and equally important reasons for choosing a particular boiler are availability and price of the proper fuel, the cost of purchasing and installing the boiler, as well as the cost and comfort of using the boiler.

Taking the design heat load of the house into consideration, the smallest biomass boiler with retort furnace fuelled with wood-pellets (which is available on the domestic market) was chosen. The nominal heating power of the boiler amounts to 9 kW, but it can work in the range of 3÷10 kW. When working with a nominal heating power, this boiler can achieve up to 86% efficiency. The technical specifications of the selected boiler are shown in Table 5.

Table 5. Technical parameters of the selected boiler.

Parameter	Value
Nominal heating power	9 kW
Range of work	3 ÷ 10 kW
Boiler working time without filling fuel (nom. heating power ÷ min. heating power)	50 ÷ 130 h
Water capacity of the boiler	45 dm^3
Min. chimney draft required	24 ÷ 34 Pa
Maximum operating pressure	1.5 bar
Diameter of the flue	159 mm

For the boiler, a central heating system consisting of plate heaters with a total heat output of 4709 W was designed. A double-walled heat exchanger with a capacity of 120 dm^3 for domestic hot water heating was also selected. The heating installation diagram is shown in Fig. 3.

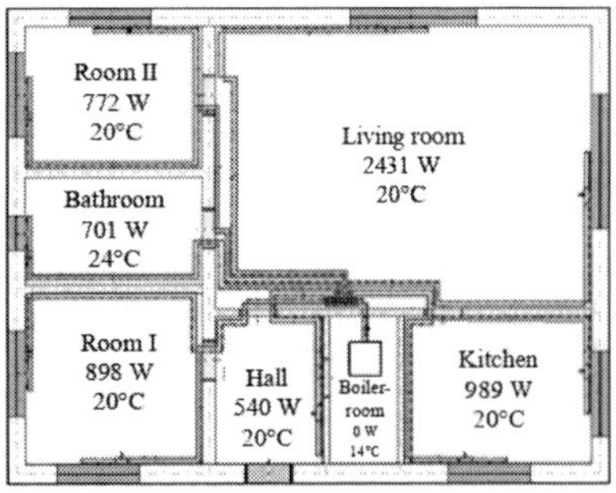

Fig. 3. The heating installation diagram.

For the heating system with biomass boiler, the equipment and operation costs have been also identified, as for the system with heat pump. The individual components of the equipment cost are given in Table 6.

Table 6. The equipment cost of the heating system with biomass boiler.

Specification	Cost [USD]
Heat source	
Biomass boiler	1616.99
Double-walled heat exchanger 120 dm^3	176.64
Total (heat source)	1793.63
An installation for heat collection	
Plate heater C11-60 (3 pcs)	220.80
Plate heater C11-30 (5 pcs)	288.27
Plate heater C21S (2 pcs)	180.56
Open expansion vessel reflex 25 dm^3, ball valve (2 pcs), manifold – 7 circuits, angle radial return valve (10 pcs), glandless circulation pump	545.62
Automatic balancing valve ASV-P (10 pcs)	1077.01
Pipe guide at the manifold (38 pcs), pipes PEAL-S10 – 191,82 m, thermostatic heads (7 pcs), flush-mounted cabinet, peschel	350.51
Total (an installation for heat collection)	2662.77
Total equipment cost	**4456.40**

In the case of a biomass boiler, it is also possible to apply for a subvention in order to cover a part of equipment cost from the same program as for a heat pump. The amount of this subsidy is also up to 15% of eligible costs, which is approx. 596.16 USD. Thus, after this subsidy is taken into account, the equipment cost for a biomass boiler system is equal to 3860.24 USD. As with heat pumps, this cost will be taken into account during calculating the payback period of investment in the biomass boiler installation.

Table 7 shows the operation cost of a pellet boiler heating system. Based on [20], the price of pellets of 208.53 USD/t and annual use of pellets (3.8 t) was assumed. The electrical power consumption of the pellets feeder, blast furnace and circulation pump is approx. 1.7 kWh/day. Operation of the biomass boiler does not require regular service, like in the case of heat pump. The annual operating time of the pellet boiler is similar to a time of the heat pump work, so the value is the same, i.e. 2426 h/yr [19]. In addition, the price of electricity is the same, i.e. 0.15 USD/kWh.

Table 7. The operation cost for pellets boiler heating system.

Specification	Cost [USD/yr]
Cost of pellet using	792.41
Annual electricity cost	93.03
Total operation cost	**885.44**

As it can be seen in Table 7, the operation cost of a pellet boiler system is slightly higher than the operation

DOI: 10.1051/e3sconf/20171901011

cost of a heat pump system. Therefore, it is necessary to set a payback period for costs incurred.

4 Results and discussion

This section presents the results of comparison of both considered heating systems. Fig. 4 shows a comparison of equipment costs, while Fig. 5 shows a comparison of operation costs.

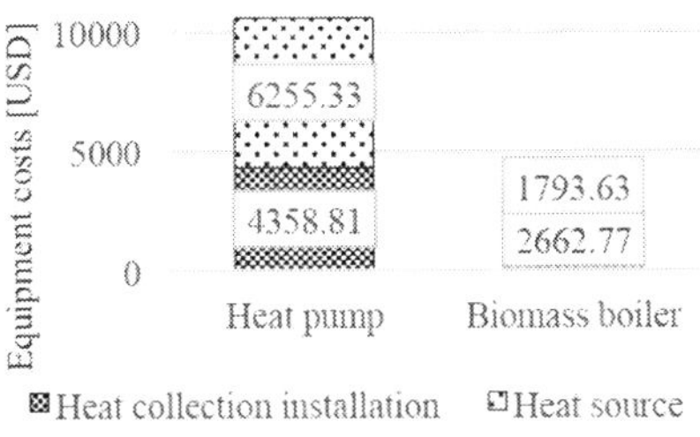

Fig. 4. Comparison of equipment cost.

The results shown in Fig. 4 indicate that the equipment costs on the heat pump is almost 3.5 times higher than on the pellet boiler. Also the costs of higher and lower heat source of heat pump is approx. 1.6 times higher than the heat collection installation of pellet boiler. The purchase costs of the installation with heat pump is calculated as the sum of costs of lower and higher heat sources. That is why these costs are significantly higher than the costs of heat collection installation working with pellet boiler. The total equipment cost of the heat pump system exceeds the total equipment cost of the biomass boiler system by approx. 2.4 times. The difference is approx. 6,158 USD. In consideration of the partial subvention [18], the cost of the heat pump over the cost of the biomass boiler will slightly decrease - to 2.3 times (the difference will be equal to approx. 4,932 USD).

Fig. 5 shows a comparison of operation costs for both heating systems. Operation costs are mainly the fuel and electricity costs, but the cost of required service inspections is also included in the case of the heat pump. The operation costs for the biomass boiler system are slightly higher than the operation costs for the heat pump system. The difference is equal to 129.14 USD/yr.

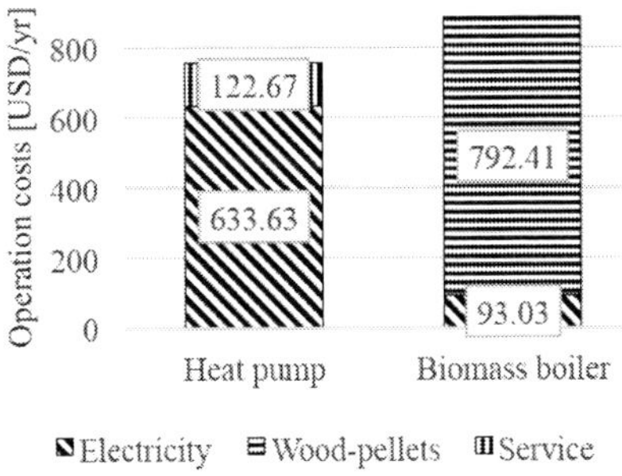

Fig. 5. Comparison of operation costs.

The simple payback period has been also calculated, i.e. after how many years the difference in operation costs between the heat pump system and the pellet boiler system equates the difference in equipment costs between those systems. This result is shown in Fig. 6. The most optimistic scenario was took into account, in which the prices of fuels and electricity are not rising (and not decreasing); also heating systems are not failing and are working with unchanging efficiency.

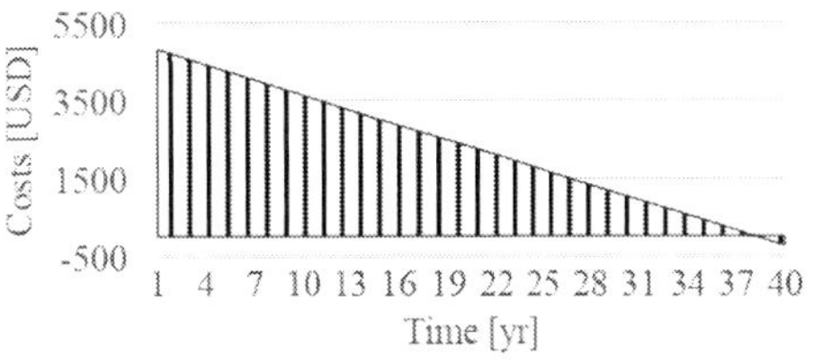

Fig. 6. The simple payback period of the heat pump system in relation to the pellet boiler system.

The results in Fig. 6 show that lower operation cost of heat pump in comparison with biomass boiler cost lets qualify heat pump as more economically justified only after 38 years of working. As stated above, the costs of potential failures are not taken into account, which in the case of heat pump may even increase the payback period. These results show that the installation with a pellet boiler is more economically justified.

For the analysed equipment, annual CO_2 emissions were also determined (Fig. 7).

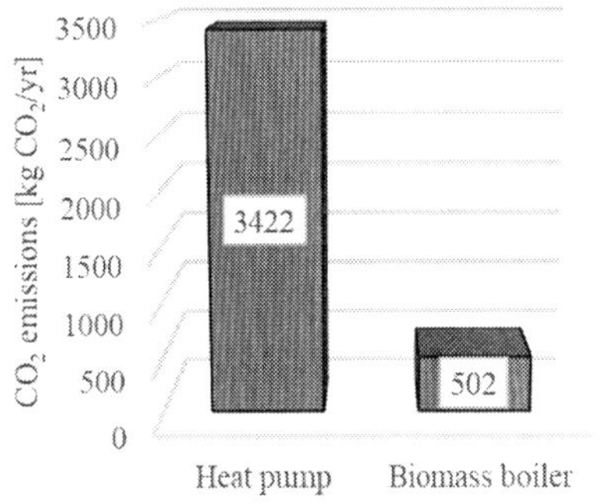

Fig. 7. Comparison of CO_2 emissions.

Although the heat pump does not burn any fuel, the electricity used by the compressor or circulating pump in Poland is obtained primarily through coal combustion. Based on the CO_2 emissivity factor for coal power generation (WE = 810 kg CO_2/MWh [21]) it was calculated that the considered heat pump produces CO_2 at the level of approx. 3,341 kg CO_2/yr. The pellet boiler also needs electricity to ensure proper operation, but it consumes much less of it what results in significantly lower CO_2 emissions (approx. 503 kg CO_2/yr). A level of CO_2 emissions from pellet combustion equal 0 kg CO_2/yr were assumed, because all CO_2 emitted during biomass combustion is assumed to be completely absorbed by plants in the process of photosynthesis [22].

Those values show that biomass boiler can be described as not only more economically justified system but also as considerably more ecological in Polish conditions. It is important to emphasise that the obtained results of CO_2 emissions, being results of the electricity used, are true only in Polish conditions. These results can be different depending on the specifics of the area. The results similar like for Poland can be obtained for Estonia, where more than 83% of electricity is made by fossil fuels burning [23]. But for example in Belgium [8] the heat pump shows lower CO_2 emissions when compared to the biomass fuelled installation, because in Belgium electricity is obtained mainly from nuclear power plants and additionally the cost of heat from pellets combustion is compared to the cost of heat from natural gas combustion. That is why in Belgium the heat pump is more ecologically and economically justified, and the results of similar analyses should be carefully taken into account.

5 Conclusions

The following conclusions can be drawn based on the performed study:

1. The total equipment costs of the HP system are over two times higher than the total costs of the biomass boiler system, even if these costs are reduced by the means of the subventions.
2. The high equipment costs of the HP based system stems from the need to make installations of lower and higher heat sources, while only a heat collection installation is required for the biomass boiler.
3. The operation costs are slightly higher for the pellet boiler but the difference is only 129.14 USD/yr.
4. The simple payback period shows that the higher equipment costs of the HP system in relation to the pellet boiler can be compensated by lower operation costs only after 38 years of using the HP system.
5. The analysis of CO_2 emissions showed that the pellet boiler is significantly more ecological.

References

1. CSO. Housing construction - inspection tables since 1991 [Internet]. 2017 [in Polish] [cited 2017 May 08]. Available from: http://stat.gov.pl/obszary-tematyczne/przemysl-budownictwo-srodki-trwale/budownictwo/budownictwo-mieszkaniowe-tablice-przegladowe-od-1991-roku,6,5.html
2. M. J. Stolarski, M. Krzyżaniak, K. Warmiński, D. Niksa, Energ. Conv. Manage. **121**, 71 (2016)
3. M. J. Stolarski, M. Krzyżaniak, K. Warmiński, M. Śnieg, Energ. Buildings **66**, 395 (2013)
4. M. Stolarski, M. Krzyżaniak, Ł. Graban, Energ. Buildings **43(2-3)**, 433 (2011)
5. T. Moore, Energ. Buildings **70**, 463 (2014)
6. A. Saari, T. Kalamees, J. Jokisalo, R. Michelsson, K. Alanne, J. Kurnitski, Appl. Energ. **92**, 76 (2012)
7. L. Gustavsson, A. Joelsson, Energ. Buildings **39(6)**, 717 (2007)
8. L. Georges, C. Massart, G. Van Moeseke, A. De Herde, Energ. Pol. **40**, 452 (2012)
9. A. Hast, T. Ekholm, S. Syri, Sustain. Cities Soc. **22**, 49 (2016)
10. L Gustavsson, Å. Karlsson, Energy **28(8)**, 851 (2003)
11. Ü. Alev, A. Allikmaa, T. Kalamees, Energy Proc. **78**, 907 (2015)
12. Å. Karlsson, L. Gustavsson, Energy Pol. **31(14)**, 1541 (2003)
13. K. Mahapatra, L. Gustavsson, Energ. Effic. **3(1)**, 1 (2010)
14. PN-EN 12831 "Heating systems in buildings. Method for calculation of the design heat load". [in Polish]
15. Ordinance of the Minister of Infrastructure of January 14, 2002 on the definition of average standards of water consumption (Dz. U. Nr 8, poz. 70) [in Polish]
16. G. Grabowski, *Heat pumps - designer's guide*, Edition I, Viessmann Academy in Poland [in Polish] (2006)
17. M. Rubik., *Heat pumps in low temperature geothermal systems.* MULTICO Publishing House, Warsaw, (2011), ISBN 978-83-7763-052-5 [in Polish]
18. National Fund for Environmental Protection and Water Management [Internet]. 2017 [in Polish] [cited 2017 May 12]. Available from: http://nfosigw.gov.pl/oferta-finansowania/srodki-krajowe/programy-priorytetowe/prosument-dofinansowanie-mikroinstalacji
19. A. Adamczewski, *Choosing the size of the heat pump* [Internet], 2017 [in Polish] [cited 2017 May 06]. Available from: http://www.instsani.pl/554 /dobor-wielkosci-pompy-ciepla
20. T. Olkowski, A. Koniecko, Ł. Przybylski, Agric. Eng. **143**, 245 (2013) [in Polish]
21. KOBiZE (2017). Emissivity factors of CO_2, SO_2, NO_x, CO and total dust for electricity based on information from domestic data base of GHG and other substances emissions in 2015 [in Polish]
22. European Commission. Commission staff working document. Impact assessment. Accompanying document to the report from the commission to the council and the European Parliament on sustainability requirements for the use of solid and gaseous biomass sources in electricity, heating and cooling. COM(2010) 11 final SEC(2010) 66.; 2010. 54p + annexes.
23. IEA Statistics, *Electricity production from coal sources (% of total)* [Internet], 2015 [cited 2017 August 01]. Available from: http://data.worldbank.org/indicator/EG.ELC.COAL.ZS?view=map&year=2015

Analysis of the efficiency of a hybrid foil tunnel heating system

Sławomir Kurpaska[1,*], and *Norbert* Pedryc[2]

[1]University of Agriculture in Krakow, Faculty of Production Engineering and Energetics, Institute of Agricultural Engineering and Informatics, Balicka st. 116B, Cracow, Poland
[2]University of Agriculture in Krakow, Faculty of Production Engineering and Energetics, Departament of Mechanical Engineering and Agrophysics, Balicka st. 120, Cracow, Poland

Abstract. The paper analyzes the efficiency of the hybrid system used to heat the foil tunnel. The tested system was built on the basis of heat gain in a cascade manner. The first step is to heat the water in the storage tank using the solar collectors. The second stage is the use of a heat pump (HP) in order to heat the diaphragm exchangers. The lower HP heat source is a cascade first stage buffer. In the storage tank, diaphragm exchangers used for solar collectors and heat pumps are installed. The research was carried out at a research station located in the University of Agriculture in Cracow. The aim was to perform an analysis of the efficiency of a hybrid system for the heating of a foil tunnel in the months from May to September. The efficiency of the entire hybrid system was calculated as the relation of the effect obtained in reference to the electrical power used to drive the heat pump components (compressor drive, circulation pump), circulation pumps and fans installed in the diaphragm heaters. The resulting effect was the amount of heat supplied to the interior of the object as a result of the internal air being forced through the diaphragm exchangers.

1 Introduction

The dynamic economic development of the country has caused a significant increase in energy demand. Energy demand should be understood as energy in the basic forms - thermal energy and electricity. In order to neutralize and reduce the amount of waste and sewage, energy expenditure is required to generate higher fuel demands. The economy of our country is mainly based on the conventional fuels such as liquid fuels, gaseous fuels and in the greatest part solid fuels in the form of hard coal. In recent years, there has been considerable interest in renewable energy sources (RES), including for the heating of gardening facilities. These studies were conducted in numerous scientific center. And so, Garcia et al. [1] analyzed the possibility of using different heat sources (traditional and renewable) for heating greenhouses located in different climatic conditions. They noted that under the European climate conditions in the Mediterranean countries the best use would be the use of solar collectors, while for countries located in the northern regions it is best to support heating by means of a heat pump. They have identified the need to store excess of energy generated by the passive heating of greenhouse interiors in energy accumulators. Rousse et al. [2] presented the results of long-term research on heat recovery through greenhouse ventilation. The bed was installed in the exchanger that condensed the vapor contained in the moist air. Dried air was injected into the object. Ossebaard et al. [3] presented an analysis of the use of particular energy sources for heating (both renewable and traditional) with regard to economic and ecological effects. Hamdan et al. [4] analyzed the use of renewable energy sources for heating greenhouses. The performance of independent sources (heat pump, solar collectors) and bivalent hybrid systems were analyzed as a combination of a pump with solar collectors. Gustavsson et al. [5] have shown the benefits of substituting traditional heat carriers for the conversion of biomass to both heating and fuel transport. The authors have shown that the need for renewable sources of energy, and the resulting benefits have counted into the reduction of harmful substances emitted into the atmosphere. Groscurth et al. [6] presented problems with the use of renewable energies for heating purposes, paying attention to the rational separation of existing renewable energy sources in order to minimize the emission of harmful substances generated during the combustion of conventional heat carriers. The financial implications of the reduction of environmental pollution have also been taken into consideration. Huang and Lee [7], based on long-term studies, have determined the consumption of electricity used to drive the heat pump. The calculations were conducted with respect to the unit increment of the temperature of the liquid stored in the buffer tank of the heat pump. Kaygusuz and Ayhan [8] described and analyzed the heat pump co-operation system in which as the bottom source was used atmospheric air, with the upper one, where energy was stored in a battery filled with a phase-changeable state.

* Corresponding author: rtkurpas@cyf-kr.edu.pl

 DOI: 10.1051/e3sconf/20171901012

Nagano et al. [9] have developed an innovative system for visualizing and analyzing the efficiency of the heat pump operation, using a ground exchanger as the bottom source. The heat pump was used for heating purposes of the prototype greenhouse. Hawlader et al. [10] studied the energy effects of a system in which a heat pump cooperated with a solar collector. Fuller [11] analyzed theoretically and performed an experimental verification of the system using heated water in solar collectors for heating the foil tunnel. The water was collected in an accumulation tank and hence was directed (in a closed system) for washing out the surface covering the object.

The energy efficiency of the system has been determined, indicating its usefulness in high radiation areas. Kurpaska [12] presented an economic and ecological analysis of the heat pump (co-operating with the ground exchanger) for heating the foil tunnel. In summary, the problem of the use of heat pumps is a current issue in various scientific institutions. Analyzing this issue will be the main aim of presented work.

2 Purpose and scope of work

The purpose of the study was to analyze the efficiency of the hybrid system for the heating of the foil tunnel in selected months. As input variables for the analyzed system were separated:

- The energy of solar radiation,
- The energy generated from solar collectors,
- The energy absorbed by the heat pump from the lower heat source,
- The energy supplied to the facility through heat exchangers.

3 Methodology of research

The research was carried out at the research station owned by the Faculty of Production Engineering and Energetics of the University of Agriculture in Cracow. Fig. 1 shows a diagram of the analyzed hybrid system.

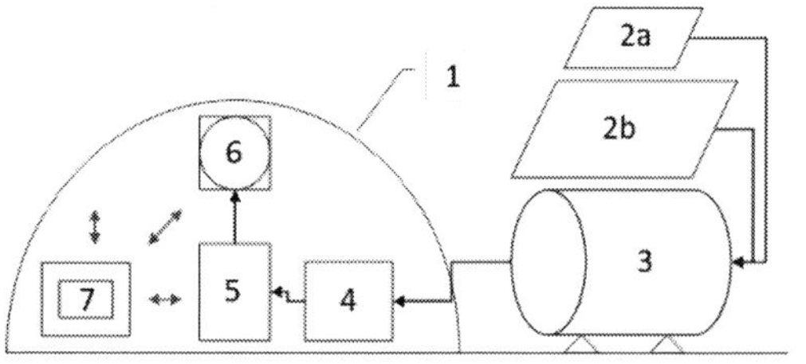

Fig. 1. Diagram of the measurement station: 1 - foil tunnel; 2a, 2b - solar collectors; 3 - accumulation tank; 4 - heat pump; 5 - buffer tank; 6 - heat source exchangers with fan; 7 - measuring and control equipment.

Heat from the compressor heat pump (type Vatra 10B1), through two diaphragm exchangers (6), was delivered to the interior of the foil tunnel (tunnel measuring 9 x 6 m and 2.5 m high, covered with 0.18 mm PE foil). Water in the accumulation tank (3) was the lower heat source of the compressor pump (4) which cooperated with the buffer tank (5). Water heating in the receiver (3) took place using solar collectors (vacuum and flat collectors). Flat collector area was 7.8 m^2, and vacuum collectors 4.3m^2.

The system was supplemented by a control and measuring system, which monitored the necessary parameters at a present frequency and controlled the operation of the heat pump components. The study was conducted for the period from April to September, and for further analysis were selected days differing in the parameters of the surrounding climate.

Calculations of significant values characteristic for the efficiency of the bivalent system in differential time $d\tau$ were made on the basis of the following equations:

- Solar radiation energy Es:

$$E_s = \sum_{i=1}^{n} R_z \cdot F_k \cdot d\tau \ \text{[J]} \quad (1)$$

- Energy gained from solar collectors Qk:

$$Q_k = \sum_{i=1}^{n} m_{cz} \cdot c_{cz} \cdot (T_p - T_z) d\tau \ \text{[kWh]} \quad (2)$$

- Efficiency of solar collectors ηk:

$$\eta_k = \frac{Q_k}{E_s} \cdot 100\% \quad (3)$$

- The energy supplied to the *Qzbd* accumulation tank:

$$Q_{zbd} = m_{wod} \cdot c_{wod} \cdot (T_{zbmax} - T_{pocz}) d\tau \ \text{[kWh]} \quad (4)$$

- The energy collected by the heat pump from the heat source Qdź:

$$Q_{dź} = \sum_{i=1}^{n} m_{cz} \cdot c_{cz} \cdot (T_z - T_p) d\tau \ \text{[kWh]} \quad (5)$$

- The energy supplied to the object through *Qgź* heat exchangers:

$$Q_{gź} = \sum_{i=1}^{2} \left(\sum_{i=1}^{n} m_{cz} \cdot c_{cz} \cdot (T_z - T_p) \right) d\tau \ \text{[kWh]} \quad (6)$$

- The demand for heat Q_{ob}:

$$Q_{ob} = \sum_{i=1}^{n} k_{osl} \cdot F_{osl} \cdot (T_{ob} - T_{ot}) d\tau \ \text{[kWh]} \quad (7)$$

- Heat pump efficiency coefficient COP:

$$COP = \frac{Q_{gź}}{P_{pc} \cdot \tau_{pc}} \quad (8)$$

where:
R_z - intensity of solar radiation, [W • m^{-2}]; F_k - surface [m^2], τ - working time [h], m_{cz} - stream of medium [kg • s^{-1}], c_{cz} – the

specific heat of the liquid [J • kg^{-1} • K^{-1}], T- temperature, where; subscript: z, p supply and return circulating medium, ob- inside the tunel, ot- ambient [°C], k_{ost}- heat transfer coefficient [W • m^{-2} • K^{-1}], τ_{PC}- heat pump working time [h], P_{pc} - heat pump power [kW].

4 Research results

Fig. 2 shows the calculated efficiency of solar collectors.

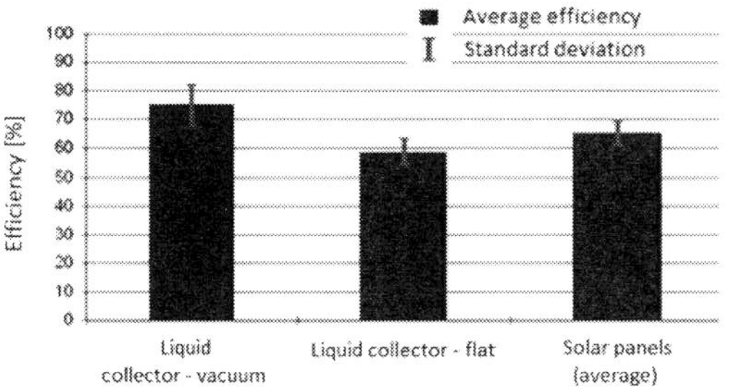

Fig. 2. Efficiency of the analyzed solar collectors.

The highest efficiency have shown vacuum collectors with an average level of about 75%, while the average performance of flat collectors was in the analyzed period, with efficiency close to 60%. The average difference between the efficiency of the collectors was 15% (standard deviation 11.67%). The calculated overall average efficiency of the solar conversion system, calculated as the arithmetic mean of the efficiency of both types of solar collectors, was approximately 65% from each of the system cycles studied.

Fig. 3 shows the effect of the sum of solar radiation on the amount of energy supplied to the storage tank during the surveyed cycles (the sum of solar radiation and energy were calculated for the period of system operation). The amount of this energy was strictly dependent on the work of the solar collectors. The energy losses that occurred when the tank was loaded were at an average level of 6.37 kWh (standard deviation of 3.09 kWh.) With the increase in solar radiation, the value of these losses increased.

Fig. 4 shows the correlation between the sum of solar radiation and the energy supplied to the accumulation tank by the solar collectors and the energy received by the heat pump from it during the tested measurement cycles (the sum of solar radiation and energy was calculated for the period of system operation cycle). It can be noticed that the amount of energy consumed from the lower heat source by the heat pump is approximately constant throughout the range (on average 31.45 kWh at standard deviation of 3.84).

Fig. 5 illustrates the effect of the heat pump operating time on the amount of energy it consumes from the accumulation tank during the measured cycle tests (working time and energy were calculated for the system cycle period).

During the longer operation of the heat pump, more energy was drawn from the bottom of the heat pump. The calculations showed that the average energy consumption from the accumulation tank, per hour of operation, was about 2.5 kWh of heat.

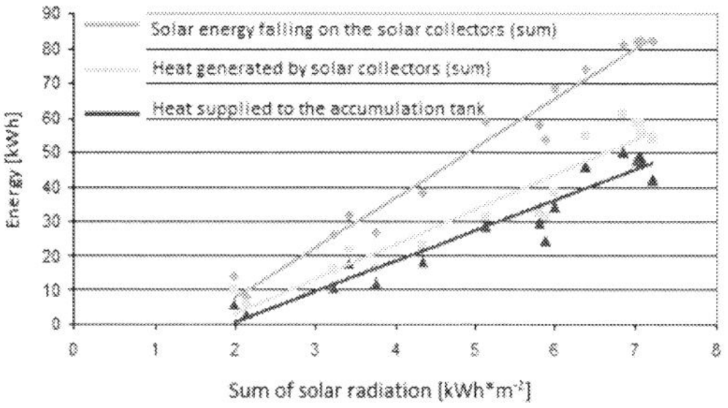

Fig. 3. The effect of total solar radiation on changes in energy components in the system.

In a given system configuration, the amount of heat supplied by the solar collectors to the storage tank, which would balance the energy taken from it by the heat pump, was only reached at a total solar radiation of approximately 5.5 kWh.

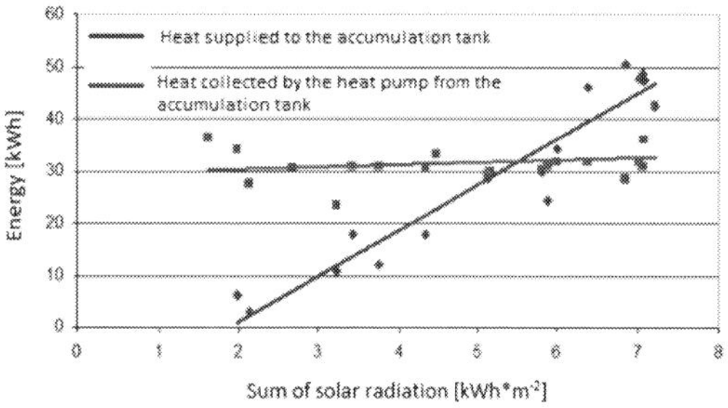

Fig. 4. The dependence of the energy supplied to the accumulation tank and the energy received from it by the heat pump as a function of the sum of solar radiation.

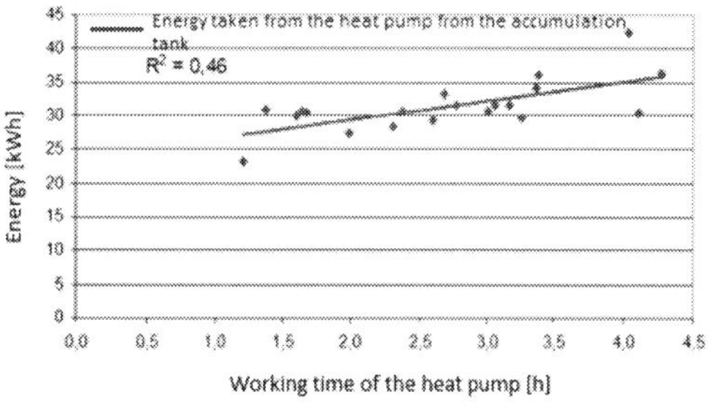

Fig. 5. Energy collected from the accumulation tank by the heat pump as a function of the system operating time.

Fig. 6 illustrates the amount of energy in the lower and upper heat sources of the heat pump with respect to the heat demand of the object (the heat demand was calculated during the operation of the liquid-air heat exchangers for the assumed temperature of 15 ° C, while the energy was calculated for the period of system operation). The heat that was collected by the heat pump from the accumulation tank could theoretically be sufficient to meet the heat demand of the object at the level of up to 0.5 kWh · m^{-2}. In calculations, the heat transfer coefficient of the object was 7 [W • m^{-2} • K^{-1}]. It should also be noted that the difference between the heat values collected by the heat pump and the heat supplied to the object was the smaller the higher the heat demand (lower outdoor temperature).

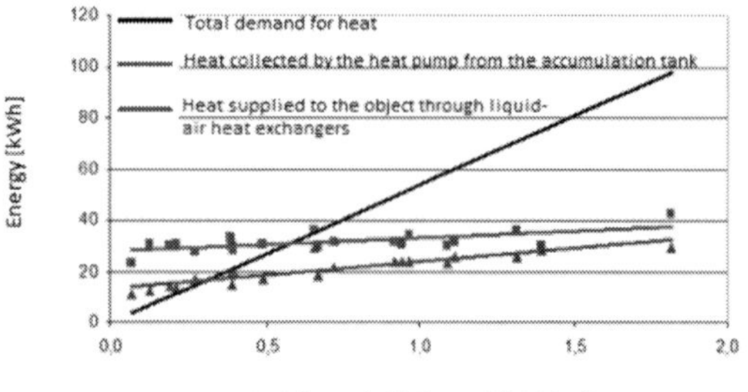

Fig. 6. Energy taken from the lower heat source by the heat pump and the energy supplied to the interior of the object by liquid-air heat exchangers as a function of the heat demand of the object.

Fig. 7 depicts the influence of the average temperature of the lower heat source on the efficiency of the hybrid system and the heat pump during the test cycles (temperature and efficiency were calculated for the system cycle period).

The temperature of the water stored in the accumulation tank had a significant effect on the heat pump's operation. Higher water temperatures increased the efficiency of the heat pump by improved heat exchange and associated with this shorter operating time (heat pumps).

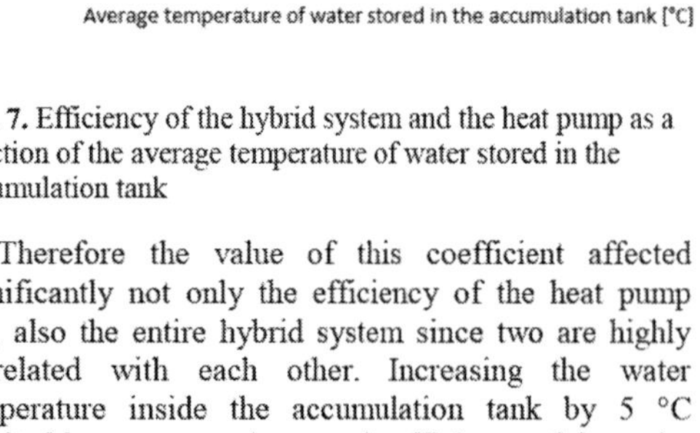

Fig. 7. Efficiency of the hybrid system and the heat pump as a function of the average temperature of water stored in the accumulation tank

Therefore the value of this coefficient affected significantly not only the efficiency of the heat pump but, also the entire hybrid system since two are highly correlated with each other. Increasing the water temperature inside the accumulation tank by 5 °C resulted in an average increase in efficiency of the entire system by 0.1. Under the tested conditions, the average efficiency of the heat pump was 1.51, while the overall system (including the power consumed by circulation pumps of diaphragm exchangers) was about 1.48.

5 Conclusions

1. The efficiency of the solar conversion system for the period under review was about 65% on average (average solar collector efficiency was 75% for vacuum collector and 60% for flat collector respectively).
2. The average value of energy losses when charging the accumulation tank was about 6.37 kWh. With the increase in the sum of solar radiation, the value of these losses increased.
3. The amount of energy consumed from the lower heat source by the heat pump is approximately constant throughout the range (average 31.45 kWh at standard deviation of 3.84). With the analyzed configuration of the system, the amount of heat supplied by the solar collectors to the accumulation tank, which would balance the energy taken from it by the heat pump, was only achieved at a total solar radiation of about 5.5 kWh.
4. Under the test conditions, the average efficiency of the heat pump was 1.51, while the overall system (including the power taken from the circulation pumps of the diaphragm exchangers) was about 1.48.

The Research was financed by the Ministry of Science and Higher Education of the Republic of Poland.

References

1. J.L. Garcia, S. De la Plaza, L.M. Navas, R.M. Benavente, L. Luna, J. Agric. Eng. Res. **69**, pp. 107-114, (1998)
2. D.R. Rousse, D.Y. Martin, R. Theriault, F. Leveillee, R. Boily, App. Therm. Eng **20**, pp. 687-706, (2000)
3. M.E. Ossebaard, A.J.M. Van Wijk, M.T. Van Wees, Energy **22**, pp. 1087- 1098, (1997)
4. M.A. Hamdan, A.I. Al-Sayeh, B.A. Jubran, App. Energy **41**, pp. 251- 264, (1992)
5. L. Gustavsson, P. Svenningsson, Energy Conv. Man. **37**, pp. 1211-1216, (1996)
6. H.M. Groscurth, T. Bruckner, R. Kummel, Energy **18**, pp. 1187- 1205, (1993)
7. B.J. Huang, C.P. Lee, Ren. Energy **29**, pp. 633- 639, (2004)
8. K. Kaygusuz, T. Ayhan, Energy Conv. Man. **40**, pp. 1377-1396, (1999)
9. K. Nagano, T. Katsura, S. Takeda, App. Therm. Eng. **26**, pp. 1578-1592, (2006)
10. M.N.A Hawlader., S.K. Chou, M.Z. Ullah, App. Therm. Eng. **21**, pp. 1049-1065, (2001)
11. R .J. Fuller, Aquac. Eng. **36**, pp. 250–260, (2007)
12. S. Kurpaska, Inż. Rol. **3**, pp. 189-197, (2013)

Possibility of reducing CO_2 emissions from internal combustion engines

Dawid Drabik[1], *Jarosław* Mamala[2,*], *Michał* Śmieja[3], and *Krzysztof* Prażnowski[4]

[1] Opole University of Technology, MEng, Faculty of Mechanical Engineering, St. Mikołajczyka 5, 45-271 Opole, Poland
[2] Opole University of Technology, Faculty of Mechanical Engineering, St. Mikołajczyka 5, 45-271 Opole, Poland
[3] University of Warmia and Mazury, Faculty of Technical Sciences, Michała Oczapowskiego 11, 10-719 Olsztyn, Poland
[4] Opole University of Technology, Faculty of Mechanical Engineering, St. Mikołajczyka 5, 45-271 Opole, Poland

Abstract. Article defines on the possibility of reduction CO_2 of the internal combustion engine and presents the analysis based on originally conducted studies. The increase in overall engine efficiency is sought after by all engineers dealing with engine construction, one of the major ways to reduce CO_2 emissions is to increase the compression ratio. The application of the compression ratio that has been increased constructional in the engine will, on one hand, bring about the increase in the theoretical efficiency, but, on the other hand, require a system for pressure control at a higher engine load in order to prevent engine knocking. For the purposes of the article there was carried out a number of studies and compiled results, and on their basis determined what have a major impact on the reducing CO_2.

1 Introduction

Modern internal combustion engines are built from the point of view of achieving the highest overall efficiency of the entire system. Those aspects that are associated with environmental protection and CO_2 reduction are very important in the production of power units. Therefore, automobile manufacturers attempt to design their cars in such a way as to ensure their cars are fuel-efficient and have good responsiveness. There are many ways to reduce CO_2, starting from those directly related to the engine and ending with the design of bodywork. Currently the race associated with automobile development is on. Internal combustion engines are still the most common units that transform fuel energy into mechanical energy. However, it should be kept in mind that traditional fuel resources are slowly being depleted; therefore, such a great emphasis is placed on using renewable power sources for cars [2].

Gasoline contains carbon and hydrogen atoms. During combustion, the carbon (C) from the fuel combines with oxygen (O_2) from the air to produce carbon dioxide (CO_2). Emission of carbon dioxide is directly connected with fuel consumption. Therefore, if fuel consumption is reduced, emission of carbon dioxide decreases. Engines are adjusted in order to minimize fuel consumption. 2.3 kg of carbon dioxide is produced from burning 1kg of fuel.

With all of the above information in mind, as well as the fact that the search process for alternative sources to power engines is very dynamic, it can be stated that a traditional four-stroke engine will be the most popular unit installed in passenger cars for a few years to come. The development of systems that enhance engine efficiency is already very advanced. New solutions to improve engines as units that transform fuel energy into mechanical energy are still being produced.

2 Efficiency of internal combustion engines

The efficiency of the internal combustion engine is an extremely important indicator of engine operation and it is connected with the possibility of reducing fuel consumption. Engineers aim at raising the efficiency since it facilitates better use of energy obtained from fuels. The main function of the internal combustion engine is to transform fuel energy into mechanical energy [4]. Overall engine efficiency is defined:

$$\eta_o = \eta_i \cdot \eta_m \cdot \eta_t \tag{1}$$

Considerations on engine efficiency should begin with the determination of theoretical efficiency η_t, which is defined as a ratio of the work of theoretical cycle L_t to the entire heat energy provided for the engine in one cycle Q:

$$\eta_t = \frac{L_t}{Q} \tag{2}$$

Efficiency that considers losses in the engine cylinder in respect of actual combustion, compression, decompression, charge exchange and other processes are called indicated efficiency. Indicated efficiency can be

* Corresponding author: j.mamala@po.edu.pl

 DOI: 10.1051/e3sconf/20171901013

presented using two equations: as a ratio of indicated work L_i to theoretical work L_t:

$$\eta_i = \frac{L_i}{L_t} \tag{3}$$

or

$$\eta_i = \frac{3600 \cdot N_i}{\eta_t \cdot G_e \cdot W} \tag{4}$$

where:
N_i – indicated power,
G_e – hourly fuel consumption in g/h or kg/h,
W – heating value of fuel kJ/kg,

Mechanical efficiency takes into account losses resulting from friction in the engine system and is defined as a ratio of useful work L_e to indicated work L_i.

$$\eta_m = \frac{L_e}{L_i} \tag{5}$$

Total efficiency of internal combustion engine is defined:

$$\eta_o = \frac{L_e}{Q_d} \tag{6}$$

The results of petrol and diesel engine efficiency are presented in Table 1.

Table 1. Results of petrol and diesel engine efficiency [4].

Efficiency	Spark ignition engine	Diesel
Theoretical efficiency η_t	0.35-0.45	0.50-0.65
Indicated efficiency η_i	0.65-0.78	0.75-0.85
Mechanical efficiency η_m	0.75-0.85	0.75-0.80
Overall efficiency η_o	0.20-0.32	0.28-0.45

One of the ways to enhance overall efficiency is to charge the compression ratio. Raising the compression ratio yields a number of positive results. Overall efficiency describes all losses in the engine.
In order to determine overall engine efficiency accurately, a complex formula considering the most important variables should be used:

$$\eta_o = \frac{1}{g_e \cdot W_d} \tag{7}$$

It can be stated that specific fuel consumption is a measure of overall engine efficiency. The research carried out by the authors showed that a direct increase in the compression ratio has an impact on the reduction of fuel consumption [4].

The CO_2 emission standards have become more strict over the years. It is connected with a considerable danger caused by emission of carbon dioxide. Fig. 1 presents changes in the CO_2 emission standards implemented over the years [3].

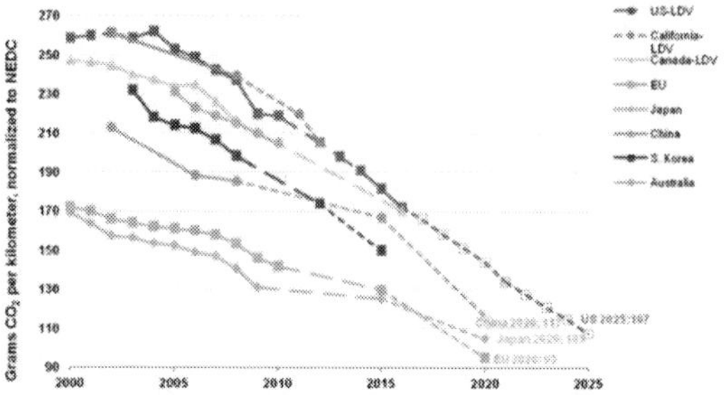

Fig. 1. The trend of carbon dioxide emission standards for passenger cars [9].

3 Solutions for dealing with emission of carbon dioxide

It is impossible and unprofitable to increase the compression ratio in an engine continuously. The possibilities of increasing the compression ratio in a petrol engine are limited mainly by a chance that a phenomenon called engine knocking occurs, which leads to serious damage to the engine design. Strength conditions and the occurrence of mechanical losses due to the compression ratio being too high cause the compression ratio to be in the range of 9 to 14 for petrol engines.

The results to be obtained by the increase in the compression ratio are so promising that many companies are trying to design an engine with a variable compression ratio. The engine built by French company MCE-5 is very promising as its compression ratio is controlled by the change of the volume of the combustion chamber with a very complex system. The presented results confirm a significant reduction of fuel consumption and carbon dioxide emission and, simultaneously, the increase in torque and power. Fig. 2 shows MCE-5 Engine VCRi.

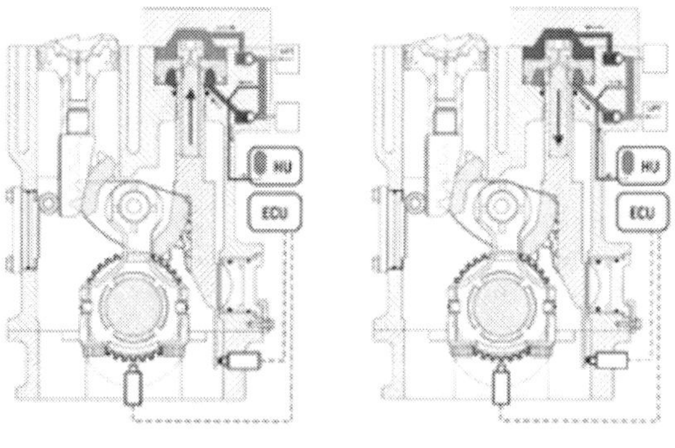

Fig. 2. MCE-5 Engine VCRi [6].

A decrease of fuel consumption in a combustion chamber is a result of these adjustments. Fig. 3 presents adjustments made in v6 engine to reduce emission of carbon dioxide.

 DOI: 10.1051/e3sconf/20171901013

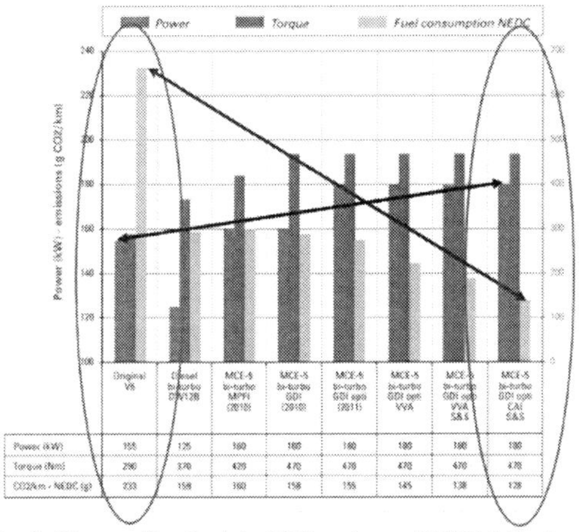

Fig. 3. Key results of original V6 engine and MCE-5 Engine VCRi bi-turbo GDI opti CAI [6].

Results:
- On a Peugeot 407 type vehicle, the most basic MCE-5 VCRi MPFI reduces CO_2 emissions by 45%.
- 16% more power.
- 62% more torque.

Other engines that are based on the idea of variable compression ratio are Scuderii Engine and SVC (SAAB Variable Compression). The compression ratio ideal for the engine in a given moment depends on road conditions. Fig. 4 shows a map of optimal compression ratio depending on the driving style.

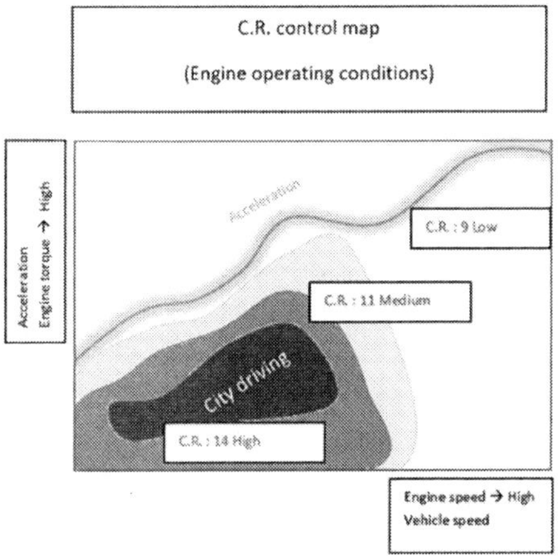

Fig. 4. Map of optimal compression ratio for an appropriate driving style [2].

The engine constructed by Infinity will be serially mounted in cars. The idea presented by Infinity engineers is based on changing the volume of a combustion chamber [1]. It adjusts compression ratio to the road conditions. A mass of a vehicle has a considerable influence on fuel consumption. Fig. 5 presents a relationship between a vehicle mass and fuel consumption.

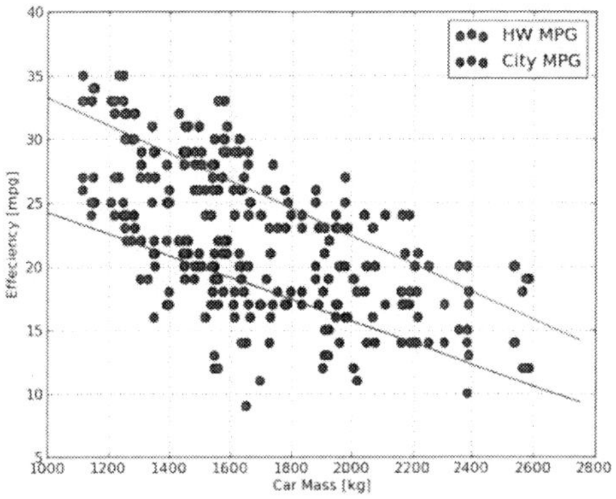

Fig. 5. Impact Car Mass on efficiency [9].

Internal combustion engines are constantly being developed. The following factors influence the development of engines:
- modern materials,
- modern oils and fuels,
- exhaust fumes emission norms,
- hybrid technology,
- modern electronic systems [4].

The internal combustion engine is a design that has evolved over the years. The first engines did not have high power or torque. Engine performance parameters say what characterises a given power unit [5].

The term engine indication mainly refers to measurements of pressure in the cylinder. Appropriate use of indication techniques gives the possibility of observing the phenomena occurring in the combustion chamber at low cost. A pressure measurement usually involves a full operating cycle of the engine [7]. Fig. 6 shows an open theoretical indicator diagram of an internal combustion engine for the two compression ratio, calculated using a computer, whereas shows an open indicator diagram for the two compression ratio. By carrying out laboratory tests, it can be stated that pressure in the combustion chamber can be modified. This translates into changes in power, torque or specific fuel consumption. Fig. 7 shows an opened actual indicator diagram

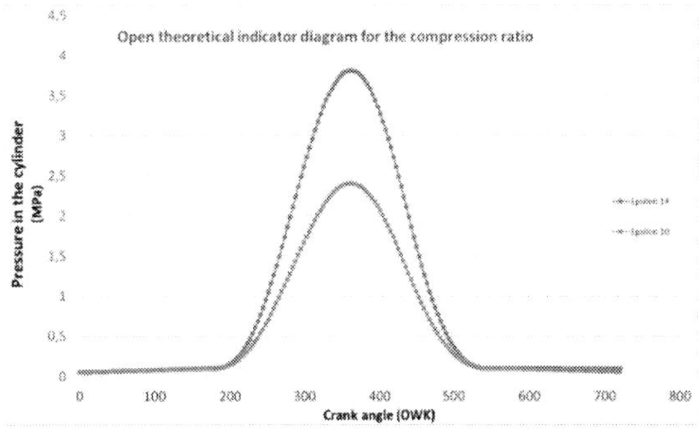

Fig. 6. Open theoretical indicator diagram for the different compression ratio.

DOI: 10.1051/e3sconf/20171901013

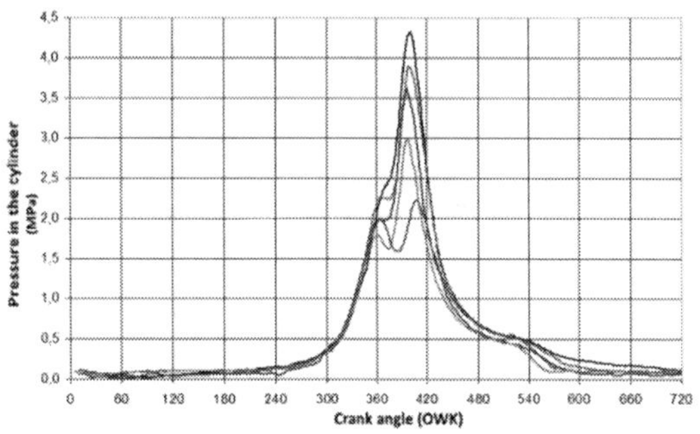

Fig. 7. Opened actual indicator diagram obtained by the change of the compression ratio.

The decrease in the compression ratio has a direct impact on the reduction of pressure in the cylinder. The increase of the compression ratio from 10 to 14 results in the raise of the pressure peak in a rate of 40%. Changes of the pressure in the cylinder enable a number of various analyses to be performed. Indication parameters can be divided into two categories:
- indirect indication parameters,
- direct indication parameters [8].
The difference between these two categories is such that direct indication parameters are determined from the pressure curve during one cycle. Indirect indication parameters require a number of calculations to be determined.

Increasing the compression ratio permanently results in high pressures without the need for an extensive charging system. However, this requires a solution that allows the operation of such a full engine load. Therefore, this work describe a system has been proposed to change the effective compression ratio of the internal combustion engine.

4 Conclusions

Overall efficiency is one of the most important engine performance parameters. It is obvious that the highest efficiency is much desired. There are many methods of increasing overall efficiency of an internal combustion engine. One of the methods, which yields very good results, is to increase the compression ratio. Unfortunately, it is road conditions that determine if the compression ratio rate. Modern engines that are installed as standard in passenger cars have a fixed compression ratio. However, there are innovative engine designs that are characterised by variable compression ratio. These engines include: MCE-5, Scuderii Engine, SVC SAAB or Infinity Engine. All tests were performed on the dynamometer. The results show that increasing the compression ratio improves the overall efficiency of the engine. The results confirm that as the compression ratio increases, the pressure in the cylinder rises.
By examining engine performance parameters, it is possible to choose such areas to be modified so that overall efficiency can be higher. Open indicator diagrams show that the higher compression ratio is assumed, the higher the pressure in the system is. This translates into the increase in overall engine efficiency. Higher compression ratio also involves lower specific fuel consumption, and lower fuel consumption means lower carbon dioxide emission into the environment.

This is an important issue that despite the growing interest in electric drive, the internal combustion engine will remain the main source of power vehicles. Car manufacturing is constantly increasing, hence the increasing production of engines. Currently around 90 million power units are produced yearly; however, it is possible that 125 million power units will be produced yearly in 2025.

The production of internal combustion engines will increase dramatically in the following years, which is an economic motivation for manufacturers. New solutions to enhance overall increase engine efficiency and reduced CO_2 are still being worked towards. Studies into the possibilities of engine development are becoming more and more expensive due to the fact that processes taking place in a power unit need to be examined more and more thoroughly. The authors are working on creating a new engine that will have a variable compression ratio. The whole purpose of the work is to show how important it is to increase the compression ratio and why compression rates cannot be high all the time.

References

1. A. Bieniek, J. Mamala, M. Graba, *Możliwości wstępnego ograniczenia emisji NOx i PM silnika wysokoprężnego w aspekcie przyszłościowych norm emisji pojazdów pozadrogowych*, ***75***, (Zeszyty Naukowe Instytutu Pojazdów, 2012)
2. D. Drabik, J. Mamala, *Effective compression ratio of combustion engine as a way of increasing efficiency* (Journal of KONES 2016)
3. M. Graba, A. Bieniek, J. Mamala, A. Lechowicz, Inżynieria Rolnicza, **75** (2011)
4. S. Luft, *Podstawy budowy silników*, 15 ***(1-3)***, (Wydawnictwa Komunikacji i Łączności, 2011)
5. J. Mamala, *Kompensacja niedostatku siły napędowej w procesie rozpędzania samochodu osobowego*, (Wydawnictwo Politechniki Opolskiej, 2011)
6. MCE-5 Development: www.mce-5.com
7. J. Mysłowski, *Doładowanie silników Pojazdy samochodowe*, ***33***, (Oficyna Wydawnictwa Komunikacji i Łączności WKŁ Warszawa, 2011)
8. M. Sobieszczański, *Modelowanie procesów zasilania w silnikach spalinowych*, **13**, (Wydawnictwo Komunikacji Łączności, Warszawa 2000)
9. www.acea.be (11.07.2017)

The efficiency of the heat pump water heater, during DHW tapping cycle

Arkadiusz Guźda[1,*], and *Norbert* Szmolke[1]

[1]Opole University of Technology, Department of Environmental Engineering, 76 Prószkowska Street, 45-758 Opole, Poland

Abstract. This paper discusses one of the most effective systems for domestic hot water (DHW) production based on air-source heat pump with an integrated tank. The operating principle of the heat pump is described in detail. Moreover, there is an account of experimental set-up and results of the measurements. In the experimental part, measurements were conducted with the aim of determining the energy parameters and measures of the economic efficiency related to the presented solution. The measurements that were conducted are based on the tapping cycle that is similar to the recommended one in EN-16147 standard. The efficiency of the air source heat pump during the duration of the experiment was 2.43. In the end of paper, authors conducted a simplified ecological analysis in order to determine the influence of operation of air-source heat pump with integrated tank on the environment. Moreover the compression with the different source of energy (gas boiler with closed combustion chamber and boiler fired by the coal) was conducted. The heat pump is the ecological friendly source of the energy.

1 Introduction

Renewable energy sources are becoming more common as a result of the increasing social awareness of the issues related to the conservation of the environment. The application of renewable sources also brings about considerable ecological and economic benefits.

It is estimated that the housing sector, regardless of the location of a country on the world map, is responsible for the use of 19 do 50 % of total energy that is required for the existence of households [1]. The use of energy needed for production of domestic hot water forms a considerable proportion in the overall energy balance; it is equal to 18 % in the USA [2], 20% in Brazil [3], 25% in the UK [4], 26% in Spain [5] and 30% in Australia [6].

Such a huge relative portion in the energy use in the housing sector leads to continuous attempts aimed at increasing the energy efficiency of equipment and heating systems, whose role is concerned with heating households and production of domestic hot water. One of the most effective ways of increasing the energy efficiency of heating installations is associated with the use of heat pumps.

Heat pumps are capable of converting energy derived from renewable sources to useful energy, which is further applied to meet the thermal demand of a building and its occupants. A system comprising a heat pump consists of a lower heat source, heat pump unit and a higher heat source, which is formed by the system applied for distribution of heating or cooling inside the building.

The operating principle of this equipment involves the use of heat with a lower temperature (i.e. lower heat source) with the purpose of generating heat with a greater temperature. This process does not lead to carbon dioxide emission into the atmosphere.

The potential lower sources include the use of ambient air, ground, water and a storage reservoir [7]). For the case of heat pumps, whose role is limited to the production of domestic hot water, the largest proportion of units apply atmospheric air to play the role of the lower heat source. In this case, we call them air heat pumps. The most common solutions of such heat pumps comprise an integrated hot water tank.

The considerable popularity of such pumps is attributable to the low investment required in them as well as the high value of the coefficient of performance (COP). The mean value of COP for heat pumps marketed in Europe was equal to 3.1 in 2016 [8].

Energy input is required to accompany the processes in which temperature of the system is raised. The type of driving energy: electrical, mechanical or chemical is dependent on the design of a heat pump [8].

Fig. 1 presents the theoretical circulation of the medium in a compression type heat pump operating according to pressure-enthalpy graph (p-h).

A dissipation of the pressure p_g of the liquid refrigerant occurs in the condenser until the medium obtains evaporation pressure p_0 in the expansion valve (line 3-4). Heat is extracted from the lower source – whose role is taken on by atmospheric air so that the phase change can occur (line 4-1). The vapor which flows from the evaporator is compressed in the

* Corresponding author: arkadiusz.guzda@gmail.com

compressor (line 1-2) until the compression pressure is obtained p_g; this is possible due to input of external energy. The next phase involves the release of heat in the condenser (line 2-3) and the cycle is repeated.

The operating performance of a heat pump is expressed by the formula

$$COP = \frac{Q}{L} = \frac{Q_O + L}{L} = 1 + \frac{Q_O}{L} \quad (1)$$

where: Q – quantity of heat absorbed by the evaporator, Q_0 – quantity of heat produced in condenser, L– work delivered to the compressor.

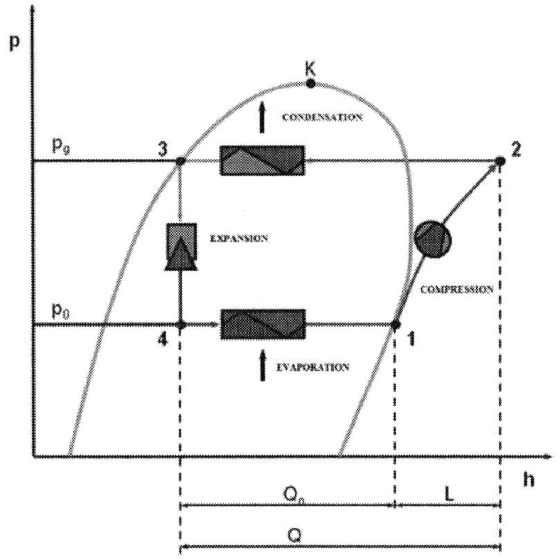

Fig. 1. Chart of the theoretical p-h cycle, where: Q_0 - the quantity of heat produced in condenser, Q - the quantity of heat absorbed by the evaporator, L - work delivered to the compressor.

The thermal efficiency of air heat pumps applied for the production of domestic how water is determined by a number of factors such as: course of tapping cycle, ambient temperature, temperature of cold water entering the pump, temperature of hot water in the storage tank, etc.

In addition, heat pumps need to be able to overheat tap water above 55°C during a thermal disinfection cycle with the purpose of preventing the growth of legionella bacteria, as they pose health threat to humans.

The authors in [4] state that for the case of heat pumps with an integrated water tank, in some conditions water overheating would be unfeasible.

In recent years, there is a number of papers concerned with the determination of the conditions affecting the performance of heat pumps. Such reports often include a description of how heat pump performance can be increased.

In paper [9], the authors proposed a model of dynamic simulation applying three types of condenser geometries, which comprises minichannels whose COP coefficient was equal to 3.85-3.93.

Concurrently, the studies reported in [10] demonstrated the existence of a dependence between the energy efficiency of the medium, quantity of the refrigerant in the circulation, length of the condensing section and the control settings of the heat pump.

It was remarked that the adequate refrigerant charge plays an important role for the effective operation of a heat pump. Other studies suggest that the adequate selection of the capacity of the water tank forms an important design consideration.

2 Experimental system and procedure

2.1 Description of the system

The study that is presented in this paper is concerned with the determination of the relation between the efficiency of an air heat pump and the use of domestic hot water. The study was performed by application of a heat pump manufactured by a Polish company named Galmet comprising a water tank with the capacity of 200 liters; the heat pump also comprises a 2 kW electric heater. The system is tested at Opole University of Technology, Faculty of Mechanical Engineering (50°41'00.3"N 17°56'37.5"E). Indoor air that is extracted from a laboratory plays the role of the lower heat source.

Table 1 contains a summary of the exploitation parameters that were measured during the operation of the heat pump.

Table 1. Summary of the measured exploitation parameters.

Domestic hot water	
▪ Temperature	°C
▪ Temperature difference	K
▪ Capacity	dm^3
▪ Volumetric flow rate	dm^3/s
▪ Thermal energy (over entire tapping cycle)	kWh
▪ Temperature of tank shell	°C
Refrigerant	
▪ Temperature of refrigerant exiting the compressor ▪ Temperature of refrigerant exiting evaporator ▪ Temperature of refrigerant exiting expansion valve ▪ Temperature of refrigerant exiting evaporator	°C
Parameters of electrical supply	
▪ Electric power	W
▪ Consumption of electricity	kWh
▪ Voltage	V
▪ Current	A
Air (lower heat source)	
▪ Air temperature	°C
▪ Relative humidity in air	%

2.2 Tapping cycles

In accordance with the PN-EN 16147 standard [11] in force now, COP coefficient is derived on basis of a daily tapping cycle, which comprises a profile of typical water usage model determined for a typical household.

The minimum duration of such a cycle is equal to 24 hours. This cycle applies energy input required for production of DHW according to definitions in standard [11] together with the energy needed for the operation of the system [12]. The decision regarding the water tapping cycle applied in the test of a particular pump is made by the manufacturer of the equipment.

The standard defines five water tapping cycles, ranging from S – XXL. These cycles vary in terms of the minimum volume of tap water use and daily profile of its tapping cycle.

The determination of the relation between the efficiency of the heat pump and use of hot water applied a study conducted for the tapping cycle (called L in the standard) as it is dedicated by manufacturer for this purpose and whose course is similar to the cycle described in the standard [11]. The statement regarding the use of hot water at a level of 199.8 liters corresponding to L cycle is given by the manufacturer.

For the cycle applied for the measurements, the equivalent of domestic hot water (volume of water used in the system) is equal to 230 dm^3 per day on average. The duration of the measurements was 18 days and they lasted from March 8th to April 5th, 2017.

The test involved three phases, whose overview is presented in Fig. 2.

Phase 1 involved the process of heating water in the tank from an initial temperature (i.e. temperature of cold water) to a temperature that was demanded for the purposes of the cycle, i.e. in the range from 45-50°C.

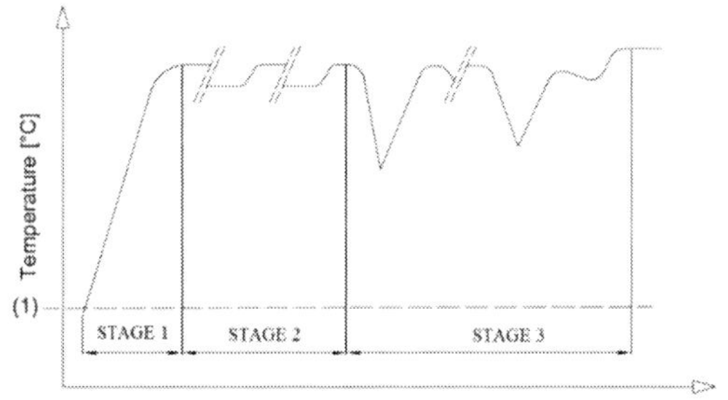

Fig. 2. Phases comprising a single tapping cycle.

Phase 2 involved the determination of energy input by the heat pump in the conditions when it is in standby mode. The final phase (phase 3) involved the determination of the energy use and coefficient of performance in the of the hot water production mode during the reference tapping cycles; throughout the duration of the cycles, various levels of water consumption were adopted.

The figure below (Fig. 3) contains the profile of the tapping cycle corresponding to the L cycle, as it was determined during the conducted study.

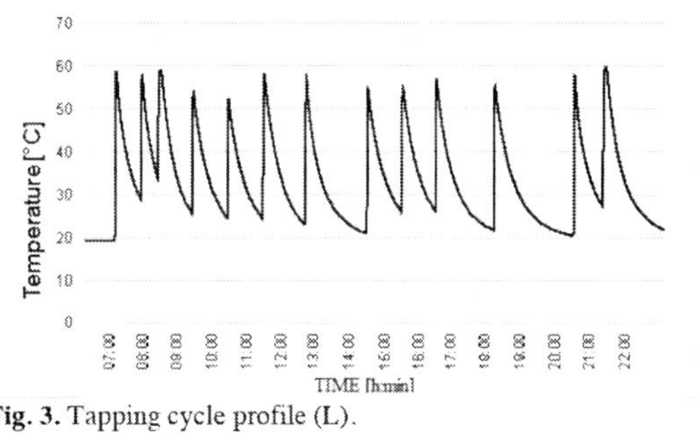

Fig. 3. Tapping cycle profile (L).

On the basis of an analysis of the above tapping profile, we can note that the greatest hot water consumption is noted for the period between 7 a.m. and 9:30 p.m. This use takes on a periodical character at constant time intervals. The longest intervals corresponding to tapping cycles are registered for the times between 12:45 p.m. and 3:30 p.m. we well as 6:30 p.m. to 8:30 p.m.

3 Test results

Fig. 4 presents a detailed temperature chart for two points: T1 and T4.

T1 is located in the vicinity of the hot water outlet nozzle whereas point T4 near the cold water inlet nozzle. The temperature peak which can be noted for the date March 20th, 2017 is associated with the increase of DHW temperature set-point at the outlet of the tank.

The temperature on the lower section of the tank was in the range from 19-30°C; whereas in the top section it was in the range from 35-48°C.

The following figure (Fig. 5) contains the chart of the compression and dissipation pressure which accompany the operation of the heat pump in the analyzed period.

The compression pressure during the operation of the heat pump was equal to 16 bar at a maximum, whereas the pressure behind the thermostatic expansion valve (TEV) was around 2.5 bar.

The subsequent chart in Fig.6 presents the values of the temperatures measured after the basic components of the heat pump, i.e. evaporator, expansion valve, before and after the condenser.

We can note that the temperature after the compressor (and before the condenser) increased to nearly 68°C, whereas the temperature of the refrigerant behind the evaporator was in the range 16-24°C.

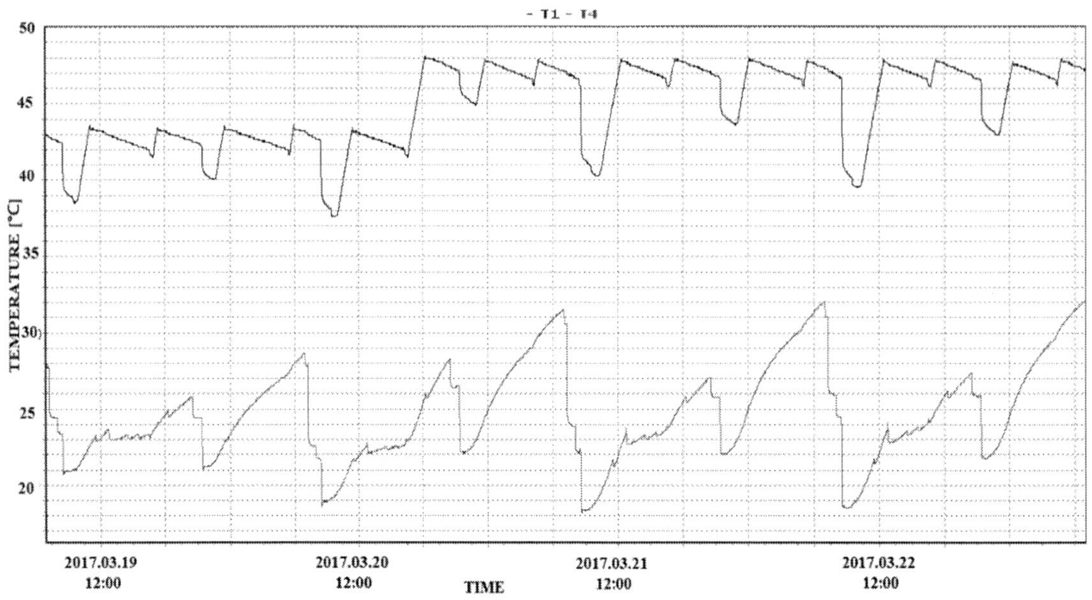

Fig. 4. Temperature chart in the water tank.

The heat flux delivered to water in the tank led to the decrease of the temperature behind condenser to a value in the range of 26-38°C. The pressure of the steam dissipation from the refrigerant resulted in the decrease of its temperature to nearly -1°C (in the initial phase) and 5°C (in the final phase).

Fig. 7 contains charts representing air relative humidity in the laboratory where the examined heat pump was located.

The air temperature in the room was in the range 16.4-19.5°C, whereas relative humidity of the air was in the range 17-40%.

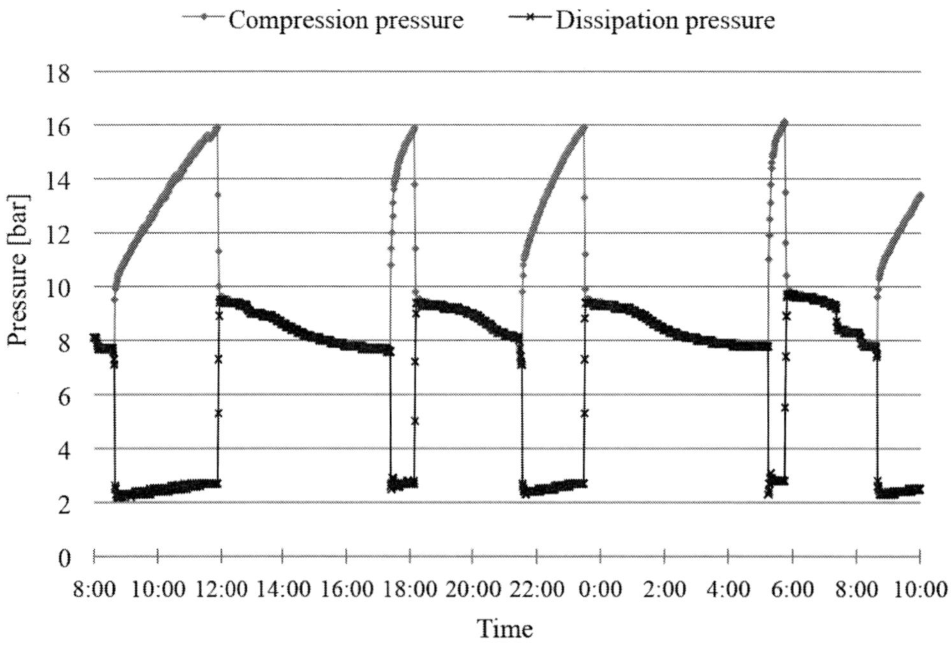

Fig. 5. Compression and dissipation pressure during the operation of the heat pump.

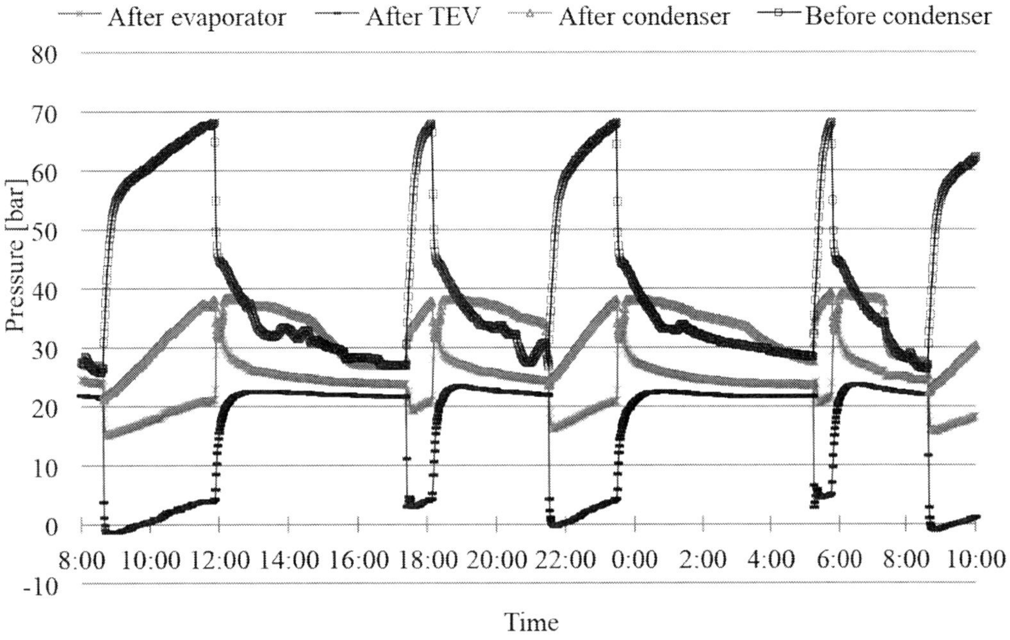

Fig. 6. Temperatures of the refrigerant.

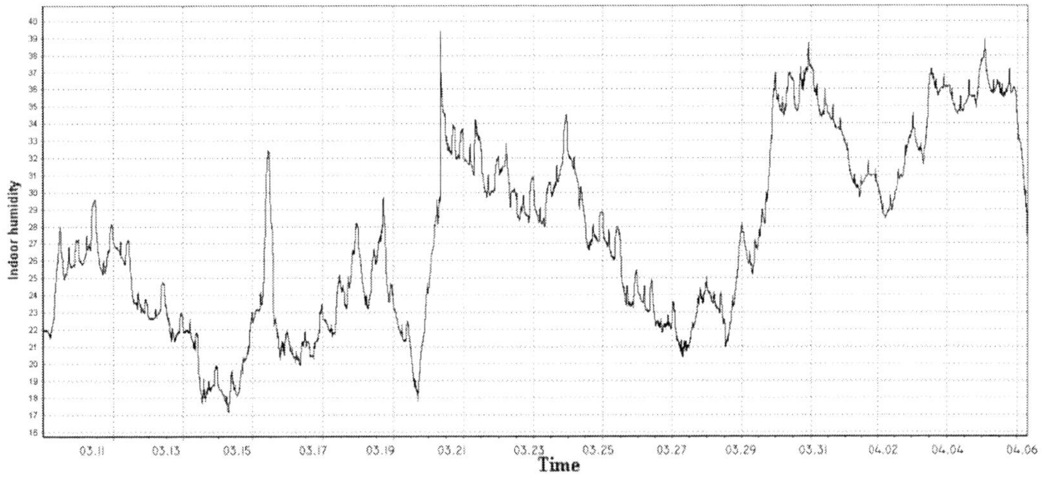

Fig. 7. Relative humidity in the laboratory.

Table 2 contains a summary of the results of the conducted research. On the basis of the analysis of the data, we can note that the heat pump produced **222.6 kWh** of thermal energy in the form of domestic hot water throughout the duration of the test cycle.

By substituting the values of thermal energy and electricity consumption throughout the test duration into formula (1), we are able to determine the value of the coefficient of performance of the heat pump, which is equal to **2.43**.

Fig. 8 illustrates the emission of pollutants into the environment, which resulted from the operation of the heat pump during the test

The calculations were performed by application of the standards in force and using software dedicated to performing ecological analysis– called Audytor EKO 1.0.

The operation of the heat pump resulted in the generation of 0.142 kg sulphur dioxide by the emitter; the amount of nitrogen dioxide was smaller and equal to 0.103 kilograms.

The emission of carbon dioxide was nearly 88 kg, and particulate matter 0.004 kg.

Table 3 presents a comparison of the results of the conducted research compared with different sources of energy (gas boiler and boiler fired by the coal).

Table 2. Test results.

	Value	Unit
Hot water use	6.3	m^3
Thermal energy generated by the heat pump over the entire test cycle	222.6	kWh
Electricity consumption by the heat pump over the entire test	91.7	kWh
COP	**2.43**	-

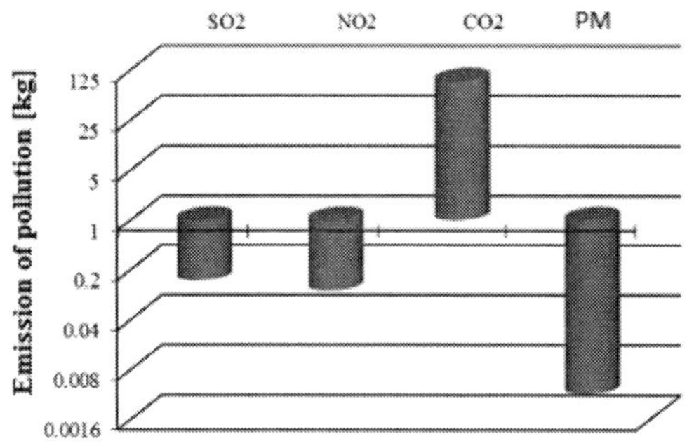

Fig. 8. Emission of pollutants in the analyzed period.

In this comparison the gas boiler with a closed combustion chamber is the most ecological source. The second place is the heat pump, the last is the boiler fired by the coal.

We can note at this point that the heat pump does not emit pollutants directly into the environment; yet, it is driven by electricity generated in Polish conditions by the power plants (emitter), which apply hard coal for the production of electricity.

Table 3. Comparison of emission different heat source.

	Air source heat pump	Gas boiler	Boiler fired by coal
SO_2 [kg/rok]	0.142	0.001	0.633
NO_2 [kg/rok]	0.103	0.037	0.073
CO [kg/rok]	-	0.007	1.484
CO_2 [kg/rok]	87.96	49.23	61.01
PM [kg/rok]	0.0037	-	0.2308
Soot [kg/rok]	-	-	0.1154
BAP [kg/rok]	-	-	0.0005

The operation of the gas boiler resulted in the generation of 0.001 kg sulphur dioxide by the emitter; the amount of nitrogen dioxide was 0.037 kilograms. The emission of carbon dioxide was 49 kg, and CO 0.007 kg. The coal fired boiler was the most unecological source of the energy. The emission of the pollutants were: SO_2 - 0.633 kg, NO_2 – 0.073 kg, CO – 1.484 kg, CO_2 – 61.01 kg, PM – 0.2308 kg, Soot – 0.1154 kg and BAP – 0.0005 kg.

4 Conclusions

On the basis of the results of the study, it was possible to determine the value of the COP coefficient with regard to the tapping cycle of domestic hot water during a cycle that is similar to the recommended one contained in [11]. The following conclusions can be made from the study:

1) Heat pump was tested in accordance with the tapping cycle L that is similar to the one found in [11].
2) The tested heat pump is characterized by a high value of the COP coefficient, whose value was 2.43.
3) The operation of the equipment resulted in the generation of 0.142 kg of sulphur dioxide, 0.103 kilograms of nitrogen, 88 kg of dioxide carbon dioxide and 0.004 kg of particulate matter
4) Heat pumps employed for the production of domestic hot water offer an ideal alternative to the traditional systems of DHW production, as they are characterized by a considerable efficiency and relatively low cost associated with the investment and exploitation.
5) The heat pump is one of the most ecological friendly source of energy, because in the compression with another sources (e.g. gas or coal fired boiler). This device does not emit pollutants directly into the environment.

In addition, it is projected that heat pumps have the potential to replace a number of traditional technologies (e.g. solid fuel boilers and gas boilers) currently applied for production of domestic hot water.

References

1. A.S. Vieira, R.A. Stewart, C.D. Beal, Energy Build. **91**, 148-162 (2015)
2. X. Liu, S. Lau, H. Li, Energy Build. **69**, 1-13 (2014)
3. E. Ghisi, S. Gosch, R. Lamberts, Energy Policy **35**, 4107-4120 (2007)
4. H. Singh, A. Muetze, P.C. Eames, Renewable Energy **35**, 873–878 (2010)
5. L. Pérez-Lombard, J. Ortiz, C. Pout, Energy Build. **40**, 394-398 (2008)
6. F. Alam, T. Theos, (Proceedings of the Fourth BSME-ASME International Conference on Thermal Engineering, 710-718, 2008)
7. M. Mironowicz, N. Szmolke, D. Skoruppa. Instal **10**, 10-15 (2014)
8. A. Gużda, N. Szmolke, J. Civil Eng. Env. Arch. **63,** 4/16, 201-208 (2016)
9. O. Ibrahim, F. Fardoun, R. Younes, H. Louahlia-Gualous, Energy **64**, 1102-1116 (2014)
10. J. Zhang, R.Z. Wang, J.Y. Wu, App. Therm. Eng. **27**, 1029-1035 (2007)
11. PN-EN 16147:2011
12. P. Lachman, Instal Reporter, **01**, 15-18 (2014)

DOI: 10.1051/e3sconf/20171901015

The microclimate parameters change in the occupied zone inside some large-volume buildings with significant influence of the heat emission from people staying inside

Paweł Albert Jelec[1,*]

[1] Cardinal Stefan Wyszyński University in Warsaw, Faculty of Biology and Environmental Sciences, Woycickiego street 1/3, 01-938 Warsaw, Poland

Abstract. The paper shows selected aspects of the microclimate change in the occupied zone of the large-volume buildings. The air temperature and some other microclimate parameters change because of heat emission from people staying inside there. There are shown exemplary methods and results of research conducted in sacral buildings and residential apartment buildings. It is based on author's research. The novelty is the research of the human heat emission in case of a big number of people staying inside sacral buildings. It can be useful for a large-volume buildings design (especially for the sacral buildings) and for microclimate shaping in many types of buildings and residential buildings also. In some buildings with a big number of people staying inside at the same time (such as the sacral buildings) heat emitted from people significantly influences on microclimate. Another situation is observed in residential apartment buildings, where heat and moisture gain are distributed in separate rooms. Human heat streams can be useful as a part of the heat balance of the room especially in case the big number of people staying in the room.

1 Introduction

The heat emitted by people can be often considered as an additional heat streams in buildings, which are useful and considerable part of the room heat balance. In sacral buildings, there are specific conditions. Volume of the room is often large, and so large is a floor area (giving the possibility of staying big number of people). Very often those buildings (especially their main nave) are tall. People are staying all time in the same place and their physical activity is not changing significantly during Mass. So, calculating the heat emitted by people in such circumstances is possible and those heat streams seem to be an important part of room heat balance and can be predicted in the design phase. So, here are shown exemplary results of research conducted in the sacral buildings and in the residential apartment buildings. It is based on author's research conducted in sacral buildings and residential apartment buildings. The main aim was to test the influence of the human heat emission on the microclimate of those types of buildings. Also, the aim was to check whether there is such an increase in internal temperature due to the human heat emission of people, which does not deteriorate the selected microclimatic parameters and is important from the point of view of energy efficiency.

The regulation of a blood supply to the skin is the basis of heat exchange between the environment and the man. Enlargement of the blood vessels results in better blood supply and increases skin surface temperature. It can also take place in the other direction, caused by artificial external heat. The blood can be transferred from the skin to the inside of the body then. Resting man emits sensible heat (79%, by convection and radiation) and latent heat (21%, by evaporation and respiration). Human emission of heat depends on physical activity, gender, clothing, and other factors. Resting metabolism occurs in the body when the person is lying down, in complete physical and mental calm, more than a 12 hours after the last meal, in neutral ambient temperature (for a man: 16÷20 °C) and it is about 1,163 W/kg (mass of the human body) [1]. At the temperature of 8 °C (possible for unheated churches) the calculated heat stream is about 370 W/m^2 (area of a body surface), with the use of the P.O. Fanger's method [1] and 193 W/m^2 with the use of the J.A. Pogorzelski's method [2] (at the air temperature of 16 °C, just for comparison, it is less value: 243 W/m^2 [1] and 150 W/m^2 [2]). In case of resting in a laying position – average internal heat stream from man it is about 70-93 W, in sitting position – 81-105 and in staying position about 93-128 W [3]. Another question is the ventilation of the room and the feedback influence of room microclimate on human heat streams emission. In case of air quality deterioration, thermoregulation metabolic processes of the human body will adapt to the ambient air temperature and humidity (also human body will react to the surfaces temperatures). In case when the big number of people is staying in the room, those processes are multiplied and they cause an increase in the temperature. So, the main

* Corresponding author: p.jelec@uksw.edu.pl

question is if it is possible to treat human heat emission as a useful heat gain in such buildings when a big number of people is staying inside and microclimate parameters is changing.

2 Materials and Methods

The sacral buildings research has started in the years 2000 at the Warsaw University of Technology (under the scientific supervision of Prof. Leszek Wolski). All tested buildings were located in the Masovian Voivodeship in Poland. The research at the residential apartment buildings has started in 2014 at the Faculty of Biology and Environmental Sciences of the Cardinal Stefan Wyszyński University of Warsaw in the Department of Environmental Engineering. The tested buildings were the residential apartment buildings and single-family residential buildings. Researches on all types of large-volume buildings mentioned above are still ongoing.

During research, there were used TROTEC's diagnostic equipment, including the EC060V thermal imaging camera and also the Infratec / Jenoptik VarioCAM HD Inspect 675 (with NETD even less than 30 mK) camera. Temperature and humidity measurements were also performed with the use of the thermohygrometers and other microclimate measurement equipment.

In the aspect of air temperature and relative humidity measurements, thermohygrometers were placed at different heights in the building at fixed measuring points or on portable tripods staying on the floor. In the case of measurements made at the points located at the significant height, there were used the author's method of using balloons with helium to hang the ropes on construction elements in the room such as tie rods on which the measuring devices later were mounted. In addition, to carry out the research surface thermometers, pyrometers and other microclimate and external climate testing equipment were also used to conduct surface temperature tests.

2.1 Objects location

The measurements in the occupied zone of buildings were performed with thermohygrometers and thermal imaging cameras. Tested buildings were located in the town of Płock and the Masovian Voivodeship in Poland. The tested buildings were 18 large-volume buildings with at least one large-volume room, and about 10 residential buildings (with separate apartments). During the research there were achieved about 300,000 data, 1,200 photos registering people staying inside buildings. Several thousand individual thermal images have been made in the last years for many different buildings, including buildings affected by the sick building syndrome in which mold occurs. Special software MAPKALK were written for data analysis also.

2.2 Tested objects characteristics

Three distinct types of sacral buildings have been identified and there have been selected 9 from group of 18 tested objects.

Below the selected sacral buildings are shown:

Object 1: Chappel „Kaplica Rzymskokatolickiej Parafii pw. Ducha Świętego", Płock, contemporary construction.

Obiekt 2 : Church „Kościół pw. św. Józefa", Płock, contemporary construction.

Object 3 : Church „Kościół pw. św. Krzyża", Płock, contemporary construction.

Object 4 : Church „Kościół pw. św. Stanisława Kostki" (lower), Płock, contemporary construction.

Object 5 : Church „Kościół farny pw. św. Bartłomieja", Płock, contemporary construction.

Object 6 : Church „Kościół pw. św. Jana Chrzciciela", Płock, 18th century construction.

Object 7: Church „Kościół rzymsko-katolicki pw. św. Stanisława Kostki" (upper), Płock, contemporary construction.

Object 8: Cathedral Basilica: „Bazylika Katedralna pw. Wniebowzięcia Najświętszej Maryi Panny", Płock, 11th century construction.

Object 9: Church „Kościół Matki Boskiej Częstochowskiej", Płock, contemporary construction.

Types of objects classified by their volumes:

Type I (small): Volume range: 1,413 – 2,760 m^3, height range in the middle part of the room: 3.7 - 8.5 m (objects 1,2,3).

Type II (medium): medium-volume objects and average height: Volume range: 3,451 - 6,000 m^3, height range in the middle part of the room: 5.8-12 m (objects 4, 5, 6).

Type III (large): large-volume and significant height objects. Volume range: 17,218-34,058 m^3, height range in the middle part of the room: 12-26.5 m (objects 7, 8, 9).

2.3 Measuring points location

Measuring points were located in different parts of the area of the church nave. It depended on its layout and the possibility of measurement equipment location. So, at objects 1, 2, 3, 5, 8 and 6 it was located along the walls and the benches (Fig. 1 a). In the objects 4, 9 and 7, because of a round shape, measuring points are located in more "dispersed" places in the space of the room (Fig. 1 b). In those points there were fixed a mounted or portable thermohygrometers - on tripods, portable poles or suspended on ropes attached to the tension rod or other existing rod. It is shown on Fig. 2 (mounting rods for the chandelier were used there to attach the ropes).

Objects, number of points and location of the measuring points over the level of the floor:

- Object 1, Points: 1-9, Heights: 0; 1; 1,8; 2,6; 5,2;
- Object 2, Points: 1-9, Heights: 0,7; 2; 4;
- Object 3, Points: 1-8, Heights: 1,2,3,4;

 DOI: 10.1051/e3sconf/20171901015

- Object 4, Points: 1-7, Heights: 1,2,3,4,5;
- Object 5, Points: 1-9, Heights: 0,1; 1; 2; 3; 4;
- Object 6, Points: 1-10, Heights: 0,1,2,3,4,6,9,12;
- Object 7, Points: 1-9, Heights: 0,1,2,3,4,5,6,7,8;
- Object 8, Points: 1-9, Heights: 0,1; 1,2,3,4,6,8,11;
- Object 9, Points: 1-8, Heights: 0,1,2,3,4.

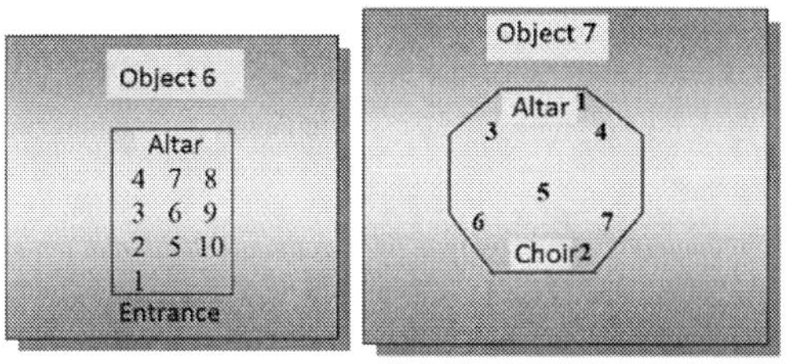

Fig. 1. Typical measuring points location basic on examples of object 6 (a) and 7 (b).

At the residential buildings, measurements were performed in the occupied zone in the different points of the rooms or there have been made thermograms using thermal imaging cameras.

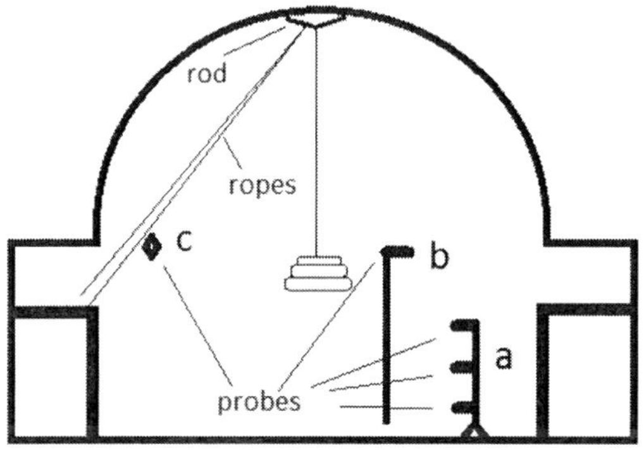

Fig. 2. The types of the measurements with the use of the thermohygrometers: probes were mounted on tripods (a), portable poles (b) or suspended on ropes fitted to the existing rods (c).

2.4 Methods of people's stay characteristics in the churches

In the churches, especially those of modern design and architectural form, there is many different constructional solutions and therefore there are different heights and volume. So, there are small churches (300 to 600 of Mass attendees) and even very large (even 3-6 thousands of Mass attendees). The differences in the impact of the human heat emission on the microclimate depend on the volume of a room, its height and also they are proportional to the area occupied by persons to the rest of the room's space and other parameters.

Regarding the specific features of the sacral buildings, there are also different proportions of seats to standing places. Because of the people standing in the churches, it is possible to obtain a relatively high density of settlement, in the order of 2.8-3.0 persons/m^2. The objects of tested types show differentiation considering such defined coefficient of the settlement. The lowest values are observed on weekdays in the morning and in the afternoon. The highest values and full use of the object space are observed at high Mass on Sundays. The settlement coefficient does not change during the Mass, which was important for assessing dependency and temperature rise. People enter the building about 15-20 minutes before the Mass and leave it about 5 minutes after the Mass. The settlement coefficient assumes the highest values in the area closest to the altar. First are occupying seating places at the benches, then standing places. The high values of settlement in Type III (large) object are the least obvious and repeatable, due to the large space available for staying as well as the different architectural features. Small objects generally have the principle of making the best use of the space of the church, and therefore the size of the human space is as large as possible in relation to the space of the room [4].

2.5 Specific formal aspects of the research methodology

During the thermography research, there were used the requirements of the polish standard PN-EN 13187 [5]. It is Polish version of the European standard EN 13187:1999 "Thermal performance of buildings. Qualitative detection of thermal irregularities in building envelopes. Infrared method".

Including for example: the difference between indoor temperature and outside temperature - at least 15 K, indoor air temperature as stable as possible, outdoor maximum test conditions, for instead before sunrise or late evening. Wind speed of not more than 1 m/s, no dampness of barriers, no measurements can be processed during bad weather, e.g. rainfall, snowfall or fog. The angle to the partitions surfaces during the measurement cannot exceed 45°. Research met the requirements of Polish civil engineering law for buildings and their location [6] and requirements of many European Standards such as PN-EN ISO 6946:1999 [7] for example.

3 Results and discussion

Below the results of the research inside sacral buildings of three types are presented. Additionally, those buildings are compared in the aspects of the specific features of the residential apartment buildings treated as the opposite cases.

3.1 Air temperature and relative humidity distribution in the large-volume rooms

Within the scope of the research, measurements have been made in the objects of the three types, described earlier in the subsection 2.3. (Type I, Type II and Type III). The space of the tested rooms was divided into zones. These zones (shown on the Fig. 3 and Fig. 4) are: "mean" - those are mean results from the whole space of the room (on the selected height over the floor), "center"

- means strictly the central part of the room, "edges" - denotes the area along the walls, "extended center" - denotes the middle part of the room, and "ext. mid. with. cent." - which means the middle part of the room, excluding its strictly central part. Then, the space of the room is divided into planes of fixed height over the level of the floor. Fig. 3 and Fig. 4 [8] concern the human occupation zone and room space just above it. Those graphs illustrate the results obtained from hundreds of measurements in many objects grouped into the three discussed types of buildings.

Buildings precise location: All selected sacral buildings were located in the city of Płock. The tested residential buildings were located in the Płock, Ciechanów, and Płońsk districts. The results show that there are a significant increases in temperature in the occupied zone in sacral buildings where many people are staying during the Masses.

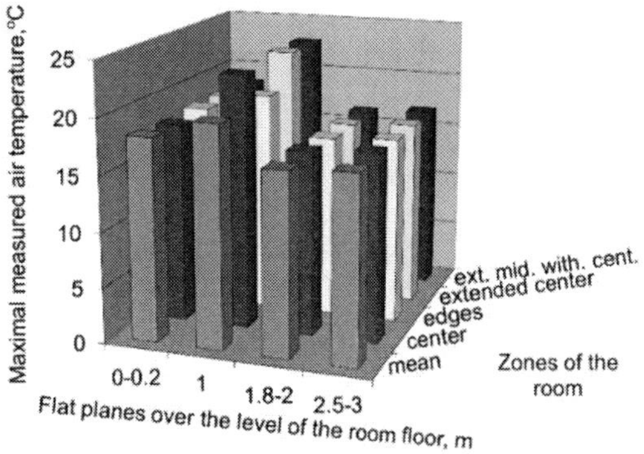

Fig. 3. The maximal air temperature after Mass inside object of Type I (small).

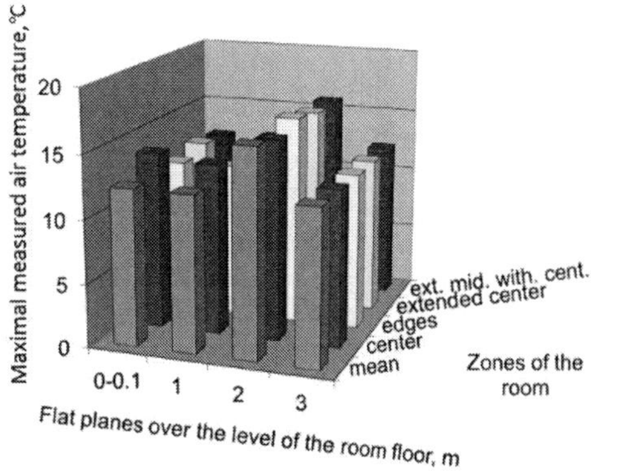

Fig. 4. The maximal air temperature after Mass inside object of Type II (medium).

It is profitable from the point of view of a thermal comfort. In addition, the highest air temperature after Mass is observed in Type I objects, which results from the smaller volume and height of objects with a relatively large number of occupants and a relatively high settlement coefficient (as a number of people referring to the area they are staying in). The tested sacral buildings of a large volume were generally unheated or poorly heated. The analyzed cases are those for which the temperature rise during people's stay is significant (not less than 0.5 ^{o}C) and the impact of external climatic conditions can be assessed as negligible. This allowed to analyze those cases which deserve attention. In those cases the heat from humans is a major cause of the increase in indoor air temperature.

In the height range of more than 3 meters (as included in Fig. 5 and 6 [8]), a convective heat transfer is observed. Fig. 5 and Fig. 6 show the situation in objects of type III (large objects).

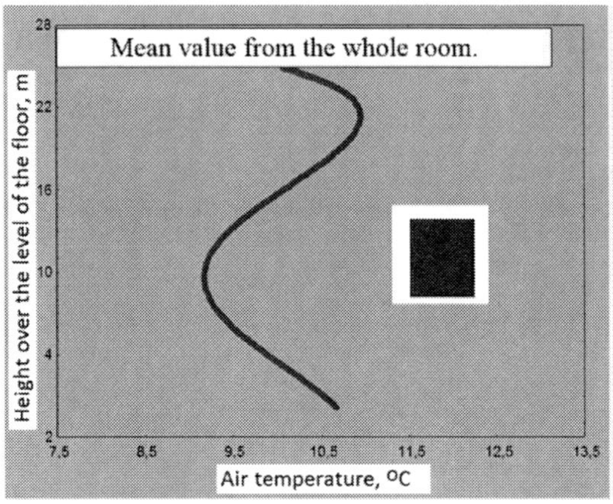

Fig. 5. The air temperature profiles after Mass inside the large objects (above the occupied zone - from 2 m to the highest level of the measurements). Black square marks area of the measuring points location.

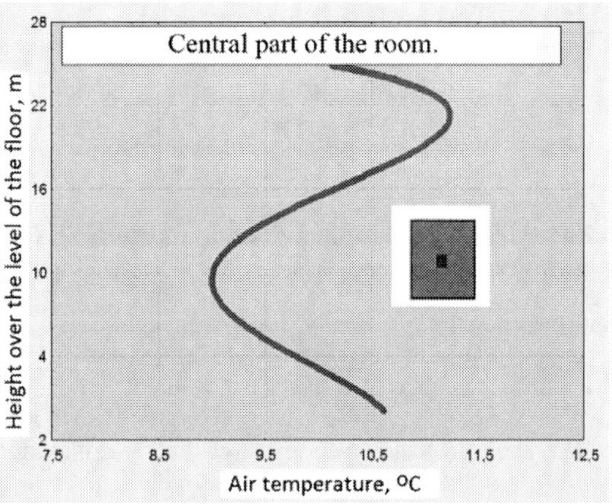

Fig. 6. The air temperature profiles after Mass inside the large objects (above the occupied zone - from 2 m to the highest level of the measurements). Small black square marks the location of the measuring points.

These objects are characterized by considerable height (the charts scales reach to 28 m, but in some cases rooms even reached the height of more than 30 m). For convective heat transfer, the curvature of the graph and the higher temperature at a certain height (from about 15 to 25 m) are observed, followed by a slight reduction in the air temperature near the ceiling (as a result of heat transfer to the outside through the partition). There is a significant increase in air temperature in the human occupation zone up to an average of about 10.5 °C in rooms where the initial temperature is around 6-8 °C.

The Table 1 shows the brief conclusions concerning changes in air temperature and relative humidity in the studied objects of the three selected types. It is interesting that the Type I objects, due to the smaller volume and height of the objects, achieve a more uniform temperature distribution in the vertical section of the room, whereas in the large objects (Type III) the most characteristic vertical profile temperature is observed. Evident convective upward flow is observed especially in this type of buildings.

Table 2 shows the mean differences between air temperature after Mass and the normative air temperature of thermal comfort.

Table 1. The horizontal and vertical heat zones.

Considered specific value	Range of the occurence inside rooms of tested types
Mean air temperature increases:	
Small objects	Inside whole room, in occupied zone, there are the highest air temperature increases.
Medium objects	Above occupied zone, there occurs gradual air temperature lowering.
Large objects	In the occupied zone, to the height of 6 m above the level of the floor, there occur the highest air temperature increases.
Mean air temperatures after Mass:	
Small objects	The higher air temperatures after Mass at the height of: from 4 m to the half of the object height.
Medium objects	Lower mean air temperaturę at the height of: lower part even to the half of the object height.
Large objects	Above the occupied zone – there occurs gradual air temperature lowering. Convectional raise of the heat stream and its flow to the top of the room.
Mean air relative humidity increases:	
Small objects	At the range of occupied zone the highest air relative humidity increases occur. Lower air relative humidity increases occur at the rest part of the room.
Medium objects	The values of the air relative humidity increase in parallel to the change of the room height.
Large objects	In the occupied zone up to the height of the half of the room there occurs similar distribution of the air relative humidity increases.

Table 2. The difference between air temperature after Mass and normative temperature of thermal comfort (mean values calculated from data achieved in the all measuring points fixed in the area of the room), °C.

Objects	Mean	Median	Minimum	Maximum
Small	3.27	2.91	-2.81	11.2
Medium	6.76	7.08	0.08	8.79
Large	5.07	5.07	2.72	7.31

Objects during tests did not always achieve the normative air temperature of thermal comfort, because of the low number of people staying in the sacral building during the test. It causes relatively low human heat gain. It doesn't affect significantly to the microclimate. Another obvious reason that the thermal comfort is often not achieved, is the very low initial air temperature in the church.

As the opposite situation, in some cases (mainly in Type I and Type II objects), there occur very high differences between the normative thermal comfort air temperature and indoor air temperature (which is higher). This is due to a large number of people and relatively huge human heat emission. Additionally, Type I and Type II objects are relatively not too high so air in room is warmed quite quickly. Also, the cumulative effect of successively air temperature increase occurs because of low intervals between the Masses (in Type I and Type II more often than in Type III objects).

Table 3 and Table 4 shows the combination of air temperature increments and air relative humidity increments in the different room zones (mean results from zones). The "mean" in the columns is the mean value for the whole room, while the "mean" in the rows is the mean value for all measurements. In addition, the maximum and minimum values are included in the tables. Particular attention could be paid to small objects, where the increase in the air relative humidity due to the presence of people in the object reaches up to about a dozen percent (in the case of long people's stay in successive Masses).

In aspect of objects comparison, there is a different situation than in the another category of objects - apartment residential buildings. There is a large number of people staying at the residential building a long time also, but the heat gains are divided into individual dwellings and distributed in the many rooms.

Table 3 also shows temperature increases in each type of objects, and the highest observed temperature increases are observed in small objects and in tall buildings where even up to several thousand people can be present inside one room at the same time. Increases in air temperature reach even 6-7 ° C.

The heat gains shown in Table 3 are important for microclimate and thermal comfort, especially if the room is unheated. In practice, this allows for self-heating of many sacral buildings. For small objects, there is often a risk of excessive air humidity, which requires adequate ventilation. Sacral objects are intensively ventilated between the Masses when the faithful leave the church.

Table 3. The mean air temperature increases, °C.

Small objects (Type I)

Type of value	Mean	Central part	Edges	Ext. middle	Ext. mid. without center
Minimum	1.593	1.525	1.571	1.365	1.325
Mean	2.144	1.908	2.146	1.958	1.946
Maximum	2.967	2.308	3.150	2.439	2.433

Medium objects (Type II)

Type of value	Mean	Central part	Edges	Ext. middle	Ext. mid. without center
Minimum	0.839	0.850	0.920	0.870	0.909
Mean	1.190	1.102	1.204	1.155	1.176
Maximum	1.519	1.450	1.458	1.470	1.700

Large objects (Type III)

Type of value	Mean	Central part	Edges	Ext. middle	Ext. mid. without center
Minimum	-0.056	0.000	-0.075	-0.033	-0.060
Mean	1.716	1.755	1.831	1.886	1.711
Maximum	6.340	5.820	6.337	7.511	6.393

Table 4. The increases in the air relative humidity during the Mass (mean values calculated for the all measuring points), %.

Small objects (Type I)

Type of value	Mean	Central part	Edges
Minimum	5.429	5.393	5.423
Mean	10.418	9.582	10.444
Maximum	15.881	15.227	15.876

Medium objects (Type II)

Type of value	Mean	Central part	Edges
Minimum	2.228	3.042	2.590
Mean	3.808	4.213	3.861
Maximum	5.059	5.569	5.073

Large objects (Type III)

Type of value	Mean	Central part	Edges
Minimum	1.280	0.665	3.352
Mean	3.702	3.003	4.397
Maximum	5.794	4.343	6.500

In the case of relatively high objects (especially in Type II and Type III), the heated air also rises upward, so it minimizes the effect of air temperature rise due to the presence of people in the room. Heat gains from humans in the considered cases (at least 0.5 ° C increase in indoor air temperature) are important for the heat balance of the buildings, especially in objects of Type I.

In the case of tested apartment residential buildings, heat gains from people were often a reason for significant humidity increases and adequate infiltration of fresh air and proper ventilation were necessary there. In some tested cases, ventilation was insufficient and led to serious troubles with inner air quality. Thus, cases of residential apartment buildings constitute partially in the opposite to the cases of sacral buildings larger in volume (Type II and III). In all mentioned buildings there is a large number of people staying inside but in sacral buildings, they stay generally in one large-volume room. However, there is also a need for adequate ventilation to prevent excessive air humidity problems. Nevertheless, the large volume of sacral objects (especially Type II and Type III) in a certain way causes that adequate air quality is provided. The ventilation needs are much smaller there than in residential apartment buildings and sacral buildings of Type I.

Nevertheless, the human-derived moisture gains are important for the microclimate of the residential buildings, often contributing (in conjunction with other factors) even to the sick building syndrome and mold (Fig. 7).

Fig. 7. Detected troubles in the tested residential apartment buildings: mold in the room corner, on the internal surface of the external wall, caused by limited infiltration of external air, poor natural ventilation, thermal bridges and significant moisture gains from people.

3.2 Specific differences in microclimate shaping between selected types of buildings

The division of tested sacral buildings into small, medium and large displays the following differences in the microclimate change between objects:

- Air temperature rises (in cases of the same occupation of people) decrease when the volume of the considered object enlarges. The greatest increase in temperature is achieved in small objects. In larger buildings, heat rises upward due to convection. Because of that, in tall buildings the effect of the air temperature increase is limited and the greatest increase is in the occupied zone.

The rise in temperature of the tested objects changes with the height above the floor level.
• The highest temperature rise is achieved in the human occupation zone (all types of tested buildings)
• In smaller sacral objects there is greater ability to design the microclimate by predicting temperature rise in case of people's presence.
• The rate of decrease in temperature depends on the use of the room after leaving it by people - whether windows and doors are opened and at how long it lasts. The sacral buildings are often ventilated between the Masses.
• The effect of raising the temperature during use of objects depends on the number of persons staying, as well as on the duration of use of the object and the density of people in the space.
• The shape of the temperature distribution caused by people staying depends on the volume of the building. In the case of smaller sacral objects, there are often several Celsius degrees of temperature increase.
• In Type I objects there is the greatest increase in air temperature and humidity in the middle zone and the lowest temperature variation between zones. In large buildings, the highest temperature rise in the middle zone and the highest temperature and humidity differences between the zones are observed.
• The distribution of the temperature that is shaping during people staying is varied in both the horizontal plane of the room and at the different heights. The presence of heat sources, the unevenness of the density of the people's occupation in the space of the room, the air movement near the windows - all those parameters influence on the formation of temperature distribution.
• It has been observed that in the lowest and smallest objects (Type I objects) buildings microclimate deterioration is most common. It depends on a large extent on the time of people's stay and their number.
• On the basis of the achieved results, it was found that in the smallest objects there are heat gains which allow to reach the thermal comfort (in winter and during whole heating season). Especially in large objects - studies do not confirm the previous views on the air temperature distribution because there is also a cooler zone in the middle part of the room. Some results suggest that further research should be carried out.

3.3 Influence of people staying in room parameters

It has been found that the considerable air temperature rise depends on some minimal number of people staying inside the room. It is possible to indicate the number of about 25 people. - Then, about 50% of cases in objects of small volume cause a noticeable increase in temperature.
• The effect of a significant temperature increase with a minimal amount of persons causing a noticeable rise in temperature occurs within about 10 minutes after entry to the object and is maintained for about 30 minutes after leaving it by Mass attendants.
• Accumulation of air temperature increase caused by the presence of people can be seen in the Sunday successive Masses. In 90% of the cases, it results in a gradual increase in temperature. It is even up to 300% higher than in the case of a single Mass.
• The effect of daytime on the temperature raise while maintaining the presence of the people is not significant and amounts 0.2 °C.
• The increase in room volume results in a reduction in possibility of the use of human heat for heating purposes. It is stated with limit of 2000 m^3. For these value there was a noticeable increase in temperature.
• Shaping a microclimate with particular attention to the effects of heat emissions from people for microclimate and energy efficiency is possible under the following conditions:
- Keeping the minimum number of occupants for the type of objects;
- Providing adequate residence time;
- Providing adequate ventilation, especially small objects, so that there is no deterioration in air quality. In larger buildings, the amount of air available may be sufficient to ensure adequate air quality in room during the people's presence.

3.4 Achieved results summary

• Considering the influence of the heat emission from people on room's microclimate, some improvement of methods of its assessment have been developed;
• In tested objects, the most considerable air temperature increases (caused by heat emission from people) are observed in the occupied zone of sacral buildings. Influence is often reduced by convectional raise but, in general, this emission is important as a useful heat gain in aspect of achieve thermal comfort and in aspect of the building heat balance;
• The influence of the human heat emission on microclimate depends on ventilation intensity. It is not so important in buildings of larger volume (especially in Type III objects).
• As an opposite case to the sacral buildings, in residential apartment buildings poor ventilation often causes significant humidity growth and sick building syndrome risk (troubles with partitions and thermal insulation, as an effect of poor ventilation, badly performed thermo-modernization, thermal bridges, etc.). The same situation can occur in sacral buildings because of the same technical problems and also because of low air temperature and high air relative humidity. It was observed in some cases in sacral buildings also and it led to the sick building syndrome, serious partitions damages and mold there.

4 Conclusions

The results of the research helped to improve some methods of assessment of the influence of the heat emission from people on the room microclimate. The preliminary assessment and conclusions are:
- There is direct dependence between number of people staying inside the sacral building and the air temperature growth.

DOI: 10.1051/e3sconf/20171901015

- The human heat emission is the considerable part of the heat balance especially in Type I objects. However, the useful heat emission should be reduced by the ventilation heat loss, but even after this reduction, there is still even a few dozens watts per person as an effective heat gain (these research results now are processed). So, almost all sacral buildings (considering Polish climatic conditions) can be designed as unheated, passive buildings.

- In the case of the largest volume buildings (Type III objects), there are less possibilities to achieve a thermal comfort conditions (especially right air temperature) only as the result of the human heat emission. But there are still significant possibilities of achieving it in the case when there are a bigger number of people is staying inside (about hundred and more).

- Higher height of some buildings causes convectional raise of the warmer air so the effect for the same number of people staying inside three tested types of buildings is different.

- Another situation is observed in the residential apartment buildings, where the heat and moisture gain from people are distributed in the separate rooms. It seems that all room has its separate microclimate, but heat emission from whole building causes microclimate changes especially in the rooms laying on the highest floor causing significant air quality deterioration.

- Inside the Type I objects, in the occupied zone, the air temperature increase was 7.76 °C, 3.77 °C in Type II objects, and in Type III objects it was 5.95 °C (as a mean value for the entire research period, considering cases when air temperature increase was at least 0,5 °C).

The achieved results indicate that thermal impact of a big number of people staying inside large-volume buildings (especially sacral buildings) can be useful as a part of heat balance of those buildings.

Research on the large-volume buildings is still ongoing and newer results will be published next year with the precise assessment of the heat gain.

References

1. PO. Fanger, *Komfort cieplny* (Arkady, Poland, 1974)
2. J.A. Pogorzelski, *Fizyka cieplna budowli* (PWN, Warszawa, 1976)
3. H. Recknagel, E. Sprenger, W. Hoffmann, E. Schramek, *Ogrzewanie i klimatyzacja. Poradnik* (EWFE, Gdańsk, 1994)
4. P.A. Jelec, The selected aspects of the microclimate change caused by the heat emission from people staying inside the large-volume buildings, *17 International Multidisciplinary Scientific GeoConference SGEM 2017 (International Multidisciplinary Scientific GeoConference SGEM ,* Varna, 2017)
5. PN-EN 13187:2001 *Właściwości cieplne budynków - - Jakościowa detekcja wad cieplnych w obudowie budynku - Metoda podczerwieni* (PKN, Warszawa, 2001)
6. Dz.U. 2002 nr 75 poz. 690 *Rozporządzenie Ministra Infrastruktury z dnia 12 kwietnia 2002 r. w sprawie warunków technicznych, jakim powinny odpowiadać budynki i ich usytuowanie* (Ministry of Infrastructure, Warszawa, 2002)
7. PN-EN ISO 6946:1999 *Komponenty budowlane i elementy budynku. Opór cieplny i współczynnik przenikania ciepła. Metoda obliczania* (PKN, Warszawa, 1999)
8. L. Wolski, P.A. Jelec, S. Grabarczyk, A. Krajewska, A. Kamiński, *Fizyka obiektów sakralnych – pod kierunkiem Leszka Wolskiego* (Polish Academy of Sciences, Warszawa, 2006)

A numerical analysis of metrological properties of Venturi tube in the air-coal particle mixture flow measurement in power industry

Grzegorz Borsuk[1*], *Bolesław* Dobrowolski[1], and *Jacek* Wydrych[1]

[1]Opole University of Technology, Mikolajczyka 5, Opole, Poland

Abstract. The main topic of the study is to evaluate the effect of particle concentration and fractional composition on the differential pressure in the Venturi tube during the measurement of flow rate of the gas-particle mixture. Gas phase flow was described by Euler method, while particles movement by Lagrange method. Calculations were performed using ANSYS Fluent Package. A series of numerical calculations were carried out on the particle loading ratio $Y \leq 2$ and their diameters of 10 µm $\leq d_p \leq$ 200 µm. It has been found that, except for the Y concentration loading ratio, the increase in the differential pressure in the venturi tube also affects the Stokes number. It depends on the flow conditions and particles diameter. A correction function has been proposed to determine the effect of particles diameter on the results of the mixture flow measurement. The results of the calculations were compared with the Lee and Crowe experimental data. A good comparison of calculation results to measurement results was found.

1 Introduction

Gas-solid particles mixture flow occurs in many industries where pneumatic conveying is used. As an example power plant industry, food industry, building materials industry and many others. Proper processing and energy conversion processes require a current knowledge of the mass flow or one of its components. This problem is particularly important in the power plant industry, where the uniform supply of burners requires the measurement of the particle coal conveyed by the gas stream. For multiphase flow measurements, a videogrammetry method seems useful, which may help to identify flow regimes [1].

One of the methods of measuring homogeneous mixture flow is Venturi tube flowmeter. It has many advantages and high accuracy when the requirements specified in the relevant normative documents are fulfilled [2]. As a result of the tests, it has been found that selected types of Venturi tubes can be used to measure mixtures flow, but the presence of an additional phase changes the characteristics of the flowmeter. The condition of using a flow meter for the measurement of gas-particle mixture flow is to calibrate it in the conditions similar to those for further use.

The purpose of this paper is to indicate the possibility of applying mathematical modelling and numerical simulation to the initial evaluation of its metrological properties during diluted gas-particle mixture flow measuring. Particular studies have been made for Venturi tube, which is often used in two-phase mixture flow measuring systems, especially in pipe system in the conventional power plant [3, 4].

2 Venturi flow meters in the gas-particle flow measurement systems

Operations on the use of gas-solid mixture flow measurements were undertaken mainly for the needs of the power industry, and more precisely – pneumatic conveying of pulverized materials such a coal dust. In 1948 Carlson and others [5] proposed a dual-reducing system, consisting of a serial ASME nozzle and an orifice. The pressure measurement on the orifice, according to authors practically independent of the solid phase presence, was the basis for calculating the gas mass flow. The pressure drop in the nozzle was used to determine the mass flow of the mixture. As Teisseyre's later studies [4, 6] show, the Carlson method is not suitable for continuous measurements due to the gradual wear of the orifice and the dependence of the orifice differential pressure on the particle concentration in the mixture.

Fabar [7] implemented the first systematic study on the Venturi tube, created with the recommendations of the ISA standard. The layout of the measuring system, also used in works Antikajn [8] and Szatil [9, 10] is shown in Fig. 1a. As a result of the study, Fabar stated the linear dependence of the pressure drop Δp_{tp}^{l} at the two-phase flow from the particle concentration in the mixture:

$$\Delta p_{tp}^{l} = (1 + E_l Y)\Delta p_g^{l} \qquad (1)$$

where:

* Corresponding author: g.borsuk@po.opole.pl

$$Y = \frac{M_s}{M_g} \quad (2)$$

is loading ratio of the particles. The value Δp_g^I corresponds to the differential pressure connected with the pure gas flow with the mass flow M_g, and E_I is the Gästerstädt number [11]. Equation (1) is used to calculate the particle concentration in the gas, assuming that the gas mass flow is known at the two-phase flow.

The Farbar method requires an additional measurement of the gas mass flow, which in essence is a two-way method. Szatil [10] has confirmed the suitability of a Venturi tube for the measurement of gas-particle mixtures and found a weak dependence of constant E_I on the solid phase concentration and gas velocity

Modification of the measurement method based on the Venturi tube, Graczyk presented [12, 13] in the form of the so-called three-signal method (Fig. 1a). This method was also used in the works of Teisseyre [4, 6], Leroch [14], Barth [15], Jung [16, 17], Szatil [10] and finally Payne and Crowe [18]. In addition to the pressure drop measurement Δp_{tp}^I (Fig. 1a), Graczyk proposed the pressure drop measurement Δp_{tp}^{II} for the whole Venturi tube. As a result of the research, this author stated that one of the important problems is the selection of proper measurement points. Δp_{tp}^{II} to provide maximum measurement sensitivity and obtain a linear dependence of the pressure drop from the particle concentration

$$\Delta p_{tp}^{II} = (1 + E_{II} Y)\Delta p_g^{II} \quad (3)$$

Pressure drop ratio is a basis for calculating the particle concentration Y and the mass flow of both phases according to the formula:

$$\pi_{tp} = {\Delta p_{tp}^{II}}\Big/{\Delta p_{tp}^{I}} \quad (4)$$

Further, using a single Venturi tube as a mass flow meter simplifies the measurement installation in a significant degree. Graczyk studies have provided the basis for the practical application of the three-signal method.

Jung's research has made a great contribution to the development of Venturi tube [16, 17]. Jung proposed modifying the three-signal method by using a nozzle with extended cylindrical part (Fig. 1b). The pressure drop combined with the flow of the mixture through the tapered cylindrical part was the basis for the calculation of the particle concentration. According to Jung, the differential pressure Δp_{tp}^I measured in the nozzle depends only on the flow of pure gas and can be the basis for calculating the gas mass flow. With the knowledge Δp_{tp}^{II} it is possible to calculate Y and the particle mass flow. The disadvantage of the Jung method is that the influence of the presence of particles on the value Δp_{tp}^I is neglected, which affects the measurement accuracy. Barth's work [15, 19] uses a different approach, by placing the measuring points at the end part of the Venturi tube (Fig. 1c).

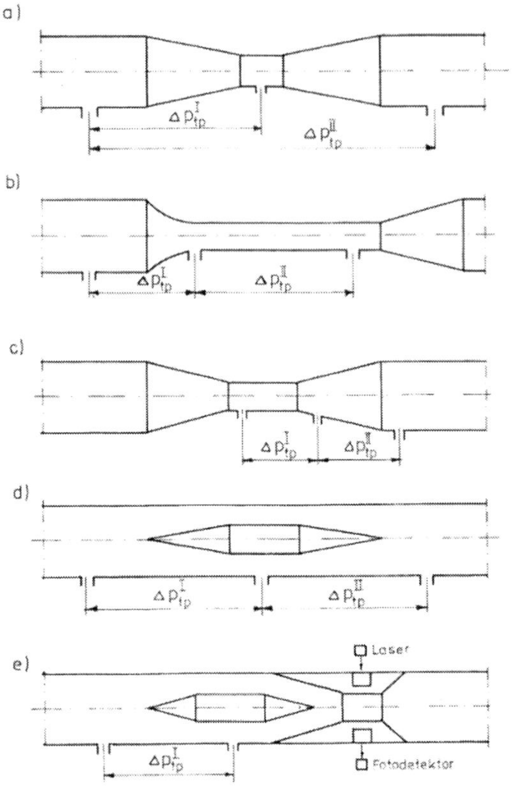

Fig. 1. Selected Venturi tube systems (described in the text).

Łapa [14, 20] has developed the concept of the Graczyk and has conducted detailed experiments on the venturi in the triple signal system. The study was conducted for the particle-air mixture. The effect of extensive empirical studies was to determine the influence of the pipe diameter and the venturi module on the Gästerstadt number value. Łapa has determined the calibration characteristics of the test specimens of Venturi tube by using the three-signal method to pipe diameters D <100 mm at Y < 2. The accuracy of the measurement method was estimated at ±8%.

For the local measurement of the gas-coal dust mixture concentration conveyed to the burners, the three-signal Venturi tube was used by Leroch [21]. As a result of experimental research, he defined the diameter and tube module to measure the actual velocity of both phases and particle concentration.

At the end of the 1970s, extensive studies on the evaluation of metrological properties of annular Venturi tube were undertaken in works [18, 22, 23, 24, 25] (Fig. 1d). In work [24] a measurement system based on the optical method of solid phase concentration measurement was proposed (Fig. 1e). Research on the development of measurement methods for two-phase mixtures are time-consuming and costly, so in this paper, it was indicated the possibility of supporting them with simulation methods.

3 Calculation methodology

The gas-particle mixture flow is described by differential equations which can be written in the generalized conservative form, with the isolation of convection, diffusion and source components [26]:

$$\frac{\partial(\rho\phi)}{\partial t} + \frac{\partial(\rho u_i \phi)}{\partial x_i} = \frac{\partial}{\partial x_i}\left(\Gamma_\phi \frac{\partial \phi}{\partial x_i}\right) + S_\phi + S_{\phi p} \quad (5)$$

In Equation (5) ϕ is generalized dependent variable, $\Gamma_\bullet$ the diffusion coefficient. The source component will include the other components of differential equations - apart from convective and diffusion. Coefficients $\Gamma_\bullet$ i $S_\bullet$ are presented in Table 1.

Table 1. Coefficients of equation (5).

Equation	ϕ	$\Gamma_\bullet$	$S_\bullet$	$S_{\bullet p}$
Continuity	1	0	0	0
Momentum equation in the x_i axis	u_i	μ_{ef}	$F_i - \frac{\partial p}{\partial x_i} + \frac{\partial}{\partial x_j}\left(\mu_{ef}\frac{\partial u_j}{\partial x_i}\right)$	$\overline{S_{ui,p}}$
Kinetic turbulence energy	k	$\frac{\mu_{ef}}{\sigma_k}$	$G_k - \rho\varepsilon$	0
Turbulent energy dissipation rate	ε	$\frac{\mu_{ef}}{\sigma_\varepsilon}$	$\frac{\varepsilon}{k}\left(C_1 G_k - C_2\rho\varepsilon\right)$	0

$$\text{where: } \mu_{ef} = \mu + \mu_t, \mu_t = C_\mu \rho \frac{k^2}{\varepsilon},$$

$$G_k = \frac{\partial u_i}{\partial x_j}\mu_{ef}\left(\frac{\partial u_i}{\partial x_j} + \frac{\partial u_j}{\partial x_i}\right)$$

$$\sigma_k = 1.00 \quad \sigma_\varepsilon = 1.30 \quad C_1 = 1.44 \quad C_2 = 1.92 \quad C_\mu = 0.09$$

Conversion of momentum between the diluted and continuous phase is described by the source term:

$$\overline{S_{ui,p}} = \frac{1}{V_E}\eta_j \int_{\delta t_j} \frac{C_D \, Re_p \, D_p}{8\pi}\left(\bar{u}_j - u_{pi}\right)dt \quad (6)$$

where V_E is the elementary cell volume and η_j describe the number of particles transported at j-trajectory in the time unit.

The particle motion was described by the Lagrange method by adding drag coefficient and gravitation:

$$\frac{d\vec{u}_p}{dt} = \frac{1}{\tau_p}\left(\vec{u} - \vec{u}_p\right) + \vec{g} \quad (7)$$

where $\vec{u}_p$ is the particle velocity vector and τ_p is the time of dynamic particle relaxation. As a result of the integration of equation (7) in the gas velocity field, trajectories of particle fractions were obtained. Knowledge of the trajectories is the basis for determining interphase interactions according to equation (6). Details of calculations can be found in [26]. ANSYS Fluent program [27] was used for numerical research.

4 Numerical research results

4.1 Boundary conditions of the research

The purpose of the study is to evaluate the degree of non-equilibrium flow phenomena on the metrological properties of the venturi tube in the dispersed flow. In particular, the calculations are carried out to set the connection between empirical constants in one-dimensional mathematical models and the Stokes number.

In this calculation, a Venturi tube with diameter ratio $\beta = \frac{d}{D} = 0.5$ was used in a pipe with a diameter D = 100 mm. The shape of the venturi tube is shown in Fig. 2, and its dimensions are given in Table 2.

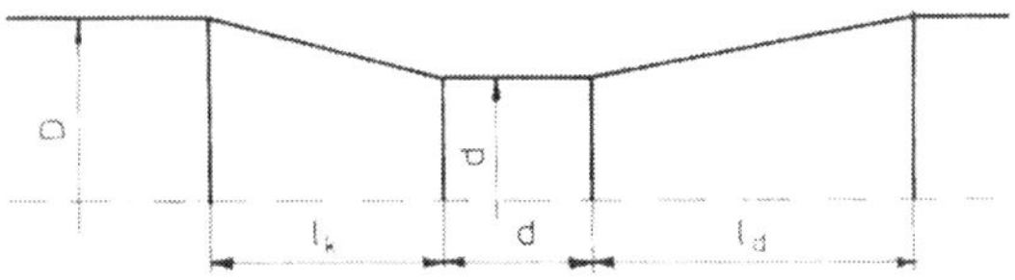

Fig. 2. Scheme of flow system with Venturi tube.

Table 2. Dimensions of Venturi tube.

D [mm]	$m=(d/D)^2$	d [mm]	l_k [mm]	l_d [mm]
100	0.25	50	135	409.8

Calculations were made for a range of Reynolds number Re = $5\cdot10^4$ – $5\cdot10^5$, with gas density of ρ_g = 1.2 kg/m^3, particles density of ρ_s = 1400 kg/m^3 and their diameter from 10 μm to 200 μm. The effect of particle concentration on differential pressure Δp_{tp} at the two-phase flow was also tested. The range of research was proper for the concentration characteristic in pneumatic conveying in a pipe system for coal particle burned in power plants ($Y \le 2$) [3, 4].

In the calculations, the influence of gravitation was omitted, which simplified the problem to two-dimensional axial-symmetrical flow. In the inlet, a fully developed turbulent flow was assumed. Fig. 3 shows the results of the pure gas test calculations, the symbol p_0 describe pressure value in the inlet cross-section.

The increase a gas velocity, as well as, the Reynolds number Re cause a significant increase the differential pressure Δp_g. The gas mass flow is calculated from the formula:

 DOI: 10.1051/e3sconf/20171901016

$$M_g = \frac{C}{\sqrt{1-\beta^4}}\pi\frac{d^2}{4}\sqrt{2\Delta p_g \rho_g} \quad (8)$$

where flow coefficient is

$$C = 4\frac{M_g}{d^2}\frac{1}{\pi}\frac{\sqrt{1-\beta^4}}{\sqrt{2\Delta p_g \rho_g}} \quad (9)$$

Flow coefficients calculated from the formula (9) were compared to the corresponding values according to the standard[2]. Differences do not exceed ±2%, which is a proof of correctness of accepted calculation methodology.

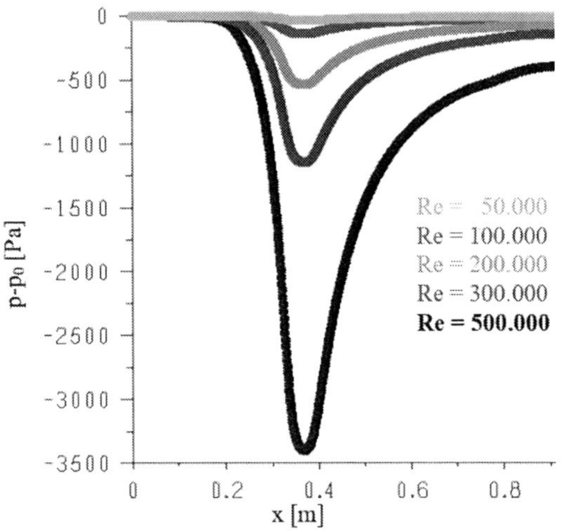

Fig. 3. Pressure distribution along the Venturi tube for different values of Reynolds number.

4.2 Calculation results for gas-particle mixture flow

A series of calculations were made for gas-particle mixtures. Fig. 4 shows pressure distribution along the Venturi tube for particle diameter d_p=10 μm with Reynolds number Re=$5\cdot10^5$ for different Y concentrations.

Similarly, results are shown in Fig. 5. Comparison, the smaller particle diameter creates a much larger pressure drop than for diameter d_p=200 μm at the same particle concentration.

The results of further calculations are shown in Fig. 6. The pressure difference measured on the venturi tube is the linear function of the particle flow ratio Y.

The angle of the characteristic inclination $\frac{\Delta p_{tp}}{\Delta p_g} = f(Y)$ depends to a significant extent on the particle diameter, from Stokes number St defined as follows:

$$St = \frac{\rho_s {d_p}^2 U}{18\mu_g D} = \frac{1}{18}\frac{\rho_s}{\rho_g}\left(\frac{d_p}{D}\right)^2 Re_G \quad (10)$$

In this case, Stokes number is in the range of 0.0324 to 129.630.

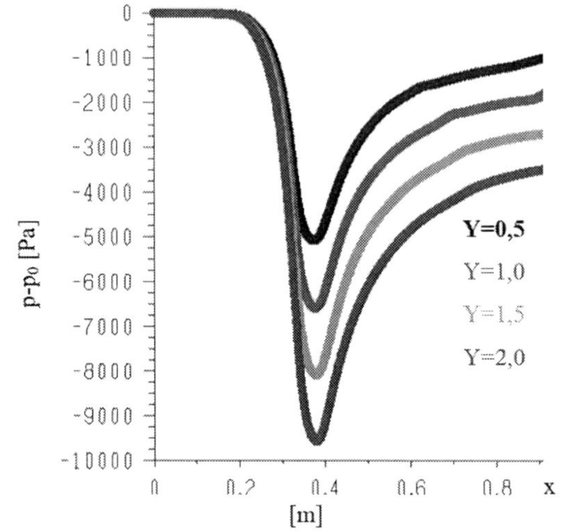

Fig. 4. Pressure distribution along the Venturi tube for Re=500 000, d_p=10 μm.

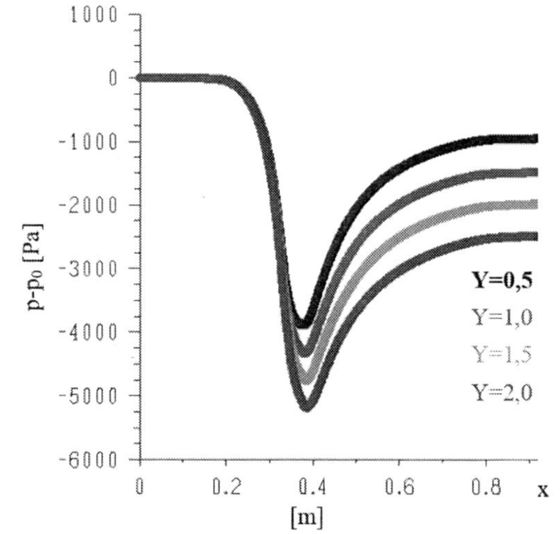

Fig. 5. Pressure distribution along Venturi tube for Re=500 000, d_p=200 μm.

When the particle diameter d_p is large, corresponding to St→∞, then $\Delta p_{tp} \cong \Delta p_g$ and the Venturi tube practically does not react to the presence of particles. Otherwise, when St→0, differential pressure $\Delta p_{tp} \cong \Delta p_h$ corresponds to the flow of a homogeneous mixture with a modified density.

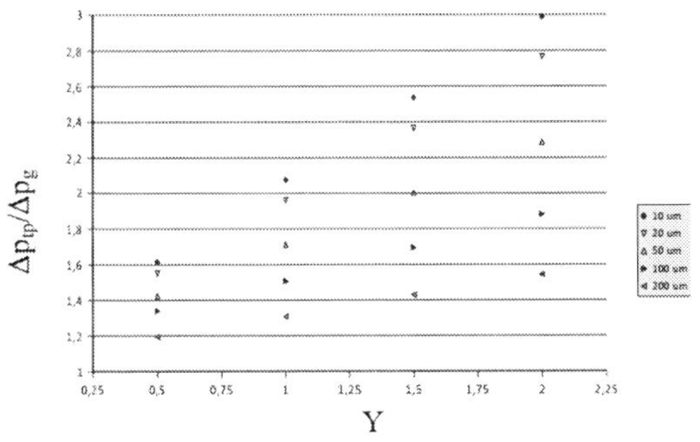

Fig. 6. Influence of particles loading ratio *Y* on differential pressure Δp_{tp} for different particle diameters.

Generally, the differential pressure measured on the Venturi tube is contained in the range

$$\Delta p_h \geq \Delta p_{tp} \geq \Delta p_g \tag{11}$$

So, if the Stokes number is not taken into account, the differential pressure Δp_{tp} does not give more precise information about the mixture mass flow. The results of a wider analysis of this problem show good agreement with numerical calculations results after the introduction of the correction function $f(\Omega)$, which is equivalent to Gästerstädt number E_I in equation (1). Then, the differential pressure Δp_{tp}^I is calculated from the formula:

$$\Delta p_{tp}^I = (1 + f(\Omega)Y)\Delta p_g^I \tag{12}$$

where

$$f(\Omega) = (1 + a\Omega^b)^{-1} \tag{13}$$

and

$$\Omega = \left(\frac{d_p}{D}\right)^2 \frac{\rho_s}{\rho_g} Re_g \beta \tag{14}$$

Values *a* and *b* in formula (13) are empirical constants depending on the fractional composition of the particles and other quantities characterizing the measurement conditions.

The function $f(\Omega)$ fulfills the asymptotic conditions:

$$\lim_{\Omega \to 0} f(\Omega) = 1 \tag{15}$$

$$\lim_{\Omega \to \infty} f(\Omega) = 0 \tag{16}$$

and ensures the fulfillment of the relation (11).

An example of the selection of empirical constants *a* and *b* is presented in Fig. 7, for the experimental data Lee and Crowe [23]. From the points distribution, it follows that the function $f(\Omega)$ good describes empirical data in a wide range of Stokes values.

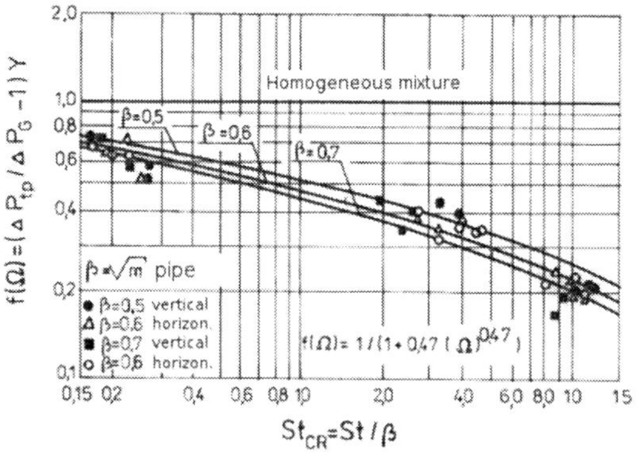

Fig. 7. The correction function *f*(Ω) compared with experimental data Lee and Crowe [16], where *a*=*b*=0.47.

Inserting Δp_g from equation (12) into equation (8) we obtain an equation for calculating the two-phase mixture mass flow M_{tp}:

$$M_{tp} = \frac{C}{\sqrt{1-\beta^4}} \pi \frac{d^2}{4} \sqrt{\frac{2\Delta p_{tp}^I \rho_g}{1 + f(\Omega)Y}} \tag{17}$$

where function $f(\Omega)$ is defined by equation (13). It follows that determining the mass flow of the gas-particle mixture requires, in addition to the particle concentration, also the knowledge of constant values a and b of a function $f(\Omega)$.

5 Conclusions

On the basis of provided analyses, the following conclusions can be drawn regarding the use of the gas-particle mass flowmeter:

1. Differential pressure Δp_{tp} measured on the Venturi tube is caused by the flow of both phases, but in general it, does not correspond to the mass flow of the two-phase mixture.

2. When the Stokes number St ⇒ 0, the differential pressure measured on the venturi corresponds to the two-phase mixture flow which can be treated as a modified density homogeneous fluid.

3. For large Stokes number, especially at St→∞, the presence of particles in the gas do not have a significant effect on the differential pressure increasing, which practically corresponds to the flow of pure gas itself.

4. With a given Stokes number, the pressure increase due to the presence of the dispersion phase is a linear function of particle loading ratio.

5. The results of the measurement of the mixture mass flow or one of its components will always be erroneous if the Stokes number value is not taken into account.

References

1. S. Anweiler, J. Environ. Manage. (2016) (https://doi.org/10.1016/j.jenvman.2017.03.040)
2. PN-EN ISO 5167-1: *Measurement of particle flow ratio using measuring nozzles*, Book of Polish Normalisation Committee, Warszawa, (2005)
3. D. Giddings, B. J. Azzopardi, A. Aroussi, Pickering SiJ. Powder Technol. **172**, 149-156 (2007)
4. M. Teisseyre, M. Mazur, *Supply of pulverized coal fired boilers – measurement issues*, Book of Wroclaw University of Technology, Wrocław (2006)
5. H. M. Carlson, P. M. Frazier, R. B. Engdahl, Trans. ASME **70**, 65-79 (1948)
6. M. Teisseyre, Books of Wroclaw University of Technology, Energetyka III **143**, 111-136 (1966)
7. L. Farbar, Trans. ASME **75**, 943-951 (1953)
8. P. A. Antikajn, Teploenergetika **12**, 35-37 (1956)
9. A. A. Szatil, Tieploenergetika **8**, 44-48 (1957)
10. A. A. Szatil, Izmieritelnaja Technika **9**, 46-48 (1961)
11. E. Gasterstädt, Zeitschrift des Vereines Deutcher Ingenieure **24**, 617-624 (1924)
12. Cz. Graczyk, *Ph.D. Thesis*, Silesian Univerisity of Technology, Gliwice (1957)
13. Cz. Graczyk, Pomiary Automatyka Kontrola **12**, 509-511 (1970)
14. J. Łapa, Measurement of the air-particle mixture flow ratio by three-signals Venturi tube – possibilities and range of application, Books of IMA i BUT, Wroclaw University of Technology, **5**, 37-50 (1971)
15. W. Barth, Chem. Ing. Tech. **3**, 164-170 (1960)
16. R. Jung, *Beiträge angenwandter Strömungs-forschung zur Entwicklung Kohlenstaubfeuerung*. Book VDI-Forschungsheft, **532** (1969)
17. R. Jung, Brennst.-Warme-Kraft **8**, 377-383 (1966)
18. A. L. Payne, C. T. Crowe, Proc. 1981, Symp. Instrum. and Control Fossil Energy Processes, ANL-81-62 (1981)
19. W. Barth, R. Nagel, K. van Wavern, Chem. Ing. Tech. **9**, 599-602 (1957)
20. J. Łapa, Pomiary Automatyka Kontrola **2**, 68-70 (1974)
21. R. Leroch, Energetyka **7/8**, 27-30 (1975)
22. J. Gołąbek, *Ph.D. Thesis*, Silesian University of Technology, Gliwice (1979)
23. J. Lee, C. T. Crowe, J. Fluids Eng. **104**, 88-91 (1982)
24. A. Payne, M. Werner, D. Plank, C. T. Crowe, The 1982 Symposium on Controls and Instrumentation of Fossil Energy Systems, Huston, June (1982)
25. L. M. Werner, C. T. Crowe, Proc. Symposium on Instrumentation and Control for Fossil Energy Processes, San Francisco, 8-9 June (1981).
26. B. Dobrowolski, J. Wydrych, JTAM **45**, 513-537 (2007)
27. ANSYS Fluent, Ansys Inc. (2016)

Visual study of capabilities of managing IC engine filling degree vs engine ecological indicators

Szymon Kołodziej[1,*], and *Grzegorz* Ligus[2]

[1]Opole University of Technology, Department of Vehicles, 5 Mikolajczyka St., Opole, Poland
[2]Opole University of Technology, Department of Environmental Engineering, 5 Mikolajczyka St., Opole, Poland

Abstract. The operation of an internal combustion engine in a transient state is related to constant changes in cylinder filling degree. Managing a steady course of engine torque demands constant corrections in the volume of air delivered to the engine. In this work, different engine throttle management strategies were analyzed. The effect of throttle velocity on emissions was depicted. The research has shown that the greatest value of work density is reached by the IC engine for relatively low values of throttle angle. For shown values, disturbances in air flow in engine intake which may adversely affect emissions due to uneven cylinder filling were researched with the use of Digital Particle Image Velocimetry. The conducted research has shown significant unevennes in air supply to individual cylinders in the engine. This effect may cause discrepancies in air excess ratio for each cylinder, and result in heightened harmful substance emission.

1 Introduction

The way in which IC engine cylinders are filled with fresh air-fuel mixture load not only determines its power, allowing to control its value, but also significantly affects the course of the combustion process. Combustion of the air-fuel mixture in wrong proportions results in obtaining a wrong value of the air/fuel equivalence ratio "λ", thusly increasing harmful substance emission into the environment. Management of quantity and flow of the air in the engine intake system is the subject of interest in research centers across the world. One of the effects of conducted research that can be observed is the electronic throttle, which is widely utilized nowadays. Forming and managing a correct correlation between the throttle and acceleration pedal is remarkably important for obtaining proper dynamic and ecological parameters. An abrupt change of throttle position by the driver with the desire to suddenly accelerate can result in fuel starvation, significantly worsening engine operation quality. The electronic throttle executed commands from the ECU, which correct engine operation parameters on the basis of information supplied by a range of sensors.

The significant majority of vehicles used nowadays is power by four-cylinder engines. An ideal state for such an engine is a situation in which all four cylinders are supplied with the same amount of air-fuel mixture, allowing each of them to reach identical operating parameters. However, in a practical situation, an unevenness in supply of fresh air into the cylinders can occur. In modern engines, the utilization of combustion sensors based on wideband lambda sensors (two for each cylinder pairs in a 4-2-1 exhaust system) partially solves this problem. However, the correction occurs after the combustion process has already taken place. Moreover, most engines utilize a narrow-band lambda sensor. The main source of information about the quality of the combustion process is the amount of free oxygen present in exhaust gasses. Fuel dosage is selected in a way that allows to obtain the stoichiometric composition of the air-fuel mixture ($\lambda=1$). Because of this, the value computed by the ECU is an averaged value for all cylinders. Despite the correct value of the average air-fuel ratio, a situation in which some of the cylinders operate on a lean air-fuel mixture, while the other ones operate on a rich air-fuel mixture, can occur. Such a situation negatively affects engine operation and its substance emissions.

One of the basic solutions of this problem is a correct design of the intake manifold, in a way which assures that the following conditions are observed:

1. The cylinder filling ratios are even for all cylinders.
2. The mass of air sucked into each of the cylinders is measured.
3. In the future, the design has to allow the utilization of separate pressure sensor for each cylinder, which will allow to introduce absolute correction in fuel dosage.

2 Throttle impact

Engine operation in transient states causes bigger difficulties in supplying the correct amount of air to the engine than operation with constant parameter values. In older vehicles, the throttle was fully connected with the acceleration pedal, meaning the driver had full control of the throttle opening angle. It is common to press the

* Corresponding author: s.kolodziej@po.opole.pl

acceleration pedal fully to gain maximum vehicle acceleration. However, such behavior adversely affects the amount of air-fuel mixture delivered to the cylinders. Limiting the possible throttle opening angle to a degree, especially at lower engine speeds, to ensure better engine performance is a more beneficial solution. This is due to the fact that most drivers do not pay attention to adverse effect of fully opening the throttle, while the electronically controlled systems choose the cylinder filling ratio more precisely and depending on the engine need for its current operating state. In Table 1, experimental data from measurements of maximum engine torque as a function of engine rotational speed and throttle opening angle was presented.

Table 1. Maximum engine torque as a function of engine rotational speed and throttle opening angle [3].

Maximum engine torque	
Engine speed [rpm]	Optimum opening of the throttle [%]
1000-1500	35
1500-1600	50
1600-2000	60
2000-2750	70
2750-6000	100

Throttle opening values in respect to engine speed presented in Table 1 were chosen with the intent of obtaining maximum engine torque in its whole range of operation. Presented data depicts the parameters of acceleration process for a popular, naturally aspirated IC engine of 1.6 liter displacement. It is apparent that at lower engine rotational speeds, maximum throttle opening angle does not affect the generated torque in a positive way. It is advisable to initially open the throttle valve only partially (as little as 35% for low engine rpm's), reaching full throttle opening as late as at 2750 rpm.

During torque measurements, specific fuel consumption was also registered. Measurement results were presented in Table 2.

Table 2. Specific fuel consumption in respect to engine rotational speed and throttle opening [3].

Minimum fuel consumption	
Engine speed [rpm]	Optimum opening of the throttle [%]
1000-2700	35
2700-2900	45
2900-3750	50
3750-4500	60
4500-6000	100

Data contained in Table 2 presents the minimum specific fuel consumption in respect to engine rpm and throttle opening. Same as for torque-related data from Table 1, full opening of the throttle at lower engine speeds does not positively affect fuel consumption.

During ecodriving the driver possesses greater consciousness about the need to correctly operate the throttle. On the other hand, when the driver desires to achieve maximum driving dynamics, it is practically impossible for him to operate the throttle in an optimum way. Too abruptly opening of the throttle at the lower range of engine rpm's results in hampering of engine dynamics, higher fuel consumption and higher CO_2 emissions.

An electronically controlled throttle provides a more advantageous way of controlling the throttle opening angle, by opening the throttle only in a certain range during full acceleration at lower engine rpms. The ECU computes the optimum throttle opening angle basing on all the operating parameters and driver input and sends the information to the throttle, where it is realized either by a stepper motor or a DC electric engine with a mechanical gear.

An excess of the driving power is needed to the correct operation of the throttle, allowing to overcome the resistances of the return spring, aerodynamic resistances associated with the air flow through the partially opened throttle plate while ensuring sufficient acceleration and deceleration in throttle movement. To examine the effect of throttle plate movement speed on engine operation indicators, a range of simulations were conducted with the use of Vehicle Road Load Simulator and a virtual engine saved in it. The vehicle acceleration phase was the subject of research, as it is in principle the most energy consuming and most substantially emissions-affecting phase of vehicle movement. Research results were presented in Figs. 1÷4.

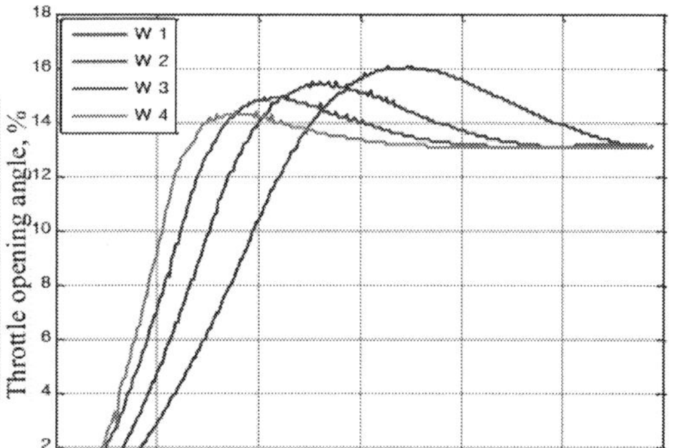

Fig. 1. Throttle opening angle vs throttle movement speed management strategy.

Throttle operation for four different variants of throttle velocity control was shown in Figs. 1 ÷ 4. The first variant depicts the slowest control variant, whereas the fourth variant (W4) is the fastest one.

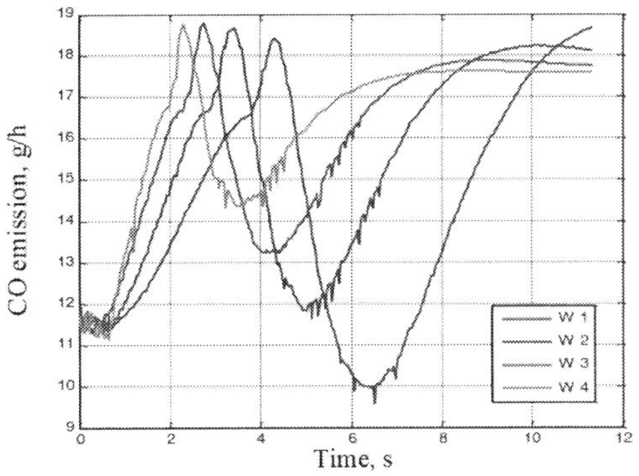

Fig. 2. Carbon monoxide emissions vs throttle opening velocity.

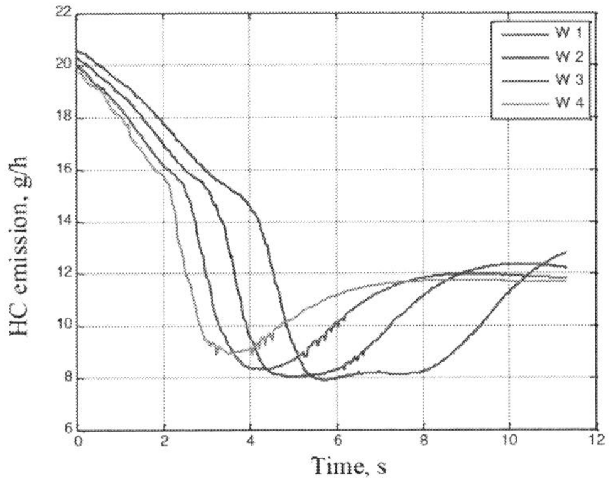

Fig. 3. Hydrocarbon emissions vs throttle opening velocity.

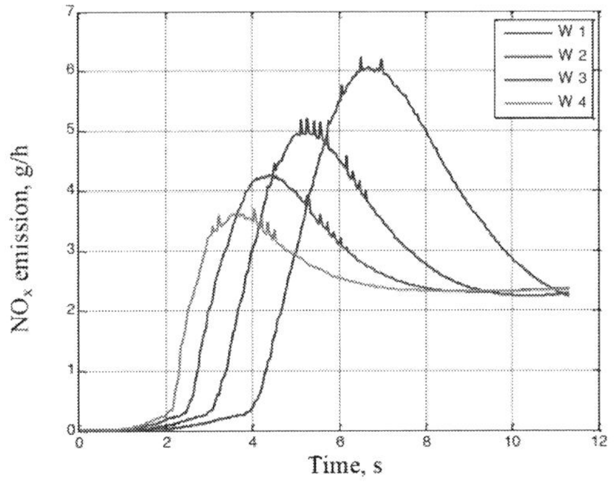

Fig. 4. NO_x emissions vs throttle opening velocity.

To showcase the differences between different throttle control variants for the chosen acceleration profile, work cycles for each researched variant were shown in fig 1. It can be seen that for the faster throttle control variant, signal overlaps faster with the velocity profile, and signal overdrive is not as significant as it is for lower throttle control speeds.

Next graphs show the results of measurements of the amount of toxic exhaust gas components (carbon monoxide, hydrocarbons and nitric oxides) for engine operation at throttle velocities depicted in Fig. 1.

The amount of carbon monoxide, which is the product of incomplete combustion at the beginning stage of the acceleration phase increases, and the intensity of said increase is dependent on the throttle velocity. In the later part of the acceleration phase the amount of emitted CO drops, its level reaching the lowest value for the slowest throttle velocity control strategy. The last phase of vehicle acceleration is related with an increase in CO emission and its stabilization for each shown case. The variant for which emissions are lowest can be determined by computing the field under each curve. For carbon monoxide, emissions are lowest in variant 3.

Hydrocarbons, which are unburned or partially burned fuel particles, are toxic substances. In Fig. 3 it was shown that HC emissions are also dependent on throttle operation velocity. The lowest HC emissions were obtained for variant 1, which is the slowest.

Nitric oxides are particularly toxic. They form as a result of reaction of the fuel with oxygen, depending on combustion pressures and peak temperatures. Nitric oxide emission shown in Fig. 4 clearly shows a lower value for higher throttle velocity. This is confirmed by calculating the field under the graph.

Only a short fragment of the acceleration was shown in Figs. 2, 3 and 4. Research for the same modifications was also conducted for a full city driving cycle, and the summary amount of emissions of mentioned exhaust gas components was the same as for presented short fragments. This confirms a previous statement that the acceleration phase is the most significant in the ecological aspect.

3 Air distribution in the intake manifold

An even cylinder filling in a multi-cylinder engine severely affects the quality of the air-fuel mixture, due to the utilization of aggregated control devices.

The literature study in the field of intake manifold research shows, that most studies rely on computer simulations [1,4]. Basing on the analysis of methods used for research of multi-phase flow phenomena, it was deduced that it is possible to experimentally investigate the flow in the intake manifold with the use of the Digital Particle Image Velocimetry (DPIV) method (Fig. 5) [2, 5].

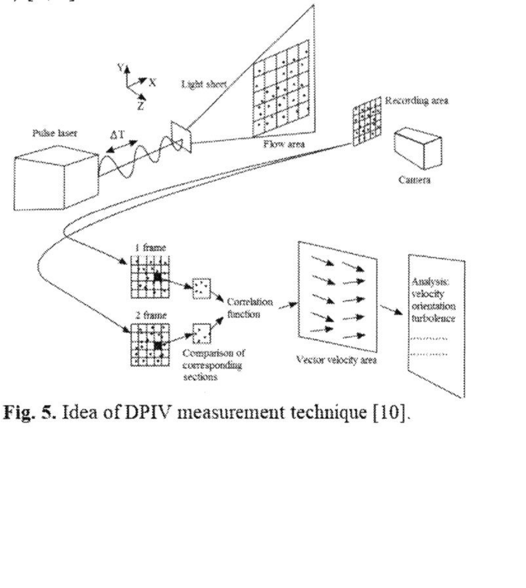

Fig. 5. Idea of DPIV measurement technique [10].

DOI: 10.1051/e3sconf/20171901017

DPIV is a noninvasive method of researching the flow phenomena with the use of a light sheet generated by laser in the inert seeding particles mixed with the medium flowing through the manifold and registered by a video camera placed perpendicularly to the light sheet. To use the described procedure, the investigated element has to have at least two transparent walls which are perpendicular to each other. Images are registered in a strictly determined time gap and analyzed in respect to changes in concentration of seeding particles suspended in the flowing gas. Consequently, placement of seeding particles from the previous picture is searched in the next one. To determine the averaged seeding particles movement a Cross Correlation Function (CCF) is used [6]. Before captured images can be subject to verification by the DPIV algorithms mentioned above, it is recommended to carry out a series of image processing with the use of graphics software. Every pair of recorded images is subject to geometrical analysis, which searched for areas of pixel concentration that are constant in position (creation of background image). The generated background image is subtracted from all raw images of the recording session. The effect of this subtraction is a new series of images which show only seeding particles flowing in interest area. Measurements, which are prepared in such a way lower the risk of occurrence of noise caused by uneven lighting etc.

Generation of image depicting the flow of seeding particles, which is free of interruptions, requires capturing a series of images, conducting full calculations for each image, and then averaging of obtained results [9].

To investigate and understand real phenomena occurring in the intake system, a Volkswagen 9A engine intake manifold was chosen as a research object. Measurements of velocity fields distribution for flow velocities corresponding to 1500 rpm and 2000 rpm and 10, 20, 30° throttle angles were conducted. The choice of parameters was backed by previously conducted simulations, which have shown that they correspond to velocity fields most often utilized in city driving. Results were shown in form of colored vectors in figures 6(a)÷12(a), whereas essential markings corresponding to those figures were applied to figure 6(a)

An initial analysis of obtained results has shown significant disproportions and unevenness in air flow through the manifold for all chosen parameter. A vortex forming above the inlet to the third cylinder may be the effect of the ejector-like operation of the air stream just above the throttle body. Its presence significantly lowers the velocity before the inlet to the fourth cylinder, thusly lowering its filling ratio. A characteristic vortex can also be seen above the inlet to the first cylinder, which is caused by the influence of manifold walls. A significant reduction in air flow velocity can be observed in this area. To eliminate the resulting effect of air particles moving against the assumed air flow direction, a transparent plastic baffle was used to reduce the manifold volume in this section. Research was repeated after modifying the manifold in such way, with results shown in Figs. 6(b)÷11(b).

The analysis of vector velocity fields presented in Fig. 6 shows significant in cylinder filling for the manifold with modified geometry.

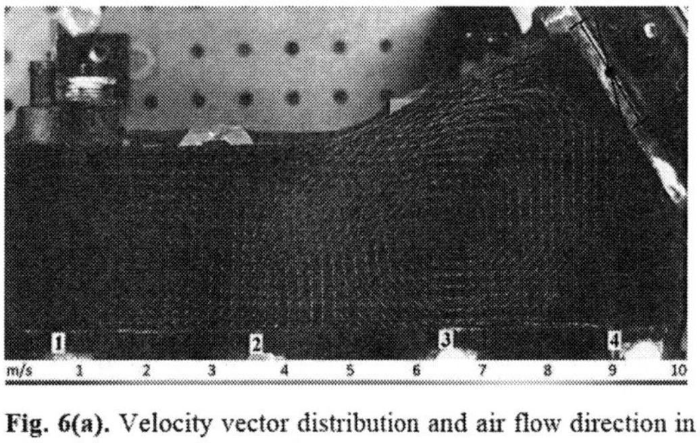

Fig. 6(a). Velocity vector distribution and air flow direction in the intake manifold for 1500 rpm and 10° throttle position angle.

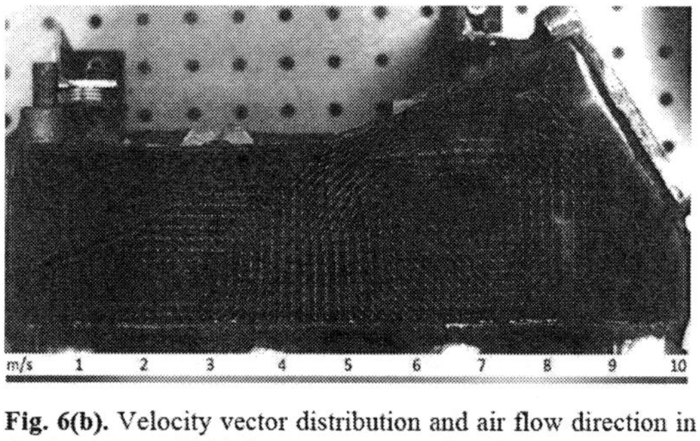

Fig. 6(b). Velocity vector distribution and air flow direction in the intake manifold for 1500 rpm and 10° throttle position angle.

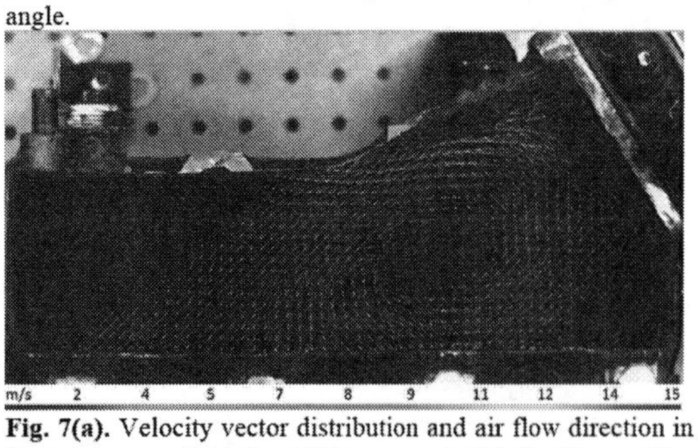

Fig. 7(a). Velocity vector distribution and air flow direction in the intake manifold for 1500 rpm and 20° throttle position angle.

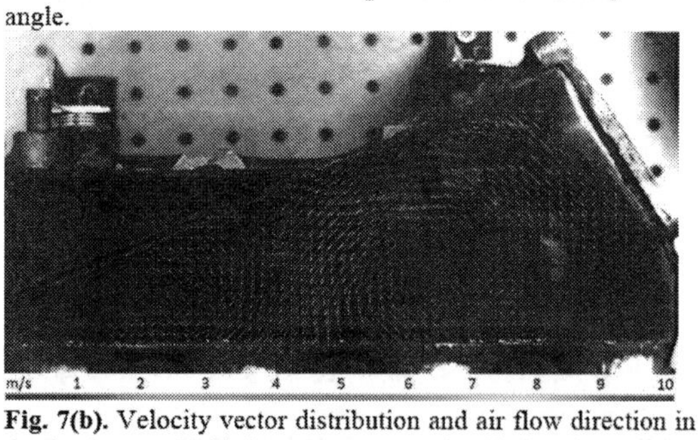

Fig. 7(b). Velocity vector distribution and air flow direction in the intake manifold for 1500 rpm and 20° throttle position angle.

DOI: 10.1051/e3sconf/20171901017

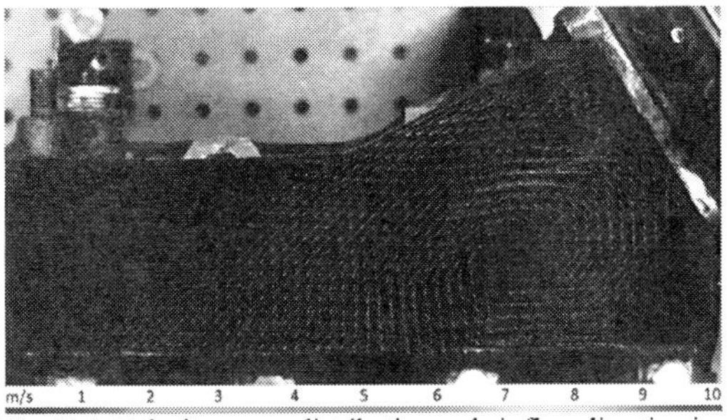

Fig. 8(a). Velocity vector distribution and air flow direction in the intake manifold for 1500 rpm and 30° throttle position angle.

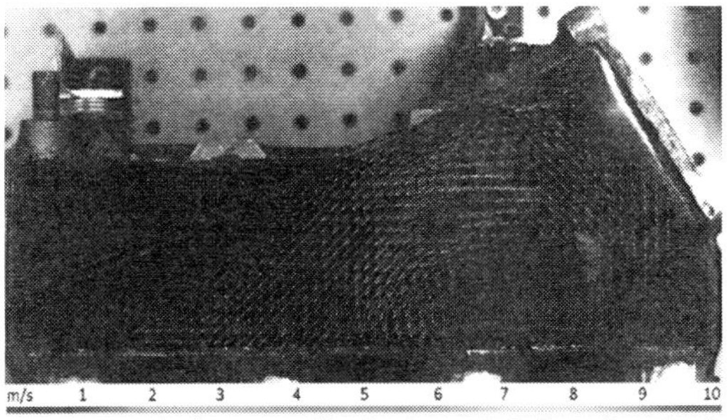

Fig. 8(b). Velocity vector distribution and air flow direction in the intake manifold for 1500 rpm and 30° throttle position angle.

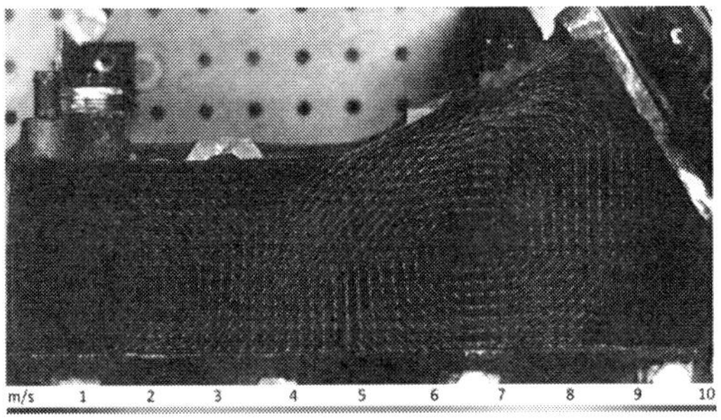

Fig. 9(a). Velocity vector distribution and air flow direction in the intake manifold for 2000 rpm and 10° throttle position angle.

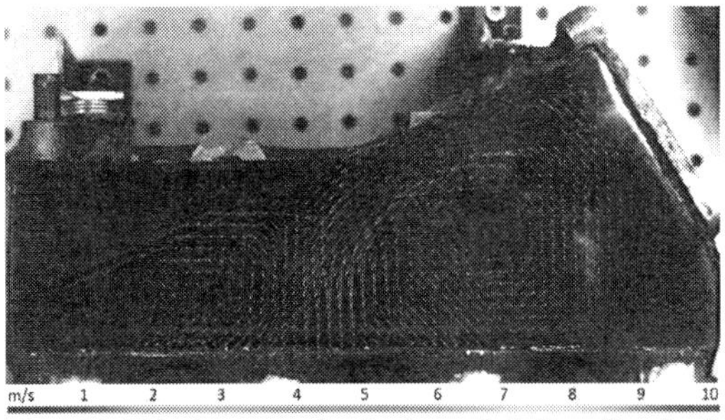

Fig. 9(b). Velocity vector distribution and air flow direction in the intake manifold for 2000 rpm and 10° throttle position angle.

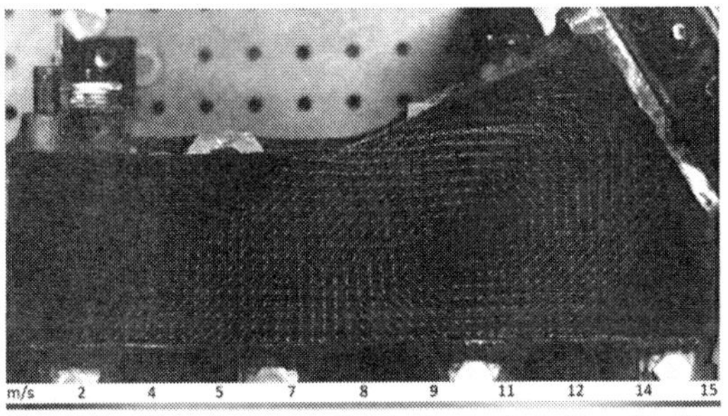

Fig. 10(a). Velocity vector distribution and air flow direction in the intake manifold for 2000 rpm and 20° throttle position angle.

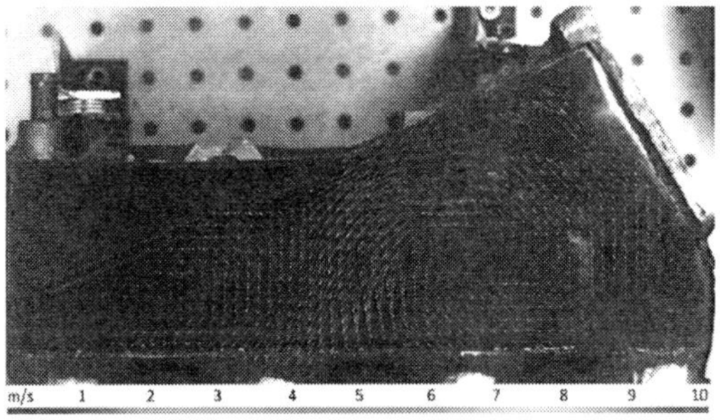

Fig. 10(b). Velocity vector distribution and air flow direction in the intake manifold for 2000 rpm and 20° throttle position angle.

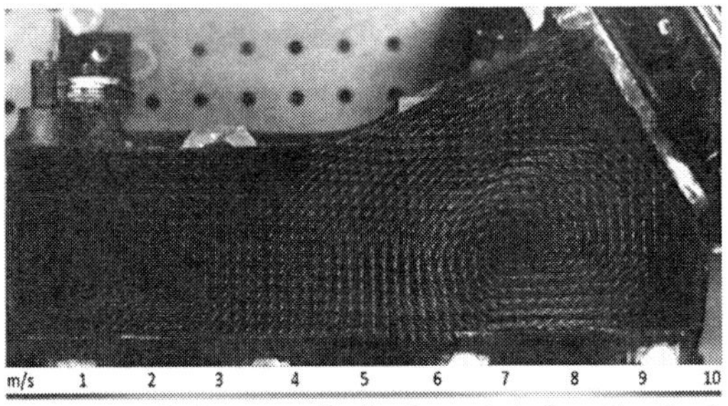

Fig.11(a). Velocity vector distribution and air flow direction in the intake manifold for 2000 rpm and 30° throttle position angle.

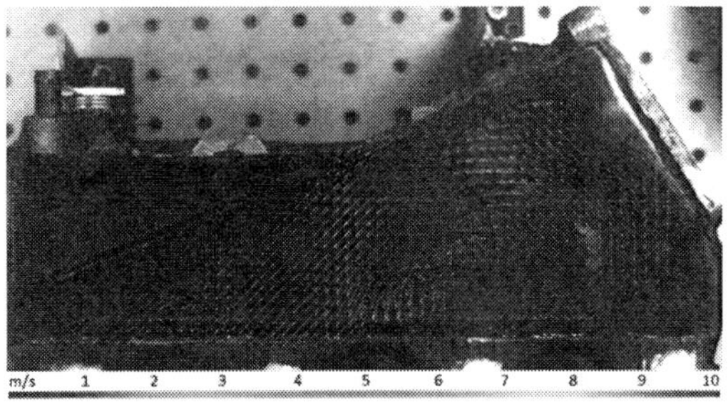

Fig. 11(b). Velocity vector distribution and air flow direction in the intake manifold for 2000 rpm and 30° throttle position angle.

Despite lack of vortex formation above cylinder 1, for each research case significantly higher flow velocities in the area of the second cylinder were noted. To precisely depict the differences in cylinder filling, a velocity profile was made along the cylinder inlet line. Results were shown in Figs. 12÷13. Position of inlet canals was schematically shown by a black line.

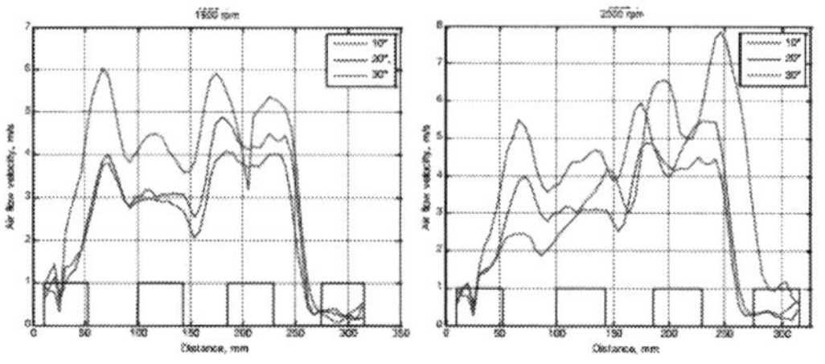

Fig. 12. Velocity profile distribution in the plane of stock manifold outlet for different throttle position angles.

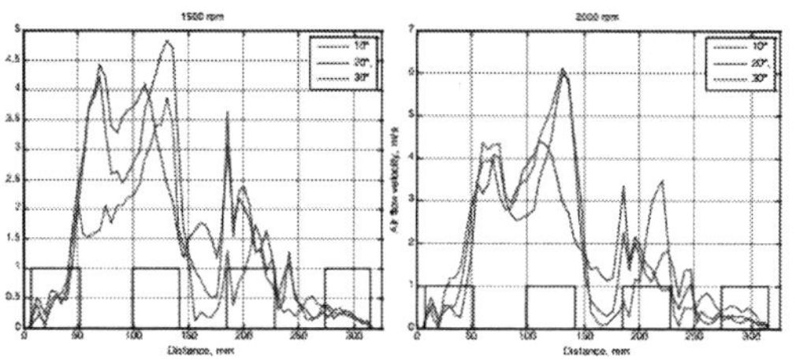

Fig. 13. Velocity profile distribution in the plane of modified manifold outlet for different throttle position angles.

A clearly visible difference in cylinder filling (Fg. 12÷13) can denote different values of the lambda parameter for each cylinder and increased harmful substance emissions. A modified manifold geometry, whose velocity profile was depicted in Fig. 13 exacerbates the disproportion in filling ratio.

4 Conclusions

Improper control of cylinder filling in an IC engine can adversely affect harmful substance emissions, as simulations have shown. The utilization of electronic throttle control systems solves this problem to a significant degree. It is vital to maintain the throttle in a good technical condition. Its contamination or wear of its components may result in an improper operation.

Unevenness of filling in a multi-cylinder engine can result in different excess air coefficient for each of the cylinders. Conducted experimental researches have shown uneven velocity field distribution at cylinder inlets of the exemplary manifold during even air extraction from every canal. Such state does not occur during engine operation - in real-life conditions, the air is being sucked out of each cylinder alternately. However, it can be assumed that the conducted research gives an outlook on flow resistances for each cylinder inlet. Discrepancies occurring during a real work cycle will affect cylinder filling to a smaller degree, however the differences in cylinder filling shown in the course of this research will occur during engine operation, albeit to a smaller degree. The research has also shown that a larger capacitive part of the manifold is essential. It not only positively affects the uniformness of the velocity field distribution, but also reduces the unwanted wave effect occurring due to the pulsating nature of valvetrain operation.

References

1. A. Nigro, A. Algieri, C.D. Bartolo, S. Bova, IJMS, **123**, 4, (2017)
2. J. Bode, J. Schorr, C. Krüger, A. Dreizler, B. Böhm, *Proceedings of the Combustion Institute* 36, (2017)
3. J. Mamala, Zeszyty Naukowe Politechniki Opolskiej, Mechanika, **87**, 309, (2005)
4. J. Mamala, Combustion engines, **4**, 139, (2009)
5. J. Rabault, J. A. Vernet, B. Lindgrenc, P. Henrik Alfredsson, IJHFF, **62**, (2016)
6. J.S. Bendat, A.G. Piersol, *Random Data: Analysis and Measurement Procedures*, (Wiley, Hoboken, 2010)
7. K. Cupiał, A. Dużyński, J. Grzelka, proceedings of KONES 2000, 26 Międzynarodowa Konferencja Naukowa Silników Spalinowych, Nałęczów (2000)
8. K. Święcicki, Modelowanie Inżynierskie, **25**, 57, (2015)
9. R.J. Adrian, J. Westerwell, *Particle Image Velocimetry*, (Cambridge University Press, Cambridge, 2011)
10. https://www.dantecdynamics.com/measurement-principles-of-piv (access 25.02.2017)

Simulation of fuel demand for wood-gas in combustion engine

Katarzyna Botwinska[1,*], *Remigiusz* Mruk[2], *Karol* Tucki[3], and *Mateusz* Wata[4]

[1,2,3,4] Warsaw University Of Life Sciences, Faculty of Production Engineering, Department of Production Organization and Engineering, Nowoursynowska 164, Warsaw, Poland

Abstract. In the era of the oil crisis and proceeding contamination of the natural environment, it is attempted to substitute fossil raw materials with alternative carriers. For many years, road transport has been considered as one of the main sources of the substances deteriorating air quality. Applicable European directives oblige the member states to implement biofuels and biocomponents into the general fuel market, however, such process is proceeding gradually and relatively slowly. So far, alternative fuels have been used on a large scale to substitute diesel fuel or petrol. Derivatives of vegetable raw materials, such as vegetable oils or their esters and ethanol extracted from biomass, are used to that end. It has been noticed that there is no alternative to LPG which, due to financial reasons, is more and more popular as fuel in passenger cars. In relation to solutions adopted in the past, it has been decided to analyse the option of powering a modern passenger car with wood gas - syngas. Such fuel has been practically used since the 1920's. To that end, a computer simulation created in SciLab environment was carried out. Passenger car Fiat Seicento, fitted with Fire 1.1 8V petrol engine with power of 40kW, whose parameters were used to prepare the model, was selected as the model vehicle. The simulation allows the determination of engine demand on the given fuel. Apart from the wood gas included in the title, petrol, methane and LPG were used. Additionally, the created model enables the determination of the engine power at the time of the indicated fuels supply. The results obtained in the simulation revealed considerable decrease in the engine power when the wood gas was supplied and the increased consumption of this fuel. On the basis of the analysis of the professional literature describing numerous inconveniences connected with the use of this fuel as well as the obtained results, it has been established that using the wood gas as alternative fuel is currently unjustified.

1 Introduction

Wood gas (german holzgaz, synthesis gas) is a gas produced in the process of gasification of biomass, in this case - wood. The process is carried out in special devices called "gas generators" and it has many stages [1]. The gasification consists in thermal transformation of fuel in solid form (wood) into the gaseous fuel (wood gas) [2]. The gas consists of the combination of such flammable substances as: hydrogen, carbon oxide, methane and non-flammable - carbon dioxide, nitrogen, water vapour. The composition of the gas depends on numerous factors, *inter alia*, on properties of the gasified material, e.g. humidity, or generator parameters, e.g. internal temperature of the hearth [1, 2].

Wood gas as fuel in passenger vehicles has already been used in the past. In 1920's a French engineer J. Imbert constructed a wood gas generator which could be used to power cars with gas fuel [1, 4]. In the following years, this technology was adopted by the automotive magnates like: General Motors, Mercedes-Benz or Ford. Due to the availability of the raw material and simple construction of the gas generator, the Europeans constructed the devices on their own by means of old gas cylinders, boilers and washing machines [1]. The wood gas became popular mainly during the World War II. In the warfare the petrol reserves were entirely used by the army primarily for the purpose of aviation. Therefore, it became necessary to use other fuels for remaining vehicles - trucks, passenger and delivery cars. It is estimated that the number of vehicles powered with wood gas in the said period equalled to 7 million [1, 4]. The use of syngas in motor vehicles entailed, however, numerous inconveniences.

Due to the low technology efficiency, arduous operation of the generator, the risk of poisoning oneself with carbon dioxide and the regained access to petrol as transportation fuel, wood gas has been superseded from the power systems of motor vehicles. The gas became more often used in stationary installations designed to heat residential units [4, 5]. It may seem that technologies based on wood gas powering of motor vehicles do not have any future. Nevertheless, legal conditions in respect of biofuels and biocomponents as well as strict environmental standards imposed on existing vehicles render it necessary to seek alternative solutions for present carriers.

* Corresponding author: katarzyna_botwinska@sggw.pl

1.1 The aims of the study

The purpose of this thesis was the evaluation of supply modern engine with wood gas.
To fulfil the formulated objective, the following values were determined:

- Demand of a petrol engine for wood gas, as well as petrol, methane and LPG;
- Reduction of power of the engine after syngas supply in relation to other fuels (petrol, methane, LPG).

1.2 Scope of the study

The scope of the thesis comprises the preparation of the simulation of internal combustion engine operation of a selected passenger car, including:

- Preparation of the simulation presenting the demand of the selected internal combustion engine for used fuels;
- Carrying out the simulation and obtaining results in the form of diagrams of temporary values monitored during the simulation and numerical values for cumulated values of particular fuel consumption, air consumption and total consumption of air and fuel mixture;
- Preparation of the simulation concerning the assessment of changes in the maximum power characteristic of the selected engine and carrying out the simulation and obtaining results on its efficiency;
- Analysis of the achieved results.

2 Materials and Methods

2.1 The object of research

To construct the simulation model, the parameters of the passenger car Fiat Seicento with a Fire 1.1 8v petrol engine with power of 40 kW were used. It is a typical urban car with low operation costs due to low fuel consumption and exhaust emission. The manufacturer provides the following values of fuel combustion for the said unit: in the mixed cycle 5.7 l/100 km, in the urban cycle 7.2 l/100 km and in the extra-urban cycle 4.8 l/100 km [6]. Table 1 below presents basic data of engine specification:

Table 1. Specification of the Fire 1.1 8v engine with the power of 40 kW [6].

Parameter	Value
Engine code	176.b2.000
Capacity	1108 cm^3
Diameter, piston stroke	70.72 mm
Maximum power	54KM (40kW) at 5500 rpm
Maximum torque	86 N·m at 3800 rpm
Arrangement of cylinders	line
Number of cylinders	4

2.2 Prepared simulation

The simulation model was constructed in the Scilab environment. The software enables the use of numbers and structures such as matrices and vectors. It has built-in graphics features which allow the creation of graphs and diagrams in two or three dimensions. It may be widely used to perform basic and advanced calculations in many fields of science. The structure of the prepared simulation is depicted below. The model was divided into a few main units containing elements necessary to carry out specific tasks.

2.2.1 NEDC test generator

Data, which was entered in the simulation unit "NEDC test generator", enabled the control of the simulated operation process of the selected vehicle and its engine according to the requirements of the "New European Driving Cycle" (NEDC) test. It is a current approval test for vehicles, which are to be considered roadworthy within the area of the European Community [7, 8]. Such data was entered into components "Signal Builder" which at the initiation of the simulation generate the following signals: Vehicle velocity [m/s] constituting the main element needed to determine a load for a vehicle, Clutch [-] this signal informs next simulation units about the connection of the drive to the engine (when it is turned on) or disconnection of the drive from the engine (turned off), Number of gear [-] through the change of gear the signal controls the transmission ratio between the engine and vehicle wheels in the next simulation elements, effort on wheels [N] is the result of calculations which include the vehicle dynamics and drags occurring when the car is in motion.

Then, within the module, calculations of a temporary value of the vehicle velocity [m/s] are carried out, and they are followed by determination of the distance covered by the vehicle with the use of integrating element and vehicle accelerations with the use of derivative term according to the correlations presented below [10]:

$$d = \int_0^{tc} v(t)\ dt\ [m] \tag{1}$$

$$a(t) = \frac{dv(t)}{dt} \left[\frac{m}{s^2}\right] \tag{2}$$

Where:
$v(t)$ - Temporary vehicle velocity in the NEDC test,
tc - Completion time of the simulation.

After that, the simulation calculates temporary load values resulting from the vehicle movement Fv [N] based on the parameters consistent with NEDC procedure and assumed for this type of vehicle (conversion mass 740 kg) according to the correlation [7, 8, 10]:

$$Fv = a + b \cdot v(t)^2\ [N] \tag{3}$$

Where:
$v(t)$ - Temporary vehicle velocity in the NEDC test,
a - Rolling resistance coefficient (5 N),
b - Drag coefficient (0.0337 $N/(km/h)^2$.

Apart from the load resulting from the vehicle motion resistance, the simulation also includes the calculation of the load resulting from the motion dynamics in the

DOI: 10.1051/e3sconf/20171901018

NEDC test according to the methodology on the basis of the correlations [7, 8, 10]:

$$Fa = m\frac{dv(t)}{dt} \text{ [N]} \quad (4)$$

Where:
$v(t)$ - Temporary vehicle velocity in the NEDC test,
m - Conversion mass of the vehicle 740 kg.

Then, the calculated load values are summed up and transferred to subsequent simulation modules. Fig. 1 shows a part of the developed simulation calculating vehicle parameters: acceleration, distance and load.

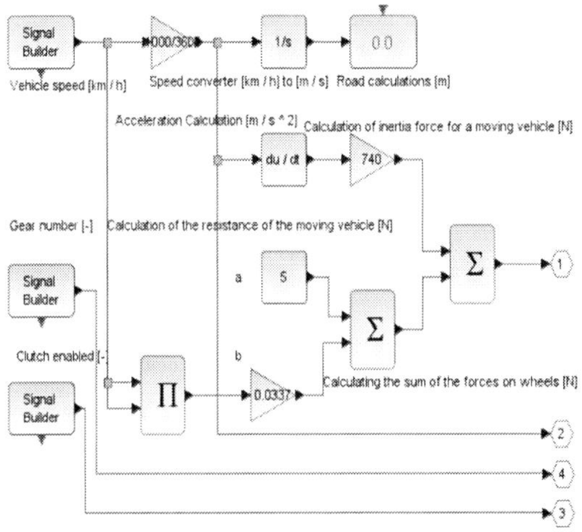

Fig. 1. Diagram of the NEDC test generator.

2.2.2 "Drive system" and "Engine" simulation units

„Drive system" unit, on the basis of temporary vehicle velocity values, vehicle wheels load, number of gear ratio and the value of signal of clutch engagement, performs calculations of the temporary values of torque of the engine Ms [N·m] and the rotational speed of the engine ω [rad/s] with the use of parameters of the selected vehicle according to the correlations [10]:

$$Ms = Fk \cdot R \cdot Pm \cdot S \, [N \cdot m] \quad (5)$$

$$\omega s = \frac{v(t)}{R} \cdot P\omega \cdot S \left[\frac{rad}{s}\right] \quad (6)$$

Where:
Fk - Temporary value of the effort on wheels [N],
R - Wheel radius [m],
Pm - Transmission ratio of the drive system for the torque (0; 0.0769; 0.137; 0.209; 0.287; 0.336 [-]),
S - Clutch engagement signal (0, 1 [-]),
$v(t)$ - Temporary vehicle velocity in the NEDC test,
$P\omega$ - Transmission ratio of the drive system for the rotational speed (0; 13.9; 7.30; 4.79; 3.48; 2.98 [-]).

Next, the "Engine" unit checks whether the calculated values are contained within the ranges of permissible changes and the values of rotational speed of the neutral gear on 83.7 rad/s (800 rpm) are determined according to the principles of engine operation in the vehicle which does not have the start/stop feature and cannot stop the engine during the longer no-load states in compliance with NEDC test.

2.2.3 "Calculations for petrol" simulation unit

Within this unit, on the basis of the assumed characteristics of the hourly fuel consumption for the selected engine in the function of the rotational speed and torque, temporary values of the fuel and air consumption necessary for operation and cumulated values are calculated. In the model was used 95 octane gasoline. In this case as well the general engine efficiency for a particular engine operation point is calculated and it constitutes a basis for further estimation of consumption of other fuels. To perform the calculations universal characteristics of the selected vehicle [9] were used which, after having been converted into digital form, were entered into the simulation component as a regular two-dimensional matrix. Such component, with the use of linear interpolation function, allows the calculation of temporary values of fuel consumption for any rotational speed values of the engine and torque. Using temporary values of the rotational speed defined previously, torque and characteristics of the hourly fuel consumption, module elements perform the calculation of the temporary value of fuel stream $\dot{pal}$ [g/s] on the basis of the following correlation [10]:

$$\dot{pal} = fu(\omega s,\ Ms)/3600 \left[\frac{g}{s}\right] \quad (7)$$

Where:
fu - Function of hourly fuel consumption depending on the rotational speed and torque [g/h],
ωs - Engine rotational speed [rad/s],
Ms - Engine torque [N·m]).

After that, including the fuel stream and a stoichiometric constant, the values of the air stream are calculated and they are essential to carry out proper process of fuel and air mixture combustion in the engine combustion chamber on the basis of the correlation [10]:

$$\dot{pow} = \dot{pal} \cdot Ste \left[\frac{g}{s}\right] \quad (8)$$

Where:
$\dot{pal}$ - fuel stream [g/s],
Ste - Stoichiometric constant for petrol (14.7 g_{pow}/g_{pal}).

Including calorific value of petrol, the energy stream $\dot{Ee}$ [J/s] provided by the fuel was calculated, and, after that, the temporary value of engine efficiency η [-] was determined with the use of the correlation [10]:

$$\dot{Ee} = Qe \cdot \dot{pal} \left[\frac{J}{s}\right] \quad (9)$$

$$\eta = \frac{Ms \cdot \omega s}{\dot{Ee}} \, [-] \quad (10)$$

Where:
$\dot{pal}$ - fuel stream [g/s],
Qe - calorific value of petrol (42 kJ/g).

DOI: 10.1051/e3sconf/20171901018

ωs - engine rotational speed [rad/s],
Ms - engine torque [N·m]).

The simulation model, based on the calculated temporary values, performs calculations of cumulated values of fuel and air consumption for the entire test, using integrating elements. Fig. 2 shows a fragment of a built-in simulation that calculates the instantaneous petrol stream to power the engine during the NEDC test, the energy flow required for the engine drive and the total fuel and air consumption values.

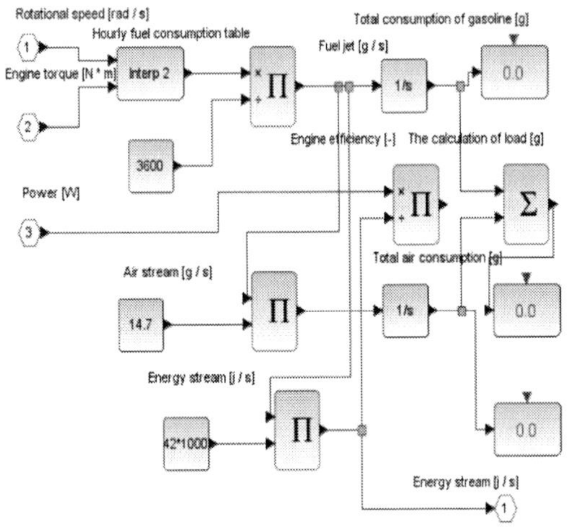

Fig. 2. Diagram of the calculations for petrol.

2.2.4 *"Calculations for methane", "Calculations for LPG" and "Calculations for wood gas" simulation units*

The said simulation units have the same structure enabling the calculation of the temporary values of demand for a particular fuel $\dot{pal}$ [g/s], including a required stream of energy provided by a given fuel determined in the previous module and calorific value of a given fuel as well as calculation of the temporary value of the air stream according to the correlation [10]:

$$\dot{pal} = \frac{\dot{Ee}}{Qe} \left[\frac{g}{s}\right] \tag{11}$$

$$\dot{pow} = \dot{pal} \cdot Ste \left[\frac{g}{s}\right] \tag{12}$$

Where:
$\dot{Ee}$ - stream of energy provided by the fuel [J/s],
Qe - Calorific value for fuels: methane, LPG, wood gas [J/g],
Ste - Stoichiometric constant for fuels: methane, LPG, wood gas $[g_{pow}/g_{pal}]$.

Table 2 below summarizes the accepted calorific values and stoichiometric steady-state combustion air requirements for individual fuels. In the developed simulation, the chemical composition of the applied wood gas was assumed:

- Nitrogen (N_2) 47% Vol.;
- Carbon monoxide (CO) 23% Vol.;
- Hydrogen (H_2) 18% Vol.;
- Carbon Dioxide (CO_2) 10% Vol.;
- Methane (CH_4) 2% Vol.

Table 2. Calorific values and stoichiometric values for fuels used in the simulation.

Fuel	Calorific values [kJ/g]	Stoichiometric values $[g_{pow}/g_{pal}]$
Petrol	42	14.7
Methane	50	17.2
LPG	46.1	15.7
wood gas	4.9	1.39

In order to depict the effects of the simulation, the elements in the discussed modules present temporary values in the form of diagrams. The cumulated values are estimated as well. Below, Fig. 3 shows the simulation elements that calculate the value of the wood gas stream in the NEDC test.

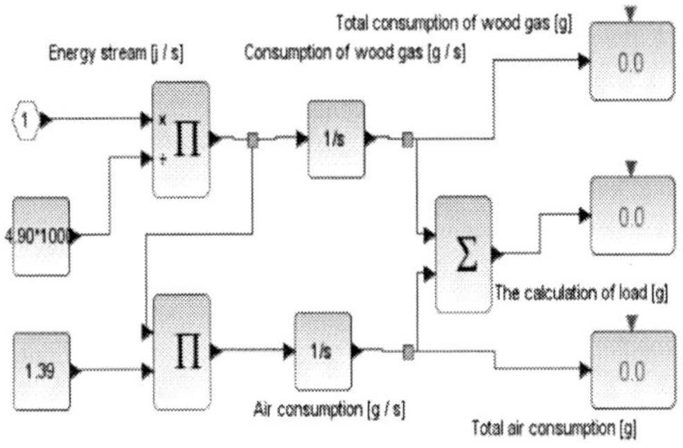

Fig. 3. Diagram of the calculations for wood gas.

2.2.5 *Simulation of the estimation of changes in maximum power characteristic*

Fig. 4 presents the reference diagram of the prepared simulation. The following simplification assumptions were applied in the simulation of the estimation of changes in maximum power characteristic of the selected engine for different fuels:

- Particular values on the maximum power curve in the function of the rotational speed for petrol in the case of a selected engine depend on the maximum load delivered to engine cylinders;
- The same values of the engine efficiency with the use of different fuels were assumed for a particular point of engine operation located on the maximum power curve;
- In the case of other fuels used, the volume of the fuel and air mixtures equal to the volumes of petrol and air mixture;
- Based on the maximum streams of mass of the used fuels which may be supplied to the engine and calorific values, theoretical energy streams are calculated, and subsequently, on the basis of the engine efficiency the progress of the theoretical curve of the maximum engine power is determined.

 DOI: 10.1051/e3sconf/20171901018

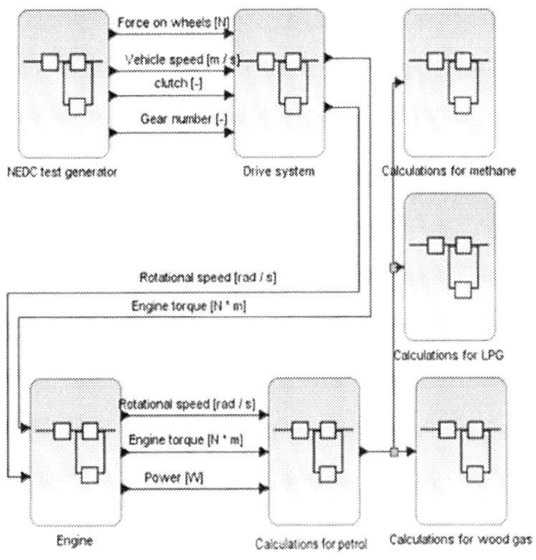

Fig 4. General diagram of the prepared simulation.

3 Results and discussion

The results were presented below as waveforms and numerical values for the included parameters.
The results of the simulation unit performance responsible for the calculation of temporary parameters of the internal combustion engine "Engine" were presented in Fig. 5 in the form of temporary waveforms of the torque values produced by the engine, engine rotational speed, including the restrictions assumed in the simulations and power produced by the engine. The parameters shown in the diagrams are key elements necessary for further calculations in the units concerning particular fuels.

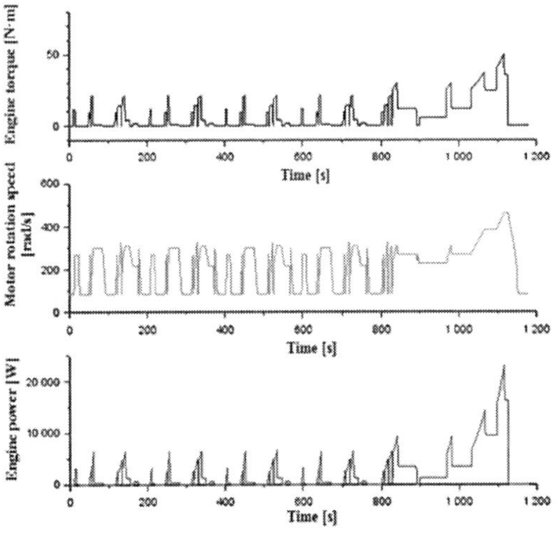

Fig. 5. The results of the simulation unit performance responsible for the calculation of temporary parameters of the internal combustion engine.

On the basis of [9], the characteristics of the hourly fuel consumption for the selected engine were adopted for the simulation. The view of the waveform of the characteristics after its conversion into numerical form is shown in Fig. 6. For technical reasons, characteristics fragments for higher maximum torque values (red line in the diagram) typical for the selected engine were extrapolated in such a manner that a regular matrix with a set of values may be obtained. It prevents calculation problems of the simulation modules for engine operation conditions near the maximum torque curve in the function of the rotational speed.

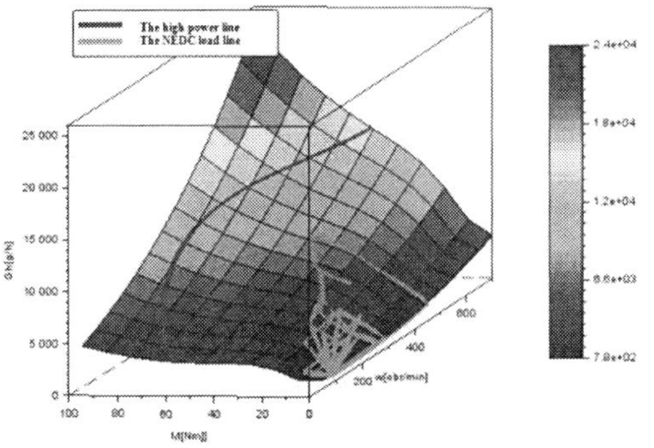

Fig. 6. Characteristics of the hourly fuel consumption in the function of the rotational speed and torque of the selected engine adopted in the simulation.

The diagram was also marked with a line (green colour) which determines the course of changes in the engine operation conditions during the simulation of the NEDC test. The form of the line confirms that the aforementioned test uses moderate ranges of the rotational speed and torque of the selected engine during the process.

Table 3 below presents the results of the cumulated values of fuel and air consumption as well as total load provided to the engine for different variants of the engine simulation in respect of the used energy sources obtained from the simulation.

Table 3. Comparison of the mass fuel and air consumption for the NEDC test for different fuels supplying the engine, calculated during the simulation.

	Petrol	Methane	LPG	Wood gas
Mass fuel consumption [kg]	0.497	0.418	0.453	4.26
Mass air consumption [kg]	7.31	7.19	7.12	5.93
Mass load consumption [kg]	7.81	7.61	7.57	10.2

In the simulation of the test, values for the engine powering with the petrol and liquefied petroleum gas are comparable due to the similarity of their energy parameters such as calorific value or stoichiometric constant. In the case of methane, the consumption of this energy source measured in grams during the test simulation is the lowest since it had the highest calorific value and stoichiometric constant.

As far as the NEDC test simulation for the wood gas is concerned, high values of its consumption were

DOI: 10.1051/e3sconf/20171901018

obtained, which results from its very low calorific value (4.9 kJ/g) in relation to other used fuels. Moreover, the results of the simulation confirm the information contained in the professional literature revealing considerable demand for this gas during the engine operation due to low energy values of the fuel. The values of the air consumption, however, in this type of engine supply, are lower than in the case of other analysed fuels. It is associated with the lower value of stoichiometric constant for the wood gas.

The next step in the prepared simulation was the estimation of the changes in maximum power characteristic for the selected engine when using the following fuels: petrol, methane, LPG and wood gas according to the methodology described in the thesis. Fig. 7 presents the results of performance of the prepared simulation in the form of diagrams of the maximum engine power in the function of the rotational speed [rad/s].

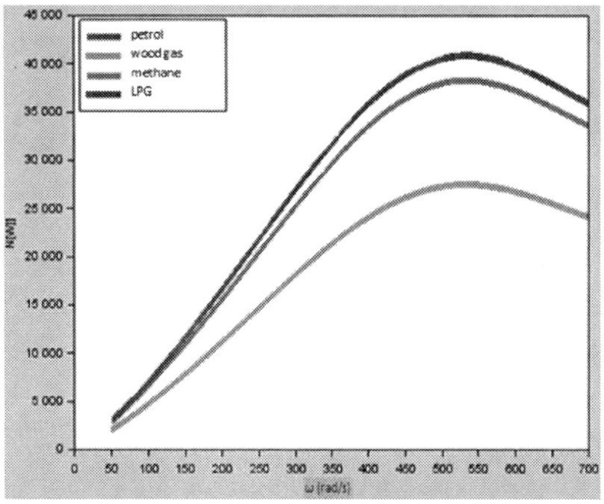

Fig. 7 Waveform of the maximum power in the function of the rotational speed for different fuels used in the simulation.

In the cases of supplying the engine with: petrol and LPG the curves of the power practically overlap. As far as methane powering is concerned, there is a slight drop of the value of the maximum power characteristic. However, a considerable decrease in the value of the maximum power characteristic was obtained in the case of powering the engine with wood gas.

4 Conclusions

The following conclusions may be drawn from the analysis of the prepared simulation model, obtained simulation results and professional literature:

- Scilab environment constitutes functional software enabling the simulation of the vehicle dynamics based on the implemented characteristics;
- The cumulated values of the engine demand for wood gas proved to be significantly higher than for other examined fuels. It results from the low calorific value of the carrier.
- The analysis of the maximum power characteristic of the engine supplied with wood gas revealed much lower values than in the case of other carriers.
- Results obtained in the simulation support data contained in the professional literature. Considering inconveniences related to the use of syngas, its low calorific value and drops in engine power, it has been established that the use of wood gas as fuel for motor vehicles is currently unjustified.

References

1. H. LaFontaine, P. Zimmerman, Złote Myśli, **2** (2008)
2. J. Taler, A. Mruk, J. Cisek, K. Majewski, Rynek Energii, **4** (2013)
3. http://www.lowtechmagazine.com/2010/01/wood-gas-cars.html
4. P. Gil, J. Wilk, RUTMech, t. XXXII z. 87 (2015)
5. I. Hetmańczyk, W. Hepner, AE, **1** (2012)
6. J. Zembowicz , WKŁ (2010)
7. T.J. Barlow, S. Latham, I.S. McCrae, & P.G. Boulter, TRL Pub. Proj. Rep. (2009)
8. Regulation No 83 of the Economic Commission for Europe of the United Nations (UNECE)
9. K. M. Romaniszyn, Combust. Engines, **43** (2004)
10. W. Serdecki, Poznan Univ. of Technol (2012)

DOI: 10.1051/e3sconf/20171901019

Possibility of direct electricity production from waste canola oil

Paweł P. Włodarczyk[1,*], *Barbara* Włodarczyk[1], and *Antonina* Kalinichenko[1]

[1]University of Opole, Department of Process Engineering, Dmowskiego Street 7-9, 45-365 Opole, Poland

Abstract. Powering high-efficiency devices, such as fuel cells, with waste products will allow for a broader development of renewable energy sources and utilisation of by- products. This publication presents the possibility of electrooxidation of the emulsion of waste rapeseed oil, prepared on the basis of the detergent Syntanol DS-10. The process of electrooxidation was carried out on platinum electrode in alkaline (KOH) and acidic (H_2SO_4) electrolyte, in the temperature range of 293-333 K. In each analysed case the process of electrooxidation took place. The maximum current density obtained was 7 mA·cm^{-2}. Thus, it has been shown that it is possible to generate electricity directly from the emulsion of the waste rapeseed oil.

1 Introduction

Today's energy production is based mainly on crude oil, coal, natural gas and nuclear energy. But in the last few years alternative energy sources have also been developing very fast. There is use of photovoltaic cells, solar collectors, heat pumps, wind turbines and more. One of these sources is also fuel cells (FC). The actual efficiency of FCs is up to 80-90% [1, 2]. Thus, the FCs are the most efficient power devices. Moreover, the main advantages of fuel cells are zero or low negative influence on the environment and silent operation [3, 4]. Most often the hydrogen is used as fuel for FCs [2-4, 5]. However, there are problems with cheap production and storage of this fuel, and that is the reason for the search of new fuels for FCs, including biofuels [3, 6-9]. Previous studies by the authors [10-15] have shown a possibility of electrooxidation (in various electrolytes) of petroleum derivatives and vegetable oils, e.g. canola oil, sun flower oil or grape seed oil. Canola seed oil is often used as additive for biodiesel fuel, especially as a blend with petroleum diesel fuel. Almost all manufacturers of diesel vehicles and engines have approved the use of B5 (a blend containing 5% of biodiesel), some approve the use of B20 blends (20% biodiesel) or higher, up to 100% biodiesel (B100) [16, 17]. Also, significant amounts of waste canola oils are produced everyday. Therefore, an attempt to use waste canola oil as fuel for fuel cells seems justified. In this case, electrooxidation of waste canola oil should be carried out in order to determine the basic ability of this substance to electrooxidate on platinum electrode. Theoretically, the current density can be calculated using the Butler-Volmer exponential function [18]

$$i = A \cdot i_0 \cdot \left(e^{\left(\frac{\alpha_a nF}{RT}(E-E_q)\right)} - e^{\left(\frac{\alpha_c nF}{RT}(E-E_q)\right)} \right) \quad (1)$$

* Corresponding author: pawel.wlodarczyk@uni.opole.pl

where i is electrode current density (A·m^{-2}); i_0 is exchange current density (A·m^{-2}); R is universal gas constant; E is electromotive force at the terminals of the real cell (V); E_{eq} is equilibrium potential (V); α_a is the so-called anodic charge transfer coefficient; α_c is the so-called cathodic charge transfer coefficient; A is electrode active surface area (m^2); n is number of electrons involved in the electrode reaction; F is faraday constant per volt gram equivalent (kJ);
or in a more compact form [18]

$$i = i_0 \cdot e^{(K \eta_{act})} \quad (2)$$

where η_{act} is activation overpotential (V); K is the factor of dependence of activation overpotential on reaction speed.
But, in real conditions the choice of fuel (or catalyst) is mainly carried out by experimental methods. The work covered the research on obtaining electricity from waste canola oil electrooxidation. This work covers the analysis of possibility of powering FCs with waste canola oil. The possibility of powering FCs with waste canola oil will allow for further development of renewable energy devices and will enable to eliminate or reduce the emissions of toxic substances.

2 Material and Methods

Tested canola oils came from five places (where it is used for frying). Oils were mixed to obtain a homogeneous sample. Next, the prepared oil was filtered in standard automotive fuel filter to eliminate solid substances and other pollution. In case of using waste canola oil as fuel for fuel cells, conducting the electric current is very important. Waste canola oil is a hydrophobic substance and does not conduct the electric current. In order to cause the conduction, an intermediate agent to solve waste canola oil in water was used. In order to cause the conduction, an intermediate agent to solve waste canola oil in water was used. Due to its excellent emulsification properties, Syntanol DS-10 was used as the detergent [19]. Syntanol DS-10 is characterized by high superficial activity, dispersion, solubilisation capabilities and emulgation [20, 21]. After the electrooxidation of the emulsion, Syntanol DS-10 can be degraded, e.g. promoted by energy transfer reactions or by bacteria [22, 23]. Investigated waste canola oil emulsion (WCOE) was obtained by mixing, in various ratios of water, waste canola oil and the detergent. WCOE was obtained with the use of a mechanical stirrer, with the speed of about 1200 rpm [10-15]. Concentration of canola oil and Syntanol DS-10 was equal to 0.0100%, 0.025% and 0.0500%. Emulsion stabilization time was about 6-8 hours (depending on the temperature). The Fig. 1 shows scheme of emulsion preparation.

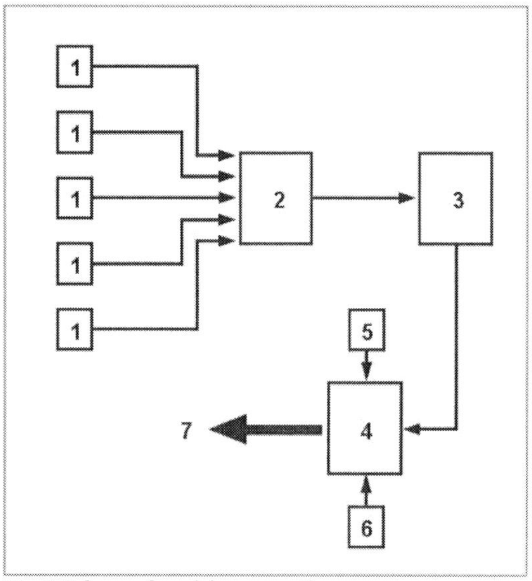

Fig. 1. Scheme of emulsion preparation
1 – points of acquisition of the waste canola oil samples, 2 – mixing of waste canola oil samples, filtration of mixed waste canola oil, 4 – WCOE preparation (by mixing), 5 – water, 6 – detergent (Syntanol DS-10), 7 – WCOE for measurements.

Measurements were done by the method of polarizing curves of electrooxidation of waste canola oil emulsion in glass vessel, on a smooth platinum electrode in aqueous solution of KOH and H_2SO_4 electrolytes. The electrode surface area was equal to 6.28 cm^2. Platinum was used as the catalyst of working electrode, due to its excellent catalytic properties [18], while saturated calomel electrode (SCE) was used as a reference electrode. The measurements were conducted at various temperatures 293, 303 and 333K. Electrochemical measurements were performed in a glass cell with potentiostat connected to a computer. Potentiostat AMEL System 5000 was used in the research. The Fig. 2 shows the scheme of glass vessel with electrodes.

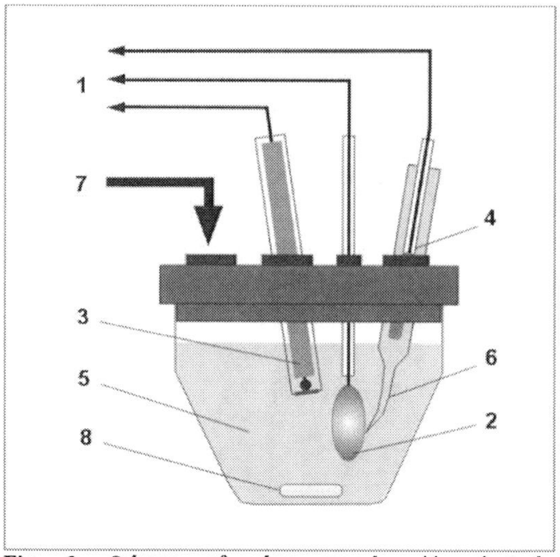

Fig. 2. Scheme of glass vessel with electrodes 1 – to potentiostat, 2 – working electrode (Pt), 3 – auxiliary electrode, 4 – reference electrode (SCE), 5 – electrolyte (KOH or H_2SO_4), 6 – Luggin capillary, 7 – WCOE, 8 – stirrer bar.

The Fig. 3 shows scheme of measurement position.

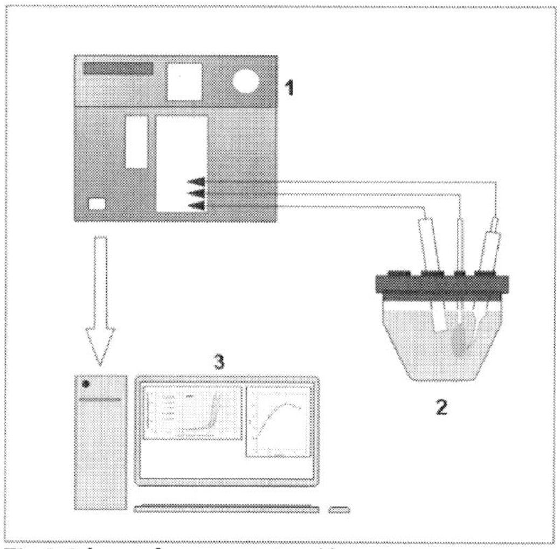

Fig. 3. Scheme of measurement position
1 – potentiostat, 2 – glass vessel, 3 – computer.

3 Results

First, the measurements were taken on electrooxidation of Syntanol DS-10 in electrolyte (aqueous solution of KOH and H_2SO_4), for various concentrations of the detergent at temperatures of 293-333K. Next, the measurements were taken on electrooxidation of WCOE in aqueous solution of KOH and H_2SO_4 electrolytes, for various concentrations of the detergent at temperatures of 293-333K. The comparison of these two processes (electrooxidation of Syntanol DS-10 and WCOE) allowed to determine whether the electricity is generated form the electrooxidation of the oil, or just from the detergent. Figs. 4-6 shows electrooxidation of WCOE in KOH at the temperature of 293-333K.

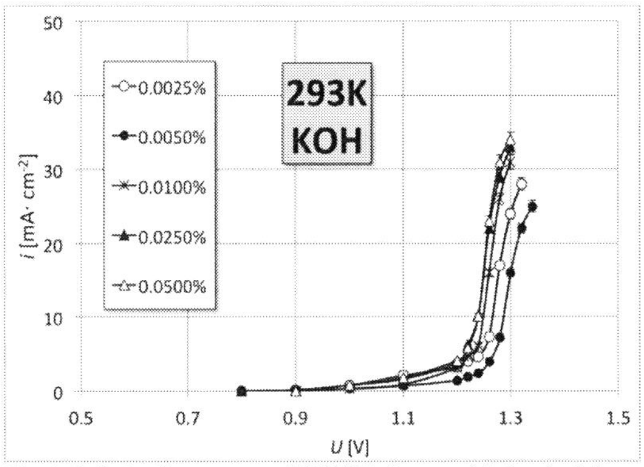

Fig. 4. Polarization curves of WCOE electrooxidation in 0,1n concentration of electrolyte KOH at 293K - concentration of canola oil was equal to 0.0100%, 0.025% and 0.0500%.

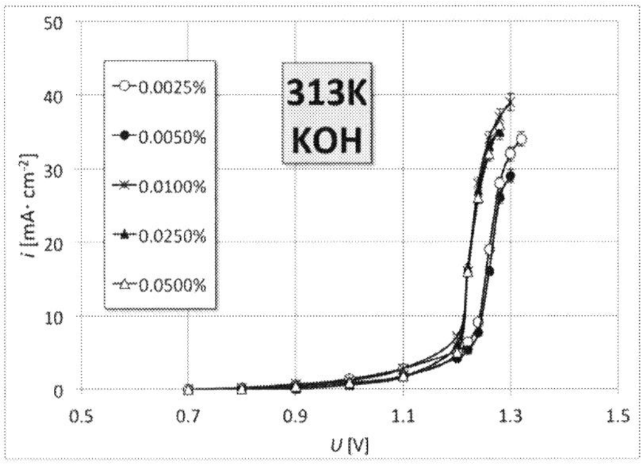

Fig. 5. Polarization curves of WCOE electrooxidation in 0,1n concentration of electrolyte KOH at 313K - concentration of canola oil was equal to 0.0100%, 0.025% and 0.0500%.

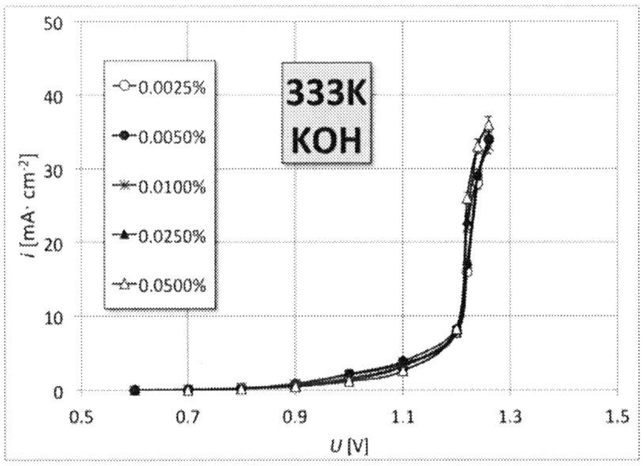

Fig. 6. Polarization curves of WCOE electrooxidation in 0,1n concentration of electrolyte KOH at 333K - concentration of canola oil was equal to 0.0100%, 0.025% and 0.0500%.

Figs. 7-9 show electrooxidation of WCOE in H_2SO_4 at the temperature 293-333K.

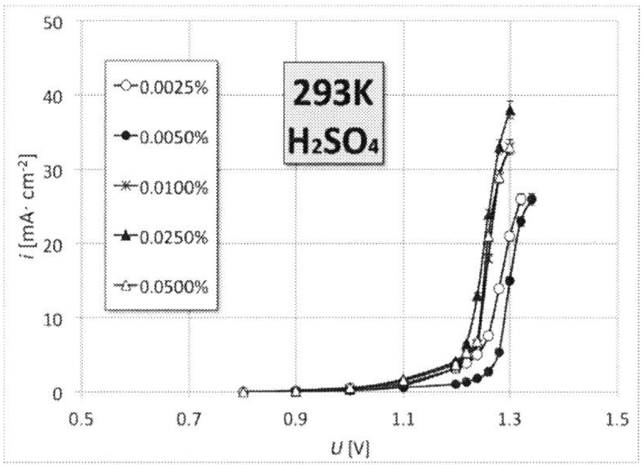

Fig. 7. Polarization curves of WCOE electrooxidation in 0,1n concentration of electrolyte H_2SO_4 at 293K - concentration of canola oil was equal to0.0100%, 0.025% and 0.0500%.

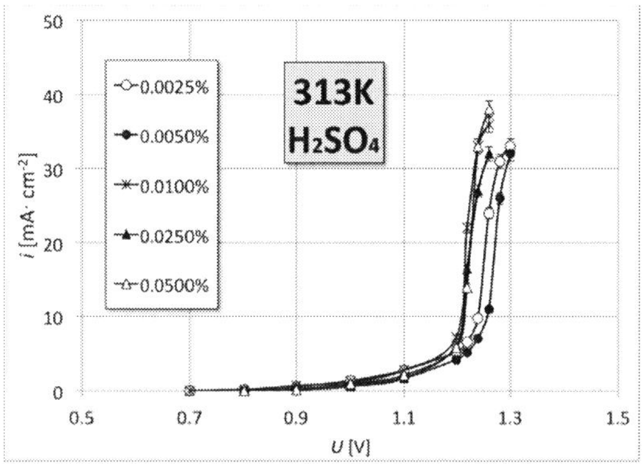

Fig. 8. Polarization curves of WCOE electrooxidation in 0,1n concentration of electrolyte H_2SO_4 at 313K - concentration of canola oil was equal to 0.0100%, 0.025% and 0.0500%.

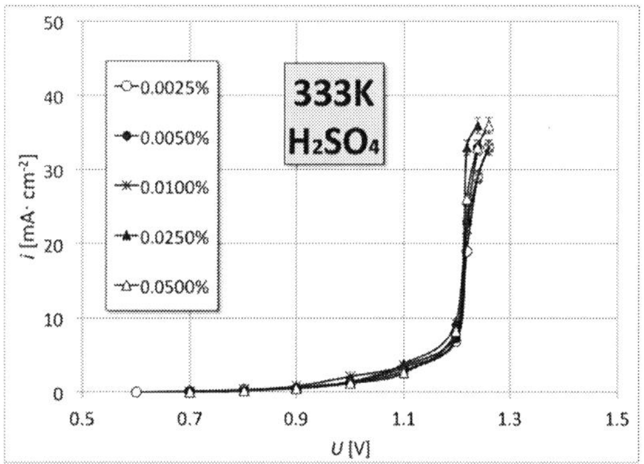

Fig. 9. Polarization curves of WCOE electrooxidation in 0,1n concentration of electrolyte H_2SO_4 at 333K - concentration of canola oil was equal to 0.0100%, 0.025% and 0.0500%.

4 Conclusions

The use of used oil to generate electricity in fuel cells would allow to develop the management of waste substances in a way that ensures low negative impact on

DOI: 10.1051/e3sconf/20171901019

the environment. Analysis of data from the measurements has shown the validity of the research undertaken in order to use the analyzed oil for energy production without the process of combustion. Potential of working electrode was establishing in about 22 minutes and was badly reproducible. Stationary, current-free real potential depends on canola oil concentration and is included in potentials range of 0.60-1.28V. These values are very close to the values obtained with pure canola oil [14]. To ascertain that the emulsion (and not the detergent) was electrooxidated, measurements of electrooxidation process run in the scope of kinetics, but the potential on the electrode is low and establishes in a long period of time. The highest results of current density (7 $mA \cdot cm^{-2}$) were obtained at the temperature of 333K for acid electrolyte. But the electrooxidation of WCOE occurred for all temperatures (293-333K), for all concentrations of waste canola oil, and for both electrolytes (KOH and H_2SO_4). Current density during the electrooxidation of WCOE in acid electrolyte (H_2SO_4) is slightly greater (0.1-0.9 $mA \cdot cm^{-2}$) than the current density during WCOE electrooxidation in alkaline electrolyte (KOH). At all measurement attempts, electrooxidation in the acid electrolyte occurred more efficiently than in the alkaline electrolyte – the measurements were more reproducible. With the temperature above 333K, the first electrooxidation of Syntanol DS-10 takes place, and only then – of canola oil emulsion. For this reason, to ensure the occurrence of the process of electrooxidation in fuel cells (with KOH or H_2SO_4 electrolyte) powered with emulsion of waste rapeseed oil, the temperature of 333K should not be exceeded.

It has been demonstrated that using the detergent Syntanol DS-10 to prepare the emulsion of waste canola oil allows for the electrooxidation of WCOE in aqueous solution of KOH and H_2SO_4, and thus enables direct conversion of waste canola oil into electrical energy (3-7 $mA \cdot cm^{-2}$). Therefore, a fundamental possibility of electrooxidation (electricity production) of waste canola oil on electrode (with platinum catalyst) in alkaline (aqueous solution of KOH) and acid (aqueous solution of H_2SO_4) electrolyte was shown in this paper. The next stage of the research should be the analysis of the composition of the emulsion after electrooxidation, as well as the selection of a suitable catalyst.

References

1. R. O'Hayre, S-W. Cha, W. Colella, F. B. Prinz, *Fuel Cell Fundamentals* (John Wiley & Sons, Hoboken, 2005)
2. D. Stolten, *Hydrogen and fuel cells. Fundamentals, Technologies and Applications*, (Wiley-VCH, Weinheim, 2010)
3. J. Larminie, A. Dicks, *Fuel cell system explained* (John Wiley & Sons, Hoboken, 2003)
4. G. Hoogers, *Fuel Cell Technology Handbook* (CRC Press, Boca Raton, 2004)
5. J. Rifkin, The Hydrogen Economy (Jeremy P. Tarcher/Penguin, New York 2003)
6. D. K. Ross, Vacuum **80** (10), 1084-1089 (2006)
7. A. Hamnett, Catalysis Today **38** (4), 445–457 (1997)
8. J. Milewski, J. Lewandowski, J. Power Technol. **93** (5), 347-353 (2013)
9. J. Milewski, K. Michalska, A. Kacprzak, J. Power Technol. **93** (3), 161-168 (2013)
10. P. P. Włodarczyk, B. Włodarczyk, J Power Technol. **93** (5), 394-396 (2013)
11. P. P. Włodarczyk, B. Włodarczyk, Infrast. Ecol. Rural Areas **4** (1), 1071-1080 (2016)
12. P. P. Włodarczyk, B. Włodarczyk, Diagnosing of the Environment Condition, Research Methods - Forecasts, Works of Ecology Commission and Environmental Protection, Scientific Society of Bydgoszcz, **10**, 205-216 (2016)
13. P. P. Włodarczyk, B. Włodarczyk, Ecol. Eng. **18** (1), 65-70 (2017)
14. P. P. Włodarczyk, B. Włodarczyk, J. Power Technol. **96** (6), 459-462 (2016)
15. P. P. Włodarczyk, B. Włodarczyk, New Trends in Management and Production Engineering - Regional, Cross-border and Global Perspectives, 188-198 (Shaker Verlag, Aachen, 2016)
16. J. Sheehan, V. Camobreco, J. Duffield, M. Graboski, H. Shapouri, *An overview of biodiesel and petroleum diesel life cycles*, National Renewable Energy Laboratory, Prepared for U.S. Department of Energy's Office of Fuels Development and U.S. Department of Agriculture's Office of Energy (1998)
17. J. Van Gerpen, Fuel Process. Technol. **86** (10), 1097–1107 (2005)
18. J.O'M. Bockris, A.K.N. Reddy, *Modern electrochemistry* (Kulwer Academic/Plenum Publishers, New York, 2000)
19. Iu. I. Sakharov, E. G. Rastiannikov, G. M. Verbitskaia, L. N. Tarasova, Vopr Pitan. **4**, 75-7 (1975)
20. A. Survila, Z. Mockus, S. Kanapeckaitė, M. Samulevičienė, Electrochimica Acta **50** (14), 2879-2885 (2005)
21. O. Paraska, S. Karvan, Technical Transactions. Mechanics, Cracow University of Technology Press, **8** (107), 203-210 (2010)
22. A. V. Kravchenko, A. G. Rudnitskii, A. F. Nesterenko, V. S. Kublanovskii, Ukrainian Chemistry Journal C/C of Ukrainskii Khimicheskii Zhurnal, **60** (11), 11-13 (1994)
23. O. V. Ignatov, Iu. V. Shalunova, L. V. Panchenko, O. V. Turkovskaia, N. M. Ptichkina, Prikl Biokhim Mikrobiol. **31** (2), 220-223 (1995)

Small photovoltaic setup for the air conditioning system

Maciej Masiukiewicz[1,*]

[1]Opole University of Technology, Department of Mechanical Engineering, 5 Mikołajczyka st., 45-758 Opole, Poland

Abstract. The increasing interest in air conditioning systems for residential applications in Poland will certainly increase the demand for electricity during the summer period. Due to this fact a growing interest in solutions that help to lower the electricity consumption in this sector is observed. The problem of increased energy demand for air conditioning purposes can be solved by transfer the consumption of electricity from the grid system to renewable energy sources (RES). The greatest demand for cooling occurs during the biggest sunlight. This is the basis for the analysis of technical power system based on photovoltaic cells (PV) to power the split type air conditioner. The object of the study was the commercial residential air-conditioning inverter units with a capacity of 2.5kW. A network electricity production system for their own use with the possibility of buffering energy in batteries (OFF-GRID system). Currently, on the Polish market, there are no developed complete solutions dedicated to air conditioning systems based on PV. In Poland, solar energy is mainly used for heat production in solar collectors. The proposed solution will help to increase the popularity of PV systems in the Polish market as an alternative to other RES. The basic conclusion is that the amount of PV energy generated was sufficient to cover the daily energy requirement of the air conditioner.

1 Introduction

The IPCC's Fifth Assessment Report-AR5 confirms unequivocally global warming and provides evidence of its substantial and wide-ranging consequences. The increase in average temperature and the prolonged periods of high temperatures are responsible for permafrost melting, heavy precipitations, floods, droughts, wildfires are some of the most significant effects caused by climate change that are likely to increase over time and can deeply affect both environment and society [1].

The study of urban energy consumption is becoming more and more important because of three main facts:

First – urban population is increasing: in 1950, 30% of the world population lived in cities, and 54% in 2014, and this ratio will reach 66% in 2050, that being around 6.5 billion of persons, i.e. 2.6 billion persons more than nowadays. Therefore, urban development is a crucial issue, in particular from an energy point of view as urban energy consumption per capita is also increasing (+32% in the last 40 years).

Second – the energy paradigm changes: the need of dramatically reducing greenhouse gas emissions as well as fossil energy issues favor the use of renewable energies, which are often decentralized and intermittent. Related polices, currently ongoing in many countries worldwide, change the previous centralized energy management scheme, which requires a better understanding and forecasting of power demand and power production, in particular in cities, where the network in dense.

Third – urban heat stress during hot seasons due to the urban heat island (UHI) effect may further intensify effects of probable more frequent heat waves in the context of climate change. This can lead to dramatic public health problems as well as energy issues due to the multiplication of active cooling devices, which would also contribute to increase urban air temperatures.

Therefore, urban energy consumption have been a critical research problem for the last 30 years (referenced 219 papers concerning only urban energy models [2]), and will certainly still remains a major issue for the following years [3].

Summer air conditioning represents a growing market in buildings worldwide, with a particularly significant growth rate observed in European commercial and residential buildings [4]. Increasing interest in residential air conditioning systems in Poland results in a simultaneous increase in demand for electricity in the summer [5]. Due to this fact, there is a growing interest in solutions that reduce the energy consumption in this sector. This problem can be solved, among others by transferring the power consumption of air conditioning systems from system sources to renewable energy sources (RES) [6].

There are no comprehensive solutions, currently available in the Polish market, dedicated to air conditioning systems based on photovoltaic cells (PV). In Poland, solar energy is mainly used for the production

* Corresponding author: m.masiukiewicz@po.opole.pl

DOI: 10.1051/e3sconf/20171901020

of heat in the solar collectors.

The use of a PV system to power air conditioning seems to be an intuitive solution. Air conditioning works most often in the summer when the sun is abundant, and photovoltaic panels provide the most electricity.

2 Problem definition

The most important problem in the design of autonomous PV systems is the selection of the power source. PV generators and battery module packs adjusted to the level of electrical power demand [7].

Calculating the required volume of PV modules to power a split air conditioner requires the daily electricity consumption estimating. The conducted analysis concerns only the summer period, where there is a need for cooling. Due to the use in the inverter air conditioner rotary compressor, consumption is variable and depends on the heat load of the air-conditioned room. The heat load depends on the number of the operating and type of the devices, the number of residential people, and the outside temperature whether condition.

The most common device in Poland is a system with a cooling capacity of 2.5kW, which roughly translates into approx. 0.85kW of electrical power demand. Due to the high cost of the PV system, we are looking for the optimum area for the modules and the size of the necessary battery capacity that will provide the power for a typical split air conditioner depending on: local sunlight, PV conversion efficiency, outdoor air temperature. These affects the amount of heat load of the residence,

When analyzing, consider periods in which the PV cells do not work (night, overcast) but there is a need for cooling. Therefore there is a need for the optimum capacity of energy battery bank adjustment to provide power during solar shortage periods to ensure uninterrupted operation of the system.

3 Measurement setup and method

The study was conducted on an autonomous system that is preferred from the seasonal power supply point of view and integrates the PV SMART-GRID system into the air conditioning system [8].

The Toshiba AvAnt residential SPLIT air conditioner inverter (RAS-107SAV-E6 outdoor unit, RAS-107SKV-E6 indoor unit) has the following characteristics: Cooling capacity – 2.5kW; Cooling range (min - max) – 1.2 ÷ 3.0kW; Power consumption (min./nominal/max.) – 0.29/0.77/1.05kW; EER – 3.25; SEER – 5.1; Energy efficiency class – A [9].

Voltcraft Energy Logger 4000 Energy Meter was used to collect instantaneous power consumption data from the air conditioner unit [10].

Analysis in real condition operation of the PV system was carried out on a research facility located on the building of the Faculty of Mechanical Engineering of the Opole University of Technology (OUT). The obtained data characterize the operation of PV cells for the Opole City region in Poland. PV installation consists of the following components: 1 – Six ETP660240WW PV modules with a power rating of 240W_p.; 2 – Sunny Island 2224 Inverter with 2200W AC power; 3 – Sunny Boy 1300TL-10 Network Inverter with 1400W DC; 4 – Sunny Home Manager; 5 – Sunny Remote Control; 5 – Battery protection Batfuse B.0100; 6 – Four 240Ah Victron AGM 6V batteries connected to give total voltage of 24V; 7 – 500W VAWT charge controller; 8 – Sunny WebBox.

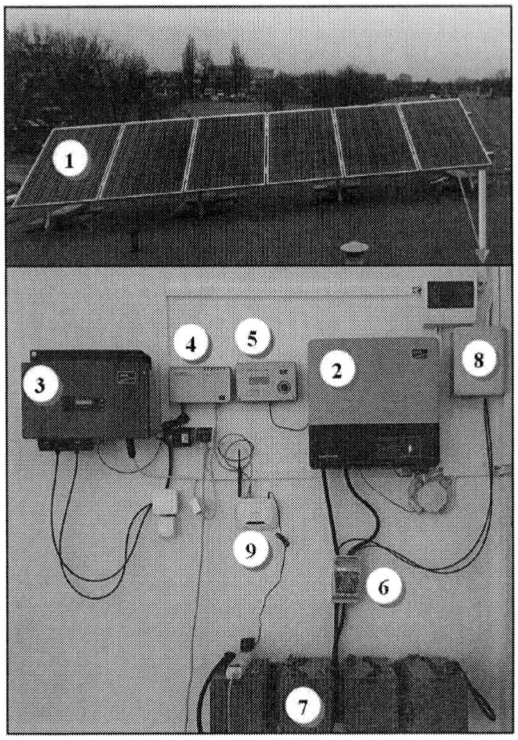

Fig. 1. PV measurement setup overview in the Department of Mechanical Engineering (Opole University of Technology).

The basic component of the PV installation are 6 polycrystalline photovoltaic modules (ET Solar ETP660240WW) serial connected with a total peak power of 1.44kWp, a single module efficiency of 14.75% and a total cell surface area of 8.761m^2 are shown in Fig. 1.

Installation archives min. electricity generation on a given day. The example of the electricity generated for the day of 2 June 2017 is shown in Fig. 2.

To present the operation of the air conditioning system supplied by the PV system, data were presented: global irradiance total solar radiation intensity for City of Opole, power output from the PV system, power consumption of the air conditioner, and outdoor air temperature.

The meteorological station of the Faculty of Mechanical Engineering of the OUT, whose geographical location is N 50°41.027' E 17°56.620', was used to record the intensity of solar radiation and outside air temperature. Data is averaged and saved with an interval of 900s.

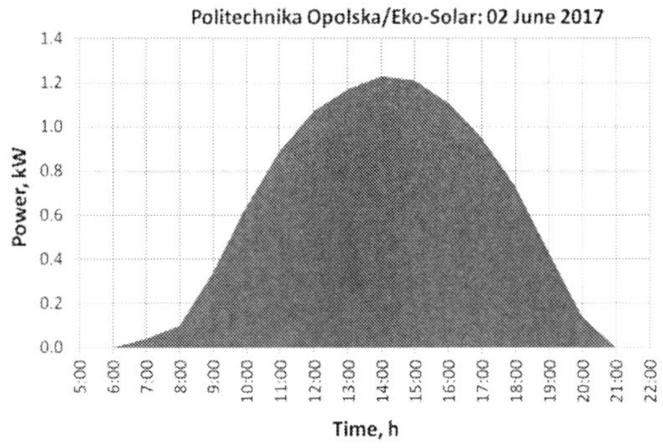

Fig. 2. Electricity generated for the day of 2 June 2017 [11].

Collection of data from PV inverters and their transmission to the server is provided by Sunny Webbox. The parameters of the PV system are constantly being saved since it is started, and access: to them is via a web page (Tab. 1.).

Data is averaged and recorded at an interval of 3600s.

Table 1. Overview of the daily report obtained from the Sunny Portal system for day of June 2, 2017 [11].

Daily Output [kWh] 6/2/2017	Monthly Output [kWh] June 2017	Yearly Output [kWh] 2017
10.15	221.97	595.38

The information about the instantaneous power consumption of the air conditioner was obtained using the Voltcraft Energy Logger 4000 energy meter. The data is written to the internal memory at interval 60s from where it can be transferred to the SD memory card.

A summary of all collected data allows to specify: First – the solar radiation with PV power output – PV system real capacity. The photovoltaic conversion efficiency was calculated from equation (1):

$$\eta = \frac{P}{E \cdot A} \cdot 100\% \qquad (1)$$

where: A – cell surface PV, m^2; E – global radiation, W/m^2; P – power, W

Second – generated power of PV installation with instantaneous power requirement by air conditioner of the actual coverage of air conditioner needs by PV system. And third – momentary power demand by air conditioner with outside temperature with the influence of outside air temperature on the heat load.

An exemplary graph of the characteristics described in the example on the day of June 6, 2017 for the city of Opole is shown in Fig. 3.

The sunshine data for the city of Opole from 2016 and 2017 were also collected to determine the availability of solar energy for PV panels in the analyzed period (Tab. 2). This data is helpful for estimating the size of batteries to compensate temporary power shortages in the PV plant.

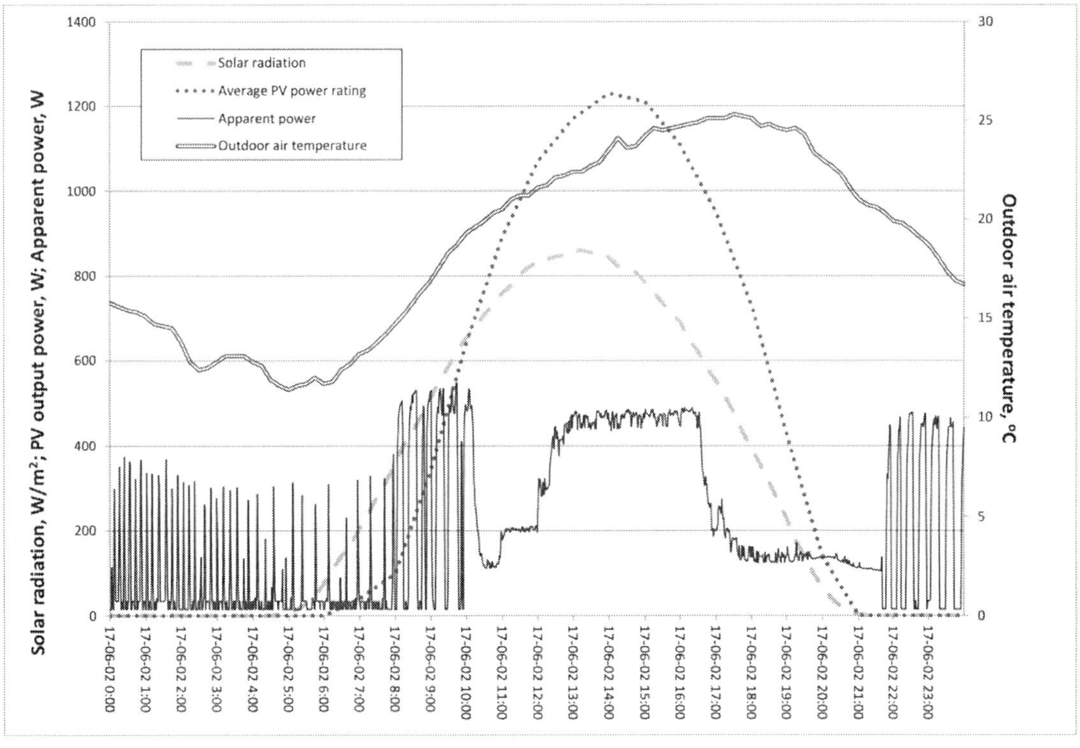

Fig. 3. Set of recorded parameters of the total measurement system for one day.

Table 2. Average sunshine hours in the city of Opole [12].

		2016		2017	
		May	June	May	June
Accumulated Sunshine Hours	h	248.3	273.3	212.8	291.3

4 Results and disussion

The study was held for 13 days, from 25.06.2017 to 06.06.2017r. On the June 7th the fault was covered by the air conditioner and switched off the cooling function. During that period, there was a demand for active power in the range of 6.7 ÷ 741.4W; Apparent power in the range of 13.7 ÷ 938.5W. The air conditioner in this period consumed 65.7kWh of energy and the daily energy demand varied between 0.74 and 11.905kWh.

During the same period, the PV installation achieved average photovoltaic conversion efficiency at 7.52%, peak at 19.16%, and generated 102.6kWh of electric energy. The highest achieved power was 1.27kWp (nominal electrical power of the system is 1.44kWp). The electrical power and efficiency of the PV system from the background of the solar radiation view is shown in Fig. 4. In the analyzed period, the outdoor air temperature variations ranged from 10.5°C to 32.2°C. All of the described characteristics are shown in Fig. 5.

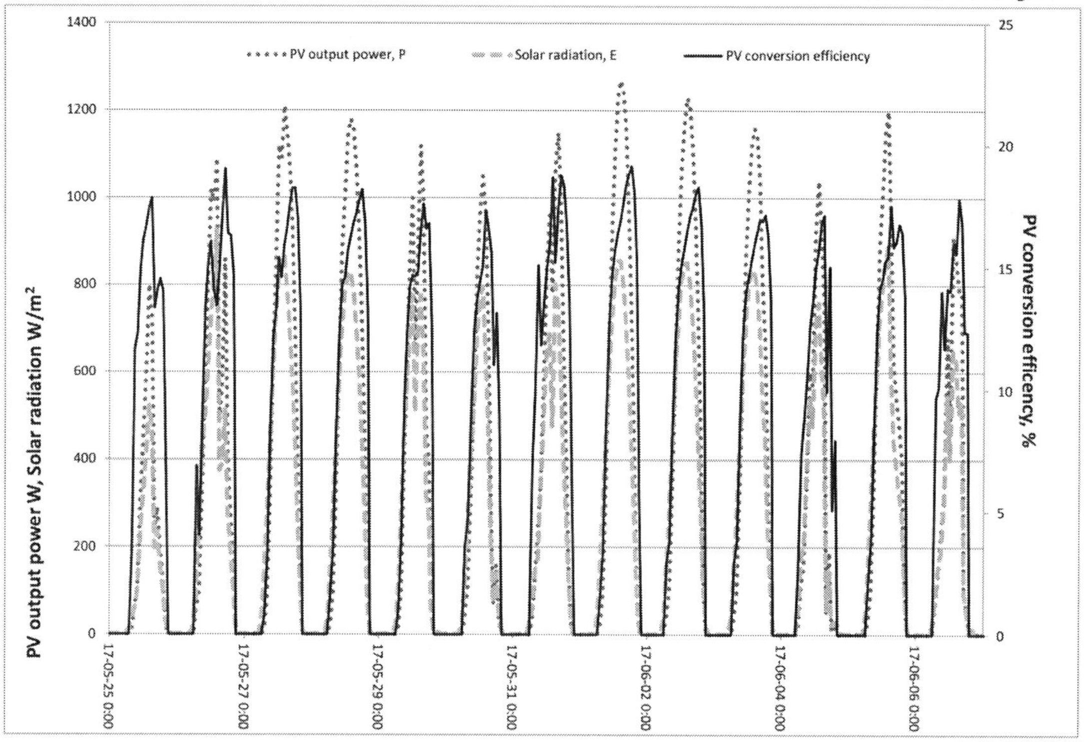

Fig. 4. Achieved electrical power and efficiency of the PV system for the analyzed period with the referenced solar irradiation.

The main task of the battery bank is the compensation of the electricity shortage supplied from the main PV source. Power shortage as well as power bank capacity depends on the day time, the radiation intensity variation and the need of energy reservation level. It is recommended to reserve the energy for 2-3 days in summer and 3-5 days in winter. In addition, in order to extend the life of the batteries, it is recommended that 50% to 100% of the power supply be used for their selection [13]. Based on the data collected, the average sunshine hours (Table 2.) and the needed for the air conditioner's undisturbed operation, the battery bank capacity was estimated. The energy reserve factor is shown in Table 3. The *F* factor was calculated by assuming the summer conditions of a minimum 2 days reserve, extending it by the number of hours without sunshine [12].

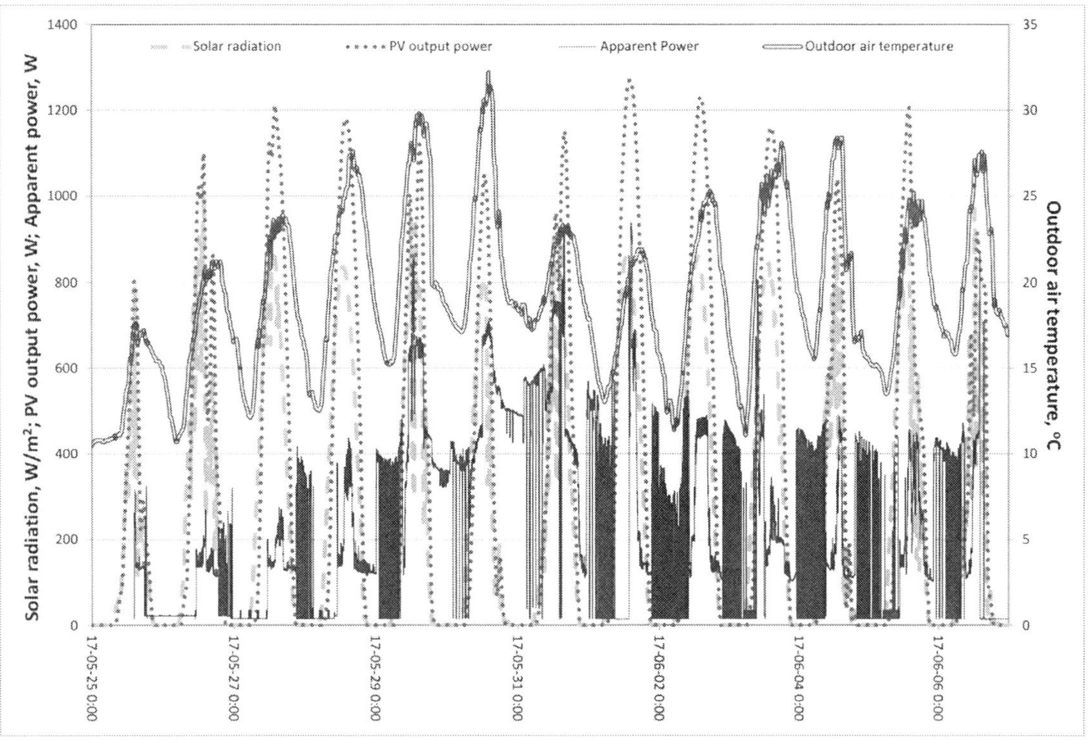

Fig. 5. The recorded parameters of the total measurement system for the analyzed period.

Table 3. Estimated *F* factor values associated with the energy reserve.

Year	2016	2017
May	3	2
June	2.7	2.5
Average	2.55	

Battery capacity was calculated from equation (2):

$$C = \frac{2 \cdot W \cdot F}{U}, Ah \quad (2)$$

where: F – factor related to energy reserve, -; W – daily energy demand, Wh; U – battery voltage, V

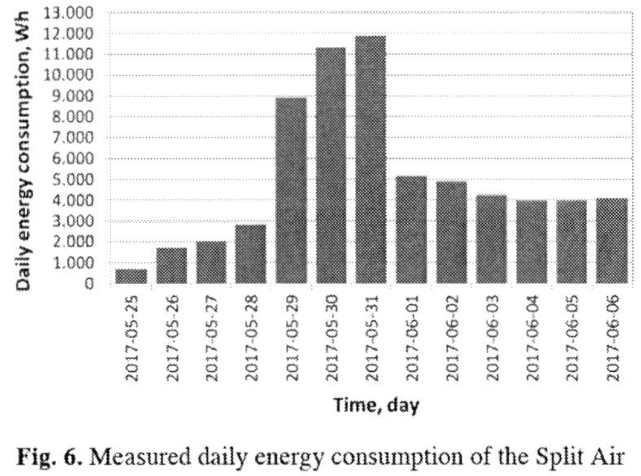

Fig. 6. Measured daily energy consumption of the Split Air Conditioning System.

The highest measured daily energy consumption was 11.9kWh is shown in Fig. 6. The required battery capacity (at 24V) was calculated according equation (2) at 2529.8Ah.

5 Conclusions

The power consumption of an air conditioner analysis, shows four characteristic modes of operation – standby mode, disturbed standby mode, work mode and disturbed work mode. During standby mode, the device exhibited a constant power requirement of about 14.0 ÷ 21.5W. The disturbed standby mode time is characterized by short time power demands of over 50W. During proper operation in the work mode, the device has a power requirement of over 50W, and the power fluctuation is fairly smooth. One of the basic conclusions, which is a promising expectation, from a commercial point of view is that the highest peaks in power demand were at maximum PV output. On the warmest days, there were periods of disturbed work mode characterized by short power drops to the standby level.

On a daily basis, there is no homogeneous configuration of four operation modes. However, there are some analogies between the following days. After a period of work mode, per day, there is a period of disturbance in the nighttime. In the warmest days (e.g. 29.05.2017-01.06.2017) instead of disturbed standby mode, work is disturbed. These effects can be caused by the building's temperature inertia, the impact of adjacent rooms, the influx of ventilation air and the household equipment operation.

Next basic conclusion is that the amount of PV energy generated was sufficient to cover the daily energy requirement of the air conditioner. However energy shortage occurred during the nighttime when the device continued to operate in the disturbed standby mode and disturbed work mode. This means that during the night, either the use batteries is necessary or give up air conditioning at all. But then there will be need to invest more energy in removing the accumulated heat the next day. However, this case will result in worse thermal comfort in the morning.

References

1. F.F. Pietrapertosa, Renew. Sust. Energ. Rev. (2017), (to be published)
2. J. Keirstead, M. Jennings, A. Sivakumar, Renew. Sust. Energ. Rev. **16**, 6 (2012).
3. L. Frayssinet, Renew. Sust. Energ. Rev. (2017), (to be published)
4. C. A. Balaras, G. Grossman, H. M. Henning, C. A. I. Ferreira, E. Podesser, L. Wang, E. Wiemken, Renew. Sust. Energ. Rev. **11**, 2 (2007).
5. Air Conditioning and Ventilation Market in Poland and Silesian Voivodeship (2013) (in Polish) http://pnt.euro-centrum.com.pl/files/post/830/Rynek-Klimatyzacji-i-Wentylacji-w-Polsce-i-wojewodztwie-slaskim.pdf [access: 12.07.2017]
6. D. Kwiecień, *Forum Wentylacja, Salon Klimatyzacja* (Warsaw, 2009) (in Polish)
7. E. Klugmann-Radziemska, *Renewable energy sources, computational examples* (WPG, Gdańsk 2013) (in Polish)
8. S. Anweiler, W. Fedak, R. Ulbrich, B. Jarosz, Instal **7-8** (2016)
9. TOSHIBA catalogues: http://www.toshiba-hvac.pl/pliki/toshiba_katalog_2015.pdf [access: 12.07.2017]
10. Voltcraft catalogues: http://www.voltcraft.pl/ (access: 15.07.2017)
11. Sunny Portal www.sunnyportal.com (access: 17.07.2017)
12. Wether reports (in Polish) http://www.weatheronline.pl (access: 14.07.2017)
13. D. Żabicki, *Batteries for photovoltaic installations* (in Polish) http://www.fachowyelektryk.pl/technologie/odnawialne-zrodla-energii/1648-akumulatory-do-instalacji-fotowoltaicznych.html

 DOI: 10.1051/e3sconf/20171901021

Heat demand mapping and district heating grid expansion analysis: Case study of Velika Gorica

Hrvoje Dorotić [1,*], *Tomislav* Novosel[1], *Neven* Duić[1], and *Tomislav* Pukšec[1]

[1]Faculty of Mechanical Engineering and Naval Architecture, Department of Energy, Power Engineering and Environment, University of Zagreb, Zagreb, Croatia

Abstract. Highly efficient cogeneration and district heating systems have a significant potential for primary energy savings and the reduction of greenhouse gas emissions through the utilization of a waste heat and renewable energy sources. These potentials are still highly underutilized in most European countries. They also play a key role in the planning of future energy systems due to their positive impact on the increase of integration of intermittent renewable energy sources, for example wind and solar in a combination with power to heat technologies. In order to ensure optimal levels of district heating penetration into an energy system, a comprehensive analysis is necessary to determine the actual demands and the potential energy supply. Economical analysis of the grid expansion by using the GIS based mapping methods hasn't been demonstrated so far. This paper presents a heat demand mapping methodology and the use of its output for the district heating network expansion analysis. The result are showing that more than 59% of the heat demand could be covered by the district heating in the city of Velika Gorica, which is two times more than the present share. The most important reason of the district heating's unfulfilled potential is already existing natural gas infrastructure.

1 Introduction

European Union (EU) has recognized energy efficiency, renewable energy sources and carbon dioxide (CO_2) reduction as the 3 main pillars of its energy and climate strategy. Specific targets have been appointed for the period until 2020: Europe will reduce its CO_2 emissions by 20%, increase the share of RES (renewable energy sources) by 20% and increase the overall energy efficiency by 20% until 2020 compared to 1990 levels [1]. According to the latest reports, EU will achieve its goals related to the CO_2 emission reduction and renewable energy sources share but additional efforts should be undertaken to achieve the planned increase in the energy efficiency [2]. This matter will have a crucial role in the upcoming strategy for the period 2020-2030, where additional increase of energy efficiency is planned, up to 27%. The energy sector is a key component in achieving said goals since it is responsible for 79% of the EU's greenhouse gas emissions [3].
Combined heat and power (CHP) units in a combination with the district heating and cooling (DHC) systems can maximize the exploitation of the waste heat which is currently inefficiently released in the atmosphere [4, 5]. Also, these systems have the ability of utilizing locally distributed renewable energy sources such as biomass, solar and geothermal energy which will additionally increase the CO_2 reduction on a local level. District heating and cooling systems will have an important role in the future energy systems with a high share of an intermittent electricity production coming from renewable energy sources such as wind turbines and solar panels. In order to balance the grid, the excess of the electrical energy could be efficiently converted and stored as a thermal, by using storage and power-to-heat technologies, which include electrical heaters and heat pumps.
In the literature, the term "Fourth generation district heating" (4DH) is reserved for DHC systems which can tackle the problems of the future energy systems. Although 4DH implies increase of the overall thermal grid efficiency by implementing lower supply temperatures, they will also have to integrate power, thermal, gas and transport grids and at the same time utilize waste heat and renewable energy sources [5]. It was already shown that Denmark could integrate up to 50% of DH with the combination of large-scale heat pumps and heat storage thus achieving great impacts on the overall energy system [6]. The analysis in this work was mostly based around the cost of the system as a whole. Similar analyses have been conducted for other countries such as China [7], Sweden [8], UK [9], Italy [10] and France [11], among others. Project Heat Roadmap Europe proved that integration of heat savings, DH in a combination with heat pumps will result in the cheapest low-carbon heat sector in Europe [12]. Analysis of the DH and RES integration has also been carried out on a city level [13], including the comparative analysis of advanced and underdeveloped systems [14]. In addition to this, hourly optimization has been proved to

* Corresponding author: hrvoje.dorotic@fsb.hr

be an important approach in the development of DH systems [15]. Other research has also demonstrated the positive impact of energy efficiency measures related to DH such as the reduction of the supply temperature [16], utilization of smart grids [17], renewables [18] and hybrid systems [19] and socio-economic impact for final consumer [20] for example.

GIS (Geographical Information System) mapping has already been proven as a valuable tool for the analysis of energy systems. In the STRATEGO project, the heat demand mapping methodology has been developed [22]. Apart from the energy demand, GIS based methods have also been used for the mapping of potential energy sources [23].

This work will demonstrate heat demand mapping method and present the results of the analysis of the economic potential for the DH grid expansion. As the case study for this work, the city of Velika Gorica in Croatia has been selected.

2 Method

The work presented in this paper can be divided in the two main steps: heat demand mapping and expansion potential analysis.

2.1 Heat demand mapping

Heat demand consists only of the space heating demand. In order to carry out the head demand mapping process, publicly available data was used. In this way, the developed method could be used for mapping of the heat demand in other cities and regions. The mapping process is handled in 3 steps:

- mapping of the locations and surface areas of buildings,
- identification of the building heights,
- identification of the types of buildings.

Each step contains one layer of information. Combining building's surface area, height (number of floors) and type, its overall heating demand can be acquired. The obtained results can be further complimented with the addition of several layers of information for example population density, energy certificates and so on.

2.2 DH expansion analysis

The analysis of the potential for the expansion of the current DH system is based on an economic assessment to check if certain areas of a city can justify the installation of the new thermal grid. In order to make this assessment, five aspects are taken into the account:

- does a DH grid already exist in the area,
- the total heat demand in the area,
- levelized cost of heat (LCOH),
- potential revenue from the supply of heat,
- cost of a new grid in the area.

According to the available data, proposed analysis can be carried out on a building level or in a grid form, as shown below. In order to calculate levelized cost of heat, method described in [21] was used. Actual heat and capacity prices were used to estimate the potential revenue.

To calculate the total cost of the DH grid expansion, two sets of data are required: the length and discounted cost of pipes. This data can be obtained by calculating the average net surface area of the buildings relative to the ground surface areas as well as the distribution length relative to the net surface area of all connected users. The calculation is shown in Equation (1).

$$L_{ng}=(L_{nu}/A_{cu})/(A_u/A_g) \tag{1}$$

Where L_{ng} represents the necessary distribution length for a given surface, L_{nu} represents the length of the existing distribution for the already connected users, A_{cu} represents the net surface area of the connected users, A_u represents the surface area of all potential users in the analysed area and A_g represents the ground surface area.

The grid expansion is considered economically feasible if the total income generated by new consumers is larger than the sum of the discounted investment in the new grid and the total cost of the thermal energy production.

Also, it is important to notice that the investment cost of the substations hasn't been taken into the account.

3 Case study

Because of the general lack of data related to the building census in Croatia, part of the process has been carried out manually. Building's location and footprint have been acquired from OpenStreetMaps (OSM) database [24] as well as the online building census GEOPORTAL [25]. Total number of floors have been collected from the city itself by using energy audits and manual inspection. This approach was also used in defining building category. In total, 7 specific heat demand categories have been used, similarly to the Sustainable Energy Action Plan (SEAP) of the city of Velika Gorica [26]:

- Households – 193 kWh/m^2 annually,
- Commercial buildings – 180 kWh/m^2 annually,
- City companies – 275 kWh/m^2 annually,
- Administrative buildings – 285 kWh/m^2 annually,
- Child care and education – 232 kWh/m^2 annually,
- Cultural buildings – 123 kWh/m^2 annually,
- Other public buildings – 128 kWh/m^2 annually.

Annual heat demand presented in SEAP was obtained by using the heat consumption data from the final users. Specific heat demands, shown above, don't include the the thermal energy needed for DHW preparation. The DH system Velika Gorica currently doesn't provide this service.

All of the mentioned data has been added and processed in qGIS [27], a free to use and open access GIS tool.

4 Results

Fig. 1 shows the heat demand map of the city with a resolution of 100 by 100 meters to demonstrate heat demand densities. This resolution has been used in all subsequent studies.

Fig. 2 shows the existing natural gas and DH distribution (existing and officially planned) within the city. As it can be seen, the gas infrastructure is spread around most of the city while DH is concentrated generally in the city centre. Normally, areas already covered by natural gas would not present suitable locations for the expansion of DH due to the competition. In this case however, gas infrastructure has been ignored in order to demonstrate the method since no expansion would be plausible if it wasn't.

Fig. 3 shows the initial results of the analysis where all areas suitable for the exploitation of DH are shown. The analysis takes into the account the potential revenue, LCOH and cost of the DH grid expansion into consideration. In cases where DH grids already exist, the results were overridden and set as suitable regardless of the costs. Since some areas identified as suitable are dislocated from the identified centres, they were excluded. The final area deemed suitable is presented in Fig. 4.

The final results show that a total of 117,182 GWh annually could be supplied by DH. This is a significant share of the total heating demand identified in the cities SEAP which equals 197,342 GWh. In the year 2008, district heating network delivered 56,242 GWh of thermal energy to the final customers, which was equal to 28% of the total heating demand [26]. This shows a great potential for a future district heating grid expansion. Thermal energy delivered by DH could be doubled, but besides a larger network, this will also require additional upgrades of the heat production facilities.

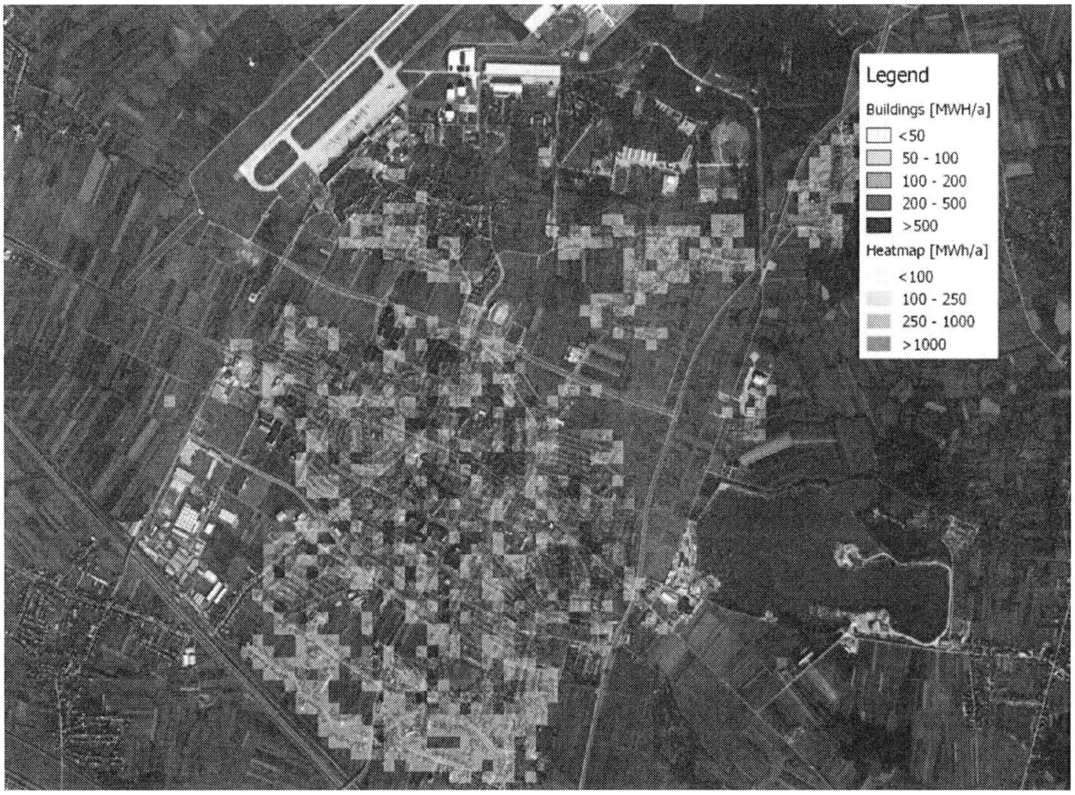

Fig. 1. Heat demand map of Velika Gorica.

DOI: 10.1051/e3sconf/20171901021

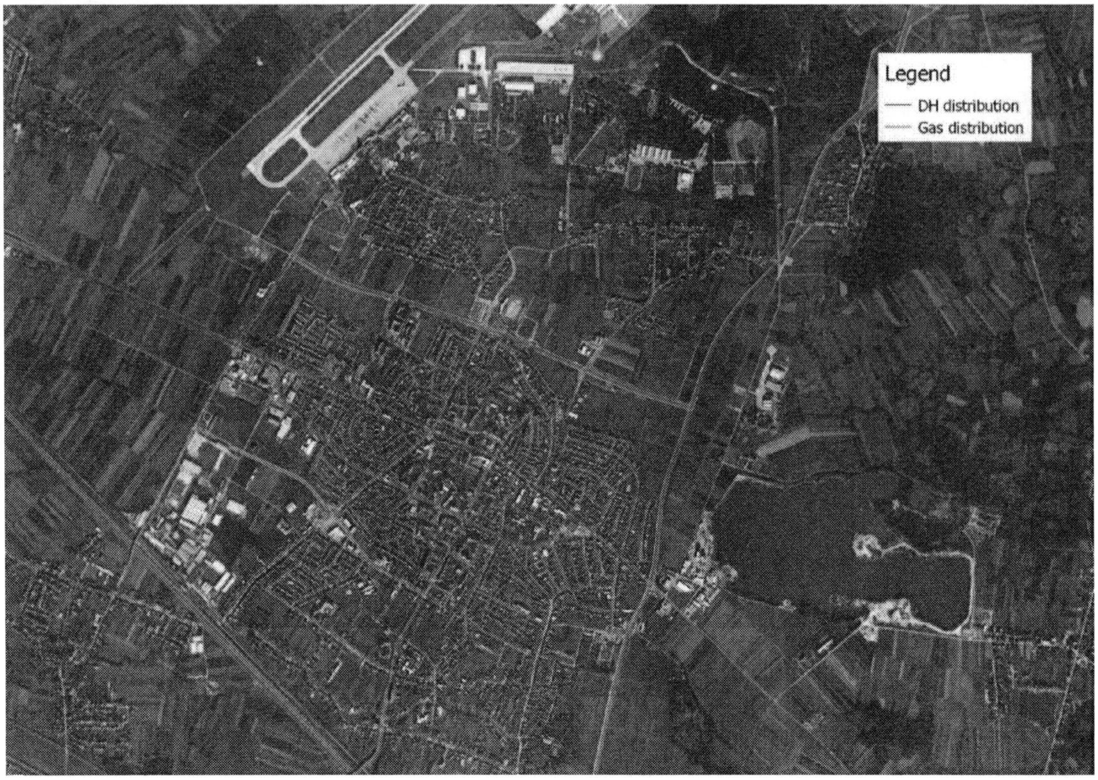

Fig. 2. Existing infrastructure in Velika Gorica.

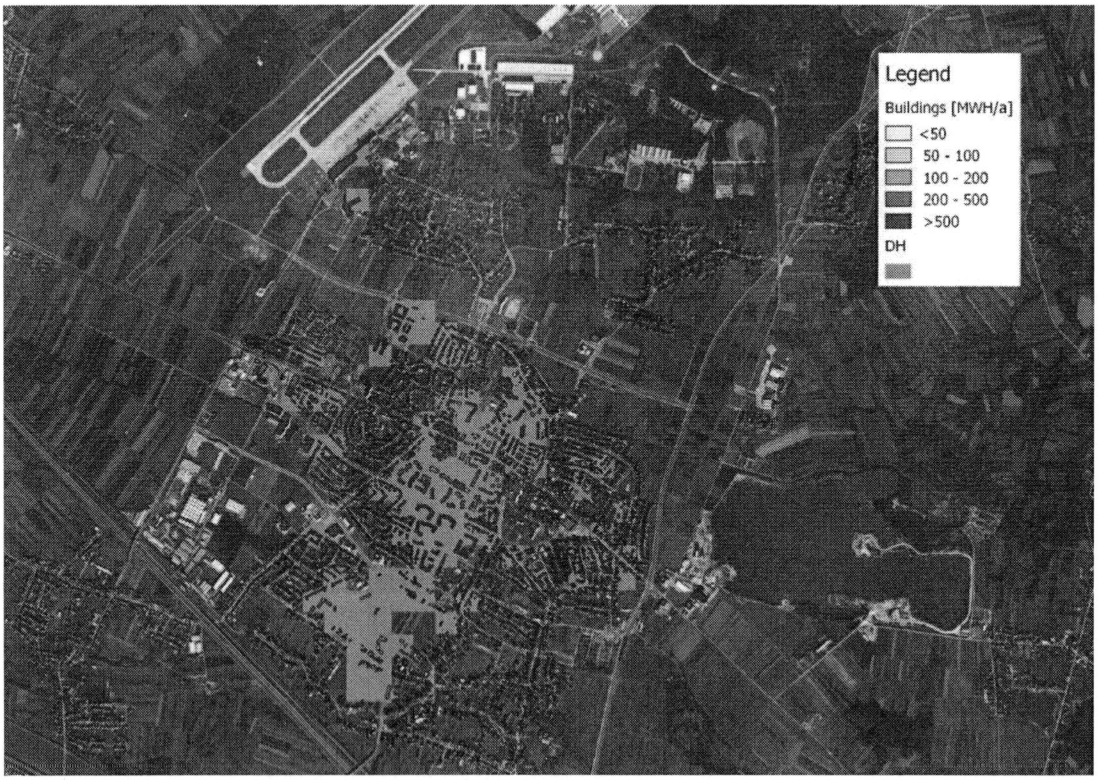

Fig. 3. Initial results.

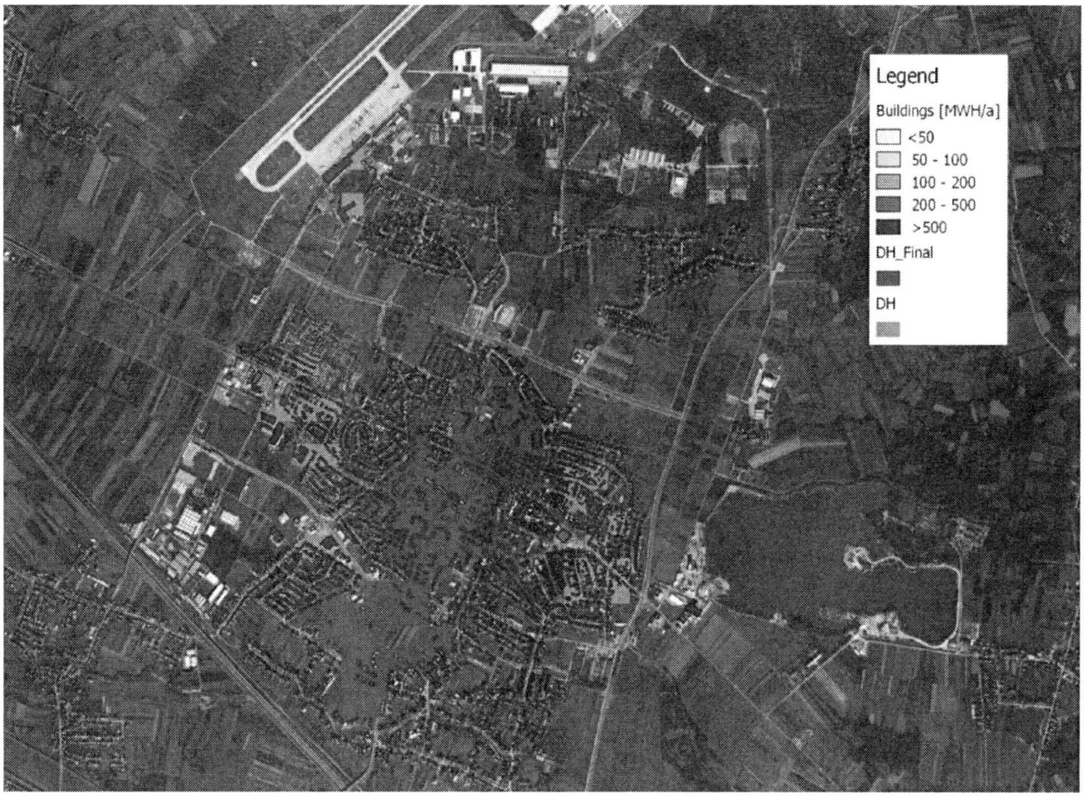

Fig. 4. Final results.

5 Conclusion and future work

GIS based methods have already been successfully used for energy demand and supply mapping. In this paper, mapped heat demand has been used for an analysis of the potential for the expansion of the district heating grid. As a case study, the city of Velika Gorica in Croatia has been chosen. To map the buildings and their heat demand, qGIS was used. The results have shown that a total of more than 117 GWh could be supplied from the DH which is more than 59% of the total heating demand identified in the city's SEAP. Those results can't be applied to the real case since the natural gas grid already covers large share of that area.

The future work of this research will include refinement and validation of the input data and results, including the sensitivity analysis of the used data. This will be supplemented with a supply side analysis in order to more precisely analyse the potentials for the utilization of the DH and the resulting integration of the power and heat sectors.

Acknowledgement

Financial support from the European Union's Horizon2020 projects CoolHeating (grant agreement 691679) and PLANHEAT (grant agreement 723757), as well as the support from the RESFlex project funded by the Environmental Protection and Energy Efficiency Fund with the support of the Croatian Science Foundation are gratefully acknowledged.

References

1. European Commission, "20 20 by 2020 Europe's climate change opportunity," Brussels (2008)
2. European Commission, "A Roadmap for moving to a competitive low carbon economy in 2050," Brussels (2011)
3. European Environment Agency, "Annual European Union greenhouse gas inventory 1990–2012 and inventory report 2014," European Environment Agency (2014)
4. R. Lund, B. V. Mathiesen, App. Energy **142**, 389-395 (2015)
5. H. Lund, S. Werner, R. Wiltshire, S. Svendsen, J. E. Thorsen, F. Hvelplund and B. V. Mathiesen, 4th Generation District Heating (4GDH): Integrating smart thermal grids into future sustainable energy systems, Energy **68**, 1-11 (2014)
6. M. Münster, P. E. Morthorst, H. V. Larsen, L. Bregnbæk, J. Werling, H. H. Lindboe and H. Ravn, Energy **48**, 47–55 (2012)
7. W. Xiong, Y. Wang, B. V. Mathiesen, H. Lund and X. Zhang, Energy **81**, 274–285 (2015)
8. L. D. Lucia and K. Ericsson, Energy Res. Social Sci. **4**, 10-20 (2014)

DOI: 10.1051/e3sconf/20171901021

9. S. J. G. Cooper, G. P. Hammond, J. B. Norman, Energy Inst. (2015)
10. N. Aste, M. Buzzetti, P. Caputo, Sustainable Cities and Society **14**, 43-55 (2014)
11. P. Gabillet, Energy Policy **78**, 189–197 (2015)
12. D. Connolly, H. Lund, B.V. Mathiesen, S. Werner, B. Moller, U. Persson, T. Boermans, D. Trier, P.A. Ostergard, S. Nielsen, Energy Policy **65**, 475-489 (2014)
13. T. Novosel, T. Pukšec, G. Krajačić, N. Duić, Energy Procedia **95**, 337-343 (2016)
14. D. Čulig-Tokić, G. Krajačić, B. Doračić, B. Vad Mathiesen, R. Krklec, J. Møller Larsen, Energy **93**, 435-443 (2015)
15. M. Pavičević, T. Novosel, T. Pukšec, N. Duić, Energy, **In press** (2017)
16. Bolonina, G. Bolonins, D. Blumberga, Environ. Climate Technol. **14**, 41-46 (2014)
17. M. A. Ancona, L. Branchini, A. De Pascale, F. Melino, Energy Procedia **75**, 1208-1213 (2015)
18. M. Elci, A. Oliva, S. Herkel, K. Klein, A. Ripka, Energy Procedia **70**, 560-567, (2015)
19. R. Mikulandrić, G. Krajačić, N. Duić, G. Khavin, H. Lund, B. V. Mathiesen, P. Østergaard, J. Sustain. Dev. Energy, Water and Environ. Syst. **3**, 282-302 (2015)
20. L. Grundhahl, S. Nielsen, H. Lund, B. Moller, Energy **14**, 1771-1778 (2016)
21. Levelized cost calculations, [Online]. Available: http://en.openei.org/apps/TCDB/levelized_cost_calculations.html [Accessed 05 May 2017]
22. Mapping the Heating and Cooling Demand in Europe, Work Package 2, Background report 5, STRATEGO project
23. R. Lund, U. Persson, Energy **14**, 1-10 (2016)
24. OpenStreetMaps, [Online] Available: http://openstreetmap.org [Accessed 10 May 2016]
25. Geoportal, [Online] Available: http://geoportal.dgu.hr/viewer/ [Accessed 10 May 2015]
26. City of Velika Gorica, "Sustainable Energy Action Plan of the City of Velika Gorica" (2011)
27. GIS, [Online] Available: https://qgis.org [Accessed 03 December 2016]

Thermal energy storage in granular deposits

Paweł Ratuszny[1,*]

[1] University of Opole, Independent Department of Process Engineering, 45-365 Opole, Dmowskiego 7-9, Poland

Abstract. Energy storage technology is crucial for the development of the use of renewable energy sources. This is a substantial constraint, however it can, to some extent, be solved by storing energy in its various forms: electrical, mechanical, chemical and thermal.
This article presents the results of research in thermal properties of granular deposits. Correlation between temperature changes in the stores over a period of time and their physical properties has been studied. The results of the research have practical application in designing thermal stores based on bulk materials and ground deposits. Furthermore, the research results are significant for regeneration of the lower ground sources for heat pumps and provide data for designing ground heat exchangers for ventilation systems.

1 Introduction

Due to the increasing demand for energy, combined with the development of energy based on renewable energy sources (RES), which, as is commonly known, generate energy in an inconsistent manner, the issue of storing energy in its various forms is becoming increasingly significant. It is usually done by using devices and media which store energy and allow to use it at different time.

Energy storage issue is not new. For many there have been various techniques and substances used for storing energy in various forms. In order to achieve the best efficiency of the system, it is most beneficial to store energy in the form it was obtained, or the form it will be used. Each transformation of energy generates losses and impairs the efficiency of the installation.

Storing energy involves a lot of technical problems associated with the storage time, the source of acquisition, and the destination. Thermal energy, regardless of time, can be stored in passive low-temperature systems, heating systems supplies, and systems used to power conventional thermal power plants where high temperatures are required [1].

1.1 Thermal energy storage

Thermal energy needs to be stored both short- and long-term, in a wide range of temperatures. It can be accumulated through a change in the internal energy of the medium, with the use of the specific heat of the substance, by applying to the phase change (using the latent heat), through physical processes, in thermo-chemical processes or through combinations of processes.

Accumulation of thermal energy with the use of the specific heat of the substance can be used for solids and liquids. Solid-liquid and liquid-gas phase changes are most common in the case of heat accumulation through phase change [2]. Methods of accumulation of thermal energy are presented in Fig. 1.

Development of effective and economical methods of storage of thermal energy obtained from RES is today a key issue for the development of renewable energy. This area requires close cooperation of both scientists and engineers, as well as investors. In order to design and execute cost-effective heat stores, extensive research is conducted throughout the world, covering all methods presented on Fig. 1. Mekhamer [4] pointed out raw saudi bantonite as an effective material to store heat energy, with a capacity of 30 kJ/g when heated to a temperature of 200° C. The material is also easily available and cheap.

A lot of research is being carried out on phase change materials, e.g. the research on using solar salt [5] as a good accumulator of solar energy in power plants CSP (concentrated solar power). The basis research into the mechanism of material melting around the heating pipe was undertaken in [6]. In paper [7] it was proposed to add PCM (Phase Change Materials) to concrete, and in the research financed by the Spanish Government [8], it was proposed to fill bulkheads with PCM. It indicated, like in this paper, that buildings should also be used as the storage of heat and that the research had a broad practical use. A lot of research is conducted on the sensible heat storage in natural materials, both in the ground and in the materials used as fillers. Paper [9] presents the results of execution of a heat store introductory project in Kerava, Finland. The subject of research presented in paper [10] is similar to the one presented in this paper, however, the storage material there is not the sand, commonly used to fill the foundation walls, but rocks which may be used for the storage of energy in the power plants CSP. The thermal properties of 5 rocks were presented. And the authors also stressed the ease of availability and low price of the material used to build the heat store.

* Corresponding author: ratuszny@uni.opole.pl

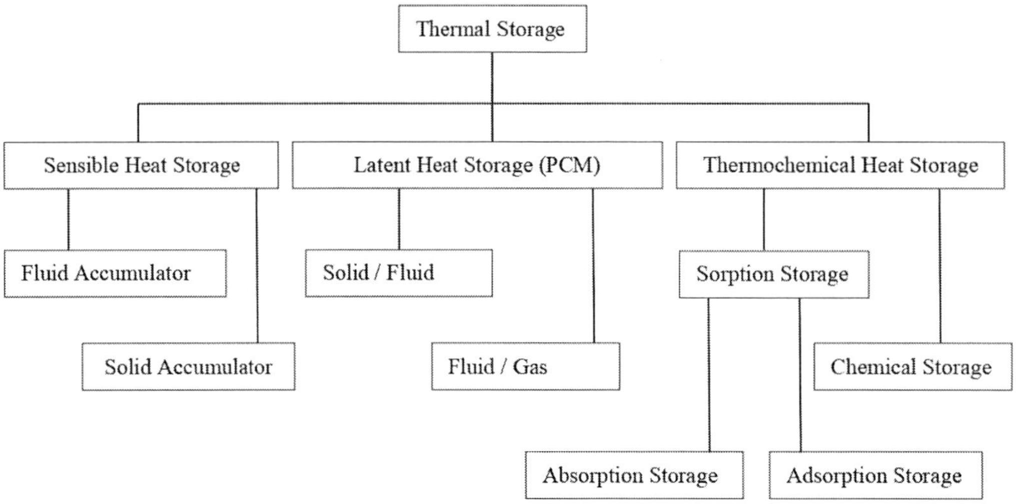

Fig. 1. The classification of thermal stores in relation to the thermodynamic properties [3].

This paper presents the results of research into bulk materials that can be used in the construction of sensible heat stores used for heating buildings. This research is part of the project, whose goal is to develop data and guidelines for designing this type of storage. In buildings without basement the space between the foundation walls is filled with compacted sand. This space can be used for sensible heat storage. The advantage of such store is the low cost of its construction. The store uses the construction elements of the building for the housing and the filling. The extra costs need to be incurred for a system of pipes to transfer heat, together with the power supply, and for padding the thermal insulation. These are small costs in comparison to the cost of the construction of a separate store, and they only slightly increase the cost of the entire house.

1.2 Accumulation of sensible heat using specific heat

This is the simplest and most common form of accumulation of thermal energy. Thermal stores that use specific heat of the substance accumulate energy by an increase in the temperature of the medium (a solid or liquid). This system operates on heat capacity and change in the temperature of the material during accumulating and discharging energy from the store [2]. The following substances can be considered as working media for sensible heat accumulation: water, air, oils (e.g. silicon), glycols, molten wax, refrigerants, as well as solids: rocks, grainy materials, soil; best suited, however, for this type of accumulation are deposits of rock, grainy deposits (soil, aggregates) and water tanks.

The amount of thermal energy accumulated in a homogeneous material depends on its specific heat, temperature changes, and the amount of heat-accumulating material according to the following formula:

$$Q = m\int_{T1}^{T2} c_w(T)dT \qquad (1)$$

The substance that has the very good storage properties using the specific heat is of course water. The downside of this medium is low temperature of accumulation and a fairly narrow temperature range (0-100°C at atmospheric pressure).

As can be observed in the picture (Fig. 2), substances with relatively good storage property are metals, natural deposits and materials - stone and soil, as well as concrete. These materials are suitable for high-temperature accumulation which means the store can accumulate considerable amounts of thermal energy, and co-operate with systems using renewable energy sources - solar collector systems for thermal energy accumulation, and heat pumps, for which the store may be the bottom source. This makes it possible for the energy stored in the summer to be used in the heating period, in the winter.

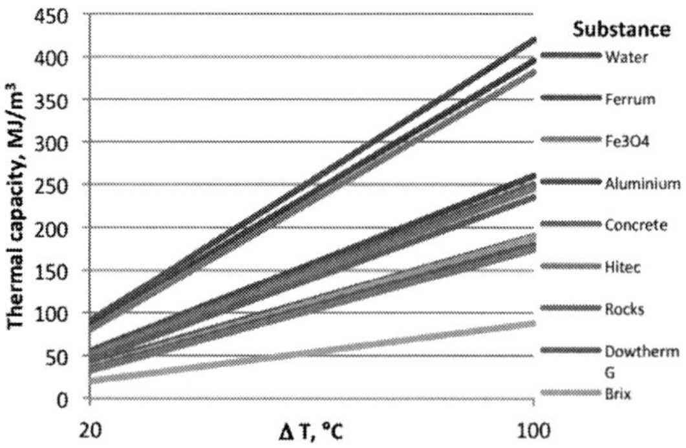

Fig. 2. Thermal capacity as a temperature rise function for certain liquids and solids - at constant specific heat [1].

Practical issues of the design and research relate primarily to the calculation of the heat capacity of the store, heat-exchange area linked to the cross-section and the length of the pipes of the heat exchanger, pressure drop, i.e. selection of circulation pumps and distribution of the heat exchanger pipes in the deposit.

1.3 Heat transfer characteristics

The basic issue, however is to determine the heat exchange method. In granular deposits it depends on the type and alignment of the particles (porosity of the deposit) and on how the heat is supplied. Different mechanisms will be applied for a deposit heated with a flowing liquid, and for a deposit with the pipe heat exchanger, in which the air is trapped in the voids between the grains. For such deposits where the particles are surrounded by stagnant fluid, heat transfer is assumed to occur in the vertical direction by the following mechanisms [11]:

1. Heat transfer through the fluid in the void space by conduction and by radiation between adjacent voids (when the voids are assumed to contain a non-absorbing gas).
2. Heat transfer through the solid phase.
   a) Heat transfer through the contact surface of the solid particles.
   b) Conduction through the stagnant fluid near the contact surface.
   c) Radiation between surfaces of solid (when the voids are assumed to contain a non-absorbing gas).
   d) Conduction through the solid phase.

Mechanisms 1 and 2 are parallel with each other. Conductivity is given by $k_\varepsilon = (\Delta T/\Delta L)$ = heat flux in void space + heat flux through solid phase.

Therfore [11]:

$$-k_\varepsilon \frac{\Delta T}{\Delta L} = -k\varepsilon \frac{\Delta T}{\Delta L} + \alpha_{rv}\varepsilon(-\Delta T) \qquad (2)$$

Allowing for the temperature drop in the particle which is the sum of temperature drop in the solid phase and temperature drop near the contact surface, and flux in the solid phase, a general heat transfer equation for a granular deposit in which the fluid is stationary can be applied: (by stagnant fluid) [11]:

$$k_\varepsilon = \varepsilon\left[k_g + \alpha_{rv}\Delta L\right] + \frac{(1-\varepsilon)\Delta L}{\dfrac{1}{k_g / l_v + \alpha_p + \alpha_{rs}} + \dfrac{l_s}{k_s}} \qquad (3)$$

Both thermal conductivity and thermal resistance for the deposit depend on the grain material properties, void fraction of deposits and the heat transfer coefficients [12]. The research results presented here show that with the increase in porosity thermal resistance decreases, regardless of the heat conduction factor for grains of the granular material. It also shows that the relationship between the thermal resistance and the thickness of the barrier for low porosity of the deposit is not linear. For example, at porosity 0.1 thermal resistance initially grows fairly quickly with the increase in the width of the barrier, until it reaches approximately 0.7 m; for such width the resistance is almost 8 times higher at porosity factor 0.1 than at factor 0.9. Furthermore, the differences in the heat conduction factors of the granules did not have a strong effect on the thermal resistance when the barrier width exceeded 1 meter. However, the impact of the ratio of penetration of heat into the grains for the air filtering the granular material to the thermal resistance decreases when the thermal conductivity rises above 40 W/mK at specific parameters.

It is clear that at the low penetration coefficient resistance decreases very quickly with the increase in conductivity coefficient [12].

These data are essential for the appropriate design of placement of the tubes of the heat exchanger in granular deposit, as well as for the selection of the material and the construction of the deposit.

The paper [13], presents the heat exchange mechanism that can be used in the design of thermal stores, in which heat is supplied by a stream of the flowing fluid. Papers [14,15] describe the heat exchange for such deposits at high temperatures, in which radiation plays a larger role.

2 Material and Methods

Temperature changes have been studied in a heated granular deposit depending on the degree of its humidity. A tests results measurement system was constructed, as shown in Fig. 3.

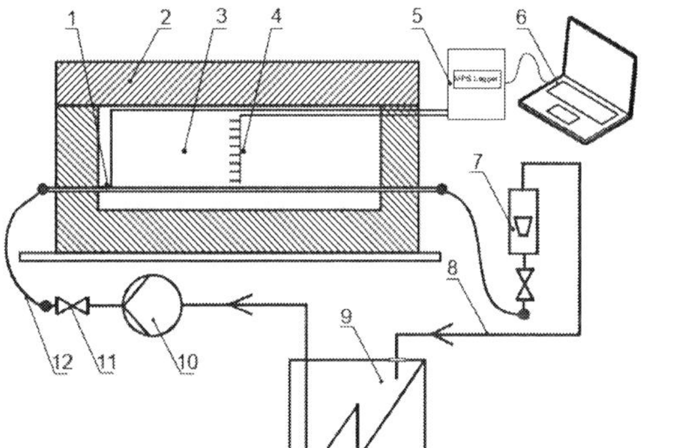

Fig. 3. The design of the laboratory workstation: 1 - pipe heating the deposit equipped with the inlet temperature sensor PT-100, 2 - thermal insulation of the measuring chamber, 3 - measuring chamber filled with loose material, 4 - temperature sensors TS-2 DM (8 pieces), 5 - measuring unit with MPS Logger, 6 - computer working with measuring module, 7 - rotameter, 8 - return of the heating water, 9 - electric thermostatic water heater, 10 - circulating pump, 11 - valve, 12 - flexible cable connecting the units of the workstation.

It consists of the measuring chamber filled with the tested granular material, the heat supply unit of a boiler with a thermostatic temperature control system for the

 DOI: 10.1051/e3sconf/20171901022

feed water, and a measuring unit for measurement of the flow of the liquid, and the temperature in various areas of the deposit.

A temperature measurement was based on a measuring and control unit equipped with a data logging function, configured for multi-point measurement and temperature monitoring. In the project, the control device was connected to 8 digital temperature sensors TS-2 DM and 1 sensor PT-100 for monitoring the temperature of feed water as it may exceed the measuring range of sensors TS-2 DM.

Tests were carried out for the deposits of quartz sand 1-1.6 mm: dry, and at about 10% moisture content. The test chamber was filled with sand, and temperature sensors were installed; sensor PT-100 was placed on the heating water inlet and digital sensors were positioned vertically at 15 mm intervals. This allowed to measure the change in temperature of the deposit depending on the distance from the place of heat supply. The heating water temperature was 65 °C.

3 Tests results and discussion

The results of the measurements for the tested deposits are shown on the charts: Fig. 4 and Fig. 5.

As evident, the temperature increased in all sensors with a very similar trend - initially it grew faster, but after a while it stabilizes at a certain level for each sensor.

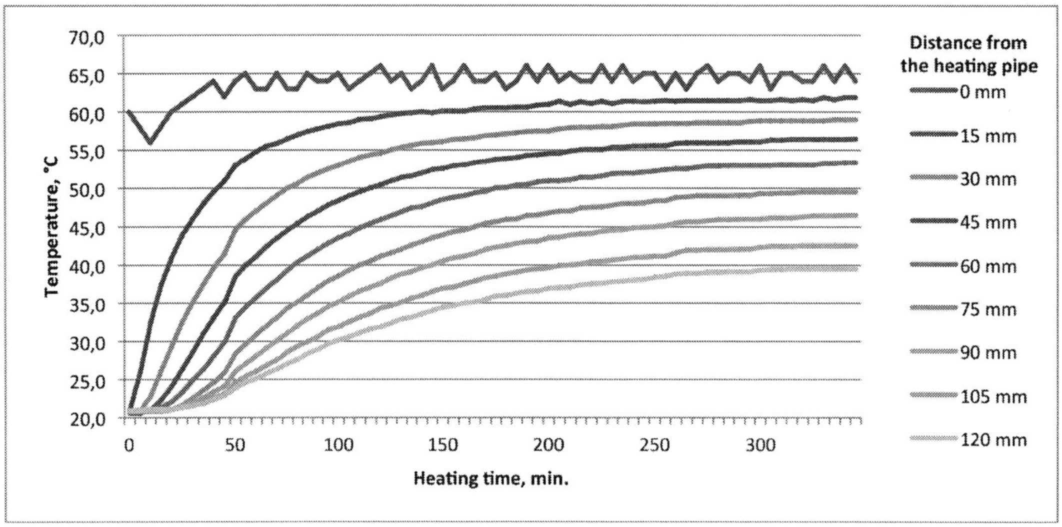

Fig. 4. Temperature chart for the deposit at different heights depending on the time of heating – quartz sand 1-1.6 mm, dry.

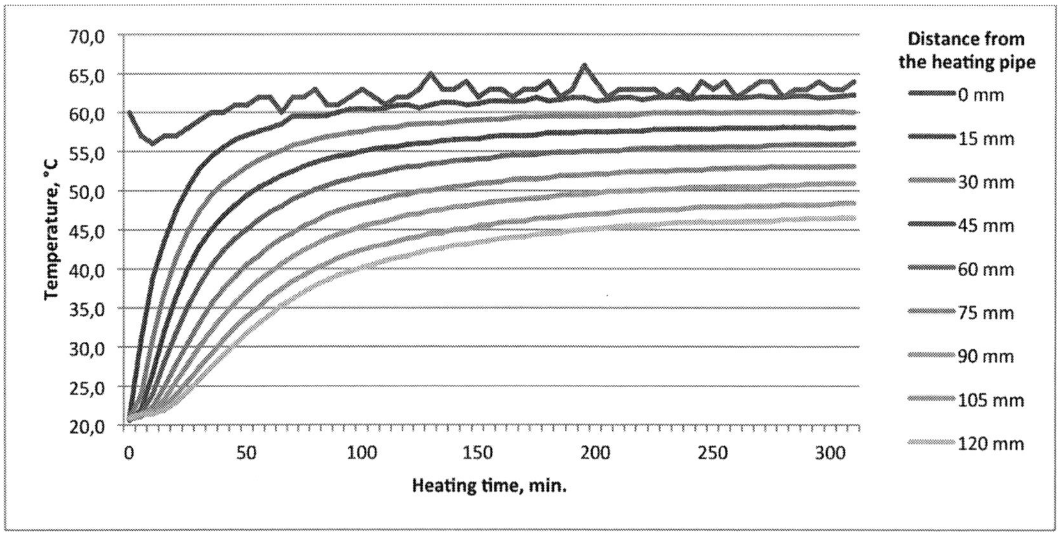

Fig. 5. Temperature chart for the deposit at different heights depending on the time of heating –quartz sand 1-1.6 mm, moist.

Temperature differences between the sensors depend on sensor placement (Fig. 6): higher temperatures were taken by sensors located closer to the heating pipes, and conversely, the further the sensor was placed (sensors were placed in succession every 15 mm) the lower temperature it showed. Sensor, PT-100, which was located directly on the heating water inlet heated up quickest and the temperature stability was achieved. Small temperature fluctuations detected by sensor PT-100 were caused by starting or stopping of the thermostat of the warming device.

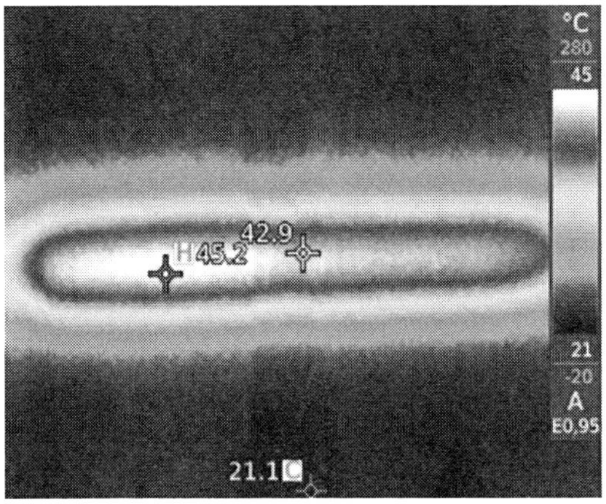

Fig. 6. Thermal photography of a granular deposit heated by a pipe exchanger. Symmetrical temperature distribution can be observed.

Charts show changes in one direction, however, it should be noted that in a granular deposit, in which the fluid flow is not possible, the temperature distribution around the heating pipe is symmetric, as is shown in the picture taken with the help of a thermal imaging camera.

4 Conclusions

The results of the tests show that for deposits of moist sand changes in temperature at various points of the deposit are faster, which is related to a higher heat transfer coefficient for radiation between the grain material and the air filling the void space in solid surface. In the deposit of 10% moisture content higher temperature is achieved faster than in the dry sand deposit; however, the temperature distribution is similar.

Areas of deposits more remote from the place of heat supply reach the balance temperature which is lower than the temperature of the supply medium. For the test material the temperature dropped by 18 Celsius degrees in the moist deposit, but more than 25 Celsius degrees for the dry deposits as detected by the sensor fitted at a distance of about 120 mm from the heating pipe. The sensor is located at a distance of 15 mm from the heating pipe recorded temperature 1.5 Celsius degrees lower. These values determine the distribution of (distance between) pipes of the heat exchanger in the proposed thermal store. Because the temperature distribution around the heating pipe is symmetric, these tubes must be distributed evenly throughout the volume of the storage and must not be placed only in the bottom area.

Tests were carried out also for other materials, such as sands of 0.5 mm grain, gravel, and garden soil. The results will be used in developing guidelines for designing sensible heat stores based on granular deposit with no liquid flow.

Finally, it should be pointed out that, taking into account the density of the stream of the stored energy, sensible heat stores are not competitive to PCM ([16]) stores using chemical reactions or sorption. There is a point, however, in doing research on sensible-heat storage because of the prevalence of granular deposits in construction - as foundation walls filling, substructure of public squares, etc. To build a sensible-heat store in such areas requires, therefore, only small investment costs, which in turn positions such stores high in terms of economic criteria. Not without significance is also the volume of the space filled with a granular material, which can easily allow of storing the appropriate amount of heat for the building. Therefore, this solution should be popularized in the construction industry.

The research results presented in this paper are part of the above mentioned area of research into using the commonly available and inexpensive materials for energy storage. It should be noted that this area of research is closely related to engineering activities, and thus it fits into the current trend of cooperation between science and industry.

Notation

Q - stored thermal energy [J],
m - mass [kg],
c_w - specific heat, J/kg·K
T - temperature, K
k_e - stagnant conductivity, that is effective thermal conductivity of porous media filled with stagnant fluid for both unconsolidated and consolidated particles, W/m·K
k_g - thermal conductivity of fluid, W/m·K
k_s - thermal conductivity of solid, W/m·K
ΔL - effective lenght between centers of two neighboring solid particles in direction of heat flow, m
l_s - effective lenght of a solid particle for heat transfer in a bed of unconsolidated particles, m
l_v - effective thickness of the fluid film adjacent to the surface of two solid particles, m
α_p - heat transfer cofficient representing the heat transfer rate through the contact surface between solid particles in a bed of unconsolidated particles or between two clogged particles in a consolidated bed, $W/m^2 \cdot K$
α_{rv} - heat transfer cofficient for thermal radiation, void space to void space, $W/m^2 \cdot K$
α_{rs} - heat transfer cofficient for thermal radiation, solid surface to solid surface, $W/m^2 \cdot K$
ε - void fraction of a packed bed of unconsolidated particles

DOI: 10.1051/e3sconf/20171901022

References

1. R. Domański, *Magazynowanie energii cieplnej*, PWN Warszawa, (1990)
2. M. Pomorski, A. Nemś, Z. Gnutek, Instal **11**, 27-32 (2015)
3. M. Danielak, Polski Instalator **10**, 38-40 (2014)
4. W. K. Mekhamer, Arab. J Chem. **9**, 264–268 (2016)
5. Q. Peng, J. Ding, X. Wei, G. Jiang, Energy Procedia **105**, 4420 – 4427 (2017)
6. T. Kousksou, M. Mahdaoui, M. Hlimi, R. Elalaiji, T. Rhafiki, Case Stud. Therm. Eng. **8,** 128–140 (2016)
7. S. Ramakrishnan, X. Wang, J. Sanjayan, J. Wilson, Procedia Eng. **180,** 1170 – 1177 (2017)
8. A. Gracia, L. F. Cabeza, Energy Build. **103,** 414-419 (2015)
9. H. Oosterbaan, M. Janiszewski, L. Uotinen, T. Siren, M. Rinne, Procedia Eng. **191,** 352 – 360 (2017)
10. H. Grirate, N. Zari, Iz. Elamrani, R. Couturier, A. Elmchaouri, S. Belcadi, P. Tochon, Energy Procedia **49**, 810 – 819 (2014)
11. D. Kunii, J.M. Smith, A.I.Ch.E. J. **6,** 1, 71-78 (1960)
12. E. Kotela, J. Leszczyński, *Analiza jednowymiarowego przepływu ciepła przez przegrodę wypełnioną materiałem granulowanym,* Mat. Konf. Debata o przyszłości energetyki, Politechnika Częstochowska (2010)
13. J.-W. Park, D. Park, D.-W. Ryu, , B.-H. Choi, E.-S. Park, Energy Geol. **181**, 142–156 (2014)
14. R. Lech, P. Szostak, H. Szeląg, Ceramic Materials **68**, 3, 213-217 (2016)
15. R. Lech, P. Szostak, H. Szeląg, Ceramic Materials, **68**, 3, 218-223 (2016)
16. M. Neugebauer, M. Wesołowski, Ciepłownictwo, ogrzewnictwo, wentylacja **44,** 9, 369-372 (2013)

Estimation of the cost of electro-mechanical equipment for small hydropower plants – review and comparison of methods

Seweryn Lipiński[1,*], and *Tomasz* Olkowski[1]

[1]Department of Electric and Power Engineering, Electronics and Automation, Faculty of Technical Sciences, University of Warmia and Mazury in Olsztyn, 11 Oczapowskiego St., 10-736 Olsztyn, Poland

Abstract. The estimate of the cost of electro-mechanical equipment for new small hydropower plants most often amounts to about 30-40% of the total budget. In case of modernization of existing installations, this estimation represents the main cost. This matter constitutes a research problem for at least few decades. Many models have been developed for that purpose. The aim of our work was to collect and analyse formulas that allow estimation of the cost of investment in electro-mechanical equipment for small hydropower plants. Over a dozen functions were analysed. To achieve the aim of our work, these functions were converted into the form allowing their comparison. Then the costs were simulated with respect to plants' powers and net heads; such approach is novel and allows deeper discussion of the problem, as well as drawing broader conclusions. The following conclusions can be drawn: significant differences in results obtained by using various formulas were observed; there is a need for a wide study based on national investments in small hydropower plants that would allow to develop equations based on local data; the obtained formulas would let to determinate the costs of modernization or a new construction of small hydropower plant more precisely; special attention should be payed to formulas considering turbine type.

1 Introduction

A proper assessment of investment cost can be regarded as crucial in any project; investment in small hydropower plant is no exception. Generally, this cost can be split into two parts. One of them is the sum of the cost of civil engineering works (land purchase, infrastructure, labour, etc.), which is extremely hard to estimate as it is highly dependent on local conditions in many aspects (e.g. difficulty of access or local material and labour costs) [1, 2]. The second component is the cost of electro-mechanical equipment, which, as indicated by the literature research [2-14], can be better estimated.

The cost of electro-mechanical equipment for new small hydropower plants most often amounts to about 30-40% of the total budget [15]. In case of modernization of existing small hydro installations, this estimation represents the main cost.

The aim of our work was to analyse mathematical formulas that allow estimating the cost of investment in electro-mechanical equipment for small hydropower plants.

2 Methodology

To achieve the aim of our work, chosen cost functions were converted into the form allowing their comparison. General rule, linking all expressions proposed in literature is their exponential form, i.e. cost (C) is given as multiplication of exponential functions where power of plant (P) and its net head (H) are bases while different constant coefficients, obtained on the basis of real data, are exponents [3-13]. It means that the generalized formula can be given in a following form:

$$C = a \cdot P^b \cdot H^c, \quad (1)$$

where a, b and c are the above mentioned constant coefficients. Only Singal and Saini [9] decided to divide the cost into four components but each of these components was still given in exponential form.

We decided to use US dollars ($) as the currency (converted with the use of the exchange rate from the year of the publication of each of the considered formulas) and not to use specific cost as this is the approach most commonly used in the literature.

3 Analysed formulas

Gordon and Penman [3] proposed first formula destined for the estimation of electro-mechanical equipment cost in 1979. Since then, many others have been developed for that purpose. The basic approach is based on the value of power of the small hydropower plant and on its net head; this approach was used in works by Gulliver

* Corresponding author: seweryn.lipinski@uwm.edu.pl

and Dotan [4], Whittington et al. [5], Voros et al. [6], Papantonis [7], Gordon [8], Kaldellis [9], Singal and Saini [10], and Aggidis et al. [11]. Ogayar and Vidal [12] took into account also turbine type (Kaplan, Semi-Kaplan, Francis or Pelton) while Cavazzini et al. [13] supplemented this approach with the use of design flow rate (but without distinction between Kaplan and Semi-Kaplan turbines).

The formulas obtained with the use of methodology described above are given in Table 1. We analysed only correlations that are indicated as those including only the cost of electro-mechanical equipment, what decreased their number from over a dozen to five only.

4 Results and discussion

Based on equations from Table 1 the costs were simulated with respect to plants' powers and net heads. Obtained results are shown in Fig. 1 and Fig. 2.

Fig. 1 shows cost with respect to plant power for three values of net heads (5, 25 and 100 [m]). Fig. 2 presents cost with respect to net head for four values of plant power (5 [kW], 100 [kW], 500 [kW] and 2 [MW]).

Table 1. Literature functions the estimation of the cost of electro-mechanical equipment for small hydropower plants.

Authors	Investment costs C [\$] in the function of power output P [kW] and net head H [m]
Gordon and Penman [4]	$C = 9000 \cdot P^{0.7} \cdot H^{-0.35}$
Gulliver and Dotan [5]	$C = 9600 \cdot P^{0.82} \cdot H^{-0.35}$
Kaldellis [10]	$C = 4125 \cdot P^{0.878} \cdot H^{-0.107}$
Singal and Saini [11]	$C = 950 \cdot P^{0.8087} \cdot H^{-0.2127} + 1180 \cdot P^{0.8145} \cdot H^{-0.2083} + 613 \cdot P^{0.8108} \cdot H^{-0.2118} + 281 \cdot P^{0.8197} \cdot H^{-0.2075}$
Aggidis et al. [12]	$C = 18552 \cdot P^{0.56} \cdot H^{-0.36}$

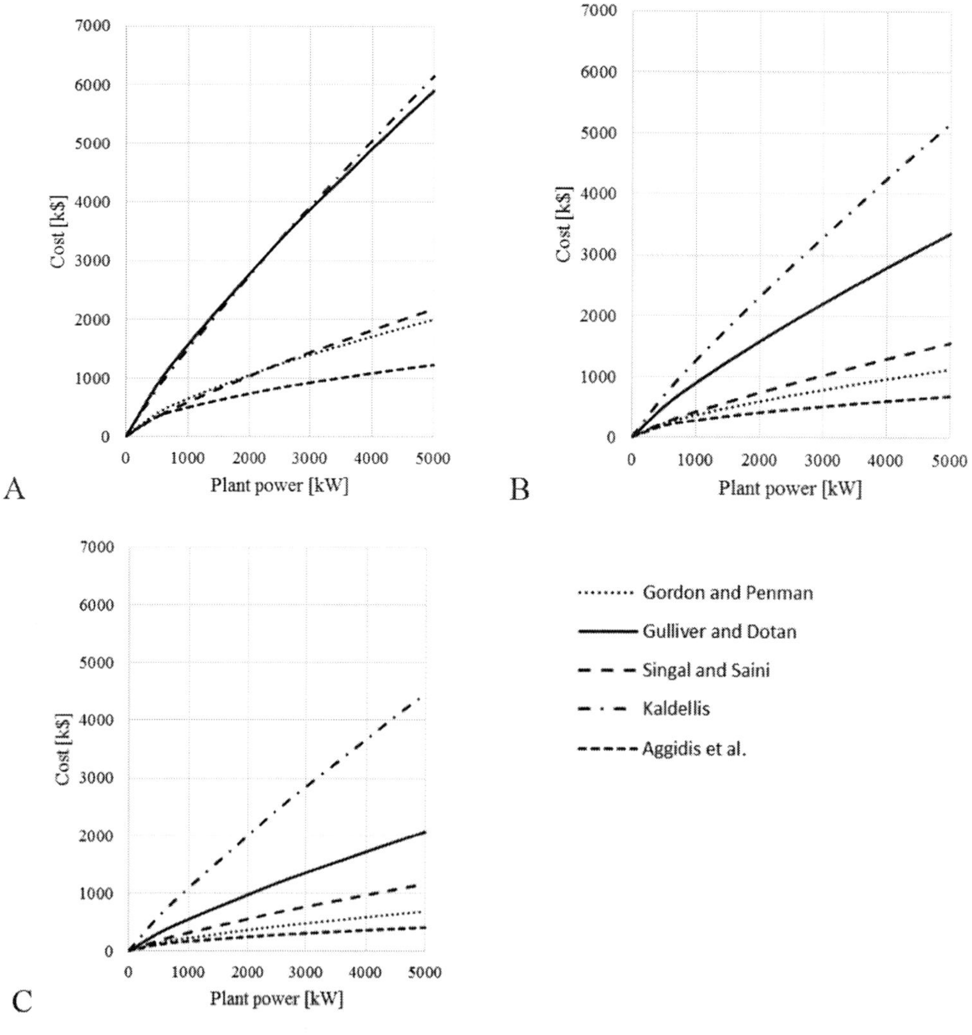

Fig. 1. Cost functions with respect to plant power for net heads equal to 5 (A), 25 (B) and 100 [m] (C).

DOI: 10.1051/e3sconf/20171901023

Fig. 1 clearly shows that results obtained with various formulas can differ greatly. Particularly formula by Kaldellis stands out - for each value of net head, it estimates cost of electro-mechanical equipment much higher than other formulas. Only for small head value, formula by Gulliver and Dotan gives similar results. Besides, formula by Gullliver and Dotan also stands out from the others, giving higher estimates than three remaining formulas also for higher net head values - not as high as Kaldellis' but still significantly higher. Formulas by Singal and Saini, Gordon and Penman, as well as Aggidis et al., gives results much closer to each other, but still considerably different. It should be noted that the most similar results were obtained using the formulas derived in 1979 and 2008, and that the formula by Aggidis et al. always gives the smallest estimates.

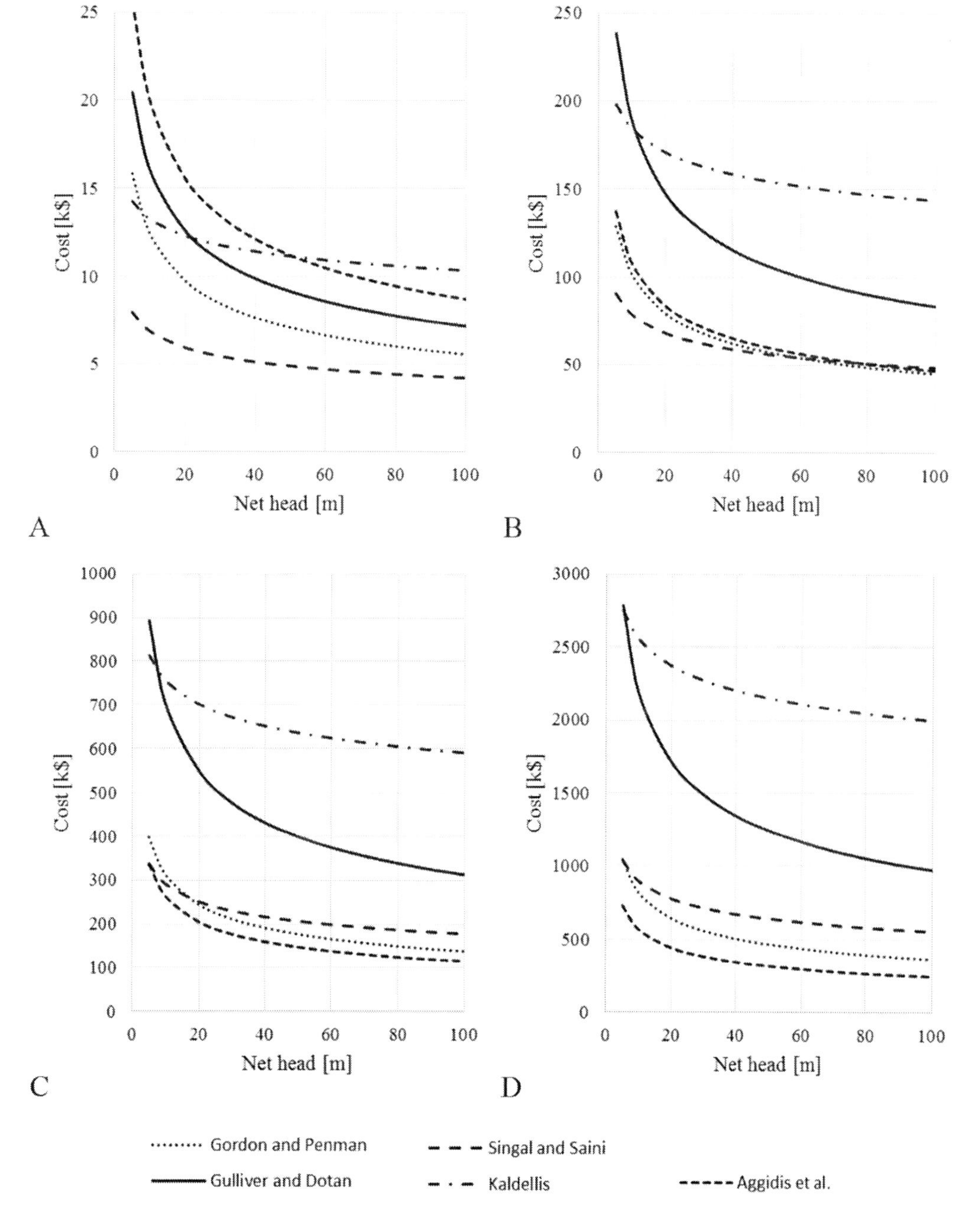

Fig. 2. Cost functions with respect to net head for plant powers equal to 5 (A), 100 (B) and 500 [kW] (C), and 2 [MW] (D).

Fig. 2 confirms conclusions that were drawn based on Fig. 1, i.e. estimates obtained with the use of various formulas differ significantly. Kaldellis' formula still gives highest estimates, except for cases of plants with small power and net heads. Intuitively expected cost reduction resulting from increase of net head has varying dynamics, depending on the approach. For example, for the same power (500 [kW]), increase of net head from 5 to 100 [m] can theoretically give approx. 65% decrease in cost (Gulliver and Dotan, Gordon and Penman, and Aggidis et al. formulas) as well as approx. 50% (Singal and Saini formula) or 27% decrease (Kaldellis formula), with a very different starting cost.

The same conclusion concerning dynamics of cost volatility results from the analysis of Fig. 1 – increase of cost being consequence of increase in plant power can have very different growth dynamics, depending on the used formula.

It is worth mentioning that the literature study shows that theoretical values obtained with the use of different methods for the actual small hydropower plants are also significantly different [12, 13]. Comparison with reals cost shows great underestimations or overestimations of the cost [12,13]. Such observation leads to the conclusion that formulas allowing to estimate the cost of investment in electro-mechanical equipment for small hydropower plants should be used very carefully, i.e. with particular regard to the considered project.

It is particularly worth to look up for the literature case studies in the context of finding an existing power plant(s) with characteristics similar to the designed one. Special attention should be payed to the formulas considering turbine type [12-14], but in our opinion, more research on that subject, considering more case studies in various geographical locations, should be performed.

The above remark applies not only to planning investments in small hydropower plants, but also to research work. The simple use of chosen formula [16, 17] raises the question of the accuracy of the obtained results and reliability of the economic analysis, which can be extremely important if the decision on the real investment is to be made.

5 Conclusions

The following conclusions can be drawn based on the performed study:

1. Significant differences in results obtained by using different formulas allowing estimation of the cost of investment in electro-mechanical equipment for small hydropower plants were observed.
2. Careful use of the analysed formulas is recommended.
3. In case of planning the investment in small hydropower plant, it is always advisable to refer to the proper case study – as similar to the one being designed as it is possible.
4. There is a need for wide studies based on national investments in small hydropower plants that would allow to develop equations based on local data and to supplement the case studies database.
5. The obtained formulas would let to determinate the costs of modernization or a new construction of small hydropower plant more precisely.
6. Special attention should be payed to formulas considering turbine type, as it can be the key to the more precise estimates of the cost of investment in electro-mechanical equipment for small hydropower plants.
7. Simulation of the cost of electro-mechanical equipment for small hydropower plants based on unification of the formulas found in literature allows deeper discussion of the problem, as well as drawing broader conclusions.
8. Such approach is novel in the context of investment in small hydropower plants and, in our opinion, it should be used more frequently, as it allows obtaining and then analysing families of curves, instead of referring only to specific cases.

References

1. L.J. Bracken, H.A. Bulkeley, C.M. Maynard, Energ. Pol. **68(C)**, 92 (2014)
2. S. Mishra, S.K. Singal, D.K. Khatod, Ren. Sust Energ. Rev **15(8)**, 3862 (2011)
3. J.L. Gordon, A.C. Penman, J. Water Power Dam. Constr. **31(9)**, 46 (1979)
4. J.S. Gulliver, A. Dotan, J. Energ. Eng. **110(3)**, 204 (1984)
5. H.M. Whittington, A.R. Wallace, D.S. Henderson, Proc. Third Int. Conf. Small Hydro, Int. Water Power & Dam. Constr., 182 (1988)
6. N.G. Voros, C.T. Kiranoudis, Z.B. Maroulis, Ren. Energ. **19(4)**, 545 (2000)
7. D. Papantonis, *Small hydro power stations* (Simeon, Athens, 2001)
8. J.L. Gordon, Hydro Rev. Wordw. **11(1)**, 37 (2003)
9. J.K. Kaldellis, Energ. Pol. **35(4),** 2187 (2007)
10. S.K. Singal, R.P. Saini, Ren. Energ. **33(12)**, 2549 (2008)
11. G.A. Aggidis, E. Luchinskaya, R. Rothschild, D.C. Howard, Ren. Energ. **35(12)**, 2632 (2010)
12. B. Ogayar, P.G. Vidal, Ren. Energ. **34(11)**, 2501 (2009)
13. G. Cavazzini, A. Santolin, G. Pavesi, G. Ardizzon, Energy **103**, 746 (2016)
14. A. Santolin, G. Cavazzini, G. Pavesi, G. Ardizzon, A. Rossetti, Energ. Convers. Manage. **52(7)**, 2533 (2011)
15. *Guide on How to develop a small hydropower plant* (ESHA, 2004)
16. D.A. Zema, A. Nicotra, V. Tamburino, S.M. Zimbone, Ren. Energ. **85**, 498 (2016)
17. D.K. Okot, Ren. Sust Energ. Rev **26**, 515 (2013)

Internal heat gain from different light sources in the building lighting systems

Dariusz Suszanowicz[1,*]

[1]University of Opole, Faculty of Natural Sciences and Technology, ul. R. Dmowskiego 7-9, 45-365 Opole, Poland

Abstract. EU directives and the Construction Law have for some time required investors to report the energy consumption of buildings, and this has indeed caused low energy consumption buildings to proliferate. Of particular interest, internal heat gains from installed lighting affect the final energy consumption for heating of both public and residential buildings. This article presents the results of analyses of the electricity consumption and the luminous flux and the heat flux emitted by different types of light sources used in buildings. Incandescent light, halogen, compact fluorescent bulbs, and LED bulbs from various manufacturers were individually placed in a closed and isolated chamber, and the parameters for their functioning under identical conditions were recorded. The heat flux emitted by 1 W nominal power of each light source was determined. Based on the study results, the empirical coefficients of heat emission and energy efficiency ratios for different types of lighting sources (dependent lamp power and the light output) were designated. In the heat balance of the building, the designated rates allow for precise determination of the internal heat gains coming from lighting systems using various light sources and also enable optimization of lighting systems of buildings that are used in different ways.

1 Introduction

Directive 2010/31/EU of the European Parliament and of the Council of 19 May 2010 on the energy performance of buildings [1] and the Construction Law of 7th July 1994 [2] require investors to prepare the energy performance of buildings. According to legislators, these acts will lead to an increase in the number of buildings with low energy requirements.

According to the energy performance of buildings preparation methodology, contained in the Minister of Infrastructure and Development Regulation of 27 February 2015 Pos. 376 [3], the annual need for non-renewable primary energy for the integrated lighting system shall be determined for buildings and commercial premises. The methods of calculating energy requirements for an integrated lighting system presented in the Regulation do not apply, however, to residential buildings and premises.

In order to be able to design passive or "zero energy" residential buildings or to plan modernization of existing ones to the standards recommended for passive buildings, it is necessary to determine precisely the energy efficiency of lighting systems used in residential buildings.

For all types of buildings (including residential buildings), the energy performance of buildings preparation methodology assumes determining internal heat gains. However, no calculations concerning heat gain from lighting systems using different sources of light have been taken into consideration. Internal heat gains from lighting affect the final energy consumption for heating public facilities as well as residential buildings [4, 5].

According to studies carried out by many researchers for various types of buildings, lighting accounts for 30% of electricity consumption in office buildings [6], from 20 to 30% in hospitals [7], from 28 to 62% in school buildings [8], whereas in residential buildings it is approximately 12% [9, 10]. The results mentioned above show that the lighting systems in public facilities is particularly important for the energy efficiency of buildings.

The heat generated in the building's interior by the lighting system is heat gain during the heating season, while in the summer ,when it is necessary to cool the rooms using air-conditioning systems, the heat from the lighting systems increases the demand for cooling in the building and thus has a negative effect on energy efficiency [11].

In order to calculate precisely the heat gains from building lighting, it is necessary to determine the energy parameters of the different light sources used in all types of buildings for which energy performance is prepared [12].

2 Materials and methods

The work analyses the consumption of electricity, luminous flux and heat flux emitted by various types of light sources most commonly used in buildings. A measuring stand was prepared, equipped with an insulated test chamber of 19 dm^3, in which different light sources

* Corresponding author: d.suszanowicz@uni.opole.pl

DOI: 10.1051/e3sconf/20171901024

were installed and the measuring instruments were connected. The diagram of the measurement station is shown in Fig. 1.

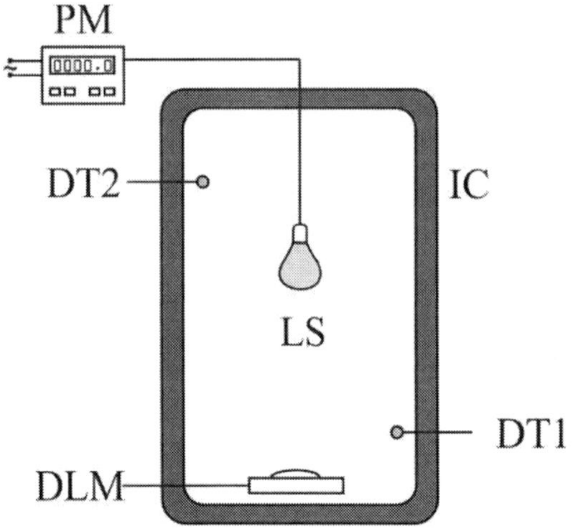

Fig. 1. Measurement station diagram: LS - light source, IC - insulated chamber, PM - multifunctional electrical power meter, DT1, DT2 - digital thermometers, DLM - digital light meter.

In a closed, insulated chamber, different light sources were placed in turn: 100 Watt incandescent light bulbs, 70 Watt halogen bulbs, 18 Watt compact fluorescent lamps and 12 Watt LED bulbs of different manufacturers with Edison E14 or E27 threads. The electricity consumption and the instantaneous power consumption were recorded using a multifunction electrical power meter. The temperature inside the measuring chamber was measured by two digital thermometers (one mounted at the top of the chamber, the other at the bottom), and the light emitted by the individual bulbs was registered using a digital light meter. The luminous flux emitted by a specific light source was recorded using a digital light meter.

All series of tests of the working parameters for individual light sources were started in identical environmental conditions (temperature inside and outside the chamber was 23°C). After switching on the light source enclosed inside the insulated chamber, the changes in the air temperature inside the chamber, the power consumption of the source and the stream of light emitted by the light source were recorded.

In order to determine the required time of observation of the operating parameters of the analysed types of light sources, it was planned to observe the temperature changes of individual types of light bulbs by using a thermographic camera.

In order to record the heat radiation emitted by the examined light sources, the images of selected light sources in the infrared medium band were taken using a thermographic camera. Sample images showing the temperature of the LED, incandescent and compact fluorescent bulbs, after 3 minutes of their operation, are shown in Fig. 2.

As can be seen from the pictures shown in Fig. 2, the surface temperatures of the different types of light bulbs during their operation are significantly different from each other. After 3 minutes of operation, the LED bulb reached a temperature of 32.6°C, a compact fluorescent lamp - 58.7°C, while the filament temperature of an incandescent light bulb reached 216°C and the surface temperature of the bulb was 134°C.

a)
b)
c)

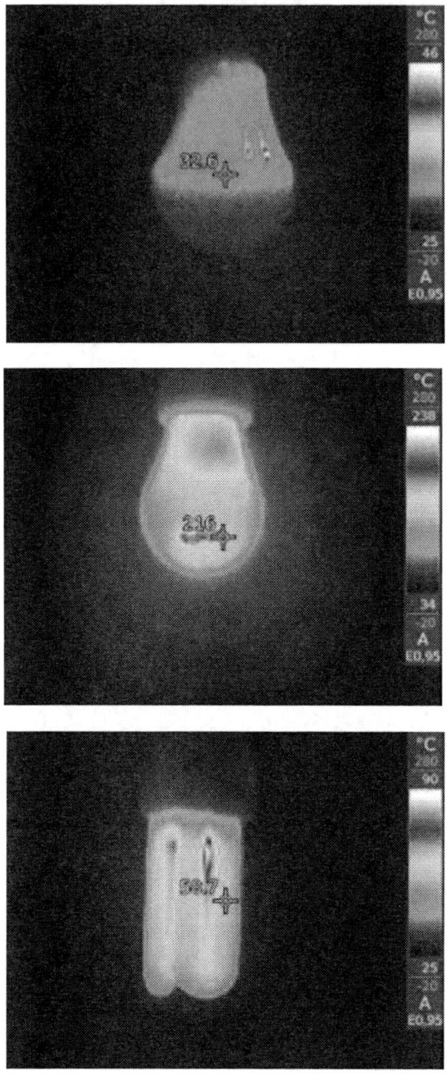

Fig. 2. Infrared images of the light sources: a) LED bulb, b) incandescent light bulb, c) compact fluorescent lamp.

Observations made by the thermographic camera allowed us to assume that the series of tests with the incandescent light source should last significantly shorter, so as not to overheat the bulb contained within a small insulated chamber from which the generated heat could not be discharged into the environment.

For each type of light source, bulbs of different manufacturers with the same nominal power values were tested by carrying five series of measurements and averaging the results.

3 Results and discussion

Sample graphs of air temperature changes at two measuring points located in the lower and upper part of the chamber recorded during the study of different light sources are shown in Fig. 3.

a)

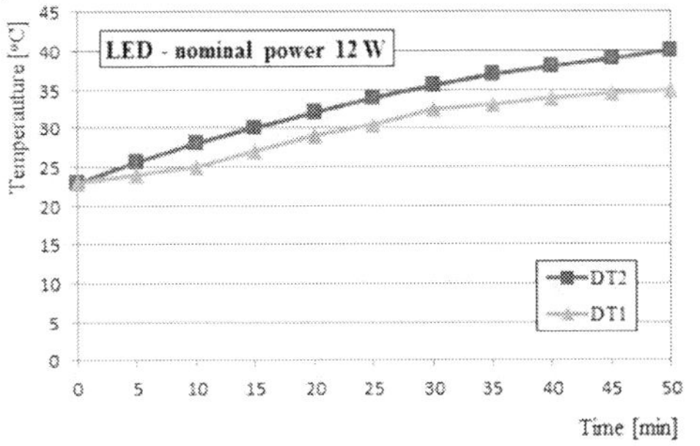

b)

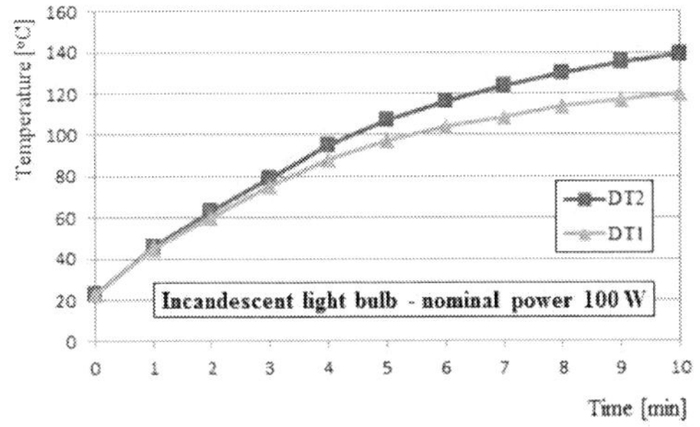

c)

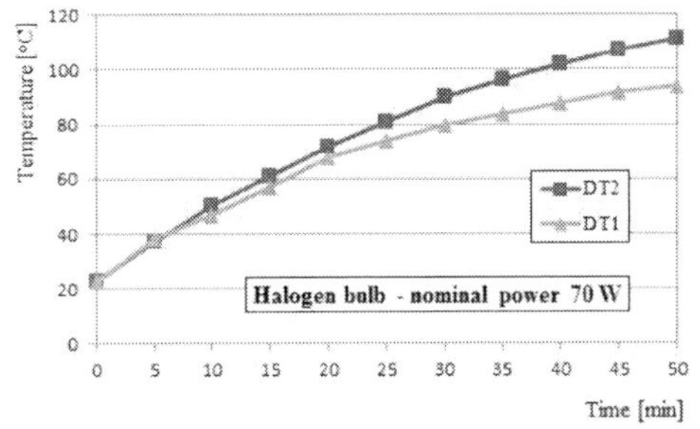

d)

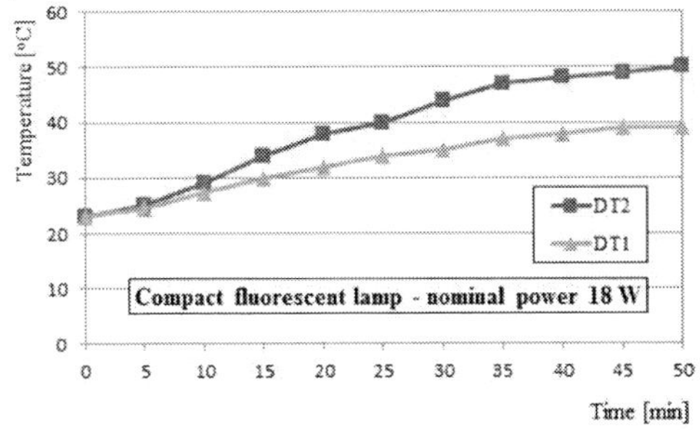

Fig. 3. Air temperature changes at two measuring points located in the lower (DT1) and upper (DT2) part of the chamber with the light sources: a) LED bulb, b) incandescent light bulb, c) halogen bulb, d) compact fluorescent lamp.

On the basis of the results obtained for the same light source across different series of measurements, we were able to determine the average temperature changes characteristic of the particular type of source.

Due to the thermal insulation of the test chamber walls, it was established that 100% of the heat emitted by the light source was consumed in order to heat the air inside the chamber.

Knowing the heat capacity of the air (dependant on the temperature inside the chamber) and the change of the average value of air temperature inside the chamber during the experiment, the heat output P_h was determined for each measurement series, i.e. the heat transferred from the light source to the air inside the chamber during the experiment. The heat output was calculated on the basis of the formula (1):

$$P_h = m \cdot c_a \cdot \Delta T \cdot t^{-1} \, [W] \tag{1}$$

where:

P_h - heat power [W],
m - mass of air [kg],
c_a - specific heat of air [J·(kg·K)$^{-1}$],
T - temperature [K],
t - time [s].

In all series of measurements, the electricity consumption of the light source was also recorded. The results of measurements of instantaneous energy consumption for incandescent bulbs, halogen bulbs, compact fluorescent lamps and LED bulbs, averaged in all series of measurements for particular source of light, were presented in Fig. 4.

The data collected from all series allowed us to establish, for each type of light source their actual electric power, which in the case of halogen bulbs and incandescent light bulbs was significantly higher (by 10%) than their nominal power. On the other hand, the actual electric power of compact fluorescent lamps and LED bulbs was lower than nominally indicated by the manufacturers.

The research allowed us to determine, for each type of light source, the unit heat flux that is the empirical heat emission coefficient (H_e), referenced to 1W of the nominal light source.

If the value of the heat emission coefficient (H_e) is 0, the source does not emit any heat energy, and if it is 1, it means that the source emits only heat. The heat emission coefficient (H_e) is determined according to the formula (2):

$$H_e = \frac{P_h}{P_e} \, \left[\frac{W}{W}\right] \tag{2}$$

where:

P_e - actual electrical power [W],

The H_e coefficient will allows to determine quickly the internal heat gains from a building lighting system equipped with a particular type of light source. The unit of the heat emission factor H_e is the $W_{\text{heat energy}}$ / $W_{\text{electric energy}}$.

 DOI: 10.1051/e3sconf/20171901024

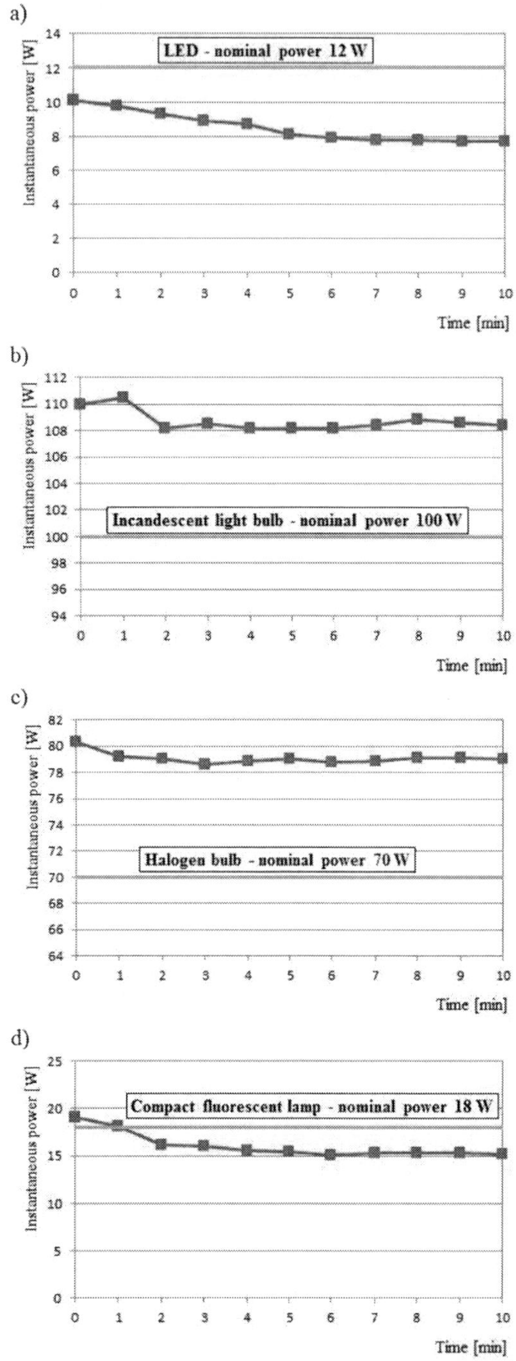

Fig. 4. Instantaneous power of the light sources: a) LED bulb, b) incandescent light bulb, c) halogen bulb, d) compact fluorescent lamp.

The luminous efficacy of each type of light source that is the quotient of the luminous flux emitted by the light source to the actual electric power (P_e) taken from the source, was also determined. The unit of luminous efficacy is lm/W. The luminous flux and the actual electrical power recorded during the experiments were used to determine the average values for each light source. The determined values of luminous efficiencies enables us to compare the energy efficiency of different types of light sources.

The average values of the heat emission coefficient (H_e) and the light efficiency determined for each light source during the analysis of test results are summarized in Table 1.

Table 1. Average values of the operating parameters of different light sources determined during the tests.

Light sources	Heat emission coefficient H_e, W/W	Luminous efficacy η, lm/W
LED bulb	0.08	140
incandescent light bulb	0.95	12
halogen bulb	0.82	16
compact fluorescent lamp	0.31	56

On the basis of experimental studies and analysis of the obtained results, we have proposed a correlation which enable us to determine the internal heat gains derived from lighting systems used in all types of buildings in the form of (3):

$$Q_{H,o} = A_f \cdot E_v \cdot \eta^{-1} \cdot H_e \cdot t_o \left[\frac{kWh}{yr.}\right] \quad (3)$$

where:
$Q_{H,o}$ - internal heat gain from lighting system [kWh/yr.],
A_f - floor area [m^2],
E_v - illuminance [lx],
η - luminous efficacy [lm/W],
H_e - heat emission coefficient [W/W],
t_o - annual light system operating hours [kh/yr.].

When determining internal heat gains according to correlation (3), it is necessary to indicate the number of operating hours of the lighting system during the year (t_o). Unfortunately, as far as residential buildings are concerned, it is very difficult to do such calculations therefore it was necessary to analyse the requirements and recommendations for lighting systems of such buildings.

On the basis of the Polish Standards [13, 14] and the recommendations included in the Minister of Infrastructure and Development Regulation on the energy performance of buildings preparation methodology or a part of a building and statement of energy performance of buildings [3], for different types of buildings the requirements concerning the average illuminance level of different types of premises and the number of lighting installation operating hours per year were developed and are summarized in Table 2.

Table 2. Requirements for lighting systems of the selected building types.

Building type	Illuminance E_v, lx	Annual light system operating hours t_o, kh/yr.
Residential	250	1.8
Office	500	2.5
Educational	300	2.0
Service	300	2.5
Store	300	5.0
Hospital	250	5.0
Production	500	2.2
Farm (chickencoop)	20	6.5

The requirements for different types of light sources, as presented in Table 2, allow us to optimize the installed power of lighting sources in both residential and non-residential buildings. Introducing the standard number of operating hours for a lighting system per year, enables us to compare the values of heat gains obtained from different buildings of the same type.

4 Conclusions

After reviewing the available literature on the subject and examining various light sources, a correlation was proposed to determine the internal heat gains $Q_{H,o}$ from lighting systems used in all types of buildings. It should be emphasized that outside of the heating season, the heat from the building lighting systems influences the temperature rise inside the building and may cause the need for cooling by means of an air-conditioning system.

Based on the results obtained, the empirical heat emission factor H_e was determined for the selected four types of lighting sources, depending on the power of the light source. The luminous efficacy η of each type of light source, i.e. the ratio of the luminous flux emitted by the light source to the energy it consumes, was also determined.

The determined working parameters of different light sources will allow for accurate incorporation of internal heat gains from lighting systems using different light sources into the building's thermal balance, and will also allow to optimize lighting systems of different types of buildings.

As demonstrated by the conducted study, only in the case of LED light sources there is no higher nominal power consumption by the bulb during the first minutes after it has been switched on. Accordingly, it can be stated that only in the case of building lighting systems equipped with motion sensors and time switches it is recommended to use only LED light sources.

References

1. Directive 2010/31/EU of the European Parliament and of the Council of 19 May 2010 on the energy performance of buildings. *Off. J. Eur. Union* **18.6.2010 L 153** (2010)
2. Act of 7 July 1994 Construction Law. *J. Laws* **pos. 290, 961, 1165, 1250** (2016)
3. Regulation Ministry of Infrastructure and Development on the methodology for determining the energy performance of a building or part of a building and energy performance certificates. *J. Laws* **pos. 376** (2015)
4. K. Flodberg, Å. Blomsterberg, M.-C. Dubois, Int. J. Energy Environ. Eng. **3**, 19 (2012)
5. E. Monstvilas, K. Banionis, V. Stankevičius, J. Karbauskaite, R. Bliūdžius, J. Civ. Eng. Manag. **16**, 3 (2010)
6. R. Zmeureanu, C. Peragine, Energy Convers. Manage. **40**, 1229-1236 (1999)
7. P. Torcellini, N. Long, S. Pless, R. Judkoff, Technical Report NREL/TP-550-34607, 1-115 (2005)
8. S. Chow, A.R. Ganji, B. 26th World Energy Engineering Congress, (Atlanta Georgia, 2003) DOI:10.1.1.576.5568
9. P. Bertoldi, B. Atanasiu, Proceedings of EEDAL'06 Conference, 267-272 (2006)
10. T. Crosbie, S. Guy, Int. J. Environ. Technol. Manage. **9**, 220-235 (2008)
11. A. Byung-Lip, P. Ji-Woo, Y. Seunghwan, K. Jonghun, L. Seung-Bok, J. Cheol-Yong, Energies **8**, 6658-6671 (2015)
12. K. Valančius, V. Lapinskienė, Environ. Eng. Conf. Proceedings, 31 (2014)
13. Polish Standard PN-EN 12464-1:2012, Light engineering (2012)
14. Polish Standard PN-EN 12665:2008, Light and lighting (2011)

DOI: 10.1051/e3sconf/20171901025

Single chamber microbial fuel cell with Ni-Co cathode

Barbara Włodarczyk[1,*], *Paweł P.* Włodarczyk[1], and *Antonina* Kalinichenko[1]

[1]University of Opole, Department of Process Engineering, Dmowskiego Street 7-9, 45-365 Opole, Poland

Abstract. The possibility of wastewater treatment and the parallel energy production using the Ni-Co alloy as cathode catalyst for single chamber microbial fuel cells is presented in this research. The research included a preparation of catalyst and comparison of COD, NH_4^+ and NO_3^- reduction in the reactor without aeration, with aeration and with using a single chamber microbial fuel cell with Ni-Co cathode. The reduction time for COD with the use of microbial fuel cell with the Ni-Co catalyst is similar to the reduction time with aeration. The current density (2.4 $A \cdot m^{-2}$) and amount of energy (0.48 Wh) obtained in MFC is low, but the obtained amount of energy allows elimination of the energy needed for reactor aeration. It has been shown that the Ni-Co can be used as cathode catalyst in single chamber microbial fuel cells.

1 Introduction

The energetic industry struggles with difficulties with producing more and more energy. Simultaneously the wastes and wastewater production increases very much, too. Both industrial and municipal wastewater is generated in huge amounts. Traditional wastewater treatment systems require lot of cost and energy to efficient work. The United States spends approximately $23 billion annually on domestic wastewater treatment, and another about $200 billion is needed for improving publicly owned treatment works [1]. So, it is important to reduce of wastewaters treatment costs. It is exist most different ways to use of wastewater [2]. At the same time new sources of energy are sought. Technical device that can combine electricity production with wastewater treatment is a microbial fuel cell (MFC) [3,4]. During wastewater treatment in MFC the activated sludge bacteria can be used for electricity production. The concept of microbial fuel cells was created the 60's of XX century [5]. Bacteria that were identified as capable of creating electricity (activated sludge is capable of producing electrons and H^+ jones) in MFCs include a wealth of genera of bacteria e.g. *Geobacter*, *Shewanella* or *Pseudomonas* [6-10]. Electrons are formed as a result of biochemical reactions catalysed by enzymes produced by bacteria. Next electrons are transferred from the anode (negative terminal) to the cathode (positive terminal) through the electric current receiver (load) [11]. In MFC organic material is oxidized on anode, and the product of oxidation are electrons (and CO_2) [12,13]. A bacterium in the anode compartment transfers electrons to the anode electrode. This occurs either through direct contact, nanowires, or mobile electron shuttles. During electron production protons are also produced in excess. These protons migrate through the proton exchange membrane (PEM) into the cathode area. The electrons flow from the anode through an external resistance to the cathode where they react with the final electron acceptor (oxygen) and protons [12-14]. So, MFCs are bio-electrochemical systems that are devices that use bacteria as catalysts to oxidize organic and inorganic matter and generate current [3,15].

The problems of using MFCs are slow process of reducing of some wastewater parameters and a low current density obtained in MFCs. But, it is possible to increase the current density by using different catalyst for electrodes (mainly for cathode). On anode the function of catalyst take over the microbes. In MFC's carbon is most often used as the electrode, but it is possible to use metal catalysts [16]. Theoretical the current density (obtained with using specific catalyst) can be calculated using the Butler-Volmer exponential function [17]. But, in real conditions the choice of catalyst is mainly carried out by experimental methods. For this reason experimental research of search a new catalysts for MFCs are still conducted [15, 16, 18-24]. Authors have attempted to demonstrate the possibilities of wastewater treatment with electricity production using the single chamber MFC (with Ni-Co alloy as cathode catalyst).

2 Material and Methods

The wastewater from municipal wastewater treatment plant was used in measurements. Research covered measurement of wastewater treatment level and electricity production. Research of wastewater treatment level covered reduction of COD, NH_4^+ and NO_3^- were conducted in the reactor without aeration, with aeration and with using a MFC [25]. Capacity of each reactor was equal $0.015 m^3$. In the first reactor the wastewater had contact with air only by wastewater mirror. In the second reactor the wastewater was aerated by air pump

* Corresponding author: barbara.wlodarczyk@uni.opole.pl

($0.27 m^3 \cdot h^{-1}$) and in the third reactor the wastewater was treatment with MFC. In test MFC the graphite brush as the anode (Chemviron Carbon) and Ni-Co alloy was used as catalyst of cathode. Ni-Co alloy were obtained by the method of electrochemical deposition and analysed by the XRD method. The alloys were deposited on mesh copper electrode. The alloys were deposited from a mixture of $NiSO_4$ and $CoSO_4$ [26, 27]. The alloys were obtained at temperature of 293K at current density of deposition id equal to 300 $A \cdot m^{-2}$ [26]. Composition of mixture for alloy deposited shows Table 1 [22].

Table 1. Mixture composition for deposited Ni-Co alloy on copper electrode.

Component	$NiSO_4$ x $7H_2O$ 195 $g \cdot l^{-1}$ $CoSO_4$ x $7H_2O$ 35 $g \cdot l^{-1}$
pH	2.0
Temperature	293 K
Current density	300 $A \cdot m^{-2}$
Co	50 %

Before the deposition of the alloy, mesh copper electrode was degreased in 25% aqueous solution of KOH (after degreasing, the surface shall be completely wettable with water), digested in acetic acid and washed with alcohol [26]. The time of the deposition was equal to 1 hour. The chemical composition of Ni-Co alloys was determined with the XRD method. The alloy with 50% of Co was selected for measurements [24, 26]. The Nafion 117 (183 μm) was used as PEM. The cathode with Ni-Co catalyst (in mesh form) was hot pressed into PEM. Cathode prepared in this way was used as one of reactor (MFC) wall (Fig. 1) [24].

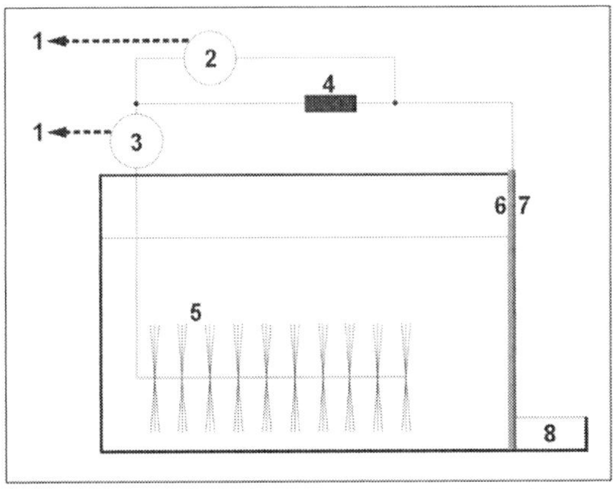

Fig. 1. Scheme of single chamber microbial fuel cell with graphite anode and Ni-Co cathode:
1 – to computer, 2 – voltmeter, 3 – ammeter, 4 – load, 5 – graphite anode, 6 – PEM, 7 – mesh cathode with Ni-Co catalyst, 8 – water (from cathode) container.

All three reactors were fed with the same wastewater with a COD concentration of 1755 ± 70 $mg \cdot l^{-1}$. The reactors were operated in batch mode till reaching 90% of COD removal (assumed reduction level) [25]. The temperature of measurements was equal 293K. During MFC operation, both voltage and current were also measured. The MFC was loaded with resistance equal 10Ω. The Fig. 2 shows scheme of measurement position.

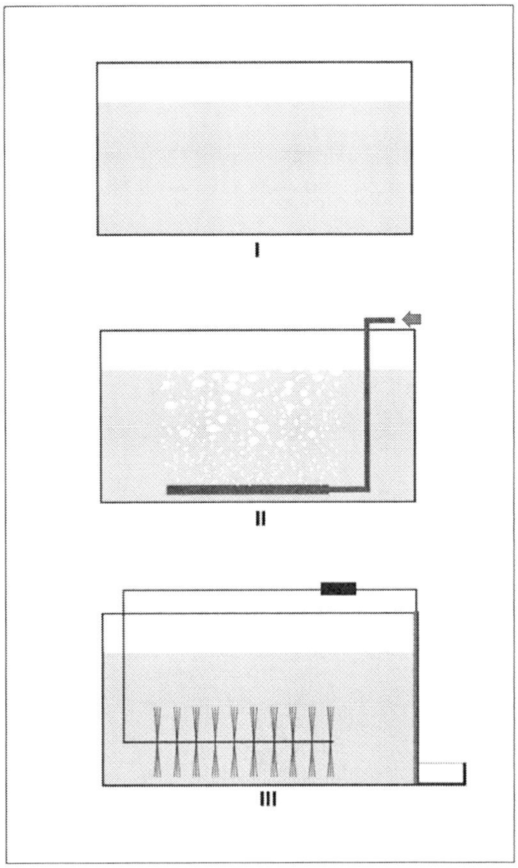

Fig. 2. Scheme of measurement position: I – without aeration, II – with aeration, II – with using MFC .

Table 2 shows the parameters of analysed wastewater.

Table 2. Parameters of analysed wastewater.

Parameter	Value
pH	6.7±0.1
COD [$mg \cdot l^{-1}$]	1,755 ± 70
NH_4^+ [$mg \cdot l^{-1}$]	11 ± 2
NO_3^- [$mg \cdot l^{-1}$]	4 ± 2

3 Results

Fig. 3 shows the decreasing the COD value during wastewater treatment without aeration, with aeration and with using a microbial fuel cell with Ni-Co cathode. Figs. 4 and 5 shows the change of concentration of NH_4^+ and NO_3^- during wastewater treatment without aeration, with aeration and with using single chamber MFC (with

 DOI: 10.1051/e3sconf/20171901025

Ni-Co cathode). All measurements were carried out at a temperature of 293K.

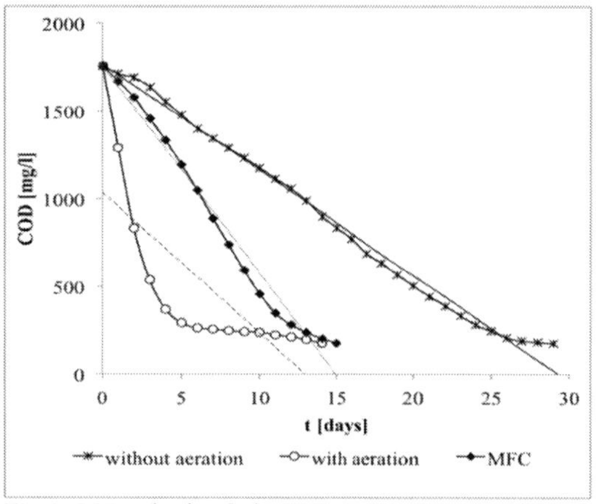

Fig. 3. COD reduction during wastewater treatment without aeration, with aeration and with using single chamber MFC with Ni-Co cathode.

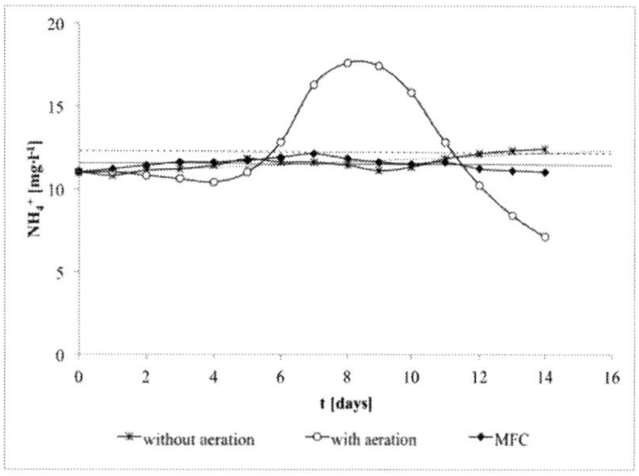

Fig. 4. Change of concentration in time of NH_4^+ during wastewater treatment without aeration, with aeration and with using single chamber MFC with Ni-Co cathode.

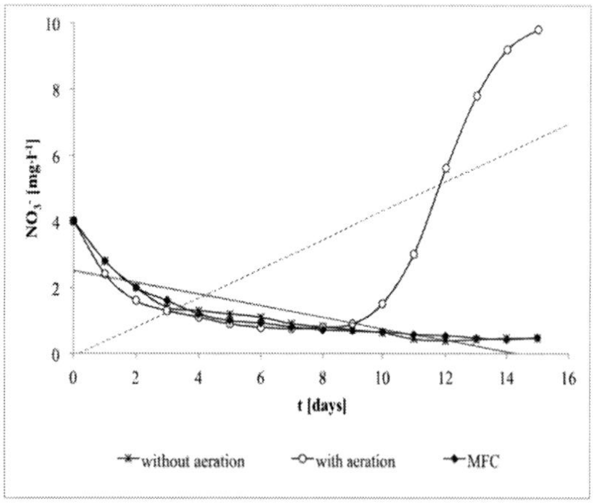

Fig. 5. Change of concentration in time of NO_3^- during wastewater treatment without aeration, with aeration and with using single chamber MFC with Ni-Co cathode.

The obtained current density was equal 0.24 mA·cm^{-2} and obtained amount of energy was equal 0.93 Wh. Based on voltage and current have been determined power curve. Figure 6 shows power curve of single chamber MFC with Ni-Co catalyst.

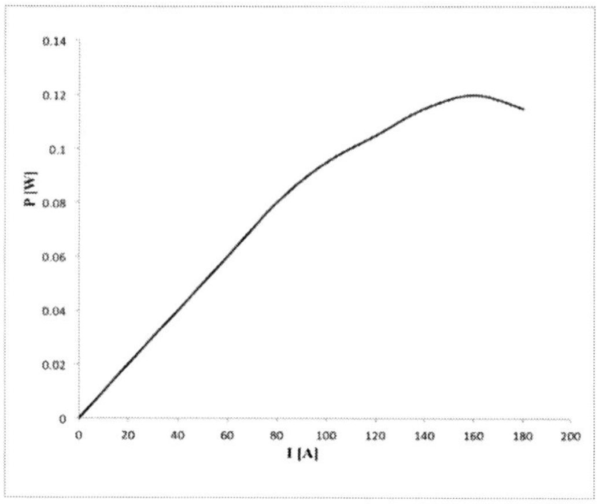

Fig. 6. Power curve of test single chamber MFC with Ni-Co catalyst.

4 Conclusions

The measurements show that in any cases (without aeration, with aeration and with single chamber MFC with Ni-Co cathode) was reached level 90% efficiency of COD reduction (Fig. 3). Measurements have shown the effectiveness of COD reduction in any cases (in any reactor) but the characteristics of curves are different. The characteristic curve of aeration is more preferred than characteristic curve of MFC (with Ni-Co cathode) because about 83% effectiveness of COD reduction after about 5 days (Fig. 3). But time for 90% (assumed reduction level) effectiveness reduction of COD with the use of single chamber MFC (15 days) is similar to the reduction time with aeration (14 days). The measurement of NH_4^+ reduction shows no changes in cases measurement in single chamber MFC (Fig. 4). The measurements (Fig. 5) have shown also the effectiveness of NO_3^- reduction (during 14 days). Increasing of NH_4^+ concentration (in II reactor - aerated) results from the attachment of hydrogen molecule to the ion ammonia (e.g. during putrefaction) (Fig. 4) [28]. The increase of NO_3^- concentration (Fig. 5) is the result of nitrification during the growth of bacteria [29,30]. The average current density (2.4 A·m^{-2}) and amount of energy (0.48 Wh) obtained in test single chamber MFC are low. But, if can accept a longer COD reduction time, the obtained amount of energy will allow elimination of the energy needed for wastewater aeration. So, use of MFC should be considered as a support traditional wastewater treatment. The obtained results are comparable to the results using e.g. carbon or Cu-B alloy as cathode catalyst [25,31].

The fundamental possibility of using of Ni-Co alloy as cathode catalyst of single chamber MFC for municipally wastewater was obtained in this paper.

References

1. US EPA. *Report. Clean watersheds needs survey overview* (2008)
2. A. Gawdzik, B. Włodarczyk, 9th ICEE, (2014). DOI:10.3846/enviro.2014.020
3. E. Logan, *Microbial fuel cell* (Wiley & Sons, Hoboken, 2008)
4. P. L. McCarty, J. Bae, J. Kim, Environ. Sci. Technol., **45**, 7100-7106 (2011) DOI:10.1021/es2014264
5. J. B. Davis, H. F. Yarbrough, Science, **137**, 615-616 (1962)
6. D.R. Bond, D.R. Lovley, Appl. Environ. Microbiol. **69**, 1548-1555 (2003)
7. S.K. Chaudhuri, D.R. Lovley, Nat. Biotechnol. **21**, 1229-1232 (2003)
8. H.J. Kim, H.S. Park, M.S. Hyun, I.S. Chang, M. Kim, B.H. Kim, Enzyme Microbiol. Technol. **30**, 145-152 (2002)
9. H.S. Park, B.H. Kim, H.S. Kim, H.J. Kim, G.T. Kim, M. Kim, I.S. Chang, Y.K. Park, H.I. Chang, Anaerobe, **7**, 297-306 (2001)
10. C.A. Pham, S.J. Jung, N.T. Phung, J. Lee, I.S. Chang, B.H. Kim, H. Yi, J. Chun, FEMS Microbiol. Lett. **223** (1), 129-134 (2003)
11. B.E. Logan, B. Hamelers, R. Rozendal, U. Schroder, J. Keller, W. Verstraete, K. Rabaey, Environ. Sci. Technol. **40** (17), 5181-5192 (2006)
12. H. Liu, R. Ramnarayanan, B. E. Logan, Environ. Sci. Technol. **38** (7), 2281-2285 (2004)
13. K. Rabaey, W. Verstraete, Trends Biotechnol. **23** (6), 291-298 (2005)
14. X. Wang X., Y. J. Feng, H. Lee, Water Sci. Technol. **57** (7), 1117-1121 (2008)
15. B. E. Logan, J. M. Regan, Trends Microbiol. **14** (12), 512-518 (2006)
16. C. Dumas, A. Mollica, D. Féron, R. Basséguy, L. Etcheverry, A. Bergel, Electrochimica Acta **53** (2), 468-473 (2006)
17. J.O'M. Bockris, A.K.N. Reddy, *Modern electrochemistry* (Kulwer Academic/Plenum Publishers, New York, 2000)
18. S. Cheng, H. Liu, B. E. Logan, Environ. Sci. Technol. **40** (1), 364-369 (2006)
19. E. Martin, B. Tartakovsky, O. Savadogo, Electrochimica Acta **58**, 58-66 (2011)
20. F. Zhao, F. Harnisch, U. Schröder, F. Scholz, P. Bogdanoff, I. Herrmann, Electrochem. Commun. **7** (12), 1405-1410 (2005)
21. B. Włodarczyk, P. P. Włodarczyk, Eng. Prot. Environ. **18** (2), 189-198 (2015)
22. P. P. Włodarczyk, B. Włodarczyk, Archives of Waste Manage. Environ. Prot. **17** (1), 111-118, (2015)
23. P. P. Włodarczyk, B. Włodarczyk, Civil Environ. Eng. Rep. **21** (2), 131-145 (2016)
24. B. Włodarczyk, P. P. Włodarczyk, Ecol. Eng. **18** (2), 210-216 (2017)
25. T. Huggins, P. H. Fallgren, S. Jin, Z. J. Ren, J. Microb. Biochem. Technol. **S6**:002 (2013)
26. P. P. Włodarczyk, B. Włodarczyk, Chinese Business Rev. **14** (3), 159-167 (2015)
27. P. P. Włodarczyk, B. Włodarczyk, China-USA Business Rev. **14** (5), 269-279 (2015)
28. J.E. McMurry, D.S. Ballantine, C.A. Hoeger, V.E. Peterson, M.E. Castellion, *Fundamentals of Inorganic, Organic and Biological Chemistry* (Pearson, London, 2009)
29. J. Łomotowski, A. Szpindor, *Modern wastewater treatment systems (in Polish)* (Arkady, Warsaw, 2002)
30. Z. Ren, H. Yan, W. Wang, M. M. Mench, J. M. Regan, Environ. Sci. Technol. **45** (6), 2435-2441 (2011)
31. B. Włodarczyk, P. P. Włodarczyk, J. Ecol. Eng. **18** (4), 224-230 (2017)

DOI: 10.1051/e3sconf/20171901026

Heat transfer coefficient as parameter describing ability of insulating liquid to heat transfer

Zbigniew Nadolny[1,*], *Przemysław* Gościński[1], and *Bolesław* Bródka[1]

[1]Poznan University of Technology, Institute of Electrical Power Engineering, 60965 Poznan, Poland

Abstract. The paper presents the results of the measurements of heat transfer coefficient of insulating liquids used in transformers. The coefficient describes an ability of the liquid to heat transport. On the basis of the coefficient, effectiveness of cooling system of electric power devices can be estimated. Following liquids were used for the measurements: mineral oil, synthetic ester and natural ester. It was assumed that surface heat load is about 2500 W·m^{-2}, which is equal the load of transformer windings. A height of heat element was 1.6 m, because it makes possible steady distribution of temperature on its surface. The measurements of heat transfer coefficient was made as a function of various position of heat element (vertical, horizontal). In frame of horizontal position of heat element, three suppositions were analysed: top, bottom, and side.

1 Introduction

Proper work of high voltage transformers depends on many factors. One of them is relatively low operating temperature, which is provided by effective cooling system [1-3]. The system consists of insulating liquid, which fills all free space and gaps in transformer. Heat transfer coefficient α determines the ability to heat transfer of the liquid. The bigger value of the coefficient, the more effective heat transfer between source of the heat and environment.

Knowledge of coefficient α of insulating liquid is essential at the design stage of the transformer (computer optimization) [4,5]. It can be also useful for transformers diagnostics [6,7]. It is assumed, that coefficient α is basically constant, and depends only on temperature. Meanwhile, reconnaissance research of the paper authors has shown that the coefficient α depends on many other factors, such as type of insulating liquids (mineral oil, synthetic ester, natural ester), the position of the heating element (vertical, horizontal), the length of the heating element and surficial heat load value. Disregard of the factors may lead to incorrect design of the transformer, which can result in increased temperature of work. Increased temperature is a reason of intensive ageing process of insulating system of transformers. The increase of temperature of several degrees results in a double reduction in the lifetime of the transformer.

According of Montsinger law (law of 8 degrees), reliability of paper insulation, impregnated with mineral oil, as a function of temperature, is described by following formula [8,9]:

$$t = 7.154 \cdot 10^4 \cdot e^{-0{,}0865 \cdot T} \qquad (1)$$

where:

t – absolute lifetime of transformer insulation [years],
T – insulation work temperature [°C].

The law is based on the initial assumption that at a temperature of 95 °C lifetime of insulation transformer is 20 years. Such a lifetime is adopted by many transformer designers. This law specifies the lifetime of transformer insulation working in any temperature. The name of "the law of 8 degrees" means that the increase or decrease in temperature of 8 °C will result in, respectively, double shortening or lengthening the lifetime of transformer insulation (Fig. 1). At temperature of 95 °C the lifetime of transformer insulation is 20 years. At a temperature of 8 degrees lower (87 °C) the lifetime is twice as long and is 40 years, and a temperature of 8 degrees higher (103 °C) the lifetime is twice less and is only 10 years. Some drawback of Montsinger law is that the law does not consider other factors such as moisture and oxygen influence [10,11].

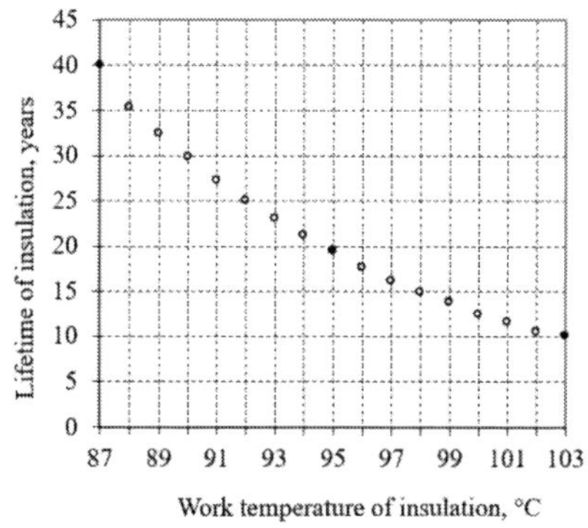

Fig. 1. Lifetime of transformer insulation *t* as a function of work temperature *T* according to the Montsinger law of 8 degrees, based on formula (1).

* Corresponding author: zbigniew.nadolny@put.poznan.pl

 DOI: 10.1051/e3sconf/20171901026

Bussing and Arrhenius-Dankin formulas, used in US to evaluate the influence of temperature on ageing process of insulation system made from organic materials, are similar in physical sense to Montsinger law [12,13].

2 Aim and range of investigation

The purpose of the research is the determination of heat transfer coefficient α of insulating liquid as a function of the liquid type and position of heating element. Following liquids were investigated: mineral oil (Nynas Draco), synthetic ester (Midel 7131) and natural ester (Envirotemp FR3). All material parameters of mentioned liquids are described in authors papers [1-3,14-17]. Vertical and horizontal positions of the heating element were analyzed. In frame of horizontal position of the element, three positions of cooling surface relative to the heating element (top, bottom, side – Fig. 2) were investigated. Each measurement was repeated three times.

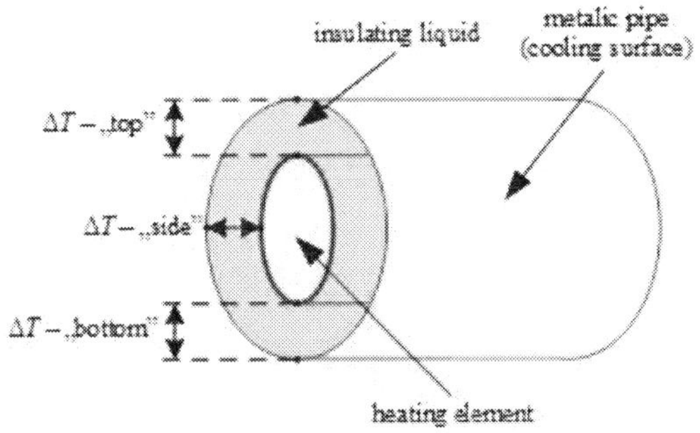

Fig. 2. Three positions of cooling surface relative to the heating element: top, bottom, and side.

3 Measurement system

Measurement system consists of heat model of transformer (Fig. 3), temperature sensors and a system to register a temperature. Heat model of transformer consists of the heating element and metallic pipe (cooling surface). Space between metallic pipe and the heating element was filled by insulating liquid.

Height of metallic pipe was 2 m, its diameter was 88.9 mm, the thickness of the pipe wall was 3.2 mm. The heating element was electric heater, its length was 1.6 m, and its diameter was 22 mm. Lateral surface of the heating element was 0.11 m^2. All mentioned data were necessary to calculate surficial heat load p (formula (2)). Space between the heating element and metallic pipe was filled by the liquid with volume 8 liters.

Two sensors to measure temperature were placed on external surface of the heating element and external surface of metallic pipe. The temperature difference ΔT, measured using the sensors, was temperature decrease necessary to calculate heat transfer coefficient α of the liquid, using following formula:

$$\alpha = p \cdot (\Delta T)^{-1} \quad (2)$$

where:
α – heat transfer coefficient of insulating liquid [$W \cdot m^{-2} \cdot K^{-1}$],
p – surface heat load [$W \cdot m^{-2}$],
ΔT – temperature decrease in insulating liquid [K].

Authors decided to place one of the sensors on external surface of metallic pipe, because it was more convenient, comparing to internal surface of the pipe. In this case, measured temperature difference ΔT consists of two temperature differences: in liquid and in the pipe. However, temperature difference in metallic pipe is much smaller comparing to the difference in liquid, what is caused by relatively big thermal conductivity λ of the metal. Finally, temperature difference in the pipe can be omitted.

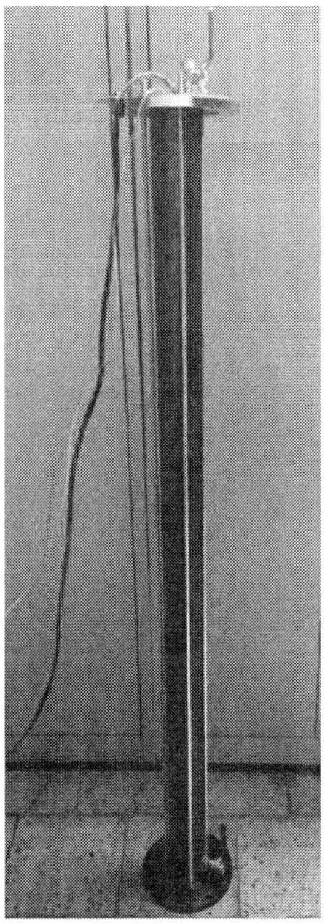

Fig. 3. Picture of measurement system: metallic pipe filled by insulating liquid, the heating element placed inside metallic pipe.

Applied voltage to the heating element was 101 V, applied current was 2.75 A. Surface heat load on lateral surface of the heating element was 2525 $W \cdot m^{-2}$.

Temperature on the heating element surface, and on metallic pipe surface, was stable after about 5 hours.
In order to increase an accuracy of measurement, registration of temperature using the sensors was made after 6 hours since the voltage was applied on the heating element.

4 Results

Table 1 presents measurement results of heat transfer coefficient α of insulating liquid as a function of liquid type and position of the heating element.

Independently from position of the heating element, mineral oil has the biggest heat transfer coefficient α. The coefficient α of mineral oil was a dozen percent bigger than the coefficient α of natural and synthetic esters. For example, for vertical position of the heating element, coefficient α for mineral oil was 115.9 $W \cdot m^{-2} \cdot K^{-1}$, and for natural ester was 93.5 $W \cdot m^{-2} \cdot K^{-1}$ – about 19% smaller. So big difference results in different temperature decrease ΔT in the liquid. The decrease ΔT for mineral oil was 21.8 °C, and for natural ester was 27.0 °C (Table 1). It means that in case of transformer filled with natural ester, temperature will be bigger by more than 5 °C, compared to transformer filled with mineral oil. So big difference of temperature will make the ageing process of transformer insulation system faster, it means it will make the ageing process of whole transformer faster. Low viscosity of mineral oil, comparing to viscosity of both types of esters, is a reason of the biggest heat transfer coefficient α of the oil.

Table 1. Heat transfer coefficient α of insulating liquid as a function of liquid type and position of the heating element, estimated using temperature decrease measurements; T_1 – temperature on cooling surface [°C], T_2 – temperature on the heating element [°C], ΔT – temperature decrease in liquid [°C], α – heat transfer coefficient of the liquid [$W \cdot m^{-2} \cdot K^{-1}$].

Liquid type	Position of the heating element	Position of cooling surface	T_1	T_2	ΔT	α
Mineral oil Nynas Draco	vertical	-	75.5	97.3	21.8	115.9
	horizontal	top	69.1	82.3	13.2	190.8
		side	64.3	80.1	15.8	160.2
		bottom	57.4	73.9	16.5	152.7
Natural ester Envirotemp FR3	vertical	-	77.4	104.4	27.0	93.5
	horizontal	top	68.4	84.2	15.8	159.5
		side	63.8	81.7	17.9	140.8
		bottom	57.3	76.1	18.8	134.3
Synthetic ester Midel 7131	vertical	-	75.3	101.5	26.2	96.6
	horizontal	top	66.3	84.1	17.8	142.1
		side	64.0	81.7	17.7	142.7
		bottom	57.5	75.7	18.2	138.5

Comparing natural and synthetic esters it is possible to say that coefficient α of synthetic ester was a few percent bigger than coefficient α of natural ester.

Horizontal position (top position of cooling surface relative to the heating element) was an exception. In this case, coefficient α of natural ester was bigger than coefficient α of synthetic ester.

On the basis of measurement results it can be concluded that type of used insulating liquid has some influence on heat transfer coefficient α of the liquid. From the point of view of the effectiveness of transformer cooling system, the use of mineral oil is more efficient, compared with synthetic or natural esters. In turn, comparing natural and synthetic esters, it can be said that there is no difference between the esters from coefficient α point of view.

Obtained results are consistent with other results of authors described in their papers [14-17]. In the papers, heat transfer coefficient α was calculated on the basis of measurements of five thermal properties of the liquid: thermal conductivity λ, specific heat c_p, viscosity υ, density ρ, and thermal expansion coefficient β.

Independently from type of the liquid, bigger heat transfer coefficient α was for horizontal position of the heating element, comparing with vertical position. For example for mineral oil, for vertical position coefficient α was 115.9 $W \cdot m^{-2} \cdot K^{-1}$, and for horizontal position (top position of cooling surface relative to the heating element) coefficient α was 190.8 $W \cdot m^{-2} \cdot K^{-1}$ – about 65% bigger. In first case, temperature decrease in mineral oil ΔT was 21.8 °C, and in second case ΔT was 13.2 °C (Table 1). Difference of temperature decreases ΔT was 8.6 °C in favor of the horizontal position of the heating element. These differences probably result of path length, which must be done by hot liquid from the heating element to cooling surface.

In case of vertical position of the heating element, the path is relatively long, because it is equal to the length of the heating element and a length of metallic pipe. In turn, for horizontal position of the heating element, the path of hot liquid from the heating element to cooling surface is relatively very short, almost equal to the thickness of the liquid.

Analyzing three variants of horizontal position of the heating element, it must be concluded that heat transfer coefficient α is the biggest in case of variant of top position of cooling surface relative to the heating element. In turn, coefficient α is the smallest in case of variant of bottom position of cooling surface relative to the heating element. In first case, transfer of heat is based on natural convection phenomenon – the move of hot liquid from the heating element up to cooling surface. In second case, transfer of heat is not based on natural convection, because hot liquid is on top (close to the heating element), and cool liquid is bottom (close to cooling surface). That is why transfer of heat in based mainly on thermal conductivity phenomenon. For example, for mineral oil, coefficient α was 190.8 $W \cdot m^{-2} \cdot K^{-1}$ for variant of top position of cooling surface relative to the heating element, and only 152.7 $W \cdot m^{-2} \cdot K^{-1}$ (about 20% smaller) for variant of bottom position of cooling surface relative to the heating element. Temperature decrease ΔT was 13.2 °C in first case, and 16.5 °C in second case. Difference of temperature

 DOI: 10.1051/e3sconf/20171901026

decreases ΔT was 3.3 °C in favor of the variant of top position of cooling surface relative to the heating element.

On the basis of mentioned results, it can be said that position of the heating element has significant influence on heat transfer coefficient α of the liquid. Horizontal position makes that coefficient α is tens of percent bigger than coefficient α for vertical position. In turn, in frame of horizontal position, heat transfer coefficient α is the biggest for variant of top position of cooling surface relative to the heating element.

5 Conclusions

Type of used insulating liquid has some influence on heat transfer coefficient α of the liquid. Mineral oil has much bigger heat transfer coefficient α compared to natural and synthetic esters. Comparing natural and synthetic esters, it must be concluded that there is no difference between the esters from coefficient α point of view.

Position of the heating element has influence on heat transfer coefficient α of the liquid. Horizontal position makes that coefficient α is tens of percent bigger than coefficient α for vertical position. In case of horizontal position, heat transfer coefficient α is the biggest for variant of top position of cooling surface relative to the heating element, comparing to other variants.

The research was financed from resources of the Ministry of Science and Higher Education for statutory activities No. 04/41/DS-PB/4293, name of the task: The Influence of Various Parameters on Thermal Properties of Dielectric Liquids Used in High Voltage Transformer

References

1. R. Lopatkiewicz, Z. Nadolny, Prz. Elektrotechniczny **84** (2008)
2. G. Dombek, Z. Nadolny, P. Przybylek, Int. Conf. High Voltage Eng. Appl. **1-4** (2014)
3. R. Lopatkiewicz, Z. Nadolny, P. Przybylek, W. Sikorski, Prz. Elektrotechniczny **88** (2012)
4. J. Jajczyk, L. Kasprzyk, Int. Conf. Heat Transfer, Therm. Eng. and Environ., 226-231 (2008)
5. L. Kasprzyk, A. Tomczewski, K. Bednarek, Prz. Elektrotechniczny **86** (2010)
6. T. Boczar, A. Cichon, S. Borucki, IEEE Trans. Dielectr. Electr. Insul. **21** (2014)
7. H. Majchrzak, A. Cichon, S. Borucki, Arch. Acoust. **42** (2017)
8. E. Jezierski, *Budowa i projektowanie* (WNT, Warszawa, 1963)
9. A Inhof, *Materiały elektroizolacyjne w technice wysokich napięć* (WNT, Warszawa, 1963)
10. F. Mosiński, B. Bocheński, Między. Konf. Transformatorowa "Transformator", 161-169 (2007)
11. F. Mosiński, Konf. Naukowo-Techniczna "Transformatory w eksploatacji", 117-119 (2005)
12. K. Kolebiński, *Materiałoznawstwo elektro-techniczne* (WNT, Warszawa, 1976)
13. M. Kaźmierski, Między. Konf. Transformatorowa "Transformator", 1.2/1-1.2/13 (2011)
14. G. Dombek, Z. Nadolny, Eksploatacja i Niezawodnosc – Maintenance and Reliability **18** (2016)
15. Z. Nadolny, G. Dombek, P. Przybylek, IEEE 2016 Conf. Electr. Insul. Dielectr. Phenom., 857-860 (2016)
16. Z. Nadolny, G. Dombek, P. Przybylek, D. Przadka, IEEE 2016 Conf. Electr. Insul. Dielectr. Phenom. 538-541 (2016)
17. G. Dombek, Z. Nadolny, Eksploatacja i Niezawodnosc – Maintenance and Reliability **19** (2017)

DOI: 10.1051/e3sconf/20171901027

Analysis of power and energy losses in power systems of electric bus battery charging stations

Jarosław Jajczyk[1,*], *Arkadiusz* Dobrzycki[1], *Michał* Filipiak[1], and *Dariusz* Kurz[1]

[1]Poznan University of Technology, ul. Piotorowo 3a, 60-965 Poznan, Poland

Abstract. Two variants of electricity transmission from the Transfer-Switching Station (TSS) to the battery charging station were analysed in the paper. The first variant under analysis referred to the transmission of electricity via a three phase a.c. line with rated voltage of 15 kV. In the second variant of electricity transmission, the d.c. line with the working voltage of 1500 V was used. For both variants, lines and other equipment such as transformers, rectifier system and voltage stabilisation system in the battery charging station were modelled. For both solutions, analysis of energy per annum was conducted depending on the distance of the charging station from the TSS. On top of this, the simultaneous operation of several chargers was taken into account, which would correspond to the case of charging many buses at the same time from a single power line. The paper demonstrates that in the case of the analysed electricity transmission systems in the electric bus battery charging systems it is possible to use a more advantageous solution, which is characterised by reduced power and energy losses.

1 Introduction

Electrically operated buses are becoming a more and more popular means of transport. The main reason for the use of the electric power in the municipal transport is the carbon efficiency which is of particular significance in densely populated areas (cities). The examples of cities, where such solutions are already in use include: Hamburg, Stockholm, Gothenburg, Vienna, Stuttgart or Geneva. On top of this, electric buses are operated at airports, i.e. places where their range may be relatively small.

The use of the electric buses does not require such high investment outlays as is the case with trams (among other things, the necessity to build tracks). In the simplest variants, it is enough to provide the depot with the charging station. However, such an approach limits the effective use of the rolling stock (the necessity to return to the depot for the charging time, the limitation of the length of lines and operating time due to the capacity of the batteries).

A solution that ensures a fuller use of the electric bus potential is to ensure the appropriate infrastructure that allows for the charging of batteries of the buses as they are in service. The solution is implemented mainly by means of fast charging stations [1-4]. The stations are situated at selected bus stops in such a way as to make it possible for the bus to cover another section of the route after the batteries are charged. One of the three following methods is used to perform the charging process: plug-in charging, by means of a pantograph or the wireless charging [1, 2, 5]. Efficient and correct performance of the battery charging process requires the appropriate powering of the charging station [6-10]. Such a station, depending on the configuration, may be powered with alternating current (a.c.) or direct current (d.c.).

The paper provides the comparison of the methods of powering of the battery charging stations in electric buses depending on the type of the supply voltage (a.c. or d.c.). The structure of both systems was presented, commencing from the TSS and ending with the charging station. The efficiencies of its respective sections were taken into consideration and on this basis, analysis of power and energy losses was carried out during the charging process with the electricity transmission by means of a.c. or d.c. lines.

2 Power supply for the battery charging station

In the case of charging of the bus batteries during short stops at bus stops, it is necessary to supply them with appropriately high power which allows for the fast charging [1, 2, 3, 4, 10]. Such a solution requires the application of a separate power line, especially in the case of locating several charging stations at a single site e.g. at intersections. Two approaches are applied in practical solutions [1, 2]. The first approach consists in the powering of the charging station by means of a three-phase alternating current line (Fig. 1). In this case, after reducing voltage in the TSS from HV to MV, electricity is transmitted via the medium voltage line. Such a solution requires the provision of the remaining elements of the power supply system within the charging station, i.e.: MV/LV transformer, rectifier system, voltage

* Corresponding author: jaroslaw.jajczyk@put.poznan.pl

DOI: 10.1051/e3sconf/20171901027

adjustment and charging control system (d.c./d.c.).

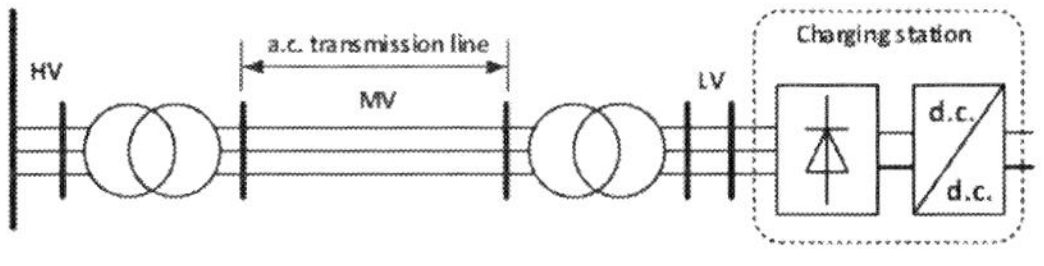

Fig. 1. A.C. line power supply.

In the second method, the direct current line (Fig. 2) is used to transmit electricity. In this case, the TSS, except the HV/MV transformer, also includes the MV/LV transformer and the rectifier system. Electricity is transmitted using the d.c. line. The level of voltages is adjusted in the charging station by means of the d.c./d.c. converter, which is also responsible for the charging process control.

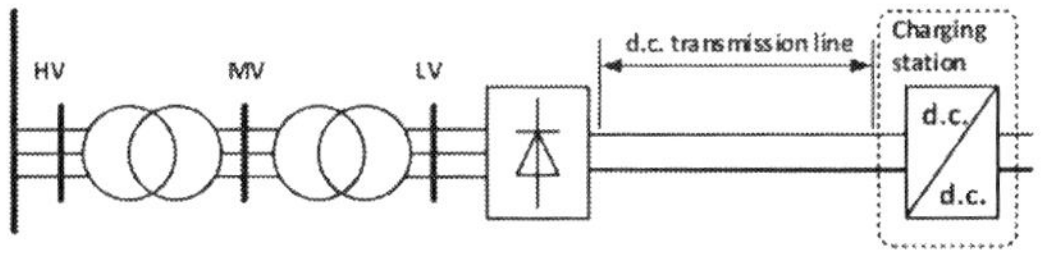

Fig. 2. D.C. line power supply.

In this solution, it is also possible to power the charging station from the tram traction. The chances are that such a solution will be applied in municipal agglomerations, where the transport network including trams, trolleybuses or subway is operative. In order to power the charging station, it is also possible to take advantage of the railway traction network, especially when the bus stop is located close to it and there is no adequate power infrastructure.

3 Analysis of power and energy losses in power supply systems for battery charging circuits

Actions related to the natural environment protection in the field of transport, and especially taking into account the issues regarding electric vehicles, cannot be understood as reduction of emissions of pollution into the air exclusively. In this case, an important issue is to strive for the highest possible efficiency of the sources and electricity transmission systems [6, 7, 9, 11]. A persistent problem is the issue of optimisation of the system [2, 12, 13, 14, 15], which consists, among other things, in the minimisation of the losses related to the transmission and conversion of electricity (rectification, adaptation of the voltage level). While balancing the amount of energy necessary to power an electric vehicle, it is also important to take into account the energy lost along the source-battery route. As well as the thermal losses related to the electricity conversion and flow of current [6, 9], in the case of the electricity transmission via lines with higher voltages, it is also necessary to consider the possible losses related to other physical phenomena [7, 9, 10, 12, 16].

The subject of analysis in this paper is such a configuration of the power supply system, in which the TSS is the source. In the case of the charging stations supplied with alternating current, the analysed elements (considered important from the point of view of the transmission efficiency) include:

- the HV/MV transformer located in the TSS,
- the MV transmission line,
- the MV/LV transformer located in the charging station,
- the rectifier system in the charging station,
- the voltage adjustment and charging control system.

In the case of the DC power supply for the charging station, the elements taken into consideration in the analysis include:

- the HV/MV and MV/LV transformers located in the TSS,
- the rectifier system located in the TSS,
- the d.c. transmission line,
- the voltage adjustment and charging control system.

The above considerations leave out the effect of the charging station equipment which consumes electricity for its own needs (e.g. for active cooling of the converter systems).

The following formula is used to calculate the total power losses in the power supply system (1):

$$\Delta P = P_{TSS} \cdot \left(1 - \prod_{i=1}^{n} \eta_i\right) \qquad (1)$$

where:
ΔP – total power losses of the transmission system,
P_{TSS} – power on output terminals of the power supply point,
η_i – efficiency of the i-th element of the power supply system,
n – number of components of the power supply system (in the issue under consideration n=5).

Power losses in transmission system ΔW were calculated using the following equation (2):

$$\Delta W = \Delta P \cdot t_r \qquad (2)$$

where:
t_r – time of electricity transmission during a year.

Efficiencies of the respective elements of the power supply system (transformers, converters) were selected on the basis of datasheets. An exception to this is the efficiency of the transmission line. As is known, power losses in the line depend both on the volume of the transmitted power and on the physical parameters of the line per se e.g.: resistivity, geometry or type of the line. In order to calculate the line efficiency (η_{line}) based on the power losses, the following formula is used (3):

$$\eta_{line} = \left(1 - \frac{P \cdot \rho \cdot l}{U^2 \cdot \cos^2 \varphi \cdot s}\right) \cdot 100 \ [\%] \qquad (3)$$

where:
P – power transmitted via the line,
U – rated voltage of the transmission system,
φ – phase angle of the receiver,
l – length of the transmission line,
ρ – resistivity of the line material,

s – cross-section of the transmission line.

An assumption was made that the power lines are cable lines, that is, they are dominant in the presently built municipal power lines.

4 Exemplary calculations of power and energy losses

Calculations of the power and energy losses were performed for cases of powering of one, two or three stations using a single power line. The assumption in the paper is that the power transmitted to one battery charging station amounts to 150 kW and 3 stations are powered by maximum one line.

To determine the power losses, formula (1) was used. The assumed efficiencies of the respective components of the charging station power supply system were listed in table 1. The ranges of efficiency were provided for the a.c. and d.c. transmission lines as efficiency value depends on, among other things, the length of the line. The actual values of these efficiencies were calculated in accordance with formula (3). For the calculation purposes, the minimum acceptable cross-section of cables was taken into account.

Table 1. Assumed efficiencies of the respective elements of the transmission system [1, 2, 5].

Element	η [%]
HV/MV transformer	99
MV/LV transformer	97
Rectifier system	99
d.c./d.c. voltage adjustment system	99
High performance a.c./d.c. charging station	95
a.c. transmission line	(99.5 ÷ 99.9)/km – depending on the transmitted power
d.c. transmission line	(97.0 ÷ 99.9)/km – depending on the transmitted power

Fig. 3 presents the exemplary relative power losses for a situation in which one station is powered. Detailed results of the calculations can be found in the paper by [17]. The obtained results suggest the consideration of energy losses related to the transport of energy to the charging station as potentially small differences in power losses may translate into significant differences in annual energy loss.

In order to analyse the energy losses, it was assumed that each station works occasionally in the 1/6- minute cycle on working days (which represents an annual average of 251 days, 16 hours each) and in the 1/12-minute cycle on holidays (which represents an annual average of 114 days, 16 hours each). On this basis, the average yearly working time t_r of one station was determined. These assumptions were made based on the statistical data and average frequency of arrival of a bus to the bus stop [18, 19]. The paper analyses the energy losses for the distance of the charging station from the TSS located within a radius of 1-20 km.

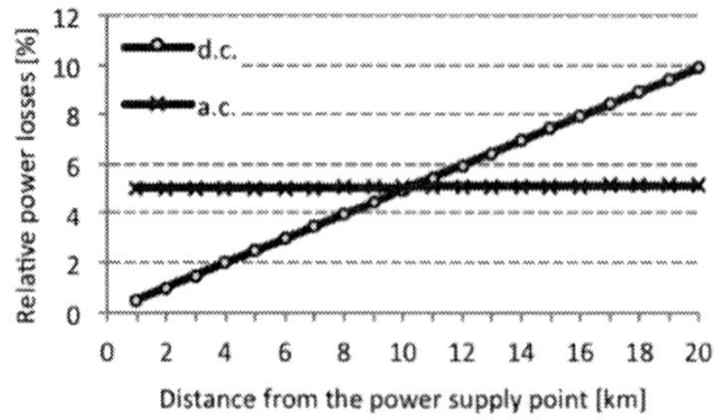

Fig. 3. Relative power losses for the power supply of one station with the capacity of 150 kW, depending on the distance from the power supply point.

The results of calculations of annual energy losses depending on the distance from the TSS for two power supply variants (a.c. line and d.c. line) are presented in Fig. 4÷6. The figures present the relationship of these losses in the case of transmission of power amounting to 150 kW (Fig. 4), 300 kW (Fig. 5) and 450 kW (Fig. 6).

Data listed in table 1 as well as the exemplary results of calculations illustrated in Figs. 3÷6 allow for a conclusion that in the case of the a.c. system, the element that generates the highest losses is the charging station and the impact of other factors is much less significant. On the other hand in the case of the d.c. system, the decisive factors include transmission losses related directly to the length of the line and the transmitted power. On top of this, in the case of reduced transmitted power, (1 or 2 charging stations powered from the line) it is possible to indicate the distance ranges for which reduced energy losses will be obtained using the d.c. systems.

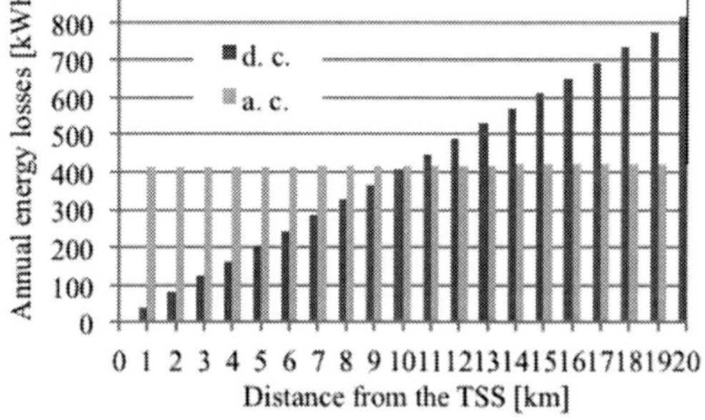

Fig. 4. Annual energy losses for the power supply of one station with the capacity of 150 kW.

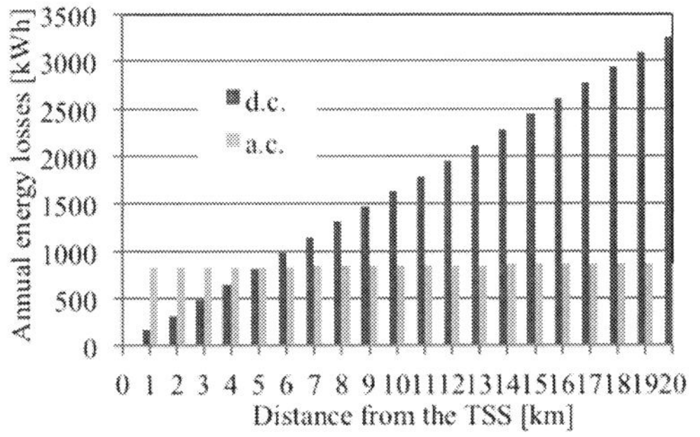

Fig. 5. Annual energy losses for the power supply of two stations with the total capacity of 300 kW (2x150 kW).

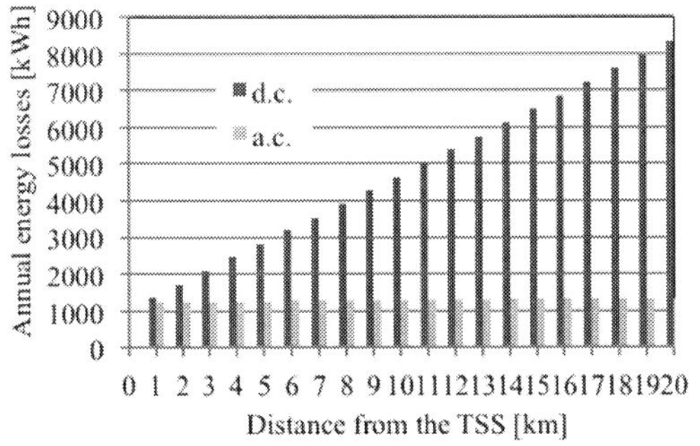

Fig. 6. Annual energy losses for the power supply of three stations with the capacity of 450 kW (3x150 kW).

5 Conclusions

The paper presents calculations of energy losses for different variants of power supply as the function of distance of the electric bus battery charging stations from the TSS. Simultaneous operation of several chargers was simulated, which corresponds to the case of charging of the batteries of many buses at the same time from a single power line. The results were presented in the graphic form. On their basis, the energy losses during the charging process were compared for the transmission of electricity via a.c. or d.c. lines.

The paper has demonstrated that for the analysed (a.c. or d.c.) electricity transmission systems in the electric bus battery charging systems, it is possible to indicate a more advantageous solution (characterised by smaller power and energy losses, i.e. higher efficiency) using direct or alternating voltage. The choice depends, among other things, on the efficiency of the elements, which are part of the electricity transmission system, the length of the line and the value of the transmitted power.

Analysis of the results of calculations allows for the conclusion that reduced losses are related to the use of d.c. systems for shorter lines and reduced power. On the other hand, in the case of longer distances, it is more advantageous to use a.c. systems. Unfortunately, it is not possible to indicate unambiguously an economically more effective solution. As well as the costs of construction of the infrastructure, it seems necessary to take into account the analysis of potential costs related to energy losses. In particular, this is noticeable in the case systems with reduced capacity (1-2 charging stations), as in the case of the long-term operation, this factor may turn out to be extremely important, considering a systematic increase in the prices of electricity. It is also necessary to remember about the main reason for use of electric vehicles, i.e. the minimization of pollution emissions. Ensuring high effectiveness of the used power supply system indirectly affects the reduction in pollution emissions.

References

1. M. Rogge, D.U. Sauer, S. Wollny, Energies **8**, (2015)
2. S. Bai, D. Yu, S. Lukic, *Energy Conversion Congress and Exposition*, 1178-1184 (IEEE, 2010)
3. L. Kasprzyk, Eksploat. Niezawodn., **19** (2017)
4. D. Burzyński, L. Kasprzyk, E3S Web of Conferences, **14**, 01041 (2017)
5. M. Filipiak, *Computational Problems of Electrical Engineering*, 1-3 (IEEE Xplore, Sandomierz, 2016)
6. K. Bednarek, Prz. Elektrotechniczny, **84/12** (2008)
7. K. Bednarek, R. Nawrowski, A. Tomczewski, *Studies in Applied Electromagnetics and Mechanics*, **22**, 363-368, (IOS Press, Amsterdam, Berlin, 2002)
8. K. Bednarek, J. Jajczyk, Prz. Elektrotechniczny **85/12** (2009)
9. Z. Piatek, T. Szczegielniak, D. Kusiak, Prz. Elektrotechniczny **87/5** (2011)
10. G. Dombek, Z. Nadolny, Eksploat. Niezawodn. **18**, (2016)
11. D. Kurz, *Computational Problems of Electr. Eng.*, 1-5 (IEEE Xplore, Sandomierz, 2016)
12. J. Jajczyk, L. Kasprzyk, *Int. Conf. Heat Transfer, Therm. Eng. Environ.*, 226-231 (WSEAS, Rhodes, 2008)
13. L. Kasprzyk, Prz. Elektrotechniczny **88/7B** (2012)
14. J. Jajczyk, *Computational Problems of Electr. Eng.*, 1-3 (IEEE Xplore, Sandomierz, 2016)
15. J. Jajczyk, Prz. Elektrotechniczny **90/4** (2014)
16. W. Opydo, A. Dobrzycki, Electr. Eng. **94** (2012)
17. A. Dobrzycki, M. Filipiak, J. Jajczyk, Poznan Univ. of Technol. Acad. J. Electr. Eng. **92** (2017)
18. Główny Urząd Statystyczny (http://www.gus.gov.pl/ 4.05.2017)
19. http://www.mpk.poznan.pl/, (4.05.2017)

Online identification of wind model for improving quadcopter trajectory monitoring

Ryszard Beniak[1], and *Oleksandr* Gudzenko[1,*]

[1]Opole University of Technology, Prószkowska Street 76, 45- 758 Opole, Poland,

Abstract. In this paper, we consider a problem of quadcopter control in severe weather conditions. One type of such weather conditions is a strong variable wind. In this paper, we ponder deterministic and stochastic models of winds at low altitudes with the quadcopter performing aggressive maneuvers. We choose an adaptive algorithm as our control algorithm. This algorithm might seem suitable one to solve the given problem, as it is able to adjust quickly to changing conditions. However, as shown in the paper, this algorithm is not applicable to rapidly changing winds and requires additional filters to smooth the impulse streams, so as not to lose the stability of the object.

1 Introduction

The development of modern technologies in the field of aerodynamics, composite materials, electronics, satellite navigation and robotics allowed to reach a qualitatively new level for the creation of unmanned aerial vehicles [1-5]. Now there are many different unmanned aerial vehicles or drones for solving various problems. One kind of such devices is a quadcopter.

Quadcopters are unmanned aerial vehicles with four rotors arranged vertically. The advantages of quadcopters include mobility, low cost and maneuverability. Such drones can reach speeds of up to 70 km/h. This fact makes difficult to control them remotely. Therefore, the best solution is to create autonomous control systems.

The priority directions of the quadcopter's development are defense, law enforcement and nature protection activities, rescue operations, environmental monitoring and scientific research [1,2]. They help solve dangerous for human problems, like penetrating the emergency nuclear reactors or plunging into the mouth of a volcano.

However, to solve these problems, it is necessary to develop a control system capable of responding to various external influences, so that the drone does not lose stability during the flight. Also one of the disadvantages of quadcopters is a limited supply of energy. For this, it is necessary that the drone overcomes the problem as quickly and efficiently as possible.

To perform these high requirements, it is necessary to develop an autonomous control system that takes into account these factors.

The purpose of this paper is to develop an adaptive control system that can take into account the wind shear during the quadcopter's movement. The term "wind shear" is used for light aircrafts to describe the impact of wind on drones' behavior. To make our reasoning more complex, we study aggressive maneuvers of the drone and allow for variable as well as instantly changing speed and/or direction of the wind. The work is a continuation of the research presented in papers [6,7].

2 Wind models

Usually, quadcopters are used at low altitudes. At these altitudes, the term "wind shear", introduced in Chapter 1, refers to a set of air movements in the lower layers of the atmosphere, from small gusts and turbulence to a powerful displacement of the air mass of one layer relative to the other. The most common definition of wind shear is the following: "change in wind speed and/or direction in space, including ascending and descending air currents". This explanation implies that any atmospheric phenomenon or even any physical obstacle in the path of the prevailing airflow, resulting in a change in the wind speed and/or direction, is essentially the cause of the wind shear. The wind shear is always present in the atmosphere [8].

The wind shear is important because of its impact on the flight characteristics of the drones and, in consequence, due to the potentially adverse effect on safety.

In general, the wind velocity $\mathbf{V}_{wind}$ in the lower layers of the atmosphere can be represented in the form (1):

$$\mathbf{V}_{wind} = \mathbf{V}_{const} + \mathbf{V}_{rand}, \qquad (1)$$

where $\mathbf{V}_{const}$ and $\mathbf{V}_{rand}$ – deterministic and stochastic components of wind velocity, respectively.

Consider the deterministic component of the wind velocity. In the lowest layers of the atmosphere, the

* Corresponding author: o.gudzenko@doktorant.po.edu.pl

direction of the wind changes slowly, and the wind velocity increases with altitude according to the logarithmic law of wind [8], represented in the form (2).

$$\mathbf{V}_{const} = \frac{\mathbf{V}_0}{k_{karman}} \ln\left(\frac{z+z_0}{z_0}\right), \tag{2}$$

where $\mathbf{V}_{const}$ – wind velocity at altitude z, $\mathbf{V}_0$ – "friction velocity", k_{karman} – Karman constant, $k_{karman} = 0.38$, z_0 – layer thickness of surface roughness.

Consider the stochastic component of wind velocity. The results of experimental studies indicate that for unsteady part of the wind, a model specified by the von Karman turbulence spectrum can be used [8]. Unfortunately, the von Karman spectrum does not lead to a rational transfer function. A suitable approximation of the von Karman model is given by the Dryden transfer functions (3) [8].

$$\begin{gathered} G_u(S) = \sigma_u \sqrt{\frac{2V_a}{L_u}} \cdot \frac{1}{\left(S + \frac{V_a}{L_u}\right)}, \\ G_v(S) = \sigma_v \sqrt{\frac{3V_a}{L_v}} \cdot \frac{\left(S + \frac{V_a}{\sqrt{3}L_v}\right)}{\left(S + \frac{V_a}{L_v}\right)^2}, \\ G_w(S) = \sigma_w \sqrt{\frac{3V_a}{L_w}} \cdot \frac{\left(S + \frac{V_a}{\sqrt{3}L_w}\right)}{\left(S + \frac{V_a}{L_w}\right)^2}, \end{gathered} \tag{3}$$

where u, v, w – unit vectors in the coordinate system associated with the quadcopter, $\boldsymbol{\sigma}$ – intensity of turbulence, $\mathbf{L}$ – spatial wavelengths, V_a – quadcopter velocity. Parameters for the Dryden model of wind gusts are defined in MIL-F-8785C [9].

Given the presented wind speed model, it is possible to record the drag force $\mathbf{F}_{res}$ as (4).

$$\mathbf{F}_{res} = \mathbf{k}_{res} \cdot \widetilde{V}_a^2 \cdot sign(\widetilde{V}_a), \tag{4}$$

where $\widetilde{V}_a$ – velocity of the quadcopter relative to air mass, $\mathbf{k}_{res}$ – coefficient of resistance.

3 Quadcopter model

In this chapter, we approach the problem of quadcopter control along a given trajectory. The considered quadcopter is assumed to perform aggressive maneuvers. An aggressive maneuver is a trajectory of motion in which the control actions of a quadcopter are close to their limitations, but do not exceed them. This will minimize the energy consumption of batteries and, hence, the maneuver finished as soon as possible.

We developed a detailed and precise mathematical model of the quadcopter. This model takes into account that the considered quadcopter is asymmetric with respect to its center of mass (5). The model is described in detail in [6, 7]. In contrast to the model presented in [6, 7], in this paper, we will examine in detail the resistance forces to the motion of a quadcopter.

$$\begin{cases} \mathbf{I}\dot{\boldsymbol{\Omega}} + \boldsymbol{\Omega} \times \mathbf{I}\boldsymbol{\Omega} + \boldsymbol{\Omega} \times \begin{bmatrix} 0 \\ 0 \\ I_m \cdot \sum_{i=1}^{4} \omega_i \end{bmatrix} = \begin{bmatrix} U_1 \\ U_2 \\ U_3 \end{bmatrix}; \\ M\ddot{\mathbf{X}} = \mathbf{G}(\boldsymbol{\eta})U_0 - \begin{bmatrix} 0 \\ 0 \\ Mg \end{bmatrix} - \mathbf{F}_{res}; \\ \dot{\boldsymbol{\eta}} = \boldsymbol{\Lambda}(\boldsymbol{\eta})^{-1}\boldsymbol{\Omega}, \end{cases} \tag{5}$$

$$\begin{cases} U_1 = k_2 \sum_{i=1}^{4} \omega_i^2 \cdot y_{i,c}; \\ U_2 = -k_2 \sum_{i=1}^{4} \omega_i^2 \cdot x_{i,c}; \\ U_3 = k_1\left(-\omega_1^2 - \omega_2^2 + \omega_3^2 + \omega_4^2\right), \\ U_0 = k_2 \sum_{i=1}^{4} \omega_i^2, \end{cases} \tag{6}$$

where $\mathbf{I}$ – inertia tensor, $\boldsymbol{\Omega}$ – angular velocity vector, I_m – motors' moment of inertia, ω_i – i^{th} motor's angular velocity of rotation, $\mathbf{U} = (U_1, U_2, U_3, U_0)^T$ – control, M – quadcopter's mass, $\mathbf{X} = (x, y, z)^T$ – quadcopter's centre of mass, $\boldsymbol{\eta} = (\phi, \theta, \psi)^T$ – Euler angles, $\mathbf{G}(\boldsymbol{\eta})$ – matrix transformation between inertial and related coordinate systems, which has the form (7), $\boldsymbol{\Lambda}(\boldsymbol{\eta})$ – transition matrix for the angular velocity, which has the form (8), k_1 and k_2 – constant coefficients, $x_{i,c}$ and $y_{i,c}$ – distances from centre of the ith motor to the quadcopter's gravity centre for axes Ox and Oy, respectively.

$$\mathbf{G}(\boldsymbol{\eta}) = \begin{bmatrix} \cos(\varphi)\sin(\theta)\cos(\psi) + \sin(\varphi)\sin(\psi) \\ \cos(\varphi)\sin(\theta)\sin(\psi) - \sin(\varphi)\cos(\psi) \\ \cos(\varphi)\cos(\theta) \end{bmatrix}, \tag{7}$$

$$\boldsymbol{\Lambda}(\boldsymbol{\eta}) = \begin{bmatrix} 1 & 0 & -\sin(\theta) \\ 0 & \cos(\varphi) & \sin(\varphi)\cos(\theta) \\ 0 & -\sin(\varphi) & \cos(\varphi)\cos(\theta) \end{bmatrix}. \tag{8}$$

The first equation (5) describes the motion around the quadcopter's center of mass. The second equation (5) describes the motion of the center of mass in an inertial coordinate system. The third equation is a kinematic equation. It connects the orientation parameters with the angular rotation velocity of the quadcopter. The system of equations (6) relates the angular rotational velocities of the quadcopter to the lifting force and torque generated by motors.

4 The control algorithm

We choose an adaptive algorithm based on inverse dynamics, described in [7], as our control algorithm. The choice of such a control algorithm will allow the quadcopter to react quickly to changes in the resistance forces and improve the control accuracy.

In this paper, the variable part in the control algorithm was the resistance force. Using the adaptive algorithm, the parameters were identified and compensated in the control algorithm. Taking into account that the speed and direction of the wind can change with time, and the quadcopter can fly over different surfaces, the parameters $\mathbf{V}_0$ and z_0 are variable in time. An algorithm for identifying these parameters was developed and can be presented in the form of (9) – (13) [10-13].

$$\mathbf{U}_k = \mathbf{\Gamma}_k \cdot \mathbf{\Phi}, \quad (9)$$

$$\mathbf{L}_k = \frac{\mathbf{P}_k \cdot \mathbf{\Gamma}_k^T}{\lambda \cdot \mathbf{E} + \mathbf{\Gamma}_k \cdot \mathbf{P}_k \cdot \mathbf{\Gamma}_k^T}, \quad (10)$$

$$\mathbf{P}_{k+1} = (\mathbf{P}_k - \mathbf{L}_k \cdot \mathbf{\Gamma}_k \cdot \mathbf{P}_k)/\lambda, \quad (11)$$

$$\mathbf{\Phi}_{k+1} = \mathbf{\Phi}_k + \mathbf{L}_k(\mathbf{U}_k - \mathbf{\Gamma}_k \cdot \mathbf{\Phi}_k), \quad (12)$$

$$\mathbf{P}_0 = 10^5 \cdot \mathbf{E}, \ \mathbf{\Phi}_0 = 0 \cdot \mathbf{E}, \quad (13)$$

where $\mathbf{\Phi}$ – auxiliary parameters vector, $\mathbf{\Phi} = (A_x, B_x, A_y, B_y, A_z, B_z)^T$, $V_{i0} = A_i \cdot k_{karman}$, $z_{i0} = e^{-B_i/A_i}$, $\mathbf{U}_k$ – control, $\mathbf{\Gamma}_k$ – matrix of the model's identification system, $\mathbf{L}_k$ – matrix of identification coefficients, $\mathbf{P}_k$ – matrix of algorithm convergence estimates, $\mathbf{\Phi}_k$ – estimation of model parameters, $\mathbf{E}$ – identity matrix, k – stands for individual iterations, λ – adaptation parameter, $\lambda = 0.95$.

The disadvantages of this algorithm include the redundancy of the data, in particular, of the parameter z_0. To select one of the three values, we needed to use additional filtering and averaging. This was because as the wind was changing sharply or there was no wind along one of the axes, the identification algorithm degenerated and was converging for a long time. In order not to lose the stability of the quadcopter, we had to impose restrictions on the identifiable parameters.

5 The simulation

For our simulations, we used an "eight-shaped" trajectory. In the beginning, the quadcopter flew up to 2 meters above the earth and carried out a maneuver. During this simulation, the velocity and direction of the wind was changing. The model parameters were as follows.

$$M = 0.7[\text{kg}], \ I_m = 0.001[\text{kg m}^2], \ g = 9.8[\text{m/s}^2],$$

$$k_1 = 0.7426 \cdot 10^{-6}[\text{kg m}^2], \ k_2 = 0.1485 \cdot 10^{-6}[\text{kg m}],$$

$$k_{x,res} = 0.0024[\text{kg/m}], \ k_{y,res} = 0.0072[\text{kg/m}],$$

$$k_{z,res} = 0.0072[\text{kg/m}],$$

$$\mathbf{I} = \begin{pmatrix} 0.0123 & -0.0012 & -0.0001 \\ -0.0012 & 0.0135 & -0.0002 \\ -0.0001 & -0.0002 & 0.0227 \end{pmatrix} [\text{kg m}^2] \quad (14)$$

Fig. 1 shows the wind velocity vector from the simulation and from the identification algorithm. In Fig. 2, we depict the parameters of the wind velocity model.

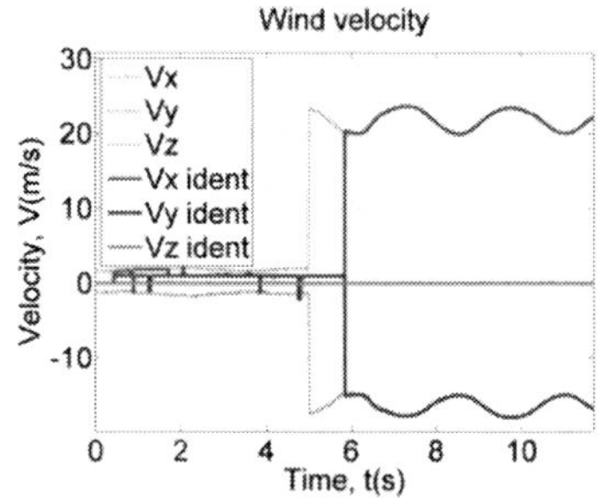

Fig. 1. Wind velocity.

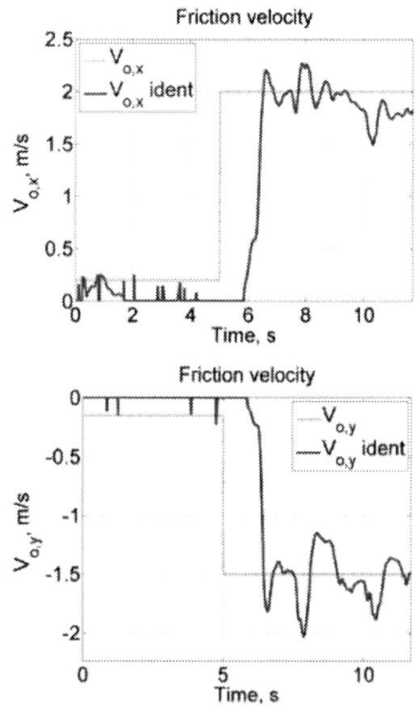

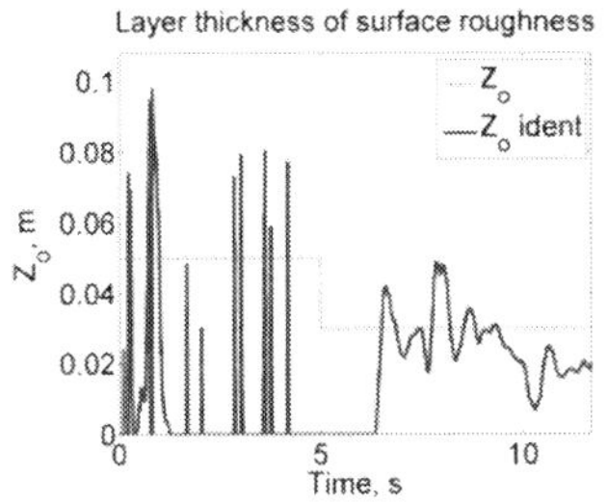

Fig. 2. Parameters of the wind velocity model.

We conclude from Figs 1 and 2 that the most critical area in this algorithm is when wind speed changes sharply (or, in other words, there is a gusty wind). The areas of wind absence are determined with a large relative error, but since it does not result from the wind's impact on the quadcopter, we may neglect it in our considerations. By imposing additional constraints on identifiable parameters, we can avoid instability of the quadcopter under sharp perturbations. Fig. 2 shows that the parameter z_0 depends on random perturbations and is poorly defined. The algorithm proved to solve the problem of quadcopter control in severe weather conditions to a satisfactory extent. This notwithstanding, in the future, we aim to improve the algorithm and add an additional filter.

Figs 3 and 4 show the motion trajectories of a quadcopter both with and without the wind's impact in the control algorithm. As you can see, both algorithms solved the problem. However, the standard deviation from the predetermined trajectory of the control algorithm without taking into account the wind was 0.3 m, while for the control algorithm taking into account the wind it was 0.21 m. The maximum deviation from the predetermined trajectory for the algorithm with the wind's impact was 0.68 m, while in case of the other algorithm it was 0.8 m.

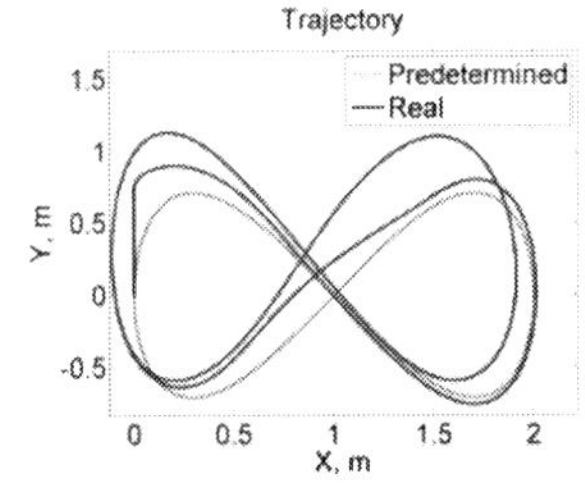

Fig. 3. The trajectory of the quadcopter motion without taking into account the wind velocity in the control algorithm.

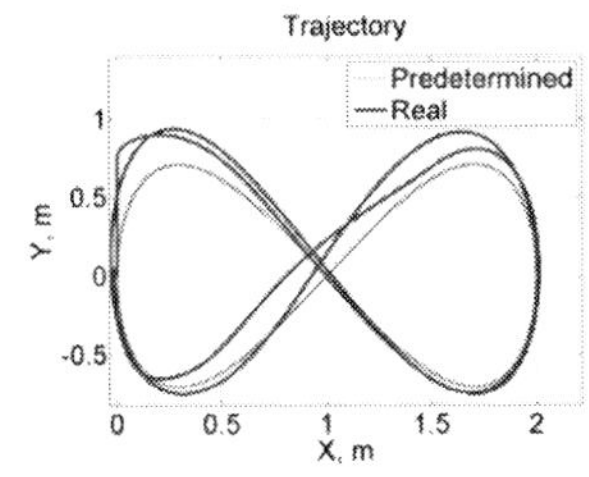

Fig. 4. The trajectory of the quadcopter motion taking into account the wind velocity in the control algorithm.

6 Conclusions

In this paper, we have approached the problem of controlling the motion of a quadcopter along a given trajectory. We chose an aggressive "eight-shaped" trajectory. The quadcopter flew under a strong variable wind. Stochastic and deterministic wind models were considered. We have solved the problem by means of the adaptive control method based on the inverse dynamics algorithm. We created an algorithm for online identification of wind model parameters. During the simulation, the algorithm proved its general effectiveness. However, it turned out not to be effective in gusty variable winds. This was because the algorithm lacked time to reconfigure and hence did not manage to take into account the influence of wind in the control algorithm. In further studies, we plan to address this problem. This notwithstanding, taking into account the influence of wind on the quadcopter's motion seems to prop up the stability of the control object and the accuracy of the quadcopter's motion along it's trajectory under varying wind conditions.

References

1. M. Sudarma, I.B. A. Swamardika, A.M. Pratama, Int. J. Electr. Comput. Eng. (IJECE) **6,** 1 (2016)
2. S. Sharma, A. Muley, R. Singh, A. Gehlot, Indian J. Sci. Technol. **9,** 43 (2016)
3. T. Boczar, A. Cichon, D. Wotzka, M. Kunicki, M. Koziol, IEEE Trans. Dielectr. Electr. Insul. **24,** 1 (2017) DOI: 10.1109/TDEI.2016.005972.
4. M. Kunicki, A. Cichon, S. Borucki, Arch. Acoust. **41,** 2 (2016) DOI: 10.1515/aoa-2016-0026
5. A. Waindok, P. Piekielny, SELM 2013 (2013)
6. R. Beniak, O. Gudzenko, J. Autom. Mobile Rob. Intell. Syst. **10,** 2 (2016) DOI: 10.14313/JAMRIS_2-2016/14
7. R. Beniak, O. Gudzenko, Proc. 21th Int. Conf. on Methods and Models in Autom. and Rob. (MMAR), Aug 2016 DOI: 10.1109/MMAR.2016.7575292
8. R.W. Beard, T.W. McLain, *Small Unmanned Aircraft: Theory and Practice* (Hardcover, 2012)
9. https://www.mathworks.com/help/aeroblks/drydenwindturbulencemodelcontinuous.html - Dryden Wind Turbulence Model
10. J.M. Selfridge, G. Tao, Syst. Sci. Control Eng., **2** (2014) DOI: 10.1080/21642583.2013.879050
11. S. Islam, P.X. Liu, A. El Saddik, Nonlinear Dyn. **78** (October 2014) DOI:10.1007/s11071-014-1425-y
12. S. Tomashevich, A. Belyavskyi, IFAC-PapersOnLine **49,** 13 (2016) DOI: 10.1016/j.ifacol.2016.07.974
13. G. Feng, R. Lozano, *Adaptive control systems* (Reed Educational and Professional Publishing Ltd, Guildford, 1999)

The computer diagnostic program (DIAG 2) for identifying states of complex technical objects

Stanisław Duer[1], and *Dariusz* Bernatowicz[2,*]

[1]Technical University of Koszalin, Department of Mechanics, 15-17 Raclawicka St., 75-620 Koszalin, Poland
[2]Technical University of Koszalin, Department of Electronics and Informatics, 2 Śniadeckich St., 75-453 Koszalin, Poland

Abstract. This paper presents a method to control an operation process of a complex technical object, with the use of trivalent diagnostic information. Also, a general diagram of the complex technical object was presented, and its internal structure was described. A diagnostic analysis was conducted, as a result of which sets of the functional elements of the object and its diagnostic signals were determined. Also, the methodology of the diagnostic examination of the technical system was presented. The result was a functional and diagnostic model, which constituted the basis for initial diagnostic information, which is provided by the sets of information concerning the elements of the basic modules and their output signals. The theoretical results obtained in the present study were verified in practice on the example of a complex and reparable technical object. It belongs to the group of technical equipment for which a short time of shutdowns is required (an ineffective use of the object).

1 Introduction

Smart systems oriented to assist human technical and technological activities are widely discussed in literature. Such systems are used to supervise and organize technical and technological processes basing upon diagnostic and quality evaluation. In case of high complexity processes being determined by large number of variables and factor such systems become very helpful as they apply to states of technical objects such as equipment, devices, production resources but also to selection of mandatory conditions and parameters of a process. Important achievements in this matter were presented in the following works [1-6] Developed ideas and particular systems consisted of many specialized modules presented in details in these works practically relate to smart and complex systems used to supervise technological processes including operational processes. Those include, among others, studies conducted by B. Buchannan, E. Shortliffe, F. Hayer-Roth, D. Waterman, W. Pedrycz, W. Kacalak, and others [7-12], describing principles of smart systems' application to solve and aid organizational processes of different technical and technological problems. It is important to note that different methods of converting human–specialist-knowledge into a form expressed by a computer programming are specified in those works. The works [13-17] provided by the author can be classified into those studies accordingly.

The paper by L. Będkowski, A. Rosiński and T. Dąbrowski [1] covers the theory of operation of technical devices. It includes a mathematical description of model of technical object considering its reliability and operation. The paper presents organization of the operation process with the use of object's models presented. The authors based their considerations upon wide experience in testing of technical devices as well as the conclusions driven during previous studies.

An early stage of realization of the renovation process is technical diagnosis of the object under operation.

It involves significant number of tasks necessary to be performed in order to determine and explicitly identify the technical states of the object in relation to a set of particular situations and events occurred or operational. Those actions are defined as identification – classification (diagnosis) of states [1,18-23].

From the very beginning, two-valued logic was applied within technical diagnosis, in which "1" is determined as Operational State and "0" as Nonoperational State. Such a diagnosis aims at identifying a nonoperational state (failure) and finding an element which is generating that state [19-21]. For the purpose of organization of the object's operational action, it is important to anticipate the states preceding nonoperational states; In such a case applying of two-valued logic has been proven insufficient [1,3, 21]. A technical diagnosis supplies the object's user with

* Corresponding author: dariusz.bernatowicz@tu.koszalin.pl

diagnostic information in order to organize the operational process. A particular example of that can be a piece of information expressed in three-valued logic. If technical objects are diagnosed by means of three-valued logic, it gives an opportunity to identify the following states of the object: *operational* – {2}, *nonoperational* – {0}, as well as *incompletely operational* – {1} [19-21]. Due to ability of recognizing an incompletely operational state of internal elements of the object the diagnostic information obtained from the object with this approach becomes, , a framework whether to make operational decisions or build operational strategies that organize new course of technical objects' operation called a state-based operation [3,4,9,13]. Where those two disciplines, that is a diagnostic theory and a theory of the technical object's operation, work together within organizing of the operational process of technical objects and optimizing the operation system.

Expected increase of available technical information is achieved by applying three-valued logic that introduces intermediate and incompletely operational states [3,7,16,18,22]. The complexity and importance of operations for modern technical object impose whole new demands to reliability and quality diagnosis. Diagnostic information provided to the user must be precise and reliable and most importantly quickly accessible. One of the methods used to counteract such problems is to apply multi-valued logic with a greater number of states [3,21]. Another reason for applying such approach is relatively low increase in calculative cost of expanding the number of distinguished states. Using multi-valued logic requires, however, many technical and economic analyses, as well as standardization of principles and rules applicable for different types of technical objects.

Research on application of four-valued logic within the diagnostic process of complex technical objects conducted by the authors resulted in creating computer diagnostic program (DIAG 2) that is a development of the intelligent system (DIAG 2). A complete multi-valued logic diagnostic process was implemented within discussed computer program. The following tasks were mandatory: creating of a diagnostic-functional diagram, defining of a set of diagnostic information, as well as classifying and concluding [3,20,21].

2 Description of the diagnostic program (DIAG 2)

The main purpose of creation of new computer diagnostic program was a necessity to use multi-states logic and to open a possibility to develop it for additional tasks in the near future. The increase of functionality or integration of additional tasks in a form of independent modules should lead to a smart system that can aid the renovation process of operational characteristics of technical states [4]. Taking the above into account, it was mandatory to apply the most accurate parameterization and highest tightness of condition as possible for tasks carried out when developing the program.

The computer diagnostic program (DIAG 2) is a "standalone" type program with Graphical User Interface (GUI), which was created using a popular object programming Java language. Using Java and Java Virtual Machine (JVM) necessary to run an application allowed to reach independence from any operating system. It relates to scientific communities, where a number of less known and popular operating systems are used. Despite the fact that JVM is the indirect runtime environment, staged operational nature of the program inconsiderably affects the performance in relation to applications created using native languages.

A similar approach was also applied when selecting a data format for both the project's structure specification and the output data of the classification process. By applying commonly used Extensible Markup Language standard (XML), it was possible to gain high flexibility and comfort while editing or saving data. Such data having a form of specified structures can be recorded both in files (documents) and assigned to XML data bases using XQuery [10]. In addition, XML standard allows to easily modify created data structures if further development is necessary.

Programme (DIAG 2) realizes a complete diagnostic process and consists of separate stages that are available in consecutive panels. Respective stages include the following tasks:

- specify a model of the technical object and based of which to create structure and diagnostic and functional diagrams,
- define diagnostic information of the examined object in a diagnostic subsystem by testing a state, analyzing a model and observing real operation processes,
- create an internal structure of the object in a form of diagnostic signals table to enable further calculations,
- classify states of the examined object using multi-valued logic (two-, three-, or four-valued) within the specified ranges of states and parameters of radial basis function networks,
- make conclusions from the diagnostic process by analyzing the gained results in a form of state maps, graphs and protocols and a comparison of classification results for different cases of multi-valued logic of states.

The diagnostic program being described is also characterized by a compact structure enabling to easily access the individual stages at any time of using it, clear GUI and an intuitive user interface. Most of the options relating to individual stages are accessible dually in the main menu or toolbars but also in the context menu with commands characteristic for a respective panel. All schemes and graphs can be separately printed out or saved on data storage devices in a form of graphic files.

3 Structure of the project and the diagnostic and functional diagram

Application of a hierarchic structure of the model in the system enabled to analyze complex technical objects

DOI: 10.1051/e3sconf/20171901029

by decomposing its components within one project. In the structure of the project depicted in Fig. 1 it can be considered more than one object, where another object can be a detailed analysis of any element of the previous one. This leads to greater depth of the object examined and tightness of its components. An internal structure of each object is divided into four levels of operational structure [4]: level one: object {O}, level two: functional units $\{E_i\}$ in an object, level three: components $\{e_j\}$ in each unit and level four: basic elements $\{e_{i,j}\}$.

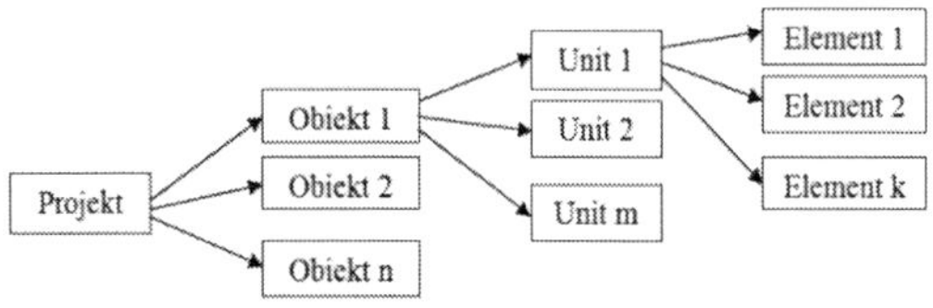

Fig.1. Structure of the project applied in (DIAG 2).

Functional units are designated in the system as units and primary elements as elements, whereas components of the 3rd level act as intermediate "elements" allowing bidirectional transformation of a hierarchic form of the object into matrix of internal structure shown in Table 1.

Table1. An example of the object's internal structure.

Object's units	Fundamental elements $\{e_{i,j}\}$ in the object				
E_1	$e_{1,1}$	...	$e_{1,i}$	...	$e_{1,J}$
⋮	⋮	...	⋮	...	⋮
E_i	$e_{i,1}$	...	$e_{i,j}$	...	∅
⋮	⋮	...	⋮	...	⋮
E_I	$e_{I,1}$	...	$e_{I,j}$	...	$e_{I,J}$

where:
$(e_{i,j})$ – j^{th} element within i^{th} assemble (module) of the object, E_i – i^{th} assemble of the object, ∅ is the element which complements the dimension of the table.

The above method to divide an operational structure of the object ensures a convenient form of describing (addressing) the elements within the structure. In addition, it also enables to change a structure of the examined object at any time, including a change of number of units or primary elements in a specific unit.

The first stage for a diagnostic elaboration is to create not only a structure of the technical object and also its functional-diagnostic diagram which can be used to determine a set of diagnostic signals $\{X_{i,j}\}$. The object's functional units being a part of the functional-diagnostic diagram shown in Fig. 2 are addressed as (E_i), where "i" stands for a number of a unit in the object. Individual elements of a unit are addressed as $(e_{i,j})$, where "j" stands for a number of an element in i-unit.

However, it is necessary to create a new project and objects to be examined prior to realizing respective stages of the diagnostic process. If any project is created or currently open, it will be closed and saved as a proper XML document. Opening a new project sets up all default parameters. If it is necessary to implement any change in parameters, it can be achieved using the option Preferences. All newly created projects are saved in the default directory "Data" in the main path of the program. When creating a project, as well as other elements of the structure such as objects, units and elements, it is required to fill in respective forms in the dialog box. Each form consists of specific information on individual elements of the structure, that is to say, full or abbreviated name, description, code, author, date, etc. If an existing project is opened, the whole structure, object's diagrams, input data and configuration variables are properly set up. (Fig. 2 and Fig. 3) [3,20-21].

The window shown in Fig. 2 divides the application into the basic parts: the left-hand part containing a hierarchic structure of the project, including its elements, and the right-hand and central one realizing consecutive stages of the diagnostic process. The left panel is the main panel being available at any time and fully integrated with particular stages of the central panel. It means that any changes in structure of the project are automatically updated in the central panel.

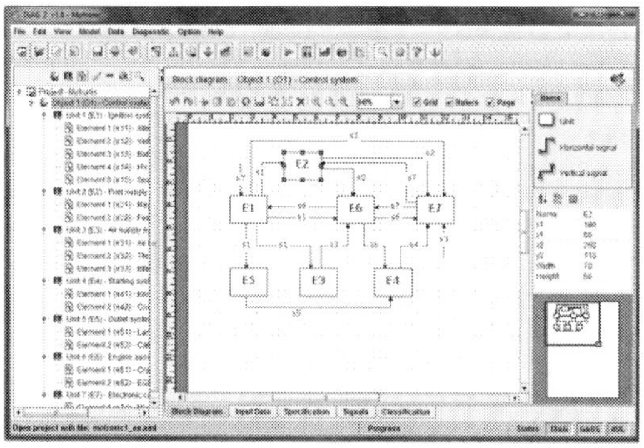

Fig. 2. Structure of the Project and a functional-diagnostic diagram of an electronic device to control an automotive engine.

A course of the diagnostic process with its preferences is considered independently for each object. It should be noted that each object, showing tight and unique features, enables to diversify a level of diagnostic depth of individual elements. If another object is chosen or a new one is created, previous results are saved and a new diagnostic process is being edited.

Stages of the diagnostic process are placed in separate panels with descriptions using tabs in the bottom of the central panel (Fig. 3).

The following tabs can be distinguished:
- Block Diagram – enables to create functional-diagnostic diagrams and specify diagnostic signals on the level of a functional unit for each object (Fig. 2),
- Input Data – enables to insert measurement data,
- nominal and diagnostic signals using a form and a data file, or a file with data generated in a random manner with preassigned parameters,
- Signals – defines a diagnostic signals table in accordance with an internal structure of the object that is used as an input structure in a classification

process. This panel also includes additional signal calculations,

- Classification – realizes the classification process of stages of the examined object depending on a type of logic used and a range of relevant, irrelevant and inadmissible variations. It enables to generate results of the classification process in a form of charts, graphs or reports.

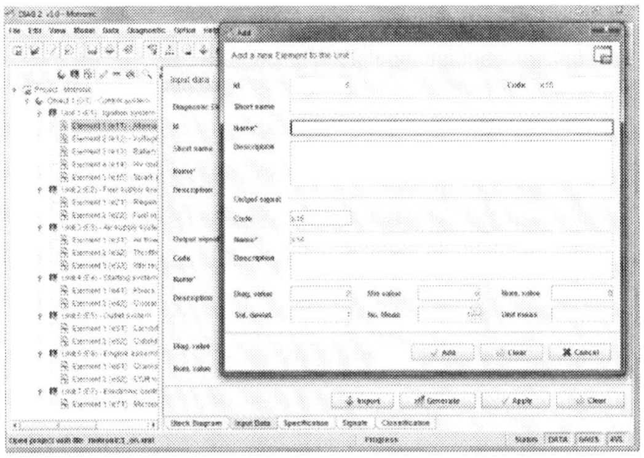

Fig. 3. Inserting input data of diagnostic signals using a form.

4 Determination of the diagnostic information on the example of a car engine

The method for the diagnosis of technical object presented was verified on the example of a reparable technical object. The analogue controller unit for combustion automotive engine with its peripheries was used for that purpose (Fig. 4). Research set-up was developed on the basis of a spark ignition engine with multi-point injection MPF1. The object was subjected to a diagnostic analysis in result of which functional-diagnostic diagram was developed. In the example, an object was used whose internal structure (Fig. 4) is composed of seven modules (E_1, E_2,..., E_7) (Table 2), and each one of them, up to five elements were distinguished [3].

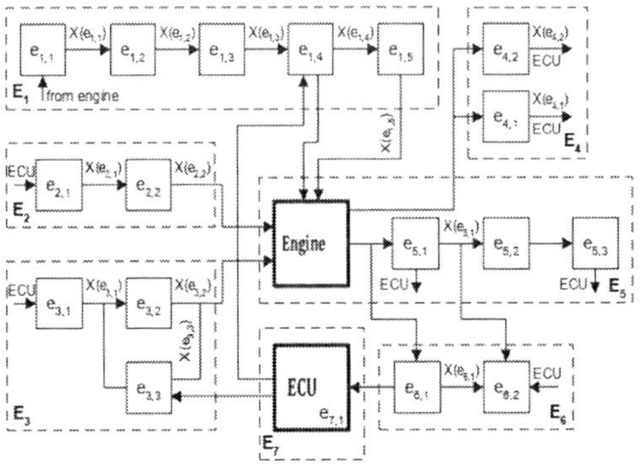

Fig. 4. Diagram of an electronic controller for an automotive engine, where: E_1 – ignition module, E_2 – fuelling module, E_3 – air-feeding module, E_4 – starting circuit, E_5 – power supply circuit, E_6 – engine block, E_7 – electronic control unit.

Table 2. Internal structure of the object.

Assembly of the object	Structure of the object $\{e_{i,j}\}$				
E_i	e_1	e_2	e_3	e_4	e_5
E_1	$e_{1,1}$	$e_{1,2}$	$e_{1,3}$	$e_{1,4}$	$e_{1,5}$
E_2	$e_{2,1}$	$e_{2,2}$	$\varnothing$	$\varnothing$	$\varnothing$
E_3	$e_{3,1}$	$e_{3,2}$	$e_{3,3}$	$\varnothing$	$\varnothing$
E_4	$e_{4,1}$	$e_{4,2}$	$\varnothing$	$\varnothing$	$\varnothing$
E_5	$e_{5,1}$	$e_{5,2}$	$\varnothing$	$\varnothing$	$\varnothing$
E_6	$e_{6,1}$	$e_{6,2}$	$\varnothing$	$\varnothing$	$\varnothing$
E_7	$e_{7,1}$	$\varnothing$	$\varnothing$	$\varnothing$	$\varnothing$

where:

E_1 – ignition module: $e_{1,1}$ – automotive alternator, $e_{1,2}$ – voltage regulator, $e_{1,3}$ – battery, $e_{1,4}$ – coil ignition, $e_{1,5}$ – sparking plug;E_2 – fuelling module: $e_{2,1}$ – fuel tank ventilation valve,$e_{2,2}$ – fuel injector; E_3 – air-feeding module: $e_{3,1}$ – air flow meter, $e_{3,2}$ – throttle position sensor, $e_{3,3}$ – idle run position controller; E_4 – starting circuit: $e_{4,1}$ – combustion knocking sensor $e_{4,2}$ – coolant temperature sensor; E_5 – power supply circuit: $e_{5,1}$ – oxygen sensor (1), $e_{5,2}$ – catalyser, $e_{5,3}$ - oxygen sensor (2); E_6 – engine block:$e_{6,1}$ – crank shaft position sensor, $e_{6,2}$ – EGR valve: E_7 – $e_{7,1}$ electronic control unit, $\varnothing$ is the element which complements the dimension of the table.

5 Classification of technical objects' states in intelligent diagnostic systems with using multi-valued logic

An external reflection of all states of the object is precisely measured values of physical variables that characterize the object. Such a variables called attributes are connected with technical parameters of the examined object. Considering above output signals of the object are highly dependent upon states of input signals and states of its primary elements (Fig. 5).

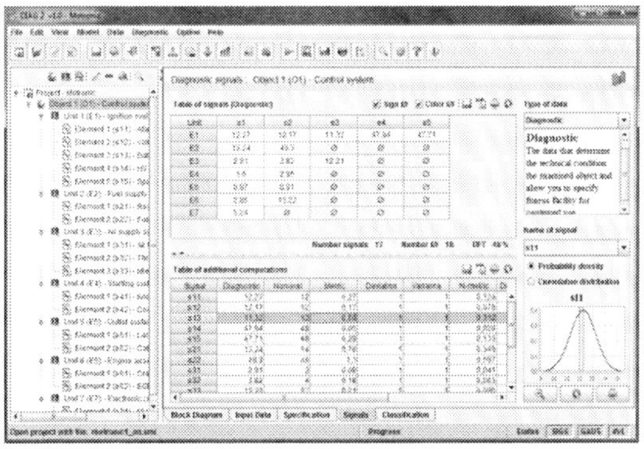

Fig. 5. A panel presenting a set of diagnostic signals and a table with additional computations.

A set of functional elements $\{e_{i,j}\}$ determined during diagnostic elaboration is a basis to put together diagnostic signals in a form of a table. An example table with diagnostic signals is shown in Fig. 5.

Before we pass to the panel "Signals", the tab "Input data" should be chosen in order to determine a set of signals as it enables to insert values of diagnostic and nominal signals.

DOI: 10.1051/e3sconf/20171901029

A diagnostic state of the object is determined on a basis of examination of a set of output signals (diagnostic signals) and it significantly depends on the results of comparison between an image of a diagnostic signal and the image of its model signal (nominal signal). A classification stage within the diagnosis process determines therefore a specified state for each primary element of the examined technical object. It is then verified whether values of output signals of particular elements are contained in a range of permissible variations. The classification of states whether for functional units or the object itself is realized basing upon the classification rules and it depends on a type of logic (a number of states) (Fig. 6).

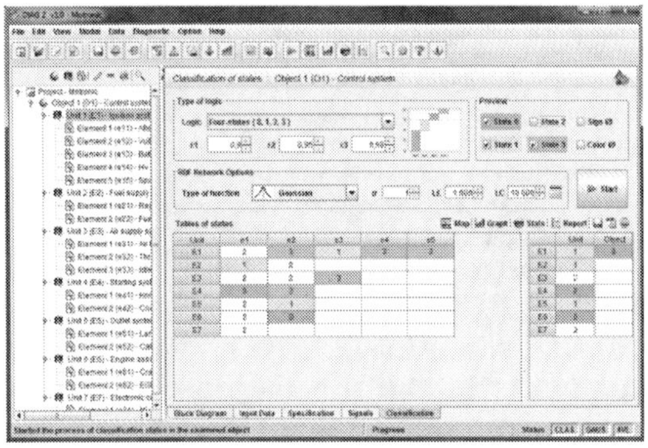

Fig. 6. The classification panel of the technical object within the diagnostic process.

The process of states classification is realized after choosing the panel "Classification". This panel contains configuration options and results tables of states (states maps). Within the configuration the following items can be defined: a type of logic of states and their ranges, an option of the graphical presentation in the states table and RBF parameters. In case of multi-valued logic acceptable states of the object are determined within the Polish standards for technical reliability and the quality of operation [3, 20].

In the described system the maximum number of states is limited to four, which therefore it enables to choose two-, three-, or four-valued logic. Symbols 0,1,2, and 3 are assigned to states as a result of classification. In case of four-valued logic, those symbols specify the following states [18]:

- Operational State marked as "3", which determines the state in which it the object performs in accordance with its destination (green) (Fig. 6),
- Incompletely Operational State marked as "2", which determines the state of the object characterized by its incomplete operation (blue) (Fig. 6),
- Critically Operational State marked as "1", which determines a state of the object characterized by extremely limited operation (yellow) (Fig. 6),
- Nonoperational State marked as "0", which determines the state in which the object is no longer functional and incapable to operate at all (red) (Fig. 6).

6 Conclusions

The paper describes the computer diagnostic program (DIAG 2). Project's structure and its individual elements such as model of the technical object, and creation of diagnostic-functional scheme are described. The paper also presents the method for determining diagnostic signals of the examined object and creating an internal structure of the object in a form of diagnostic signals tables. A significant part of the paper is a description of the classification process of the examined object's states in multi-valued logic realized with specified ranges of states and RBF parameters. An innovative part of the work is that diagnostic conclusions making (analysis) realized in multi-valued logic and ways of presenting of the results gained.

References

1. L. Bedkowski, T. Dabrowski, *Basic of the maintenance theory p. 2.* WAT, Warsaw, p. 187, (2006)
2. S. Duer, Def. Sci. J. **60**, 5, (2010)
3. S. Duer, *Intelligent system of supporting the process renewal of operating characteristics in complex technical objects,* Technical University of Koszalin, Koszalin, p. 242 (2012)
4. Z. Palkova, Warsaw U. Lif. Sci. **55**, (2010)
5. A. Rosiński, Reliability, R. Saf. Theory App. **3**, (2010)
6. A. Rosiński, Adv. in Saf. Reli. R. Man. (2012)
7. B. Buchannan, E. Shortliffe, *Rule – Based expert systems*, Addison – Wesley Publishing Company, p. 387, (1985)
8. F. Hayer-Roth, D. Waterman, D. Lenat, *Building expert systems*, Addison – Wesley Publishing Company, p. 192, (1983)
9. W. Kacalak, M. Majewski, J.M. Zurada, Lecture Not. in Art. Int. 6114, (2010)
10. W. Kacalak, M. Majewski, Lecture Notes in Computer Science 7666, Part IV, Springer 2012, 115-122, (2012)
11. W. Pedrycz, *Fuzzy Control and fuzzy systems*. John Walley End Sons, Inc. p. 432, (1993).
12. D. Waterman, *A guide to export systems*, Addison – Wesley Publishing, (1986)
13. S. Kobayashi, K. Nakamura, *IEEE Expert*, Oct. (2011)
14. P. Linz, R. Wang, *Exploring Numerical Methods: An Introduction to Scientific Computing using Matlab,* University of California, Davis, p. 172, (2003)
15. P. Linz, *An Introduction to Formal Languages and Automata,* University of California, Davis, (2002)
16. Z. Palkova, T. Rodny, Slovak U. of Agri. (2010)
17. Tresp, R. Molitar, European Con. on Art. Int. (1998)

18. M. Bialko, *The Basic properties of neural networks and hybrid expert system*, Tl University of Koszalin, Koszalin, p. 341, (2000)
19. S. Duer, R. Duer, Neural Comp. App. **19**, 5, (2010)
20. S. Duer, Neural Comp. App. **21**, 1, (2012)
21. S. Duer, R. Duer, S. Mazuru, Romanian Ass. Nonc. Tech (2016)
22. M. Majewski, W. Kacalak, Polish J. Envir. Stu. **18**, 3B, (2009)
23. L. Pokorádi, S. Duer, Int. Sci. Con. Adv. Mech. Eng. (2016)

Minimisation of the LCOE for the hybrid power supply system with the lead-acid battery

Leszek Kasprzyk[1,*], *Andrzej* Tomczewski[1], *Karol* Bednarek[1], and *Artur* Bugała[1]

[1]Poznań University of Technology, Faculty of Electrical Engineering, 60-965 Poznan, Piotrowo 3A, Poland

Abstract. The paper presents the methodology of minimisation of the unit cost of production of energy generated in the hybrid system compatible with the lead-acid battery, and used to power a load with the known daily load curve. For this purpose, the objective function in the form of the LCOE and the genetic algorithm method were used. Simulation tests for three types of load with set daily load characteristics were performed. By taking advantage of the legal regulations applicable in the territory of Poland, regarding the energy storing in the power system, the optimal structure of the prosumer solar-wind system including the lead-acid battery, which meets the condition of maximum rated power, was established. An assumption was made that the whole solar energy supplied to the load would be generated in the optimised system.

1 Introduction

In the recent years, a rapid increase in the number of renewable energy sources has been noticeable. The quantity of electricity produced in the European Union from renewable sources has been doubled in the period of the recent 15 years. This increase is caused, on the one hand, by the vision of exhaustion of fuels used in conventional power plants, and on the other hand, a decrease in the prices of renewable energy sources with the simultaneous increase in their effectiveness [1, 2]. It is not without significance that renewable sources do not have such a negative impact on the natural environment as the conventional solutions (particularly in terms of emission of harmful substances into the atmosphere).

The most popular renewable sources include wind farms and photovoltaic farms [3]. Unfortunately, these sources are characterised by high dependence on weather conditions and small efficiency [4,5]. For this reason, while designing the power supply systems built of such type of sources, the selection of their optimum configuration, which concerns the type and number of used sources and systems that store the energy and improve quality of the supplied energy is absolutely important [6-11]. On a global scale, what may also be the subject of optimisation is the location of the generation system and transmission lines [12-18]. It is also necessary to emphasise that in the process of solving technical problems, the more and more frequently decisive criteria include economic ratios. This refers to various engineering issues [14, 19-20].

For this reason the paper deals with the subject of optimisation of hybrid power supply systems based on the example of the solar-wind system with the lead-acid battery pack, designed to power loads with the known daily load curve. The economic ratio has been selected as the criterion of optimization. Attention is focused on prosumer systems, taking into account the applicable legal regulations in Poland [21].

2 Characteristics of the hybrid power supply system

The application of power supply systems which take advantage of independently operating sources of solar or wind type to ensure the continuity of operation of loads, usually requires the replenishment of energy from the power grid [22]. This is the result of the stochastic nature of changes in wind and solar energy and the occurrence of deterministic components which do not match with typical daily load curves of objects. The joint use of PV modules and wind turbines, that is the hybrid system, allows for the reciprocal compensation of shortages of electricity generated by both sources during a year. Momentary differences between generated and installed capacities are, however, an important problem in this type of power supply systems. With the assumption that the load is supplied only with electricity generated in the hybrid system, power interruptions may only be eliminated using one of two solutions: significant oversizing of the installed capacity of the system or the incorporation of energy storages into the system.

The present paper takes into consideration the solar-wind systems with the lead-acid battery pack, whose general structure is presented in Fig 1. Based on the legal regulations related to renewable prosumer systems, which are applicable in Poland, the energy storage is divided into two sections: the classic battery pack and the power system (an energy storage with the efficiency of 0.7 or 0.8 depending on the system's capacity) [21].

As well as PV modules (solar section) and wind turbines (wind section) the system incorporates the lead-

* Corresponding author: leszek.kasprzyk@put.poznan.pl

acid battery pack intended for deep discharge as well as network inverter intended for synchronisation of electricity parameters with the power system. Because of the necessity to settle the amounts of energy collected from and discharged into the power grid, a bidirectional electricity meter is used in the system. The load is characterised by daily load curve $P_o(t)$.

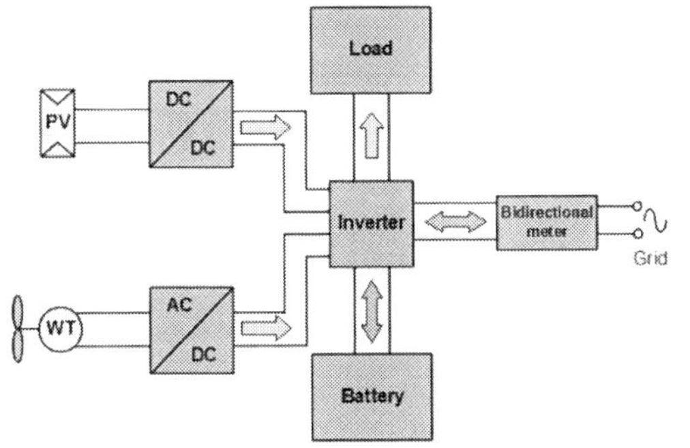

Fig. 1. Structure of the hybrid power supply system of solar and wind type, including the lead-acid battery pack and exchange of energy with the power system
(PV – photovoltaic modules, WT – wind turbines).

The assumed algorithm of operation of the system covers the powering of the load directly from the solar and wind system, the additional charging of the energy storage from the PV panels and wind turbines, the drawing of energy from the energy storage in order to supplement the energy generated by the PV modules and wind turbines, the discharge of energy into the power system (production exceeding the demand) and its recovery in the case of the energy deficit from the solar-wind system and lead-acid energy storages. In Fig. 1 grey arrows mean the direction of flow of electricity. An assumption was also made that the charging of the storages with powers which exceed the permissible values causes a change in life L_C of the batteries in accordance with the following formula:

$$L_C = \begin{cases} \text{dla } (I_{AVG} > 0) \ 1000 - 0{,}34 \cdot \exp\left(26{,}2 \dfrac{I_{AVG}}{C_{20}}\right) \\ \text{dla } (I_{AVG} < 0) \ 1000 - 0{,}5 \cdot \exp\left(-37{,}23 \dfrac{I_{AVG}}{C_{20}}\right) \end{cases} \quad (1)$$

where:
C_{20} – rated capacity (20 hours),
I_{AVG} – average battery charging current (the negative value means discharge).

3 LCOE

The comparison of systems intended for electricity generation is possible, using many criteria. One of the most frequently used indicators is the unit cost of its generation [23]. It takes into account all the outlays incurred in the electricity production process (investment, operation, costs of fuel, etc.), also considering the variability of the value of money in time, credits, inflation, etc. Owing to the diversified approach to the method of defining of a unit cost of energy generation, several methods are used commonly of which the most important ones include: UNIPEDE developed by the International Union of Producers and Distributors of Electrical Energy and LCOE (Levelized Cost of Electricity) [10, 24, 25]. For the purposes of this paper, the second of the listed methods, popular in many countries of Europe, the U.S. and in international analyses is used.

The LCOE method uses the cost model in which the aspects of revenues related to the sale of energy and various forms of financial support for investments are omitted. The value of the LCOE represents the average cost of energy production converted into 1 kWh of energy generated in the system. This is the minimum price of energy, at which the sum of discounted revenues is equal to the sum of discounted costs spent on construction and operation, as well as financial costs considered most frequently during the lifetime of a unit. The overall relationship which describes the LCOE method does not take into account the systems in which energy storages were used, as well as the cases of use of the power system as a storage for the previously generated energy. In consideration of the above, it is necessary to modify the basic relationship [10, 24, 25], which assumes the following formula for the described system:

$$LCOE = \frac{\sum_{y=0}^{N} \dfrac{\left(I_{PV(y)}P_{PV} + I_{WT(y)}P_{WT} + I_{ES(y)}A_{ES}\right)}{(1+r)^y}}{\sum_{y=0}^{N} \dfrac{A_{(y)}}{(1+r)^y}} + \quad (2)$$

$$+ \frac{\sum_{y=0}^{N} \dfrac{\left(M_{PV(y)}P_{PV} + M_{WT(y)}P_{WT} + M_{ES(y)}A_{ES}\right)}{(1+r)^y}}{\sum_{y=0}^{N} \dfrac{A_{(y)}}{(1+r)^y}}$$

where:
y – year index,
N – number of years of analysis (lifetime of the system),
$I_{PV(y)}$, $I_{TW(y)}$, I_{ES} – unit investment outlays (PLN/kW) for the solar and wind sections respectively,
$I_{ES(y)}$ – unit investment outlays (PLN/kWh) for the energy storage,
$M_{PV(y)}$, $M_{WT(y)}$ – unit operational costs (PLN/kW) for the solar and wind sections per year y respectively,
$M_{WT(y)}$ – unit operational cost (PLN/kWh) of the energy storage per year y,
P_{PV}, P_{WT} – power (kW) of the solar and wind sections,
A_{ES} – battery pack energy capacity (kWh),
$A_{(y)}$ – amount of energy (kWh) generated in the hybrid system, intended to supply the load with the known daily load curve during year y,
r – discount rate.

4 Mathematical model of the hybrid power supply

The development of the analytical model of the hybrid system characterised in section 2 requires the application of partial models: the PV module, the wind turbine and the energy storage, and their linking to the control algorithm. Such a model will be used to simulate the operation of the system. The values of wind speed v_w and irradiance G_r (measuring samples) from a period of one year, determined as mean values from the period of 36 seconds will be used as the excitation. The measuring data was made available by Mr. Krzysztof Markowicz from the Faculty of Geophysics at the Warsaw University and comes from the radiation transfer measurement station situated in the vicinity of Rzeszów. The said data will be used to establish the values of energy generated in the system, the energy transferred into the load, the storage and to or from the power system. The model requires complexity which allows for obtaining of accurate results of analysis in a short time.

4.1 PV modules

One of the standard electric models of a photocell [3, 22, 25-27, 28, 29] is used in the mathematical modelling of PV modules. In the analysed case, the double-diode model (DDM) was used; its structure is presented in Fig. 2. Its use guarantees significant accuracy of the mapping of the real behaviour of the PV installation during changes in irradiance and temperature, however it does not extend the time of analysis significantly.

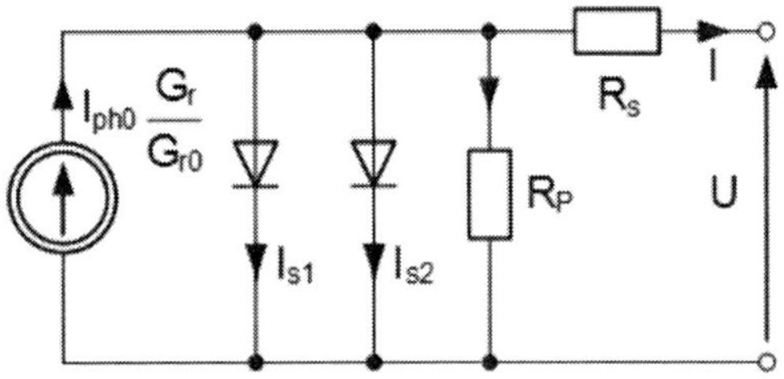

Fig. 2. Double-diode equivalent circuit of a photocell [28].

Power generated in the photocell P, with the set irradiance G_r, is determined upon solving of a non-linear equation that describes the dependence of current I on voltage U on terminals of element [10, 30, 31]:

$$I = I_{ph0}\frac{G_r}{G_{r0}} - I_{S1}\left[e^{\left(\frac{q(U+R_sI)}{N_1k_BT}\right)} - 1\right] - I_{S2}\left[e^{\left(\frac{q(U+R_sI)}{N_2k_BT}\right)} - 1\right] - \frac{(U+R_sI)}{R_p} \tag{3}$$

where:
I, U – current and voltage on the terminals of the photocell,
I_{ph0} – solar current generated at irradiance G_{r0},
G_r – real irradiance on the photocell surface,
I_{s1}, I_{s2} – saturation currents of the first and second diode,
N_1, N_2 – quality factors of the first and second diode,
R_s – series resistance,
R_p – parallel resistance,
k –Boltzmann constant,
T – photocell temperature,
q – elementary electron charge.

Knowing the number and method of connection of the PV module cells, the establishment of its current-voltage characteristics for irradiance G_r, requires the solving of equation (3). Determination of the family of characteristics I=f(U) for irradiance ranging between 50 W/m^2 and 1500 W/m^2 allows for the establishment of the characteristics of maximum power points as the function of irradiance MPP=f(G_r) being the numerical realization of the maximum power point tracking system. One of the most frequently used methods to solve equation (3) is the Newton-Raphson method.

4.2 Wind turbines

A wind turbine is a complicated mechanical-electrical system and the modelling of its operation in dynamic states belongs to complex and time-consuming tasks [14]. However, the use of the full mathematical model of the turbine is not always required. This also refers to the case under consideration in which it is only necessary to determine the amount of electricity generated in the wind section of the hybrid system during a period of one year [32]. In connection with the above, the simplified model of the wind turbine which covers only the modelling of its power curve P=f(v_w) was used in the conducted analyses. For this purpose, interpolation or approximation is used. For the purposes of this paper, splines were used. This allows for the indication of the corresponding wind turbine power for the measuring samples of wind speeds recorded to one decimal point.

4.3 Lead-acid batteries

Lead-acid batteries are most frequently modelled using circuit models, i.e. equivalent circuits built of electrical components, which reflect chemical phenomena that take place in an electrochemical cell. Such models consist of the source of voltage serially connected with the group of resistive-capacitive blocks (in parallel connection with each other, with time constants that reflect the inertia of electrochemical phenomena), the resistor (that reflects the resistance of electrodes, electrolyte and connections between the respective conductive elements inside the battery) and the coil (that reflects the inductance of internal circuits of the battery). Furthermore, such models take into account parasitic phenomena that take place in the cell (e.g. heat resulting from gasifying during electrolysis) by means of resistance connected in parallel [33].

Depending on the use of the battery, there is no need to take into account some of its elements in the circuit model. Because of low variability of the flowing current in the case under consideration, the circuit inductance may be omitted and one RC block is left in the main branch (Fig. 3) [34]. All the listed elements of the equivalent circuit are non-linear and change to a great extent depending on many factors such as the battery charge level, electrolyte temperature, or current which flows through the battery.

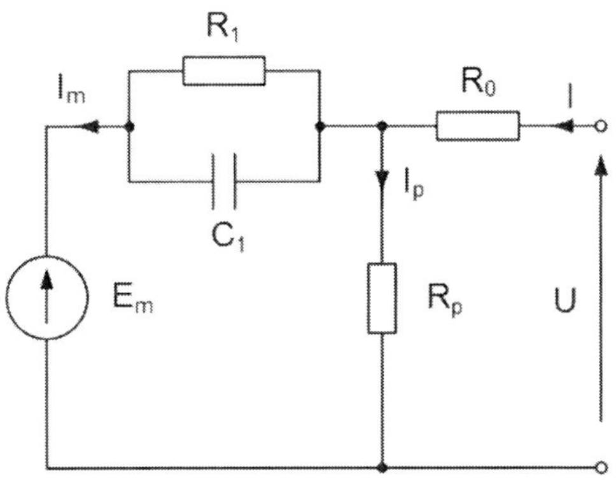

Fig. 3. Equivalent circuit of the lead-acid battery [33].

Their values are determined based on the formula presented in papers of [33-35], while the battery capacity is determined from the following formula:

$$C(I,T) = \frac{K_C C_{0^*}\left(1+\frac{T}{-T_f}\right)^{\varepsilon}}{1+(K_C-1)\left(\frac{I}{I_n}\right)^{\delta}} \tag{4}$$

where:
K_C, C_{0^*}, ε, δ – constants determined on the basis of battery specifications,
I – battery load current,
I_n – battery rated current.

The values of load and current on battery terminals during the charge and discharge process are determined by analysing the transient states in the circuit, using standard circuit methods.

5 Minimisation of the unit cost of generation of electricity from the solar and wind system with the electrochemical energy storage

The basic aim of the paper is to present the optimisation algorithm intended for establishment of the structure of the hybrid power supply system (described in chapter 2), which guarantees the continuous supply of electricity to the load with the set daily load characteristics and minimises the cost of energy generated in the system without drawing the energy from the power grid. At the same time, the sale of the surplus electricity is not taken into account (prosumer installation), except its "storage" in the power system in accordance with the applicable law [21].

The average unit costs of electricity generation determined in accordance with the standard of the so called levelized cost of energy (*LCOE*) were taken as the objective function [PLN/kWh] – formula (2). Based on the detailed analysis of this issue, it was found that the objective function was multimodal and the character of the task is that of optimization with restrictions. Five parameters which have a direct effect on the value of the objective function and which create the vector of decision variables **x** were indicated: x_1 – type of used PV modules, x_2 – number of PV modules, type x_1, x_3 – type of used wind turbines, x_3 – number of wind turbines, type x_2, x_5 – energy capacity of lead-acid battery pack. All the variables are of integer type and the restrictions result from the content of the database of turbines and PV modules, assumed rated power P_n of the system (e.g. structural limitations of the roof) and maximum time T_{MAX}, during a period of one year, in which the system may fail to supply energy to the load. The rated power of the optimised hybrid system P_{no} is determined by variables from x_1 to x_4.

Taking into account the restrictions, modified objective function $J_m(\mathbf{x})$ is the sum of the *LCOE* and the factor associated with standard penalties related to structural and functional restrictions [25,36,37]. In the case of the considered minimisation task, the modified objective function looks as follows:

$$J_m(\boldsymbol{x}) = LCOE(\boldsymbol{x}) + \sum_{i=1}^{m+n} F_{k(i)}(\boldsymbol{x}, \rho_{K(i)}) \tag{5}$$

where:
m, n – number of structural and functional restrictions respectively,
$F_{k(i)}$ – penalty function for the i-th restriction,
$\rho_{K(i)}$ – scalar penalty factor of the i-th restriction.

Taking into account the nature of the task and the properties of the objective function, the genetic algorithm metaheuristics, which is effective for multi-modal tasks with restrictions was used to minimise the objective function (5) [29,35,36].

6 Results of optimisation and simulation tests

Based on the mathematical model of the hybrid power supply system presented in section 4 and the optimisation method proposed in section 5, the IT system designed for minimisation of the *LCOE* (2) was implemented in the MATLAB and MS Visual Studio 2015 environments. The genetic algorithm method implemented in the MATLAB environment allows for extensive modification of the method of operation of its respective blocks (selection, mutation, crossing, scaling, selection pressure, etc.) in relation to the basic algorithm. The performed calculations included the use of the elitist strategy (transfer of 1 individual) and random selection based on odds without repetitions. The restrictions are taken into account using the Augmented Lagrangian Genetic Algorithm (ALGA) which described in detail in the paper by [1]. Additionally, the selection probability was assumed in the conducted calculations at the level of 0.7 while the mutation probability amounted to 0.05. Based on the conducted preliminary tests, an assumption was made that a generation consists of 40 individuals, while 60 generations are altogether under analysis. Three loads with the average daily power of 3 kW and daily load profiles $P_o(t)$ presented in Fig. 4 (variants: I, II and

DOI: 10.1051/e3sconf/20171901030

III) were selected for the calculations. An assumption was made that the maximum rated power of the solar-wind system may amount to $P_n = 20$ kW. It was also assumed that the optimised power supply system guaranteed the supply of electricity to the load all year round, and the power system is the second storage for energy generated in photovoltaic modules and wind turbines right after lead-acid batteries.

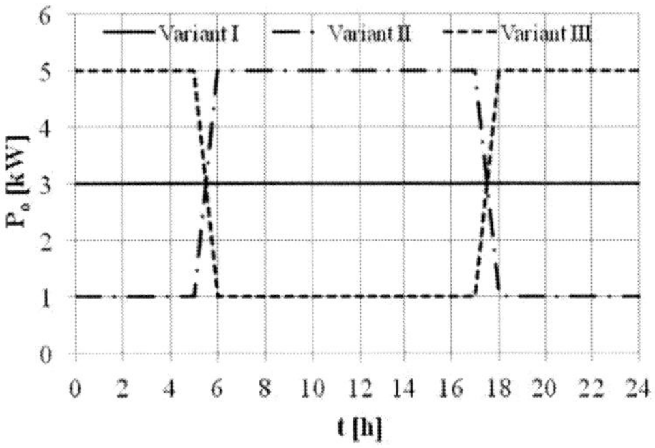

Fig. 4. Daily load patterns for the variants of the load analysed in the paper.

In order to establish the value of the *LCOE* the average unit investment and operational costs converted into 1 kW of installed power (solar and wind system) as well as the costs of one kWh of installed energy capacity of batteries given in the report [38] were used.

For each type of the load, a series of 5 optimisation calculations were carried out using the developed IT system. For each of them the rated power of the optimized system, the capacity of the solar and wind sections as well as the value of the *LCOE* and the energy capacity of the battery pack were determined.

Fig. 5 presents the pattern of the average value of the *LCOE* as a function of the number of a generation for five algorithm implementations and the case of the load with profile II. On the other hand, Table 1 lists the values of optimal solutions and the *LCOE* obtained for three daily load curves given in Fig. 4.

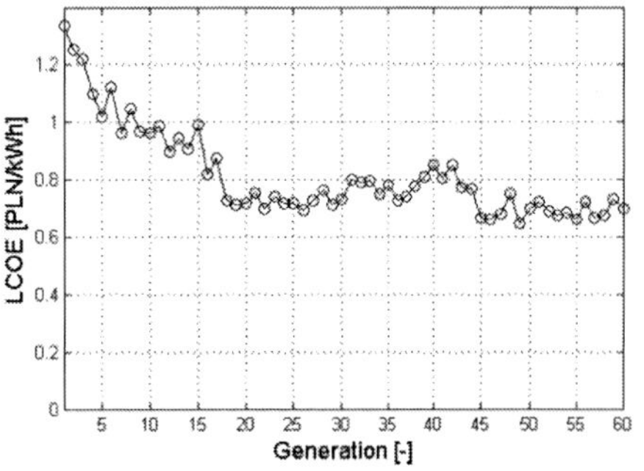

Fig. 5. Average values of the *LCOE* (from 5 algorithm) as the function of the number of a generation for load variant II.

No solution, which would fulfil the assumptions made, was found for variant III. The reason for this is the high demand for electricity at night when the power supply system is not supported by any photovoltaic sources.

Table 1. Parameters of optimal solutions for the analysed types of loads (Fig. 4) and $LCOE_{AVG}$ determined for 5 genetic algorithm implementations.

Parameters	Load variant		
	I	II	III
P_{no} [kW]	19.1	16.85	no solution
P_{PV} [kW]	4.1	1.85	no solution
P_{WT} [kW]	15.0	15.0	no solution
P_{PV}/P_{no} [%]	21.5	11.0	no solution
P_{WT}/P_{no} [%]	78.5	89.0	no solution
A_{EM} [kWh]	1	1	no solution
A_{EM}/P_{no} [kWh/kW]	0.05	0.06	no solution
LCOE [PLN/kWh]	**0.516**	**0.508**	no solution
$LCOE_{AVG}$ [PLN/kWh]	**0.519**	**0.512**	no solution

7 Conclusions

The developed algorithm and the calculation system enable the establishment of the optimal structure of the hybrid power supply system, for which the cost of generation of electricity used to power the load with the set daily load characteristics is minimal.

The used model of the hybrid system and the optimisation algorithm (genetic algorithm) are effective and enable the establishment of the global minimum for the multimodal objective function with restrictions despite the complexity of the task and the range of the area of acceptable solutions. Owing to the applied modifications of the genetic algorithm, high repeatability of results was obtained. The differences between the best and average values (from five algorithm implementations) are insignificant and amount to less than 1% (Table 1).

The optimisation calculations were carried out for three load variants (Fig. 4) with the average daily power of 3 kW. The hybrid system with the maximum rated power equal to 20 kW, the storage capacity with the maximum value of 100 kWh and assumption of the lack of purchase of energy from the network were searched. In the cases of load variants I and II, the system structure which fulfils the set requirements was found (table 1). In the case of variant III, a solution, which meets the assumed power supply criteria, does not exist.

In the analysed cases, the inclusion of the energy storage into the structure of the hybrid power supply system slightly increases the unit cost of electricity generation - *LCOE* (by about 3%), however, it allows the powering of the load all year round without drawing energy from the power grid. Owing to the use of the power system as additional energy storage, the capacity of the selected battery pack is small and amounts to 1 kWh (for variants I and II). Its annual operating time is short, which results in rare replacements of energy storages and reduction in operational costs. It also confirms the legitimacy of using electrochemical energy storages in the form of relatively cheap lead-acid batteries.

References

1. R. Conn, N. I.M. Gould, L. Toint, Math. Comput. **66** (1997)
2. G. Sinden, Energy Policy **35** (2007)
3. R. Kumar, R.A. Gupta, A.K. Bansal, Swarm Evol. Comput. **8** (2013)
4. R. Prasad, E. Natarajan, Energy **31** (2006)
5. M. B. Shadmand, M. Pasupuleti, R. Balog, *IEEE Energy Conversion Congress and Exposition*, 3626-3632, (2011)
6. L. Kasprzyk, A. Tomczewski, K. Bednarek, Prz. Elektrotechniczny **87/12b** (2011)
7. M. Manbachi, F. Mahdloo, M.R. Haghifam, A. Ataei, C. Yoo, Eksploat. Niezawodn. **4** (2010)
8. A. Zarinchang, N. Faghih, J. Zarinchang, Eksploat. Niezawodn. **14** (2012)
9. A. Jiakai, H. Yan, W. Wei, Eksploat. Niezawodn. **17** (2015)
10. M. Bortolini, M. Gamberi, A. Graziani, Energy Conv. Manag. **86** (2014)
11. S. Diaf, M. Belhamel, M. Haddadi, A. Louche, Energy Policy **36** (2008)
12. K. Bednarek, J. Jajczyk, Prz. Elektrotechniczny **85/12** (2009)
13. J. Jajczyk, L. Kasprzyk, *Int. Conf. on Heat Transfer, Therm. Eng. and Environ.* (WSEAS, 226-231, 2008)
14. A. Tomczewski, Sci. World J. **2014**, 643769 (2014)
15. J. Jajczyk, *Computational Problems of Electrical Engineering* (IEEEXplore, 1-3, 2016)
16. L. Kasprzyk, R. Nawrowski, A. Tomczewski, *Intelligent computer techniques in applied electromagnetics* (Springer, 2008)
17. K. Bednarek, R. Nawrowski, A. Tomczewski, *Int. Conf. on Parallel Comput. in Electri. Eng.* (Parelec, 2000)
18. Z. Nadolny, G. Dombek, P. Przybylek, *IEEE 2016 Conf. Electr. Insul. Dielectr. Phenom.*, 857-860 (2016)
19. G. Dombek, Z. Nadolny, Eksploat. Niezawodn. **19**, (2017)
20. J. Rymaszewski, M. Lebioda, E. Korzeniewska, Prz. Elektrotechniczny **88/12B** (2012)
21. Polish Government Act dated 22.06.2016 amending the Act on Renewable Energy Sources and some other acts (in polish)
22. A. Kaabeche, M. Belhamel, R. Ibtiouen, Solar Energy **85** (2011)
23. R. Hongbo, G. Weijun, R. Yingjun, Renew. Energy **34** (2009)
24. K. Branker, M. J.M. Pathak, J.M. Pearce, Renew. Sust. Energ. Rev. **15** (2011)
25. Ould. Bilal, V. Sambou, C.M.F Kébé, P.A. Ndiaye, M. Ndongo, Energy Procedia **14** (2012)
26. K. Abdelhamid, R. Ibtiouen, Solar Energy **103** (2014)
27. H. Belmili, M. Haddadi, S. Bacha, M. F. Almi, B. Bendib, Renew. Sust. Energ. Rev. **30** (2014)
28. A. Bugała, G. Frydrychowicz-Jastrzębska, Z. Zbytek, J. Dach, D. Janczak, Int. Conf. Front. Sensors Technol. (MATEC Web of Conferences, Hong Kong, 2016)
29. M. Kolhe, IEEE Trans. Energy Convers. **24** (2009)
30. J. I. Rosell, M. Ibáñez, Energy Conv. Manag. **47** (2006)
31. A. Bugała, G. Frydrychowicz-Jastrzębska, Z. Zbytek, J. Dach, D. Janczak, *International Conference on Frontiers of Sensors Technologies* (MATEC Web of Conferences, Hong Kong, 2016)
32. T. Petru, T. Torbjörn, IEEE Trans. Power Syst. **17** (2002)
33. Burzyński, L. Kasprzyk, Poznan Uni. of Technol. Acad. J. **92** (2017)
34. Burzyński, L. Kasprzyk, E3S Web of Conferences **14**, 01041 (2017)
35. L. Kasprzyk, Eksploat. Niezawodn. **19/2** (2017)
36. M. Maleki, F. Pourfayaz, Solar Energy **115** (2015)
37. M. Maleki, F. Ameri, F. Keynia, Renew. Energy **80** (2015)
38. G. Wiśniewski, Instytut Energetyki Odnawialnej, Warszawa (2013)

DOI: 10.1051/e3sconf/20171901031

New approach to wireless data communication in a propagation environment

Wojciech P. Hunek[1,*], and *Paweł* Majewski[2]

[1,2] Opole University of Technology, Institute of Control Engineering, Prószkowska 76, 45-758 Opole, Poland

Abstract. This paper presents a new idea of perfect signal reconstruction in multivariable wireless communications systems including a different number of transmitting and receiving antennas. The proposed approach is based on the polynomial matrix *S*-inverse associated with Smith factorization. Crucially, the above mentioned inverse implements the so-called degrees of freedom. It has been confirmed by simulation study that the degrees of freedom allow to minimalize the negative impact of the propagation environment in terms of increasing the robustness of whole signal reconstruction process. Now, the parasitic drawbacks in form of dynamic ISI and ICI effects can be eliminated in framework described by polynomial calculus. Therefore, the new method guarantees not only reducing the financial impact but, more importantly, provides potentially the lower consumption energy systems than other classical ones. In order to show the potential of new approach, the simulation studies were performed by author's simulator based on well-known OFDM technique.

1 Introduction

As we all know the environment in which we live and move has a huge impact on our behavior. In fact, without proper air conditions we cannot to function normally. That is why it is so important to be able to determine the positive or negative impact of the environment on our actions. In this paper, a special attention has been paid to the parasitic influence of the environment on the wireless data transmission process. Through an application of different tools we can effectively identify and reduce the noise impact under various engineering tasks including a multivariable real perfect signal reconstruction problem. A recently introduced new method of signal recovery is presented and confirmed by simulation examples here.

Today's wireless data communication standards are already well developed. There are a number of solutions that allow for balanced transmission of information in order to achieve a significant bandwidth throughput (for example WiMAX, WiFi 802.11n, DVB-T or LTE/LTE Advanced standards), the several of them implement the well-known OFDM mechanism [1,2]. The use of multi-variability, which is a response to the paradigm of achieving a significant throughput, entails additional problems such as inter-channel (ICI) and inter-symbol interference (ISI). Notwithstanding, the basic method of independent separate between channels is the usage, for instance, the commonly practiced tool in form of SVD factorization [3-7]. However, all of those methods suffer from the parasitic effect of the environment in the form of noise. This paper presents the results of simulation studies using a new signal reconstruction method dedicated to propagation environments described by nonsquare polynomial matrices. A new approach implies the use of appropriate tools associated with systems described by matrix polynomial framework. A key role plays here the Smith-factorized the polynomial matrix *S*-inverse, which implements the so-called degrees of freedom [8-15]. As is shown by simulation studies, the new idea can restrain the negative influence of the propagation environment on the process of signal reconstruction/recovery.

The paper is organized in the following manner. A system representation is given in Section 2. Next unit includes the notion of *S*-inverse. A new approach to problem of signal recovery in MIMO telecommunications systems is presented in Section 4. A simulation study of Section 5 confirms a big potential of new method. Finally, the conclusions are shown at the end of the paper.

2 System representation

In this paper the simulation studies of wireless data communications were carried out using a system with N_T-transmitting antennas and N_R-receiving antennas described by the polynomial matrix of the form [16]

$$\underline{\mathbf{C}}_{N_R \times N_T}(q^{-1}) = \begin{bmatrix} \underline{\mathbf{c}}_{11}(q^{-1}) & \cdots & \underline{\mathbf{c}}_{1n}(q^{-1}) \\ \vdots & \ddots & \vdots \\ \underline{\mathbf{c}}_{m1}(q^{-1}) & \cdots & \underline{\mathbf{c}}_{mn}(q^{-1}) \end{bmatrix}, \quad (1)$$

where operator q^{-1} is the background shift operator. And polynomials describing individual cells of $\underline{\mathbf{C}}(q^{-1})$ matrix are finite ipulse response type (FIR). For such described

* Corresponding author: w.hunek@po.opole.pl

environment model, the signal transition through the propagation path is illustrated in the Fig. 1. Therefore, the signal transition equation can be written according to the formula [16]

$$\mathbf{R}'(t) = \underline{\mathbf{C}}(q^{-1})\mathbf{S}(t) + \zeta(t), \qquad (2)$$

where $\mathbf{R}'(t)$, $\mathbf{S}(t)$ and $\zeta(t)$ are the vectors of the noisied received signal, transmitted signal and noise signal, respectively. In simulation studies we also assumed that the analyzed disturbance is an uncorrelated white noise with zero-mean value.

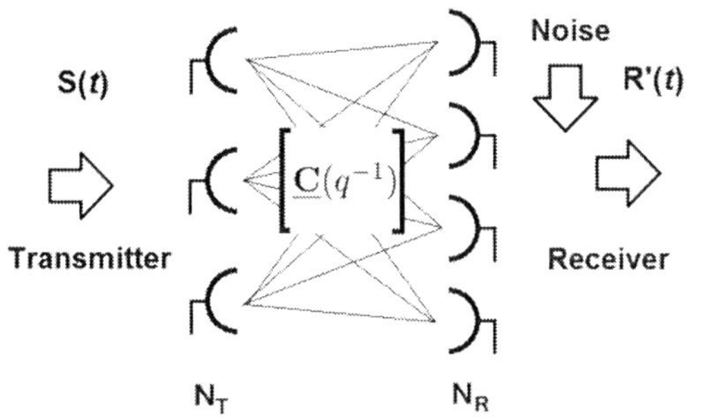

Fig. 1. Model of wireless signal path [16].

Because fo the new method is based on a polynomial approach, the next section presents alternative tools that are equivalent to those commonly used in the classical reconstruction processes, where the propagation environments are described by parametric matrices [17-19].

3 Polynomial matrix *S*-inverse

Due to the fact that in the new approach the signal reconstruction process uses the polynomial matrices to described a propagation parameters, the well-known SV decomposition for channel separation cannot be applied. On the other hand, in the literature we can find a several solutions related to SVD factorization in form, e.g., P-SVD (Polynomial-SVD) [20-22]. In this paper a new approach to perfect signal recovery is presented, which is based on so-called polynomial matrix *S*-inverse. This tool does not only provide a channel separation using the so-called Smith factorization, but penetrates the whole process of signal reconstruction. The Smith factorization for the matrix $\underline{\mathbf{C}}(q^{-1})$ being of the full normal rank with FIR-polynomials is the following

$$\underline{\mathbf{C}}(q^{-1}) = \underline{\mathbf{U}}(q^{-1})\underline{\Sigma}(q^{-1})\underline{\mathbf{V}}(q^{-1}), \qquad (3)$$

where the unimodular matrices $\underline{\mathbf{V}}(q^{-1})$ and $\underline{\mathbf{U}}(q^{-1})$, as equivalent to the parameter signal reconstruction approach, are precoder and equalizer, respectively [20], whilst matrix $\underline{\Sigma}(q^{-1})$ includes eigenvalues of matrix $\underline{\mathbf{C}}(q^{-1})$. Taking into account this factorization we obtain right and left polynomial matrix *S*-inverses [8]

$$\underline{\mathbf{C}}^{\mathrm{R}}(q^{-1}) = \underline{\mathbf{V}}^{-1}(q^{-1})\underline{\Sigma}^{\mathrm{R}}(q^{-1})\underline{\mathbf{U}}^{-1}(q^{-1}), \qquad (4)$$

$$\underline{\mathbf{C}}^{\mathrm{L}}(q^{-1}) = \underline{\mathbf{V}}^{-1}(q^{-1})\underline{\Sigma}^{\mathrm{L}}(q^{-1})\underline{\mathbf{U}}^{-1}(q^{-1}). \qquad (5)$$

Due to the fact that in the presented in the paper the simulated studies consider the telecommunications systems with a greater number of receiving antennas than transmitting ones ($N_T < N_R$), our attention is focused only on nonunique left inverses. This results that the matrix $\underline{\Sigma}(q^{-1})$, which is depicted in Eqn. (5), is of the following form [23]

$$\underline{\Sigma}^{\mathrm{L}}(q^{-1}) = \left[\underline{\mathbf{D}}_{N_T\times N_T}(q^{-1}) \quad \underline{\mathbf{M}}_{N_T\times(N_R-N_T)}(q^{-1})\right], \qquad (6)$$

where matrix $\underline{\mathbf{M}}(q^{-1})$ contains the crucial degrees of freedom. As has already been confirmed by formal proof [16], in a classical approach, where the Moore-Penrose inverse associated with SVD factorization is commonly used for nonsquare systems, a corresponding to the matrix $\underline{\Sigma}^{\mathrm{L}}(q^{-1})$ structure has zero – degrees of freedom [20]. Therefore in this case we cannot influence the signal perfect recovery through an application of special selected degrees of freedom. The next section presents a new issue concerning the multivariable signal perfect reconstruction including Smith factorization of $\underline{\mathbf{C}}(q^{-1})$.

4 New approach to wireless signal reconstruction in polynomial matrix environment

Having the background and the necessary tools needed in the new approach to the reconstruction process we can finally go through the description of the main issue of the paper. Let the signal pass through propagation environment described by Eqn. (1) defines the formula (2). After using the Smith factorization, for the purpose of the ICI cancellation, we obtain

$$\mathbf{R}'(t) = \underline{\mathbf{U}}(q^{-1})\underline{\Sigma}(q^{-1})\underline{\mathbf{V}}(q^{-1})\mathbf{S}(t) + \zeta(t). \qquad (7)$$

After applying the left polynomial *S*-inverse of Eqn. (5), we receive the reconstructed signal form

$$\mathbf{S}(t) = \underline{\mathbf{V}}^{-1}(q^{-1})\underline{\Sigma}^{\mathrm{L}}(q^{-1})\underline{\mathbf{U}}^{-1}(q^{-1})\mathbf{R}'(t) - \underline{\mathbf{V}}^{-1}(q^{-1})\underline{\Sigma}^{\mathrm{L}}(q^{-1})\underline{\mathbf{U}}^{-1}(q^{-1})\zeta(t), \qquad (8)$$

or, after some assumptions [16], as follows

$$\begin{aligned}\mathbf{S}(t) &+ \underline{\mathbf{V}}^{-1}(q^{-1})\underline{\Sigma}^{\mathrm{L1}}(q^{-1})\underline{\mathbf{U}}^{-1}(q^{-1})\zeta(t)\\ &= \underline{\mathbf{V}}^{-1}(q^{-1})\underline{\Sigma}^{\mathrm{L}}(q^{-1})\underline{\mathbf{U}}^{-1}(q^{-1})\mathbf{R}(t)\\ &+\underline{\mathbf{V}}^{-1}(q^{-1})\underline{\Sigma}^{\mathrm{L2}}(q^{-1})\underline{\mathbf{U}}^{-1}(q^{-1})\zeta(t),\end{aligned} \qquad (9)$$

where $\mathbf{R}(t)$ is the perfect obtained signal, whilst $\underline{\Sigma}^{\mathrm{L1}}(q^{-1}) \neq \underline{\Sigma}^{\mathrm{L2}}(q^{-1})$ and $\underline{\Sigma}^{\mathrm{L}}(q^{-1})$ can be any. It is crucial, that trough an application of Eqn. (9) we would like to delete any noises in terms of selection of appropriate degrees of freedom $\underline{\mathbf{M}}(q^{-1})$. Therefore, the performance index is determined in the following form [24,25]

$$\underline{\mathbf{M}}_{\mathrm{opt}}(q^{-1}) = \arg \min_{\underline{\mathbf{M}}(q^{-1})} \sum_{t=0}^{N-1} \{[\mathbf{S}'(t)$$

 DOI: 10.1051/e3sconf/20171901031

$$- \mathbf{S}(t)]^{\mathrm{T}}[\mathbf{S}'(t) - \mathbf{S}(t)]\}, \tag{10}$$

which was the basis for genetic algorithm to obtain the optimal degrees of freedom. The goal of the optimization task is receive the smallest difference between the transmitted $\mathbf{S}(t)$ and reconstructed $\mathbf{S}'(t)$ signals. The big potential of the new approach shows next section.

5 Simulation study

In the simulation studies, a system containing two transmiting antennas and three receiving antennas ($N_T = 2$, $N_R = 3$) is considered. The parameters describing the propagation properties of the environment are described by the hypothetical polynomial matrix of the form

$$\underline{\mathrm{C}}(q^{-1}) = \begin{bmatrix} 0.8 + 0.4q^{-1} & 0.8 \\ 1 - 0.6q^{-1} + 0.2q^{-2} & 1 - 0.6q^{-1} \\ 1 + 0.4q^{-1} & 0.8 \end{bmatrix}. \tag{11}$$

After applying the new method to the process of signal reconstruction, the matrix (11) is factorized and inversed finally to obtain

$$\underline{\mathbf{U}}^{-1}(q^{-1}) = \begin{bmatrix} 1 & 0 & \cdots \\ 1 & 0 & \cdots \\ -1.25 - 1.75q^{-1} + 2.5q^{-2} & 1 & \cdots \end{bmatrix}$$

$$\begin{matrix} \cdots & 0 \\ \cdots & -1 \\ \cdots & 2.5q^{-1} - 2.5q^{-2} \end{matrix}\Bigg], \tag{12}$$

and

$$\underline{\mathbf{V}}^{-1}(q^{-1}) = \begin{bmatrix} 0 & -5 \\ 1.25 & 5 + 2.5q^{-1} \end{bmatrix}, \tag{13}$$

whilst crucial in the new approach a block matrix containing the degrees of freedom is a form

$$\underline{\Sigma}(q^{-1}) = \begin{bmatrix} 1 & 0 & \underline{\mathbf{M}}_{11}(q^{-1}) \\ 0 & 1 & \underline{\mathbf{M}}_{21}(q^{-1}) \end{bmatrix}. \tag{14}$$

For the purpose of simulating studies of wireless data transmission an authors' simulator, which implements the new signal reconstruction method, was developed in the Matlab environment. This tool, besides using the new method, takes into account the widely used mechanism OFDM for increasing the transmission of data. However, due to space limitation reasons the simulator is not discussed here in detail (for more information see Ref. [16]). The procedure allows to select the optimal degrees of freedom, which were determined using a genetic algorithm based on the performance index (10). Moreover, said tool also give us the ability to graphical presentation the transmitted bits in the form of constellation maps.

In order to show the obtained results, the basic information related to data transmission, should be given (Table 1).

The simulation studies were performed as follows. For the considered matrix (11) describing the dynamics of the environment, the transmission of individual bits was simulated.

Table 1. Data transmission parameters.

Parameter	Value
Number of transmitting antennas	2
Number of receiving antennas	3
Number of subcarriers	16
Type of constellation	64-QAM
Number of transmitted bits	103776

In this scenario the different levels of noise were taking into account, which, according to Eqn. (2), blurred the received signal. Basis on the selected degrees of freedom the Bit Error Rate (BER) index was calculated. The BER index results of new method and classical one (i.e., the approach where the degrees of freedom of the matrix $\underline{\Sigma}(q^{-1})$ of Eqn. (14) are zero) were compared separately. For such studies, the comparative results are shown in Fig. 2. The constellation points obtained from the signal reconstruction process for the classical approach and the new method are presented in the Figs. 3 and 4, respectively.

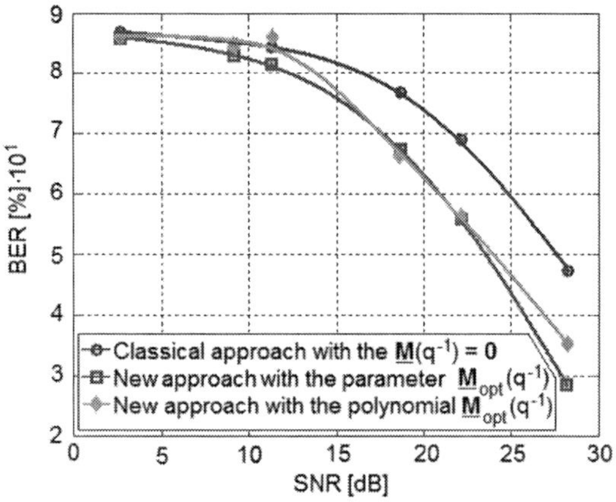

Fig. 2. BER vs. SNR.

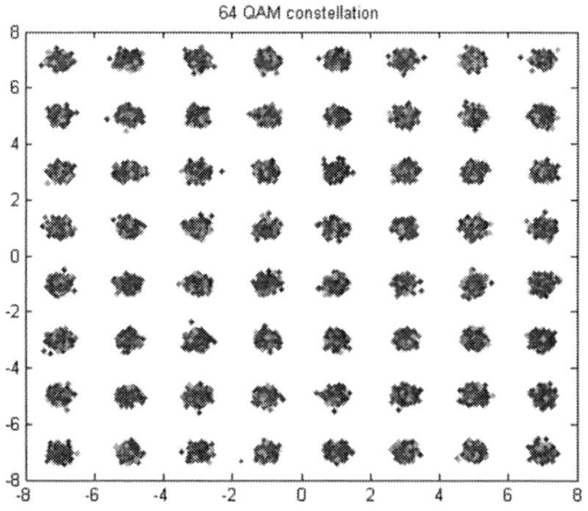

Fig. 3. Classical approach with SNR = 28 dB, BER 47%.

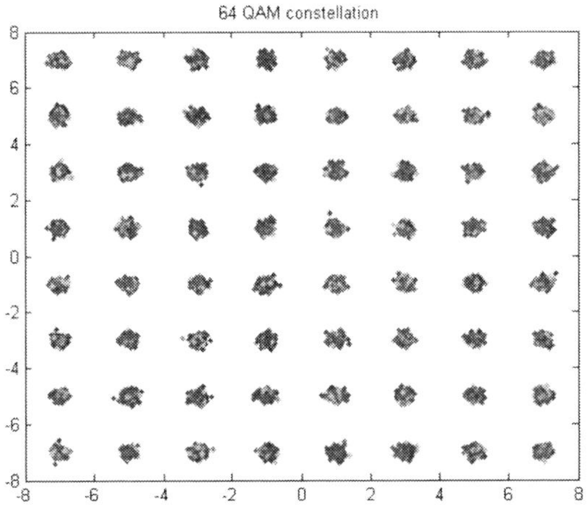

Fig. 4. New approach with SNR = 28 dB, BER 27 %.

6 Conclusions

This paper presents a new method for signal reconstruction process. A signal recovery technique based on polynomial matrix description of propagation properties of the natural environment has been proposed. A new approach implements the polynomial matrix *S*-inverse, which contains the crucial degrees of freedom. The superiority of the new method over classical solutions was proven by simulation studies.

References

1. H. Shu, E. P. Simon, L. Ros, Sig. Proc.**117** (2015)
2. D. Mattera, M. Tanda, Dig. Sig. Proc. **49** (2016)
3. O. Nibouche, J. Jiang, Dig. Sig. Proc. **23** (2013)
4. M. Nagahara, IEEE Sig. Proc. Let. **22** (2015)
5. Q. H. Spencer, A. Lee Swindlehurst, M. Haardt, IEEE Tran. on Sig. Proc. **52** (2004)
6. W.P. Hunek, *Contribution of a new matrix H-inverse to stabilizing the perfect control for LTI MIMO state-space systems*, Proc. of the 21st Int. Conf. on Met. and Mod. in Auto. and Rob. (MMAR), 64–69 (2016)
7. W.P. Hunek, *New SVD-based matrix H-inverse vs. minimum-energy perfect control design for state-space LTI MIMO systems*, Proc. of the 20th IEEE Int. Conf. on Sys. The Cont. and Comp. (ICSTCC), 14–19 (2016)
8. W.P. Hunek, *Towards a General Theory of Control Zeros for LTI MIMO Systems* (Opole Univ. of Tech. Press 2011)
9. W.P. Hunek, K.J. Latawiec, Int. J. Appl. Math. Comp. Sci. **21** (2011)
10. W.P. Hunek, K.J. Latawiec, Cont. Cyb. **38** (2009)
11. W.P. Hunek, Bul. Pol. Aca. Sci. – Tech. Sci. **59** (2011)
12. W.P. Hunek, K.J. Latawiec, P. Majewski, P. Dzierwa, *An application of a new matrix inverse in stabilizing state-space perfect control of nonsquare LTI MIMO systems*, Proc. of the 19th IEEE Int. Conf. on Met. and Mod. in Auto. and Rob. (MMAR), 451–455 (2014)
13. W.P. Hunek, K.J. Latawiec, R. Stanisławski, M. Łukaniszyn, P. Dzierwa, *A new form of a σ-inverse for nonsquare polynomial matrices*, Proc. of the 18th Int. Conf. on Met. and Mod. in Auto. and Rob. (MMAR), 282–286 (2013)
14. W.P. Hunek, *An application of polynomial matrix σ-inverse in minimum-energy state-space perfect control of nonsquare LTI MIMO systems,* Proc. of the 20th Int. Conf. on Met. and Mod. in Auto. and Rob. (MMAR), 252–255 (2015)
15. W.P. Hunek, P. Dzierwa, ITC **43** (2014)
16. P. Majewski, *Research towards increasing the capacity of wireless data communication using inverses of nonsquare polynomial matrices*, Doc. diss. (to appear) (2017)
17. D. Sellathambia, J. Srinivasanb, S. Rajaramc, *Hardware implementation of adaptive SVD beamforming algorithm for MIMO-OFDM systems,* Proc. of Int. Conf. on Des. and Man, (IConDM'2013), Proc. Eng. 84–93 (2013)
18. M. Sadek, A. Tarighat, A. H. Sayed, IEEE Tran. on Wire. Com., **6** (2007)
19. E.K.S. Au, S. Jin, M.R. McKay, W.H. Mo, IEEE Tran. Wire. Com. **7** (2008)
20. N. Moret, A. Tonello, S. Weiss, *MIMO precoding for filter bank modulation systems based on PSVD*, Proc. of the 73rd IEEE Veh. Tech. Conf. (VTC'2011), 1–5 (2011)
21. I.A. Akhalaghi, H. Khoshbin, Int. J. Tomo. Stat. **11** (S09) (2009)
22. H. Zamiri-Jafarian, M. Rajabzadeh, *A polynomial matrix SVD approach for time domain broadband beamforming in MIMO-OFDM systems*, Proc. of the 67th IEEE Veh. Tech. Conf. (VTC'2008), 802–806 (2008)
23. P. Majewski, Sci. Bul. Opole Univ. of Tech. **69** (2013)
24. W.P. Hunek, P. Majewski, *Perfect reconstruction of signal – a new polynomial matrix inverse approach,* EURASIP J. Wir. Com. and Net., (under review) (2017)
25. W.P. Hunek, P. Majewski, Przegląd Elekrotechniczny **93** (2017) (in Polish)

 DOI: 10.1051/e3sconf/20171901032

Application of numerical model of the photovoltaic module to the practical comparative analysis

Damian Krzesimowski[1,*], and *Artur* Pawelec[2]

[1]Kielce University of Technology, Faculty of Management and Computer Modelling, al. 1000-lecia PP 7, Kielce, Poland
[2]Kielce University of Technology, Faculty of Environmental, Geomatics and Energy Engineering, al. 1000-lecia PP 7, Kielce, Poland

Abstract. This paper includes a model of photovoltaic module based on a basic catalogue data, taking into account the analysis of series and shunt resistance. These resistances are searched with accuracy 0.001 Ohms, but their influence is taken into account in the electrical parameters for each iteration. In addition, the model takes into account changes in environmental parameters: solar radiation intensity and the operating temperature, for which is sought the optimal operating point of the module. This allows to perform a step towards practical applications of the model, including comparative analysis. In the literature numerical models are used to verify the functioning of the PV modules as a function of voltage or current. In the paper is included description of searching the voltage at which the operating point will achieve maximum power for given environmental parameters. This approach to the issue allows to determine the power plane, on which the inverter should move, and this was the first aim of the work. Moreover, it is therefore possible to compare different PV modules without necessity to build the test stands. The output numerical data for different PV modules can be compared with each other and the best model can be chosen to fulfil the condition of the maximum power per day for the desired geographical location. The ability to perform this kind of analysis is the second aim of the work.

1 Introduction

Nowadays designers of photovoltaic installations have a number of modules to choose from many manufacturers. They have to select the module for specific working conditions. They are supported by data sheets, laboratory tests and numerical tests. The data sheets include basic information about the PV module. The results of laboratory tests are detailed and may contain redundant information that is unnecessary from the investor's point of view. In contrast, computer simulations can be tailored to the investor's expectations. The most important thing is to match the module and its number to the conditions in which the installation is to be operated. The engineer can use laboratory data, but these data, though accurate, should be treated as the output parameters of the module under ideal conditions. It is much more convenient and faster to use a computer program, to input parameters of the selected PV module and to check the results for the given environmental conditions. However, these data are burdened with the modelling error of the PV module. For computational reasons computer models are subject to simplification. This results in differences in the results for laboratory and numerical data. If a single module is considered, the differences will be negligible. However, if the results are extrapolated to the entire system, it may be found that the differences go beyond the statistical error. For this reason, it is necessary to develop numerical models that will generate results on the one hand within the tolerance limits, on the other hand, to ensure that these results are sufficiently consistent with laboratory results.

The most commonly used are single diode equivalent and dual diode equivalent models of photovoltaic cells. Single diode models may, but does not have to, take into account the serial and shunt resistances in the equivalent circuit. The number of photovoltaic cells connected in series to obtain a module is one of the input parameters necessary for the calculation. The number of tools used in the calculation is not large. For analysis purposes, the software MATLAB/Simulink is usually used [1–5]. The proposed models of PV modules included in the papers mentioned above can be compared with the use of this particular software environment. On the other hand, the graphical and numerical results presented in the publications allow to verify own solutions, not necessarily realized with the use of the same tool.

The equivalent circuit model of PV cell is needed in order to simulate its real behaviour. Using the physics of p-n junctions, a cell can be modelled as DC current source in parallel with one diode that represent currents escaping due to diffusion and charge recombination mechanisms. The consideration of the recombination loss leads to more precise model known as one-diode

* Corresponding author: d.krzesimowski@tu.kielce.pl

DOI: 10.1051/e3sconf/20171901032

model shown in Fig. 1, used in the described research. Two resistances R_s and R_{sh} are including to the model the contact resistances and the internal PV cell resistance respectively.

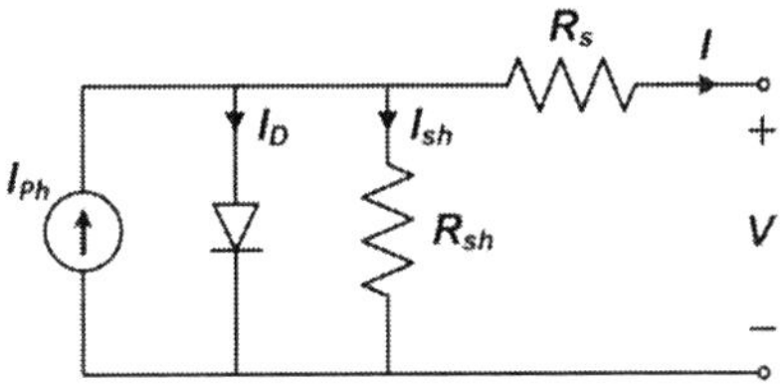

Fig. 1. Single diode equivalent circuit for the silicon photodiode.

The reason for this choice is that this model is well resolved mathematically and sufficiently precise for the purposes of comparative analysis. In addition, only the data sheet of the PV module is required for calculations. However, such sheets do not contain all the necessary data that must be calculated before the final results are obtained. In the study the Scilab software (similar to MATLAB) has been used, and the results are scripts and text files with environmental input data. At the moment no visualization of the proposed mathematical model has been made with the use of the xcos module, but it is possible to present the model as a block diagram or as a description of applied mathematical methods. For the purposes of the paper the last solution had been chosen.

2 Materials and methods

The numerical model proposed in this article is based on the following parameters of the PV module:

- typical peak power P_{mp},
- number of cells N_s,
- short-circuit current I_{sc},
- open-circuit voltage V_{oc},
- current at peak power I_{mp},
- voltage at peak power V_{mp},
- temperature coefficient of short-circuit current K_I,
- temperature coefficient of short-circuit voltage K_V,
- ideality (material) factor A,
- series resistance R_s,
- shunt resistance R_{sh}.

The last three parameters are not usually listed in data sheets. The ideality factor is defined by the production technology of the PV cell. For the purposes of calculations, the values contained in the table I of the paper [6] has been used. However the values, which in this case are arbitrarily imposed, are a kind of simplification. Semiconductor manufacturing processes may vary from factory to factory, so the exact value of the ideality factor is never known.

The resistances can be accurately determined, it is sufficient to find a series resistance to the purposes of calculation the shunt resistance. In the presented model they are determined with accuracy of 0.001 Ohm.

As the environmental conditions STC (Standard Test Condition) values had been used. Additional input data are:

- temperature of the PV module T_c,
- irradiance G,
- Boltzmann constant k,
- elementary charge q.

The first step towards described model is to determine the modified ideality factor, also called thermal voltage. It can be calculated by using formula (1) based on [4].

$$a = (N_s \cdot A \cdot k \cdot T_c) / q \qquad (1)$$

Next step is to determine the reverse saturation current under STC conditions. According to [7] the reverse saturation current of diode at the reference temperature is given by formula (2) with the diode ideality factor added.

$$I_0 = I_{sc} / \{exp[(q \cdot V_{oc}) / (A \cdot k \cdot T_c)] - 1\} \qquad (2)$$

The reverse saturation current is temperature dependant but for the purposes of finding series and shunt resistance this fact is not important.

The values of series resistance in an ideal situation tend to zero, while the value of shunt resistance tend to infinity. In fact, the serial resistance is in the range of 0 to 1 Ohm and the shunt resistance is several hundred Ohm. It has already been mentioned that one of the input parameters for determining the shunt resistance is the serial resistance. In the presented model a table of values of serial resistance is created from 0.001 to 1.000 Ohm with an accuracy of 0.001 Ohm, therefore consists of 1000 values. Each of these values is substituted with formula (4), according to [4], to determine the corresponding shunt resistance.

$$AuxEq3 = V_{mp} + I_{mp} \cdot R_s - V_{oc} \qquad (3)$$

$$R_{sh} = (V_{mp} + I_{mp} \cdot R_s) / \{I_{sc} - I_{sc} \cdot [exp(AuxEq3 / a)] + I_{sc} \cdot [exp(-V_{oc} / a)] - (P_{mp} / V_{mp})\} \qquad (4)$$

The iteration starts at $R_s = 0$ which must increase in order to move the modelled Maximum Power Point until it matches with the P_{mp}. There is only one pair of resistances that satisfies this condition. To numerically determine this power output current I should be calculated. One of the essential parameters at this stage is the photocurrent I_{pv} for STC conditions at the currently tested pair of R_s and R_{sh}. In many papers the calculation is simplified by assuming $I_{pv} = I_{sc}$. Presented model is improved by taking advantage of the iterative solution of R_s and R_{sh}. Each iteration updates R_s and R_{sh} toward the best model solution, so according to [4] formula (5) is introduced in the model.

$$I_{pvSTC} = I_{sc} \cdot [(R_s + R_{sh}) / R_{sh}] \qquad (5)$$

Formula (5) uses the resistances R_s and R_{sh} to determine $I_{pv} \neq I_{sc}$. The values of resistances are initially unknown but as the solution of the algorithm is refined along successive iterations the values of resistances tend to the best solution and (5) becomes valid and effectively determines the light-generated current taking into account the influence of the series and shunt resistances of the array.

Only at this point can be determined the output current and output power. However, it is necessary to prepare an iterative table containing triggering voltage values V, in fact generated by the inverter. In the presented model the voltage is determined from 0 to V_{oc} in iteration level calculated by using formula (6).

$$VstepR = (V_{oc} + 0.1 \cdot V_{oc}) / 100 \quad (6)$$

The margin $0.1 \cdot V_{oc}$ is intended to improve readability of charts. For the purpose of determining the resulting current, formula (8) had been used according to [1].

$$AuxEq7 = V + I \cdot R_s \quad (7)$$

$$I = I_{pv} - I_0 \cdot [exp(AuxEq7 / a) - 1] - AuxEq7 / R_{sh} \quad (8)$$

Iterative numerical solution of formula (8) is impossible. In the presented model a patch has been introduced that enables the generation of resistance pairs and the corresponding calculated maximum power, bypassing the problem of the presence of the current I on both sides of the formula (8). It consists of three elements: the first is the accuracy of the determination of the serial resistance. The second is the number of iterations for determining the power characteristics. The third and most important is the determination of the current I with using the formula (8) based on the value of this current determined in the previous iteration. Therefore the formula (7) has been redefined in (9).

$$AuxEq7(i) = V(i) + I(i\text{-}1) \cdot R_s \quad (9)$$

In order to minimize the error thus introduced into the final result, an experimentally determined value of 1.0004 was applied in the formula (10) which describes the power for the current iteration.

$$P_{mp,e}(i) = V(i) \cdot I(i) \cdot 1.0004 \quad (10)$$

The final result is the calculated power graph plotted on the P_{mp} power diagram from the data sheet and the pair table R_s, R_{sh} and $P_{mp,e}$. The solution is the pair of resistance for which $P_{mp,e} \cdot P_{mp}$. The pairs for which $R_{sh} < 0$ or $P_{mp,e} < 0$ should be rejected.

It is worth noting that the parameter A is taken onto account in calculations, and is the only arbitrarily determined parameter. Because it cannot be sure whether the selected value is correct, the obtained resistance values cannot be treated as true. This means that in the calculations not only pair of R_s and R_{sh} is sought but the combination of three values: A, R_s and R_{sh}. If for example instead of $A = 1.2$ it is assumed to be $A=1.3$ then the values of resistance pairs will be different.

Based on the conclusions of one of the paper on this issue [8] it can be assumed that a change in the ideality factor entails a change in the resistances, that no significant differences in the power characteristics of a specific PV module can be found. The condition is the consistent use in the calculations predetermined value A and calculated on the basis of values R_s and R_{sh}. For the purpose of coarse error determination in the final results, a value for the test PV module had been assumed as $A = 1.2$ and $A = 1.3$, then the characteristics of $I(V)$, $P(V)$ and $FF(V)$ has been obtained. The Fill Factor (FF) indicator had been introduced to the calculations for the purposes of evaluating the performance of the test module, based on the solution presented in [9]. This indicator is described by the formula (11)

$$FF(i) = P_m(i) / (V_{oc} \cdot I_{sc}) \quad (11)$$

The results of searching the resistances are presented in Fig. 2. For the purposes of this paper to the calculation module Bruk-Bet Solar BEP300 had been used with catalogue parameters listen in Table 1.

Table 1. Catalogue parameters of the tested PV module.

type	poly-Si
P_{mp}	300 W
N_s	72
I_{sc}	8.95 A
V_{oc}	44.2 V
I_{mp}	8.4 A
V_{mp}	35.8 V
K_I	0.049 %/°C
K_V	-0.32 %/°C

On the assumption $A = 1.2$ value $R_s = 0.267\ Ohm$, $R_{sh} = 900.124\ Ohm$, but if $A = 1.3$ then the result of calculations is: $R_s = 0.22\ Ohm$, $R_{sh} = 3143.122\ Ohm$. The next step was to compare $I(V)$ and $P(V)$ characteristics and value of FF for the STC conditions. Results are presented in Fig. 3 and Fig. 4. For ideality factor equal to 1.2 the FF point is obtained for $V = 36.66\ V$ and for ideality factor equal to 1.3 the FF point is obtained for $V = 36{,}71\ V$. The difference is only $0.05V$ and the absolute difference is equal to 0.14%. The same can be claimed about the value of FF, the values for $A = 1.2$ and $A = 1.3$ are almost the same. The result of finding a FF point is presented in Fig. 5. There is an insignificant change in shape of the characteristics marked in red and blue, but the most important point on the graph is always in the same place.

 DOI: 10.1051/e3sconf/20171901032

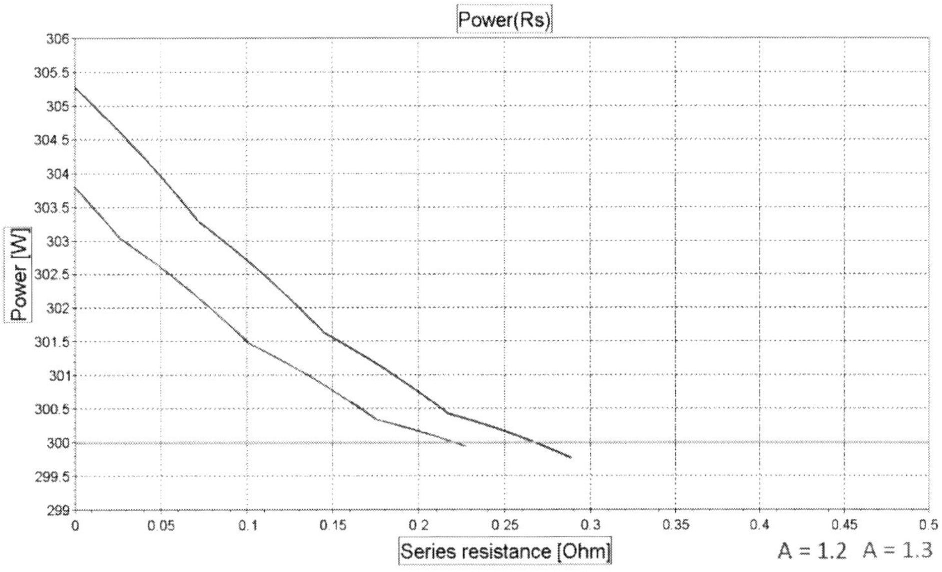

Fig. 2. Results of searching the series resistance for ideality factor (A) equal to 1.2 and 1.3.

3 Results and discussion

To verify the numerical model the power characteristics of a specific PV module should be experimentally determined over given interval of time and then the results should be compared with the obtained by using mathematical method. There are few such verifications in literature, the presented model allows it. The ability to use measured environmental parameters and use them to simulate continuous operation of the PV module with known catalogue parameters had been implemented. This is undoubtedly a great advantage of the presented solution, it allows to simulate PV modules without the need to build test stands.

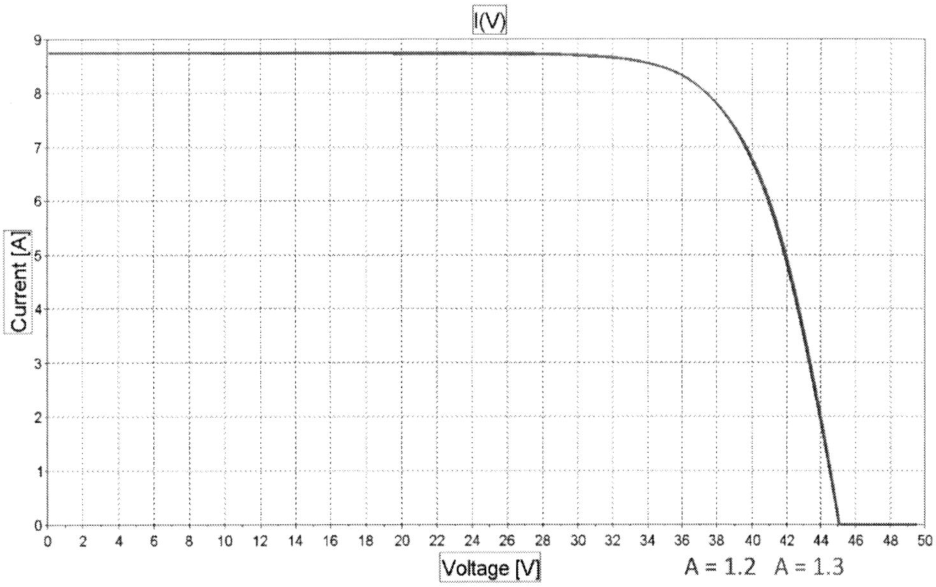

Fig. 3. I(V) characteristics for ideality factor equal to 1.2 and 1.3 under STC conditions.

DOI: 10.1051/e3sconf/20171901032

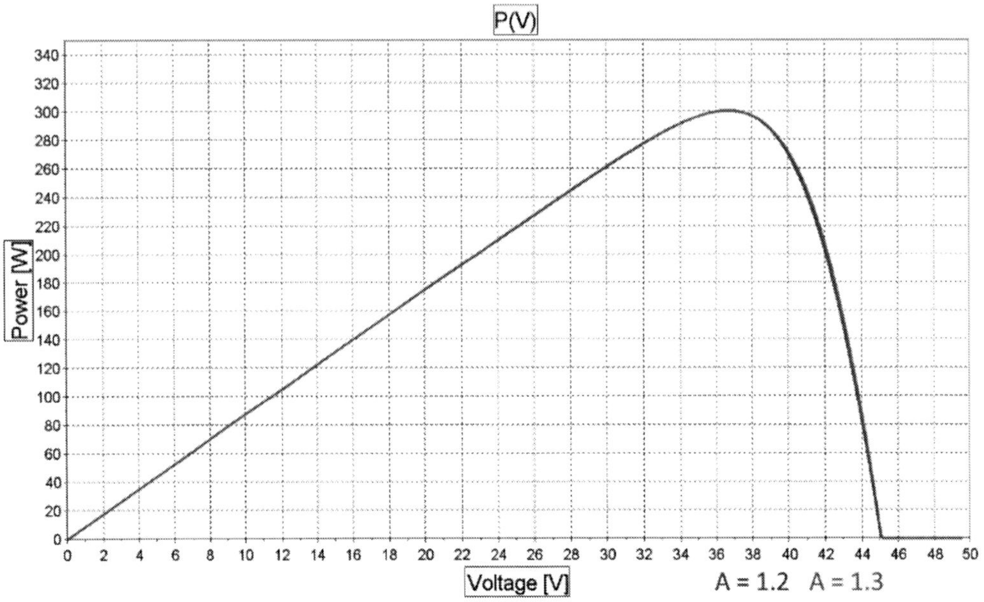

Fig. 4. P(V) characteristics for ideality factor equal to 1.2 and 1.3 under STC conditions.

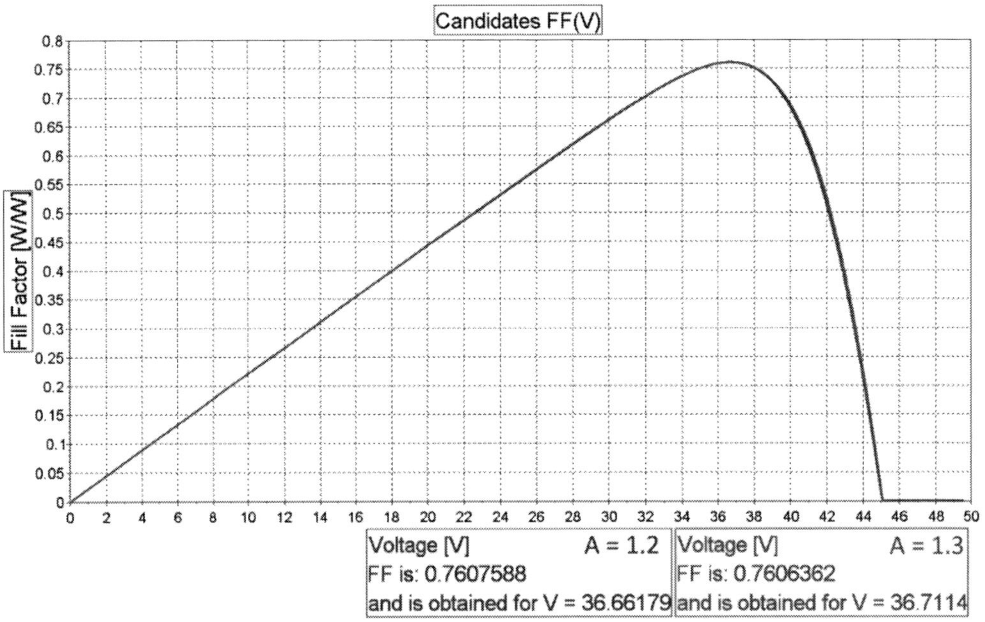

Fig. 5. FF(V) candidates characteristics for ideality factor equal to 1.2 and 1.3 under STC conditions.

The key to achieving a satisfactory final solution is the self-regulation mechanism of the voltage applied to the module according to the maximum power criterion.
The search of this power is an iteration method and is carried out with every change of work parameters of the module. The iterative control parameter is voltage *V*, calculations start with *0* with the iteration level specified by the formula (12). The iteration loop is performed 1000 times.

$$VstepP = (V_{oc} + 0.1 \cdot V_{oc}) / 1000 \quad (12)$$

The modified ideality factor is also calculated

according to (1). The saturation current is calculated according to [10] with using the formula (14), which for the STC conditions assumes the value of the formula (2).

$$\Delta T = T_c - T_{STC} \quad (13)$$

$$I_0 = (I_{sc}+K_I \cdot \Delta T) / \{exp[(V_{oc}+K_V \cdot \Delta T) / a] - 1\} \quad (14)$$

This improvement aims to match the open-circuit voltages of the model with the experimental data for a very large range of temperatures.

Formula (14) is obtained from (2) by including in the equation the current and voltage coefficients K_I and K_V.

The saturation current I_0 is strongly dependent on the temperature and (14) proposes a different approach to express the dependence of I_0 on the temperature so that the net effect of the temperature is the linear variation of open-circuit voltage according to the practical voltage/temperature coefficient.

For calculation purposes it is also necessary to determine the photocurrent for the STC conditions with using the formula (5) and the photocurrent taking into account the temperature, which according to [1] can be expressed by the formula (15).

$$I_{pv} = (I_{pvSTC} + K_I \cdot \Delta T) \cdot (G/G_{STC}) \quad (15)$$

At this stage it is possible to determine the output current with the iterative method using formulas (8) and (9). 1000 values of $I(i)$ are obtained, which subsequently multiplied with the appropriate results of $V(i)$ allow to determine the power $P(i)$. Also in this case the problem of numerical determination of the current $I(i)$ had been solved by using values of $I(i-1)$ and the correction of resultant value of power with the formula (10).

It remains to find such a pair $V(i)$ and $I(i)$ for the resulting power to be the greatest. Since the voltage iteration starts at 0 and ends at $1.1 \cdot V_{oc}$ maximum is sought after all possible power characteristics. This is done at the expense of some computing power, which is not a problem as long as changes in the operating parameters of the module occur rarely enough.

The Fill Factor (F) indicator is calculated in each iteration and the maximum value of *FF* corresponds to the most optimal voltage *V* that should be applied for the PV module.

For verification of the model and its future commercial uses, the presented solution will generate a power graph over time. Sample graphs from data collected in every 5 minutes are shown in Figs. 6 and 7. These graphs can be compared with the experimental data and firstly can help to determine the errors of the model and secondly can help to make the necessary corrections in the form of constants used in the formulas.

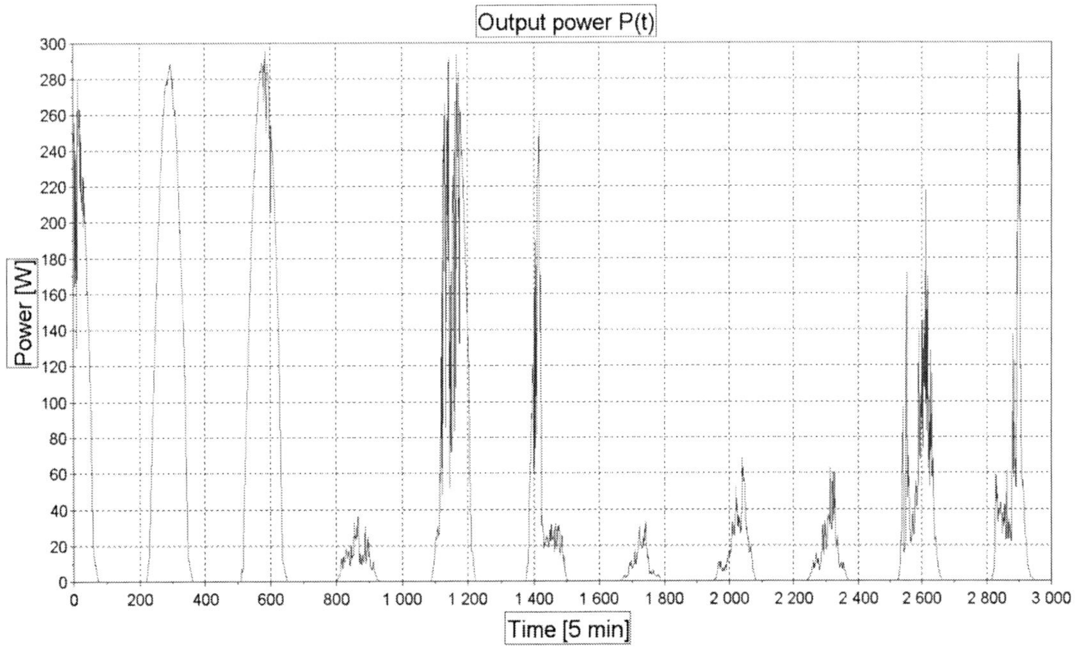

Fig. 6. Characteristics of output power of tested PV module based on the real work conditions. Sampling time is 5 min.

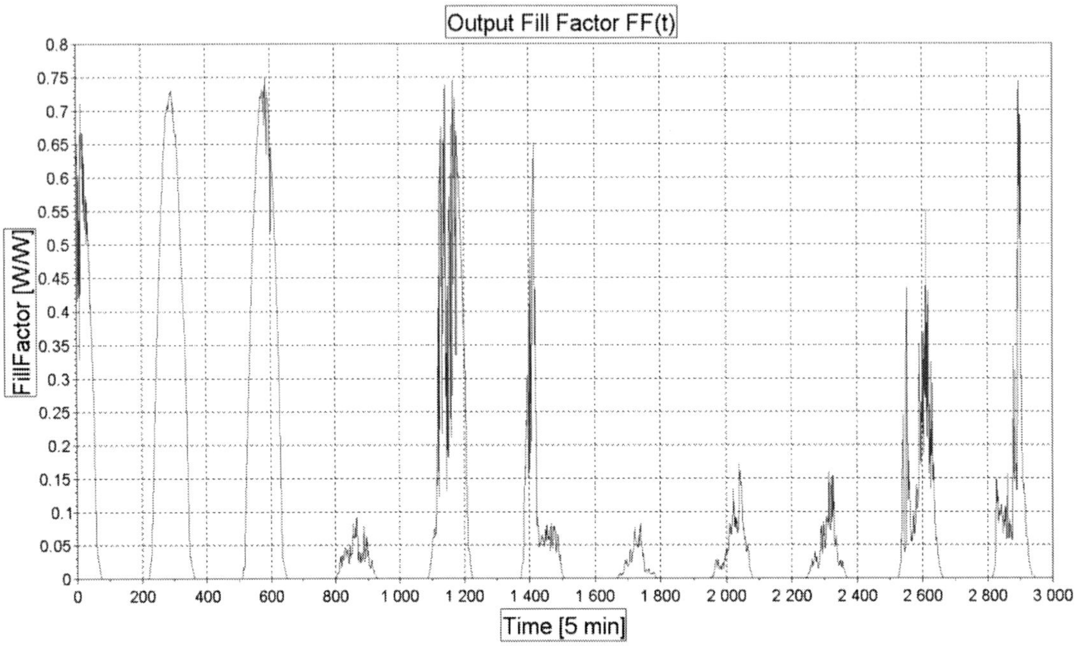

Fig. 7. Characteristics of output Fill Factor of tested PV module based on the real work conditions. Sampling time is 5 min.

4 Conclusions

The numerical model of the photovoltaic module presented in this paper is one of the most accurate in the literature. It is based on commonly known mathematical formulas that have been slightly processed for the purpose of increasing the accuracy of the model. The problem of being unable to find a numerical solution of the equation to the output current in a single diode equivalent model with series and shunt resistance had been solved. This had been achieved at the expense of time to obtain solutions and negligible mistake compensated numerically. The result is a new mathematical tool that can be used to compare PV modules with different parameters under the same operating conditions. The novelty is that there is no need to build test stands composed of different PV modules, as well as the ability to get results in minutes, not in a dozen days or so.

The method of comparing the results of undertaken in such way calculations remains to solve. The most important parameter from the investor's point of view is the output power. If modules of different typical peak power are tested, standardisation is required but only after the final results are obtained. In a sense, a parameter that can be helpful is determined from the formula (11) Fill Factor. It allows to determine the performance of a particular module and compare that value with those obtained for other modules without having to use the module's power parameter. Another way to compare modules can be the sum of the output power calculated after the iteration of the operating parameters. In this situation, the investor obtains a graph allowing to evaluate the dynamics of the module's performance under the same conditions. However, this parameter requires mathematical expression, easy to modify and practical implementation. Third way is to use a standardisation of the output power. The output power value for each of the tested PV modules may be reduced to an arbitrarily determined value, e.g. 1kW. This method would allow to plot the normalised power characteristics of a given module in the irradiance function. This chart could be the most important information for the investor on the data sheet, if it would be placed there.

References

1. D. Bonkoungou, Z. Koalaga, D. Njomo, Int. J. Emerg. Technol Adv. Eng. **3**, 3, 493-502 (2013)
2. J. Park, H-g.K. Yongyun Cho, C. Shin, Adv. Sci. Technol. Lett. **73**, 147-155 (2014)
3. B. Alsayid, Int. J. Comput. Technol. Electron. Eng. **1**, 3, 6-11 (2012)
4. H. Bellia, R. Youcef, M. Fatima, J. Astron. Geophys. **3**, 53-61 (2014)
5. X.H. Nguyen, M.P. Nguyen, Springer Environ. Syst. Res. **4**, 24 (2015)
6. H.-L. Tsai, C.-S. Tu, Y.-J. Su, *Proceedings of the World Congress on Engineering and Computer Science* (2008)

DOI: 10.1051/e3sconf/20171901032

7. A. El Shahat, J. Power Technol. **94,** 1, 50-66 (2014)
8. V.V. Ramana, D. Jena, D.N. Gaonkar, Int. J. Renew. Energy Res. **6,** 3, 970-974 (2016)
9. W. Abd El-Basit, A.M., El-Maksood, F., El-Moniem Saad Soliman, Leonardo J. Sci. **23**, 13-28 (2013)
10. M.G. Villalva, J.R. Gazoli, E.R. Filho, IEEE Trans. Power Electron. **24,** 5, 1198-1207 (2009)

An innovative hybrid insulation switch to enable/disable electrical loads without overvoltages

Konrad Zajkowski[1,*]

[1]Koszalin University of Technology, DP of Energy, 15-17 Raclawicka St. 75-620 Koszalin, Poland

Abstract. This article describes an innovative project for the design of a low voltage switch to enable/disable electrical devices from an AC source, which is characterised by the absence of arc ignition. When powered off, this device is a galvanically separated of the receiver from the power source. The proposed hybrid interaction has two switches: the insulating switch made of mechanical contacts and the current switch made of elements from power electronics. The absence of arc ignition increases the equipment uptime and improves the comfort-powered receiver and equipment located nearby.

1 Introduction

The generation of overvoltage, which ignites the arc on the switch contacts, is accompanied by disconnection of the load from the electrical network. The problem of overvoltages is up-to-date and often discussed in the literature [1-4]. The immediate cause of the overvoltage is a disconnected non-zero current, which flows through the inductance [5-7]. This is a consequence of the continuation of the energy function as described by equations of current through the coil.

A model consisting of an AC power source, power line, and energy receiver includes several inductances connected in series. They represent the nature of the line, the receiver and the source [8-10]. The resultant inductance is equal to the sum of all inductances in a closed circuit. This means that the electric arc generation will be accompanied by the character for any receiver, including the resistive.

For the contacts in the switch, while operating an AC current circuit switch with a receiver of any type, the electric arc impacts can be observed on its surface. A repeated interaction electrical arc reduces the ability of the working device, manifested by increasing resistance, arcing and sticking contacts. When using a conventional installation of low voltage switches, there is a destruction of the working layer of these contacts. This results in a deterioration of the electrical conductivity in the "ON" position. Moreover, the generation of the electric arc on the connector contacts adversely affects the operation of the other devices in the vicinity of the switch [11]. Frequent commutation causes harmonics in the power supply circuit of other technical receivers [12].

The construction of conventional electrical switches is based on mechanical contacts that are exposed to all of the negative effects of the electric arc. The electric arc is not observed, when commutation is made at the instantaneous value of current equal to $i(0) = 0$. The power electronic switch is built with a triac.

Switching by power electronics has been used in systems of higher voltages to date. They support the switching process but do not provide the isolation of the receiver from the phase conductor in a disconnected state.

Switches are used at higher voltages [13]. Thyristor stacks are connected in parallel to the mechanical switches (Fig. 1).

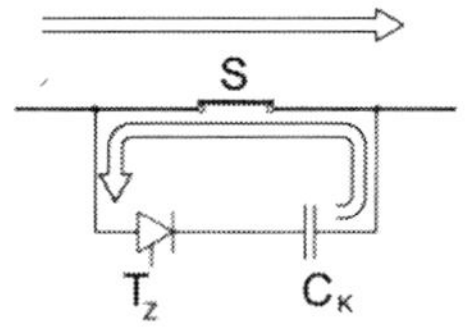

Fig. 1. An Electrical diagram of an exemplary thyristor switch; S – mechanical switch, T_Z – thyristor stacks, C_K – commutation capacitor [13].

This configuration results in no electric arc, but to fulfil the required isolation from the source of the receiver in the OFF state, it must be liberated from an additional switch, which is connected in series with the thyristor stack. To achieve the primary shock protection, the isolation should be in the "OFF" state.

2 Technical description of the solution

The procedure for eliminating the phenomena of electric arc ignition tends to improve the switching capacity of the contacts, extends the uptime and improves the comfort facilities located nearby. Additionally, is possible the work of this low-voltage switch in an environment exposed to the appearance of explosive vapours.

The reduction of the electric arc is possible by using

* Corresponding author: konrad.zajkowski@tu.koszalin.pl

one or several methods simultaneously to reduce the overvoltages:

- A changing environment that surrounds the switch contacts (neutral gases or vacuum),
- Use of non-linear surge arresters,
- A change in the nature of the receiver by reducing the impact of induction,
- Commutation at the time when the current value is zero.

It is possible to reduce the arc ignition when commutation is performed at a time when the sine wave current passes through zero. A power electronic component that causes the disconnection of the sinusoidal current at a natural zero crossing of the current is a triac with an RC suppressor. The triac is turned ON by the gate current and is turned OFF when the anode current is zero.

Each commutation in a non-zero current introduces an additional harmonic distortion to the mains. The ON switch was held at the time of transition for the sinusoidal current through zero and requires a specialised system to start the gate current in the triac, which is synchronised with the detection of the zero anode current.

Electronic power components that are in a disconnected state do not cause the separation of the receiver from the phase conductor. To meet this requirement, the triac must cooperate with the mechanical contacts. This hybrid switch is the subject of this study.

The advantages of this method are:

- Switching operations, i.e., ON and OFF are held at the instantaneous current value equal to zero,
- No generation of additional harmonic distortion in the current waveform,
- Total absence of an electric arc during switching,
- Extension time without a failure switch,
- Increase the switching ability of the device,
- Improve the comfort of work devices that operate in the vicinity,
- Separation of the receiver from the phase conductor in the OFF state,
- Increasing the life of the receiver by reducing the inrush current during cold start,
- This device does not receive energy from a source in the OFF state.

The disadvantages are:

- Increasing the complexity of the switch and the cost of its production,
- Reduction in the amplitude of the sine wave voltage that powers the receiver.

The separation of the receiver in the OFF state and the absence of an electric arc in commutation are feasible in a hybrid system as shown in Fig. 2.

The five-layer thyristor serves as an element which turning OFF the current flowing through the receiver when the instantaneous value of current is equal to zero. However, the ON function is a thyristor with the detectors for the zero current crossing. The function for the circuit isolator is a SB mechanical switch. The condition for the proper interaction of the two switches is the correct sequence of actions.

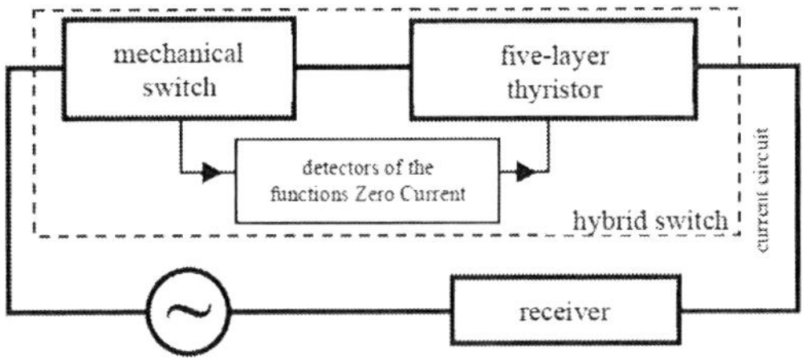

Fig. 2. A block diagram of the hybrid switch to connect the source and receiver.

The switching ON procedure should be performed in the following sequence (Fig. 3):

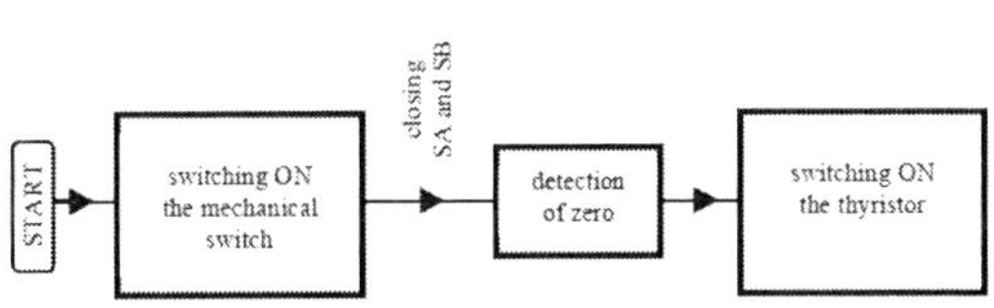

Fig. 3. The algorithm procedure for switching ON the hybrid switch.

The switching OFF procedure should be performed in the following sequence (Fig. 4):

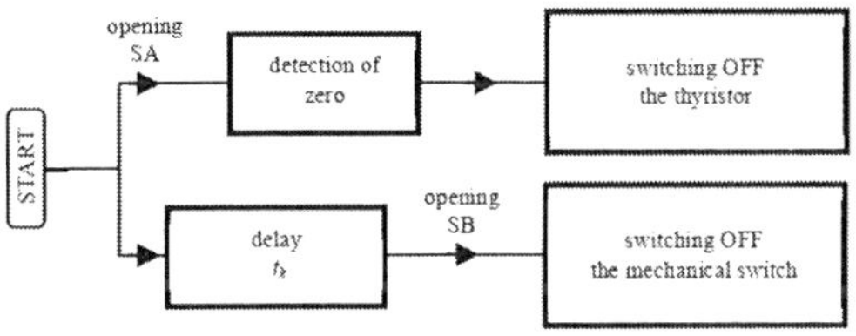

Fig. 4. The algorithm procedure for switching OFF the hybrid switch.

Disconnecting the SB mechanical switch should occur after the moment of time in which the sine wave current passes through zero (there will be an activation switch thyristor). For safety reasons, the delay time t_k should be in the range:

$$2s \geq t_k \geq 1.1 \cdot t_{\min}, \tag{1}$$

where the t_{min} delay value depends on the frequency power source, and it is determined from the relationship:

$$t_{\min} = \frac{1}{2f} = \frac{1}{2 \cdot 50Hz} = 0.01s. \tag{2}$$

2.1 Implementation of the mechanical system

The mechanical system is composed of an assembly of two bistable mechanical switch, current switch SB and auxiliary switch SA, which initiates the procedure for

DOI: 10.1051/e3sconf/20171901033

commutation (Fig. 5a).

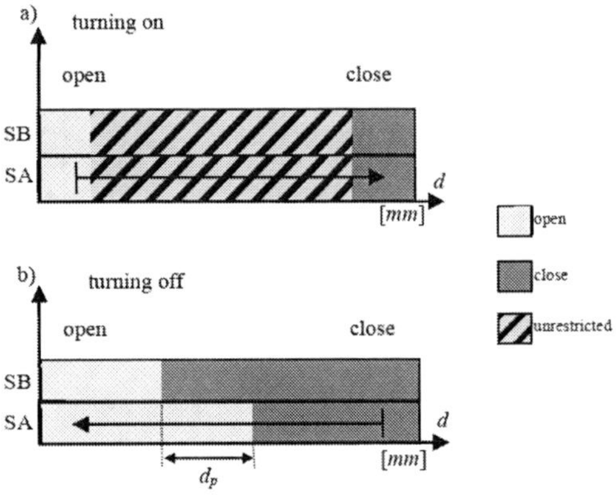

Fig. 5. Draft of the mechanical switch contacts: on the turning ON process a), on the turning OFF process b).

In the process of switching OFF, both mechanical contacts are moved using a common mechanism that provides the t_k time offset between the opening of the contact SB and contact SA (Fig. 5c). In the process of switching ON (Fig. 5b), a special time offset is not required between commutations SA and SB. The distance d_p must be chosen so that at a certain speed of movement, the mechanism has met condition (1).

2.2 Implementation of the electronic part of the device

The Triac switch (4 in Fig. 6) was built with the power electronic element T_1 type BTA16/600 that works with control triac T_2 type MOC3063.

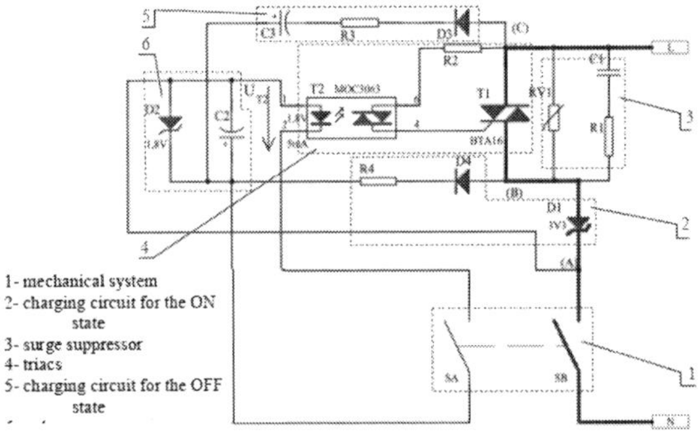

Fig. 6. Scheme of system for the hybrid electronic switch.

The procedure for turning ON this system begins after switches SA and SB are closed. The sequence and time shifts between the activation of SA relative to SB (according to Fig. 5b) do not affect the action of the system. The start of the switching ON procedure is possible only after the closure of two components: SA and SB. After closing SA and SB, switch T_1 is opened, the current flowing through the diode D_1 is equal to zero and the U_{BA} voltage drop on the diode is also equal to zero.

In this case, the power source (2 in Fig. 6), which is built with the elements D_1, D_4 and R_4, is deactivated. The current flowing through D_4 and R_4 is equal to zero. The D_1 diode does not have the flowing energy needed to charge the voltage source (6 in Fig. 6) that is constructed with diode D_2 and capacitor C_2. In the absence of the U_{C2} voltage on capacitor C_2, both the T_1 and T_2 triacs are in an open state. Opening element T_1 allows the observation of the entire power supply voltage U_p between points C and A $U_{CA} = U_p$. Through diode D_3, which acts as a rectifier, the current begins to flow in one direction. The value of this current depends on the values of the resistor R_3 and the capacitor C_3. This current begins charging capacitor C_2 up to the zener voltage for diode D_2. When the contact SA is closed, the U_{C2} source voltage activates the starting procedure for the triac T_2. Optotriac MOC3063 is equipped with the zero crossing detection of the sine wave of current, and the triac T_1 will be activated only at the time of the zero crossing for the sine wave of current flowing through the receiver. Because of the activation of the triac T_1, the voltage U_{CB} starts to drop to approximately 1.5 *V*, which corresponds to the voltage drop at the closed thyristor. With such a small voltage value, the current flowing through diode D_3 is zero. With the active key T_1, the current flowing through the diode D_1 will determine the voltage drop U_{BA} at the level of the zener voltage of the diode. The nonzero voltage U_{BA} causes a unidirectional flow of current through diode D_4 and resistor R_4. These components are responsible for the power supply voltage source (6 in Fig. 6) when the key T_1 is in the active state. As a result, diode D_1 is included in series with the receiver to reduce the voltage drop on the receiver (in the negative half of the sinusoid for a value of 0.7 *V* and in the positive half of the sinusoid for a value equal to the zener voltage of the diode).

The shutdown procedure begins after opening contact SA (according to Fig. 5c) - that is, after the t_a time that is required for its opening procedure, counting from the start of the shutdown procedure (Fig. 7).

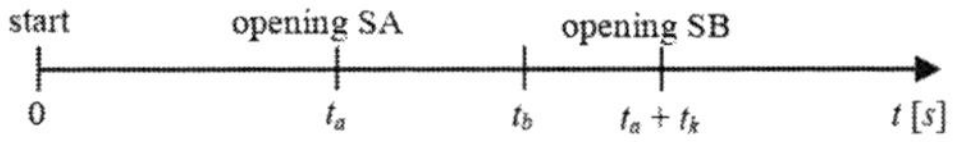

Fig. 7. Procedure for turning OFF the mechanical switch.

In the period between the opening of contacts SA and SB (when $t_a < t < t_a + t_k$), the disconnect operation on triac T_1 is realised. This operation occurs because of the power interruption on the transmitting diode in optotriac T_2. The main circuit is disconnected using triac T_1 only at the time t_b. This time is the moment in which the sine wave of the current flowing through T_1 is equal to zero. Subsequently at $t = t_a + t_k$, contact SB is opened. This causes the isolation of the receiver on the phase conductor L from the power source.

3 Observations of electrical signals on an experimental model

The purpose of these measurements in the test device is determine the voltage waveforms as a function of time. The device according to Fig. 6 has been assembled. Resistive element has been used as a receiver. Waveforms were measured: the power supply voltage U_p, the logical state of the contacts SA and SB, and the voltage at the receiver U_0. AC / DC measuring card was used to record the waveforms. The team of mechanical contacts SA and SB was made using two electromagnetic relays.
The required t_k delay time is the time between the start of the current flow through the coil of the relay and the closing of the relay contacts (approx. 4 *ms*). This time depends on the value of the current that flows through the coil. Resistive type receiver was used. The observed results are shown in Fig. 8.
Activation of triac T_1 is done when three conditions are met (Fig. 8):

- The wave of *Up* signal passes through zero,
- SA and SB is in the attached state.

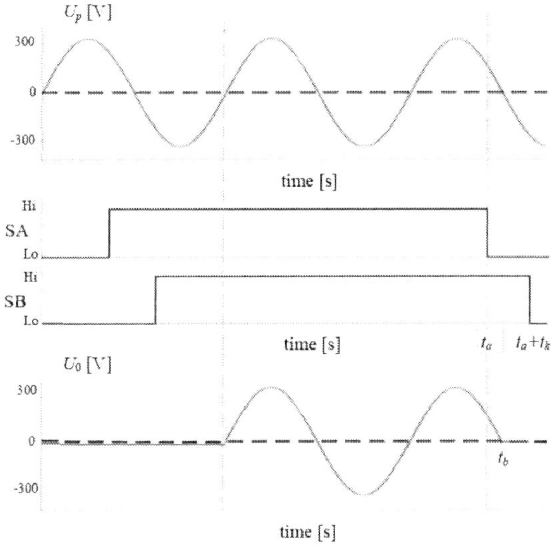

Fig. 8. The selected electrical waveforms in the circuit, where: t_a, t_b, t_a+t_k points according to Fig. 7, U_p – voltage power source, SA & SB – logical state of the contacts, U_0 – voltage at the receiver

Deactivation of triac T_1 is done when (Fig. 8):

- SA switch are transfered in the open state,
- Sinusoidal waveform of the current flowing through the triac (see U_0, when the receiver is resistive) reaches a value equal to zero.

After disconnection of triac T_1 followed by opening the contact SB. This operation occurs at zero current.

4 Cooperation of the device with a receiver

The hybrid switch is installed in a circuit with low voltage AC power and with the phase conductor in series with the receiver. The value for the minimum and maximum power of the receiver depends on the type of installed diode D_1.

In place of the diode D_1, it is possible to use a linear resistance with a suitably selected resistance value and power. It is a solution that should be dedicated to a specific electrical device.

5 Conclusions

The discussed hybrid switch [14,15] is a series connection of the mechanical switch (SB) and triac switch (T_1). It enables the realisation of the commutation process under the conditions required for an insulating switch. No ignition of an electric arc during the commutation occurs only when the conditions are right: Fig. 3 - when switching ON, Fig. 7 - when switching OFF. Switching ON and OFF is performed at the time when the transition sinusoidal current goes through zero. This results in an improved switching process, resulting in an increased uptime hybrid switch and improving the working conditions of the electrical equipment powered from a common main.

Diode D_1 is plugged in series with the power circuit receiver and causes a slight decrease in the supply voltage. The energy needed to power this switch is taken directly from the current path. Therefore, the third wire (N in Fig. 6) connecting the power supply with the switch is not required. The hybrid switch is coupled by using two terminals in series with the receiver and the source similar to the embodiment of a conventional switch (e.g., switch placed in the wall of the house to the light switch). The hybrid switch is a device that can be easily replaced, and it can replace conventional switches that are recessed without modifications of the domestic electrical installation. When powered OFF, the switch does not receive power from the source.

The failure of the hybrid switch does not cause problems for the user. In the event of damage to diode D_1 or triac T_1, the switch will continue to function as a conventional switch without limiting the overvoltages.
Increasing the complexity of the hybrid switch will result in increased costs of production. According to the preliminary estimates, the increase will be approximately 50% of the sales in the first phase and will gradually decrease with an increase in demand for this product.

References

1. Y. Lizhi, L. Guishu, S. Haifeng, L. Xin, L. Xixiao, Lightning (APL), 7th Asia-Pacific International Conference, 727-730 (2011)

DOI: 10.1051/e3sconf/20171901033

2. J. Magnusson, R. Saers, L. Liljestrand, G. Engdahl, IEEE Trans. Power Electron. **29**, 6, 2715–2722 (2014)
3. H. Shareef, S.N. Khalid, M.W. Mustafa, A. Mohamed, IEEE Int. Conf. Power and Energy Conference, p.357–361 (2008)
4. Z. Tao, S. Liu Xue, Z. Yin, S. Peng, IEEE Int. Conf. Applied Superconductivity and Electromagnetic Devices, 430–431 (2013)
5. K. Zajkowski, Neural Comput. Appl. **24**, 2, 431-439 (2014) DOI: 10.1007/s00521-012-1239-0
6. S. Duer, K. Zajkowski, R. Duer, J. Paś, Neural Comput. Appl. **23**, 3-4, 913-925, (2013) DOI: 10.1007/s00521-012-1016-0
7. S. Duer, K. Zajkowski, Neural Comput. Appl. **23**, 7-8, p.2185-2197, (2013) DOI: 10.1007/s00521-012-1169-x
8. F. Wu, B. Sun, K. Zhao, and L. Sun, IEEE Trans. Ind. Electron. **60**, 10, 4450-4457 (2013)
9. M. Siergiejczyk, J. Paś, A. Rosiński, IET Intell. Transp. Syst. **10**, 9, 587-593 (2016)
10. J. Paś, M. Siergiejczyk, Diagnostyka, **17**, 1, 49-55, (2016)
11. W. Choromański, J. Dyduch, J. Paś, *Archives of Transport*, Polish Academy of Sciences, **23**, 2, 137-152, Warsaw (2011)
12. S. Duer, K. Zajkowski, I. Płocha, R. Duer, Neural Comput. Appl. **22**, 7-8, 1581-1590, (2013) DOI: 10.1007/s00521-012-1052-9,
13. W. Deluga, Annu. Set Environ. Prot. **15**, 944-965, Koszalin (2013)
14. K. Zajkowski, patent for invention No. P.406646, Polish Patent Office, Protection for: Koszalin University of Technology (2016)
15. K. Zajkowski, patent for invention No. P.406647, Polish Patent Office, Protection for: Koszalin University of Technology (2016)

E3S Web of Conferences **19**, 01034 (2017)
EEMS 2017
DOI: 10.1051/e3sconf/20171901034

The assessment of exploitation process of power for access control system

Michał Wiśnios[1,*], and *Jacek* Paś[1]

[1]Military University of Technology, Faculty of Electronic, Institute of Electronic Systems, 00-908 Warsaw, 2 Gen. S. Kaliskiego Street, Poland

Abstract. The safety of public utility facilities is a function not only of effectiveness of the electronic safety systems, used for protection of property and persons, but it also depends on the proper functioning of their power supply systems. The authors of the research paper analysed the power supply systems, which are used in buildings for the access control system that is integrated with the closed-circuit TV. The Access Control System is a set of electronic, electromechanical and electrical devices and the computer software controlling the operation of the above-mentioned elements, which is aimed at identification of people, vehicles allowed to cross the boundary of the reserved area, to prevent from crossing the reserved area and to generate the alarm signal informing about the attempt of crossing by an unauthorised entity. The industrial electricity with appropriate technical parameters is a basis of proper functioning of safety systems. Only the electricity supply to the systems is not equivalent to the operation continuity provision. In practice, redundant power supply systems are used. In the carried out reliability analysis of the power supply system, various power circuits of the system were taken into account. The reliability and operation requirements for this type of system were also included.

1 Introduction

The basic criterion providing the power supply continuity for the access control system includes high reliability of all power supply devices forming the power grid [1-3]. The quality of the supplied electricity, which the proper functioning of a number of receivers depends on, does not remain irrelevant [2,4,5]. The quality parameters of electricity – supplying the electric receivers, including the biometric access control system, were determined in the relevant standards – e.g. PN-EN 50160 or regulations of the Journal of Laws No. 93/2007 [6]. The standards and regulations strictly define the electricity parameters, however, there is a lack of executive regulations, which would constitute a basis for enforcement against the receivers of interference introduced to the grid [7-9]. For the purposes of the devices forming the access control system, the most dangerous are short power interruptions and their specific cases defined as voltage dips [10,2,11]. In order to continue the access control system operation, in case of voltage losses longer than 10 ms, it is important to use the emergency source of electricity, e.g. accumulators' batteries or UPS devices [12,13]. The Access Control System is a set of electronic, electromechanical and electrical devices and the computer software controlling the operation of the above-mentioned elements, which is aimed at:

• identification of people or vehicles allowed to cross the boundary of the reserved area [14,15];
• prevention from crossing the boundary of the reserved area by unauthorised people or vehicles [16];
• generation of an alarm signal informing about an attempt of crossing by an unauthorised entity.

In Fig. 1, a simplified diagram of the access control system used in the large facility was presented.

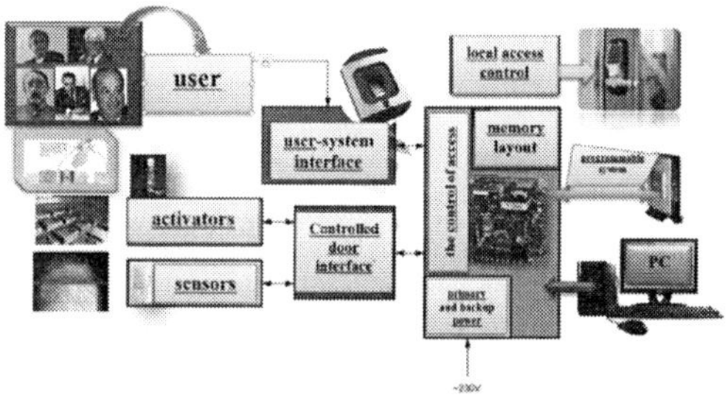

Fig. 1. Block diagram of the access control system.

Currently, in the controlled facilities, where the access control system was applied, a technique of verifying the user, which uses biometrics, is applied [17,18,19]. This technique allows for precise identification of individuals with the use of their unique characteristics. Initially (due to high costs), the biometric readers were installed only in the systems designed for special-purpose facilities, e.g. secret offices. The rapid development of microprocessor technologies and electronic programmable systems resulted in the situation that the price of biometric readers decreased, however, the

* Corresponding author: michal.wisnios@wat.edu.pl

operation precision and reliability increased [20-22]. In order to control the processes of passenger and transport traffic in large areas and facilities, the distributed access control systems are used [23-25] – Fig. 2. The integration of safety systems with access control systems makes it possible to introduce new functionalities – e.g. control over a moving object within the zone, possibility of blocking specific premises, creation of interactive documentation of the users' stay in a given facility, remote monitoring, archiving, overview and diagnosing of the system with the use of TCP/IP network, counting the number of objects at the entrance and exit, blocking the entrances to the zones in case of a breach of e.g. perimeter protection of the facility, etc. [25,26].

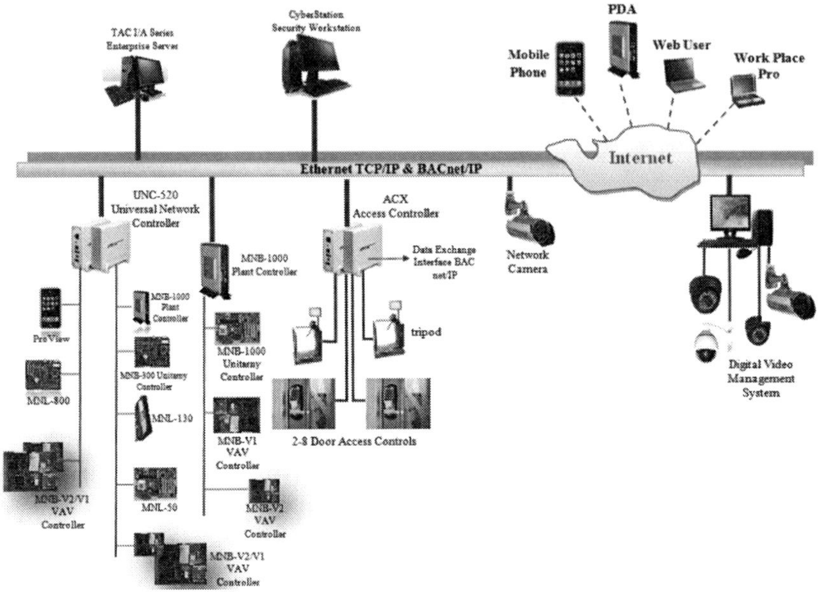

Fig. 2. Separate integrated access control system.

2 Methods for implementation of access control systems' power supply

The power supply of access control systems can be implemented in three different ways, which must take into account, among others, the areal extent, on which the system is placed, requirements specified in the standards and recommendations, and also the economic balance – Fig. 3.

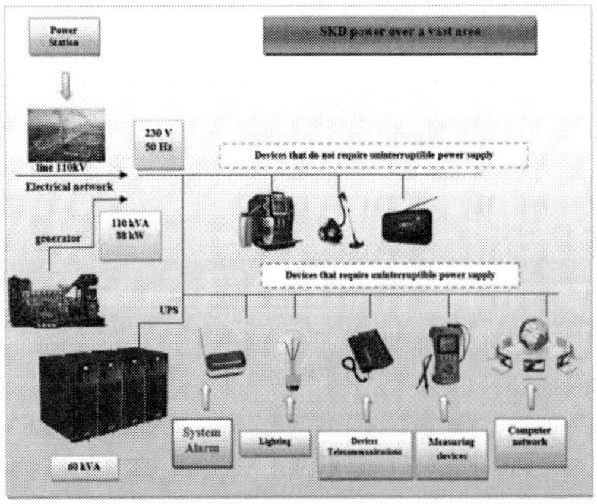

Fig. 3. Implementation of the access control systems' power supply.

The access control system's power supply can be implemented as follows:

- concentrated (power supply is connected to the mainboard – control panel, all the system's elements and devices are supplied from surveillance lines or transmission buses);
- distributed (in the system, there are many power supplies due to high power consumption by individual elements and devices, which are located at a considerable distance from the control panel – voltage drops in the surveillance lines and transmission buses from quiescent current and alarm current, all the power supplies occurring in the system should be monitored by the control panel due to sabotage and technical parameters of the output voltage – Fig. 4);
- mixed (system, due to the large areal extent, on which it is installed [27], uses at least two control panels, the concentrated system is supplied from the power supply with the use of surveillance and transmission lines, which are connected to the control panel – Fig. 5, control subpanels are supplied from other power supplies, or modules 1,2,…,n can be supplied in the system by independent power sources, like in the distributed system, all the power supplies should be monitored by the control panel due to sabotage and technical parameters of the output voltage.

The selection of a type and a method of power supply in the access control system should be preceded by drawing

 DOI: 10.1051/e3sconf/20171901034

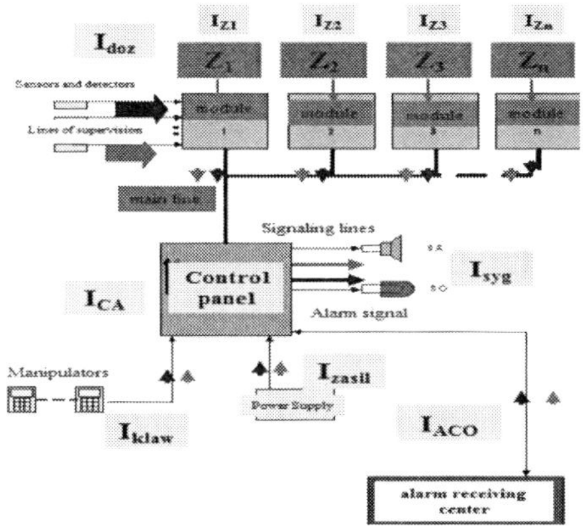

Fig. 4. Power supply implementation in the distributed access control system.

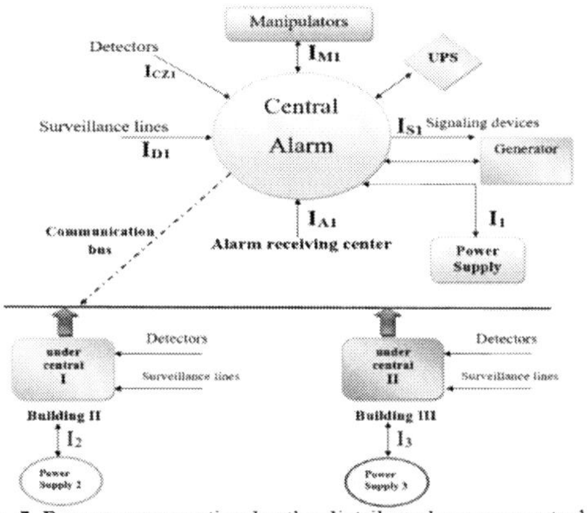

Fig. 5. Power consumption by the distributed access control system's device.

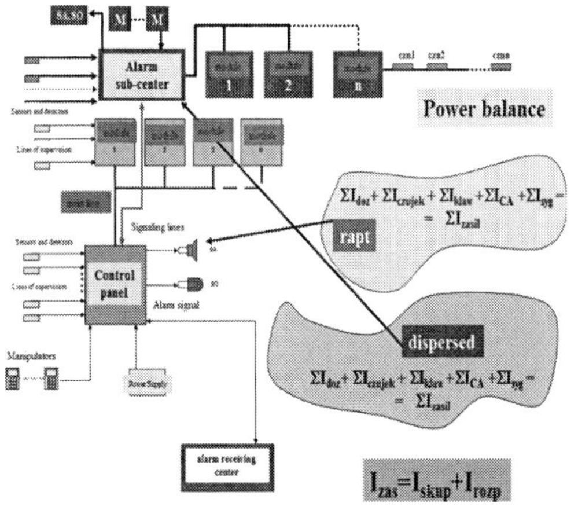

Fig. 6. Power supply implementation in the mixed access control system.

up the power balance for two types of the system operation – the state of monitoring and alarming. Taking into account the alarm system class – access control system (from 1 to 4), it is possible to determine the value of the capacity of backup power supply sources – accumulators' batteries are the most commonly found in the systems.

3 Reliability analysis of the power supply system used in the access control system in the vast area

By analysing the operation of the power supply system used in the access control system, which is operated in the vast area, it can be stated, on the basis of the presented figures 3 - 6, that it has a mixed reliability structure [22,28,29]. The damage to the main power supply of the system supplying the control panel results in the transition to the backup power supply. It results in the transition of the access control system from the state of complete usability $R_O(t)$ to the state of unreliability of safety $Q_B(t)$. During the main power supply failure, some functions of the access control system may be suspended, however, it does not affect safety in the vast area. The system's operation time on the backup power supply is determined by the internal battery capacity, which is specified for given values of the quiescent and alarm currents of sensors and actuators. If the power supply from the electrical grid is not reactivated at that time (main power supply failure), the access control system changes to the state of unreliability of safety S_B[30]. The access control system is used in the vast area – Fig. 5 has three control panels, mutually configured and monitored by the alarm centre. The power supply of individual control subpanels is implemented in a similar manner – the main power supply, the accumulators' battery. However, in case of the main power supply failure, the connections that can be used for mutual electricity exchange were not foreseen in the system. Fig. 7 illustrates the relationships in the considered access control system in terms of safety.

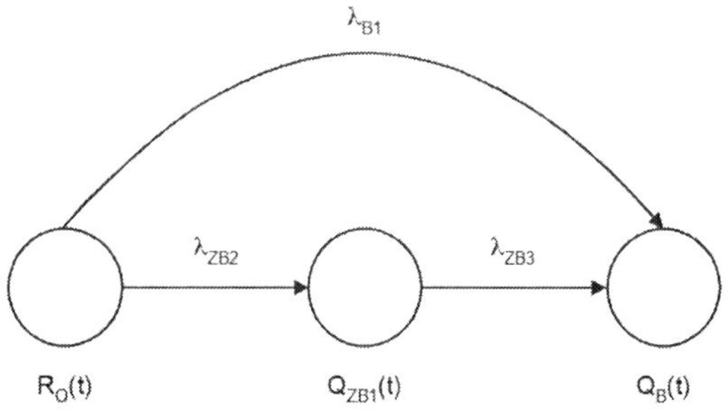

Fig. 7. Relationships in the power supply system of electronic safety systems. Markings in Fig.: $R_O(t)$ – probability function of the system's staying in the state of complete usability S_{PZ}, $Q_{ZB}(t)$ – probability function of the system's staying in the state of impendency over safety S_{ZB}, $Q_B(t)$ – probability function of the system's staying in the state of unreliability of safety S_B, λ_{B1} – intensity of transitions of the automatic transfer switching equipment, λ_{ZB2} , λ_{ZB3} – intensity of transitions of external power lines and uninterruptible power supply.

DOI: 10.1051/e3sconf/20171901034

The system presented in Fig. 7 can be described by the following Chapman–Kolmogorov equations:

$$\begin{aligned} R_0'(t) &= -\lambda_{B1} \cdot R_0(t) - \lambda_{ZB2} \cdot R_0(t) \\ Q_{ZB1}'(t) &= \lambda_{ZB2} \cdot R_0(t) - \lambda_{ZB3} \cdot Q_{ZB1}(t) \\ Q_B'(t) &= \lambda_{B1} \cdot R_0(t) + \lambda_{ZB3} \cdot Q_{ZB1}(t) \end{aligned} \quad (1)$$

Assuming baseline conditions:

$$\begin{aligned} R_0(0) &= 1 \\ Q_{ZB1}(0) &= Q_B(0) = 0 \end{aligned} \quad (2)$$

and using certain mathematical transformations[31] (including the Laplace transform), the following relationships are obtained:

$$R_0(t) = e^{-(\lambda_{B1}+\lambda_{ZB2})\cdot t} \quad (3)$$

$$Q_{ZB1}(t) = \lambda_{ZB2} \cdot \left[\frac{e^{-(\lambda_{B1}+\lambda_{ZB2})\cdot t} - e^{-\lambda_{ZB3}\cdot t}}{\lambda_{ZB3} - \lambda_{B1} - \lambda_{ZB2}} \right] \quad (4)$$

$$Q_B(t) = \frac{\lambda_{B1}}{\lambda_{B1} + \lambda_{ZB2}} \cdot \left[1 - e^{-(\lambda_{B1}+\lambda_{ZB2})\cdot t}\right] + \lambda_{ZB2} \cdot \lambda_{ZB3} \cdot \left[\frac{e^{-(\lambda_{B1}+\lambda_{ZB2})\cdot t}}{(\lambda_{B1}+\lambda_{ZB2})(\lambda_{B1}+\lambda_{ZB2}-\lambda_{ZB3})} - \frac{e^{-\lambda_{ZB3}\cdot t}}{(\lambda_{B1}+\lambda_{ZB2}-\lambda_{ZB3})\cdot\lambda_{ZB3}} + \frac{1}{(\lambda_{B1}+\lambda_{ZB2})\cdot\lambda_{ZB3}} \right] \quad (5)$$

The received relationships make it possible to determine the probabilities of the power supply system's staying n the state of complete usability R_O, impendency over safety Q_{ZB} and unreliability of safety Q_B.

By applying the relationships (3÷5), it is possible to determine the probability values of the power supply system's staying in the distinguished states. Such a procedure is shown in the following example.

Example

Assuming the following values describing the analysed system:

- operation time – 1 year:

$$t = 8760\ [h]$$

- intensity of transitions from the state of complete usability S_{PZ} to the state of impendency over safety S_{ZB1}:

$$\lambda_{ZB2} = 0.0000003429\left[\frac{1}{h}\right]$$

- intensity of transitions from the state of impendency over safety S_{ZB1} to the state of unreliability of safety S_B:

$$\lambda_{ZB3} = 0.0000002285\left[\frac{1}{h}\right]$$

- intensity of transitions from the state of complete usability S_{PZ} to the state of unreliability of safety S_B:

$$\lambda_{B1} = 0.0000001142\left[\frac{1}{h}\right]$$

Transition rate value $\lambda = 0.0000003429\left[\frac{1}{h}\right]$ corresponds to reliability $R = 0.997$.

Transition rate value $\lambda = 0.0000002285\left[\frac{1}{h}\right]$ corresponds to reliability $R = 0.998$.

Transition rate value $\lambda = 0.0000001142\left[\frac{1}{h}\right]$ corresponds to reliability $R = 0.999$.

For the above input values using the equations (3÷5), it is possible to obtain:

$$\begin{aligned} R_0 &= 0.997002 \\ Q_{ZB1} &= 0.001996 \\ Q_B &= 0.001002 \end{aligned}$$

The obtained values allow for reliability assessment of the analysed power supply system.

4 Conclusions

The issue presented in the article concerned the reliability analysis of the power supply system of the access control systems, which are operated in a large facility. The power supply system, which consists of two dependent power supply systems, was analysed. The first system is the power supply, which implements its function by the control panel (subpanel).

The second system includes the accumulators' batteries located in individual subsystems of the access control system – Fig. 5. The technical parameters of the power supply and accumulators' batteries (current, output voltages, etc.) are controlled by individual control panels, and the information on its technical condition is transmitted to the alarm receiving centre. The individual subsystems of the access control system can be treated as independent due to the implemented power supply function.

According to legal regulations, the power supply of individual subsystems should be implemented from separate electric circuits (main cable connector) appropriately protected by overvoltage and short-circuit. As a result of the adopted assumptions, the calculations, which allow to determine the probability values of the power supply system's staying in the distinguished states, were carried out. The calculations can be useful for assessing various types of power supply systems' solutions and rational selection, with assumed initial conditions and selection criteria.

In the further studies, it is planned to conduct further reliability analyses that take into account other types of the power supply systems – e.g. UPS.

References

1. J. Paś, Prz. Elektrotechniczny **10**, 91 (2015)
2. E. Pilo, (editor), *Power supply, energy management and catenary problems* (WIT Press, 2010)
3. L. Wang, (editor), *Modeling and control of sustainable power systems* (Springer-Verlag, 2012)
4. L. Będkowski, T. Dąbrowski, *The basis of exploitation, part II: The basis of exploational reliability* (Military University of Technology, Warsaw, 2006)
5. J. Dyduch, J. Paś, A. Rosiński, *The basic of the exploitation of transport electronic systems* (Publishing House of Radom University of Technology, Radom, 2011)
6. M. Krzykowski, Polityka Energetyczna – Energy Policy J. **17**, 3, (2014)
7. M. Siergiejczyk, J. Paś, A. Rosiński, IET Intell Transp Sy. **10**, 9, DOI: 10.1049/iet-its.2015.0183 (2016)
8. M. Siergiejczyk, J. Paś, A. Rosiński, Modeling of process of exploitation of transport telematics systems with regard to electromagnetic interferences, in *Tools of Transport Telematics*, given as the monographic publishing series – „Communications in Computer and Information Science", 531 (Springer-Verlag, pp. 99-107, 2015)
9. T. Dabrowski, M. Bednarek, K. Fokow, M. Wisnios, Prz. Elektrotechniczny **88,** 11A (2012)
10. J. Paś, M. Siergiejczyk, Diagnostyka, **17**, 1 (2016)
11. J. Wiatr, A. Boczkowski, M. Orzechowski, *Overvoltage protection, selecting wires and their protection in low voltage wiring systems* (MEDIUM Publishing House, Warsaw, 2010)
12. M. Siergiejczyk, J. Paś, A. Rosiński, Kones. **23**, 3, (2016)
13. J. Wiatr, M. Miegon, *UPS power supplies and accumulator batteries in emergency power systems.* (MEDIUM Publishing House, Warsaw, 2008)
14. P. Łubkowski, D. Laskowski, The selected issues of reliable identification of object in transport systems using video monitoring services, in the monograph Telematics – support for transport, given as the monographic publishing series – „Communications in Computer and Information Science", vol. 471, Springer-Verlag, pp. 59-68 (2014)
15. J. Paś, *Operation of electronic transportation systems* (Publishing House University of Technology and Humanities, Radom, 2015)
16. R. Burdzik, Ł. Konieczny, T. Figlus, Concept of on-board comfort vibration monitoring system for vehicles, in *Activities of Transport Telematics*, vol. 395, Springer, pp. 418-425 (2013)
17. M. Wiśnios, T. Dąbrowski, M. Bednarek, Prz. Elektrotechniczny **91**, 10 (2015)
18. M. Wiśnios, T. Dąbrowski, J. Konbin **29**, 1 (2016)
19. M. Kuchta, M. Siergiejczyk, J. Paś, Vibroeng. **18**, 7 (2016)
20. D. Laskowski, P. Łubkowski, E. Pawlak, P. Stańczyk, Anthropotechnical systems reliability, in *Safety and Reliability: Methodology and Applications - Proceedings of the European Safety and Reliability Conference ESREL 2014*, CRC Press/Balkema, London, pp. 399-407 (2015)
21. M. Stawowy, P. Dziula, Comparison of uncertainty multilayer models of impact of teleinformation devices reliability on information quality, in *Proceedings of the European Safety and Reliability Conference ESREL 2015*, CRC Press/Balkema, pp. 2685-2691 (2015)
22. M. Stawowy, Model for information quality determination of teleinformation systems of transport, in *Proceedings of the European Safety and Reliability Conference ESREL 2014*, CRC Press/Balkema, pp. 1909–1914 (2015)
23. M. Siergiejczyk, K. Krzykowska, A. Rosiński, Reliability assessment of cooperation and replacement of surveillance systems in air traffic, in *Proceedings of the Ninth International Conference Dependability and Complex Systems DepCoS-RELCOMEX*, given as the monographic publishing series – Advances in intelligent systems and computing", (Springer, pp. 403–411 (2014)
24. M. Siergiejczyk, K. Krzykowska, A. Rosiński, Reliability assessment of integrated airport surface surveillance system, in *Proceedings of the Tenth International Conference on Dependability and Complex Systems DepCoS-RELCOMEX*, given as the monographic publishing series – „Advances in intelligent systems and computing", vol. 365, Springer, pp. 435-443 (2015)
25. A. Rosiński, *Modelling the maintenance process of transport telematics systems* (Publishing House Warsaw University of Technology, Warsaw, 2015)
26. A. Rosiński, Reliability analysis of the electronic protection systems with mixed – three branches reliability structure, in *International Conference European Safety and Reliability*, ESREL 2009, pp. 1637–1641 (2009)
27. M. Kłodawski, K. Lewczuk, I. Jacyna-Gołda, J. Zak, Arch Transp. **41**, (2017)
28. A. Rosiński, Design of the electronic protection systems with utilization of the method of analysis of reliebility structures, in *Nineteenth International Conference On Systems Engineering (ICSEng 2008)*, Las Vegas, USA, pp. 421-426 (2008)
29. A. Rosiński, Reliability analysis of the electronic protection systems with mixed m–branches reliability structure, in *International Conference European Safety and Reliability*, ESREL 2011, pp. 2064-2071. (2011)
30. R. Billinton, R.N. Allan, *Reliability evaluation of power systems* (Plenum Press, New York, 1996)
31. Zajkowski K., Neural Comput Appl. **24**, 2, (2014)

DOI: 10.1051/e3sconf/20171901035

Comparison of capacitive and inductive sensors designed for partial discharges measurements in electrical power apparatus

Michał Kunicki[1,*]

[1]Opole University of Technology, ul. Prószkowska 76, 45-758 Opole, Poland

Abstract. In the paper results of simultaneously conducted measurements achieved using capacitive and inductive sensors are presented according to different PD model sources immersed in a mineral transformer insulation oil. All measurements are preceded under laboratory conditions using typical measurement set up commonly applied for on-site PD detection: measuring impedance and capacitor and high frequency current transformer (HFCT) are used respectively. Measuring frequency and voltage level influence as well as phase resolved PD patterns analysis are investigated in the research. Various fundamental PD signal descriptors assigned for selected frequencies are also proposed and compared for chosen sensors. The main purpose of the presented research is to compare PD measurement results achieved using selected type of sensors during laboratory measurements and to point the best application areas in fields of PD detection in high voltage apparatus under normal operating conditions. Furthermore a proper measurement results interpretation coming from different sensors as well as measurement conducting problems and achieved patterns disparities are also discussed in the paper.

1 Introduction

Monitoring of electrical power apparatus insulation system condition is a crucial item of fleet maintenance policy according to their reliability as well as providing a continuous electrical power supply for costumers [1–3]. Due to insulation system ageing process some degradation of dielectrics may occur which usually leads to local partial discharge (PD) generation. PD detection is proved to be one of the most effective insulation condition monitoring tools. There are many well-known and widespread PD detection methods but only electrical method (EM) supports a calibrated apparent charge measurement [4–9]. Measurement methodology and procedures are described in IEC 60270 standard. A phase resolved PD pattern (PRPD) has been widely known as an essential tool for a PD analysis provided by the EM. Measured apparent charge values and phase angle of the power voltage cycle correlation have been supported by the PRPD. Not only the highest effectiveness but also analysis capabilities of the measured phenomena have been delivered by a PRPD pattern tool [10–14]. A stochastic nature of a PD phenomena has been the highest challenge for all of PD measurement methods so far. Most of the physical quantities values registered during a measurement tightly depend on the PD source nature as well as on environment conditions. Any modification of the environment or the PD generation conditions radically affect the final measurement results. Thereunder in order to provide an adequate measurement results interpretation an objective comparison of achieved data with representative database need to be supported. In most cases an individual, relative measurement result is not an explicit and its interpretation may differ with reference to different apparatus or environment conditions [15-17]. Generally there are two sensor types designed for PD detection in EM method: inductive (Fig.1) and capacitive (Fig. 2) ones.

Fig. 1. General view of an inductive sensor (HFCT).

Some of capacitive sensors main fields of applications are e.g. power transformers, motors, generators and switchgears, while inductive sensors are usually used for cable applications [11,18,19]. One of the most significant advantages of inductive sensor is relatively high safety level during measurements yielded for the testing team – it does not require a direct high voltage connection: HFCT is to be clamped around the tested

* Corresponding author: m.kunicki@po.opole.pl

DOI: 10.1051/e3sconf/20171901035

cable or grounding. Although a direct high voltage connection need to be supported during capacitive coupling PD measurements a HFCT cannot be applied for not insulated active parts, such as power lines or live clamps. Furthermore, different voltage supply phase occurred PD sources discrimination is also impeded while single HFCT grounding measurement is applied [20,21].

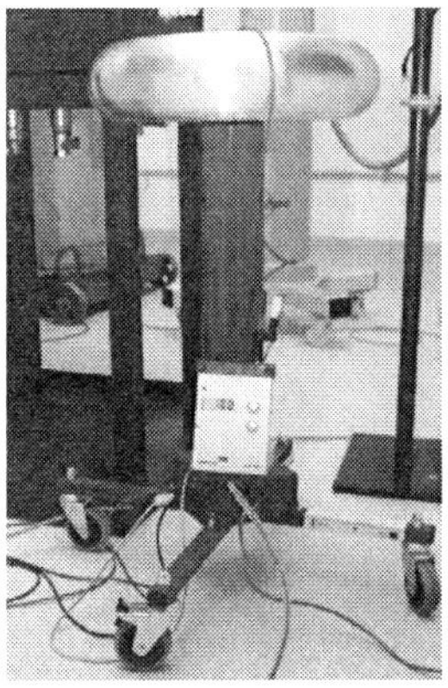

Fig. 2. General view of a capacitive sensor.

PD measurement results achieved using selected type of sensors during laboratory measurements comparison and the best application areas in fields of PD detection in high voltage apparatus under normal operating conditions recommendations have been the main purpose of the presented research.
Furthermore a proper measurement results interpretation coming from different sensors as well as measurement conducting problems and achieved patterns disparities have also been discussed in the paper.

2 Research methodology

All presented research has been preceded under laboratory conditions. Two electrode configurations have been selected for PD source modeling: point to point (8 mm gap) and surface type. A PD model source has been immersed in steel tank filed with brand new mineral insulation oil, commonly used for paper-oil insulation systems in contemporary electrical power transformers. A high voltage (HV) has been supplied by the test transformer with a ratio 220/110000.
A HV level has been adjusted using the automatic voltage control unit, connected to the primary winding of the test transformer (Fig. 3). In case of the surface type configuration a plate electrode has been grounded and 8 mm thick pressboard plate has been used as a solid dielectric. Two-track layout MPD600 system from Omicron has been applied for the research.
The first – capacitive – measuring track has consisted of a coupling capacitor MCC210 with 1 nF of capacity, a CPL542A quadripole with 30 μF of capacity (used for impedance measuring also), a MPD600 module equipped with battery power supply and a MCU504 control unit.

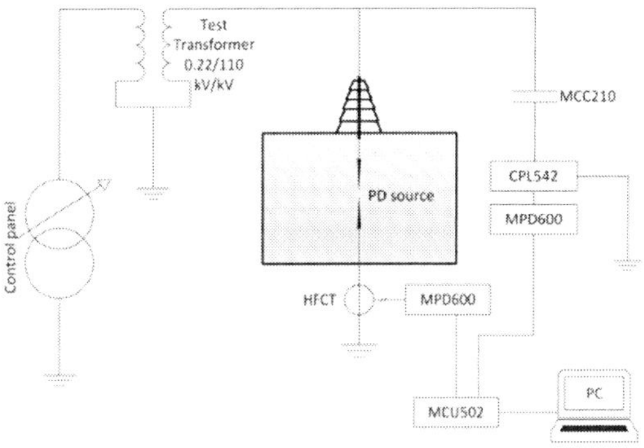

Fig. 3. Measuring system layout.

The second – inductive – measuring track has consisted of HFCT – Rogowski coil 994C.T.Split from M&B Systems Ltd, with its sensitivity of 4.5 V/A, connected to the other MPD600 module and MCU504. A MPD software has been used for phase resolved PD patterns (PRPD) acquisition. All PD data have been recorded by a PC in order to further post-measurement analysis. According to a noise free laboratory environment no supplementary filtering has been applied during measurements. All measurements have been proceeded with two types of sensors simultaneously. A voltage level influence on measurements results as well as measuring frequency band have been investigated in the research. Three voltage levels between 23 and 31 kV have been applied and three frequencies: 0.4, 2, 6 MHz for each electrode configuration. The integration window width has been set to 400 kHz for both channels and has not been adjusted during all measurements. Capacitive as well as inductive measuring channels have been calibrated with CAL542 charge calibrator using 1 nC charge injection.

3 Results and discussion

Survey analysis based on amplitude spectra comparison during PD generation achieved for selected electrode configurations has been the first step of the research. Exemplary amplitude spectra results have been presented in Fig. 4. High convergence of signals registered using different sensors may by noticed below approx. 10 MHz in case of both applied electrode configurations, and a relative signals variation has not exceeded 5 dB. Results achieved for surface type electrode configuration have been found relevant up to approx. 20 MHz. In case of point-point configuration a high disparity has been observed for frequencies over 10 MHz where relative signals variation reached even 20 dB. Similar situations have been indicated for all applied voltage levels.
In Fig. 5 there have been presented some exemplary PRPD patterns obtained for 24 kV powered point-point configuration and a frequency of 0.4 MHz. A high similarity might be observed according to PRPD shape as well as phase domain position of PD occurring and apparent charge levels of signals registered by selected sensors.

DOI: 10.1051/e3sconf/20171901035

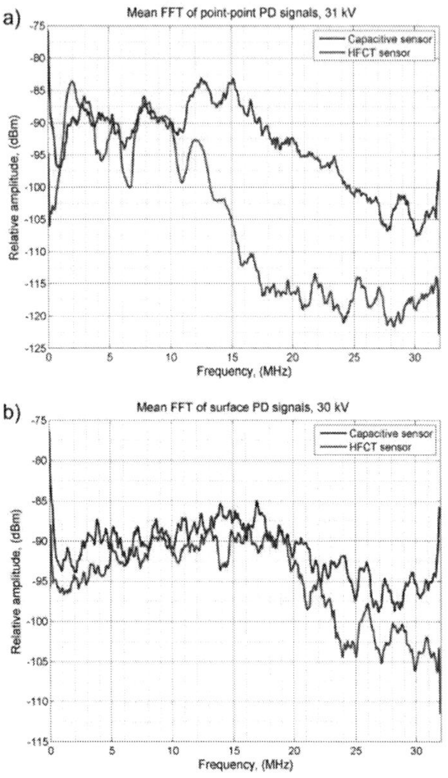

Fig. 4. Exemplary amplitude spectra of signals generated by PDs registered using selected sensors: a) point to point electrode 31 kV, b) surface type electrode 30 kV.

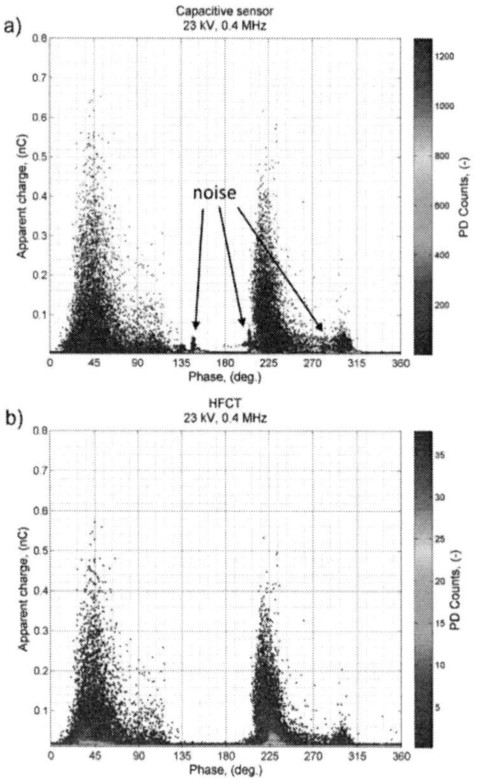

Fig. 5. Exemplary PRPD patterns registered using selected sensors for 23 kV, 0.4 MHz (point to point electrode): a) capacitive sensor, b) HFCT sensor.

Nevertheless some differences have been noticed. Capacitive sensor captured apparent charge values have been found approx. 10% higher than inductive ones. Also a grind derived interference noise significantly visible in case of capacitive sensor results has not appeared on HFCT patterns. Form the power grid point of view an electrode has been read as a circuit discontinuity, so interferences have been cut off the grounding and a PD signal has been the only registered by the sensor. The capacitor has been connected by the HV side so it has captured PD signals as well as grind derived interference.

Point-point configuration exemplary PRPD results achieved for 31 kV and a frequency of 6 MHz have been presented in Fig. 6. The most significant disparities have been recognized during that measurements, especially according to apparent charge dynamics and registered values. A mean apparent charge value when capacitive sensor had been applied has amounted approx. 15 pC while result for HFCT sensor has been 170 pC. Also a max apparent charge values comparison has resulted in the same disparities: about 60 pC for capacitive and over 1 nC for HFCT one sensors respectively.

The described observation has not depended on an applied voltage level, and has repeated during all measurements for 6 MHz and point-point configuration. Despite charge dynamics differences a phase domain PD activity has been found convergent: PD occurring phase angle range according to the first and second supply voltage half-periods (10°-80° and 190°-260°) as well as phase angles related to max apparent charge values within both half-periods (40° and 220°). Also a PD density results have been identified as very similar. No additional noise interferences have been noticed.

Surface type electrode configuration comparative measurements have been the next stage of the presented research. Some exemplary PRPD patterns registered using selected sensors for 24 kV and 0.4 MHz have been illustrated in Fig. 7.

Conclusion has been similar to the one presented according to the previous electrode and a frequency of 0.4 MHz: selected sensors registered signals high similarity might be observed; capacitive sensor captured apparent charge values have been found approx. 10% higher than inductive ones; a grind derived interference noise significantly visible in case of capacitive sensor results has not appeared on HFCT patterns. The described situation has not been influenced by the voltage level adjustments. Comparing PRPD patterns registered for selected electrode configurations some obvious differences have been observed: different PRPD shapes have been associated with each electrode, also a different share of PD activity within each half-period has been pointed.

DOI: 10.1051/e3sconf/20171901035

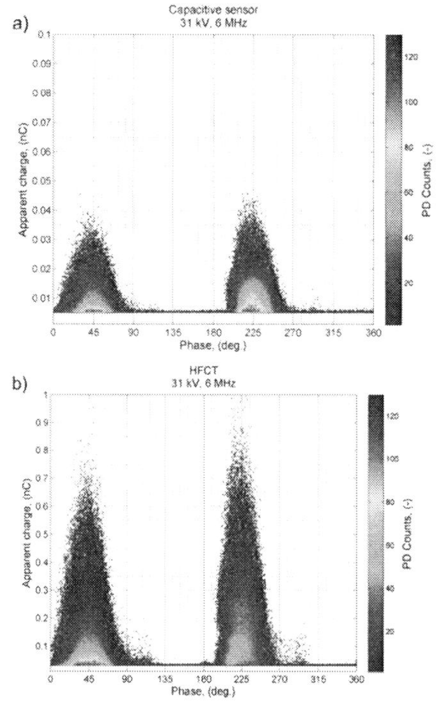

Fig. 6. Exemplary PRPD patterns registered using selected sensors for 23 kV, 6 MHz (point to point electrode): a) capacitive sensor, b) HFCT sensor.

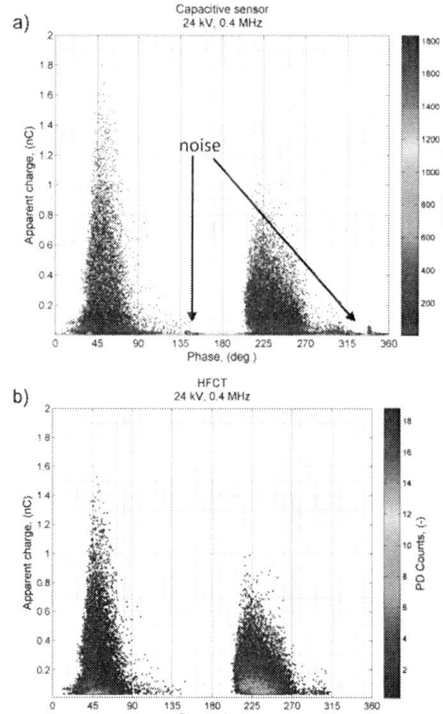

Fig. 7. Exemplary PRPD patterns registered using selected sensors for 24 kV, 0.4 MHz (surface type electrode): a) capacitive sensor, b) HFCT sensor.

In Fig. 8 there have been presented measurement results captured for surface type configuration with a voltage of 30 kV and a frequency of 2 MHz. Selected sensors registered signals high similarity might be observed according to PRPD shape as well as phase domain PD occurring. Also a PD density results have been found as very similar. No additional noise interferences have been noticed according to all previous results achieved for all frequencies above 1 MHz. Nevertheless some significant mismatches have been pointed in relation to apparent charge amplitudes while different sensors results had been compared. Generally approx. 40% of amplitudes variations have been observed: max apparent charge values within the first half-period have reached about 2.3 nC and over 3 nC in case of capacitive and inductive sensors applications respectively. The same trend has been found according to the second half-period: max apparent charge values have reached about 1.5 nC and over 2.5 nC in case of capacitive and inductive sensors applications respectively. However a PD density results have been identified as very convergent. No voltage level influence on described properties has been noticed during research in case of surface type configuration.

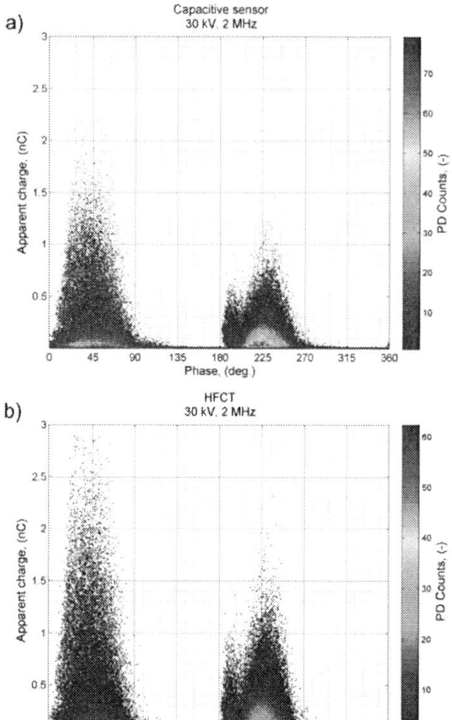

Fig. 8. Exemplary PRPD patterns registered using selected sensors for 30 kV, 2 MHz (surface type electrode): a) capacitive sensor, b) HFCT sensor.

4 Conclusions

Nowadays a PD diagnostics for sure has been found one of the crucial nexus of electrical power distribution systems. Despite there are various commonly known and widespread PD testing methods an improvement of every of them has still been a current issue, especially in the aspects of measurement results interpretation as well as a testing methodology, simplifying and calibration, above all, an on-site application during an apparatus normal service. Furthermore, advanced PD testing tools purchase costs need also to be considered. In the presented paper various exemplary representative measurement results as well as results analysis have been showed. It has been confirmed that PD measurement results achieved using capacitive and inductive sensors simultaneously have not been influenced by supply voltage level, no matter what electrode configuration had been applied. Nevertheless a significant dependency on measuring frequency has been indicated during research. Only result achieved in measuring conditions related with IEC60270 (0.4 MHz) have showed an accepted consistence level.

On the grounds of the presented conclusion, above described research on-site verification as well as a more detailed calibration issue have seemed to be the next aim of the investigation that are to be proceeded by the author as a further study and development on the described problem.

The work was co-financed from funds of the National Science Centre, Poland as a part of the Preludium Research Project No. 2015/19/N/ST8/03909.

References

1. T. Boczar, A. Cichoń, S. Borucki, IEEE Trans. Dielectr. Electr. Insul. **21,** 854–865 (2014)
2. S. Borucki, IEEE Trans. Power Deliv. **27,** 670–676 (2012)
3. J. Buchacz, A. Cichon, J. Skubis, IEEE Trans. Dielectr. Electr. Insul. **24**, 552–558 (2017)
4. D. Zmarzly, L. Nagi, S. Borucki, T. Boczar, Acta Phys. Pol. A **125**, 1377–1379 (2014)
5. Ł. Nagi, D. Zmarzły, T. Boczar, P. Frącz, IEEE Trans. Dielectr. Electr. Insul. **23**, 2036–2041 (2016)
6. M. Kozioł, D. Wotzka, T. Boczar, P. Frącz, J. Spectrosc. **1814754**, (2016)
7. D. Wotzka, A. Cichoń, T. Boczar, Arch. Acoust. **37,** 19–22 (2012)
8. T. Boczar, S. Borucki, A. Cichoń, D. Zmarzły, IEEE Trans. Dielectr. Electr. Insul. **16**, 214–223 (2009)
9. T. Boczar, A. Cichoń, D. Wotzka, M. Kunicki, M. Kozioł, IEEE Trans. Dielectr. Electr. Insul. **24**, 120–128 (2017)
10. M. Majidi, M. Oskuoee, Electr. Power Syst. Res. **119,** 100–110 (2015)
11. R. Sarathi, I. Merin Sheema, J. Rajan, IEEE Trans. Dielectr. Electr. Insul. **21**, 674–682 (2014)
12. A. Kraetge, S. Hoek, M. Koch, W. Koltunowicz, IEEE Trans. Dielectr. Electr. Insul. **20**, 2043–2051 (2013)
13. M. K. Chen, J. M. Chen, C. Y. Cheng, IEEE Trans. Dielectr. Electr. Insul. **21**, 1405–1414 (2014)
14. S. Coenen, S. Tenbohlen, IEEE Trans. Dielectr. Electr. Insul. **19**, 1934–1940 (2012)
15. P. Piekielny, A. Waindok, in *International Symposium on Electrodynamic and Mechatronic Systems (SELM), Zawiercie, 15-18 May 2013* 51–52 (2013)
16. M. Kunicki, A. Cichoń, S. Borucki, Arch. Acoust. **41**, 265–276 (2016)
17. G. Dombek, Z. Nadolny, Eksploat. i Niezawodn. – Maint. Reliab. **19**, 62–67 (2017)
18. F. Álvarez, F. Garnacho, J. Ortego, M. Á. Sánchez-Urán, Sensors **15**, 7360–7387 (2015)
19. M. Shafiq, G. A. Hussain, L. Kutt, M. Lehtonen, Measurement **49**, 126–137 (2014)
20. J. Seo, H. Ma, T. Saha, IEEE Trans. Dielectr. Electr. Insul. **22**, 1105–1117 (2015)
21. A. Cavallini, X. Chen, G. C. Montanari, IEEE Trans. Power Deliv. **25**, 814–824 (2010)21.

New Approaches to Minimum-Energy Design of Integer- and Fractional-Order Perfect Control Algorithms

Wojciech P. Hunek[1,*], and *Łukasz* Wach[1]

[1] Opole University of Technology, Institute of Control Engineering, Prószkowska 76, 45-758 Opole, Poland

Abstract. In this paper the new methods concerning the energy-based minimization of the perfect control inputs is presented. For that reason the multivariable integer- and fractional-order models are applied which can be used for describing a various real world processes. Up to now, the classical approaches have been used in forms of minimum-norm/least squares inverses. Notwithstanding, the above-mentioned tool do not guarantee the optimal control corresponding to optimal input energy. Therefore the new class of inverse-based methods has been introduced, in particular the new σ- and H-inverse of nonsquare parameter and polynomial matrices. Thus a proposed solution remarkably outperforms the typical ones in systems where the control runs can be understood in terms of different physical quantities, for example heat and mass transfer, electricity etc. A simulation study performed in Matlab/Simulink environment confirms the big potential of the new energy-based approaches.

1 Introduction

The issues involving the stability and robustness of MV/perfect control for LTI MIMO integer-order discrete-time systems in state-space framework become the field of intense scientific research [1-7]. Taking into account the integer-order perfect control the quest for advance takes the form of utilization different types of parameter of polynomial right inverses. Until now, the mostly used and widely presented in the literature is unique minimum-norm right T-inverse, regarded as minimum energy one [1,4,5]. However, application the unique T-inverse produce only one inverse model control (IMC) system, hence it cannot be used to wide class of cases where the control inputs remain unstable under perfect control law. To overcome such obstacle there is a need for finding nonunique right inverse letting us generate an infinite set of IMC systems, amongst which we can look for stable and robust ones. The first method can be application of newly introduced set of nonunique parameter and polynomial right inverses, in particular σ- and H-inverse [3-5,8-9]. The crucial features of above-mentioned inverses are so-called degrees of freedom which effectively impact the robustness, especially in terms of minimum-energy perfect control design. Aforementioned exploration of new ways of enhancement the robustness of discrete-time integer-order perfect control systems, in authors opinion, can be also widen by implementation fractional-order model considered as the model closer to the real world objects, giving the prospect to develop more efficient i.e. lower energy consuming and more accurate control strategies. The method of employment fractional model is based on entering slight modification into the order α of well-known Grünwald–Letnikov discrete-time fractional difference operator Δ^{α} of non-integer order α [10-11], which for classical perfect control systems is equal to α=1. The gist of approach is such modification of the perfect control system order that do not distort the original system i.e. leaves parameter α value still close to α=1 but influence its stability and robustness. In order to evaluate the outcome the energy-based performance indices necessary to assess the robustification of the perfect control, is introduced. The presented methodology indicates the direction to develop the new algorithms for minimizing the energy of fractional-order MV/perfect control inputs. Of course, the new method is mainly dedicated to nonsquare systems i.e. systems having different number of input and output variables.

2 System representation

Linear time-invariant (LTI) discrete-time fractional-order system $S(\mathbf{A_d},\mathbf{B},\mathbf{C})$ described by the Grünwald–Letnikov fractional difference operator is defined by the formula [10-11]

$$\begin{cases} \Delta^{\alpha}\mathbf{x}(k+1)=\mathbf{A_d}\mathbf{x}(k)+\mathbf{B}\mathbf{u}(k);\ \mathbf{x}(0)=\mathbf{x_0} \\ \mathbf{y}(k)=\mathbf{C}\mathbf{x}(k) \end{cases}, \qquad (1)$$

where k denotes the discrete time, whereas Grünwald–Letnikov fractional difference operator Δ^{α} of an order α ($0<\alpha<2$, $\alpha\in\mathbb{R}$) is defined by following formula:

$$\Delta^{\alpha}\mathbf{x}(k)=\sum_{j=0}^{k}(-1)^{j}\binom{\alpha}{k}\mathbf{x}(k-j), \qquad (2)$$

whilst

[*] Corresponding author: w.hunek@po.opole.pl

$$\binom{\alpha}{j} = \begin{cases} 1 & j = 0 \\ \frac{\alpha(\alpha-1)\ldots(\alpha-j+1)}{j!} & j > 0 \end{cases}. \quad (3)$$

The system dimensions is defined as follows: n_u-inputs $\mathbf{u}(k)$, n_y-outputs $\mathbf{y}(k)$ and n-state vector $\mathbf{x}(k)$.

Remark 1. For classical state-space systems with α=1 we have

$$\begin{cases} \mathbf{x}(k+1) = \mathbf{A}\mathbf{x}(k) + \mathbf{B}\mathbf{u}(k);\ \mathbf{x}(0) = \mathbf{x}_0 \\ \mathbf{y}(k) = \mathbf{C}\mathbf{x}(k) \end{cases}, \quad (4)$$

where $\mathbf{A} = \mathbf{A}_d + \mathbf{I}_n$ and $\mathbf{I}_n$ is the identity n-matrix.

3 Integer-/fractional-order perfect control

To consider the discrete-time perfect control for fractional-order system denoted by formulas (1-3) it is essential to determine and minimize the performance index described by the following formula

$$J(\mathbf{u}(k)) = \sum_{k=0}^{\infty}[\mathbf{y}(k+1) - \mathbf{y}_{\mathbf{ref}}(k+1)]^{\mathrm{T}} \times [\mathbf{y}(k+1) - \mathbf{y}_{\mathbf{ref}}(k+1)], \quad (5)$$

in which $\mathbf{y}(k+1)$ represents one step deterministic output predictor and $\mathbf{y}_{\mathbf{ref}}(k+1)$ indicates the one-step output reference/setpoint. This enables defining the fractional-order perfect control law or right-invertible systems [10]

$$\mathbf{u}(k) = (\mathbf{CB})^{\mathrm{R}}\,[\mathbf{y}_{\mathbf{ref}}(k+1) - \mathbf{CA}_d\mathbf{x}(k) + \mathbf{C}\sum_{j=1}^{k+1}(-1)^j\binom{\alpha}{k}\mathbf{x}(k-j-1)]. \quad (6)$$

For classical state-space systems with parameter $\alpha = 1$ the Eqn. (6) reduces to the multivariable integer-order perfect control

$$\mathbf{u}(k) = (\mathbf{CB})^{\mathrm{R}}\,[\mathbf{y}_{\mathbf{ref}}(k+1) - \mathbf{CA}_d\mathbf{x}(k)]. \quad (7)$$

Remark 2. In further consideration are taken into account only state-space LTI MIMO discrete-time systems with product of **CB** being right invertible, where symbol 'R' denotes ant right inverse. This imposes that number of inputs of the object is higher than number of its outputs ($n_u > n_y$). For such systems the perfect control exists [1]. For left-invertible systems ($n_u < n_y$) the mentioned control strategy cannot be obtained. The issue concerning the implementation of nonsquare right inverses into perfect control laws is discussed in next section

4 Inverses of nonsquare matrices

At the beginning the definition of classical unique minimum-norm T-inverse of $\underline{\mathbf{B}}(q^{-1})$ with dimension $n_y \times n_u$, dedicated to full normal rank systems in the backward shift operator domain, is presented [1].

Definition. Let the polynomial matrix $\underline{\mathbf{B}}(q^{-1}) = \underline{\mathbf{b}}_0 + \underline{\mathbf{b}}_1 q^{-1} + \cdots + \underline{\mathbf{b}}_m q^{-m}$ be of full rank n_y (or n_u). The (unique) minimum-norm right (or least square left) T-inverse of $\underline{\mathbf{B}}(q^{-1})$ is defined as $\underline{\mathbf{B}}_0^{\mathrm{R}}(q^{-1}) = \underline{\mathbf{B}}^{\mathrm{T}}(q^{-1})[\,\underline{\mathbf{B}}(q^{-1})\underline{\mathbf{B}}^{\mathrm{T}}(q^{-1})]^{-1}$ (or $\underline{\mathbf{B}}_0^{\mathrm{L}} = [\underline{\mathbf{B}}(q^{-1})\underline{\mathbf{B}}^{\mathrm{T}}(q^{-1})]^{-1}\underline{\mathbf{B}}^{\mathrm{T}}(q^{-1})$).

For the purpose of further analysis the parameter system is given due consideration which forms a special instance of polynomial one under $\underline{\mathbf{B}}(q^{-1}) = \underline{\mathbf{b}}_0$. However, the unique T-inverse generates exactly one solution, which in many cases do not meet the requirements of conducted study.

An enhancement of aforementioned unique right T-inverse is nonunique right σ-inverse involving so-called degrees of freedom. The recent formalization of σ-inverse is presented in Ref. [4] in form of Corollary 1.

Corollary. Let the polynomial matrix $\underline{\mathbf{B}}(q^{-1}) = \underline{\mathbf{b}}_0 + \underline{\mathbf{b}}_1 q^{-1} + \cdots + \underline{\mathbf{b}}_m q^{-m}$ be of full normal rank n_y (or n_u) and let $z^s\underline{\boldsymbol{\beta}}(z^{-1}) = \boldsymbol{\beta}(z) \in \mathbb{R}^{n_y \times n_u}[z]$ be arbitrary, including an arbitrary order s. Assume additionally that the product $\underline{\mathbf{B}}(q^{-1})\underline{\boldsymbol{\beta}}^{\mathrm{T}}(q^{-1})$ is of full normal rank n_y (or n_u). Then an old form of σ-inverse (see [1], p. 57, Eqn. (5.3)) can be equivalently redefined as $\underline{\mathbf{B}}^{\mathrm{R}}(q^{-1}) = \underline{\boldsymbol{\beta}}^{\mathrm{T}}(q^{-1})[\,\underline{\mathbf{B}}(q^{-1})\underline{\boldsymbol{\beta}}^{\mathrm{T}}(q^{-1})]^{-1}$.

The polynomial form of σ-inverse can be downgraded to parameter one for $\underline{\mathbf{B}}(q^{-1}) = \underline{\mathbf{b}}_0$, and have the parameter σ-inverse formula simplified do the following form

$$\underline{\mathbf{B}}^{\mathrm{R}}(q^{-1}) = \boldsymbol{\beta}^{\mathrm{T}}[\,\underline{\mathbf{B}}(q^{-1})\boldsymbol{\beta}^{\mathrm{T}}]^{-1}, \quad (8)$$

with parameter degrees of freedom $\boldsymbol{\beta}$. Some issues concerning $\boldsymbol{\beta}$ having polynomial matrices can be found in Ref. [5].

Remark 3. For $\underline{\boldsymbol{\beta}}(q^{-1}) = \underline{\mathbf{B}}(q^{-1})$ the σ-inverse specializes to T-inverse.

Remark 4. In Eqn. (8) it should rather be written $\mathbf{B}^{\mathrm{R}}$ instead of $\underline{\mathbf{B}}^{\mathrm{R}}(q^{-1})$, as well as **B** instead of $\underline{\mathbf{B}}(q^{-1})$.

Having the notion of polynomial inverses and their parameter representations, we are able to implement the T-, σ-inverses into the fractional-order perfect control law as in Eqn. (6), which is needed to run the simulation study. In the simulation study T-inverse is used, σ-inverse is applied only for cases in which T-inverse generates unstable control inputs. An H-inverse, basing on SVD decomposition, with parameter degrees of freedom **L** [3]

$$\mathbf{A}^{\mathrm{R}} = (\mathbf{V}^{\mathrm{T}})^{-1}\boldsymbol{\Sigma}^{\mathrm{R}}\mathbf{U}^{-1}, \quad (9)$$

where $\boldsymbol{\Sigma}^{\mathrm{R}} = \begin{bmatrix}\boldsymbol{\Sigma}_{\mathrm{m}}^{-1} \\ \mathbf{L}\end{bmatrix}$, can be used instead. However, application of σ-inverse with parameter degrees of freedom $\boldsymbol{\beta}$ is more convenient solution, specifically.

5 Energy-based improvement in fractional-order perfect control

The aim of the study is to present that even minor change in the order of the system may lead to the increase of perfect control robustness. The manipulation of parameter α, where α = 1 for classical systems i.e. integer-order systems, can directly affect the perfect control counteracting the undesirable rumbling. To accomplish the objectives of the study, the measures enabling evaluation how the alteration of the parameter α influence the robustness of the perfect control in terms of energy, have to be establish. Moreover the value for which the perfect control energy and state vector is close to the minimum, has to be determined. The introduced below indices are basing on the perfect control and state vector energy. To measure the impact of perfect control energy $E(\mathbf{u}(k))$ and the state vector energy $E(\mathbf{x}(k))$ the following performance indices are introduced, respectively

$$E(\mathbf{u}(k)) = \sum_{k=0}^{\infty}[\mathbf{u}(k) - \mathbf{u}_{\mathbf{stab}}(k)]^{\mathrm{T}}[\mathbf{u}(k) - \mathbf{u}_{\mathbf{stab}}(k)], \quad (10)$$

$$E(\mathbf{x}(k)) = \sum_{k=0}^{\infty}[\mathbf{x}(k) - \mathbf{x}_{\mathbf{stab}}(k)]^{\mathrm{T}}[\mathbf{x}(k) - \mathbf{x}_{\mathbf{stab}}(k)]. \quad (11)$$

The $\mathbf{u}_{\mathbf{stab}}$ and $\mathbf{x}_{\mathbf{stab}}$ are the stable prefect control input and state vectors, accordingly.

6 Simulation study

Let us take into account the integer order system $\mathrm{S}(\mathbf{A}, \mathbf{B}, \mathbf{C})$ in state-space, where $\mathbf{A} = \begin{bmatrix} 0.25 & -0.75 \\ -0.05 & -0.80 \end{bmatrix}$, $\mathbf{B} = \begin{bmatrix} -0.75 & 0.90 \\ -0.10 & 0.30 \end{bmatrix}$, $\mathbf{C} = [0.90 \quad -0.90]$ and initial state vector $\mathbf{x_0} = [-4.00 \quad 9.00]$. The system is chosen because of rumblings in control, which make it difficult to stabilize. Simulation study is carried out in Matlab/Simulink environment with the time horizon $k_{max} = 200$ embracing the subsequent simulations for several values of fractional order difference parameter α modified with step 0.05. For every simulation the output remains at the reference/setpoint for $k \geq d = 1$, under the stabilizing fractional-order perfect controls (and $\mathbf{x}(k)$ runs). The $E(\mathbf{u}(k))$ and $E(\mathbf{x}(k))$ indices are computed for each simulation and garnered in the Tables 1 and 2.

Tables 1 and 2 present results of simulation for values of parameter $\alpha \geq 1$, for $\alpha < 1$ the perfect control become unstable. The results of simulations show that slight increase in the order of perfect control (between $1 < \alpha \leq 1.15$) affect the improvement in perfect control robustness in terms of control energy needed for its stabilization. Some results of simulation study are presented in Figs. 1-6. It can be observed that significant increase in perfect control robustness is obtained for $\alpha \approx 1.15$, where the value $E(\mathbf{u}(k))$ is close to minimum. There is also decline in $E(\mathbf{x}(k))$ value, although for mentioned α does not reach a minimum value. Moreover the Fig. 5 may suggest that for $\alpha = 1.20$ there is still improvement in perfect control robustness. In fact the value $E(\mathbf{u}(k))$, interpreted as the energy of perfect control, increases, which means that system go beyond its optimum point.

It should also be strongly indicated that the robust fractional-order perfect control strictly dedicated to non-integer order plants described by the Grünwald–Letnikov fractional difference operator can be obtained in a different way. Another example could be an application of various nonunique right inverses of CB product (for example see Eq. (6)). Due to special selected degrees of freedom we can significantly improve the robustness of this control strategy. The issue corresponds with the perfect control design of multivariable integer-order systems.

Tab. 1 – summary of the simulation study results for $E(\mathbf{u}(k))$.

α	$\mathbf{u_1}$	$\mathbf{u_2}$	$E(\mathbf{u}(k))$	$\mathbf{u_{1stab}}$	$\mathbf{u_{2stab}}$
1.00	46.1374	39.3123	85.4497	-0.6182	0.5707
1.05	14.2739	12.1624	26.4363	-0.6180	0.5705
1.10	8.6567	7.3761	16.0328	-0.6179	0.5704
1.15	8.2402	7.0213	15.2615	-0.6179	0.5704
1.20	10.4085	8.8688	19.2772	-0.6179	0.5704
1.25	14.4960	12.3516	26.8476	-0.6179	0.5704

Tab. 2 – summary of the simulation study results for $E(\mathbf{x}(k))$.

α	$\mathbf{x_1}$	$\mathbf{x_1}$	$E(\mathbf{x}(k))$	$\mathbf{x_{1stab}}$	$\mathbf{x_{2stab}}$
1.00	562.7609	614.9315	1177.6924	1.2070	0.0959
1.05	179.6233	231.7935	411.4168	1.2070	0.0959
1.10	96.0203	148.1910	244.2113	1.2070	0.0959
1.15	62.0046	114.1754	176.1800	1.2070	0.0959
1.20	45.5737	97.7445	143.3183	1.2070	0.0959
1.25	37.7172	89.8878	127.6050	1.2070	0.0959

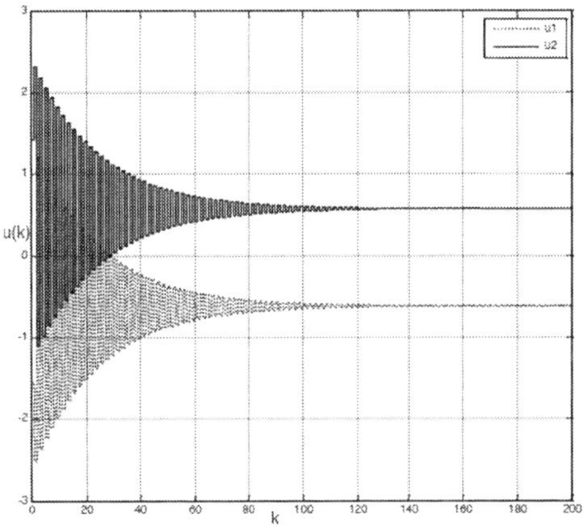

Fig. 1 – $\mathbf{u}(k)$ for $\alpha = 1, \mathrm{k} = 200$.

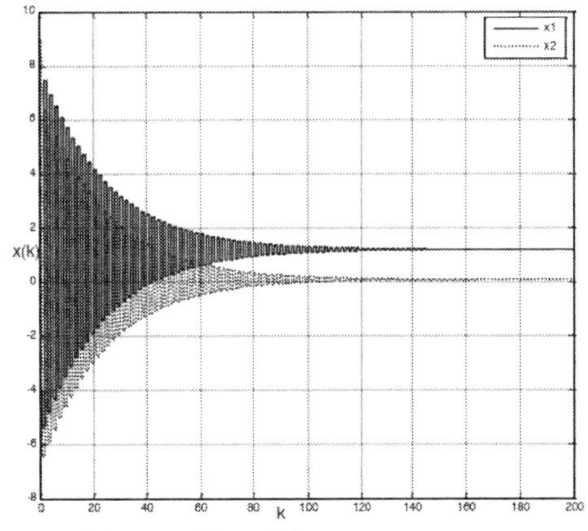

Fig. 2 – $\mathbf{x}(k)$ for $\alpha = 1, \mathrm{k} = 200$.

DOI: 10.1051/e3sconf/20171901036

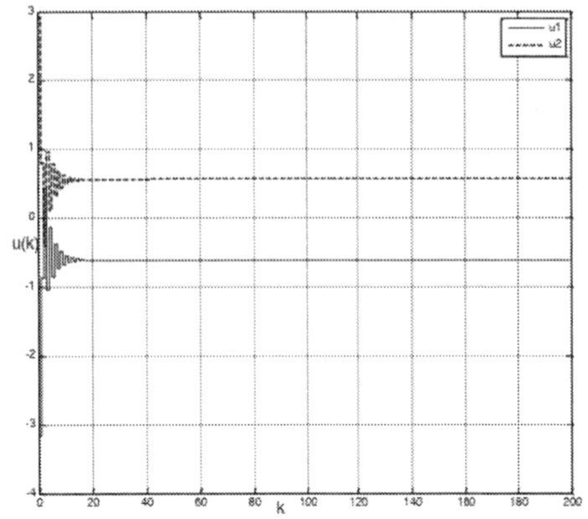

Fig. 3 – $\mathbf{u}(k)$ for $\alpha = 1.15, \mathrm{k} = 200$.

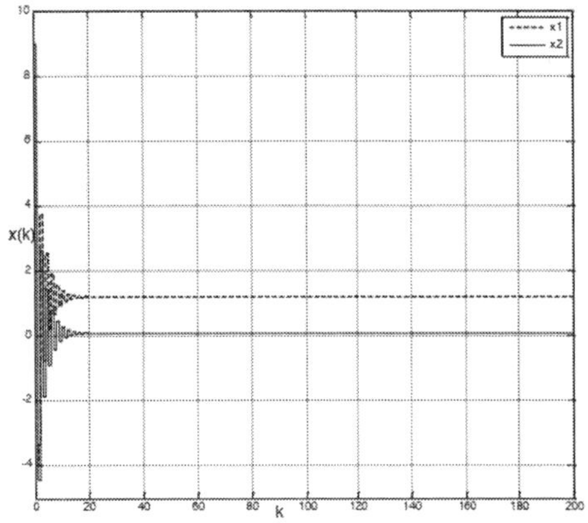

Fig. 4 – $\mathbf{x}(k)$ for $\alpha = 1.15, \mathrm{k} = 200$.

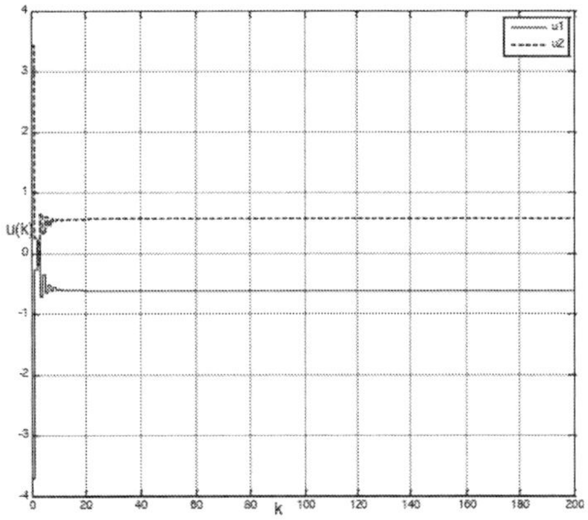

Fig. 5 – $\mathbf{u}(k)$ for $\alpha = 1.20, \mathrm{k} = 200$.

Note that for fractional-order systems of Eqns. (1) the output remains at the reference/setpoint for $k \geq d = 1$ even under unstable perfect control, this is a main feature of this control strategy. Nevertheless, due to an application of different tools we can immediately obtain the minimum phase behavior of non-integer order plants. It can only be done for nonsquare systems having different number of input and output variables in contrast to the square ones where the mechanism of nonuniqueness does not exist.

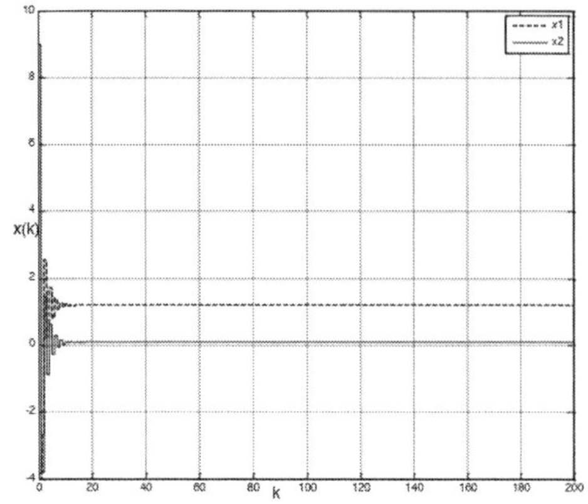

Fig. 6 – $\mathbf{x}(k)$ for $\alpha = 1.20, \mathrm{k} = 200$.

7 Conclusions

In the paper the new approach to minimize the energy of perfect control and growth of the system robustness by application of fractional-order model has been presented. The simulation examples of multivariable discrete-time system in state-space framework have been carried out in Matlab/Simulink environment. The results of simulation study, evaluated using the introduced authors' indices, show the decrease of energy required to stabilize the system inputs as well as decline in state vector energy. That confirms the correctness of the presented strategy and demonstrates the potential of proposed method in reducing control energy. The future study can be focused on search of the analytical formula that proofs the results obtained in the simulation tasks.

References

1. W.P. Hunek, *Towards a General Theory of Control Zeros for LTI MIMO Systems*, Opole Univ. of Tech. Pre. (2011)
2. W.P. Hunek, K.J. Latawiec, P. Majewski, P. Dzierwa, *An application of a new matrix inverse in stabilizing state-space perfect control of nonsquare LTI MIMO systems,* Proc. of the 19th IEEE Int. Conf. on Met. and Mod. in Auto. and Rob. (MMAR), 451-455 (2014)
3. W.P. Hunek, *New SVD-based matrix H-inverse vs. minimum-energy perfect control design for state-space LTI MIMO systems.* Proc. of the 20th IEEE Int. Conf. on Sys. Theory, Cont. and Comp. (ICSTCC'16), Sinaia, Romania, pp. 14-19, (2016)
4. W.P. Hunek, *An application of polynomial matrix σ-inverse in minimum-energy state-space perfect control of nonsquare LTI MIMO systems,* Proc. of the 20th Int. Conf. on Met. and Mod. in Auto. and Rob. (MMAR), 2521–255 (2015)
5. W.P. Hunek, *An application of new polynomial matrix σ-inverse in minimum-energy design of robust minimum variance control for nonsquare LTI*

MIMO systems. Proc. of the 8th IFAC Symp. on Robust Cont. Design (ROCOND'2015), Bratislava, Slovakia, pp. 149–153, (2015)

6. K. J. Latawiec, *The Power of Inverse Systems in Linear and Nonlinear Modelling and Control*. (Opole Univ. of Tech. Press, Opole, 2004)
7. W. P. Hunek, Ł. Wach, *Towards a New Stability Criterion for Fractional-Order Perfect Control of LTI MIMO Discrete-Time Systems in State-Space* - Proc.of 3rd IEEE Int. Conf. on Cyb. CYBCONF-2017, Exeter, UK, pp. 134-139, (2017)
8. Ł. Wach Ł., W.P. Hunek, JAMRIS **10**, (2016)
9. A. Ben-Israel, T.N.E. Greville, *Generalized Inverses, Theory and Applications* (Springer-Verlag, 2 ed., New York, 2003)
10. Ł. Wach, W.P. Hunek, *Perfect control for fractional-order multivariable discrete-time systems, Theoretical Developments and Applications of Non-Integer Order Systems, Lecture Notes in Electrical Engineering* **357**, Proc. of the 7th Conf. on Non-int. Ord. Calc. and its Appl. (RRNR'2015), Szczecin, Poland, pp. 233-237, (2015)
11. D. Sierociuk, *Estimation and control of discrete fractional order state space systems* Ph.D. dissertation, Warsaw University of Technology, Warsaw, Poland, 2007 (in Polish)

 DOI: 10.1051/e3sconf/20171901037

Settlement of reactive power compensation in the light of white certificates

Konrad Zajkowski[1,*]

[1]Koszalin University of Technology, DP of Energy, 15-17 Raclawicka St. 75-620 Koszalin, Poland

Abstract. The article discusses the problem of the determination of savings on active energy as a result of a reactive power compensation. Statutory guidance on the required energy audit to obtain white certificates in the European Union was followed. The analysis was made on the basis of the Polish Law. The paper presents a detailed analytical method and an estimation method taking into account the impact on the line, the transformer and the generator. According to the relevant guidelines in the European Union, the reduction of CO_2 emissions by calculating the saving of active power should be determined.

The detailed method and an estimation method proposed for the determination of savings on active energy as a result of the reactive power compensation carried out possess some errors and inconvenience. The detailed method requires knowledge of the network topology and a determination of reactive power Q at each point of the network. The estimation method of analysis is easy in execution, especially if the consumer of energy is the main or the most significant purchaser of electricity in the network. Unfortunately, this latter method can be used only for activities that do not require high computational accuracy. The results obtained by this method are approximate values that can be used for the calculation of economic indicators. The estimation method is suitable for determining the number of white certificates when a power audit concerns a recipient of electricity, the structure of which is a large number of divisions scattered at many different locations in the power system.

1 Introduction

In Europe, the white certificates are a market tool for improving energy efficiency. This is document demonstrating energy savings as a result of investments to improve energy efficiency.
The investments that reduce the production of electricity, which reduces the negative impact on the environment is one of activities. An energy audit is an element that makes the settlement of the actions undertaken possible. The relevant guidelines are set out in the following regulations: [1-6].

Among the many energy efficiency improvement measures [7] that make it possible to apply for white certificates, is to limit the flow of reactive power [2]. This action consists in performing size electrical measurements and analysing the location of the compensation devices.
The subject of energy savings is active energy, which is saved by the action of reducing the flow of reactive energy. The effect of reducing the reactive power is to reduce the active power in the form of energy losses on the power grid.

An engineering approach in this area will involve a determination of savings on primary energy (individually on each grid element) being the result of reducing reactive power flows.

2 Methodology

An analysis in a detailed and simplified audit should be prepared by means of the balance method [2]. It is important to analyze the components in the energy distribution system by selecting the shortest route connecting the generator to the power compensation location.

Two calculation methods will be presented: detailed (sections 2.1 - 2.4) and estimation (sections 2.5 - 2.6). In section 3, a computational example using the estimation method proposed by the author is presented.

2.1 Transmission line section

Reactive power flows through a line segment cause:
- a limitation of the active power transfer,
- an increase of active power losses,
- an increase of current and voltage drops,
- difficulties in the work of grid elements.

The transmission line can be described with impedance $R + jX$. This approach was presented by the authors in the following positions: [8-13].
The active power P is absorbed by the receiver (Fig. 1), and for the reactive power Q, the receiver is a prosumer (receiver or supplier). Both the active and reactive power generated by the receiver affect

* Corresponding author: konrad.zajkowski@tu.koszalin.pl

DOI: 10.1051/e3sconf/20171901037

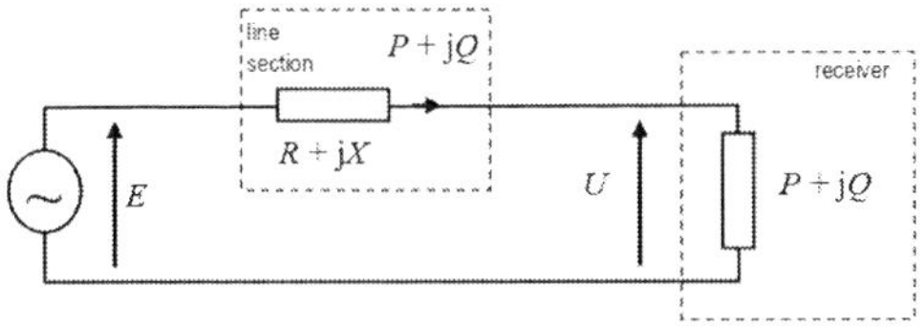

Fig. 1. Mathematical model of the section of the power line.

the ΔP power losses on the $R + jX$ impedance:

$$\Delta P = \frac{P^2 + Q^2}{U^2} R. \tag{1}$$

When the compensator is added (Fig. 2), the reactive power passing through the line is reduced.

$$Q' = Q - Q_C. \tag{2}$$

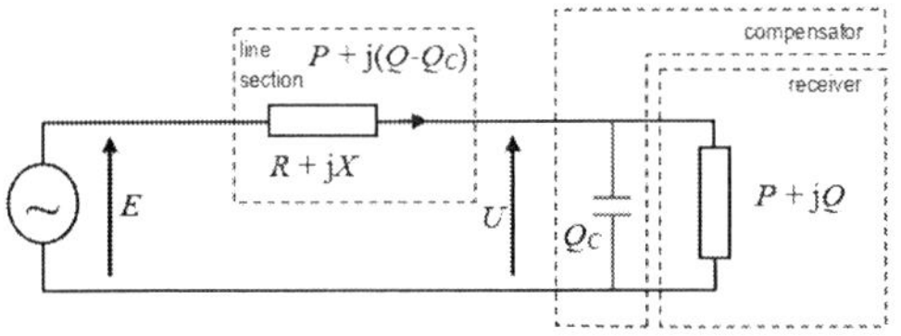

Fig. 2. Mathematical model after compensation.

Typically, this power is not reduced to zero. With partial compensation, the remaining reactive power continues to load the flow. The difference between the active power loss (1) determined prior ΔP and after $\Delta P'$ compensation is a profit on δP_P active power:

$$\delta P_P = \Delta P - \Delta P' = \frac{2QQ_C - Q_C^2}{U^2} R, \quad \delta A_P = \delta P_P \cdot T_p, \tag{3}$$

where: Q_C – power of compensator, δA_P – active energy, T_p – work time of compensator Q_C.

In the mth section of the transmission line, the loss of active power is dependent on the power transmitted by the next line section (Fig. 3):

$$\delta P_{P(m)} = \frac{Q_{(m+1)}^2 - Q_{(m+1)}'^2}{U_{(m)}^2} R_{(m)}, \tag{4}$$

where: $Q_{(m+1)}$ – reactive power transmitted from $m+1$ system, $Q'_{(m+1)}$ – reactive power from $m+1$ system after switching on the compensator in the receiver.

The reactance in the $m+1$ section changes the reactive power acting on the mth section. Analogically, the reactive power is changed after the compensation Q'.

Due to the variability of the parameters over time, the values given in formulas (3) and (4) are values averaged over the T_p time interval. To determine the economic profit resulting from the compensation should be known:

- network topology,
- the length of the mth section of the line,
- the conductor cross-section area,
- the conductivity of the conductive material,
- reactive power before $Q_{(m)}$ and after $Q'_{(m)}$ the compensation.

Unfortunately, for all network elements this data is often not available.

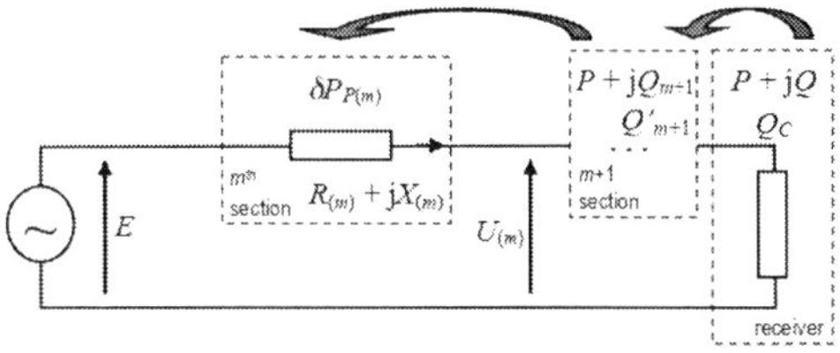

Fig. 3. Impact between network segments.

2.2 The transformer

The transformer for the power line is the R_T resistance at which the loss of active power takes place. This loss is described in formula (1). Reactive power flows through the transformer cause an increase of active power losses in the transformer. Load losses ΔP_{oT} in a three-phase transformer are the result of the flow of current I through the winding:

$$\Delta P_{oT} = 3I^2 R_T. \tag{5}$$

When the current is equal to the I_{nT} nominal current, the load losses will be:

$$\Delta P_{onT} = 3I_{nT}^2 R_T. \tag{6}$$

From equations (5) and (6), a relationship is obtained which allows the loss to be measured at any transformer load:

$$\Delta P_{oT} = \Delta P_{onT}\left(\frac{I}{I_{nT}}\right)^2 = \Delta P_{onT}\left(\frac{S}{S_{nT}}\right)^2. \tag{7}$$

where from the transformer's datasheet: ΔP_{onT} – nominal load power dissipation, S_{nT} – nominal power of the transformer.

The savings on the active power losses in the mth transformer using the reactive power compensation in the receiver equals the loss difference before and after the compensation:

$$\delta P_{T(m)} = \Delta P_{onT(m)}\left(\frac{Q_{(m+1)}^2 - Q_{(m+1)}'^2}{S_{nT(m)}^2}\right). \tag{8}$$

Thus, the gain on active energy in the mth transformer is:

$$\delta A_{T(m)} = \delta P_{T(m)} \cdot T_p. \tag{9}$$

The parameters $Q_{(m+1)}$ and $Q'_{(m+1)}$ in formula (8) is to be understood as reactive power at the output of the

DOI: 10.1051/e3sconf/20171901037

preceding segment, which may be another transformer or preceding a power line segment (same as shown in Fig. 3).

In chapters 2.1 and 2.2, the change in active power caused by the impact of individual sections of the power system (transformer or line segment) has been omitted. These chapters also omit the change in the voltage value $U_{(m)}$ that occurs when the compensator is connected.

2.3 The generator

The effect of reactive power from the segment preceding the generator, causes a reduction the ability to produce active power. When the power factor is less than nominal, it is impossible to use the full power of the turbine due to the thermal load of the generator. An increased reactive power flow causes a demagnetization of the rotor and a reduction of the electromotive force of the generator.

In the case of asynchronous generators, reactive power must be supplied from the network or from the compensator with an individual compensation.

In practice it is difficult to estimate the demand for reactive power of each segment in the power system. Due to the existence of the reactance of individual section in the system, reactive power is required and can't be completely zero.

To determine the influence of power in the generator is the use of similarities in electrical properties between the transformer and the generator.

$$\delta A_G = \delta P_G \cdot T_p = \Delta P_{onG} \left(\frac{Q_{(m+1)}^2 - Q_{(m+1)}'^2}{S_{nG}^2} \right) T_p \cdot \qquad (10)$$

where: ΔP_{onG} – load losses at a nominal current flow through the generator, S_{nG} – nominal power of the generator.

The production costs of reactive power in generators are substantially smaller than in receivers. However, the production of reactive power in the generator increases the transmission costs of reactive power in the network. Therefore, from an economic point of view, the production of reactive power in the generator and its transmission to customers is unprofitable.

2.4 Energy savings in all system segments

Energy saving δA that results from the reduction of active power losses in all the parts of the network by using the reactive power compensation of the receiver requires knowledge of the network topology and the determination of (4), (8) and (10):

$$\delta A = \delta A_G + \sum_m \left(\delta A_{P(m)} + \delta A_{T(m)} \right). \qquad (11)$$

Determination of value (11) requires:
- knowledge of network topology,
- measurement of $Q_{(m)}$ power at each point of the network,
- knowledge of the change in reactive power $Q'_{(m)}$ caused by the Q_C power compensation in the receiver.

This information is not available for every point in the network. In practice, the reactive power $Q'_{(m)}$ can't be measured. The powers $Q_{(m)}$ and $Q'_{(m)}$ can be determined analytically, but this requires detailed information about the entire network.

2.5 The energy equivalent of reactive power

Inclusion in the calculations of additional current flows, ie: a complete power grid structure with several generators and multiple receivers, results in additional computational complexity calculations.

A solution to this problem might be the use of the ratio (Fig.4) which was determined in the mid twentieth century: *the energy equivalent of reactive power* [14-16]. This ratio is defined as the differential of the active power losses in relation to reactive power.

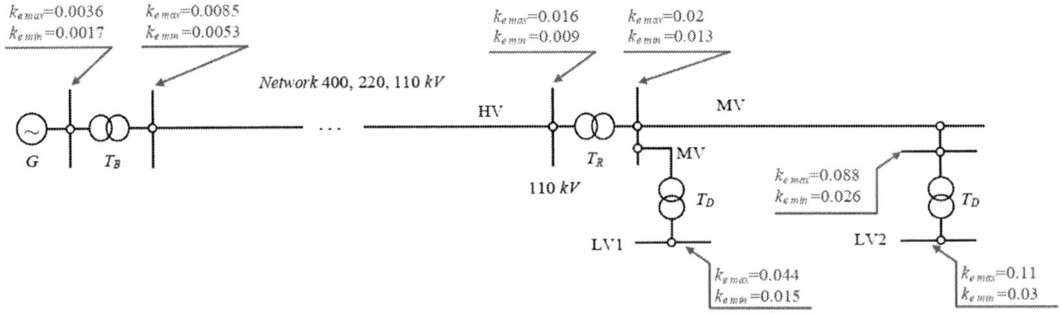

Fig. 4. The values of the energy equivalent of reactive power estimated in 2007 [7].

$$k_e = \frac{\delta}{\delta Q}\Delta P = \frac{2Q}{U^2}R \quad \left[\frac{kW}{kVAr}\right]. \tag{12}$$

The value of coefficient k_e strongly depends on reactive power, and it varies with changes to this power. This quantity permits one to calculate the change of active power losses caused by changes to reactive power that is collected at a given point in the network. The value of this factor depends from the value of reactive power and the distance of the point analyzed to source, whose measure is resistance R. According to the literature [9] and [12], the value of this coefficient is determined for the minimum and maximum reactive power loads (Fig.4).

Due to the variability of the load, the average value of the reactive power energy equivalent in a given network node is accepted in calculations:

$$k_{e\,avg} = \frac{k_{e\min} + k_{e\max}}{2}. \tag{13}$$

In order to estimate how much power losses are reduced in the grid that supplies a given network point as a result of the installation of a battery of condensers, it is enough to multiply the value of the reactive power energy equivalent for this point by the power of the battery installed.

The reduction amount of losses is calculated from the following formula:

$$\delta P = k_{e\,avg} \cdot Q_C, \quad \delta A = \delta P \cdot T_p. \tag{14}$$

The equivalent of reactive power k_e is an economic tool. It is an indicator approximating many of the phenomena discussed in sections 2.1 – 2.4 (only the most important has been discussed). Using this factor, the calculations are error-prone, the value of which depends on the accuracy of the determination of k_e.

The value of the energy equivalent of reactive power k_e depends from the following:

- the distance of the point analyzed in the network (the resistance path from the generator to the point analyzed in the network),
- the load of the network - an adoption of the arithmetic average of the minimum and maximum values is only a partial correction of an error in estimation,
- network topology and its complexity.

This means that an estimation of the reduction of losses based on the energy equivalent of reactive power is burdened with significant errors. The latest k_e values are the result of random testing for a limited part of the power system and were performed in 2007 [7].

Due to considerable computational uncertainty, this coefficient allows you to know only the range of power values. However, there are situations where the use of this indicator is necessary. For example, when you need to designate economic indicators that account for the installation of compensator that are connected to the network and:

- the exact structure of the network is unknown,
- it is not possible to count the currents and energy losses on individual network segments (see section 2.4),
- an economic task involving a very large number of recipients.

Approximate results are obtained when the k_e coefficient is used, so this calculation will be referred to as *estimation method*. Comprehensive calculations for each network member (see section 2.1 – 2.4) will be referred to as *detailed analytical method*.

2.6 Determination of primary energy savings

Determination of primary energy savings is needed to account for the economic effects of improving energy efficiency. It is possible to apply for white certificates, which are awarded by tender. They are documents that weigh actions to improve the energy efficiency of company's or state.

The saved active energy is prepared for one calendar year by making adequate measurements and calculations [2]. The values of the average energy determined in billing invoices can be used as input data. This can be determined when using advanced electronic energy meters, which possess separate individual records with a division into inductive and capacitive reactive power. For this purpose, it is best to accept the entire billing year in an analysis, including all the seasons, and which consists of i-billing periods.

When the estimation method is used, only inductive reactive power can be used, since the k_e coefficient is determined only for such power. Prior to compensation, the average power factor for inductive reactive power is:

$$\tan\varphi_{(i)} = \frac{A_{Q(i)}}{A_{P(i)}}, \tag{15}$$

where: A_Q - total inductive reactive energy read from the i th invoice, A_P - consumed active energy during this period.

The value of $\tan\varphi'$ after the compensation can be determined in two ways:

a) when there are invoices from the period after the compensation, this coefficient is calculated analogously to (15).
b) when there is no invoice after the compensation (settlement made before the compensator is enabled), the minimum value (e.g. 0.4) from contract, should be taken.

Assuming invariable active energy, after partial compensation there is still reactive energy, which is:

$$A'_{Q(i)} = \tan\varphi' \cdot A_{P(i)}. \tag{16}$$

Therefore, reactive power reduced by the compensator in the ith billing period is equal to:

$$A_{C(i)} = Q_C \cdot T_p = A_{Q(i)} - A'_{Q(i)}. \tag{17}$$

It is reactive inductive energy reduced by the battery of capacitors.

Taking the energy equivalent of reactive power determined for the point being the closest to compensation according to (14), the value of active power reduction in the network caused by compensation A_C in the ith billing period is as follows:

$$\delta A_{(i)} = k_{e\,avg} \cdot A_{C(i)}. \tag{18}$$

Taking invoices for the whole year, the total final energy savings (within the meaning of the Act [1]) constitute the sum of active energy savings resulting from compensation:

$$\delta A = \sum_i \delta A_{(i)} \quad [kWh/year]. \tag{19}$$

In accordance with the guidelines of the International Energy Agency (IEA) and the Organisation for Economic Co-operation and Development (OECD), the value of (19) expressed in the tons of oil equivalent (*toe*) is:

$$\delta A_{toe} = \frac{\delta A}{11630} \quad [toe/year]. \tag{20}$$

The average annual primary energy savings referred to in [1] in accordance with [2] are equal to:

$$\delta A_p = \delta A \cdot w_i, \quad \delta A_{p\,toe} = \delta A_{toe} \cdot w_i, \tag{21}$$

where: δA_p – the amount of primary energy savings expressed in primary fuel $[kWh\ /\ year]$ or $[GJ\ /\ year]$, δA – final energy savings, expressed in $[kWh\ /\ year]$ or $[GJ\ /\ year]$, w_i – effort ratio of non-renewable primary energy which depends from the final energy carrier (Tab.1).

The w_i factor is defined according to the fuel or energy source used in a given country, which according to [4] should be one of the values:

Table 1. Effort ratio of non-renewable primary energy for the production and supply of electric energy.

Way of producing electricity	w_i
mixed-mode production	3.0
production of electricity from solar energy	0.7

For a specified value of δA_p expressed in $[kWh\ /\ year]$, the reduction of CO_2 emissions in a year is determined according to [5]:

$$\Delta E_{CO2} = \delta A_p \cdot 3.6 \cdot 10^{-3} \cdot we_{CO_2} \quad [kg/year], \tag{22}$$

where: we_{CO2} – CO_2 emission factor for fuels.

The quantity of the reduction of CO_2 emissions is calculated using CO_2 emission factors contained in [5] and under the EU Emissions Trading System (EU ETS) for a given year (Table 2).

Table 2. Example we_{CO2} reference CO_2 emission factors for fuels [6] in accordance with the guidelines of the Intergovernmental Panel on Climate Change (IPCC) from the year 2006.

Fuel type	we_{CO2} $MgCO_2/TJ$	Fuel type	we_{CO2} $MgCO_2/TJ$
petroleum	73.3	coking coal	94.5
shale oil	73.3	lignite	101.1
gas / diesel oil	74.0	bituminous coal	94.2
ethane	61.6	natural gas	56.1
solvent naphtha	73.3	industrial waste	142.9
petroleum coke	97.5	waste oils	73.3
refinery gas	51.3	peat	105.9

A project aimed at an improvement of energy efficiency, as a result of which energy savings in (21) are obtained in an amount being an equivalent to at least 10 *toe* on average over the year, can be included in a tender [1]. Several projects of the same type aimed at an improvement of energy efficiency, as a consequence of which the total energy savings are achieved in the amount being an equivalent to at least 10 *toe* on average over a year, may be brought together.

3 Results

The proposed calculation method using k_e factor, which the author called the estimation method, was used to designate white certificates for one Polish company. This economic settlement covered some 350 branches subordinate to the company, in which the reactive power compensation was made. These calculations has been checked and reviewed. White certificates have been awarded to this company.

Table 3. Summary values from invoices and calculations.

i	days	$A_{P(i)}$ kWh	$A_{Q(i)}$ kVArh	$\tan\varphi_{(i)}$ VAr/W	$A'_{Q(i)}$ kVArh	$A_{C(i)}$ kVArh	$\delta A_{(i)}$ kWh
1	61	204862	174510	0.852	81945	92565	6480
2	60	184128	171452	0.931	73651	97801	6846
3	61	189452	185423	0.979	75781	109642	7675
4	61	208512	160214	0.768	83405	76809	5377
5	61	210245	182134	0.866	84098	98036	6863
6	61	195428	181247	0.927	78171	103076	7215
Σ	365	1192627	1054980			577929	40455

The computational example presented here concerns one point where reactive power compensation has been made. The settlement was based on 6 invoices that covered the entire calendar year. The $\tan\varphi'$ coefficient (after the compensation) was assumed to be equal to the contract value, that is 0.4. This receiver has been connected to a low voltage network (LV2 in Fig.4). The energy equivalent of reactive power is equal: $k_{e\,min}$= 0.03, $k_{e\,max}$= 0.11. It is assumed that the energy consumption in the year in which the company is applying for white certificates will be at a similar level as in the year when meter measurements were conducted. Statements and calculations were made on the basis of invoices from a representative calendar year. The total final energy savings are as follows:

$$\delta A = 40\ 455\ kWh/year.$$

This corresponds to the value expressed in tons of the oil equivalent (20):

$$\delta A_{toe} = 3.48\ \ toe/year.$$

The electricity consumer is connected to the power grid, which is characterized by the production of electricity from renewable and non-renewable sources. In this case (according to Tab.1), w_i factor is equal to 3. According to Equation (21), the average annual primary energy savings are:

$$\delta A_p = \delta A \cdot w_i = 121\ 365\ kWh/year,$$

in *toe*: $$\delta A_{p\,toe} = \delta A_{toe} \cdot w_i = 10.44\ \ toe/year.$$

As a result of the foreseen reactive power compensation, the threshold of 10 *toe* will be exceeded, so the entity can apply for white certificates.
The CO_2 emission factor for power plants and combined heat and power plants that use bituminous coal is as follows [5]:

$$we_{CO2} = 93.8\ kg/GJ.$$

The reduction of CO_2 emissions per year (22) is:

$$\Delta E_{CO2} = \delta A_p \cdot 3.6 \cdot 10^{-3} \cdot we_{CO_2} = 40.98\ \ tons_{CO_2}/year.$$

4 Discussion and Conclusions

The proposed two methods of determining the savings of active energy due to the compensation of reactive power have some faults and difficulties. Where the network topology is simple and the measured power values are known at each point in the network, the detailed analytical method may be used. In other cases, a estimation method can be used, which requires the knowledge of the values of k_e factor: the energy equivalent of reactive power. The disadvantage of this method is the inaccuracy of calculations, which is dependent from the error in the determination of k_e coefficient, but in economic calculations, which the purpose is to know the approximate value of the power saved, such a calculation is acceptable.

When considering a recipient of electricity with a large number of divisions scattered in many different places in the power system, so far no computational energy audit has been developed. In these cases, no electricity audits were made. The author of this article has consulted experts from several countries in Europe. Only Italians performed such audits, but in calculations they limited themselves to analysing the last segment of the electricity network - between the meter and the receiver.

By using the estimation method, the author performed two power audits in 2014 and 2016. The calculations concerned one of Polish companies dealing with the sale of automotive fuels. Both audits have passed a positive review in the Ministry and have been granted white certificates.

References

1. Journal of Laws of the Republic of Poland (Dz.U. 2011 r. Nr 94, poz. 551 z późn.zm.), *Ustawa o efektywności energetycznej*
2. Journal of Laws of the Republic of Poland (Dz.U. 2012 r. poz. 962), *Rozp. Ministra Gosp. w sprawie szczegółowego zakresu i sposobu sporządzania audytu efektywności energetycznej*
3. Journal of Laws of the Republic of Poland (Dz.U. z 2012 r. poz. 1039), *Rozp. Ministra Gosp. w sprawie sposobu obliczania ilości energii pierwotnej odpowiadającej wartości świadectwa efektywności energetycznej*
4. Journal of Laws of the Republic of Poland (Dz.U. Nr 201 z 2008r. poz. 1240), *Rozp. Ministra Infrastr. w sprawie metodologii obliczania charakt. energet. budynku i lokalu mieszkalnego lub części budynku*
5. Journal of Laws of the Republic of Poland (Dz.U Nr 183 z 2008 r. poz.1142), *Rozp. Ministra Środow. w sprawie sposobu monitorowania wielkości emisji substancji objętych wspólnotowym systemem handlu uprawnieniami do emisji*
6. Ministry of Environment, *Wartości opałowe (WO) i wskaźniki emisji CO2 (WE) w roku 2009 do raportowania w ramach Wspólnotowego Systemu Handlu Uprawnieniami do Emisji za rok 2012*, http://www.mos.gov.pl/
7. R. Kołodziej, Elektroenergetyka **11-12** 1-2, 119-127 (2012)
8. W. Szpyra, *Straty Energii Elektrycznej w Sieciach Elektroenergetycznych*, p. 59-69, Poland (2008)
9. S. Duer, K. Zajkowski, I. Płocha, R. Duer, Neural Comput. Appl. **22**, 7-8, 1581-1590 (2013) DOI: 10.1007/s00521-012-1052-9
10. S. Duer, K. Zajkowski, Neural Comput. Appl. **23**, 7-8, p. 2185-2197 (2013) DOI: 10.1007/s00521-012-1169-x
11. K. Zajkowski, Neural Comput. Appl. **24**, 2, p.431-439 (2014) DOI: 10.1007/s00521-012-1239-0,
12. K. Zajkowski, S. Scaticailov, Nonconventional Technol. Rev. **6 XX**, 2, 54-61, (2016)
13. J. Smyczek, K. Zajkowski, EUROSIS-ETI, p. 278-281, Univ. of Malaga, Spain (2004)
14. M. Kuchta, M. Siergiejczyk, J. Paś, JVE Int. Ltd. J. Vibroeng. **16**, 8, 4039-4044 (2014)
15. W. Szpyra, Przegląd Elektrotechniczny **2**, 144-147, (2011)
16. I. Coroama, M. Gavrilas, I. Ovidiu, Politehnica University of Timisoara, Romania (2010)

E3S Web of Conferences **19**, 01038 (2017)
EEMS 2017
DOI: 10.1051/e3sconf/20171901038

Creating of structure of facts for the knowledge base of an expert system for wind power plant's equipment diagnosis

Stanisław Duer[1,*], *Paweł* Wrzesień[2], and *Radosław* Duer[3]

[1]Technical University of Koszalin, Department of Mechanics, 15-17 Raclawicka St., 75-620 Koszalin, Poland
[2]Vortex Energy Polska sp. z o.o., ul. Malczewskiego 26, 71-612 Szczecin, Poland
[3]Technical University of Koszalin, Department of Electronics and Informatics, 2 Śniadeckich St., 75-453 Koszalin, Poland

Abstract. This article describes rules and conditions for making a structure (a set) of facts for an expert knowledge base of the intelligent system to diagnose Wind Power Plants' equipment. Considering particular operational conditions of a technical object, that is a set of Wind Power Plant's equipment, this is a significant issue. A structural model of Wind Power Plant's equipment is developed. Based on that, a functional – diagnostic model of Wind Power Plant's equipment is elaborated. That model is a basis for determining primary elements of the object structure, as well as for interpreting a set of diagnostic signals and their reference signals. The key content of this paper is a description of rules for building of facts on the basis of developed analytical dependence. According to facts, their dependence is described by rules for transferring of a set of pieces of diagnostic information into a specific set of facts. The article consists of four chapters that concern particular issues on the subject.

1 Introduction

Expert systems are programs that can help or replace human experts in a specific field. Such systems can provide pieces of advice, recommendations or diagnoses relating to problems appearing in a specific field. Computer programs developed in that way are of great importance and use for fields that are poorly formalized (without having mathematical backgrounds, which could be created algorithms for) e.g. diagnostics, medicine, etc. Expert systems are characterized not only by their capability to solve non-algorithmically defined problems but also have other advantages e.g. collected expert knowledge can be easily transmitted and available if experts are not present at time (due to their sick health or retirement) and coded knowledge can be easily delivered (it is easier to copy a computer program than teach another human) and a response can be acquired fast, and there are no human symptoms like fatigue nor stress.

In literature, supporting and intelligent systems are presented in detail. Such systems are used for supervision or organization of technical – technological problems in terms of their diagnosis and quality evaluation. Those systems are particularly useful when a number of variables are analyzed and it is required to consider complex factors having influence on those variables relating to: a technical condition of technical equipment, tools, manufacturing means, or selection of conditions and parameters of processes. Important achievements are brought by works of B. Buchannan, E. Shortliffe, W. Kacalak, and others [1-13]. Those works help to build up a set of conclusion rules, create knowledge representation, knowledge collection and analysis. Their works considerably help to practically solve those problems when creating expert knowledge, including its application in diagnostics or operation of technical equipment. There should be also considered works by F. Hayer-Roth, D. Waterman and others in this issue since their works [4,5,14-19] cover mathematical background in terms of development and organization of expert systems, including rules for creation of expert knowledge bases, creation of conclusion rules and means of their analyses (look through and sort out).

The paper by F. Hayer-Roth, D. Waterman, D. Lenat [5,19] included descriptive and mathematical grounds for designing of expert systems. The rules are quite well developed for the creation of an expert database including a set of inference rules and ways of their analysis (review and sorting).

In papers of M. Białko, W. Kacalak, W. Pedrycz and others [2,5-8,10,15,19,20], presented a wide approach to the issue of neural networks, machine teaching of neural networks, genetic algorithms as well as sets and fuzzy knowledge. The paper includes well-developed issues of machine teaching and training of neural networks. Such problems were solved as the following: an introduction to neural networks, teaching of layered neural networks as well as teaching through competition between neurons in a network. The paper also solves in a good manner the problems of the construction and functioning of genetic algorithms. The paper covers extensively fuzzy knowledge systems, ways to represent this knowledge and to use it in the organization of artificial neural networks and expert systems.

* Corresponding author: stanislaw.duer@tu.koszalin.pl

The papers by (L. Będkowski, T. Dąbrowski, A. Rosiński and L. Pokoradi, [1, 16-18] presents the theoretical background of the reliability of technical devices in the operation process. The author presented mathematical basis of the policy rules of the organization of repairs (replacement of components (constructional elements)) of devices. Another important practical problem which was solved in the study is the representation of the ways and methods to set periods of repairs (replacements of functional parts) of devices and their optimization. The paper also covers preventative maintenance procedures including drawbacks of this type of maintenance. A large part of the paper deals with the issues of the development and verification of the maintenance policy strategy. The study is also of a large practical importance as regards the organization of the development of theoretical models of technical devices' maintenance processes as well as the conditions and rules of their modification.

The paper by (Duer 2010-2016) [4,14,20-25], makes the use more specific (and improved) of the results of diagnostic testing in the organization of a technical object's maintenance system. The paper presents the mechanism of a negative change of states in a technical object, as a result of which there occurs in the object a reduction of its operational properties: a change of the state. The author also presented a diagram and a description of artificial neural network structures and mathematical dependencies which express the idea of the functioning of the network in compliance with the algorithm developed for it. The paper also presents theoretical grounds for diagnosing of technical objects in trivalent logic with the use of an artificial neural network. The results of the study were supported with an example of a diagnostic information database for the device tested.

This paper aims at developing a set of rules for a knowledge base of the expert system that will allow identifying defects of the object under examination.

2 Functional – diagnostic structure of wind power plant's equipment

Wind Power Plants (WPPs) can be divided into two types in terms of localization, that is onshore and offshore (WPPs). Onshore (WPPs) are usually characterized by irregular localization of particular Wind Turbine Generators (WTGs) due to topographical conditions [4, 20, 24] (Fig. 1). On the other hand, offshore (WPPs) are characterized by a regular localization structure of particular (WTGs). For offshore, is used (DC) connection into the land. Onshore (WPPs) feed the grid by means of (MV) or (HV) power transformers. Fig. 1 shows a structure of an onshore (WPP), which consists of three (WTGs) connected by means of MV step-up power transformers and (MV) cable lines into the (HV) Substation that is used to output power into the National Distribution System by (HV) power line at the Point of Common Coupling (PCC) [4, 5,13,20].

A functional structure of the (WPP) was created (Fig. 1) for a developed system for supervision and safety usage of a wind power plant. A designed knowledge base relates to the (WPP) consisting of five (5) unit (WTGs) of type Vestas V90 2.0 MW connected to the (HV) Substation by means of two 20-kV cable lines, which are an internal Electrical Infrastructure of the Wind Power Plant (Fig. 1). The (WPP) was designed to be located in the north of Poland, that is to say, in the 1st grade wind energy zone in Poland. The (HV) Substation was designed as power output into the grid [11,20-23].

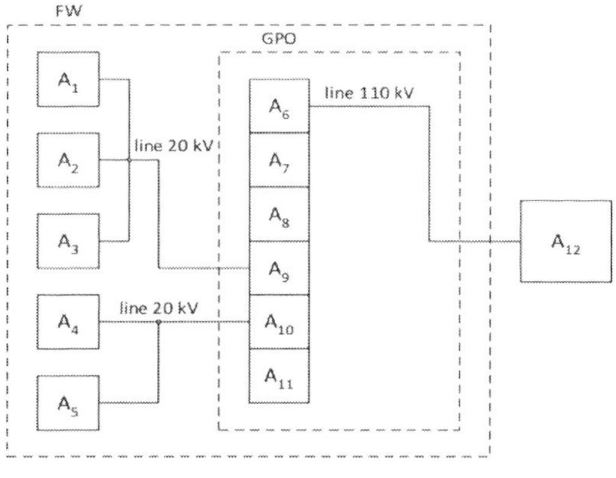

Fig. 1. Block diagram of the WPP, where: $A_1 \div A_5$: no. $1 \div 5$ – Wind Turbine Generators, A_6 – Power Transformer Field, A_7 – Voltage Measurement Field, A_8 – Shunt Reactor Filed, A_9 – Line no. 1 for Wind Turbine Generators no. $1 \div 3$, A_{10} – Line no. 2 for Wind Turbine Generators no. $4 \div 5$, A_{11} – General Signalization, A_{12} – Point of Common Coupling, GPO – Wind Farm Substation, GPZ – Grid Operator's Substation, FW – Wind Power Plant, PCC – Point of Common Coupling.

The Wind Farm Substation is connected into the Grid Operator's Substation at the (PCC) by means of a (HV) cable power line. A 20/110-kV power transformer is fed by two 20-kV cable power lines connected to (WTGs) (Fig. 2). The issue concerning creating of expert knowledge bases is a specialized field of science that relates to development of methods and construction principles, and organization of sets of pieces of engineering, diagnostic and operational expertise information on a specific technical and complex object under examination. A basis to create a set of engineering knowledge on a specific object under examination is knowledge, which is possessed by a user of the specific object in terms of its construction, usage and operation. Engineering knowledge on the specific object is constantly provided with (expanded) diagnostic data of the object determined by diagnostic evaluations.

A diagnostic evaluation of the technical object is an analytical – creative process for acquiring sets of pieces of diagnostic information on a specific object.

A set of pieces of diagnostic information that is determined through that process consists of sub-sets of pieces of information on: making of a functional –

DOI: 10.1051/e3sconf/20171901038

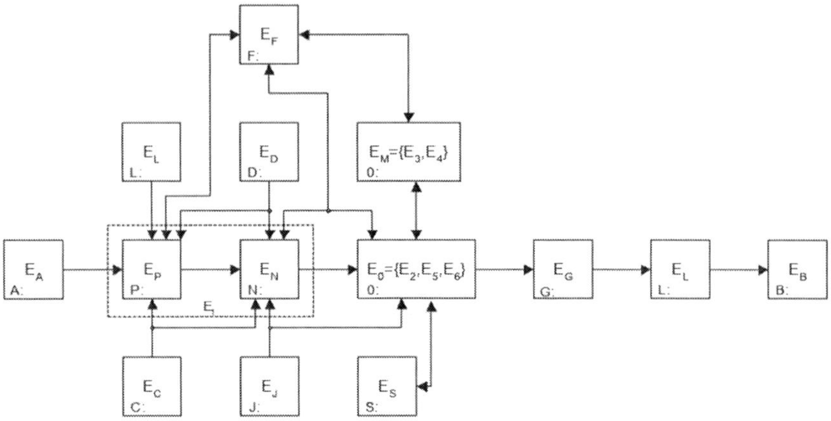

Fig. 2. Functional – diagnostic diagram of the Wind Power Plant's equipment, where: A – Environmental and Ambient, B – Grid, C – Brake Systems, D – Hydraulics, E – Yaw System, F – Communications, G – Connectors, H – Protection Relays, I – Pitch System, J – Cooling and Heating Systems, K – Transmission, L – MV Unit Power Transformer, M – Controller, N – Gearbox, O – Generator, P – Rotor.

diagnostic model, determining of a sub-set of basic elements, determining of a sub-set of diagnostic signals and their reference signals (Fig. 2).

To make diagnostic evaluation of the technical object, it requires possessing diagnostic and engineering knowledge on construction, operation, and diagnosis of such technical objects [11,20-23]

This section presents information by its stages due to a multilevel way of transformation of information on the object into a form of an expert knowledge. Stage 1 is a presentation of a fundamental theory relating to functioning and operation of (WTG) and (WPP's) equipment. Stage 2 is a description of issues relating to diagnostic evaluation of the examined equipment. As a result of this approach, engineering and diagnostic knowledge on (WTG) and (WPP's) equipment is acquired. Issues relating to creation of expert knowledge bases will be presented in the next section (Fig. 2).

A basis to carry out diagnostic research on technical equipment is functional – diagnostic models of such objects [4-10]. Thus, a functional – diagnostic model of the (WPP) was developed (Fig. 2 and Table 1).

Table 1. Internal structure of the Wind Power Plant model.

Object's units	Fundamental elements $\{e_{i,j}\}$ in the Wind Power Plant model				
E_1	$e_{1,1}$	$e_{1,2}$	$e_{1,3}$	$e_{1,4}$	$e_{1,5}$
E_2	$e_{2,1}$	$e_{2,2}$	Ø	Ø	Ø
E_3	$e_{3,1}$	$e_{3,2}$	$e_{3,3}$	Ø	Ø
E_4	$e_{4,1}$	$e_{4,2}$	Ø	Ø	Ø
E_5	$e_{5,1}$	$e_{5,2}$	Ø	Ø	Ø
E_6	$e_{6,1}$	$e_{6,2}$	Ø	Ø	Ø
E_7	$e_{7,1}$	Ø	Ø	Ø	Ø

where:

$e_{1,1}$, $e_{1,2}$, $e_{1,3}$, $e_{1,4}$, $e_{1,5}$ – Wind Turbine Generators no. 1 ÷ 5, $e_{2,1}$, $e_{2,2}$ – MV cable power lines no. 1 & 2, $e_{3,1}$ ÷ $e_{3,3}$ – MV switchgear no. 1 ÷ 3, $e_{4,1}$ – a measurement and diagnostic system of electrical parameters in specific elements of the wind power plant, $e_{4,2}$ – a diagnostic system of the wind power plant, $e_{5,1}$ – a shunt reactor of the wind power plant $e_{5,2}$ – an auxiliary power transformer of the wind power plant, $e_{6,1}$ – HV power transformer, $e_{6,2}$ – HV Switchgear, $e_{7,1}$ – Substation.

As a result of the functional – diagnostic evaluation, seven i- functional sets were determined in the wind power plant model. A sub-set of j-fundamental (functional) elements was determined in each i-set. A determined set of fundamental elements $\{e_{i,j}\}$ in the wind power plant model, which determines its internal structure, is shown in the table 1.

The developed functional – diagnostic model of the Wind Power Plant shown in Fig. 2 was also a basis to determine a set of diagnostic signals, as well as their reference signals. The determined set of diagnostic signals $\{X(e_{i,j})\}$ is shown in the Table 2.

Table 2. Set of diagnostic signals for an object.

Object's units	Set of diagnostic signals in the Wind Power Plant $\{X(e_{i,j})\}$				
E_1	$X(e_{1,1})$	$X(e_{1,2})$	$X(e_{1,3})$	$X(e_{1,4})$	$X(e_{1,5})$
E_2	$X(e_{2,1})$	$X(e_{2,2})$	Ø	Ø	Ø
E_3	$X(e_{3,1})$	$X(e_{3,2})$	$X(e_{3,3})$	Ø	Ø
E_4	$X(e_{4,1})$	$X(e_{4,2})$	Ø	Ø	Ø
E_5	$X(e_{5,1})$	$X(e_{5,2})$	Ø	Ø	Ø
E_6	$X(e_{6,1})$	$X(e_{6,2})$	Ø	Ø	Ø
E_7	$X(e_{7,1})$	Ø	Ø	Ø	Ø

An effective measurement and registry system for a set of diagnostic measuring signals of a wind turbine generator $\{X(e_{i,j})\}$ is designated by means of a diagnostic model of the object (see Fig. 2). Example measurement values of diagnostic signals of a wind turbine generator and a power transformer of the specific electrical substation are presented in Fig. 3.

By analyzing values of measurement signals, a set of reference diagnostic signals $\{X_w(e_{i,j})\}$ of wind turbine generators is designated and shown in Table 3.

 DOI: 10.1051/e3sconf/20171901038

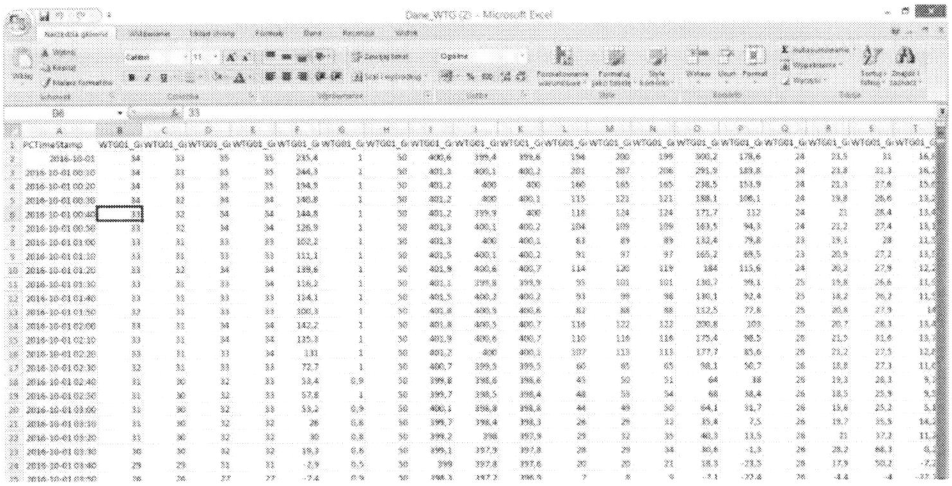

Fig. 3. Measurement values from a wind turbine generator.

Table 3. Set of measurement diagnostic signals and reference signals $\{X_w(e_{i,j})\}$ for one wind turbine generator in a Wind Power Plant as an example.

Object Units		Signals of Basic Object Components	
Symbol	Description	Symbol	Description of example diagnostic signals
E_1	Nacelle with drivetrain	$X(e_{1,1})$	Wind speed and direction measurement system - **18 reference value 18**
		$X(e_{1,2})$	Yaw system - **240 referencevalue 240**
		$X(e_{1,3})$	Drivetrain - **239 reference value 240**
		$X(e_{1,4})$	Rotor - **57 reference value 60**
		$X(e_{1,5})$	Pitch system - **12,5 reference value 13**
E_2	LSS	$X(e_{2,1})$	LSS stabilisation system - **0,0038 reference value 0,0035**
		$X(e_{2,2})$	LSS - **746 reference value 750**
		$X(e_{2,3})$	LSS grease system- **28 reference value 30**
E_3	Gearbox	$X(e_{3,1})$	Gearbox temperature regulation system - **38 reference value 40**
		$X(e_{3,2})$	Gearbox- **1487 reference value 1500**
		$X(e_{3,3})$	Clutch - **1487 reference value 1500**
		$X(e_{3,4})$	Mechanical brake- **1487 referencevalue 1500**
E_4	Generator	$X(e_{4,1})$	Generator temperature regulation system - **38 reference value 40**
		$X(e_{4,2})$	Generator- **690 reference value 690**
E_5	Magnetizer	$X(e_{5,1})$	Magnetizer- **3,35 reference value 3,40**
		$X(e_{5,2})$	Generator current regulation system - **1,38 reference value 1,40**
E_6	Converter	$X(e_{6,1})$	Power regulator system- **1,67 reference value 1,80**
		$X(e_{6,2})$	Converter- **678 reference value 690**
E_7	MV Transformer	$X(e_{7,1})$	MV transformer- **1,75 reference value 1,80**

3 Structure of facts in the expert knowledge base for wind power plant's equipment

When creating a system for supervision and safety usage of the wind power plant, 306 facts were developed, that is parameters, defined in groups of elements, equipment, and units required to specify a technical and functional condition of the whole wind power plant, including wind turbine generators, (MV) equipment and networks, (HV) equipment and networks, and the substation [23]. 109 rules describing specific functional states of the wind power plant were developed on the basis of facts (parameters). Fig. 2 shows a block diagram of the wind power plant's structure in the developed expert system. Diagnostic tests on technical equipment are based on functional – diagnostic models for those objects [8-16, 19]. Thus, a functional – diagnostic model of the wind power plant (see Fig. 2) was developed for this paper. The developed model of the object under examination is a basis to determine a set of diagnostic signals and their reference signals. A functional – diagnostic diagram of the object under examination is shown in Fig. 2. As a result of the functional – diagnostic analysis, fourteen functional units were identified in the object. A sub-set of functional elements was determined in each unit and diagnostic output signals being a set of diagnostic signals $\{X(E_i)\}$ were determined. Fig. 2 shows a graphical representation of the block structure of the wind power plant in the developed expert system.

Fundamental elements of the wind power plant were analyzed in order to determine facts (parameters and working units) for the wind power plant (Fig. 2). As a result of this, two fundamental functional – structure units were determined: a wind turbine generator and a substation (Fig. 2). Then, basic blocks consisting of characteristic parameters of equipment, systems and networks were determined for each functional – structure unit. Wind turbine generators were divided into the blocks A - P, where: A – Environmental and Ambient, B – Grid, C- Brake Systems, D - Hydraulics, E -Yaw System, F - Communications, G - Connectors, H - Protection Relays, I - Pitch System, J - Cooling and Heating Systems, K – Transmission, L - MV Unit Power Transformer, M - Controller, N - Gearbox, O - Generator, P-Rotor.

Facts for the Substation were categorized into the Blocks Q-W, where: R- Feeder Cable Line Field, R - Shunt Reactor Filed, S - Power Transformer Field, T - Auxiliaries Field, U - Voltage Measurement Field, W- General Signalization.

Each block of facts for wind turbine generators and the substation (Fig. 2) consists of several parameters, where an exact number of parameters depends on specific functions and a number of equipment in a specific block [4]. The Block A- Environmental and Ambient consists of 12 parameters, including 3 parameters PA010, PA011, and PA012 defined as reserve and without any assigned values / characteristics in order to ensure a possibility for future expanding the expert system if necessary. The Block A- Environmental and Ambient consists of the following facts:

- Cut_{in} wind speed,
- Wind speed for WTG ready,
- Wind speed for WTG nominal power,
- Wind speed for WTG non-nominal power,
- Cut_{out} wind speed,
- Wind speed for WTG survive conditions,
- Wind direction,
- Ambient temperature,
- Air pressure,
- Reserve.

Threaten (13) defining categories were developed in order to identify characteristics of facts (parameters) as follows (Fig. 4):

1. ***Object*:** specifies place assigned to a specific fact e.g. wind turbine generator or substation.
2. ***Voltage:*** specifies voltage assigned to a specific fact e.g. 20-kV or 110-kV.
3. ***Direction*:** where signals come from e.g. WTG or substation.
4. ***Block*:** specifies a group which a specific fact was assigned to, e.g. Block A.
5. ***Field*:** specifies a field in the MV switching room, which a specific fact is assigned to, e.g. feeder cable line field.
6. ***Type*:** specifies a type of a specific fact e.g. measurement or states.
7. ***Symbol:*** all parameters (facts and rules) have their own symbol, which is made as follows:

$$for:\ X = (P\ or\ R) \rightarrow If\ Yes\ Y\ then\ ZZZ \quad (1)$$

where:
X – specifies a parameter's type: X = (P or R), P – facts, R – rules, Y – specifies a specific block, where: Y = A (where: A – Environment and Ambient) or B etc., ZZZ – three digit number assigned to a specific parameter.

8. ***Abb.*:** a letter abbreviation is assigned to all parameters of the system in order to make identification easier;
9. ***Default*:** default value;
10. ***Minimum*:** min. values are determined for facts;
11. ***Maximum*:** max. values are determined for facts;
12. ***Unit*:** measurement units are given for each fact;
13. ***State*:** state 0 or state 1 for two-bit facts.

Lp.	Obiekt	Napięcie	Kierunek	BLOK	Pole	Parametr	Rodzaj	Symbol	Skrót Pl.	Default	min	max	Jedn.	Stan 0	Stan 1
								Środowisko i warunki otoczenia							
1.	Turbina	20kV	EW		Liniowe odpływowe	Prędkość wiatru załączania turbiny	pomiary	PA001	PWZT	3.5	3.5	-	m/s		
2.	Turbina	20kV	EW		Liniowe odpływowe	Prędkość wiatru postoju turbiny w gotowości do pracy	pomiary	PA002	PWPTGP		0	3.5	m/s		
3.	Turbina	20kV	EW		Liniowe odpływowe	Prędkość wiatru pracy znamionowej turbiny	pomiary	PA003	PWPZT		12	20	m/s		
4.	Turbina	20kV	EW		Liniowe odpływowe	Prędkość wiatru pracy nie znamionowej turbiny	pomiary	PA004	PWPNZT		3.5	12	m/s		
5.	Turbina	20kV	EW		Liniowe odpływowe	Prędkość wiatru wyłączania turbiny	pomiary	PA005	PWWT	20	20	-	m/s		
6.	Turbina	20kV	EW	A	Liniowe odpływowe	Prędkość wiatru przetrwania turbiny	pomiary	PA006	PWPT	60	-	60	m/s		
7.	Turbina	20kV	EW		Liniowe odpływowe	Kierunek wiatru	pomiary	PA007	KierWiatr		0	360	stopień		
8.	Turbina	20kV	EW		Liniowe odpływowe	Temperatura otoczenia	pomiary	PA008	Temp		-10	50	C		
9.	Turbina	20kV	EW		Liniowe odpływowe	Ciśnienie barometryczne	pomiary	PA009	CiśBar		0	1500	hPa		
10.	Turbina	20kV	EW		Liniowe odpływowe	Rezerwa		PA010							
11.	Turbina	20kV	EW		Liniowe odpływowe	Rezerwa		PA011							
12.	Turbina	20kV	EW		Liniowe odpływowe	Rezerwa		PA012							
13.	Turbina	20kV	EW					Sieć							
14.	Turbina	20kV	EW		Liniowe odpływowe	Napięcie znamionowe nN	pomiary	PB001	NapZnam_nN		0	690	V		
15.	Turbina	20kV	EW		Liniowe odpływowe	Prąd znamionowy nN	pomiary	PB002	PrądZnam_nN		0	2200	A		
16.	Turbina	20kV	EW		Liniowe odpływowe	Moc znamionowa czynna P	pomiary	PB003	MocCzynZnam		0	2000	kW		
17.	Turbina	20kV	EW		Liniowe odpływowe	Moc znamionowa bierna Q	pomiary	PB004	MocBierZnam		0	2000	kVA		
18.	Turbina	20kV	EW		Liniowe odpływowe	Nastawa częstotliwości f	pomiary	PB005	NastCzęst		47.5	51.5	Hz		
19.	Turbina	20kV	EW		Liniowe odpływowe	Nastawa współczynnika mocy	pomiary	PB006	NastWspMocy		-1	1	stopień		
20.	Turbina	20kV	EW		Liniowe odpływowe	Prąd fazowy IL1	pomiary	PB007	PrądFaz1		0	1800	A		
21.	Turbina	20kV	EW		Liniowe odpływowe	Prąd fazowy IL2	pomiary	PB008	PrądFaz2		0	1800	A		
22.	Turbina	20kV	EW		Liniowe odpływowe	Prąd fazowy IL3	pomiary	PB009	PrądFaz3		0	1800	A		
23.	Turbina	20kV	EW		Liniowe odpływowe	Napięcie fazowe UL1	pomiary	PB010	NapFaz1		0	400	V		
24.	Turbina	20kV	EW	B	Liniowe odpływowe	Napięcie fazowe UL2	pomiary	PB011	NapFaz2		0	400	V		
25.	Turbina	20kV	EW		Liniowe odpływowe	Napięcie fazowe UL3	pomiary	PB012	NapFaz3		0	400	V		
26.	Turbina	20kV	EW		Liniowe odpływowe	Moc czynna P	pomiary	PB013	MocCzyn		-120	2000	kW		
27.	Turbina	20kV	EW		Liniowe odpływowe	Moc bierna Q	pomiary	PB014	MocBiern		-1000	1000	kvar		
28.	Turbina	20kV	EW		Liniowe odpływowe	Częstotliwość f	pomiary	PB015	Częstotl		0	60	Hz		
29.	Turbina	20kV	EW		Liniowe odpływowe	Temperatura inwertera - faza 1	pomiary	PB016	TempInwt1		-30	150	°C		
30.	Turbina	20kV	EW		Liniowe odpływowe	Temperatura inwertera - faza 2	pomiary	PB017	TempInwt2		-30	150	°C		
31.	Turbina	20kV	EW		Liniowe odpływowe	Temperatura inwertera - faza 3	pomiary	PB018	TempInwt3		-30	150	°C		
32.	Turbina	20kV	EW		Liniowe odpływowe	Temperatura - szyny zbiorcze	pomiary	PB019	TempSzynZb		-30	150	°C		
33.	Turbina	20kV	EW		Liniowe odpływowe	Rezerwa		PB020							
34.	Turbina	20kV	EW		Liniowe odpływowe	Rezerwa		PB021							
35.	Turbina	20kV	EW		Liniowe odpływowe	Rezerwa		PB022							

Fig. 4. Table of the Facts for the Block A – Environment and Ambient and the Block B – Grid (example).

Interpretation of dependence (1). If a fact or a rule (X = P or R) with a symbol A or B, etc. exists in the Block Y, then a signal of specific parameters (characteristics) is assigned. An example of the description of the structure of a developed fact or rule is as follows:

- **PA001** – means a fact from the Block A and of the number of 001,

- **RC003** – means a rule from the Block C and of the number 003, etc.

If the fact **PA001** exists, then it means that it is a fact from the Block A and of the number 001. Thus, the following information can be found in the facts' table (Fig. 4): ***Object*** – wind turbine generator, ***Voltage*** – 20-kV, ***Direction*** - WTG, ***Block***- A (where: A-Environment and Ambient), ***Field*** - feeder cable line field, ***Parameter*** - Cut_{in} wind speed, measurement type, Abb.- PWZT, parameters: average - 3.5, min. - 3.5, Unit - m/s. The implemented method of defining facts, which is above described, allowed to present facts in a simple and clear form of the table that enables to find parameters easily e.g. through characteristics or block division.

Fig. 4 shows example facts for a wind turbine generator based on the Block A – Environment and Ambient and the Block B - Grid. Fig. 5. shows facts for the substation based on the Block Q – feeder cable line filed.

Facts for all Blocks of the developed system for supervision and save usage of the wind power plant were developed in accordance with the above specified method.

4 Conclusions

A basic concept of the expert system depends on transferring expert's knowledge into a computer program equipped with a knowledge base, specific conclusion rules, and a user or graphical interface used for communication. All knowledge collected in the system can be used many times by many users searching for help or support. The computer gives users the best pieces of advice, according to computer's logic, and explains logic used for making conclusions. This term can be also proper to systems used in situations where a human, though they have a great knowledge on a specific field, is not able to use it optimally.

If a user of the object must take a rapid decision in a situation where there is many pieces of information but it is not confirmed, then they may act under stress, which would lead into taking wrong decisions. Thus, a machine is sometimes better than a human. When it goes for operating wind power plants where decisions must be taken fast and optimally under conditions of thousands of signals and pieces of information coming, expert systems seem to much better than humans. Thus, real time expert systems have been developed, which ensure more effective strategies of monitoring and control, where lots of data is coming e.g. monitoring of signals in nuclear power plants, space programs, interpretation of radar signals, monitoring of patients in medicine. Due to expert systems used in those fields, a number of specialists can be reduced, quality of systems can be improved, and operational cost of systems can be reduced.

References

1. L. Bedkowski, T. Dabrowski, *Basic of the maintenance theory p. 2.* WAT, Warsaw, p. 187, (2006)
2. M. Bialko, *The Basic properties of neural networks and hybrid expert system*, Technical University of Koszalin, Koszalin, p. 341, (2000)
3. B. Buchannan, E. Shortliffe, *Rule – Based expert systems*. Addison – Wesley Publishing Company, p. 387 (1985)
4. S. Duer, *Intelligent system of supporting the process renewal of operating characteristics in complex technical objects*, Technical University of Koszalin, Koszalin, p. 242 (2012)
5. F. Hayer-Roth, D. Waterman, D. Lenat, *Building expert systems*. Addison – Wesley Publishing Company, p. 192 (1983)
6. W. Kacalak, M. Majewski, J. M. Zurada, Lecture Not. in Art. Int. **6114** (2010)
7. W. Kacalak, M. Majewski, Lecture Not. in Com. Sci. **7666** (2012)
8. S. Kobayashi, K. Nakamura, *IEEE Expert*, Oct. (2011)
9. P. Linz, R. Wang, *Exploring Numerical Methods: An Introduction to Scientific Computing using Matlab*. University of California, Davis, p. 172, (2003)
10. P. Linz, *An Introduction to Formal Languages and Automata*. University of California, Davis, (2002)
11. M. Majewski, W. Kacalak, Polish J. Env. Stu. **18**, 3B (2009)
12. Z. Palkova, Warsaw Uni. of Lif. Sci. Pre. **55** (2010)
13. Z. Palkova, T. Rodny, *Modeling of object-related database structures in the development of artificial irrigation system*, November 2010, Slovak University of Agriculture in Nitra (2010)
14. S. Duer, Defence Sci. J. **60**, 5 (2010)
15. W. Pedrycz, *Fuzzy Control and fuzzy systems*, John Walley End Sons, Inc, p. 432, (1993)
16. L. Pokorádi, S. Duer, International Sci. Con. on Adv. in Mech. Eng. (2016)
17. A. Rosiński, Reliability, Ris. and Saf. Theory and App., **3** (2010)
18. A. Rosiński, Advances in Saf. Rel. and Ris. Man. (2012)
19. D. Waterman, *A guide to export systems*, Addison – Wesley Publishing Company (1986).
20. S. Duer, R. Duer, S. Mazuru, Romanian Ass. of Nonc. Tech., **XX**, 2 (2016)
21. S. Duer, R. Duer, Neural Com. & App. **19**, 5 (2010)
22. S. Duer, Neural Com. & App. **1** (2012)
23. S. Duer, K. Zajkowski, I. Płocha, R. Duer, Neural Com. & App., **22**, 7 (2013)
24. Duer S., Zajkowski K.: Neural Com. & App. **23**, 7, (2013)
25. S. Duer, K. Zajkowski, R. Duer, J. Paś, Neural Com. & App. **23**, 3-4, (2013)

Analysis of the impact of thermal resistance of the roof on the performance of photovoltaic roof tiles

Dariusz Kurz[1,*], and *Ryszard* Nawrowski[1]

[1]Poznan University of Technology, Faculty of Electrical Engineering, Institute of Electrical Engineering and Electronics, St. Piotrowo 3a, 60 – 965 Poznań, Poland

Abstract. The paper explores the issues related to the impact of thermal resistance of the roof on the electrical parameters of photovoltaic roof tiles. The methodology of determination of the thermal resistance and thermal transmittance factor was presented in accordance with the applicable legal regulations and standards. A test station was presented for the purpose of measurement of the parameters of photovoltaic roof tiles depending on the structure of the roof substrate. Detailed analysis of selected building components as well as their impact on the design thermal resistance factor and thermal transmittance factor was carried out. Results of our own studies, which indicated a relation between the type of the roof structure and the values of the electricity generated by photovoltaic tiles, were presented. Based on the calculations, it was concluded that the generated outputs in the respective constructions differ by maximum 6%. For cells with the highest temperature, the performance of the PV roof tiles on the respective roof constructions fell within the range between 0.4% and 1.2% (depending on the conducted measurement) and amounted to 8.76% (in reference to 9.97% for roof tiles with the lowest temperature).

1 Introduction

A photovoltaic roof tile is an example of a BIPV (Building Integrated Photovoltaics) element which integrates the roof covering with PV cells. The BIPV elements, except the generation of electric current in the process of photovoltaic conversion, must meet the requirements set for building materials. In the case of solar roof tiles installed in the roof, they must ensure appropriate water, thermal and sound insulations. Additionally, the solar roof tiles, as compared with the traditional PV panels installed on the supporting structure over the roof, as a consequence of poorer wind cooling, are subject to the faster heating of PV cells and are characterised by lower performance [1, 2]. On top of this, the photovoltaic tiles cannot be placed on solar tracking systems which would increase the generated power outputs [3]. Thermal processes occurring within the roof space (such as heat exchange by convection and radiation) necessitate the performance of complex analytical calculations (using criterial numbers, principles of probability or numerical calculations) by means of specialist computer software or parallel calculations [4-6].

Continuously growing demand for electricity (largely, also the demand of individual recipients) entails the need to obtain it from other sources. However, photovoltaic micro-installations constitute a certain danger for the power system, which is related to the hardly predictable quantity of produced electricity, its transmission to power plants or other recipients through power lines or the electromagnetic impact on the surrounding environment [7, 8].

Growing requirements set for newly built buildings, which refer to their energy-efficiency, necessitate the application of building materials with the lowest thermal transmittance factor U [W/m^{2}K] (i.e. with a large thermal resistance R [m^{2}K/W]). In the 90-ties in Poland, the thermal transmittance factor for a building roof was established at the level of 0.45 W/m^{2}K, while at present, pursuant to the Regulation of the Minister of Infrastructure of 1 January 2017, it may not exceed 0.23 W/m^{2}K [9, 10]. Hence, the roof insulation significantly affects a decrease in efficiency of electricity generation from photovoltaic roof tiles, which is of no importance in the case of performance of traditional PV panels [1,2].

2 Methodology of determination of thermal resistance and thermal transmittance factor of the roof.

2.1 Thermal transmittance factor *U*

Determination of the thermal insulation of roofs must be carried out in accordance with the applicable legal acts, i.e. at present, PN-EN ISO 6946:2008. The thermal transmittance factor U for a building partition, with the

* Corresponding author: dariusz.kurz@put.poznan.pl

assumption of the one-dimensional heat flow must be determined on the basis of the following equation [10]:

$$U = \frac{1}{R_T} \leq U_{max} \tag{1}$$

where:
R_T – total thermal resistance.

The value of the thermal resistance is determined in accordance with the following equation [10]:

$$R = \frac{d}{\lambda} \tag{2}$$

where:
d – thickness of a layer of material in a component [m],
λ – design thermal conductivity coefficient of the material calculated in accordance with ISO 10456, or coefficient taken from values given in the table [W/mK].

2.2 Total thermal resistance of the building component consisting of homogenous layers

Total thermal resistance R_T of a flat building component which consists of homogenous layers thermally perpendicular to the heat stream must be calculated from the following formula [10]:

$$R_T = R_{si} + R_1 + R_2 + \cdots + R_n + R_{se} \tag{3}$$

where:
R_{si} – heat transfer resistance on the inner surface,
$R_{1...n}$ – design thermal resistance of each layer,
R_{se} – heat transfer resistance on the outer surface.

2.3 Total thermal resistance of the building component consisting of homogenous and inhomogeneous layers

The hipped roof end is an inhomogeneous partition, therefore, it is necessary to calculate supremum R_T' and infimum R_T'' of the total thermal resistance for the repeatable section of a partition (the so called component). Such a repeatable section for the hipped roof end is the component with the width equal to the axial spacing of a rafter. Further calculations are carried out in an identical manner as is the case with homogenous partitions. The total thermal resistance of the component which consists of homogenous and inhomogeneous layers parallel to each other can be determined from the following formula [10]:

$$R_T = \frac{R_T' + R_T''}{2} \tag{4}$$

Prior to this, the partition must be divided into sections consisting of thermally homogenous layers and their relative areas must be determined in accordance with the formula given below [10]:

$$f_{a...n} = \frac{A_{a...n}}{A} \tag{5}$$

where:
$f_{a...n}$ – relative areas of each section [-],
$A_{a...n}$ – area of the section perpendicular to the direction of heat flow [m²],
A – total area of the partition perpendicular to the direction of heat flow [m²], whereby

$$f_a + f_b + \cdots + f_n = 1 \tag{6}$$

Fig. 1 presents the cross-section of the roof surface with a division into sections and layers (based on the example of a roof with boards, top-cover membrane and mineral wool). Symbol x marks a counter-batten, and symbol y stands for a rafter. The component with the width from the half of one rafter to the half of the other rafter was assumed as the repeatable section. It is possible to distinguish three sections in it (two identical sections a and one section b). On the other hand, the roof consists of 6 layers (A – H). The specified layers comprise:

- A – pine rafters and mineral wool,
- H – only pine rafters (without wool),
- B – pine boards,
- G – vapour-permeable film,
- C – top-cover membrane,
- D – pine counter-battens and air void,
- E – pine battens,
- F – photovoltaic roof tile.

The respective sections (visible in Fig.1) include:

- a – pine rafters and pine boards with top-cover membrane or vapour-permeable film (depending on the type of the roof structure), the width of the section is g_1 = 0.04 m,
- b – mineral wool, boards and top-cover membrane or vapour-permeable film or air void (depending on the type of the roof structure), the width of the section is g_2 = 0.72 m.

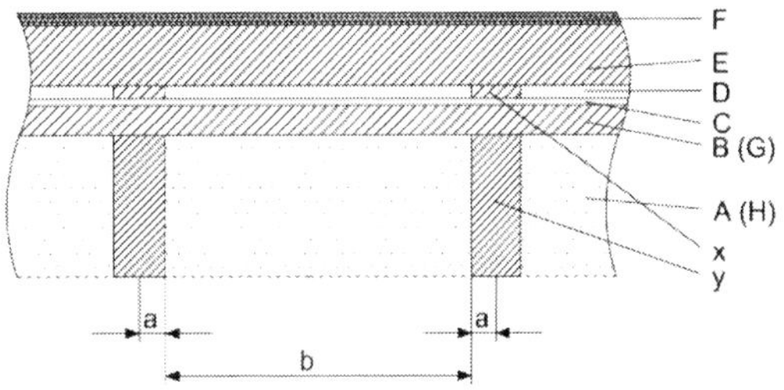

Fig. 1. Cross-section of the roof surface with a division into sections (a, b) and layers (A – H).

During the determination of the thermal resistance factors of the roof structure, layers A, B, C as well as G and H, which will replace layers A and B respectively in other types of roof structures subjected to consideration in this paper, will be of particular importance. Details regarding the layers are given below in the paper.

2.4 Supremum of total thermal resistance R_T'

The supremum of total thermal resistance R_T' is determined with the assumption of a one-dimensional thermal stream perpendicular to the surface of the component in accordance with the formula below [10]:

 DOI: 10.1051/e3sconf/20171901039

$$\frac{1}{R_T'} = \frac{f_a}{R_{Ta}} + \frac{f_b}{R_{Tb}} + \cdots + \frac{f_n}{R_{Tn}} \quad (7)$$

where:
R_{Ta}, R_{Tb}, ..., R_{Tn} – total thermal resistances from the environment to the environment of each section, calculated in accordance with the formula (3).

2.5 Infimum of total thermal resistance R_T''

The infimum of total heat resistance R_T'' is determined with the assumption that all the surfaces parallel to the surface of the component are isothermal. Equivalent thermal resistance R_j of each thermally inhomogeneous layer is calculated from the following equation [10]:

$$\frac{1}{R_j} = \frac{f_a}{R_{aj}} + \frac{f_b}{R_{bj}} + \cdots + \frac{f_n}{R_{nj}} \quad (8)$$

where:
j – consecutive component layer in the section.

The standard [8] also allows for an optional method of determination of R_j in accordance with the following formula:

$$R_j = \frac{d_j}{\lambda_j''} \quad (9)$$

where:
d_j – thickness of the thermally inhomogeneous layer [m],
$\lambda j''$ – equivalent thermal conductivity coefficient of the inhomogeneous layer, whereby:

$$\lambda_j'' = \lambda_{aj} f_a + \lambda_{bj} f_b + \cdots + \lambda_{nj} f_n \quad (10)$$

The infimum of the total thermal resistance is (based on relation 3) equal to [8]:

$$R_T'' = R_{si} + R_1 + R_2 + \cdots + R_n + R_{se} \quad (11)$$

If the air layer is part of the inhomogeneous layer, it can be treated as a material with equivalent thermal conductivity coefficient for which parameters are determined by the standard [10].

3 Types of roof structures

There are several types of roof structures in single-family buildings depending on the preferences of the house owners, their financial capacities and the functions of the attic. The space under the roof can be used as a habitable attic (with the possibility of creation of accommodation spaces) or inhabitable attic (unheated premises). The function of the attic determines its temperature, therefore it will influence the process of heat exchange through the roof covering and roof insulation, thus, the performance conditions of the photovoltaic roof tiles. The working PV roof tiles will transfer the heat to the roof structure, whereby the heat will be accumulated and transferred to the attic to an extent that will depend on the thermal resistance of the structure. The most frequently encountered roof substrates include:

- D1 – vapour-permeable film insulated with mineral wool (for habitable and inhabitable attics),
- D2 – vapour-permeable film without insulation (for inhabitable attics),
- D3 – boards and top-cover membrane insulated with mineral wool (for habitable attics).

The paper analyses the three most frequent roof structures mentioned above in terms of estimation of their thermal resistance and impact on the performance of photovoltaic roof tiles, related to an increase in the temperature and a decrease in the efficiency of electricity generation.

In certain roof structures, some layers are absent or are replaced by others (given in brackets). The detailed specification of the materials occurring in the respective types of roof structures is listed in Table 1. (where "+" means the presence and "–" the absence). The data of the structural (construction) materials applied in the built fragments of roofs (visible in Fig. 2) was collected in Table 2 (based on the datasheet of materials or the standard [10]).

Table 1. Specification of materials in the respective types of roof structures.

Roof	Layer							
	A	H	B	G	C	D	E	F
D1	+	-	-	+	-	+	+	+
D2	-	+	-	+	-	+	+	+
D3	+	-	+	-	+	+	+	+

Table 2. Values of parameters of building materials [10].

Material	d [m]	λ [W/mK]	R [m²K/W]	U [W/m²K]
1. Mineral wool	0.1	0.042	2.3810	0.42
2. Rafter (0.08 m wide)	0.1	0.3	0.3333	3
3. Board	0.025	0.3	0.0833	12
4. Membrane	0.002	0.18	0.0111	90
5. Film	0.0002	0.04	0.0050	200
6. Counter-batten	0.01	0.3	0.0333	30
7. Batten (0.06 m wide)	0.04	0.3	0.1333	7.5
8. PV roof tile	0.0055	0.8545	0.0064	155.5

The design values of physical properties for the wooden elements (rafter, board, counter-batten and batten – reported in table 2) were assumed on the basis of PN-EN

 DOI: 10.1051/e3sconf/20171901039

6946:1999 for the pine wood cut along the medium density 550 kg/m^3 in semi-humid conditions [10].

The values of heat transfer resistances from the inner side R_{si} = 0.17 m^{2}K/W, outer side R_{se} = 0.04 m^{2}K/W and for the PV tile were assumed or calculated in accordance with the standard [10].

4 Determination of roof parameters

In accordance with the applicable regulations and methodology reported in section 2 of this paper, thermal resistances and thermal transmittance factors were determined for three different roof structures (described in detail in section 3). The detailed calculations were presented only for roof D3, and the values of the obtained parameters for each type of roof were listed in table 3.

4.1 Determination of supremum R_T' for sections *a* and *b*

In order to determine supremum R_{Ta} and R_{Tb} for the respective sections, it is necessary to take into account the layers occurring in them (in accordance with the numeration of materials in table 2):

$$R_{Ta} = R_{si} + R_2 + R_3 + R_4 + R_{se} \tag{12}$$

$$R_{Tb} = R_{si} + R_1 + R_3 + R_4 + R_{se} \tag{13}$$

Values R_i are determined in accordance with equation (2). The total area of partition A was determined as the product of thickness of the roof structure and the width of the repeatable component section:

$$A = (2g_1 + g_2) \cdot (d_2 + d_3 + d_4) \tag{14}$$

The relative area of each section f_a, f_b was determined as follows:

$$f_a = \frac{g_1 \cdot (d_2 + d_3 + d_4)}{A} \tag{15}$$

$$f_b = \frac{g_2 \cdot (d_2 + d_3 + d_4)}{A} \tag{16}$$

The supremum was determined in accordance with equation 7:

$$\frac{1}{R_T'} = \frac{f_a}{R_{Ta}} + \frac{f_b}{R_{Tb}} + \frac{f_a}{R_{Ta}} \tag{17}$$

4.2 Determination of infimum R_T'' for layers *A*, *B* and *C*

In each of the layers, there is one material for which it is necessary to determine the equivalent thermal conductivity coefficient and then the equivalent thermal resistance in accordance with equations (8) – (11):

$$\lambda_A'' = 2\lambda_2 f_a + \lambda_1 f_b \tag{18}$$

value of the total infimum:

$$R_T'' = R_{si} + R_A + R_B + R_C + R_{se} \tag{19}$$

4.3 Determination of the value of R_T and U for the roof

Having determined the values of the infimum and supremum of a component, it is possible to calculate the value of total thermal resistance R_T for it, consisting of homogenous and inhomogeneous layers parallel to each other, using equation (4), and the thermal transmittance factor U in accordance with formula (1).

4.4 Specification of determined parameters for the tested roofs

By using the presented relations and values of structural materials of the roof, the parameters of thermal resistances and thermal transmittance factors were determined for the roof structures under consideration. (Table 3).

Table 3. Values of parameters determined for the tested roofs.

Parameter		Values		
Symbol	Unit	Roof		
		D1	D2	D3
R_{Ta}	[m^{2}K/W]	0.5483	0.5483	0.6378
R_{Tb}	[m^{2}K/W]	2.5960	0.2150	2.6854
A	[m^2]	0.0802	0.0802	0.1016
f_a	[-]	0.05	0.05	0.05
f_b	[-]	0.9	0.9	0.9
R_T'	[m^{2}K/W]	1.8901	0.2289	2.0328
λ_A	[W/mK]	0.0678	-	0.0678
λ_B	[W/mK]	-	-	0.3
λ_C	[W/mK]	-	-	0.18
λ_G	[W/mK]	0.04	0.04	-
λ_H	[W/mK]	-	0.4391	-
R_T''	[m^{2}K/W]	1.6899	0.4427	1.7794
R_T	[m^{2}K/W]	1.79	0.3358	1.9061
U	[W/m^{2}K]	0.5587	2.9777	0.5246

While analysing the values of the thermal transmittance factor obtained by way of analytical calculations, it is possible to notice that none of them meets the requirements set for roofs in the applicable legal acts. In real-world solutions, roof rafters are about 18 cm thick (and not 10 cm as in built fragments) and the thickness of the applied thermal insulation in the form of mineral wool is about 30 cm (and not 10 cm as in the analysed

 DOI: 10.1051/e3sconf/20171901039

case). It is also possible to use materials with better (lower) thermal conductivity coefficient λ. After a change in the thickness and quality of the materials used, it would be necessary to carry out the whole analysis again (in a manner identical to the one presented before), until obtaining the satisfactory values of parameter U. However, from the point of view of the performance of the photovoltaic roof tiles, a general principle that indicates the impact of the thermal transmittance factor (thermal resistance of the roof) on the reduction in the efficiency of generation of the electricity due to an increase in the temperature of the PV cells is important.

5 Measurement of electrical parameters of photovoltaic roof tiles placed on different roof structures

Fig. 2 presents the measuring station which serves the purpose of performing measurements in environmental conditions and electric parameters of the solar roof tiles installed in three different roofs D1 – D3 (described above). The PV cells are directed towards the south and inclined at an angle of 37° in relation to the ground surface (optimal conditions for Poznan) [11-13]. Three photovoltaic roof tiles were serially connected with each other on each of the roofs, and as a result of this the total installed power on a single roof amounted to 156 W.

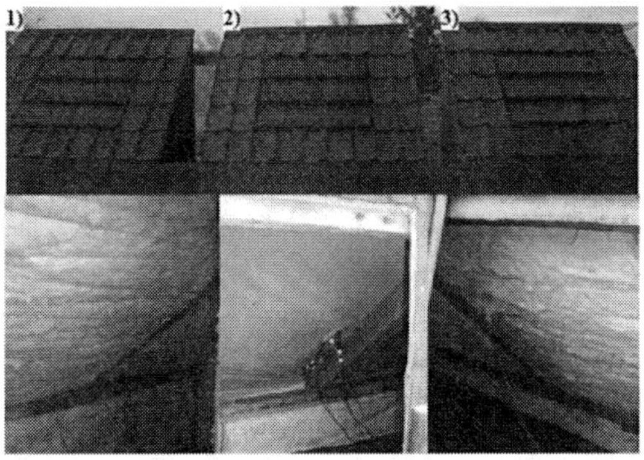

Fig. 2. View of roofs built on the basis of the presented three material structures of a different kind.

Table 4 lists the specifications of the PV roof tile in STC conditions (E = 1000 W/m^2, t_o = 25°C, AM = 1,5). It occupies the area of 0.5 m^2, however, in view of the required overlapping of roof tiles (laying in a way similar to the traditional roof covering) its active area containing PV cells is smaller by 0.1 m^2.

The measurements of parameters of the presented solar installations were conducted on a selected sunny day in the month of August. The values of the respective measured parameters obtained during the measurements were collected in Table 5. It can be noticed that the installation built into roof D2 (which only consists of film without insulation), i.e. with the highest thermal conductivity coefficient, generated the highest power at the level of slightly less than 110 W (at environmental conditions - E = 993.34W/m^2, t_{pv} = 28,69°C), which is lower by about 30% than that declared by the manufacturer in the datasheet for the STC conditions.

Analysis of the measurement results demonstrates unambiguously the impact of the roof structure (used building materials) and their design thermal resistances on the values of temperatures of PV cells and air within the roof space (t_{gap} in layer D). Higher temperatures of the PV cells determine the reduction of voltages and generated power.

Table 4. Values of parameters of the PV roof tile in STC [14].

U_m [V]	I_m [A]	P_{max} [W]	ν_{Isc} [%/°C]	η [%]
9.8	5.3	52	0.027	14.9
U_{oc} **[V]**	I_{sc} **[A]**	ν_P **[%/°C]**	ν_{Uoc} **[%/°C]**	S **[m^2]**
12	5.55	– 0.38	– 0.285	0.4

Taking roof D2, which has the lowest thermal resistance, as the base roof (as in the case of this structure, there is free exchange of heat from layer D with the air on the building attic) the measured values were compared with the values determined analytically on the basis of temperature coefficients given by the manufacturer in the roof tile datasheet. The maximum differences in the obtained values amounted to ±1.5%.

Table 5. Measured values of the installation parameters with PV roof tiles placed on various structures.

E [W/m^2]		**993.34**	**567.37**	**360.37**
D1	U_{oc} [V]	29.98	30.22	29.93
	I_{sc} [A]	5.33	2.91	1.62
	U_m [V]	22.51	24.19	24.65
	I_m [A]	4.71	2.57	1.43
	P_{max} [W]	106.02	62.17	35.25
	η [%]	8.89	9.13	8.15
	t_{pv} [°C]	33.61	31.48	30.94
	t_{gap} [°C]	36.31	35.42	34.35
D2	U_{oc} [V]	30.31	30.53	30.24
	I_{sc} [A]	5.33	2.76	1.60
	U_m [V]	23.06	24.96	24.01

cont. Table 5.

D2	I_m [A]	4.75	2.72	1.45
	P_{max} [W]	109.54	67.89	36.26
	η [%]	9.19	9.97	8.39
	t_{pv} [°C]	28.69	27.73	27.21
	t_{gap} [°C]	29.32	28.91	28.74
D3	U_{oc} [V]	29.57	29.76	29.44
	I_{sc} [A]	5.12	2.75	1.57
	U_m [V]	22.20	23.87	24.37
	I_m [A]	4.65	2.50	1.42
	P_{max} [W]	103.23	59.68	34.61
	η [%]	8.66	8,76	8.00
	t_{pv} [°C]	34.44	32.84	31.14
	t_{gap} [°C]	37.42	37.85	36.28

The differences between the power generated by PV roof tiles, which have the highest temperature (D3) and the lowest one (D2), range between 0.5 and 12%, depending on the performed measurement and the prevailing environmental conditions. On the other hand, the performance of PV roof tiles ranged between 8 and 9.97%, that is, between 4 and 12%, depending on the structure and the performed measurement.

6 Conclusions

Photovoltaic roof tiles constitute an option for persons who wish to have their own photovoltaic installation, maintaining the aesthetic appearance of their houses. However, the stricter and stricter requirements of the civil engineering industry regarding the energy-efficiency of buildings and quality of building materials have an adverse effect on the performance of photovoltaic systems integrated with the building. It is necessary to conduct an accurate analysis of thermal resistances of the roof structure and to estimate the real power outputs from the installed solar installation in order to make the investor fully aware of the achieved profits and time of return on investment.

Analysis of the measuring data demonstrates unambiguously the link between the type of the roof structure and the values of idle voltages, voltages at maximum power point and generated electric power. Provision of a better roof insulation (with lower thermal transmittance factor) causes faster and stronger heating of the PV cells (resulting from worse heat exchange) and thus a decrease in efficiency and quantity of the electricity generated from photovoltaic roof tiles - from a few to several percent.

References

1. M. D'Orazio, C. Di Perna, E. Di Giuseppe, Renewable Energy **68**, 378 (2014)
2. A. Gunawan, P. Hiralal, G. A. J. Amaratunga, K. Tan, S. Elmes, *Photovoltaic Specialist Conf.*, (PVSC, 776-781, 2014)
3. G. Frydrychowicz - Jastrzębska, A. Bugała, Energies **8**, 1025 (2015)
4. D. Kurz, *Computational Problems of Electr. Eng.*, 1-5 (IEEE Xplore, Sandomierz, 2016)
5. L. Kasprzyk, Prz. Elektrotechniczny, **12**, 128 (2007)
6. J. Jajczyk, *Computational Problems of Electr. Eng.*, 1-3 (IEEE Xplore, Sandomierz, 2016)
7. K. Bednarek, R. Nawrowski, A. Tomczewski, *Studies in Applied Electromagnetics and Mechanics*, **22**, 363-368, (IOS Press, Amsterdam, Berlin, 2002)
8. K. Bednarek, Int. J. Occup. Saf. Ergon. (JOSE), **16**, 357 (2010)
9. Notice of Minister of Infrastructure and Development of 17 July 2015 on the Homogenous Text of the Regulation of the Minister of Infrastructure on the Technical Requirements to be Met by Buildings and Their Location, Journal of Laws of 2015, item 1422
10. PN-EN ISO 6946:2008 Building components and building elements – Thermal resistance and thermal transmittance
11. D. Głuchy, D. Kurz, G. Trzmiel, Prz. Elektrotechniczny **89/6**, 281 (2013)
12. Haberlin H., *Photovoltaics. System Designed and Practice* (John Wiley & Sons Ltd., 2012)
13. Jastrzębska G., *Ogniwa słoneczne* (Wydawnictwa Komunikacji i Łączności, 2013)
14. FTDS52 photovoltaic tile data sheet [www.fotton.eu], date:14.02.2017

DOI: 10.1051/e3sconf/20171901040

Thermal properties of mixtures of mineral oil and natural ester in terms of their application in the transformer

Zbigniew Nadolny[1,*], and *Grzegorz* Dombek[1]

[1]Poznan University of Technology, Institute of Electrical Power Engineering, 60965 Poznan, Poland

Abstract. The article describes research results of thermal properties of mineral oil and natural ester. Percentage proportions of both the liquids were as follows: 100/0, 95/5, 80/20, 50/50, 20/80, 0/100. The authors present measurement results of thermal conductivity, viscosity, specific heat, density, and thermal expansion of the created mixtures. The measurements were taken in a relatively wide temperature range: 25 °C, 40 °C, 60 °C, and 80 °C. On the basis of the measurement results, convection heat transfer coefficient α was calculated and the most advantageous proportion of both the components of the mixture was pointed in terms of cooling effectiveness of the transformer.

1 Introduction

Insulating liquids are an integral part of the insulating system of most power transformers in the world. Due to their good properties, they have a role of an electric insulation and cooling medium [1-3]. Keeping in mind that heat transport in the transformer proceeds the following way: heat source → paper impregnated with liquid → insulating liquid → tank → air; it is liquid properties that will substantially decide about the efficiency of heat transfer [2, 4-6].

The properties which determine the ability of a liquid to heat transfer are: thermal conductivity λ, kinematic viscosity υ, specific heat c_p, density ρ, and thermal expansion β. These properties determine the ability of a liquid to cool the device, i.e. coefficient α [7-9]. The greater this ability is, the higher the value of coefficient α. In turn, coefficient α is greater when thermal conductivity, specific heat, density, and liquid thermal expansion are higher and its viscosity is lower [1, 2, 5].

The most frequently used insulating liquid in power engineering devices is mineral oil [1, 10, 11]. The popularity of applying mineral oil results from its low price and properties which have been well known for decades. Lately, stricter and stricter requirements concerning reliability of power engineering devices and a prospect of a real deficit of petroleum have contributed to search for insulating liquids alternative to mineral oil [12-16].

In the early 1990s, research on transformer liquids of plant origin was initiated. After less than ten years later, commercial application of natural esters was started in distribution transformers. Now we are observing their wider and wider application, including power transformers [11]. The advantages of applying natural esters instead of mineral oil are very good fire properties (high flash and fire point, and low net calorific value) and environmental ones, which include first of all full biodegradability and low toxicity [17-19]. Undoubtedly, another advantage resulting from applying natural esters is also retarding the process of paper insulation ageing of the transformer [12, 20, 21]. Unfortunately, these esters, in comparison to mineral oil, are also characteristic of much higher viscosity and higher solidification temperature [2, 11].

Recently research on properties of insulating liquid mixtures has been conducted in many scientific centers in the world. These mixtures can be obtained in two ways. The first one is the so-called retrofilling, which consists in intentional replacing one insulating liquid (most often mineral oil) by another (natural or synthetic esters). This allows, among others, prolonging the remaining operation time of the power device [22]. This process, however, does not guarantee full removing the liquid which filled the transformer before. A small amount of it (not exceeding 8%) can be deposited in, for example, saturated paper insulation, windings, the core, and other hardly accessible places [23]. Thus as a result of filling the modernized unit with a new insulating liquid, we unintentionally obtain a mixture with the liquid which previously filled the transformer. The other way consists in applying an intended mixture in the new transformer, which consists of two or more insulating liquids of a known proportion. This results in obtaining a mixture, which is characteristic of improvement of some, selected properties in reference to the base liquids [10, 13-15, 24, 25].

Independently of the method how the mixture is created, the purpose of the application is to improve properties of the insulating liquid, which consequently affects the operation period of the transformer. Research on mixtures of insulating liquids which is conducted in many scientific centres all over the world refer mainly to electric properties. The remaining properties, including

* Corresponding author: zbigniew.nadolny@put.poznan.pl

thermal properties, attract much less attention, which is not a proper approach. Thus we should take into consideration that applying mixtures of insulating liquids which are characteristic of better electric properties and worse thermal properties in reference to the base liquids, will result in accelerating ageing processes taking place in the insulating system of the transformer and reducing the time of its operation. Therefore, this article presents research results concerning properties that influence the efficiency of transformer cooling properties.

2 Purpose and range of the research

The purpose of the undertaken research was to determine the influence of proportions of mineral oil with natural ester on thermal properties of a created mixture. The research range covered the measurement of thermal conductivity λ, kinematic viscosity υ, specific heat c_p, density ρ, and thermal expansion β. On the basis of the measured properties and the equation presented below, coefficient α was determined [8, 9]:

$$\alpha = \sqrt[n+1]{c \cdot \lambda^{1-n} \cdot g^{n} \cdot \delta^{3n-1} \cdot \beta^{n} \cdot \rho^{n} \cdot c_p^{\,n} \cdot \upsilon^{-n} \cdot q^{n}} \quad (1)$$

where: α – convection heat transfer coefficient of the liquid [$W \cdot m^{-2} \cdot K^{-1}$], c, n – constants dependent on the character of the flow, λ – thermal conductivity [$W \cdot m^{-1} \cdot K^{-1}$], g – acceleration of gravity [$m \cdot s^{-2}$], δ – characteristic dimension connected with liquid flow [m], β – thermal expansion [K^{-1}], ρ – density [$g \cdot l^{-1}$], c_p - specific heat [$J \cdot kg^{-1} \cdot K^{-1}$], υ – kinematic viscosity [$mm^2 \cdot s^{-1}$], q – surface thermal load [$W \cdot m^{-2}$]. The characteristic dimension δ was equal to 1 m. By contrast, the values of constants c and n were assumed according to the figures listed in Table 1.

Table 1. Values of constants c and n dependent on the flow type [8,9].

Flow type	*Gr·Pr*	*c*	*n*
No flow	$< 10^{-3}$	0.45	0
Laminar flow	$10^{-3} - 5 \cdot 10^{2}$	1.18	0.125
Transitional flow	$5 \cdot 10^{2} - 2 \cdot 10^{7}$	0.54	0.25
Turbulent flow	$>2 \cdot 10^{7}$	0.135	0.333

For measurements of thermal properties, the authors used mineral oil named Nytro Taurus manufactured by Nynas and natural ester named Envirotemp FR3 made by Cargil. Both liquids in investigations were new and unused. Water content was 2 ppm in case of mineral oil, and 34 ppm in case of natural ester. Acid number was less than 0,010 $mg_{KOH} \cdot g^{-1}{}_{liquid}$ for mineral oil, and 0,020 $mg_{KOH} \cdot g^{-1}{}_{liquid}$ for natural ester. Measurements of thermal conductivity of the mixtures were done using authorial measurement system, described in [26]. Measurements of kinematic viscosity, density, and thermal expansion of the mixtures were done using authorial measurements systems according to obligatory standards [27-29]. Measurement of specific heat of the mixtures were done using the differential scanning calorimeter Mettler Toledo DSC1. In order to do the investigations, the authors prepared mixtures of mineral oil and natural ester of the following percentage concentrations of both the liquids: 100/0, 95/5, 80/20, 50/50, 20/80, 5/95, and 0/100. The measurements were taken at the temperature of 25 °C, 40 °C, 60 °C, and 80°C. The results of the investigations are presented in the following section.

3 Results and discussion

3.1 Thermal conductivity of the mixture of mineral oil and natural ester

Table 2 and Fig. 1 contain measurement results of thermal conductivity λ of a mixture of mineral oil (MO) and natural ester (NE) depending on ester concentration. As we can observe, with increase of natural ester content, thermal conductivity λ considerably increases.

Table 2. Measurement results of thermal conductivity λ of the mixture of mineral oil (MO) and natural ester (NE).

Proportion of mineral oil (MO) and natural ester (NE)		Temperature *T*			
		25 °C	40 °C	60 °C	80 °C
		Thermal conductivity λ [$W \cdot m^{-1} \cdot K^{-1}$]			
100% MO	0% NE	0.133	0.130	0.128	0.126
95% MO	5% NE	0.136	0.134	0.132	0.130
80% MO	20% NE	0.144	0.141	0.139	0.137
50% MO	50% NE	0.158	0.156	0.153	0.152
20% MO	80% NE	0.173	0.171	0.168	0.166
5% MO	95% NE	0.179	0.177	0.174	0.172
0% MO	100% NE	0.182	0.180	0.178	0.175

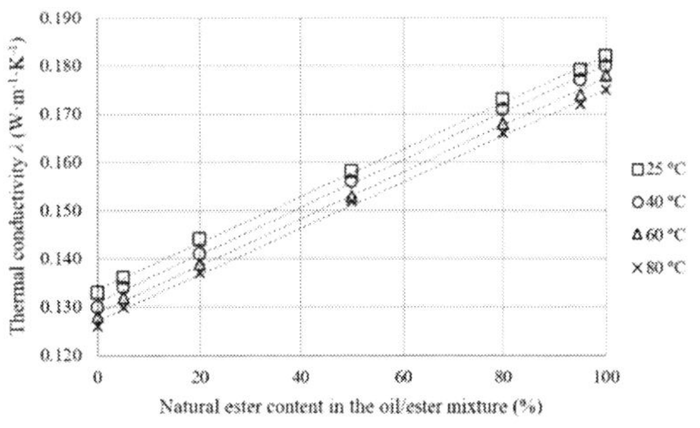

Fig. 1. Thermal conductivity λ of the mixture of mineral oil and natural ester.

This increase is practically linear. The increase of λ was equal to 36.8% (for 25 °C), 38.5% (for 40 °C), 39.1% (for 60 °C), and 38.9% (for 80 °C). As we can see, this increase was at a constant level and it practically did not depend on temperature. Summing up, we can say that natural ester increases thermal conductivity of the mixture of mineral oil and natural ester, and consequently they can improve cooling properties of such a mixture.

3.2 Viscosity of the mixture of mineral oil and natural ester

Table 3 and Fig. 2 present research results of kinematic viscosity υ of the mixture of mineral oil and natural ester, depending on the ester concentration.

Table 3. Measurement results of kinematic viscosity υ of the mixture of mineral oil (MO) and natural ester (NE).

Proportion of mineral oil (MO) and natural ester (NE)		Temperature T			
		25 °C	40 °C	60 °C	80 °C
		Kinematic viscosity υ [mm^2·s^{-1}]			
100% MO	0% NE	17.08	9.59	5.37	3.43
95% MO	5% NE	17.67	9.93	5.44	3.59
80% MO	20% NE	21.34	12.25	6.89	4.46
50% MO	50% NE	31.68	18.48	10.28	6.50
20% MO	80% NE	45.81	26.23	14.67	9.25
5% MO	95% NE	53.42	30.78	17.24	10.88
0% MO	100% NE	56.29	32.66	18.29	11.50

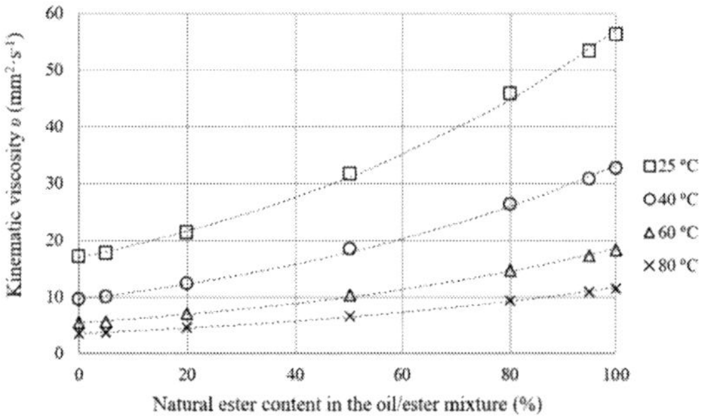

Fig. 2. Kinematic viscosity υ of the mixture of mineral oil and natural ester.

As it was shown, increase of natural ester content increases viscosity of the mixture very significantly. Initially, this increase is tiny (up to 5% of ester content) and next it is very clear (over 5% of ester content). Increase of viscosity υ was equal to 230% (for 25 °C), 241% (for 40 °C), 241% (for 60 °C), and 235% (for 80 °C). This increase, caused by increase of natural ester content in the mixture, was practically independent of temperature. Summing up, we can state that natural ester clearly increases viscosity of the mixture. This fact can result in substantial deterioration of thermal properties of the mixture.

3.3 Specific heat of the mixture of mineral oil and natural ester

Table 4 and Fig. 3 present measurement results of specific heat c_p of the mixture of mineral oil and natural ester depending on ester concentration. Increase of the natural ester content caused a certain increase of the specific heat of the mixture. This increase was linear. The increase of c_p was equal to 6.6% (for 25 °C), 5.4% (for 40 °C), 4.3% (for 60 °C), and 5.7% (for 80 °C). As we can see, the increase of specific heat caused by the increase of the natural ester content in the mixture practically did not depend on temperature. Summing up, we can say that natural ester can slightly improve cooling properties of the mixture of mineral oil and natural ester.

Table 4. Measurement results of specific heat c_p of the mixture of mineral oil (MO) and natural ester (NE).

Proportion of mineral oil (MO) and natural ester (NE)		Temperature T			
		25 °C	40 °C	60 °C	80 °C
		Specific heat c_p [J·kg^{-1}·K^{-1}]			
100% MO	0% NE	1902	1974	2077	2187
95% MO	5% NE	1904	1975	2078	2192
80% MO	20% NE	1925	1995	2094	2212
50% MO	50% NE	1961	2025	2119	2226
20% MO	80% NE	1980	2036	2123	2233
5% MO	95% NE	2024	2080	2160	2257
0% MO	100% NE	2028	2082	2166	2259

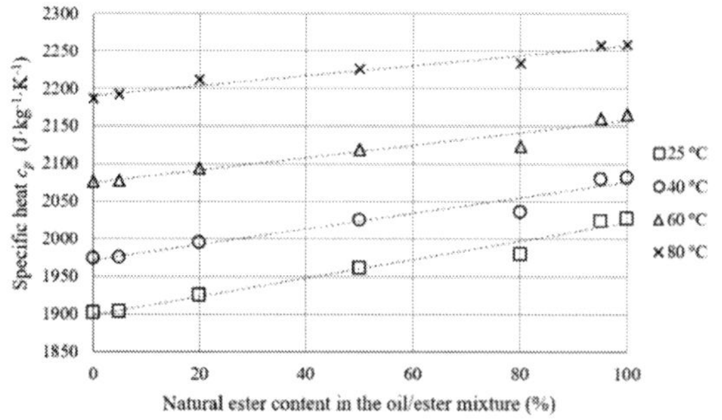

Fig. 3. Specific heat c_p of the mixture of mineral oil and natural ester.

3.4 Density of the mixture of mineral oil and natural ester

Table 5 and Fig. 4 present measurement results of density ρ of a mixture of mineral oil and natural ester, depending on ester concentration. As it was shown, the increase of natural ester content caused the linear increase of mixture density. Increase of density ρ was equal to 5.8% (for 25 °C), 6.0% (for 40 °C), 5.6% (for 60 °C), and 5.8% (for 80 °C). This increase by a few percent, caused by the increase of natural ester content in the mixture, was practically independent of temperature. Summing up, we can say that increase of natural ester content can cause a slight improvement of cooling abilities of the mixture of mineral oil and natural ester.

Table 5. Measurement results of density ρ of the mixture of mineral oil (MO) and natural ester (NE).

Proportion of mineral oil (MO) and natural ester (NE)		**Temperature *T***			
		25 °C	40 °C	60 °C	80 °C
		Density ρ [g·l^{-1}]			
100% MO	0% NE	867	857	845	832
95% MO	5% NE	870	860	847	834
80% MO	20% NE	877	867	854	842
50% MO	50% NE	890	880	869	856
20% MO	80% NE	906	896	883	871
5% MO	95% NE	915	904	891	878
0% MO	100% NE	917	908	892	880

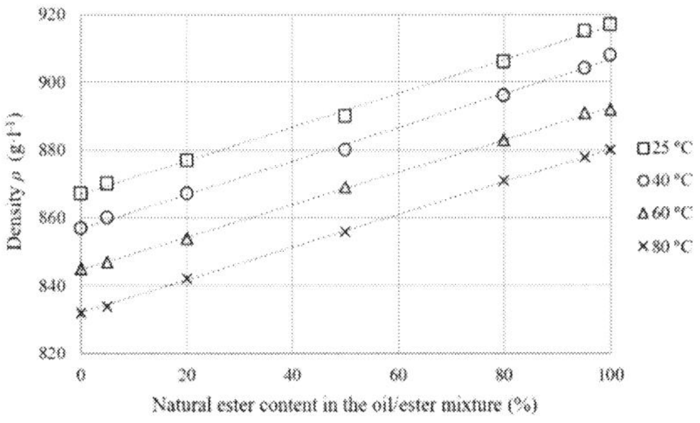

Fig. 4. Density ρ of the mixture of mineral oil and natural ester.

3.5 Thermal expansion of the mixture of mineral oil and natural ester

Table 6 and Fig. 5 show measurement results of thermal expansion β of a mixture of mineral oil and natural ester depending on ester concentration. As we can see, increase of natural ester content practically did not cause any changes of thermal expansion of the mixture. The lack of expansion changes was noted for all the temperature values. Thus we can say that increase of natural ester content does not result in any changes of cooling abilities of an analyzed mixture.

Table 6. Measurement results of thermal expansion β of the mixture of mineral oil (MO) and natural ester (NE).

Proportion of mineral oil (MO) and natural ester (NE)		**Temperature *T***			
		25 °C	40 °C	60 °C	80 °C
		Thermal expansion β [K^{-1}]			
100% MO	0% NE	0.00075	0.00076	0.00078	0.00080
95% MO	5% NE	0.00075	0.00076	0.00078	0.00081
80% MO	20% NE	0.00076	0.00077	0.00078	0.00081
50% MO	50% NE	0.00075	0.00077	0.00079	0.00080
20% MO	80% NE	0.00075	0.00076	0.00078	0.00081
5% MO	95% NE	0.00075	0.00076	0.00079	0.00081
0% MO	100% NE	0.00074	0.00076	0.00078	0.00080

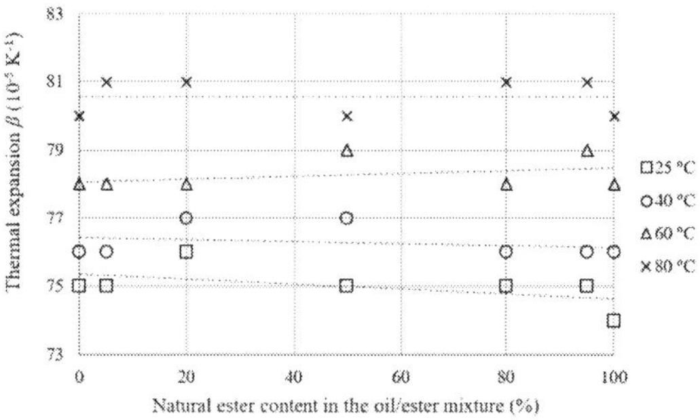

Fig. 5. Thermal expansion β of the mixture of mineral oil and natural ester.

3.6 Convection heat transfer coefficient of the mixture of mineral oil and natural ester

Table 7 and Fig. 6 present calculation results of the convection heat transfer coefficient α of a mixture of mineral oil and natural ester depending on ester concentration. The calculations of coefficient α were done on the basis of the measurement results presented in the previous subsections (3.1-3.5).

As it results from the table and the figure, with increase of natural ester content in the mixture, its coefficient α was initially rising a little (up to 5% of ester content) and next it was clearly decreasing (starting from 5% of ester content). This trend could be observed independently of temperature. The initial increase of coefficient α (up to 5% of ester content) was very small and it was merely 0.4% (for 25 °C), 0.8% (for 40 °C), 1.3% (for 60 °C), 0.9% (for 80 °C). In turn, the drop of coefficient α (starting from 5% of ester content) was

 DOI: 10.1051/e3sconf/20171901040

more significant and it was 4.5% (for 25 °C), 4.8% (for 40 °C), 5.4% (for 60 °C), and 8.1% (for 80 °C).

Table 7. Measurement results of convection heat transfer coefficient α of the mixture of mineral oil (MO) and natural ester (NE).

Proportion of mineral oil (MO) and natural ester (NE)		Temperature T			
		25 °C	40 °C	60 °C	80 °C
		Convection heat transfer coefficient α [$W{\cdot}m^{-2}{\cdot}K^{-1}$]			
100% MO	0% NE	93.56	107.89	125.70	141.65
95% MO	5% NE	93.92	108.70	127.32	142.86
80% MO	20% NE	92.94	106.63	123.67	139.56
50% MO	50% NE	88.65	101.98	118.64	134.15
20% MO	80% NE	85.18	97.98	113.91	129.43
5% MO	95% NE	84.04	96.60	112.44	127.12
0% MO	100% NE	83.46	96.12	111.80	126.17

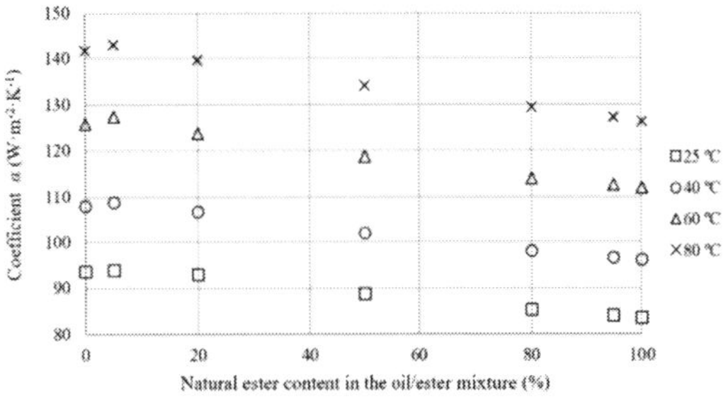

Fig. 6. Convection heat transfer coefficient α of the mixture of mineral oil and natural ester.

The initial increase of coefficient α caused by increase of ester content in the mixture (up to 5%) resulted from many effects. These include the increase of thermal conductivity, specific heat, and density, despite a slight increase of viscosity. In turn, the drop of coefficient α (over 5% of ester content) was caused by a violent viscosity increase, despite resulting increase of thermal conductivity, specific heat, and density. We can say on this basis that the best combination of the mixture components is 95% of mineral oil and 5% of natural ester.

4 Conclusions

The increase of natural ester content in the mixture of mineral oil and natural ester resulted in increase of thermal conductivity by almost 40%, viscosity increase by over 200%, specific heat increase by about 5%, density increase by about 5%, and no changes of thermal expansion. In contrast, coefficient α has its maximum for a mixture of 95% of mineral oil and 5% of natural ester.

Coefficient α was 141.65 $W{\cdot}m^{-2}{\cdot}K^{-1}$ in case of pure mineral oil and 142.86 $W{\cdot}m^{-2}{\cdot}K^{-1}$ in case of mixture with maximal convection heat transfer coefficient, for 80 °C (Table 7). Temperature decrease ΔT (between windings and tank) for pure oil is 24.7 °C, and for mixture with maximal α is 24.5 °C (surface heat density equals 3500 $W{\cdot}m^{-2}$). It means that difference of ΔT between pure oil and mixture whit maximal α is just 0.2 °C.

From the practical point of view we can say that it makes no sense to apply the mixture (95% of mineral oil and 5% of natural ester) of the maximum coefficient α. This results from the fact that for such a mixture, this coefficient is only by 1-2% higher than for the coefficient α of pure mineral oil.

In the case of retrofilling, i.e. replacing mineral oil with ester (e.g. natural ester), we can expect that a few percent of oil will remain in the transformer in its hardly available parts (paper insulation, crevices). In this way, a mixture of oil and ester is created of the proportion 5/95. The coefficient α of such a mixture (Table 7, Figure 6) is much lower than the coefficient α of pure mineral oil or the mixture of the maximum value of α. Thus we can expect that after retrofilling the transformer, the created mixture will have worse cooling properties. On the other hand, the coefficient α of the mixture is a slightly higher than the coefficient α of pure natural ester. Thus we can expect that after retrofilling the transformer, the created mixture will have better cooling properties than natural ester.

Measured values of thermal properties of mixtures (thermal conductivity, kinematic viscosity, specific heat, density, and thermal expansion) were intermediate values of pure oil and pure natural ester. But coefficient α values, calculated using formula (1), were not intermediate values of pure oil and pure natural ester. The coefficient α had some unexpected maximum, what is important results.

Future investigations of a mixture with used mineral oil and unused natural ester should be made in order to better understanding of thermal consequence of retrofilling the transformer.

The research was financed from resources of the Ministry of Sciene and Higher Education for statutory activities No. 04/41/DS-PB/4293, name of the task: The Influence of Various Parameters on Thermal Properties of Dielectric Liquids Used in High Voltage Transformer.

References

1. T.O. Rouse, IEEE Electr. Insul. Mag. **14** (1998)
2. G. Dombek, Z. Nadolny, Eksploatacja i Niezawodnosc – Maintenance and Reliability **18** (2016)
3. Z. Nadolny, G. Dombek, P. Przybylek, IEEE 2016 Conf. Electr. Insul. Dielectr. Phenom., 857-860 (2016)
4. R. Lopatkiewicz, Z. Nadolny, Prz. Elektrotechniczny **84** (2008)

5. Z. Nadolny, G. Dombek, P. Przybylek, D. Przadka, IEEE 2016 Conf. Electr. Insul. Dielectric Phenom., 538-541 (2016)
6. R. Lopatkiewicz, Z. Nadolny, P. Przybylek, W. Sikorski, Prz. Elektrotechniczny **88** (2012)
7. F.B. Waddington, GEC J. Sci. Technol. **49** (1983)
8. F.P. Incropera, D.P. DeWitt, T.L. Bergman, A.S. Lavine, *Fundamental of heat and mass transfer* (Willey, 2007)
9. R. Rudramoorthy, K. Mayilsamy, *Heat and mass transfer* (Pearson, 2011)
10. R. Liao, J. Hao, S. Grzybowski, IEEE 2011 Int. Conf. Dielectric Liq., 1-4 (2011)
11. Cigre WG A2.35, *Experience in service with new insulating liquids. Brochure 436* (Cigre, Paris, 2010)
12. Y. Bertrand, L.C. Hoang, Int. Conf. Prop. Appl. Dielectric Mater., 491-494 (2003)
13. I. Fofana, V. Wasserberg, H. Borsi, E. Gockenbach, IEEE Electr. Insul. Mag. **18** (2002)
14. I. Fofana, V. Wasserberg, H. Borsi, E. Gockenbach, IEEE Electr. Insul. Mag. **18** (2002)
15. C. Perrier, J.L. Beroual, L. Bessade, IEEE Trans. Dielectr. Electr. Insul. **13** (2006)
16. T. Boczar, A. Cichon, S. Borucki, IEEE Trans. Dielectr. Electr. Insul. **21** (2014)
17. K. Longva, Nordic Insul. Symp., 293-295 (2005)
18. T.V. OOmmen, IEEE Electr. Insul. Mag. **18** (2002)
19. C.P. McShane, J.L. Corkran, K.J. Rapp, J. Luksich, Conf. Electr. Insul. Dielectric Phenom., 124-127 (2003)
20. L. Ruijin, H. Jian, C. George, M. Zhigin, Y. Lijm, IEEE Trans. Dielectr. Electr. Insul. **18**, 1626-1637 (2011)
21. P. Trnka, V. Mentlik, M. Svoboda, IEEE Int. Conf. Dielectric Liq., 1-4 (2014)
22. C.P. McShane, J. Luksich, K.J. Rapp, IEEE-IAS/PCA Cem. Ind. Tech. Conf., 141-147 (2003)
23. I. Fofana, V. Wasserberg, H. Borsi, E. Gockenbach, IEEE Electr. Insul. Mag. **17**, 17-30 (2001)
24. R. Liao, J. Hao, L. Yang, S. Liang, J. Yin, Int. Conf. High Voltage Eng. Appl., 588-591 (2010)
25. C.P. McShane, IEEE-IAS/PCA Cem. Ind. Tech. Conf., 31-40 (2001)
26. G. Dombek, Z. Nadolny, P. Przybylek, Prz. Elektrotechniczny, 148-151 (2014)
27. ASTM D 1903-08
28. ISO 3675:1998
29. ISO 3104:1994

DOI: 10.1051/e3sconf/20171901041

Influence of the power supply parameters on the projectile energy in the permanent magnet electrodynamic accelerator

Andrzej Waindok[1], and *Paweł* Piekielny[1,*]
[1]Opole University of Technology, Department of Electrical Engineering and Mechatronics, 45-920 Opole, ul. Prószkowska 76, Poland

Abstract. The main objective of the research is to investigate, how the power supply parameters influence the kinetic energy of the movable element, called commonly a projectile or bullet. A calculation and measurement results of transient characteristics for an electrodynamic accelerator with permanent magnet support were presented in the paper. The calculations were made with using field-circuit model, which includes the parameters of the power supply, mass of the bullet and friction phenomenon. Characteristics of energy and muzzle velocity verso supply voltage (50 V to 350 V) and capacitance value (60 mF to 340.5 mF) were determined, as well. A measurement verification of selected points of calculation characteristics were carried out for investigated values of muzzle velocity. A good conformity between calculation and measurement results was obtained. Concluding, presented characteristics of the muzzle velocity and energy of the projectile vs. power supply parameters indicate, that accelerators could be used for fatigue testing of materials.

1 Introduction

Electrodynamic accelerators (EA), commonly called railguns, are investigated for many years [1, 2]. Mostly these researches were dedicated to military applications [3-5]. However, in recent years, the application of the accelerators in the material testing becomes more popular [6]. Studies on dynamic defragmentation of planetary materials [7] and hypervelocity material testing [8] by using an accelerator are carried out, especially. The use of accelerators for fatigue testing of materials has many advantages: wide measuring range, high velocity impact tests and the possibility of using a ballistic part with different face geometry. These advantages allow to use accelerators in the material testing in industry, in military and especially in the area related with exploration of the space [9].

The fatigue testing of materials can also include study of rails and projectile materials which work under extreme conditions. The shape of rails and projectile is also very important and influences the system efficiency significantly [10-13].

For fatigue testing of materials, an important aspect is the precise determination of the impact energy over a large range of its changes. The value of the projectile energy depends on its mass and velocity. Through an appropriate choice of the power supply parameters (voltage and capacity) it is possible to control the muzzle velocity. Thus, the prediction of the projectile energy, based on the parameters of the power supply system, mass and geometry of the projectile face, is very important. This can be achieved by appropriately formulated computational model of the accelerator.

In order to increase the efficiency of the railgun there is a need to conduct the experimental work and create a proper calculation model of the system. The appropriate choice of calculation method and variables influencing the work of the whole system is very important for analysis of the device operation [14-17]. The model could be used, for example, in order to increase the thrust, without increasing the excitation current [18]. The main aim of this study is the prediction of the projectile energy with using the presented calculation model.

2 Physical model

The analysis was done for the hybrid accelerator with ferromagnetic core and permanent magnets (Fig. 1). For this solution, the magnetic field has two sources: a current flowing through the circuit and permanent magnets. This type of accelerators allows to increase the value of the magnetic field in the projectile area for the same power supply configuration.

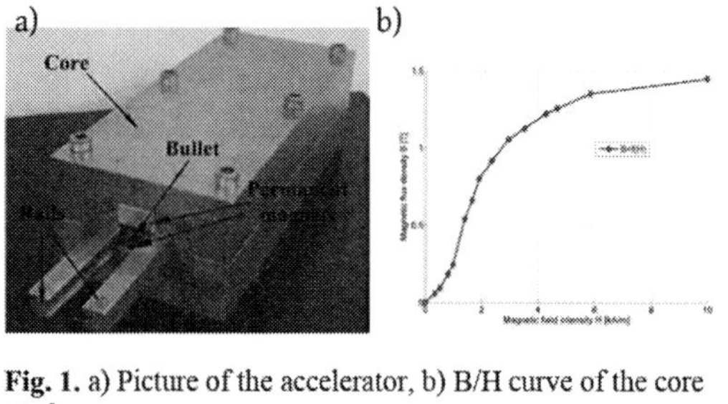

Fig. 1. a) Picture of the accelerator, b) B/H curve of the core steel.

* Corresponding author: p.piekielny@po.opole.pl

For the construction with permanent magnets, it is important to take care about the circuit connections in the power system. There is a possibility of demagnetization of permanent magnets in case of power supply reverse polarity. The blocking diode between the thyristor and the ground of the capacitors bank has been used to prevent capacitor charging in the negative direction.

The 3D visualization of the laboratory stand used in the measurement verification is presented in Fig. 2. As an energy source a capacitors bank with a total capacity of 340.5 mF and nominal voltage of 250 V was used.

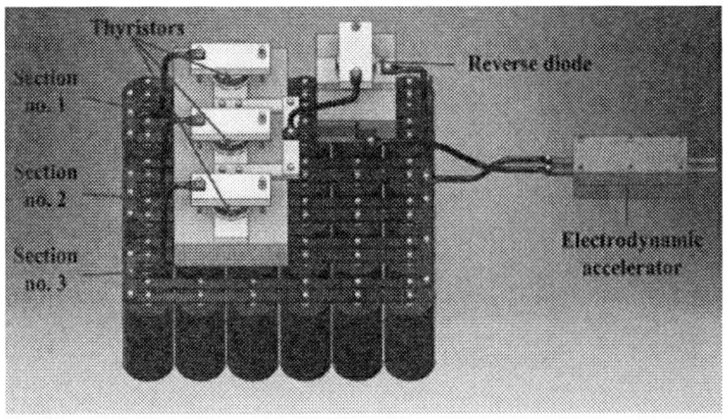

Fig. 2. 3D visualization of the investigated system.

In order to control the power source energy more precisely, the capacitor battery is divided into three sections of the same capacity. Each section is triggered with a separate thyristor (model T95-1900 from Kubara LAMINA Company). Current measurement was made in common ground of thyristors by recording the voltage drop on a 0.517 mΩ resistance. Velocity of the projectile was determined by measuring the flight time through the optical gate at the end of rails.

During the shot the following signals were recorded with using an oscilloscope: the voltage on the capacitor bank u and the excitation current i.

3 Calculation model

The calculations have been made with using a field-circuit method. The field parameters, which are magnetic flux density and thrust, have been determined with using finite element method (FEM) in Maxwell software. The voltage boundary conditions were assumed on the rails ends and on the outer boundaries the zero Dirichlet condition was assumed. The nonlinear B-H curve has been included in the model (Fig. 1b). Maxwell's equations in the following form have been solved:

$$\nabla \times \mathbf{H} = \mathbf{J}\,;\ \nabla \times \mathbf{E} = 0\,;\ \nabla \bullet \mathbf{B} = 0 \tag{1}$$

where:

- **H** – magnetic intensity vector [A/m],
- **J** – current density vector [A/m²],
- **E** – electric intensity vector [V/m],
- **B** – magnetic flux density vector [T].

An exemplary magnetic flux and current distribution is presented in Fig. 3 (U=100 V, z=0.1 m). The highest value of the magnetic flux is observed in the projectile area (about 5.14 T). The largest value of the current density (Fig. 3b) is observed in the area of the bullet, on the inner edges of the connecting elements (over 3.9 kA/mm^2).

a)

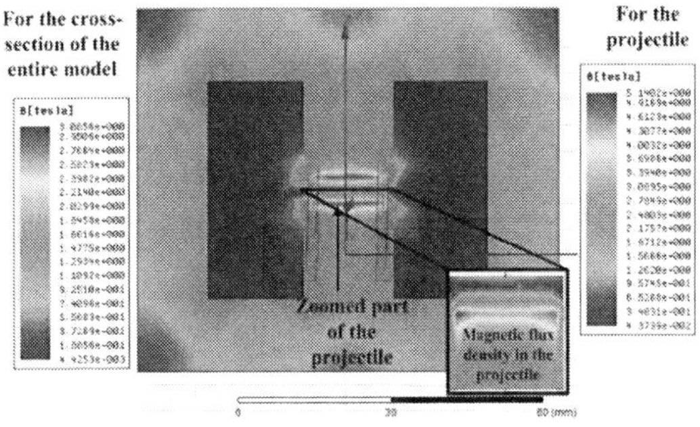

b)

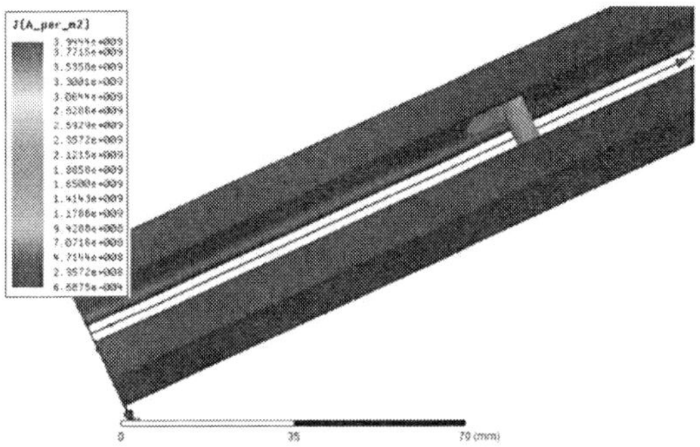

Fig. 3. a) Magnetic flux distribution in the cross section of the railgun, b) Excitation current density distribution in the rails and projectile (U=100 V, z=0.1 m).

Using the model, the current and magnetic flux distributions were determined, which are needed to calculate the magnetic flux and Lorentz force. The magnetic flux Φ is calculated with using the integral:

$$\Phi = \int_S \mathbf{B} \bullet \mathbf{n} dS \tag{2}$$

where:

- **n** – unit vector normal to surface S,
- S – surface limited by the rails and projectile, parallel to the plane YZ (Fig. 3).

The source of the thrust is a Lorentz force **F**:

$$\mathbf{F} = \int_{\Omega_m} (\mathbf{J} \times \mathbf{B}) d\Omega_m \tag{3}$$

where:

- Ω_m – volume of the projectile [m³].

In Fig. 4 (U=100 V, z=0.1 m) the volume force density in rails and projectile is presented. The highest value of force density, directed along z-axis, is observed in the area of the projectile. In the case of rails the value of acting force density is lower than in the projectile and is directed along y-axis out of the rails.

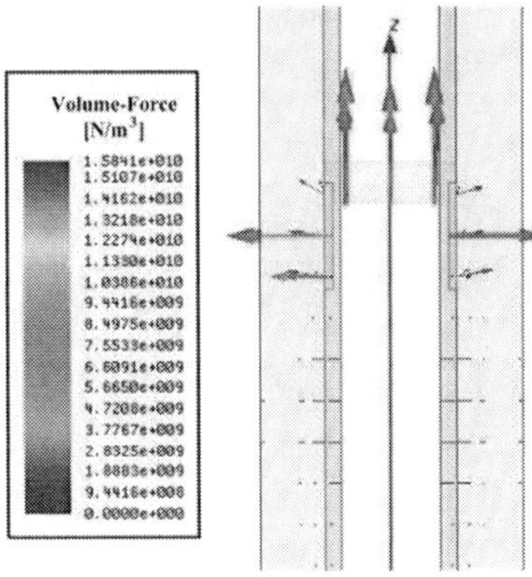

Fig. 4. Volume force density in rails and projectile.

In the transient calculations, the dynamic inductance L_d is an important parameter

$$L_d = \frac{\partial \Phi}{\partial i} = \frac{1}{i}\frac{\partial W_m}{\partial i} \tag{4}$$

where:

i – current flowing through the rails [A],
W_m – magnetic energy [J].

The circuit model is described by the ordinary differential equations in the form:

$$\frac{dv}{dt} = \frac{F(i,x) - Dv}{m} \; ; \quad \frac{dx}{dt} = v \tag{5}$$

$$\frac{di}{dt} = \frac{-Ri - \frac{d}{dx}(\Phi(i,x))v - \frac{q}{C}}{\frac{d}{di}\Phi(i,x)} \tag{6}$$

where:

R – resistance of the accelerator circuit [Ω],
v – projectile velocity [m/s],
q - capacitor charge [C],
C - capacitance of the capacitor banks [F].

The circuit parameters (Table 1) have been obtained based on measurements and analytical calculations.

Table 1. Main circuit parameters of the model.

Quantity	Unit	Value
Projectile mass m	[g]	4.6
Friction coefficient D	[Ns/m]	0.4
Drag coefficient C_x	-	1.05
Accelerator circuit resistance R	[mΩ]	1.82-2.2
Accelerator circuit inductance L	[μH]	0.84-0.89

The field-circuit model was implemented in Matlab/Simulink software (the block diagram is presented in Fig. 5). In order to use the model, the thrust and magnetic flux values vs. excitation current and projectile position have been determined (Figs. 6 and 7) and included in the model with using Look-up tables.

In the mechanical part of the model, in order to increase the calculation precision, the drag coefficient is taken into account. The air resistance P is calculated based on equation:

$$P = 0.5 C_x g A v^2 \tag{7}$$

where:

g – air density (1.2 kg/m³),
A – area of the projectile front part (120 mm²),
v – muzzle velocity.

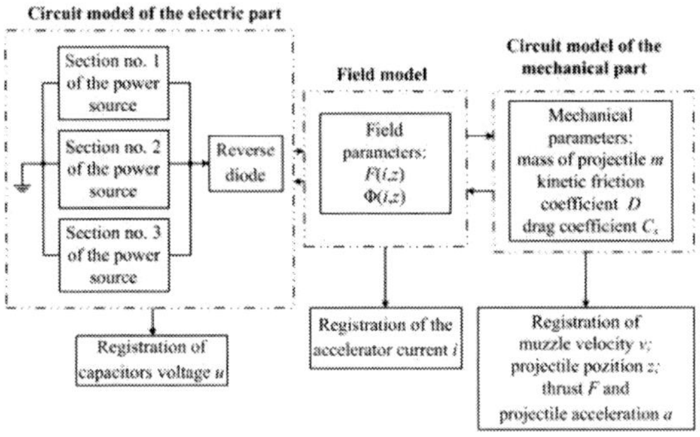

Fig. 5. The field-circuit model.

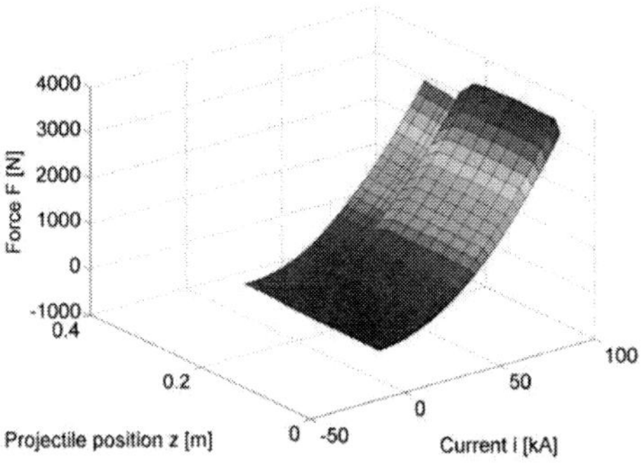

Fig. 6. Thrust vs. current and projectile position values.

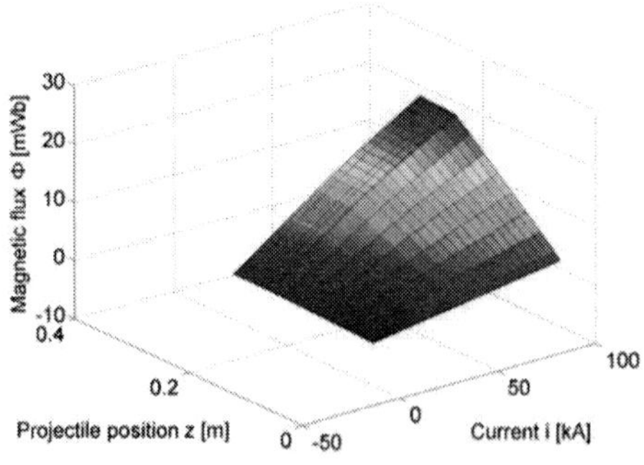

Fig. 7. Magnetic flux vs. current and projectile position values.

3 Measurement verification

The mathematical model was verified for different values of capacitances and voltages. In Figs. 8-11 some transients are presented.

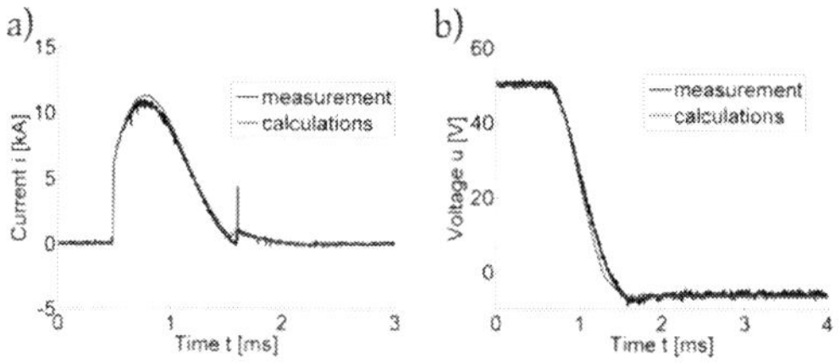

Fig. 8. Excitation current (a) and capacitor voltage (b) transients for U=50 V and C=114 mF.

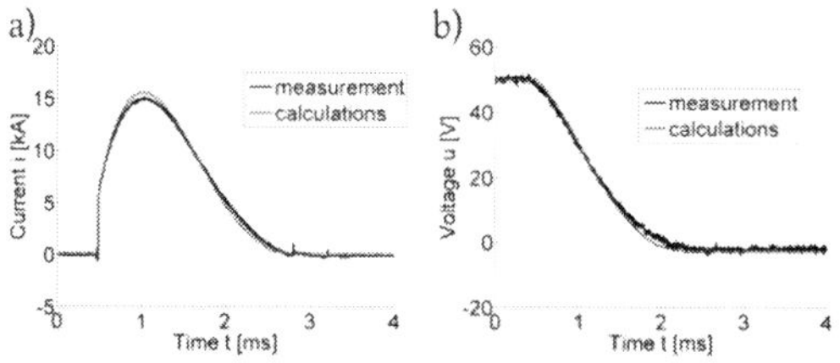

Fig. 9. Excitation current (a) and capacitor voltage (b) transients for U=50 V and C=340.5 mF.

The increasing of capacitance by a factor 3, causes only a slightly increasing in the peak current value (from 12 kA to 15 kA). The duration of the impulse increases by almost 70%. The increasing of voltage by factor 2 causes proportional increasing of peak current value. The duration time does not change vs. voltage value. The shapes of the measured waves are very similar to the calculated ones.

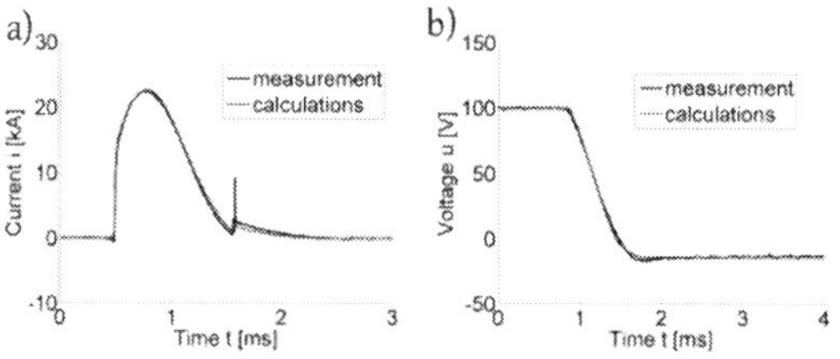

Fig. 10. Excitation current (a) and capacitor voltage (b) transients for U=100 V and C=114 mF.

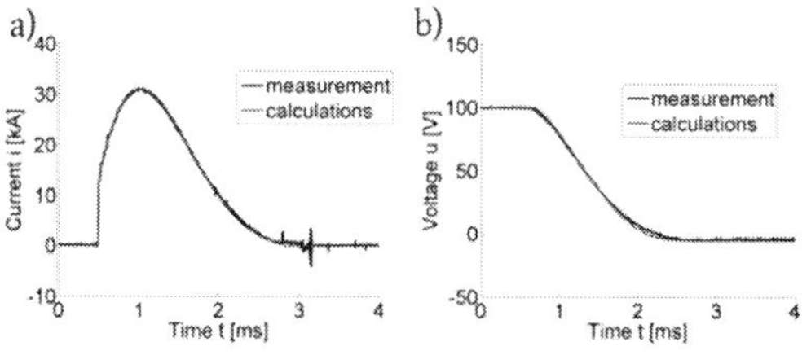

Fig. 11. Excitation current (a) and capacitor voltage (b) transients for U=100 V and C=340.5 mF.

The exit muzzle velocity and peak current value were verified as well. Some measurement verification values have been given in Table 2. There is a very good conformity visible in the measured and calculated peak current values. It means, that the electrical part of the model is quite correct. Some problems are observed in the measurement verification of muzzle velocity. For low energies there is a huge difference between them. For higher energies the differences between measurements and calculations of muzzle velocity are lower. It means, that there are some problems with the mechanical part of the model. The main problem is the proper consideration of the friction and drag force. The experiments show, that these forces changes vs. source energy. It could be due to force acting on the rails, which changes slightly the distance between rails, and consequently the friction coefficient. Additionally, the drag coefficient could be different for higher velocities (more than 100 m/s).

Table 2. Measurement verification of the exit muzzle velocity and peak current value

Initial capacitor voltage and capacitance value	**Velocity [m/s]**		**Peak current [kA]**	
	Measu -red	Calcul-ated	Measu -red	Calcul-ated
U=50 V, C=114 mF	13.37	4.41	11.03	11.29
U=50 V, C=340.5 mF	48.64	41.45	15.15	15.58
U=100 V, C=114 mF	44.80	40.73	22.63	22.75
U=100 V, C=340.5 mF	126.26	140.53	31.33	31.19

4 Analysis of the projectile energy vs. supply parameters

The control of the projectile energy is very important from the fatigue testing point of view. In this chapter calculation and measurement results of muzzle velocity and energy vs. capacitance and initial voltage have been presented (Figs. 12-15). It should be mentioned, that there is a very high repeatability of shots for this kind of railgun [19, 20].

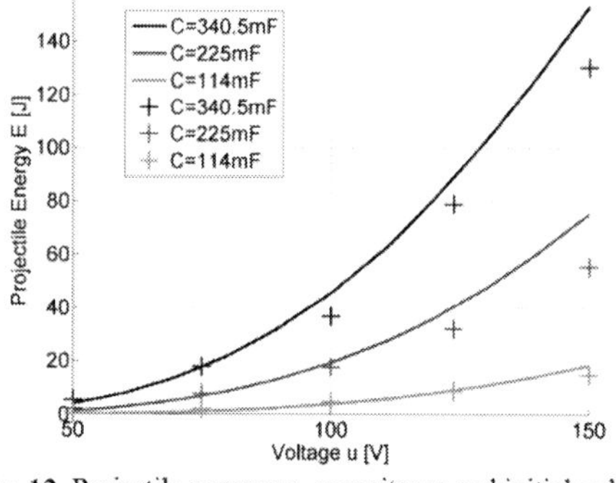

Fig. 12. Projectile energy vs. capacitance and initial voltage values (measurements are denoted by +)

The projectile energy increases, as expected, exponentially to the capacitor voltage (Fig. 12). The

DOI: 10.1051/e3sconf/20171901041

muzzle velocity changes almost linear vs. capacitor voltage (Fig. 13). The calculation model works better for lower energies. The differences between calculations and measurements are increasing vs. capacitance value. It means, that the mathematical model needs some improvements, especially in the area of mechanical phenomenon.

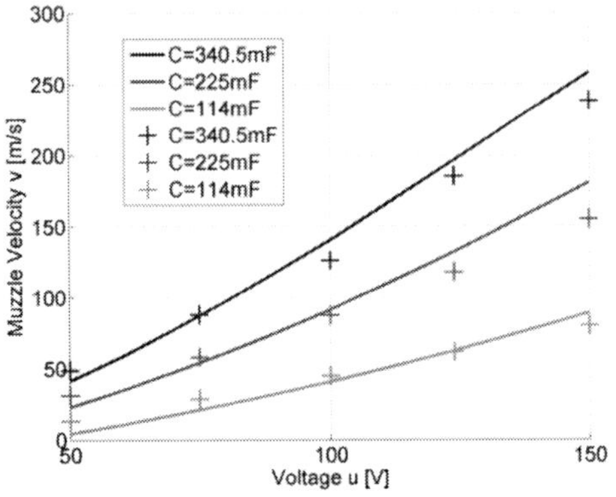

Fig. 13. Muzzle velocity vs. capacitance and initial voltage values (measurements are denoted by +).

The mathematical model allows to estimate the projectile energy for higher voltages, than presented in Figs. 12 and 13. Thus, there has been carried out analysis for initial voltages, which are presently not possible to set in the measurement stand. In Figs. 14 and 15 some results have been given.

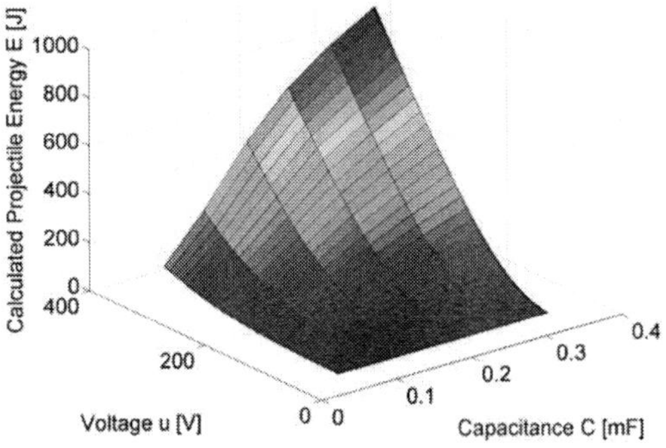

Fig. 14. Calculated projectile energy vs. capacitance and initial voltage values.

For very high energies, not analysed in the experiments, the friction and drag force are playing an important part. Thus, for voltage values higher than 170 V and capacitances higher than 250 mF, the projectile energy increases almost linearly vs. capacitor voltage (Fig. 14). It means, that the muzzle velocity increases slowly than the voltage (Fig. 15).

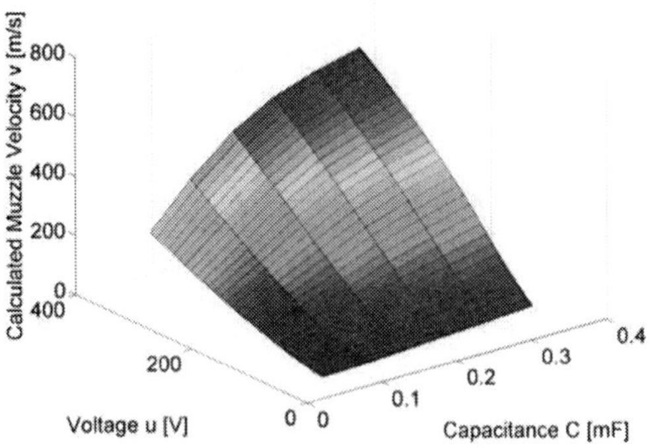

Fig. 15. Calculated muzzle velocity vs. capacitance and initial voltage values.

5 Conclusions

Concluding, presented characteristics of the muzzle velocity and energy of the projectile vs. power supply parameters indicate, that there is a capability to use accelerators for fatigue testing of materials. However, in order to predict the energy of the projectile, the proper mathematical model has to be formulated. The presented model is quite precise and could be used in prediction of projectile energy. Although, in relatively high energy range, the model should be improved, what will be done in future investigations.

Generally, for relatively low velocities (below 200-300 m/s) the projectile energy increases exponentially vs. voltage and nearly linear vs. capacitance value. For higher velocities, the drag force becomes an important factor. In this case, the energy increases linearly vs. supply voltage.

From Figs. 14 and 15 it is visible, that with using the presented laboratory stand, it is possible to reach energies of almost 800 J. Further optimization of the electrodynamic accelerator construction and supply system should make possible improving of this value.

The increasing demand for fatigue testing of materials in high energy range (e.g. impacts in the space), satisfies the importance of presented investigations.

References

1. S. Katsuki, H. Akiyama, N. Eguchi, M. Sued, M. Soejima, T. Maeda, K.M. Sato, IEEE Trans. Magn. **31**, 183-188 (1984)
2. D.A. Sink, L.J. Krzastek, IEEE Trans. Magn. **27**, 266-271 (1991)
3. M. Tower, C. Haight, IEEE Trans. Magn **20**, 298-301 (1984)
4. S. Hundertmark, M. Schneider, D. Simicic, G. Vincent, IEEE Trans. Plasma Sci. **42**, 3180-3185 (2014)
5. J. McFarland, I.R.A. McNab, IEEE Trans. Magn. **39**, 289-294 (2003)
6. M. Schneider, G. Vincent, J.D. Hogan, J.G. Spray, IEEE Trans. Plasma Sci. **43**, 1162-1166 (2015)

DOI: 10.1051/e3sconf/20171901041

7. J.D. Hogan, J.G. Spray, R.J. Rogers, G. Vincent, M. Schneider, Int. J. Impact Eng. **62**, 219-228 (2013)
8. S.A. Poniaev, S.V. Bobashev, B.G. Zhukov, R. O. Kurakin, A.I. Sedov, S.N. Izotov, S.L. Kulakov, M.N. Smirnova, Acta Astronaut. **109**, 162-165 (2015)
9. P. Lehmann, B. Reck, M.D. Vo, J. Behrens, IEEE Trans. Magn. **43**, 480-485 (2007)
10. L. Tang, J. He, L. Chen, S. Xia, D. Feng, IEEE Trans. Plasma Sci. **43**, 1585-1591 (2015)
11. L. Jin, B. Lei, Q. Zhang, R. Zhu, IEEE Trans. Plasma Sci. **43**, 1220-1224 (2015)
12. W. Xu, W. Yuan, Y. Sun, P. Yan, J. Li, IEEE Trans. Plasma Sci. **41**, 1542-1546 (2013)
13. L. Chen, J. He, Z. Xiao, Y. Pan, IEEE Trans. Plasma Sci. **39**, 417-421 (2011)
14. T. Boczar, A. Cichon, D. Wotzka, M. Kunicki, M Koziol, IEEE Trans. Dielectr. Electr. Insul. **24**, 120-128 (2017)
15. R. Beniak, O. Gudzenko, MMAR, 1104-1109 (2016)
16. J. F. Gieras, Z. J. Piech, B. Z. Tomczuk, *Linear synchronous motors*, (CRC Press, Taylor & Francis Group, 2011)
17. J. Domin, K. Kluszczyński, Prz. Elektrotech. **89**, 21-25 (2013)
18. A. Waindok, P. Piekielny, SELM, 51-52 (2013)
19. A. Waindok, P. Piekielny, Prz. Elektrotech. **92**, 246-249 (2016)
20. A. Waindok, P. Piekielny, Prz. Elektrotech. **93**, 152-155 (2017)

Mathematical model of optical signals emitted by electrical discharges occuring in electroinsulating oil

Michał Kozioł[1,*]

[1] Opole University of Technology, Prószkowska 76, 45-758 Opole, Poland

Abstract. The article presents a parametric model describing the registered distributions spectrum of optical radiation emitted by electrical discharges generated in the systems: the needle- needle, the needle-plate and in the system for surface discharges. Generation of electrical discharges and registration of the emitted radiation was carried out in three different electrical insulating oils: fabric new, operated (used) and operated with air bubbles. For registration of optical spectra in the range of ultraviolet, visible and near infrared a high resolution spectrophotometer was. The proposed mathematical model was developed in a regression procedure using gauss-sigmoid type function. The dependent variable was the intensity of the recorded optical signals. In order to estimate the optimal parameters of the model an evolutionary algorithm was used. The optimization procedure was performed in Matlab environment. For determination of the matching quality of theoretical parameters of the regression function to the empirical data determination coefficient R^2 was applied.

1 Introduction

One of the most important areas of research activities is elaboration of mathematical models which can be used for mathematical description of objects, technical and technological process and physical phenomena.

The application of mathematical modeling requires thorough knowledge of the analyzed physical phenomena and the appropriate selection of mathematical tools that will allow description of the phenomenon and estimation of the parameters of the applied equation [1].

Generation of electrical discharges in the insulating oils is accompanied by physical phenomena. The most important of these are: ultra-high frequency (UHF) [2,3], acoustic wave generation (AE) [4–7], optical radiation [8–10] and high-energy X-rays [11,12]. Based on these phenomena, measurement methods have been applied, which are used in diagnostics of insulation systems in power transformers [13–15].

As part of the work a parametric model has been developed, which describes the recorded spectral distributions of the emitted optical radiation for all investigated forms of electrical discharges (ED). The emitted optical radiation was recorded during the ED generation in the following systems: point-point, point-plate, surface discharges.

Measurements of the optical radiation were carried out using the high-resolution optical spectrophotometer HR 4000 manufactured by Ocean Optics that enables registration of optical signals in the radiation range UV-VIS-NIR (200 nm – 1100 nm). The light signal was delivered to the spectrophotometer using optic-fibre made of polymer fibre - POF type (Polymer Optical Fibre)[16–19].

The envelope of the optical spectrum was determined by separating the maximum values that were then interpolated with the nonlinear function to obtain a set of data corresponding to the recorded characteristics.

In the process of estimation model sigmoidal-gaussian regression function has been used. The dependent variable was the intensity of the recorded optical signals. The independent variable was the wavelength component. The model was based on a literature review [20,21].

Curently in research works to determine the function parameters evolutionary algorithms (EA) are commonly used. They are based on a stochastic search. EA as opposed to other methods are characterized by the following features [22,23]:

- possibility of solving tasks with complicated structure,
- possibility to use for tasks requiring complex parallel computing,
- allow to take into account limitations occurring for the analyzed space,
- simple implementation of a genetic algorithm that does not require extensive mathematical procedures.

Evolutionary methodology unfortunately is not devoid of defects, which can be counted [24]:

- lower efficiency of genetic algorithms compared to traditional gradient algorithms,
- complicated genetic algorithms require precise definition of a number of parameters that need to be implemented individually for each task,

* Corresponding author: m.koziol@po.opole.pl

- when each time the algorithm is run in the task, it usually causes some other solutions.

Process of creating a model consists of:

- analysis of the construction of an object or phenomenon,
- observation of mechanisms of occurrence of the analyzed phenomenon,
- selection of appropriate mathematical tools,
- verification of the created model by comparing its response with the input value.

2 Model of sigmoidal-gaussian regression function

Model of sigmoidal-gaussian regression function is described by equation (1):

$$MSG(\lambda) = \frac{A}{1 + ke^{-(\lambda-\mu)}} \sum_{i=1}^{n} a_i e^{-\left(\frac{\lambda-\mu_i}{c_i}\right)^2} \quad (1)$$

where:

$MSG(\lambda)$ – approximation of the sigmoidal-gaussian regression function,
$\lambda \in \mathbb{R}$ - wavelength,
$(A, a_i) \in \mathbb{R}$ – estimates of structural parameters, intensity,
$k > 0$ – estimation of the structural parameter, rising,
$c_i > 0$ – estimation of structural parameter, scale,
$(\mu, \mu_i) \in \mathbb{R}$ – estimates of structural parameters, location,
n – number of dominant components in spectral distribution.

The genetic algorithm available in the Matlab environment has been used to estimate optimal parameters of the optical spectrum model [25].

Based on the trial simulations and computing capabilities of the available hardware ,the following parameters of the genetic algorithm were adopted:

- number of designated variables: 15 – 18,
- number of generations: 1000,
- population size: 300,
- initial values: selected globally,
- Scope of the search: globally defined.

To determine the measure of matching regression function parameters with empirical values the determination coefficient R^2 was used (2) [26–28]:

$$R^2 = \left(\frac{\sum_{\forall i}(x_i - \mu)(\tilde{x}_i - \tilde{\mu})}{\sqrt{\sum_{\forall i}(x_i - \mu)^2 \sum_{\forall i}(\tilde{x}_i - \tilde{\mu})^2}} \right)^2 \quad (2)$$

where:

x_i – the value of the empirical dependent variable,
$\tilde{x}_i$ – the theoretical value (estimated) of the distribution,
μ – arithmetic mean of the empirical dependent variable,
$\tilde{\mu}$ – the theoretical arithmetical mean of the distribution,
$i = (1,2,...,n)$, $n = 3648$ – number of components that fall within one spectral distribution.

3 Applicability analysis of MSG model

3.1 Measuring conditions

Optical spectral recording was performed for three spark gaps systems: point-point, point-plate and surface discharges. Electrical discharges was generated for the alternating voltage (AC) and for three technical conditions of mineral oil: pure mineral oil, operated (used) mineral oil, operated mineral oil with air bubbles. During the measurements for point-point and point-plate type systems, the distance between the electrodes was constant and amounted to 6 mm. The electrodes for surface discharges type system were separated by a dielectric, which was a glass plate of 10 mm thickness.

The value of breakdown voltage (U_b) was defined for each spark gap system, before starting the measurements. Values of breakdown voltage for the applied spark gap are presented in Table 1.

Optical signals were recorded using a high-resolution optical spectrophotometer HR4000. To transmit the optical signals from spark gap systems to the spectrophotometer, optical fiber was used.

Table 1. Values of breakdown voltage for the applied spark gap systems.

Spark gap systems	Values of breakdown voltage U_b [kV]		
	Brand new mineral oil	Operated (used) mineral oil	Operated mineral oil with air bubbles
point-point	48.5	45.6	44.8
point-plate	47.5	45.2	46.3
Surface discharges	49.8	49.8	49.8

3.2 Application of the MSG model

The adopted model (MSG) was used to describe the optical emission distribution emitted by the ED generated in the point-point system for the three electrolytic insulating oils studied: pure mineral oil, exploited mineral oil, exploited mineral oil with air bubbles.

For comparison and reproduction of the results of measurements and analysis of the optical signals generated by the tested ED forms, the relative voltage values was applied. The relative voltage value was referred to brekdown voltage (U_b), which was determined for each of the spark gap systems.

Degree of matching models to the registered spectrum of optical radiation were analyzed for a relative supply voltage in the range of 0.60 U_b to 0.95 U_b, with a graduation rate of 0.05 U_b separately for each of the three tested insulating oils.

Fig. 1-3 show exemplary spectrum distributions of optical radiation, Recorded by optical spectrophotometry (blue color) and models (red color), emitted by ED in the tested insulating oils: pure (fig. 1), operated (fig. 2) and operated with air bubbles (Fig. 3). For comparative purposes, the numerical values of determinants of R^2 for the MSG model are summarized in Tables 2-4.

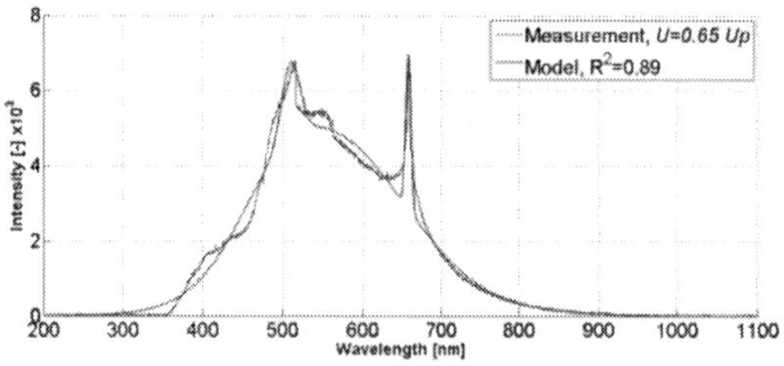

Fig. 1. Distribution of the spectrum of recorded optical radiation (blue color) and modeled (red) emitted by the ED generated in the point-point system in pure mineral oil for a relative supply voltage value of 0.65 U_b.

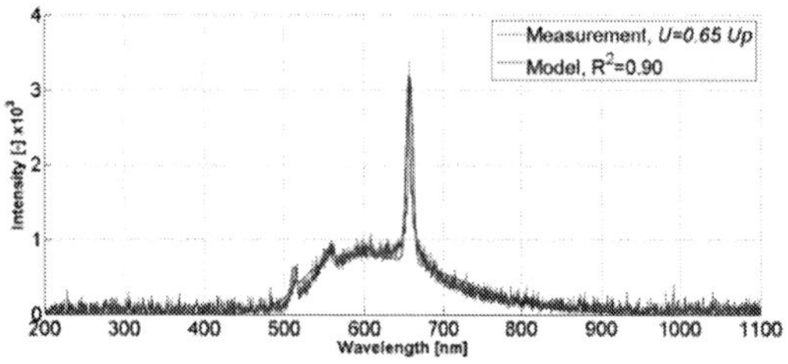

Fig. 2. Distribution of the spectrum of recorded optical radiation (blue color) and modeled (red) emitted by the ED generated in the point-point system in operated mineral oil for a relative supply voltage value of 0.65 U_b.

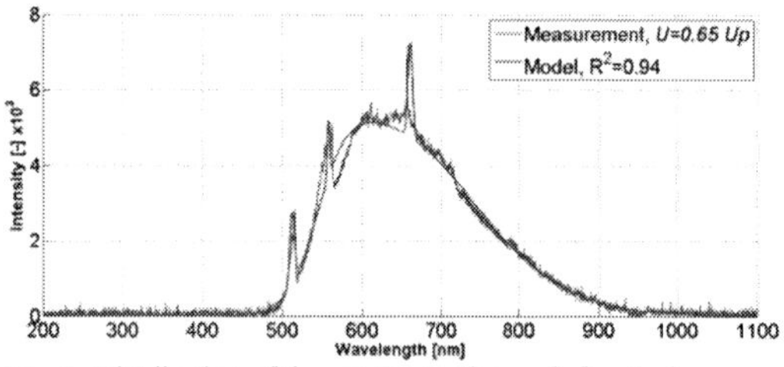

Fig. 3. Distribution of the spectrum of recorded optical radiation (blue color) and modeled (red) emitted by the ED generated in the point-point system in operated mineral oil with air bubbles for a relative supply voltage value of 0.65 U_b.

Table 2. Mean values of the determination factor R^2 calculated for the model (MSG) used in the point-point system, for the three tested insulating oils.

Relative voltage value	Value of determination coefficient R^2		
	Brand new mineral oil	**Operated (used) mineral oil**	**Operated mineral oil with air bubbles**
0.60	0.84	0.82	0.91
0.65	0.87	0.91	0.95
0.70	0.86	0.96	0.84
0.75	0.95	0.93	0.93
0.80	0.92	0.90	0.90
0.85	0.96	0.89	0.87
0.90	0.93	0.94	0.96
0.95	0.92	0.89	0.88
Median	**0.92**	**0.90**	**0.90**
Arithmetic average	**0.91**	**0.90**	**0.90**
Standard deviation	**0.04**	**0.04**	**0.04**

The determined R^2 values for individual power supply values were averaged (arithmetic mean) from the 15 analyzed measurements for each case.

In an analogous manner, the adopted model (MSG) was used to describe the distribution of emitted optical radiation generated by ED in the point-plate system and in the surface discharge system for all tested insulating oils. Examples of spectral distributions of optical radiation are presented in Fig. 4-9.

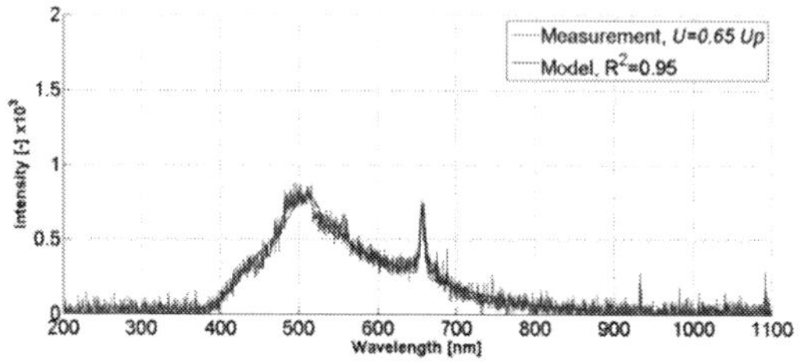

Fig. 4. Distribution of the spectrum of recorded optical radiation (blue color) and modeled (red) emitted by the ED generated in the point-plate system in pure mineral oil for a relative supply voltage value of 0.65 U_b

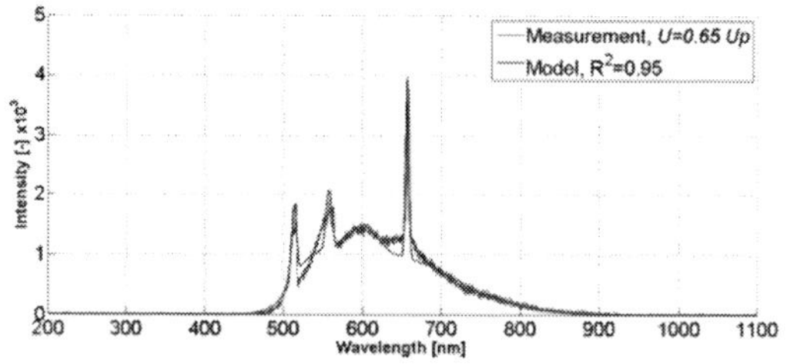

Fig. 5. Distribution of the spectrum of recorded optical radiation (blue color) and modeled (red) emitted by the ED generated in the point-plate system in operated mineral oil for a relative supply voltage value of 0.65 U_b.

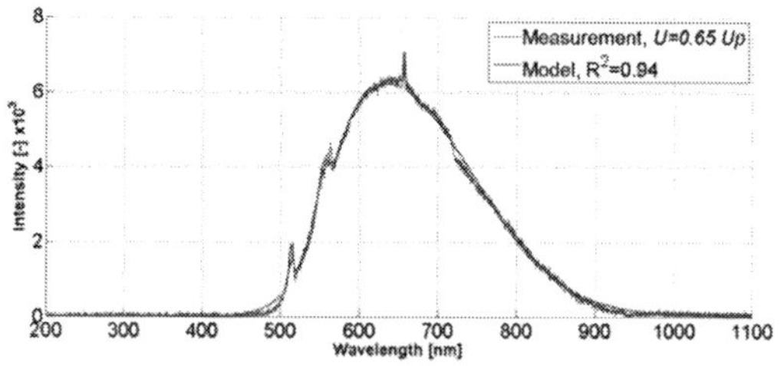

Fig. 6. Distribution of the spectrum of recorded optical radiation (blue color) and modeled (red) emitted by the ED generated in the point-point system in operated mineral oil with air bubbles for a relative supply voltage value of 0.65 U_b.

Table 3. Mean values of the determination factor R^2 calculated for the model (MSG) used in the point-plate system, for the three tested insulating oils.

Relative voltage value	Value of determination coefficient R^2		
	Brand new mineral oil	**Operated (used) mineral oil**	**Operated mineral oil with air bubbles**
0.60	0.92	0.89	0.91
0.65	0.91	0.94	0.93
0.70	0.95	0.91	0.96
0.75	0.94	0.93	0.94
0.80	0.96	0.92	0.95
0.85	0.91	0.90	0.97
0.90	0.89	0.87	0.96
0.95	0.90	0.94	0.95
Median	**0.91**	**0.91**	**0.95**
Arithmetic average	**0.92**	**0.91**	**0.95**
Standard deviation	**0.02**	**0.02**	**0.02**

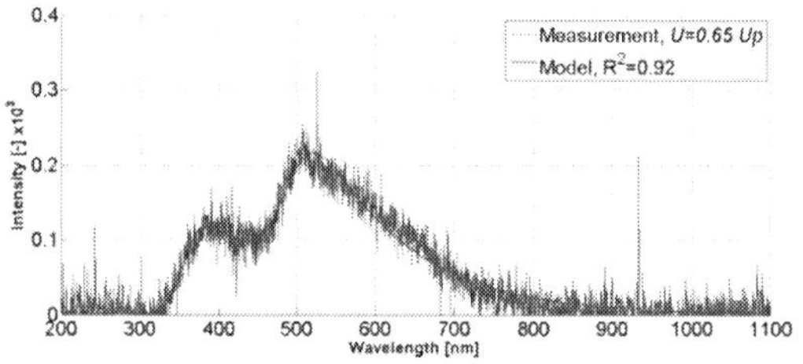

Fig. 7. Distribution of the spectrum of recorded optical radiation (blue color) and modeled (red) emitted by the ED generated in the surface discharge system in pure mineral oil for a relative supply voltage value of 0.65 U_b.

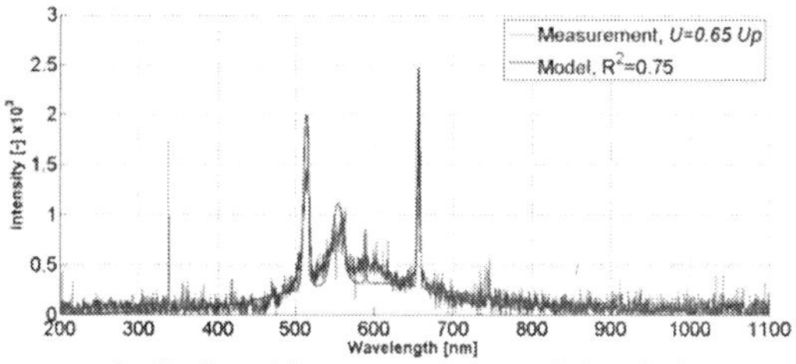

Fig. 8. Distribution of the spectrum of recorded optical radiation (blue color) and modeled (red) emitted by the ED generated in the surface discharge system in operated mineral oil for a relative supply voltage value of 0.65 U_b.

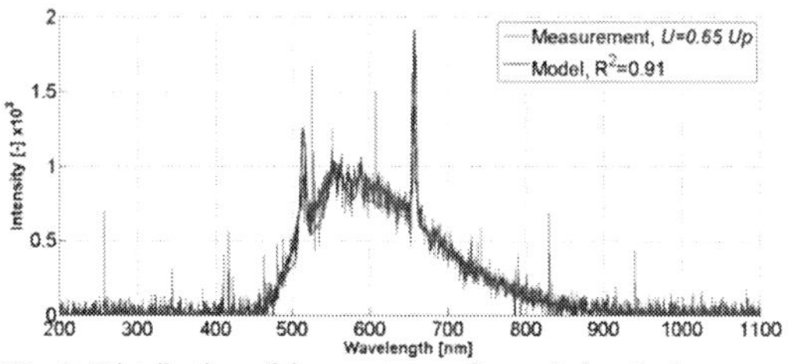

Fig. 9. Distribution of the spectrum of recorded optical radiation (blue color) and modeled (red) emitted by the ED generated in the surface discharge system in operated mineral oil with air bubbles for a relative supply voltage value of 0.65 U_b.

Table 4. Mean values of the determination factor R^2 calculated for the model (MSG) used in the surface discharge system, for the three tested insulating oils.

Relative voltage value	Value of determination coefficient R^2		
	Brand new mineral oil	**Operated (used) mineral oil**	**Operated mineral oil with air bubbles**
0.60	0.91	0.79	0.83
0.65	0.97	0.85	0.96
0.70	0.98	0.87	0.91
0.75	0.97	0.90	0.89
0.80	0.96	0.90	0.87
0.85	0.97	0.94	0.91
0.90	0.95	0.85	0.89
0.95	0.94	0.84	0.90
Median	**0.96**	**0.86**	**0.89**
Arithmetic average	**0.96**	**0.87**	**0.89**
Standard deviation	**0.02**	**0.05**	**0.04**

4 Comparative analysis of the obtained results

For comparative purposes, Figs. 10-12 show arithmetic mean and median values for determinants R^2, separately for each of the three tested insulating oils.

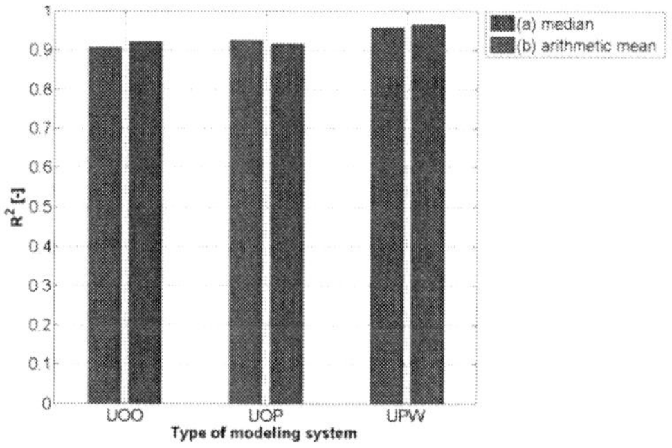

Fig. 10. Median (a) and arithmetic mean (b) of R^2, calculated for ED generating systems: point-point (UOO), point-plate (UOP) and surface discharge system (UPW); generated in pure mineral oil.

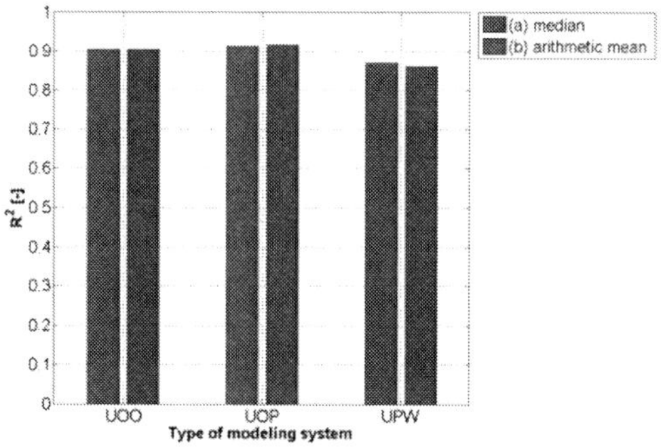

Fig. 11. Median (a) and arithmetic mean (b) of R^2, calculated for ED generating systems: point-point (UOO), point-plate (UOP) and surface discharge system (UPW); generated in operated mineral oil.

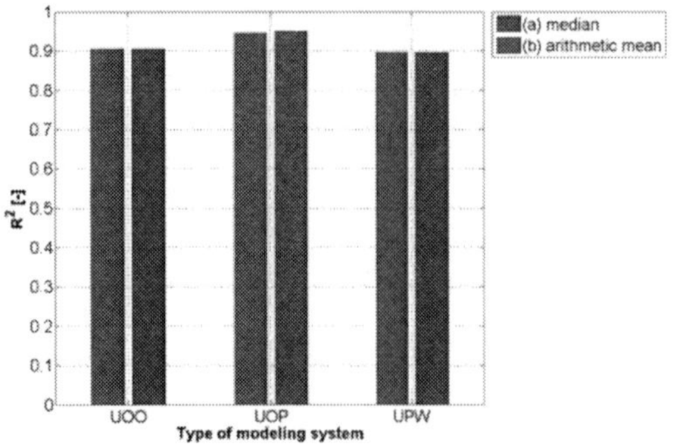

Fig. 12. Median (a) and arithmetic mean (b) of R^2, calculated for ED generating systems: point-point (UOO), point-plate (UOP) and surface discharge system (UPW); generated in operated mineral oil with air bubbles.

Based on the comparative analysis of the determinants of R^2, it can be concluded that the applied mathematical model (MSG) is characterized by good and very good matching of parameters to empirical values, which is confirmed by arithmetic meanings within the range of 0.8 - 1.0.

Mathematical model for the entire range of the emitted optical signal consists of the models components sum of optical spectrum structures. These structures consist of: continuous spectrum, which is described by the sigmoid function and band spectrum, described by the Gaussian function.

The model obtained can be finally described using function (3). The most important parameter of this dependence of coefficient values are: μ_{11}, μ_{12}, μ_{13}. Based on them, it is possible to determine the dominant component wavelengths present in the analyzed spectral range.

$$MSG(\lambda) = \left(\frac{A_1}{1+k_1 e^{-(\lambda-\mu_1)}}\right)\left(\frac{A_2}{1+k_2 e^{-(\lambda-\mu_2)}}\right) + \left(a_{11} e^{-\left(\frac{\lambda-\mu_{11}}{c_{11}}\right)^2}\right) + \left(a_{12} e^{-\left(\frac{\lambda-\mu_{12}}{c_{12}}\right)^2}\right) + \left(a_{13} e^{-\left(\frac{\lambda-\mu_{13}}{c_{13}}\right)^2}\right) \quad (3)$$

Values of equation (3) were determined in the optimization process using a genetic algorithm (GA) implemented in a Matlab environment. Tables 5-6 shows the sample values of parameters that were calculated for the spectral distribution of the optical radiation emitted in the point-point system.

Table 5. Examples of mathematical model parameters (MSG) values determined for the entire spectral range of emitted radiation generated in the point-point system.

Parameter type	Parameter value	Parameter type	Parameter value
A_1	3.35	a_{11}	5.70
k_1	0.03	**μ_{11}**	**514.96**
μ_1	510.54	c_{11}	10.11
A_2	3.60		
k_2	0.01		
μ_2	722.89		

Table 6. Examples of mathematical model parameters (MSG) values determined for the entire spectral range of emitted radiation generated in the point-point system.

Parameter type	Parameter value	Parameter type	Parameter value
a_{12}	1.50	a_{13}	1.02
μ_{12}	**562.05**	**μ_{13}**	**656.10**
c_{12}	7.72	c_{13}	4.99

Based on the obtained values of determinants of R^2, it can be concluded that the proposed model (MSG) and the procedures used to determine its parameters were chosen correctly. The model can be used to determine the dominant values of wavelength components. This makes it possible to apply an adopted mathematical model to a data block that may be a part of an expert system.

5 Conclusions

The purpose of creating mathematical models is to describe the behavior of a model object as an influencing factor. Such description allows to predict the behavior of an object with changes in influencing factors. As a result, is possible to analyze and design objects using only their mathematical description or simulation model.

Proposed in this paper mathematical model for describing optical signals emitted by electrical discharges in insulating oil can be used to build the knowledge base.

Obtained results with optical signal modeling can also be used to build a database consisting of comparative patterns of emited optical spectra. Such a solution can be a component for designing an expert system that will be dedicated to diagnostics of high voltage insulation systems.

References

1. J.M. González-Barrios, S. Ruiz-Velasco, Metrika **61** (2005)
2. S.M. Hoek, A. Kraetge, M. Koch, O. Kessler, M. Heindl, in Proc. of 2012 IEEE Int. Conf. Condition Monit. and Diagnosis, *CMD 2012* (2012) doi:10.1109/CMD.2012.6416438
3. S. Coenen, S. Tenbohlen, IEEE Trans. Dielectr. Electr. Insul. **19** (2012)
4. T. Boczar, A. Cichoń, S. Borucki, IEEE Trans. Dielectr. Electr. Insul. **21** (2014)
5. M. Kunicki, A. Cichoń, S. Borucki, Arch. Acoust. **41** (2016)
6. G. Kil, D. Park, I. Kim, S. Choi, WSEAS Trans. Power Syst. **3** (2008)
7. H. Majchrzak, A. Cichoń, S. Borucki, Arch. Acoust. **42** (2017)
8. P. Frącz, T. Boczar, D. Zmarzly, T. Szczyrba, in Acta Phys. Pol. A **124** (2013)
9. P. Rózga, P. Tabaka, in 2015 IEEE 11th Int. Conf.

Prop. Appl. of Dielectr. Mater. (ICPADM, 2015)

10. D. Siebler, P. Rohwetter, R. Brusenbach, R. Plath, Procedia Eng. **120** (2015)
11. Ł. Nagi, D. Zmarzly, T. Boczar, P. Frącz, IEEE Trans. Dielectr. Electr. Insul. **23** (2016)
12. S. Adili, L. G. Herrmann, C. M. Franck, in Annual Report - Conf. Electr. Insul. Dielectr. Phenom., CEIDP (2012) doi:10.1109/CEIDP.2012.6378737
13. Z. Nadolny, G. Dombek, P. Przybylek, in *IEEE* Conf. Electr. Insul. Dielectr. Phenom. (2016)
14. P.M. Mitchinson, P.L. Lewin, P. Jarman, IEEE Electr. Insul. Mag. **29** (2013)
15. T. Boczar, A. Cichon, D. Wotzka, M. Kunicki, M. Koziol, IEEE Trans. Dielectr. Electr. Insul. **24** (2017)
16. M. Kozioł, D. Wotzka, T. Boczar, P. Frącz, J. Spectrosc. **2016** (2016)
17. T. Boczar, D. Zmarzły, P. Frącz, IEEE 11th Int. Conf. Prop. Appl. Dielectr. Mater. (2015) doi:10.1109/ICPADM.2015.7295378
18. T. Boczar, D. Zmarzly, in IEEE Int. Conf. Dielectr. Liq. (2005) doi:10.1109/ICDL.2005.1490036
19. S. Biswas, C. Koley, B. Chatterjee, S. Chakravorti, IEEE Trans. Dielectr. Electr. Insul. **19** (2012)
20. J. Saragih, in Proc. of the IEEE Comput. Soc. Conf. Comput. Vision and Pattern Recognit. (2011) doi:10.1109/CVPR.2011.5995618
21. D. Wotzka, A. Cichoń, T. Boczar, Arch. Acoust. **37** (2012)
22. T. Bartz-Beielstein, J. Branke, J. Mehnen, O. Mersmann, Wiley Interdiscip. Rev. Data Min. Knowl. Discovery **4** (2014)
23. C. Gagn, J. Mach. Learn. Res. **13** (2012)
24. L. Dioşan, M. Oltean, Genet. Program. Evolvable Mach. **10** (2009)
25. M. Kumar, M. Husian, N. Upreti, D. Gupta, Int. J. Inf. Technol. Knowl. Manag. **2** (2010)
26. O. Harel, J. Appl. Stat. **36** (2009)
27. R. Anderson-Sprecher, Am. Stat. **48** (1994)
28. R. C. Quinino, E. A. Reis, L. F. Bessegato, Teach. Stat. **35** (2013)

DOI: 10.1051/e3sconf/20171901043

Noise variance estimation for Kalman filter

Ryszard Beniak[1], *Oleksandr* Gudzenko[1,*], and *Tomasz* Pyka[1]

[1]Opole University of Technology, Prószkowska 76, 45-758 Opole, Poland

Abstract. In this paper, we propose an algorithm that evaluates noise variance with a numerical integration method. For noise variance estimation, we use Krogh method with a variable integration step. In line with common practice, we limit our study to fourth-order method. First, we perform simulation tests for randomly generated signals, related to the transition state and steady state. Next, we formulate three methodologies (research hypotheses) of noise variance estimation, and then compare their efficiency.

1 Introduction

Noise variance estimation is one of key unresolved issues associated with signal filtering [1-4]. If the noise variance is estimated incorrectly, and the estimation output is then used for signal filtering, real time measurement results might be distorted. To estimate the measurement noise one may calculate, for instance, measurement variance, as in [1]. Estimation of the process noise, however, is a way more complicated. The latter might be carried out by means of Bayesian, maximum likelihood, correlation and covariance matching [1, 4, 5]. Another acceptable approach is to assume that the process noise is constant, while taking into account the varying measurement noise.

Kalman filter is one of the most frequently used filters [6-8]. It is an extended version of recursive filtering [1, 6]. The latter is applied only for the measurement noise, while the former can be used also for the process noise. However, to use Kalman filter we need to have precise information about the object (in other words, we might encounter problems resulting from the observer effect). However, due to varying degree of the object's complexity, this is not always possible. Having complete information about the object, we can select the elements of the positive definite covariance matrix after they have been modified. If we lack more specific information about the object, we might define the elements of the covariance matrix incorrectly, and, in extreme cases, the matrix may turn out not positive definite. In consequence, the measurement results might prove imprecise.

This paper presents a method for noise variance estimation based on verification of three hypotheses, with Krogh's [9] method for numerical integration applied in each case. Small degree of complexity allows to apply these methods in real time measurement. We use values corresponding to individual hypotheses to explain the measurement noise **R**. The process noise is assumed to be constant and equal to **Q**.

2 Krogh algorithm

The Krogh algorithm [9] was first described in 1960s and is usually applied for numerical integration. One of its unique features is that it allows for changes in values of determinants in each timestep. At the same time, in spite of the change in determinants, their convergence of the individual processes of differential equation solutions is preserved. The Krogh algorithm is defined as follows [9].

May w determine a polynomial given for time points t_n, $t_{n-1}, \ldots, t_{n-q+1}$. The polynomial is defined by Newton forward differences formulae:

$$M_{q-1,n}(t) = w[t_n] - (t - t_n)w[t_n, t_{n-1}] + \ldots \\ (t - t_n)(t - t_{n-1})\ldots(t - t_{n-q+2})w[t_n, t_{n-1}, \ldots, t_{n-q+1}], \quad (1)$$

where

$$w[t_n, t_{n-1}, \ldots, t_{n-i}] = \begin{cases} w(t_n) & i = 0 \\ \dfrac{w[t_n, \ldots, t_{n-i+1}] - w[t_{n-1}, \ldots, t_{n-i}]}{t_n - t_{n-i}} & i = 1,2\ldots \end{cases} \quad (2)$$

Consider a following polynomial $M^*_{q,n}(t)$

$$M^*_{q,n}(t) = M_{q-1,n}(t) + \ldots \\ (t - t_n)(t - t_{n-1})\ldots(t - t_{n-q+1})w[t_{n+1}, t_n, \ldots, t_{n-q+1}], \quad (3)$$

where M defines the predictor, while M^* determines the corrector. By inserting additional symbols

* Corresponding author: o.gudzenko@doktorant.po.edu.pl

 DOI: 10.1051/e3sconf/20171901043

$$\begin{aligned} &\Delta t_i = t_i - t_{i-1} \\ &\tau = (t - t_n)/\Delta t_{n+1} \\ &\xi_i(n+1) = \Delta t_{n+1} + \Delta t_n + \ldots \Delta t_{n+1-i} \\ &\beta_0(n+1) = 1 \\ &\beta_i(n+1) = \left[\xi_0(n+1)\xi_1(n+1)\ldots\xi_{i-1}(n+1)\right]/\left[\xi_0(n)\ldots\xi_{i-1}(n)\right] \\ &\varphi_0(n) = w(t_n) \\ &\varphi_i(n) = \xi_0(n)\xi_1(n)\ldots\xi_{i-1}(n)w[t_n, t_{n-1}, \ldots t_{n-i}] \end{aligned} \tag{4}$$

where i denotes the concurrent sample numbers from equations (1) to (4), we can yield the dependence which defines the range of the polynomial coefficients

$$\varphi_{i+1}(n+1) = \varphi_i(n+1) - \beta_i(n+1)\varphi_i(n). \tag{5}$$

3 Algorithm

Below we present the algorithm for the measurement variance and the hypotheses we aim to test.

The algorithms consist of five main steps, which three hypotheses are based on:

1. Calculating coefficients ϕ according to (5).
2. Calculating the variance for each of the coefficients ϕ.
3. Estimating the mutual dependencies between the calculated variances and ordering them from the largest (r_{max}) to the smallest one (r_{min}); the number of variances are denoted with l.
4. Calculating the arithmetic mean $\bar{c}$ from the mutual dependencies from the third step.
5. Calculating the variance of the signal measured with noise as σ_{sn}.

Hypothesis 1:

$$h_1 = \frac{1}{l}\bar{c}\,\sigma_{sn}, \tag{6}$$

Hypothesis 2:

$$h_2 = \begin{cases} (r_{\max} - r_{\min})/r_{\max} & r_{\max} \geq 1 \\ (r_{\max} - r_{\min})\sigma_{sn} & r_{\max} < 1 \end{cases} \tag{7}$$

Hypothesis 3:

$$h_3 = \begin{cases} \sqrt{\left(\frac{1}{l}\sigma_{sn} - \sigma_{10}\right)^2} & n \geq 10 \\ \frac{1}{l}\sigma_{sn} & n < 10 \end{cases} \tag{8}$$

where σ_{10} denotes the variance of the first 10 samples.

In this paper, we use a modified form of Kalman filter, which can be written in short as follows [6, 8]

$$\hat{\mathbf{x}}_{k/k-1} = \mathbf{A}\hat{\mathbf{x}}_{k-1/k-1} + \mathbf{B}\mathbf{u}_k, \tag{9}$$

$$\mathbf{P}_{k/k-1} = \mathbf{A}\mathbf{P}_{k-1/k}\mathbf{A}^T + \mathbf{Q}, \tag{10}$$

$$\hat{\mathbf{x}}_{k/k} = \mathbf{K}_k\left(\mathbf{z}_k - \mathbf{H}\hat{\mathbf{x}}_{k/k-1}\right), \tag{11}$$

$$\mathbf{K}_k = \mathbf{P}_{k/k-1}\mathbf{H}^T\left(\mathbf{H}\mathbf{P}_{k/k-1}\mathbf{H}^T + \mathbf{R}\right)^{-1}, \tag{12}$$

$$\mathbf{P}_{k/k} = (\mathbf{I} - \mathbf{K}_k\mathbf{H})\mathbf{P}_{k/k-1}, \tag{13}$$

where **Q** matrix of the process noise, **R** matrix of the measurement noise depending on the hypothesis, **K** denotes the Kalman gain, **A** – state transition matrix, **B** – control matrix, **P** – state variance matrix, **H** – measurement matrix, **z** – measurement variables, **I** – identity matrix, $\hat{\mathbf{x}}$ – estimated state, index k denotes a sample in k-th time point.

Next, in order to verify the proposed hypotheses, we have carried out tests for randomly generated signals, both for steady state and transition states.

4 Simulation results

In this Chapter, we present simulation results for a signal generated randomly for the transition state. The signal was generated by means of the Taylor series

$$x(k) = \sum_{n=0}^{m} \frac{f^{(n)}(a)}{n!}(k-a)^n. \tag{14}$$

We restricted our calculations to the fourth order derivative (m=4). The value of the parameter a was randomly drawn from the interval <-100, 100>. Following that, a white noise is superimposed on the randomly generated signal.

Next, we estimated the noise using the algorithm from Chapter 3 and the proposed hypotheses. The results are presented in Fig. 1. In addition, in this Figure we show the variance of the noise with and without the source signal – σ_{sn}, σ_n, respectively.

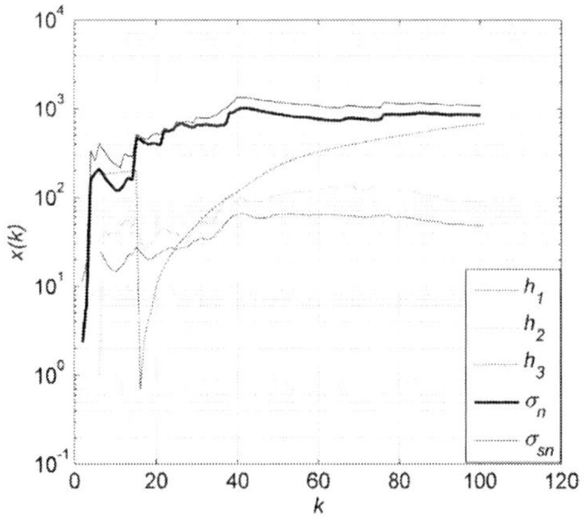

Fig. 1. Verification of the proposed hypotheses for the transition state.

From Fig. 1 we can conclude that, based on Hypothesis 3, we can obtain the most accurate estimation of the noise variance.

 DOI: 10.1051/e3sconf/20171901043

Next, based on Fourier series, we generate the signal for the steady state

$$x(k) = a_0 + \sum_{n=1}^{m} \left(a_n \cos nk + b_n \sin nk\right), \qquad (15)$$

where coefficients a_0, a_n and b_n are drawn randomly from the interval <-100, 100> and then, consistently, we restricted the series to a fourth order (m=4).

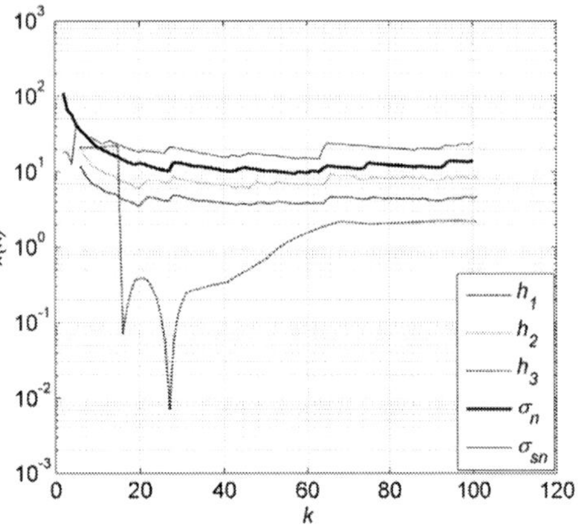

Fig. 2. Verification of the proposed hypotheses for the transition state.

As the former case, we imposed a white noise on the randomly generated signal. The results, along with the variance of the distorted source signal and the variance of the noise itself, are depicted in Fig. 2. The best estimation of the noise variance in the steady state was obtained when Hypothesis 2 was applied.

To confirm that the above conclusions are correct, we carried out additional verification, for distorted signal of sinusoidal, sawtooth, rectangular and constant waveforms. A white noise was superimposed on the sinusoidal signal. Next, we checked the proposed hypotheses. The result is shown in Fig. 3.

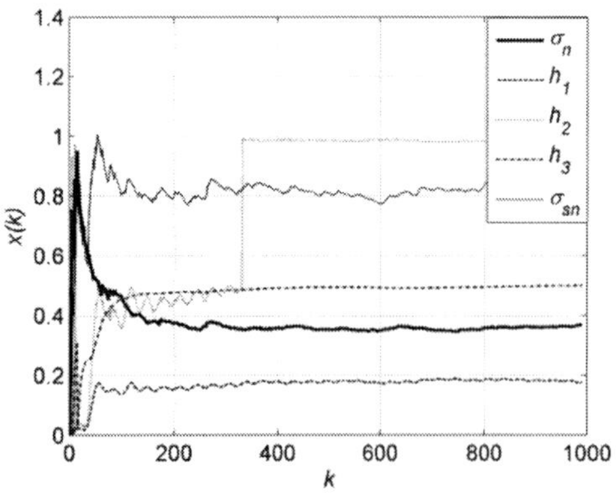

Fig. 3. Verification of the proposed hypotheses for the sinusoidal signal with a superimposed white noise.

We repeated the above procedure for the sawtooth signal with a superimposed white noise. The result is presented in Fig. 4. We conclude that by applying either Hypothesis 2 or 3, we yielded the most accurate estimation for sinusoidal and sawtooth shape. We also verified the rectangular and constant signal. We imposed a white noise on both and checked if the algorithm estimates the noise variance correctly.

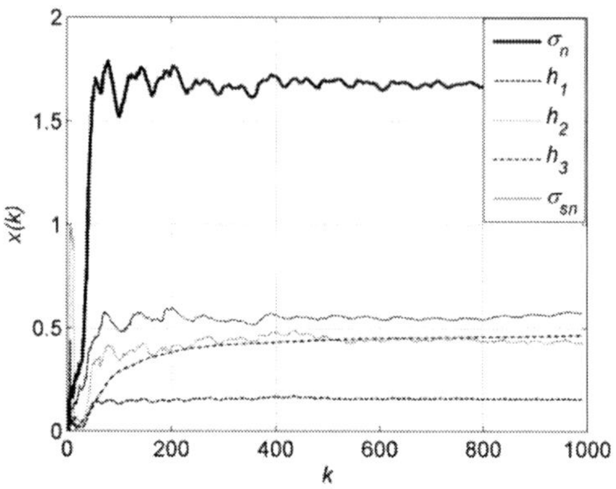

Fig. 4. Verification of the proposed hypotheses for the saw signal with a superimposed white noise.

The results are presented in Fig. 5 (for the rectangular signal) and Fig. 6 (constant signal). Both Hypotheses, 2 and 3, estimated the noise variance best.

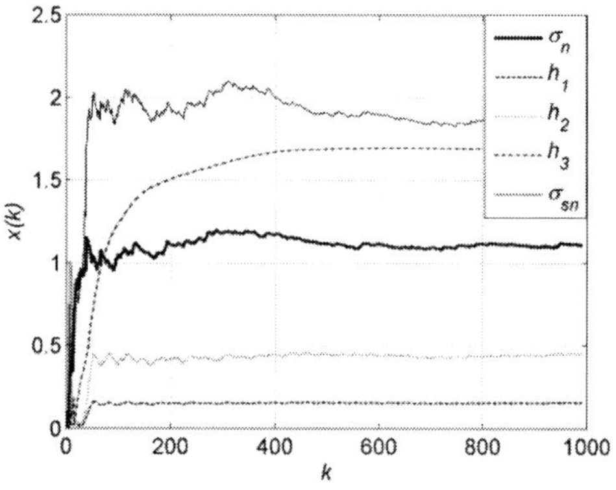

Fig. 5. Verification of the proposed hypotheses for the rectangular signal with a superimposed white noise.

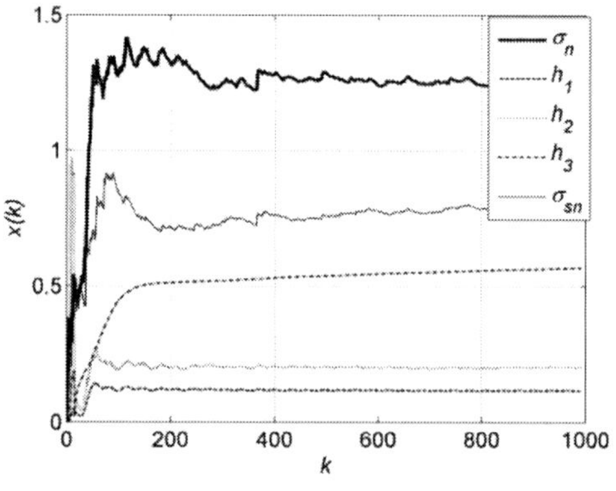

Fig. 6. Verification of the proposed hypotheses for the constant signal with a superimposed white noise.

We can observe that in most cases the proposed hypotheses allow for an accurate estimation of the noise variance, with the best results obtained for Hypotheses 2 and 3. The noise was estimated particularly well in case of randomly generated signals, both for steady state (Hypotheses 2 and 3) and transient state.

Next, we confronted the above results with the results obtained from measurements performed for a real medium-power industrial drive with an SEE355ML4BS frequency inverter. The characteristics of the drive are as follows. U_Δ=400 V, I_Δ=523 A, P_n=315 kW, T_n=2020 Nm, n_n=1489 rpm, η=96.6 %, I_d=6.2 kgm^2, where U_Δ denotes line-to-line voltage, I_Δ line current, P_n nominal power, T_n nominal torque, n_n nominal rotary velocity, I_d moment of inertia on the motor shaft, η efficiency. The considered object also comprised a 20W39MX3GV pomp of capacity equal to 550m^3/h, liquid column of 130 m and rotary velocity equal to 1490 rpm.

Next, we performed measurement of the converter drive described above. We used a measurement device consisting of two NI 6133 measurement cards, which enable 16-channel measuring with a maximum sampling rate of 1 MS/s. We applied CWT30LF Rogowski coils from PEM with measurement error of 0.2%. The sampling rate was lowered to 400 kS/s. We yielded a very distorted result. Hence, we applied linear Kalman filter, assuming that the object's model is completely unknown. Therefore, the filtering depending exclusively on the assumed noise variance and measurement results.

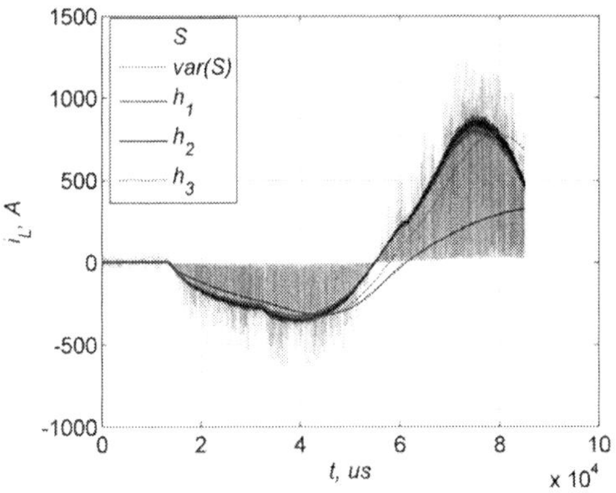

Fig. 7. Measurement results of the line current i_{L1}.

Figs. 7 and 8 show the measurement results S of the line current i_{L1} and i_{L2}, both raw and after Kalman filter has been applied. We assumed that the process noise **Q** is constant. The measurement noise **R** is estimated based on the three hypotheses (h_1, h_2 and h_3, respectively) and the variance of the measurement results - $var(S)$. Such an assumption is frequently applied by engineers; yet, as Figs. 7 and 8 show, it proved to be an oversimplification, and to yield inaccurate results. In extreme cases, such an assumption might distort measurement results. However, from Figs. 7 and 8 we may also conclude that the hypotheses, particularly Hypothesis 2, proposed in this paper estimate the measurement noise precisely. For Hypotheses 1 and 3, the results are quite smooth. However, in case of these hypotheses, we can observe a phase shift, which is unwelcome.

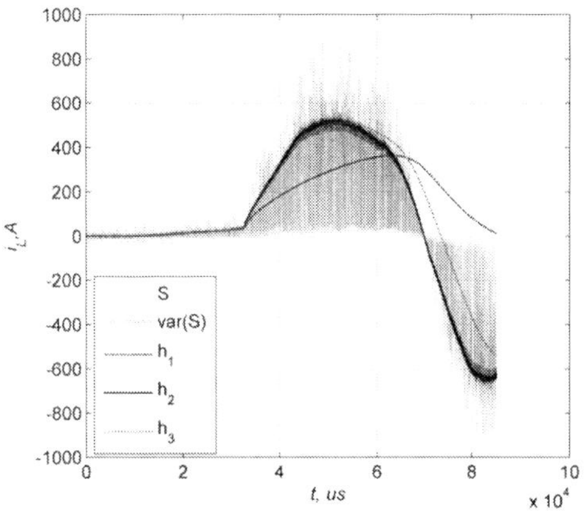

Fig. 8. Measurement results of the line current i_{L2}.

Thus, both hypotheses should not be applied in practice. Therefore, our results indicate that Hypothesis 2 may prove a handy tool when estimating the measurement noise by means of Kalman filter.

5 Conclusions

In this paper, we compared three hypotheses, which facilitate estimating the measurement noise. We applied the estimation results in Kalman filter. Next, confronted the results from theoretical considerations with the results from measurements, carried out in working conditions, and obtain a satisfactory outcome, particularly for one of the hypotheses. Taking into account the above as well as simplicity of our approach, it may be recommend for real-time measurements.

References

1. S. M. Mohan, N. Naik, R.M.O. Gemson, M.R. Ananthasayanam, TR/EE2015/401, (2015)
2. R. Bos, X. Bombois and P. M. J. van der Hof, *Proc. of the* 16th *IFAC World Congress*, Czech Republic, (2005)
3. Y. Bulut, D. Vines-Cavanaugh, D. Bernal, *Proc. of the IMAX*-XXVIII, USA, (2010)
4. M. Nilsson, *Proc. of the Reglermöte*, Sweden (2006)
5. S. Kosanam and D. Simon, *Proc. of the IASTED Int. Conf. Intelligent Syst. Control* (2004))
6. T. Zielinski, *Digital signal processing from theory to application* (WKiŁ, Warsaw, 2005) (in Polish)
7. B. M. Åkesson J. B. Jørgensen, N. K. Poulsen, S. B. Jørgensen, *Proc. of the* ESCAPE17, (2007)
8. K. H. Eom, S. J. Lee, Y. S. Kyung, Ch. W. Lee, M. Ch. Kim and K. K. Jung, Sensors **11**, (2011)
9. F. T. Krogh, *Numerical Solution of Ordinary Differential Equations* (Springer, New York, 1974)

Comparative analysis of cellulose pressboard and aramid paper used in air insulation systems of high-voltage devices

Tomasz Turba[1,*], and *Paweł* Frącz[1]

[1]Opole University of Technology, Faculty of Electrical Engineering, Automatic Control and Computer Science, Prószkowska 76, 45-758 Opole, Poland

Abstract. The paper presents results of a comparative analysis of parameters of two kinds of solid dielectrics used in air insulation systems to prevent occurring partial discharges. The research works regarded materials made of: cellulose pressboard and aramid paper. All measurements were performed under laboratory conditions by changing the value of partial discharges generation voltage until breakdown occurred in the inhomogeneous environment that was simulated using needle-plate (made of copper) electrode system. The main contribution which resulted from studies is a statement that potential use of aramid paper as a dielectric can extend the life of a high voltage electric device as compared to standard cellulose pressboard usage due to higher electric resistances to breakdown or detection of corona voltage. Results shown that the aramid paper has greater electric resistance to breakdown in comparison to cellulose with no difference between both on detecting corona of partial discharge.

1 Introduction

The research concerns a field of science which is related to generation of electric discharges on surfaces of high-voltage (HV) air insulation systems. Important problem is ageing process which progress during operation of equipment causing deterioration of insulating properties. Main ageing factors include: UV radiation, temperature fluctuations, ozone and nitrogen oxides, rime deposition and partial discharges (PD) [1]. Natural insulation of most air insulation systems provides atmospheric air having dielectric strength equal to 32 kV / cm in normal conditions. High voltage lines (e.g. 110 kV), distributions line along with power plants with substations form the key elements of a power system, which purpose is to produce and transmit energy to consumers. Interruptions in electricity supply can lead to financial losses, so all components of the power system should work reliably, safe and at the lower possible cost. Thus, to improve the dielectric strength of high voltage equipment it is necessary to use barriers of solid dielectric materials by placing them between electrodes in perpendicular or parallel manner [2]. The main objective of research conducted within this scope is to compare insulating properties to PD resistance of cellulose pressboard and aramid paper to determine as accurately as possible with the lowest installation cost whether aramid paper could succeed cellulose in normal operations. The studies involved measurements of electrical discharges occurring on the surface of both dielectrics. Based on the gathered data the initial voltage and breakdown values were estimated and compared to confirm the hypothesis.

2 Materials and methodology for comparative analysis

2.1 Metrological parameters of the measuring system

There are many measurement methods of detecting partial discharges occurring in insulation systems of electrical power machines. Methods can be divided into two groups: invasive and non-invasive methods. Invasive methods can detect and record PD currents with help of measuring probes expressing the PD value. This involves shutting off power supply for the device during connection/disconnection of probes which is a huge disadvantage [3]. Non-invasive methods include: acoustic emission [4,5], optical emission [6,7], electromagnetic or radio-frequency detection [8,9]. This paper concerns the use of invasive, classical method. Fig. 1 presents circuit diagram of measuring system using the experimental methodology which involved construction of a needle-plate electrodes system supplied by a 110 kV test transformer to simulate inhomogeneous environment. Between the needle and the plate a double bakelite plate with a circular hole of diameter s equal to 30 cm inside is placed. Between the plates the dielectric sample is installed as can be seen on Fig. 2. Consecutive measuring of electric strength is done by drifting apart the distance a [cm] between dielectric sample and edge electrode.
Next step after breakdown is to extend distance from plate electrode to new sample series (d [cm]). Incrementation by 1 cm after breakdown voltage is set

* Corresponding authors: tturba@gmail.com

 DOI: 10.1051/e3sconf/20171901044

for extending the distance from the edge electrode with constant value of d for a new series of measurements.

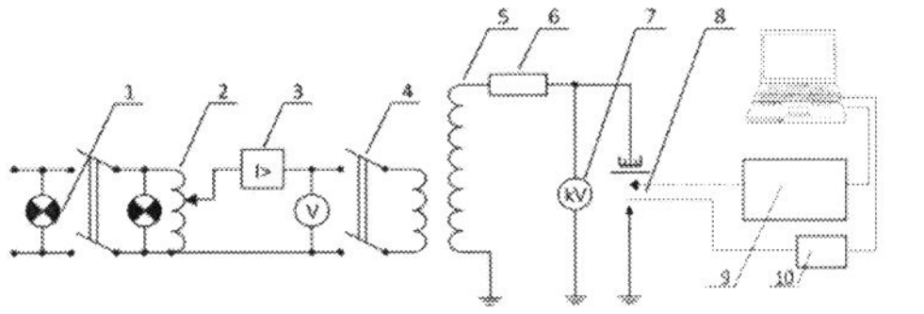

Fig. 1. Circuit diagram of the measuring system: 1 – pilot light, 2 – voltage regulator, 3 – overcurrent protection, 4 – circuit breaker, 5 – test transformer, 6 – protective resistor, 7 – electrostatic voltmeter, 8 – edge-plate configuration.

Next, the distance d is incremented by 1 cm and new measurement series is processed with extending the needle electrode 1 cm for each trial beginning again from 1 cm. Three measurements for each distance were processed due to increase the probability of correct results. Each voltage for distance is a sum of arithmetic average value.

Fig. 2. Measurement system (needle-plate configuration) with aramid paper installed. Distances a, d, s diameter and r radius are shown.

The distance a is in range from 1 to 15 cm and the distance d from sample to plate varies from 1 to 12 cm. When the distance a is fully measured (every centimetre had its breakdown voltage value measured), the d is being incremented by 1 cm. If the breakdown voltage of 110 kV power transformer cannot be achieved on a particular distance a, the measurement is finished and the distance d is also incremented (distance a after that is restarted from 1 cm for new group of measurement). The algorithm is shown in Fig. 3. In the research besides breakdown voltage, value also measured was corona voltage to verify whether the material type and parameters has any influence for observing phenomena. The Uo measurement took place during the measurement of Ub by generating voltage from 0 V to nearest possible visual descriptor seen by human eye in darkened lab room which corresponds to a value shown by electrostatic voltmeter. The value is being saved to database, and the generation of higher voltage continues until breakdown is achieved. It is possible to measure earlier (lower voltage value) and more accurate by using UV corona discharge camera like Ofil DayCor SuperB or spectrophotometer [6,7]. Using the OE method is part of future research of this study.

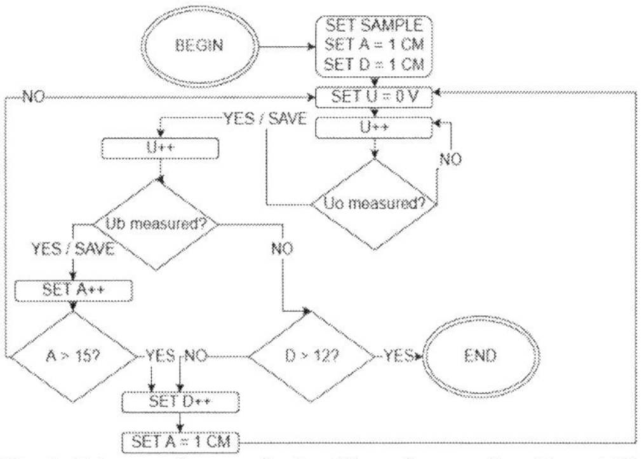

Fig. 3. Scheme of research algorithm of measuring Uo and Ub with increasing values of a and d distances in correct order.

2.2 Characteristic of materials applied in the study

Dielectrics can be classified according to different criteria. The most commonly used is the division from point of view of chemical properties. Therefore, there are three main categories: non-organic (e.g. glass, ceramic), organic (e.g. cellulose, resin) and polymers (e.g. aramid paper, polyvinyl chloride) [10]. Pressboard is a type of pressed cellulose cardboard produced by wet felting of vegetable fiber and flattening. For cellulose preparation, there is a need to apply two processes: acidic and alkaline, where it is less damaging for chemical molecules. Pressboard cellulose can be simply formed, trimmed or crushed so it is a commonly used material for insulating electrical conductors and transformers of high power [10]. It is a composite material like paper with the variation of sheet thickness ranged from 0.1 mm to plates of 50 mm. It is characterized as highly resistant to discharge and can withstand short-term thermal load up to 350 ° C.

The aramid paper is a polymer from the group of amides used in the production of fiber and sheets having both high mechanical strength and thermal resistance. It is made from synthetic fiber belonging to a group of polymers having amide linkages. The characteristic feature is existence of aromatic moieties in their main chains. If there is more in the structure of moieties, the greater mechanical strength, heat and fire resistance applies, but decreases the solubility (difficult in processing – more expensive). The material is akin to Kevlar in view of preparation by polymerizing 1,3-diaminobenzene and isophthalic acid [11]. Yarns of aramid fibbers are formed by performing the wet liquid crystalline polymer solution in sulfuric acid. The fibbers are pressed into the slot with blowing air and then immersed in the bath to solidify. Sheets are produced on a paper machine sieve with a flat head by combining fibrids and flocs in aqueous solution. Finally, it is subjected to hot calendering giving it higher density resulting better mechanical and electrical properties [11].

Both cellulose pressboard and aramid can withstand high temperature impact and are similar in winding on a transformer coil of high power. Thus, they can be

DOI: 10.1051/e3sconf/20171901044

compared with the difference of cost. The aramid is far more expensive to produce than cheap cellulose, but initial research shows [12] that if probability of prolonging life of a high power equipment with aramid as insulation is greater than the cellulose – the risk of investment could be returned. The damage or investigation of shortcoming fault can be far more expensive in the mean of years of exploitation [13].

Materials used in research are both 0.1 mm thick and the size is 40 x 40 cm^2. The cellulose is from Tetrans company, model N125 and aramid paper is Nomex DuPont, model 410. When the sample is outdone after electrical breakdown, it is changed to a new, fresh one without changes in chemical structure because of short-thermal activity and ozone influence.

3 Comparative analysis of measurement results

Figs. 4-8 presents quadratic polynomial curve fitting models of aggregated values of *Uo* corona voltage, *Ub* breakdown voltage for cellulose and aramid papers.
It was respectively measured according to algorithm from Fig. 3 for various distances between electrodes *a* [cm] and constant value of distance to dielectric barrier *d* [cm].

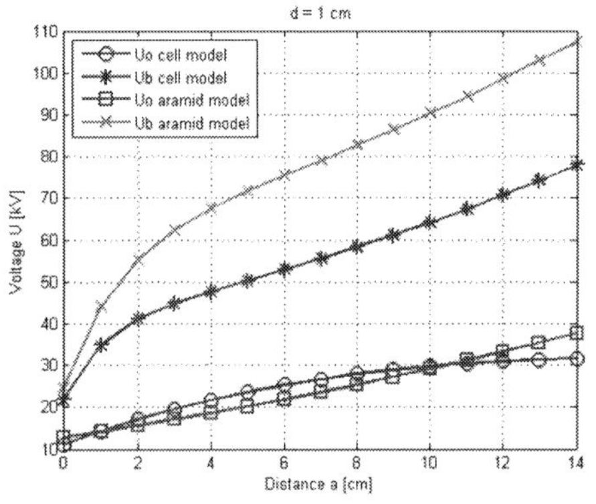

Fig. 4. Dependency of the voltage *Uo* and *Ub* to the distance *a* 1 cm ÷ 14 cm between electrodes at the constant distance *d* = 1 cm for both materials.

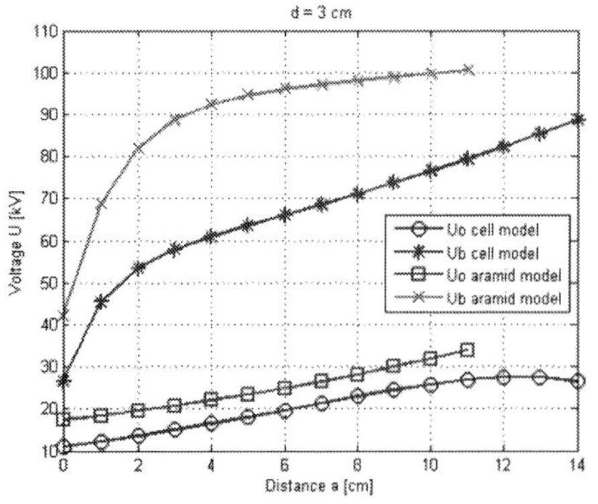

Fig. 5. Dependency of the voltage *Uo* and *Ub* to the distance *a* 1 cm ÷ 14 cm between electrodes at the constant distance *d* = 3 cm for both materials.

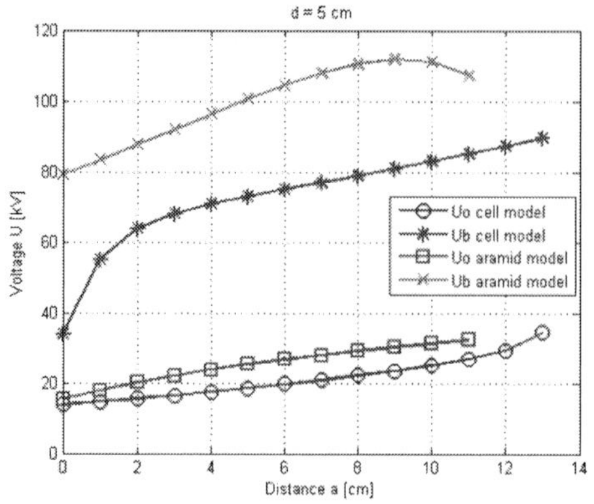

Fig. 6. Dependency of the voltage *Uo* and *Ub* to the distance *a* 1 cm ÷ 14 cm between electrodes at the constant distance *d* = 5 cm for both materials.

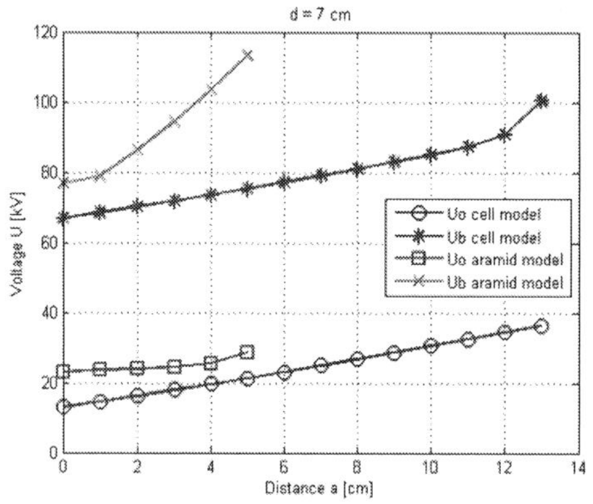

Fig. 7. Dependency of the voltage *Uo* and *Ub* to the distance *a* 1 cm ÷ 14 cm between electrodes at the constant distance *d* = 7 cm for both materials.

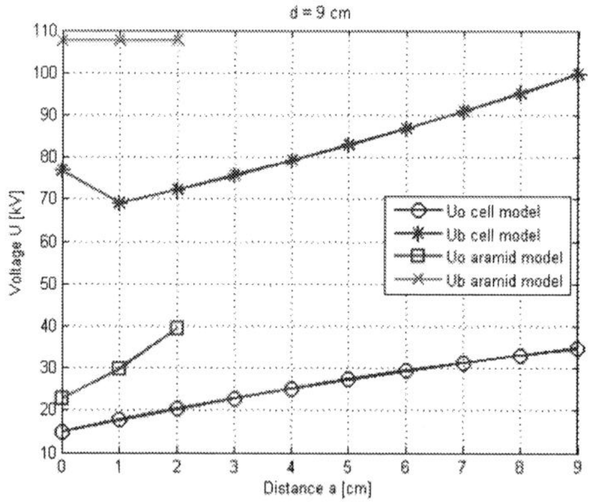

Fig. 8. Dependency of the voltage *Uo* and *Ub* to the distance *a* 1 cm ÷ 14 cm between electrodes at the constant distance *d* = 9 cm for both materials.

4 Discussion of analysis results

Based on the analysis, it was conducted that the measured corona voltages values has a low diversity coefficient value to assess that aramid paper is more resistant dielectric compared to cellulose paper. In the opposition stands the case of breakdown voltage, where the high difference was shown, when used various barriers as insulation. Furthermore, under laboratory environment, the research must have been finished when the aramid paper dielectric barrier of thickness 0.1 cm could not be penetrated. Voltage of 110 kV was maximum possible value for transformer to generate on the power line connected to plate electrode. As can be seen on Figs. 4-6, the corona voltage U_o of barriers are in similar range of values. On Fig. 7 the U_o value for aramid paper is measured only when there was an electric penetration of barrier through generation of breakdown short-circuit. The lowest possible to measure value of corona voltage for cellulose pressboard of thickness 0.1 cm was 12.33 kV. In case of aramid paper of thickness 0.1 cm the lowest corona voltage value was 12.87 kV. The highest possible to measure value of corona voltage for cellulose pressboard was 23.90 kV and 28.77 kV for aramid paper respectively. The maximum possible distance between electrodes to create a breakdown voltage with cellulose was 12 cm which had given the full spectrum of measurement as was assumed for study. Aramid paper was far more resistant to short-circuit, thus the distance was lower in comparison – 9 cm with only two possible measurement distances between electrodes (a = 2 cm). Power transformer of 110 kV electric voltage was limited to destroy the sample with higher breakdown voltage value. Value of corona voltage increases almost linearly for both dielectric types. Value of breakdown voltage increases gradually with the increase of corona voltage for the lower values. However, this is not a linear relation. The distance between electrodes is directly proportional to higher voltage values generated in both dielectric types. Research have shown that aramid paper is far more resistant to electric breakdown in comparison to cellulose paper of same thickness. For PD detection purposes to observe U_o it has no significance. Both values of corona voltage of PD on different materials are almost the same irrespectively which material was used.

5 Conclusions

Results of comparative analyses of corona and breakdown voltage in relation to the distance between HV and ground electrodes were presented in this paper. In particular two kinds of solid dielectrics used as barriers in air insulation of high voltage systems were considered: pressboard cellulose and aramid paper. The measured relations were registered under laboratory conditions using electrostatic voltmeter – in order to determine the influence of additional parameters as e.g. the distance between electrodes and the barrier, on the achieved results. The conducted analysis showed differences in values of corona and breakdown discharge occurrence for different solid dielectrics. The results obtained in the studies on aramid paper may help to improve diagnostics of power equipment, and succeed cellulose as a universal, insulation material for e.g. dry transformers (if risk of investment is calculated). In the future work, we aim to perform similar measurements for higher voltage values and for various thicknesses of the barrier. For corona discharge measurement it could be observed by using special UV spectrum camera. It has better accuracy than human eye and results can be obtained quicker on lower voltage value where human cannot see the UV emission of flowing electrons [6].

References

1. M. Kozioł, D. Wotzka, T. Boczar, P. Frącz, J. Spectro. **2016** (2016)
2. Ch. Liu, W. Zhang, M. Xu, L. Shi. Y. Zhao, IEEE ICD, **2,** 772-775 (2016)
3. P. Frącz, T. Boczar, D. Zmarzły, T. Szczyrba, Acta Phys. Pol. A **124**, 413-416 (2013)
4. H. Majchrzak, A. Cichoń, S. Borucki, **42,** 29-35 (2017)
5. T. Boczar, D. Zmarzły, Insight **45**, 488-492 (2003)
6. S. Krzewiński, P. Frącz, I. Urbaniec, T. Turba, J. Spectro. **2017** (2017)
7. P. Frącz, I. Urbaniec, T. Turba, S. Krzewiński, J. Spectro. **2016** (2016)
8. A. Reid, M. Judd, C. Johnstone, UPEC **44**, 1-4 (2009)
9. M. Siegel, M. Beltle, S.Tenbohlen, S. Coenen, IEEE Trans. Dielectr. Electr. Insul. **24**, 331-339 (2017)
10. J. Gielniak, A. Graczkowski, H. Moranda, P. Przybylek, K. Walczak, Z. Nadolny, H. Mościcka-Grzesiak, IEEE Trans. Dielectr. Electr. Insul. **20**, 982-987 (2013)
11. J. Song, M. Wen, L. Li, A. Isenberg, IEEE Int. Conf. on Dielectr. **2**, 625-629 (2016)
12. S. Wolny, M. Lepich, IEEE Int. Conf. on Dielectr. **2**, 1163-1166 (2016)
13. T. Suwnansri, iEECON, 1-4 (2014)

Detection of ionizing radiation generated by electrical discharges in the air using sphere-sphere system

Łukasz Nagi[1,*]

[1] Opole University of Technology, Prószkowska 76, 45-758 Opole, Poland

Abstract. The article presents the results of studies on the detection of ionizing radiation generated during electrical discharges. The obtained energy spectra for a system of sphere spark gaps, and the dependence of the obtained energies on the geometric arrangements of the detection devices and of the electrical discharge generating devices are presented. Additionally, the recorded signals from the scintillation detector, and the energy spectrum received for the examined system are also presented. It is significant that the field distribution related to the shape of the spark gaps is of key importance for the received measurement data. The completed studies for the sphere - shaped spark gaps are a part of studies on recording ionizing radiation generated by full and partial discharges in the air and in mineral oil.

1 Introduction

Both full and partial discharges are a significant threat not only for human health and life, but also for electrical supply security. Nowadays, effective diagnostics of power equipment and networks is one of the most important tools for safe work and for ensuring a reliant power system, without the risk of e.g. uncontrolled electric power shortages. Electrical discharge detection is the main subject of the studies at the Institute of Electric Power Engineering and Renewable Energy (IEIEO). Laboratory experiments precede the application of innovative methods in the industry. Many methods of partial discharge detection developed at IEIEO have a practical application. The problem of the occurrence of electrical discharges and their effect on high voltage technology have resulted in developing numerous methods to detect these physical phenomena not only at the Opole University of Technology. The main reason for the world-wide works on this issue is that the sooner discharges are detected the faster the hazards can be reduced. On the contrary, the reduction of energy losses and economic losses is important for the development of technology, science, and the society's quality of life, therefore diagnostics of electrical engineering devices and facilities is very crucial. The existing methods can be divided into invasive and non-invasive methods. The invasive methods use a direct connection to the facilities being tested, and thus partial discharge currents can be detected with great precision. The non-invasive methods offer lower accuracy, but faster and more comfortable traceability of PD (partial discharges). At present, power network and device diagnostics are based on several electrical discharge detection methods. These are methods for detecting mostly PD that precede failures. The most effective methods include UHF [1–3], the acoustic emission method (EA) [4–6],], the optical method [7–10]. Another result of the studies on electrical discharges is defining the power balance of the phenomenon. This would require determining energies coming from each type of energy that is released during a discharge. From light energy through heat, sound, to high frequency electromagnetic radiation. Electrical discharges generate ionizing radiation [11,12] to, both in natural atmospheric discharges [13–15] and in laboratories [16–18]. Studies on partial discharges in a laboratory proved the existence of such radiation also at a voltage between the electrodes not higher than 120kV [19,20]. When preparing effective ionizing radiation detection methods on actual facilities, the energy balance of an electrical discharge phenomenon can be developed. The article presents the results obtained in the course of laboratory tests on a model spark gap.

2 Measurement procedure and results

A system that consists of the following three systems has been designed to generate and to measure highly-energetic radiation:
-a system generating electrical discharges,
-a system to control the measurement desk,
-a system to measure highly-energetic radiation.

The first part is comprised of a set of spark gaps powered with high voltage from a transformer, and the third part of a scintillation detector placed at the negative electrode or at a certain distance from this electrode. The detector is fitted on an arm that can be positioned at any place in the space with a precision of down to 0.1 mm. During the measurement, the voltage was increased until the system's breakdown.

The tests with the sphere gap system with a sphere diameter of 20 mm were performed for 10 various

* Corresponding author: l.nagi@po.opole.pl

distances between the electrodes. For the research 172 measurements were performed to take a dataset. Fig. 1 characterizes the dependence between the maximum recorded energy and the distance between the spark gaps. With the increase in the distance, and hence the increase in the system's breakdown voltage, the maximum recorded ionizing radiation energy increases until it reaches the plateau for a given system. Figs. 2 - 4 present average scintillation energy in this system. Each diagram presents different models that match the fitting. A similar tendency can here again be observed, when along with an increase in the breakdown voltage (caused by an increase in the distance between the spark gaps) the maximum energy recorded by the detector increases. The presented models adjusted to the obtained data have high R^2 coefficients, the best one is for the exponential model. Figs. 5 and 6 show the sum of scintillation energy in this system. Each chart reflects different models that match the fitting. As in previous characteristics concerning the maximum and the average energy recorded in the tests, it can be observed here that, along with an increase in the distance between the spark gaps, the energy recorded by the detector increases. The linear model is the best fitting for the total energy measured.

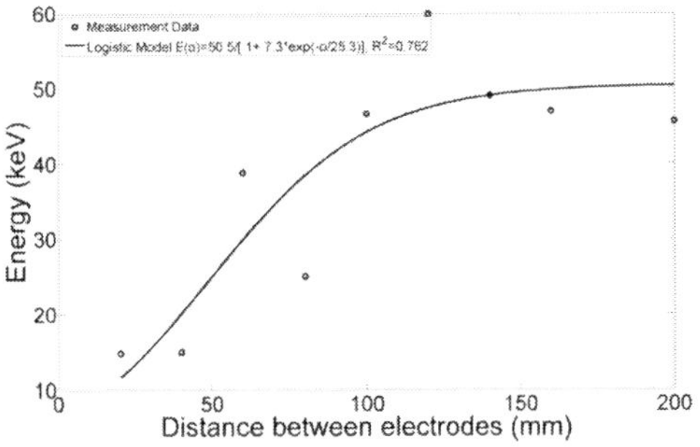

Fig. 1. The dependence between the maximum scintillation energy and the distance between the electrodes in a sphere gap system with a sphere diameter of 20 mm in a test performed in the air.

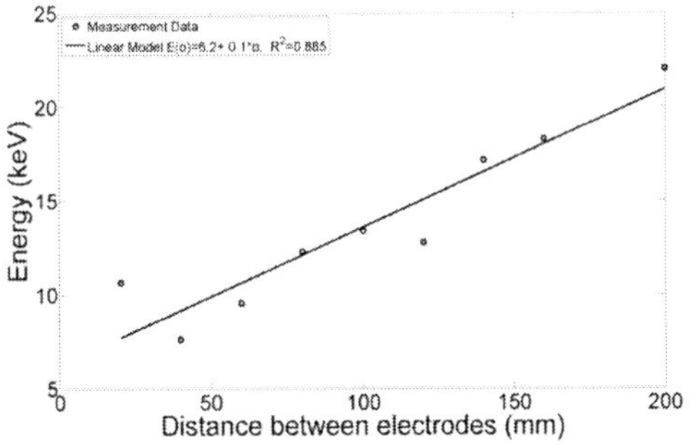

Fig. 2. The dependence between the average scintillation energy and the distance between the electrodes in the sphere gap system with a sphere diameter of 20 mm in a test performed in the air.

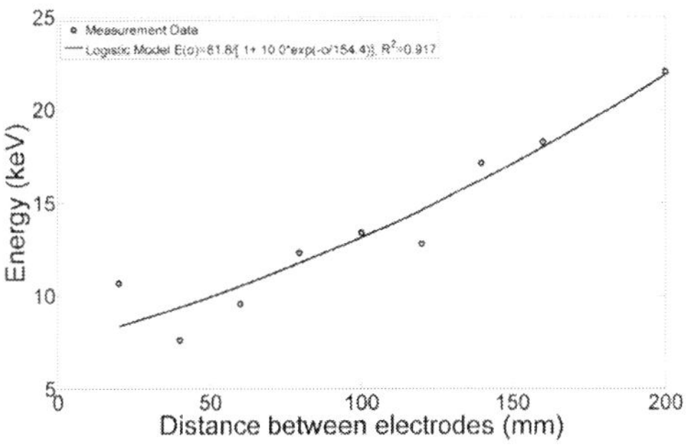

Fig. 3. The dependence between the average scintillation energy and the distance between the electrodes in the sphere gap system with a sphere diameter of 20.

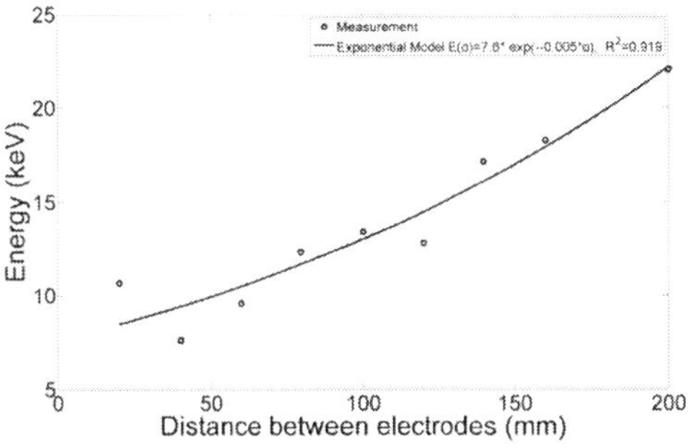

Fig. 4. The dependence between the average scintillation energy and the distance between the electrodes in the sphere gap system with a sphere diameter of 20.

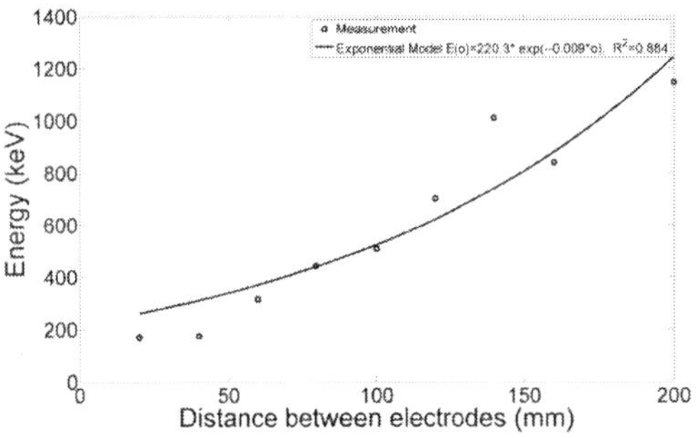

Fig. 5. The sum of the scintillation energy depending on the distance between the electrodes in the sphere gap system with a sphere diameter of 20.

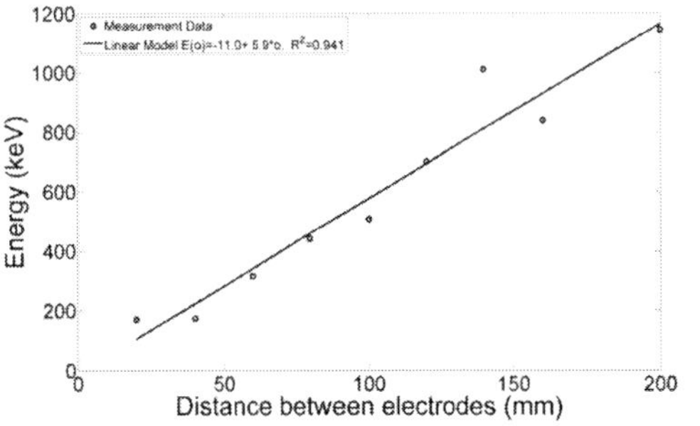

Fig. 6. The sum of the scintillation energy depending on the distance between the electrodes in the sphere gap system with a sphere diameter of 20 mm.

Fig. 7 presents the energy spectrum for the selected distances between the electrodes, while Fig. 8 shows the energy spectrum recorded in this electrode arrangement. With an increase in the distance between the electrodes, the recorded spectrum shifts towards higher energies, also, the total count number is higher for a greater distance between the spark gaps. The energy spectrum shows that for this test system, the highest energies recorded occur at higher voltage, and the highest count number in one energy interval falls between 60-120 mm of the distance between the electrodes.

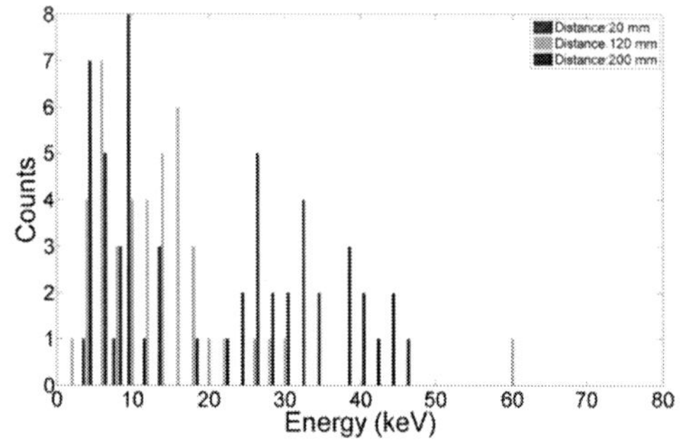

Fig. 7. The energy spectrum for the selected distances between the electrodes in the sphere gap system with a sphere diameter of 20 mm in a test performed in the air.

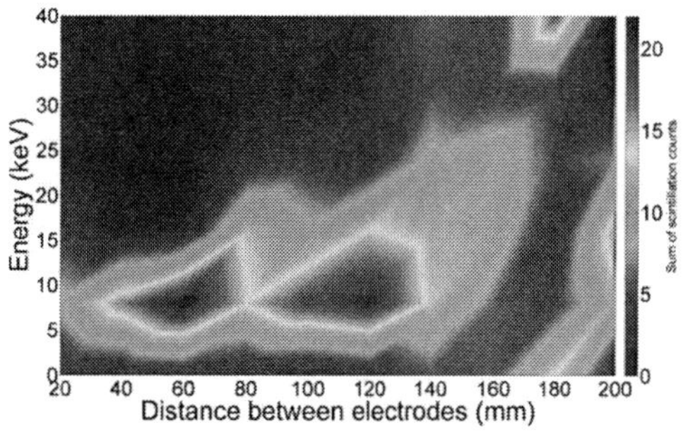

Fig. 8. The scintillation energy spectrum in the sphere gap system with a sphere diameter of 20 mm in a test performed in the air.

During the measurements, the voltage, the breakdown current, and the shape and number of scintillating bolts at the time of a breakdown in the system were recorded at the same time. Sample partial data registered in the test are presented in Figs. 9 and 10. They reflect the recorded scintillation peak, and the signal with noise and interference removed, respectively. Figs. 11 and 12 present exemplary scintillation models for a ϕ 20 mm sphere system in the air. Fig. 13 shows the time courses of the signals from the synchronized measuring voltage and current intensity systems, and from the ionizing radiation measuring system.

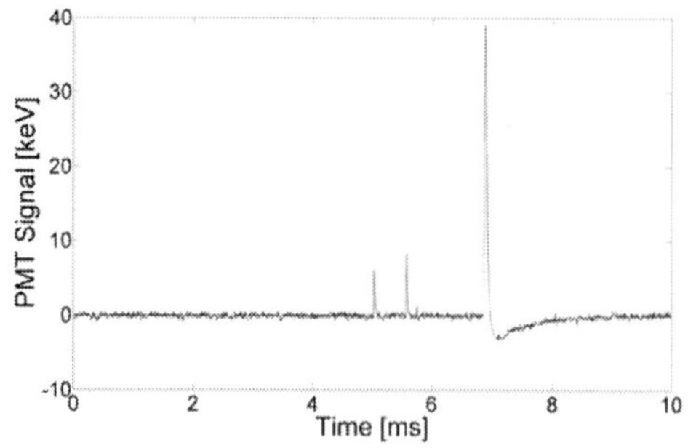

Fig. 9. Sample scintillation registered for the system. The distance between the 60 mm electrodes. The distance between the detector and the negative 1 mm electrode.

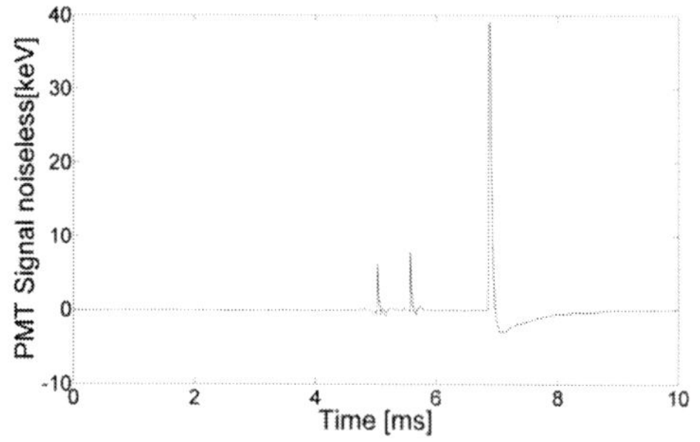

Fig. 10. Sample scintillation from figure 9 with noise and interference removed.

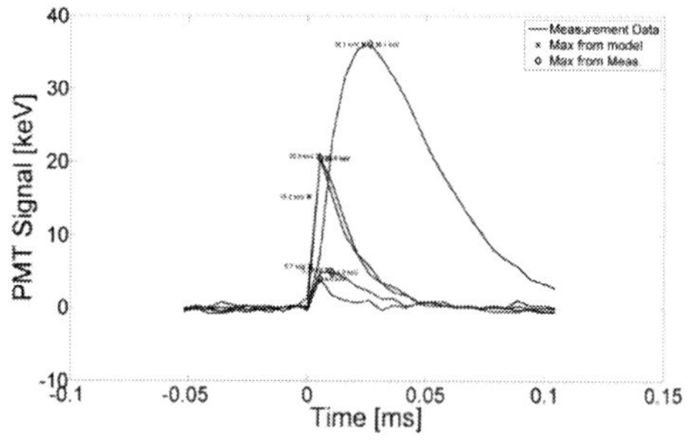

Fig. 11 Different mathematical models for a sample scintillation in the sphere gap system with a sphere diameter of 20 mm in a test performed in the air.

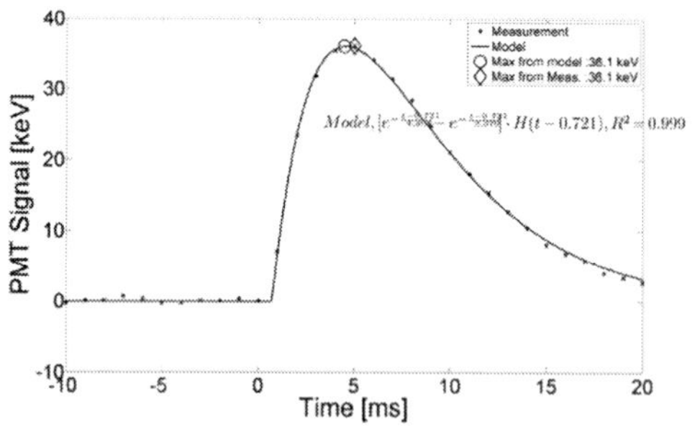

Fig. 12. The best fitted model for a sample scintillation in the sphere gap system with a sphere diameter of 20 mm in a test performed in the air.

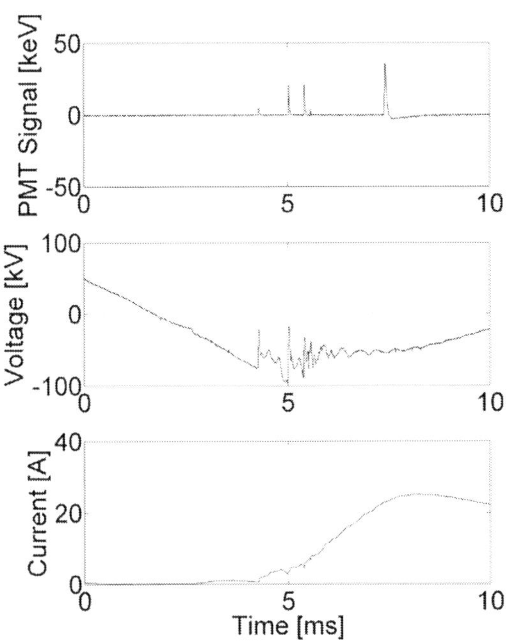

Fig. 13. The time courses of the signals from the multiplier photo-tube, and of the current and voltage signal on the primary winding in the transformer at breakdown.

3 Conclusions

The test in the sphere-sphere gap system demonstrated a high dependence of the data recorded on the distances between the electrodes. The maximum scintillation energy, the total energy in the measurement, and the average energy per scintillation in the measurement increase along with the increases in the distance between the spark gaps.

An increase in the distance increases the breakdown voltage, which, in turn, may result in an increase in the ionization energy of air molecules, and further to higher bremsstrahlung energies when excited molecules return to their primary state. The energy spectrum in this system suggests that the greatest number of scintillations is recorded at a distance between the electrodes equal to 120 mm. An increase in the distance of more than 120 mm lowers the count of recorded scintillations, and the energy of individual excitations increases. Along with an increase in the distance, the radiation spectrum shifts towards higher energies, reducing the number of recorded scintillations at the same time.

ACKNOWLEDGEMENT

The work was co-financed from funds of the National Science Centre (NCN) as part of the PRELUDIUM programme, project No. 2014/15/N/ST8/03680

References

1. W. Gao, D. Ding, W. Liu, X. Huang, IEEE Trans. Power Deliv. **29**, 38–47 (2014)
2. M. Kunicki, A. Cichoń, Meas. Autom. Monit. **61**, 12–15 (2015)
3. M. Siegel, M. Beltle, S. Tenbohlen, IEEE Trans. Dielectr. Electr. Insul. **23**, 1580–1588 (2016)
4. A. Hekmati, Int. J. Electr. Power Energy Syst. **77**, 250–255 (2016)
5. D. Wotzka, T. Boczar, D. Zmarzly, Acta Phys. Pol. A - Opt. Acoust. Methods Sci. Technol. **116**, 428–431 (2009)
6. M. Kunicki, A. Cichoń, S. Borucki, Arch. Acoust. **41**, 265–276 (2016)
7. T. Boczar, D. Zmarzly, P. Fracz, IEEE 11th Int. Conf. Prop. Appl. Dielectr. Mater. 2015 740–743 (2015). doi:10.1109/ICPADM.2015.7295378
8. P. Frącz, T. Boczar, S. Borucki, A. Cichoń, D. Zmarzły, Acta Phys. Pol. A **122,** 814–817 (2012)
9. P. Fracz, IEEE Trans. Dielectr. Electr. Insul. **20**, 1909–1914 (2013)
10. S. Biswas, C. Koley, B. Chatterjee, S. Chakravorti, IEEE Trans. Dielectr. Electr. Insul. **19**, 18–28 (2012)
11. P. O. Kochkin, C. V Nguyen, a P. J. van Deursen, U. Ebert, J. Phys. D. Appl. Phys. **45**, 425202 (2012)
12. P. O. Kochkin, A. P. J. van Deursen, U. Ebert, J. Phys. D. Appl. Phys. **48**, 25205 (2015)
13. J. R. Dwyer *et al.*, J. Geophys. Res. Sp. Phys. **117**, n/a-n/a (2012)
14. J. R. Dwyer, J. Geophys. Res. **113**, D10103 (2008)
15. P. Kochkin, A. P. J. van Deursen, A. de Boer, M. Bardet, J.-F. Boissin, J. Phys. D. Appl. Phys. **48**, 425202 (2015)
16. J. R. Dwyer *et al.*, J. Geophys. Res. **113,** D23207 (2008)
17. J. R. Dwyer *et al.*, Geophys. Res. Lett. **32**, 1–4 (2005)
18. P. O. Kochkin, a P. J. van Deursen, U. Ebert, A. Van Deursen, U. Ebert, J. Phys. D. Appl. Phys. **47**, 145203 (2014)
19. D. Zmarzly, L. Nagi, S. Borucki, T. Boczar, Acta Phys. Pol. A **125**, 1377–1379 (2014)
20. Ł. Nagi, D. Zmarzły, T. Boczar, P. Frącz, IEEE Trans. Dielectr. Electr. Insul. **23**, 2036–2041 (2016)

 DOI: 10.1051/e3sconf/20171902001

Air purification from a mixture VOCs in the pilot-scale trickle-bed bioreactor (TBB)

Rafał Sarzyński[1], *Agnieszka* Gąszczak[1], *Daniel* Janecki[2,*], *Grażyna* Bartelmus[1]

[1]Institute of Chemical Engineering, Polish Academy of Sciences, ul. Bałtycka 5, 44-100 Gliwice, Poland
[2]Department of Process Engineering, University of Opole, ul. Dmowskiego 7-9, 45-365 Opole, Poland

Abstract. The efficiency of the air bio-purification from the mixture of two volatile organic compounds (styrene and p-xylene) was studied. The process was carried out in a pilot-scale trickle-bed bioreactor installation designed to purify ~200 m^3h^{-1} of the polluted air. The bioreactor operated at concurrent flow of gas and liquid (mineral salt solution) through packing (polypropylene Ralu rings) covered with a thin layer of microorganisms (bacterial consortium of *Pseudomonas* sp. E-022150 and *Pseudomonas putida* mt-2). The experiments, carried out for various values of a reactor load with pollutant, confirmed the great efficiency of the investigated process. At the tested bed load with pollution (inlet specific pollutant load was changed within the range of 41 – 84 gm^{-3} h^{-1}), styrene conversion degree changed within the range of 80-87% and p-xylene conversion degree within the range of 42-48%.

1 Introduction

Air pollution caused by human activity is an international problem since harmful substances emitted into the atmosphere are transmitted across the state borders both in the ground zone, in the rhythm of meteorological changes, and through the stratosphere. Air pollution, first of all, with volatile organic compounds (VOCs) is becoming a more and more serious problem since they are mostly toxic substances, aromatically troublesome, and they are also tropospheric ozone precursors.

There is no one universal method of air purification due to a complex chemical composition of VOCs and various process conditions. Classical physicochemical methods such as absorption, adsorption, condensation, incineration or catalytic combustion are expensive and often generate secondary pollution. Besides the above mentioned, biological methods of VOCs elimination are becoming more and more popular since they make it possible to purify big streams of air containing even very small and changeable in time amounts of harmful substances. Moreover, the bio-purification methods require the lowest financial outlay connected with VOCs utilization per 1 m^3 of the purified air. By comparison, depending on the used method, the expenditure equals: 30-200 $ - condensation, 40-190 $ - absorption, 30-600 $ - combustion, 25-155 $ - adsorption and 13-90 $ - biodegradation methods [1].

It is worth noticing that biological methods of gas purification can be used only if the removed substance is biodegradable and water soluble (even slightly).

The subject of research in the presented study is the process of air bio-purification from a mixture of volatile organic compounds (styrene/ p-xylene) carried out in a pilot-scale trickle-bed bioreactor (TBB) unit. In such apparatuses the column packing is made of an inert material on which microorganisms are immobilized.

The pollution removed from the air is the only source of carbon and energy for the microorganisms. The purified air and mineral salts solution flow concurrently down the bed making it possible to use high flow rates of both phases without causing the onset of flooding in the column – a phenomenon which considerably limits the operating range of the counter-current flow. Their advantages are also both low investment and operational costs and larger potential of optimal operating parameters control.

The substances removed from the air were a mixture of styrene/p-xylene.

Styrene is one of very important raw materials for industry connected with plastics production. The world production capacity of styrene amounts to about 32.5 million tons and keeps growing by 2-3% a year on average. Such common and quantitatively large styrene consumption considerably influences the quantity of its emission into the atmosphere estimated at 25 000 Mg per year [2]. In general, the concentration of styrene in industrial waste gases may reach 1 gm^{-3} [3] and average range varies between 0.15 and 0.4 gm^{-3}. Styrene and its metabolites are known to have serious negative effects on human health [4].

The other component of the mixture polluting the air was one of xylene isomers. Xylenes (o-, m- and p-dimethylbenzene) are harmful chemicals that can be found in many consumer products such as paints, lacquers, varnishes, adhesives, inks and dyes, cleaners and degreasers, aviation gasoline, etc. They are also used

* Corresponding author: zecjan@uni.opole.pl

to produce a lot of plastics and synthetic fibres, insecticides and pesticides and other chemicals [5].
Crude oil and coal tar are their natural sources. Therefore, practically the total production of these substances is localized in refinery and petrochemical industry. The industrial emission of xylenes into the atmosphere is estimated at about 34 000 Mg per year in the USA, 7 300 Mg per year in Canada, out of which 58% was connected with the use of solvents and 39% with car exploitation [6]. The industrial emission of xylenes in Japan is estimated at about 48000 Mg per year [7]. About 10 – 22% of the produced xylene is used for fuel enrichment. However, due to weak solubility in water and slight adsorption in soil only about 1% of xylenes introduced into the environment is to be found in water or soil, 99% goes into the atmosphere. Therefore, there is an ongoing need for the development of better treatment methods to eliminate or reduce the content of xylene in gaseous effluents prior to their release into the atmosphere. Both styrene and xylene, due to their toxic action and lowering the standard of living, were placed on the list of 189 harmful air pollutants by the US EPA (Environmental Protection Agency) – Clean Air Act 1990 [8].
In the literature one can find a lot of studies whose authors examined the process of purifying the air from styrene or xylene in biofilters using various microorganisms in the research. Experiments performed in trickle-bed bioreactors are rather limited. Extensive analysis of literature data on air biopurification from styrene/xylene was presented in our earlier papers [9,10].

2 Experimental installation and experimental procedures

The research into the efficiency of air purification from the mixture of styrene/xylene was carried out in a pilot installation shown schematically in Fig. 1. Its main element was a stainless steel reactor having inner diameter 1.084 m filled with polypropylene Ralu rings (Raschig GMBH company) 50.8 mm in diameter (□ = 0.95; a=110 m^{-1}). The height of the packing layer was about 1.8 m. The biological material immobilized on packing was a mixed population of microorganisms: *Pseudomonas sp.* E-93486 strain showing 97% similarity to *Pseudomonas putida* and 97% similarity to *Pseudomonas stutzeri* from 16SrRNA analysis, which does not have the ability to use xylene as a source of carbon but actively degrades styrene and *Pseudomonas putida* mt-2 strain degrading p-xylene. Both strains, coming from VTT Culture Collection (Finland), were carefully adapted to degrade the tested compounds and the kinetics of the bodegradation reactions was determined (tests in batch reactor).
The pilot-scale installation was automated in such a way that the control and changes of the operational parameters were possible from the operator's cabin by changing the data on the computer control screen. The following parameters were determined during the operation of the installation once a day:
-styrene and p-xylene concentration in the gas stream at the inlet and outlet of the bioreactor (Varian 450 GC chromatograph equipped with a CP-WAX 52CB column and a FID detector)
- carbon dioxide concentration in the air leaving the bioreactor (Varian 450 GC chromatograph),
- microorganism concentration in the recirculated liquid (absorbance measurement at λ=550 nm),

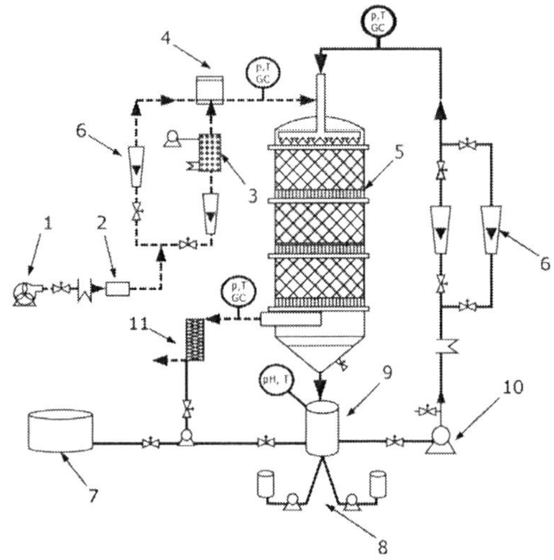

Fig. 1. Schematic diagram of experimental set-up.
- - - - gas phase, —— - liquid phase, p - pressure measurememt, T - temperature measurement, GC - gas chromatography analysis, pH – pH measurement
1- compressor, 2 - air filter, 3 - VOC's dosing system, 4 - gas mixing chamber, 5 - packed bioreactor, 6 - flow meter, 7- mineral salt solution container with dosing pumps, 8 - pH regulation system with dosing pumps, 9 - liquid recirculation tank, 10 - liquid pump, 11- droplet separator.

- oxygen and nitrogen concentrations in the circulating liquid (ISE max CAS40D probe and tests by HACH company)
- VOCs concentration in the circulating in the bioreactor installation liquid (Varian 3800 chromatograph equipped with an auto-sampler, a CP-WAX column and a FID detector).
The following were recorded 'on line' in the computer memory: flow rate of both phases, the temperature in the marked points of the installation, pH and the amount of the oxygen dissolved in the solution, gas pressure drop in bed, and the level of liquid in the containers.
Prior to immobilization the reactor was sterilized with water having the temperature of 95°C for 24 hours and next with air having the temperature of 110°C for 48 hours.

3 Experimental results

The process was carried out at optimal for bacteria strains conditions: the temperature of 30°C and pH 7 maintained by batching the solutions of KH_2PO_4/NaOH according to the indications of the pH electrode.

The system adapted itself to the utilized pollution with difficulty, especially to p-xylene. Only after 70 days did the bed start to operate in a stable way. During 250 days of continuous operation of the installation, out of which 70 days should be regarded as the period of bed adaptation, 6 measurement series were carried out which differed in the styrene and p-xylene concentration in the air supplied to the reactor. Volumetric flow rate of air was kept at the level of 100 m^3h^{-1} which gives the mean gas residence time in bed of t_g = 60 s. The flow rate of liquid spraying the packing, which was mineral salts and microelements solution securing good condition of microorganisms immobilized on packing, was kept at the level of 8 m^3h^{-1}. The composition of mineral medium was the same as given in the study [11]. The changes in styrene/p-xylene concentration in influent and effluent gas as well as conversion degree are shown in Fig. 2 and Fig. 3.

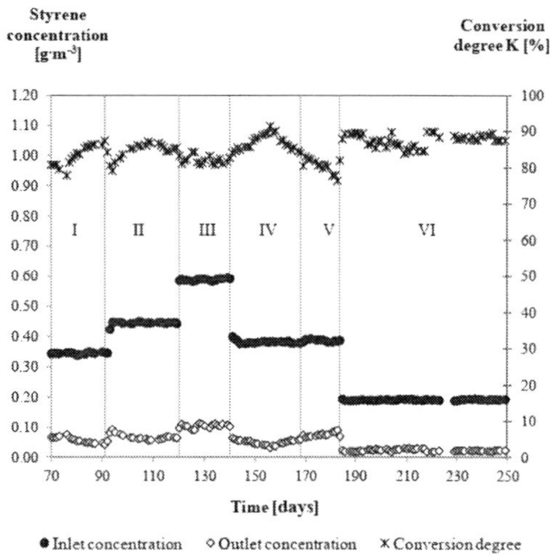

Fig. 2. Changes in the concentrations and conversion degree of styrene during the process of air purification from styrene/p-xylene mixture.

The average values of inlet concentration and conversion degrees (K) of both components for 6 series of experiments were listed in Table 1.

The conversion degrees were calculated from the dependence:

$$K = \left(\frac{C_g^0 - C_g^1}{C_g^0} \right) \cdot 100\% \qquad (1)$$

where:

C_g^0 and C_g^1 are inlet and outlet styrene/p-xylene concentrations (gm^{-3}).

During the whole period of tests the styrene conversion degree was relatively high and, depending on the bed load with pollution, it varied in the range of 80-87%. For p-xylene the conversion changed within the range of 42-48%.

The effectiveness of the process was evaluated by calculating the specific elimination capacity (EC) ($gm^{-3}h^{-1}$) defined as:

$$EC = \frac{C_g^0 - C_g^1}{t_g} \qquad (2)$$

where:

$t_g = V_{bed}/V_g$ is average gas residence time (h);
V_{bed} is bed volume (m^3_{bed});
V_g is volumetric gas flow rate (m^3h^{-1}).

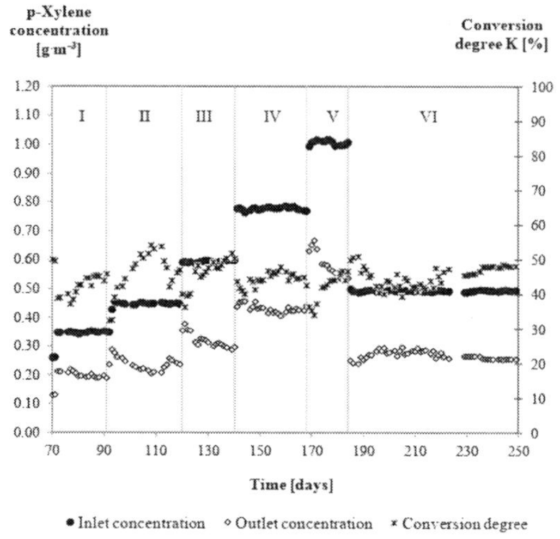

Fig. 3. Changes in the concentrations and conversion degree of p-xylene during the process of air purification from styrene/p-xylene mixture.

Table 1. Average values of pollutants inlet concentrations and conversion degrees obtained in experiments (subscripts: s- refers to styrene, x- refers to p-xylene).

No. series Days	C_{gs}^0 gm^{-3}	C_{gx}^0 gm^{-3}	K_s %	K_x %
I **70-90**	0.346	0.346	84.0	41.8
II **91-120**	0.447	0.449	85.0	47.6
III **121-140**	0.383	0.778	87.0	44.3
IV **141-168**	0.579	0.605	82.2	46.4
V **169-183**	0.388	1.007	80.5	42.0
VI **184-250**	0.191	0.493	87.1	44.2

No clogging of the column, due to excess growth of biomass on the packing material, was observed. The changes of biofilm thickness were signalled by the increase of pressure drop in the column. After termination of each measurement series the stream of liquid spraying the bed was doubled for about 3 h to remove the excess biofilm.

The experimental results were plotted in Fig. 4 as a dependence of a specific elimination capacity, EC

($gm^{-3}h^{-1}$) vs. inlet specific pollutant load, PL^0 ($gm^{-3}h^{-1}$) which was determined from the relationship:

$$PL^0 = \frac{C_g^0}{t_g} \quad (3)$$

In Fig. 4 the experimental results obtained for the air purification process from styrene only (for the same volumetric flow rates of both phases) were also presented for comparison. As it can be seen, similar styrene conversion degrees were achieved in both processes. In V series only, in which the concentration of p-xylene in the air fed to the bioreactor was increased to ~1 gm^{-3}, a systematic decrease in time of styrene conversion was observed (see Fig. 2). Therefore, fearing the poisoning of the styrene degrading strain, this measurement series was ended after 14 days.

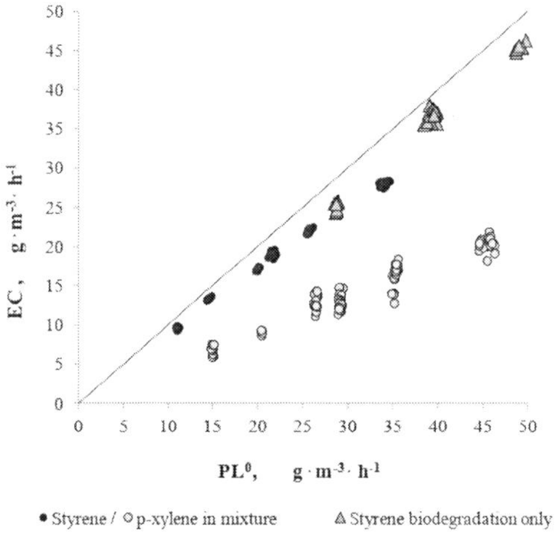

Fig. 4. Elimination capacity (EC) vs. inlet specific pollutant load (PL^0) for air purification from the mixture styrene/p-xylene and from styrene only.

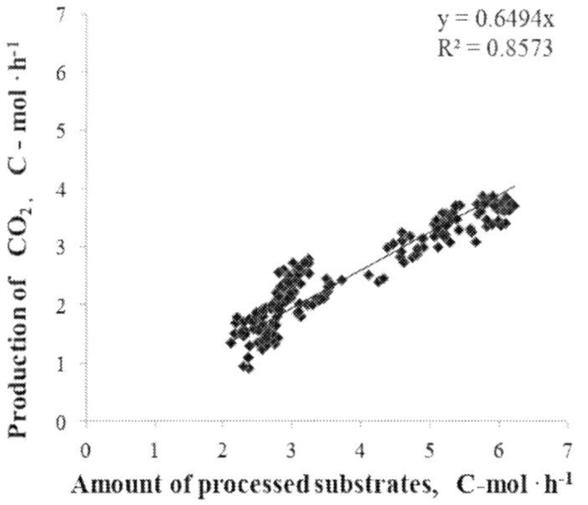

Fig. 5. Production of carbon dioxide vs. the amount of processed substrates.

In the investigated process, substrates were biodegraded in aerobic conditions to carbon dioxide, water and biomass. Hence, monitoring the carbon dioxide concentration in the gas phase can provide valuable information on the operation of the trickle-bed bioreactor. For example, the global biomass yield coefficient equaling 0.649 was calculated from the measurements of the amount of carbon dioxide produced in the process (Fig. 5).

4 Conclusions

The carried out research confirmed the possibility of using a trickle-bed bioreactor to purify big streams of air from a mixture of volatile organic compounds (styrene/p-xylene). At the specific pollutant load exceeding 70 $gm^{-3}h^{-1}$ about 2000 g of VOCs was removed from the air every day. The conducted experiments provided a lot of crucial information connected with conducting of the process and made it possible to work out procedures ensuring the effective and long-term operation of the installation. Moreover, the experiments showed good activity of the examined bacterial strains in the process; due to such feature the examined microorganisms may be recommended for technical application.

References

1. R. Munoz, S. Villaverde, B. Guiyesse, S. Revah, Biotechnol. Adv. **25,** 410 (2007)
2. H. Jorio, L. Bibeau, M. Heitz, Environ. Sci. Technol. **34,** 1764 (2000)
3. B. Bina, R. Dehghanzadeh, H. Pourmoghadas, H. Kalantary, A. Torkian, J. Res. Med. Sci. **6**,280 (2004)
4. P. Vodicka, M. Koskinen, A. Naccarati, B. Oesch-Bartlomowicz, L. Vodickova, K. Hemminki, Drug Metab. Rev. **38**, 805 (2006)
5. O. Liu, A. Babajide, P. Zhu, L. Zou, Chem. Eng. Technol. **29,** 320 (2006)
6. E. Elmrini, N. Bredin, Z.Shareefdeen, M. Heitz, Chem. Eng. J. **100**, 149 (2004)
7. E. Jeong, M. Hirai, M. Shoda, J. Biosci. Bioeng. **102**, 281 (2006)
8. J.R. Robledo-Ortiz, D. Ramirez-Arreola, A. Perez-Fonseca, C. Gomez, et al. , Int. Biodeter. Biodegr. **65**, 539 (2011)
9. A. Gąszczak, G. Bartelmus, I. Greń, Appl. Microbiol. Biotechnol. **93**, 565 (2012)
10. K. Klepacka, Prace Naukowe IIChPAN **14**, 41 (2010).
11. I. Greń, A. Gąszczak,. Szczyrba, S. Łabużek, Polish J. of Environ. Stud. **18,** 383 (2009)

Influence of calcination process on the formation of selected air pollutants

Jacek Wydrych[1*], *Bolesław* Dobrowolski[1], *Grzegorz* Borsuk[1*], *Sławomir* Pochwała[1]

[1]Opole University of Technology, Mikolajczyka 5, Opole, Poland

Abstract. The subject of the study is to analyze the phenomena of thermal flow in the precalcinator chamber of the exchanger's furnace tower including the combustion of coal dust and decarbonisation of raw lime powder. During the research were provided development of a mathematical model of particulate solid fuels combustion, calcining the raw material, NO_x and CO_x formation. Moreover conducting the number for the current and the upgraded design of the precalcinator and analysis of the results. In this study, a mathematical model based on Euler's method to describe the motion of the gas phase and the Lagrange method to describe the motion of particles [1-4]. In the calculations there were assumed fractional particles raw material and fuel, and the following processes: flow of exhaust gases from the rotary kiln through the precalcinator chamber, heat exchange between the particles of raw material and exhaust gases, the additional fuel combustion in the precalcinator, the process of raw material calcination, transformation of gaseous substances, effect of the additional (tertiary) air delivery on the processes in the chamber.

1 Introduction

With certainty, cement factories to some extent have cause environmental pollution. Cement industry releases carbon dioxide into the air directly by the heating of limestone, as well as the blasting in quarries causes of dust and pollutes the air [5-7]. For this reason, the cement industry is one of the primary producers of carbon dioxide, a major greenhouse gas. Studies related to the generation of carbon dioxide and other pollutants during the decarbonisation of limestone in the precalcinator chamber [8-10]. The calculations considered several different options. In particular alternatives of modified flow system by supplying tertiary air and points location of raw material to the precalcinator chamber. Completed numerical calculations have shown that the design of the precalcinator and conditions of its operation (presence of tertiary air) affects of the heat-flow and thermochemical processes in the chamber. It was found that the increase in the concentration of NOx in the furnace precalcinator chamber in all alternatives is smaller than the emission limit regulated by law.

2 Modeling the calcination process of clinker meal particles

Modeling the calcination process of clinker meal is a multi-stage issue. During moving the particles of raw material through the precalcinator chamber, the five stages of the calcination process can be distinguished, which is accompanied by:

1. In the pre-heating zone, the limestone is heated from room temperature to about 800°C by hot gases (for example: combustion products, CO_2 from the calcination process and additional air).
2. From about 800°C, the carbon dioxide pressure resulting the dissociation of limestone is equalized with the partial pressure CO_2 in the gas of the precalcinator chamber.
3. When the limestone temperature exceeds 900°C so the distribution temperature and the partial pressure exceeds 1 atmosphere, the dissociation process can proceed from the surface toward the particle center.
4. If the calcium carbonate contained in the molecules dissolves before the particles leave the calcination zone, the lime begins to sinter.
5. Calcium particles that may contain limestone leave the calcination zone and are cooled by the air supplied for combustion.

The reaction of thermal decomposition of calcium carbonate can be described by the following formula [11]:

$$CaCO_3 + \xrightarrow{178\ kJ/kg} CaO + CO_2 \uparrow \quad (1)$$

$$100\,g \xrightarrow{\qquad\qquad} 56\,g + \quad 44\,g$$

The shrinking core model was adopted to study the distribution of spherical limestone particles in the calcination process. At a given time step there is a centrally located core of unconverted carbonate surrounded by a calcium oxide coating with a reaction phase occurring on the surface between the core and the coating [12,13]. On the reaction surface, the bass balance

* Corresponding author: j.wydrych@po.opole.pl

for the calcining material can be written in the following form:

$$\frac{dm_{p,L}}{dt} = -r_c \frac{Mw_{CO_2}}{Mw_{CaCO_3}} \quad (2)$$

where

$$r_c = {}^r k_s' \cdot e^{-E_2/RT_L} \cdot 4 \cdot \pi \cdot r_p^2 \cdot Mw_{CaCO_3} \quad (3)$$

while mp,L, $MwCO_2$, Mw_{CaCO3} are the mass of the reacting limestone, the atomic mass of carbon dioxide and the calcium carbonate. r_c is the calcination coefficient of calcium carbonate and r_p is the shrinking radius of the part of the non-calcined particle. The value ${}^r k_s'$ is calculated from the formula:

$${}^r k_s' = {}^r k_s \quad \text{for} \quad p_{CO_2} < 10^{-2} p_{eq}$$

$${}^r k_s' = {}^r k_s \left(\frac{p_{CO_2} - p_{eq}}{p_{eq}} \right) \quad (4)$$

$$\text{for} \quad 10^{-2} p_{eq} < p_{CO_2} < p_{eq}$$

where

$$p_{eq} = 1.826 \cdot 10^7 \exp\left(-\frac{19680}{T_L} \right) \quad (5)$$

The ${}^r k_s$ component is the calcium carbonate calcination constant, while T_L is the $CaCO_3$ particle temperature. The presence of CO_2 in gases affects the calcination process, but it is difficult to determine the effect of the partial pressure exerted by CO_2 on this process [14]. The energy balance for the calcareous non-calcined core of the particle can be written as follows:

$$\frac{d(m_{p,L} C_{p,s} T_L)}{dt} = h_{c,L} \cdot {}^L A_p \cdot (T_g - T_L) + \sigma \cdot \varepsilon_L \cdot {}^L A_p \cdot (T_g^4 - T_L^4) + C_{p,s} \cdot T_L \cdot \frac{dm_{p,L}}{dt} \quad (6)$$

where T_L is the particle core temperature, εL is the emissivity of the particle, ${}^L A_P$ means the outer core of the particle [15]. Fig. 1 shows temperature and pressure distributions in the particle volume during the course of the calcination process.

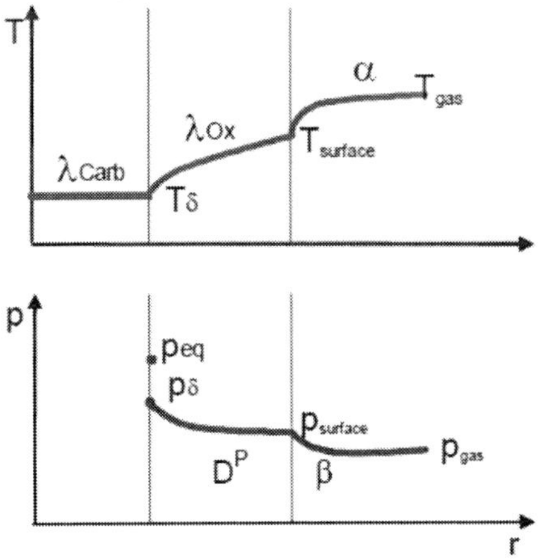

Fig. 1. Temperature and pressure distribution in the particle volume during the course of the calcination process.

The specimen comprises a dense carbonate core surrounded by a porous oxide layer. In the calcination reactor at temperature T_{gas}, heat is transferred by radiation and convection (symbolized by α) to the solid surface at a temperature of $T_{surface}$. By means of thermal conduction (λ_{Ox}) heat penetrates through the porous oxide layer to reach the reaction front, where the temperature is T_δ. Fig. 2 shows the $CaCO_3$ calcination model.

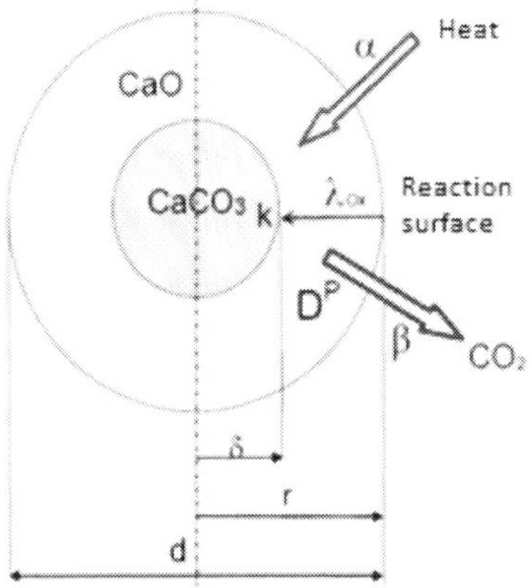

Fig. 2. $CaCO_3$ calcination mechanism model.

As the reaction enthalpy is many times greater than the internal energy, the heat flowing further into the core is negligible during the reaction. Therefore the core temperature is only slightly lower than the front temperature. Once heat is supplied, the chemical reaction (k) then takes place, for which the driving force is the deviation of CO_2 partial pressure from the equilibrium (p_{eq} - p_δ). The released CO_2 diffuses (D^P) through the porous oxide layer to the surface and finally passes by convection (β) to the surroundings where the CO_2 partial pressure p_{gas} exists.

3 Modeling the NOx formation process

The NO_x forming mechanisms included in the ANSYS Fluent package allow modeling of thermal, prompt and fuel nitrogen oxides as well as their reduction during post-combustion [16]. For solving the predicted NO_x emissions, the transport equations for nitric oxide concentration are solved. For thermal and prompt NO_x estimation, only the transport equation for the NO component is required in the following form:

$$\frac{\partial}{\partial t}(\rho Y_{NO}) + \nabla \cdot (\rho u Y_{NO}) = \nabla \cdot (\rho D \nabla Y_{NO}) + S_{NO} \quad (7)$$

where: Y_{NO} is the mass fraction of the NO component in the gas phase, D is the effective diffusion coefficient, while S_{NO} is the source component.

In order to estimate fuels NO_x, additional transport equations are required for the following gaseous components:: HCN, NH_3, and N_2O [17]. The formation of thermal NO_x is described by a group of high-temperature chemical reactions, so-called extended Zeldovich mechanism [18]. The basic reactions describing the formation of such compounds from elemental nitrogen can be written as:

$$\begin{aligned} O + N_2 &\Leftrightarrow N + NO \\ N + O_2 &\Leftrightarrow O + NO \\ N + OH &\Leftrightarrow H + NO \end{aligned} \qquad (8)$$

The rate of NO formation by reaction (8) can be written as follows:

$$\begin{aligned} \frac{d[NO]}{dt} &= k_{f,1}[O][N_2] + k_{f,2}[N][O] + k_{f,3}[N][OH] \\ &- k_{r,1}[NO][N] - k_{r,2}[NO][O] - k_{r,3}[NO][H] \end{aligned} \qquad (9)$$

where the concentration of components in parentheses is in g mol/m^3.

Combustion of hydrocarbon fuels can result in prompt NO_x formation, especially in low-temperature zones with increased fuel content, where the fuel residence time is short. Such conditions may be present on the burner surfaces, in gradual combustion or in gas turbines [19-21]. Prompt NO_x outweigh the flames. The basic reactions describing the formation of such compounds can be written as:

$$\begin{aligned} CH + N_2 &\Leftrightarrow HCN + N \\ N + O_2 &\Leftrightarrow NO + O \\ HCN + OH &\Leftrightarrow CN + H_2O \\ CN + O_2 &\Leftrightarrow NO + CO \\ CH_2 + N_2 &\Leftrightarrow HCN + NH \end{aligned} \qquad (10)$$

The first of the presented reactions is the most important. Research of the probability density distribution of the NO_x concentration point with the CH concentration point shows a close correlation. This proves that the majority of NO_x produced in the flame zone is the prompt NO_x produced by the CH reaction. It is assumed that this reaction controls the rate of prompt NO_x formation according to the correlation:

$$\frac{d[NO]}{dt} = k_0[CH][N_2] \qquad (11)$$

When the fuel is in solid form, transient compounds are formed at relatively high temperatures: HCN and NH_3. These compounds cause the formation of NO_x from elemental nitrogen through nitrous oxide N_2O. Nitrogen enters the combustion and air supply. For certain specific conditions, i.e. oxygen-rich, high-pressure areas, this transient combustion mechanism can contribute up to 90% of all NO_x. The simplest form of this mechanism can be written as two reversible basic reactions:

$$\begin{aligned} N_2 + O + M &\Leftrightarrow N_2O + M \\ N_2O + O &\Leftrightarrow 2NO \end{aligned} \qquad (12)$$

where M is generally the third component. Because the first reaction engages the third component, the formation of this mechanism is possible at higher pressure. Both reactions fix the oxygen so the occurrence of the mechanism is possible in the areas rich in this component. According to the laws on the reactions rate, the formation of NO_x by means of transition mechanism can be written as:

$$\frac{d[NO]}{dt} = 2\left(k_{f,2}[N_2O][O] - k_{r,2}[NO]^2\right) \qquad (13)$$

It is assumed that N_2O is in a quasi-stationary state, which implies dependence:

$$[N_2O] = \frac{k_{f,1}[N_2][O][M] + k_{r,2}[NO]^2}{k_{r,1}[M] + k_{f,2}[O]} \qquad (14)$$

4 The results of numerical calculations

In this work mathematical model based on Euler's method to describe the motion of the gas phase and the Lagrange method to describe the motion of particles [22-26]. Numerical calculations were made on pre-generated computational meshes, as shown in Fig.3. In order to take into account the combustion process, it was necessary to add carbon fuel data as used in the precalcinator in the program. The following parameters were used:

- calorific value – 24.6 MJ/kg,
- chemical composition as molar proportions: C – 0.409, H – 0.573, O – 0.014, N – 0.004,
- 10 fractional RRS distributions were used for: minimum diameter 10^{-6} m, maximum diameter $250 \cdot 10^{-6}$ m, main $100 \cdot 10^{-6}$ m and separation factor 3.5.

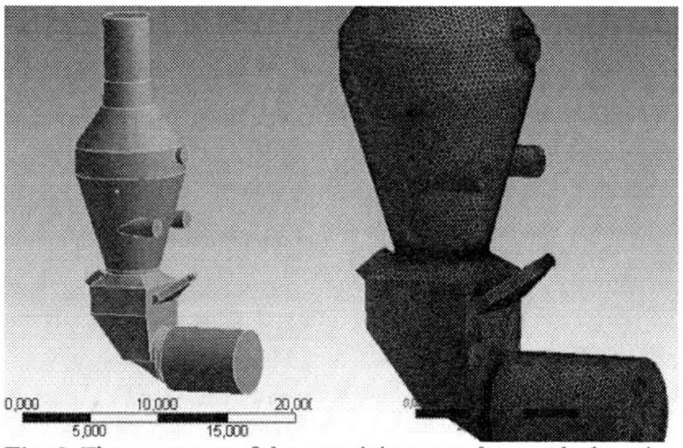

Fig. 3. The geometry of the precalcinator and numerical mesh.

4.1 Boundary conditions

The scheme analyzed in the calculation of the precalcinator is shown in Fig. 4.

It was assumed that a gas from the rotary kiln with weight composition 64% N_2, 29% CO_2 and 7% O_2. Individual components of the mixture were further treated separately due to the combustion process and the calcination process. Several different variants were considered in the calculation:

- W0 – Simulation for the current design of the precalcinator chamber with typical furnace operating conditions (Fig. 4).
- W0_3p – Simulation for the actual design of the precalcinator chamber at the III air intake at the intended operating conditions for the furnace (Fig. 4).

- W0_3p_Ex – Simulation for the current design of the precalcinator chamber at the third stage of air delivery at the intended operating conditions of the furnace and when the fuel in the calciner is increased (Fig. 4).

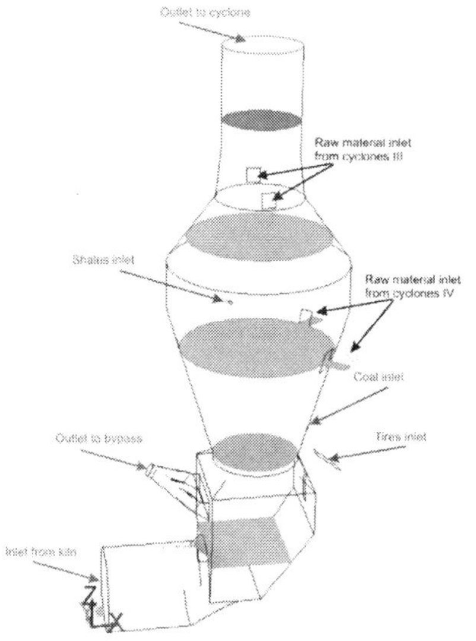

Fig. 4. Computing area.

Table 1. summarizes the accepted boundary conditions for the inlet and outlet sections. Fig. 5 shows the RRS distribution for clinker meal.

Due to the difficulty to determine actual particle size, the fractional composition of the clinker meal was scaled so as to obtain the observed separation of the meal into the rotary kiln and the heat exchangers tower. This required the use of an iterative method where the particle diameter was gradually increased. The calculations were carried out, the meal distribution was analyzed until the desired chamber/kiln ratio of 35%/65% was reached.

Table 1. Boundary conditions

Section	v [m/s]	T [K]	p_s [Pa]
Inlet from kiln	15	1370	-400
Outlet to bypas	-30	1100	-400
Outlet to cyclones	From the balance of the incoming mass	1112	-800
Coal inlet	100	300	0
Shales inlet	45	300	0
Tire inlet	Locked – treated as a wall		
Raw material inlet III	From the particle flow rate	1000	0
Raw material inlet IV		1100	0

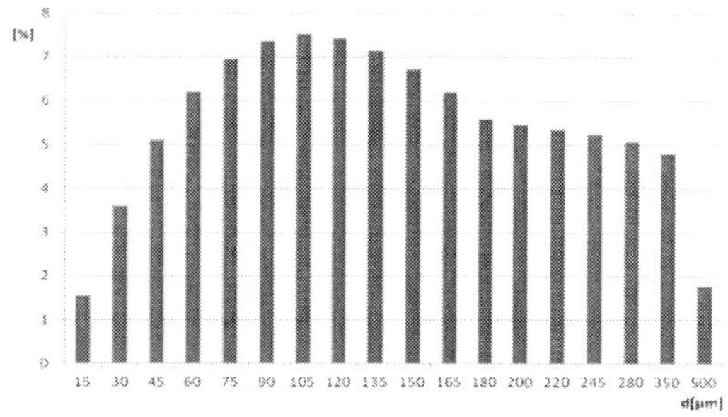

Fig. 5. Mass distribution of individual fraction obtained from the R-R-S method.

4.2 Calculation results for gas-particle mixture flow

Fig. 6 shows the velocity sections in the lengthwise midsection for the analyzed variants. Fig. 7 presents the temperature profile for proper conditions. Fig. 8 shows the velocity profile in cross-sections. Fig. 9 shows the particle concentration distribution in lengthwise section.

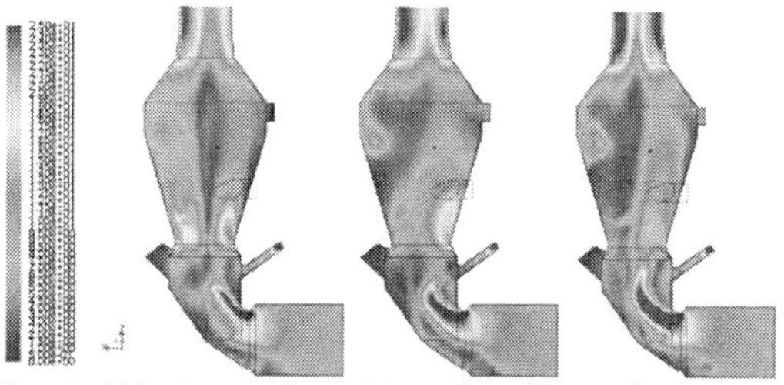

Fig. 6. Velocity profile (m/s): variant W0, variantW0_3p, variant W0_3p_ex

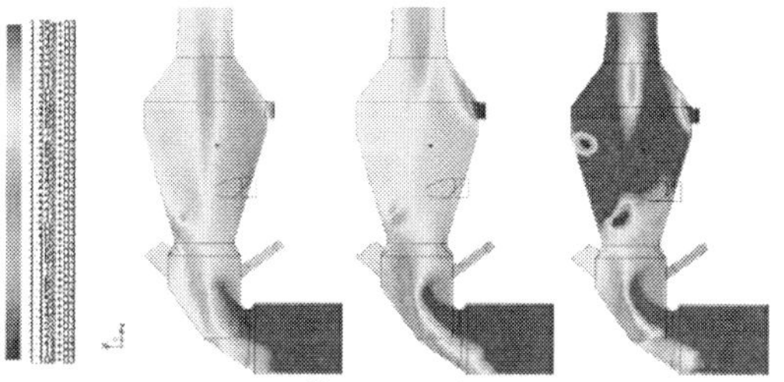

Fig. 7. Temperature profile (K): variant W0, variant W0_3p, variant W0_3p_ex.

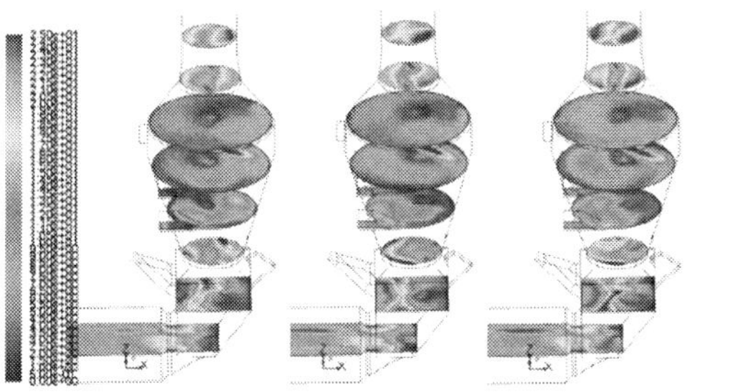

Fig. 8. Velocity profile (m/s), variant W0, variantW0_3p, variant W0_3p_ex.

DOI: 10.1051/e3sconf/20171902002

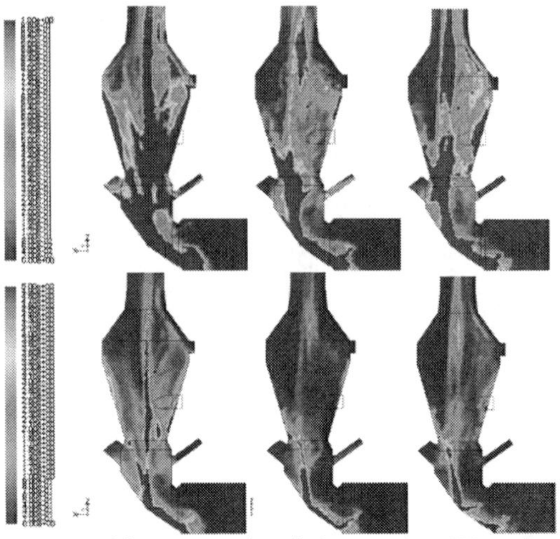

Fig. 9. Particle concentration (kg/m3): variant W0, variant W0_3p, variant W0_3p_ex.

Fig. 10 shows the temperature profile in cross-sections. Fig. 11 presents the CO_2 concentration distribution in crosswise sections and NO_x mass distribution in crosswise sections is shown in Fig. 12.

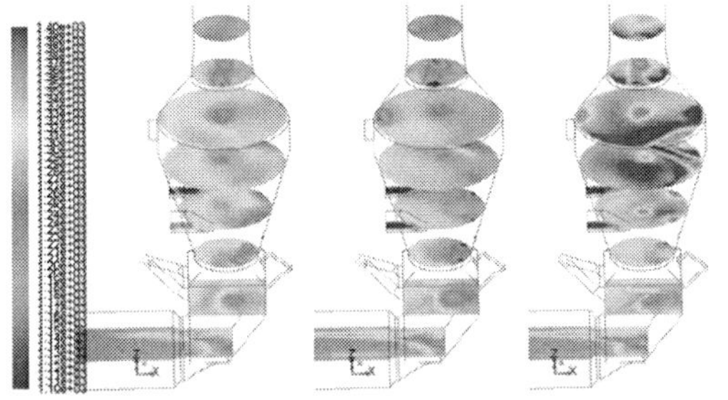

Fig. 10. Temperature profile (K), variant W0, variantW0_3p, variant W0_3p_ex.

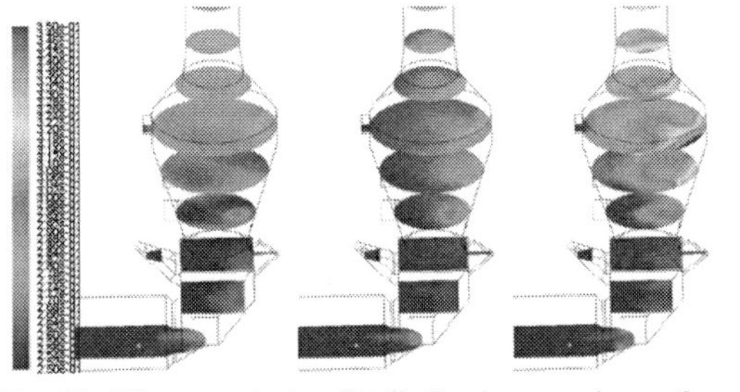

Fig. 11. CO_2 concentration distribution in crosswise sections: variant W0, variant W0_3p, variant W0_3p_ex.

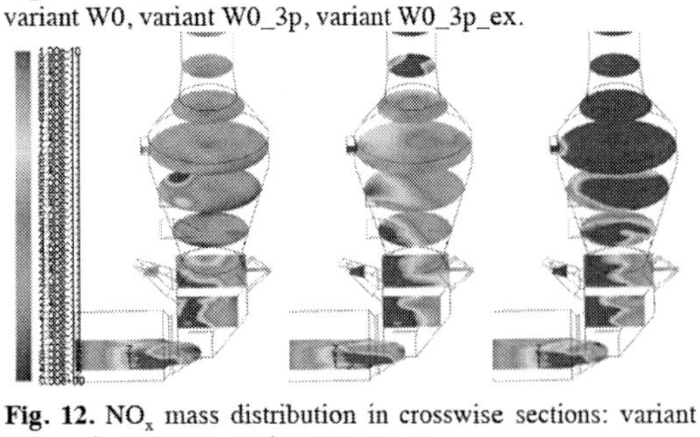

Fig. 12. NO_x mass distribution in crosswise sections: variant W0, variant W0_3p, variant W0_3p_ex.

Table 2 and 3 show selected values in the outlet section of the chamber. Compiled data are average values in the cross-section of the discharge from the precalcinator. They can be useful for determining the size of derivatives and also for choosing the optimal variant. Table 4 shows the percent distribution of the meal between the inlet to the kiln and the outlet from the chamber.

Table 2. Average values in outlet section

Variant	Temperature [K]	Velocity [m/s]	Particle concentra-tion [kg/m^3]
W0	1,269	17.23	0.3207
W0_3p	1,254	19.25	0.2978
W0_3p_Ex	1,351	20.92	0.2922

Table 3. Average values in outlet section.

Variant	Pressure drop between inlet and outlet [Pa]
W0	515.8 (-233.5 282.3)
W0_3p	345.6 (-186.5 159.1)
W0_3p ex	365.3 (-185.8 179.5)

Table 4. The sum of mass percentages of meal coming to both outlets of cyclones III and IV for three different flow variants.

Variant	From Cyclone IV		From Cyclone III	
	to cyclones	to kiln	to cyclones	to kiln
W0	14.05%	85.86%	56.35%	43.44%
W0_3p	5.82%	94.16%	55.01%	44.30%
W0_3p_Ex	9.26%	90.74%	53.85%	46.12%

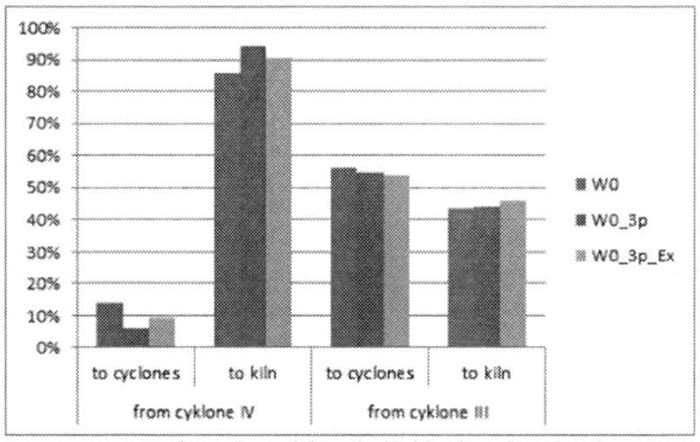

Fig. 13. Separation of meal into individual outlets for three different computational variants.

Based on Table 3 and Fig. 13, it can be concluded that, with respect to the basic variant W0, the variant with additional air W0_3p causes an increase in the number of particles coming from the III and IV row of the cyclones to the bottom outlet in the kiln. Table 5 and Fig. 14 show the aggregate flow rate of the scalded pulp reaching the funnel and kiln of both cyclones. The presented data show that in the case of III row cyclones the particle mass reaches the kiln approximately 2.5 kg/s. The particle mass flowing out the kiln is the largest for variant W0 (4.5 kg/s).

Table 5. Scaled meal mass flow rate [kg/s] reaching the individual outlets of cyclones III and IV row for tested variants.

Variant	From Cyclone IV		From Cyclone III	
	to cyclones	to kiln	to cyclones	to kiln
W0	1.7981	2.3425	4.5994	2.4340
W0_3p	0.8565	1.9468	2.6828	2.8216
W0_Ex	0.9007	2.1653	2.6424	2.4619

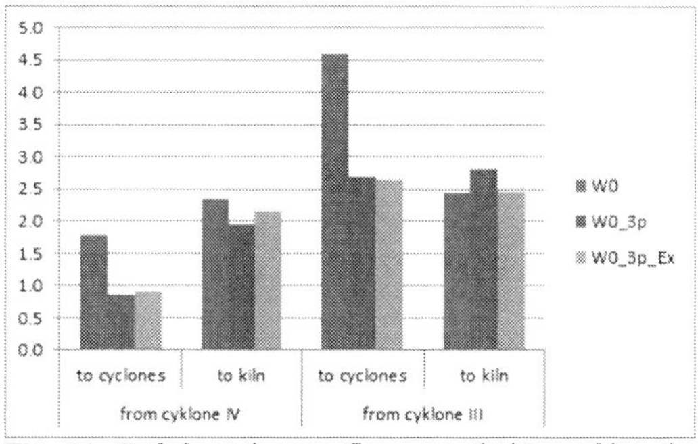

Fig. 14. Scaled meal mass flow rate [kg/s] reaching the individual outlets.

5 Conclusions

Completed numerical calculations have shown that the construction of the precalcinator and its operating conditions (the presence of the tertiary air, the combusted fuel flow rate) affect the heat-flow and thermochemical processes in the chamber. It has been found that the increase of NO_x concentration in the kiln heater chamber in all variants is three orders less than the permitted emission of 500 mg/Nm^3. Lack of empirical data on NO_x concentration at the outlet of the rotary kiln makes it difficult to analyze in-depth reductions in the precalcinator in detail.

It is advisable to carry out further detailed numerical studies on the optimization of the precalcinator chamber construction (burner location, air supply, a method of feeding the meal). It is also necessary to support numerical results with measurements on the object and detailed analysis of fuel and raw materials to refine the values of mathematical model coefficients.

References

1. B. Dobrowolski B., J. Wydrych, JTAM, Warszawa, **45**, 25 (2007)
2. D.K. Fidaros, C,A. Baxevanou, C.D. Dritselis C.D., Vlachos N.S., Powder Technol. **171**, 15 (2007)
3. Z. Hu, J. Lu, L. Huang, S. Wang, Commun. Nonlinear Sci. Numer. Simul. **11**, 12 (2006)
4. L. Huang, J. Lu, F. Xia, W. Li, H. Ren, Chem. Eng. Process. **45**, 10 (2006)
5. E. Worrell, L. Price, N. Martin, C. Hendriks, L. Ozawa Meida, Annu. Rev. Energy Env. **26**, 27 (2001)
6. S. Zhang, E. Worrell, Crijns-Graus W., Appl. Energy,**147**, 22 (2015)
7. B.M. Al Smadi B, K.K. Al-Zboon, K.M. Shatnawi, Jordan J. Civ. Eng. **3**, No. 3, 18 (2009)
8. K.T. Kaddatz, M.G. Rasul, A. Rahman, Procedia Eng. **56**, 8 (2013)
9. A. Rahman, M.G. Rasul, M.M.K. Khan, S. Sharma, Energy Procedia **61**, 6 (2014)
10. M.T. Javeda, N. Irfana, B.M. Gibbs, J. Environ. Manage. **83**, 39 (2007)
11. P. Dai, B. González, J.S. Dennis, Chem. Eng. J. **304**, 11 (2016)
12. I. Martínez, G. Grasa, R. Murillo, B. Arias, J.C. Abanades , Chem. Eng. J. **215–216**, 9 (2013)
13. A. Teklay, C. Yina, L. Rosendahl, L.L. Køhler, Appl. Clay Sci. **103**, 10 (2015)
14. H. Mikulcic, E. Berg, M. Vujanovic, P. Priesching, L. Perkovic, R. Tatschl, N. Duic, Chem. Eng. Sci. **69**, 9 (2012)
15. A. Teklay, C. Yina, L. Rosendahl, M. Bøjer, Cem. Concr. Res. **61–62**, 9 (2014)
16. ANSYS Fluent, Ansys Inc. (2016)
17. C. Schonenbeck, R. Gadiou, D. Schwartz, Fuel **83**, 8 (2004)
18. Y.B. Zeldovich, G.I. Barenblatt, Combust. Flame **3** (1959)
19. S. Bhunia S., A.K. Sadhukhan, P. Gupta, Fuel Process. Technol. **158**, 12 (2017)
20. W.P. Adamczyk, R.A. Bialecki, M. Ditaranto, P. Gladysz, N. Erland, L. Haugen, A. Katelbach-Wozniak, A. Klimanek, S. Sladek, A. Szlek, G. Wecel, Energy (to be published in 2017)
21. J. McConnell, B. Goshayeshi, J.C. Sutherland, Fuel **201**, 12 (2017)
22. D.K. Fidaros, C.A. Baxevanou, C.D. Dritselis, N.S. Vlachos, Powder Technol. **171**, 15 (2007)
23. W.K.H. Ariyaratne, A. Malagalage, M.C. Melaaen, L.A. Tokheim, Chem. Eng. Sci. **123**, 13 (2015)
24. G. Borsuk, B. Dobrowolski, J. Wydrych, Chem. Process. Eng. **27**, 12 (2006)
25. G. Borsuk, B. Dobrowolski, Chem. Process. Eng. **27**, 3/2, 15 (2006)
26. R. Saidur, M.S. Hossain, M.R. Islam, H. Fayaz, H.A. Mohammed, Renew. Sustainable Energy Rev. **15**, 14 (2011)

E3S Web of Conferences **19**, 02003 (2017) DOI: 10.1051/e3sconf/20171902003
EEMS 2017

Simulation of diesel engine emissions on the example of Fiat Panda in the NEDC test

Katarzyna Botwinska[1,*], *Remigiusz* Mruk[2], *Jacek* Słoma[3], *Karol* Tucki[4] and *Mateusz* Zaleski[5]

[1,2,3,4,5] Warsaw University Of Life Sciences, Faculty of Production Engineering, Department of Production Organization and Engineering, Nowoursynowska 164, Warsaw, Poland

Abstract. Road transport may be deemed a strategic branch of modern economy. Unfortunately, a rapid increase in the number of on-road motor vehicles entails some negative consequences as well, for instance, excessive concentration of exhausts produced by engines which results in deterioration of air quality. EURO emission standards which define acceptable limits for exhaust emissions of power units is an example of an activity performed in attempt to improve air quality. The EURO standard defines permissible amount of exhausts produced by a vehicle. Presently new units are examined through NEDC test. For the purpose of this thesis, a virtual test stand in a form of a computer simulation of a chassis dynamometer was used to simulate emission of a diesel engine (compression-ignition engine) in the NEDC test. Actual parameters of the 1.3 MultiJet engine of the Fiat Panda passenger car of 2014 were applied in the model. The simulation was carried out in the Matlab Simulink environment. The simulation model of the Fiat Panda passenger car enables the designation of the emission waveform for all test stages which corresponds to the values received during an approval test in real-life conditions.

1 Introduction

Air pollution caused by excessive exhaust emission from vehicle engines constitutes a serious problem, which has been growing for years. Transport-related contamination is the most severe within the area of urban agglomerations where the traffic density is definitely the highest [1]. The improvement of road infrastructure, wide availability of vehicles and relatively low prices of used cars result in a sharp increase in the number of on-road cars both in Poland and within the area of the European Community [1, 2]. The problem lies in the age and technical condition of motor vehicles present on roads. Unfortunately, low prices and average level of wealth of society contribute to the increase of the number of vehicles in poor technical condition which frequently fail to meet emission standards provided for their years of production [3]. Internal combustion engines, powered with petroleum-derived fuels, emit a lot of harmful substances having negative, direct and indirect, influence on live organisms. Harmful substances produced by internal combustion engines of power units include: hydrocarbons, nitrous oxides, carbon oxides, sulphur oxides, particulates, aldehydes and heavy metals. Such substances are toxic to the natural environment, therefore, the need to improve air quality is more and more emphasised [2].

European emission limits are legal regulations, which define acceptable amounts of particular components of exhausts. The standards apply both to petrol and diesel engines. The first emission standard, Euro 1, was introduced in 1993 and it established the emission level for passenger cars as well as passenger and light trucks [1]. The currently binding standard is Euro VI which was introduced by the Regulation (EC) no 715/2007 of the European Parliament and of the Council of 20 June 2007. NEDC (New European Driving Cycle) test is closely related to the standard. It is a binding approval test detecting average fuel consumption of the vehicle and the emission of carbon dioxide and harmful substances (CO, NOx, HC, PM). It is the Euro VI standard, which determines permissible exhaust limits. Currently, every new vehicle which is to be accepted as roadworthy has to undergo such test and receive an approval certificate. This certificate is a prerequisite for roadworthiness [4].

The test lasts 18 minutes and 40 seconds in total and is divided into two parts: urban driving cycle and extra-urban driving cycle. The urban driving cycle includes four repeated cycles in which the vehicle is accelerated to the speeds of 15, 32 and 50 km/h, then it is stopped for a specific time. The distance of approximately 4 km is covered in this cycle, and the average speed amounts to 18.35 km/h. The total pausing time of the car is over 3 minutes, during which the manual gearbox is in the neutral position with clutch engaged. This stage is supposed to represent movement of the vehicle in urban agglomerations. After that, the next, extra-urban stage begins in which the vehicle is accelerated in turn to the speeds of 70, 100 and 120 km/h. The entire stage lasts 6 minutes and 40 seconds, the vehicle is not stopped but it is decelerated to 50 km/h. The distance covered in this

* Corresponding author: katarzyna_botwinska@sggw.pl

part is about 7 km [5, 6]. The test is performed in laboratory conditions on a specially prepared chassis dynamometer. It allows the simulation of specific engine loads resulting from the stages of the test. The stand is fitted with control and measurement instruments, which among other functions, enable the measurement of the amounts of substances in produced exhausts [4].

1.1 The aims and scope of the study

This thesis is aimed at recreating the NEDC test of a selected passenger car. For that purpose, a computer simulation was used which reflects the cooperation between a chassis dynamometer and a model vehicle according to the assumed research cycle. Therefore, the objective of the thesis was to prepare a simulation model of a chassis dynamometer and simulation of exhaust emission of a diesel engine of a selected Fiat Panda passenger car. The scope of the project included the construction of the aforementioned model and the analysis of the obtained results.

2 Materials and methods

2.1. The object of research

In the simulation the parameters of the 1.3 MultiJet II 16 v, 70 kW engine of Fiat Panda Easy passenger car of 2014 were applied. It is a power unit which conforms to the highest level of the emission standard Euro VI. According to manufacturing specifications the values of fuel combustion for the said engine are: in the urban cycle 4.7 l/100 km, in the extra-urban cycle 3.5l/100 km and in the mixed cycle 4,7l/100 km. CO_2 emission in the mixed cycle amounts to 104 [g/km] [7]. Table 1 demonstrates specifications of the model engine.

Table 1. Specification of 1.3 MultiJet II 16v, 70 kW engine.

Parameter	Value
Capacity	1248 cm3
Maximum power	55 kW (75 KM) at 4000 [rpm]
Maximum torque	190 N·m at 1500 [rpm]
Generation	MultiJet II fitted with diesel particulate filter (DPF)
Arrangement of cylinders	line
Number of cylinders	4
Injections	Direct MultiJet "Common Rail" controlled electronically with a turbocharger and intercooler

2.2 Prepared simulation

Simulation models were created in Matlab Simulink environment. They were developed on the basis of documentation containing requirements for the NEDC test [5, 6] and they allowed to obtain information about the options for chassis dynamometer selection and to carry out simulation of exhaust emission of the Fiat Panda car with the MultiJet engine.

To ensure that the simulation is run properly, standard data for the NEDC test was supplemented with information allowing the control of the signal of the engaged gear, the signal of engagement / disengagement of the clutch and signal of the time controlling the simulation which allows the program to perform all key activities according to the test schedule. Thus, in order to obtain a set of gear values in the gearbox in the form of integers, characteristic points for the manipulation of the signal of the gear number and clutch engagement were established for particular stages of the NEDC test for urban and extra-urban cycles. The prepared data controlling the simulation for the NEDC test was entered into the simulation environment as data sets for the tool enabling the creation of continuous waveforms based on characteristic points [8, 9]. Fig. 1 below depicts the view of waveforms.

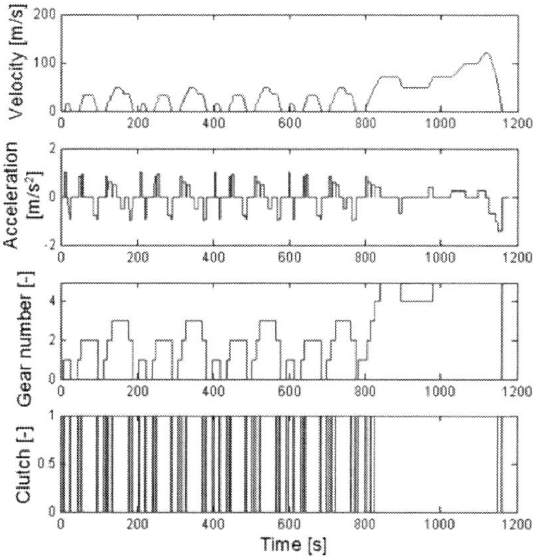

Fig. 1. Component for entering a data set with the view of the introduced control signals in the simulation project of the NEDC test.

The component was entered into the simulation model from which the signals of particular control values for further model elements were then derived. Using the vehicle velocity signal, the "Derivative" component was used to generate acceleration signal necessary for further calculations of the inertia forces of the chassis dynamometer. For that purpose, the following mathematical correlation was used:

$$a(t) = \frac{dv(t)}{dt} \left[\frac{m}{s^2}\right] \quad (1)$$

Where:
$a(t)$ - acceleration (m/s²),
$v(t)$ - car velocity (m/s).

Apart from calculation components, the fragment of the model contains "NEDC diagram" element depicting the generated control waveform. The individual parts of the simulation model were built basis on the mathematical reliance contained in the literature sources [10 - 11].

DOI: 10.1051/e3sconf/20171902003

Fig. 2 below presents the fragment of the model responsible for the generation of signals controlling the test.

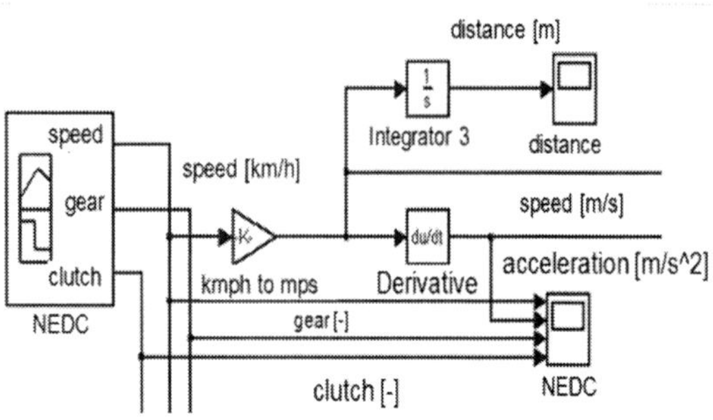

Fig 2. Fragment of the model responsible for the generation of signals controlling the test.

Then a fragment of the model responsible for the determination of the internal friction value of the chassis dynamometer was constructed according to the following correlation:

$$F_T(t) = a + bv(t)^2 \qquad (2)$$

Where:
$F_T(t)$ - load from the friction process absorbed by the chassis dynamometer (N),
a - value equal to the rolling resistance (N),
b - value equal to the air drag coefficient (N/(km/h)^2),
$v(t)$ - velocity (km/h).

Forces generated by the dynamometer related to the required mass inertia during the vehicle velocity changes were determined pursuant to the following correlation [10]:

$$F_m(t) = m\frac{dv(t)}{dt} \qquad (3)$$

Where:
$F_m(t)$ - inertia load absorbed by the chassis dynamometer (N),
m - Equivalent inertia (kg),
$v(t)$ - velocity (km/h).

The simulation involved matrix calculations which enable the simultaneous generation of many cases of the dynamometer configuration for the selected car mass. In the above correlation specific parameters of the chassis dynamometer included in [6] were used. Table 2 presents the applied correlations.

Table 2. Configuration set of the chassis dynamometer selected for the simulation.

Vehicle reference mass (RM) (kg)		Equivalent inertia	
965 < RM ≤ 1 080		1 020	
Power and load absorbed by the dynamometer at 80 km/h		Road load coefficient	
6.0	270	6.0	270

In order to acquire information allowing the selection of chassis dynamometer elements responsible for generation of load in the simulation, components were used to calculate temporary value of the power from the car wheels or provided to the system to obtain characteristics of velocity changes in compliance with the guidelines. Temporary power value was calculated from the following correlation which also included a matrix calculus:

$$P(t) = F(t)v(t) \qquad (4)$$

Where:
$P(t)$ - temporary power value (W),
$F(t)$ - force generated on the dynamometer (N),
$v(t)$ - velocity (km/h).

Fig. 3 below shows the fragment of the model determining the above values:

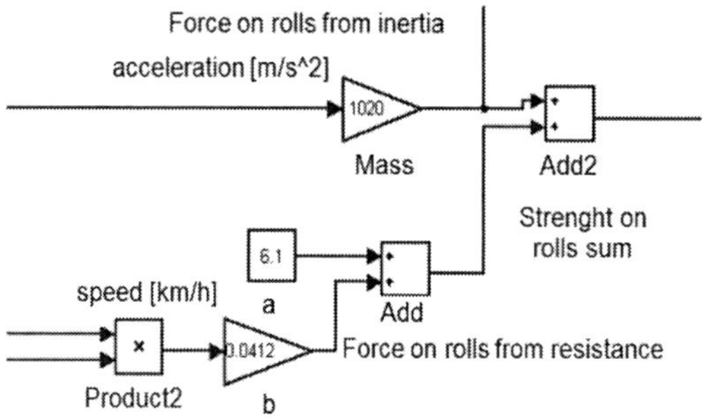

Fig. 3. Fragment of the simulation model responsible for the calculation of the internal friction forces, inertia forces, temporary power of the chassis dynamometer in the NEDC test.

The next step was to create a simulation model of the NEDC test for the configuration of the chassis dynamometer compliant with the parameters of Fiat Panda with the 1.3 MultiJet engine. Fig. 4 below depicts a fragment of the simulation model responsible for the calculation of the rotational speed of vehicle wheels, gearbox ratio, engine rotational speed ωe and engine torque Me. Those parameters are necessary to establish temporary emission rates of selected exhaust components. For that purpose, the following mathematical correlation was used:

$$Me = (F_T + F_m) \cdot R \cdot Tr \cdot C\ [N \cdot m] \qquad (5)$$

$$\omega e = \frac{v(t)}{R} \cdot T\omega \cdot C \left[\frac{rad}{s}\right] \qquad (6)$$

Where:
R - wheel radius [m],
Tr - Transmission ratio of the drive system for the torque (0 1/13.46; 1/7.05; 1/4.55; 1/3.24; 1/2.42] [-]),
C - Clutch engagement signal (0, 1 [-]),
$v(t)$ - Temporary vehicle velocity in the NEDC test,
$T\omega$ - Transmission ratio of the drive system for the rotational speed (0; 13.46; 7.05; 4.55; 3.24; 2.42] [-]).

Below Fig. 4 shows a fragment of the simulation responsible for the calculation of the rotational speed of vehicle wheels, gearbox ratio, engine rotational speed and engine torque in the NEDC test.

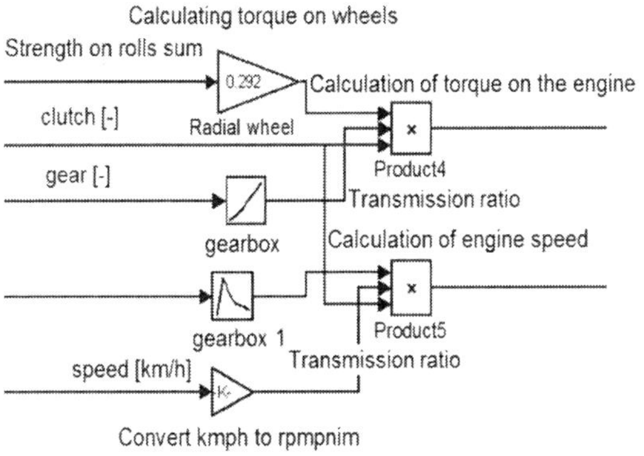

Fig. 4. Fragment of the simulation responsible for the calculation of the rotational speed of vehicle wheels, gearbox ratio, engine rotational speed and engine torque in the NEDC test.

Simulation based on instantaneous values: engine speed, engine torque and fuel and air consumption characteristics of the engine as a function of speed and torque, calculates the instantaneous fuel and air streams fed to the engine in the NEDC test. The simulation uses the mathematical formulas presented below:

$$\dot{air} = f_{air}(\omega e,\ Me)\ \left[\frac{g}{s}\right] \quad (7)$$

$$\dot{fuel} = f_{fuel}(\omega e,\ Me)\ \left[\frac{g}{s}\right] \quad (8)$$

where:

f_{air}, f_{fuel} - Function of air, fuel consumption depending on the rotational speed and torque [g/s]

Below Fig. 5 shows a fragment of the model that calculates instantaneous air and fuel streams to the engine in the NEDC test.

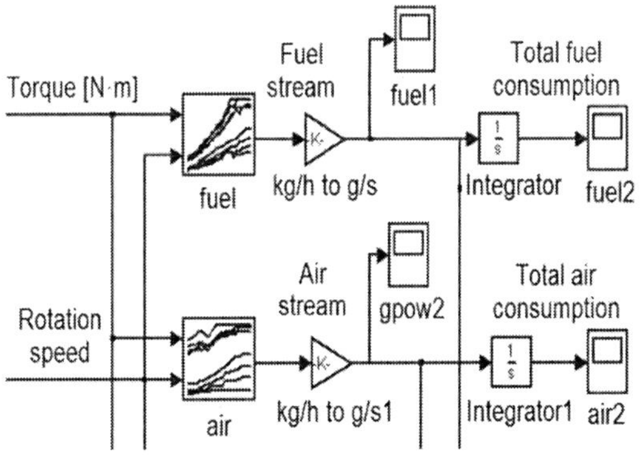

Fig. 5. Fragment of the model responsible for the calculation of the temporary values of the air and fuel streams in the NEDC test.

Additionally, the fragment of the model responsible for the calculation of the temporary values of the streams of selected exhaust components is presented below. It includes fuel consumption $\dot{fuel}$ [g/s] and air consumption $\dot{air}$ [g/s] in the function of the engine rotational speed ωs [rpm] and torque produced by the engine Ms [N·m], however, for the diagram legibility purposes, the mentioned calculation fragments were omitted in the below fragment.

$$\dot{CO} = f_{CO}(\omega s,\ Ms)\cdot(\dot{air} + \dot{fuel})\ \left[\frac{g}{s}\right] \quad (9)$$

$$\dot{CO2} = f_{CO2}(\omega s,\ Ms)\cdot(\dot{air} + \dot{fuel})\ \left[\frac{g}{s}\right] \quad (10)$$

$$\dot{NOx} = f_{NOx}(\omega s,\ Ms)\cdot(\dot{air} + \dot{fuel})\ \left[\frac{g}{s}\right] \quad (11)$$

Where:

f_{CO}, f_{CO2}, f_{NOx} - Function of the carbon monoxide, carbon dioxide, nitrogen oxides content of the exhaust gases depending on the rotational speed and torque [m%],

ωs - Engine rotational speed [rad/s],

Ms - Engine torque [N·m]),

$\dot{air}, \dot{fuel}$ - Air, fuel stream [g/s].

Fig. 6 shows a fragment of the model responsible for the calculation of the temporary values of the streams of selected exhaust components in the NEDC test.

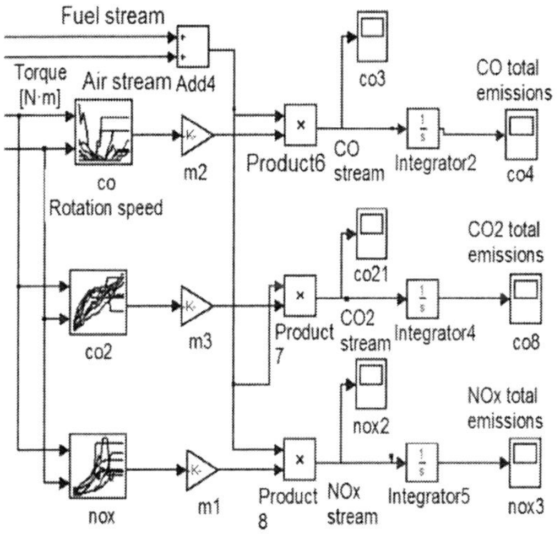

Fig. 6. Fragment of the model responsible for the calculation of the temporary values of the streams of selected exhaust components in the NEDC test.

3 Results and discussion

As a result of the simulation of operation of the 1.3 MultiJet engine within the NEDC test, the waveforms of stream [g/s] and consumption [s] of the following media were obtained: fuel and air. The waveforms are depicted below. Fig. 7 shows changes in the fuel stream [g/s] and fuel consumption [s] during the simulation.

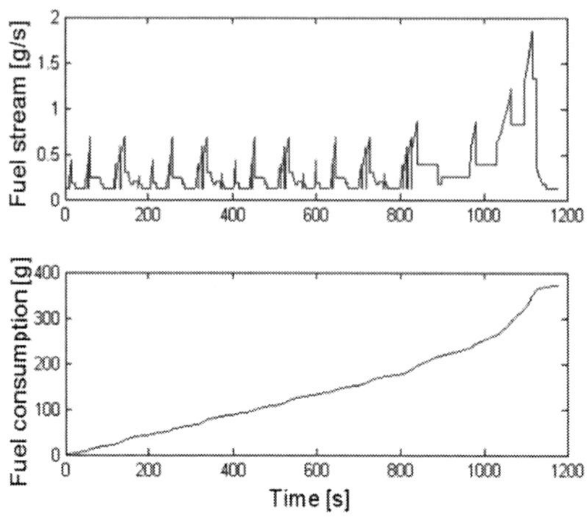

Fig. 7. Changes in the fuel stream [g/s] and fuel consumption [s] during the simulation.

Fig. 8 shows changes in the air stream [g/s] and air consumption [s] during the simulation.

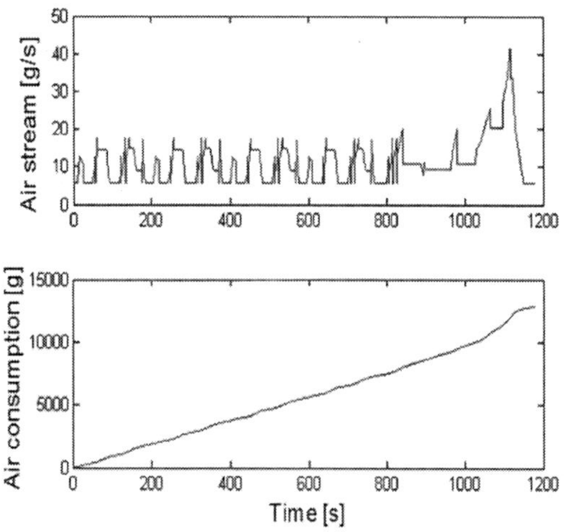

Fig. 8. Changes in the air stream [g/s] and air consumption [s] during the simulation.

As a result of the simulation of operation and emission of the 1.3 MultiJet engine within the NEDC test, the waveforms of stream [g/s] and emission [g] of the following substances were obtained: carbon oxide CO, carbon dioxide CO_2 and nitrogen oxides NO_x. The obtained streams of fuel and air streams converge to the engine work during the NEDC test. There were analogies between these parameters and vehicle speed at a given test point. Based on the similarities with scientific sources about the NEDC test, waveforms can be considered valid. The course of the carbon monoxide stream changes with the speed of the vehicle. In the first phase of the 'urban' test, a cyclical increase in the carbon monoxide content can be observed. Maximum values characterize the fastest vehicle speed at any given time for the urban phase. The value of the parameter close to zero corresponds to the stopping the vehicle. In the second extra-urban phase, two points with the highest concentration can be seen, which means the maximum speed for this phase. In the extra-urban cycle, the carbon monoxide content is lower than in the urban cycle. Fig. 9 shows changes in the stream [g/s] and emission [g] of carbon oxide CO during the simulation.

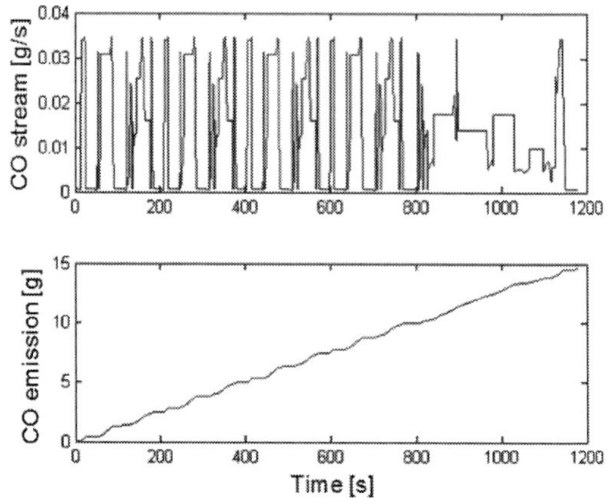

Fig. 9. Changes in the stream [g/s] and emission [g] of carbon oxide CO during the simulation.

The course of carbon dioxide also changes with the speed of the vehicle. The increase in concentration is proportional to the increase in speed. Compared to the course of carbon monoxide, however, there is a dramatic increase in the concentration for the In the extra-urban cycle phase. Fig. 10 shows changes in the stream [g/s] and emission [g] of carbon dioxide CO2 during the simulation.

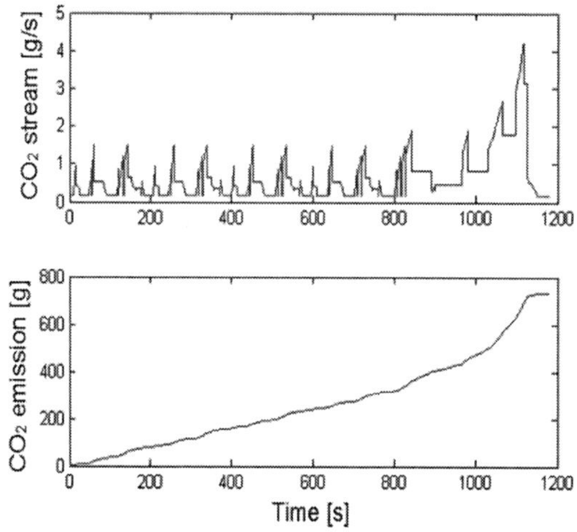

Fig. 10. Changes in the stream [g/s] and emission [g] of carbon dioxide CO_2 during the simulation.

Based on the results of the emission of nitrogen oxides, their concentration was highest with sudden increase in vehicle speed. The content of nitrogen oxides in the exhaust gas is, however, much lower than that of carbon mono and dioxide. Fig. 11 shows changes in the stream [g/s] and emission [g] of carbon oxide CO during the simulation.

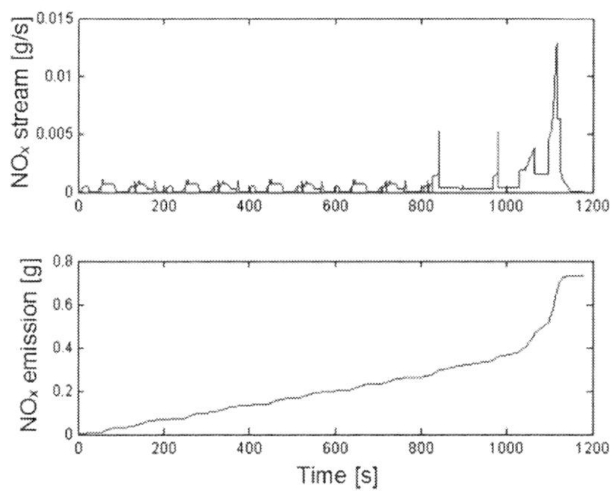

Fig. 11 Changes in the stream [g/s] and emission [g] of nitrogen oxides NO_x during the simulation.

4 Conclusions

The following conclusions may be drawn from the analysis of the prepared simulation model, obtained simulation results and professional literature:

- Matlab Simulink environment facilitates the creation of models and simulation of dynamic elements operation without the need to use complicated mathematical correlations;
- The developed simulation model provided temporary waveforms the total power for the simulation process. Such waveforms enable the determination of the maximum power a vehicle should be loaded with to have proper waveforms of velocity and may constitute the basis for the selection of the brake (in the case of the set with the highest vehicle mass for the NEDC test the value equals to 54 kW);
- Simulation model for Fiat Panda 1.3 MultiJet provides not only the temporary waveforms of emission for particular test stages but also cumulated waveforms the value of which corresponds to the real processes of the approval tests;
- Developed simulation models may be the basis for the creation of real systems for chassis dynamometer control with the simulation of mass inertia.

References

1. I. Czapliński, R. Smolec, Autobusy: technika, eksploatacja, systemy transportowe **17** (2016).
2. E. Trzensik, M. Świątłoń, Zesz. Nauk. Polit. Śl. (2011).
3. M. Witaszek, K. Witaszek,. Zesz. Nauk. Polit. Śl. Transport (2015).
4. J. Merkisz, M. Andrzejewski, J. Pielecha, Combust. Engines **50** (2011).
5. Barlow, T. J., Latham, S., McCrae, I. S., & Boulter, P. G., TRL Pub. Proj. Rep. (2009)
6. Regulation No 83 of the Economic Commission for Europe of the United Nations (UNECE).
7. http://cenniki.fiat.pl/fiat/download/fiat/katalog/panda/katalog/nowa_panda_katalog.pdf
8. B. Mrozek, Z. Mrozek, PLJ (1998)
9. T. Kumar, OUP (2011)
10. W. Serdecki, Poznan Univ. of Technol (2012)
11. A. Ambrozik, T. Ambrozik, D. Kurczyński, Post. Nauk. i Tech., (2012)

Particulate matter in the rural settlement during winter time

Tomasz Olszowski[1,*]

[1]Department of Thermal Engineering and Industrial Facilities, Opole University of Technology, Mikołajczyka 5, 45-271 Opole, Poland

Abstract. The objective of this study was to analyzed the variability of the ambient particulates mass concentration in an area occupied by rural development. The analysis applied daily and hourly $PM_{2.5}$ and PM_{10} levels. Data were derived on the basis of measurement results with the application of stationary gravimetric samplers and optical dust meter. The obtained data were compared with the results from the urban air quality monitoring network in Opole. Principal Component Analysis was used for data analysis. Research hypotheses were checked using U Mann-Whitney. It was indicated that during the smog episodes, the ratio of the inhalable dust fraction in the rural aerosol is greater than for the case of the urban aerosol. It was established that the principal meteorological factors affecting the local air quality. Air temperature, atmospheric pressure, movement of air masses and occurrence of precipitation are the most important. It was demonstrated that the during the temperature inversion phenomenon, the values of the hourly and daily mass concentration of $PM_{2.5}$ and PM_{10} are very improper. The decrease of the PM's concentration to a safe level is principally relative to the occurrence of wind and precipitation.

1 Introduction

During the cold months of the year, anthropogenic activity leads to a considerable deterioration of air quality [1]. This situation is particularly relevant for geographical the area of the former Eastern Bloc, where household heating is based on useful energy supply from outdated systems. This problem is however, not limited only to urban areas [2, 3]. In the areas of compact rural settlement, the ratio of solid fuels combustion in the total production of thermal energy is in the range from 70 to 95% [4]. Unfortunately, papers discussing the results of research on air quality in rural areas are scarce [5].

It's widely known, that in Poland, the air quality is worse during the cold season than in warm [6]. In the cold season, the main source of ambient aerosol is uncontrolled emission from local individual heating systems. These systems are commonly outdated, which leads to the larger degree of enriching the troposphere with a load of hazardous substances.

Among the air pollutants, particulate matter (PM) has the greatest impact on the quality of the environment. This effect can be attributed to the large load and chemical properties of the emitted particles. As it contains a variety of hazardous substances deposited on its surface, suspended particulate matter is known to have a negative impact on the health of humans and local conditions of the microclimate. The group of the most hazardous includes fine particles, formed by the particles of up to 2.5 micrometers in diameter ($PM_{2.5}$). Such fine particles can penetrate into the blood vessels and human organs following the processes of inhalation and gas exchange. As given in [6]: *knowledge of the possible origin of $PM_{2.5}$ is still incomplete. There is a limited number of appropriate $PM_{2.5}$ sampling points in Central and Eastern Europe which substantially restricts accurate identification of PM sources.*" Equally, the information regarding the ratio of fine particles in PM_{10} (i.e. ones with the aerodynamic diameter <10 μm) is scarce. This ratio is often considered as the principal indicator of the air quality in terms of the particulate matter mass concentration alongside with heavy metal and PAH's.

Due to the scarcity of reports into the condition of troposphere in the areas of Central and Eastern Europe in the cold season, a study into the variability of the concentrations of $PM_{2.5}$ and PM_{10} was undertaken. The main objective of this study was to analyzed the variability of the ambient particulates mass concentration in an area occupied by rural development. The objective of this study was to determine the ratio of the fine fraction in PM_{10} and a comparison between air quality in the rural and urban areas.

The scope of a research project in this area involved the verification of a hypothesis regarding the equal ratios of $PM_{2.5}$ in PM_{10} in the areas with different areas of urban development. The hypothesis was verified for the conditions of smog occurrence and during the periods when it was absent. In addition, the study led to the substantiation of the hypothesis regarding the impact of meteorological parameters on the air quality in the areas of rural development.

2 Materials and methods

2.1 Measurement site

* Corresponding author: t.olszowski@po.opole.pl

The main research was conducted in rural area over 3 cold seasons (2015–2017). The registration of meteorological parameters and concentration of $PM_{2.5}$ and PM_{10} in the rural area were realised in the centre of village Kotórz Mały (Poland, 50° 43′ 50″ N; 18° 02′ 36″ E; 1025 inhabitants). The measurement point was located in direct neighborhood of compact rural building development area.

The objective of a comparison of two areas with different levels of urban development, the work applied data from a town measurement station managed by The Voivodeship Inspectorate of Environmental Protection [7]. The meteorological data representative of a urban area and concentration of PM's were determined in provincial town of Opole (Poland, 50° 40′ 13″ N; 17° 55′ 12″ E, 122,000 inhabitants). The monitoring station was situated in the centre of the city, near low and high residential housing and commercial development. Measurement point was situated 12 km south-west from the observation point in Kotórz Mały.

The tested zone in the village is characterized by low buildings, which predominantly use obsolete individual heating systems (coal, in a proportion of 91%). In the urban area, the ratio of the individual household heating systems is considerable smaller. Within the radius of 1 km from the measurement spot, over 60% of buildings apply district heating.

2.2 Measurements procedure and results analysis

Simultaneous 24-hour aspiration of the $PM_{2.5}$ and PM_{10} was carried out with two PNS3D15 low-volume dust meters. Aspirations were conducted by 91 days of each winter. The aspiration headers were installed 2 m above ground. The apparatus were located 5 meters apart. The PM separators were applied with Whatman GF/A fibreglass air filters with a diameter of 47mm. Prior to and after aspiration, the filters were seasoned over 24 h in conditions of constant temperature and humidity and, subsequently, their weight was determined by a differential scale (RADWAG XA 52/2X). The procedure for estimating the aerosol mass concentration was conducted in accordance with guidelines of the European Standard [8]. The expanded concentration measurement of uncertainty (U) did not exceed 3.6%.

Hourly aspirations of aerosols in rural area were provided with using the DustTrak DRX 8533 Aerosol Monitor TSI®. The measurement procedure was realized by application of the guidelines given by the manufacturer [9]. Hourly registrations were conducted only in January 2017.

To assess the meteorological conditions, two portable weather stations were used (Davis Vantage®). Values of main parameters (temperature (T), atmospheric pressure (P), relative humidity (RH), wind speed (W), precipitation (R) and duration of sunshine (S)) were collected with data-logger.

The raw data was verified by the Shapiro-Wilk test with the purpose of determination of the distribution of the measurement data. The data did not demonstrate normal distribution, hence, the analysis of the registered data was performed by the use of a non-parametric test. The comparison of the results applied Mann-Whitney U test. The data applied for statistical analysis was rescaled in accordance with the algorithm (1), which reduced the impact of the absolute value of a given variable and enabled all variables to be limited within the range {0;1}.

$$X' = \frac{X - X_{min}}{X_{max} - X_{min}} \quad (1)$$

The relationships between variables were examined using Spearman correlation [10]. Significance level of 0.05 was adopted. Finally, the study used Principal Component Analysis (PCA). The PCA is a technique used to emphasize variation and bring out strong patterns in a dataset. It's often used to make data easy to explore and visualize. A detailed description of the procedures for implementing the PCA is shown at work [11].

3 Results and discussion

Table 1 contains a summary of data regarding selected meteorological parameters registered during the 3-seasons observations in rural area. Throughout three successive winters, the mean temperature was slightly above the multi average measured over many years. The atmospheric pressure and relative humidity were characterized by the lowest variability coefficients (0.01 and 0.12, respectively) and their mean values were characteristic for the analyzed period. The wind speed and hydrometeor precipitation was associated with the occurrence of variable weather conditions. The meteorological condition was determined by the occurrence of the high pressure area with a center located in Russia (considerable insolation, windless weather, lack of precipitation – on average 23% of the winter period) and western recirculation (area associated with occurrence of rainfall, higher temperature, lack of insolation, low pressure, considerable movement of air masses). During the examined period, the values of meteorological parameters registered in the weather station located in the city were statistically almost identical to the ones registered in the rural areas (*p-values* for the Mann-Whitey test: 0.11 (T); 0.09 (RH); 0.18 (W); 0.19 (P); 0.17 (R); and 0.44 (S)).

Table 1. Meteorological parameters characterising the conditions during the observations.

parameter	MIN	MAX	AVG	SD
$PM_{2.5}$ [$\mu g \cdot m^{-3}$]	8.00	142	31.9	29.8
PM_{10} [$\mu g \cdot m^{-3}$]	11.0	195	44.6	34.8
T [°C]	-12.2	11.7	1.75	4.22
RH [%]	59.1	96.4	80.0	9.65
W [m/s]	0.00	16.1	3.06	9.71
P [hPa]	978	1041	1016	11.4
R [mm]	0.00	14.0	1.39	2.63
S [h]	0	9.3	2.33	2.95

Despite the fact that throughout the seasons in 2015 and 2016 most meteo conditions were associated with the effect of western circulation, the mean values of the daily mass concentration of particulate matter measured throughout three seasons in the rural areas (C) were found to be alarming. The mean value of the mass concentration of $PM_{2.5}$ in the rural areas was at a similar level to the concentration of $PM_{2.5}$ determined not only in Opole (32.4 μg/m^3, *p-value* = 0.21), but also in the towns with the highest level of pollution in Poland. At the same time, these values considerably exceeded the mass concentrations measured in the rural areas in Central and Northern Europe [6]. A statistically insignificant difference (*p-value* = 0.38) was also noted during the comparison of mean mass concentrations of PM_{10} (in Opole, it was 46.2 μg/m^3).

The adverse aerosanitary conditions expressed in terms of the number of days when the admissible daily mass concentration of PM_{10} is exceeded (C >50 μg/m^3) occurred in the rural areas on average throughout 30 days of the winter season. At the same time, the arbitrarily determined condition accompanied by smog occurrence, i.e. for $C_{PM2.5}$>40 μg/m^3 occurred on average for the duration of 23.7 days. The data for the urban areas does not demonstrate statistically significant differences.

Figs. 1 and 2 show the values of the ratio of $PM_{2.5}$ in PM_{10} in the urban and rural areas for the entire range of the daily data (Fig. 1), which arbitrarily confirm the condition involving smog occurrence (Fig. 2), respectively. Since the study reported here deals with mass concentrations of particulate matter in the air, the ratios registered in the two measurement spots are considered as very high. From the point of view of human health, the ratio of $PM_{2.5}$ in PM_{10} is unacceptable. In both cases, the location of the median can suggest that the higher ratio of the fine fraction can be noted in the rural area. The value of the probability excluded this hypothesis for comparison with the data in Fig. 1 (p-value = 0.46).

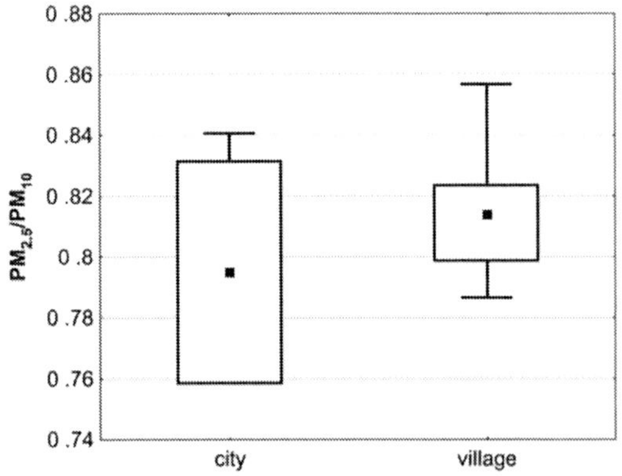

Fig. 1. Contribution of $PM_{2.5}$ to PM_{10} (all number of observation). Boxes show the range between the 25th and 75th percentiles. The whiskers extend from the edge of the box to the 5th and 95th percentiles of the data. The squares inside indicate median values.

The phenomenon of a local concentration of particulate matter in the ambient air is very adverse. This fact is particularly attributable to the most hazardous smog component formed by fine ambient particulate matter. The results demonstrate that the human exposition in the ambient air to the most adverse aerosanitary conditions is associated with the considerably greater hazard in the rural areas. In this case, the mean ratio of $PM_{2.5}$ in PM_{10} is nearly 10% greater and is statistically distinct from the ratio that was measured for the urban air (*p-value* = 0.03). The occurrence of the smog phenomenon is correlated with the increase of the respirable aerosol fraction in the suspended PM. The analysis of the 24-hour concentration of the fractions of the particulate matter demonstrates that during the episodes of smog, the ratio of $PM_{2.5}/PM_{10}$ was above >0.85. For the mean daily mass concentrations of PM_{10} smaller than 50 μg/m^3, the mean value of this ratio was equal to 0.79. The differences are confirmed by the results of the Mann-Whitney test (*p-value* = 0.002).

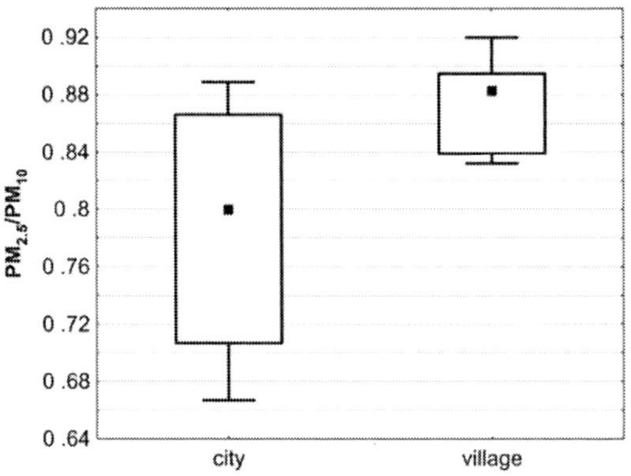

Fig. 2. Contribution of $PM_{2.5}$ to PM_{10} (during smog occurrence).

The occurrence of the circumstances when the admissible levels of the particulate matter are exceeded was directly linked to the characteristics of the emission sources, and primarily with the intensity of the application of the thermal energy sources. This increased intensity of domestic boiler use was determined by the weather conditions.

Table 2 contains the results of the Spearman correlation between the PM's mass concentration measured at village and the remaining registered meteorological data. The analysis of correlation demonstrates that only the length of insolation does not correlate with the values of the mass concentration of particulate matter suspended in the ambient air. The results suggest that the aerosanitary conditions improve during the occurrence of precipitation similar to unstable atmosphere. An increase in the velocity of air advection, as well as the occurrence of precipitation and temperature increase affect both indirectly and directly the mass concentrations of aerosols. With the exception of wind speed and temperature, the remaining correlations are weak, though they are statistically significant. The results of statistical analysis correspond to literature data referred to urban area [12, 13] and show the main meteorological conditions that affect the modification process of an aerosanitary situation. The result of the basic analysis of the relation between he variables reveals the parameters that should be considered in the further study.

Table 2. Spearman correlation coefficient values for considered parameters.

parameter	$PM_{2.5}$	PM_{10}
T [°C]	-0.50*	-0.50*
RH [%]	0.22*	0.21*
W [m/s]	-0.64*	-0.62*
P [hPa]	0.33*	0.36*
R [mm]	-0.33*	-0.34*
Sun [h]	-0.01	-0.02
PM_{10}	0.97*	1.00

*significant with $p < 0.05$.

The scope of the experiment also involved the verification of the variability of the mass concentrations of the examined aerosol fractions over 60 minute intervals. The test took two weeks, including 6 days accompanied by the occurrence of a permanent temperature inversion accompanied by occurrence of smog phenomenon. Both gravimetric tests (24 hours) as well as optical ones (1 hour) demonstrate that enhanced by an anticyclone, the three-day period of temperature inversion occurrence caused a situation in which local air immission levels did not meet the standards required for the protection of human health.

In general, the mean daily values of C for $PM_{2.5}$ and PM_{10} registered on the basis of hourly measurements were greater by 11-14% than the results that were concurrently obtained by the use of on the basis of the gravimetric analysis. By analogy, the mean ratio of the fine particles in the PM_{10} fraction was greater as well (0.96). The difference that was noted stems from the different approaches adopted during the measurements methodology and the determined quantities. The gravimetric analysis is based on the determination of the volume of particles in the air and measurements of the mass of particulate matter that is collected during aspiration and extracted from the filter. The mass concentration concerns only the solid particles dispersed in the air. For the case of the optical measurements, the aerosol mass is calculated on the basis of the he light beam diffusion proportionally to the particle mass passing with the air through the measurement section. The mass concentration in this case involves both the solid and liquid phase that is dispersed.

The results in Fig 3 illustrate the data derived from 36 subsequent hourly observations with regard to the parameters, which define the quality of the troposphere in the rural areas. The selected period illustrates the impact of the temperature inversion (accompanying smog, from hour 1 to 25-26 during the observation) as well as the impact of the variability of the most relevant weather parameters. This figure demonstrates the existence of two maximums of particulate matter concentrations. It can be note, however, that these occurrences are not affected by the variations in the ambient air temperatures and wind speed. The course of the lines reflects the morning and afternoon increase of the intensity of exploitation of individual heating systems in residential buildings. Throughout the duration of the smog, the levels of hazardous substances in the air were very high. Nevertheless, the results demonstrate that the outdoor activities should be particularly avoided during the seasons corresponding to the increased exploitation of the heating systems.

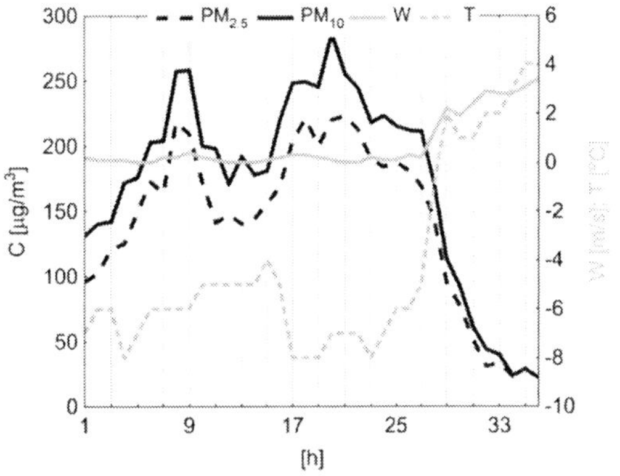

Fig. 3. Variability of selected parameters in time during and after the occurrence of temperature inversion.

Fig. 3 demonstrates that after 3 hours from the occurrence of advection movement accompanied by a concurrent considerable increase in temperature, the aerosanitary conditions considerable improve and the air quality can be considered as good as a result of the dissipation of pollutants. The change in the weather conditions determines the decrease of the intensity of the use of local, point emission sources resulting from the combustion of solid fuels, which also contributes to the improvement of ambient air quality.

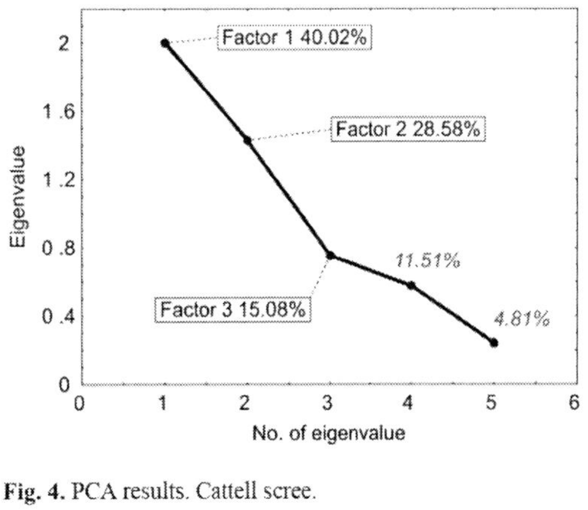

Fig. 4. PCA results. Cattell scree.

The results of the experiment reflect the relevance of the principal components in the explanation of the relevance of the input factors (per cent in the variability in the database). The Keiser criterion was used as the principal test of the selection of the relevance parameters, which meant that the factors investigated had an eigenvalue of >1 [14]. The selection of the factors was also supported by the Cattell scree test [15,16] (Fig. 4). The graphical interpretation indicates that in the analyzed case, three of the principal components had a decisive role. Cumulative percentage of variation explained by three components reach almost 84%. Although factor no. 3 has the eigenvalue of <1 its impact on the justified variability of the primary data is relevant.

The relations between the primary and experimental principal components are presented in a graphical form in Fig. 5 and 6. Each of the variables is represented by a vector. The sense and length of the vector decides on the degree in which the particular variables have an impact on the values of the principal components. In the analyzed example, nearly all of the input variables are located in the vicinity of the values 1 or -1. This means that the larger part of information contained in these variables is carried by the principal components. The fact of adjacent location of two variables indicates a strongly positive correlation. The variables located on the opposite sides are negatively correlated.

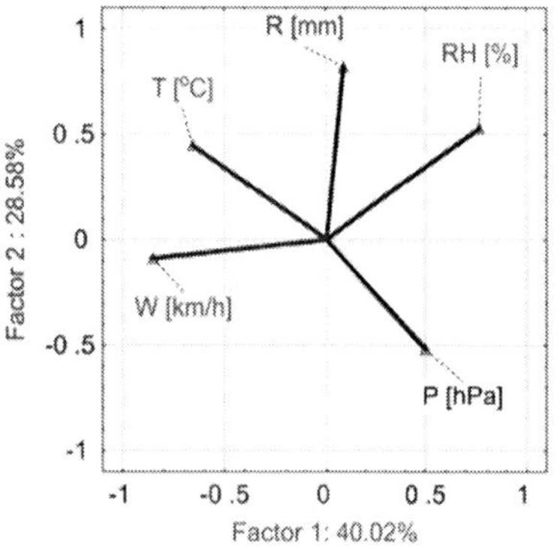

Fig. 5. The PCA results. Configuration of examined variables in the realm of factors 1-2.

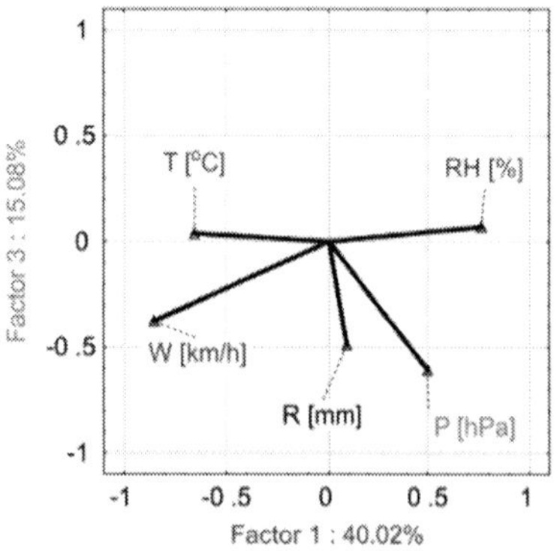

Fig. 6. The PCA results. Configuration of examined variables in the realm of factors 1-3.

The Principal Component Analysis has successfully extracted three principal components which illustrate the influence of meteorological parameters on $PM_{2.5}$ and PM_{10} mass concentration. For the examined case, the first component is responsible for explaining 40% of information regarding the concentrations used as part of the input variables.

One can note that the principal component No.1 explains only the specific variables responsible for meteorological parameters connected with advection, air temperature and water vapor content in the air. The remaining principal components account for the variables which also describe the weather parameters; however, all of them can be considered as individual components (amount of precipitation - factor 2 and atmospheric pressure – factor 3). All of factors (common and separate components) account for over 83% of information regarding the mass concentration of considered PM's.

4 Conclusions

The conducted research demonstrated that throughout the winter season, the ground-level troposphere in the rural areas is considerably enriched by particulate matter emission and the values of the mass concentration of PMs does not statistically differ from the levels recorded in the urban areas. The values of the daily and hourly mass concentration of $PM_{2.5}$ and PM_{10} are very improper. The ratio of the fine particles in the particulate matter suspended in the air, whose hazard to human health is known to be high, is greater in the rural areas than in the towns for the case of the occurrence of smog which is boosted by the temperature inversion phenomenon. The analysis of the main components revealed that the metrological factors play a critical role on the ambient air quality. In the cold season, the control of air quality is performed by the temperature, which determines the level of pollution. An increase of this parameter leads to the smaller utilization of the domestic boilers. The occurrence of the advection and precipitation as a result of the changes in atmospheric pressure leads to the local dilution of particulate matter in the ground-level ozone even in the areas characterized by the considerable emission of pollution of anthropogenic origin.

It seems, that the results cannot be supra-local dimension. But they can find confirmation of areas with similar characteristics of emission and climatic-meteorological conditions. At the local scale, the results of the experiment show that it is necessary to replace old heating systems with new ones that do not generate significant amounts of dust. As a consequence, there is the necessity of implementing immediate change in the Polish legislation.

References

1. J. Schwarz J., M. Cusack, J. Karban et al., Atmos Res **176-177** (2013)
2. J. Krynicka, A. Drzeniecka-Osiadacz, Pol. J. Environ. Stud. **22 (4)** (2013)

DOI: 10.1051/e3sconf/20171902004

3. Z. Mijić, A. Stojić, N. Perisić, M.,Rajsić, S. Tasić, M. Radenković, J. Joksić, Atmos. Environ. **44** (2010)
4. T. Olszowski, Proscince **1** (2014)
5. K.M. Nam, N.E. Selin, J.M. Reilly, S. Paltsev, Energy Policy **38** (2010)
6. W. Rogula-Kozłowska, G. Majewski, B. Błaszczak, K. Klejnowski, P. Regula-Kopiec, Int. J. Environ. Res. Pub. Heal. **13** (2016)
7. http://www.opole.pios.gov.pl:81 (acces 10. 07. 2017)
8. EN 12341:2014.
9. www.tsi.com (acces on 10.07.2017)
10. M. John, C. E. Priebe, Comput. Stat. Data Anal. **51**(9) (2007)
11. I.T. Jolliffe, Principal Component Analysis (Springer-Verlag, New York, USA,1986)
12. S. Hormann, B. Pfeiler, E. Stadlober, Austrian J. Stat. **34**(307) (2005)
13. E. Maraziotis, L. Sarotis, O Marazioti, P. Marazioti, Glob NEST J. **10** (2008)
14. A.R. Ferro, R.J. Kopperud, L.M. Hildemann Environ Sci Technol. **38** (2004)
15. H.F. Kaiser, Edu Psych Meas. **20** (1960)
16. R.B. Cattell, Multivar Beh Res. **1** (1966)

E3S Web of Conferences **19**, 02005 (2017) DOI: 10.1051/e3sconf/20171902005
EEMS 2017

Unmanned Aerial Vehicles for Environmental Monitoring with Special Reference to Heat Loss

Stanisław Anweiler[1,*], *Dawid* Piwowarski[1], and *Roman* Ulbrich[1]

[1]Department of Mechanical Engineering, Opole University of Technology, 5 Mikołajczyka st., 45-758 Opole, Poland

Abstract. This paper presents the design and implementation of device for remote and automatic monitoring of temperature field of large objects. The project aimed to create a quadcopter flying platform equipped with a thermal imaging camera. The object of the research was district heating installations above ground and underground. The results of the work on the implementation of low-cost (below 750 EUR) and efficient heat loss monitoring system. The system consists of a small (<2kg) multirotor platform. To perform thermal images micro camera FlirOne with microcomputer Raspberry Pi3 was used. Exploitation of UAVs in temperature field monitoring reveals only a fraction of their capabilities. The fast-growing multirotor platform market continues to deliver new solutions and improvements. Their use in monitoring the environment is limited only by the imagination of the user.

1 Introduction

Growing share of the municipal sector in energy consumption develops many activities and requirements for energy consumption reduction and the observation of trends necessity. For individual objects there is an energy audit available and the building's energy performance analysis is possible, which significantly contributes to the energy intensity reduction of the economy. In the case of actions relating to larger areas, such as complexes of buildings, districts or even entire cities, such procedure would require a huge financial investment and time. For some time, there are proposals to use the techniques of aerial laser scanning for the urban heat island (UHI) effects [1]. In recent years, attempts are being made to use the unmanned aircraft with thermal imaging techniques. Recent advances in UAVs for civilian use make it possible to regularly monitor environments from the thermophysical point of view, at spatial and temporal scales that would be difficult to achieve using conventional methods. Previous aerial monitoring of thermophysical environments has been expensive and time consuming [2].

This paper presents the design and implementation of device for remote and automatic monitoring of temperature field of large objects. The project aimed to create a quadcopter flying platform equipped with a thermal imaging camera. For this commercially available UAV model was used – 3DR-Iris+ [3]. The object of the research was district heating installations above ground and underground. A series of images to locate the places of heat loss and thus damage to the hull insulation. Additionally a group of public buildings heat loss have been studied.

The aim of the work is the design and implementation of low-cost and efficient heat loss monitoring system. The system consists of a small (<2kg) multirotor platform. To perform thermal images micro camera FlirOne with microcomputer Raspberry Pi3 was used.

In recent years there has been a noticeable increase in the municipal sector's consumption of energy. This is a prerequisite for introducing actions to limit this tendency. One such method is the visualization of heat loss. This is a serious and problematic topic of contemporary engineering when dealing with whole complex of buildings or extensive district heating networks. For individual objects, the subject seems simpler for energy audits and energy performance of buildings.

Taking on the subject of visualization of heat loss in large objects, we face such problems as the difficulty of monitoring due to the surface of the examined objects, the area in which they are located - often industrial objects in hazardous areas with difficult access. Another problem is the cyclical nature of such research.

Periodic thermal imaging of studied objects allows for observation of changes and their tendencies, allows defining defect spots - defective insulation, subassembly or leakage of heating medium. This allows for the development of systematic reviews and periodic repairs.

Another difficulty in monitoring large objects is the way of conducting the study itself. During manual monitoring, by trained staff, it generates high costs and prolongs the time of testing. It also poses a risk to the crew itself because of the area in which the object may be located.

All these aspects create a need for a new method for identifying heat losses in large facilities.

* Corresponding author: s.anweiler@po.opole.pl

2 Materials and methods

In last years laser scanning has been used - LIDAR. This is a modern data mining technique for the numerical model of terrain [4]. The accuracy of the measurement is independent of the lighting conditions. The accuracy range is in the range of 0.1 - 0.2m, which is undoubtedly one of the advantages. The disadvantage is the absorption of laser pulses by clouds, fog, water, asphalt and tar.

Research uses a manned aircraft equipped with the necessary equipment. The necessity of having the appropriate permits and licenses to fly in the study areas is one of the first difficulties. Another is the cost of the study itself. The need for a temporary reservation of the airspace in which the flights are to take place is not a matter of delaying the study. If the weather is not proper for fly or the measuring device fails, the flight must be cancelled. It generates losses and high maintenance costs for systematic research.

When analysing the difficulty of visualization of heat loss in large objects, it is concluded that the best method is to take photos from the air. However, the use of manned flying unit in this area may outweigh the costs of research alone. A good unmanned aerial vehicle operated by a terrestrial operator or a pre-defined route can be a good solution - based on GPS coordinates.

The use of unmanned aircraft in environmental monitoring is a relatively new issue. They are used with a positive effect in the monitoring of the animal population, monitoring of areas, in search of missing persons. Even their use in monitoring geothermal areas has been described [2,5]. There are many flying models on the market. In addition to price, they distinguish between their capabilities and performance. Flying platform used for study of heat lost should be able to fly on the planned route - e.g. along the heat pipe or overhead line, the possibility of flight in hover mode. This will allow to inspect chimneys co to jest, chimneys, GSM towers and provide a series of images of studied object. The flying model must be able to take off from small and uneven surfaces. Investigated objects may be in the vicinity of other large objects, high vegetation, etc. Its design should allow it to fly in difficult conditions such as high dustiness, high humidity. The great advantage is the possibility of transporting the equipment by one person and carriage in a passenger car, possibly small VAN. It requires the necessity of small size of the platform. In thermal imaging, it is necessary to make them at a constant height to not generate a measurement error. This dictates another criterion when selecting a measuring device.

In recent years, the technology of multirotor models has developed considerably, resulting in stable flight and relatively low maintenance cost [6]. Two families of unmanned aircraft were considered in this research. These were planes, and in it an airplane and a so-called flying wing. The second group were the copters, which included helicopters and multicopters (quadrocopters, hexacopters, etc.). The fundamental difference between them is that in the multicopter there is no precise flight mechanics, only a number of independent electric motors whose rotation is independently changed. Only the rotation speed is responsible for stable flight [3]. Condition of model selection is shown in Table 1. The cost of proposed method is lower than 750 EUR.

Table 1. Condition of model selection

Platform type	Small/uneven terrain launch	Dusty/humid environment operation	Mobility/one person transportation	On-board GPS	Hove ring	Constant altitude flight
Airplane	No	Yes	Yes	Yes	No	Yes
Flying wing	No	Yes	Yes	Yes	No	Yes
Multicopter	Yes	Yes	Yes	Yes	Yes	Yes
Helicopter	Yes	Yes	Yes	Yes	Yes	Yes

All the criteria are met by the second group of unmanned flying objects - the copter. Due to the presence of precision mechanics on board of the helicopter, multicopters were chosen. Two models were considered. First was the popular DJI Phantom version 3 Standard. However, it has been ruled out for problematic adaptation of non-DJI components. The next model was 3DR Iris +. It has very similar capabilities and performance to the first model. This is an open platform. This means that the user has full access to the source code that dictates the behaviour of the platform on the fly and thus to edit. Iris + allows to connect additional modules such as video accessories, thermal imaging cameras, sensors that may not necessarily have been issued by 3DR.

The Iris + 3DR platform is built on the Pixhawk flight controller, which incorporates sensors such as an accelerometer, a gyroscope, and a barometer. The barometer allows you to fly at a constant height at an accuracy of up to one meter.

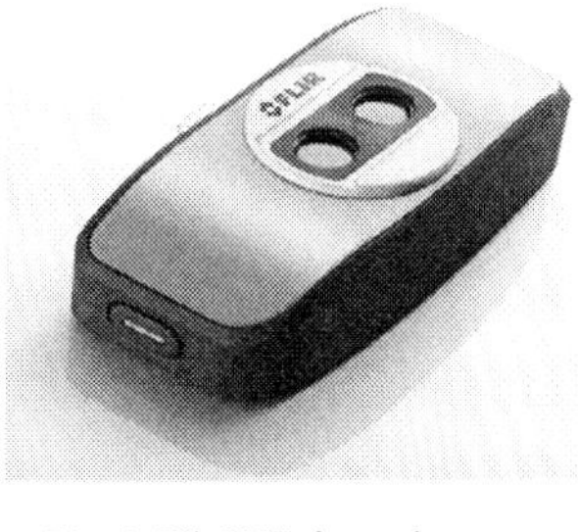

Fig. 1. FlirONE thermal camera

DOI: 10.1051/e3sconf/20171902005

The thermal imaging camera used, shown in Fig. 1, is a small Flir One device. Its measuring range is -20 to 120°C. The camera captures two images simultaneously - the first one from the standard camera, the other from the thermal camera, resulting in a 640x480 image. The measurement resolution is 0.1°C but the accuracy is ± 3°C. It is dictated by the fact that the camera's optics are able to capture a temperature difference of up to 0.1°C, but the data processed further by the processor generate error within ± 3°C. The image is refreshed with rate 8.7 Hz.

The camera requires Android or iOS system to work properly. This was resolved by using the popular Raspberry Pi microcomputer. The Android software was installed on the computer and then the camera was attached. At the given stage of the survey, the camera takes pictures at the specified interval of time, e.g. every 5 or 10 seconds. Photos are saved to a memory card in RP3. This provides a collection of photos for further process. A set of described devices located on the test platform is shown in Fig. 2.

Fig. 2. Quadcopter infrared vision platform assembly.

3 Results and discussion

To confirm the usefulness of the camera for the purposes of the research, a series of photographs were made. One of the examined elements was a fragment of ECO Opole heat district. Air temperature during research was 2°C. Obtained material found that the heat pipe is well insulated, however, we managed to find a few weak points. One was the concrete support through which the heat pipe passed - Fig. 3.

The thermal images captured the difference in temperature at the spot where the heat exchanger passes through the concrete block. This generates some loss in transmitted heat, which can only be visualized with the help of a thermal camera.

Next examined object was photovoltaic farm - Fig 4 a) and b).

The photos made it possible to state that the devices are working properly. There is no such thing as hotspots which could suggest that the efficiency of photovoltaic farm is reduced. The presence of hot spots is highly unwanted as it affects the efficiency of the photovoltaic farm [7]. The image that researchers wanted to capture was illustrated by the presence of hot spots in Fig. 4 c).

a)

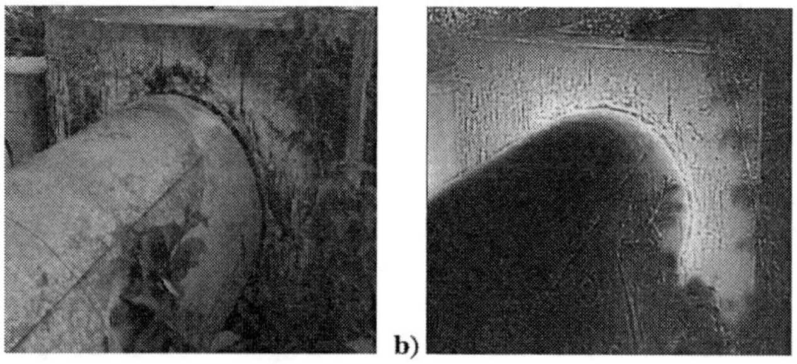

b)

Fig 3. ECO heating network fragment - concreate support A-general view; B-IR view.

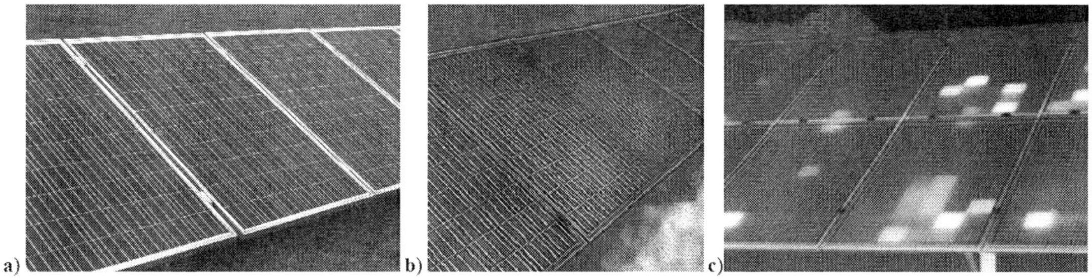

Fig. 4. Examined photovoltaic farm and damaged PV panels. [8]

Photo shows exactly the previously mentioned hot spots, which can only be revealed with thermal camera. As a preliminary study, attempts were made to take pictures of the underground heating pipes. The examined section is part of the ECO Opole heating network. The object consists of two pipes with a diameter of 125mm in double insulation. The temperature in the feed pipe is 91°C and the return pipeline is about 50°C. The heating system is shown in Fig. 5

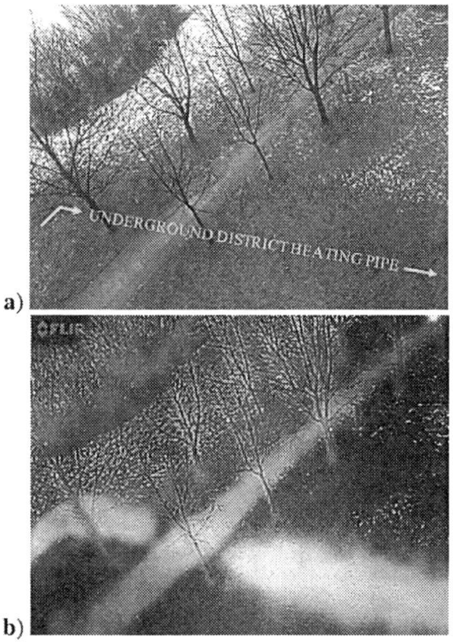

Fig. 5. Underground district heating pipe a-VIS and b-IR footage.

The study was conducted on 05.02.2017r. Under the following atmospheric conditions: temperature -1°C, pressure 1020hPa. The weather was windless.

Thanks to taken pictures, are easy to trace the underground heating route and compare it with the GIS map and verify its correctness.

The next stage of the research will be the interpretation of acquired thermal imagery. This will require the introduction of linear temperature field markers, tests for conditions simulated with a heat source of known power and surface temperature, actual conditions and sequences for periodic monitoring of manufacturing plants, building complexes and district heating networks.

4 Conclusions

The purpose of the research will be to develop a method of detecting heat losses in large objects - with a great emphasis on industrial objects, heat networks, complexes of buildings. Preliminary results confirm the usefulness of the selected 3DR Iris + platform together with the thermal imaging camera to identify heat losses in the tested objects.

The low cost of the platform maintaining allows periodic measurements. By compiling the collected data, it is possible to indicate defective or wearable elements in advance, planning and systematize periodic inspections and repairs. All of these activities lead to a reduction in the risk of breakdowns, lower maintenance costs for the lines and, consequently, a reduction in the negative impact on the environment.

Using UAVs in temperature field monitoring reveals only a fraction of their capabilities. The fast-growing multirotor platforms market continues to deliver new solutions and improvements. Their use in monitoring the environment is limited only by the imagination of the user.

References

1. W. Su, Y. Zhang, Y. Yang, G. Ye, Sustainability, **6**, 10 (2014)
2. A. Nishar, S. Richards, D. Breen, J. Robertson, B. Breen, Renew. Energ. **86**, (2016)
3. S. Anweiler, D. Piwowarski, J. Clean. Prod. **155**, 1 (2017)
4. A. Strzałka, *PhD Thesis*, Opole University of Technology, (2013)
5. M. C. Harvey, J. V. Rowland, K. M. Luketina, J. Volcan. Geoth. Res. **325** (2016)
6. S. Yahyanejad, B. Rinner, ISPRS J Photogramm, **104** (2015)
7. J. Carlson, D. Menicucci, P. Vorobieff, A. Mammoli, H. He,. Appl. Therm. Eng. **70**, 1 (2014).
8. Aerial thermal infrared in Melbourne Victoria www.uaviation.com.au/thermal [access: 10.07.2017]

Aerosol in selected laboratories at Faculty of Mechanical Engineering, Opole University of Technology

Tomasz Olszowski[1,*]

[1]Department of Thermal Engineering and Industrial Facilities, Opole University of Technology, Mikołajczyka 5, 45-271 Opole, Poland

Abstract. The paper contains the results of a study into mass concentration of the dispersed aerosol fraction with the aerodynamic diameter of up to 2.5 and 10 micrometers. The study was conducted during classes with students participating in them in two laboratories located at Faculty of Mechanical Engineering, Opole University of Technology as well as outdoor outside the building. It was demonstrated that the values of the mass concentration of $PM_{2.5}$ and PM_{10} measured in the laboratories differ considerably from the levels measured in the ambient air in the outdoor areas surrounding the faculty building. It was concluded that the diversity of $PM_{2.5}/PM_{10}$ ratio was greater in the laboratories. Direct correlation was not established between the concentrations of the particular PM fractions in the two investigated environments. It was demonstrated that there is a statistically significant relation between the concentration of $PM_{2.5}$ and PM_{10} and the number of people present in the laboratory. The conducted cluster analysis led to the detection of the existence of dominant structures determining air quality parameters. For the analyzed case, endogenic factors are responsible for the aerosanitary condition. The study demonstrated that the evaluation of air quality needs to be performed individually for the specific rooms.

1 Introduction

Epidemiological studies have shown that there is a strong correlation between the air pollutants concentration and increase the prevalence of respiratory diseases and finally mortality of people [1, 2]. People spend 85–90% of their time in enclosed spaces, including 40% in buildings with public access [3]. For this reason, it is important to pay attention to the parameters of air which they used to breathe. One of the parameters of air quality is associated with the mass concentration of particulate matter suspended in the air.

Over the years, it was taken that indoor air quality is directly relative to the parameters of the external air penetrating the buildings. Nevertheless, around the turn of last century, it was observed that the concentration of aerosols inside the buildings is often considerably higher in comparison to the quality of external air infiltrating the area [4].

The recent studies indicate that the air quality indoors is relative to both the factors associated with the external environment as well as endogenic factors. The studies reported in [5] deal with the characteristics of the factors affecting the quality of air inside buildings, which were found to include air pollution, microclimate conditions, level of air ionization, exploitation conditions of the buildings as well as their design. The factors deciding about the quality of the inside environment include the ones associated with the presence as well as animal and human activity indoors. The activities inside buildings can considerably affect not only on the level of the concentration of classical aerosol particles as well as the level of bioaerosols [6-8]. Air quality is also considerably influenced by the effect resulting from the existence of ventilation and air conditioning systems. The concentration of aerosol particles is also dependent on the season. In the moderate climate, higher aerosol concentrations are registered during winter [9].

This paper focuses on the verification of the level of mass concentration of $PM_{2.5}$ and PM_{10} in the selected laboratory rooms of a university during the classes with students participating in them. It also aims to determine the relation between the particulate matter concentration and the parameters which characterize the quality of air outdoors.

2 Materials and methods

2.1 Measurement sites

The study was conducted in the laboratories in the conditions when students participated in the classes. The laboratories are situated at the Faculty of Mechanical Engineering, Opole University of Technology (50°40'59.9"N; 17°56'42.62" E). It was ensured that the rooms selected for the testing were similar in size and layout. For the purpose of minimizing the impact of direct insolation on the potential air movement indoors, the two selected laboratories were located adjacent and

* Corresponding author: t.olszowski@po.opole.pl

protected from the direct exposition to sunlight. The first of them, Atmospheric Pollution Control laboratory (APC) has a cubic capacity of 98m^3, whereas the other one, Thermodynamic laboratory (TD) comprises 118 m^3. The rooms are used periodically (only during classes) and are regularly cleaned up. The rooms comprise the same resources in terms of the material, design etc. The air flow resulting from leakages through drafty windows and doors is characterized by similar infiltration parameters. Both rooms receive ventilation air through outdated gravitational ventilation.

In addition, with the purpose of determining the potential impact of the external conditions on the air quality indoors, the measurements of the dedicated quantities were also performed outdoors. The equipment, including aerosol aspirator and temperature sensor was exposed outside the rooms on an observation platform situated 3 meters away from the windows in the horizontal plane.

2.2 Measurements procedure and results analysis

The mass concentration of the aerosol fractions of up to 2.5 and 10 micrometers in size was measured concurrently by application of Environmental *DustTrak* Aerosol Monitor (Models 8335 DRX TSI®). Every day, prior to the measurements, the apparaus was calibrated. It was done with the purpose of setting the value of the calibration coefficient to match to the current parameters of the indoor and outdoor air. Concurrently, prior to the specific measurements, the appatus was set to zero. The flow of air through the measurement cells of the apparatus was always equal to 1 [dm^3·min^{-1}] during the observations.

The aspiration of $PM_{2.5}$ and $PM_{10}$was performed for the time constant of 10 minutes and the registration constant of 10 seconds. The data acquisition and storage occurred automatically. The registration involved 960 independent observations indoors and 96 outdoors.

Throughout the aerosol aspiration, data regarding outdoor and indoor air temperature was registered. In addition, the number of persons present in class was also recorded.

The raw data was verified by the Shapiro-Wilk test with the purpose of determination of the distribution of the measurement data. The data did not demonstrate normal distribution, hence, the analysis of the registered data was performed by the use of a non-parametric test. The comparison of the results applied Mann-Whitney U test.

Another verification of the dependencies applied the cluster analysis (CA) [10]. It was done with the aim of obtaining a uniform group of the tested objects, thus facilitating the identification of the principal characteristics and structure of the analyzed data, consequently enabling the researcher to classify the standard objects.

As a result of the objective analysis of the data cluster, it will be possible to establish the mutual relations between the elements forming a given set. This is particularly relevant in the aspect of modeling processes occurring in the ambient air. The relations between the parameters characterizing the ambient air are particularly important during the estimation of the parameters regarding probability distribution, offering the further assessment of the variability of air quality in rooms.

The analysis applied all investigated variables. The data was rescaled on the basis of the algorithm (1), which reduced the impact of the absolute value of a given variable and enabled all variables to be limited within the range {0;1}. The analysis applied the procedure of agglomeration – by determining the measure of the distance between the objects (Euclidean distance).

$$X' = \frac{X - X_{min}}{X_{max} - X_{min}} \quad (1)$$

3 Results and discussion

Table 1 summarizes the data on the number of observations of selected parameters of air and data on the use of classrooms. The results suggest that the physical parameters and the number of students present in the laboratores were almost identical in both rooms. A comparison undertaken by means of the Wilcoxon test demonstrated a lack of the differences both in terms of the aerosanitary conditions and number of students present in class (p-value 0.18 accompanied by a materiality level a = 0.05). This, in turn, implies that the further study should adopt observations with regard to the environment that are discussed without classification of the data according to the location of the rooms.

Table 1. Selected variables collection.

Site	No of observ.	$PM_{2.5}$ [µg·m^{-3}]	PM_{10} [µg·m^{-3}]	T [^{O}C]	No of people
APC	640				
Min		32.5	40.0	19.1	3.00
Max		179	215	23.3	14.0
AVG		64.5	89.5	22.1	8.80
SD		34.6	53.8	1.69	3.90
TD	320				
Min		31.2	39.9	19.0	4.00
Max		186	219	23.1	15.0
AVG		63.5	89.9	22.0	9.00
SD		35.8	51.4	1.71	3.95
O	96				
Min		30.1	42,0	-4.00	n/a
Max		95.3	151	19.0	n/a
AVG		49.6	69.6	8.80	n/a
SD		29.9	24.5	8.09	n/a

Fig. 1 demonstrates the results of the mass concentration of aerosol with the fractions of up to a 2.5 and 10 micrometers in size. Fig. 1A presents the results of the mass concentration recorded during the measurements in the indoor (I) and outdoor air (O) during the episodes of the unstable weather (temperature 16-18 ^{O}C, light wind, good dispersion of pollution). Fig.

1B presents the results accompaying the occurrence of smog episodes in the urban area (temperature inversion, T in the range from -4 to -1 °C, windless weather). In both cases, the values of the mass concentrations of $PM_{2.5}$ and PM_{10} established in the rooms signifcantly differ from the levels that were determined in the outdoor air in the vicinity of the faculty building.

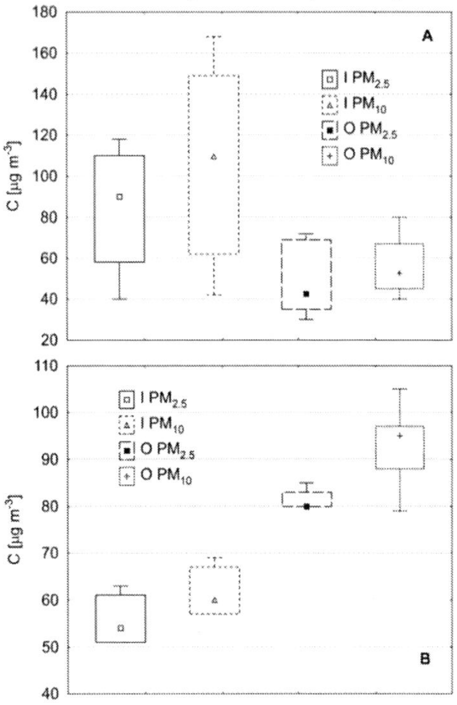

Fig. 1. Mass concentration of $PM_{2.5}$ and PM_{10} for different weather conditions. Boxes show the range between the 25th and 75th percentiles. The whiskers extend from the edge of the box to the 5th and 95th percentiles of the data. The squares inside indicate median values.

The results of the Wilcoxon test unequivocally confirm the differences. For the adopted materiality level of α=0.05, p-value was equal to 0.004, which clearly confirms the existance of the statistically significant differences between the concentrations in the two environments. It is noteworthy that the concentrations of aerosols indoors do not seem to be relative to the outdoor parameters. The indoor temperature was virtually constant (coefficient of variability Cv was lower than 0.1%). The outdoor temperature was characterized by a greater variability (CV $\approx$ 0.92%). Within the range of similar temperatures (results derived from 600 observations (I) and 60 (O), when the difference between the indoor temperature (I T) and the outdoor temperature (O T) was around 20%), we observed from 1.5 to 4-times higher concentrations of the particular fractions in the rooms. This demonstrates the impact of the local sources of pollutants as well as inadequate ventilation. Concurrently, the faulty ventilation aided in the maintenance of higher standards of the indoor air quality during the smog episodes, when the measured concentrations of $PM_{2.5}$ and PM_{10} were, respectively, 25 and 35 [µg m^{-3}] lower in the indoor air compared to the outdoor air. Unfortunately, after a comparison with the 24-hour values, the PM_{10} levels recorded in the laboratories were found to exceed the admissible levels.

The comparison of the outdoor concentrations of PM_{10} with the results gathered in the laboratories [11] demonstrates that the indoor air quality is more adverse. The mass concentrations of the aerosols in the laboratories is nearly 40 [µg·m^{-3}] higher than the mass concentration of the particles with the same fraction. The ratio of the dispersed phase in the aerosol can be assessed to be negligible. The considerable differences in the PM_{10} values are attributable equally to the poor ventilation, as well as to the character of the space utilization during the classes. The activity in the laboratories, in contrast to lecture rooms, is associated with a considerable degree of physical movement of the students in class. In turn, this results in the escalation of the PM_X emission from the clothes and skin, which is accompanied by a permanent phenomenon of the resuspension of solid particles from the surface. Consequently, this contributes to the enrichment of the air with the solid particles.

The study did not involve the measurements of total suspended particles (TSP). The ratio of PM_{10} in the TSP is in the range from 0.4 to 0.8 [12]. Under the assumption that this ratio is low in the laboratories, we can assume that during the cold season, and in the conditions marked by the insufficient quality of ventilation and air filtration, aerosanitary conditions in the laboratories are extremely adverse. In accordance with data provided by DECOS [13], a safe level of total suspended particles is equal to 200 µg·m^{-3} in the consideration of the maximum anticipated ratio of the fine particle fraction. For the lowest ratio of PM_{10}, the value of TSP is exceeded.

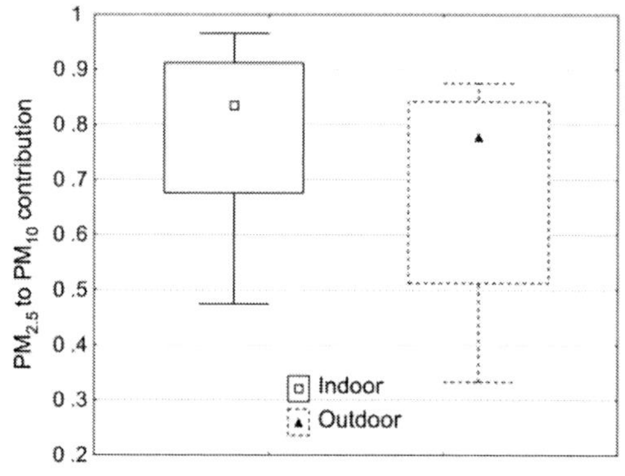

Fig. 2. Ratio of $PM_{2.5}$ in PM_{10}.

Fig. 2 presents the ranges of the ratios of the respirable fraction in PM_{10} in the two environments. A basic statistical analysis yieds that a greater diversity of the ratio of $PM_{2.5}$ in PM_{10} is noted in the rooms. The Cv value was equal to 0.11% for (I) and 0.05% for (O). The statement regarding the greater diversity of the ratio of $PM_{2.5}$ in PM_{10} noted in the rooms means that the variety and the number of the emission sources in the rooms have a considerable impact on the aerosol concentration.

Only for the case of the outdoor aerosol, the sources of emission are not that much diversified. The difference that is clearly visible in the figure is also confirmed by the results of the Wilcoxon test (p-value = 0.039, for $\alpha = 0.05$).

Separately for the internal and external environment, there is a heavy dependence between the mass concentration of $PM_{2.5}$ and PM_{10}. The Spearman's *rho-* was equal to 0.92 and 0.98 for (I) and (O), respectively.

Table 2 contains the results for the Spearman's rank correlation between the registered parameters.

Table 2. Spearman correlation coefficient values for selected parameters.

I $PM_{2.5}$ to O $PM_{2.5}$	**I PM_{10} to O PM_{10}**	**O T to I $PM_{2.5}$**	**O T to I PM_{10}**
-0.188	-0.298	0.205	0.237
I T to I $PM_{2.5}$	**I T to I PM_{10}**	**I P to I $PM_{2.5}$**	**I P to I PM_{10}**
0.374	0.456	0.678	0.728

On the basis of the Guilford scale [14], we can conclude about the lack of a correlation between the concentrations of the particular particle fractions in both environments. The relation between the indoor concentration of $PM_{2.5}$ and PM_{10} and the temperature (and therefore also weather) does not seem to be statistically material. A weak, yet significant, positive correlation was established for the indoor temperature and mass concentration of aerosol in the laboratories. A statistically significant and strong dependence ($PM_{2.5}$) and certain relation (PM_{10}) regarding the mutual impact of the two parameters was established for the mass concentration of $PM_{2.5}$ and PM_{10} and the number of students present in the laboratory.

A graphical representation of the cluster analysis is presented in Fig. 3. The conducted classification offered the possibility to classify the groups of the analyzed variables.

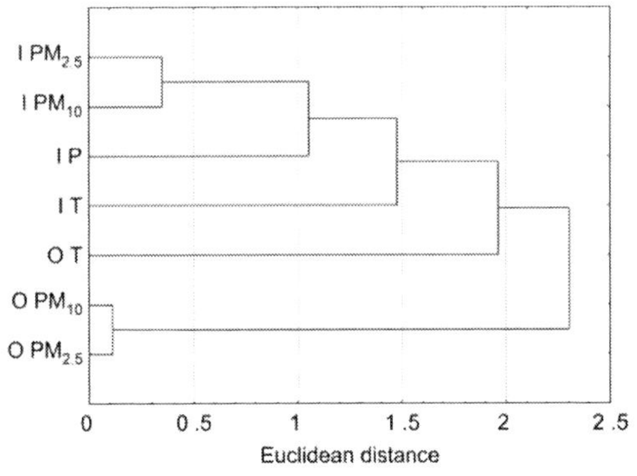

Fig. 3. Dendrogram for the verified objects.

Fig. 3 demonstrates that the most homogenous group is formed by the variables characterizing the parameters measured in the laboratory. The computed Euclidean distance between I $PM_{2.5}$ and I PM_{10} was equal to 0.35, whereas between I PMx and I P and I T, it was 1.07 and 1.57, respectively. For the remaining parameters, the distances exceeded the indicative value of 2, which clearly demonstrates the divergences with regard to the homogeneity of the parameter types (I) and (O). The performed analysis confirms that the values of the parameters determined for the outdoor environment are considerably distinct from the ones that were established for the laboratory environment. This result can confirm the impact of the particular parameters on the value of the mass concentration of $PM_{2.5}$ and PM_{10} in the laboratories.

4 Conclusions

In the analyzed case, the aerosanitary conditions are primarily determined by the endogenic factors. Undoubtedly the results are affected by the physical containment of the laboratory space resulting from the poor ventilation in them. The considerable degree of limitation of air exchange with the ambient air result in the maintenance of qualitative parameters in the rooms only during the smog episodes. In general, during the heating season, the aerosanitary conditions have to be considered as unsatisfactory. The study also demonstrated that the assessment of indoor air quality should be performed individually for every case. Concurrently, it is known that the modernization of the ventilation systems in the examined laboratories is an urgent necessity.

References

1. ECA. **22** (2000)
2. US EPA. **EPA/600/P-99/002bF**; (2004)
3. L. Morawska, C.R. He, G. Johnson, H., E. Uhde, G. Ayoko, Environ Sci Technol. **43** (2009)
4. W.J. Fisk, D. Faulkner, D. Sullivan, M.J. Mendell, Aerosol Sci Tech. **32(6)** (2000)
5. H. Fromme, J. Diemer, S. Dietrich, J. Cyrys, J. Heinrich, W. Lang, M. Kiranoglu, D. Twardella, Atmos Environ. **42** (2008)
6. A. Afshari, U. Matson, L.E. Ekberg, Indoor Air. **15(2)** (2005)
7. T. Hussein, K. Hameri, M.S.A. Heikkinen, M. Kulmala M. Atmos Environ. **39** (2005)
8. H. Guo, L. Morawska, C.R. He, D. Gilbert, Atmos Environ. **42** (2008)
9. L. Morawska, C.R. He, G. Johnson, H. Guo, E. Uhde, G. Ayoko, Environ. Sci Technol. **43** (2009)
10. P.J. Rousseeuw, J. Comput App Math **20** (1998)
11. M. Lenart, T. Olszowski, Ecol Chem Eng A **23(2)** (2016)
12. C. Alves, T. Nunes, J. Silva, M. Duarte, Aerosol Air Quality Res **13** (2013)
13. B. Kromhourt, H.J. Boleij, Ann Occup Hyg. **39** (1995)
14. J.P. Guilford, *Psychometric methods* (New York, McGraw-Hill, 1954)

Calibration of parameters of water supply network model using genetic algorithm

Tomasz Boczar [1,*], *Norbert* Adamikiewicz[2], and *Włodzimierz* Stanisławski[3]

[1] Opole University of Technology, ul. Prószkowska 76, 45-758 Opole, Poland
[2] Miejskie Wodociągi i Kanalizacja w Kędzierzynie-Koźlu Sp. z o.o., ul. Filtrowa 14, 47-200 Kędzierzyn-Koźle, Poland
[3] State Higher Vocational School in Nysa, ul. Armii Krajowej 7, 48-300 Nysa, Poland

Abstract. Computer simulation models of water supply networks are commonly applied in the water industry. As part of the research works, results of which are presented in the paper, OFF-LINE and ON-LINE calibration of water supply network model parameters using two methods was carried out and compared. The network skeleton was developed in the Epanet software. For optimization two types of dependent variables were subjected: the pressure on the node and volume flow in the network section. The first calibration method regards to application of the genetic algorithm, which is a build in plugin - "*Epanet Calibrator*". The second method was related to the use of function *ga*, which is implemented in the MATLAB toolbox Genetic Algorithm and Direct Search. The possibilities of application of these algorithms to solve the issue of optimizing the parameters of the created model of water supply network in both cases: OFF-LINE and ON-LINE calibration was examined. An analysis of the effectiveness of the considered algorithms for different values of configuration parameters was performed. Based on the achieved results it was stated that application of the *ga* algorithm gives higher correlation of the calibrated values to the empirical data.

1 Introduction

Water supply networks are one of the main components of the technical infrastructure of cities and settlements. Due to water losses there is a necessity for operational cost reduction, which may be performed through continuing development of new supporting methods for monitoring and diagnostics of these facilities. Apart from the extensions and narrowing of the pipe cross section, sediment causing changes in surface roughness, etc., cracks, causing water leaks, are the most common damage to the water supply infrastructure. But only some of them are manifested by the outflow of water to the surface, the rest are invisible and difficult to locate [1,2]. The problem is growing in the areas of cities, where as a result of the exploitation of deposits, e.g. coal or calcium, located below the city, the occurrence of the failures is more common. This issue will have an impact also after the cessation of mining. Current methods of locating failures in water supply systems use monitoring devices. These methods are, however, less effective in case of increasingly common PVC and PE pipes [3,4,5]. For that reason, Authors performed research studies aimed at implementation of a device for signaling of occurrence and localization of a defect in the water supply network. The main component of the system for detecting and locating leaks in a water supply network is a computer model that simulates network operation. This paper presents the results of research works dealing with the calibration procedure of the proposed mathematical water supply network.

2 Methodology

The object under the study was the water supply network of Kedzierzyn-Kozle, Poland. It is build from the following elements: two pumping stations (water treatment plant and water treatment station), two zonal pumping stations located in Slawiecice and Cisowa, transmission pipelines with valves, junction valves and the water recipients. The system of pipelines consists of buses of a total length of ~153 km and of junction valves of total length of ~74 km. The network covers an area of 6392 x 12322 m (excluding wholesale customers in the neighboring communities). The city has an area of 123.4 km^2 where ~64000 residents live. Fig. 1 shows the general structure of the network that was created in the EPANET environment.

The water supply network simulation model before application was calibrated using empirical data, obtained during experimental measurements. Within carried out research works, the OFF-LINE parameters were adjusted using two methods. The first method was to use a genetic algorithm implemented in EPANET using the "EPANET Calibrator" plugin, while the second method involved the use of the MATLAB toolbox for Genetic Algorithm and Direct Search, using the *ga* function.

* Corresponding author: t.boczar@po.opole.pl

Both OFF-LINE and ON-LINE adjustment use the ready-to-use features available in the environment and the applicability of these algorithms to solve the optimization problem of the water supply model was investigated.

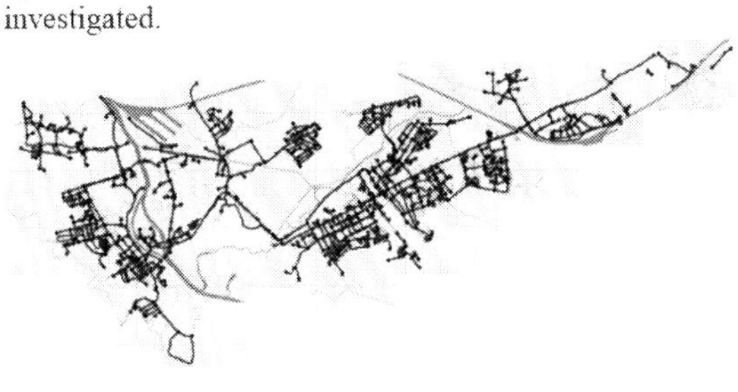

Fig. 1. Screenshot of the network created in the EPANET with the map in the background.

The "EPANET Calibrator" plug-in used during the OFF-LINE adjustment process was created by the "Resan Labs" team from Brazil. This solution makes it possible to correctly and repeatedly estimate the optimal parameters of model parameters. The calculation result is saved to a file that can then be further analyzed in the EPANET environment.

The use of the "Epanet Calibrator" plug-in requires the network to be split into separate regions (sectors). In each node or pipeline of a given sector, the value of the sought parameter is the same. In case of the roughness parameter, the areas in which the type and age of the infrastructure are similar should be selected. Fig. 2 shows the network divided into sectors.

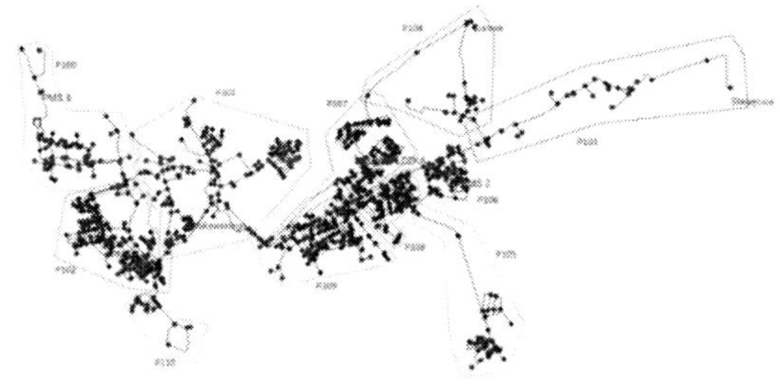

Fig. 2. Screenshot of the network created in the EPANET.

After establishing the sectors, reference values should be entered. The "EPANET Calibrator" optimizes model parameters for two types of dependent variables: water pressure in a given node and volume flow in a given network segment.

With a dedicated interface, one can enter the value of each parameter in successive time units manually, or one can import a time sequence directly from a CSV file. After the optimization process is started, the algorithm will modify the model parameters so that the resulting dependent variables correspond to the reference values. As part of the carried out works values measured on the object that were entered using CSV files were used.

In the next step, one determines which model parameters will be optimized. The EPANET Calibrator allows one to choose from three options:

- *roughness* of the pipe walls,
- the amount of water *leakage*, which is determined by the emitter coefficient,
- *roughness* and *leakage* at the same time. In the considered case, this option was applied.

The algorithm used in the plug-in seeks to minimize the error between measured (reference) values and the simulation obtained by both of the following values:

- network pressure values in a strictly defined network nodes, which are: p_{ZUW}, p_{SUW}, p_{PMS1}, p_{PMS2}, p_{PMS3}, p_{PMS4}, p_{PMS5}, p_{PMS6} and suction pressure at two pumping stations: p_{CIS}, p_{SLA},
- volumetric flow in strictly defined pipes, which are: v_{ZUW}, v_{SUW}.

The EPANET Calibrator allows one to set values for four selected genetic algorithm parameters. The drawback of this plug-in is that there is no information on the progress of the calculation except for the successive iteration error.

Parameters of the genetic algorithm applied in the EPANET Calibrator were defined as follows:

- population size = 200,
- number of generations = 500,
- probability of mutation = 0.01,
- probability of crossing = 0.9,
- the optimized parameter = *roughness*,
- dependent variable = *pressure* (network pressure).

3 Calibration results

For the mentioned settings, the mean square error (RMSE) values in all ten sectors were calculated and the obtained results are presented in Table 1 as compared to data before optimization. Further, exemplary graphics representations of the results of the algorithm are presented in Figs. 3 and 4.

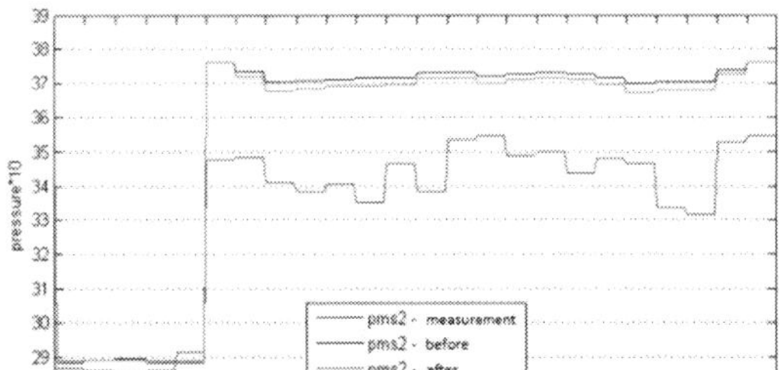

Fig. 3. Summary of genetic algorithm for PMS 2- pressure in measurement point no 2: the measured values (red), simulation result before optimization (blue), result of optimization (green).

In the next step, performance analysis of the algorithm for different values of the configuration parameters was performed. The obtained profit was compared by increasing the size of initial population or number of generations. Fig. 5 shows the evolution of the genetic algorithm for an initial population of 200 individuals and 500 generations.

Fig. 6 shows the opposite situation: 500 individuals and 200 generations. By analyzing the curves in Figs. 5 and 6, it can be stated that a greater number of population does not imply better or faster calculations. For 200 individuals, a satisfactory result was reached already after forty minutes of calculations, while for 500 individuals the algorithm achieved optimum results already in the sixteenth minute of operation.

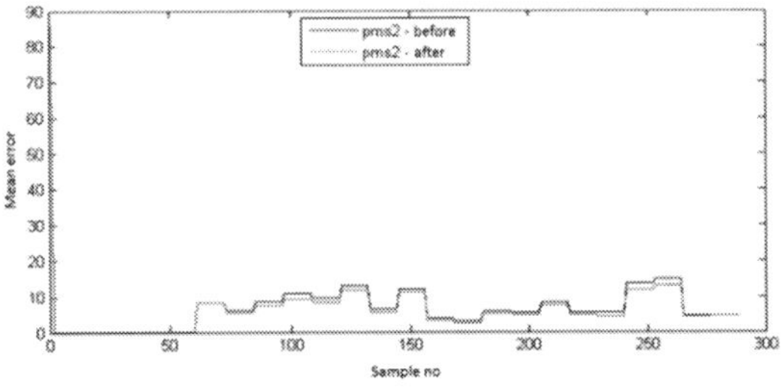

Fig. 4. Summary of genetic algorithm for PMS 2- pressure in measurement point no 2: error before optimization (blue), error after optimization (green).

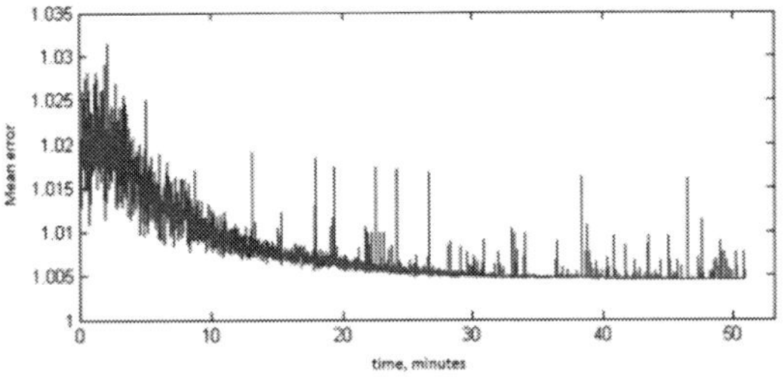

Fig. 5. The process of optimization in time. Population of 200 individuals, number of generations: 500.

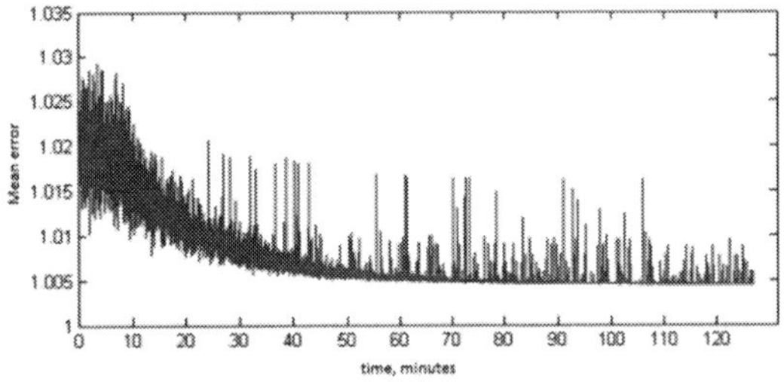

Fig. 6. The process of optimization in time. Population of 500 individuals, number of generations: 200.

The undoubted advantage of a large number of populations is the widening of the scope of the search while maintaining the convergence of the algorithm to the final result. Modification of mutation parameters would give a similar or better extension effect, but it could cause problems with maintaining convergence to the desired optimum value.

This phenomenon shows the results of optimization attempts shown in Figs. 7 and 8. The algorithm searching a larger solution space significantly minimizes the error. However, single individuals cannot keep their genetic information long enough for the analysis of a narrowed neighborhood to enable finding the best result.

The results of calculations performed using the EPANET calibration tool were not satisfactory and the and the mean squared error was not significantly reduced.

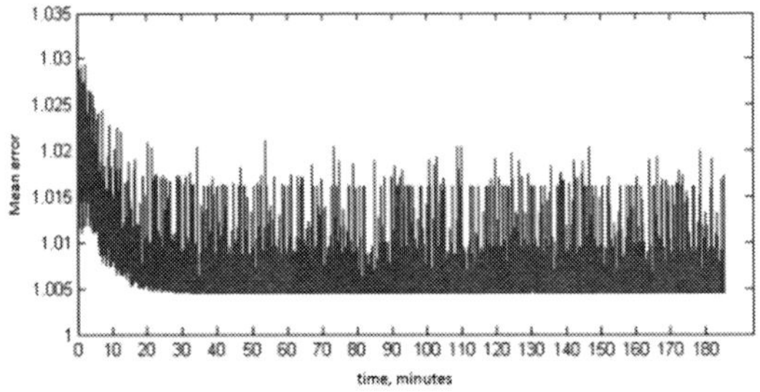

Fig. 7. The process of optimization in time. Population of 200 individuals, number of generations equal to 400, modified probability of mutation = 0.1.

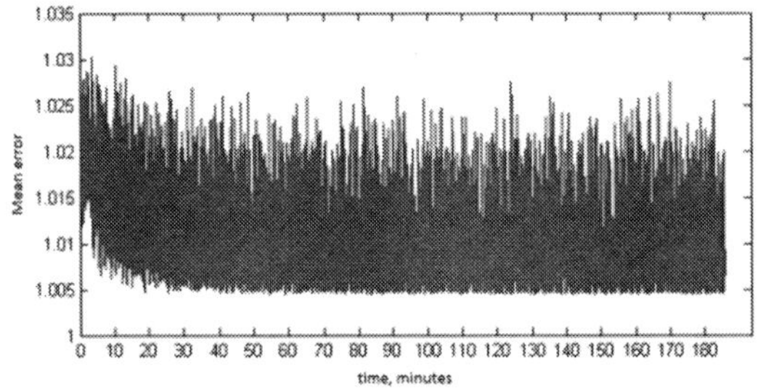

Fig. 8. The process of optimization in time. Population of 200 individuals, number of generations equal to 400, modified probability of mutation = 0.5.

Thus, in order to enhance the calibration results, the genetic algorithm implemented in MATLAB within the function *ga* was applied in combination with the epanet2.dll dynamic link library. For optimization the emitter parameter value was subjected in ten defined network sectors. On the prepared network, using the defined sectors, optimization was started with the algorithm configured according to the following settings:

- population size = 200,
- number of generations = 400,
- probability of mutation = 0.01,
- probability of crossing = 0.9,
- optimized parameter - node leakage = *emitter*,
- dependent variable: network *pressure* in 10 sectors,
- initial values: *roughness* = 0.1, *emitter* = 0.04.

As result of optimization the mean squared error value was significantly reduced, as shown in Table 1. It should be noted that successful optimization of the parameters using the "EPANET Calibrator" plug-in, is not possible when using automate diagnostic processes e.g. in an expert system, since there exist no correct interfaces to the system.

Table 1. Summary of optimization results – RMSE error values before optimization and values obtained using EPANET build in genetic algorithm and MATLAB tollbox.

Location – sector name	RMSE **before optimization**	RMSE **after optimization with EPANET**	RMSE **after optimization with MATLAB**
p_{ZUW}	0.3104	0.3083	0.2767
p_{SUW}	0.7968	0.7777	5.4206
p_{PMS1}	1.5688	1.4049	0.4892
p_{PMS2}	6.5923	5.8241	1.0097
p_{PMS3}	1.1115	0.9184	0.5122
p_{PMS4}	10.7789	9.7080	1.9766
p_{PMS5}	11.2838	10.609	0.9033
p_{PMS6}	8.4838	8.0646	0.9485
p_{CIS}	10.1014	9.3956	2.7899
p_{SLA}	15.5286	14.248	1.7254
Average	**6.6556**	**6.1250**	**1.6052**

4 Conclusions

The paper depicts the procedure of calibration of water supply network model using two methods: a build-in EPANET calibrator and an algorithm using MATLAB toolbox. The process considers optimization of the *roughness* and *leakage* parameters, which are estimated based on measurement data by using genetic algorithm. The motivation behind the performed comparative analysis was to show the difference between the accuracy of the model parameters calculated using the two considered methods.

Based on the research task performed it was shown that it is crucial to optimize the model parameters in order to gather a better fitted simulation model to the empirical data. The calibration and optimization tasks are necessary for achieve a precise simulation model, which may be applied as an module in an expert system for detection and localization of leakages in water supply networks. Example of such an expert system an interested reader may find in [5].

Acknowledgment

The work was co-financed by the Polish National Center for Science and Development within the project no PBS3/B3/33/2015.

References

1. H. Jia, W. Wei, K. Xin, Front. Environ. Sci. Eng. **2**, 57-62 (2008)
2. R. Schmid, *Review of modelling software for piped distribution networks. 2nd Ed.*, Swiss Centre for Development Cooperation in Technology and Management (2002)
3. B. Ulanicki, A. Zehnpfund, F.M. Alzamora, *Proc. 2nd Intl. Conf. on Hydroinformatics*, Zurich, (1996)
4. A. Preis, A.J. Whittle, A. Ostfeld, L. Perelman, J. Water Resour. Plan. Manage. **137,** 343-351 (2010)
5. D.A. Savic, Z.S. Kapelan, P.M.R. Jonkergow, Urban Water J. **6**, 3-22 (2009)
6. T. Boczar, D. Zmarzły, P. Frącz, *E-proceedings of the 37th IAHR World Congress*, Kuala Lumpur, Malaysia (2017)

The removal of chloramphenicol from water through adsorption on activated carbon

Joanna Lach[1,*], *Agnieszka* Ociepa-Kubicka[2]

[1] Czestochowa University of Technology, Faculty of Infrastructure and Environment, ul. Brzeźnicka 60a, 42-200 Częstochowa, Poland
[2] Czestochowa University of Technology , Faculty of Management, Armii Krajowej 19 B, 42-200 Częstochowa, Poland

Abstract. The presented research investigated the removal of chloramphenicol from water solutions on selected activated carbon available in three grades with different porous structure and surface chemical composition. Two models of adsorption kinetics were examined, i.e. the pseudo-first order and the pseudo-second order models. For all examined cases, the results of tests with higher value of coefficient R^2 were described by the equation for pseudo-second order kinetics. The adsorption kinetics was also investigated on the activated carbons modified with ozone. The measurements were taken from the solutions with pH values of 2 and 7. Chloramphenicol was the most efficiently adsorbed on the activated carbon F-300 from the solutions with pH=7, and on the activated carbon ROW 08 Supra from the solutions with pH=2. The adsorption of this antibiotic was in the majority of cases higher from the solutions with pH=2 than pH=7. The modification of the activated carbons with ozone enhanced their adsorption capacities for chloramphenicol. The adsorption is influenced by the modification method of activated carbon (i.e. the duration of ozonation of the activated carbon solution and the solution temperature). The results were described with the Freundlich and Langmuir adsorption isotherm equations. Both models well described the obtained results (high R^2 values).

1 Introduction

One of the new types of contaminants that can be found in water and wastewater are pharmaceuticals. The presence of drugs in treated wastewater was reported as early as in 1976 (clofibric acid - USA) [1]. Since that time, a wide range of almost all pharmaceutical types have been found in untreated and treated wastewater, surface water and even drinking water [2].

The presence of pharmaceuticals in wastewater is due to their common use in the prophylaxis and treatment of people and animals as well as in agriculture (including fish farming) [3] and therefore, the pharmaceuticals are found in municipal wastewater. The highest concentrations of pharmaceuticals are observed in wastewater from hospitals and pharmaceutical manufacturing plants [4].

One of the major groups of pharmaceuticals are antibiotics. The invention of penicillin by Alexander Fleming in 1928 marked a breakthrough in medicine. It is considered one of the greatest inventions of the 20th century. Unfortunately, very soon, a drug-resistance phenomenon was started to be observed (in the mid-30s of the 20th century for sulphonamides and in the second half of the 40s for penicillin and streptomycin). Currently, drug resistance has become a global problem [5]. In Europe, it is estimated that 25 000 of patients die because of infections caused by drug resistant bacterial strains. In the USA, as many as 70% bacteria involved in hospital infections are resistant to at least one antibiotic [6, 7]. The currently observed presence of antibiotics in surface water and, consequently, in drinking water is dangerous for the functioning of the environment and for human health [8].

Prolonged exposure, even to low antibiotic concentrations can: reduce the human immunity, have carcinogenic teratogenic or mutagenic properties or, having an effect similar to that of hormones, disturb the proper functioning of the human organism [9]. An unavoidable and, probably, the key result is the emergence of super-bacteria resistant to all antibiotics [10].

Removal of various contaminants through adsorption on activated carbon can be used for a wide variety of applications. It allows the removal of both inorganic and organic compounds from water and wastewater [11]. Different types of antibiotics can be effectively removed by this method [12-14].

The paper reports the results obtained from the study on the adsorption rate of chloramphenicol on three different grades of activated carbon that are widely used in water treatment. The study aimed to compare the selected activated carbons and to determine the effect of pH and temperature on the adsorption efficiency. The paper also presents the efficiency of chloramphenicol adsorption on the activated carbon modified with ozone.

* Corresponding author: jlach@is.pcz.czest.pl

2 Materials and methods

The water adsorption tests were conducted under static conditions by adding 1 g of activated carbon to 250 cm^3 of the solution, while stirring it on a mechanical shaker at the rotating speed of 160 rpm. The adsorption kinetics tests were conducted for the initial concentration of 0.5 $mmol/dm^3$, whereas the adsorption isotherm tests for the concentrations of 0.05; 0.1; 0.2; 0.5 and 1.0 $mmol/dm^3$ from the solutions with pH=2; 4; 6; 8 and 10, respectively. The pH value was regulated using the solution of HCl or NaOH. Because of high efficiencies of adsorption on the modified carbons, the adsorption efficiency tests were carried out for the concentration of 1.5 $mmol/dm^3$.

The influence of temperature on the adsorption process was determined for the solutions at the temperature of 20, 30 and 40°C, respectively, using a shaking water bath.

The activated carbons were modified with ozone in a water solution at different temperatures (20, 40 and 60°C). The 150 g sample of activated carbon was transferred into each Erlenmeyer flask and 300 cm^3 of distilled water was added. Then the flask was placed in the water bath. This mixture was purged with ozone at the concentration of 3 mg/dm^3 and the flow rate of 40 mg/dm^3. This modification process was continued for 60 and 120 minutes, respectively. The tests were conducted for three activated carbon grades with the characteristics given in Table 1 [13].

Table 1. Physical and chemical properties of activated carbons (PN-83/C-97555) [15].

Parameter	Activated carbons		
	ROW 08	F-300	WG-12
Bulk density,, g/dm^3	417	542	450
Surface area, m^2/g	890	859	1098
Water absorption, cm^3/g	0,97	0,72	0,61
Mechanical strength, %	98	97	98
pH of water extract	8,6	6,8	6,8
Methylene blue number, LM	34	31	28
Iodine adsorption, LI, mg/g	760	1055	1117

Chloramphenicol (Chloromycetin) used in this study was manufactured by Sigma-Aldrich (Fig. 1).

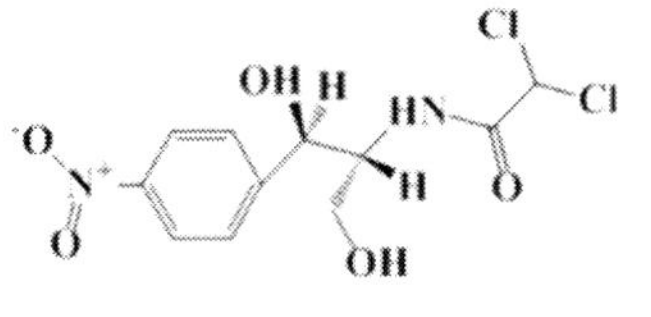

Fig. 1. The structural formula of chloramphenicol.

The molecular weight is 323.14 g/mol and the chemical formula is $C_{11}H_{12}Cl_2N_2O_5$, pKa=5.5. The concentration of chloramphenicol was determined spectrophotometrically at the wavelength of λ=278.
The adsorption isotherms were described with the Langmuir and Freundlich equations (Table 2).

Table 2. Isotherms and their linear forms.

	Isotherm	Linear form
Lang-muir	$q=\frac{q_m K_L C_e}{1+K_L C_e}$	$\frac{1}{q}=\left(\frac{1}{K_L \cdot q_m}\right)\frac{1}{C_e}+\frac{1}{q_m}$
Freun-dlich	$q=K_F C_e^{\frac{1}{n}}$	$\ln Q=\frac{1}{n}\ln C_e+\ln K_F$

where:

q is the amount of metal ions adsorbed per unit mass of adsorbent (mg/g), C_e - equilibrium concentration of the solute in the bulk solution (mg/dm^3), q_m – the solid phase concentration corresponding to the complete monolayer coverage of adsorption sites, K_L - the constant related to the free energy of adsorption, K_F - Freundlich isotherm constants, n - adsorption intensity.

Various models have been suggested to describe the kinetics of adsorption of the solute molecules on the sorbent. The Pb adsorption kinetics data were correlated with the adsorption kinetic models [16]:

- the pseudo-first-order kinetic model

$$\log(q_e - q_t) = \log q_e - \frac{k_1}{2.303}t \quad (1)$$

- the pseudo-second-order kinetic model

$$\frac{t}{q_t} = \frac{1}{k_2 q_e^2} + \frac{1}{q_e}t \quad (2)$$

where:
k_1 is the rate constant for the pseudo-first-order kinetic model,
k_2 is the rate constant for the pseudo-second-order kinetic model,
q_e is the amount of solute adsorbed et equilibrium and q_t is the amount of solvent adsorbed at time t.

3 Results and discussion

At the first testing stage, the kinetics of chloramphenicol adsorption from the solution with the concentration of 0.5 $mmol/dm^3$ on three activated carbon grades (WG-12, F-300 and ROW 08 Supra) was examined (Fig. 2). The equilibrium was first reached (after 7 hrs) for the activated carbon with the grade ROW 08 Supra. For the remaining two activated carbons the time to reach the equilibrium was much longer and amounting to 9.5 hrs. The equilibrium constants of

DOI: 10.1051/e3sconf/20171902008

adsorption kinetics were determined and presented in Table 3. The chloramphenicol adsorption kinetics followed the pseudo-second order model, as indicated by higher values of the correlation coefficients (Table 3).

The effect of pH on the efficiency of chloramphenicol adsorption on the activated carbon WG-12 was examined (Fig. 3). The highest efficiencies were achieved for the process proceeding from the solution with pH=2, while slightly poorer efficiencies were reported for the solution with pH=4. The rates of chloramphenicol adsorption from the solutions with pH= 6.8 and 10 were similar, but lower than for pH=2 and 4. This is most likely due to the fact that chloramphenicol undergoes dissociation at higher pH values and non-dissociated molecules (predominating at pH=2 and pH =4) are more efficiently adsorbed compared to dissociated ones. The obtained isotherms were described with the Langmuir and Freundlich model (Table 4). Both model types with high values of the correlation coefficient R^2 described the obtained test results.

Also, the effect of the solution temperature (in the range from 20 to 40°C) on the efficiency of chloramphenicol adsorption was determined in this study (Fig. 4, Table 5). It was observed that the higher the adsorption temperature was, the higher the adsorption efficiency was reported. The differences were not large.

Taking into consideration unmodified activated carbons it was found that chloramphenicol was best adsorbed on the activated carbon F-300 from the solutions with pH=7, and on the activated carbon ROW 08 Supra from the solutions with pH=2. The efficiency of adsorption on the carbons modified with ozone was examined. In the case of all investigated activated carbons with different grades, the efficiencies of chloramphenicol adsorption were higher on modified activated carbons (Table 6). For the activated carbons WG-12 and ROW 08 Supra it was noted that the higher the temperature and the longer the modification time, the more efficient the adsorption of chloramphenicol was. In the case of the activated carbon F-300, the optimal modification temperature was 20°C. The obtained results can be associated with the increase in the specific surface area of the activated carbons due to the modification with ozone [17]. The modification of activated carbon enables also the change of the chemical structure of the activated carbon surface. It was found that ozone led to the increase in the number of acid groups (capable of exchanging anions) as well as basic groups (capable of exchanging anions) [18]. The highest adsorption efficiency (94.7%) was observed when the process proceeded from the solution with pH =2 on the activated carbon F-300 modified with ozone at 20°C for 120 minutes. The most significant differences in the adsorption efficiency were observed for the process proceeding from the solution with pH=7. The efficiency of adsorption on the unmodified activated carbon WG-12 was 72%, whereas on the activated carbon modified at 60°C for 120 minutes, it amounted to 89.3%.

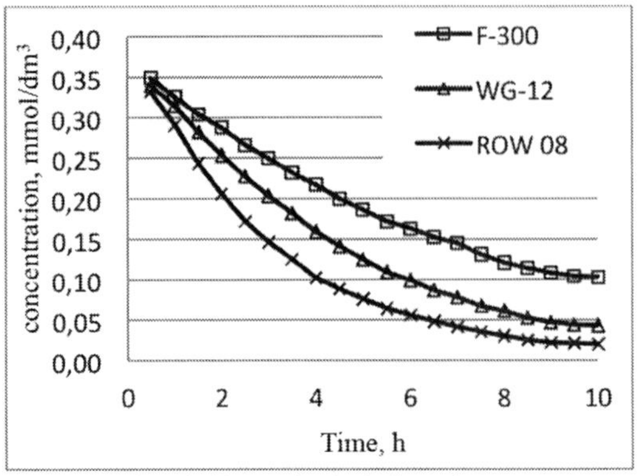

Fig. 2. Kinetics of chloramphenicol adsorption.

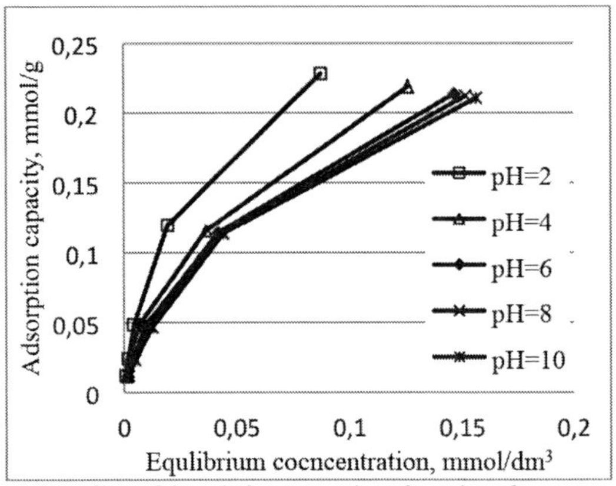

Fig. 3. Effect of the solution pH on the adsorption of chloramphenicol.

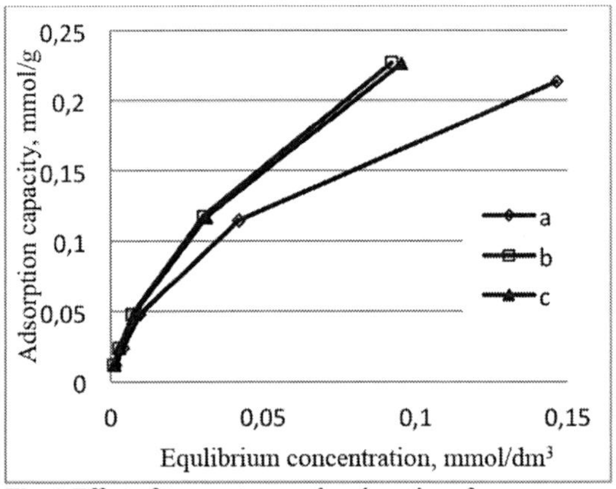

Fig. 4. Effect of temperature on the adsorption of chloramphenicol: a – 20°C, b - 30°C, c - 40°C.

Table 3. Pseudo first- and pseudo second-order rate constants for the adsorption of chloramphenicol on the activated carbons.

Activated carbon	q_e (exp) mg/g	First-order kinetic model		Second-order kinetic model	
		k_1 1/h	R^2	k_2 g/(mmol h)	R^2
ROW 08	0.120	0.051	0.8991	9.799	0.9978
F-300	0.099	0.046	0.8110	12.245	0.9889
WG-12	0.114	0.052	0.9648	10.773	0.9832

Table 4. The values of Freundlich and Langmuir constants for the adsorption of chloramphenicol.

Isotherm	Constants	pH of the solution				
		2	4	6	8	10
Freundlich	K_F, mmol/kg	1.22	0.82	0.81	0.81	0.81
	1/n, -	0.63	0.60	0.64	0.66	0.67
	R^2	0.9815	0.9853	0.9906	0.9930	0.9935
Langmuir	q_m, mmol/g	0.203	0.194	0.165	0.167	0.150
	K_L, dm^3/mmol	74.08	57.53	46.33	40.43	42.48
	R^2	0.9920	0.9951	0.9957	0.9965	0.9915

Table 5. The values of Freundlich and Langmuir constants for the adsorption of chloramphenicol.

Isotherm	Constant	Process temperature, °C		
		20	30	40
Freundlich	K_F, mmol/kg	1.07	1.14	1.10
	1/n, -	1.32	0.66	0.65
	R^2	0.9464	0.9979	0.9981
Langmuir	q_m, mmol/g	0.165	0.138	0.145
	K_L, dm^3/mmol	46.33	87.66	80.71
	R^2	0.9959	0.9865	0.9911

Table 6. The efficiency of chloramphenicol adsorption from the solutions with the concentration of 1.5 mmol/dm^3, %.

Activated carbons	pH solution	
	2	7
Activated carbon type WG-12		
WG-12	83.3	72.0
WG-12–O_3-120 min 20^0C	92.0	88.7
WG-12 –O_3-60 min 40^0C	92.0	87.3
WG-12 –O_3-120 min 40^0C	90.0	85.3
WG-12 –O_3- 60 min 60^0C	89.3	84.7
WG-12 –O_3-120 min 60^0C	93.3	89.3
Activated carbon type ROW 08 Supra		
ROW 08	90.0	88.7
ROW 08–O_3- 60 min 20^0C	93.3	94.0
ROW 08–O_3-120 min 20^0C	93.3	92.0

 DOI: 10.1051/e3sconf/20171902008

Table 6 (cont.)

ROW 08–O_3- 60 min 40^{0}C	94.0	92.7
ROW 08–O_3-120 min 40^{0}C	94.7	92.0
Activated carbon type F-300		
F-300	88.0	90.7
F-300–O_3- 60 min 20^{0}C	92.0	90.7
F-300–O_3-120 min 20^{0}C	94.7	90.7
F-300–O_3-60 min 40^{0}C	90.7	88.0
F-300–O_3-120 min 40^{0}C	91.3	86.7

4 Conclusions

The adsorption of chloramphenicol on the investigated activated carbons WG-12, F-300 and ROW 0 followed the pseudo-second order kinetics equation. Considering the unmodified activated carbons it was found that chloramphenicol was the most efficiently adsorbed on the activated carbon F-300 from the solutions with pH=7, and on the activated carbon ROW 08 Supra from the solutions with pH=2. When examining the effect of the solution pH on the adsorption rate it was found that the lower the solution pH was, the larger the capacity of the activated carbon was reported. This is the result of more favorable adsorption of non-dissociated molecules which predominated in acid solutions. Higher adsorption efficiency was observed when the process proceeded from the solution at the temperature of 40°C, compared to 30°C or 20°C, but the differences were not large. The modification of the activated carbons with ozone resulted in the increase in their adsorption capacities with respect to chloramphenicol. The optimal modification temperature and duration depended on the activated carbon type.

Acknowledgements: This research was financed from the statutory grant No. BS/PB – 401/301/12.

References

1. K. Sosnowska, K. Styszko-Grochowiak, J. Gołaś, Krakowska Konferencja Młodych Uczonych, 395, (2009)
2. A. Szymonik, J. Lach, K. Malińska, Chem. Eng. S. **24**/1, 65 (2017)
3. F. Yu, Y. Li, S. Hans, J. Ma, Chemosphere 153, 365 (2016)
4. A. Koszowska , M. Ebisz, T. Krzyśko-Łupicka, Environmental Medicine **18/1**, 62 (2015)
5. Z. Markiewicz, Z. A. Kwiatkowski, *Bakterie, antybiotyki, lekooporność*. (Wyd. Naukowe PWN, Warszawa, 2012)
6. D. Żabicka, E. Literacka, K. Bojarska, Aktualności NPOA **3**, 1 (2012)
7. L. Bruton, J.S. Lazo, K.L. Parker., *Farmakologia Goodmana & Gilmana* (Wyd. Czelej, Lublin, 2007)
8. L. Wollenberger, B. Halling-Sørensen, K. Kusk, Chemosphere **40,** 723 (2000)
9. O.A. Jones. J.N. Lester, N. Voulvoulis Trends Biotechnol. **23**, 163 (2005)
10. J.L. Martínez, Environ. Pollut. **157**, 2893 (2009)
11. J. Lach, E. Okoniewska, L. Stępniak, A. Ociepa-Kubicka, Desalin. and Water Treat. **52** 3979 (2014)
12. E. Bocos, E. Alfaya, O. Iglesias, M. Pazos, M. Á. Sanromán, Sep. Purif. Technol. **151,** 243 (2015)
13. X. Zhang, W. Guo, H.H. Ngo, H. Wen, N. Li, W. Wu, J. Environ. Manage. **172**, 193 (2016)
14. L. Qin, Z. Zhou, J. Dai, P. Ma, H. Zhao, J. He, A. Xie, C. Li, Y. Yan, J. Taiwan Inst. Chem. E. **62,** 228 (2016)
15. J.Lach, E. Okoniewska, L. Stępniak, E. Ociepa, Ann. Set Environ. Prot. **15/3**, 2142 (2013)
16. L. Largitte, R. Pasquier, Chem. Eng. Res. Des. **109**, 495 (2016)
17. J. Rivera-Utrilla, M. Sánchez-Polo, Chemosphere **65,** 1131 (2006)
18. M. Sa´nchez-Polo, U.von Gunten, J. Rivera-Utrilla, Water Res. **39**, 3189 (2005)

Analysis of factors affecting the ecological status of the large water bodies on the basis of monitoring and integrated 3D models

Ireneusz Zbiciński[1,*], *and Aleksandra* Ziemińska-Stolarska[1]

[1]Lodz University of Technology, Faculty of Process and Environmental Engineering, Wolczanska 213 Street, 90-924 Lodz, Poland

Abstract. The paper presents water monitoring methods used to determine and predict the ecological status of large water bodies based on calibrated 3D water quality models. State monitoring system of quality of water was discussed and critically reviewed. Mobile measurements technique to collect vast number of measuring points across the large water body during one day cruise to make maps of pollutants distributions was described. Maps of water quality parameters show dynamic changes of spatial distributions of pollutants both in time and along the length of the lake which proves that one point measurements cannot be representative for the large water body. Critical evaluation of calibration principles and accuracy of the most frequently used water quality models, WASP and GEMSS are presented. On the basis of CFD calculations, close correlation between water quality and hydrodynamics in large water bodies regarding the physicochemical aspects and biological production was proved. Neglecting the effect of hydrodynamic on eutrophication process introduce large element of uncertainty in determination of nutrients flow in the large water bodies. An example of scenarios for improving the ecological potential of dam reservoirs to obtain significant reduction of selected nutrients concentration in the lake is discussed.

1 Introduction

Declining water quality has become a global problem due to human populations grow, expansion of industrial and agricultural activities and climate change which alter the hydrological and biogenic cycles. Globally, the most prevalent water quality problem is eutrophication, a result of high-nutrient loads, which substantially impairs beneficial uses of water. The main sources of the pollution are emissions to water of sewage and pollutant effluents, which increase load of biogenic and other substances causing e.g. seasonal cyanobacterial blooms.

Anthropogenic activities effecting water quality include also deepening and widening of waterways, the construction of port basins causing disturbance of bottom sediments. Wastewater delivered to water bodies is mainly produced through agricultural activities (natural and artificial fertilization, animal production, leakages from fertilizer containers, manure, plant protection products) and municipalities (lack of sewage system, leakage of septic tanks, runoff of sewage from the streets).

Significant contribution in the eutrophication of waters is caused also by emission from point sources produced by municipal and industrial wastewater treatment plants or discharges from sanitary and storm sewers.

To collect data necessary for any kind of water management decisions, state program of monitoring of quality of water bodies has been developed [1].

Currently, the following types of water quality of monitoring are implemented:

Diagnostic monitoring; carried out every 6 years: covers around 100 indicators: biological and physico-chemical, salinity, acidification, biogenic substances, aerobic conditions, particularly harmful substances-priority substances, etc.

Operational monitoring: carried out every 3 years controls biological elements and physico-chemistry of waters. The goal of operational monitoring is to determine the state of the JCWP (Uniform Part of Surface Waters) and to track changes resulting from the action programs dedicated to improve the state of the waters.
Additional requirements are set for recreational areas, including bathing areas and areas sensitive to eutrophication caused by pollution from municipal sources.

Research monitoring is used to determine the effect of accidental pollution on environment or to explain the reason of not achieving the environmental goals attained to given JCWP.

Due to high costs of monitoring, even for a single measuring point (costs of measuring system, installation of the buoy, maintenance), standard measurements are carried out from 3 to 6 years. To use the data from years

* Corresponding author: ireneusz.zbicinski@p.lodz.pl

between the measurements, the Mark Heritage Rule [1] was introduced. "Mark (Factor) heritage rule" says that data collected in previous years can be used to estimate current ecological status of the water body according to the following principles: biological elements can be used on the level of single factor for 6 years for fish species, the other for 3 years. Physico-chemical and hydromorfological elements must be taken from the last year of biological evaluation whereas chemical elements, from the last 6 years.

Question arises how representative are the data collected from single measuring point for the large water body? Can we assume uniform distribution of physico-chemical, biological and hydromorfological parameters in large reservoirs?

The question is crucial for accurate description of ecological status of water bodies and correct management decisions regarding current and future actions towards maintenance and improvement of environment quality.

2 Measurements of water quality parameters

Problem of quality and quantity of data for correct description of ecological status of large water bodies was considered in the frame of MONSUL project (2015-2017) devoted to determination of ecological status of Sulejow Reservoir [2]. The ultimate goal of the project was to develop a universal tool for critical and balanced decisions concerning environmental protection and planning related to the environment of the reservoir waters.

In the MONSUL project mobile measurements technique to collect large number of measurements of water quality parameters together with the exact time and location across the entire reservoir during one day cruise was developed. The measurement were made using EXO 2 multi-parameter probe (YSI, USA) with data collection and transmission to provide continuous measurements of the following parameters: water temperature, pH, dissolved oxygen concentration, conductivity, chlorophyll and blue green algae (BGA) concentration and ammonium ion concentration.

Traditional, stationary measurements were also carried out using the buoy equipped with the EXO 2 multi-parameter probe and a meteorological station measuring air temperature, humidity, wind speed and direction, atmospheric pressure and insolation. For the purpose of measurements in motion, a special extension boom was installed on the motorboat to attach the EXO2 probe. The construction of the boom enabled to control probe immersion depth and allow access to the probe for observation and periodic cleaning. After tests of motorboat measurements and comparing results from mobile and stationary measurements in the same location point (reference value), optimal speed of the boat (around 5 km/h) at which the results of both measurements matched with good accuracy was set. A typical route for one day motorboat measurements tracked with a GPS system is shown in Fig. 1. Elaborated methodology of mobile measurements enabled to collect around 700 during one cruise.

Data collected during mobile measurements, after transformation into a spatial datum located on the surface of the reservoir (GIS processing systems) allow to make precise maps of water quality parameters in the water body.

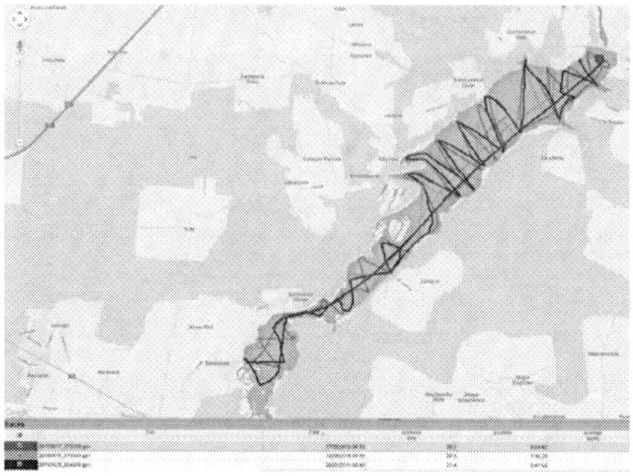

Fig. 1. An example of the route of the motorboat for one day mobile measurements determined by a GPS system.

A series of mobile measurements of the water quality parameters carried out in June/September 2015 shown dramatic changes of spatial distributions of pollutants both in time and along the length of the lake, Fig. 2 and 3.

Spatial distributions of chlorophyll and blue green algae (BGA) concentration between June and September 2015 are shown in Figs. 2 and 3.

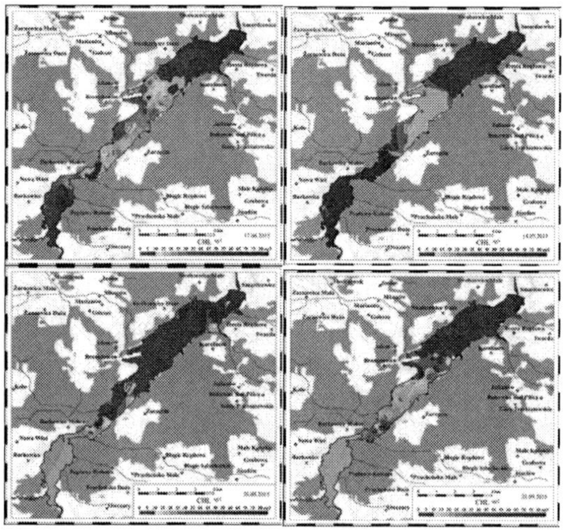

Fig. 2. Spatial distribution of chlorophyll concentration in June/September 2015.

Analysis of Figs. 2 and 3 proves that one point measurements cannot be representative for the entire water body.

Differences in concentrations of chlorophyll and blue green algae are even tenfold in the same measuring day, which confirms that not only quality but also quantity of the data are important for all actions to improve of ecological status of large water body [3].

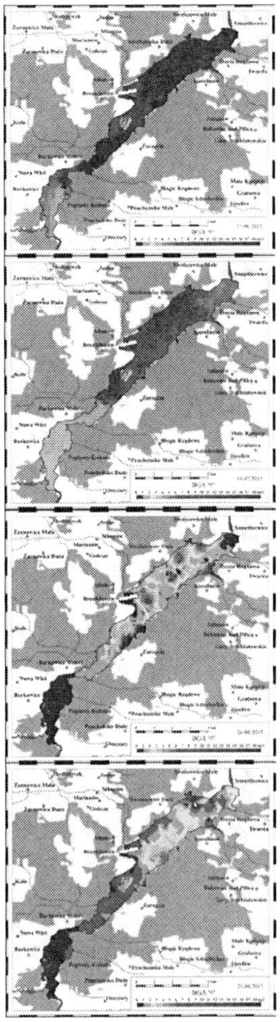

Fig. 3. Spatial distribution of blue green algae (BGA) concentration June/ September 2015.

To support water management decisions and study relationships in the ecosystem surface water quality models can be used.

3 Water quality models

The objective of water quality modeling is to study the characteristics of the spatial and temporal distribution of water quality variables under different conditions and various external factors. There is a big diversity of water quality models, from a single factor of water quality indicators to multifactor, from steady-state models to dynamic, from point source models to the coupling models of point and nonpoint sources, and from zero-dimensional (input-output models) to one-, two- and three-dimensional models [4,5].

Significant efforts were expended in developing mathematical models to predict eutrophic status of the natural and man-made water bodies, as a function of nutrient inputs [6,7,8,9]. With understanding of relationship between the eutrophication processes and hydrodynamics, as well as with more advanced computing capability, multidimensional models of lakes and reservoir have been developed to study water quality problems [10,11,12,13].

Cao and Zhang [14] classified of water-quality models based on water body types, model-establishing methods, water quality coefficients, components, model property, spatial dimension, and reaction kinetics. Selected water quality models with a brief description are presented below:

Streeter-Phelps models, 1-D steady-state models focus on oxygen balance and one-order decay of BOD [15].

QUAL, 1-D river and stream water quality models

suitable for dendritic river and non-point source pollution including steady-state or dynamic models [14].

Water Analysis Simulation Program, WASP, Dynamic compartment-modeling program for aquatic systems, including water column and the underlying benthos which allows to investigate 1, 2, and 3-D systems, and a variety of pollutant types. Can be linked with hydrodynamic and sediment transport models that can provide flows, depths velocities, temperature, salinity and sediment fluxes [16,17].

MOHID, 3-D water modeling system which integrates modules for the simulation of hydrodynamics, water quality and sediment processes. The integration of MOHID different tools, (MOHID Water, MOHID Land and MOHID Soil) can be used to study the water cycle in an integrated approach [18].

Generalized Environmental Modeling System for Surface Waters, GEMSS, 3-D hydrodynamic and transport models embedded in a geographic information and environmental data system (GIS). Compute time-varying velocities, water surface elevations, and water quality constituent concentrations in rivers, lakes, reservoirs, estuaries, and coastal water bodies [19].

MIKE, A finite-difference model in an orthogonal grid system for free surface flows. Includes modeling components for advection-diffusion, water quality, heat exchange with the atmosphere, heavy metals, eutrophication, flooding and drying of intertidal areas, and sediment processes. The model can simulate the fate and transport of conservative or linearly decaying constituents, including: nutrient cycling; dissolved oxygen levels; exchange of metals between the bed sediments and the water column; and sediment transport, deposition, and entrainment [20].

The most frequently used models of water quality, with vast number of practical applications are WASP and GEMSS.

Critical evaluation of principles and accuracy of simulations for both models is presented below.

3.1 WASP

The Water Quality Analysis Simulation Program, WASP, is a general dynamic mass balance framework for modeling contaminant fate and transport in surface waters. To date WASP has been employed in over one hundred modeling applications that have included rivers, lakes, estuaries and ocean environments [16,21].

WASP Transport Options: A body of water is represented in WASP as a series of discrete computational elements or segments. Environmental properties and chemical concentrations may vary spatially among the segments. Each variable is advected and dispersed among water segments, and exchanged with surficial benthic segments by diffusive mixing.

WASP Toxicant Modules: The primary WASP toxicant code, TOXI, is combined with separate databases to support four modules—simple toxicant, nonionizing toxicants, organic toxicants, and mercury.

WASP Eutrophication Module: The WASP eutrophication code, EUTRO, EUTRO output reports state variable concentrations, along with key forcing functions and process rates for every segment at user-specified print intervals. Major processes and variables included in the EUTRO module, Fig. 4.

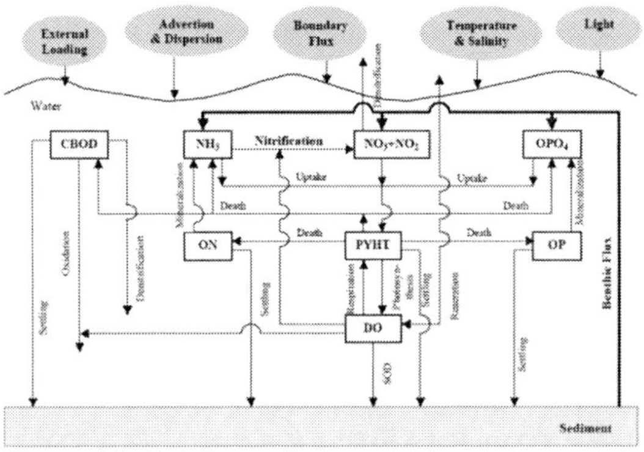

Fig. 4. Major processes and variables included in the EUTRO module [21].

3.2 GEMSS

Model GEMSS® (Generalized Environmental Modeling System for Surface waters) produced by Environmental Resources Management (ERM® Inc. Exton, PA, USA) [19] is a three-dimensional and multifunctional mathematical model designed to simulate the surface water environment (GEMSS® software). GEMSS, like WASP consist of different modules used for specific applications. The main modules of GEMSS® family are: hydrodynamic, sediment transport, differential thermal, chemical and oil spills, toxic spread, construction of wetlands, water quality. WASP's module EUTRO is directly used in GEMSS.

The basic principle of the water-quality models is based on the conservation of mass. The water volume and water-quality constituent masses being studied are tracked and accounted for over time and space using a series of mass balance equations assuming vertical and lateral homogeneity.

There are two main challenges deciding about accuracy of water quality models; correct determination of kinetics of nutrients cycle and correct hydrodynamics of water flow. The first challenge is satisfied by calibration technique to adapt kinetic model parameters to a particular water body by achieving the smallest discrepancies between simulation results and measurement data.

As an example, calibration principle on the basis of phytoplankton growth modeling is described below.

Phytoplankton growth kinetics assume a central role in prediction of the eutrophication process. Changes the amount of phytoplankton during the day in the particular segment is defined by the concentration of chlorophyll „a" (mg/L/day). Quantity of phytoplankton in the water column depends on the growth and sedimentation rate as well as mortality. Phytoplankton concentration in the water body is described by the equation:

$$\frac{\partial C_P}{\partial t} = (R_G - R_D - R_S)C_p \tag{1}$$

where: C_p - phytoplankton concentration (mg/L), R_G - growth rate constant (day^{-1}), R_D - death rate constant (day^{-1}), R_S - settling rate constant (day^{-1}).

The specific growth rate R_G is defined via the following equation:

$$R_G = G_{max} X_T X_L X_N \tag{2}$$

where: G_{max} - the maximum growth rate in 20°C, at optimum light and nutrients conditions (day^{-1}), X_T - the temperature adjustment factor, dimensionless, X_L - the light limitation factor (function of incident solar radiation (ly/day)), X_S - the nutrient limitation factor as a function of dissolved inorganic phosphorus and nitrogen.

The total biomass reduction rate R_D for the phytoplankton in the segment due to the death rate is expressed by the equation:

$$R_D = k_{1R}(T) + k_{1D} + k_{1G} Z(t) \tag{3}$$

k_{1R} = endogenous respiration rate constant, day^{-1}, k_{1D} = mortality rate constant, day^{-1}, k_{1G}= grazing rate constant, day^{-1}, Z(t) = herbivorous zooplankton population grazing on phytoplankton, mg C/L.

The settling of phytoplankton cells makes an important contribution to the overall mortality of the phytoplankton population, particularly in lakes, reservoirs and coastal oceanic waters. Sedimentation of phytoplankton is described by the relationship:

$$R_S = \frac{v_S A_S}{V} \tag{4}$$

where: v_S = settling velocity, m/day, A_S = surface area, m^2, V = segment volume, m^3.

 DOI: 10.1051/e3sconf/20171902009

Table 1 shows the variables and parameters in the net phytoplankton growth equations. In the second column an average range of constants in kinetics equations for phytoplankton growth is displayed.

Table 1. Phytoplankton Growth Terms (WASP).

Description	Average range	Sulejow Reservoir
Light extinction coefficient	0,1 - 5	1, m^{-1}
Segment depth	0,1 - 30	0,3-7, m
Water temperature	0 - 35	0-25, oC
Fraction of day that is daylight	0,3 - 0,7	0,3 -
Average Daily Surface Solar Radiation	200 - 750	500 Ly/day
Zooplankton Population - Rate Constants	0 - 1	0,5 mg C/l
Endogenous Respiration Rate at 20°C	0,05	0,05
Settling Velocity	0,01 - 0,2	0,15
Death Rate	0 - 0,25	0,05
Grazing Rate	0 - 5	1,5

In the next step, in all water quality models, specific value of constant for a given water body should be determined. This step is called calibration of the model; the constants are determined by trial and error method using experimental data for analyzed water body.

Selected constants for phytoplankton kinetics calculated by WASP for the Sulejow Reservoir are presented in Table 1. Such an approach gives reliable and exact values of the simulated parameters of the eutrophication process assuming high quality and sufficient quantity of the data.

Similar approach is applied for phosphorus cycle, nitrogen cycle and oxygen balance. An example of GEMSS simulations (both validation and calibration) for water temperature and nitrates concentration in the Sulejow Reservoir is shown in Fig. 5 and 6.

a)

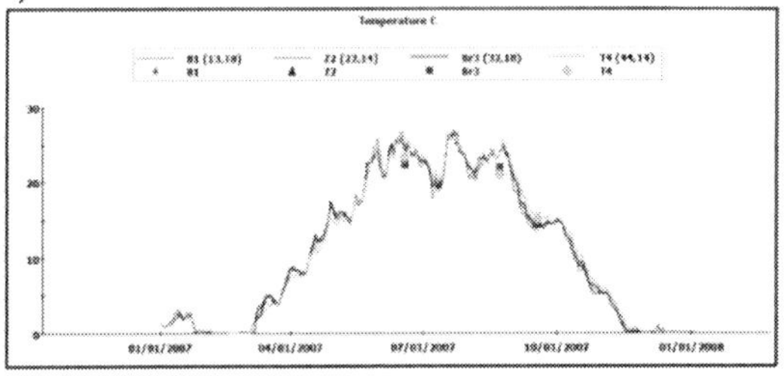

b)

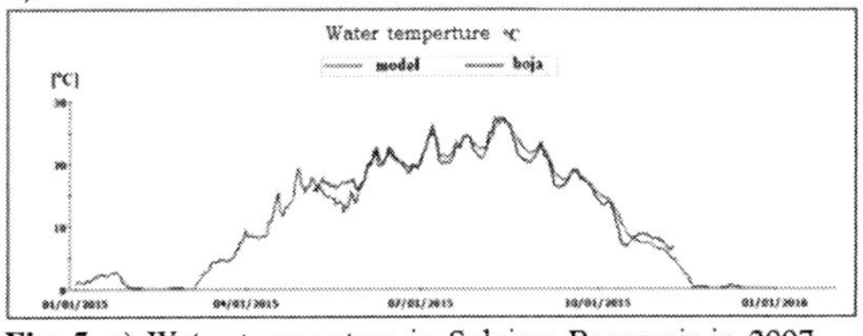

Fig. 5. a) Water temperature in Sulejow Reservoir in 2007 – model calibration b) Water temperature in Sulejow Reservoir in 2007 – model verification.

a)

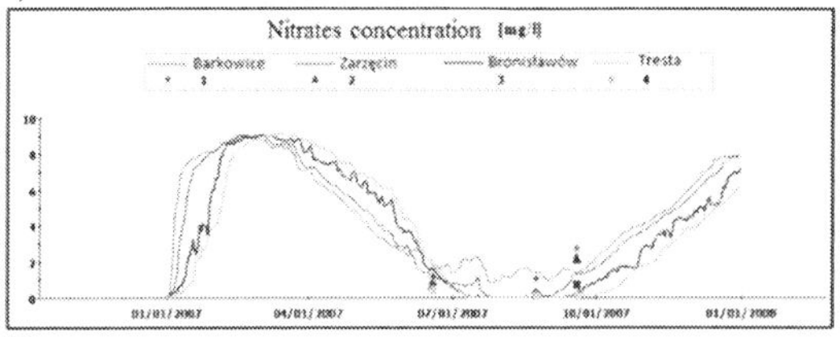

b)

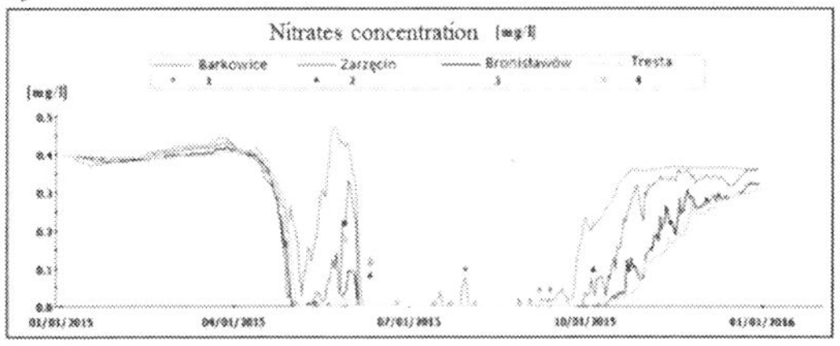

Fig. 6. a) Nitrates concentration in the Sulejow Reservoir in 2007 – model calibration **b)** Nitrates concentration in the Sulejow Reservoir in 2015 – model validation.

GEMSS model for the Reservoir was calibrated using data from 2007 [22]; good match of theoretical calculations with experimental measurements for prediction of both parameters in 2015 proves accuracy and versatility of GEMSS methodology. [23].

On the bases of calibrated water quality model scenarios for improving the ecological potential of the Reservoir can be determined, Fig. 7.

Analysis of Fig. 7 shows that 50 % reduction of nutrients input results in threefold concentration of nitrogen in the lake.

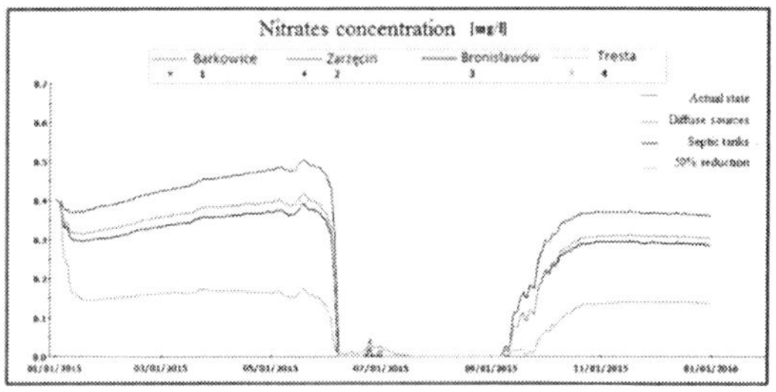

Fig. 7. Scenarios for improving the ecological potential of the Reservoir.

The results of simulations might help decision maker to undertake actions to reduce inflow of nutrients to obtain significant reduction of concentration of biogenic substances in the lake.

4 Effect of hydrodynamics on nutrients distribution

There is a number of literature suggestions that essential factor affecting phytoplankton community structure and cyanobacteria appearance is the hydrodynamics in the lake [13, 24, 25, 26, 27].

All water quality models contain hydrodynamic modules but owing to large size and complexity of the computational grid, none of existing water quality

DOI: 10.1051/e3sconf/20171902009

models can deliver reliable picture of hydrodynamics in a big water bodies. Important issue is how simplified hydrodynamics might affect final predictions of nutrient flow in the large water reservoirs.

For accurate determination of hydrodynamics Computational Fluid Dynamics (CFD) technique is commonly used. CFD technique is complex, requires experience and knowledge in elaboration of geometry of the object, optimization of numerical grid, selection of proper turbulence model (for big water bodies flow is turbulent because of the low water viscosity and large transverse dimensions) and, always verification of calculated flow patterns with experimental data [27].

Additionally, especially for shallow, polymictic lake, the impact of wind might be important for the distribution and mixing of the water masses.

In Fig. 8 results of CFD calculations (Ansys Fluent 14.0) for 17th-kilometer long Sulejow Reservoir show strong effect of wind on surface streamlines [17]. Significant differences in the course of the path lines and velocity fields are noticeable when the effect of the wind is considered. The simulations show, that when steady flow pattern develops in the basin, a large recirculation zones are formed bellow the outlet of the reservoir.

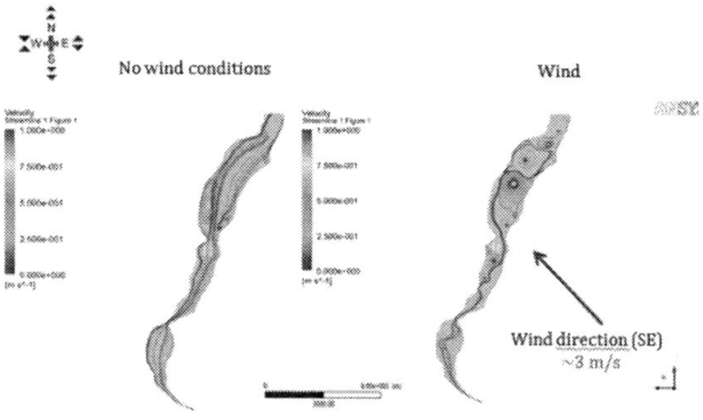

Fig. 8. Effect of wind on surface streamlines in the Sulejow Reservoir [17].

Fig. 9 shows the surface velocity field below the outlet of the reservoir. The length and color of the vectors depict the magnitude of the velocity, as well as the direction of flow. Such circulating structures, reached the size of even half kilometer and increase the water retention time. Results of CFD simulations were verified by field measurements in selected cross-sections of the lake [17].

Recirculation zones have also a clear impact on the distribution of phytoplankton cells during the water blooms. Lacustrine area of the reservoir is characterized by significant slowing the flow < 0,1 m/s, which develops favorable conditions for the growing of algae species.

Hydrodynamics obtained from CFD calculations cannot be directly used in water quality models. Manual transfer of data, however, is possible which increase accuracy of determination of nutrients concentrations in the lake. Fig. 10 shows comparison of chlorophyll „a" concentrations during the vegetation season, combined with the hydrodynamic conditions.

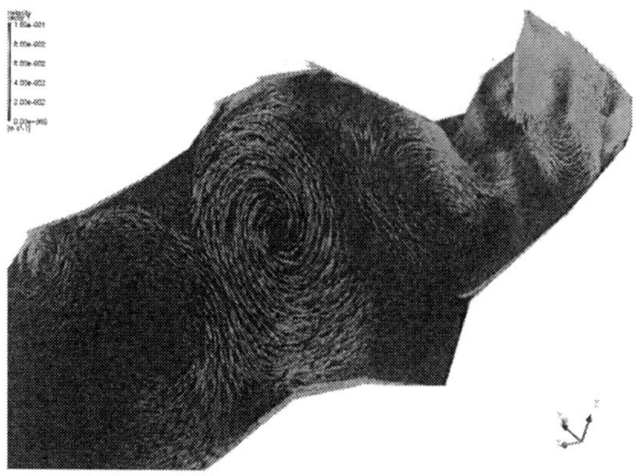

Fig. 9. Surface velocity field below the outlet of the reservoir, with arrows indicating direction and color scale indicating magnitude (velocity in m/s).

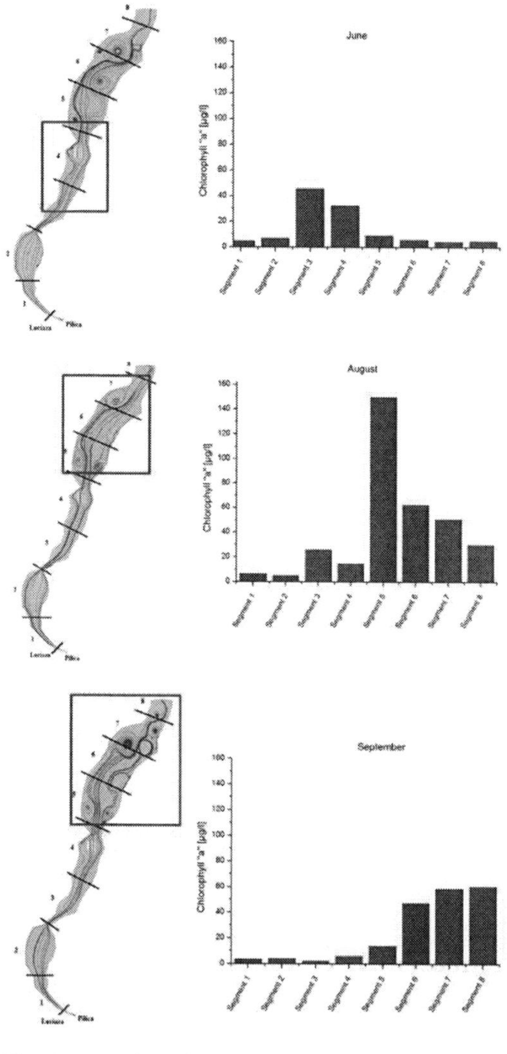

Fig. 10. Comparison of chlorophyll „a" concentrations during the vegetation season, combined with the hydrodynamic conditions. Red squares present areas with the highest concentration of the analyzed parameter.

Analysis of the results shows the movement of plankton biomass from segments 2, 3 and 4 in June to segments 6, 7 and 8 in August and September. Results of simulations show that chlorophyll "a" content is strongly affected by hydrodynamics; in the recirculation zones higher concentrations of the parameter are due to higher retention time [28,29].
Analysis of Fig. 10 shows that accurate chlorophyll "a" concentration can not be obtain without determination of hydrodynamics in large water bodies.

Grid with 17 million elements was necessary to predict correctly hydrodynamics in Sulejow Reservoir whereas typical Hydrodynamic Module in water quality models (WASP, GEMSS, etc.) offers space for about 60000 numerical cells (around 300 times less than in CFD model) only which introduce large element of uncertainty in calculations of hydrodynamic and further in determination of nutrients flow in the large lakes.

Identification of mechanisms which control water quality in dam reservoirs is a scientific basis for undertaking practical actions, whose aim can be both to retain the status of the aquatic environment and also to improve it in the defined time horizon, depending on changing anthropogenic pressure on the natural environment.

5 Conclusions

State monitoring of water quality is rare and delivers sparse data which might not be representative for the large water body.

The application of mathematical modelling to water quality has proved to be a powerful support in water resources management. As a predictive tool, mathematical model allows forecasting and evaluation of effects of changes in the surrounding environment on the ecological status of the analyzed water bodies. As a diagnostic tool, permits a description of highly complex, real conditions. WASP model was used to determine the water quality in the Sulejow Reservoir and GEMSS model was used to determine the percent reduction in discharges derived from supplying rivers, necessary to improve the water quality standard in the Sulejow Reservoir.

One of the most important challenge controlling accuracy of water quality models is satisfied by calibration technique to adapt kinetics model parameters to a particular water body by achieving the smallest discrepancies between simulation results and measurement data. Such an approach can give a reliable description of the eutrophication process assuming high quality and sufficient quantity of the measured data.

The methodology of mobile monitoring measurements of water quality parameters which is compliant with the standards used in surface water monitoring studies developed in a frame of MONSUL project may deliver large number of high quality data.

Hydrodynamic characteristics, such as residence time, surface circulation pattern and turbulent mixing are affecting quality of surface water bodies. In order to predict areas of algal blooms, the hydrodynamics should be correlated with the nutrient kinetics growth.

Application of calibrated and verified water quality models, can be used to identify mechanisms controlling water quality and to elaborate control strategies and scenarios for improving the ecological potential of water body with reduced cost of monitoring.

References

1. http://www.gios.gov.pl/pl/stan-srodowiska/pms (accessed August 05. 2017)
2. www.monsul.wipos.p.lodz.pl (accessed August 05. 2017)
3. H.B. Glasgow, J.M. Burkholder, R.E. Reed, A.J. Lewitus, J.E. Kleinmann, J. Exp. Mar. Biol. Ecol. **300**, (2004)
4. Q. Wang, S. Li, P. Pia, Ch.Qi, F. Ding, Sci World J, **231768**, 7 (2013)
5. Z.X. Xu, S.Q. Lu, J Hydrodyn. **15**, 2 (2003)
6. S.E. Jorgensen, Lake Reserv Manage J. **3** (2001)
7. D.M. Di Toro, J.P. Connolly, Mathematical Models of Water Quality in Large Lakes, (USEPA, 1980)
8. R. V. Thomann, J.A. Muller, *Principles of surface water quality modeling and control,* (Harper & Row Publishers, 1987)
9. W.S Lung, *Water quality modeling for waste load allocations and TMDLs*, (John Wiley & Sons Inc., 2001)
10. G.T. Orlob, *Mathematical Modeling of Water Quality: Streams, Lakes, and Reservoirs* (John Willey & Sons, 1983)
11. J.L. Martin, S.C. McCutcheon, *Hydrodynamics and transport for water quality modeling* (CRC Press, Inc., 1999)
12. S.D. Lau, V.R. Stovin, I. Guymer, Water Sci. Technol. **55**, 4 (2007)
13. R. Pastres, S. Ciavatta, Environ. Modell Softw. **20**, 8 (2005)
14. X.J. Cao, H. Zang, J. Water Resour. Archit. Eng. **4**, 4 (2006)
15. D.J. O'Connor, Water Resour Res, **3**, (1967)
16. D.M. Di Toro, J.J. Fitzpatrick, R.V. Thomann, Water Quality Analysis Simulation Program (WASP) and Model Verification Program (MVP) - Documentation, (Hydroscience Inc., Westwood, NY, 1983).
17. A. Zieminska-Stolarska, Analysis of the flux of biogenic substances on water eutrophication in the large water bodies on the example of the Sulejow Reservoir, PhD Thesis, Lodz University of Technology, (2014)
18. R. Miranda, F. Braunschweig, P. Leitão, R. Neves, F. Martins, A. Santos, Ecol. Environ **40** (2000)
19. GEMSS® Workshop Guide, ERM Inc., (Exton, PA, USA, 2009)

20. Danish Hydraulic Institute, MIKE21: User Guide and Reference Manual (Danish Hydraulic Institute, Horsholm, 1996)
21. T.A. Wool, R.B. Ambrose, J.L. Martin, E.A. Comer, Water Quality Analysis Simulation Program (WASP) - version 6.0. USEPA, Atlanta, 2001)
22. Voivodeship Inspectorate For Environmental Protection in Lodz. Report on the state of water purity in the Sulejow Reservoir in 2007, (Lodz 2008)
23. B. Rakowska, M. Sitkowska, E. Szczepocka, B. Szulc, Oceanol. Hydrobiol. Stud. **1** (2005)
24. J. Kubrak, E. Nachlik, Hydraulic bases of flow capacity calculation for river beds, (SGGW, Warsaw, 2003)
25. V.A. Verkhozina, O.M. Kozhova, Y.S. Kusner, Aquat. Ecosyst. Health **3**, 2 (2000)
26. Deus, R., Brito D., Kenov I.A., Lima M., Ecol. Modell. **253** (2013)
27. A.G. Andersson, P. Andreasson, T.S. Lundström, Eng. Appl. Comp. Fluid **7**, 1 (2013)
28. K. Izydorczyk, M. Tarczynska, Ecohydrol. Hydrobiol. **5** (2005)
29. A.K.M. Kabzinski, H. Grabowska, Gospodarka Wodna, **5** (2008)

Model of geothermal water deironing on the basis of a triangulation

Iwona Klosok-Bazan[1,*], and *Robert* Olszewski[2]

[1]Opole University of Technology, Mechanical Department, Opole Mikolajczyka Str. 5, Poland
[2]Warsaw University of Technology, Faculty of Geodesy and Cartography, Warsaw, Pl. Politechniki 1, Poland

Abstract. The main of this research was to recognize the efficiency of geothermal water deironing processes. To this, studies on a laboratory scale were conducted. In the paper the modeling of geothermal water deironing processes using new tool basis on triangulation has been used. Carrying out the analysis of potential possibilities of using geothermal water and taking economic aspects of water deironing processes into account, the authors of this article conducted research into the geothermal water deironing processes by adjusting existing knowledge on the removal of iron from cold groundwater. The conducted research confirmed the possibility of adjusting such processes to the treatment of geothermal water, but their effectiveness differs depending on the temperature and salinity of water, but this relation is not linear. The analyzes indicated that the best deironing effect was obtained for water whose salinity does not exceed 10 g/L and the temperature 30°C.

1 Introduction

Natural hot waters have been used since earliest time, and warm springs occur in many region of the world. Reservoirs of geothermal water are found in locations where the earth's heat is near enough to the surface so that water or steam can reach the top [1-3]. Geothermal waters quality is very diverse. Iron is one of the solute available as pyrite and pyrrholite from rock-forming constituents which usually is found at very low concentration levels [3]. Iron is not considered hazardous to humans health but it causes technological and esthetic problems. Iron present in the water at concentrations exceeding the level of 0.2mg/l, the following problems can occur: iron precipitates give water a reddish color when exposed to air; deposition of iron precipitates in the distribution system can reduce the effective pipe diameter and eventually clog the pipe; additionally aesthetic value of the equipment used in balneology and recreation lower because of the reddish residue [2-11]. Therefore, the water with iron should be treated in deironing processes. Iron removal by aeration and separation is a common method for removal of iron as from groundwater going into the public water supply system as well from geothermal water using for balneology [7-11].

Alternative processes, such as ion-exchange, oxidation with oxidizing agents including chlorine and potassium permanganate, filtering with activated carbon and other filtering materials, supercritical fluid extraction, bioremediation and treatment with limestone, have been proposed in order to facilitate the operation and allow the removal of high amounts of iron [7, 9].

The quality of the deironing process is a function of many factors changing in time [3,4]. Deironing processes of so-called cold water (water temperature up to 20° C) are rather well known and broadly described in literature [7-9]. However, geothermal water deironing processes have not been the subject of so many studies, which induced the authors of this article to become interested in this issue. To bring up the problem, laboratory tests were conducted and the model of the process on the basis of a triangulation was made.

Triangulation is one of the methods of interpolation enable conversion of discrete data to a continuous form. It allows to test the properties of the mathematical field which is the model of continuous distribution of the quantitative feature of the phenomenon. Appropriately selected interpolation methods allow for development of the correct model of spatial distribution of a given phenomenon on the basis of the limited amount of source data. Interpolation is a numerical method which allows to define a so-called interpolation function in a given range which accepts the nominal values in set points called nodes. Interpolation is often applied in the experimental circumstances where you have a limited amount of data for definition of value relationships [12]. The interpolation algorithms can be applied both in two-, three- and multi-dimensional space [13]. The easiest example is the application of the linear function (Fig. 1) for interpolation of the value of dependant variable (y) for a given value of independent variable (x). The knowledge of the value of the (endogenous) variable

* Corresponding author: i.klosok-bazan@po.opole.pl

DOI: 10.1051/e3sconf/20171902010

explained for two values of the explanatory (exogenous) variable is, in this case, sufficient to define y value for any x in a set field of considerations.

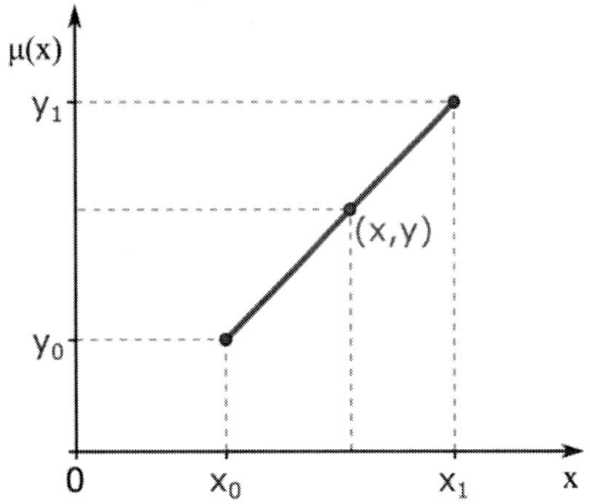

Fig. 1. Linear interpolation.

This easiest linear approach can also be applied in case of three-dimensional space. Having at disposal a set of irregularly distributed points with coordinates (xi, yi) and known values of attributes (zi), (e.g. height of the point above sea level, temperature, etc.), we can determine (interpolate) the demanded value in the external point of the area in coordinates defined in any manner (x,y) (Fig. 2).

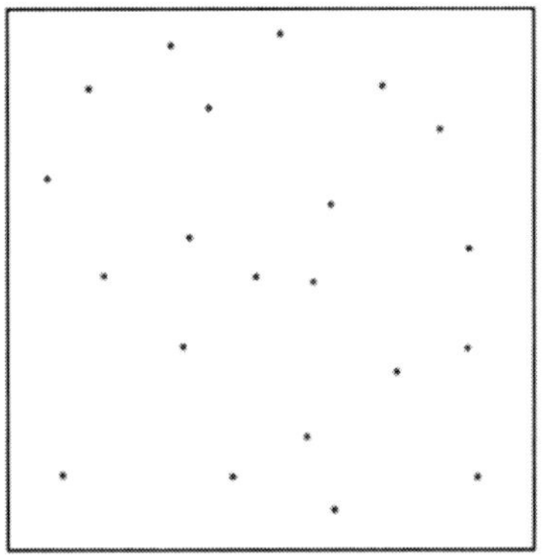

Fig. 2. The set of irregularly distributed measurement points.

Spatial interpolation with the linear method requires to set the network of Delaunay triangles or, in other words, determine Thiessen polygons on the basis of the set of measurement points. When they are determined in this manner, they fulfil the following conditions (Fig. 3): sums of triangle edges are the lowest and the sum of triangle heights is the lowest (the sum of length of two longest triangle edges is the lowest).

Delaunay triangles set (in X, Y, Z space) can be the basis for creation of the numerical space model - a so-called statistical surface [12] based on the set of distributed points (e.g. numerical model of topographical relief [13-15].

A so-called triangle irregular network (TIN) created in the process of Delaunay triangulation is, as a rule, transformed to the form of a regular network (most often with square mesh) which significantly facilitates the interpolation of spatial data (Fig. 4).

Determination of the interpolated value ZP in a given point P is performedby the application of a simple formula (Fig. 5) [12-15].

A)

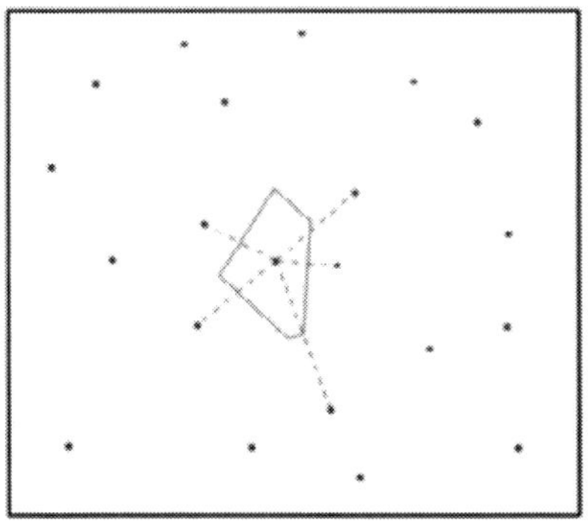

B)

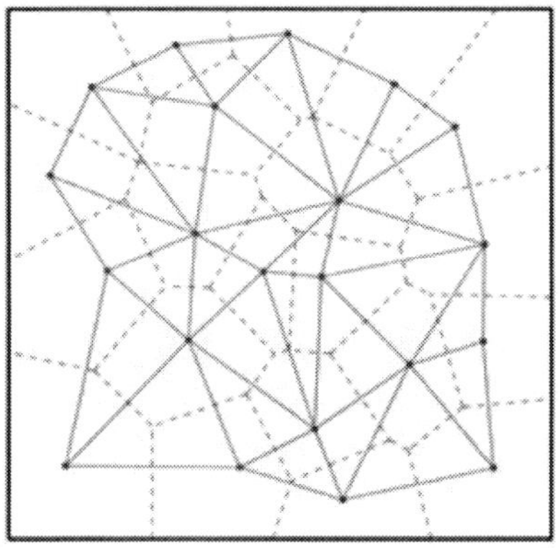

Fig. 3. Thiessen polygons (A) and Delaunay triangulation (B).

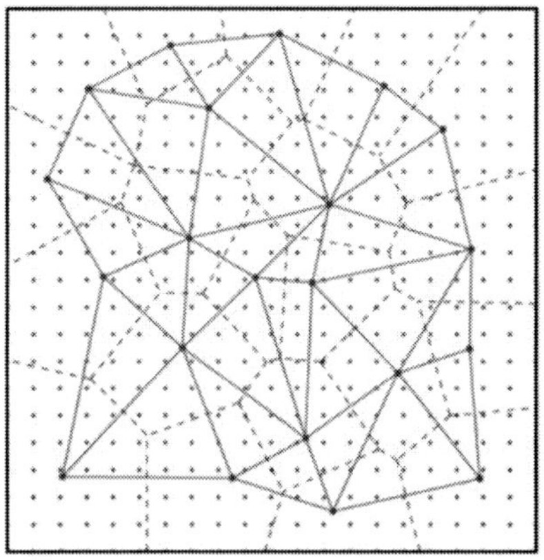

Fig. 4. The transformation of TIN model into a set of regularly distributed points.

 DOI: 10.1051/e3sconf/20171902010

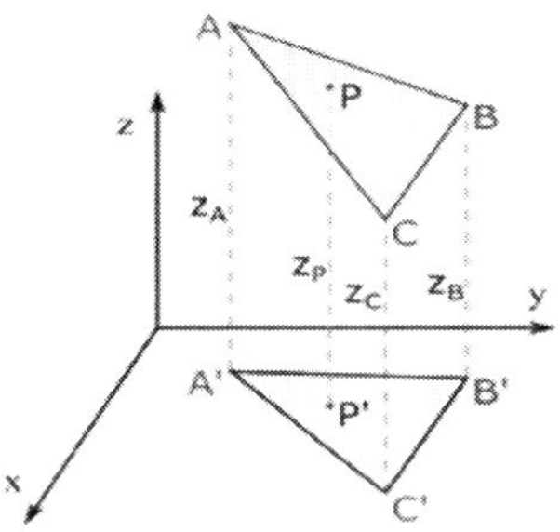

Fig. 5. Determination of the value of dependent variable in TIN model (interpolation linear function).

$$z_P = \frac{z_A S_A + z_B S_B + z_C S_C}{S_A + S_B + S_C} \quad (1)$$

where:
S_A- P'B'C' triangle field;
S_B- P'A'C' triangle field;
S_C- P'A'B' triangle field.

2 Materials and methods for laboratory tests

The fundamental aim of this article was to develop the model of effectiveness of geothermal water deironing processes for balneological purposes on the basis of trangulation. For this reason the laboratory tests were conducted.

Geothermal water was prepared on the basis of distilled water and pure reagents added proportionally. Due to the composition of the analyzed geothermal water and brines, it was assumed that the salinity of model water will be obtained by dosing sodium chloride (NaCl) and sodium sulphate (Na_2SO_4) in range 4 – 150 g/L. After the preparation of saline water and the adjustment of pH to the level of natural geothermal water, i.e. pH = 7.0 ± 0.2, water was heated until it reached the desired temperature.

The tests were conducted at constant, high concentration of iron, which was added to water in the form of pure iron sulphate in the amount 0.05 g/dm^3. It corresponded to the concentration of iron at the level of 18.42 mg/dm^3. Bivalent iron in the form of $FeSO_4$ was added to water with the assumed level of salinity, after the adjustment of pH, to the sample in the water bath.

The temperature range applied in the test was 20-50 ° C. Samples prepared in such a way were subject to the aeration and, then, filtration processes after which iron was measured according to the standard PB-16/PL, edition no. 1 of 02.01.2014 Colorimetric determination of the total iron and divalent iron content in water containing small amounts of organic compounds using thiocyanate test.

The results obtained in the laboratory were used to create a model based on the previously described tool.

3 Results and discussion

The determination of optimum process zones for specific ranges of salinity and temperatures is extremely important (Fig. 6). It is necessary to provide an appropriate amount of oxygen so that the oxidation process could proceed correctly. In case of deironing of standard groundwater, this process should proceed effectively as 8-11°C is the temperature at which the oxygen solubility is relatively high [7, 9].

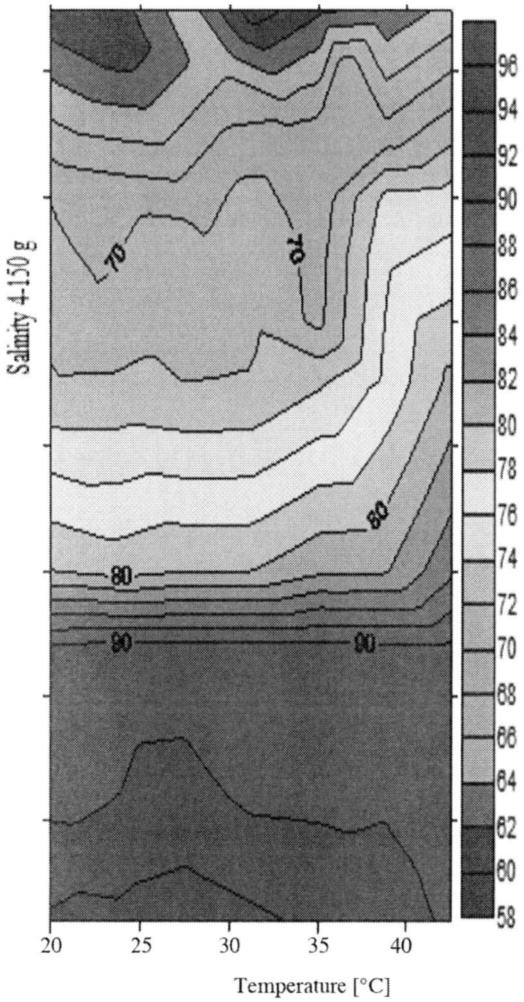

Fig. 6. Deironing process effectiveness model developed using triangulation model – thematic map.

The provision of an appropriate amount of oxygen dissolved in geothermal water with the temperature significantly exceeding 20°C is much more difficult. That is why the experimental determination of the time of reaction for the least optimum conditions (the highest salinity and the highest temperature) was the first, very important task to perform. As a result of the conducted analysis, it was stated that the aeration time should be about 30 minutes. The presented relation makes (Fig. 6) it possible to determine expected trends in the scope of changes in the effectiveness of deironing processes depending on changes in the temperature and salinity of treated water. From the presented relation it appears that the water deironing effectiveness decreases as the salinity increases, but this relation is not linear. The effectiveness of the deironing process is also affected by the second analyzed parameter, i.e. temperature. Figure 6 draws attention to the peculiar peninsula that was occurred when temperature range to 30°C.

4 Conclusions

The conducted research confirmed the possibility of adjusting such processes to the treatment of geothermal water, but their effectiveness differs depending on the temperature and salinity of water. The removal of iron from water of salinity of 120-140 g/dm^3 at water treatment temperature 20-25 °C is the least effective while the highest effectiveness of the process was recorded for water of the lowest salinity level. The optimum process was reached at the temperature 25-35 °C. The use of triangulation model based on quantity equations describing the relation between process variables allowed to predict the degree of iron removal in various cases depending on the process temperature rather precisely.

References

1. W. Górecki and another, *Atlas of Geothermal Resources of Paleozoic Formations in the Polish Lowlands*, Kraków (2006)
2. A.J. Ellias, W.A Mahon, *Chemistry and Geothermal Systems*. Academic Press, New York (1977)
3. J.L. Burguera, M. Burguera, C.E. Rondon, Anal. Chim. Acta 366 (1998)
4. I. Klosok-Bazan, Ochrona Srodowiska **36**, 2 pp. 43-46 (2014)
5. S. Armannson, J Pet. Sci. Technol. **33**, 1 pp. 181-188 (1981)
6. I. Klosok-Bazan, A. Sowa, Operational problems associated with the high iron concentration in thermal waters used for balneology. *Proceedings of 15th International Multidisciplinary Scientific GeoConference SGEM,* Albena, pp. 469-476 (2015)
7. A. Kowal, M. Świderska-Bróż, *Oczyszczanie wody*, Wydawnictwo Naukowe PWN, Warszawa (1997)
8. I. Klosok-Bazan, Modeling of geothermal water deironing processes using artificial neural networks. *Proceedings of 16th International Multidisciplinary Scientific Geo Conference SGEM*, Albena, pp. 187-193 (2016)
9. R.D. Letterman, *Water quality and treatment. A handbook of community water supplies*. McGraw-Hill Book Company (1999)
10. Ch. Bong-Yeon, Process Biochem. **40** pp. 3314–3320 (2005)
11. I. Klosok-Bazan, Analisys of Recreational Use and Treatment Possibilities of Geothermal Water Taken from Sandstone Layer in Opole City. *Proceedings of 16th International Multidisciplinary Scientific Geo Conference SGEM*, Albena, pp. 27 - 33 (2016)
12. A.H. Robinson, *Elements of Cartography*, John Wiley&Sons (1995)
13. Z. Kurczyński, R. Preuss, *Podstawy fotogrametrii*, Oficyna Wydawnicza Politechniki Warszawskiej, Warszawa (2000)
14. A. Buczek, R. Olszewski, Studium możliwości koherencji komponentów TOPO i NMT Bazy Danych Topograficznych, Archiwum Fotogrametrii, Kartografii i Teledetekcji **17b**, (2000)
15. M-J. Krak, F. Ormeling, *Kartografia. Wizualizacja danych przestrzennych*, Wydawnictwo Naukowe PWN, Warszawa (1998)

Viscosity of diluted suspensions of vegetal particles in water

Adriana Szydłowska*, and *Jerzy* Hapanowicz

Department of Process Engineering, Faculty of Mechanical Engineering, Opole University of Technology, Poland

Abstract. Viscosity and rheological behaviour of sewage as well as sludge are essential while designing apparatuses and operations employed in the sewage treatment process and its processing. With reference to these substances, the bio-suspensions samples of three size fractions ((i) 150÷212 μm, (ii) 106÷150 μm and (iii) below106 μm) of dry grass in water with solid volume fraction 8%, 10% and 11% were prepared. After twenty four hours prior to their preparation time, the suspension samples underwent rheometeric measurements with the use of a rotational rheometer with coaxial cylinders. On the basis of the obtained results, flow curves were plotted and described with both the power model and Herschel-Bulkley model. Moreover, the viscosity of the studied substances was determined that allowed to conclude that the studied bio-suspensions display features of viscoelastic fluids. The experimentally established viscosity was compared to the calculated one according to Manley and Manson equation, recommended in the literature. It occurred that the measured viscosity values substantially exceed the calculation viscosity values, even by 105 times. The observations suggest that it stems from water imbibition of fibrous vegetal particles, which causes their swelling and decreases the amount of liquid phase in the suspension.

1 Introduction

Sewage treatment is a very complex process because it consists of many operations, e.g. piping transport, mixing, aeration or sedimentation [1]. In turn, sewage which is not subjected to purification is sometimes applied as a heat source forced through heat pumps [2]. In order to achieve a correct design of the apparatuses within the process of sewage processing (in terms of effectiveness and profitability), knowledge of these substances viscosity in various process conditions is essential. The viscosity influences mainly the suspensions flow pattern, their flow resistances, the phenomena of heat transfer and also the velocity of solid particles settling or gas bubbles lifting. The viscosity is a parameter depending not only on the temperature. Its value may also result from the solid concentration, the solid properties and also the shear time and shear rate bound up with the flow velocity in a pipe or a mixing intensity in a tank. Fluctuations of suspensions viscosity can be verified due to the rheometeric measurement results of these fluids. The rheometeric measurements are the most state-of-the-art and desirable in the waste-water treatment sector as well as in many other fields of industry what is mentioned by the authors of the papers [3,4].

Sewage rheology became an important research subject in the early eighties of the last century and it resulted from increasing population of people and, thus an increasing amount of sewage which could not be stored in any way [4]. Since then, a stronger emphasis has been placed on increased efficiency of the sewage treatment process while keeping costs as low as possible.

The fact that rheology belongs to one of the most important fields of knowledge, which allows to design the sewage treatment process correctly, is confirmed by numerous articles concerning this issue, inter alia [1,3-8]. The authors of these articles recognize unanimously the sewage as non-Newtonian fluids which do not obey the Newton's law of viscosity

$$\eta = \frac{\tau}{\dot{\gamma}}, \tag{1}$$

in which η means a coefficient of dynamic viscosity [Pa·s], τ is shear stress [Pa] and $\dot{\gamma}$ is shear rate [s^{-1}]. However, as far as determination of the rheological behaviour of sewage is concerned, the scientists' opinions are divided. In articles [2,7] it has been proven that sewage can be treated as shear-thinning substance whereas the authors of the article [6] revealed that sewage belongs to viscoelastic fluids with yield stress. Due to the difficulties with elaboration of the correct measurement methodology, the establishment of the rheological stability of sewage has not been successful yet [5]. Unequivocal rheological classification of sewage may be considered burdensome due to its very complex composition and structure. Sewage constitutes a multiphase system formed by water, long-chain polymers, bacteria, colloids, flocculants and also particles of organic origin including vegetal origin [5]. In addition, the designation of the sewage rheological

* Corresponding author: a.szydlowska@doktorant.po.edu.pl

DOI: 10.1051/e3sconf/20171902011

behaviour may sometimes seem inconclusive as a consequence of its continuously changing inner structure.

Usually, although the rheometeric measurements of sewage are extensive, they are limited to the establishment of the rheological characteristics of different kinds of sewage sludge [9] e.g. activated sludge [3] or digested sludge [7] which is partly recirculated. Additionally, changes of the physical properties of the studied sludge are indicated. The changes result from the impact of the subsequent physical and chemical processes of successive stages within the waste-water treatment processes. The impact of the solid concentration and the grain geometry of the solid in the sludge on the rheological properties is also investigated [8,10]. However, the literature of the subject reveals insufficient information about spontaneous changes of sewage physical properties, which should interplay with their rheological properties. Hence, the primary aim of this paper is to determine the viscosity of the bio-suspensions samples which were prepared in advance. In order to investigate whether the time of existence of bio-suspension affects their rheological properties, suspensions samples of vegetal particles in water were studied.

2 Materials and methods

The aqueous suspensions of dry grass (of hay) were measured. Such a solid indubitably occurs in sewage sludge and furthermore, a combination of water (as well as an aqueous solution) with a vegetal solid is found in various industry sectors.

The solid was milled and sieved and, as a result, three solid size fractions were obtained: (i) 150÷212 μm, (ii) 106÷150 μm and (iii) below 106 μm. It was assumed that the particles of the applied vegetal material constitute elements of fibres with size that suits a given fraction. The suspensions with the solid volume fraction of each of the fractions equal to 8%, 10% and 11% (the solid mass fraction of the samples equals 9%, 11% and 13% respectively) underwent the rheometeric measurements. The samples were prepared according to the gravimetric method what required earlier determination of the solid real density. The physical properties of the suspensions components and the suspensions notations which were used for the needs of the results documentation are compiled in Table 1.

The volume of the prepared samples corresponded to the measuring gap capacity of the rheometer and it was equal to 25 cm^3. Operating with the solid mass fraction, the earlier calculated masses of the solid were measured out and appropriate amounts of water were added.

The viscosity measurements were conducted with the rotational rheometer with coaxial cylinders similarly to the other researchers [7,8,11,12]. A set of cylinders with the width of a measuring gap equal to 2 mm was applied. The measurements were performed at a temperature of 21°C ± 1°C within the range of shear rate from 1 s^{-1} to 437 s^{-1} which corresponds, among others, to such processes like forcing in a pipe or mixing in a tank. In order to consider the phenomena occurring in sewage sludge which is processed for tens of hours, the prepared suspensions underwent the measurements after twenty four hours since their preparation time. It allowed to consider the changes of the solid physical parameters resulting from the contact between the solid and water.

Table 1. The suspensions physical properties and their accepted notations

Material	Real density [kg/m^3]		
Water	998.5		
Hay	1175.0		
Accepted notations	Size fraction [μm]	Solid fraction	
		vol. %	wt. %
S-I-8	(i) 150÷212	8	9
S-I-10		10	11
S-I-11		11	13
S-II-8	(ii) 106÷150	8	9
S-II-10		10	11
S-II-11		11	13
S-III-8	(iii) below 106	8	9
S-III-10		10	11
S-III-111		11	13
The notation example: S-II-10 – the suspension with solid size fraction 106÷150 μm and solid volume fraction 10%.			

The samples were prepared in leakproof vessels which prevented from evaporation of water. However, a fluidity loss of some samples occurred during the experiments, which caused limitations within the range of shear rate.

3 Results and discussion

The results of rheometeric measurements allowed to plot the flow curves of the studied suspensions in the established range of the shear rate changes. The studied samples displayed features of non-Newtonian fluids, therefore their flow curves were primarily established for determination with the use of the power model

$$\tau = k\dot{\gamma}^{n}, \qquad (2)$$

which is depicted in the literature of the subject accordingly, since employing the equation (2) to describe the flow curves of the sewage sludge is frequently used. In model (2) symbols k and n constitute its parameters which correspond to the substance consistency and the tendency of its viscosity changes in function of the shear rate, respectively. Values of the parameters k and n for each of the studied samples together with the values of the coefficient of determination R^2 are compiled in Table 2.

The obtained values of the model (2) parameters for each of the suspensions suggest very high viscosity of the suspensions and a strong downward trend. Simultaneously, substantially high values of the parameter k can indicate the studied substances are endowed with the yield stress. The best way to verify that assumption would be an experiment which would analyse relationships between the shear stress in function of very low values of shear rates. However, due to a more qualitative than quantitative character of the conducted analysis, the occurrence of the yield stress of the studied substances was verified in a theoretical way.

Table 2. Parameters of models (2) for the studied bio-suspensions.

Size fraction [μm]	vol. %	k [Pa·s^n]	n [-]	R^2 [-]
(i)	8	97.99	0.14	0.10
150÷21	10	127.38	0.36	0.99
2	11	346.33	0.12	0.99
(ii)	8	91.65	0.15	0.92
106÷15	10	194.52	0.11	0.97
0	11	431.97	0.08	0.96
(iii)	8	10.45	0.36	0.97
below	10	74.33	0.25	0.98
106	11	126.21	0.17	0.94

Consequently, Herschel-Bulkley model was employed

$$\tau = \tau_y + \eta_p \dot{\gamma}^m, \tag{3}$$

because if the substance has the yield stress, the model (3) will be more accurate (successful) than the model (2). Model (3) is rather dedicated to concentrated suspensions, than diluted ones, however, its application in this case is justified by parameters values of model (2) which turned out to be peculiar for the studied diluted suspensions. In the equation (3) the symbol τ_y is a yield stress [Pa], while η_p means the so-called plastic viscosity [Pa·s^m]. The obtained values of model (3) parameters for individual bio-suspensions are compiled in s. Due to an insufficient number of the measuring points for the suspensions with the solid volume fraction equal to 11%, a description of the points distribution with the model (3) was impossible.

Table 3. Parameters of models (3) for the studied bio-suspensions.

Size fraction [μm]	vol.%	τ_y [Pa·s]	η_p [Pa·s^m]	m [-]	R^2 [-]
(i)	8	129.4	6.28	0.48	1.00
150÷	10	110.72	32.95	0.76	0.99
212	11	-			
(ii)	8	100.2	1.16	1.48	0.99
106÷	10	137.37	61.58	0.26	0.97
150	11	-			
(iii)	8	24.13	1.81	0.61	0.98
below	10	92.86	18.19	0.43	0.99
106	11	-			

The evaluation of the obtained values corresponding to model (3) parameters allows to state the substances subjected to the rheometeric measurements may be characterized by the yield stress. It constitutes a substantial peculiarity due to the fact that the prepared bio-suspensions should be classified as dilute where water constitutes their liquid phase that is Newtonian liquid with low viscosity. It is also worth highlighting that a slight increase of solid fraction results in a significant increase in the studied bio-suspensions viscosity.

The research results assessment also based on a comparison of the suspension viscosity values and the trend of their changes against the viscosity of a pure liquid phase as well as suspension viscosity calculated according to Manley and Manson equation

$$\eta_z = \eta_c\left(1 + 2{,}5\phi + 10{,}05\phi^2\right), \tag{4}$$

which is recommended in the literature [13] as a typical equation for this kind of two-phase systems. In the equation (4): η_z –delineates suspension viscosity [Pa·s]; η_c – is the viscosity of a liquid phase [Pa·s]; ϕ - is solid volume fraction [-]. This equation is recommended to calculate suspensions viscosity with up to 30% of solid volume fraction. Lack of parameters requiring extra measurements such as e.g. maximum solid fraction is an advantage of this equation. In Fig. 1-3 the viscosity of the studied bio-suspensions resulting from the measurements is shown against the background of the values calculated according to (4) and the viscosity of the liquid phase.

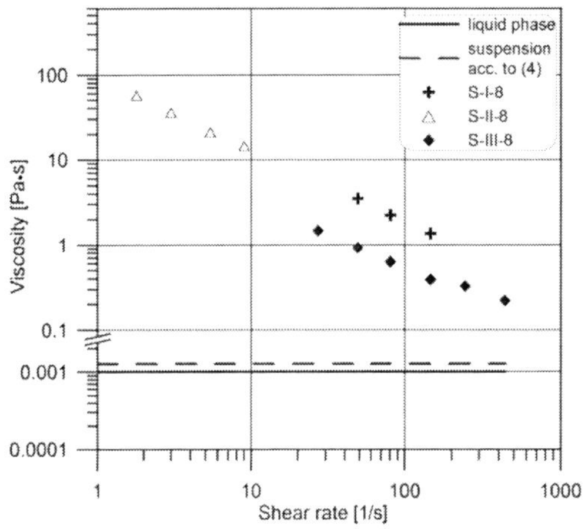

Fig. 1. Viscosity of bio-suspension with 8 vol.% solid fraction.

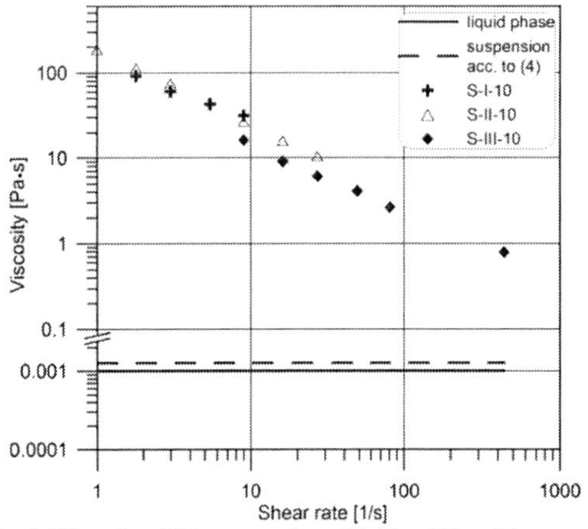

Fig. 2. Viscosity of bio-suspension with 10 vol.% solid fraction.

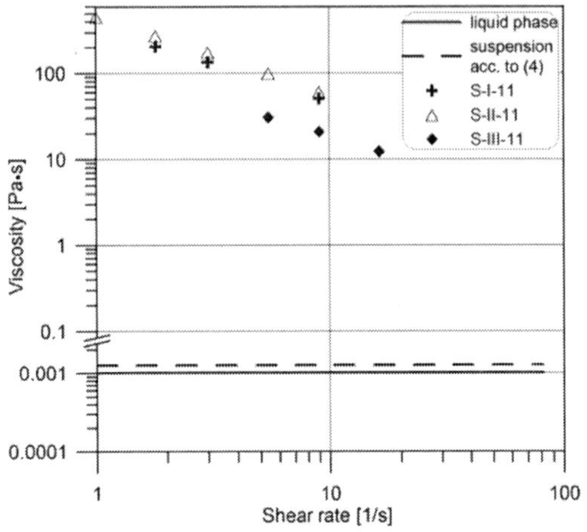

Fig. 3. Viscosity of bio-suspension with 11 vol.% solid fraction.

Basing on the graphs pinpointed in the Figs. 1-3, it is possible to conclude that the trend of the bio-suspensions viscosity fluctuations is strongly downward. Yet, the fact that the measured viscosity values are either slightly over or five times higher than the viscosity values calculated according to (4) is even more significant. Apart from that, such a model (4) (like many other models dedicated to suspensions) cannot average the real results as it does not predict substance viscosity changes in a function of the shear rate. Investigating the points of distribution in the graphs, the impact of the solid size fraction on the values of the bio-suspension viscosity is easy to notice.

Yet, analysing the conditions of the rheometeric research it has to be added that a distinct increase of solid sizes and a partial disappearance of the liquid phase occurred in the samples of the suspensions prepared in advance. This effect corresponds to vegetal particles imbibition due to partial water absorption by the particles and it results in the increase of the solid volume fraction in the suspension. Thereby, the discussed measurement results can be treated as concerning concentrated suspensions what should exclude a surprise factor.

4 Conclusions

The major conclusion stemming from the conducted analysis concerns the duration of vegetal particles presence in a given multiphase system, which can affect the viscosity of the sewage (and any other aqueous bio-suspension) and their rheological behaviour. The contact of the phases results in a partial imbibition of the liquid by the particles and, consequently, it leads to both an increase in their size and a decrease of the volume of the unbounded liquid phase. As a result, the bio-suspension with the specified concentration of the dry matter and the specified original size of the grains comes into existence. Though, in fact it is the multiphase system, which exhibits problematic parameters for an unequivocal quantitative adjudication. Such a substance would be characterized with completely different rheological properties other than the typical suspension containing grains which do not undergo the imbibition. Thereby the establishment of the bio-suspensions viscosity seems impossible while basing on the conventional dependencies, such as employing Manley and Manson equation in this paper. These equations take into account neither the contact time of the phases and resulting in the imbibing effect of vegetal particles, nor the non-Newtonian behaviour of the dilute suspensions.

Due to the aforementioned complex sewage composition, the rheological behaviour of the liquid phase itself, which forms sewage, becomes a separate issue for consideration. The articles [14,15] highlighted the complexity of the rheological properties of the suspensions consisting of the non-Newtonian liquid. It is recommended to conduct the quantitative assessment of the imbibition effect of vegetal particles, which would further enable to determine the impact of the imbibing effect on the bio-suspension viscosity.

References

1. N. Eshtiaghi, F. Markis, S.D. Yap, J.C. Baudez P. Slatter, Water Res. **47**, 5493-5510 (2013)
2. X-h. Wu, F. Wang, D-X. Sun, W-H. Yang, Exp. Therm. Fluid Sci. **35**, 612-617 (2011)
3. N. Ratkovicha, W. Hornc, F.P. Helmusc, S. Rosenbergerc, W. Naessensd, I. Nopensd, T.R. Bentzen, Water Res. **47**, 463-482 (2013)
4. I. Seyssiecq, J.H. Ferrasse, N. Roche, Biochem. Eng. J. **16**, 41-56 (2003)
5. J.C. Baudez, J. Environ. Manage. **78**, 232-239 (2006)
6. J.C. Baudez, F. Markis, N. Eshiaghi, P. Slatter, Water Res. **45**, 5675-5680 (2011)
7. N. Eshtiaghi, S.D. Yap, F. Markis, J.C. Baudez P. Slatter, Waster Res. **46**, 3014-3022 (2012)
8. M. Mori, I. Seyssiecq, N. Roche, Process Biochem. **41**, 1656-1662 (2006)
9. V. Lotito, A.M., Lotito, J. Environ. Manage. **137**, 189-196 (2014)
10. T. Li, Y. Wang, Y. Dong, J. Evnviron. Sci. **24**, 1917-1922 (2012)
11. X. Cao, Z. Jiang, W. Cui, Y. Wang, P. Yang, Procedia Environ. Sci. **31**, 113-121 (2016)
12. B.J. Konijn, O.B.J. Sanderink, N.P. Kruyt, Powder Technol. **266**, 61-69 (2014)
13. M. Dziubiński, J. Prywer, *Mechanika płynów dwufazowych* (WNT, Warszawa, 2009)
14. J. Hapanowicz, Inż. Ap. Chem. **55**, 136-138 (2016)
15. A. Szydłowska, J. Hapanowicz, Inż. Ap. Chem. **56**, 26-28 (2017)

 DOI: 10.1051/e3sconf/20171902012

Long-term trends in river flow: a case study of the Soła River (Polish Carpathians)

Mariola Kędra[1,*]

[1]Cracow University of Technology, Department of Environmental Engineering, Institute of Engineering and Water Management, Warszawska 24, 31-151 Cracow, Poland

Abstract. This study analyses potential trends in river flow for the most recent 60 years (1956–2015). The study area is situated in the Soła catchment in the Polish Carpathians. The focus of the study was to evaluate long-term trends in mean monthly river flow for each season of the year as well as to compare the direction of these trends for sites located a distance upstream and downstream from a cascade of three dams built on the Soła River. Moreover, potential long-term trends in seasonal precipitation were also studied. The data indicate a significant increase in minimum discharge during the winter for the upstream site. Increases in mean and maximum discharge are significant for spring at that site as well. In contrast, a significant decrease in minimum discharge was identified for the downstream site during the spring. Moreover, significant decreases in mean and minimum discharge were noted for the summer season at the downstream site only. No trends in discharge were identified for the autumn at either site. Significant trends in precipitation were noted for each season of the year, with increases during the winter, spring, and autumn (3.6–9.8 mm per decade) in the vicinity of the upstream site, but a decrease in minimum summer precipitation (–5.0 mm per decade) at the downstream site. The revealed differences in the direction of seasonal trends for the upstream and downstream sites studied suggest an anthropogenic impact on river flow downstream from the cascade of dams. Overall, the identified decrease in summer discharge at the downstream site, accompanied by a decrease in summer precipitation, indicate the need for adaptive water management in the studied catchment to ensure water availability for the summer season.

1 Introduction

In mountain regions of mid-latitudes, stream and rivers are typically rich in water owing to substantial orographic precipitation. Unpolluted rivers with well oxygenated water of relatively low temperatures throughout the year and dynamic flow create suitable environmental conditions for cold water species [1]. Mountain rivers are also affected by human activity, which directly (water diversion, dam and reservoir construction, channel engineering works) or indirectly (urbanization, removal of vegetation, road construction) [2-3], alters flow and thermal conditions, impairing river ecosystems [4]. Of direct interventions in river environments, the construction of dams and the changes to fluvial environments that they introduce, have been well documented [5-16]. Dams essentially alter the downstream flow of water and sediment, modifying biochemical cycles and the habitat structure [17]. Dams with hypolimnetic (bottom layer) water releases can weaken the synchronous behaviour of air and water temperatures and significantly change water temperature dynamics in the downstream river section [18-19]. Moreover, fragmentation of free-flowing rivers by dams causes disadvantageous ecosystem isolation [17].

The impact of a changing climate, mainly atmospheric warming, on freshwater resources and the functioning of aquatic ecosystems has been extensively studied over the last few decades [20-27]. The warming trend in global mean surface air temperature of 0.12 °C per decade (1951–2012) [28], accompanied by likely changes in precipitation [29], may worsen flow and thermal conditions in fluvial environments, with serious implications for both aquatic life and humans. Therefore, ensuring sufficient freshwater resources for domestic, agricultural, and industrial use, while maintaining environmental flows for ecological health of rivers [30], is one of the most important management concerns in a changing climate. In this context, the identification of significant changes and long-term trends in hydro-climatic variables such as river flow and precipitation may give rise to a more realistic planning and management effort for existing water resources.

The aim of this study is to: (1) assess potential long-term (60-year) trends in river flow for each season of the year; (2) compare the direction of these trends for sites

* Corresponding author: mariola.kedra@iigw.pk.edu.pl

 DOI: 10.1051/e3sconf/20171902012

located a distance upstream and downstream from a cascade of three dams built on a mountain river in the Polish Carpathians; (3) compare potential trends in seasonal values of river flow (Q) and precipitation (P) for nearby sites. Detection of statistically significant trends in Q and P can facilitate future water management and planning in the studied mountain river catchment; it may also serve as the basis for informed decisions that mitigate adverse effects of a changing climate in key water source areas, especially in mountain river basins.

2 Materials and methods

2.1 Study area

The research was carried out on the Carpathian river Soła, the right-bank tributary of the upper Vistula River in southern Poland (Fig. 1). From its source in the Żywiec Beskids 1070 m above sea level (a.s.l.) to its mouth 227 m a.s.l., the Soła River is 89 km in length [31].
The Soła catchment covers an area of 1,391 km^2 [31], with the highest peak (Pilsko: 1,557 m) being located in the Żywiec Beskids, and the lowest elevation at the Soła River mouth. The rock layer underlying the Soła catchment is mainly composed of sandstone, loam, shale, and gravel [32].

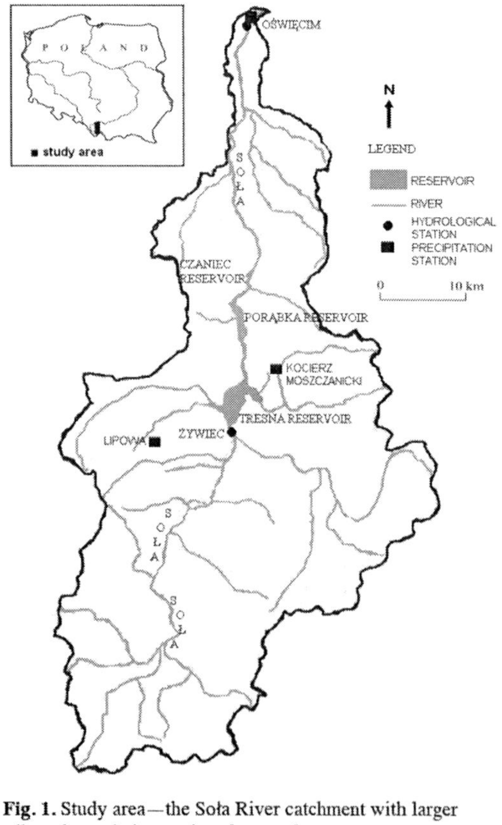

Fig. 1. Study area—the Soła River catchment with larger tributaries and observation sites used.

Climate conditions in the study area vary depending on location. In the lower-lying parts of the catchment, the air temperature is higher, with an annual average of 6–8 °C, and precipitation totals of 600–800 mm a year. The highest peaks in the Silesian and Żywiec Beskids are characterised by a cool climate with the annual mean air temperature below 4 °C and annual precipitation totals reaching 1,400 mm [33].

According to CLC mapping [34], the Soła catchment area is 51.5% forested, while agricultural and developed lands comprise 36.9% and 10.0%, respectively, while waters cover 1.6%. A cascade of three dams (Tresna, Porąbka, and Czaniec) was built between 1936 and 1967 on the Soła River [35]. Physical characteristics of the studied reservoirs are listed in Table 1.

Table 1. Physical characteristics of the studied reservoirs [35].

Statistic	Tresna	Porąbka	Czaniec
River length (km)	41.9	34.6	31.0
Drainage area (km^2)	1037	1091	1120
Capacity (mln m^3)	94.6	28.4	1.2
Depth (m)	23.0	23.8	7.5
Surface area (km^2)	10.00	3.67	0.45
Year of construction	1967	1936	1967
Functions[a]	F, S, E, L(6.7)	F, S, E, L(8.5)	B, E

[a] F = flood control; S = water supply; E = electricity production; L(q) = increasing base flow values to q m^{3}s^{-1}; B = balancing reservoir.

2.2 Data and methods

2.2.1 Data

The considered river flow series consist of monthly means for two gauging sites (Żywiec and Oświęcim) situated along the middle and lower course of the Soła River, respectively (Table 2). Monthly precipitation totals for nearby sites (Lipowa, Kocierz Moszczanicki, and Oświęcim) were also taken into account (Table 3). The available flow data for the Żywiec site begin in November 1955; this yields the longest feasible time period studied (60 years), with consecutive seasons from winter (December 1955, January and February 1956) to autumn 2015. Monthly flow and precipitation data for the years 1956–2015 (December 1955 – November 2015) were made available for research purposes by the Polish Institute of Meteorology and Water Management—the National Research Institute; precipitation data for the town of Oświęcim for the years 1956–1980 were obtained from Polish state precipitation yearbooks. The Żywiec site is adjacent to the upper part of the Tresna Reservoir (TR), and Oświęcim is located 28 km downstream from the Czaniec Reservoir (CR).

Table 2. Physical characteristics of the study sites [31] and descriptive statistics of mean monthly discharge data for the 60-year period (1956–2015).

Statistic	Żywiec site	Oświęcim site
Latitude (N)	49°41'11"	50°02'24"
Longitude (E)	19°11'36"	19°13'07"
River length (km)	50.2	3.0
Drainage area (km^2)	785	1386
Median elevation (m)	683	576
Mean catchment slope (–)	0.034	0.018
Mean discharge (m^{3}s^{-1})	15.56	20.61
Minimum discharge (m^{3}s^{-1})	1.45	1.44
Maximum discharge (m^{3}s^{-1})	108.02	193.00
Land use (%) [34]		
Forest	62.2	51.5
Agricultural land	31.3	36.9
Developed land	6.5	10.0
Waters	0.0	1.6

Table 3. Details of precipitation sites used. Mean annual precipitation was calculated using monthly data for 1956–2015.

Statistic	Lipowa site	Kocierz Moszcza-nicki site	Oświęcim site
Latitude (N)	49°39'58"	49°44'25"	50°02'23"
Longitude (E)	19°04'36"	19°15'17"	19°14'50"
Altitude (m a.s.l.)	530	430	243
Mean annual *P* (mm)	1040.5	1083.1	751.8

2.2.2 Methods

After visual inspection of monthly data compiled in a Microsoft® Excel 2007 spreadsheet, a quality check was performed in order to detect spurious values. Subsequently, each 720-element series was arranged as the matrix $\boldsymbol{M}_{ik}$, where the index i (for row) represents a year (i = 1, ..., n), n = 60, and the index k (for column) represents a month (k = 1, ..., 12). For each season (winter–autumn) of the year (1956–2015), that is, for the relevant three columns of the matrix (1–3, 4–6, 7–9, 10–12), arithmetic mean, minimum, and maximum values were computed as time series $\{x_i, i = 1, ..., n\}$. Next, trend analysis was performed on each seasonal series $\{x_i\}$ separately in order to assess the strength of potential monotonic trends in seasonal values of Q and P. The linear model of time trend was assumed in the form: $x(t) = a \cdot t + b$, where a is the slope coefficient, b is the intercept, and t is time. The non-parametric, 2-sided Mann-Kendall test (MK)[36-38] was used. The null hypothesis of no trend (H_0: $a = 0$) was tested equivalently (H_0: $\tau = 0$), using Kendall's rank correlation coefficient τ [36-38] at the significance level α: 0.05 (medium evidence) and 0.01 (strong evidence). The τ is expressed as:

$$\tau = \frac{2S}{n(n-1)} \quad (1)$$

$$S = \sum_{i=1}^{n-1} \sum_{j=i+1}^{n} \text{sign}(x_j - x_i) \quad (2)$$

where: S is Kendall's S statistic, and sign(x) is the sign function. The magnitude of trends was determined using Theil-Sen's slope [38-40] as a robust estimate of the slope, with the slope and intercept defined as:

$$\hat{a} = \text{median}\left(\frac{x_j - x_i}{t_j - t_i}\right), i = 1, \cdots, n-1, j = i+1, \cdots, n \quad (3)$$

$$\hat{b} = \text{median}(\{x_i\}_{i=1}^{n}) - \hat{a}\text{median}(\{t_i\}_{i=1}^{n}) \quad (4)$$

No assumption of normality, but serial interdependence is required for the MK test [38]. Therefore, the trend-free pre-whitening procedure (TFPW) [41] was applied to each data series.

3 Results and discussion

For the river discharge studied, significant trends (p<0.05) were identified for each season of the year (Table 4) except autumn (September–November). For winter (December–February, Fig. 2) and spring (March–May, Fig. 3), increasing trends were noted for the upstream site (Żywiec) only; an increase in minimum Q (0.67 m^{3}s^{-1} per decade, strong evidence) during the winter was followed by an increase in mean and maximum Q during the spring (1.35 and 2.03 m^{3}s^{-1} per decade, respectively, Figs. 3a–b). In contrast, a significant decrease in minimum Q (–1.29 m^{3}s^{-1} per decade, Fig. 3c) was identified for the downstream site (Oświęcim) during the spring season. Moreover, significant decreasing trends in mean and minimum Q were identified for summer (June–August) at the downstream site only, with a rate of change of 2.18 and 1.05 m^{3}s^{-1} per decade, respectively (Fig. 4), while remaining trends in summer Q were found to be decreasing as well, but remained statistically insignificant (at p<0.05).

For precipitation, significant trends (p<0.05) were identified for each season of the year (Table 5). For winter, an increase in mean, minimum, and maximum P (3.6–5.5 mm per decade) was identified at one site (Lipowa, approximately 8.4 km W from Żywiec). For spring, an increase in mean and maximum P (3.7–9.0 mm per decade) was found at one site as well (Kocierz Moszczanicki, approximately 8.0 km NE from Żywiec).

 DOI: 10.1051/e3sconf/20171902012

On the other hand, a decrease in minimum P (–5.0 mm per decade) was identified during the summer at the downstream site only (Oświęcim). Moreover, increasing trends in autumn P (5.2–9.8 mm per decade) were found at the two sites (Lipowa and Kocierz Moszczanicki) in the vicinity of Żywiec.

Table 4. Rate of change $\hat{a}$ (m^3s^{-1} $decade^{-1}$) for river flow (Q) in each season of the year (1956–2015) based on Theil-Sen's slope estimation [39–40] with trend free pre-whitening [41]. Statistical significance of the trends was evaluated via the 2-sided Mann-Kendall test [36-37].

Q statistic	Winter (Dec–Feb)	Spring (Mar–May)	Summer (Jun–Aug)	Autumn (Sep–Nov)
Mean (Żywiec)	0.74	1.35**	–1.02	0.58
Minimum (Żywiec)	0.67**	0.11	–0.40	0.11
Maximum (Żywiec)	0.72	2.03*	–1.82	1.00
Mean (Oświęcim)	0.34	0.57	–2.18*	0.12
Minimum (Oświęcim)	0.03	-1.29*	–1.05**	–0.02
Maximum (Oświęcim)	0.03	1.06	–3.15	–0.08

* data significant at $p < 0.05$; ** data significant at $p < 0.01$.

Table 5. Rate of change $\hat{a}$ (mm $decade^{-1}$) for precipitation (P) in each season of the year (1956–2015) based on Theil-Sen's slope estimation [39–40] with trend free pre-whitening [41]. Statistical significance of the trends was evaluated via the 2-sided Mann-Kendall test [36-37].

P statistic	Winter (Dec–Feb)	Spring (Mar–May)	Summer (Jun–Aug)	Autumn (Sep–Nov)
Mean (Lipowa)	4.6**	2.2	–4.8	5.2*
Minimum (Lipowa)	3.6*	1.0	-2.5	1.2
Maximum (Lipowa)	5.5*	2.8	-5.2	7.2*
Mean (Kocierz Moszcz.)	1.3	3.7*	-3.3	5.6**
Minimum (Kocierz Moszcz.)	1.7	0.5	–0.4	1.8
Maximum (Kocierz Moszcz.)	0.2	9.0*	-3.8	9.8*
Mean (Oświęcim)	0.5	1.0	-3.0	1.7
Minimum (Oświęcim)	0.5	0.4	-5.0*	1.7
Maximum (Oświęcim)	0.9	2.2	-2.2	1.0

* data significant at p<0.05; ** data significant at p<0.01.

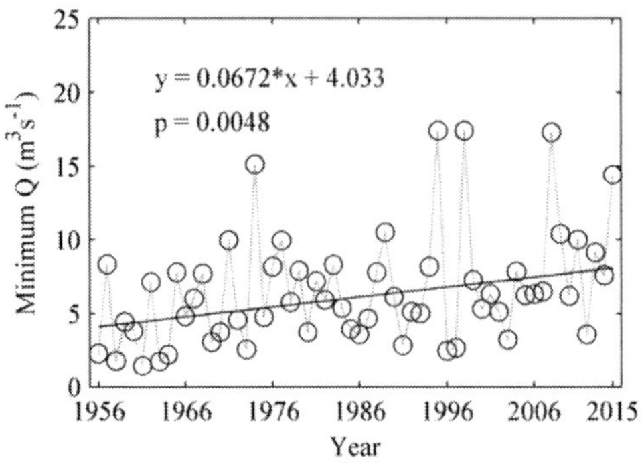

Fig. 2. Winter flow rates (Q) at Żywiec (upstream from the TR dam) for the 60-year period (1956–2015) with a significant trend shown in minimum Q.

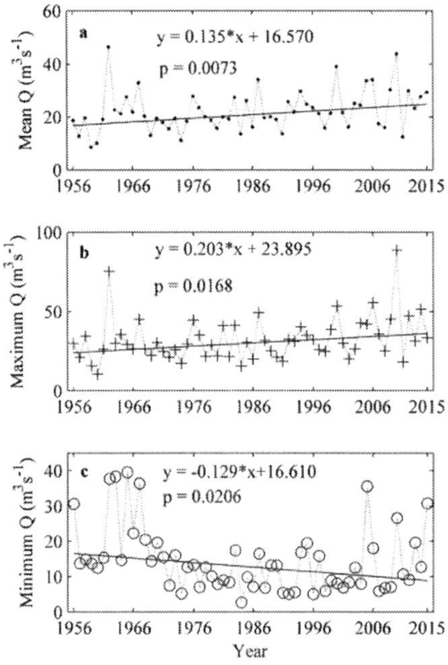

Fig. 3. Spring flow rates (Q) for the 60-year period (1956–2015) with significant trends shown: at Żywiec (upstream from the TR dam) in mean Q (a) and maximum Q (b);
at Oświęcim (downstream from the TR dam) in minimum Q (c).

Direct effects of climate change include changing river temperature and flow patterns [2]. It is also clear that river flow is regarded a master variable, which shapes essential characteristics of fluvial ecosystems [42]. In order to undertake adequate decisions mitigating adverse climate-change impacts on fluvial environments and on existing freshwater resources, an assessment of ongoing changes in flow, especially long-term tendencies, appears crucial and indispensible. In this study, we are focused on a mountain river with long-term flow series pertaining to the most recent 60 years (1956–2015), and on concurrent precipitation from neighbouring sites.

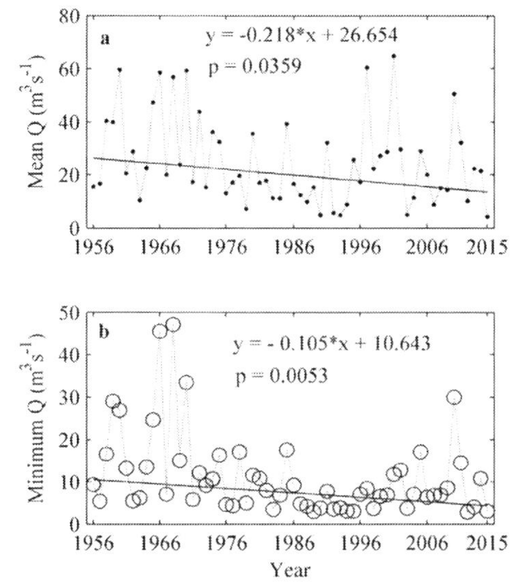

Fig. 4. Summer flow rates (Q) at Oświęcim (downstream of the CR dam) for the 60-year period (1956–2015) with significant trends shown in: (a) mean Q; (b) minimum Q.

The obtained results suggest that river flow tends to be progressively higher during the winter (minimum values of Q shown for the upstream site, Fig. 2) as well as during the spring (mean and maximum values of Q shown for the upstream site, Figs. 3a-b). An increasing trend in winter Q (at Żywiec) is consistent with a similar trend in winter P (mean, minimum, and maximum values of P at a nearby site, Lipowa). An increasing trend in spring Q can be associated with increased spring precipitation regionally, especially for March and May, as revealed for the neighbouring catchments of the Skawa, Raba, and Dunajec rivers [27, 43]; moreover, it corresponds with an increase in spring P found at Kocierz Moszczanicki. However, such an increasing tendency in spring Q is absent for the downstream site (28 km downstream from the CR dam), and even a decreasing tendency is present (minimum values, Fig. 3c). Moreover, another difference in revealed changes for the upstream and downstream sites pertains to a significant decreasing trend in summer flow rates (mean and minimum values, Fig. 4) for the downstream site only, which is accompanied by a decreasing trend in P (minimum values of P at Oświęcim). On the other hand, summer is the only season when all remaining trends in Q and P (Table 4 and 5, respectively) show a decreasing, yet statistically insignificant, tendency consistent with revealed, significant decreases in summer Q and P for the mentioned nearby catchments [27, 43]. No trend in autumn Q was identified, while increasing trends in autumn precipitation were found (at Lipowa and Kocierz Moszczanicki).

It seems also relevant to mention that differences in magnitude and/or direction of hydro-climatic trends depend on changes in dynamic interactions between large numbers of factors acting upon mountain river basins. In particular, ongoing land use and land cover changes (LULCC) are widespread in the Carpathians due to a shift in land use after 1990, reflected in farm abandonment and forest expansion [44]. According to CLC mapping, which is now available for the years 1990, 2000, 2006, and 2012, the same direction of LULCC occurred in the Soła catchment over the period 1990–2012. For the Żywiec and Oświęcim subcatchments, increases in forest cover (5.9 and 4.7%, respectively) and developed land (1.0 and 3.0%, respectively) occurred at the expense of agricultural land. This implies further differences in land cover between the studied subcatchments (Table 2), suggesting different impacts of LULCC on discharge and precipitation changes during the studied time period, but a detailed analysis is needed to investigate this issue.

Overall, the noted decrease in summer river flow volume at Oświęcim, accompanied by a decreasing trend in summer precipitation, should be carefully studied by the Regional Water Management Board in Cracow, responsible for water management in the Soła catchment. The revealed trends suggest a possible water deficiency during the summer season, indicating the need for adaptive water management of existing water resources in the studied catchment in order to provide an adequate water supply for local residential communities, while preserving environmental flows for the ecological health of rivers [30].

4 Conclusions

In the conducted analysis pertaining to changing flow conditions in a mountain river, long-term (60-year) trends in river flow for each season of the year (1956–2015) except autumn were found. The direction of these trends differs for sites located a distance upstream and downstream from a cascade of three dams. For the upstream site (Żywiec), a significant increase in river flow (minimum values) during the winter is followed by significant increases in flow (mean and maximum values) during the spring. For the downstream site (Oświęcim), significant decreases in flow (mean and minimum values) during the spring and summer are observed. The identified differences suggest an anthropogenic impact on river flow downstream from the studied cascade of the Tresna, Porąbka, and Czaniec reservoirs.

Concomitant, long-term (60-year) trends in precipitation are significant for each season of the year (1956–2015). Increasing tendencies in winter, spring, and autumn P were found in the vicinity of the upstream site, while a decrease in summer P was identified at the downstream site.

In general, the identified decrease in summer discharge at the downstream site, accompanied by a decrease in summer precipitation, suggest the need for adaptive water management in the Soła catchment, due to a likely water deficiency during the summer season.

This work was supported by the Polish Ministry of Science and Higher Education (grant no. Ś1/194/2017/DS task 4).

References

1. K. Bieniarz, P. Epler, *Ichthyofauna*, in: *The Upper Vistula Basin, part II* (PWN Warszawa-Kraków, pp. 69–81, 1991) (In Polish)
2. J.D. Allan, M.M. Castillo, *Stream ecology. Structure and function of running waters* (2nd ed., Springer, 2007)
3. G.C. Poole, C.H. Berman, Environ. Manage. **27**, 787–802 (2001)
4. B. Babbitt, BioScience **52**, 656–658 (2002)
5. R.M. Baxter, Annu. Rev. Ecol. Syst. **8**, 255–283 (1977)
6. M.E. Power, W.E. Dietrich, J.C. Finlay, Environ. Manage. **20**, 887–895 (1996)
7. G.M. Kondolf, Environ. Manage. **21**, 533–551 (1997)
8. N.L. Poff, J.D. Allan, M.B. Bain, J.R. Karr, K.L. Prestegaard, B.D. Richter, R.E. Sparks, J.C. Stromberg, BioScience **47**, 769–784 (1997)
9. F.J. Magilligan, K.H. Nislow, B.E. Graber, Geology **31**, 569–572 (2003)
10. F.J. Magilligan, K.H. Nislow, Geomorphology **71**, 61–78 (2005)
11. G.E. Petts, A.M. Gurnell, Geomorphology **71**, 27–47 (2005)
12. D. Caissie, Freshw. Biol. **51**, 1389–1406 (2006)
13. W.L. Graf, Geomorphology **79**, 336–360 (2006)
14. N.L. Poff, J.D. Olden, D.M. Merritt, D.M. Pepin, Proc. Natl. Acad. Sci. USA **104**, 5732–5737 (2007)
15. R. Soja, Ł. Wiejaczka, Water Environ. J. **28**, 473–482 (2014)
16. M. Kędra, Ł. Wiejaczka, K. Wesoły, River Res. Applic. **32**, 561–571 (2016)
17. N.L. Poff, D.D. Hart, BioScience **52**, 659–668 (2002)
18. M. Kędra, *Reservoir-induced changes in dynamics and synchrony of river water temperatures revealed by RQA and CRQA*, in: *Recurrence plots and their quantifications: expanding horizons* (Springer Proceedings in Physics, vol. 180, Springer, pp. 289–300, 2016)
19. M. Kędra, Ł. Wiejaczka, Water Environ. J. **30**, 31–39 (2016)
20. B.A. Sinokrot, H.G. Stefan, J.H. McCormick, J.G. Eaton, Clim. Change **30**, 181–200 (1995)
21. R.D. Ries, S.A. Perry, Clim. Res. **5**, 197–206 (1995)
22. B.W. Webb, Hydrol. Processes **10**, 205–226 (1996)
23. O. Mohseni, T.R. Erickson, H.G. Stefan, Water Resour. Res. **35**, 3723–3733 (1999)
24. O. Mohseni, H.G. Stefan, J.G. Eaton, Clim. Change **59**, 389–409 (2003)
25. M. Daufresne, M.C. Roger, H. Capra, N. Lamouroux, Glob. Change Biol. **10**, 124–140 (2004)
26. R.E. Hari, D.M. Livingstone, R. Siber, P. Burkhardt-Holm, H. Güttinger, Glob. Change Biol. **12**, 10–26 (2006)
27. M. Kędra, Water **9**, 22 (2017)
28. D.L. Hartmann, A.M.G. Klein Tank, M. Rusticucci, L.V. Alexander, S. Brönnimann, Y. Charabi, F.J. Dentener, E.J. Dlugokencky, D.R. Easterling, A. Kaplan, B.J. Soden, P.W. Thorne, M. Wild, P.M. Zhai, *Observations: Atmosphere and Surface*, in: *Climate Change 2013: The Physical Science Basis. Contribution of Working Group I to the Fifth Assessment Report of the Intergovernmental Panel on Climate Change* (Cambridge University Press, 2013)
29. K.E. Trenberth, Clim. Res. **47**, 123–138 (2011)
30. M. Acreman, M.J. Dunbar, Hydrol. Earth Syst. Sci. **8**, 861–876 (2004)
31. W. Chełmicki, *Location, classification and characteristics of the basin*, in: *The Upper Vistula Basin, part I* (In Polish, PWN Warszawa-Kraków, pp. 15–29, 1991)
32. S. Węcławik, *Geological structure*, in: *The Upper Vistula Basin, part I* (In Polish, PWN Warszawa-Kraków, pp. 30–41, 1991)
33. T. Niedźwiedź, B. Obrębska-Starklowa, *Climate*, in: *The Upper Vistula Basin, part I* (In Polish, PWN Warszawa-Kraków, pp. 68–83, 1991)
34. CLC 2012, http://land.copernicus.eu/pan-european/corine-land-cover (accessed 4 May 2017)
35. J. Hennig, I. Hennig, A. Roszkowski, *Retention reservoirs*, in: *The Upper Vistula Basin, part II* (PWN, Warszawa-Kraków, pp. 121–143, 1991) (In Polish)
36. H.B. Mann, Econometrica **13**, 245–259 (1945)
37. M.G. Kendall, *Rank correlation methods* (4th ed. Charles Griffin, London, 1975)
38. D.R. Helsel, R.M. Hirsch, *Statistical methods in water resources. Techniques of water-resources investigations* (Book 4, U.S. Geological Survey, 2002) https://pubs.usgs.gov/twri/twri4a3/ (accessed 17 July 2017)
39. H. Theil, Proc. R. Neth. Acad. Arts Sci. **53**, 386–392, 521–525, 1397–1412 (1950)
40. P.K. Sen, J. Am. Stat. Assoc. **63**, 1379–1389 (1968)
41. S. Yue, P. Pilon, B. Phinney, G. Cavadias, Hydrol. Process. **16**, 1807–1829 (2002)
42. M.E. Power, A. Sun, G. Parker, W.E. Dietrich, J.T. Wootton, BioScience **45**, 159–167 (1995)
43. M. Kędra, Clim. Res. **72**, 251–265 (2017)
44. A. Björnsen Gurung, A. Bokwa, W. Chełmicki, M. Elbakidze M. Hirschmugl, P. Hostert, P. Ibisch, J. Kozak, T. Kuemmerle, E. Matei, K. Ostapowicz, J. Pociask-Karteczka, L. Schmidt, S. van der Linden, M. Zebisch, Mt. Res. Dev. **29**, 282–288 (2009)

Examination of coagulant additives on qualitative composition of selected thermal waters

Ewelina Łukasiewicz[1,*], and *Mariusz* Rząsa[1]

[1] Department of Thermal Engineering and Industrial Facilities, Faculty of Mechanical Engineering, Opole University of Technology, Poland

Abstract. On the territory of Poland occur rich deposits of thermal waters. Although the utilisation of these waters is continuously extending, Poland is not exploiting their full geothermal potential due to high investment costs. Thermal waters industry in Poland to date indicates operations within mainly balneology as well as recreation objectives. Higher temperature values of these waters foster a washout in the surrounding rocks resulting in a high concentration of diluted substances which must be often removed. The following study investigates thermal waters from three intakes for which coagulation processes were conducted. Research clearly shows that coagulant additives not impact on the qualitative composition of thermal water, what is very important according to medicinal properties of water. The study results may be further applied as a valuable piece of information for further exploitation in balneology or within the heating sector and other installations.

1 Introduction

On the territory of Poland are rich deposits of thermal waters which are gradually more exploited as a national trend. Pursuant to the provisions of the Geological and Mining Act of June, 9, 2011 (Journal of Laws of 2011, No 163 item 981) thermal water is referred to as an underwater type with a temperature exceeding 20°C at the outflow of the source, and it does not originate from the drainage of mine headings [1].

According to the development of the scientific Polish Geothermal Association (PGA), at least 6600 km^2 of geothermal water with temperatures of 27-125 °C is accumulated within the area of Poland. These resources are present in almost all countries in certain geothermal provinces and districts. The vast majority of the area is the Polish Province of Central-European sedimentary swimming pools with geothermal water, including tanks: the Cambrian, Devonian-Carboniferous, Permian, Zechstein, Triassic, Jurassic and Cretaceous [2].

In Europe and throughout the world, water and geothermal energy are used primarily in heating systems and as domestic hot water and in individual installations, recreational ones and spa centers, in agriculture (for heating greenhouses, crops under foil cover and heating the soil). Any other uses of the thermal waters include water fish cultures, algae, industrial processes such as e.g. drying or pasteurization, melting snow and ice from sidewalks and roadways or airport runways heating. Carbon dioxide, edible salt and other chemicals can be recovered from the geothermal water. They are applied in cosmetics production or they are bottled as a curative and mineral [3] supplement. Frequently, the geothermal water can be used as the drinking type of water [4]. The pores, cracks, fissures and fault rocks of the Earth's crust also contain the geothermal waters. In Poland, geothermal waters (or thermal types) are characteristic for their temperature exceeding 20°C. With regard to the fact that in Poland the geothermal waters are exploited in balneology and recreation sectors, their quality shall comply with the limit values specified in the Regulation of the Minister of Health dated November 9, 2015 concerning the requirements to be fulfilled by the swimming pool water [5]. The selected physico – chemical requirements to be fulfilled by the swimming pool water are presented in the Table 1.

Table 1. Selected physico-chemical requirements to be fulfilled by the water in the swimming pools [5].

Parameter	Unit	Water supplied into a swimming pool basin from the circulation system		Water inside the swimming pool basin	
		min	max	min	max
pH of fresh water		6.5	7.6	6.5	7.6
pH of salt water		6.5	7.8	6.5	7.8
turbidity	NTU	-	0.3	-	0.5
aluminium	mg/L		-		0.2
total iron	mg/L		-		0.2
nitrates	mg/L		20	-	20
oxidisability	mg/L	-	-	-	4

* Corresponding author: e.lukasiewicz@po.opole.pl

The thermal water in the intake frequently does not fulfil such requirements, therefore it shall be subjected to the purification process. Coagulation is one of the purification methods concerning waters with an increased level of iron and manganium content [6,7].

2 Coagulation

Coagulation is a process commonly applied for removal of the surface colloidal contaminants both of organic and inorganic origin as well as hardly falling suspensions. It is dramatically important to assure high effectiveness of coagulation due to its further purification stages. It results from a possibility of diminishing a demand for application of oxidising agents or disinfectants.

A non-hydrolised aluminium sulphate (VI) $Al_2(SO_4)_3$ belongs to one of the most widely used coagulants in the Polish water treatment plants. However, it is more frequently substituted with the initially hydrolised coagulants which respond with increased efficiency to the removal of organic contaminants, disintegration of colloids with negative electric charge and disinfectants causing water turbidity. Moreover, they display greater resilience to the temperature as well as pH values fluctuations. The initially hydrolised coagulants include polyaluminium chlorides and sulphates. The increased values of alkalinity among such types of substances arise from the presence of hydroxyl groups. The alkalinity indicator serves as a polymerisation degree measure [7,8].

The thermal water composition is very diverse as a consequence of a contact with rocks and minerals and a high degree of salinity. Before using such water in the geothermal systems and heat exchangers it must undergo a process of purification. The process of removing iron ions seems to be of key importance here. The most common forms of the iron element are hydrogen carbonate, iron (II) sulphate or iron (IV) sulphate. For the sake of insoluble forms precipitation of iron hydroxide, $Fe(OH)_3$ and methods such as aeration, liming and coagulation [9,10,11] are commonly applied. The suspended substances referred to as colloidal reveal such common features as a small size, a large surface area and a relatively high electric charge of particles. A destabilization of a colloidal system is often possible only due to a chemical treatment.

3 Methodology

The objective of the following research is assessment of the coagulation process in the course of de-ironing of the thermal waters as well as determination of the coagulant additive impact on the ionic water content. Here, the coagulation process was performed with the use of initially hydrolised aluminium coagulants known under the trading names as Flokor 1,2A and Flokor DM17H. The selected physico–chemical data of the applied coagulants is plotted in the Table 2. The quality analysis was exerted on the water from three thermal water intakes: Uniejów, Wołczyn and Lądek-Zdrój. In Uniejów, the water temperature at the level of intake is 68°C, in Wołczyn it reaches 37°C, and in Ladek-Zdrój it is as high as 44°C. The source of Uniejów intake water is PIG/AGH-2 drilling well and this water fulfils the criteria of mineral healing water 0.63% chloro – sodium, hiperthermal water. Thus, it may be used either for bathing or for supplementary drinking treatment.

Table 2. Physico – chemical data of the applied coagulants.

	Flokor 1,2A	**Flokor DM17H**
State	light grey or transparent liquid	light grey or transparent liquid
Turbidity in 20°C [g/cm^3]	1.290 ± 0.05	1.275 ± 0.05
pH	4.20 ± 0.50	3.50 ± 0.50
Aluminium Al^{3+} [%]	12.00 ± 0.50	8.50 ± 0.50
Al_2O_3 [%]	22.60 ± 0.90	16.00 ± 0.90
Chlorides Cl- [%]	6.50 ± 0.50	7.50 ± 0.50
Alkalinity [%]	85.00 ±5.00	75.00 ± 5.00
Solidification temperature [°C]	about - 5°C	about - 22°C

Geotermia Uniejów Partnership was founded in Q3 of 1999 under the initiative of Uniejów Municipality as well as the Provincial Environmental Protection and Water Management Fund in Łódź. The distribution of heat to the recipients began in September 2001. 180 power connections have been established till the present date, out of which 30 are connections with a large – scale customers who collect 80% of the total amount of the heat output. The total heating plant power is 5.6 MWt, out of which 3.2 MWt is generated by the geothermal spring and 2.4 MWt originates from the peak-load source basing on two oil-fuelled boiler stations.

The water from Wołczyn intake comes from a VII A drilling well and possesses 23.3 g/dm^3 of solids. It is a chloride-sodium-calcium-bromide bringe also rich in iron and boron. It may be successfully applied either for bathing or for inhalations. So far, the waters have not been covered by any investment.

Lądek-Zdrój water is extracted from L-2 „Zdzisław" drilling well and sulphide-fluoride baths are among its principal applications.

The waters from Wołczyn and Uniejów revealed an increased colour intensity as well as iron and manganium content, thus they were subjected to removal with the use of a coagulant additive. Lądek-Zdrój water did not require any treatment and therefore the only experimental studies were led for this water to verify the interplay of a coagulant additive with the ionic water content.

Tests on iron, manganium and aluminium content were performed with a spectrophotometric method with the use of a UV-VIS spectrophotometer. The ionic analysis was run with the use of a IC chromatograph. Ph-metry was done in accordance with PN-90 C-04540/01 norm [12], conductivity test - PN-EN 27888 norm [13], colour test - PN-EN ISO 7887 norm [14], turbidity test - PN-EN ISO 7027 norm [15]. The volume coagulation was conducted in the ambient temperature

 DOI: 10.1051/e3sconf/20171902013

and within the first 24 hours since the collection time of water samples.

The study results presented beneath are divided according to the three intakes. Uniejów water was subjected to coagulation with the use of Flokor 1,2A in the following doses: 10, 20 and 40g/m^3. Wołczyn water underwent coagulation with the use of Flokor 1,2A and Flokor DM17H, in the following doses: 10, 20, 50 and 100 g/m^3 and 30g/m^3, respectively. The water sample was also submitted for the ionic analysis. Lądek- Zdrój water was not prone for treatment and in this case the impact assessment was run exclusively of the coagulant additive against the ionic water content. Flokor 1,2 A coagulant was submitted in a dose of 40 g/m^3.

4 Results and discussion

4.1 Thermal water - Uniejów

Table 3 is a collection of the test results for Uniejów water. This water is characterised with an increased colour intensity and iron content. Consequently, the coagulation was arranged with the use of Flokor 1,2A in the following doses: 10 g/m^3, 20 g/m^3 and 40 g/m^3. Ionic analysis was not carried out for this water.

Table 3. Quality of Uniejów water intake.

Index	Unit	Water			
		Raw	Flokor 1,2A doses		
			10g/m^3	20g/m^3	40g/m^3
pH		7.18	7.22	7.25	7.24
σ	mS/cm	11.57	11.58	11.63	11.61
Total Fe	mg/L	0.418	0.110	0.0246	0.0057
Mn	mg/L	0.128	0.0623	0.0563	0.0552
Colour	mg Pt/L	34	6	4	2

4.2 Thermal water - Wołczyn

The desired testing was performed on the raw water subjected to coagulation and Flokor 1,2 A coagulant in doses of 20 and 50 g/m^3 and Flokor DM17H in a dose of 30g/m^3 (Table 4 and Table 5).

Table 4. Anions concentration in water intake of Wołczyn.

Peak name	Raw	20g/m^3 1,2A	50g/m^3 1,2A	30g/m^3 DM17H
	[mg/L]	[mg/L]	[mg/L]	[mg/L]
Chlorides	14,850	14,750	15040	15594
Bromides	76.38	78.62	78.84	76.75
Sulphates	2573.1	2597.4	2629.5	2641.9

Table 5. Cations concentration in water intake of Wołczyn.

Peak name	Raw	20g/m^3 1,2A	50g/m^3 1,2A	30g/m^3 DM17H
	[mg/L]	[mg/L]	[mg/L]	[mg/L]
Lithium	6.39	6.35	6.46	6.70
Sodium	8300	8250	8410	8690
Ammonium	6.8	6.87	7.35	6.95
Magnesium	369.4	367.9	374.1	388.9
Calcium	1850	1911	1870	1950

The temperature in the intake was 37 °C. The ionic analysis was carried out and the coagulant additive was evaluated according to its impact on the ionic content of water .

In accordance with Pourbaix's graph, i.e. the graph of potential/pH for the iron compounds, the solid forms of iron occur when the medium reaches pH values of > 7. The tests performed for the model waters with increased salinity and temperatures indicated that the applied coagulants accrue the highest effectiveness levels (de-ironing) within the pH range of 7.9 – 8.1 [5,9].

Due to the fact that the water sample had a pH = 6.51, it was essential to increase the pH level up to the required range (Table 6).

Table 6. Quality of Wołczyn water intake.

Index	Raw	DM17 H 30 g/m^3	1,2A 10 g/m^3	1,2A 20 g/m^3	1,2A 50 g/m^3	1,2A 100 g/m^3
pH	6.51	7.28	-	7.16	7.31	-
σ mS/cm	42.9	43.00	-	43.10	43.11	-
Total Fe (decantate) mg/L	1.42	0.927	22.50	0.847	0.891	21.60
Total Fe (filtrate) mg/L	0.80	0.006	<0.005	<0.005	<0.005	<0.005
Mn (filtrate) mg/L	1.13	1.09	-	1.08	1.04	-
Colour (decantate) mg Pt/L	94	58	-	54	22.50	-
Colour (filtrate) mg Pt/L	2	2	-	2	2	-

The amendment of pH was generated with the use of 0.1m NaOH. Wołczyn intake water was yellowish-brown in colour >1000 mg Pt/dm^3, thereby it shall become less intensive. Moreover, it included an increased amount of total iron i.e. 65,30 mg/dm^3.

4.3 Thermal water - Lądek Zdrój

The water in Lądek-Zdrój is reveals a high quality what is manifested by its determination as potable water, thus it has not required any treatment. The ionic analysis for that water was presented below. For the sake of the investigation and control, it was decided to conduct coagulation in order to assess the impact of a coagulant aluminium additive on the content of water. (Table 7 and Table 8).

Table 7. Cations concentration in water intake of Lądek-Zdrój.

Peak name	Raw	40 g/m^3 1,2A
	[mg/L]	[mg/L]
Lithium	0.0257	0.0267
Sodium	49.1891	50.7335
Potassium	0.7197	0.7859
Calcium	3.0555	2.3571

Table 8. Anions concentration in water intake of Lądek Zdrój.

Peak name	Raw	40 g/m^3 1,2A
	[mg/L]	[mg/L]
fluorides	10.7757	10.7655
chlorides	5.1649	12.0046
bromides	0.0339	0.0363
nitrates	-	0.0174
phosphates	0.4807	0.4169
sulphates	19.4341	19.9799

5 Conclusions

Uniejów intake water is characterised with an increased level of iron content, in raw water its content equals 0.418 mg/L. Even the lowest dose of Flokor 1,2 A, i.e. 10g/m^3 enabled one to reduce its content up to 0.110 mg/dm^3, thus the norm 0.2 mg/dm^3 was ensured. The colour intensity and turbidity were heavily reduced, however it shall be here highlighted that in case of the thermal waters and healing or mineral waters, the colour is induced by the desired components of these waters. Taking the potable waters into account, their colour intensity and turbidity shall stay at a consumer-acceptable level and they shall be devoid of any inappropriate alterations. The coagulant developed as a result of reduction of the total iron as well as manganium amount, it also decreased the colour intensity and turbidity up to the quality norms for the potable water (Journal of Laws 2015, item 1989).

Considering the water from Wołczyn, a coagulant additive has not affected significantly the ionic content of water. In this case, chlorides content constituted a value which rose substantially i.e. by circa 0.75 grams for Flokor DM17H and a dose of 30g/m^3. De-ironing with the use of both Flokor 1.2A and DM17H resulted in a satisfactory outcome. The iron content in a decanter dropped by circa 99%. Flokor 1,2A doses, i.e. 10 and 100g/m^3 have not caused any dramatic reduction of the total iron amount. The amount of manganium compounds has not been reduced. The filtration ensured a high degree of colour intensity decrease – up to 2 mg Pt/dm^3. Lądek-Zdrój water ionic content has not undergone a substantial modification after submitting it with a coagulant. The only parameter where an ample difference might be noticed concerns chlorides as their content more than doubled.

Prior to application of the thermal waters in balneology, water treatment is not essential due to a high value of medicinal effects of such waters. Nevertheless, in the heating sector and energetics that is in installations where e.g. heat exchangers would be applied, water treatment would also be necessary as the compounds might be responsible for the encrustation of pipes, corrosion or other technological issues.

References

1. Act from 9 June 2011 Geological and Mining Law (Journal of Laws 2011 No. 163 Item 981)
2. A. Zimny, M. Karch, K. Szczotka, PGA, Kraków (2008) www.pga.org.pl (in Polish)
3. Y. Noorollahi, H. Y. Sahzabi, Proceedings World Geothermal Congress (2005)
4. B. Kępińska, Przegląd Geologiczny **58**, 7 (2010)
5. Regulation of the Minister of Health dated November 9, 2015 concerning the requirements to be fulfilled by the swimming pool water (Journal of Laws of 2015, item 2016)
6. A. Nowacka, M. Włodarczyk-Makuła, Ann. Set Environ. Prot. **16** (2014)
7. E. Podgórni, M. R. Rząsa, Pol. J. Environ. Stud. **23**, 6 (2014)
8. J. Gumińska, M. Kłos, Gaz, woda i technika sanitarna **5**, 195 (2011)
9. H. Alçiçek, A. Bülbül, M. C. Alçiçek, J. Volcanol. Geotherm Res. **309** (2016)
10. M. Goosen, H. Mahmoudi, N. Ghaffour, Energies **3** (2010)
11. Z. Zięba, E. Grzesiak, Infrast. Ecol. Rural Areas **8**, 1 (2010)
12. Electrometric determination of pH in water and wastewater having the electrolytic conductivity of 10 microsimens/cm and above: Polish Standard PN-90/C-04540/01
13. Determination of electrolytic conductivity: Polish Standard PN-EN 27888:1999
14. Examination and determination of colour (ISO 7887:2011)
15. Determination of turbidity - Part 1: Quantitative methods (ISO 7027-1:2016)

Analysis of Wastewater Treatment Efficiency in a Soft Drinks Industry

Joanna Boguniewicz-Zabłocka[1,*], *Andrea G.* Capodaglio[2] and *Daniel* Vogel[1]

[1] Opole University of Technology, ul. Mikołajczyka 5, 45-271Opole, Poland
[2] DICAr,University of Pavia, via Ferrata 3, 27100 Pavia, Italy

Abstract. During manufacturing processes, most industrial plants generate wastewater which could become harmful to the environment. Discharge of untreated or improperly treated industrial wastewaters into surface water could, in fact, lead to deterioration of the receiving water body's quality. This paper concerns wastewater treatment solutions used in the soft drink production industry: wastewater treatment plant effectiveness analysis was determined in terms of basic pollution indicators, such as BOD, COD, TSS and variable pH. Initially, the performance of mechanic-biological systems for the treatment of wastewater from a specific beverages production process was studied in different periods, due to wastewater flow fluctuation. The study then showed the positive effects on treatment of wastewater augmentation by methanol, nitrogen and phosphorus salts dosed into it during the treatment process. Results confirm that after implemented modification (methanol, nitrogen and phosphorus additions) pollution removal occurs mostly with higher efficiency.

1 Introduction

The beverage production sector belongs to the agro-food industry. This industry is currently under consumer pressure, implementing significant changes as the result of increasing public interest in the so-called "healthy lifestyle". Products well-fitting to such a style have essentially much less sugar content, thus reducing the calorie content of drinks. Significant increase in sale volumes was recently observed in the so-called "functional waters" segment, that is, products based on ordinary spring water enriched with vitamins, minerals and/or flavoring extracts [1]. The beverage industry accounts for a notable 11.2% of the entire food sector income in Poland. Bottled water has been driving the industry for a few years, and in 2015, the public has spent nearly 12% more on it than in the previous year, with the highest increase in isotonic drinks alone. Economic forecasts predict a faster rising pace of water sales in the near future, than the industrial sector average, with the value of sales in this special category drinks likely to grow even faster through innovation. Juices, nectars and fruit drinks account for about one third of the beverage market in Poland. Their sales increased by 11.3% and 6.4%, respectively, over the previous year, while customer expenditure increased by 8.4% and 1.7%, respectively [1]. The observed increase in production is associated with higher industrial water consumption, implying higher volumes of waste water is being discharged into the environment [2].

In order to limit possible deterioration of water quality in receiving water bodies, the efficiency of treatment processes adopted in this industry should be consistently improved as the industry is expected to produce increasing wastewater quantities every year [3]. The most potentially harmful component of this flow is technological wastewater formed during washing of equipment and facilities, containing residuals of detergents, as well as other cleaning products [4], and is characterized by a great variability of pollutant concentrations over time. Its characteristics depend largely on the type of specific production, and by the adopted production cycle. For these reasons, industrial wastewater treatment plants (IWWTPs) are not designed according to classic process schemes adopted from municipal sewage treatment plants (MSTPs).

IWWTPs are characterized by an individual specificity deriving directly from quantity and quality (initial and expected) of treated wastewater, technology used in the industry, as well as specific operating conditions. Thus, the identification of a suitable solution for an IWWTP is relatively complex. The basic problem is to correctly determine amounts of incoming sewage, as well as pollutant loads, so that the plant will not be overly under or over-sized with respect to their foreseeable extended variability [5]. A proper solution should be resilient to such variability and perform consistently under any foreseeable circumstance.

Sewage treatment plants based on activated sludge technology face various operational problems. A main problem may be bulking of the sludge caused by excessive growth of filamentous bacteria [6]. Other

* Corresponding author: j.boguniewicz@po.opole.pl

problems could be due to the lack of sufficient amounts of organic compounds for the proper progress of the denitrification process [7]. As a result, deterioration of outflow quality may occur in treatment plants based on these processes. The solution may lay in the use of methanol or other chemicals dosed in the original wastewater order to improve treatment [8].

Inflow fluctuations and changes in pollutant concentration are additional problems in conventional activated sludge facilities. Under such conditions, Sequencing Batch Reactor (SBR) characterized by cyclic operation could be desirable. The major differences between SBR and conventional, continuous-flow, activated sludge system is that SBR tanks carry out equalization, aeration and sedimentation functions sequentially over time, rather than in the conventional space sequence of continuous-flow systems. Many studies [9,10] demonstrate the efficiency of SBRs for organic matter and nutrients removal from wastewater, as well as the present one. Other growing technologies include Membrane BioReactors (MBRs) that are more frequently been used for both industrial and domestic wastewater treatment [11,12].

Wastewater from the soft drink industry may include wasted soft drinks and syrup, water from bottles and cans washing, detergents and caustics, and traces of lubricants used in the production process machinery. These pollutants contain organic and inorganic substances described by the 5-day biochemical oxygen demand (BOD_5), and chemical oxygen demand (COD) [3]. Since wastewater from the soft drinks industry generally contains organic substances that are readily biodegradable [13], biological processes are mainly used for its treatment. The effectiveness/efficiency of the treatment depends on many factors, as described by Anielak [14] in a review of solutions and technologies for the treatment of food industry wastewater.

This paper presents an example of activated sludge-based IWWTP. Purpose of the study is to analyse the process of wastewater treatment in conditions of irregular inflow and varied composition. Treatment effectiveness was determined in terms of basic pollution indicators: COD, BOD_5 and Total Suspended Solid (TSS).

2 Materials and methods

2.1 Characteristics of the specific IWWTP

The sewage treatment plant object of this study has been thoroughly modernized, and put into operation back in 2010. Its maximum capacity is 1635m^3/d, and the average flow of wastewater in 2015 was 805m^3/d. Wastewater treatment is carried out with mechanical, biological and chemical processes to reach the desired quality objectives imposed by law [15] allowing its disposal into the environment.

Wastewater is pumped from a holding tank with a volume of 1500 m^3 by two pumps with capacity of 55 m^3/h to bar-and-mesh screens to hold large solids.

The so-called "sketches" are removed, transported by conveyor, dewatered and compacted in its final section. Wastewater then flows into the inner ring of the grit chamber, there sand settles to the bottom of the chamber. From this chamber the water-sand mix is pumped to a free-standing separator, where dehydration of the mix occurs, followed by the sand being dumped at a landfill. Quantitative and qualitative equalization of wastewater then occurs in a tank (active volume 175 m^3) equipped with 3 pumps, feeding waste water from this to the biological block [16].

This consists of separation chamber, selector, activated sludge reactor, secondary clarifier and stabilization chamber. The biological processes block is divided into 3 identical technological lines. The system's scheme is illustrated in Fig. 1. Sewage aeration and degree of internal recirculation are controlled by an online oxygen meter. The recirculation system consists of a tank and two pipes equipped with shut-off valves. Separation of effluents from the activated sludge mixture occurs in secondary clarifier equipped with external recirculation pumps, feeding the settled solids to the selector. Waste sludge removed from the system is pumped to an aerobic stabilization chamber, where final stabilization of the sludge occurs by continued aeration, and stabilized sludge is then forwarded to a dewatering press.

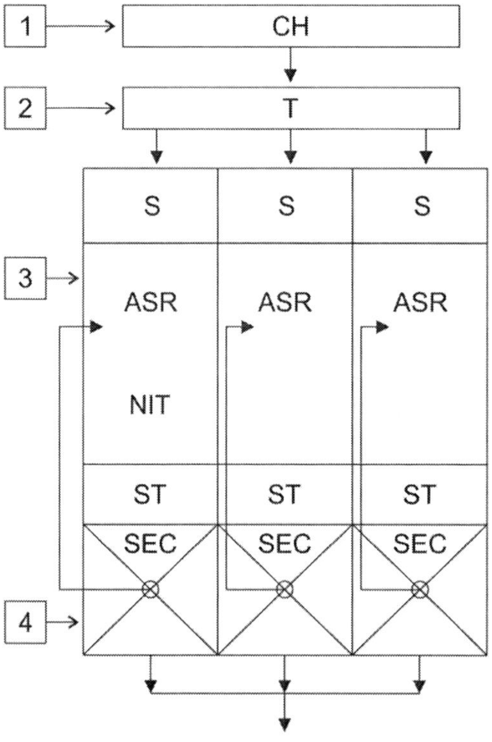

Fig. 1. Diagram of the biological part of wastewater treatment facilities where S- separation tank, ASR – activated sludge reactor (NIT- nitrification sector), ST – sludge thickener, SEC - secondary clarifier, CH – chamber, T- neutralisation tank and Chemical addition: 1 - urea, 2- ammonium phosphate ,3 – methanol, and 4 - FLOKUAT TB54HC.

 DOI: 10.1051/e3sconf/20171902014

The capacity of individual sections is: 40 m^3 separation chamber, selector 170 m^3, activated sludge reactor 1285 m^3, secondary clarifier 216 m^3 each (total 648 m^3), sediment thickener 20m^3 (total 60 m^3), sludge stabilization chamber 500 m^3.
In biological part of IWWTP the wastewater flow is too low to ensure proper use of the tree activated sludge lines, at the same time the sludge loading with organic contaminants would be too low. Therefore there are only two aeration tanks in use. Internal recirculation is carried out separately for each tank.

2.2 Characteristics of the receiving water body

The treated wastewater is discharged to the Młynówka ditch, which reaches the Biała River after about 500 meters, a left tributary of Osobłogi R., which later flows into the Oder River in Krapkowice. The control point located on the Biała River, shows good ecological status and satisfies the requirements for protected waters [17].

2.3 Methodology

The procedures for wastewater sampling and the instrumentation used for water quality parameter measurements were determined according to recommendations of Polish State Sanitary Inspection. The spectrophotometer manufactured by HACH Co. model DR 2000 (Hach, Loveland, CO, USA) was applied for analysis of ammonium, nitrate, nitrite, and phosphorus (spectrophotomety VIS/UV). Hydrogen ion concentration (pH) was measured with a pH meter. BOD_5 was determined by dilution of samples and dissolved oxygen measurements by Winklers' method, and measurement of COD was conducted via a simplified dichromate method. TSS was filtered, placed into a 105°C oven to dry, and weighed. Some data was obtained directly from the company's.
To assess the effectiveness of the treatment facility, all monitoring data were used. The reduction ratio of individual pollutants was calculated using formula 1:

$$\eta = (1 - C_e / C_o) \times 100\% \qquad (1)$$

where C_0 is the concentration of the indicator in the raw wastewater and C_e concentration in the treated effluent.

Flow values determined on the basis of measurements were: Q_{avd} = 805.3 m^3/d; amount of raw sewage produced: 293 934 m^3/year. It should be emphasized that the plant has a water permit for the discharge of treated industrial waste water to the Młynówka ditch in total amounts of Q_{avd} = 1150 m^3/d and Q_{maxd} = 1635 m^3/d.

3 Results and discussion

An important factor influencing the efficiency of wastewater treatment processes is the intensity and variability of inflows and their characteristics. Due to frequent changes of organic compounds and nitrogen loads, problems with proper operation of the activated sludge process have been observed in the recent past, affecting treatment efficiency. The most frequently observed problem was an improper ratio of nitrogen and carbon in the influent solution, that was an impediment to the achievement of efficient biological treatment. This could have been due to the high percentage of "light" (or "diet") drink products, with lower sugar content compared with traditional drink products.
In this case, after some testing, it was determined that the best option was to introduce supplementary chemicals from the outside. Urea (46%N) in dosage of 90 kg/d and ammonium phosphate (40% P_2O_5 in diluted solution) at a dose of 50 kg/d were therefore added to the neutralisation tank before biological treatment.
Furthermore, periodically, biological treatment presented poor sludge settling properties due to sludge bulking or poor floc aggregation (pinpoint sludge), losing biomass in the effluent with a deteriorating effect on the process. A loss of biomass implies also greater effluent organic loads, as 1mg of escaping biomass is approximately equivalent to 6 mg additional COD in the effluent. The obvious solution for this, in the absence of a complete process revamping, is an occasional addition of coagulating chemicals, as seen required. In this facility, considerable efficiency improvements in removal of BOD_5, COD were observed when the synthetic coagulant FLOKUAT TB54HC (proprietary name) was added at the effluent section of the aeration tank.

Fig. 2 shows the amount of sewage directed to the treatment plant compared to the volume of production, illustrating the fluctuation and ratios of sewage inflow to the treatment plant, following characteristic changes in the production of beverages. It can be seen that up to a drink production of about 10000 m^3/month, the amount of treated wastewater is roughly double that value, while for production levels > 10000 m^3/month, the volumes of treated wastewater is less than proportional to that quantity. In month VIII, when production of drink beverages was 20000 m^3, water treatment stopped at 25000 m^3, or just 25% more of the former. It seems that, at the lower production levels, the amount of process (and washing) water has an almost invariant generation pattern. It will be interesting to study if these volumes can further be optimized by enhanced internal operations, recycling or internal reuse.

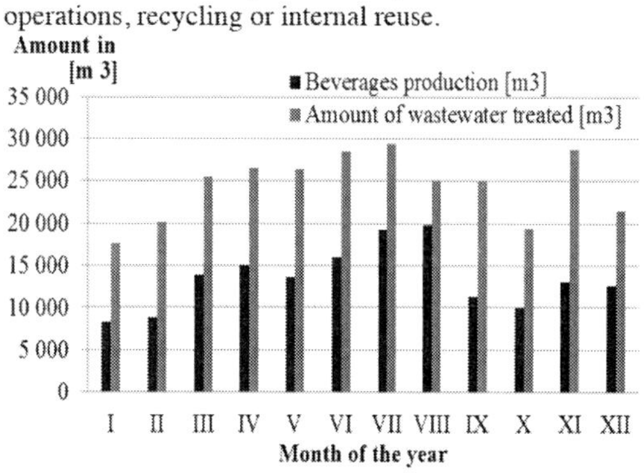

Fig. 2. The volumes of beverages production and amount of wastewater treated.

DOI: 10.1051/e3sconf/20171902014

Analyzing data presented in Fig. 2, significant fluctuations in the amount of monthly generated sewage, reaching up to 10 000 m^3, or 330 m^3/d can be observed. The mean value of the inflow was 805 m^3/d during the year in question. Minimum values of daytime inflows ranged from 500 to 600 m^3/d in January and October, maximum value was about 900 m^3/d. A fluctuation of one third to one half of the volume of wastewater treated could warrant the study of alternative treatment strategies (i.e. temporary re-opening of the third line currently out of operation).

Table 1 shows pollutant concentration of sewage before biological treatment (after equalization tank with chemicals addition) and treated. The permissible values of the effluent parameters are in accordance with the Ordinance of the Minister of the Environment of 18 November 2014 [15] concerning the conditions to be met when discharging wastewater to water or land. The WWTP is obliged by permit to ensure the following quality of effluent wastewater: pH 6.5 – 9,0; $BOD_5 \leq 25$ mg O_2/l; $COD_{Cr} \leq 125$ mg O_2/l; TSS ≤ 35 mg /l; Total Phosphorus ≤ 30 mg P/l; Total nitrogen ≤ 15 mg N/l; Ammonia nitrogen ≤ 0.124 mg N-NH_4 /l; or in alternative, to reduce the pollutants by: BOD_5 min. 90%, COD_{Cr} min. 92%, TSS min. 60%.

Table 1. Mean values of wastewater pollutants concentrations before biological treatment (Inflow) and treated (Outflow).

Sample		1	2	3	4
BOD mg O_2/l	In	304	321	626	408
	Out	3.8	12	5.4	6.6
COD mg O_2/l	In	778	838	1216	1076
	Out	18	98	30	75
TSS mg /l	In	27	27	21	22
	Out	17	14	7	16
P mg P/l	In	6	-	3	2
	Out	4.8	5.2	1.8	5.4

Removal efficiencies for BOD_5, COD, and total suspended solids (TSS) are shown in Fig. 3-5, in both situations on no chemicals support (A: before modification) and with chemicals additional dosages (B: after modification). BOD_5 removal is improved in the latter case by at least 50%, COD by at least 25%.

BOD_5 in treated wastewater ranges from 304 to 626 mg O_2/l during the year, according to the quantity and type of beverages produced. These fluctuations result in a high variations in total activated sludge biomass. The maximum BOD_5 concentration occurs in September, at the same time the highest removal efficiency is reached, after process adjustment, at approximately 99%. Therefore, the BOD_5 parameter in the purified effluent never exceeds even half the allowable value. Without chemical addition, effluent values were consistently around the discharge limit.

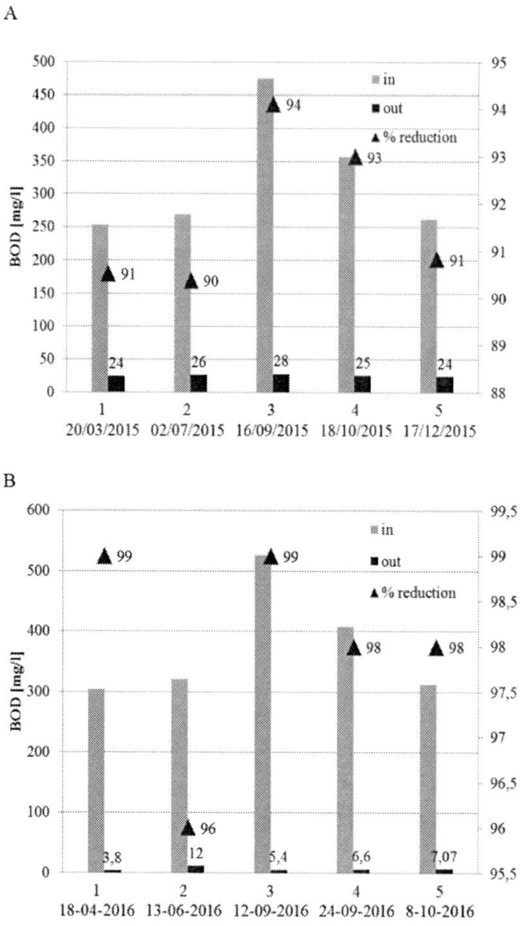

Fig. 3. Efficiency of BOD_5 removal: (A) before modification; (B) after modification.

Similarly to BOD_5, COD is also changing, the concentration before biological treatment of wastewaters ranging from 778 to 1216 mgO_2/l. Its removal efficiency varies from 88% to 97%, always fulfilling discharge limitations. Before chemical addition to the process, effluent values were mostly around, or above, effluent limits.

By comparing BOD_5 and COD removals, it is easily appreciated that wastewater contains easily biodegradable organic substances. BOD_5 to COD ratio, ranging between 0.3 and 0.6, indicates susceptibility of sewage to biological treatment, although, as indicated by many papers [9,10], knowledge of the participation of individual COD fractions allows for a more accurate assessment of the susceptibility of sewage to biological treatment than the overall quotient.

With regard to suspended solids, their removal efficiency is much smaller, but the limit values for outflows are not exceeded due to the low concentrations in the post-production effluents.

 DOI: 10.1051/e3sconf/20171902014

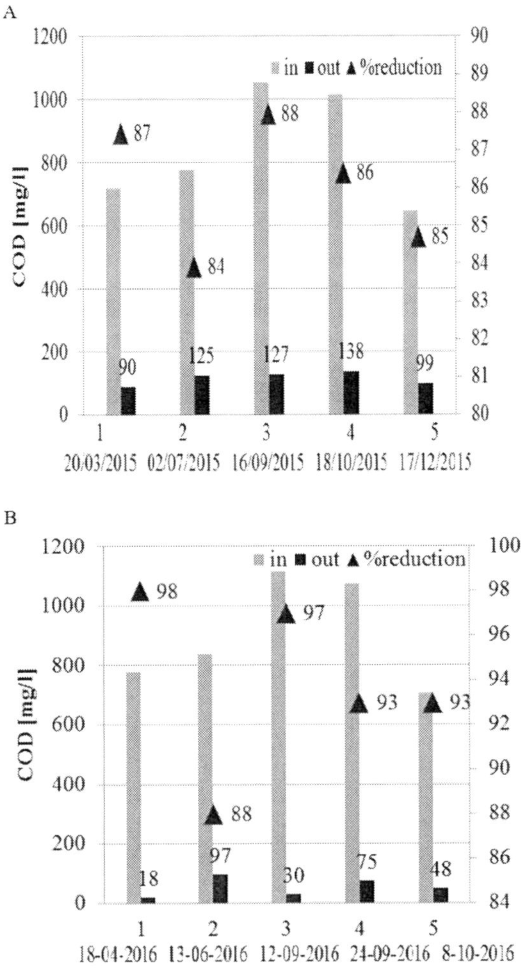

Fig. 4. Efficiency of COD removal: (A) before modification; (B) after modification.

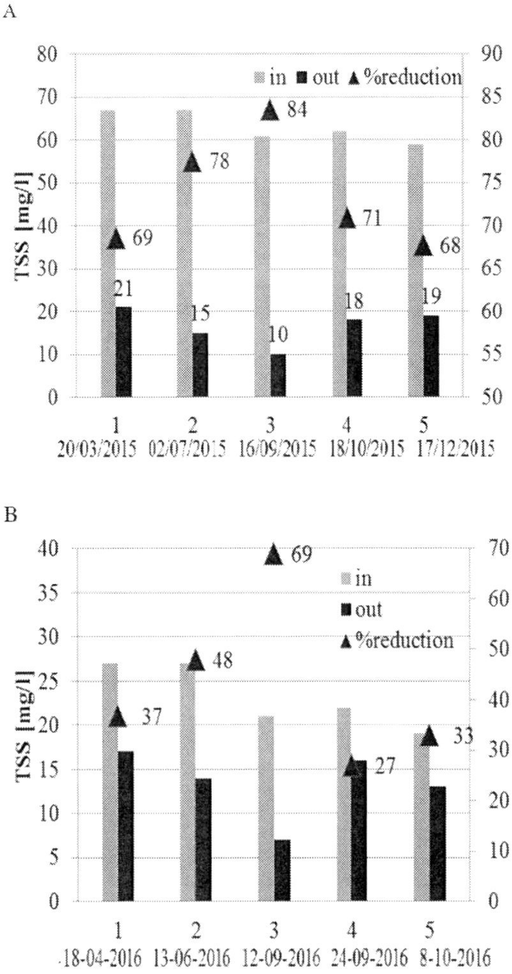

Fig. 5. Efficiency of TSS removal : (A) before modification; (B) after modification.

This treatment facility is characterized by high removal efficiencies, after chemicals addition had been introduced, in agreement with values required by regulations. BOD_5 on the outflow does not exceed 10 mgO_2/l, with values in inflow sewage up to 600.0 mgO_2/l, and efficiency between 96-98%.

TSS effluent concentration is on average 15 mg/l, with overall removal efficiency of 50% due to low inlet values at the treatment plant. COD was observed to decrease from 1216.0 mg/l to 30 mg/l. Concentration of phosphorus in the effluent was up to 5 mgP/l, with biological phosphorus removal periodically assisted by chemical precipitation with coagulant dosed into the effluent of the activated sludge reactor at a dose of 40-50 l/d of PIX (ferric chloride) coagulant.

This is most important in the spring when excessive effluent P concentrations can cause euthrophication in receiving waters. Similar threats have already been described by many authors including Capodaglio [18,19], Zhang [20], and others [13,21]

While in this case, biological treatment was assisted by external dosage of chemicals, as required to improved treatment efficiency, an alternative solution to chemicals dosing, could gave consisted in configuration changes of wastewater treatment process sequence. Such changes could involve replacing one aeration chamber with denitrification chamber.

However, since frequent problems had been observed due the lack of organic carbon in the influent, achievement of good biological denitrification could have not been assured. This would have required also the addition of supplemental carbon, which was considered an un-appropriate choice. There are many different compounds that could be used to aid in denitrification. Among the most commonly used, and best documented is methanol [9,22].

Another solution could also have been the adoption of the pre-denitrification stage. Such a scheme will be tested in this industrial wastewater treatment plant to evaluate effectiveness and cost-efficiency.

4 Conclusions

It is usually cheaper, technically simpler and more efficient to clean industrial effluents at the source, rather than attempting their treatment after mixing/dilution with household effluent. However, due to the fluctuation of pollutant loads from industrial plants, and especially in smaller facilities, constant monitoring of process parameters and careful efficiency evaluation of individual steps is required. Production changes significantly affect the removal efficiency of the main parameters: organic compounds, nitrogen and phosphorus. Biological treatment could be assisted by the dosage of chemicals, as required, to avoid effluent deterioration. The addition of chemicals to assist treatment processes in this case was considered an optimal choice to support industrial wastewater treatment in sub-optimal conditions. A pre-denitrification process option was however considered as an alternative solution and will soon be tested as it could effectively eliminate the necessity of chemicals addition. The wastewater treatment plant object of this evaluation can provide complete treatment of wastewater in accordance with the requirements of the Regulation of the Minister of the Environment of 18 November 2014. The purified effluent parameters are lower than the required values and their quantity and quality do not adversely affect the receiver.

References

1. PAIiIZ. Food industry in Poland. Polish Agence of Information and Investments S.A. (2013) (in Polish)
2. L.L. Fillaudeau, P. Blanpain-Avet, G. Daufin, J. Cleaner Prod. **14**, 5 (2006)
3. J. Chen, S. Seng, Y. Hong, *Wastewater Treatment,* 7, pp. 255–269, (AG-Metal Zenica, Taylor & Francis Group, LLC, 2006)
4. E. Podgórni., J. Boguniewicz-Zabłocka, Aplikacyjne i teoretyczne problemy w przemyśle rolno-spożywczym - postęp naukowo-techniczny, Politechnika Opolska, pp. 205-214 (2016) (in Polish)
5. B. Bartkiewicz, *Industrial Wastewater Treatment,* (WN PWN, Warszawa, 2006) (in Polish)
6. A.G. Capodaglio, H. Jones, V. Novotny, X. Feng, Water Res **25**(10), 1217-1224, (1991)
7. A.G. Capodaglio, P. Hlavínek, M. Raboni, Ambiente & Áqua - Interdisciplinary J. Appl. Sci. (2016)
8. Z. Heidrich, *Water and wastewater management guide,* (Verlag Dashofer, Warszawa, 2012) (in Polish)
9. A. Anielak, Unconventional methods of nutrients removal in SBR reactors. Gaz, woda i technika sanitarna (in Polish) (2006)
10. W. Dąbrowski, M. Puchlik, Ochrona Środowiska, **12**, 735-746 (2010)
11. A.G. Capodaglio, A. Callegari, Water Practice Technol. **10**,3 (2015)
12. A.G. Capodaglio, A. Callegari, J. Water, Sanitation Hyg. Dev. **6**, 3 (2016)
13. T. Imoobe and P. Okoye, J. Pure Appl. Sci. **4**, 1, (2011)
14. A. Anielak, *Dairy industry and its water-sewage economics*, Agro Przemysł **2**, 57–59 (in Polish), (2008)
15. Ordinance of the Minister of the Environment of Poland of 18 November 2014 (Journal of Laws no1800)
16. D. Vogel, Diploma Thesis, Politechnika Opolska (2017) [in Polish]
17. State of the environment in Poland. Main Inspectorate for Environmental Protection, Library of Environmental Monitoring, (2015) Access 12.12.2016,http://www.gios.gov.pl//zalaczniki/artykuly/Sygnaly%20 calosc_pol2015.pdf. [in Polish]
18. A.G. Capodaglio, J. Boguniewicz, F. Salerno, E. Lorents, D. Copetti, E. Legnani, G. Tartari, *River Basin Management* – Lawson (ed.) (ICE, London, ISBN 0 415 392004, 2005)
19. A.G. Capodaglio, A. Muraca, G.Becchi, Water Sci. Technol. **47**, p. 291 – 298 (2003)
20. R.B. Zhang, X. Qian, X.C. Yuan, R. Ye, B.S. Xia, and Y.L. Wang, Int. J. Environ. Res. Public Health **9**, 4504-4521 (2012)
21. A.Azzellino, R.Salvetti, R.Vismara, L.Bonomo, Sci. Tot. Environ. **371** 214-222 (2006)
22. K.R. Pagilla, M. Urgun-Demitras, R. Ramani, Water Sci. Technol. **53**, 3, 165-17 (2006)

Water Loss Reduction as the Basis of Good Water Supply Companies' Management

Agnieszka Ociepa-Kubicka[1*] and *Krzysztof* Wilczak[2]

[1]Czestochowa University of Technology , Faculty of Management, Armii Krajowej 19 B, 42-200 Częstochowa , Poland
[2]Water Supply and Sewerage Company in Gliwice, Rybnicka 47, 44-100 Gliwice , Poland

Abstract. Companies using water distribution systems to reduce the operating costs and increase the reliability of water supply systems, as well as to protect disposable water resources, must search for ways to reduce water losses. The article points out the economic and environmental aspects of water losses. The possibilities of using international water loss assessment standards have been analysed. The reflections presented in the paper refer to the current trends and world standards in the field of water distribution systems management. The article presents the results and analysis of water losses for the water supply network operated by the Water Supply and Sewerage Company in Gliwice (Przedsiębiorstwo Wodociągów i Kanalizacji w Gliwicach, PWiK). The losses were determined on the basis of numerous indicators and compared with other distribution systems. At present, most indicators of water loss are at a very good or good level. The Infrastructure Leakage Index (ILI), as one of the most reliable loss indicators for the surveyed distribution system, assumed values from 3.33 in 2012 to 2.06 in 2015. The recent drop in ILI values indicates the effectiveness of the Company's strategy for water leakage reduction. The success comprises a number of undertakings, such as ongoing monitoring, pressure reduction and stabilisation, repairs and replacement of the most emergency wires.

1 Introduction

Water losses occur in every water distribution system but only in some of them their levels can be considered as acceptable. Pipeline failure can be a result of design, engineering and operating mistakes, or natural wear and tear of materials. Among many factors, the most important influence on the actual water losses has: the incorrect technical condition of the wiring network and the level of network pressure and its fluctuations in the daily cycle. It should be stressed that the work of the waterways in the workplace has the effect of mining. Rock mass movements, land subsidence, bumps and corrosivity of groundwater are major causes of damage to water supply systems.

Due to the optimisation of water production costs and the need to protect water resources, rational reduction of water losses must be a priority ask for water supply companies. Poland, with a resource index of 1460 m3/M a year, has been seen in the World Water Council report since 1990 as a country with a water deficit.

The International Water Organisation (IWA) plays an important role in the implementation of sustainable water management, including the reduction of water losses [1,2]. The IWA guidelines for determining, analysing and evaluating water losses in the distribution systems will enable comparing the regularity of the operation of water supply networks by different plants. Many countries like Germany, Switzerland, Austria, United Kingdom, Denmark, Spain have been provided by IWA and have developed professional programs in economic documentation of technical, economical and reliability with investigation systems [1-4].

The purpose of this article is to analyse and evaluate water losses in the water supply network operated by the Water Supply and Sewerage Company in Gliwice (PWiK). These losses were considered in the context of system reliability and water supply to customers. The percentage water loss indicator and the standardised international methodology based on indicators recommended by IWA were used to determine the water losses. The obtained values of indicators for the discussed system were compared to the Polish and international standards.

The results and evaluations presented in this paper are based on the data obtained from the PWiK Gliwice and literature analysis.

2 Economic and environmental aspects of water loss

Water supply companies are required to supply the right amount of water, at the right pressure and of the right quality. By implementing these tasks, water companies must strive to optimise the costs and revenues of their operations. The financial policy of the plants should protect the recipients of services against undue,

*Corresponding author: agnieszkaociepa22@wp.pl

DOI: 10.1051/e3sconf/20171902015

excessive increase in the rates of water charges. Calculation of the prices of water and sewage disposal must be based on the costs incurred for the provision of services [5]. Each m^3 of the produced and pressed into the system water, entails various costs of, among others, energy costs and water treatment. It is estimated that the removal of each cubic meter of water is a gain of approximately 1.0-1.5 kWh of energy needed to produce and transport water in the distribution system [6]. According to the regulations, the costs of activities resulting from the production of unsold water, including losses, are charged to all users of the water supply network. It should be emphasised that in Poland, average prices for water consumption and wastewater discharge systematically increase. As compared to 2008, the average price in 2011 was by 27%, and as of January 1, 2016 by 70% higher (Price formation 2016). Based on the analysis of water and wastewater prices for households in Polish cities of over 50 thousand, in 2016, the lowest gross prices were PLN 6.56 (Zamość and Biała Podlaska) and the highest PLN 18.0 (Mysłowice and Dąbrowa Górnicza).

The results of the NIK (The Supreme Audit Office) audit, indicate that in 2013, 2014 the share of household expenditure on water purchase and sewage disposal in the income of a given farm was for large Polish cities from about 2.50 to 4.00%. Studies indicate that the upper social acceptability index should not exceed 3%. In Poland, these indicators are significantly higher than in the Netherlands, Italy, Great Britain, Germany and Portugal where the index is significantly lower than 1.0%.

Therefore, limiting water losses should be the key task of water supply companies. In this way, the plants can save on the costs of running the network, as well as on long-term capital costs, as the smaller the amount of water injected into the network affects the investment costs. It is also obvious that reducing losses is not only financial savings. There are also benefits of the use of the water supply system improvement, and also the image of an enterprise whose primary responsibility is to provide reliable water supply.

Also taking into account the need to protect water resources, the rational reduction of water losses by distribution system operators must be a priority. Controlling and limiting water losses not only affects the cost of water supply, but also protects the world's drinkable water resources, of which there are only around 3% in the world. Losses should include not only the costs of waste and the environmental costs of lost water, but other costs resulting from leaks, such as subsidence or road crashes, and even the costs of street crashes caused by the removal of water supply breakdowns [7]. It should also be stressed that water shortages are a problem not only for African countries or other desert areas but also for our country. Poland, with the resource index of 1460 m^3M^{-1} a year, has been noted in the World Water Council Report since 1990 as a country with a water deficit. The unfavourable situation is deepened by the small amount of water retained in water bodies and the degree of certain water bodies' pollution [8].

3 Materials and methods

The analysis covered the data from the years 2012-2015, obtained from the Water Supply and Sewerage Company in Gliwice, which included: water supplied to the network, used for social welfare purposes, non-production and production purposes, network length, quantity and length of water supply connections, material and age structure of the network, the number of recipients. On this basis the Percentage Water Loss Indicator(PLI) andUnitary Water Loss per Capita Indicator(Qlos) were estimated, as well as the IWA recommended indicators [9]:

- Unique Real Leakage Balance Indicator (RLB1 and RLB2)
- Non-Revenue Water Basic Index (NRWB)
- Unavoidable Annual Real Losses Indicator (UARL)
- Infrastructure Leakage Index (ILI).

The above ratios are calculated on the basis of the formulas presented in Table 1 and on the basis of data obtained from the waterworks.

4 Water losses on the example of the Water and Sewerage Company in Gliwice

4.1 Operating range

The Waterworks and Sewerage Company in Gliwice covers the towns of Gliwice and Pyskowice and the communes of Kozłów, Smolnicka, Taciszów, Kleszczów, Rzeczyce and Ligota Łabędzka in the municipality of Rudziniec. Water is supplied to about 200.000 people and the total length of the water supply system is 813.59 km.

The results indicate that 49% of the treated water meets the quality requirements for water intended for human consumption. The remaining 51% is treated.

4.2 The material structure and the age of the water supply network

In the material structure of the network, the grey cast iron dominates, which is currently mainly referred to as polyethylene, and, in the range of large diameters – spheroidal cast iron. Exchanges are necessary because of the internal incrustation of wires or significant deterioration of pipe strength due to the graphitisation of old grey cast iron, often even more than 100 years old.

Approximately 50% of the water supply network has been operated for over 30 years. There are less than 30% of the lines built in the last decade. Cables operated for over 50 years are mostly made of grey cast iron and have a small share of steel. On the other hand, wires of the age range of 31-50 years are mostly made of steel, whereas in the last 20 years, the networks have been built mainly of polyethylene and nowadays also in a small share of ductile cast iron.

4.3 Water balance and indicators characterising water loss

Balancing water in the network is the basis for evaluating the operation of the water distribution system in terms of water consumption and loss. Based on its results, the companies decide to initiate actions aimed at liquidation of water losses [10-12]. The amount of water loss in the water supply network is due to the difference between the water introduced into the grid and the water sold and consumed for the needs of the water supply. This can be expressed by the formula:

$$Vlos = Vsup - Vts - Vsol \quad (1)$$

where:
Vlos – water loss in the distribution system, m^3
Vsup – water supplied to the network, m^3
Vts – volume of water consumed on so-called own needs of the water supply, i.e.: rinsing of water and sewage networks, tanks, for business purposes etc., m^3
Vsol – water sold, m^3.

The PLI water loss percentages are the most commonly used water loss indicators. The indicator allows to determine the share of water losses in relation to the volume of water entering the water supply (Table 1). For the reliable and objective assessment of water losses for PWIK Gliwice, the loss indicators recommended by IWA were also calculated, which are summarised in Table 1

All indicators presented in Table 1 are described in detail in numerous papers [4,12,13].

Tables 2 and 3 summarise the distribution system data that are needed to calculate the IWA recommended water loss rates.

Table 4 presents the water loss indicators for the water supply system analysed, calculated from the formulas 2-8.

Table 1. Compilation of water loss indicators [4,12,13].

Water loss indicator	Indicator formula
Percentage Water Loss Indicator **PLI**	$PLI = (V_{los}/V_{sup}) \cdot 100 \quad [\%]$ (2) V_{los} – water loss in the distribution system, m^3/year V_{sup} – water supplied to the network, m^3/year
Unique Real Leakage Balance Indicator– **RLB**	$RLB_1 = V_{los}/[(M+R) \cdot 365] \quad [m^3 \cdot km^{-1} \cdot d^{-1}]$ (3) M – main network's length, km R – ramification network's length, km $RLB_2 = (V_{los} \cdot 1000)/(L_{wc} \cdot 365) \quad [dm^3 \cdot supply^{-1} \cdot d^{-1}]$ (4) L_{wc} – number of water connections
Non-Revenue Water Basic Index – **NRWB**	$NRWB = [(V_{sup} - V_{sol})/V_{sup}] \cdot 100 \quad [\%]$ (5) V_{sol} – water sold, m^3/year
Unavoidable Annual Real Losses Indicator– **UARL**	$UARL = [18 \cdot (M+R) + 25 \cdot Wc + 0,8 \cdot L_{wc}] \cdot 0.365 \cdot p \quad [m^3 \cdot year^{-1}]$ (6) Wc– length of water connection, km p – average pressure in the tested network 46 m H_2O 0.365 – conversion factor per year and m^3
Infrastructure Leakage Index– **ILI**	$ILI = V_{los}/UARL \quad [-]$ (7)
Unitary Water Loss per Capita Indicator $\mathbf{Q_{los}}$	$Q_{los} = (V_{los} \cdot 1000)/(LM \cdot 365) \quad [dm^3 \cdot d^{-1} M^{-1}]$ (8) LM – number of inhabitants using the supply

Table 2. Compilation of water balance for 2012 – 2015.

Year	Supplied water [m^3]	Own post-supply use [m^3]	Water to be sold [m^3]	Sold water [m^3]	Loss [m^3]	Loss [%]
2012	11 452 004	113 692	11 338 312	9 625 983	1 712 329	14.95
2013	11 072 530	153 521	10 919 009	9 417 694	1 501 315	13.56
2014	10 363 660	125 296	10 238 364	9 304 428	933 936	9.01
2015	10 700 283	118 805	10 581 478	9 513 710	1 067 768	9.98

Table 3. Compilation of the network length, quantity and length of water connections.

Year	Length of main network M [km]	Ramification networks length R [km]	Length of water connections [km]	Number of connections [items]	Total [km]
2012	103.19	465.85	234.85	181334	803.89
2013	103.19	469.90	234.54	18123	807.63
2014	102.69	472.70	235.62	18159	811.01
2015	102.69	475.49	235.41	18150	813.59

Table 4. Compilation of water loss indicators for 2012-2015.

Year	PLI [%]	Q_{los} [$dm^3 \cdot d^{-1} M^{-1}$]	RLB_1 [$m^3 \cdot km^{-1} \cdot d^{-1}$]	RLB_2 [$dm^3 \cdot supply^{-1} \cdot d^{-1}$]	NRWB [%]	UARL [$m^3 \cdot year^{-1}$]	ILI [-]
2012	14.95	23.5	8.2	258.7	15.95	514129	3.33
2013	13.56	20.8	7.2	227.0	14.95	515076	2.91
2014	9.01	13.1	4.5	140.9	10.22	516708	1.81
2015	9.98	15.0	5.0	161.2	11.09	517342	2.06

The ILI leakage rate infrastructure assumes a value of between 3.33 in 2012 and 2.06 in 2015. According to the American Water Association (AWWA), the state of the water supply network is rated as very good for values lower than 3.0 or weak for values greater than 5.0. The average ILI leakage index values in the Polish cities ranged from 3.16 to 16.62 [6], which may lead to the conclusion that the state of the analysed network is currently very good.

5 Discussion of the results

Percentage loss of water, still widely used in Poland and in the world, is unreliable because it does not take into account such important factors as: the length of the network, the number of connections and their length, and the pressure and hydraulic load of the water supply network. However, it can be used to assess the variability of water losses in a multi-year distribution system. It is therefore recommended that the use of the classical method of calculating the percentage of losses be supplemented by the IWA standardised methodology, which will allow us to obtain more complete information on the amount of wasted water lost from the network. It should be emphasized that PWiK Gliwice has complete information allowing for a comprehensive assessment of water losses according to the recommended international standards.

It should be emphasised, however, that more and more water losses are being analysed in line with international standards. Bergel's comprehensive water loss analysis for 334 small group waterworks is an example. The author sets the range of percentages in the analysed water groups at 18.8-23.6%, but also determines the per capita net loss per unit and determines all indicators recommended by IWA [14]. Comparison of the percentage water loss indices for different systems can be unjust to plants. For example, in 2014, the percentage loss index for the distribution system operated by the Water Supply and Sewerage Company in Częstochowa was 9.0% and ILI 1.54 [13]. In the same year, the wastage rate for the Water Supply and Sewerage Company in Myszków was 17% at ILI 1.48 [15]. For the above water supply systems, with a clearly differentiated percentage of water loss index, the ILI values are comparable.

The analysis of loss indicators for the distribution system of PWiK Gliwice shows that the plant has taken effective measures to reduce the leakage. In recent years, losses have been reduced to a level that demonstrates good technical condition of the water supply network. The percentage of water start-up has decreased from about 15 in 2012 to about 9.0% in recent years. The infrastructure leakage rate during the period under review has been systematically reduced to ca. 2.0. It should be noted that in the literature a slightly different scales of interpretation of ILI can be found in the literature. According to the more rigorous state of the network, it is a good level [16] according to others at very good level [17]. Also, the value of the unit real RBL loss at 160 $dm^3 \cdot dl^{-1} \cdot d^{-1}$, against the background of Polish systems, indicates that the losses are maintained at a good level. Currently in Western countries, permissible actual losses are assumed at 100 $dm^3 \cdot dl^{-1} \cdot d^{-1}$ [14]. The next indicator recommended by IWA for comparison of the state of water distribution systems is the NRWB, Non-Revenue Water Basic Indicator, which is defined in Poland as a volume of unsold water. Similarly to previous indicators, it has decreased in recent years from about 16% in 2012 to about 11% in 2015. It shows a good technical state of the network against the national data where the weighted average for this system size is about 24% [18] The net water loss per capita is also a good indicator of water quality, with a value of 15.0 $dm^3 \cdot d^{-1} \cdot M^{-1}$ in 2015. In Bergel group water systems, this indicator ranged from 24.0 to 39.9 $dm^3 \cdot d^1 \cdot M^{-1}$ on average, while the Hotloś links analysis for 10 municipal waterworks indicated the average range of this indicator from 16 to 35 $dm^3 \cdot d^{-1} \cdot M^{-1}$ [13]. With a hydraulic loading indicator of about 50.0 $m^3 d^{-1}$ km^{-1}, the unit losses for the system under review are 5.0 $m^3 \cdot km^{-1} \cdot d^{-1}$, which gives approx. 0.2 $m^3 \cdot km^{-1} \cdot d^{-1}$. In accordance with German requirements, a loss of 0.2 $m^3 \cdot km^{-1} \cdot d^{-1}$ is allowed at such hydraulic system load.

6 Conclusions

There are several positive trends in water intake in Poland, consisting in the reduction of the overall water consumption especially for production purposes, the reduction of underground water consumption by industry, and more efficient use of water from the collective water supply. Important factors influencing water saving should be the implementation of water-saving technologies and the reduction of losses in water supply systems.

In this respect, reducing water losses should be a key task for plants operating water supply systems. In addition, water companies must strive to optimise the costs of their operations to protect their service recipients from undue and excessive increases in water charges. By reducing the losses, plants can reduce the costs of the current network exploitation and save on the investment costs. However, it should be noted that the complete elimination of water losses is not often possible and unprofitable. Plants operating water distribution systems in Poland and in other European countries must implement and apply modern standards for water loss management.

The calculated water loss rates allowed for the analysis and assessment of the technical condition of the water supply system, and it can be concluded that the plant has succeeded in successfully reducing water losses. Based on the analysis of water losses in the distribution system operated by the Water Supply and Sewerage Company in Gliwice, it can be stated that the process of loss reduction takes place in a multidirectional plant, progressively and systematically achieving more and more satisfactory results. At present, most indicators of water loss are at a very good or good level. The success comprises a number of undertakings, such as ongoing monitoring, pressure reduction and stabilisation, repairs and replacement of the most emergency wires. Monitoring allowed, among other things, for the monitoring and pressure management, which made greater pressure stability possible, and thus, less failure and loss. Active network control was implemented to quickly detect failures by providing operational teams with specialist equipment such as noise loggers and correlates.

References

1. G. Ferrari, G, D. Savic, Procedia Eng. **119** (2015) DOI: 10.1016/j.proeng.2015.08.874.
2. J.A, Zambrano, M.Gil-Martinez, M.Garcia-Sanz, I. Irizar, Water Sci. and Technol. **60,** 2 (2009)
3. M. Clarke, P. Boden, A. McDonald, Water Environ. J. **26** (2012)
4. M. Kwietniewski, Ochr Środ. **35**, 4 (2013)
5. Ordinance of the Minister of Infrastructure and Construction of 11 August 2016 amending the regulation on tariffs, model of application for tariff approval and settlement conditions for collective water supply and collective sewage disposal (Journal of Laws 2016, item 1301)]
6. H. Hotloś, Environ Prot Eng. **36,** 3 (2010) http://epe.pwr.wroc.pl/2010/hotlos_3-2010.pdf.
7. C.H. Ashton, V.S. Hope, Urban Water **3** (2001)
8. W. Kuczyński, W. Żuchowicki, Annual Set the Environment Protection, Environment Protection Yearbook, **12** (2010)
9. A. Lambert, W. Hibner, IWA **10** (2000)
10. F. Piechurski, Napędy i Sterowania, **2** (2013)
11. F. Piechurski, Napędy i Sterowania **1** (2014)
12. J. Rak, D. Trojan, J. Civil Eng., Environ. Archit. **245** (2014)
13. E. Ociepa, K. Merc, Gaz Woda i Tech. Sanitarna **90, 7** (2016), DOI:10.15199/17.2016.7.3
14. T. Bergel, Gaz Woda i Tech. Sanitarna **8** (2012)
15. W. Kędzia, E. Ociepa, Inż i Ochr. Środ. **18** (2015)
16. J. Rak, A. Tunia, Instal. **5** (2012)
17. A. Lambert, Non-Revenue Water Conf., Malta (2012)
18. T. Bergel, Gaz Woda i Tech. Sanitarna **10** (2012)

Long-term trends in river water temperature: a case study of the Raba River (Polish Carpathians)

Mariola Kędra[1,*]

[1]Cracow University of Technology, Department of Environmental Engineering, Institute of Engineering and Water Management, Warszawska 24, 31-151 Cracow, Poland

Abstract. This study analyses potential trends in river water temperature (Tw) over a period of 50 years (1960–2009). The study area is located in the Raba catchment in the Polish Carpathians. The aim of the study was to evaluate long-term trends in daily Tw for each season of the year as well as to compare the direction of these trends for sites located some distance upstream and downstream from the Dobczyce Reservoir (DR), built in 1986 on the Raba River. For autumn and winter, increases in mean, median, and minimum Tw are significant for the downstream site (39 km from the DR), with a rate of change of 0.18–0.26 °C per decade. Contrasting results were obtained for the spring and summer seasons; significant increases (0.22–0.47 °C per decade) in mean, median, and maximum Tw were found for the upstream site, in concordance with seasonal air temperature trends, but a significant decrease in maximum Tw (–0.42 and –0.35 °C per decade, respectively) was identified for the downstream site. The revealed discrepancies in the direction of seasonal trends for the upstream and downstream sites studied suggest an anthropogenic impact on Tw downstream from the DR. Overall, a significant increase in summer Tw suggest the need for appropriate flowing water management that would mitigate adverse effects of climate warming on the fluvial environment.

1 Introduction

Water temperature (Tw) in streams and rivers is one of the most important determinants of aquatic life conditions. Tw regulates the rate of biological and chemical reactions affecting the growth, development, and survival of aquatic organisms as well as their distribution within a fluvial environment and across geographic regions [1]. Tw depends on heat exchange processes between a water body and its surroundings, that is, at the surface and bed of rivers [2]. The exchange of heat across the water surface is determined by short-wave and long-wave radiation, evaporation, and conduction between water and ambient air; therefore, changing meteorological conditions such as air temperature, solar radiation, cloud cover, relative humidity, and wind speed induce changes in water temperature [2-3] to an extent dependent on insulating and buffering processes, which operate within a stream system [4].

The impact of climatic factors, mainly air temperature, on the functioning of aquatic ecosystems is extensively discussed in the context of progressive climate change [5-13]. The world-wide warming trend in mean surface air temperature of 0.12 °C per decade (1951–2012) [14] together with likely changes in precipitation [15] may create flow and thermal conditions unfavourable for native aquatic biota. As stream temperature measures the amount of heat energy per unit volume of water [4], hence even relatively short-term (weekly, monthly) decreases in flow volume (induced by climatic or anthropogenic factors such as water withdrawal) can temporarily enhance a potential increase in Tw, with detrimental consequences for aquatic organisms. In particular, fish, aquatic ectothermic vertebrates [16], can tolerate only a certain temperature range [1] and respond rapidly to the temperature of their habitat; temperature can therefore be viewed as an ecological resource shared between three thermal guilds (cold, cool, and warm water fishes) [16]. A progressive increase in Tw will result in warmer fluvial conditions unsuitable for those native fish species which prefer colder water. In that case, even temperature variation produced by habitat heterogeneity [17] may be insufficient refuge for them.

The impact of anthropogenic factors, including land use change and in-river impoundments, on the thermal regime of rivers has been well documented [18-23]. In particular, removal of vegetation (riparian or upland) makes small streams more prone to a rise in Tw due to their relatively low thermal capacity [4,24]. Dams alter downstream thermal conditions depending on their size, and operational mode and water release mechanism [20, 22]. Downstream of dams with hypolimnetic (bottom layer) releases, the thermal regime is cooler in summer and warmer in winter, compared with the predam

* Corresponding author: mariola.kedra@iigw.pk.edu.pl

conditions, affecting the populations and types of species present [22]. Hypolimnetic releases can deteriorate the synchronous behaviour of air and water temperatures and significantly modify water temperature dynamics in the downstream river reach [25-26]. Modelling results indicate that climate change impact will be more pronounced downstream of dams with epilimnetic (surface layer) water releases than with hypolimnetic releases [6].

The aim of this study is: (1) to assess potential long-term (50-year) trends in river water temperature for each season of the year; (2) to compare the direction of these trends for sites located some distance upstream and downstream from a relatively large reservoir with hypolimnetic water releases, which was built on a mountain river in the Polish Carpathians. As water temperature is regarded a fundamental ecological variable [27] an assessment of potential changes in Tw, especially long-term trends, appears necessary and vital. This should facilitate appropriate management of the river thermal regime – an integral part of environmental flow [27], with emphasis on remedial action mitigating adverse effects of climate warming on natural fluvial environments. Moreover, important differences in the direction of trends in water temperature for upstream and downstream sites can further underscore the usefulness of river reaches downstream from hypolimnetic dams (tailwaters) for desirable coldwater trout fisheries [28].

2 Materials and methods

2.1 Study area

The research was conducted on the Carpathian river Raba, the right-bank tributary of the upper Vistula River in southern Poland (Fig. 1). From its source in the Gorce Mountains 780 m above sea level (a.s.l.) to its mouth 180 m a.s.l., the Raba River is 132 km in length [29]. The Raba catchment covers an area of 1,537 km^2 [29], with the highest peak (Turbacz: 1,310 m) being located in the Gorce Mountains, and the lowest elevation at the Raba River mouth. The rock layer underlying the Raba catchment is primarily composed of sandstone, shale, marl, loam, and gravel [30].

Climate conditions in the study area vary depending on location. In the lower-lying parts of the catchment, the air temperature is higher, with an annual average of 6–8 °C, and precipitation totals of 600–800 mm a year. The highest peaks in the Gorce Mountains are characterised by a cool climate with the annual mean air temperature below 4 °C and annual precipitation totals reaching 1,200 mm. Precipitation totals increase as the elevation increases; however, this tendency is disturbed by the influence of mesorelief landforms [30].

The Raba catchment area is 57.6% agricultural, while forest and developed land comprise 35.4% and 6.2%, respectively, while water covers 0.8% [31]. A retention reservoir with a holding capacity of 125 mln m^3 was built in 1986 on the Raba River [32]. The dam is 30.5 m high, located at the 60.1 river kilometre, is equipped with a gated spillway and fixed, low-level intakes. The Dobczyce Reservoir (DR) is the main source of drinking water for the city of Cracow. Its function is flood prevention, and increasing base flows in the Raba River. The dam is also used for electric energy generation [32].

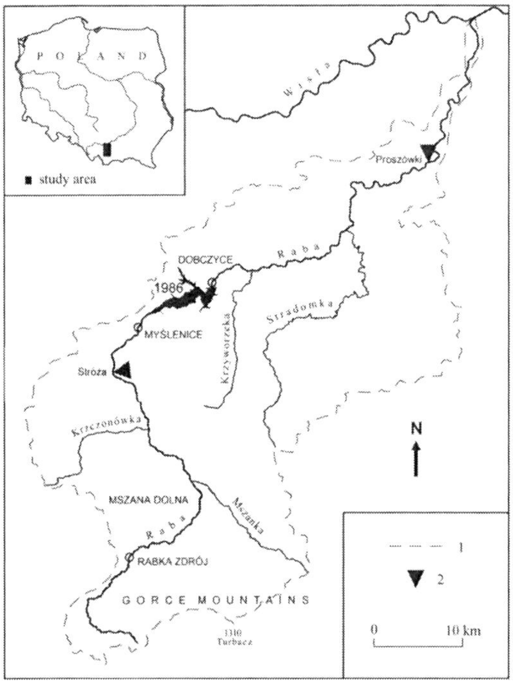

Fig. 1. Study area—the Raba River catchment with larger tributaries: (1) catchment boundary, (2) stream gauging sites.

2.2 Data and methods

2.2.1 Data

The examined water temperature series consist of daily observations (at 6:00 Coordinated Universal Time, or CUT) made consecutively at two gauging sites (Stróża and Proszówki) situated on the Raba River (Table 1). The data for the years 1960–2009 (December 1959 – November 2009) were made available for research purposes by the Polish Institute of Meteorology and Water Management—the National Research Institute, and in some cases were obtained from Poland's state-published surface water yearbooks. The available water temperature series for Proszówki ends in 2009; this determined the time period studied. The sites Stróża and Proszówki are located along the upper and lower course of the Raba River, approximately 21 km and 39 km from the DR dam, respectively. Moreover, gridded data series of daily mean air temperature for Stróża and Proszówki for 1962–2009 (December 1961 – November 2009) were extracted from the high quality, homogenised CARPATCLIM Database © European Commission JRC [33] for comparison purposes. The data are available for the years 1961–2010 (January 1st to December 31st) only.

DOI: 10.1051/e3sconf/20171902016

Table 1. Physical characteristics of the study sites [29] and descriptive statistics of daily water temperature (Tw) data for the 50-year period (1960–2009).

Statistic	Stróża site	Proszówki site
Latitude (N)	49°47'49"	49°59'47"
Longitude (E)	19°55'29"	20°26'18"
River length (km)	80.6	21.5
Drainage area (km^2)	644.1	1470.4
Median elevation (m)	581	423
Mean catchment slope (–)	0.030	0.016
Mean discharge (m^3s^{-1})	10.10	17.02
Minimum Tw (°C)	0.0	0.0
Maximum Tw (°C)	22.2	24.8
Median Tw (°C)	7.2	8.9
Mean Tw (°C)	7.6	9.1
Standard deviation of Tw (°C)	5.9	6.7

2.2.2 Methods

After visual inspection of daily temperature data, compiled in a Microsoft® Excel 2007 spreadsheet, a quality check was performed in order to identify spurious values. Next, each series was arranged as the matrix $\boldsymbol{T}_{ki}$, where the index k (for row) represents a day of the year (k = 1, ..., 366), and the index i (for column) represents a year (i = 1, ..., n), and n = 50. For the relevant rows of the matrix (1–91, 92–183, 184–275, 276–366), arithmetic mean, median, minimum, and maximum values were calculated as n-element series $\{x_i\}$ for each season (winter–autumn) of the year (1960–2009). Next, trend analysis was performed on each seasonal series separately in order to evaluate the strength of potential monotonic trends in seasonal temperature. The linear model of time trend was assumed in the form: $x(t) = a{\cdot}t + b$, where a is the slope coefficient, b is the intercept, and t is time. The non-parametric, 2-sided Mann-Kendall (MK) test [34-36] was used. The null hypothesis of no trend (H_0: $a = 0$) was tested equivalently (H_0: $\tau = 0$), using Kendall's rank correlation coefficient τ [36-38] at the significance level α: 0.05 (medium evidence) and 0.01 (strong evidence). The τ is expressed as:

$$\tau = \frac{2S}{n(n-1)} \quad (1)$$

$$S = \sum_{i=1}^{n-1} \sum_{j=i+1}^{n} \operatorname{sign}(x_j - x_i) \quad (2)$$

where S is Kendall's S statistic, and sign(x) is the sign function. The magnitude of trends was determined using Theil-Sen's slope [36-38] as a robust estimate of the slope, with the slope and intercept defined as:

$$\hat{a} = \text{median}\left(\frac{x_j - x_i}{t_j - t_i}\right), i = 1, \cdots, n-1, j = i+1, \cdots, n \quad (3)$$

$$\hat{b} = \text{median}(\{x_i\}_{i=1}^{n}) - \hat{a}\text{median}(\{t_i\}_{i=1}^{n}) \quad (4)$$

No assumption of normality, but serial interdependence is required [36]. Therefore, the trend-free pre-whitening procedure (TFPW) [39] was applied to each data series.

3 Results

For water temperature, significant trends (p<0.05) were identified for each season of the year (Table 2). For autumn (September–November) and winter (December–February), increases in mean, median, and minimum Tw were found for the downstream site only (Proszówki), with a rate of change of 0.22–0.26 °C per decade for autumn (Fig. 2), and 0.18–0.24 °C per decade for winter (Fig. 3). For spring (March–May), a significant increase in maximum Tw (0.47°C per decade) was identified for the upstream site (Stróża, Fig. 4a); in contrast, a significant decrease in maximum Tw (–0.42 °C per decade) was found for the downstream site (Fig. 4b), together with an increase in minimum Tw (0.03 °C per decade, Fig. 4c). For summer (June–August), increases in mean, median, and maximum Tw are significant for the upstream site (0.22–0.29 °C per decade, Figs. 5a–c); in contrast, a significant decrease in maximum Tw was identified for the downstream site (–0.35°C per decade, Fig. 5d).

Table 2. Rate of change $\hat{a}$ (°C $decade^{-1}$) for water temperature (Tw) in each season of the year (1960–2009) based on Theil-Sen's slope estimation [37-38] with trend free pre-whitening [39]. Statistical significance of the trends was evaluated via the 2-sided Mann-Kendall test [34-35].

Tw statistic	Winter (Dec–Feb)	Spring (Mar–May)	Summer (Jun–Aug)	Autumn (Sep–Nov)
Mean (Stróża)	0.04	0.21	0.25*	–0.08
Median (Stróża)	0.00	0.14	0.22*	0.00
Minimum (Stróża)	0.00	0.00	0.24	–0.04
Maximum (Stróża)	0.30	0.47**	0.29**	0.06
Mean (Proszówki)	0.24**	–0.13	–0.10	0.22
Median (Proszówki)	0.18**	–0.21	–0.07	0.26*
Minimum (Proszówki)	0.00	0.03*	0.00	0.22*
Maximum (Proszówki)	0.29	–0.42*	–0.35*	0.00

* data significant at p<0.05; ** data significant at p<0.01.

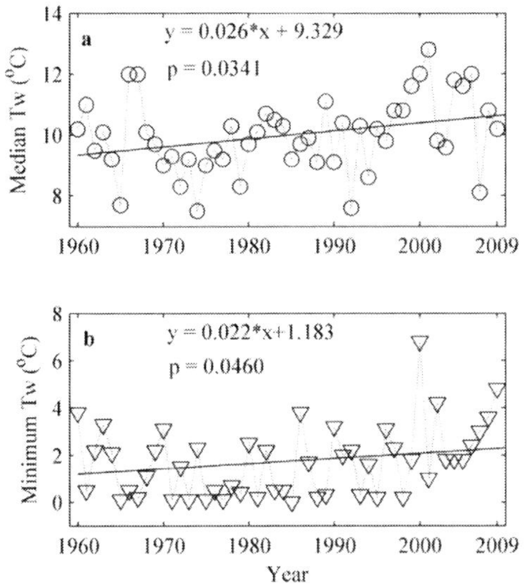

Fig. 2. Autumn water temperatures (Tw) at Proszówki (downstream of the DR dam) for the 50-year period (1960–2009) with significant trends shown in: (a) median Tw; (b) minimum Tw.

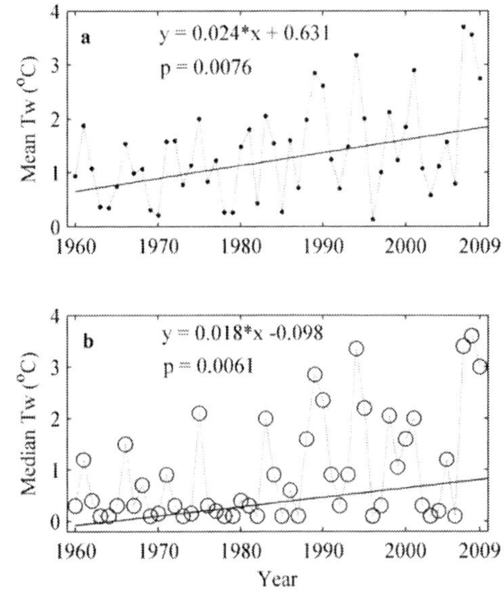

Fig. 3. Winter water temperatures (Tw) at Proszówki (downstream of the DR dam) for the 50-year period (1960–2009) with significant trends shown in: (a) mean Tw; (b) median Tw.

In order to explain the detected differences in seasonal Tw, analogous trend analysis was performed for seasonal air temperature (Ta) for the period 1962–2009. As listed in Table 3, significant trends (p<0.05) were found for each season of the year except autumn. All seasonal trends in Ta were increasing, and highly consistent for both the upstream and downstream sites.

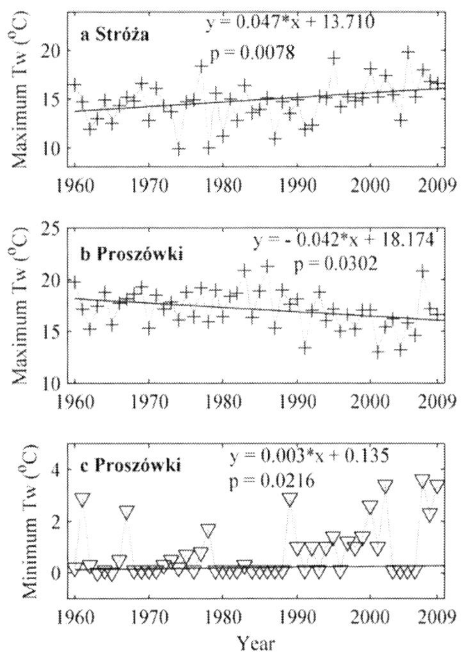

Fig. 4. Spring water temperatures (Tw) for the 50-year period (1960–2009) with significant trends shown in: (a) maximum Tw at Stróża (upstream of the DR dam); (b)–(c) maximum and minimum Tw at Proszówki (downstream of the DR dam), respectively.

Table 3. Rate of change $\hat{a}$ (°C decade^{-1}) for gridded air temperature (Ta) in each season of the year (1962–2009) based on Theil-Sen's slope estimation [37-38] with trend free pre-whitening [39]. Statistical significance of the trends was evaluated via the 2-sided Mann-Kendall test [34-35].

Tw statistic	Winter (Dec–Feb)	Spring (Mar–May)	Summer (Jun–Aug)	Autumn (Sep–Nov)
Mean (Stróża)	0.48*	0.37**	0.44**	0.10
Median (Stróża)	0.45	0.30*	0.47*	0.19
Minimum (Stróża)	0.87*	0.68	0.42*	–0.13
Maximum (Stróża)	0.70*	0.30	0.53**	0.12
Mean (Proszówki)	0.52*	0.38**	0.46**	0.12
Median (Proszówki)	0.42	0.37**	0.48**	0.22
Minimum (Proszówki)	0.75	0.52	0.45*	–0.11
Maximum (Proszówki)	0.79*	0.18	0.52*	0.06

* data significant at p<0.05; ** data significant at p<0.01.

DOI: 10.1051/e3sconf/20171902016

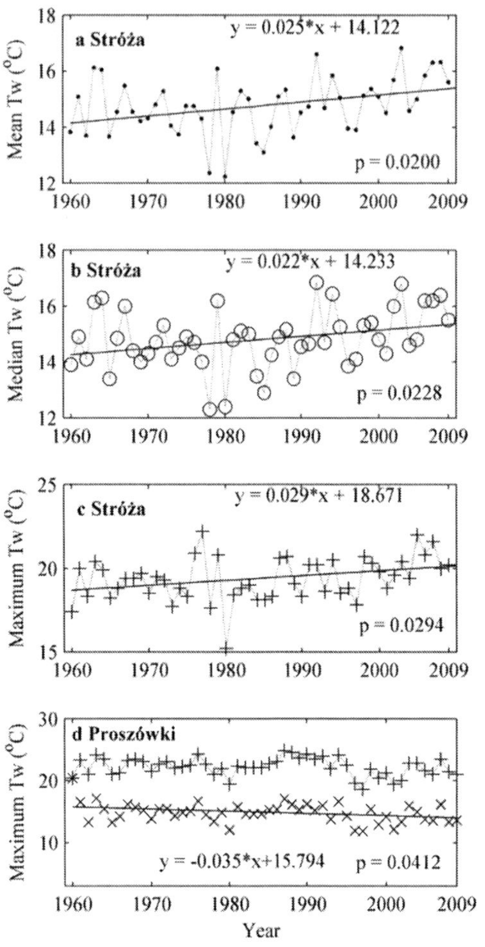

Fig. 5. Summer water temperatures (Tw) for the 50-year period (1960–2009) with significant trends shown in: (a)-(c) mean, median, and maximum Tw at Stróża (upstream of the DR dam); (d) maximum Tw (+original series, x pre-whitened series) at Proszówki (downstream of the DR dam), resp.

Increases in mean, minimum, and maximum Ta during the winter (0.48–0.87 °C per decade) were followed by increases in mean and median Ta during the spring (0.30–0.37 °C per decade). For summer, increasing trends were found for each of the investigated series, that is, for mean, median, minimum, and maximum air temperature, with a rate of change of 0.42–0.53 °C per decade. Furthermore, monotonic trend analysis was repeated for seasonal water temperature relating to two consecutive time periods before and after the construction of the DR complex (1960–1985, and 1986–2009). At the upstream site, an increase in minimum winter Tw (0.05 °C per decade, strong evidence) as well as a decrease in maximum autumn Tw (–0.98 °C per decade) were found for 1960–1985, while a considerable increase in maximum spring Tw (1.44 °C per decade, strong evidence) was followed by increases in mean and median Tw (0.65–0.70 °C per decade) for the summer of 1986–2009 (Table 4). At the downstream site, no trend in Tw was found for 1960–1985 (Table 5);

Table 4. Rate of change $\hat{a}$ (°C decade^{-1}) for water temperature (Tw) at Stróża. See details in caption of Table 2.

1960–1985				
Tw statistic	Winter (Dec–Feb)	Spring (Mar–May)	Summer (Jun–Aug)	Autumn (Sep–Nov)
Mean	–0.21	–0.36	–0.29	–0.57
Median	–0.08	–0.67	–0.30	–0.25
Minimum	0.05**	0.00	0.00	–0.22
Maximum	–0.80	0.00	–0.38	–0.98*
1986–2009				
Tw statistic	Winter (Dec–Feb)	Spring (Mar–May)	Summer (Jun–Aug)	Autumn (Sep–Nov)
Mean	–0.05	0.56	0.65*	0.36
Median	–0.10	0.58	0.70*	0.39
Minimum	0.00	0.00	0.62	0.00
Maximum	0.00	1.44**	0.53	0.49

* data significant at p<0.05; ** data significant at p<0.01.

Table 5. Rate of change $\hat{a}$ (°C decade^{-1}) for water temperature (Tw) at Proszówki. See details in caption of Table 2.

1960–1985				
Tw statistic	Winter (Dec–Feb)	Spring (Mar–May)	Summer (Jun–Aug)	Autumn (Sep–Nov)
Mean	0.07	–0.03	–0.07	–0.31
Median	0.00	–0.23	–0.13	0.00
Minimum	0.00	0.00	–0.25	–0.38
Maximum	0.00	0.81	–0.25	0.17
1986–2009				
Tw statistic	Winter (Dec–Feb)	Spring (Mar–May)	Summer (Jun–Aug)	Autumn (Sep–Nov)
Mean	0.21	–0.08	–0.25	0.85**
Median	0.13	0.30	–0.04	0.83
Minimum	0.00	0.42	0.27	0.85
Maximum	–0.23	–0.91	–1.32**	0.47

* data significant at p<0.05; ** data significant at p<0.01.

on the other hand, a decreasing trend in maximum summer Tw (–1.32 °C per decade, strong evidence) was followed by an increase in mean autumn Tw (0.85 °C per decade, strong evidence) for 1986–2009.

4 Discussion

On the basis of an exemplary mountain river with a long-term (50-year) record of temperature measurements (for the period 1960–2009), the obtained results suggest that water temperature tends to be progressively higher during the spring (maximum values of Tw shown for the upstream site, Fig. 4a), as well as during the summer (mean, median, and maximum values of Tw shown for the upstream site, Figs. 5a–c). Such increasing tendencies are consistent with the increasing trends found in spring and summer air temperature (Table 3) for 1962–2009 at both studied sites, and for spring and summer Tw for 1986–2009 at the upstream site (Table 4). Increasing trends in spring Tw are in line with results obtained for a 30-km^2 catchment in Scotland over the period 1968–1997 [40]. Moreover, increasing summer Tw can also be associated with diminished summer river flow in the Raba catchment for 1985–2014 [41]. If these tendencies continue, the created thermal conditions will likely be unsuitable for native, coldwater fish (e.g. salmonids) inhabiting the Raba River [42]. Conversely, increasing trends in spring and summer Tw are absent for the downstream site (approx. 39 km from the DR dam with hypolimnetic releases), and even decreasing trends are present for 1960–2009 (maximum values, Fig. 4b and 5d, respectively) and 1986–2009 (maximum summer values, Table 5). Other differences in seasonal Tw at the upstream and downstream sites exist as well, including significant increases in autumn and winter Tw (mean, median, and minimum values, Figs. 2, 3) for 1960–2009, as well as for 1986–2009 (mean autumn Tw, Table 5) at the downstream site only. While increasing trends in winter Tw (mean and median values, Fig. 3), in line with [40], could partly be explained by increasing trends in winter Ta (Table 3), no trends in autumn air temperature were noted for either site (Table 3) for 1962–2009. On the other hand, these discrepancies can be explained by the well-documented impact of dams with hypolimnetic releases, which tend to create warmer conditions in autumn and winter, but cooler conditions in spring and summer [22, 43]. It appears that cooler conditions shaped by the DR dam compensate for an increase in spring and summer Tw downstream.

It is also apparent that the magnitude and/or direction of hydro-climatic trends depend on dynamic interactions between numerous factors acting on river catchments. Specifically, land use and land cover changes (LULCC) are common in the Carpathians due to a major shift in land use after 1990, leading to farm abandonment and forest expansion [44]. According to CLC mapping (data available for the years 1990, 2000, 2006, and 2012), the same direction of LULCC occurred in the Raba catchment over the period 1990–2012. For the Stróża and Proszówki subcatchments, increases in the surface area of developed land (4.0 and 4.2%, respectively) and forest cover (4.4 and 3.3%, respectively) occurred at the expense of agricultural land; both change factors are of a similar magnitude, but detailed analyses are needed in order to quantify their effects on water and air temperatures as well as river flow.

On the whole, the identified increasing trends in spring and summer water temperature should facilitate adaptive management of the thermal regime of the Raba River upstream from the DR dam in order to mitigate adverse effects of climate warming on this natural fluvial environment. In addition, important differences in the direction of trends in water temperature for the upstream and downstream sites indicate the usefulness of the 39-km downstream river section of the DR dam for coldwater trout fishery production [28].

5 Conclusions

In the herein described analysis pertaining to changing thermal conditions in a mountain river, long-term (50-year) trends in river water temperature for each season of the year (1960–2009) were found. The direction of these trends differs for sites located some distance upstream and downstream from a relatively large manmade reservoir with hypolimnetic water releases. For the upstream site, Tw tends to be progressively higher in spring (maximum values) and summer (mean, median, and maximum values), in concordance with seasonal air temperature trends. This implies the need for appropriate flowing water management that would mitigate adverse effects of climate warming on the fluvial environment.

For the downstream site, Tw tends to be higher in autumn and winter (mean, median, and minimum values), but lower in spring and summer (maximum values). The identified differences point to an anthropogenic impact on river water temperature downstream from the Dobczyce Reservoir. On the other hand, this creates opportunities for using the 39-km river reach below the DR dam as a natural trout fishery, provided that the DR dam is managed adaptively for changing climate conditions.

This work was supported by the Polish Ministry of Science and Higher Education (grant no. Ś1/194/2017/DS task 4).

References

1. J.D. Allan, M.M. Castillo, *Stream ecology. Structure and function of running waters* (2nd ed., Springer, 2007)
2. J.E. Edinger, D.W. Duttweiler, J.C. Geyer, Water Resour. Res. **4**, 1137–1145 (1968)
3. B.A. Sinokrot, H.G. Stefan, Water Resour. Res. **29**, 2299–2312 (1993)
4. G.C. Poole, C.H. Berman, Environ. Manage. **27**, 787–802 (2001)
5. J.D. Meisner, Trans. Am. Fish. Soc. **119**, 282–291 (1990)
6. B.A. Sinokrot, H.G. Stefan, J.H. McCormick, J.G. Eaton, Clim. Change **30**, 181–200 (1995)

7. R.D. Ries, S.A. Perry, Clim. Res. **5**, 197–206 (1995)
8. B.W. Webb, Hydrol. Processes **10**, 205–226 (1996)
9. O. Mohseni, T.R. Erickson, H.G. Stefan, Water Resour. Res. **35**, 3723–3733 (1999)
10. O. Mohseni, H.G. Stefan, J.G. Eaton, Clim. Change **59**, 389–409 (2003)
11. M. Daufresne, M.C. Roger, H. Capra, N. Lamouroux, Glob. Change Biol. **10**, 123–140 (2004)
12. R.E. Hari, D.M. Livingstone, R. Siber, P. Burkhardt-Holm, H. Güttinger, Glob. Change Biol. **12**, 10–26 (2006)
13. A.C. Johnson, M.C. Acreman, M.J. Dunbar, S.W. Feist, A.M. Giacomello, R.E. Gozlan, S.A. Hinsley, A.T. Ibbotson, H.P. Jarvie, J.I. Jones, M. Longshaw, S.C. Maberly, T.J. Marsh, C. Neal, J.R. Newman, M.A. Nunn, R.W. Pickup, N.S. Reynard, C.A. Sullivan, J.P. Sumpter, R.J. Williams, Sci. Total Environ. **407**, 4787–4798 (2009)
14. D.L. Hartmann, A.M.G. Klein Tank, M. Rusticucci, L.V. Alexander, S. Brönnimann, Y. Charabi, F.J. Dentener, E.J. Dlugokencky, D.R. Easterling, A. Kaplan, B.J. Soden, P.W. Thorne, M. Wild, P.M. Zhai, *Observations: Atmosphere and Surface*, in: *Climate Change 2013: The Physical Science Basis. Contribution of Working Group I to the Fifth Assessment Report of the Intergovernmental Panel on Climate Change* (Cambridge University Press, 2013)
15. K.E. Trenberth, Clim. Res. **47**, 123–138 (2011)
16. J.J. Magnuson, L.B. Crowder, P.A. Medvick, Am. Zool. **19**, 331–343 (1979)
17. C.C. Coutant, Environ. Biol. Fishes **18**, 161–172 (1987)
18. G.W. Brown, J.T. Krygier, Water Resour. Res. **6**, 1133–1139 (1970)
19. N.H. Ringler, J.D. Hall, Trans. Am. Fish. Soc. **104**, 111–121 (1975)
20. B.W. Webb, D.E. Walling, Appl. Geogr. **16**, 211–227 (1996)
21. C.L. Lowney, Water Resour. Res. **36**, 29473–2955 (2000)
22. N.L. Poff, D.D. Hart, Bioscience **52**, 659–668 (2002)
23. J.L. Lessard, D.B. Hayes, River Res. Appl. **19**, 721–732 (2003)
24. D. Caissie, Freshw. Biol. **51**, 1389–1406 (2006)
25. M. Kędra, *Reservoir-induced changes in dynamics and synchrony of river water temperatures revealed by RQA and CRQA*, in: *Recurrence plots and their quantifications: expanding horizons* (Springer Proceedings in Physics, 180, Springer, pp. 289–300, 2016)
26. M. Kędra, Ł. Wiejaczka, Water Environ. J. **30**, 31–39 (2016)
27. J.D. Olden, R.J. Naiman, Freshw. Biol. **55**, 86–107 (2010)
28. C.W. Krause, T.J. Newcomb, J.D. Orth, River Res. Applic. **21**, 581–593 (2005)
29. W. Chełmicki, *Location, classification and characteristics of the basin*, in: *The Upper Vistula Basin, part I* (In Polish, PWN Warszawa-Kraków, pp. 15–29, 1991)
30. S. Kurek, A. Misztal, J. Pawlik-Dobrowolski, *Natural environment as a factor controlling runoff and chemical composition of water in the Raba catchment*, in: *The Raba catchment as an area of recharge of water and pollutants for the reservoir in Dobczyce* (In Polish, Politechnika Krakowska, Monogr. 145, pp. 13–31, 1993)
31. CLC 2012, http://land.copernicus.eu/pan-european/corine-land-cover (accessed 4 May 2017)
32. J. Hennig, I. Hennig, A. Roszkowski, *Retention reservoirs*, in: *The Upper Vistula Basin, part II* (In Polish, PWN, Warszawa-Kraków, pp. 121–143, 1991)
33. S. Szalai, I. Auer, J. Hiebl, J. Milkovich, T. Radim, P. Stepanek, P. Zahradnicek, Z. Bihari, M. Lakatos, T. Szentimrey, D. Limanówka, P. Kilar, S. Cheval, Gy. Deak, D. Mihic, I. Antolovic, P. Nejedlik, P. Stastny, K. Mikulova, I. Nabyvanets, O. Skyryk, S. Krakovskaya, *Climate of the Greater Carpathian Region* (Final Technical Report, 2013, www.carpatclim-eu.org)
34. H.B. Mann, Econometrica **13**, 245–259 (1945)
35. M.G. Kendall, *Rank correlation methods* (4th edn., Charles Griffin, London, 1975)
36. D.R. Helsel, R.M. Hirsch, *Statistical methods in water resources. Techniques of water-resources investigations* (Book 4, U.S. Geological Survey, 2002) https://pubs.usgs.gov/twri/twri4a3/ (accessed 17 July 2017)
37. H. Theil, Proc. R. Neth. Acad. Arts Sci. **53**, 386–392, 521–525, 1397–1412 (1950)
38. P.K. Sen, J. Am. Stat. Assoc. **63**, 1379–1389 (1968)
39. S. Yue, P. Pilon, B. Phinney, G. Cavadias, Hydrol. Process. **16**, 1807–1829 (2002)
40. S.J. Langan, L. Johnston, M.J. Donaghy, A.F. Youngson, D.W. Hay, C. Soulsby, Sci. Total Environ. **265**, 195–207 (2001)
41. M. Kędra, Clim. Res. **72**, 251–265 (2017)
42. K. Bieniarz, P. Epler, *Ichthyofauna*, in: *The Upper Vistula Basin, part II* (In Polish, PWN Warszawa-Kraków, pp. 69–81, 1991)
43. R. Soja, Ł. Wiejaczka, Water Environ. J. **28**, 473–482 (2014)
44. A. Björnsen Gurung, A. Bokwa, W. Chełmicki, M. Elbakidze M. Hirschmugl, P. Hostert, P. Ibisch, J. Kozak, T. Kuemmerle, E. Matei, K. Ostapowicz, J. Pociask-Karteczka, L. Schmidt, S. van der Linden, M. Zebisch, Mt. Res. Dev. **29**, 282–288 (2009)

Assessment of application of selected waste for production of biogas

Monika Pawlita-Posmyk[1], and *Małgorzata* Wzorek[1,*]

[1]Department of Process Engineering, Faculty of Mechanical Engineering, Opole University of Technology, Opole, Poland

Abstract. Recently, the idea of biogas production has become a popular topic in Poland. Biogas is a valuable source of renewable energy with a potential application in electricity and heat production. Numerous types of technological solutions of biogas production are closely linked to the availability of substrates in the area, as well as their quantity and their properties. The paper presents the assessment of application in biogas production selected wastes such as communal and household sewage sludge and waste from a paper production in Opole region (Poland). The annual productions of methane, biogas and electricity were estimated. Chosen physico-chemical properties important in fermentation process were taken into consideration in the assessment. The highest value of potential energy was obtained using waste from the paper industry but the most appropriate parameters for this process has sewage sludge from the municipal sewage treatment plant. The use of sewage sludge from domestic and municipal sewage and waste from the paper industry creates the opportunity to reduce the amount of waste materials.

1 Introduction

In Poland, the production of biogas derived from waste materials gains in popularity and the number of biogas plants is rising [1-3]. In 2016, in Poland, the register of the President of the Agricultural Market Agency (ARR) included 94 agricultural biogas installations on the list of biogas manufacturers. Their total installed electrical capacity was equal to 101.234 MW (this means a 21% increase in relation to 2015) [4].

The methane fermentation in a biogas plant involves biogas production through a process involving four phases (hydrolysis, acidogenesis, acetogenesis and methanogenesis) [5] on the basis of resources with various volume ratios of hydrocarbons, lipids and proteins [6-7]. The composition of such biogas is different depending on the applied substrates, however, typically it consists mostly of CH_4 (40-70%) and CO_2 (15–60 %) as well as other compounds in small amounts: H_2O (2-7%), N_2 (2-5%), O_2 (0-2%) < 1% H_2, NH_3 (0-1%) and H_2S (0.005–2%) [5, 8-11]). The principal sources of biomass include: agricultural waste, communal solid waste, sewage sludge, forestry waste as well as industrial and animal remains, etc. [12]. According to [13] another interesting solution suitable for biogas production from waste is digestion of poultry manure or litter.

Agriculture forms the principal source of biodegradable waste. On the basis of data as of 2016, the ratios of the resources applied in the production of agricultural biogas are as follows [4]:

- slurry 24.04%,
- fruit and vegetable remains 20.64%,
- distillery stillage 14.81%,
- maize silage 13.62%,
- beet pulp 6.89%,
- sludge from food and agricultural industry 3.88%,
- sludge obtained from dairy production 2.76%,
- sewage 2.67%,
- green forage 1.76%,
- other 8.93%.

These resources were applied in the production of 250 million m^3 of biogas in 2016 (i.e. an increase of 21% in relation to 2015). In total, the production of electricity derived from agricultural biogas was equal to 524.6 GWh [4]. Nevertheless, the production of biogas in an anaerobic digestion is small in relation to the available technical potential [14].

The production of biogas is an important option in managing the existing wastes, as this process applies organic matter, which is difficult in application by other methods. In addition, biogas plants can aid in the utilization of waste produced in a given area, as the waste can be used locally.

Sludge from communal wastewater treatment plants and domestic treatment plants form one group of waste whose application can be potentially problematic. Although communal sludge obtained in wastewater treatment plants has been considered to be an adequate substrate for the production of biogas in large wastewater treatment plants for a long time, some areas, such as Opole province face a problem associated with its utilization.

Communal sewage sludge forms a considerable hazard to the environment due to its large volume and the

* Corresponding author: m.wzorek@po.opole.pl

chemical composition.

In the last years, the dynamic development of domestic sewage treatment plants has led to a considerable increase in the volume of waste being produced in such facilities. On the basis of data obtained from the Central Statistical Office [15,16], the number of domestic sewage treatment plants increased from 181,000 in 2014 to around 203,000 in 2015.

The majorities, i.e. almost 85% of wastewater treatment facilities applied in individual households are located in rural areas (around 84% of underground containers without drainage systems and 92% of domestic sewage treatment plants). The variability of the physico-chemical properties of waste from domestic sewage treatment plants leads to the refusal to accept this type of waste by the communal wastewater plants due to the potential destabilizing effect of this waste in the normal treatment process. Hence, other measures are sought to further process this type of waste.

Waste obtained from the paper industry forms another group of waste posing a problem with its utilization. A large volume of waste in the form biomass originates from the processes in the cellulose and paper industry and its storage poses a hazard to the environment. The problems are associated with the high moisture content in this type waste, which is coupled with the large volume and variability in its composition. The methods applied to process this waste are often too expensive and in some cases their impact on the environment can be problematic. In 2011, the overall production of this waste was equal to 932,900 tons, and in 2013, the production and processing of cellulose mass, paper and cardboard produced 1,085,000 tons of waste [19].

The objective of this paper is to assess the potential for the use of substrates formed by waste derived from communal and household sewage treatment plant and waste from a paper production plant applied in the production of biogas. The use of these substrates in the methane fermentation process could lead to the solution of a problem associated with the utilization of waste in the area of the Opole province (Poland).

2 Materials and methods

2.1 Substrates

The assessment of the properties that are important in the context of the application in biogas production involved a study involving four types of waste derived from facilities located in the Opole province.

The sludge that was identified for the study came from two domestic sewage treatment plants (DSS I, DSS II), sewage sludge from a communal wastewater treatment plant (SS) and post-processing waste from paper production (PW).

The assessment of the properties of sludge obtained from domestic wastewater treatment applied from two facilities. The first of them is a domestic sewage treatment plant comprising a dual biological zone with a fluidized bed. This facility consists of an initial settling tank and two independent zones applied for an aerobic conditioning and a secondary tank. This plant is installed as an onsite facility for a single-family house occupied by a family of four and has been constantly applied for a period of 4 years. The other sewage treatment plant applied in the research comprises a two-chamber tank coupled with a drainage system. A single family of three uses this facility. It has been in use for 5 years.

Another type of the waste used in this study was formed by communal sewage sludge originating from a mechanical-biological wastewater treatment plant capable of increased treatment of biogens (WWTP). This plant was designed for a load of 225,000 population equivalent (PE) of waste, i.e. 45 000 m^3 per day.

The annual volume of waste conditioned by the plant is equal to around 96,966,000 m^3 per year (including 12% of wastewater from industry) [18]. The sewage sludge samples were extracted from waste mixing chamber (i.e. sludge was obtained from the compactors that were dehydrated in the compactor and the initial settling tank). The post-processing waste from the paper industry made up the third type of the waste. The waste samples were extracted following the dehydration process (i.e. the composition of the waste was 40% dry mass) from the prisms where it is stored prior to its treatment or disposal.

2.2 Analysis methods

The sewage sludge and the waste from the paper industry were extracted and analyzed in accordance with the standards in force. The analysis involved the list of the following indicators:

- water content in accordance with PN-EN 12880:2004,
- pH in accordance with PN-EN 12176:2004,
- dry mass in accordance with PN-EN 12879:2004,
- total phosphorus in accordance with PN-EN 13346:2002, PN-EN ISO 11885:2009,
- total Kjeldahl nitrogen in accordance with PN-EN 13342:2002,
- ammoniacal nitrogen, in accordance with PN-75/C-04576/15,
- heavy metal concentration: Cd, Cr, Cu, Ni, Pb, Zn, in accordance with PN-EN 13346:2002, PN-EN ISO 11885:2009 and Hg in accordance with PN-EN 13346:2002, PN-EN 1483:2007,
- magnesium in accordance with PN-EN 13346:2002, PN-EN ISO 11885: 2009,
- total organic carbon (TOC) in accordance with PB/FCH/17/ B:10.04. 2013.

3 Results and discussion

The analysis was initiated by assessing the total volume of the production of the waste in the area of the Opole province as well as in Poland with the purpose of determining the potential for biogas production from the waste.

The rough calculations were performed on the basis of the following relations:

- methane production (MP) [19]

$$MP = AW \times DM \times OM \times PCH_4 \quad (1)$$

- biogas production (BP) [19]

$$BP = MP \times \%CH_4 \quad (2)$$

- electricity production (EP) [20]:

$$EP = MP \times GCV_{CH4} \times \eta_{EL} \quad (3)$$

where: AW – amount of waste production per year, DM – dry mass content, OM – organic matter content in the waste, PCH_4 – potential for methane production (on the basis of an assumption that 1 ton of waste gives 100 m^3 of biogas), $\%CH_4$ methane content in the biogas (on the basis of an assumption it is 60%), GCV_{CH4} – calorific value of methane (9.17 kWh/m^3) η_{EL} – efficiency of electricity production (35% [20]).

Table 1 contains a summary of the input data taken for calculations and the estimated production of biogas and electricity.

Table 1. Production of waste, methane, biogas and electricity from selected waste substrates.

	Unit	DSS[1]	DSS[2]	SS[1]	SS[2]	PW[1]	PW[2]
Dry mass (DM)	% d.m.	2.65		4.00		33.5	
Organic matter (OM)	% d.m.	83.4		75.4		31.3	
Amount of waste (AW)	t/year	467,712[3]	11.520[3]	555,982 (d.m.)	16,487 (d.m.)	1,085,500	47.207
Biogas production (BP)	m^3/year	1,722,817.15	42,433.92	2,794,736.2	82,874.65	18,970,017.08	824,981.65
Methane production (MP)	m^3/year	1,033,690.29	25,460.35	1,676,841.7	49,724.79	11,382,010.25	494,988.99
	m^3/h	118.00	2.91	191.42	5.68	1,299.32	56.51
Electricity production (EP)	MWh/ year	3.317.63	8,171.45	5,381.82	15,959.30	36,530.56	1,588.67

[1] – on the national scale, [2] – on the scale of the Opole province, [3] – calculations under the assumption that there are 203,000 domestic sewage treatment plants in Poland [15]; 5,000 – number of domestic sewage treatment plants in the Opole province; of which sewage sludge forms about 1-2% by volume; 2.4 m^3/person· month- standardized water use per inhabitant in a household [21]; households comprising a domestic sewage treatment plant consists of 4 people.

The waste from the paper industry (PW) has the greatest potential for biogas production. This is due to the amount of the waste generated both locally and on the national scale. The potential for biogas production of the waste originating from the communal wastewater treatment plant and domestic sewage plants can give the production of 8,699.45MWh/year of electricity on the national scale and 24,130 MWh/year on the scale of the Opole province. If we were to use all available waste for biogas production, we can reduce its amount by 2,109,194 tons per year on the national scale and by 75,214 tons per year on the scale of the Opole province. The applied waste can yield 23,487,570.42 m^3 of biogas per year on the national scale and 950,290.22 m^3 per year on the scale of the Opole province.

We need to note that the waste applied in the methane fermentation must be characterized by adequate properties. The results of the analysis involving physico-chemical properties of the waste are presented in Table 2 and the data is compared with the results with regard to the most common substrates used in biogas production.

One of the parameters having a considerable impact on the course of the methane fermentation, and in particular on the stability of the process is associated with the maintenance of waste is pH. The optimum value of pH should be in the range from 6.5 - 8.0, which guarantees the beneficial environment for the growth of anaerobic bacteria [26]. An increase of pH above 8.5 makes the environment toxic for them [27-28]. The examined waste had a pH values that were slightly acid, (DSS I, DSS II, SS), and even neutral one (PW).

All waste materials taken for comparison purposes demonstrated a high water content (i.e. from 66.5% to 98.5% with a note that the greatest value was recorded for household sewage treatment plant DSS II). The high water content is important in the aspect of the digestion process conducted by bacteria, as well as for their longevity and for obtaining gas that is rich in CH_4.

Matheri et al. [28] recommend that the moisture content in the process of methane fermentation should be equal to 90%.

One of the key parameters in terms of the efficiency of biogas production is associated with the high content of the organic matter in the waste. It determines the course of the fermentation process and the volume of the biogas. During the course of the microbiological processes, the amount of organic matter in the waste decreases in time and, as a consequence, a decrease of the amount of methane production occurs, and the profitability of the process drops with the smaller potential for application for energy production. Various substances undergo disintegration at a different pace.

Among the analyzed materials, the waste from the paper industry demonstrates the lowest content of organic substances (31.3% d.m.). This value is also two times lower from the water content in both the domestic sewage plant and communal wastewater treatment plants. The study did not also establish considerable differences between the properties of the sludge from various domestic sewage treatment plants. The total phosphorus concentration is marginally present in the analyzed waste. In the sludge from WWTP the concentration of phosphorus is at a level of 1.81% d.m., whereas for DSS I and DSS II the values are equal to 0.586% d.m. and 0.504% d.m., respectively (i.e. a

Table 2. Physico-chemical properties of waste and other substrate tested by the authors.

Parameter		Unit	Waste				Cattle slurry [25,26]	Cattle manure [25-27]	Maize silage [26,28]
			DSS I	DSS II	SS	PW			
Water content		%	96.2	98.5	96.0	66.5	88.0–95.0	67.0–87.0	-
pH		-	6.8	6.2	6.3	7.0	-	7.7	-
Content of organic substances		% d.m.	84.4	82.4	75.4	31.3	75.0–85.0	68.0–76.0	95.6
Total phosphorus		% d.m.	0.586	0.504	1.81	0.058	-	-	0.24
Total nitrogen		% d.m.	4.47	3.44	5.70	0.47	2.6-6.7	1.1-3.4	1.76
Ammoniacal nitrogen		% d.m.	0.35	<0.01	0.36	0.06	-	-	0.16
Magnesium		% d.m.	0.18	0.27	0.33	0.35	0.3-0.7	1.3	0.12
TOC		% d.m.	55.0	50.1	44.9	10.0	-	-	-
C:N		-	12.3:1	14.6:1	7.9:1	21.3:1	6–20:1	11–30:1	32.7:1
Heavy metals	Cd	mg/kg d.m.	0.759	0.910	0.827	0.186	0.30	0.29	0.07-0.20
	Cr	mg/kg d.m.	8.83	21.10	48.20	12.30	7.30	12.90	0.50-0.78
	Cu	mg/kg d.m.	360.0	136.0	107.0	76.4	44.5	39.0	2.59-5.0
	Ni	mg/kg d.m.	9.13	9.13	18.90	4.30	5.90	5.20	0.30-5.00
	Pb	mg/kg d.m.	18.9	20.8	20.0	21.5	7.7	30.0	0,7-2.0
	Zn	mg/kg d.m.	1161	614	1078	98.7	270	190	15.1-56
	Hg	mg/kg d.m.	<0.05	0.07	0.42	<0.05	0.06	0.03	0.01

d.m. – dry mass

difference of 0.082 % d.m.).

The content of the total phosphorus in the waste from the paper industry is lower in comparison to the remaining results (i.e. over 10 times lower in relation to the sludge from the domestic sewage treatment plants (DSS I and II), and 31 times lower in comparison to the sludge from the communal wastewater treatment plant (in comparison to the result recorded for maize silage).

The total content of the organic carbon forms an important consideration, as its content in the substrate affects the activity of the bacteria.

The sludge from the domestic sewage treatment plant DSS I had 5 times greater content of the organic carbon from the waste from the paper industry (10 % d.m.) and was characterized by the smallest concentration of the total nitrogen (0.47 % d.m.) among the analyzed waste. The sludge from the communal wastewater treatment plant also demonstrated over 4 times greater TOC content in comparison to the waste from the paper industry. The sludge from both domestic sewage treatment plants had a similar content of the organic carbon (a difference of 4.9 % d.m. was noted). The content of the total nitrogen in the examined waste was in the range from 0.47 to 6.70 % d.m. The lowest concentration of the ammoniacal nitrogen was registered in the waste from the domestic sewage treatment plant DSS II, whereas the concentration of the ammoniacal nitrogen in DSS I and SS was almost at the same level. An optimum ratio of carbon, nitrogen and phosphorus (C:N:P) of 100:3:1 is an important condition for the gaining [29-30] a high efficiency of CH_4 production. In order to ensure that the anaerobic fermentation is performed in the conditions of an optimum C:N ratio, the biodegradable material should demonstrate a high C:N ratio [25,31]. The insights from the literature offer various ranges of the C:N ratio to achieve the adequate course of the methane fermentation process. Abbasi et al. [32] give that this ratio should be 20-30 :1, whereas the work [24,33] describes an experiment conducted in the conditions when the C:N ratio was 15-30:1. However, these authors found that microorganisms should have a ration of C:N 20-30:1 to ensure their optimum growth [24,31]. Another range of the C:N that was found to be optimum for the composition of the input for biogas production is equal to 10-25:1. The ratio of C:N in the waste, including DSS I, DSS II, PW, cattle slurry and cattle manure was adequate in this respect, whereas for SS it was equal to 7.9:1, which is far below the optimum value for the anaerobic process. For the case when the C:N ratio is too low (i.e. <10:1), nitrogen is released in the form of annonia, and this leads to the increase of the environment pH [25].

Another important condition, which affects the course of biogas production, is associated with the heavy metal content in the waste. It is necessary to reduce the amount of heavy metals in the substrates to a minimum by applying a process of selection of substrates used in the fermentation process.

However, not all metals have a negative impact on this process. The growth of microorganisms requires an adequate content of magnesium and trace elements such as copper and zinc [34].

The content of magnesium has a particular impact on the growth rate of microorganisms and this impact is similar to the role of basic nutrients. While the moderate concentrations stimulate the growth of the bacteria, excessive concentrations reduce this growth and even greater can play a considerably inhibitory effect on the process and even lead to toxicity [35]. The content of magnesium in the range from 1000 to 1500 mg/l can have a moderately slowing effect, whereas the range starting from 3000 mg/l can stop the growth of microorganisms altogether [36]. The content of Mg in the analyzed waste was only marginal, as we recorded 0.12-1.3 mg/kg d.m.

The waste applied for the testing has a various content of heavy metals. From the analysis of heavy metals, we learned that DSS I and II have a low content of heavy metals, just as for the case reported with regard to another domestic sewage treatment plant [37]. Such small volumes should not considerably affect the fermentation process. Heavy metals can occur in higher concentrations in communal waste and sludge [38, 39,40]. The diversity of the substrates, bacteria strains and environmental factors offer the explanation to the variability of the heavy metal concentrations, as well as their relative toxicity [35,41]. The inhibitory effect accompanied by the reduction of the production of biogas was reported for the cases when: Cu-40mg/l, Ni-10mg/l, Cd-70mg/l, Pb-340mg/l, Zn-400mg/l, Cr^{3+}-130mg/l, Cr^{6+}-110mg/l [36]. The concentration of heavy metals in the waste from the paper industry is similar to the result gained for the cattle manure, yet with a note that the latter contains more Cu. The tested waste, have a greater Zn concentration (614-1161 mg/kg d.m.) compared to the cattle slurry, cattle manure and maize silage (15.1-270 mg/kg d.m.). Among all analyzed types of waste, we learned that DSS I, DSS II, SS and PW were characterized by the highest values of Cu concentration (76.4-360 mg/kg d.m.). The concentration of Cr in DSS II (21.10 mg/kg d.m.) and SS (48.20 mg/kg d.m.) was higher in comparison to other types of waste (0.50-12.90 mg/kg d.m.). In addition, SS had a considerable level of Ni content (18.90 mg/kg d.m.).

4 Conclusions

Poland is a country with a considerable potential for biogas production, which can be applied in the methane fermentation process, among others based on waste. This process can involve the use of a wide range of organic waste, which is considered as noxious waste in terms of its utilization potential.

The examined waste is characterized by a considerable potential for use in biogas production. In particular, it can be applied jointly in co-fermentation with other substrates. It can form a cheap and easily accessible substrate for the production of biogas in the facilities that are already in operation.

The tested sewage sludge derived from domestic sewage treatment plants and a communal wastewater treatment plant demonstrates adequate physico-chemical properties for this purpose. This in particular refers to the adequate content of organic parts, TOC and nitrogen concentration. In addition, the concentration of heavy metals was low, which suggests that the amount available in the substrate should not have a negative impact on the fermentation process.

The application of sewage sludge in the production of biogas obtained from domestic sewage treatment plants, communal wastewater treatment plants and waste from the paper industry can offer an opportunity to solve the existing problem associated with utilization of this type of waste in the Opole province. The use of biogas for energy production can have a positive effect on the overall condition of the environment as well on the energy balance on the scale of the domestic economy. A considerable advantage associated with the use of selected substrates is associated with their accessibility, low cost of acquiring it and providing a solution to its utilization.

References

1. E.S. Energimyndigheten, ES **3** (2015)
2. M. Svensson, D. Baxter. IEA Bioenergy Task 37 (2015)
3. R. Feiz, J. Ammenberg, Resour. Conserv. Recycl. **122** (2017)
4. Agricultural Market Agency, www.arr.gov.pl, (18.06.2017)
5. A. Fontana, V. Patrone, E. Puglisi, L. Morelli, D. Bassi, M. Garuti, L. Rossi, F. Cappa, Bioresour. Technol. **218** (2016)
6. I. Angelidaki, D. Karakashev, D. J. Batstone, C. M. Plugge, A. J. M. Stams, *Methods in Methane Metabolism*, (Part A. Academic Press, 2011)
7. J. Ammenberg, R. Feiz, Resour. Conserv. Recycl. **122** (2017)
8. M. Franco-Morgado, C. Alcántara, A. Noyola, R. Muñoz, A. González-Sánchez, Sci. Total Environ. **592** (2017)
9. A. Akbulut, Energy **44** (2012)
10. Y. Shen, J. L. Linville, M. Urgun-Demirtas, M. Mintz, S. W. Snyder, Renew. Sust. Energ. Rev. **50** 346–362 (2015)
11. A.S. Mehr, M. Gandiglio, M. Mosayeb Nezhad, A. Lanzini, D.M.S. Mahmoudi, M. Yari, M. Santarelli, Appl. Energ. **191** (2017)
12. Y. Santosh, T.R. Sreekrishnan, Koli S., Rana V. Bioresource Technol. **95**, 1 (2004)
13. M. Tańczuk, R. Junga, S. Werle, M. Chabiński, Ł. Ziółkowski, Renewable Energy (2017) http://dx.doi.org/10.1016/j.renene.2017.05.074
14. M. Börjesson, E. Ahlgren, Energy **48** (2012)
15. Central Statistical Office, Warsaw 2016 www.stat.gov.pl (18.06.2017)
16. Central Statistical Office, Warsaw 2015 www.stat.gov.pl (18.06.2017)
17. National Waste Management Plan - update 2014, Warsaw (2015) [in Polish]

18. Statement of WIOŚ Opole nb. 1/W/2016, Opole 2016 [in Polish]
19. J. J. Chávez-Fuentes, A. Capobianco, J. Barbušová, M. Hutňan, Waste Biomass Valorization, **8**,5 (2017)
20. M. Tańczuk, W. Kostowski, M. Karaś, Energy Convers. Manage. **125** (2016)
21. Ordinance of the Minister of Infrastructure of January 14, 2002 on the definition of average standards of water consumption, (J. Act No 8, item 70)
22. S.O. Dahunsi, S. Oranusi, V.E. Efeovbokhan, J. Cleaner Prod. **156** (2017)
23. R. Kigozi, AO Aboyade, E. Muzenda, *Proceedings of the World Congress on Engineering and Computer Science* (2014)
24. A.N. Matheri, S.N. Ndiweni, M. Belaid, E. Muzenda, R. Hubert, Renewable Sustainable Energy Rev. **80** (2017)
25. A. Jędrczak, *Biological waste processing* (PWN, Warsaw, 2008) [in Polish]
26. www.agengpol.pl (25.07.2017)
27. L. Zhu, Ch. Yan, Z. Li, Bioresour. Technol. **220** (2016)
28. M. Fugol, H. Prask, Inż. Rolnicza **9** (2011)
29. R. Conrad, F. Bak, H.J. Seitz, B. Thebrath, H.P. Mayer, H. Schütz, FEMS Microbiol. Lett. **62**, 5 (1989)
30. E. Kovács, R. Wirth, G. Maróti, Z. Bagi, K. Nagy, J. Minárovits, G. Rákhely, K. L. Kovács, Bioresour. Technol. **178** (2015)
31. A.H. Igoni, M.J. Ayotamuno, C.L. Eze, S.O.T. Ogaji, S.D. Probert, Appl. Energy **85**, 6 (2008)
32. T. Abbasi, D.M. Tauseef, S.A. Abbasi, *Biogas Energy* (Springer, New York, 2012)
33. L. Deressa, S. Libsu, R.B. Chavan, D. Manaye, A. Dabassa, Environ. Ecol. Res. **3** (2015)
34. M. Bożym, I. Florczak, P. Zdanowska, J. Wojdalski, M. Klimkiewicz, Renew. Energy **77** (2015)
35. Y. Chen, J. J. Cheng, K. S. Creamer, Bioresource Technol. **99**, 4044–4064 (2008)
36. G. D. Zupančič, V. Grilc, *Managemet of Organic Waste,* (InTech, 2012)
37. M. Pawlita-Posmyk, M. Wzorek, Chemik, **10** 623-625 (2016)
38. A. Król, K. Mizerna, Chemik, **10** (2015)
39. M. Wzorek, Ecol. Chem. Eng. S, **19**, 4 (2012), doi: https://doi.org/10.2478/v10216-011-0044-5
40. M. Wzorek, Waste Manage. Res. **32**, 5 (2014), doi.org/10.1177/0734242X14529618
41. E. Abdelsalam, M. Samer, Y.A. Attia, M.A. Abdel-Hadi, H.E. Hassan, Y. Badr, Renewable Energy **87** (2016)

The study of heavy metals leaching from waste foundry sands using a one-step extraction

Marta Bożym[1,*]

[1]Opole University of Technology, Faculty of Mechanical Engineering, 76 Prószkowska Street, 45-758 Opole, Poland

Abstract. There are a number of leaching test which are used to evaluate the effect of foundry waste disposal on the environment (TCLP, SPLP, ASTM at al.). Because the spent foundry sand are usually deposited at the ground level and they have a similar structure to the soil, survey mobility of metals using the same methods seems appropriate. One-step extraction allows for the evaluation of the mobility and bioavailability of metals in soil and waste. Waste foundry sands have been successfully used as a component in manufactured soils in U.S., but concern over metal contamination must be eliminated before considering this direction of use. The study evaluated the leaching of heavy metals (Cd, Pb, Cu, Zn, Cr, Ni) from deposited waste foundry sands. The overall, as well as heavy metals were extracted by different type of extractants: H_2O, CH_3COOH, HCl, EDTA, $MgCl_2$ and NaCOOH. These extractants are most commonly used to study the mobility and bioavailability of metals in soil and waste. In the present study applicable standards and methodology described in the literature in analysis were used. The results allowed to evaluate the bioavailability of metals leached from those wastes.

1 Introduction

After Poland's accession into the European Union, it took an obligation upon itself to pass and implement the environmental laws of the EU. With regard to industrial activity, *Council Directive 96/61/EC* was passed on industrial emissions (*Integrated Pollution Prevention and Control - IPPC*) [1,2]. One of the provisions of this directive stipulates the necessity of limiting the volume of waste production in industry. The implementation of this directive in foundries needs to involve waste regeneration to the greatest extent that is possible. The list of the solid waste from foundries includes molding and core masses, dusts and slimes from gas treatment activities, smelter slags, waste refractory materials and spent grinding materials.

The greatest ratio of the waste is formed by foundry sands, which are commonly further reused after their regeneration [3,4]. The waste foundry and core sands can be successfully applied in other industrial branches as well. Among others, they can be used in road building and production of asphalt, as well as in building engineering in the production of concrete mortars. In addition, they are applied in the production of ceramic produce in the place of sand. An interesting solution which is applied in the US is associated with the application of waste foundry sands in gardening [5,6]. Another solution is postulated by the Finns. Waste foundry sands are applied in composting stations there. However, regardless of the actual application, all waste needs to be tested with regard to leachability parameters prior to their reuse. The assessment of heavy metal leaching usually applies immersion in water extracts. In addition, an adequate assessment of the impact of the waste on the environment can apply dedicated extraction methods, such as ASTM, TCLP, SPLP along with others [7-11]. For the case when molding sands are applied in agriculture and gardening, it is recommended to conduct a study of the soil as well. This is done with the purpose of determining the mobility of heavy metals and potential toxic impact of waste on plant and soil organisms [12-14]. In 2007, US EPA confirmed that benefits resulting from the application of waste foundry sands in agriculture and gardening as well as in geotechnical applications. The advantages of this application include the protection of natural resources as a result of reducing the demand for the raw materials, saving of energy and reduction of greenhouse gas emissions as a result of the reduction of mining activities and decrease of storage costs. Lindsay and Logan [15] state that the application of waste foundry sands in agriculture needs to be supported by research, which has to confirm their applicability for the designated purpose and the ones which provide means of monitoring their impact on the environment. Up to this date, there are no long-term studies into long-standing use of waste foundry sands in agriculture and into the ecological outcomes of this application. Dayton et al. [16] explain that the purpose of assessing the impact of the waste foundry sands in agriculture need to involve studies into leaching of the pollutants and its impact on the plants. It is often emphasized that waste foundry sands can be applied in the environmental after the materials comprise natural binders, e.g. bentonite or materials following

* Corresponding author: m.bozym@po.opole.pl

their thermal regeneration, which involves a thermal decomposition of the organic binders.

The assessment of heavy metal mobility in environmental samples applied extraction techniques. Such means of analysis provide information on the ratio of forms with various bioavailability, which is later converted into the total content. This proves useful for instance in the conditions when assessment of metal ion migration from soil into plant is undertaken. The assessment of bioavailability of microelements in solid samples applies leaching methods by application of various extracting agents. The mobility and bioavailability of metals is relative to their physical form and type of binding with the matrix. There are various methods applied for this purpose. The methods involving chemical metal leaching can be classified into single- and multi-stage. The advantage of the single-stage methods are associated with their short duration and the capability to obtain the results quickly. In contrast, the multi-stage (speciation) methods apply techniques of separation of the individual metals form in a given matrix through a number of stages. The advantage of these methods involves a precise manner of assessing the chemical forms of metals and the ability to easily predict their bioavailability. The drawbacks are connected with the time-consuming process and the necessity to use a number of extraction agents.

Heavy metals and other elements of the environment are bound to the main fractions in soil. In a paper dating back from 1979, Tessier et al. [17] identified 5 fractions. Fraction 1 is formed by metals bound in an active form (soluble fraction) and exchangeably bound metals (exchangeable fraction). This is a fraction comprising metals with the greatest mobility and bioavailability. Fraction 2 is formed by metals bound to carbonates, phosphates and sulfates (carbonate fraction). Metals in this fraction are potentially mobile and bioavailable in certain environmental conditions, usually in the conditions accompanied by a low pH. Fraction 3 is formed by metals bound to Fe and Mn oxides. These structures are unstable in the condition marked by a low oxidation and reduction potential (Eh), and the reduction of the Mn and Fe oxides under anaerobic conditions and their further decomposition can lead to the adsorption of heavy metals. Fraction 4 is formed by metals bound to organic matter and sulfides (oxidizable fraction). The metals bound to this fraction can be released as a consequence of mineralization in the conditions of an extremely oxidized environment. The last one, i.e. residual fraction is made up by metals bound to the mineral matrix, and this fraction is virtually unavailable for plants. The literature of the area contains a number of further modifications of the Tessier's method [17]. Due to a variety of procedures involving metal extraction, there is an issue associated with their harmonization and standardization. As a consequence, a study was initiated in 1987 within the framework of the EC Standards, Measurement and Testing Program (SM&T, formerly BCR). As a result, standardized procedures of metal separation techniques in soils and environmental were introduced and a three-stage fractioning method was developed [18]. However, several European countries apply more simplified techniques used for assessing bioavailability of metals in soils, as they involve just a single stage. For instance, 1M HCl solution is applied for the purpose in assessment of bioavailability in the Chemical and Agricultural Stations in Poland [19-21]. In other countries, this purpose is served by the use of salt solutions with the concentration that is similar to the one in soils. Other solutions include the use of 0.01 M $CaCl_2$ in the Netherlands [22], 0.1 M $NaNO_3$ in Switzerland [23,24], and 1M NH_4NO_3 in Germany [25]. The legal regulations in France provide for the use of the DTPA (NFX31-121) and 1M NH_4CH_3COO buffered solutions [26].

The aim of this study was a research involving waste foundry sands in terms of the leaching of heavy metals from them. The research was undertaken by application of selected extraction agents, which define the mobility and bioavailability of metals present in the waste.

2 Materials & Methods

The research reported here applied waste foundry sands that were derived from waste deposited in an industrial waste landfill located in one of Polish foundries. The waste was extracted from various spots in the landfill in order to test waste stored across various periods in time: 1) mixed waste landfilled in the 1980s utilized in the production of road-building aggregates, 2) waste foundry and core sands landfilled in the 1990s that were not mixed with other types of waste, 3) and 4) mixed molding waste put to landfill throughout 2012 throughout various seasons, 5) mixed foundry waste landfill in the 1980s and remaining there in an unchanged form. The samples were collected from several spots located in the landfill. A laboratory sample with the weight of around 1 kg was separated by the quartering technique. After they were taken to a laboratory, the samples were dried, homogenized, ground in a mortar and passed through a sieve. The samples were subjected to extraction by application of selected extraction agents and the level of heavy metal content was determined in them by application of Flame Atomic Absorption Spectrometry (FAAS) by means of Thermo M6 *SOLAAR* spectrometer. The determination of the total metal content in the waste samples, they were mineralized with aqua regia by application of the Milestone Start D Microwave digestion system. The following extraction agents were applied for the analysis:

- Water (H_2O) in accordance with the procedure provided in the Polish PN-EN 12457-4 standard [27], metals in the forms with greatest mobility are subjected to leaching by this method;
- 0.11M CH_3COOH (0.11M Ac) in accordance with the procedure provided in SM&T [18]: 40 mL of a 0.11 mol/LCH3COOH solution was added to 1 g of dry material and all was shaken for 16 h in room temp. Tis method is applied for leaching of metals in the ion-exchangeable and carbonate forms. Acetic acid is a non-specific acid and can be applied for leaching metals bound to silicates and carbonates;

DOI: 10.1051/e3sconf/20171902018

- 0.1M HCl [26,28], this method is used for leaching metals bound to the majority of mobile fractions, this also includes the ability of leaching selective sulfates and metal silicates to a certain extent. Methodology: dry material was extracted with 0.1M HCl with 1:10 ratio and shacked 1hr in room temp.;
- 1M $MgCl_2$ in accordance with Tessier et al. [17], it is used for leaching ion-exchangeable fractions, however, it does not impact metals bound to alumino-silicates and hydroxides due to the neutral pH of the salts. Methodology: 1g of material was extracted at room temperature for 1hr with 8 ml of MgCl2 solution at pH 7;
- 1M $NaCH_3COO$ (NaAc) with the pH value of 5.0 in accordance with Tessier et al. [17], it is used for leaching metals bound to ion-exchangeable and carbonate fraction, i.e. mobile fractions. Methodology: 1g of material was extracted with 8 ml of 1M sodium acetate at pH 5 for 5 hrs at room temperature;
- 0.05M Na_2EDTA (EDTA) in accordance with McGrath et al. [29]. Methodology: 5g of material was extracted with 50 ml of 0.05 M Na2EDTA solution (pH 7) for 1 hr. This technique is applied for leaching carbonate and organic fractions of metals. It indicates the occurrence of metals in the forms that are best assimilated by plants. For the case of this extraction agent, there is a possibility of isolating metals from organic complexes and insoluble organo-metallic compounds [30].

All the tests that are reported here were repeated at least three times. The purpose of the confirmation of the analytical correctness applied the conducted analysis of certified material „Metals in soil" CRM028-050 Tusnovics.

3 Results and discussion

No considerable differences between samples of waste (No. 1-5) could be observed with regard to heavy metal leaching. Therefore the results were mediated and reported as the mean of the five analyzed samples. The examined waste was characterized by a various heavy metal content. Fig. 1 contains the cumulative information regarding heavy metal concentrations in the examined samples.

The highest concentration of heavy metals in sample no. 5 was analyzed, i.e. sample collected from the waste put to landfill in 1980s and remaining there in an unchanged form, as it was not processed further. It was established that chromium has the highest concentration in this type of waste (264 mg kg^{-1}). A similar level of chromium concentration was recorded in sample no. 3, which is formed by the waste landfilled in 2012. The high concentration of chromium in comparison to other metals was found by the special characteristics of the production process. The plant from which the waste was derived manufactures cast iron and cast steel components. In this case, the high concentration of chromium in comparison to other metals was affected by the special characteristics of the production process.

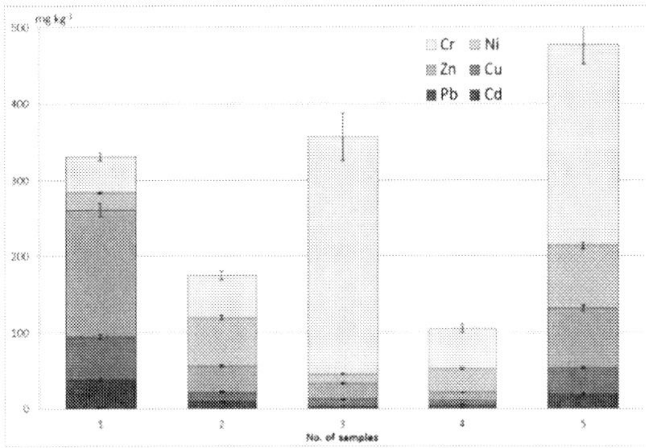

Fig. 1. Total heavy metals concentration in foundry waste samples.

The plant from which the waste was derived manufactures cast iron and cast steel components. Such production often applies additions of other metals, including nickel, zinc and other, which provides the end production with relevant product parameters. Hence, among all metals, the zinc and nickel concentration was increased. The examined waste demonstrated a very small cadmium concentration (0.5-1 mg kg^{-1}). The lowest metal concentration was established in sample no. 4, which is formed by waste landfilled in 2012 and is derived from the most recent production. The differences between the metal concentrations in samples originating from 2012 are noteworthy, in the consideration of the fact that they were put to landfill at different times. This could also be attributable to special characteristics of the production process. This also demonstrates that the composition of the foundry waste varies depending on time and type of production. The composition of from waste foundries is affected by the use of waste foundry sands, ashes, slag, refinery materials and others. Besides, the composition of the cast iron and type of binders used in the molding sands also impacts the waste composition. A common hazard is associated with waste foundry sands containing organic binders, which can release hazardous compounds including phenol, formaldehyde, PAHs and others [2-4].

The degree of pollutant leaching impacts the harmful effect of landfilled waste. This aspect is taken into consideration during the designed further application of this waste. Table 1 contains the results of heavy metal leaching from the examined waste by application of six extraction agents. The metal leaching for a each fraction was converted into the percentage ratio in relation the total content.

The study demonstrates that the leaching in the environment when H_2O was used was marginal. This type of analysis is performed with the purpose of determining the impact of the landfilled waste on the environment [31]. As a result of the fact that the recorded results of metal leaching from the waste was low, the waste can be classified as neutral. The extraction by use of deionized water usually results in leaching of metal ions and ion complexes of the soluble organic matter. Such metals are actively bound (soluble fraction). It contains metal forms with the highest mobility and availability [30,32].

The extraction by application of neutral salts of organic acids and strong alkalis leads to the leaching of the fractions with the highest mobility, as they can be released from waste during storage, whereas metals in chemically bound forms are not susceptible to leaching [30]. This fraction comprises metals with weak sorption characteristics, principally the ones, which are blocked on the surface of mineral particle by relatively weak electrostatic forces and the ones that can be released during ion exchange. In the examined waste, the ratio of metals included in this fraction is relatively small (1-12%) (Table 1). The greatest leaching capability was noted for the case of zinc. This result is also gained with regard to soils. Zinc occurs predominantly in the mobile form in soils.

Table 1. The percentage of elution compared to total concentration of heavy metals calculated for each sample and total content of heavy metal.

solution	Cd	Pb	Cu	Zn	Ni	Cr
	%					
H_2O min-max average SD	<LOQ	<LOQ	<LOQ	<LOQ	<LOQ	<LOQ
Ac						
min-max	9-26	0-11	8-13	15-22	2-6	1-3
average	19	4	10	18	3	2
SD	7	4	3	3	2	1
HCl						
min-max	27-37	0-12	8-15	17-23	4-7	2-5
average	32	6	12	20	6	3
SD	4	5	2	3	1	1
$MgCl_2$						
min-max	5-12	0-8	2-8	10-15	1-4	0-1
average	9	3	6	12	2	1
SD	3	3	3	2	1	1
NaAc						
min-max	24-30	0-7	4-9	13-20	1-4	1-4
average	27	2	7	16	2	2
SD	2	3	2	3	1	1
EDTA						
min-max	25-52	0-14	13-18	20-25	6-9	2-5
average	42	10	15	23	7	3
SD	11	6	2	2	1	1
	Total content [mg kg^{-1}]					
min	0.553	2.4	6.1	10.6	11.6	47.1
max	0.974	37.7	56.2	166	81.6	311
average	0.777	14.3	23.7	62.1	41.9	146
SD	0.189	14.6	21.3	63.5	29.3	130

<LOQ – limit of *quantitation* (Cd 0.05; Pb, Cu, Zn, Ni, Cr 0.5 mg kg^{-1} d.m.)

The extraction performed by application of 0.11M acetic acid [18] leads to leaching of the ion-exchangeable metal fraction and carbonate fractions, as these forms are mobile. These forms can be released from waste. Acetic acid was leached the metals from the samples in a wide range. The greatest level of leaching was recorded with regard to cadmium (9-26%) and zinc (15-22%). The content of copper in the mobile fraction was equal to as little as 10% of the total metal content in the waste. Concurrently, the level of Pb, Ni and Cr leaching was low(2-4%).

According to Rao et al. [30] and Tessier et al. [17], the testing concerned with leaching of metal bound to carbonate fraction should apply acetic acid salts, of which sodium and ammonium acetate are most commonly used. The extractions applying NaAc, similar as in the case of 0.11M Ac, cadmium (24-30%) and zinc (13-20%) are predominantly used (Table 1). As for the level of leaching results of the remaining metals by the application of NaAc, it was found to be small (2-7%). This is due to the fact that cadmium and zinc in the tested samples occur in the form bound to carbonates, and these forms are likely to be leached from landfills and in the conditions when they are applied in road construction and building engineering. Complexing agents (EDTA, DTPA) and inorganic acids form the strongest extraction agents [30,33,34]. Complexing reagents are capable of dissolving both the exchangeable fraction, as well as metals bound to the organic matter and hydroxides in soil. The metal content in these extract is significantly correlated with its ratio in the plants. Hence, the chelating factor is most commonly applied for the assessment of bioavailability of metals in soils [18,30,33]. EDTA forms a more powerful extraction agent than the remaining chelating substances, such as DTPA, HEDTA, NTA and EGTA. It was demonstrated that EDTA can lead to leaching greater quantities of metals compared to inorganic acids [28]. Acid extractions are applied with the purpose of leaching metals with stronger bonds, e.g. ones in the form of carbonates, Fe-Mn oxides and the ones that are bound to organic substances. Hydrochloric acid is capable of dissolving large quantities of microelements, which considerably exceed the nutritional demand of plants. As a result of this, its use is often criticised [35]. The procedures applied in Poland apply a method using 1M HCl for the assessment of the potential availability of metals in soils [21]. Therefore, it is recommended to substitute 1M HCl solution with the one at a lower concentration: 0.1M HCl [26,28]. For the case of the tested waste, the recorded percentage ratios of metals after leaching by HCl and EDTA agents was the highest. The use of these extraction agents resulted in leaching the greatest quantities of cadmium and nickel, equal to: Cd 27-37% (HCl) and 25-52% (EDTA) vs. Zn 17-23% (HCl) and 20-25% (EDTA), respectively (Table 1). Concurrently, the ratio for the remaining metals was considerably lower. These studies suggest that the 0.05M Na_2EDTA proved to be the most effective extraction agent (in comparison to HCl).

Throughout the assessment concerned with bioavailability of certain metals, some authors remark that salt solutions, such as 0.01M $CaCl_2$, 0.1M $NaNO_3$ and 1.0M NH_4Ac, prove much more applicable in the determination of metal availability for plants in comparison to complexing extraction agents (EDTA, DTPA) and acids (0.1M HCl, 1M HCl) [28,36]. The insights in the area suggest that the examined waste was characterized by a low percentage concentration of mobile and bioavailable fractions. Only for the case of cadmium and zinc, higher levels of leaching were registered in comparison to other metals. However, this fact does not disqualify their application in agriculture, since the registered cumulative cadmium concentration was very low, at a level, which normally occurs in soils. Aside from that, zinc has a very low toxicity and is considered to be indispensable for the plant growth. A greater hazard can be posed by organic compounds leached from landfilled waste. For these reason, the application of foundry waste is recommended only with regard to foundry sands not containing organic binders and the ones that are subjected to thermal regeneration prior to their use again.

4 Conclusions

1. Due to the low level of heavy metal leaching by application of water extracts, according to standard [26], the examined material formed by foundry sands was classified as a neutral waste. In general, the heavy metal content in water extracts measured in accordance with the PN-EN 12457-4 standard [26] was determined to be at a level below the limit of quantification.
2. The examined waste was characterized by various levels of heavy metal content. The greatest level was registered in sample no. 5, which involved the waste landfilled in the 1980s. The lowest concentration of pollutants was recorded in sample no. 4, which was landfilled in 2012. Besides, a considerable variability of the samples was observed with regard to the heavy metal content in them.
3. Chromium was the dominant metal in the examined waste. Slightly lower levels were registered with regard to zinc. This is attributable to the composition of the steel and iron in the produced casts. The waste had the lowest level of the pollution with cadmium.
4. It was established that cadmium forms the metal with the greatest mobility. However, due to the low cumulative content of this metal, the waste does not pose a serious hazard to the environment. The remaining metals occurred primarily in the immobile forms in the examined waste.
5. The highest ratio of heavy metal leaching was recorded as a result of application of HCl and EDTA eluents. These eluents can be used to demonstrate the quantities of mobile and potentially bioavailable forms. However, these metal forms can be leached from waste after a longer exposition period. The lowest leaching was observed by application of water, followed by $MgCl_2$ salt. The use of the last eluent on this list results in the leaching of metals bound to exchangeable and carbonate fraction, which demonstrates the greatest mobility.
6. In general, the results indicate the low mobility and bioavailability of metals contained in the examined waste.

References

1. E. Bagińska, Arch. Foundry **2**(3), 11-15 (2002)
2. M. Holtzer, Foundry J. Polish Foundryman's Assoc. **4**, 314-321 (2004)
3. M. Holtzer, Arch. Foundry Eng. **7**, 83-88 (2007)
4. M. Holtzer, Foundry J. Polish Foundryman's Assoc. **12**, 950-955 (2004)
5. M. Bożym, I. Dąbrowska, *Problems in environmental protection in the Opole Voivodeship - waste and sewage*, edited by K. Oszańca, Atmoterm, 24-41 (2012)
6. E. Smoluchowski, M. Zgut, Foundry Rev. Sci. Pract. **5**, 31-33 (2005)
7. M. Bożym, K. Zalejska, Arch. Waste Manag. Environ. Prot. **16**(3), 1-6 (2014)
8. A. Król, Environ. Prot. Eng. **38**(4), 29-40 (2012)
9. A. Król, Chem. Industry **10**, 971-973 (2007)
10. J. Kuterasińska, A. Król, Architecture Civil Eng. Environ. **3**, 61-67 (2015)
11. M. Wzorek, Ecol. Chem. Eng. S, **19**, 4, 617-627 (2012), doi: https://doi.org/10.2478/v10216-011-0044-5
12. R.S. Dungan, J.S. Kim, H.Y. Weon, A.B. Leytem, Ann. Microbiol. **59**(2), 239-246 (2009)
13. H. Zhang, L. Su, X. Li, J. Zuo, G. Liu, Y. Wang, Front. Environ. Sci. Eng. **8**(1), 89–98 (2014)
14. K.C. Bastian, J.E. Alleman, Waste Manag. **18**, 227-234 (1998)
15. B.J. Lindsay, T.J. Logan, J. Resid. Sci. Techn. **2**(1), 3-12 (2005)
16. E.A. Dayton, S.D. Whitacre, R.S. Dungan, N.T. Basta, Plant Soil **329**, 27–33 (2010)
17. A. Tessier, P. G. Campbell, M. Bisson, 1979. Anal. Chem. **51**(7), 844–851 (1979)
18. A.M. Ure, P. Quevauviller, H. Muntau, B. Griepink, Int. J. Env. Anal. Chem. **51**, 135–151 (1993)
19. M. Bożym, A. Rajmund, Environ. Prot. Natural Res. **26**(1), 1-6 (2015)
20. A. Rajmund, M. Bożym, Engin. Environ. Prot. **17**(2), 231-241 (2014)
21. E. Ociepa, A. Ociepa-Kubicka, E. Okoniewska, J. Lach, Ann. Set Environ. Prot. **15**, 1772–1786 (2013)
22. V.J.G. Houba, T.M. Lexmond, I. Novozamsky, J.J. van der Lee, Sci. Total Environ. **178**(1–3), 21–28 (1996)
23. H. Hani, S. Gupta, Commission of the European Communities, Report 10361, 157–167 (1996)
24. VSBo, V.S. Verordnung über Schadstoffgehalt im Boden, Swiss ordinance on polluttants in soils. Nr. 814.12, 3000 Bren, Switzerland (1986)
25. DIN, Deutsches Institut fur Normung, Bodenbeschaffenheit. Extraktion von Spurenelemente mit Ammonium-nitratlosung. Vornorm DINV 19730, DIN (ed.) Boden-Chemische Bodenuntersuchungsverfahren, Berlin, Germany (1995)
26. E. Meers E., G. Du Laing, V. Unamuno, A. Ruttens, J. Vangronsveld, F.M.G. Tack, M.G. Verloo, Geoderma **141**, 247–259 (2007)
27. PN-EN 12457-4:2006 - Characterization Of Waste - Leaching - Compliance test for leaching of granular waste materials and sludges - Part 4: One stage batch test at a liquid to solid ratio of 10 l/kg for

DOI: 10.1051/e3sconf/20171902018

materials with particle size below 10 mm (without or with size reduction).
28. S. Chen, L. Sun, L. Chao, Q. Zhou, T. Sun, Bull. Environ. Contam. Toxicol. **82**, 43–47 (2009).
29. S.P. McGrath, J. Cegarra, 1992. J. Soil Sci. **43**, 313–321
30. C. R.M. Rao, A. Sahuquillo, J.F. Lopez Sanchez, Water Air Soil Poll. **189**, 291–333 (2008)
31. Regulation of the Minister of Economy of 16 July 2015r. on the acceptance of waste for landfill disposal (OJ 2015, item 1277)
32. V. Seguin, C. Gagnon, F. Courchesne, Plant Soil **260**, 1–17 (2004)
33. M. Anju, D.K. Banerjee, Environ. Monit. Assess. **176**, 67–85 (2011)
34. A. Bermond, I. Yousfi, J.P. Ghestem, Analyst **123**,785–789 (1998).
35. S. Mercik, *Agricultural chemistry Theoretical and practical basics*, SGGW Warsaw, Poland (2004)
36. N.W. Menzies, M.J. Donn, P.M. Kopittke, Environ. Pollut. **145**, 121-130 (2007)

Leaching of heavy metals from cementitious composites made of new ternary cements

Justyna Kuterasińska-Warwas[1,*], and *Anna* Król [2]

[1] Institute of Ceramics and Building Materials, Opole, Poland
[2]Opole University of Technology, Opole, Poland

Abstract. The paper presents a comparison of research methods concerning the leaching of harmful substances (selected heavy metal cations ie. Pb, Cu, Zn and Cr) and their degree of immobilization in cement matrices. The new types of ternary cements were used in the study, where a large proportion of cement clinker was replaced by other non-clinker components - industrial wastes, ie. siliceous fly ash from power industry and granulated blast furnace slag from the iron and steel industry. In studied cementitious binders also ground limestone was used, which is a widely available raw material. The aim of research is determining the suitability of new cements for neutralizing harmful substances in the obtained matrices. The application of two research methods in accordance with EN 12457-4 and NEN 7275 intends to reflection of changing environmental conditions whom composite materials may actually undergo during their exploitation or storing on landfills. The results show that cements with high addition of non-clinker components are suitable for stabilization of toxic substances and the obtained cement matrices retain a high degree of immobilization of heavy metals at the level of 99%.

1 Introduction

The present trend in the use of alternative fuels and waste materials in cement and concrete production may increase the risk of contamination of final products (Fig.1 1) [1-4].

Such activities raise some doubts due to the impact of such materials on the environment. This risk particularly concerns the increased content of certain harmful substances, such as heavy metals, which, as a result of contact with the water, can leach into the surrounding environment [5].

Many environmental studies carried out on cementitious materials with wastes from other industries have shown that these materials have a high degree of immobilization of heavy metals [3, 6-15].

This is due to the properties of hardened cement matrix, which has been used for many years as an effective and safe material in the process of permanent solidification of hazardous waste [10,12,16].

The immobilization properties of cement matrix (i.e. mortar and concrete) correspond to hydrated calcium silicates and in particular C-S-H phase (calcium silicate hydrate). Hydrated calcium silicates are the dominant constituent of hardened cement pastes and can stand for up to 70% of the volume.

The amount and morphology of the C-S-H phase depends on the selection of binder components [17]. Some authors believe [18, 19] that the major factors influencing the immobilization potential of hydrated calcium silicates are the ability to incorporate foreign elements into the structure, the effect of hydrated calcium silicates as ion exchangers, low permeability of the CSH phase, acting as a barrier to particles migration and its large specific surface area (which gives the so-called sorption potential).

In addition, the C-S-H phase remains in the state of equilibrium with the liquid phase present in the pores and, depending on the concentration of calcium hydroxide at this stage, emits or takes up Ca^{2+} ions from the one.

The high (> 12.4) pH of the liquid present in the pores favours the precipitation of hydroxides and other insoluble heavy metal compounds [3,18,20-21].

The evaluation of hereby paper assessed the immobilization properties of mineral matrices, made of selected types of composite cements of non-standard composition. Applied cements are the object of standardization work on the revision of the current cement standard (EN 197-1) [22-23]. These matrices solidified heavy metals (Cu, Cr, Pb and Zn) previously introduced into them as soluble salts.

The research is to show that matrices made of new types of composite cements will be capable of immobilizing contained heavy metals and what will be the amount of metals leaching out, depending on the application of a different leaching scenario, i.e. the test method used.

The effect of addition of heavy metals on the strength of cement mortars (after 7, 28 and 90 days of hardening) was also determined.

* Corresponding author: j.kuterasinska@icimb.pl

DOI: 10.1051/e3sconf/20171902019

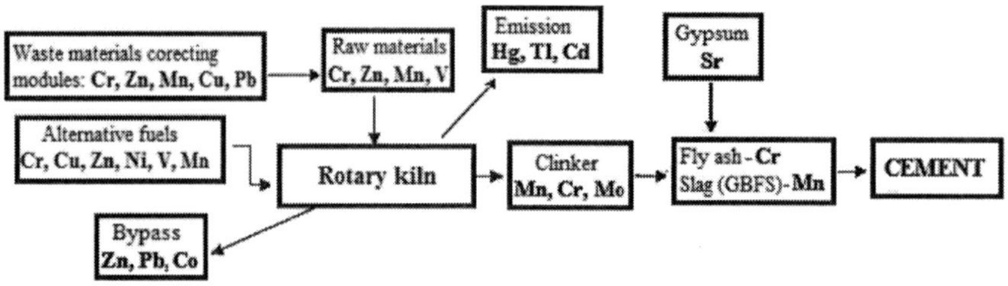

Fig. 1. Sources of heavy metals in cement production [4, 12].

2 Selection oftest methods – reflecting the two leaching scenarios

There could be several scenarios according to which the material can be leached. The impact on the amount of leachates is primarily affected by the environmental conditions where the material is exposed (temperature variables, leaching liquid and its pH, time and type of exposure, mechanical influences i.e. erosion, frost). Especially important is the pH of the leaching medium, which the lower, the more aggressively affects the exposed cement matrices, accelerating their dissolution and release of toxic elements [3,8, 24-27]. Moreover, the expression of the material content (particulate or monolithic) and its parameters such as porosity, expansion area or reactivity are critical [25-27]. The necessity to assess the release level of harmful substances from construction products is required by the Construction Products Directive [12, 28]. Particularly, when building materials become matrices solidifying waste from other industries, they should be subjected to appropriate environmental tests. Composites that solidify hazardous waste often end up in nature through practical use, such as civil engineering and communication works, but may also be deposited in landfills [12]. Since they are thus exposed to external factors, they may be subject to damage and loss of utility. Consequently, they should have good physical-mechanical properties that will ensure longevity in the long run. Durability is essential due to the fact of heavy metals being leached into the aquatic or soil environment which the matrices are in contact with. In the study the leaching teste were performed using 2 test methods. The first one is a dynamic one, batch-based test (according to EN 12457: 2006) and is used to characterize waste destined for landfill of a given type (I scenario - fragmented material) [29-30]. The second one is a method performed according to the Dutch standard NEN 7375: 2004 [31] and is considered as a static one. This method allows to reflect the environmental conditions in which the material is constantly immersed in the leaching liquid in monolithic form (II scenario).

3 Materials and test methods

3.1 Materials used for testing

Four types of mortars made of new types of composite cements were subjected to leaching tests. The increased content of a mixture of two non-clinker main components in their composition is a key feature of these cements. These binders are present in the three-component combinations: K-S-L/LL, K-S-V and K-V-L/LL (K-clinker, S-granulated blast furnace slag, V-siliceous fly ash, L/LL - limestone). Cements were obtained by mixing finely ground non-clinker components with a semi-finished product acting as clinker carrier. The semi-finished product (C) used was characterized by an increased content of SO_3 (5.1%) and a high specific surface (6200 cm^2/g). Binders were mixed in a laboratory homogenizer in appropriate proportions - each mixture was homogenized for 30 min (Table 1).

Table 1. The composition of tested cements

No.	Cement type	Percentage content of component, [%]			
		C	S	LL	V
1.	CEM II/ C(30%S+10%LL)	60	30	10	-
2.	CEM II/ C(30%V+10%LL)	60	-	10	30
3.	CEM VI (45%S+10%LL)	45	45	10	-
4.	CEM VI (45%S+10%V)	45	45	-	10

The mortars were prepared according to the formula given in PN-EN 196-1 [32] by mixing 450 g of selected composite cement, 1350 g of standard sand (maximal grain size d ≤ 2mm) and 225 ml of water.

Prepared mortars were characterized by a water/cement ratio w/c = 0.5. Samples in the shape of rectangular beams were formed with dimensions 40x40x160 mm. Samples were stored under conditions specified in PN-EN 197-1 [22], i.e. at t = 20 ± 1 ° C, and humidity RH> 90%.

Heavy metal ions Cr^{6+}, Cu^{2+}, Zn^{2+} and Pb^{2+}) were introduced into the prepared mortars as their soluble salts: $Na_2Cr_2O_7 \cdot 2H_2O$, $Cu(NO_3)_2 \cdot 3H_2O$, $Pb(NO_3)_2$, $Zn(NO_3)_2 \cdot 6H_2O$. Heavy metals were added to mortars in the amount of 0.2% calculated on the metal cation with respect to the binder mass (heavy metal salts dissolved in mixing water).

3.2 Research methodology

Scope of the study consisted of the compressive strength of mortars after 7, 28 and 90 days of hardening (with and without heavy metals addition), additionally, water extracts from their debris were made (I leaching scenario). In parallel, identically prepared mortars, freshly formed after being demoulded, were placed in containers and poured with deionized water in a ratio of water to a sample volume of 5: 1 (II scenario). Eluates were collected after the specified time intervals, i.e. 0.25, 1, 2.25, 4, 9, 16, 36 and 64 days. The mortars were made and tested according to the procedure stated in the standard for cement testing PN-EN 196-1: 2006 [32]. The water extracts for the I leaching scenario were constructed according to the methodology described in the standard PN-EN 12457-4: 2006 (maximal particle size diameter d $\leq$ 10mm, water extract made at a liquid to solid ratio of 1/10, shaking time 24 h) [30] Water extracts for the monolithic sample, according to the second leaching scenario, were made in compliance with the proceedings of the Dutch standard NEN 7375: 2004 [31]. An Inductively Coupled Plasma Mass Spectrometer (ICP MS) was used to read the concentrations of heavy metals in the water extracts. The microstructure of pastes with the addition of selected heavy metals was also investigated using the scanning electron microscope (SEM) in combination with the X-ray diffraction microscope (EDS).

4 Test results

The compressive strength of mortars made of new types of composite cements with previously introduced soluble heavy metal salts was studied. The results were referred to comparative mortars (without added metal). In order to determine the effect of the composition of individual cements on the degree of immobilization of heavy metals in formed matrices, leaching of heavy metal was investigated using the two test methods.

4.1 The effect of heavy metal addition on the strength of mortars

Fig. 2 shows the results of the compressive strengths after 7, 28 and 90 days of hardening of mortars prepared from the analyzed cements. The results of mortars with heavy metals addition were compared with mortars without the additives.

On the basis of mortar strength results obtained after the analyzed periods, it can be seen that 0.2% addition of heavy metal salts slightly decreases the strength level. For mortars that had reached the age of 90 days, the addition of copper in two cases (CEM II/C (30% S + 10% LL) and CEM VI (45% S + 10% V)) marginally increased strength (1% and 4% accordingly). The addition of zinc and lead usually resulted in a mild decrease in strength over all periods (from about 1% to 12%). The lowest decrease in strength was observed for CEM VI cement (45% S+10%V) , and even an increase of 1.6% for chromium and 3.2% for lead addition was

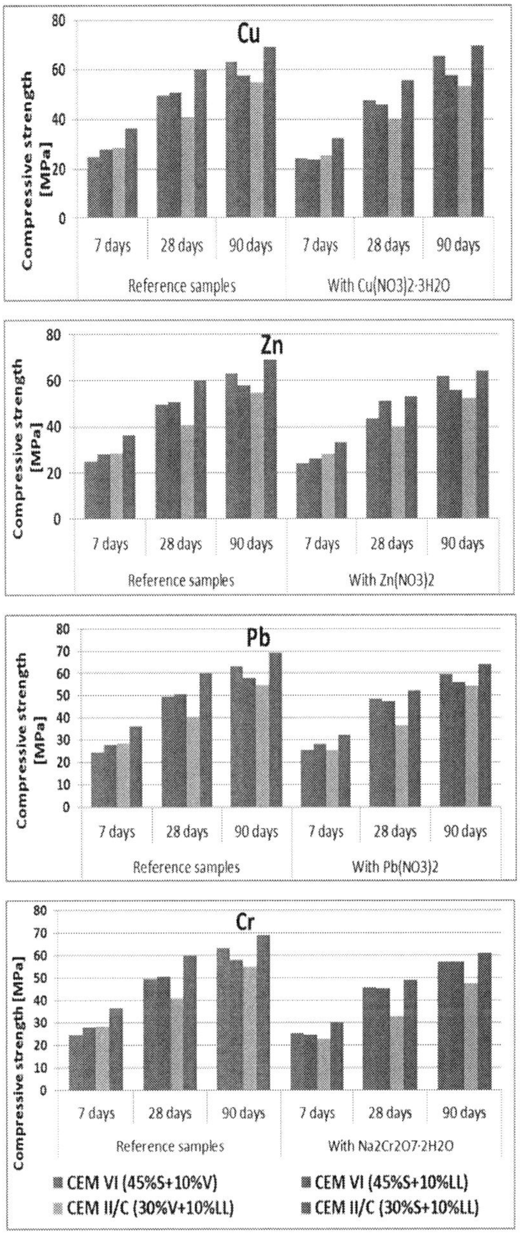

Fig. 2. Comparison of the strengths of 7, 28 and 90-day mortars made of composite cements without and with the addition of heavy metal (Cr, Cu, Zn, Pb salts).

observed in strength after 7 days of hardening. Nevertheless, for the remaining samples of composite cements, the addition of chromium salts had the most impact on the strength reduction (for 7 and 28-day mortars CEM II/C (30% S+10% LL) the drop exceeded 20%. Such a slight impact of 0.2% of heavy metal addition on the mechanical properties of the mortars indicates the good quality of tested matrices, which will certainly be beneficial for the degree of immobilization of heavy metals in their structure.

4.2 Leaching of heavy metals from crushed mortars (I leaching scenario)

Concentrations of individual heavy metal ions in the eluates obtained from crushed cement mortars after 7, 28 and 90 days of hardening are shown in Table 2. It has been found that leaching of metals for mortar at early stages (7 days) is significantly higher than mortars after 90 days. This is particularly evident on the example of chromium and lead. This is due to the lesser degree of cement matrix hydration over a shorter period of time, the consequence of which is relatively less compact microstructure of the paste. This has a positive influence on the leaching of heavy metals, which should in principle be adsorbed by the C-S-H phase. This phase, at early hardening period already occupies a substantial part of it, however, its amount increases with time, which is due to the continuous hydration of active phases of cement binder. After 28 days of mortar sample curing process, a significant decrease in leaching of analyzed heavy metals can be observed, which is undoubtedly due to the greater part of C-S-H phase.

Table 2. Concentration of heavy metals in eluates in 7, 28 and 90-day mortars dosed with their salts.

Sample age	Cement type	Concentration of heavy metal ions in eluat [μg/dm^3]			
		Cu^{2+}	Zn^{2+}	Pb^{2+}	Cr^{6+}
7 days	CEM II/C (30%S+ 10%LL)	5.40	13.40	18.75	2151.5
28 days		0.18	2.78	2.84	1166.1
90 days		0.38	0.01	0.21	726.7
7 days	CEM II/ C (30%V+ 10%LL)	1.27	13.60	309.75	2951.0
28 days		1.29	0.05	3.44	2333.4
90 days		1.22	0.02	0.22	2312.7
7 days	CEM VI (45%S+ 10%LL)	2.90	14.00	109.54	2112.5
28 days		<0.05	<0.05	1.60	1126.8
90 days		<0.05	0.01	0.17	561.6
7 days	CEM VI (45%S+ 10%V)	1.96	14.60	252.43	2275.0
28 days		<0.05	<0.05	1.60	650.4
90 days		1.26	0.01	0.17	371.8

Considering the dependence of heavy metals on the composition of tested cements, it can be seen that most metals are leached out of the matrix on fly ash-lime cement (CEM II/C (30% V + 10% LL). Thus, it may indicate that the fly ash-lime cement matrix has the least compact microstructure.

The compressive strength results, which are the lowest for this type of cement described in hereby study may reflect the situation (Fig 2).

Since a specific amount of individual heavy metal cations was added to the mortar, their total content in the mortar was known. Hence, their level of immobilization in the matrices examined was calculated (Table 3).

The results show that analyzed heavy metals are effectively retained inside mortars on new types of composite cements. Such matrices are characterized by a high level of heavy metal immobilization, mainly of

Table 3. Immobilization level of heavy metals in cement pastes.

Sample age	Cement type	Heavy metals immobilization level [%]			
		Cu^{2+}	Zn^{2+}	Pb^{2+}	Cr^{6+}
7 days	CEM II/ C (30%S+10%LL)	99.99	99.99	99.99	98.34
28 days		100.00	99.99	99.99	99.10
90 days		100.00	100.00	100.00	99.44
7 days	CEM II/ C (30%V+10%LL)	99.99	99.99	99.84	97.73
28 days		99.99	100.00	99.99	98.20
90 days		99.99	100.00	100.00	98.22
7 days	CEM VI (45%S+10%LL)	99.99	99.99	99.94	98.38
28 days		100.00	100.00	99.99	99.13
90 days		100.00	100.00	100.00	99.56
7 days	CEM VI (45%S+10%V)	99.99	99.99	99.87	98.25
28 days		100.00	100.00	99.99	99.50
90 days		99.99	100.00	100.00	99.71

99.8% -100% for Cu, Zn and Pb - depending on the age of the sample. The least effective cement composition for immobilization of heavy metals has been performed by the combination of siliceous fly ash and limestone in CEM II/C cement (30%V + 10% LL), which clearly shows the chromium immobilization ratio of approximately 97.73% for mortars after 7 days of hardening and 98.22% for mortars after 90 days.

Despite the fact that tested samples were crushed and subjected to dynamic testing, the level of immobilization of selected heavy metals in the analyzed cement matrices turned out to be high. The results obtained therefore prove, that there are the possibilities to apply these cement types in the stabilization of industrial waste containing high concentrations of heavy metals.

4.3 Immobilization of heavy metals from monolith (II leaching scenario)

In order to reflect the environmental conditions in which fresh cement composites are subjected to constant water stress (e.g. concrete retention tanks, drainage wells and drainage water receivers), monolithic samples were tested. The amount of leachates was observed at fixed intervals, i.e. after 0.25, 1, 2.25, 4, 9, 16, 36 and 64 days [31].

Table 4 gives the concentrations of heavy metals in the eluates for the various test periods.

The leachability test carried out on monolithic samples showed an identical trend in the leaching of individual metals from the test matrices.

Chromium from the matrix on fly ash – lime cement CEM II/C (30% V + 10% LL) was leached in the largest amount, the least was copper.

Table 4 Aggregated concentration of heavy metals in water extracts made from mortars dispensed with their salts after a specified period of time.

Cement types	Concentration of heavy metal ions in eluates [μg/dm^3] after specified time [days]							
	0.25	1	2.25	4	9	16	36	64
	Cu^{2+}							
CEM II/ C(30%S+10%LL)	2.35	2.54	3.1	5.56	6.09	6.28	6.33	9.14
CEM II/ C(30%V+10%LL)	1.51	2.2	3.16	3.71	4.24	5.0	7.66	10.37
CEM VI (45%S+10%LL)	2.43	2.59	3.13	5.64	6.09	6.38	6.67	8.85
CEM VI (45%S+10%V)	2.06	2.13	2.63	2.94	3.59	4.03	4.96	10.11
	Zn^{2+}							
CEM II/ C(30%S+10%LL)	6.71	22.25	22.3	23.51	23.56	23.61	23.66	36.83
CEM II/ C(30%V+10%LL)	264.79	305.55	305.6	324.18	324.23	324.28	324.33	337.34
CEM VI (45%S+10%LL)	32.51	65.46	65.51	244.32	244.37	244.42	245.31	277.19
CEM VI (45%S+10%V)	130.51	211.74	211.79	211.84	211.89	212.16	215.29	237.55
	Pb^{2+}							
CEM II/ C(30%S+10%LL)	27.01	44.32	44.55	44.61	45.07	46.17	46.39	50.84
CEM II/ C(30%V+10%LL)	35.54	68.17	68.48	68.54	68.82	68.87	69.03	75.6
CEM VI (45%S+10%LL)	41.21	63.89	64.08	64.58	64.61	64.66	65.03	74.52
CEM VI (45%S+10%V)	35.61	62.43	62.67	62.79	62.81	62.94	63.05	69.1
	Cr^{6+}							
CEM II/ C(30%S+10%LL)	4569.5	6357.0	6713.2	6813.3	6886.75	6979.05	7148.05	7393.75
CEM II/ C(30%V+10%LL)	4602.0	7267.0	8391.5	8695.05	8808.15	8877.7	9083.75	9319.05
CEM VI (45%S+10%LL)	4114.5	5603.0	7663.5	7716.8	7803.25	7898.8	8045.7	8290.1
CEM VI (45%S+10%V)	741.0	3549.0	8365.5	8465.6	8572.85	8702.85	8909.55	9103.25

As the mortar was used right after the demoulding process, the level of metals leached was significantly different compared to the results of the dynamic test [30]. In general, this level is considerably higher. The only exception is the amount of lead that leached from about 51 to 75 μg/dm^3 until the last day of the test, and in the dynamic method for 7-day samples, its concentration was 109.54 to 309.8 μg/dm^3 (except for the matrix on cement CEM II/C (30% S + 10% LL)). Looking at individual cement matrices, it can be stated, that matrix on CEM II/C (30% S + 10% LL) cement is the best retainer for zinc and lead, yet from the very beginning of the test. However, the greater Portland clinker content plays a significant role here - favors a greater intensity of hydration reaction and the formation of more C-S-H phase, a key factor in the heavy metals binding process. The highest amount of heavy metals leached within 6 to 24 hours (0.25-1 day) of the test.

4.4 The influence of new cement type compositions on the immobilization of heavy metals.

Referring to the composition of individual cements, it can be seen that the higher content of mineral additives (in favor of the decreasing amount of Portland clinker), especially blast furnace slag, results in reduced leaching of chromium the first day of stationary test [31]. This is particularly evident on the example of CEM VI cement (45%S-10%V). However, upon completion of the study, the total amount of washed chromium in this cement is higher when compared to CEM VI slag-lime cement (45% S-10% LL) containing the same proportion of mineral additives (55%).

The observation shows that the addition of fly ash to cement with respect to limestone deteriorates the immobilization properties of the cement matrices in

relation to all heavy metals analyzed. This is clearly indicated by the concentration of these elements in cement eluates with 30% of fly ash addition CEM II/C (30%V + 10% LL) (despite the higher content of clinker).
The pozzolanic reaction of fly ash in the cement system proceeds slowly, hence, such high content of fly ash in cement effectively slows down the formation of C-S-H phase, so crucial for binding of heavy metals. The addition of blast furnace slag affects the sealing of the cement matrix due to the reaction of portlandite ($Ca(OH)_2$) formed from the hydrolysis of clinker calcium silicates of with silica anions from the slag hydration process.
Aforementioned reaction results in an increased number of C-S-H phases in paste and a decrease in the capillary porosity of the hardened mixture, which hinders the diffusion of various substances [12].
Observing the increased leaching of chrome against other heavy metals, it should also be noted that chrome has a higher affinity for ettringite and calcium monosulphates than to the C-S-H phase alone.
The chemical structure of calcium sulfate enables the attachment of mono- or divalent ions (both anions and cations). Sulphoalumiantes enable chromium to be incorporated into their structure by replacing the sulphate group [1, 12, 19].
The worst immobilization properties for both test methods are performed by the composition of fly ash-lime cement (CEM II/C 30% V+10% LL). Limestone in the cement composition acts as a microfiller, which in consequence increases the tightness of the cement matrix. Unfortunately, it is a raw material, considered as chemically inert in the cement system, which reduces the immobilization potential in relation to heavy metals. Furthermore, the correlation of limestone with siliceous fly ash (due to its high fineness, fly ash also acts as a micro aggregate), which as a pozzolanic additive, reacts slowly (depends on the amount of portlantide of the clinker phase reaction), will result in slower synthesis of the C-S-H phase, which, judging by the results, negatively affects immobilization and strength properties of the fly ash-limestone matrix (Table 3 and Fig. 2). Other slag-lime cements, containing more Portland clinker in their composition, perform better immobilization properties of heavy metals.

4.5 Microstructure – SEM image

Figs. 3-6 show the microstructure of selected cement pastes with heavy metals addition.
Microstructural observations were made after 90 days of hydration (until the tests were implemented they had been stored in the air conditioning cabinet at 20 ± 2° C and relative humidity > 95% according to [32]).
Hydrating cement pastes without the addition of heavy metals show mainly the presence of the gel, compact

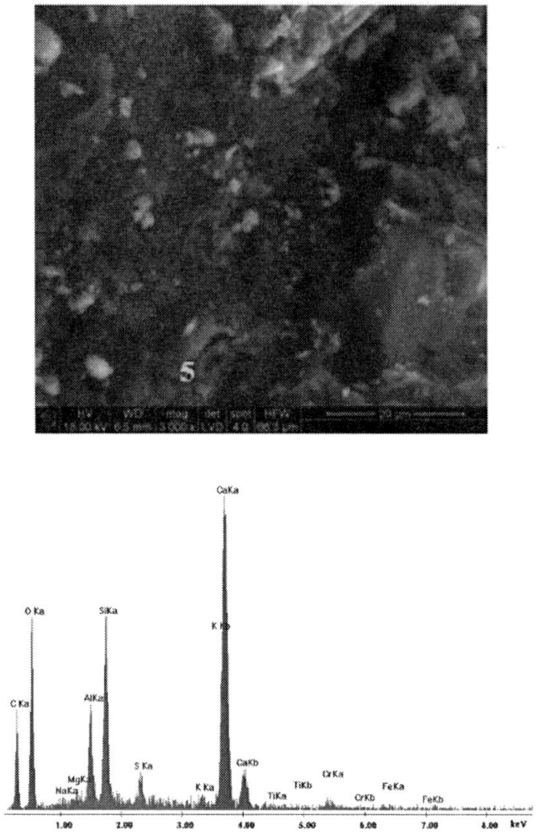

Fig. 3. Microstructure of paste on CEM VI (45%S+10%V) binder with the addition of chromium salt. Visible areas filled with C-S-H phase on slag particles and ash particle. The EDS analysis at point 5 indicates the presence of chromium ions embedded in the C-S-H phase.

C-S-H phase, but small areas of its fibrous, variant could also be observed.
In case of the addition of zinc, copper and lead salts, no significant differences were observed in pastes microstructure.

Only in the case of the addition of chromium ions to the slag cement paste (CEM VI (45%S+10%LL)), a greater amount of fibrous C-S-H was found.
Chromium was also the only metal possible to identify by X-ray microanalysis. Probably the amount of metals introduced was too small to prove their presence by chemical spot analysis (EDS).

All the pastes showed also not fully reacted particles of individual components (fly ash, slag or limestone particles covered with small amounts of C-S-H phase – (Figs. 3, 5 and 6).
Small amounts of ettringite and monosulfate also proved their presence.

 DOI: 10.1051/e3sconf/20171902019

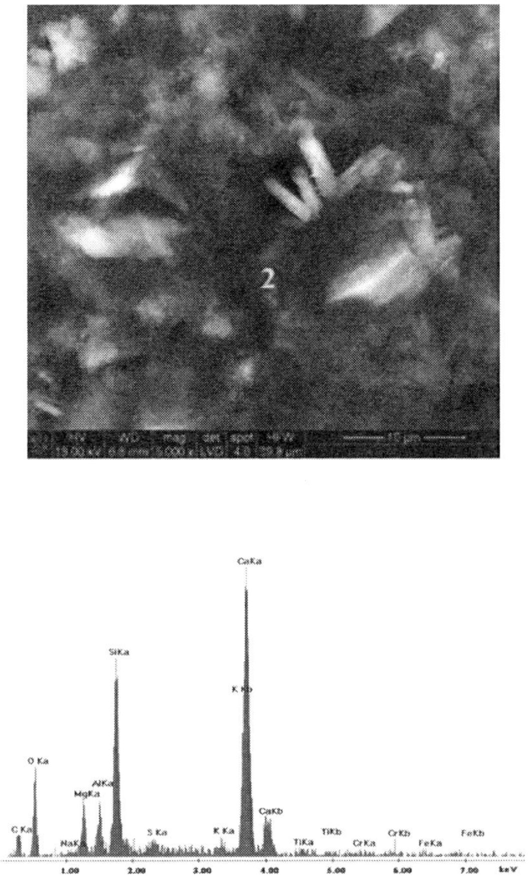

Fig. 4. Microstructure of paste on CEM VI (45%S+10%LL) binder with the addition of chromium salt. Gel phase C-S-H and calcium aluminate (point 1). The EDS analysis at point 2 indicates the presence of chromium ions embedded in the ettringite phase.

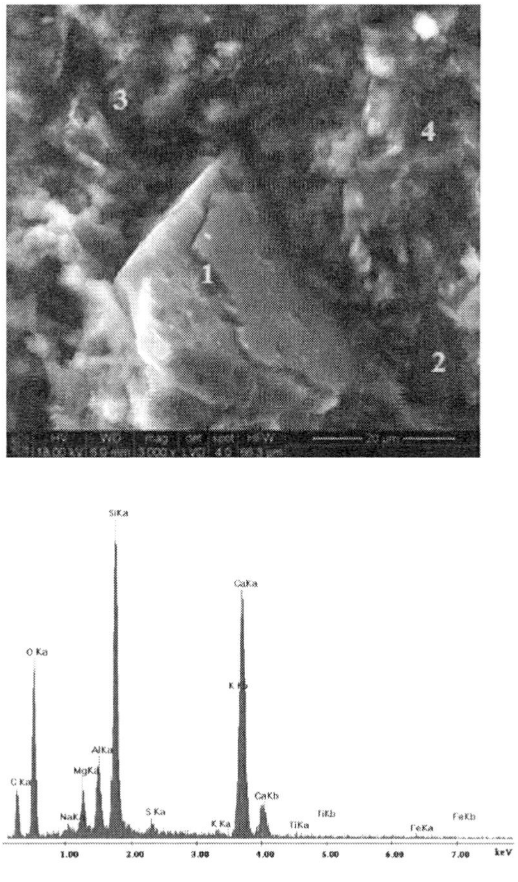

Fig. 5. Microstructure of paste on CEM VI (45%S+10%LL) binder with the addition of zinc salt. Visible unreacted slag particle (EDS- point 1,point 2), compact C-S-H phase (point 4) and portlandite (point 3).

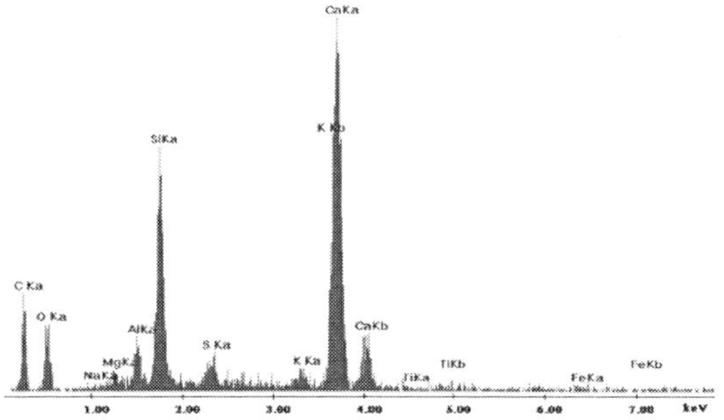

Fig. 6. Micrustructure of paste on CEM II/C (30%V+10%LL) binder with the addition of copper salt. Visible ash particle (point 1), limestone (point 2), C-S-H phase (EDS - point 3).

5 Conclusions

The use of mineral additives replacing Portland cement clinker has many environmental and economic benefits. Decreasing the content of Portland clinker, the element of cement composition expensive in the production, is favorable for the reduction of CO_2 emissions and allows the management of waste from other industries. In addition, the use of cement with significantly reduced clinker content for waste stabilization, including those containing heavy metals, is a cheap and effective method. The research has shown that the new CEM II/C and CEM VI composite cements, consisting of even more industrial waste, are suitable to use them as matrices for the immobilization of heavy metals. Batch-based tests proved that Cu, Zn and Pb immobilization levels were on the average for these matrices respectively from 99.87% for 7-day mortars to 100% for 28 and 90-day matrices. It was only chromium that its immobilization level lowered to 97.73% for 7-day samples on CEM II/C fly ash-lime cement (30% V + 10% LL) to about 99.71% for 90-day matrices on CEM VI slag-fly ash cement (45% S-10% V). The studies showed the correlation between the composition of the applied composite cements and the degree of immobilization of heavy metals in the mineral matrices made thereof. Moreover, depending on the purpose of the material/object, the appropriate test method should be selected, as the results obtained are influenced by the type of materials exposure, environmental factors or form of the tested material (monolith or disintegrated material).

References

1. A. Vollpracht, W. Brameshuber, Cem. Concr. Res. **79**, 76-92 (2016)
2. H. Lu, F. Wei, J. Tang, J. P. Giesy, Jour. Environ. Manag. **169**, 319-327 (2016)
3. W. Gwenzi, N. M. Mupatsi, Waste Manag. **49**, 114-123 (2016)
4. D. Kalarus, T. Baran, M. Ostrowski, Sc. Works of ICBM. **25**, 7-17 (2016),
5. C. Rosik-Dulewska, U. Karwaczyńska, Środkowo-Pom. Tow. Nauk. Ochr. Śr. **10**, 2005-2019(2008)
6. A. Król, Przem. Chemiczny **10**, 971-973 (2007)
7. F. Song, L. Gu, N. Zhu, H. Yuan, Chemosphere **92**, 4, 344-350 (2013)
8. A. Król, D. Jagoda, Cem. Wap. Bet. **2**, 90-101 (2012)
9. W. Müllauer, R. E. Beddoe, D. Heinz, Cem. Concr. Comp. **58**, 129-139 (2015)
10. M. A. Cinquepalmi, T. Mangialardi, L. Panei, A. Evangelista Paolini, L. Piga, Jour. Hazard. Mat. **151**, 2–3, 585-593 (2008)
11. Md. A. Habib, N. Md. Bahadur, A. J. Mahmood, Md. Anarul Islam, Jour. Saud. Chem. Soc. **16**, 3, 263-269 (2012)
12. A. Król, *Uwalnianie metali ciężkich z kompozytów mineralnych z uwzględnieniem oddziaływania środowiska* (Wydawnictwo Politechniki Opolskiej, Opole, 2012)
13. I. Buj, J. Torras, M. Rovira, J. de Pablo, Jour. Hazard. Mat. **175**, Issues 1–3, 789-794 (2010)
14. A. Król, Archit. Civil Eng. Environ. **4**, 3, 71-76 (2011)
15. J. Kuterasińska, Cem.Wap., Bet. **4**, 250-257 (2012)
16. A. Król, Environ.l Protect. Eng. **38,** no 4, 29-40 (2012)
17. A. Akyıldız, E. T. Köse, A. Yıldız, Constr. Build. Mat. **138**, 326-332 (2017)
18. X.D Li, C.S Poon, H Sun, I.M.C Lo, D.W Kirk, Jour. Hazard. Mat. **82**, 3, 215-230 (2001)
19. J. Androniuk, C. Landesman, P. Henocq, A. G. Kalinichev, Phys. Chem. Earth, **99**, 194-203 (2017)
20. Q.Y. Chen, M. Tyrer, C.D. Hills, X.M. Yang, P. Carey, Waste Man., **29**, 1, 390-403 (2009)
21. I. Navarro-Blasco, A. Duran, R. Sirera, J.M. Fernández, J.I. Alvarez, J. Hazard. Mat. **260**, 89-103 (2013)
22. EN 197-1 *Cement – Part 1: Composition, requirements and conformity criteria for common cements* (2012)
23. pr EN 197-1*Cement - Part 1: Composition, specifications and conformity criteria for common cements* (rev. June 2014).
24. M. Dell'Orso, T. Mangialardi, A. Evangelista Paolini, L. Piga, Jour.Hazard. Mat. **227–228**, (2012)
25. Y. J. Du, M. Li Wei, K. R. Reddy, Z. Peng Liu, F. Jin, Jour.Hazard. Mat. **271**, 131-140 (2014)
26. A. P. Galvín, J. Ayuso, J. Ramón Jiménez, F. Agrela, Waste Man. **32**, 1, 88-95 (2012)
27. A. Król, K. Mizerna, Conference (SEED) Krakow, POLAND Book Series: E3S Web of Conferences, Volume: 10, Article Number: UNSP 00050, (2016)
28. Construction Products Directive – CPD 89/106/EEC, (1989)
29. K.. Mizerna, A. Król, Ecol. Eng. **47**, 151-156 (2016),
30. EN 12457-4 *Characterization of waste. Leaching. Conformity test for leaching of granular waste materials and sludge. Part 2: Single-stage portion test at a liquid to solid ratio of 10 l/kg for materials with particle size less than 10 mm (with or without size reduction)* (2006)
31. EA NEN 7375 *Leaching characteristics of moulded or monolithic building and waste materials. Determination of leaching of inorganic components with the diffusion test. The tank test.* (2004)
32. EN 196-1 *Methods of testing cement. Part 1. Determination of strength.* (2006)

Environmental assessment of applicability of mineral-organic composite for landfill area rehabilitation

Kamila Mizerna[1,*], *Anna* Król[1], and *Adrian* Mróz[2]

[1] Opole University of Technology, Department of Environmental Engineering, 5 Mikołajczyka Str., 45-271 Opole, Poland
[2] Gogolin Municipal Multi-branch Enterprise, 15 Ligonia Str., 47-320 Gogolin, Poland

Abstract. This paper undertakes an assessment of the impact of a mineral-organic composite on the environment as well as the potential for its application for land rehabilitation purposes. The analysis involves the release of the leachable contaminations from the material subjected to testing. This material was formed by a composite manufactured on the basis of communal bottom ash and stabilized sewage sludge. The sludge resulting from wastewater treatment was subjected to stabilization and dehydration in waste pounds at the phase of pre-watering until 20% of dry mass is obtained. Subsequently, they were mixed with bottom ash, which was obtained from selective waste collection, in a 1:1 mass ratio. The analysis involved the leaching of inorganic contaminants in the form of heavy metals, sulphates (VI), chlorides, and fluorides as well as organic compounds in the form of organic carbon solution under the effect of leachant with a various level of pH. The analysed components were characterized by various leaching behaviour depending on the leachant pH. On the basis of the results, it was able to assess the potential hazard posed by the examined material on the environment as a consequence of its application for landfill area rehabilitation.

1 Introduction

The landfill rehabilitation process is the restoration or re-allocation of the utility value to degraded land as a result of the deposition of waste, through proper land formation, improvement of chemical and physical properties, restoration of soil and regulation of proper water relations [1]. Rehabilitation of the landfill can be carried out in the technical, biological or target site management. Excavation areas require reclamation by initiating soil-forming processes in reclaimed substrate, but above all, the restoration of plant cover. Technical reclamation consists of neutralizing the substrate acidity by liming and mineral fertilization, also biological reclamation through organic fertilization and the introduction of properly selected species of pioneer plants [2, 3].

The principal goal of the reclamation ought to be the minimization of potential destructive impacts on the environment [4]. According to the Regulation [5], landfill site rehabilitation should be performed in a way that protects the landfill from harmful effects on surface water and groundwater and air. Moreover, it is implemented in a way that integrates the area of the landfill with the surrounding environment and allows to observe the impact of the landfill on the environment. Reclamation works are the final stage of the operation and the day of completion of rehabilitation process is the same day that the landfill is closed. Non-waste materials or well-defined waste, including stabilized sludge and bottom ashes, are used for reclamation. The ash should be mixed prior to application in a 1: 1 ratio with dehydrated stabilized municipal sewage sludge. The use of sewage sludge as a reclamation material gives the municipal utilities the ability to manage this type of waste by own means. Likewise, this action is most desirable in waste management, where direct storage of unprocessed sewage sludge is not permitted [6].

In order to assess the potential environmental impact of materials used in reclamation, a leaching compliance test [7] shall be performed to provide information on the release of a given component from the material into the aquatic environment. The most dangerous contaminants are the heavy metals (mainly mercury, thallium, arsenic, cadmium and lead). Therefore, many studies on the content and leaching of heavy metals from waste materials of various origins are carried out worldwide [8-15]. Waste materials deposited in the natural environment are exposed to various physical, biological and chemical factors affecting the mobility of heavy metals. Contaminants in waste, exposed to water, can be leached into the soil and into surface and groundwater [16]. Thus, before the introduction of waste into the environment, the entrepreneur is obliged to carry out a leaching compliance test to assess the extent of their mobility to the environment.

The aim of the study was the assessment of leaching of mineral-organic composite produced on the basis of two communal wastes, i.e. stabilized sewage sludge and bottom ash from small scale boilers from households.

* Corresponding author: mizerna.kamila@gmail.com

 DOI: 10.1051/e3sconf/20171902020

The tested material shall be the subject of the reclamation of communal landfills. For this purpose, the leaching of inorganic contaminants: heavy metals, chlorides (Cl^-), fluorides (F^-) and sulphates (VI) (SO_4^{2-}) were analysed. The content of organic compounds in the form of dissolved organic carbon (DOC) was also evaluated. A leachant of varying pH was used to potentially assess the change of the environmental reaction to the level of contaminant release. The role of applying composite is to create a humus layer and create a suitable habitat for vegetation development. Used for rehabilitation of land requires mixing with the top layer of soil. Applied for macro levelling, cover layer formation and bulk filling should be properly applied and scraped to achieve an effect consistent with the land development project.

2 Materials and methods

The research material was a mineral-organic composite produced on the basis of communal bottom ash and stabilized sewage sludge. The waste came from one of the municipality in Poland in the Opole province. The sludge formed during the extraction and compaction stages were subjected to a stabilization process based on the oxygen distribution of the organic mass. It was carried out in a sludge stabilization chamber. The sludge was then aerated by aeration grate with membrane diffusers. Subsequently, it was dehydrated on the filter press and settling plots to 20% of dry mass. Stabilized sewage sludge was mixed in a 1:1 mass ratio with household furnace ash derived from selective waste collection. The mineral-organic composite is shown in Fig. 1. The samples for research were collected by quartering from the wastewater treatment plant and transferred immediately to the laboratory. A representative laboratory sample of about 2 kg was prepared from which a number of test portions were formed afterwards.

Fig. 1. Mineral-organic composite at waste management plant.

Leaching compliance test was performed on composite based on the basic test in accordance with EN 12457-2: 2002 [7]. A sample of <4mm was arranged from which water extracts were prepared at a liquid to solid ratio of L/S = 10 dm^3/kg.

Liquid with different pH (4, 7, 12) was the leachant to assess the effect of liquid reaction change on the release level of contaminants. The pH of water was lowered by using nitric acid (V) and raised with sodium hydroxide. The extracts were shaken on a laboratory shaker for 24 hours, and the slurry was filtered. Analysis of heavy metal concentrations was carried out by flame atomic absorption spectrometry (FAAS) and inductively coupled plasma atomic emission spectrometry (ICP-OES). The content of anions (F^-, Cl^-, SO_4^{2-}) was determined by ion chromatography.

3 Results and discussion

Sewage sludge for many wastewater treatment plants and municipal facilities become a problem in terms of their disposal [17]. New ways of their utilization are constantly being sought. Hence, attempts have been made to use sludge in combination with communal bottom ash as a material for landfill rehabilitation (mineral-organic composite). Stabilized sewage sludge being a component of the test material was characterized by an annual average content of heavy metals within the range given in Table 1. The waste was characterized by the highest content of Cu and Zn. These elements can be derived from domestic wastewater, as well as industrial wastewater and surface runoff. They can also be present in wastewater as a result of corrosion of sewer pipes. Sewage sludge meets the requirements of Regulation [6] for heavy metals content in the sewage sludge used in land rehabilitation. The next part of hereby paper presents the results of the chemical composition and the leachability of contaminants from the composite in order to assess the possibility of its use as a reclamation material.

Table 1. Annual average content of heavy metals in the stabilized sewage sludge.

Heavy metal	Total content [mg/kg d.m.]
Cadmium (Cd)	3.63
Copper (Cu)	1900
Nickel (Ni)	12.7
Lead (Pb)	36.8
Zinc (Zn)	1167
Mercury (Hg)	0.15
Chromium (Cr)	17.6

Table 2 shows the content of oxide forms of components present in the material. Results of the analyses indicate a significant content in the dry matter of nutrients, mainly silicon, as well as iron, magnesium, aluminium, sulphur, calcium and sodium and less potassium. Additionally, a high ignition loss of 18.6% and a total organic carbon (TOC) content of 13.0% were also determined.

Water extracts were prepared from the test composite to determine the level of leaching of contaminants that may adversely affect the soil and water environment. A leachant of different pH was used for their preparation.

The results obtained were compared with the values approved for non-hazardous and inert landfills (Table 3), which include communal waste disposal sites, according to the Regulation of the Minister of Environment dated 16 July 2015 on the acceptance of waste for landfill disposal [18]. The leaching criteria were determined at pH of leachant around 7. The tested material is used for rehabilitation of non-hazardous and inert landfills, hence the results are compared with these criteria.

Table 2. Chemical composition of mineral-organic composite.

Properties	Content [wt. %]
Loss on ignition	18.6
TOC	13.0
SiO_2	37.5
Fe_2O_3	5.57
MgO	4.67
Al_2O_3	2.15
SO_3	1.93
CaO	1.44
Na_2O	1.10
K_2O	0.95

Table 3. Concentrations of the released components of the composite at a pH of the leachant of 4, 7 and 12 compared to the limit values according to the Regulation [18]

Component	Content in water extract [mg/kg]			Limit value [18] of leaching [mg/kg]
	pH 4	pH 7	pH 12	
Dissolved organic carbon (DOC)	1370	1410	1150	800
Fluorides (F^-)	1.20	1.20	1.00	150
Chlorides (Cl^-)	1110	1140	1100	15000
Sulphur (SO_4^{2-})	17500	18000	17500	20000
Barium (Ba)	0.80	0.90	0.70	100
Zinc (Zn)	0.26	0.25	0.72	50
Copper (Cu)	3.06	3.00	3.17	50
Lead (Pb)	<0.50	<0.50	<0.50	10
Nickel (Ni)	0.20	0.19	0.17	10
Cadmium (Cd)	<0.05	<0.05	<0.05	1
Arsenic (As)	0.12	0.12	0.13	2
Chromium (Cr_{total})	0.08	0.08	0.07	10
Molybdenum (Mo)	2.44	2.36	2.36	10
Antimony (Sb)	0.18	0.19	0.20	0.7
Selenium (Se)	0.07	0.08	0.07	0.5
Mercury (Hg)	<0.008	<0.008	<0.008	0.2

Among the anions at the highest level of leaching were sulphates (17500-18000 mg/kg), chlorides (1100-1140 mg/kg) and fluorides (1.00-1.20 mg/kg). Regarding heavy metals, the leaching was in the following decreasing series: Cu> Mo> Ba> Zn> Sb> Ni> As> Cr. Very toxic metals such as Pb, Cd and Hg are below the limit of determination. The content of non-metal Se, which the high concentration in the organism could be dangerous was also analysed in water extracts. No exceedance of permissible leaching values in Regulation [18] was observed for all concentrations of elements. The only component not meeting the leaching criteria was dissolved organic carbon (DOC), which acceptable concentration in water extract prepared with a pH 7 leachant was exceeded by 610 mg/kg. However, the high value of this parameter should not become an obstacle when applying this material in landfill rehabilitation. Organic substances occur also naturally in the environment, and any leachate from the composite upon entry into ground or surface waters may be diluted. Moreover, the organic matter contained in water can complex heavy metals and limit further migration in the environment [19].

There were slight differences in the majority of concentrations of the test components depending on the pH of water used. However, the use of leachant with pH 12 resulted in a 18% reduction in DOC leaching, 17% in F^-, and Ba in 22%, relative to pH 7. While it contributed to the increase in the Zn concentration of 188%. Scientific research of heavy metal leaching conducted in Poland and worldwide prove [12-14, 20, 21] that heavy metal cations, including zinc, are generally characterized by an increased tendency to release in acidic environment. Still, based on the research results, it is clear that this correlation is not close and leaching of heavy metals can be influenced by many factors, including the chemical composition of the material. The increase in the release level of a given contaminant to the aqueous phase, influenced by a pH change, may be essential during the determination of its high concentrations. At this point, lowering or increasing the liquid pH, due to the influence of various environmental factors, may result in exceeding the allowable storage values.

4 Conclusions

The obtained test results allow to state that the mineral-organic composite is an environmentally safe material when used for rehabilitation purposes. Concentrations of components (heavy metals, chlorides, fluorides, sulphates) which could pose a risk to the aquatic environment in the water extracts did not exceed the normative limit values. The use of leachant with acidic and base character did not contribute to a significant increase in leaching of the analysed components. Zinc was the only element, during the process of leaching at pH 12, with nearly 3-times increased concentration when compared with the use of neutral water. Due to the composition of the tested material, its introduction into the top layer of the landfill site may contribute to the increase of the organic matter content and to the growth of the nutrient components, mainly silicon, iron and magnesium. Increasing the amount of organic matter is conducive to the formation of humus levels and increasing the content of assimilable forms of nutrients. These processes in turn, create conditions for the development of vegetation in degraded land.

References

1. K. d'Obyrn, W. Wójcik, AUNC Ekonomia **XLVI**, 2, 225–237 (2015) doi: dx.doi.org/10.12775/AUNC_ECON.2015.014
2. S. Góral, Ecol. Eng. **3**, 161–178 (2001), http://www.archive.ineko.net.pl/pdf/IE-3.pdf
3. K. Klimont, Probl. Inż. Rol. **2**, 165-176 (2011), http://www.itp.edu.pl/wydawnictwo/pir
4. J. Kurian, S. Esakku, R. Nagendran, C. Visvanathan, 2005. A decision making tool for dumpsite rehabilitation in developing countries. *Proceedings of Sardinia 2005*, Tenth International Waste Management and Landfill Symposium, S. Margherita di Pula, Italy (2005)
5. Regulation of the Minister of Environment of 30 April 2013 on landfills, Poland's Journal of Laws, 2013, item 523
6. Regulation of the Minister of Environment of 6 February 2015 for municipal sewage sludge, Poland's Journal of Laws, 2015, item 257
7. EN 12457-2:2002 Characterisation of waste. Leaching. Compliance test for leaching of granular waste materials and sludges. Part 2
8. M. Bożym, I. Florczak, P. Zdanowska, J. Wojdalski, M. Klimkiewicz, Renew. Energy, 77, 467-472 (2015), doi: http://dx.doi.org/10.1016/j.renene.2014.11.010
9. A. Król, Przem. Chem. **10**, 971-973 (2007)
10. A. Król, Environ. Prot. Eng. **38**, 4, 29-40 (2012), doi: 10.5277/EPE120403
11. T. Olszowski, M. Bożym, Atmos. Environ. **94**, 22-27 (2014), doi : 10.1016/j.atmosenv.2014.05.011
12. M.J. Quina, J.C.M. Bordado, R.M. Quinta-Ferreira, Waste Manage. **29**, 2483-2493 (2009), doi: 10.1016/j.wasman.2009.05.012
13. M. Vitková, V. Ettler, O. Šebek, M. Mihaljevič, Mineral. Mag. **72**, 1, 521–524 (2008), doi: 10.1180/minmag.2008.072.1.521
14. V. Cappuynus, R. Swennen, J. Hazard. Mater. **158**, 1, 185-195 (2008), doi: 10.1016/j.jhazmat.2008.01.058
15. M. Wzorek, T. Baran, M. Ostrowski, Cement Wapno Beton, **18**, 4, 207-215 (2013)
16. H.A. van der Sloot, A. van Zomeren, Mine Water Environ. **31**, 92-103 (2012), doi: 10.1007/s10230-012-0182-8
17. M. Wzorek, Ecol. Chem. Eng. S, **19**, 4, 617-627 (2012), doi: https://doi.org/10.2478/v10216-011-0044-5
18. Regulation of the Ministry of Economy of 16 July 2015 on the acceptance of waste for landfill, Poland's Journal of Laws, 2015, item 1277
19. W.H. Błaszczyk, Zesz. Prob. Nauk Rol. **476**, 61-68 (2001)
20. K. Mizerna, A. Król, Appl. Mech. Mat. **797**, 408-414 (2015), doi: 10.4028/www.scientific.net/AMM.797
21. H.A. van der Sloot, Horizontal standardization and harmonization of leaching test methods for waste, secondary raw materials, construction materials and (contaminated) soil, *Proceedings of the Wascon Conference*, San Sebastian (2003), https://www.ecn.nl/docs/society/leaching/wascon_005.pdf

Environmental, technical and technological aspects of hazardous waste management in Poland

Justyna Pyssa[1,*]

[1]AGH University of Science and Technology, Faculty of Energy and Fuel, Al. Mickiewicza 30, 30-059 Cracow, Poland

Abstract. The issue of recovery and disposal of hazardous waste is not a new concern. The waste comes from various processes and technologies and therefore the bigger emphasis should be placed on reducing quantities of generated hazardous waste (which is often connected with changes in the technology of manufacturing a given product) and limitation of their negative influence on natural environment. Plants specializing in waste processing processes should meet the so-called cardinal triad of conditions deciding on the full success of investment, and namely: economic effectiveness, ecological efficiency and social acceptance. The structure of generation of hazardous waste in EU-28 has been presented in the paper. Methods of hazardous waste disposal in Poland have been discussed. Economic and ecological criteria for the selection of technology of hazardous waste disposal have been analyzed. The influence of the hazardous waste on the environment is also presented. For four groups of waste, which are currently stored, alternative methods of disposal have been proposed.

1 Introduction

It was nearly till the end of the 70's of the previous century that the storage was the most commonly used and legally accepted method of waste disposal in Poland. The policy conducted in such manner, in which the emphasis was placed first of all on the increase of production capacity and not on ecological effects, contributed to the devastation of natural environment. For that reason actions were taken aimed not only at changes in legislation, but also in mentality and ecological awareness of the society. The idea of sustainable development started to penetrate the ecological politics of countries implying that criteria of sustainability would gain a position equivalent to economic criteria.

Currently the primary objective of natural environment protection is not so much the control of degree of environmental pollution and waste disposal, but the prevention of their generation.

2 Hazardous waste in the EU and Poland

Hazardous waste is generated in all sectors of the economy. The structure of produced waste results from the industrial specificity of a given country. In all of the European Union (EU-28) a total of 95.02 million Mg of hazardous waste was generated in 2014 [1]. In comparison with 2012, the quantity decreased by 5.67 percent. The largest decreases in the amount of produced hazardous waste one can observe in Ireland, Belgium, the United Kingdom and Bulgaria. The biggest increases in the number of generated hazardous waste one can see in Cyprus, Denmark, Estonia and Finland. In Fig. 1 the structure of production of hazardous waste in countries of the European Union (EU-28) in the years 2010 to 2014 is presented.

Changes in the quantity and quality of generated hazardous waste depend on the development of particular industry branches. In Poland in the year 2014 1,876,577 Mg of hazardous waste was produced. The largest part of hazardous waste underwent the processes of recovery other than the energy recovery and except backfilling – 1,492,430 Mg – which comprises 79.53 percent. Another process in terms of the quantity of managed waste is the deposit onto or into land – 244,786 Mg (13.04%). 135,443 Mg (7.22%) of hazardous waste underwent the process of incineration/disposal (D10). 2,492 Mg (0.13%) of hazardous waste underwent the processes of incineration and energy recovery (R1), while the land treatment and release into water bodies were applied to 1,426 Mg (0.08%) of produced hazardous waste. In Fig. 2 the percentage of particular manners of managing hazardous waste in Poland is illustrated.

3 Methods of waste disposal

Before making the decision as the choice of technology of disposal of a given group of waste it is necessary to familiarize oneself with many issues of both legal and technical and economic nature. In the majority of countries legal provisions have been introduced which precisely define processes connected with sourcing, storage, transportation and technologies of treatment of given groups of waste. The knowledge in this area is provided by directives of the European Union and

* Corresponding author: jpyssa@agh.edu.pl

DOI: 10.1051/e3sconf/20171902021

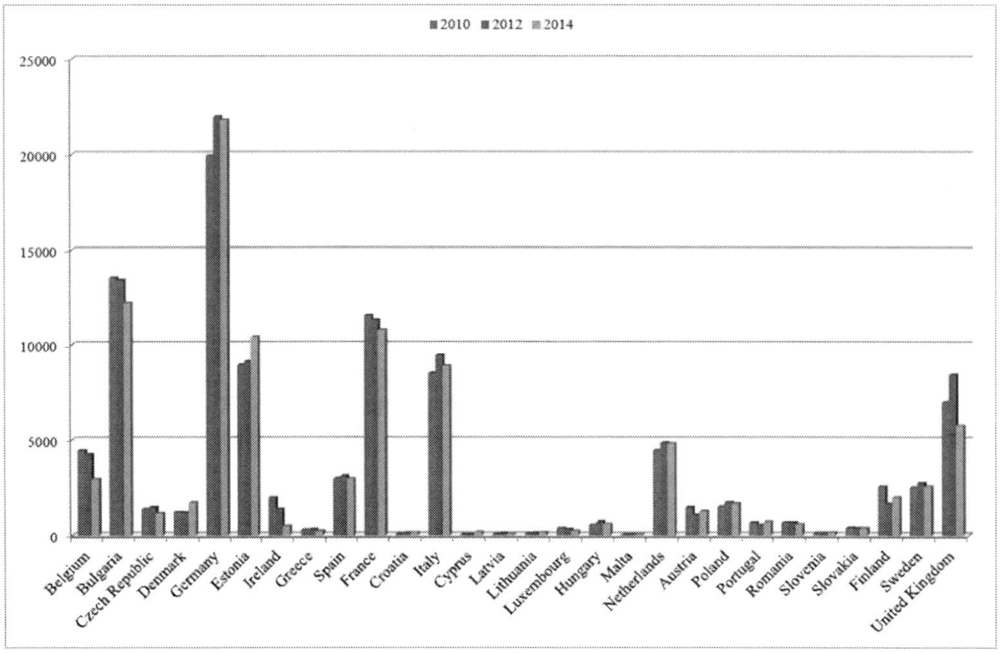

Fig. 1. Structure of production of hazardous waste in EU-28 in the years 2010 to 2014 (author's own research based on [1]).

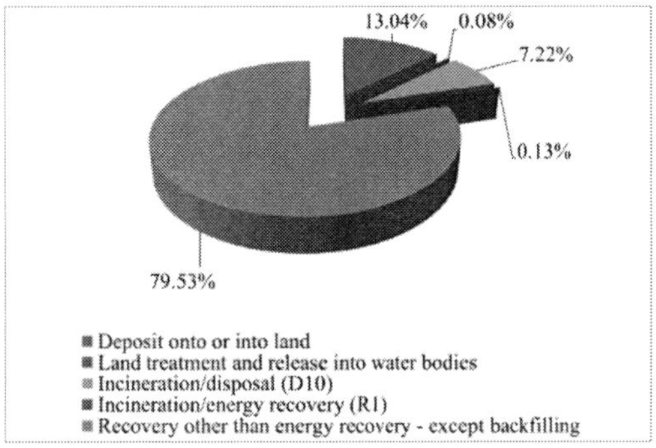

Fig. 2. Proceeding with hazardous waste generated in Poland in 2014 (author's own research based on [1]).

provided by directives of the European Union and changes in legal provisions of the Polish law.

The manner of waste disposal is selected according to properties of substances which are going to be the subject for the process. There are no universal methods thanks to which all and any waste can be disposed. This is why the selective collection of hazardous waste is so important [2]. The division of technologies of hazardous waste disposal is presented in the Table 1.

In the classification of hazardous waste disposal technologies one adopts the rule of grouping wastes according to their properties and possibilities of disposal. Depending on the chemical composition waste can be disposed in identical conditions. In Table 2 there is a comparison of particular groups of hazardous waste along with applied methods of its disposal. The choice of optimal technology of economicutilization, disposal and storage depends on many factors [2-5], and namely:

Table 1. Technologies of disposal/recovery of waste.

Method of disposal/recovery	Description
Physico-chemical	The aim is the transformation of hazardous waste to such a state which will allow for final disposal non-burdensome to the environment through energy and raw material utilization or which will allow safe storage as an alternative
Biological	It applies to organic waste and the aim is the improvement of agricultural condition of farmlands (natural circulation of elements in nature)
Thermal	The aim is the energy recovery and reduction of the mass of waste – energy recycling
Transformation of waste into form of primary and secondary raw materials	It leads to saving in natural resources and energy (material recycling)
Depositing waste on ordered landfills	Final way of managing residual waste further disposal of which is not justified either in technical and economic terms

— characteristics of waste (physical, chemical and technological properties, raw material composition, nuisance to the environment),
— quantity of waste deciding about the capacity of facilities (absorbability of landfills, efficiency

Table 2. Methods of disposal of hazardous waste (author's own research based on [3]).

Type of waste	Recovery	Incineration	Physico-chemical or biological treatment	Solidification	Land forming
Liquid waste similar to sewage			X		
Waste containing acids and alkalis			X		
Waste containing heavy metals			X	X	After chemical fixation
Waste containing inorganic toxic chemical compounds			X	X	After chemical neutralization
Waste of high chemical reactivity			X		
Waste containing neutral inorganic substances	X				
Used oils and solvents	X	X			
Used resins, paints, solvents, glues	X	X			
Waste containing organic chemical substances	X	X	X		
Out-dated plant protection products		X	X		
PCBs, chlorinated hydrocarbons		X			
Waste containing organic biodegradable substances		X	X		

of installations for commercial exploitation waste disposal), frequency of waste generation and spatial distribution of sources of formation,

- possibility of continuous reception of products which emerged during the treatment of waste such as secondary raw materials, compost and waste heat,
- field conditions such as: availability of sites for building, location of facilities, hydro-geological conditions of the surface, meteorological conditions, preserving protection zone, influence of waste on ground waters, soil and air,
- technical (availability of proven technologies) and financial (investment and operating costs) possibilities of execution of construction of facilities,
- social acceptance for the location and proposed solutions.

Plants, in which hazardous waste is treated, must fulfill three essential conditions which are: economic efficiency, environmental efficiency and social acceptance. As far as the ecological factor is indisputable, two remaining elements of the balance configuration are burdened with strong financial and social interaction. This is the result of- among others – high investment costs and the phenomenon described in the literature as the 'NIMBY' (*not in my backyard*) syndrome [2, 6].

In Fig. 3 existing installations of hazardous waste recovery and disposal in Poland are indicated. The biggest number of installations is present in southern Poland (voivodships: silesian (24), malopolska (17) and opolskie (8)). In central Poland most of facilities are located in masovian (9), lodzkie (9) and kujawsko-pomorskie (7) voivodships. And in the north of Poland the number is 11. It is connected with the location of industrial plants producing hazardous waste in those voivodships.

Fig. 3. Location of installations of hazardous waste recovery and disposal in Poland [8].

4 Conditions of waste management

Recovery of raw materials from waste, waste disposal as well as storage requires not only proper technologies, but expensive installations and equipment first of all. One also cannot forget about costs related with temporary warehousing, transportation and loading. Additionally an important factor to which the attention should be drawn is burdening the environment with by-products (sewage, waste gases, secondary waste). Building of new facilities for waste management enables the introduction of modern technologies and ensures the environment protection at a proper level. This is why it is so important to familiarize oneself with the structure of hazardous waste generation in a given area and application of optimal organizational and technological solutions codified in a form of BAT– *Best Available Techniques* [6].

4.1 Environmental conditions

Inconsistent waste management is an indication of the ineffective utilization of resources in terms of the environment protection. Incorrect waste management affects all components of the environment, including the state of ecosystems and the health and living quality of people. Highly-consumptive lifestyle and industry development has influence on generating huge quantities of waste, hazardous waste in particular. All investments in the area of waste management – regardless of nature, scale and functions they are supposed to perform in the future – influence the environment. The direct or indirect effect can be building and construction work, exploitation of newly-created and modernized facilities and installations as well as processes occurring in them.

Properly designed facilities of hazardous waste management reduce the pressure onto the environment which has influence on the improvement of quality of abiotic atmospheric components (air status/quality, water and soil quality). It has a positive impact on biological diversity of a region and proper functioning of ecosystems.

Both in the phase of investment realization and later exploitation interactions connected with the emission of noise, dust and air pollution emerge. The emissions depend on the sort of waste treated, applied technology and processes and location characteristics of a given facility. The exploitation of waste management facilities such as incineration plants, waste treatment plants, recycling plants and landfills result in emergence of secondary waste apart from typical contaminants that is dust and gaseous substance emission, thermal energy, radiation, noise and odor emission and discharge of pollutants into the water. It is therefore necessary to seek such technologies which, in addition to being very efficient in removing hazardous waste, will emit low volumes of secondary pollutants.

In general, the majority of waste treatment plants cause the emission of carbon dioxide, ammonia and volatile organic compounds. Other polluting chemical compounds are hydrogen chloride, hydrogen sulphide and dioxins.

In Table 3 processes of waste treatment along with accompanying emissions into the atmosphere are compared.

Another by-product/by-effect from the activity of hazardous waste treatment plants is contaminated sewage. The main components causing pollution are heavy metals, chloro-organic compounds and nitrogen- and phosphorus-compounds.

In Table 4 the main processes of waste treatment along with generated contaminants, introduced to the water, are compared.

Table 3. Major contamination emitted as a result of hazardous waste treatment (author's own research based on [2, 8]).

Process	Chemical compound
Biological processes	• Carbon oxides • Nitrogen oxides (N_2O, NO, NO_2) • Volatile organic compounds • Ammonia • Microbial contaminants
Physico-chemical processes	• Acids (HCL, HF i.a.) • Ammonia
Energy treatment of waste	• Carbon oxides • Sulphur oxides • Nitrogen oxides (N_2O, NO, NO_2) • Volatile organic compounds • Particulate pollutants (include metals)
Incineration	• Carbon oxides • Sulphur oxides • Nitrogen oxides (N_2O, NO, NO_2) • Volatile organic compounds • Particulate pollutants (include metals) • Acids (HCL, HF i.a.)
Storage and landfilling	• Particulate pollutants (include metals)
Treatment of waste oils	• Volatile organic compounds
Treatment of solvents	• Volatile organic compounds
Processes of separation of water contaminated with hydrocarbons	• Volatile organic compounds
Storage and warehousing of organic substances	• Volatile organic compounds

Table 4. Main contaminants of water resulting from treatment of hazardous waste (author's own research based on [2, 8].

Process	Chemical compound
Biological processes	Phosphorus; Nitrogen; Metals (e.g. As, Cd, Cu, Hg, Ni, Sn, Zn)
Physico-chemical processes	Phosphorus; Nitrogen; Metals (e.g. As, Cd, Cu, Hg, Ni, Sn, Zn)
Energy treatment of waste	Organic compounds (benzene, toluene, ethylbenzene, xylene, phenol)
Treatment of solvents	Chlorinated aromatic hydrocarbons; Organic compounds (benzene, toluene, ethylbenzene, xylene, phenol)
Treatment of waste oils	Metals (e.g. As, Cd, Cu, Hg, Ni, Sn, Zn); Organic compounds (benzene, toluene, ethylbenzene, xylene, phenol)
Storage and landfilling	Metals (e.g. As, Cd, Cu, Hg, Ni, Sn, Zn)

4.2 Economic conditions

Total investment costs of facilities depend on their capacities, location and technical solutions. Costs of facilities increase along with the capacity growth, whereas capital intensity indicator, unit exploitation cost and economic efficiency indicator decrease [2, 5].

Because of high investment costs it is necessary to prepare a detailed analysis of a group of waste produced in a region before making a decision as to the building of a plant treating and disposing hazardous waste. Such analysis should take into account first of all the waste which will be generated in a given region in a longer time perspective. It may turn out repeatedly that it is cheaper to store the waste and then transport it to existing installations instead of building a waste treatment plant.

In Table 5 unit costs of processes of recovery/ disposal of particular types of hazardous waste are compared.

Table 5. Unit costs of processes of recovery/disposal of hazardous waste (author's own research based on [9].

Method of recovery/disposal	Cost of recovery/disposal PLN/Mg
Thermal disposal	3,050
Immobilization in concrete	1,220
Individual processes	12,200
Separation of emulsion	366
Utilization in industrial processes	488
Chemical sewage treatment	976
Physico-chemical treatment	1,830
Disposal of waste from explosives	2,440
Storage until development of technologies	976
Storage in landfills for hazardous waste/secure chemical landfill	345.2 – 2,761.3
Storage of asbestos	170.3 – 732
Recovery of mercury	4,880
Recovery of solvents	3,050
Recovery of metals	610

The given values have been calculated from the original data (stated in PLN for the year 2002) with the use of GDP Deflator (annual percent) Poland 1991-2016 (The United Nations Industrial Commodity Statistics Database). The year 2015 has been adopted as the reference year.

Within certain groups of waste one can note very significant differences in unit costs of disposal/recovery. It arises undoubtedly from the fact that waste treatment plants are of different sizes which is directly connected with their capacities (the more waste treated, the lower unit costs). The cost of waste treatment is associated directly with the cost of storage and delivery to the treatment plant. The price is often the result of individual agreements between a waste-generating facility and waste disposal plant and can be different for various producers.

4.3 Technical and technological conditions

Recovery of raw materials from waste, waste disposal and storage require expensive installations, equipment and proper technologies. It involves costs related with waste transportation and reloading as well as investment and operating costs for facilities [10]. One should not also forget about burden to the environment through by-products such as sewage, waste gases and others [11, 12]. Nevertheless, building of new facilities for waste management enables the introduction of modern technologies and ensures the environment protection at a proper level. This is why it is so important to familiarize oneself with the structure of hazardous waste generation in a given area and application of optimal organizational and technological solutions codified in a form of BAT – *Best Available Techniques*.

Table 6 illustrates the division of used and proven technologies of hazardous waste disposal.

Table 6. Division of methods of disposal of hazardous waste taking account groups of disposed waste [11].

Method	Group of waste	Application
Thermal	02, 03, 04, 05, 07, 08, 10, 12, 13, 14, 16, 18, 20	In general thermal methods are applied where there is a possibility of mineralization of dangerous organic substances included in waste. Thermal decomposition can take place in classic chamber furnaces (containing special barriers causing turbulences), rotary kilns, fluid chambers and plasma furnaces [10]. Many methods of thermal utilization of dangerous waste consists in its total incineration together with other fuels and use of energy in processes of production, for example cement. In order for a given installation to be suitable for dangerous waste disposal, it must meet specific technical conditions determined in relevant regulations. The regulations define limit values for the temperature, oxygen content, time for maintaining exhaust gases in the incineration chamber [13].
Biological	02, 05, 06, 07, 08, 11, 12, 13, 16, 17, 18, 19	Biological methods are generally used for cleaning of plant protection products, heavy metals and oil derivative waste in the soil. They consist in selective actions by relevant bacterial strains decomposing toxic substances [14].
Chemical, physical, physico-chemical	06, 07, 09, 10, 11, 12, 13, 16, 17, 19, 20	The most commonly applied chemical methods consist in the neutralization of acids and alkalis. It is the precipitation of sparingly soluble chemical compounds of heavy metals included in waste in order to

DOI: 10.1051/e3sconf/20171902021

Table 6 (cont.)

		lower its solubility in water and processes of immobilization (sealing in concrete mass, vitrification). Another method is de-halogenation of chloro-organic compounds from used transformer oils, solvents and plant protection products. Other methods are also: distillation, gas stripping, extraction, adsorption as well as other physical processes [10].
Storage	10, 17, 19	Exceptional way of hazardous waste disposal is storage [15].

5 Case study – waste management options

The integrated approach for the environment protection against the negative influence of industrial activity places the emphasis first of all on the prevention, reduction and elimination of contamination. At the same time it is supposed to ensure the rational management of natural resources [16, 17]. Such approach has to take into account economic aspects too, so the consideration of costs and benefits. Using this kind of measures is not an easy task. It is necessary to combine several indicators and namely – technical and economic and ecological ones [18].

In order to make an analysis, it is worth taking a look at four selected instances of hazardous waste for which the manner of disposal to date was mostly the storage (Table 7).

Table 7. Proposal of alternative proceeding with chosen kinds of hazardous waste.

Waste[1]	Current condition[2]	Proposal
010505·	D5 (92%), D2 (8%)	In case of waste containing hydrocarbons 010505*, permissible methods are thermal processing of waste, bioremediation and solidification with use of pozzolana.
060403·	D5 (80%), D9 (10%), D8 (10%)	Solid waste containing arsenic should undergo processes of solidification, while liquid waste containing arsenic should be subject to processes of precipitation and then solidification after dehydration.
120118·	D5 (100%)	Sludge from treatment of metals containing oils should be solidified.
150202·	D5 (100%)	Waste from 150202* should undergo thermal degradation.

[1] – numbers of codes in accordance with the Regulation of the Minister for the Environment as of the 9th of December 2014 concerning the catalogue of waste (Journal of Laws 2014, item 1923) [19]
010505· – oil-containing drilling muds and wastes
060403· – wastes containing arsenic
120118· – metal sludge (grinding, honing and lapping sludge) containing oil
150202· – absorbents, filter materials (including oil filters not otherwise specified), wiping cloths, protective clothing contaminated by hazardous substances
[2] – numbers of codes in accordance with the Act on waste (Journal of Laws 2013, item 21) [20]
D2 – Land treatment (e.g. biodegradation of liquid or sludgy discards in soils, etc.)
D5 – Specially engineered landfill (e.g. placement into lined discrete cells which are capped and isolated from one another and the environment, etc.)
D8 – Biological treatment not specified elsewhere in this Annex which results in final compounds or mixtures which are discarded by means of any of the operations numbered D1 to D12
D9 – Physico-chemical treatment not specified elsewhere in this Annex which results in final compounds or mixtures which are discarded by means of any of the operations numbered D1 to D12 (e.g. evaporation, drying, calcination, etc.)

The main feature which makes the utilization of extractive waste difficult is the high content of water and the consistency of waste related to that (colloidal and sludgy in case of drilling waste and liquid in case of flow-back fluid). Hindrance is also the changeable and hard-to-predict chemical composition of waste (e.g. high variables of contents of heavy metals and salts of some alkaline elements (usually K and Na) in a form of chlorides, sulphates and hydrogen carbonates [21]. Biological treatment methods rely on the natural metabolic processes of living organisms (e.g. bacteria, fungi and higher plants) to destroy contaminants or convert them to a less toxic or less available form. The processes usually involve metabolisation of carbon-bearing substrates by organisms to generate energy for growth and reproduction. Bioremediation is commonly more effective for water-soluble contaminants since most biological activity takes place in this phase. The cost of biological treatment is generally lower than other treatment methods and many biological treatment techniques do not give rise to process residues requiring further treatment or disposal. Some of the techniques require large process areas and long time periods to achieve acceptable concentrations [22]. One of methods of recycling of drilling waste is solidification (immobilisation) of output. It is most commonly used in case of betonite drilling fluids [23]. The method consists in adding binding materials in a form of Portland cement, gypsum, limestone, fly ashes, blast furnace slag and hydraulic and puzzolans binder [21]. It causes hydraulic binding of soluble compounds (chlorides, heavy metals, petroleum-derived substances). This also increases physicochemical stability and mechanical resistance of the obtained material [21, 23].

The environmental contamination by arsenic is incidental in nature – it is a result of conduct of high-temperature processes on a large scale with the use of organic and non-organic natural resources as reagents and energy factors. Liquid and solid waste containing arsenic is the outcome of those processes. In the vast majority of cases the extraction of arsenic contained in

 DOI: 10.1051/e3sconf/20171902021

the waste is not economically viable because of too low concentration. For the disposal of liquid waste one utilizes precipitation methods consisting in binding removed ions containing arsenic into compounds insoluble in water. Obtained sediments are dehydrated in order to decrease their volume. The solid waste is subject to processes of solidification in a matrix of Portland cement with an addition of calcium, iron compounds and fly ashes [24-30].

Difficulties in complete management of metallurgic waste, especially for metallurgical purposes, arise mainly from the lack of proper economically viable technologies of its treatment for further use (e.g. concentration and dehydration of sludges, de-oiling of scale, removal of harmful elements) [31]. Metallurgic waste is especially difficult to dispose due to high hydration and high content of oils. This is why more and more frequently research is conducted on the possibility of use of the waste as additions for concrete coatings against ionizing radiation. As it arises from the performed examinations [32] the addition of the waste increases mechanical and adsorption properties of concrete coatings.

Materials such as used sorbents, oil filters and textile materials contaminated with hazardous substances should be disposed with use of thermal methods. Remains after the incineration (slag and ash) amount to 20-60 percent by weight and 10-20 percent by volume of the original mass of waste [5].

Each strategy of development of hazardous waste recovery/disposal plant requires taking investment initiatives executed in specific enterprises. One should bear in mind, however, that the primary objective of company operation is generating profits. The development strategy may involve not only modernization and innovative investments, but also ventures regarding the public interest connected with specific legal regulations – as in the described cases – which serve the environment protection.

On the basis of data provided in the literature [6, 8, 33] which includes the structure of costs related to construction of facilities, facility maintenance and costs of treatment of particular waste, discount costs have been calculated – both investment and operating ones– for three installations: hazardous waste combustion plant, hazardous waste solidification plant and bioremediation facility. The aim of analysis was calculating the value of unit cost (PLN/Mg) with the use of dynamic (discount) methods. In this work [34] NPV (*Net Present Value*) and IRR (*Internal Rate of Return*) indicators have been calculated. Only those two indicators have been calculated as others are static and rather useless for the analysis of facilities functioning for more than a dozen of years. The calculation was supplemented by the count of an average cost of waste treatment which should be the basis for determining the price. Such cost is defined as a quotient of discounted total costs and a discounted amount of treated waste. Indicators for prices listed in the literature sources have also been compared (Table 8).

The price of disposal of 1 Mg of hazardous waste for the technology of immobilization is 469 PLN. Calculated costs of combustion of 1 Mg of waste are at 1,758 PLN, while the cost of bioremediation of 1 Mg of waste is 310 PLN.

Table 8. Listing of calculated costs of hazardous waste disposal (author's own research based on [9, 22, 34, 35].

Technology	Calculated disposal cost, [PLN/Mg]	Unit costs of disposal quoted in literature, [PLN/Mg]
Immobilization	469	132[1] – 1,220[2]
Incineration of solid hazardous waste	1,758	150 – 3,667[1] 3,050[2] 1,000 – 2,400[3]
Bioremediation	310	150 – 3,667[1] 315[3]

The given values have been calculated from the original data (quoted in USD for the year 2005[1]) with the use of US Dollar Implicit Price Deflators for Gross Domestic Product 1929-2013. The year 2010 has been adopted as the reference year. The value has been calculated from the original data in USD[1] and Euro[3] with the NBP (National Bank of Poland) average exchange rate (year 2010) at 3.0157 PLN/USD and 4.03 PLN/€.
The given values have been calculated from the original data (stated in PLN for the year 2002[2]) with the use of GDP Deflator (annual percent) Poland 1991-2016 (The United Nations Industrial Commodity Statistics Database). The year 2010 has been adopted as the reference year.
[1] – prices from the year 2005 [35]
[2] – prices from the year 2002 [9]
[3] – prices from the year 2010 [22]

The calculated values are within the range of actual prices. Economic criteria, although significant, cannot still prevail in deciding about the method of disposal or recovery. The most important is till the proper neutralization of waste. The selected methods do not generate unreasonable costs and therefore they do not pose a threat to the economic development and they reduce significantly the contamination of the environment at the same time.

6 Summary

The hazardous waste poses a serious threat to all components of the environment (soil, water, air). This is why the selection of proper process of disposal or recovery (consistent with engineering knowledge) is so important. Due to its properties, the stage of proper selective collection and storage of waste (before the actual disposal/recovery) is very important. Any emission or irregularity may result in serious pollution of both soil and water.

One should ascertain that:

1. The structure of generated waste is a result of industrial specificity of a given state. In all of the European Union UE-28 a total of 95,02 million Mg of hazardous waste was produced in 2014. Poland is in a group of those countries, which generated the least (2 percent) of hazardous waste in 2014. The most of hazardous waste was produced in France (11%), Bulgaria (13%) and Germany (23%).
2. The biggest part of dangerous waste underwent processes of recovery other than energy recovery –

except backfilling – 79,53%. Unfortunately, some 13% of hazardous waste is still deposited.

3. In the classification of technologies of hazardous waste disposal one accepts the rule of grouping waste according to its properties and possibilities of disposal (Table 1). Depending on the chemical composition the waste can be disposed in identical conditions. In Table 2 particular groups of hazardous waste along with applied methods of their disposal are presented.
4. In the work all aspects of dangerous waste management (both environmental, technical and economic), which should be followed while choosing modern techniques of management of hazardous waste, have been discussed.
5. Alternative methods of proceeding with four groups of waste for which the most frequently applied method of disposal is the storage have been proposed (Table 7).
6. Calculated values of prices of waste disposal placed in Table 8 are in a range of actual prices.
7. Proposed changes in the manner of proceeding with waste collected currently on hazardous waste landfills contribute significantly to limiting this practice which is the least favorable to the environment.

The work is financed within the frames of statutory research no 11.11.210.373 at the Faculty of Energy and Fuels, AGH University of Science and Technology.

References

1. Eurostat (2017) http://ec.europa.eu/eurostat/web/environment/waste/d atabase
2. J. Pyssa, Gospod. Surowcami Min. **24, 1/1,** 113-125 (2008)
3. Cz. Rosik-Dulewska, *Fundamentals of Waste Management* (Polish Scientific Publishers PWN, Warsaw, 2015) (in Polish)
4. J. Bendkowski, M. Wengierek, *Logistics of Waste – Logistic Processes in Waste Management* (Publisher Politechnika Śląska, Gliwice, 2002) (in Polish)
5. J. Bendkowski, M. Wengierek, *Logistics of Waste – Facilities for Waste Management* (Publisher Politechnika Śląska, Gliwice, 2004) (in Polish)
6. J. Pyssa, Przemysł chemiczny/Chemical industry **89/7**, 927–934 (2010)
7. http://www.access.zgwrp.org.pl/materialy/dokument y/PlanyGospodarkiOdpadami/1/kpgo/Rozdzial_3.3.p df
8. Best Available Techniques (BAT) Reference Document for Waste Treatment. Industrial Emissions Directive (2010/75/EU) (Integrated Pollution Prevention and Control). European Union, December 2015
9. Monitor Polski (the Official Journal) no 11, item 159, Resolution no 219 of the Council of Ministers of the 29th of October 2002 on National waste management plan, Warsaw, of the 28th of February 2003
10. M.N. Rao, R. Sultana, S.H. Kota, *Solid and Hazardous Waste Management. Science and Technology* (Butterworth-Heinemann, 2017) p. 159–207.
11. J. Pyssa, IOP Conference Series: Earth and Environmental Science (2017) (to be published)
12. J. Wu, Y. Guo, Ch. Li, H. Qi, J. Cleaner Prod. **149,** 49–59 (2017)
13. Regulation of the Minister of Economic Development of the 21st of January 2016 (Journal of Laws 2016, item 108) on requirements concerning the conduct of process of thermal transformation of waste and ways of proceeding with waste resulting from that process
14. T.C. Zhang, R.Y. Surampalli, R.D.Tyagi, S.K. Benerji, *Current Developments in Biotechnology and Bioengineering. Solid Waste Management* (Elsevier B.V., 2017) p. 311–340
15. I. Pecorini, F. Baldi, D. Bacchi, E.A. Carnevale, A. Corti, Waste Manage. **63,** 96–106 (2017)
16. *National Hazardous Waste Management Plan 2014 – 2020* (Environmental Protection Agency, Ireland, 2014)
17. Resolution no 88 of the Council of Ministers of the 1st of July 2016 on National waste management plan 2022 (M.P. 2016, item 784)
18. M. Gollinger-Tarajko, *Recovery of Waste – Technologies and Possibilities* (Publisher IGSMiE PAN, Kraków, 2005) p. 14–22.
19. Regulation of the Minister for the Environment as of the 9th of December 2014 concerning the catalogue of waste (Journal of Laws 2014, item 1923)
20. Act on waste (Journal of Laws 2013, item 21) of the 14th of December 2012 (uniform text Journal of Laws 2016, item 1987, 1954)
21. J. Pyssa, E3S Web of Conferences **10**, 00076 (2016)
22. Technical and Economic Aspects of developing a National Difficult Waste Facility (NaDWaF) (SKM Enviros, July 2010)
23. T. Steliga, M. Uliasz, Nafta-Gaz, **5**, 273-283 (2012)
24. C. Visvanathan, Resour. Conserv. Recycl. **16**, 201–212 (1996)
25. C. Vandecasteele, V. Dutre, D. Geysen, G. Wauters, Waste Manage. **22/2**, 143–146 (2002)
26. T.S. Singh, K.K. Pant Journal of Hazardous Materials, **131/1–3,** 29–36 (2006)
27. Ch.D. Johnson, J. Feldmann, D.E. Macphee, Waste Manage. **27,** 375–379 (2007)
28. J. Kumpiene, A. Lagerkvist, Ch. Maurice, Waste Manage. **28,** 215–225 (2008)
29. I.H. Yoon, D. H. Moon, K.W. Kim, K.Y. Lee, J.H. Lee, M. G. Kim, J. Environ. Manage., **91/11**, 2322–2328 (2010)
30. X. Wang, F. Zhang, Z. Nong The Scientific World Journal, **2014,** Article ID260504 (2014)

31. A. Frackowiak, M. Pacholewska J. Buzek, Rudy i Metale Nieżelazne **51/1**, 325-332 (2006)
32. M. Alwaeli, *Fuel from waste V* (Publisher Helion S.A. Gliwice, 2005) p. 391-394 (in Polish)
33. Best Available Techniques (BAT). Reference Document on Waste Incineration. May 2017
34. J. Pyssa *Selection of technologies for the disposal of hazardous waste in the aspect of environmental protection on the example of the Małopolska province.* Doctoral thesis (Akademia Górniczo-Hutnicza, Kraków, 2008) p. 1-191.
35. M. G. Puder, J. A. Veil, *Offsite Commercial Disposal of Oil and Gas Exploration and Production Waste: Availability, Options and Costs.* ANL/EVS/R-06/5. Environmental Science Division, August 2006

Sustainable agriculture and protection of the environment

Ewa Siemianowska[1], *Andrzej* Wesołowski[1], *Krystyna A.* Skibniewska[1*], *Józef* Tyburski[2] and *Marcin* Gurzyński[1]

[1]UWM in Olsztyn, Faculty of Technical Sciences, 10-724 Olsztyn, Poland
[2]UWM in Olsztyn, Faculty of Environmental Sciences and Agriculture, 10-724 Olsztyn, Poland

Abstract. The economic, environmental and social development should not degrade the environment but it should leave it for the next generations in the state that it is presently or even better. The principle of sustainable agriculture is to cover the human needs for food without damage to the environment. The aim of the article was to research the farmers' awareness of the principle of sustainable agriculture and balanced fertilization and their influence on the environment. Among 100 farmers of the Tczew district (Poland) there was done questionnaire research on the determination rates of nitrogen fertilizers and on the regulation of fertilizers usage in Poland. Most of farmers declared a good knowledge of good agricultural practices and of balanced fertilization and the awareness of threats issuing from their activities. At the same time in Poland since the announcement of the Nitrate Directive of the former European Common Market (1992) up till now (2013) the application of nitrogen fertilizers doubled and the yield of wheat increased only by 15%, which means the increase of environmental burden with this chemical element.

1 Introduction

The aim of sustainable growth is to ensure the present and the future generations the access to the environment and the keeping it in non-degraded state. The issue of sustainable development is very important from the point of view of society, but also from the economic point of view. Though the policy of sustainable growth the EU tries to bring the environment to a state which prevents it from more damage [1, 2]. No implementation of the policy of sustainable growth, which serves the recognition and solving of the present environmental, social and economic problems, would mean that achieving these goals is impossible. Thanks to the policy of sustainable growth we obtain more and more energy from renewable sources, we reduce the emission unwholesome ashes and gases to the atmosphere. Progress in environmental protection can be made through responsible economic, environmental and social policy, which is accepted by society and enterprises [3 – 5]. People are very much aware today of threats produced by irresponsible usage of environment [6–9].

In the case of surface and ground (drinking) waters, their contamination with nitrogen compounds (nitrates) has mainly an agricultural origin. In Poland (similarly to other EU countries) over 50% of nitrogen as well as phosphorus contaminating water come from agriculture [10]. One of the measures taken by the EU was an introduction of so-called Nitrate Directive and the principles of so-called sustainable agriculture.

1.1 Sustainable agricultural development

A definition of sustainable agriculture is based on the concept of sustainable growth, according to which the sustainable agriculture development is such a use of environment which combines the use and maintenance of earths' resources for the present and future generations in the present or better state, but not worse [11,12]. It will be possible to maintain the present state of soil fertility and the genetic diversity of animals and crops, keeping at the same time the environment in a good condition, by implementation of rules of sustainable development in agriculture, forestry and fishery.

In today's agriculture, fertile soils are cultivated in most cases, and cultivation of low quality soils requires covering high expenditures on means of production. The effectiveness of such an approach is not sufficient on low quality soils, which results among others from the higher expenditures on fertilization and its harmful influence on the environment [2, 13]. Królczyk et al. [14] demonstrated that yields of wheat and rapeseed can even be doubled in Poland due to broad implementation of rules of sustainable agriculture.

1.2 Balanced fertilization versus water quality

Fertilization of crops is the basic yield producing factor, whereas it is the most threatening to the water environment. Especially important is a risk of penetration of biogenes coming from nitrogen and phosphorus fertilizers to ground waters and water reservoirs, which are used in large doses, because of a very high demand of crops for above mentioned macro elements [15, 16]. Excessive/unbalanced fertilization results in disturbing the macro elements equilibrium in

* Corresponding author: kas@uwm.edu.pl

waters, which in turn leads to degradation of biodiversity in surface waters and contamination of surface and ground waters. Aware of these processes, the EU issued so–called Nitrate Directive – Council Directive 91/676 of the European Common Market as early as a quarter of century from Dec. 12. 1991 in order to prevent water pollution with nitrates of agricultural origin. The Nitrate Directive of the EU is to protect the environment from penetration of biogenes from fertilizers to waters, among this to prevent using too high doses of nitrogenous fertilizers in accordance with the principles of sustainable growth [17, 18].

The Nitrate Directive obliges member states of the EU to take a number of measures, among them to delimitate areas of member states that are especially susceptible from which:

- run-offs / washing out take place to the surface and/or ground waters, which contain or may contain 50 mg/l of nitrates, if measures described in the Directive will not be taken
- run-offs take place to waters which results or may result in their eutrophication.

These areas are zones susceptible to contamination with nitrogen compounds and described according to the Council Directive 91/676 European Common Market as threatened zones (Nitrate Vulnerable Zones - NVZ), whereas according to the Polish parliamentary act from July 18, 2001 Water Law as nitrate vulnerable zones or NVZ.

Countries which join the EU are obligated to implement the provisions of the Nitrate Directive and the transposition the directive to the domestic law. The member states are obliged to monitor the concentration of nitrates in surface and ground waters and to assess the state of eutrophication of fresh surface waters and the estuary and seashore waters with the aim of reviewing and completion NVZ. They should also determine an agenda for NVZ survey as well as review agendas and additional measures at least every 4 years. The duty of member states is to inform the European Council within 6 months about all changes or completion of NVZ zones and of agenda, and to make reports for the European Council containing information on each 4-year period of implementation of the Nitrate Directive, within 6 months since the termination of the period to with it pertains [19, 20].

All these measures and abiding the legal regulations are to conduct a rational management of fertilizers on farms. An important aspect of applying a balanced fertilization is a recognition of the abundance of the soil with micro– and macro elements, as well as of other chemicals properties of the soil. Thanks to analyses of the soil, the appropriate rates of nutrients for each crop can be counted [21 – 23]. It allows to limit the biogenes contamination of surface and ground waters. Moreover, a balanced fertilization leads to a financial savings.

1.3 Transposition of the Nitrate Directive to the domestic law in Poland

In Poland the transposition of the Nitrate Directive to the national law began before the access to the EU in 2004 and introduces a number of parliamentary acts and decrees starting from 2000, among others the parliamentary act from July 26, 2000 *On fertilizers and fertilization*, the parliamentary act from July 18, 2001 *Water Law*, the decree of the Minister of Environment from Dec. 2002 (Journal of Laws No 241, position of 2093) and (Journal of Laws No 4, from 2003, position of 44), the decree of the Minister of Agriculture and Rural Areas Development from June 1, 2001 *On the detailed applications of fertilizers and on conducting trainings of their use* (Journal od Laws No 60, position of 616), 32 decrees of the directors of Regional Boards of Water Management on the *determination of waters which are susceptible to the contamination with nitrogen compounds of agricultural origin* and of *areas especially susceptible* as well as *On introduction of the agenda aiming to limit of running off of nitrogen of agricultural origin for the especially susceptible areas* (National Board of Water Management - RBofWM). Moreover Poland as a member state undertook the realization of the Council Directive 91/676/ECM from Dec. 12, 1991 on protection of waters against the contamination with nitrates of agricultural origin, on the date of the EU accession that is since May 1, 2004.

In the framework of the implementation of the Nitrate Directive areas especially susceptible to nitrate of agricultural origin, from which the outflow of nitrogen of agricultural origin to waters should be limited were delimitated. These areas composed barely 2% of the national territory and they were legally defined by the means of 11 decrees of the directors of respective regional boards of water management, issued mostly at the end of 2003 and in the 1st quarter of 2004.

For all NVZ 21 programs were introduced by the directors of RBofWM. The decrees were published in the Journal of Laws of Provinces and thus they became local law. The introduction of the schedules began on May 1, 2004 and lasted for 4 years till April 30, 2008. The tasks were addressed especially to farmers and partially also to institutions professionally related to agriculture and environment and to the circles cooperating with farmers. To the most important tasks were following the principles of GAP (Good Agriculture Practice) by farmers and realization of investments of building tanks and platforms for collection and storage natural fertilizers (originated from animal husbandry). Farmers were provided with organizational and technical help in the realization of the water protection investments on the farms (i.e. in the building of tanks and platforms for collection and storage natural fertilizers and of devices for treating of household sewage) and agricultural advising in preparing fertilization plans on the farms. The whole activities were controlled according to the schedules and the monitoring of the quality of surface and ground waters was done in the areas of special susceptibility, with the aim of assessment of the result of the action plans. Also, the training of the farmers on abiding law and the principles of GAP was carried out. An improved Code of

 DOI: 10.1051/e3sconf/20171902022

GAP was prepared which was adjusted to the recommendations of the Nitrate Directive [23, 24].

Realization of the article10th of the Council Directive 91/676 ECH obliged Poland to submit to the European Commission a report containing information on the 4–year period of introduction of this Directive, within 6 months since the termination of the period to which it pertains (in the case of Poland to the end of October 2008). The Report of the Minister of Environment on the realization of the regulation of the Council Directive from December 12, 1991 on the protection of waters from the contamination with nitrates of agricultural origin (91/676/ECM) deals with verification of the areas in Poland which are especially susceptible to the nitrates of agricultural origin and introduction of the plans of action aiming at the limitation running off of nitrates. Also, assessment of the contamination of waters with nitrates of agricultural origin and of the state of agriculture the borders of areas of special susceptibility lying within the administration of each RBofWM were verified by reducing by 26.2% in comparison to the state from 2004, therefore, in the period between May 2008 and April 2012 there were 19 NVZ (earlier 21 zones). The total NVZ area amounted to 4 623.14 sq km, which makes ca. 1.48 % of the national area, 0.5% less than in 2004 (earlier 2%). The reason for shrinking of the territory of the areas of special susceptibility in 2008 was, among others, their delimitation in the river drain basins according to geodetic units and not to commune borders, which was the case in 2004 [23-26].

New plans of action were prepared for the verified and newly-delimitated NVZ areas. At the end of 2012 Poland submitted to the European Commission a successive report entitled *A report of the realization of the provisions of the Council Directive from Dec. 12, 1991 on the protection of waters against the contamination with nitrates of agricultural origin (91/676/ECH) in the period of May 1, 2008 – April 30, 2012*, which was a contribution to the Report of the European Commission for the 27 EU. The European Commission was not satisfied with the acreage of the areas of special susceptibility, which functioned in the period of May 1, 2008 and April 30, 2012 and made barely 1.48% of the national area. So, in the period of January – June 2012 each RBofWM made a new verification of the special susceptibility areas on the basis of the expert appraisement submitted by the Ministry of Agriculture and Development of Rural Areas entitled: *An assessment of the agricultural pressure on the state of surface and ground waters and the indication of the areas of special susceptibility of the contamination of the state of surface agricultural origin* made in 2011 by the Institute of Soil Science and Plant Cultivation in Puławy, Poland, containing scientific evidences and analyses on the state of agriculture in Poland with the assessment of agricultural pressure on the state of surface waters (rivers, lakes, transmission waters and seashore waters) and ground waters, with the additional aid of the work of the Warsaw University entitled *Delimitation of areas under real pressure of agricultural activities considering the water contamination with nitrogen compounds* instructed by the National Board of Water Management. As a result of this verification the special susceptibilities areas expanded from 4 603.47 sq km (1.48% of the national area) to 13 935.06 sq km (4.48 % of national area), which makes a 3-fold (ca. 200%) increment compared to the state of 2008 [24].

Since the moment of the access of Poland to the EU our farmers have been given so-called direct supplements to each hectare of agricultural lands. From the very beginning it was clear that there is a well-founded apprehensions that at least a significant part of this money will be used to intensify production, among which to increase fertilization [27]. The degree to which this apprehension came true is shown in Table 1.

Table 1. The changes of rates of mineral fertilizers in Poland in the period of 1992-2013 and of the average yields of wheat.

Years	Rates of mineral fertilizers, kg ha^{-1}				Wheat yields, tha^{-1}
	N	P	K	Ca	
1991	39.9	22.3	32.9	117.2	3.80
2004	56.3	20.4	25.7	91.5	4.28
2013	80.7	25.6	26.7	40.3	4.44

Source: The Statistical Yearbook of Poland, GUS 1992, 2005, 2014

Contrary to the intention of the EU legislation, and particularly to the Nitrate Directive, the use of nitrogen fertilizers increased significantly in Poland from ca. 40 kg per 1 ha in the year of passing the Directive in the then countries of the EU (1991), through 56.3 kg in the year of Polish accession to the EU (2004), up to 80.7 kg N per 1 ha in 2013. However the doses of phosphorus fertilizers increased beyond the balance needs, but potassium fertilization decreased. It should be emphasized that the reduplication of the doses of nitrogen fertilizers was not accompanied with the proportional increase of crop yielding. In Poland wheat is the most often cultivated cereal and also of the highest nitrogen demand.

This species has heightened its yield by ca. 15%. It indicates the increasing disproportion between the growing fertilization and growing yields. In other words the decreasing effectiveness must be accompanied with the larger losses of N to the environment, among these to the ground and surface waters. What is worse, the growing doses of nitrogen fertilizers was accompanied with the decreasing calcium fertilization – in the period at issue it diminished almost 3-fold. Meanwhile nitrogen is one of more important factors that acidify soil, (it is assumed that each kg of mineral nitrogen causes soil acidity, which should be compensated by usage of 2 kg of CaO). In the period under discussion the 2-fold increased doses of N and 3-fold diminished of Ca had to influence negatively the state of soils and waters (decreasing productivity of fertilizers, increase of washing out of nutrients). It is doubtful whether Polish farmers are really aware of such goings-on of the situation and are mature to a reflection and among this the modification of way fertilization. To the deteriorating state of the water habitat in Poland in the period under discussion testifies among others the result of research, documenting the increasing surplus of N (lack of balance) and the growing contamination of ground waters with nitrates and pesticides [27]. The balance surplus of nitrogen in Poland amounts to ca. 60 kg of N per 1 ha a year, whereas the recommended quantity stands at 25-35 kg of N [28]. In this context, it should be noted that such mobile nutrient as nitrogen can be utilized in 100% in creating a yield. And the higher its doses the greater damage to the environments on very intensive farms the yield – producing effectiveness of N drops even below 50% [27].

In some regions of Poland the content of nitrates in ground waters has approach the top admirable norm – 50 mg per 1 litre. The realistic method of water protection against the contamination with nitrates (and at the same time with pesticides) more and more often used in practice in Germany, is a conversion of farms in the water protection zones to the organic system [29]. In this context it is worth mentioning that on the neighboring Germany in the period of 1992-2013, the average rates of nitrogen fertilizers were subject to small changes, amounting at the level of ca. 100 kg ha^{-1} (Table 2). In the same time the rates of phosphorous fertilizers were halved, what considerable diminished the risk of eutrophication of waters with this element (reduction of surplus of P to zero). The rates of potassium fertilizers were also diminishing by more than 40%, and at the same time the yield of wheat increased from 6.31 to 8.00 tha^{-1}. Thus undoubtedly the effectiveness of utilization of fertilizers was improved and their surplus was diminished, which testifies to the better balance of fertilization in that country than before the passing of the Nitrate Directive. Hence it is justifiable to claim that in Germany the progress was made in the protection of waters in the period of functioning of the Directive, whereas in Poland the situation worsened considerably.

Table 2. The changes of rates of mineral fertilizers (N, P, K) in Germany in the period of 1991-2013 and of the average yields of wheat [30].

Years	Rates of mineral fertilizers, kg ha^{-1}			Wheat yields, t ha^{-1}
	N	P	K	
1991	99,2	33,8	48,5	6,31
2004	105,3	19,3	28,3	8,17
2013	98,9	17,1	27,8	8,00

The introduction of the Nitrate Directive to the national law and to the Code of Good Agricultural Practice (GAP) did not stop the intensification of agricultural production or the increase of expenditures on chemical pesticides. We are witnessing the process of creation of large-area farms, on which the recommendations of appropriate crop protection and that of proper fertilization are not followed (among them the Nitrate Directive) and the balance between crops and animal production is unsettled. The non-observance of GAP causes that agriculture has still a negative impact on the natural environment of rural areas [31].

In 2007-2013 Poland attended Baltic Deal programme aiming to rise competence of farmers in sustainable agriculture [32]. Hundreds of farmers and agricultural advisers attended seminars, trainings, courses, workshops etc. learning how to reduce nutrients losses not to do harm to production effectiveness and farm competitiveness. The aim of the study was to research the farmers' awareness of the principle of sustainable agriculture and balanced fertilization and their influence on the environment.

2 Materials and methods

The research was done in 2016 by the means of pioneer study by the method of sounding measurement, applying a direct technique. The study group was made up to 100 farmers, running farms in Tczew district Pomeranian Province, Poland. It is a region of ones of the best soils in the country, where agriculture is one of the most intensive, and the wheat is a dominate crop. The study was conducted at a shop of pesticides, using a questionnaire of 22 questions. The selection of research objectives was made at random.

During the collection of research materials, the system analysis of information was applied which was oriented at the analysis of secondary sources [33], among them the works on the knowledge of counting of fertilizers' rates and the familiarity of the regulation of fertilizers' application in Poland. The research material was worked out with the aid of the horizontal and comparative economic analysis.

3 Results and discussion

In the researched group of farmers, the age was very various. The majority of the inquired were male. The most numerous group of the inquired people was composed of farmers, whose farms' area ranged from 10 to 20 hectares (36%), afterwards below 10 ha (30%), in the interval of 20-50 ha (20%) and 50-100 ha (14%).

Among the inquired only 38% had the knowledge on soil composition determined by a chemical analysis in a chemical-agricultural laboratory (Table 3). Such information allows to balance fertilization and to optimize the expenditures spent on fertilizers, resulting at the same time in decreasing environmental damage. Most farmers (62%) did not get their soil chemically analyzed, which may be the reason for applying inappropriate, unbalanced (often too high) rates of fertilizers and deficiency of nutrients in soil and in crops.

This fact is all the more alarming that only a part of the inquired was aware of the need of analysis of soil samples for the content of nutrients with recommended frequency, i.e. every 2 (12%) or 3 years (18%). Farmers which underwent examinations with the frequency of every 5 year (8%) should consider more frequent analyses because of rather fast changes of nutrients content in soil (Table 4).

DOI: 10.1051/e3sconf/20171902022

Table 3. Declared by farmers sources and needs of knowledge on the macro- and micro- elements content in soil.

The entire group	0-10 ha	10-20 ha	20-50 ha	50-100 ha
Analysis of soil in a chemical-agricultural laboratory				
38%	4%	14%	14%	6%
Own practice of many years'				
44%	10%	22%	4%	8%
Unnecessary information				
18%	16%	0%	2%	0%

Table 4. Declared by farmers frequency of soil chemical composition analyzing.

The entire group	0-10 ha	10-20 ha	20-50 ha	50-100 ha
Every 2 years				
12%	0%	6%	2%	4%
Every 3 years				
18%	4%	6%	8%	0%
Every 5 years				
8%	0%	2%	4%	2%
I do not examine				
62%	26%	22%	6%	8%

The farmers applied in most cases only mineral fertilizers (68%). It issues from the dominating trend in Poland of reducing animal husbandry, especially that of pig and cattle (Table 5).

Table 5. Declarative kind of applied fertilizers.

The entire group	0-10 ha	10-20 ha	20-50 ha	50-100 ha
Mineral				
68%	20%	28%	14%	6%
Organic				
0%	0%	0%	0%	0%
Mineral-organic				
32%	10%	8%	6%	8%
Natural				
0%	0%	0%	0%	0%

A half of the farmers under examination (50%) declared the knowledge of the regulations on application of organic/natural fertilizers. According to law these fertilizers can be applied from March 1 to November 30. The majority of inquired (58%) was aware of advantages of using organic fertilizers, though there was also a group of farmers (10%) convinced that this kind of fertilization did not supply nutrients to the soil (Table 6).

Table 6. Declared time-limits of application of organic natural fertilizers.

Time-limits of application of organic fertilizers				
The entire group	**0-10 ha**	**10-20 ha**	**20-50 ha**	**50-100 ha**
From March 1 to November 30				
50%	14%	22%	8%	6%
From February 1 to October 30				
24%	8%	4%	6%	6%
From April 1 to November 30				
26%	8%	10%	6%	2%
Entire year				
0%	0%	0%	0%	0%
Knowledge on impact of organic fertilization				
The entire group	**0-10 ha**	**10-20 ha**	**20-50 ha**	**50-100 ha**
It shapes the content of nutrients and organic matter in soil				
58%	16%	26%	8%	8%
It influences only the content of organic matter in soil				
30%	10%	6%	8%	6%
I inethink that organic fertilization is useless				
10%	4%	4%	2%	0%

Most farmers (88%) knew the principles of application of nitrogen fertilizers as provided by parliamentary acts. This knowledge allows their proper usage and the limitation of environmental damage. Unfortunately, the majority of the inquired (54%) made allowances for a nutrition demand of crops only by visual examination while determining the rates of nitrogen fertilizers. Solely a part of the farmers allowed the nutrients content in soil and the requirements of crops (18%). The most distressing phenomenon was the fact that there were farmers who did not even consider the crops' needs while fertilizing (Table 7).

Basing on the knowledge of determining the rates of nitrogen fertilizers by the farmers it should be concluded that a half of them (50%) considered only the nutrition requirements of crops to fit the planned yield. The residues of N from the previous growing season should be also taken into account, which was a case in 42% of the inquired.

An avoidance of the present content of N in a soil in determining the doses of nitrogen for the planned yield, may result in its excessive washing out to the ground and surface waters. The majority (70%) of inquired was conscious what a dose of nitrogen they may have supplied in natural fertilizers in a year (Table 7).

Table 7. Declared by the farmers knowledge on the principles of N fertilizers application and practice.

Knowledge on the principles of application of nitrogen fertilizers				
The entire group	**0-10 ha**	**10-20 ha**	**20-50 ha**	**50-100 ha**
Possibility of application all year round				
0%	0%	0%	0%	0%
Application to soil in appropriate condition				
88%	24%	32%	20%	12%
In spring				
12%	6%	4%	0%	2%
Declared knowledge on the principles of determining of rates of nitrogen fertilizers				
The entire group	**0-10 ha**	**10-20 ha**	**20-50 ha**	**50-100 ha**
Exclusively on the basis of the content of N in soil				
0%	0%	0%	0%	0%
On the basis of the content of N in soil and crops requirements				
22%	2%	14%	2%	4%
On the basis of the content of N in soil and crops requirements for the planned yield				
18%	4%	0%	12%	2%
Exclusively on the basis of crops requirement				
54%	18%	22%	6%	8%
I do not consider the above mentioned factors				
6%	6%	0%	0%	0%
Declared ways of determining the rates of nitrogen fertilizers				
The entire group	**0-10 ha**	**10-20 ha**	**20-50 ha**	**50-100 ha**
I allow the content of N in soil and crops requirements for the yield				
42%	6%	14%	14%	8%
I allow only crops requirements for the yield				
50%	8%	22%	6%	6%
I always apply 300-350 kg N per 1 ha				
8%	8%	0%	0%	0%
Observance of permissible rates of nitrogen in natural fertilizers in a year				
The entire group	**0-10 ha**	**10-20 ha**	**20-50 ha**	**50-100 ha**
Yes				
30%	12%	8%	6%	4%
No				
70%	18%	28%	14%	10%

The farmers under examination demonstrated the knowledge on the influence of souring of soils on the assimilation of nutrients by the crops. The inquired were (96%) conscious of the importance of liming and the correlation between the pH of soil and the assimilation of macro- and microelements. The majority of the farmers (70%) cares of the appropriate level of the soil pH on their farms (Table 8).

Table 8. Declared knowledge and practice of liming.

Knowledge on the impact of liming on the assimilation of macronutrients				
The entire group	**0-10 ha**	**10-20 ha**	**20-50 ha**	**50-100 ha**
Yes				
96%	26%	36%	20%	14%
No				
4%	4%	0%	0%	0%
Frequency of application of lime				
The entire group	**0-10 ha**	**10-20 ha**	**20-50 ha**	**50-100 ha**
On the basis of the needs issuing from an analysis of the pH of soil				
70%	14%	24%	18%	14%
Once in a 4 years				
24%	10%	12%	2%	0%
When I recall to mind				
6%	6%	0%	0%	0%

It has been proved that the farmers know the principles of fertilization in the proximity of water reservoirs and watercourses. Most farmers (84%) declared the necessity of observing a special caution during application of fertilizers in the neighborhood of water reservoirs (Table 9).

The greatest part of the inquired farmers (62%) had also the knowledge of the issues of sustainable agriculture and they had proper associations with sustainable farming. The majority (86%) gave a right answer, which proves the good awareness of the farmers of the issues of sustainable growth (Table 9). Moreover most of the farmers (66%) claims that they observed GAP. Fortunately enough, the most numerous group of farmers (22%) that do not observe these rules, are the owner of small farms up to 10 ha (Table 10).

Table 9. Observing caution during application of fertilizers by water reservoirs.

The entire group	**0-10 ha**	**10-20 ha**	**20-50 ha**	**50-100 ha**
Yes				
84%	24%	28%	18%	14%
No				
16%	6%	8%	2%	0%

Table 10. Declared knowledge on the goals of sustainable farming knowledge on the goals of sustainable farming.

Knowledge on the goals of sustainable farming knowledge on the goals of sustainable farming				
The entire group	**0-10 ha**	**10-20 ha**	**20-50 ha**	**50-100 ha**
Maximization of yield, does not make allowances for the environment needs				
4%	4%	0%	0%	0%
Optimization of yield, minimization of inputs, allowing the environmental needs				
34%	14%	12%	6%	2%
Optimization of yield and inputs, allowing the environmental needs				
62%	12%	24%	14%	12%
Associations with sustainable agriculture				
The entire group	**0-10 ha**	**10-20 ha**	**20-50 ha**	**50-100 ha**
Social, economic and environmental responsibility				
86%	26%	30%	16%	14%
Social and environmental responsibility				
12%	2%	6%	4%	0%
No associations with sustainable agriculture				
4%	4%	0%	0%	0%
Applying the principles of GAP				
The entire group	**0-10 ha**	**10-20 ha**	**20-50 ha**	**50-100 ha**
Yes				
66%	8%	32%	14%	12%
No				
34%	22%	4%	4%	2%

4 Conclusions

In the questionnaire study most of the inquired declared the knowledge of the principles of GAP and need of applying a balanced fertilization. The farmers declared that they were aware of the influence of agricultural activities on the environment. Anyway the statistical data pertaining the application of mineral fertilizers in Poland prove that declared knowledge on the above–mentioned matter bears no relation to the fertilization practice. There is a need of greater and greater indoctrination of agricultural society of the necessity of putting into practice the principles of sustainable growth, and particularly of balanced fertilization [34-36].

The presented statistical data testify the duplication of the Nitrate Directive of the EU (1991) to 2013, with barely a 15% increase of wheat yields. It indicates the lack of a proper fertilizers' balance and the augmentation of environment threat – in other words the nitrate Directive was implemented in Poland in such a way that it has not worked so fair. In the same time in the neighboring Germany, a considerable increase of yields was obtained white keeping the same size of N rates. In that country the Nitrate Directive was properly implemented and yielded the assumed result, i.e. the improvement of the effectiveness of fertilization and the state of the environment.

In the above context it can be concluded that in Poland the common and at higher level agricultural advisory is indispensable in the field of preparing fertilization plans and of increasing of the pro-environmental awareness among farmers.

References

1. S. Baker, *Politics of Sustainable Development* (Routledge, London and New York, 2012)
2. K. Palanisamy, K. Parthasarathy, Probl. Sust. Dev. **12**, 1 (2016)
3. N. Dempsey, G. Bramley, S. Power, C. Brown, Sust. Dev. **19**, 5 (2011)
4. B. Giddings, B. Hopwood, G. O'brien, Sust. Dev. **10**, 4 (2002)
5. E. Mieszajkina, Probl. Sust. Dev. **12**, 1 (2016)
6. N.A. Ashford, Int. J. Sust. Higher Educ. **5**, 3 (2004)
7. M. Kowalska, W. Knapik, M. Bogusz, Farm. Probl. Sust. Dev. **11**, 2 (2016)
8. M. Kruszyński, M. Golinowska, M. Borkowska, T. Wiciak, Prog. Plant Prot. **55**, 1 (2015)
9. D. Savić, V. Jeremi, N. Petrović, Probl. Sust. Dev. **12**, 1 (2016)
10. A. Granstedt, J. Tyburski, W. Kooker, J. Stalenga, Fragm. Agron. **3(95)** (2007)
11. P.R. Hobbs, K. Sayre, R. Gupta, Philos. Trans. R. Soc. Lond. B. Biol. Sci. **363**, 1491 (2008)
12. E. Lichtfouse, M. Navarrete, P. Debaeke, V. Souchère, C. Alberola, J. Ménassieu, *Sustainable Agriculture (*Springer Netherlands, 2009)
13. L.C. Ponisio, C. Kremen., Proc. R. Soc. B **283**, 1824 (2016)
14. J.B. Królczyk, A.E. Latawiec, M. Kuboń, Pol. J. Environ. Stud. **23** (2014)
15. K. D'haene, J. Salomez, S. De Neve J. De Waele, G. Hofman, Agric. Ecos. Environ. **192** (2014)
16. H.J. Van Grinsven, L. Bouwman, K.G. Cassman, H.M. Van Es, M.L. Mccrackin, A.H. Beusen, J. Environ. Qual. **44**, 2 (2015)
17. M. Arauzo, M. Valladolid, Agric. Ecosys. Environ. **179**, (2013)
18. G.L. Velthof, J.P. Lesschen, J. Webb, S. Pietrzak, Z. Miatkowski, M. Pinto, J. Kros, O. Oenema, Sci. Total Environ. **468** (2014)
19. H.J.M. Van Grinsven, M. Holland, B.H. Jacobsen, Z. Klimont, M.A. Sutton, W.J. Willems, Environ. Sci. Technol. **47**, 8 (2013)

20. H.J.M. Van Grinsven, H.F.M Ten Berge, T. Dalgaard, B. Fraters, P. Durand, A. Hart, G. Hofman, B.H. Jacobsen, S.T.J. Lalor, J.P. Lesschen, B. Osterburg, K.G. Richards, A.-K. Techen, F. Vertès, J. Webb, W.J. Willems, Biogeosciences, **9**, 12 (2012)
21. M.L. Cayuela, L. Van Zwieten, B.P. Singh, S. Jeffery, A., Roig M.A. Sánchez-Monedero, Agric. Ecosys. Environ. **191** (2014)
22. U. Stockmann, M.A. Adams, J.W. Crawford, D.J. Field. N. Henakaarchchi, M. Jenkins, I. Wheeler, Agric. Ecosys. Environ. **164**, (2013)
23. X. Xu, P.E. Thornto, W.M. Post, Global Ecol. Biogeogr. **22**, 6 (2013)
24. Krajowy Zarząd Gospodarki Wodnej (National Board of Water Management), http://www.kzgw.gov.pl/pl/Informacje-na-temat-dyrektywy-azotanowej.html (access 18.07.2016)
25. S. Pietrzak, Pol. J. Agron. **11** (2012)
26. K. Pulikowski, I. Małecka, Z.J. Małecki, V. Moszinskij, Zesz. Nauk.. Inż. Ląd. Wod. Kształt. Środ. **8-9** (2013)
27. A. Granstedt, J. Tyburski, K. Glińska-Lewczuk, *Rolnictwo ekologiczne jako metoda ochrony środowiska (*UWM Olsztyn, Poland, 2007)
28. A. Harasim, Studia i Raporty IUNG-PIB **29(3)** (2012)
29. J. Suchardt, J. Tyburski, *Rolnictwo ekologiczne jako metoda ochrony środowiska, UWM Olsztyn (*UWM Olsztyn, Poland, 2013)
30. *The Statistical Yearbook of Poland*, (GUS 1992, 2005, 2014)
31. G. Pe'er, L.V. Dicks, P. Visconti, R. Arlettaz, A. Báldi, T.G. Benton, S. Collins, M. Dieterich, R.D. Gregory, F. Hartig, K. Henle, P.R. Hobson, D. Kleijn, R.K. Neumann, T. Robijns, J. Schmidt, A. Shwartz, W.J. Sutherland, A. Turbé, F. Wulf, Sci. **344**, 6188 (2014)
32. Baltic Deal, http://www.balticdeal.eu/about-baltic-deal/ (access 18.07.2016)
33. Z. Kędzior, *Badania rynku. Metody zastosowania* (PWE, Warszawa, Poland, 2005)
34. B.J. Karatzoglou, J. Cleaner Prod. **49** (2013)
35. A. Kotecki, Fragm. Agron. **32** (2015)
36. B.T. van Zanten, P.H. Verburg, M Espinosa, S.Gomez-y-Paloma, G. Galimberti, J. Kantelhardt, M. Kapfer, M. Lefebvre, R. Manrique, A. Piorr, M. Raggi, L. Schaller, S. Targetti, I. Zasada, D. Viaggi, Agron. Sustain. Dev. **34**, 2 (2014)

Environmental costs resulting from the use of hard coal to electricity generation in Poland

Katarzyna Stala-Szlugaj[1,*], and *Zbigniew Grudziński* [2]
[1]Ph.D. Eng., The Mineral and Energy Economy Research Institute, Polish Academy of Sciences, Wybickiego 7A, 31-261 Cracow, Poland
[2]D. Sc., Eng., Associate Professor; The Mineral and Energy Economy Research Institute, Polish Academy of Sciences, Wybickiego 7A, 31-261 Cracow, Poland

Abstract. In the world's fuel mix used for generating electricity, the most common fossil fuel is coal. In the EU, coal combustion and electricity generation entail the need to purchase emission allowances (EUA) whose purchase costs affect the costs of electricity generation significantly. The research described in the article shows how current market conditions shape the profitability of generating electricity from coal and how Clean Dark Spread (CDS) changes as a function of changes in energy and coal prices at the assumed levels of emission and prices of EUA allowances. The article compares the results of CDS calculations in two variants. Areas have been highlighted where prices of both coal and EUA allowances cause CDS to assume values at which the prices of generated electricity do not cover the costs of fuel (i) and CO_2 emission allowances, cover all costs (ii), or constitute positive prices (iii), but still do not cover all fixed costs. With higher power plant efficiency, CO_2 emissions are lower (0.722 t/MWh). The costs of purchasing fuel required to generate 1 MWh of electricity are also lower. In such case—even with relatively high prices of coal—a power plant can achieve profitability of electricity generation.

1 Introduction

In the world's fuel mix, for years, coal has had the biggest share among fuels used for generating electricity. According to [1], in 2015, the share of coal was 40%, and coal was used to generate 9,639 TWh of electricity (Fig. 1). The second place is occupied by natural gas (22%) and the third one—by water (16%). To compare, in the 6,131 TWh of electricity produced in 1973 [2], the share of coal was lower by 3 percentage points, the second place went to petroleum and petroleum products (25%), and the third one—also to water (21%). Fig. 2 shows the share of coal power industry (both hard coal and brown coal) in the energy mix of selected countries in 2015.

The following countries (Fig. 2) have the highest share of coal power generation in the world: Kosovo (97% in 2014), South Africa (90%), and Poland (79%). On a global scale, coal power generation (including both hard coal and brown coal) amounted to 40% of total electricity production.

In the Polish power system, coal (both hard and brown) is the primary solid fuel used to generate electricity. Between 2010 and 2015, electricity generation in Poland increased by 5% (from 157.7 to 164.9 TWh [3]). In coal-fired power plants (including both hard coal and brown coal), there has been a 4% drop (from 136.5 to 130.5 TWh). In the case of hard coal-fired power plants, the share decreased by 8.6 percentage points; production dropped by 10.1 TWh and amounted to 77.7 TWh in 2015. Brown coal-fired power plants achieved a growth (by 1.2 percentage points). Their electricity generation increased to 52.8 TWh.

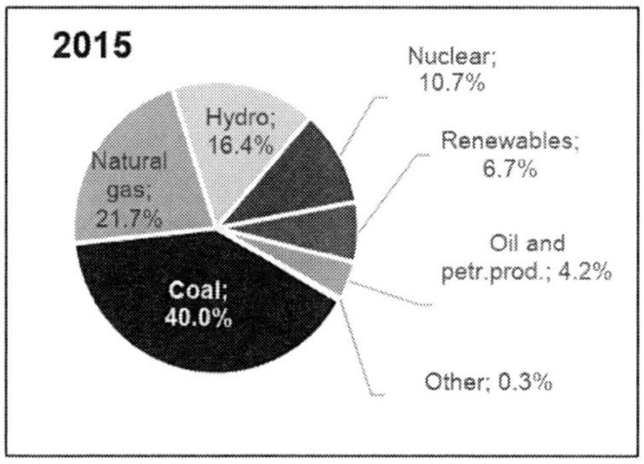

Fig. 1. Worldwide breakdown of electricity generation by fuel in 2015 [1].

However, coal combustion is often associated with the emission of various pollutants: e.g. CO_2, SOx, NOx, particular matter (PM), and fly ash [4, 5, 6]). In the case of electricity generation, it is often necessary to carry out many simulations to calculate fuel and environmental costs [7, 8].

The article aims is to present simulation calculations that will show changes in the theoretical producer's margin (CDS—Clean Dark Spread) at the assumed prices of coal and electric energy.

* Corresponding author: kszlugaj@min-pan.krakow.pl

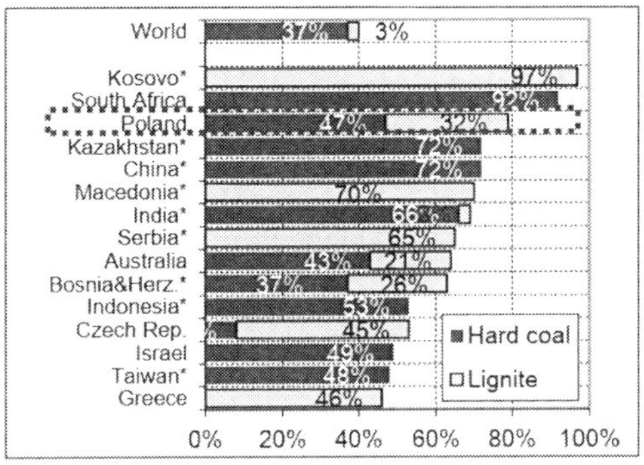

Fig. 2. Share of coal in the energy mix of selected countries of the world in 2015 (* in 2014) [1].

2 CO_2 emissions in the world

In 2015, worldwide emissions of CO_2 amounted to 36.2 billion t (Fig. 3). Even though this number is higher than in 1990 by 59% (or 13.5 billion tCO_2), it was the first time when a decline in comparison with the previous year occurred (by -0.3%, i.e. 0.1 billion tCO_2). A slowdown in the growth of emissions has been observed since 2013.

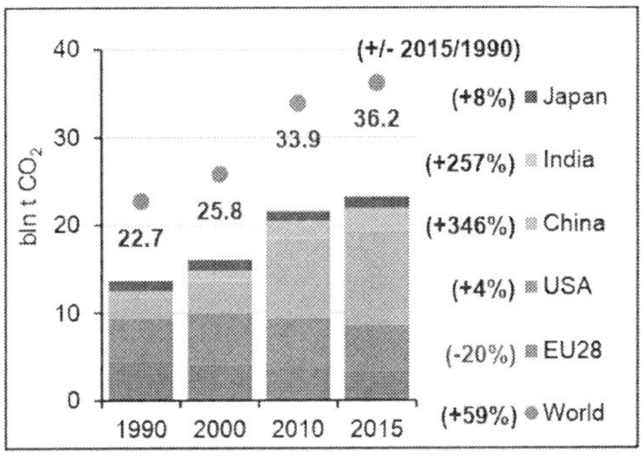

Fig. 3. CO_2 emissions by selected country, 1990-2015 (own work based on [9]).

In the case of EU28 countries, emissions decreased by 20% (0.9 billion tCO_2) compared to 1990 and amounted to 3.5 billion tCO_2 in 2015 (Fig. 4).

Asia's strong industrial growth and transferring energy-intensive industry to this part of the world by global corporations contribute to a large increase in CO_2 emissions. For example, in China, it increased by as much as 346% (by 8.3 billion tCO_2), and in India—by 257% (1.8 billion tCO_2).

Worldwide CO_2 emissions per capita amounted to 4.9 t in 2015, which constitutes an increase by 14% compared to 1990. The highest increase of emissions was observed in: China (285%, 7.7 tCO_2/capita in 2015) and Malaysia (170%, 8.1 tCO_2/capita in 2015). In the EU28 countries, it decreased by 25% compared to 1990 and amounted to 6.9 tCO_2/capita. In Poland, this emission is lower than in the Netherlands and Germany. In 2015, it amounted to 7.6 tCO_2/capita, which constitutes a decrease by 20% compared to 1990 and by 10% compared to 2010.

One of the major sources of CO_2 emissions is heat and electricity generation (Fig. 5). This industry contributed to 42% of the global CO_2 emissions (13.6 billion t) in 2014.

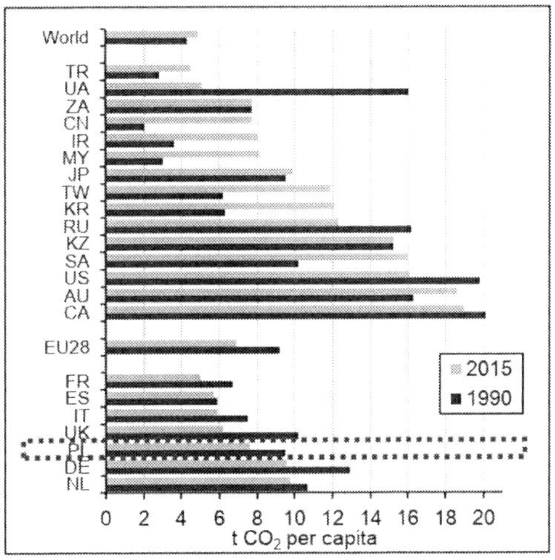

Fig. 4. CO_2 emissions by selected country, tCO_2/capita (own work based on [9]).

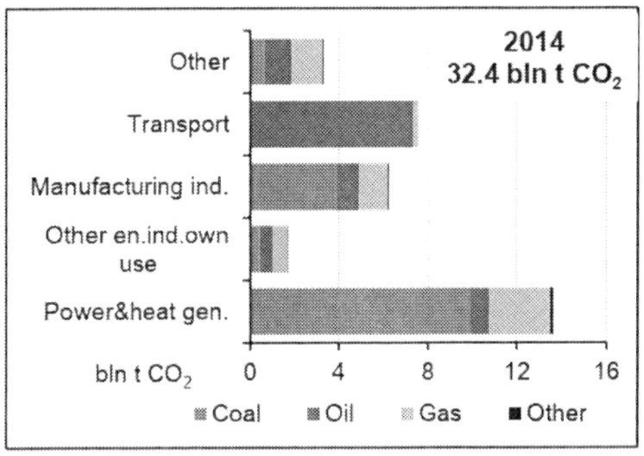

Fig. 5. World — CO_2 emissions by sectors, 2014 [10].

Coal is one of the fuels that are used to generate electricity and have the largest impact on CO_2 emissions. In 2014, coal-fired power plants accounted for 31% of worldwide CO_2 emissions (9.9 billion tCO_2).

3 Assessment of the influence of the prices of CO_2 emission allowances on the prices of electricity

The costs related to CO_2 emissions are an important factor influencing electricity prices. The costs of purchasing CO_2 emission allowances will have an increasing share in the costs of electricity generation. This is caused by the high CO_2 emissions associated with generating electricity from coal compared to other energy carriers. For the EU, reducing CO_2 emissions is the main priority of the climate policy [11,12,13,14]. This makes the events related to the CO_2 emission allowances trading market extraordinarily important for the power industry in Poland. These events affect the assessment of the competitiveness of coal compared to other fuels.

In Europe, an emissions trading system has been introduced following the implementation of Directive 2003/87/EC (the so-called ETS Directive) and Directive 2009/29/EC that improves the EU ETS system. This market has been functioning since January 2005, currently within the so-called third trading period (2013–2020). The participation in the system is mandatory for the types of business activities specified in the directive. Emission allowances are tradable both on exchanges and on the over-the-counter market. They can also be sold or purchased on the basis of commercial agreements with any company [15].

The most liquid market of derivatives for CO_2 emission allowances is the European Climate Exchange (ICE-ECX) based in London. Fig. 6 presents monthly average (spot) price changes of EUA emission allowances (ICE-ECX exchange) in EUR/tCO_2 in the years 2006–2017. The diagram also shows the dates of individual emission allowance trading periods (2005–2007, 2008–2012, 2013–2020). Companies subject to the emissions trading system must acquire allowances that decrease in number in each trading period. The price level of emission allowances in the third trading period is approx. 4.3–8.5 EUR/t.

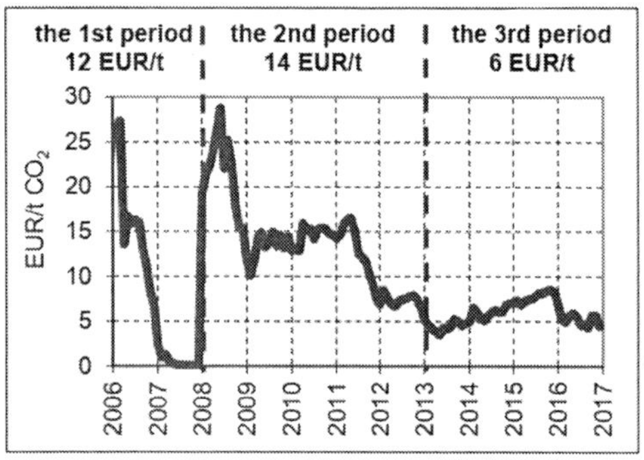

Fig. 6. Average monthly prices of EUA emission allowances (ICE-ECX), EUR/tCO_2—futures contracts as of December of each year [16]).

In January 2017, Thomson Reuters (TR) analysts presented forecasts for the pricing of EUA allowances in the years 2017–2030, at nominal and real prices. Real (using 2015 as the base year) and nominal prices (Tab.1) [16].

According to the data presented, the prices of the allowances will increase. By 2020, there will be an increase of approx. 50%. This will mainly be an effect of the Market Stability Reserve (MSR). The extent of its influence will depend on the allowance transfer ratio from auctions to the MSR reserve. The planned level of this indicator will be either 12% or 24%.

As a result, the level of CO_2 emission allowance prices will increasingly affect the level of electricity prices. This cost element can also significantly affect the price competitiveness between individual energy carriers. It may occur especially after it becomes necessary to acquire all (100%) of the required CO_2 emission allowances at market prices.

In order to investigate the impact of CO_2 emission allowance prices on electricity generation costs, an analysis of the theoretical energy producer's margin (spread) has been carried out.

Table 1. Forecast for the pricing of EUA allowances in the years 2017–2030 by Thomson Reuters (* using 2015 as the base year) [16].

Kind of Prices	2017	2020	2025	2030
Nominal prices, EUA	5.9	9.0	14.3	23.8
Real prices*, EUA	5.8	8.3	12.1	18.7

Spread is defined as the difference between the market price of electricity and the price of fuel used to generate it, with due regard for the efficiency of the process (both figures must be expressed in the same units, e.g.: EUR/MWh, USD/GJ, PLN/MWh, etc.).

For coal, the term used is "Dark Spread"; for gas, it's "Spark Spread". If, when calculating the spread, the cost of CO_2 emissions (related to a given fuel's combustion) is taken into account additionally, the word "Clean" is added to the spread name.

Therefore, "Clean Dark Spread" (CDS) is the theoretical producer's margin for generating energy from coal at market prices for energy, fuel, and emission allowances, calculated as a simple relation [11, 18]:

$$CDS = C_{EE} - C_W - C_{CO2} \quad [PLN/MWh] \qquad (1)$$

where:

- C_{EE} – electricity energy price,
- C_W – fuel purchase price,
- C_{CO2} – emission allowances price.

It can be assumed that the theoretical producer's margin should cover fixed costs of energy generation. Fig. 7 shows the changes in fixed costs in Polish hard coal-fired power plants in the years 2010–2016 as well as the changes in the PLN/USD exchange rate during this period. On the basis of this data, it can be assumed that the recent level of fixed costs is approx. 12.5 USD/MWh (50 PLN/MWh).

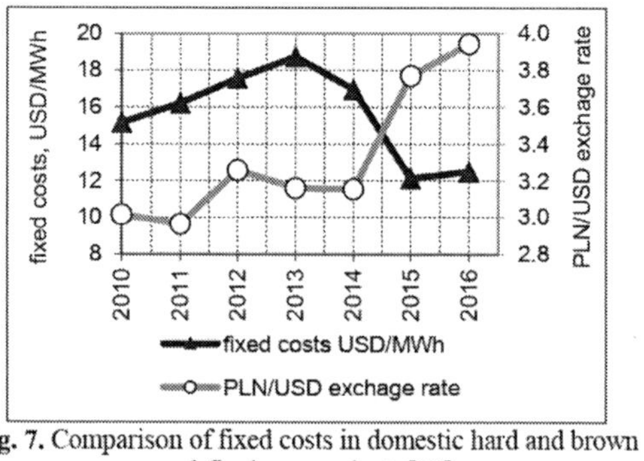

Fig. 7. Comparison of fixed costs in domestic hard and brown coal-fired power plants [19].

Table 2. Input data for CDS calculations—options 1 and 2.

Parameter	Input data for calculation	
Electricity price	35-50 USD/MWh (150-200 PLN/MWh)	
Coal price	1.72-3.50 USD/GJ (7-15 PLN/GJ)	
Calorific value	5 500 kcal/kg (23 GJ/kg)	
Prices of allowances for CO_2 emissions	Calculations were made at two levels: 5 and 10 EUR/ton CO2	
Exchange rates	4.0 PLN/USD and 4.2 PLN/EUR	
Power plant efficiency	36%	46%
CO_2 emission factor (WE) - according to [17]	92.3 kg/GJ= 0.923 tCO_2/MWh	92.3 kg/GJ= 0.722 tCO_2/MWh

Table 2 presents input data used for CDS calculations. The variability of the market prices of electricity in the range of 35–50 USD/MWh (140–200 PLN/MWh) has been assumed. The range of these prices is close to the average monthly spot prices at TGE (Polish Power Exchange) in the last 2 years. Coal prices vary in the range of 2–3.50 USD/GJ (8–15 PLN/GJ). The net as received (NAR) calorific value of this coal is 5,500 kcal/kg (approx. 23 MJ/kg). This number is close to the average parameters of coal combusted by Polish power plants.

Recently, in international trade, a number of price indexes appeared for coal with decreased quality compared to the so-called standard coal (6,000 kcal/kg, NAR). One of them is 5,500 kcal/kg coal traded in such countries as Australia or South Africa [20].

Calculations have been made for two pricing levels of emission allowances: 5 and 10 EUR/tCO_2, assuming that a power plant must buy 100% of required

Table 3. Results of simulation calculations of CDS at EUA price of **5 EUR/tCO_2** and power plant efficiency of 36% and 46%.

Coal Price, USD/GJ	Electricity price USD/MWh						
	35.0	37.5	40.0	42.5	45.0	47.5	50.0
	Power plant efficiency: 36%						
2.00	10.2	**12.7**	**15.2**	**17.7**	**20.2**	**22.7**	**25.2**
2.25	7.7	10.2	**12.7**	**15.2**	**17.7**	**20.2**	**22.7**
2.50	5.2	7.7	10.2	**12.7**	**15.2**	**17.7**	**20.2**
2.75	2.7	5.2	7.7	10.2	**12.7**	**15.2**	**17.7**
3.00	0.2	2.7	5.2	7.7	10.2	**12.7**	**15.2**
3.25	-2.3	0.2	2.7	5.2	7.7	10.2	**12.7**
3.50	-4.8	-2.3	0.2	2.7	5.2	7.7	10.2
	Power plant efficiency: 46%						
2.00	**14.5**	**17.0**	**19.5**	**22.0**	**24.5**	**27.0**	**29.5**
2.25	**12.5**	**15.0**	**17.5**	**20.0**	**22.5**	**25.0**	**27.5**
2.50	10.6	**13.1**	**15.6**	**18.1**	**20.6**	**23.1**	**25.6**
2.75	8.6	11.1	**13.6**	**16.1**	**18.6**	**21.1**	**23.6**
3.00	6.7	9.2	11.7	**14.2**	**16.7**	**19.2**	**21.7**
3.25	4.7	7.2	9.7	12.2	**14.7**	**17.2**	**19.7**
3.50	2.8	5.3	7.8	10.3	**12.8**	**15.3**	**17.8**

 DOI: 10.1051/e3sconf/20171902023

allowances. Two levels of power plant efficiency have been assumed: 36% (variant I) and 46% (variant II). The results of the CDS calculations for these variants have been summarized in Tables 3 and 4.

Each of the result tables (Tables 3 and 4) for the two price levels of EUA emission allowances, 5 and 10 EUR/MWh, consists of two parts with different power plant efficiency: 36% in the upper part and 45% in the lower part. Calculation results are given in USD/MWh. The calculations have been carried out for coal prices given in USD/GJ.

The figures in the individual table fields express the calculated CDS values in USD/MWh. The highlighted coal price level in the range of 2.25–3.0 USD/GJ (9–12 PLN/GJ) is the estimated range of prices of steam coal delivered for the commercial power industry in the coming years.

It can be assumed that the minimum theoretical producer's margin that will cover the costs of electricity generation at the assumed prices of fuel and CO_2 emission allowances is approx. 12.5 USD/MWh in the case of hard coal-fired power plants.

In the result tables, the fields containing calculated CDS values of ≥12.5 USD/MWh are highlighted in bold. Such values mean that, at given prices of coal, electricity, and CO_2 emission allowances, the producer will get a margin covering at least the fixed costs.

Negative figures mean that energy generation is completely unprofitable. With such parameters (input data), energy prices will not even cover the costs of fuel and emission allowances.

Table 4. Results of simulation calculations of CDS at EUA price of **10 EUR/tCO₂** and power plant efficiency of 36% and 46%.

Coal Price, USD/GJ	Electricity price USD/MWh						
	35.0	37.5	40.0	42.5	45.0	47.5	50.0
	Power plant efficiency: 36%						
2.00	5.3	7.8	10.3	**12.8**	**15.3**	**17.8**	**20.3**
2.25	2.8	5.3	7.8	10.3	**12.8**	**15.3**	**17.8**
2.50	0.3	2.8	5.3	7.8	10.3	**12.8**	**15.3**
2.75	-2.2	0.3	2.8	5.3	7.8	10.3	**12.8**
3.00	-4.7	-2.2	0.3	2.8	5.3	7.8	10.3
3.25	-7.2	-4.7	-2.2	0.3	2.8	5.3	7.8
3.50	-9.7	-7.2	-4.7	-2.2	0.3	2.8	5.3
	Power plant efficiency: 46%						
2.00	9.7	12.2	**14.7**	**17.2**	**19.7**	**22.2**	**24.7**
2.25	7.7	10.2	**12.7**	**15.2**	**17.7**	**20.2**	**22.7**
2.50	5.7	8.2	10.7	**13.2**	**15.7**	**18.2**	**20.7**
2.75	3.8	6.3	8.8	11.3	**13.8**	**16.3**	**18.8**
3.00	1.8	4.3	6.8	9.3	11.8	**14.3**	**16.8**
3.25	-0.1	2.4	4.9	7.4	9.9	12.4	**14.9**
3.50	-2.1	0.4	2.9	5.4	7.9	10.4	**12.9**

A comparison of the results of CDS calculations in variants I and II shows how big an impact has power plant efficiency on the achieved result: with higher efficiency, CO_2 emissions are lower (emission factor: 0.7680 t/MWh). The costs of purchasing fuel required to generate 1 MWh of electricity are also lower with higher power plant efficiency. It is also noteworthy that a change in the price of coal by only 1 USD/GJ causes a change in CDS in the range of 2–2.5 USD/MWh.

4 Conclusions

Despite the considerable increase in CO_2 emissions since 1990 (59%), in recent years (since 2012), the dynamics of this increase has weakened significantly. In 2015, for the first time, worldwide CO_2 emissions decreased slightly compared to the previous year. Three large areas: the U.S., China, and the EU (EU28) are responsible for 52–55% of global emissions. This share has been constant since 1990.

Coal combustion is responsible for 40% of worldwide CO_2 emissions, and the share of energy generation in this emission is 42%. The second place in this rating is occupied by transport with a share of 23%.

In the EU, coal combustion and electricity generation entail the need to purchase emission allowances. This situation causes these purchase costs to affect the costs of electricity generation significantly. Emission allowance prices depend on a large number of factors that often have opposite effects. These factors destabilize the market. Among these factors are:

• Increased emission reduction targets in the EU by 2030 and 2050.
• Changes in energy resource prices, especially coal and gas.
• Price relationships between major energy resources (coal/gas).
• Uncertainty of the economic development of the EU and the global economy.
• Planned reform of the EU ETS system.
• Functioning of the Market Stability Reserve (MSR).
• Implementation of the energy efficiency directive by the European Commission.

CO_2 emissions caused by coal combustion in power plants entail the need to purchase EUA allowances. The research described in the article shows:

• How current market conditions (variable prices of coal and electricity) shape the profitability of generating electricity from coal.
• How CDS changes as a function of changes in energy and coal prices at the assumed levels of emission and prices of EUA allowances.

The result tables highlight three areas of results where the CDS values are:

• Negative—the prices of electricity do not even cover the costs of fuel and CO_2 emission allowances.
• Over 12.5 USD/MWh (minimum value of the theoretical producer's margin), which allows for covering all electricity generation costs.
• Positive, but below 12.5 PLN/MWh—prices do not cover all fixed costs associated with electricity generation.

A comparison of the results of CDS calculations in two variants shows how big an impact has power plant efficiency on the achieved result: with higher efficiency, CO_2 emissions are lower (emission factor: 0.722 t/MWh). The costs of purchasing fuel required to generate 1 MWh of electricity are also lower with higher power plant efficiency. In such case, a power plant can achieve profitability of electricity generation, even with relatively high prices of coal.

Publication prepared within the statutory research of The Mineral and Energy Economy Research Institute, Polish Academy of Sciences.

References

1. European Association for Coal and lignite, *Coal industry across Europe*, 6th edition, (2017)
2. Key World Energy Statistics 2016 (Publ. OECD/IEA, Paris, 2016)
3. *Statystyka Elektroenergetyki Polskiej* (in Polish). (Publ. Agencja Rynku Energii SA, Warsaw, Yearbooks from 2010–2016)
4. B. Klojzy-Karczmarczyk, J. Mazurek, Mineral Resour. Manage. **29**, 4, 95–106 (2013)
5. B. Klojzy-Karczmarczyk, J. Mazurek, K. Paw, Mineral Resour. Manage. **32**, 3, 111–134 (2016)
6. B. Klojzy-Karczmarczyk, J. Mazurek, J. Mucha, E3S Web of Conferences **10**, 00036 (2016)
7. J. Kamiński, P. Saługa, Mineral Resour. Manage. **30**, 1, 39–52 (2014)
8. E.J. Sobczyk, J. Kicki, W. Sobczyk, M. Szuwarzyński, Resour. Policy **3**, 94–99 (2017)
9. Janssens-Maenhout G., Muntean M. and Peters J.A.H.W., *Trends in global CO_2 emissions; 2016 Report* (Publ. PBL Netherlands Environmental Assessment Agency The Hague, 2016)
10. *CO_2 emissions form fuel combustion, Highlights 2016 edition* (Publ. OECD/IEA, Paris, 2016)
11. Z. Grudziński, *Metody oceny konkurencyjności krajowego węgla kamiennego do produkcji energii elektrycznej* (in Polish) (Studia, Rozprawy, Monografie 180, 2012)
12. T. Olkuski, Annual Set The Environment Protection **15**, 1474–1488 (2013)
13. G. Wojtkowska-Łodej, Polityka Energetyczna – Energy Policy J. **17**, 3, 269–280 (2014)
14. M. Malec, J. Kamiński, P. Kaszyński, Polityka Energetyczna – Energy Policy J. **19**, 1, 21–34 (2016)
15. J. Preś, P. Jankowski, W. Hofman, *Analiza rynku handlu uprawnieniami do emisji CO_2* (in Polish) (Monthly reports from years 2013–2014, www.handel-emisjami-co2.cire.pl, on-line: 28-04-2017)
16. www.theice.com, on-line: 28-04-2017)
17. Raport z rynku CO_2 (in Polish) (KOBIZE, 58, 2017)
18. Z. Grudziński, K. Stala-Szlugaj, Ann. Set Environ. Prot. **17**, 1389–1402 (2015)
19. *Sytuacja techniczno–ekonomiczna sektora elektroenergetycznego* (in Polish), (Publ. Agencja Rynku Energii SA, Warsaw, quarterly from 2010–2017)
20. U. Lorenz, Polityka Energetyczna – Energy Policy J. **19**, 3, 19–33 (2016)

DOI: 10.1051/e3sconf/20171902024

The influence of shale gas mining activities on the natural environment in Poland

Justyna Pyssa[1,*]

[1]AGH University of Science and Technology, Faculty of Energy and Fuel, Al. Mickiewicza 30, 30-059 Cracow, Poland

Abstract. In the paper the main threats to the environment arising from works related to the exploration of gas from unconventional sources have been discussed. The influence of exploration works on the atmosphere (the emission of gases to the atmosphere as a result of fuel combustion, the emission of noise by drilling equipment) has been explained. Threats to the hydrological balance associated with the intake of water for performing the process of fracturing have been discussed. Legal regulations connected with waste management, water resource management and protected areas (Nature 2000) have been presented.

1 Introduction

The exploration for shale gas in Poland has been lasting for seven years now. The locations of Polish gas fields range from the middle-east coast of the Baltic sea through the central part of Poland to the Lubelskie region (the Lublin basin). Additionally, four boreholes were drilled in the south-west part of Poland (in the Silesian Lowland and the Opole region) [1].

All stages of works connected with the exploration and extraction of hydrocarbons from unconventional sources affect the natural environment.

The consequence of exploitation is taking large pieces of land for the need of a drill site along with necessary infrastructure.

The extent to which drilling works influence particular components of the environment depends on many factors. The most important are the degree of urbanization of the area, sensitivity of individual elements of the environment to pollution, type of drilling equipment, depth of boreholes, sort of drilled rocks, and type and scope of works stimulating the inflow of hydrocarbons to the hole.

2 Concessions and area of exploration in Poland

The possibility of exploitation of energy raw materials from new sources always raises expectations especially in the area of economic advantages. As of the 28th of February 2017, the number of concessions for the exploration and prospection of shale gas deposits in Poland amounted to 20 [2].

Until the 28th of February 2017 – 27 exploration boreholes were drilled in Poland, 54 of which are vertical holes and 18 bent-horizontal ones [2]. In a part of boreholes special treatments were conducted – 28 procedures of fracturing and 9 measures of micro-fracturing (DFIT – Diagnostic Fracture Injection Test). In Table 1 a summary of drilled holes along with applied treatments in the years 2010-2016 is presented.

Table 1. Summary of drilled exploratory boreholes for shale gas along with the type of applied special treatments (data as of the 28th of Feb 2017) [author's own research based on 2].

Year	Number of boreholes	Type of treatments conducted in borehole	
		Fracturing	Micro-fracturing/ Diagnostic Fracture Injection Test (DFIT)
2010	3	2	1
2011	12	7	0
2012	24	8	7
2013	14	4	1
2014	15	5	0
2015	4	0	0
2016	0	2	0

Locations of Polish deposits range from the middle-east Baltic coast through central Poland to the Lubelskie region [3].

One can distinguish three basins geologically and these are: baltic, podlaskie and lubelskie. Moreover, 4 boreholes were drilled in south-west part of Poland (in Silesian Lowlands and Opolskie region) [1].

In Fig. 1 areas of exploration for shale gas are presented (as of the 30th of April 2016). Areas in which concessions for the natural shale gas exploration apply are marked in red, while areas for which requests for the exploration of natural shale gas have been submitted are marked in pink.

* Corresponding author: jpyssa@agh.edu.pl

DOI: 10.1051/e3sconf/20171902024

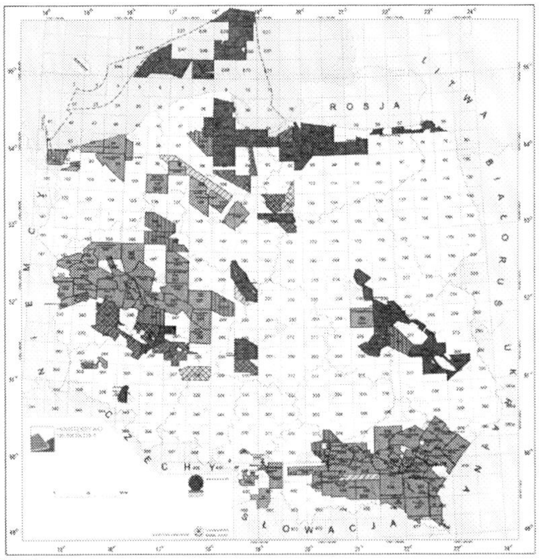

■ – areas in which concessions for the natural shale gas exploration apply,
□ – areas for which requests for the exploration of natural shale gas have been submitted

Fig. 1. Areas of exploration for shale gas [4].

3 Environmental aspects of exploration and exploitation of shale gas deposits

Particular phases of exploration and exploitation of shale gas deposits have an influence on the condition of natural environment. In case of failure to comply with procedures, human errors and random incidents, the pollution of water (surface and ground) and the emission of contaminants to the air may occur. There is also a significant risk of adverse influence of performed work on the biodiversity and landscape around a drilling rig.

3.1 Preparatory work

The degradation of soil environment occurs most often during the preparatory and assembly work. They include activities connected with the preparation of access roads, area of drilling rig and also during laying foundations for technological equipment. The area intended for the drilling rig usually takes from one to several hectares [5-7]. Still, taking into account the fact that there can be several bore-holes and each one of them requires bringing access roads, the total extent of land surface requiring the reclamation later can be significant.

Within the scope of work connected with the preparation of area for exploitation one also performs sealing of sites intended for warehousing of fuels, materials for drilling fluid and surface system for drilling fluid cleaning with the aid of geo-membrane [6].

3.2 Drilling work

The degradation of soil and water environment most often occurs during warehousing and distribution of diesel oil, warehousing of materials for drilling fluid, compiling and cleaning of drilling fluid, temporary storage of sewage and drilling waste in the area of a rig as well as during preparation for and performing fracturing procedures [6]. Cleaning of drilling fluids takes place in the area of a drilling rig with the use of vibrating, linear and double-deck sieves, desanding equipment, desludgers, centrifuges, hydrocyclones, flocculating stations. Those measures enable the re-use of drilling fluid both for boring another hole and applying as groundwork for compiling a new drilling fluid [1].

During extraction work sometimes there may happen an emergency discharge of drilling fluid, brine, natural gas and hydrogen sulphide which accompanies them. Most of the time it is a result of human error, equipment failure, hard-to-predict deposit conditions [6].

3.3 Noise emission

The main sources of noise in the area of drilling rig are power generators, internal combustion engines of drilling equipment and drilling fluid pumps, the drilling fluid pumps and vibrating sieves [6]. Additionally, an important source of sound is the vehicle traffic connected with deliveries of materials and equipment [7]. The research conducted in Poland in the area of drilling rig [5] revealed that the activity of the drilling rig influenced the acoustic environment in the immediate vicinity of workshop. The key factor limiting the level of noise pollution is the distance between the location of workshop and residential areas. The averaged level of noise in the immediate vicinity of workshop amounted to 81 dB, at a distance of 200 meters – 50.6 dB, while at the distance of 500 m – 37.9 dB. The norm for farm building is 55 dB in a day and 45 dB at night.

3.4 Pollution emission to atmosphere

The exploration and production of shale gas may include a variety of potential air emission sources that change depending on the phase of operation. In the early phases of operation, emissions may come from such sources as drilling rigs whose engines may be fueled by either diesel or natural gas and from fracturing operations where multiple diesel-powered pumps are often used to achieve the necessary pressure [8, 9].

Other sources may include the well completion process, which may involve the venting or flaring of some natural gas, and vehicular traffic with engine exhaust and dust from unpaved roads [5, 8]. In Table 2 values of maximum concentrations emitted at particular stages of work in relation to the valid Polish legislation are presented.

As it arises from the data placed in Table 2 at all stages of conducting work related with the shale gas exploration in Poland no excess in the emission of NO_2, benzene and BTEX was recorded. Among measured indicators of contamination, significant excesses were recorded in case of hydrocarbons – C_2 to C_{12}, especially in case of fracturing and reception of procedure fluid.

Table 2. Values of maximum concentrations examined at particular stages of work in relation to the Polish legislation (author's own research based on [5, 10, 11]

Indicator	SO_2 [μg/m^3]	NO_x [μg/m^3]	Methane [μg/m^3]	Hydrocar-bons C_2-C_{12} [μg/m^3]	VOCs [μg/m^3]	Benzene [μg/m^3]	BTEX [μg/m^3]
Reference value- averaged per 1 hour [10]	350	200	n/a	3,000	n/a	30	850
Reference value- averaged per year [10]	20	40	n/a	1,000	n/a	5	63
Reference value- averaged per 1 hour [11]	350	n/a	n/a	n/a	n/a	n/a	n/a
Reference value averaged per year [11]	20	30	n/a	n/a	n/a	5	n/a
Neutral state/ current state	n/a	88,7	1,000–1,400	2,800–3,500	6,500–6,600	< 1	n/a
Drilling	18–133	24–47	n/a	3,920	32,714	< 1	635
Fracturing	119–815	62–105	1,300–3,470	2,900–8,544	5,500–15,400	3.2	33–485
Reception of post-procedure fluid	196	119	10,108	7,620	11,177	< 1	23.5

n/a – not applicable
VOCs – volatile organic compounds
BTEX – benzene, toluene, ethylbenzene and total xylenes emissions

The level for SO_2 was also exceeded to a large extent during the fracturing process. One should also pay attention to the emission of hydrocarbons at the reception of post-procedure fluid. In order to counteract the emission it is necessary to use gas separators and techniques of afterburning (when the concentration is too low for self-ignition).

3.5. Pollution of ground and surface water

The contamination of ground water and useful aquifer layers may occur during the leakage of chemical substances (included in the composition of fracturing liquid) to the ground surface and as a result of breakages and leaks of pipes – in case of incorrect cementing casing strings while drilling.

Nowadays a frequently examined problem is the potential influence of hydraulic fracturing on underground aquifer layers. The current research conducted in Poland does not indicate that crevices resulting from procedures of hydraulic fracturing can reach the zone of underground water [12].

In north Poland in Pomerania rock formations, which are drilled because of the exploration of shale gas, are located at a depth of 3,000 – 4,500 m below the earth surface and the range of crevices is contained within a perimeter of 100 to 200 meters vertically [5, 12]. Aquifer layers lying at a depth of 100 to 300 meters are isolated from gas-bearing shales by a bundle of impenetrable sediments in a form of clays and marls stemming from Upper Silurian of a thickness of 2,000 – 3,000 m [12]. They are also isolated by a cover of locally present Zechstein evaporites (rock salt, calcium sulphate – anhydrite) of a thickness of 280 to 500 m [5] and local Triassic clays and shales of a thickness of 200 meters [12]. Therefore the sealing of shale complexes in Pomerania, as well as tightness of fault zones, is unquestionable. On the basis of conducted research [5] no threat of migration of fracturing fluids and gas to the surface and the level of service water has been noted.

In the Lublin area there are fault zones of complex tectonic evolution which makes that sealing complexes do not maintain full continuity. The Upper Silurian complex is pointed as the main sealing [5] (Ludlow and Pridoli). It is situated above fractured complexes of Lower Silurian. In the Syczyn OU-2K hole the thickness of that complex is 857 meters and the location of bearing series is at 2,700 m, while in the Zwierzyniec-1 hole the thickness of Upper Silurian is 1,387 m at a depth of bearing series at 3,100 meters [5].

3.6 Seismic hazard

In Poland the seismic hazard is not present. The occurrence of seismic events has not been noted in areas covered by licenses for the shale gas exploration. The area of Pomerania is the calmest in Europe in terms of seismic activity. The minimum risk of occurrence of seismic tremors concerns the Lublin region as well [5].

3.7 Waste arising during drilling

* Corresponding author: jpyssa@agh.edu.pl

As a result of works connected with the exploration and recognition of deposits of natural gas from unconventional accumulations 2 sorts of waste are generated [1, 5, 6, 13]:

1. Waste connected with drilling works – so-called extractive and drilling waste. Here one can classify used drilling fluids, flow-back fluids, reservoir fluids and solid waste produced during the purification of flow-back fluid (the sludge from separators and reservoirs, flow-back proppant etc.),
2. Waste resulting from the functioning of drilling rig (plastics, packages, filters, sorbents, used oil and grease, communal waste, etc.)

The main sources of contaminants present in the drilling waste are chemicals used for compiling and regulation of parameters of technological drilling fluids, biocides, petroleum substances, corrosion inhibitors, chemicals used for completion of deposits and stimulation of inflow of hydrocarbons and reservoir fluids – which are most often brines and crude oil [1, 6]

Each drilling fluid has an increased pH value, increased amounts of undissolved and dissolved solid substances (*Total dissolves solids* – TDS) [1, 14, 15], surfactants, chloride and sulphate ions and dissolved organic carbon (*Total organic carbon* – TOC). It also includes small amounts of petroleum-derived hydrocarbons [1,15, 16], heavy metals (such as: arsenic, cadmium, barium, chromium, copper, mercury, nickel, lead, antimony, selenium, zinc, tin, cobalt) [1, 14, 15, 17] and also trace amounts of radioactive substances which come from shale formations [1, 12, 14].

Drill cuttings (output) are crumbled fragments of rocks existing in the geological profile of the hole (clays, sands, loam, sandstones, shales, limestones, marlstones, claystones and dolomites) which are contaminated with remains of the drilling fluid, dissolved salts and heavy metals and – in trace quantities – radioactive elements (such as uranium, thorium and their decay products – radium and radon) [1, 12].

In Poland several types of drilling fluids are applied for drilling exploratory holes. The decision on the selection of sort of drilling fluid is taken after taking into consideration the geological, technical and reservoir conditions [1].

3.8 Water intake

One of important elements of the influence on environment is the intake of both surface and ground water used in the process of hydraulic fracturing. The water utilized in the procedure of hydraulic fracturing can be brought from available waterworks, rivers and lakes and nearby waste treatment plants [12]. The optimum solution in obtaining water for the procedure of hydraulic fracturing is the re-use of flow-back fluid [1].

The available resources of aquiferous formations in Poland amounts to 13,300 million m^3/year at an average consumption of 11 % [12]. Locations of Polish shale gas deposits range from the middle-east Baltic coast through Mazovia to the east part of the Lubelskie region, which are areas of lower-than-average water consumption. In Table 3 there is a comparison of volumes of groundwater resources along with the utilization rate in a given area, including testing grounds in Pomerania and the Lubelskie region.

4 Legislation

The extraction of shale gas in Poland requires taking into consideration both the European and domestic legislation as well as directions in which it may evolve. On the one hand it has to do with the care about the natural environment, on the other hand it is a concern about creating incentives to produce the gas from unconventional resources which would influence the country's energy security to a large extent.

Rules of law which have an effect on undertakings connected with the exploration and exploitation of shale gas are presented below. To the most important ones one can rate acts dealing with the potential contamination of ground water by chemicals, soil pollution, waste disposal, influence on the area and landscape and effects of gas emission.

4.1 Water management

Significant element of the influence on the environment is the utilization of ground and surface water for the need of technological processes – hydraulic fracturing in particular. It is necessary to explore in what way performed work can affect the condition of ground water resources available for management and if the work can cause a potential reduction in the availability of those resources for other users – especially the human population.

The Water Framework Directive – the European Parliament and Council directive 2000/60/WE of the 23rd of October 2000 establishing the framework for community action in the area of water policy [23] and its implementation to the Polish legislation through provisions of the Water Law (Journal of Laws of 2016, item 352, 1250 and 1948) [24] constituted the *necessity of assessment of the qualitative status of ground water and its chemical condition...performed within the frames of evaluation of the status of uniform bodies of ground water* (JCWPd), according to the article 38.5b. The data is placed in Table 3, while on Fig. 2 the rate of utilization of ground water resources in Poland in February 2017 is presented. In the areas of shale gas exploration the light-blue colour dominates which appears to confirm that the rate of utilization of ground water resources available for management, so the ratio of actual ground water intake to resources, is less than 15 percent. It is only the area where two exploratory holes - Lubocino and Gapowo (pomorskie voivodship) – are located that the colour of marking is green which means that the rate of utilization of ground water resources available for management is from 30 – 60%.

The directive on ground water – European Parliament and Council Directive 2006/118/WE of the 12th of December 2006 on protection of the ground water against pollution and status deterioration [25] – which

 DOI: 10.1051/e3sconf/20171902024

Table 3. Volume of groundwater resources along with their utilization, data for the year 2012 [author's own research based on 18-22]

Uniform Bodies of Ground Water (JCWPd) number	Unit	11[18]	13[19]	19[20]	28[21]	90[22]
Testing ground		Łebień	Lubocino, Gapowo	Stare Miasto	Wysin	Syczyn, Zawada
Area	(km^2)	3,969.0	2,856.0	3,917.4	4,057.4	4,901.0
Resources available for development (ZDZP)	m^3/day	708,387.0	384,261.0	375,026.0	423,733.0	675,140.0
Rate of resource use	%	6.9	37.7	6.0	8.4	11.8

introduces the obligation of monitoring of ground and surface water and identification of potential sources of pollution. It determines standards of quality for ground water (appendix I) and threshold values for pollutant concentrations and indicators of contaminants (appendix II).

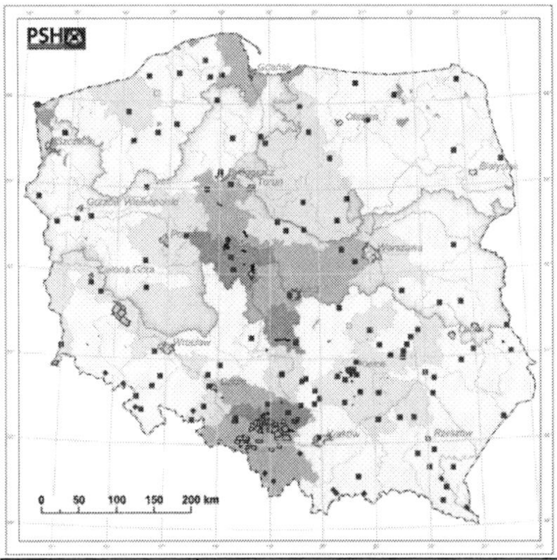

Evaluation of level of reserves:	Rate of utilization of resources of groundwater available for management (ratio of actual ground water intake to resources [%])
> 20% < 20% 0%	< 15 15 – 30 30 – 60 60 – 75 75 – 90 90 – 100 > 100

Fig. 2. Rate of utilization of ground water resources in Poland (data for February 2017) [26].

4.2 Waste management

The basic legislative act concerning the management of extractive waste is the Act as of 10th of July 2008 on extractive waste (Journal of Laws No 138, item 865) [28] which transposes the provisions of the Directive 2006/21/WE [29] into the national legislation.

The primary objective of the act is the prevention of waste generation in the extractive industry and reduction of the adverse impact of waste on people and the environment. Those assumptions are supposed to be executed through the implementation of:

- rules of management of extractive wastes,
- rules of administration of facility for neutralizing extractive wastes,
- procedures related with obtaining permits and licenses connected with the extractive waste management,
- procedures related with the prevention of major functional failure in waste neutralization facilities of the category A.

According to the act, the stage of extractive waste management commences at the phase of planning of mining activities – so at the stage of design of the method of mineral extraction. Environmental aspects have become the crucial issue in the extractive waste management and they are the very key condition determining the selection of recovery process. Technological and economic issues ceased to be the priority [29]. The research on characteristics of the extractive waste is conducted on the basis of regulation of the Minister for the Environment of the 30th of June 2013 on characteristics of the extractive waste (Journal of Laws 2013 item 759) [30]. The detailed scope of characteristics is presented in the appendix to the regulation. Outcomes of the research determine the choice of technology for further management of waste. Details regarding results of the study can be found in publications [1,5].

4.3 Environmental management

Proceeding regarding the assessment of influence of planned undertaking on the environment is regulated by the act of the 3rd of October 2008 on providing information on the environment and its protection, contribution of society to the environment protection and environmental impact assessment (Journal of Laws of 2008 item 199, no 1227) [31]. The latest changes of provisions entered into force on the 1st of January 2017. The act transposes the provisions of the European Parliament and Council Directive 2011/92/UE of the 13th of December 2011 [32] on assessment of effects produced by certain private and public undertakings on the environment and the Council Directive 92/43/EWG of the 21st of May 1992 [33] on protection of natural habitats and wild fauna and flora.

Table 4 The forms of environmental protection in Poland [34]

Specification	Number of the objects	Area		Rectrictions on hydrocarbon deposits exploitation
		[thousand ha]	In % of total area of the country	
Natural parks	23	314.5	1.0	Exploitation of deposits is not possible
Nature reserves	1,469	164.2	0.5	Exploitation of deposits is not possible
Landscape parks	121	2,529.0	8.1	Exploitation is only possible after consultation with the parc authorities, as such activity should be taken into account in managements plans
Protected landscape areas	386	6,992.5	22.4	Economic activity is only slightly limited
Natura 2000 areas- areas of special birs protection	144	557	17.6	Network does not introduce bans, protection system if flexible
Natura 2000 areas- areas of special habitat protection	823	3,795	12.4	Network does not introduce bans, protection system if flexible

The act [31] imposes on enterprisers the obligation of submitting reports on the influence of undertaking on the environment made for the needs of issuing a concession, adopting decisions on environmental conditions for realization of undertaking – so-called environmental decisions, influence of operations of a plant on transformation of morphology and landscape in the area of drilling rig and the influence of procedure of fracturing.

According to the article 3.1.8, under the term of environmental impact one should *understand proceeding in the field of assessment of influence of planned undertaking on the environment, which includes in particular*:

- *a) verification of report on the influence of undertaking on the environment*
- *b) obtaining opinion and agreements required by the act*
- *c) ensuring that society can participate in the proceeding.*

Within the frames of assessment of the influence of undertaking on the environment one determines, analyzes and evaluates (art. 62.1):

1. direct and indirect influence of a given undertaking especially on the environment, human population and health and living conditions, ..., landscape, ..., availability of deposits,

1a. risk of occurrence of major accidents and natural and construction disasters,

2 possibilities and ways of limiting and preventing the negative influence of undertaking on the environment,

3 required range of monitoring;

and (art. 62.2.) one analyzes and assesses the influence of undertaking on the area of Natura 2000.

Exploration, documentation and exploitation of shale gas should be carried out without negative

Approximately 32% of Poland is covered by various forms of protection, which represent a diverse set of measures to protect nature. Within the protected areas, restrictions on exploitations of hydrocarbon deposits – depending on the type of area – may occur (Table 4) [34].

In Fig. 3 protected areas in Poland are presented. National parks with their buffer zones are marked in dark-green and landscape parks with their buffer zones are marked in light-green, while areas of Natura 2000 (Birds Directive and Habitats Directive altogether with a surface area of above 500 ha) are marked in grey.

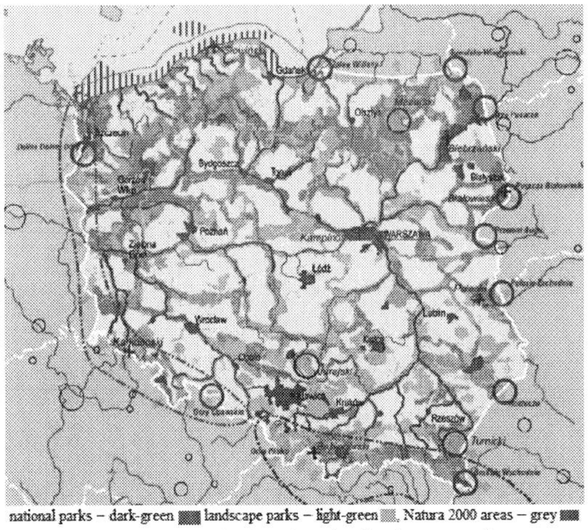

national parks – dark-green ■ landscape parks – light-green ■, Natura 2000 areas – grey ■

Fig. 3. Protected areas (national parks, landscape parks, Natura 2000 areas) [35]

In Poland all plants carrying out drilling for the purpose of exploration of shale gas deposits are located at a distance of 0,1 to 15 km from protected areas. Two of them – Lubocino (pomorskie voivodship) and Syczyn (lubelskie voivodship) – are located within the boundaries of area of protected landscape. As it appears

DOI: 10.1051/e3sconf/20171902024

from the data listed in the study [5], work conducted in those plants can have a potential short-term negative influence on the environment. It particularly concerns: air (connected with the direction of prevailing winds), water (related to the direction of water runoff and its pace) and noise (connected with transportation).

4.4 Chemical Management

Issues related with the application of chemical substances in the process of hydraulic fracturing are the subject to legislative actions of the European Union. In accordance with the European Parliament and Council Regulation 1907/2006 of the 18th of December 2006 [36] (with further amendments) – REACH Regulation – there is a legal obligation of registration of chemical substances introduced to the market taking into account all realistic scenarios of exposure. For substances CAS (Chemical Abstracts Service) numbers were defined in order to facilitate their identification. The regulation was implemented to the Polish legislation through provisions of the Act on chemical substances and their mixtures (Journal of Laws of 2015 item 1203 and Journal of Laws of 2016 item 2003) [37].

Because of the potential risk of contamination of the environment with chemical additions used in fracturing, the obligation of showing composition and percentage of all substances used in fracturing fluids by operators was implemented. The detailed composition of fracturing fluid in particular boreholes in Poland is listed on the website of the Polish Exploration and Production Industry Organization [38]. In Table 5 there is a summary of composition of a selected fluid applied in the shale gas exploration in the Lublin region.

Table 5. Hydraulic Fracturing Fluid Composition (Zwierzyniec1) [38]

Ingredients	Chemical Abstracts Service Number (CAS)	Maximum Ingredient Concentration in Hydraulic Fracturing Fluid (% by mass)
Water		92.44%
Potassium chloride	7447-40-7	3.330%
Crystalline silica	14808-60-7	2.039%
Silica crystalline-cristobalite	66402-68-4	1.653%
Hydrochloric acid	7647-01-0	0.176%
Polysaccharide	Proprietary	0.144%
Acetic anhydride	108-24-7	0.073%
Acetic acid	64-19-7	0.044%
Proprietary Component		0.025%
Terpenes and terpenoids, sweet orange oil	68647-72-3	0.025%
Isopropanol	67-63-0	0.025%
Modified alkane	—	0.008%
Sodium hypochlorite	7681-52-9	0.008%
Diethylene glycol	111-46-6	0.007%
Methanol	67-56-1	0.004%
Sodium hydroxide	1310-73-2	0.001%

5 Summary

Thanks to the technical development it is possible to limit the adverse influence of oil industry on the natural environment. However, one should be fully aware of threats created at all stages of obtaining gas from shale formations.

1. The research conducted in Poland in the area of drilling rig revealed that the activity of the drilling rig influenced the acoustic environment in the immediate vicinity of workshop. The key factor limiting the level of noise pollution is the distance between the location of workshop and residential areas. Results presented in the article showed that emissions of noise from the area of workshop did not affect the excess of valid maximum levels and did not constitute any threat to the comfort of residents.
2. In the article the influence of particular stages of work connected with the exploration of shale gas on the ambient air has been analyzed with respect to the current and compulsory legislation. No excess of NO_x, benzene and BTEX was recorded. Reference values were exceeded for hydrocarbons – C_2 to C_{12} (fracturing and reception of post-treatment fluid). The level for SO_2 was also exceeded to a large extent during the fracturing process. The data is presented in Table 2.
3. The problem of potential influence of hydraulic fracturing on groundwater is frequently examined. The research to-date conducted in Poland does not show that crevices resultant during procedures of hydraulic fracturing may reach the zone of groundwater both in the Pomerania and Lubelskie region.
4. Another danger, to which the attention of many researchers is attached, can be the excessive intake of water used in treatments related with fracturing. From the data presented in Table 3 it appears that the utilization of resources of ground waters in the areas of shale gas exploration ranges from 6 to 36,7 %. Therefore, at the present stage of exploration work there is no threat of depletion of those resources for other users.
5. Shale gas exploration areas in Poland are located at a relatively long distance from legally-protected areas. Only two of them – Lubocino (Pomeranian voivodeship) and Syczyn (Lublin voivodeship) are situated within boundaries of a protected landscape area. The negative influence of exploration facilities on protected areas can be of short-term character connected with emissions of gases and noise.

The analysis of issue presented in the paper concerns the phase of exploration and identification of gas from shale formations currently realized in Poland. In case of the deposit exploitation phase it is going to vary in the scale of venture. It pertains to the water and waste management in particular.

The work is financed within the frames of statutory research no 11.11.210.373 at the Faculty of Energy and Fuels.

References

1. J. Pyssa, E3S Web of Conferences **10**, 00076 (2016)
2. http://infolupki.pgi.gov.pl/pl/stan-prac-poszukiwawczych/aktualnosci/stan-prac-poszukiwawczych-za-gazem-z-lupkow-w-polsce-marzec-0
3. *EIA/ARI World Shale Gas and Shale Oil Resource Assessment* (Advanced Resources International, Inc. June, 2013)
4. http://infolupki.pgi.gov.pl/pl/mapy-koncesji-shale-gas
5. *The environmental and shale gas exploration. Results of studies on the soil-water environment, ambient air, acoustic climate, process fluids and wastes.* (Directorate General for Environment Protection, Warsaw 2015)
6. J. Macuda, Prz. Geol. **58**, 266–270 (2010)
7. PGI-NRI, *Environmental impact of hydraulic fracturing treatment performed on Łubień LE-2H well. Final Report* (Warsaw, November 2011).
8. *Modern Shale Gas Development in the United States. A Primer* (U.S. Department of Energy. Office of Fossil Energy. National Energy Technology Laboratory, 2009)
9. D. Pakuła, Medycyna Pracy, **66(1)**, 99-117 (2015)
10. Regulation of the Minister for the Environment of the 26th of January 2010 on reference values for certain substances in the air (J. Laws of 2010, no 16, item 87)
11. Regulation of the Minister for the Environment of the 20th of August 2012 on levels of certain substances in the air (Journal of Laws of 2012, item 1031)
12. T. Steliga, M. Uliasz, Nafta-Gaz, **5**, 273-283 (2012)
13. J. Macuda, A. Starzycka, Zeszyty Naukowe IGSMiE PAN, **85**, 259-271 (2013)
14. L. Torres, O.P. Yadav, E. Khana, Sci. Total Environ. **539**, 478–493, (2016)
15. D. Zhang, T. Yang, Petrol. Explor. Develop. **42(6)**, 876-883, (2015)
16. W. Orem, C. Tabu, M. Varonka, H. Lerch, A. Bates, M. Engle, L. Crosby, J. McIntosh, International J. Coal Geology, **126**, 20-31 (2014)
17. S. L. Brantley, D. Yoxtheimer, S. Arjmand, P. Grieve, R. Vidic, J. Pollak, G. T. Llewellyn, J. Abad, C. Simon, Int. J. Coal Geology, **126,** 140-156 (2014)
18. https://www.pgi.gov.pl/logowanie-0/docman-dokumenty-pig-pib/docman/psh/zadania-psh/jcwpd/jcwpd-1-19/4543-karta-informacyjna-jcwpd-nr-11/file.html
19. https://www.pgi.gov.pl/logowanie-0/docman-dokumenty-pig-pib/docman/psh/zadania-psh/jcwpd/jcwpd-1-19/4505-karta-informacyjna-jcwpd-nr-13/file.html
20. https://www.pgi.gov.pl/logowanie-0/docman-dokumenty-pig-pib/docman/psh/zadania-psh/jcwpd/jcwpd-1-19/4494-karta-informacyjna-jcwpd-nr-19/file.html
21. https://www.pgi.gov.pl/logowanie-0/docman-dokumenty-pig-pib/docman/psh/zadania-psh/jcwpd/jcwpd-20-39/4444-karta-informacyjna-jcwpd-nr-28/file.html
22. https://www.pgi.gov.pl/logowanie-0/docman-dokumenty-pig-pib/docman/psh/zadania-psh/jcwpd/jcwpd-80-99/4401-karta-informacyjna-jcwpd-nr-90/file.html
23. European Parliament and Council Directive 2000/60/WE of the 23rd of October 2000 establishing frames for community action in the area of water policy
24. Water Law (J. Laws of 2016, item 352, 1250 and 1948)
25. European Parliament and Council Directive 2006/118/WE of the 12th of December 2006 on protection of ground water against pollution and status deterioration
26. http://www.pgi.gov.pl/psh/psh-2/aktualna-sytuacja-hydrogeologiczna/4714-komunikat-o-biezacej-sytuacji-hydrogeologicznej-w-okresie-01-02-2017-28-02-2017/file.html
27. Act as of 10th of July 2008 on extractive waste (J. Laws No 138, item 865) and (J. Laws 2016, item 1579)
28. European Parliament and Council Directive 2006/21/WE of the 15th of March 2006 on management of waste deriving from extractive industry
29. Pyssa J., E3S Web of Conferences **14**, 02024 (2017)
30. Regulation of the Minister for the Environment of the 20th of June 2013 on characteristics of extractive waste (J. Laws of 2013, item 759)
31. Act of the 3rd of October 2008 on providing information on the environment and its protection, contribution of society to the environment protection and environmental impact assessment (J. Laws of 2008 item 199, no 1227) (uniform text Journal of Laws of 2016, items 353, 831, 961, 1250, 1579, 2003)
32. European Parliament and Council Directive 2011/92/UE of the 13th of December 2011 on assessment of effects produced by certain private and public undertakings on the environment
33. Council Directive 92/43/EWG of the 21st of May 1992 on protection of natural habitats and wild fauna and flora
34. B. Uliasz-Misiak, A. Przybycin, B. Winid, Energy Policy **65**, 68-77 (2014)
35. Monitor Polski – Official Gazette, (2012, item 252),
36. European Parliament and Council Regulation (WE) 1907/2006 of the 18th December 2006 on registration, evaluation and granting authorization and applied limitations in the field of chemicals (REACH).
37. Act of the 25th of February 2011 on chemical substances and their mixtures (Journal of Laws of 2015 item 1203 and Journal of Laws of 2016 item 2003)
38. http://www.opppw.pl/en/fracturing_fluid_composition/23

DOI: 10.1051/e3sconf/20171902025

Analysis of municipal waste generation rate in Poland compared to selected European countries

Beata Klojzy-Karczmarczyk[1*], and *Said* Makoudi[2]

[1]Mineral and Energy Economy Research Institute, Polish Academy of Sciences, Wybickiego 7A, 31-261 Krakow, Poland
[2]Mineral and Energy Economy Research Institute, Polish Academy of Sciences, Wybickiego 7A, 31-261 Krakow, Poland

Abstract. The generated municipal waste rates provided in the planning documents are a tool for forecasting the mass of waste generated in individual waste management regions. An important issue is the decisive separation of two concepts: waste generated and waste collected. The study includes analysis of the generation rate for Poland with division into urban and rural areas. The estimated and projected rate of municipal waste generation for Poland provided in subsequent editions of National Waste Management Plans (KPGO) changed since 2000 within wide range from about 300 to more than 500 kg per capita in an individual year (kg/pc/year). Currently, the National Waste Management Plan for the years 2017-2022 estimates municipal waste generation rate at approx. 270 kg/per capita/year with a projected increase to 330 kg/per capita/year in 2030. Most European countries adopt higher municipal waste generation rate, often exceeding 600 kg/per capita/year. The objective of the paper is therefore to analyze the causes of this difference in the declared values. The morphological composition of municipal waste stream in Poland and in selected European countries (e.g. France, Belgium, Switzerland) was analyzed. At present it is not possible to balance the value of the generation rate with the rate of waste collection in Poland. The conducted analyzes allow for determining a number of reasons for variation of the rate value in particular countries, mostly morphological composition of municipal waste, inclusion of household-like waste from infrastructure facilities or not and amount of waste collected in rural areas. The differences in the generation rates and provided possible reasons indicate the need to harmonize the methodology for estimating rates of municipal waste generation in various countries, including Poland.

1 Introduction

The planning of municipal waste management is closely linked to the need to develop reliable rates of waste generation with the share of individual morphological fractions throughout the stream. The mass and quality of municipal waste generated by the inhabitants vary for urban and rural areas. The ecological awareness of the community is also differentiated, which significantly influences the condition of waste management, mainly the volume of generation, collection, management of waste and issues related to the possibility of creating illegal waste disposal sites [ie. 1–4].

The composition of municipal waste is dominated by raw materials (paper and cardboard 5-19%, glass ca. 10%, plastics 10-15% and metals) and biodegradable waste (29-33%), collected selectively or separated in mechanical-biological processing installations for various purposes [5-11]. Due to the nature of heating individual households and using hard coal for this purpose [12, 13], the waste from coal combustion is also important in the stream of municipal waste, and their quantity is strictly related to the quality of the burned raw material [14]. These waste contributes to the fine fraction <10 mm and to the mineral waste. Particularly high share of these waste in the morphological composition is observed for rural areas, where fine fraction reaches 17% and mineral waste reach 6% [15, 20, 21].

The study includes an analysis of the rate of municipal waste generation since the beginning of planning activities in the waste management sector in Poland, ie. since 2000. Literature analysis has shown that rates for municipal waste generation vary for European countries. The morphological composition of municipal waste stream in Poland and selected European countries (e.g. France, Belgium, Switzerland) was therefore analyzed. For Poland, a stream of generated municipal waste collected both selectively and as mixed waste was analyzed in division into three groups: a large city (over 50,000 inhabitants), a small town (less than 50,000 inhabitants) and rural areas. The generated municipal waste rates provided in the planning documents are a tool for forecasting the mass of waste generated in individual waste management regions and thus for the design of the demand for municipal waste processing facilities and their processing capacity.

The objective of the presented work is to show the existing differences in the values of generation indicators for Poland and selected European countries and to search

* Corresponding author: beatakk@min-pan.krakow.pl

for the reasons for such a situation. The essence of the work is to show differences in the morphological composition of the municipal waste stream despite the same definitions adopted in each country.

2 Generation rate in Poland in the years 2000-2016

The conducted morphological investigations of municipal waste and their collection rates in individual years are the basis for the estimation of municipal waste generation rates. Over the years, these rates have been developed using a variety of methods, taking into account the waste collection volume, the guidelines of relevant standards and mainly concerned the large urban agglomerations [16, 17]. This way, the estimated national rates together with the morphology of waste are further used to develop the voivodship waste management plans.

In the subsequent editions of National Waste Management Plans (KPGO), the municipal waste generation rate has changed since 2000 between 353 and 280 kg/per capita/year (kg/pc/year) (Table 1). In addition, the national plans adopted by the resolutions from 2003 [18] and 2010 [20] separately provide rates for urban areas (424-346 kg/pc/year) and rural areas (223-234 kg/pc/year). Generation rates in particular years have been compiled on the basis of the volume of waste collection in the so-called base year (Table 1), preceding the development of a specific plan. Moreover, based on the current situation, at the beginning of the planning activities the projected generation rate values over the years even exceeded 500 kg/per capita/year (Table 1). In subsequent editions of national plans, however, the projected rates were lower. As a result, the projected generation of municipal waste was equated to the collected waste (Fig. 1, Table 1, 2) both in the base year in 2014 and in the years for which the forecast was prepared.

Table 1. Generation of municipal waste in Poland.

Data source	Base year	Generation [Gg]	Estimated generation rate [kg/pc/year]	Projected generation rates in years [kg/pc/year]
KPGO 2006 [18]	2000	13,503	353	2005: 407 2010: 458 2014: 522
KPGO 2010 [19]	2004	11,802	309	2010: 289 2014: 301 2018: 313
KPGO 2014 [20]	2008	12,101	314	2008: 317 2014: 339 2020: 377
KPGO 2022 [21]	2014	10,804	280	2025: 313 2030: 330

KPGO - the National Waste Management Plan

The lowest generation rates are provided in the latest national plan for the years 2017-2022 [21]. Currently, the National Waste Management Plan for the years 2017-2022 estimates municipal waste generation rate at approx. 270-280 kg/per capita/year (depending on the source of the data) with a projected increase to 330 kg/per capita/year in 2030. The information provided by the Central Statistical Office [22] in some cases differs slightly from the numbers reported in national planning documents, however, the tendency of change is maintained (Fig. 1). The National Waste Management Plan 2022 [21] assumes a unified municipal waste generation rate, regardless of the nature of the area, although the shares of the individual morphological fractions differ. It also assumes a practical equalization of the volume generated and collected and the increase of the rate in individual years according to two variants: low hypothesis (annual increase in the amount of generated municipal waste per inhabitant will be at 0.6%) and high hypothesis (annual increase will be at 1%). Table 1 provides forecasts of generation according to the high hypothesis.

Table 2. Collection of municipal waste in Poland

Data source	Base year	Collection [Gg]	Calculated collection rate [kg/pc/year]
KPGO 2006 [18]	2000	12,226	319
KPGO 2010 [19]	2004	9,800	256
KPGO 2014 [20]	2008	10,036	263
KPGO 2022 [21]	2014	10,804	280

KPGO - the National Waste Management Plan

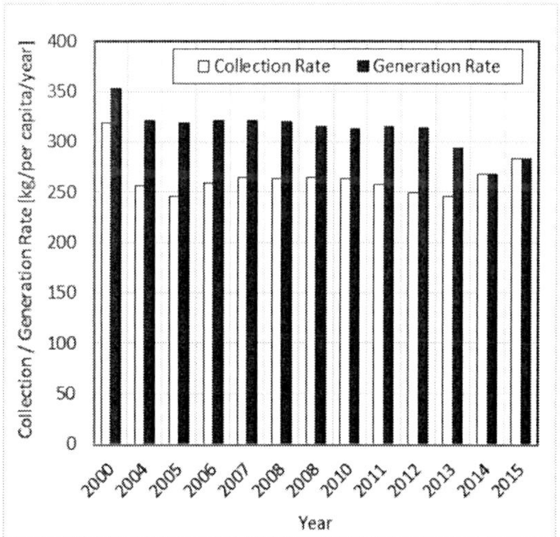

Fig. 1. Comparison of generation and collection rates for Poland in the years 2000-2015 [21, 22].

3 Generation rate in the European countries

DOI: 10.1051/e3sconf/20171902025

The average rate of municipal waste generation in Europe, calculated on the basis of 27 countries excluding Poland, was 475 kg per capita in 2013 [23]. In the same year, the generation rate for Poland was 294 [21], which was significantly lower than the EU average. Such a trend has been maintained for many years (Fig. 2). The analysis prepared and published in 2012 by E. Kaca & G. Kaca [24] highlights these discrepancies in municipal waste generation in Poland and other EU countries.

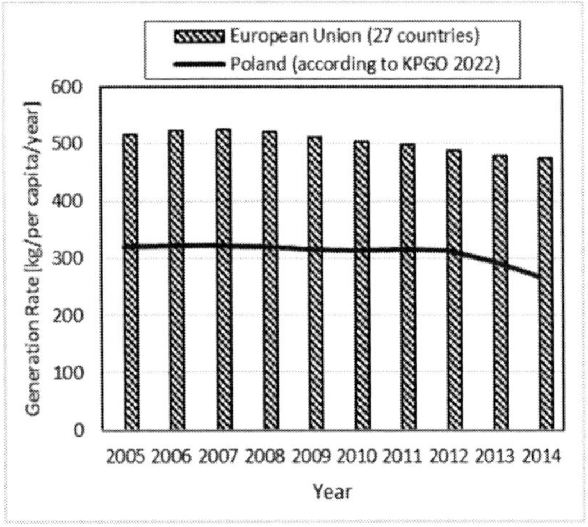

Fig. 2. Comparison of rates of municipal waste generation in EU countries and Poland in the years 2005-2014 [21, 23].

Poland is among the countries with the lowest rate of municipal waste generation, together with Romania and Estonia (Table 3). According to Eurostat data [23], the generation of municipal waste in EU countries was in the range of 254 to 752 kg/per capita/year in 2013. The highest generation of municipal waste for many years has been recorded in Denmark and the lowest for Romania (Table 3).

Table 3. Rates of municipal waste generation in European countries [23, 29].

Waste generation in 2013 [kg/per capita/year]	European countries
<300	Poland, Estonia, Romania
300-400	Czech Republic, Latvia, Slovakia, Hungary
400-500	Belgium, Bulgaria, Croatia, Finland, Spain, Lithuania, Portugal, Slovenia, Sweden, United Kingdom, Italy
500-600	Austria, France, Greece, Netherlands, Ireland, Malta
600-700	Cyprus, Luxembourg, Germany
>700	Denmark, Switzerland [29]

Rates of municipal waste generation in different European countries are undoubtedly dependent on the actual volume of generation and collection, and thus consumption and wealth of the population. They may also be the result of a varied level of ecological awareness of the inhabitants, and consequently the process of minimizing waste generation, according to the hierarchy of waste management. In search for reasons for differentiation of generation rates in individual countries, however, attention should be paid to the interpretations of the meaning of expressions concerning waste generated and collected and to the range of morphological fractions that feed the municipal waste stream.

4 Municipal waste stream in selected European countries

4.1 Methodology of analysis

French-speaking countries (Switzerland, France and Belgium) were selected for the analysis of the generation rate and the municipal waste stream. Alongside Denmark, one of the highest generation rates may be found in Switzerland (729 kg/pc/year in 2014 [29], Denmark - 758 kg/pc/year in 2014 [23]). On the other hand, France and Belgium show modest generation rate values, close to the EU average [23] (Table 3). The values of the generation indicators were reported in statistical and planning documents for each country. The methodology for estimating these indicators was not analyzed, but the morphological stream feeding the municipal waste was. Such procedure allowed to identify the possible causes of such different values.

The following section (section 4.3) contains a simulation showing how the generation rate for Poland will change if municipal waste collected in infrastructure facilities is included in the estimation process and the collection of waste in rural areas is increased. It was assumed that the generation rate currently announced for Poland is based only on waste collected in individual households. The volume of generated waste calculated on this basis was increased by the mass that could be collected from the infrastructure facilities and by the mass allegedly generated (but presumably not collected and utilized by the households) in rural areas.

4.2 Analysis of morphological composition of waste

The European Environment Agency, in spite of the different definitions of municipal waste used in each country, uses the definition of municipal waste management [23, 25], which states that municipal waste is waste primarily generated by households and similar waste from such sources like trade, offices and public institutions (household-like waste). The amount of municipal waste generated consists of waste collected by the municipal authorities or on their behalf and disposed of in accordance with the adopted waste management system. It should be noted that according to this definition, the terms waste generated and waste collected are identical and expressed in the same way.

Poland. According to the Waste Act [26], municipal waste is household waste and non-hazardous waste from other waste producers which, by its nature or composition, is similar to household waste. According to the National Waste Management Plan 2010 [19], municipal waste in Poland in 2/3 consists of household waste and in 1/3 of waste from infrastructure. Municipalities receive waste from all residential properties, while non-residential waste (including infrastructure) is not routinely received by the municipality, so it may not be included in the reporting. In addition, the morphological studies from Poland poorly take into account rural areas in the stream of municipal waste. This areas, which are inhabited by 40% of the country's population, are not representative regarding both the quantity and morphological composition of municipal waste. The majority of waste generated in rural areas is managed on one's own account. Generally, the following waste is collected, selectively or in a mixed manner: glass, metals, plastics, multi-material waste, textiles, hazardous waste, large-size waste, accounting for approximately 30% of the total mass of waste generated in rural areas. It is estimated that the remaining waste is managed on one's own account to the extent of about 70%. Self-management of waste definitely lowers generation rates, which are based on the size of waste collection.

France. Municipal waste is all waste (household and waste from infrastructure – similar to municipal waste, collected in the same manner as household waste) and management of these waste is within the competence of a local government unit [27]. Municipal waste includes mixed household waste, household waste selectively collected, business waste similar to municipal waste, large-size household waste, waste collected at selective waste collection points, household hazardous waste, street and square cleaning waste, green waste from households and infrastructure, and additionally sediments from sewage treatment plants. In France, according to Eurostat data [23], the municipal waste generation rate was 535 kg per capita in 2009, according to data from the French Environmental Protection Agency - Ademe [27] 374 kg/per capita/year in households alone, while 318 kg per capita was collected from these households in the analyzed year.

Switzerland. In Switzerland, waste is subdivided into four basic categories: construction waste, municipal ("urban") waste, hazardous waste and sewage sludge. Urban (municipal) waste is household waste and similar waste generated by handicrafts and the service sector, unless these entities employ more than 250 employees [28]. These is waste consisting of waste from households, administrative buildings, small farms, squares and gardens, public utility buildings and streets [29]. The analysis of data shows that urban waste is composed of household waste and mixed municipal waste from the infrastructure. Generation (115,000 - 118,000 Mg) of urban waste expressed in kg per capita per year show similar values between 600 and 660 kg per capita in particular years between 1990 and 2005. The rate of generation of household waste in 2005 alone was slightly less than 480 kg per capita [29, 30].

Belgium. Municipal waste is household and household-like waste. Municipal waste is waste collected by municipal services, in the selective collection stations, sweeping machines, excluding construction materials [31]. In Belgium, 436 kg/per capita/year of municipal waste was generated in 2014, composed of mixed waste and selectively collected waste [23, 31]. In addition, the Wallonia Government Regulation of 5 March 2008 on household waste management and incurred costs defines gross household waste as residual household waste after subtraction of the fractions sorted by residents [32]. The amount of gross domestic waste generated in urban areas is 130 kg/per capita/year, whereas urban-rural areas generate only 102 kg/per capita/year. Gross household waste generation in kg/per capita/year is 27% higher in urban areas [32].

In reference to the above interpretations of the definition of municipal waste in different countries, it should be noted that the term "municipal waste" is used in different ways in European countries, depending on the adopted waste management system. Differences between countries to some extent result from the inclusion or exclusion of certain types of waste in municipal waste streams (eg. sewage sludge, household-like waste from infrastructure). The actual and provided rates of municipal waste generation and management are also affected by the amount of waste generated from the infrastructure and the share of rural areas in the countries where the volume of generation is significantly underestimated.

4.3 Simulation of the generation rate for Poland

On the basis of the analysis one can notice that the generation rates for Poland are among the lowest in Europe. It could be assumed that the generation rates estimated for Poland could rise and become equal to the average value of rates for the EU when infrastructure waste and actual amounts of waste from rural areas, underestimated by the current collection and forecasting system are included in the municipal waste stream. In order to estimate the value, a forecast has been prepared in this paper, taking into account the increase in generation in these two areas, which, in real terms, undoubtedly necessitates an increase in the efficiency of collection in these areas. The expected results of these activities are presented in Table 4.

The simulation was conducted for 2014, which is the base year in the current KPGO 2022. It is possible to assume that additional 20% of the mass of waste corresponding to the mass of waste generated in the infrastructure should be taken into account (the waste generated from the infrastructure is 30% of the total [19]) and 70% of the mass from rural areas (in the case of Poland this means 40% of the country's population). Under such assumptions, the generation rate could increase even up to 540 kg/per capita/year (option I). In the other variant (variant II of the simulation), lower mass values of waste included in the municipal waste stream were adopted, reflecting waste generation from the infrastructure and rural areas at 15% and 50%,

 DOI: 10.1051/e3sconf/20171902025

respectively. Such a procedure allowed to obtain the value of generation rate for Poland at 432 kg/per capita/year. The average of these two variants provides a waste generation rate of 486 kg/per capita/year. This way, the value of generation rates is comparable to the average value obtained for other countries at 475 kg/per capita/year.

Table 4. Simulation of the rate of municipal waste generation for Poland, taking into account the assumed generation volumes from infrastructure facilities and from rural areas.

Simulated values for 2014 (base year for KPGO 2022)	Variant I		Variant II	
	(1) 20%	(2) 70%	(1) 15%	(2) 50%
Assumed (additional) generation: infrastructure and rural areas [Tg]	4.16	5.82	2.49	3.32
Total generation (taking into account infrastructure and rural areas) [Tg]	20.78		16.62	
Population [million] [22]	38.46			
Simulated generation rates (the expected values) [kg/pc/year]	540		432	

(1) infrastructure
(2) rural areas

5 Summary and conclusions

To a large extent the value of the generation rate adopted in the parent planning documents and then transferred to the downstream documents is the result of an interpretation of the amount of collected waste. In order to determine the demand for municipal waste processing plants, the amount of collected waste is significant, but the volume of waste generation is an important issue for the prediction of potential changes in the waste management system in each region of each country. The estimated and projected rate of municipal waste generation for Poland provided in subsequent editions of the National Waste Management Plans (KPGO) changed since 2000 within wide range: from about 300 to more than 500 kg per capita in an individual year (kg/pc/year). Most European countries adopt higher municipal waste generation rate, often exceeding 600 kg/per capita/year. The conducted analyzes allow for determining a number of reasons for variation of the rate value in particular countries:

- to a large extent the generation rates reflect the amount of waste collection, which in practice provides underestimated values;
- the share of morphological fractions in municipal waste streams is closely tied to the adopted municipal waste collection system in a given country;
- the morphological composition of the municipal waste is different in analyzed countries;
- the share in the municipal waste stream of waste from the infrastructure is variable and depends on the adopted and implemented system;
- the inclusion of household-like waste from infrastructure facilities in estimating the value of the generation rate is a necessary element;
- the amount of waste collected in rural areas is probably much lower than the volume actually produced;
- underestimation of generation in rural areas is a particularly important problem in countries with a high share of such areas, such as Poland;
- in estimating the rate of municipal waste generation in Poland, the mass of waste generated by the infrastructure and the predicted mass of waste to be collected in rural areas has a significant impact on increase of its value.

Reliable generation rates depend on the size and location of new investments related to the processing of municipal waste. The differences in the generation rates identified in the paper and provided possible reasons indicate the need to harmonize the methodology for estimating rates of municipal waste generation in various countries, including Poland.

An important issue in the management of municipal waste is the decisive separation of the meaning of concepts, including generated waste and collected waste. At present it is not possible to balance the value of the generation rate with the rate of waste collection in Poland, which is particularly important in rural areas. In order to obtain reliable indicators of municipal waste generation, it is necessary to conduct an up-to-date survey of the amount and composition of municipal waste in waste management regions, in line with the unified methodology developed jointly with other European countries.

It is difficult to identify the approach to the morphological composition of each country's municipal waste stream based on the Eurostat data provided, and caution should be exercised when comparing European countries, including those in the EU.

The publication was carried out in the framework of the statutory works of MEERI PAS.

References

1. W. Sobczyk, *Aspekty społeczne i środowiskowe gospodarki odpadami* (Wyd. AGH Kraków, 2016) (in Polish)
2. W. Sobczyk, A. Biedrawa-Kozik, A. Kowalska, Rocznik Ochrona Środowiska (Annu. Set Environ. Prot.), **14**, 262–273 (2012)
3. M. Mamak, A. Kicińska, Uniwersytet Zielonogórski Zeszyty Naukowe, **162**, Inżynieria Środowiska **42**, 138-154 (2016) (in Polish)
4. M. Żygadło, *Gospodarka odpadami komunalnymi* (Wyd. Politechniki Świętokrzyskiej, Kielce 2002) (in Polish)
5. B. Klojzy-Karczmarczyk, S. Makoudi, J. Staszczak, Rocznik Ochrona Środowiska (Annu. Set Environ. Prot.), **17**, 1162–1177 (2015) (in Polish)
6. B. Klojzy-Karczmarczyk, J. Staszczak, Polityka Energetyczna – Energy Policy J. **20(2)**, 143-154 (2017) (in Polish)

7. B. Klojzy-Karczmarczyk, S. Makoudi, J. Staszczak, Zeszyty Naukowe IGSMiE PAN, **87**, 149-162 (2014) (in Polish)
8. S. Makoudi, J. Staszczak, Zeszyty Naukowe IGSMiE PAN, **84**, 89–101 (2013) (in Polish)
9. Cz. Rosik-Dulewska, *Podstawy Gospodarki Odpadami* (Wyd. Naukowe PWN, 2011) (in Polish)
10. Cz. Rosik-Dulewska, Gospodarka Surowcami Mineralnymi – Mineral Resources Manage. **21**, 57–72 (2005) (in Polish)
11. P. Łach, K. Kiprian, Prace Instytutu Ceramiki i Materiałów Budowlanych, **7(18),** 59-70 (2014) (in Polish)
12. K. Stala-Szlugaj, Gospodarka Surowcami Mineralnymi – Mineral Resources Manage. **32(3),** 5-22 (2016)
13. K. Stala-Szlugaj, Rocznik Ochrona Środowiska (Annu. Set Environ. Prot. **13(2)**, 1877-1889 (2011) (in Polish)
14. K. Stala-Szlugaj, Rocznik Ochrona Środowiska (Annu. Set The Environ. Prot. **14**, 909-918 (2012) (in Polish)
15. A. Jędrczak, *Analiza dotycząca ilości wytwarzanych oraz zagospodarowanych odpadów ulegających biodegradacji* (Zielona Góra, 2010) (in Polish)
16. L. Sieja, Ochrona powietrza i problemy odpadów, **40(1)**, 28-33 (2006) (in Polish)
17. M. Malinowski, A. Woźniak, Infrastruktura i Ekologia Terenów Wiejskich, **13**, 29-39 (2010) (in Polish)
18. M.P. [Official Gazette of the Republic of Poland] of 2003, No. 11, Item 159 (http://isap.sejm.gov.pl)
19. M.P. [Official Gazette of the Republic of Poland] of 2006, No. 90, Item 946 (http://isap.sejm.gov.pl)
20. M.P. [Official Gazette of the Republic of Poland] of 2010, No. 101, Item1183 (http://isap.sejm.gov.pl)
21. M.P. [Official Gazette of the Republic of Poland] of 2016. Item 784 (http://isap.sejm.gov.pl)
22. GUS, Central Statistical Office of Poland (http://stat.gov.pl/, on-line, accessed: 21.03.2017)
23. EUROSTAT, Eurostat Statistics Explained (http://ec.europa.eu/eurostat/, http://ec.europa.eu/eurostat/statistics-explained/, on-line, accessed: 25.04.2017)
24. E. Kaca, G. Kaca, Problemy Inżynierii Rolniczej, **3(77),** 123–134 (2012)
25. European Environment Agency, *Managing municipal solid waste, a review of achievements in 32 European countries*, EEA Report 2/2013 (https://www.eea.europa.eu/, on-line, accessed: 21.03.2017)
26. Dz.U. [Journal of Laws] of 2013, Item 21 (http://isap.sejm.gov.pl)
27. ADEME, French Environment & Energy Managenet Agency (http://www.ademe.fr/, on-line, accessed 24.04.2017) (in French)
28. République et canton de Genève, *Plan de Gestion des Déchets du Canton de Genève 2014 -2017* (https://demain.ge.ch/document/plan-gestion-dechets.../, on-line, accessed: 24.04.2017) (in French)
29. OFEV, Federal Office for the Environment (https://www.bafu.admin.ch/, on-line, accessed: 25.04.2017) (in French)
30. République et canton de Neuchatel, SCPE, *Plan cantonal de gestion des déchets* (http://www.ne.ch/, on-line, accessed: 24.04.2017) (in French)
31. Economie, Statistics Belgium (http://statbel.fgov.be/, on-line, accessed : 25.04.2017) (in French)
32. Service Public de Wallonie, *Bilan du Plan wallon des déchets Horizon 2010, Volet déchets ménagers et assimilés* (http://environnement.wallonie.be, on-line, accessed: 25.04.2017) (in French)

Effect of replacement of tin doped indium oxide (ITO) by ZnO: analysis of environmental impact categories

Aleksandra Zieminska-Stolarska[1*], *Magda* Barecka[1], and *Ireneusz* Zbiciński[1]

[1]Lodz of University of Technology, Faculty of Process and Environmental Engineering, Wolczanska 213 Street, 90-924 Lodz, Poland

Abstract. Abundant use of natural resources is doubtlessly one of the greatest challenges of sustainable development. Process alternatives, which enable sustainable manufacturing of valuable products from more accessible resources, are consequently required. One of examples of limited resources is Indium, currently broadly used for tin doped indium oxide (ITO) for production of transparent conductive films (TCO) in electronics industry. Therefore, candidates for Indium replacement, which would offer as good performance as the industrial state-of-the-art technology based on ITO are widely studied. However, the environmental impact of new layers remains unknown. Hence, this paper studies the environmental effect of ITO replacement by zinc oxide (ZnO) by means life cycle assessment (LCA) methodology. The analysis enables to quantify the environmental impact over the entire period of life cycle of products—during manufacturing, use phase and waste generation. The analysis was based on experimental data for deposition process. Further, analysis of different impact categories was performed in order to determine specific environmental effects related to technology change. What results from the analysis, is that ZnO is a robust alternative material for ITO replacement regarding environmental load and energy efficiency of deposition process which is also crucial for sustainable TCO layer production.

1 Introduction to LCA

Life cycle assessment (LCA) is a technique for evaluating the environmental effects associated with a product or service over the entire period of its life cycle ("from cradle to grave") by:

- Compiling an inventory of relevant energy and material inputs and environmental releases [Fig. 1]
- Evaluating the potential environmental impacts associated with identified inputs and releases
- Interpreting the results to help decision-makers make a more informed decision.

A complete LCA consistent with ISO 14040 series standards is composed of four interrelated phases:

1. Goal Definition and Scoping - define and describe the product, process or activity. Establish the context in which the assessment is to be made and identify the boundaries and environmental effects to be reviewed for the assessment.
2. Inventory Analysis - identify and quantify energy, water and materials usage and environmental releases (e.g., air emissions, solid waste disposal, waste water discharges).
3. Impact Assessment - assess the potential human and ecological effects of energy, water, and material usage and the environmental releases identified in the inventory analysis.
4. Interpretation - evaluate the results of the inventory analysis and impact assessment to select the preferred product, process or service with a clear understanding of the uncertainty and the assumptions used to generate the results [1].

Vast development in LCA methodologies in last 15 years, elaboration of new LCA techniques, and development of huge databases allows obtaining valuable information about environmental load produced by a product within its life cycle.

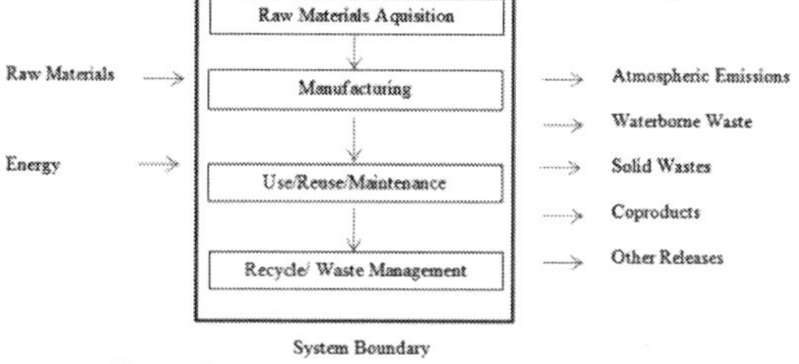

Fig. 1. Life Cycle Stages [2]

1.1 INREP project

INREP "Towards Indium free TCOs" is a Horizon 2020 collaborative research project with the objective to develop and deploy valid and robust alternatives to

* Corresponding author: aleksandra.zieminska-stolarska@p.lodz.pl

indium (In) based transparent conductive oxides (TCO) materials as electrodes. In-based materials, mainly ITO (Indium Tin Oxide), are technologically entrenched in the commercial manufacture of components like LEDs, solar cells, touch-screens, so replacing them with In-free transparent conducting oxides (TCOs) will require holistic approach. INREP looks for development of new TCOs and their deposition technologies with the optimum opto-electrical properties for the economic and safe manufacture of opto-electronic components. The INREP philosophy aimed at developing solutions for the targeted applications. This action will also bring innovative deposition equipment closer to an industrial level in order to accelerate the deployment of indium free TCOs and achieve cost reduction for large area applications. INREP includes also life cycle assessments of the environmental impact of the developed TCO materials and cost of ownership analyses of their formation technologies over the entire period from application in manufacturing, through component operation into waste management.

1.1.1 LCA in the INREP project

LCA analysis can be performed with different calculation techniques and a broad variety of approaches was reported so far. However there is no standard methodology, or even no legally accepted guidelines facilitating the choice of LCA method. The only aspect, which can be systematically verified, is accordance with ISO standards ISO 14040 and ISO 14044. Therefore, analysis of each of existing methods should be performed prior to the selection of LCA tool for a given process.

Within the scope of INREP project, LCA is performed as a comparative analysis enabling final ranking of environmentally friendly TCO layers. Hence, the following criteria for subsequent choice of the LCA approach were selected:

- The method must follow ISO standards and have a broad number of applications reported, to ensure that the approach was extensively tested
- The database of materials and utilities related to the method must be very broad to ensure availability of coefficients for all compounds used in the TCO layer deposition process (e.g. Zinc Oxide)
- The method should enable a deep LCA analysis and include all categories of impact on environment and human health
- The result of the calculation should be a single score, enabling easy comparison and interpretation
- LCA calculation tool must enable clear and transparent analysis.

Based on aforementioned criteria, all available methods were screened for availability of coefficients for compounds used in the TCO layer deposition process. As we found, none of the available LCA methods has the coefficients for all of the compounds used in INREP. Consequently, we selected technique based on availability of coefficients for the key compounds (such as Zinc Oxide). Finally, SimaPro methodology was selected for further LCA analysis.

1.1.2 SimaPro for the LCA analysis

SimaPro® methodology, which allows easy modelling and analyzing of complex life cycles in a systematic and transparent way, has been chosen as a basic tool for LCA analysis in INREP

The first steps for setting an inventory analysis are the definition of the functional unit (which is the environmental cost of deposition of 1 cm^2 of TCO) and scope of the analysis. Since the main goal of the INREP project is to develop and bring to the market efficient Indium- free TCOs, the functional unit of analysis is a layer of new TCO.

The next step of LCA analysis consists of collection of detailed data on the process. Based on this data, additional mass and energy balances can be calculated and subsequently all impacts on environment are quantified. Within the scope of INREP project, physical/chemical vapour deposition (PVD) techniques and materials being the promising candidates for ITO replacement e.g. ZnO were analysed.

1.1.3 Life cycle assessment for different layers deposited with PVD technique

For the physical vapour deposition technique (PVD), a solid source of ITO or ZnO for the deposited compound is used, what facilities the exact quantification of materials used for deposition and waste generated by the process [3]. Therefore, only the units of consumption needed to be recalculated. Additionally, the impact of the transport of raw materials was assessed. All recalculated impacts are shown for the ITO layer deposited by the PVD process (Table 1).

Table 1. Data obtained for ITO-based layer, PVD technique.

Inputs/outputs	Unit	Amount
Cooling water	m^3/cm^2	2,80E-05
ITO target	g/cm^2	1,01E-04
Argon gas	m^3/cm^2	1,21E-07
Oxygen gas	m^3/cm^2	9,88E-10
Electricity	kWh/cm^2	7,53E-05
Waste for reuse, recovery, recycling	g/cm^2	6,01E-03
Waste for final disposal	g/cm^2	1,54E-05
Waste water without treatment	m^3/cm^2	1,33E-10

What is crucial in LCA analysis is the availability of impact coefficients in the selected database. In case of PVD processes and related source of ITO or ZnO such data was available, what ensured exact analysis without need for further assumptions. Resulting process tree for the ZnO layer, showing different stages of the deposition

 DOI: 10.1051/e3sconf/20171902026

process and related impacts is shown in Fig. 2. The values of ecopoints, characterizing the environmental impact, are summarized in Table 2. For the analysed process, the electricity consumption was clearly the dominant contribution to the total environmental impact. Replacement of ITO by ZnO layer proved to a good strategy towards minimization of the environmental impact produced by TCO layer deposition process. The impact of ZnO layer, expressed as a single score accounted for 58% of the ITO layer impact (deposited with the same technique). Consequently, the ZnO material usage is responsible only for a minor part of the environmental impact (ca. 2%).

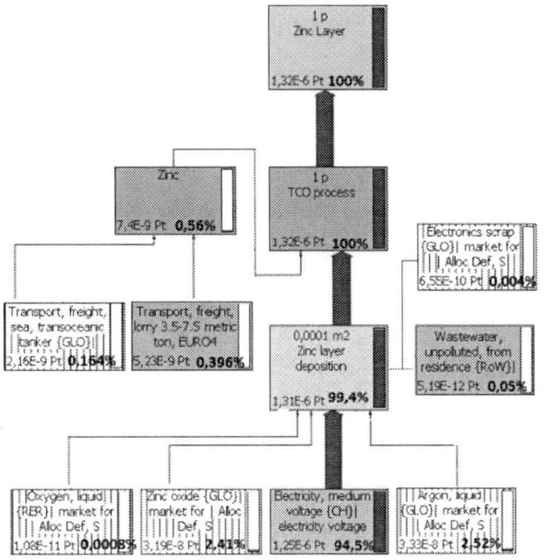

Fig. 2. Exemplary process tree for ZnO, PVD technique, SimaPro®.

Table 2. LCA results for TCO layers deposited with the PVD technique.

No.	Compound	Deposition Technique	Single Score [Pt]	Main impact [%]
1	ITO (reference case)	PVD	2,29E-6	Indium tin oxide 93,3
2	ZnO	PVD	1,32E-6	Electricity 94,5

2 Impact categories

The impact categories are defined and selected in order to describe the impacts caused by the emissions and the consumption of natural resources during the production, use and disposal of the product or processes. In most of Life Cycle Impact Assessment (LCIA) methodologies, the emissions and consumption of resources are attained to three main areas: ecosystem quality, human health and natural resources depletion (Fig. 3).

ReCiPe Endpoint H/A method was applied in INREP project to end up life cycle analysis with single score, SimaPro®.

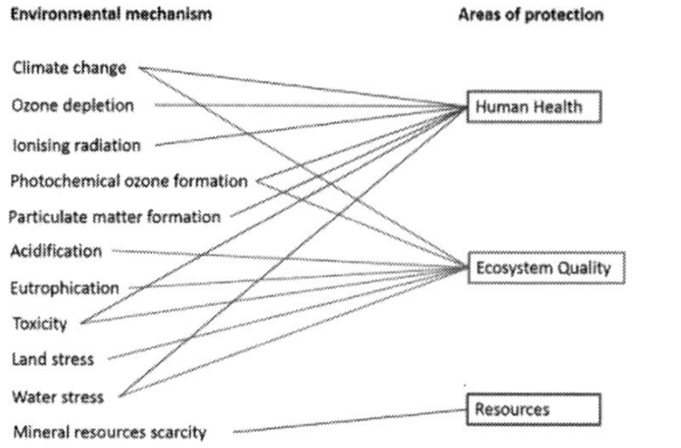

Fig. 3. An example of overall scheme of the environmental life cycle impact assessment framework linking life cycle impact results to midpoint categories to damage categories (areas of protection) [4].

A single score methodology requires subjective opinions in order to weight the relative importance of different environmental impact categories [5].

The subjective views are represented by (quantitative) weighting factors, which represent the relative importance of different environmental impacts and are used to calculate one single, aggregated value (single score). Results of LCA in a form of a single score makes decision making process clear and easier, especially when it is difficult to have an overall conclusion merely based on midpoint results [6-8].

Results of the LCA analysis indicate that for damage categories, the highest environmental impact is connected with human health, than resources depletion and finally ecosystem quality for both ZnO and ITO (Fig. 4 and 6). However specific values of ecopoints for ZnO are lower than for ITO. For human health category, ZnO has 6E-7 Pt while ITO: 1,1E-6 Pt, in case of resources: ZnO has 4.51e-7, ITO – 9,4E-7 Pt, for ecosystem both values are similar – 2,5E-7 Pt.

Total single score for ZnO is 1,32E-6 Pt whereas for ITO is 2,29E-6 Pt.

In terms of impact categories (Fig. 5 and 7) fossil depletion is mostly effected for production of ZnO layer, and also climate changes due to the gas emissions and transportation (Fig. 5). For the production of Indium layer, the biggest environmental effect is for human toxicity and metal depletion.

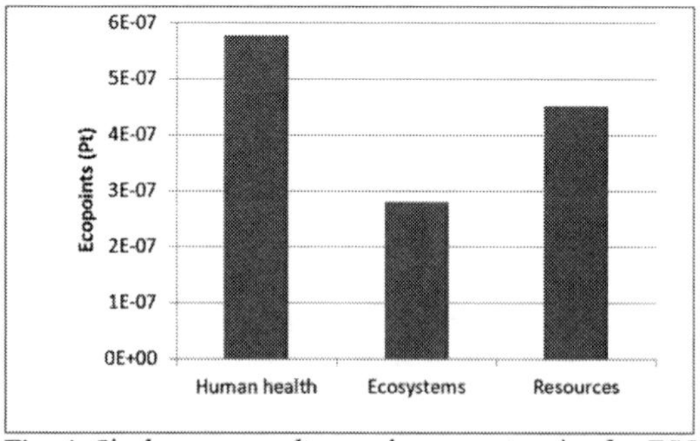

Fig. 4. Single score results per damage categories for TCO production process: ZnO, PVD technique.

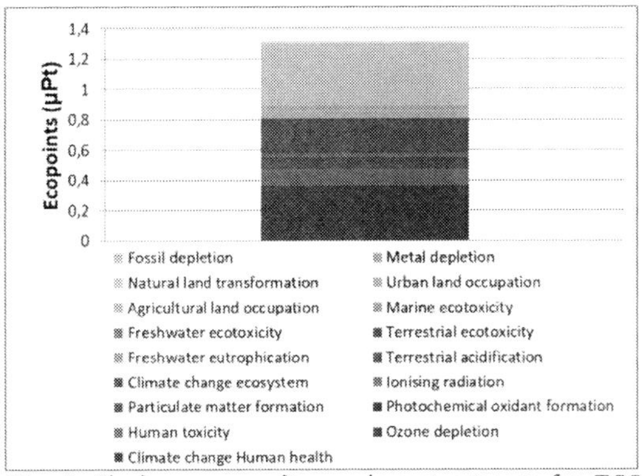

Fig. 5. Single score results per impact category for TCO process: ZnO, PVD technique.

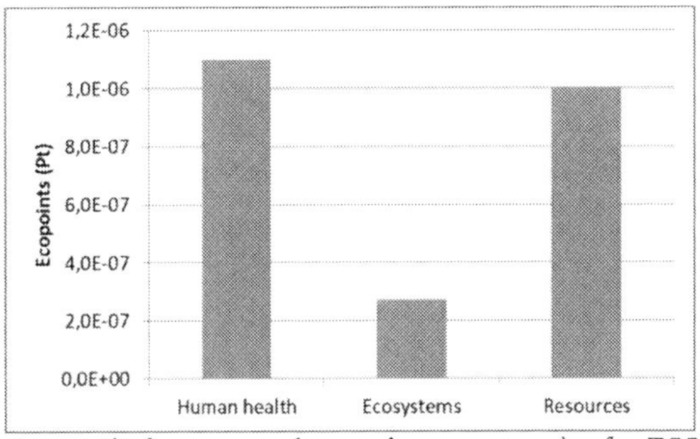

Fig. 6. Single score results per damage categories for TCO production process: ITO, PVD technique.

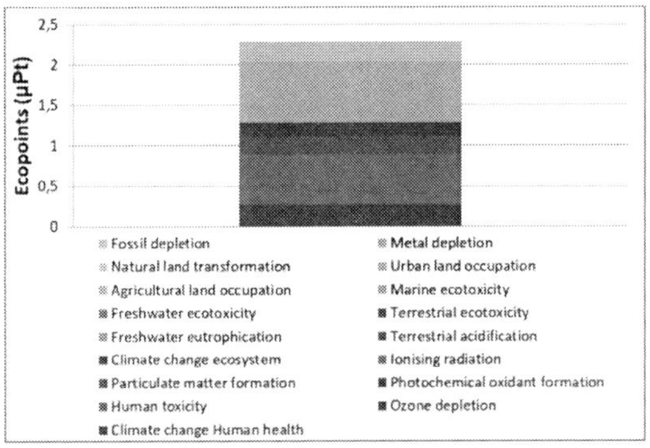

Fig. 7. Single score results per impact category for TCO process: ITO, PVD technique.

Subsequently, effects of process scale-up were studied based on a sensitivity analysis. We found that deposition technique and related energy efficiency were also crucial for sustainable TCO layer production.

3 Conclusions

Within the INREP project, robust alternatives for broadly used ITO-based TCO layers are tested in terms of environmental impact. To this end, life cycle assessment (LCA) was carried out to assess environmental impact over the entire period of its life cycle ("from cradle to grave", during manufacturing, operation and waste management) for each component of the layer. Results of LCA, expressed as single score (Ecopoints), were benchmarked towards the current industrial standard TCO layer composed from ITO. The LCA calculations, based on data from lab scale devices, showed that ZnO was a proper material for ITO replacement.

For the PVD technique, replacement of ITO by ZnO reduces environmental load by factor of 58%.

Results obtained from LCA support further process improvement and help to determine possibilities of environmental impact minimization.

Acknowledgements

"This project has received funding from the European Union's Horizon 2020 research and innovation programme under grant agreement No 641864"

References

1. International Organization for Standardization 14040. Environmental Management — Life Cycle Assessment — Principles and Framework (2006)
2. J.C. Bare, G.A. Norris, D.W. Pennington, T. McKone, J. Ind. Ecol. 6, 3-4 (2003)
3. O. G. Griffiths, J.P. O'Byrne, L. Torrente-Murciano, M.D. Jones, D. Mattia., M. C. McManus, J. Clean Prod. 42, (2013)
4. O. Jolliet, A. Brent, M. Goedkoop, N. Itsubo, R. Mueller-Wenk, et al. Final report of the LCIA definition study. (2003)
5. ISO 14044: 2006, Environmental management - Life cycle assessment Requirements and guidelines, International Standardisation Organisation (2006)
6. R. Heijungs, J. Guinée, R. Kleijn, V. Rovers, Int. J. Life Cycle Assess. **12**, 4 (2007)
7. P.B. Shantheyanda. Characterization of aluminum doped zinc oxide thin films for photovoltaic applications, Master Thesis, University of Central Florida, Orlando, Florida, (2009)
8. R. Hischier, S. Hellweg, C. Capello, A. Primas, Int. J. Life Cycle Assess. 10, 1 (2005)

DOI: 10.1051/e3sconf/20171902027

Eco-Balance analysis of the disused lead-acid-batteries recycling technology

Ewa Kamińska[1*], and *Tomasz* Kamiński[2]

[1]MT Institute, Economic Research Department, 80 Jagiellońska Str.,03-301 Warsaw, Poland
[2]MT Institute, Transport Management and Telematics Centre, , 80 Jagiellońska Str.,03-301 Warsaw, Poland

Abstract. The article presents the results of the eco-balance analysis of the disused lead-acid batteries recycling process. Test-dedicated technology offers the possibility to recover other elements, for example, polypropylene of the battery case or to obtain crystalline sodium sulphate. The life cycle assessment was made using ReCiPe and IMPACT2002 + methods. The results are shown as environmental points [Pt]. The results are shown in the environmental categories, specific for each of the methods grouped in the impact categories. 1 Mg of the processed srap was a dopted as the functional unit. The results of the analyses indicate that recycling processes may provide the environmental impact of recycling technology less harmful. Repeated use of lead causes that its original sources are not explored. Similarly, the use of granule production-dedicated polypropylene extracted from battery casings that are used in the plastics industry, has environmental benefits. Due to the widespread use of lead-acid batteries, the attention should be paid to their proper utilization, especially in terms of heavy metals, especially lead. According to the calculations, the highest level of environmental benefits from the use of lead from secondary sources in the production of new products, was observed in the refining process.

1 Introduction

In Poland, there are two methods of processing battery scrap. The first one allows the processing of non-broken up batteries. Technological gases which contain substances requiring high-temperature post-combustion in the dust extraction system, can be considered a technological disadvantage. This complicates the process of exhaust purification and practically prevents the reduction of chlorine emissions.

The second method allows only lead metallurgical fractions to be introduced into metallurgical processing. Plastics are, in turn, processed to produce commercial products. Some of the plastic fractions that do not find an industrial application is stored as waste. Metallurgical processing of lead fractions is performed in electric, vertical, rotary or rotary-pendulum furnaces. The study analysed the recycling of disused batteries using the first method.

Fig. 1 shows a schematic diagram of a recycling system for lead acid batteries for which an eco-balance analysis has been carried out.

The first stage in the recycling of disused lead-acid batteries involves the mechanical break up the batteries, then filtering out the electrolyte and disassembling the battery components. At this stage the hydrodynamic method of separation is used and metallic, plastics, sulphated lead paste, polyethylene and electrolyte fractions are formed.

The lead paste along with the electrolyte is directed to the desulfurization line, where sodium method is used to convert sodium sulphate to lead carbonate using sodium carbonate.

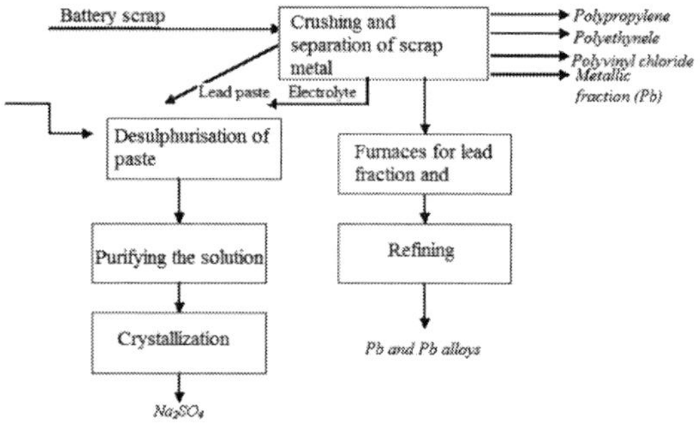

Fig. 1. The scheme of scrap metal recycling of lead-acid batteries by using the desulphurization process. [1,2]

As a result, a 2% desulphurized paste and a sodium sulphate solution are formed. This solution, following pre-filtration and purification, is directed to the evaporation and crystallization line. The result is anhydrous sodium sulphate, used in the cosmetic and metallurgical industries [3,4]. The next stage is production of raw lead obtained with the fire processing of desulphurized lead paste together with the metal

* Corresponding author: ewa.kaminska@its.waw.pl

fraction in the tilt-and-turn furnaces. The final stage is the refining process in which the raw lead is melted in the refining boilers and subjected to the process of removing impurities of foreign metals. During the refining process, the slags are formed from which lead is recovered in the tilt-and-turn furnaces [5,6].

The management of disused lead-acid batteries is realized in compliance with by Directive 2006/66/EC of the European Parliament and of the Council [7]. According to the provisions of the Directive, the battery and accumulator management system includes, inter alia, is formed by the recycling companies. Currently, there are 5 companies operating in Poland with such business profile. There are battery manufacturers, users and the entities involved in collecting waste batteries and accumulators. The system also includes the entities introducing batteries and accumulators (3064 entities) onto the market, processing plants (27) and the ones combining both forms of activity (3) [8]. According to the information in the report on the functioning of the recycling system in the country for the year 2015, 81958 Mg of car batteries were put on the market [9].

In the paper [10], the authors describe the results of the eco-balance analysis of the disused lead-acid batteries recycling technology based on typical data from the Italian market. In this paper, similar boundaries have been adopted. This is important as some people believe that the LCA should not cover only the recycling phase.

Unterreiner et al. [11] describes recycling technology for three types of batteries. Authors presented the results of the LCA analysis for three types of batteries, including lead-acid batteries (LCiA) where the evaluation was performed using ReCiPe method. Information about lead-acid-batteries recycling derives from the literature. The authors point out the importance of the lead-acid batteries recycling which can lessen the environmental impact by 50%.

Alistair et al. [12] carried out an LCA analysis among the others for lead-acid batteries. It was found that the manufacture and production stages do not play a dominant role in the environmental impact. On the other hand, high recycling indicators significantly reduce LAB's environmental impact.

Tian et al. [13,14] compared the recycling technology of the five types of lead-acid batteries using national life cycle databases. In addition, they have modelled the costs of these technologies. They highlighted the issue of the sulphur dioxide emission which causes the very high level of the environmental acidification. The result of the economic analysis of recycling carried out for different batteries is similar. The authors point out the need to seal the recycling system of disused batteries in China, since only a portion of the lead is recovered. The share of recovered lead in the lead production in China is only 30%. The authors propose to implement the American recycling system model (e.g. recycle through the network of sulphuric acid producers) and supporting Best Available Technique (BAT) in this area.

Ferget. et. al. [15] point out the high content of lead, rutile and iron compounds in the products derived from waste polypropylene obtained from the recycled lead-acid batteries and the presence of harmful airborne compounds generated by the combustion of polypropylene. These are chlorine and bromine compounds. Despite the possibility of recovery of this material, the level of contamination in the resulting material should be monitored. The tests should be carried out to determine the presence of calcium carbonate, rutile and iron oxide and lead in the form of dioxides and sulphate. The papers [16,17] describe the situation in China concerning the lead-acid batteries recycling system and the level of lead pollution. The need of recycling used up lead has been brought the attention, which will be possible when the battery collection and utilization system is sealed.

2 Assumptions for the study

The data used to carry out the eco-balance analysis of the energy materials flow and pollutants emissions is both a real data from the company involved in the lead-acid batteries recycling process and the literature data. It was assumed that only lead-acid batteries from passenger cars are used in the technological process. 1Mg of recycled battery scrap was adopted as a functional unit. A simplified analysis was carried out. It was conducted taking into account the following processes:

1. Processing of battery scrap, desulfurization of lead paste and sodium sulphate crystallization, designated as PI,
2. Re-melting of the desulphurized lead paste to a form of raw lead, labelled as PII process,
3. Refining of raw lead and re-melting into the form of raw lead from other lead materials, labelled as process PIII.

The boundary of the testing system is the recycling of disused lead-acid batteries, from the moment they are delivered to the processing plant until the end-product – refined lead - is received. The additional processes such as: polypropylene recovery, sodium sulphate production, landfill of polyvinyl chloride waste and slag were also included in the eco-balance. The company has the necessary environmental permits and a number of awards related to its pro-ecological undertakings. The technology for which the analysis was carried out is referred to as "Best Available Technique" (BAT) and is included in the guidelines for non-ferrous metal management in Poland.

The quantitative data defining the mass and type of raw materials, wastes and emissions obtained for 1 Mg of disused lead acid batteries will be the basis for the study.

3 Methodology

The SimaPro v. 7.3.3. software, dedicated to carrying out eco-balance analysis, was used to estimate the potential environmental impact of lead-acid batteries recycling technology. This tool has a built-in database of seventeen methods used for analyses. The data specific for the European conditions include such methods as: CML2 Baseline 2000 and 2002, ReCiPe, IMPACT2002 +, Ecoindicator99. The study used ReCiPe and IMPACT2002 + methods which were created by combining two eco-balance methods: CML and Ecoindicator99 (Table 1 and Table 2).

Table 1. Categories of damage and impact in the ReCiPe method [19].

Damage category	Unit	Impact category
Human Health (HH)	DALY	Ozone depletion Human toxicity Particulate matter formation Photochemical oxidant formation Climate change Ionising radiation
Ecosystem Quality (EQ)	Species / Year	Terrestrial acidification Freshwater eutrophication Terrestrial and freshwater ecotoxicity Agricultural/urban land occupation Natural land transformation
Resources (R)	$	Metal depletion Fossil depletion

DALY – Disability Adjused Life Years; Species/Year – extinction of the species during the year; $ - increase in costs resulting from the extraction o of resources.

Table 2. Categories of damage and impact of the Impact2002+ method [19]

Damage category	Unit	Impact category
Human health	DALY	Carcinogens Non-carcinogens Ozone layer depletion Human toxicity Respiratory organics Respiratory inorganic Photochemical oxidation Ionising radiation
Ecosystem quality	PDF*m2 *year	Aquatic ecotoxicity Terrestrial ecotoxicity Terrestrial acidification/nitrification Aquatic acidification Aquatic eutrophication Land occupation
Climate change	Kg CO_2 eq	Global warming
Resources	MJ primary	Non-renewable energy Mineral extraction

DALY - Disability Adjused Life Years; PDF*m^2*year – Potentially Disappeared Fraction; MJ – megajoule

They differ in how they represent the damage category – resources. In the ReCiPe method, the category is expressed by the increase in costs caused by the extraction of resources. In the second method, the category is determined by the amount of primary energy [18].

Unfortunately, SimaPro software does not have built-in methods and databases that take into account Polish conditions. Table 1 shows the damage categories and impact categories that are specific to the ReCiPE method.

The IMPACT2002 + method has identified 14 environmental impacts and 4 damage categories, including: human health, ecosystem quality, climate change and resource exploitation [20].

Table 2 shows the categories specific to the second IMPACT2002 + method used.

4 Results and discussion

The results of the analysis presented in Fig. 2 show that the PIII is the most advantageous process in terms of environmental impact.

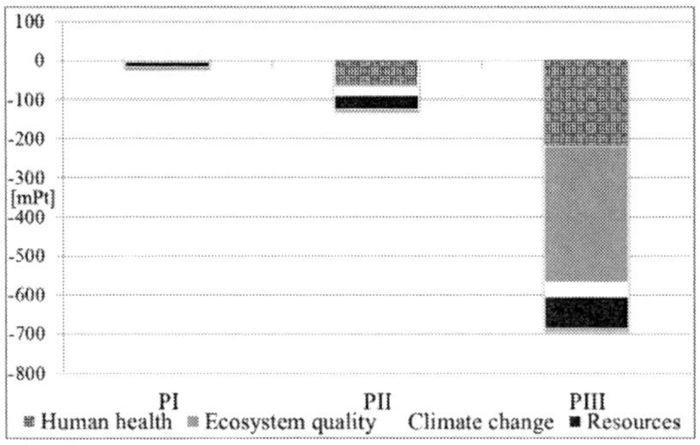

Fig. 2. The results of the analysis of the Eco-balance sheet for the processes of PI-PII-PIII in terms of the 4 categories of impact (impact categories). IMPACT2002+ methods.

This is related to the so-called production avoidance. This is due to the use of waste materials to manufacture new items and avoid the use of primary materials. The total level of environmental benefits for the refining process is -700 mPt.

Fig. 3 shows the results of comparative analysis of PI-PII-PIII processes for the damage category - ecosystem quality.

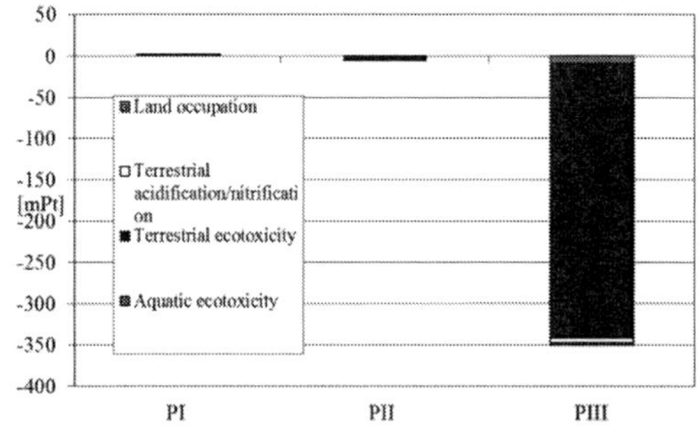

Fig. 3. Impact category related to the ecosystem quality for the PI-PII-PIII processes. IMPACT2002+ Method.

Almost 95% of all identified environmental benefits in this group are related to the Terrestrial eco-toxicity

 DOI: 10.1051/e3sconf/20171902027

category and are (-336.23) Pt. Environmental benefits in this area were identified for PII and PIII processes.
In the category related to acidification and eutrophication of the aquatic environment, no environmental impact was identified. Global warming in the category of climate change is (-44.34) Pt. For each process received values (10.31; -20; -34) Pt. In the group of resource-related factors, the environmental benefits for all non-renewable energy (-13.92; -35.06; -87.49) Pt. On the other hand, for the mineral extraction category, environmental impacts were identified as low, for all processes (-0.05) Pt. Fig. 4 shows the results of the eco-balancing analysis for the storage of polyvinyl chloride (PVC), slag, and the utilizing processes of the waste materials for the production of new materials. This includes waste polypropylene (PP) from the battery casing from which the commercially produced granular material as well as crystalline sodium sulphate resulting from the lead paste cleaning treatment.

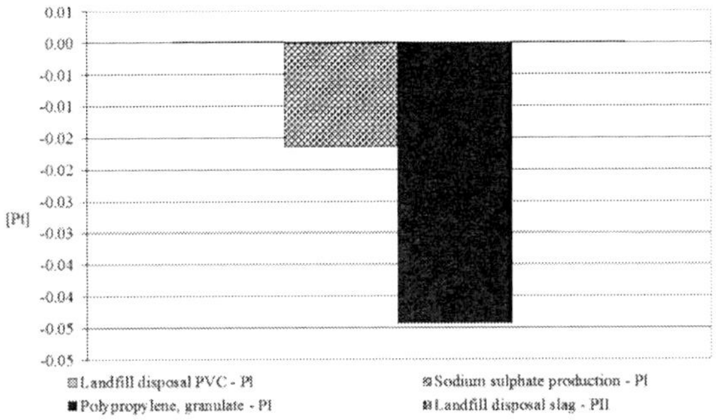

Fig. 4. The comparison of the effect of selected waste management in terms of environmental impact categories selected characteristic for the IMPACT2002+ method (PI – PII).

Fig. 5. shows the results of the eco-balance analysis of PI-PII-PIII processes carried out by ReCiPe method. Results for individual processes are as follows: (-2.93 Pt; -29Pt; -87Pt).

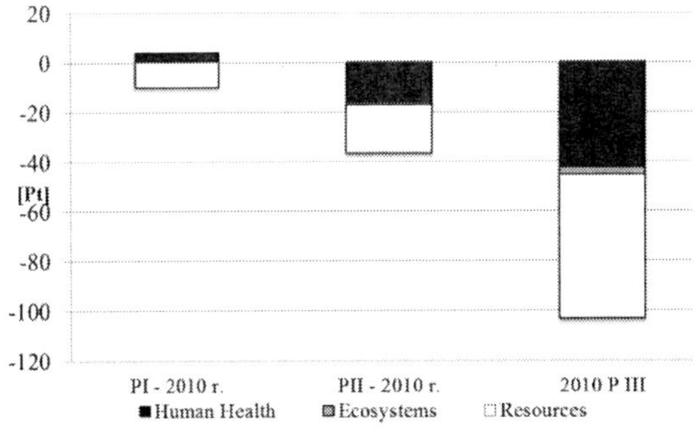

Fig. 5 The results of the analysis of the Eco-balance sheet for the processes of PI-PII - PIII in terms of the 3 categories of impact.(impact categories). ReCiPe methods.

The environmental benefits have been identified primarily in the *resources* category which represent 55.18% of all identified impacts. The impacts on human health account for 54% of the calculated environmental impacts.
Fig. 6. shows in detail the results of the eco-balance analysis for the categories related to human health. The environmental benefits have been identified for most of the identified processes.
In the particulate matter formation category, negative values have been obtained for PII and PIII (-5.4 and -13.58) Pt. In total, this represents 52.81% of all identified environmental benefits associated with human health.

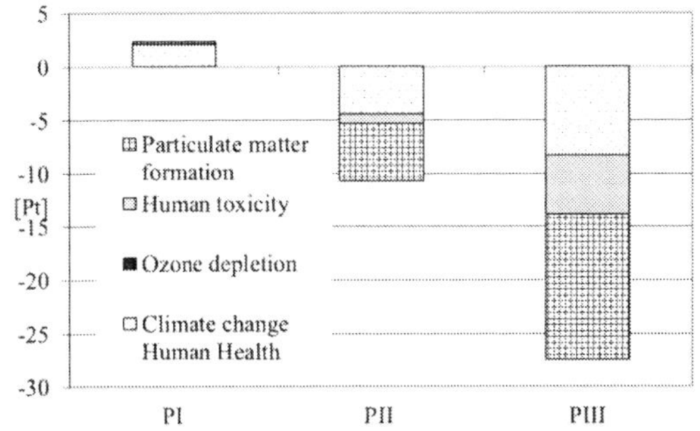

Fig. 6. Impact category related to human health for the PI-PII-PIII processes. ReCiPe methods.

The level of total environmental damages associated with dumping the waste in the landfill has reached the following values: for polyvinyl chloride (Pt 0,03) Pt, while the storage of slag which is waste from re-melting the lead paste reaches (0.04) Pt.
The environmental benefits were identified for (PP) production processes (-14) Pt and sodium sulphate production (-5) Pt.
The potential environmental impact of PIII storage processes has not been assessed because no information has been obtained about this.
The results in Fig. 7 show that no environmental benefits can be identified in the waste landfilling process.

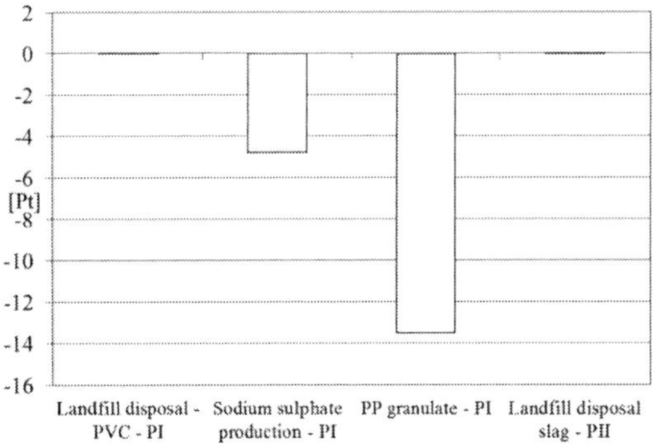

Fig. 7. The comparison of the effect of selected waste management in terms of environmental impact categories selected characteristic for the ReCiPe method (PI –PII).

The result of the analysis may be due to the lack of economic use of waste and, the need to obtain raw materials for their production.

DOI: 10.1051/e3sconf/20171902027

5 Conclusions

The total environmental burden estimated for recycling technology of 1 Mg of lead-acid batteries is (-119.60) Pt for ReCiPe method and (-817) Pt for IMPACT2002 + method. The negative outcome indicates the environmental benefits that result from the so-called avoided emissions. This means that, thanks to the use of recycled lead for the production of refined lead, the use of polypropylene from battery casings in a form of granules i.e. base product for production or the production of sodium sulphate, it is possible to utilize potentially harmful components of the battery. Despite the storage of polyvinyl chloride or the non-use of slag, the processes that positively affect the environment are superior to those generating environmental damage. The differences in the results of the analyses carried out by two different methods may be due to distinctions in weighing techniques used in individual methods as well as the use of different characterization parameters for the same impact categories.

The paper highlights the possibility of using environmental analyses to obtain information on the level of environmental damage that could potentially influence the assessment of whether the object under analysis will be perceived as environmentally friendly. Conducting the analyses in these areas will enable to take action aiming at improving the solution to the battery recycling system in the country as well as emphasize the need for extreme caution before actually implementing organizational and technological changes which could have a negative environmental impact.

It should be stressed that there are no papers in Poland regarding eco-balance analyses applied to the lead-acid batteries recycling technology. It should be emphasized that the analyzes carried out were based on the real data, which quantitatively determine the realization of the processes of recycling and utilization of raw materials as well as the waste recovered from disused batteries. The analyses presented in the paper do not reflect all possible technologies of recycling lead-acid batteries in Poland. However, they do allow to identify the most characteristic environmental burdens for recycling such waste. Based on the research conducted on the actual process of processing and disposal of battery scrap, the following conclusions can be drawn:

- the recycling process of 1 Mg of the battery scrap conducted in 2010, brings about environmental benefits of (-119.53) Pt in the ReCiPe method, for the European conditions and (-0.81822) Pt in the IMPACT2002 + method,
- reuse of materials and substances is a source of environmental benefits, while the waste disposal processes (landfill) have a negative impact on the environment,
- in respect to the waste (reuse of lead paste, plastics and sulphuric acid) is an environmentally beneficial activity,
- the process of landfilling non-recyclable waste (polyvinyl chloride) causes environmental damage.

As far as the impact categories are concerned, the dominant effect on the overall level of environmental benefits achieved through the implementation of the actual battery scrap utilization scenario is achieved by the effects of the reduction of resources. This is mainly due to the limitation of the use of primary raw materials for the production of refined lead, polypropylene and sodium sulphate.

References

1. Reference Document of the Best Available Techniques in the Non-Ferrous Metals Industry, access. 10.05.2017 https://ippc.mos.gov.pl/ippc/custom/BAT_met_niez_r5.pdf (access: 6.05.2017)
2. A. Chmielarz, W. Wężyk, K. Kamiński, Ł. Bratek, W. Malec, IMN Report No. 6406/07 (not published)
3. E. Kamińska, PhD thesis (2013)
4. J. Bendkowski, M. Wengierek, *Waste Logistics Waste Management Facilities* (Silesian Technic. Univ. 2004)
5. http://www.gios.gov.pl/images/dokumenty/prtr/poradnik_metale_20101210.pdfM (acess 10.04.2017)
6. E. Kamińska, A Skarbek-Żabkin, Logistyka **6** (2014)
7. Directive 2006/66/EC of the European Parliament and of the Council of 6 September 2006, repealing Directive 91/157/EEC. on batteries and accumulators (rechargeable batteries) and waste batteries and accumulators. [O. J. L 266 of 26.09.2006]
8. http://rzseie.gios.gov.pl (access 3.04.2017)
9. Report on the functioning of the batteries and accumulators management and used batteries and accumulators for 2015, Waste Management Division, Chief Inspectorate of Environmental Protection (2016)
10. R. Salomone, F. Mondello, F. Lanuzzo, G. Micali, Envir. Asses. **35**, 2 (2005) 206-219
11. L. Unterreiner, V., Jülch, S. Reith, Energy Procedia **99** (2016)
12. A. Davidson, S. Binks, J. Gediga, Int. J. Life Cycle Assess. **21** (2016)
13. X. Tian, Y. Gong, Y. Wu (et al.), Conserv. Recycl. **93** (2014)
14. X. Tian, Y. Wu, P. Hou (et al.) J. Clean. Prod. **144** (2017)
15. E. E Ferg, N. Rust, Polym. Test. **26** (2007)
16. Z. Sun, H. Cao, X. Zhang (et. al), Waste Manage. **64** (2017)
17. W. Zhang, J. Yang, X. Wu (et. al), Renew. Sustain. Energy Rev. **61** (2016)
18. A. Mark, J. Huijbregts, J. Zoran J. N. Steinmann (et al.), Int J. Life Cycle Assess. **2** (2017)
19. SimaPro v. 7.3.3 Analyst, computer program (2012)
20. E. Kamińska, J. Merkisz, T. Kamiński, Chemik **67**, 10 (2013)

Environmental impact of geothermal power plants in Aydın, Turkey

Ersel Yilmaz[1,*], and *Mustafa Ali* Kaptan[2]

[1]Department of Biosystems Engineering, Adnan Menderes University, Aydın, Turkey
[2]Department of Soil Science and Plant Nutrition, Adnan Menderes University, Aydın, Turkey

Abstract. Geothermal energy is classified as a clean and sustainable energy source, like all industrial activities, geothermal energy power plants (GEPP) technology has also some positive and negative effects on the environment. In this paper are presented by attent not only on environmental impacts of GEPP onto Büyük Menderes River and fresh water sources, which ere used for irrigation of agricultural fields from tousands of years in basin, but also on water quality contents like heavy metals and gases emition due to drilling and electricity producing technology of GEPP's. Aydın province is located in the southwestern part of the region and its city center has around 300000 population. The high geothermal potential of this region became from geographical location, which is held on active tectonic Alpine-Himalaya Orogen belt with active volcanoes and young faults. Since 1980's to 2016 there is about 70.97% (662.75 MW) of installed capacity by according to the Mineral Research and Exploration General Directorate, there are totally 290 well licensed (540 explore licenses and 76 business licenses), and 31 geothermal powerplants purposely installed. Topic is important because of number of GEPP increased rapidly after 2012 to now a days to 36 in whole basin.

1 Introduction

Now a days Geothermal energy is called as clean and sustainable energy source although some bad samples in the world. Like all industrial activities in the world it has also positively and negatively effect on environment. The environmental effects may be temporary or irreversible and by include changes onto landscape and land use, by gases emissions into the atmosphere, on surfaces and subsurfaces water, noise, land subsidence, seismicity and solid-liquid rests and wastes.

2 Geothermal Power Plants

Energy is one of the fundamental and enticing requirements of the economic and social development of a country. It is one of the driving requirements. In this respect, "Energy Security" is a vital element of economic security and national security. Now a days, the energy consumed in the world is obtained from a large number of energy sources; fossil resources such as gasoline, natural gas and coal make up 87% of these resources [1]. In Fig. 1. is shown the destribution of types of energy consuption in hole countries in the world.

Geothermal power plants are one of the most important renewable energy sources listed in green energy status in the world. Geothermal power stations built in the framework of respect for nature, with nature's ability to provide closed-loop equipment is operating without damaging the structure of nature. Turkey is primary energy consumer, depends on energy sources such as oil, lignite, hard coal, and natural gas, which are import from other countries. That's why renewable energy potential can be seen as a great opportunity to substitute natural sources from an economic, environmental and social perspective [2].

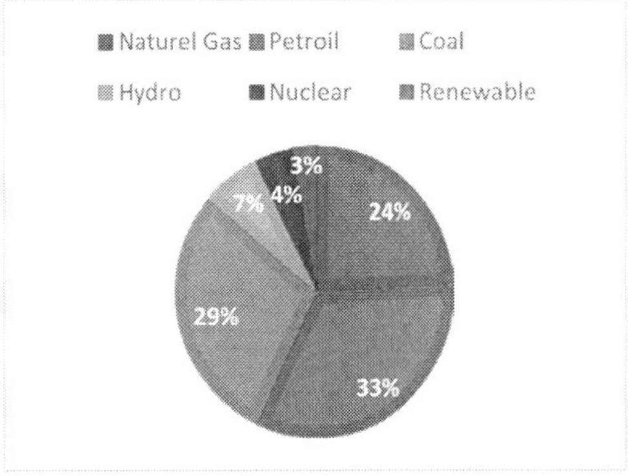

Fig. 1. Global primary energy consumption rates in 2016.

Total geothermal potential of Turkey is estimated around 30000 MW. In geothermal, Turkey has a proven 4000 MW of resources over 20°C of which 600 to 1000 MW are high temperature fields able to produce electricity [3].

Turkey has huge potential as renewable energy. For

*Corresponding author: eyilmaz@adu.edu.tr

 DOI: 10.1051/e3sconf/20171902028

example, geothermal potential is estimated to be 315 GWt and approximitly 5743 GWe, with the total installed power generation capacity in 2016 of 820.86 MWt (directly thermal used) and 100 MWe (converted to electricity). The high geothermal potential is the unique because of geographical location of the country, especially these region are located on the active tectonic Alpine-Himalaya Orogen belt with active volcanoes and young faults. Mostly of them are located in the Aegean and Central Anatolian region [2].
After 2009, numbers of geothermal energy power plants started to increase in Turkey by depending on electric energy needs of industry. Nowadays geothermal industry has achieved around 5% annual growth rate in Turkey especially into west part and Central Anatolia. Especially in the Büyük Menderes and Gediz Graben's geothermal systems have been explored new geothermal fields by MTA and Turkish private sector [4].

Turkey initially in 2014 in the field of geothermal energy while in Kenya, including Indonesia and the Philippines, have entered into 21 new plants commissioned in about 610 MW. Worldwide, 23.5 billion kWh/year of electricity was produced in 2014 [2]. But Turkey has more devoloped then the others as produced electricity by using Geothermal Energy. This production 0.3% of the 73 million MWh / year, with a total capacity of 12800 MW, operating in 24 countries was conducted from geothermal energy. According to this indicator in 2020 14500-17600 MW of geothermal energy sector, in 2030 is estimated to reach 27000-30000 MW. 27% of total electricity production in the Philippines, 7% in California, 86% of the total heat energy needs are met from geothermal sources in Iceland [1].
Geothermal areas and power plants can be classified by different point of view, as according to temperature can be classified like below:

Table 1. Comparative description of different decentralized technologies [6].

Technology	Annual generation (TWhel y^{-1})	Capacity factor (%)	Mitigation Potential ($GtCO_2$)	Energy requirements [(kWhth $(kWhel)^{-1}$]	CO_2 emissions [g $(kWhel)^{-1}$]
Small hydro	≈250	≈50	≈100	n.a.	45
Wind	260	24.5	≈450–500	0.05	≈65
Solar-photovoltaic	12	15	25–200	0.4/1–0.8/1	40/150 – 100/200
Concentrating Solar	≈1	20–40	25–200	0.3	50–90
Geothermal	60	70–90	25–500	n.a.	20–140
Biomass	240	60	≈100	2.3–4.2	35–85

-Low enthalpy contains area (20-70 °C)
-Central enthalpy pitches area(70-150 °C)
-High enthalpy contains area (>150 °C)
High Enthalpy for the production of electricity from geothermal water by using flash and binary methods, but problem is difficulties on re-injection because of rocks type and geography.
The environmental effects may be temporary or irreversible and include changes to landscape and land use, emissions into the atmosphere, surface and subsurface waters, noise, land subsidence, seismicity and solid waste.
Air and chemical pollution can be:

- Hydrogen Sulphide and its derivatives (H_2S) are also emitted, causing a bad smell and forming the acid rain causing compound SO_2 in the atmosphere.
- Emitted carbon dioxide (CO_2) and H_2S are heavy gases and can lingering valleys causing increased pollution for local populations who lives closer to geothermal power plants
- Including radioactive element radon, toxic elements arsenic, mercury, ammonia, boron (highly toxic to plants) and other polluting heavy metals.
- Waste steam is sprayed over surrounding vegetation (usually rare species in geothermal areas), while waste water is either rejected (inducing earthquakes), or pumped into streams and lakes untreated, as treatment is considered too costly to be economically viable.

Dissolved minerals (e.g., boron, mercury and arsenic) contained in the liquid streams generated during the exploration, stimulation and production phases may poison surface or groundwater and harm local vegetation. It is therefore important to monitor wells during drilling and subsequent operations so that any leakage can be detected and managed rapidly.
Local weather changes leading to:

- influence on agriculture products,
- loss of rare thermophillic plants and algal growth,
- loss of scenery and tourism.

In Table 1 is presented comparative description of different decentralized technologies and special attention was paid on CO_2 emission [6].
As of 2016, Turkey has 861 MW of installed capacity of geothermal power by numbers of 31 Geothermal power plants and produced 4213 GWh energy and 70.97% of

 DOI: 10.1051/e3sconf/20171902028

this capacity (662.75 MW) is located in Aydın Region. Today by 36 geothermal power plants produce approximitly 5743 GWh energy in Turkey [7]. Geothermal power plants in international area had reached the 12800 MW installed power capacity at the end of 2014 and Turkish geothermal sector which 4.79% of the total is, power originating from geotermal energy. In Fig. 2 shown the temperatures contains of geothermal area of europe by help of determined of rocks temperature which are 5000 m depth [3].
In Fig. 3 there is shown that the geothermal areas and locations of geothermal power plants in hole Turkey for give information. Aydın province, with electricity generated from geothermal energy is to meet its own energy needs. The geothermal basin found in Aydın Province is the largest geothermal resource in Turkey and as a consequence the district is home for a number of geothermal electric power companies (Fig. 5).

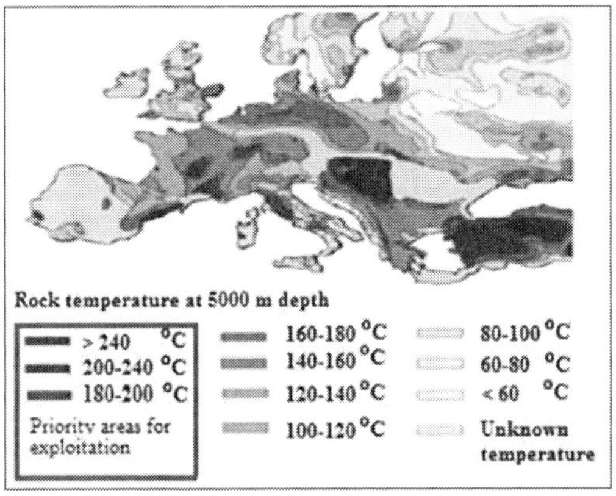

Fig. 2. Map of temperatures of geothermal area of European [3].

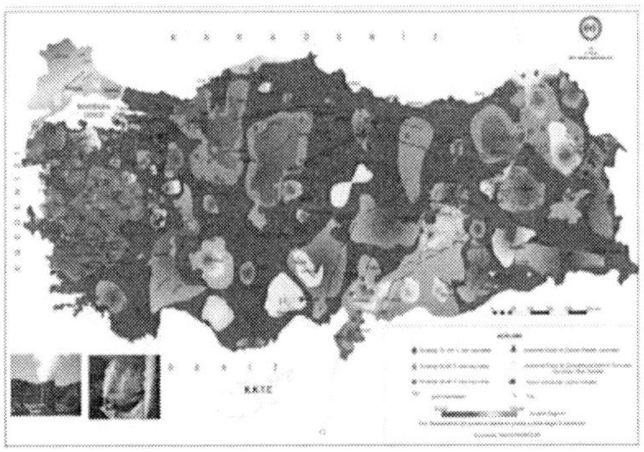

Fig. 3. Geothermal areas and locations of geothermal energy power plants in Turkey [8, 5].

The geothermal reservoir is large and the heat extraction limits for each plant are set at a level where the reinjected geothermal fluid is re-heated within the ground before being "used again". In this way, the hot water can be considered to be "cultivated" beneath the ground rather than "mined" [4]. In Fig. 4 is presented geothermal source areas in western Anatolia.

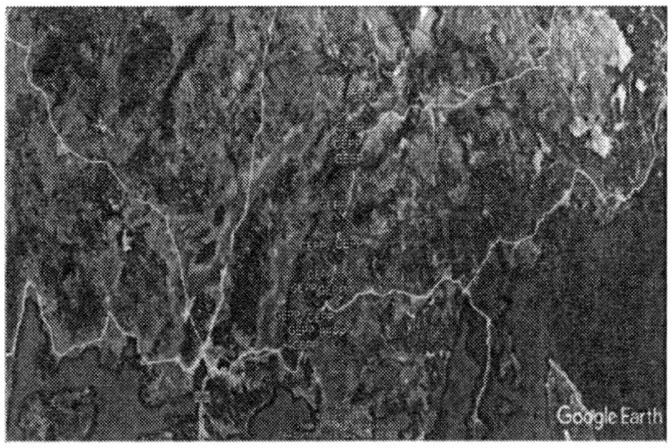

Fig. 4. Main geothermal areas of western Anatolia (Büyük and Küçük Menderes, Gediz basin).

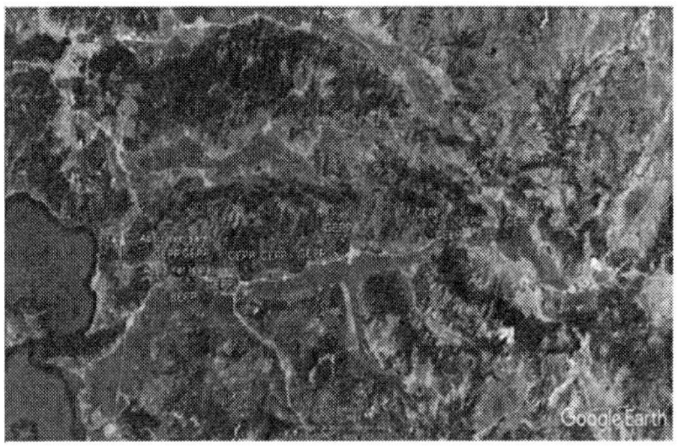

Fig. 5. Geothermal power plants locations which are produce electricity in Büyük Menderes Basin since 1986.

3 Environmental aspects of geothermal power plants

3.1 Boron toxicity

The many geothermal water sources were formed in the region as a result of the formation of the Büyük Menderes graben and high concentration of boron (B) in some water resources was reported [9]. The contamination of the hot-water resources with high boron consantration is one of the most important source the high concentration of B in irrigation water resources like well and surface water especially Büyük Menderes River. Also as presented in Table 2, average ratio of boron level is around 80 mgl^{-1} in samples which is belongs to GEPP's wells, and there is a sign of one of source of boron in water resources. Therefore, this situation has become an important problem not only in Aydın region, but also for whole basin in the future. Boron is one of the micro nutrients required for normal growth of plants. Furthermore, boron is the only one micronutrient element which is non metal. Boron uptake greatly depends on transpiration by plant. It is known that the Boron toxicity showes negative effect on metabolic functions of the plant. The boron toxicity showes reduction on yield and yield components of plants owing to the plant tissues death, the active leaf area and photosynthetic activity reduction due to the high level of boron. The typical symptoms of boron

 DOI: 10.1051/e3sconf/20171902028

toxicity show itself as chlorosis on the older leaves tips, and then following by the necrosis, finally the last phase show itself as a burnt leaves because of the boron is immobile in the plants (Fig. 6).

Table 2. Some chemical properties of geotermal waters in Germencik project research area, (avarage of last drilled geothermal wells, were analised in Lab. of ADU Agriculture Faculty) (mg l^{-1}).

EC (μmho cm^{-1})	**Potassium (K^{+})**	**Nitrite (NO_2^{-1})**	**Arsenic (As)**	**Iron (Fe)**	**Aluminium (Al)**
4650	105.9	nd	0.520	0.1060	0.256
pH	**Lithium (Li^{+})**	**Nitrate (NO_3^{-1})**	**Zinc (Zn)**	**Boron (B)**	**Copper (Cu)**
8.15	5.99	33.21	0.044	80.41	0.028
Eh (mV)	**Carbonate (CO_3)**	**Bromide (Br^{-1})**	**Lead (Pb)**	**Manganese (Mn)**	**Cobalt (Co)**
-64.1	nd	nd	0.033	0.038	0.018
Magnesium (Mg^{+2})	**Bicarbonate (HCO_3^{-1})**	**Phosphate (PO_4^{-3})**	**Cadmium (Cd)**	**Platinum (Pt)**	**Mercury (Hg)**
5.90	2320	nd	0.038	0.025	nd
Calcium (Ca^{++})	**Chloride (Cl^{-1})**	**Fluoride (F^{-1})**	**Nickel (Ni)**	**Chromium (Cr)**	**Strontium (Sr)**
7.26	300.30	11.70	0.025	0.014	2.670
Sodium (Na^{+})	**Sulphate (SO_4^{-2})**	**SiO_2**	**Barium (Ba)**	**Titanium (Ti)**	
1300	130.65	215.70	0.145	0.028	

nd: not determined

B toxicity sources are listed as follows soil B content, unconscious industrial waste management, thermal waters which are rich in soluble B and discharge of this thermal water to river, lake, irrigation waters and soils [12, 13]. It is known that in some locations of northern Aegean and the territory of the Central and Eastern Anatolia have toxic levels of Boron in soils. The boron content in the soil of Aydın seems to be enough and high in some places [14]. According to data from the Mineral Research & Exploration General Directorate, there are totally well 290 license (214 explore license and 76 business license) and 46 geothermal water resource purposely in Aydın region in 2012 [15]. The many of these resources drain its waste water to the Büyük Menderes River. Thus waste water contaminates ground water and the boron content of the water sources for agricultural purposes is increasing [16]. Recent studies [17, 18, 24] have shown that boron toxicity is a major problem at Büyük Menderes valley because of activity of geothermal power plants and their discharge of geothermal waste water due to drilling well to have effiancy and take it under control and into next stage due to holding electrical energy by using flash methode. Fluids emerging from some of these areas contain high boron concentrations and cause environmental problems for irrigation waters in agricultural areas where boron contaminates aquifers and soils [9]. Boron has accumulated in the plants more than required in the area [19].

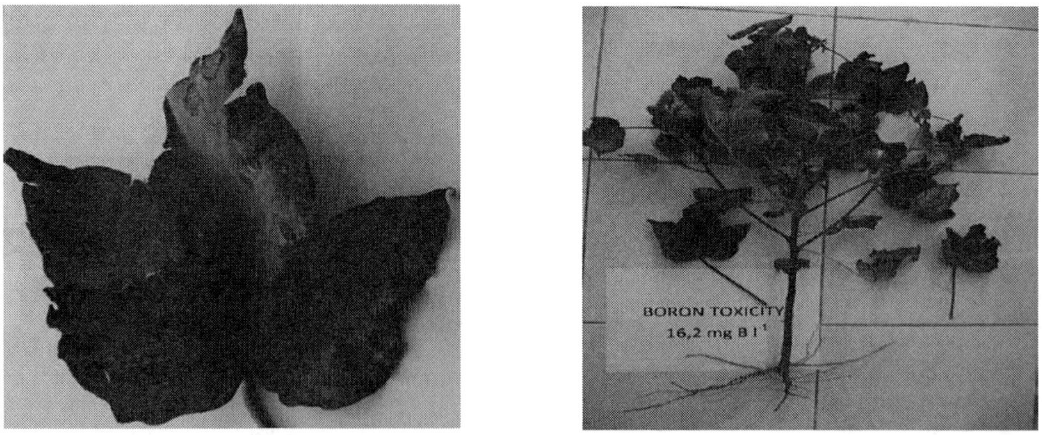

Fig. 6. Boron toxicity symptoms on cotton plants [17].

Table 3. Soil Boron Sufficiency Levels [20].

B (mg kg^{-1})	Status
0-0.4	Severely Deficient
0.5- 0.9	Deficient
1.0- 1.4	Adequate
1.4- 4.9	High
>5.0	Toxic

Table 4. Classifications of Boron for Irrigation Waters of Different Degree of Boron Tolerance [21].

Classes	Non tolerant Plants	Semi Tolerant Plants	Tolerant Plants
	mg l^{-1}		
Excellent	< 0.33	< 0.67	< 1.00
Good	0.33 – 0.67	0.67- 1.33	1.00 - 2.00
Permissible	0.67 - 1.00	1.33 - 2.00	2.00 - 3.00
Doubtful	1.00 - 1.25	2.00 - 2.50	3.00 - 3.75
Unsuitable	> 1.25	> 2.50	> 3.75

Boron toxicity can easily occurs especially arid and semi-arid regions in worldwide owing to the fact that the range of B deficiency and B toxicity is narrow (Table 3) [10, 11].

Kanber [22] in order to determine the pollution levels of some underground and surface waters in Aydın province, it has been reported that B contents of irrigation waters are between 0.474 and 8.234 mg^{-1}. In other study, Aydın and Seferoğlu [16] in order to determined boron contents of irrigation water, in Germencik region, between 0.33 and 6.41 mg l^{-1} , where, underground hot water sources were located close to irrigation water.

3. 2 Heavy metal toxicity

Heavy metals are not only natural components of the Earth's crust, but also important things for healthy of human and for sustainablity of nature and environment. They cannot be degraded or destroyed.

As In Table 2 point out in same table, there are heavy metals in water samples of drilled wells. These will be effected as cumulativly also on heavy metal toxicities in soils of basin and may be possible transfer to culture plants by irrigation water, which is also contamined, these rest-waste water can be contamine fresh water sources or surface water more than natural contamination before, which is used for irrigation, because of type of rocks and soil structures of region. To a small extent they enter our bodies by growing up of contaminated water's irrigated vegetables food, and via orginated animal food products, via drinking water and air. As trace elements, some heavy metals (e.g.copper, selenium, zinc) are essential to maintain the metabolism of human body, however, at higher concentrations they can be toxic [25].

There are point out in scientific stadies some determination of effect of heavy metals (e.g. Al, As, Cr, Cd, Pb, Hg, Ni) on living bodies in nature, about it's higher concentrations how they can be toxic and determined that heavy metals poisoning could result, for instance, from drinking-water contamination (e.g. lead pipes), high ambient air concentrations near emission sources, or intake by the food chain [26, 28].

3.3 Microclimatic changes

Because of GEPP's does not burn fuel like fossil fuel plants, they release virtually no air emissions. On emissions level have also influent technology. Influence of technologies on environmental impact is shown in Table 5.

Steam may have an impact on environment, ifluencing on microclimate. Local weather changes influence on agriculture products, and also loss of rare thermophillic plants such as olive trees.

The development of a geothermal areas can involve the removal of trees and brush to facilitate the installation of the power house, substation, well pads, piping, emergency holding ponds, etc. However, once a geothermal plant is built, reforestation and plantings can restore the area to a semblance of its original natural appearance, and can serve to mask the presence of buildings and other structures.

 DOI: 10.1051/e3sconf/20171902028

Table 5. Influence of technologies on environmental impact [23].

Description	Dry system	Flash steam	Binary cycle
Emissions	Highest (low pressure exhaust steam is vented to atmosphere	Moderate	No emissions because of secondary cycle loop is closed
Fluid waste	Moderate e.g. 65 kg $(kWh)^{-1}$	Highest e.g. 8 kg $(kWh)^{-1}$	Lowest
Application	Very basic one, dry steam natural hydrothermal reservoirs available	Wherein geothermal fluids contain low levels of impurities	Wherein geothermal fluids contain low levels of impurities

Fig. 7. Environmental impact of steam on local skale.

Fig. 8. An image of the cooling system of GEPP's.

Geothermal power plants shall give importance to geothermal district cooling. 70-80% of hot geothermal resources are located in Aydın and this area is extremely hot during the summer season and cooling is an important need. In Fig 7, 8 is shown about how impact of vapor of cooling systems on environment however to utilize cooling geothermally would be important in terms of the environment and decrease the dependance to fosil fuels and foreign countries.

4 Conclusions

This study was carried out in order to draw attention to the increasing problem of boron toxicity in the soil of Aydın region affected by wastes opened for authorized or unauthorized use for agricultural purposes and geothermal plants every day. Although geothermal energy is one of a clean and renewable resource, but it leads to B contaminated to irrigation water and soil during the electricity production or other process, by these way it may contain to drink water resources also, and it contains some gases which are important for gas emition for healthy of all living body. It is known that geothermal water contain not only boron but also heavy metals at different levels and concentrations. Although this study refers to the toxicity and pollution of boron caused geothermal water, it is also necessary to determine its heavy metal concentrations and pollutions on the other hand. Thats why in all stages should be organised (e.g. by lows, directives, articles, etc.) to controlled all wastes and restless in each countries which has geothermal sources for cover future of earth and life. Many of geothermal plants drain their waste water to the Büyük Menderes River, which is used to irrigation of agricultural area for thousands of years. Geothermal wastewater contaminated to ground water and boron content of the water sources for agricultural purposes opened is increasing. Some data shows concentration of boron level is 0.43–2.34 $mgkg^{-1}$ in soils and its toxicity is becoming increasingly an importance situation while the cotton growth in this region is very common and that's why it is believed that the severity and expand of the toxicity will become higher in the future.

Heavy metals are also dangerous like boron for human life, because they tend to bioaccumulate. As it known bioaccumulation means; that it's increase in the concentration of a chemical material which is in a biological organism over time, and compared to the chemical's concentration in environment. Compounds cumulatively increased in living things any time, they are taken up and stored faster than they are broken down as time (metabolized) or excreted. Still, food poisoning from heavy metals is very rare and in most cases only occurs after environmental pollution. As we known the most important example of such an environmental pollution occurred in Japan between 1932 and 1955 [26, 27, 28]. As we categorized general damage of heavy metals can be say about its damages, cause of their high concentration in human body like below;

-They damage to the central nervous system,

-Dementia
-Loss of memory
-Listlessness
-Severe trembling
-Alzheimer's disease
-Lung embolism
-Respiratory failure
-Birth defects
-Asthma and chronic bronchitis
-Allergic reactions such as skin rashes, mainly from jewelry
-Heart disorders, and more [25-28].

As a result, human should change GEPP's electricity producing technology, from the type of flash steam system, however its called green and sustainable energy to the other tyeps of clean and green systems, for to not have more gain by reselling old technologies to the developing or undeveloped countries and should have produce new technologies for its cooling needs to not cooling by thrue out vapor to atmosphere and should cover natural environment as possible as for new generations.

References

1. BP Energy outlook to 2035, **96**, (2016)
2. TP crude oil and natural gas sector report, May, **17**, (2016)
3. T. Kaya, TPIC and METU, An overview on Drilling and Geothermal Energy Projects in Turkey, GRC Annual Meeting International Luncheon Reno, NV, USA, **16** (2012)
4. O. Mertoglu, S. Simsek, N. Basarir, Proceedings World Geothermal Congress 2015, 19-25 April Melbourne, Australia, **16** (2015)
5. R. Ilgar, *Ekolojik Bakışla Jeotermal Kaynaklara Dualist Yaklaşım*, (Elektronik Sosyal Bilimler Dergisi, **4**, 13 88-98, Çanakkale, 2005)
6. A.A. Bazmi, G., Zahedi, Renewable Sustainable Energy Rev. **15**, 8, 20 (2011)
7. www.enerjiatlası.com/jeotermal/geotermal energy powerplants, (07.05.2017)
8. www.mta.gov.tr (10.03.2017)
9. Ü. Gemici, G. Tarcan, J. Volcaol. Geotherm. Res **116**, 215-233 (2002)
10. F.M. Eaton, J. Agric. Res. **69**, 237–277 (1944)
11. U.C. Gupta, Y.W. Jame, C.A. Canadian J. Soil Sci. **65**, 381-409 (1985)
12. B. Cartwright, B.A. Zarcinas, A.H. Mayfield, Aust. J. Soil Res. **22**, 261-272 (1984)
13. R.O. Nable, G.S. Banueuels, J.G. Paull, Plant Soil **193**, 181-198 (1997)
14. Anonymous, Tarım-Bor Araştırma ve Uygulama Programı, http://www.boren.gov.tr/tr/ isbirlikleri/ tarim-bor-arastirma-ve-uygulama-programi, (20.02.2015)
15. Anonymous, Yazılı Soru Önergesi (7/8210). TBMM Başkanlığı, 18.06.2012 tarih ve A.01.0.GNS.0.10. 02.00-72340 sayılı yazısı (2014) http://www.tbmm.gov.tr/develop/owa/ yazili_soru_sd. sorgu _baslangic (13.08.2014)
16. G. Aydın, S. Seferoğlu, Aydın Yöresinde Kullanılan Bazı Sulama Sularının Bor konsantrasyonunun Bitki Beslenmesi ve Toprak Kirliliği Açısından İncelenmesi. TOGTAG-1767, Aydın (1999)
17. M.A. Kaptan, Pamukta (*Gossypium hirsutum* L.) Bor Toksisitesi ve Humik Madde Uygulamasının Etkileri. Adnan Menderes Üniversitesi, Fen Bilimleri Enstitüsü, Doktora Tezi, Aydın (2013)
18. M. Aydın, M.A. Kaptan, Z. Dalkılıç, Relationship between fruit cracking and nutritional status of fig (*Ficus carica*L. cv. sarilop) plantations in the lowland of Aydın, Turkey. International Soil Fertility Congress, Germany (to be published).
19. C. Koç, Environ. Prog. Sustainable Energy **30**, 347-357 (2011)
20. B. Wolf, Soil Sci. Plant Anal. **2**, 263-374 (1971)
21. M.D. Thorne, H.B. Peterson, *Irrigation Soil* (2nd Ed., The Blackinston Co. Inc, NewYork, 1954)
22. P. Kanber, *Aydın İli Bazı Yeraltı ve Yerüstü Su Kaynaklarının Kirlilik Durumlarının Belirlenmesi* (ADÜ. FBE. Yüksek Lisans Tezi, Aydın, 2007)
23. R. Banos, F. Manzano-Agugliaro, F.G. Montoya, C. Gil, A. Alcayde, J. Gomez, Renewable Sustainable Energy Rev. **15**, 1753–1766 (2010)
24. M.A. Kaptan, M. Aydın, Z. Dalkılıç, Agro Life Sci. J. **1**, 3, 75-80 (2014)
25. http://www.lenntech.com/periodic/elements, (03.06.2017)
26. http://www.food-info.net/tr/index.htm, (03.06.2017)
27. http://www.who.int/emergencies/famine/en/, (03.06.2017)
28. C. Bakar, A. Baba, 1.Tıbbi Jeoloji Çalıştayı, J. **1**, 162-185, (2009)

DOI: 10.1051/e3sconf/20171902029

Health-risk assessment based on an additive to paints made from isobutyric aldehyde condensation products

Wilhelm Jan Tic[1,*]

[1]Opole University of Technology, Department of Environmental Engineering, Mikołajczyka 5 Street, 45-227 Opole, Poland

Abstract. Solvents are primarily used for making protective coatings. Considering their chemical nature, there are a great variety of coatings, including those based on liquid hydrocarbons and organic chloroderivatives. These products are a serious load to the environment because of their physicochemical properties, therefore, they have for some time been replaced with more-environmentally friendly, new generation products. One of them is the hydroxyester HE-1: made from isobutyric aldehyde condensation products, it is an alternative to those coalescents for paints and varnishes which are intended to be replaced or their use restricted.
The results of selected toxicological tests relating to the human health risk effect of the hydroxyester HE-1 – environmentally-friendly additive to paints and varnishes are presented. The test results indicate that HE-1 causes skin irritation in rabbit only when used at its maximum concentrations. No lesions in the cornea or iris were observed in any of the test rabbits after the application of the hydroxyester HE-1. In the mutagenic effect test of HE-1 on the bacteria *Salmonella typhimurium,* the result was negative. Based on the test results, it was found that the hydroxyester HE-1 may only have a human health risk effect when used at its maximum concentrations.

1 Introduction

Nearly one-half of the volume of globally used solvents is in the production of protective coatings. These solvents are volatile products, made by the processing of natural gas and petroleum. They include liquid hydrocarbons and organic chloroderivatives which are a serious load to the environment [1, 2]. This is regulated in the Decopaint Directive (2004/42/EC) which came into force on 1 January 2007 [3].
It relates to the reduction of emissions of solvents from decopaints and automotive varnishes, other than those covered by the limitations of Directive No. 1999/13/EC concerning solvent emissions [4]. According to the Decopaint Directive, any substance having its initial boiling point at 250°C or lower, as measured at a pressure of 101.3 kPa, is a Volatile Organic Compound (VOC).
Lower VOC levels in paints can be obtained by developing paint formulations which contain coalescents not classified as VOC, or which contain coalescents and glycols classified as VOC in amounts not higher than the limits referred to in the applicable laws [5].

Phthalates were the prior choice as additives to paints and varnishes. Considering their physicochemical properties, there are two groups of phhthalates: high-molecular and low-molecular products. The high-molecular phthalates, which include di-isononyl phthalate (DINP) and di-isodecyl phthalate (DIDP), represent 80% of the consumption of phthalates in Europe alone. The low-molecular phthalates include dibutyl phthalate (DBP), benzyl-butyl phthalate (BBP), and di-2-ethyl-hexyl phthalate (DEHP), which are classified as products with high human health risk [1]. Based on animal tests, it was found that the low-molecular phthalates are toxic, therefore, they may not be used for manufacturing toys, articles for children, cosmetics, and medical devices [6].
The adverse effect of phthalates on human health has been confirmed in a number of other research reports [2,7,8].

Most of the conventional coalescents, which compete with one another on the market, have their boiling points below 250°C – the criterion for being classified as VOC – thus falling into the category of VOC with applicable quantitative limitations. Some new products on the market have their boiling points above the criterion for VOC (Table 1). The products with boiling points below 250°C are approved for use on the condition that their total VOC emissions are not higher than the limits set out in the Decopaint Directive [3].
The withdrawals in the sector of coalescents for paints and varnishes or products with limited use can be replaced, as an alternative, with the hydroxyester HE-1 (Fig. 1).

The aim of the study was to determine a human health risk effect of hydroxyester HE-1 and another advanced additives to paints and varnishe.

* Corresponding author: w.tic@po.opole.pl

 DOI: 10.1051/e3sconf/20171902029

Table 1. Structural formulae and boiling temperatures of advanced additives to paints and varnishes [5,9].

Compound	CAS Number	Boiling point [°C]	Structural formula
Propane-1,2-diol	57-55-6	187.4	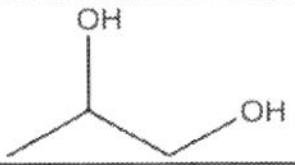
Oxydipropanol	25265-71-8	231	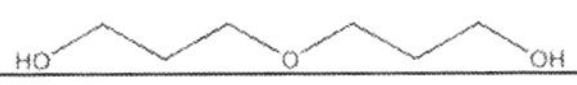
2-(2-Butoxyethoxy)ethanol	112-34-5	225-234	
(2-Methoxymethylethoxy)propanol	34590-94-8	184-197	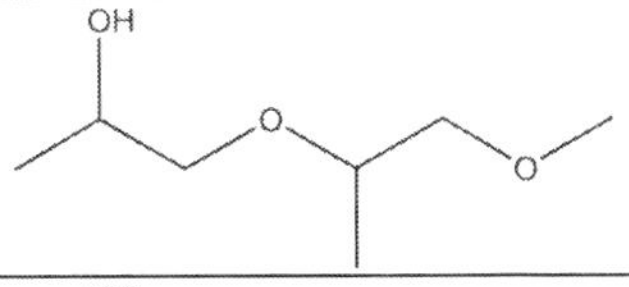
1-Butoxypropan-2-ol	5131-66-8	171	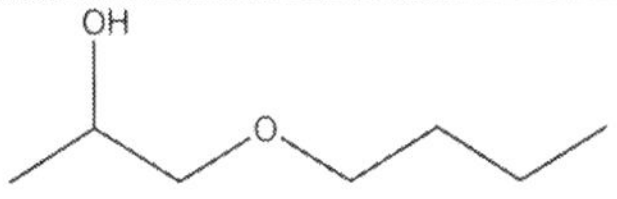
[(Butoxymethylethoxy)methylethoxy] propan-1-ol	55934-93-5	275	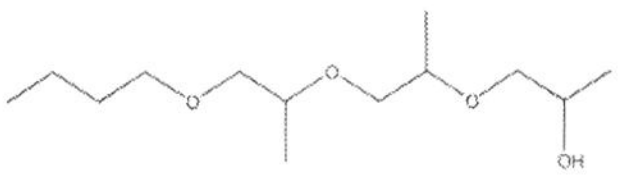
1-Isopropyl-2,2-dimethyltrimethylene diisobutyrate	6846-50-0	281.5	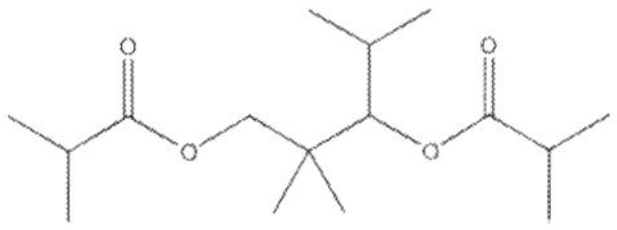

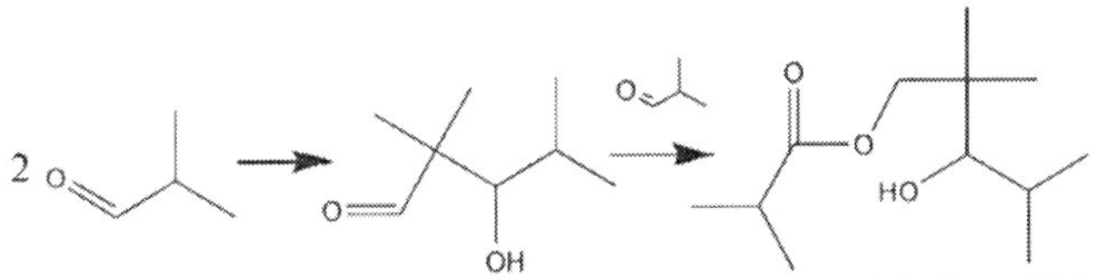

isobutyl aldehyde   3-hydroxy-2,2,4-trimethylpentanal   3-hydroxy-2,2,4-trimethylpentyl isobutyrate (Hydroxyester HE-1)

Fig. 1. General diagram of condensation of isobutanal to obtain hydroxyester HE-1.

2 Material and methods

2.1. Material

Hydroxyester HE-1 is obtained in a sequence of chemical reactions, where isobutyric aldehyde is the basic starting material in the process of aldol condensation with the subsequent Cannizaro and Tiszczenko reaction, as shown in the diagram in Fig. 1. HE-1 is hydrophobic solvent and its major application is a coalescent in water-based architectural paints.

The product has a boiling point of 255°C and is not classified as a VOC [10, 11]. Since a HE-1 production plant of a capacity of more than 100 Mg per year is going to be launched, an assessment of the product in terms of its toxicological and ecotoxicological properties is required [12]. In this paper, the results of the human health risk tests of HE-1 are discussed. The toxicological tests were carried out in accordance with Good Laboratory Practices at the Institute of Industrial Organic Chemistry, Pszczyna, Poland.

2.2 Methods

The results of three selected toxicological tests of the hydroxyester HE1 are presented in this paper. They include, among other things: dermal irritation/corrosion test, eye irritation test, and bacterial reverse mutation test. Later in this paper, the hydroxyester HE-1 is compared, in terms of toxicological properties, with other advanced additives to paints and varnishes of which the CAS numbers, boiling temperatures and structural formulae are provided in Table 1.

* Corresponding author: w.tic@po.opole.pl

2.2.1 Acute dermal irritation/corrosion test in rabbit according to OECD Test Guideline No. 404 [13]

Grading the effects of acute dermal irritation/corrosion is an important step in the toxicity assessment of test materials, including the hydroxyester HE-1. The test results indicate the potential hazards of dermal exposure to the test material.

The objective of the test was to obtain information on the health-risk effect of potential dermal exposure to the test material. An initial test was carried out using one animal. A single 0.5 cm^3 dose of the test material was applied to a hairless skin site of one animal (rabbit 1) and secured with asuitable tape. The exposure time was 4 hours.

After the examination of the exposed skin, the test material was applied to the skin of two more animals (rabbits 2 and 3) for a period of 4 hours in order to confirm the presence or absence of irritant effect. The procedure was the same as in the application of the test material to the skin in rabbit 1.

For the duration of the experiment, all the test animals were subjected to general clinical observations for morbidity and mortality. Detailed clinical observations of the exposed skin in the test animals were conducted at 1, 24, 48 and 72 hours after exposure as well as at 7 and 14 days after exposure [14].

The results of the detailed clinical observation of the test animals were reported using the classification referred to in the OECD Test Guideline No. 404/ Method B.4. (Table 2).

Table 2. Grading of dermal reactions: erythema and eschar formation, oedema.

Erythema and eschar formation	
No erythema	0
Very slight erythema (barely perceptible)	1
Well defined erythema	2
Moderate to severe erythema	3
Severe erythema (beef redness) to eschar formation preventing formation of erythema	4
Oedema	
No oedema	0
Very slight oedema (barely perceptible)	1
Slight oedema (edges of area well defined by definite raising)	2
Moderate oedema (raised 1mm)	3
Severe oedema (raised more than 1 mm and extending beyond area of exposure)	4

According to [15], a substance or mixture applied to the skin in rabbit is classified as irritant after developing a clearly inflammatory condition which persists for at least 24 hrs after exposure.

The skin condition is clearly inflammatory if:
- mean grading for either the erythema and eschar formation or for the oedema is at least 2,
- both the erythema and eschar formation and the oedema formation were graded individually in two or three test animals as being equivalent to the mean score of at least 2 for the respective animals.

In both of the above cases, the mean values were calculated from all the reactions graded after 24, 48 and 72 hrs.

A clearly inflammatory condition of the skin will persist in at least two animals until the end of the observation. The following specific effects are taken into consideration: cell growth, desquamation, discoloration, cracking, eschar formation, hair loss [15].

2.2.2 Acute eye irritation/corrosion test in rabbit according to OECD Test Guideline No. 405 [16]

The objective of the acute eye irritation/corrosion test was to obtain information on the health-risk effect of potential exposure of the eyes to the hydroxyester HE-1.

According to the test procedure, 0.1 cm^3 of the hydroxyester HE-1 was placed in the conjunctival sac of one eye of each test animal whereas the other eye, which remained untreated, served as control. In order to confirm the presence or absence of irritation, the procedure was applied to three test animals. For the duration of the experiment, all the test animals were under general clinical observation, conducted daily for morbidity and mortality. Detailed clinical observations of the exposed eyes, regarding lesions of the cornea, iris, and conjunctiva were conducted at 1, 24, 48 and 72 hours after exposure.

For grading the acute eye irritation/corrosion effect, the following classification of lesions in the eye was applied (Table 3). The grading relates to the lesions in the cornea, iris and conjunctiva [17].

Table 3. Grading of lesions in the eye.

Cornea (opacity: degree of density)	
No ulceration or opacity	0
Scattered or diffuse areas of opacity, details of iris clearly visible	1
Easily discernible translucent area; details of iris slightly obscured;	2
Necrous area; no details of iris visible; size of pupil barely discernible	3
Opaque cornea; iris not discernible through the opacity	4
Iris	
Normal	0
Markedly deepened rugae, congestion, swelling, moderate circumcorneal hyperaemia; or injection; iris reactive to light	1
Hemorrhage, gross destruction, or no reaction to light	2
Conjunctivae – redness (refers to palpebral and bulbar conjunctivae; excluding cornea and iris)	
Normal	0
Some blood vessels hyperaemic (injected)	1
Diffuse, crimson color; individual vessels not easily discernible	2
Diffuse, beefy red	3
Conjunctiva – swelling (refers to lids and/or nictitating membranes)	
Normal	0
Some swelling above normal	1
Obvious swelling, with partial eversion of lids	2
Swelling, with lids about half closed	3
Swelling, with lids more than half closed	4

According to [15], a substance or preparation has irritant effect on the eye if, following test substance application in the conjunctival sac of the test animal, it produces obvious lesions arising within 72 hours following application, which persist for at least 24 hours.

Ocular lesions are graded as obvious if mean scores correspond to any one of the following values:
- corneal opacity – grade 2 or higher than 2 but lower than 3;
- iridial lesion – grade 1 or higher than 1 but not higher than 1.5;
- circumcorneal hyperaemia – grade 2.5 or higher;
- conjunctival swelling – grade 2 or higher.

If three test animals were used in the test, the substance or preparation is considered to be irritant to the eye if ocular lesions in two or three test animals correspond to one of the grades referred to above, except for iridial lesion, where grade higher than 1 and lower than 2 applies, and for circumcorneal hyperaemia, where grade 2.5 or higher applies [15].

2.2.3 Bacterial reverse mutation test according to OECD Test Guideline No. 471 according to [18]

It was the objective of the test to determine the potential genotoxic effects produced by HE-1 in a standard Ames test – short-term mutagenicity test, recommended as a screen for genotoxic activity [19]. The bacterial reverse mutation test uses amino-acid requiring strains of *Salmonella typhimurium* to detect point mutations, which involve substitution, addition, or deletion of one or a few DNA base pairs. This test detects chemical substances producing mutations which revert mutations present in the test strains and restore the functional capability of the bacteria to synthesize an essential amino-acid. The revertant bacteria are detected by their ability to grow in the absence of the amino acid required by the parent test strain.

Two essential mutagenicity tests were used in the tests described in this paper: plate incorporation (without metabolic activation) and preincubation (with metabolic activation). In the plate incorporation method, components are mixed with an overlay agar and plated immediately onto minimal medium. In the preincubation method, the test mixture (bacteria + test substance) is incubated before being added to an overlay agar and plated onto minimal medium. In both techniques, after 2 or 3 days of incubation, revertant colonies are counted and compared with the number of spontaneous revertant colonies on solvent control plates. Mutagenicity of the test substance is shown in the growing count of the revertant colonies [20].

The test substance has a mutagenic effect if the concentration-related (over the range tested) and reproducible increase at one or more concentrations in the number of revertant colonies per plate with or without metabolic activation, $M_f > 2$ [20].

In order to determine mutagenicity, the result was also compared with the negative control, dimethyl sulfoxide (DMSO).

Initially, the dose range was selected by testing HE-1 at concentrations up to 5 mg/plate – the maximum test concentration according to the OECD Test Guideline No. 471. This preliminary experiment was performed to determine the HE-1 concentrations for the proper test. The following HE-1 test concentrations were used in the proper test: 0.0001, 0.001, 0.01, 0.1, 0.5, 1 mg HE-1/plate. The proper test assessed HE-1 for mutagenicity using the bacterial strains of *Salmonella typhimurium* TA 100, TA 98, TA 97, TA 1535 and TA 102. The tests were carried out in the absence of an external metabolic activation system – in the standard Ames plate test (without modification and with preincubation), and with a S9 fraction derived from the Sprague-Dawley rat and treated with Aroclor 1254 – in the standard plate test without modification. Three experiments without metabolic activation and two experiments with metabolic activation were performed.

3 Results and discussion

3.1 Acute dermal irritation/corrosion in rabbit according to OECD Test Guideline No. 404

Table 4 shows the results of the acute irritation/corrosion of the skin in two rabbits, graded from 0 to 4. The reaction of the test animals to the hydroxyester HE-1 was assessed by grading the skin reaction (Table 2). The acute irritation of the skin was assessed based on mean scores at 24, 48 and 72 hours.

At 1 hour after exposure, observation of the skin in the test site in rabbit 1 and rabbit 2 indicated a very slight erythema (barely perceptible). At 24 hours after exposure, the erythema was more intense and was graded as moderate to severe in the two rabbits. Moreover, a very slight (barely perceptible) oedema was observed in rabbit 1.

At 48 and 72 hours after exposure, the erythema in rabbit 1, which had lost some of its intensity by that time, was graded as well defined. Moreover, no oedema was found on the exposed skin in rabbit 1. During both observations, rabbit 2 continued to have a moderate to severe erythema and a very slight (barely perceptible) oedema. Moreover, epidermal desquamation was detected in that animal (Table 4). At 7 days after exposure, the skin of the two rabbits showed a very slight (barely perceptible) erythema and epidermal desquamation in the test site. The skin of rabbit 2 showed no oedema in the test site. 14 days after exposure, both rabbits (1 and 2) showed no pathological dermal changes in the test site (Table 4). At 24, 48 and 72 hours after exposure, mean value for erythema was 2.3 for rabbit 1 and 3.0 for rabbit 2. For oedema, mean values were 0.3 and 0.7, respectively (Table 4). Therefore, on the basis of the test results and according to [15], it is justified to say that the hydroxyester HE-1 causes skin irritation in rabbit.

3.2 Acute eye irritation/corrosion in rabbit according to OECD Test Guideline No. 405

The results of acute eye irritation/corrosion in rabbit are shown in Table 5. The effect of the hydroxyester HE-1 on the cornea, iris and conjunctiva was evaluated by scoring the degree of lesions in the eye in rabbit (Table 3). Following test substance application, lesions were observed in the cornea, iris and conjunctiva of the eye in rabbit.

Table 4. Grading of acute dermal irritation/corrosion in rabbit.

Rabbit No.	Reaction	Values at						Mean score at 24, 48 and 72 hours
		1 hour	24 hours	48 hours	72 hours	7 days	14 days	
1	Erythema	1	3	2	2	1 ED	0	2.3
	Oedema	0	1	0	0	0	0	0.3
2	Erythema	1	3	3 ED	3 ED	1 ED	0	3.0
	Oedema	0	0	1	1	0	0	0.7

ED – epidermal desquamation

Table 5. Grading of acute eye irritation/corrosion in rabbit.

Rabbit No.	Part of eye		Values at					Mean score at 24, 48 and 72 hours
			1 hour	24 hours	48 hours	72 hours	7 days	
1	Cornea		0	0	0	0	0	0
	Iris		0	1	0	0	0	0.3
	Conjunctiva	erythema	2	3	3	1	0	2.3
		swelling	1	1	1	0	0	0.7
2	Cornea		0	0	1	0	0	0.3
	Iris		0	1	0	0	0	0.3
	Conjunctiva	erythema	2	3	1	1	0	1.7
		swelling	2	2	0	0	0	0.7
3	Cornea		0	0	1	0	0	0.3
	Iris		0	1	0	0	0	0.3
	Conjunctiva	erythema	2	3	1	1	0	1.7
		swelling	1	1	0	0	0	0.3

At 1 hour following test substance application, no changes were observed in the cornea or iris of the eye in the three test rabbits. The conjunctiva of the eye in the three test rabbits showed a diffuse, crimson redness with individual vessels not easily discernible. Some swelling of the conjunctiva was observed in rabbits 1 and 3 and swelling with partial eversion of lids in rabbit 2.
The three test rabbits were observed to have hyperaemia with swelling of the nictitating membrane, circumcorneal injection and an exudate on the eye lids and lid hairs.

At 24 hours following test substance application in the three test rabbits, no changes were detected in the cornea, the iris was hyperaemic, and reaction of the pupil to light was correct. Diffuse beefy redness was observed in the conjunctiva in the three test rabbits. The test rabbits were observed to have the same conjunctival swelling as at 1 hour, that is, some swelling above normal in rabbits 1 and 3 and obvious swelling with partial eversion of the lids in rabbit 2. Moreover, each of the three test animals continued to have a hyperaemic and swollen nictitating membrane, and circumcorneal injection. An exudate on the eye lids and lid hairs was detected in the test rabbits 2 and 3 and a small amount of discharge in the test rabbit 1.

Based on the test results and [15], it is safe to say that the hydroxyester HE-1 is not an eye irritant/corrosive in rabbit.

3.3 Bacterial reverse mutation test according to OECD Test Guideline No. 471

For the bacteria *Salmonella typhimurium* TA 100, mean range of spontaneous mutations was 184-297 revertant colonies per plate. The ratio between the number of revertant colonies per plate and the number of revertant colonies per control plate was not observed to exceed 2. At a ratio of 1 mg/plate in the experiments with and without activation system, the number of revertant colonies was significantly reduced ($M_f < 0.8$).

For *Salmonella typhimurium* TA 1535, mean range of spontaneous mutations was 19-25 revertant colonies per plate. In one experiment without metabolic activation, a concentration of 1 mg/plate statistically significantly reduced the number of revertant colonies ($M_f < 0.6$). Another experiment, with metabolic activation, a concentration of 0.01 mg/plate produced a statistically significant decrease in the mutation frequency in comparison with the positive control, though without biological significance ($M_f = 0.78$).

For the bacteria *Salmonella typhimurium* TA 97, mean range of spontaneous mutations was 114-193 revertant colonies per plate. In one experiment without metabolic activation at a concentration of 1 mg/plate, a statistically significant increase in the mutation frequency in respect of the controls was detected, though without biological significance ($M_f = 0.66$).

In the experiment with metabolic activation at a maximum concentration of 5 mg/plate, the level of frequency of revertant colonies was reduced in comparison with the negative control (M_f = 0.35), and concentrations of 1 and 0.5 mg/plate significantly reduced the number of revertant colonies, though without biological significance.

For *Salmonella typhimurium* TA 102, mean range of spontaneous mutations was from 248 to 350 revertant colonies per plate. In the experiments without metabolic activation, the frequency range of revertant colonies was not higher in comparison with the negative control. In all the experiments, the number of revertant colonies was statistically significantly reduced at a concentration of 1 mg/plate (M_f < 0.77), comparably to 0.5 mg/plate in one experiment. It was observed that the toxicity of the concentrations 0.5 and 1 mg/plate was related to the duration of contact of the test substance with the test bacterial strain before plating it onto the Petri dishes. In the experiments with metabolic activation, HE-1 was not observed to increase the number of revertant colonies above the frequency range of spontaneous mutations.

For the bacteria *Salmonella typhimurium* TA 98, mean range of spontaneous mutations was 22-53 revertant colonies per plate. In the experiments without metabolic activation, the frequency range of revertant colonies was not higher in comparison with the negative control. In the experiments with metabolic activation, the mutagenic potential of HE-1 was not confirmed in the experiment with the fraction derived from the rat liver.

In the dose range from 0.0001 mg/plate to 1 mg/plate, HE-1 was not observed to increase the number of revertant colonies above the frequency range of spontaneous mutations. Figures 2 and 3 show the number of revertant colonies per plate for 5 different strains of *Salmonella typhimurium* with and without metabolic activation. All the results were compared with the negative control (DMSO). For the strain TA 102, at a HE-1 dose of 0.0001 mg/plate in the tests with and without metabolic activation, the number of revertant colonies was not determined.

From the data shown in Fig. 2 and Fig. 3, it was observed that the number of revertant colonies per plate was lower than that for the negative control regardless of the HE-1 dose in a majority of cases. For the bacterial strains TA 100 and TA 97 in the experiments without metabolic activation for a HE-1 dose of 0.001 mg/plate, the numbers of revertant colonies determined for the negative controls were exceeded, although not very highly: the percentages were 0.5 and 3%, respectively. A similar observation was made in the experiments with metabolic activation for a HE-1 dose of 0.0001 mg/plate and the strain TA 97 and for a HE-1 dose of 0.1 mg/plate and the strain TA 102, where the numbers of revertant colonies determined for the negative controls were slightly exceeded, by 1.4 and 1.8%, respectively.

The mutagenic effect test results indicate that HE-1 causes no statistically significant, dose-related increase in the number of revertant colonies and no statistically significant reproducible positive response to any of the test points. Based on the test results, the hydroxyester HE-1 was not found to produce a mutagenic effect in the test system.

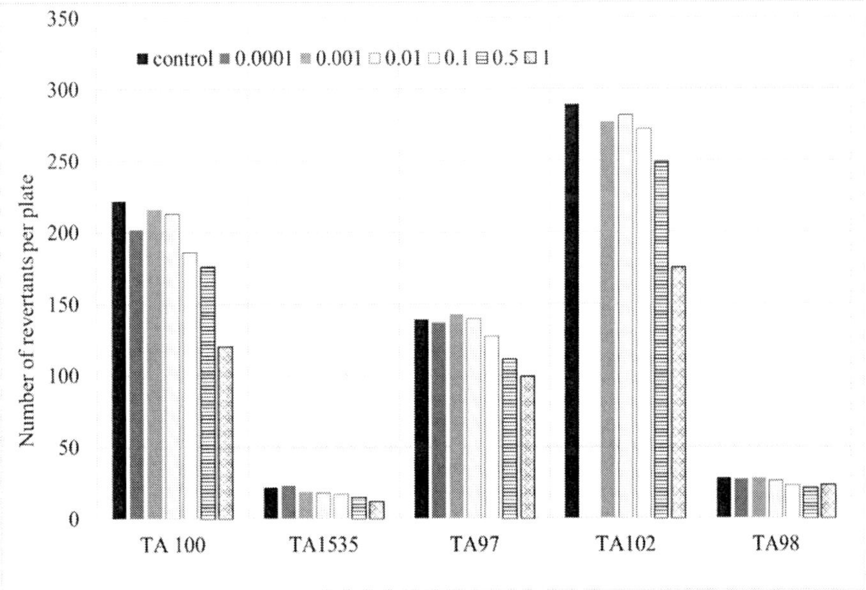

Fig. 2. Number of revertants per plate for various strains of *Salmonella typhimurium* without activation.

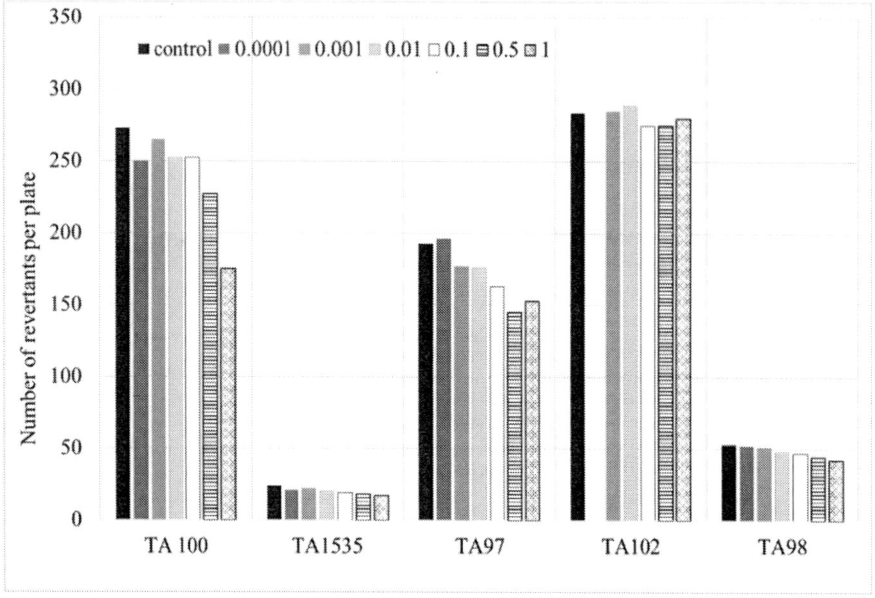

Fig. 3. Number of revertants per plate for various strains of *Salmonella typhimurium* with activation.

Table 6. Comparison of essential toxicological properties of advanced additives to paints and varnishes [6].

Compound	Acute skin irritation/corrosion	Acute eye irritation/corrosion	Mutagenic effect
57-55-6	no irritation/corrosion	no irritation/corrosion	no mutagenic effect
25265-71-8	no irritation/corrosion	no irritation/corrosion	no mutagenic effect
112-34-5	no irritation/corrosion	tests not according to OECD guidelines	no mutagenic effect
34590-94-8	no irritation/corrosion	no irritation/corrosion	no mutagenic effect **
5131-66-8	irritation	Irritation	no mutagenic effect
55934-93-5	no irritation/corrosion	no irritation/corrosion	no mutagenic effect
6846-50-0	no irritation/corrosion	no irritation/corrosion	no mutagenic effect **
HE-1*	irritation	no irritation/corrosion	no mutagenic effect

* - results of own tests
** - test on mammalian cells according to OECD 476

3.4 Comparison of toxicological properties of hydroxyester HE-1 and advanced additives to paints and varnishes

Table 6 shows the results of tests of the essential toxicological properties of HE-1 and advanced additives to paints and varnishes which are described in the literature and of which the names and structures were provided in Table 1. Information about the toxicological properties of advanced additives to paints and varnishes was collected from the ECHA website [9]. The following properties were compared: dermal and ocular irritation and mutagenic effect. For the compounds with the CAS numbers: 34590-94-8 and 6846-50-0, the mutation test was carried out using mammalian cells according to the OECD Test Guideline No. 476 [21]. Only three of the advanced additives to paints and varnishes shown in Table 6, namely those with the CAS numbers: 55934-93-5 and 6846-50-0 and HE-1, are not classified as VOC. When comparing their toxicological properties, it can be observed that HE-1 may have an irritant effect only when used at high concentrations. None of the compounds produces a mutagenic effect and none is an irritant. Although the other additives to paints and varnishes in Table 6 do not show any human health risk effect, their boiling temperatures are below 250°C, therefore they are classified as VOC.

4 Conclusions

The results of toxicological tests are discussed on the basis of three selected examples: acute dermal irritation test, acute eye irritation/corrosion test, and mutagenic effect test.

For the acute dermal irritation/corrosion test in rabbit, mean score for erythema was 2.3 and 3.0, respectively, in rabbit 1 and rabbit 2. For oedema, mean score was 0.3 and 0.7, respectively, in rabbit 1 and rabbit 2. Therefore, based on the skin reaction grading, test results and [15], it is justified to say that the hydroxyester HE-1 is a skin irritant in rabbit at its high concentrations only.

In the acute eye irritation/corrosion test in rabbit, no lesions were detected in the cornea or iris in any of the test animals. Only erythema was observed in the conjunctiva in the three test rabbits, which manifested itself as injected blood vessels and congestion of the nictitating membrane. Based on the test results and [15], it is safe to say that the hydroxyester HE-1 does not produce eye irritation in rabbit.

In the mutagenic effect tests, HE-1 did not produce any statistically significant, dose related increase in the number of revertant colonies, and no statistically significant, reproducible positive response to any of the test points. The results of the mutagenic effect tests indicate that HE-1 is not mutagenic in the test system used.

When comparing the results of toxicological tests for the hydroxyester HE-1 and advanced additives to paints and varnishes, it is safe to say that HE-1 is an environmentally friendly alternative to those additives which either are classified as VOC or are involve human health risk.

Grant number 13856. This work was supported by Poland National Centre for Research and Development in the years 2007-2010.

References

1. P. Ventrice, D. Ventrice, E. Russoa, G. De Sarro, Environ. Toxicol. Phar. **36**, 88 (2013)
2. D. Lithner, Å. Larsson, G. Dave, Sci. Total Environ. **409**, 3309 (2011)
3. Directive 2004/42/CE of the European Parliament and of the Council (2004)
4. Council Directive 1999/13/EC (1999)
5. W.J. Tic, W. Hreczuch, Przem. Chem. **87**, 1126 (2008)
6. J.H. Kim, J. Yun, J.K. Sohng, J.M. Cha, B.C. Choi, H.J. Jeon, S.H. Kim, C.H. Choi, 2007. Environ. Toxicol. Pharmacol. **23**, 272 (2007)
7. E.S. Beach, B.R. Weeks, R. Stern, P.T. Anastas, Pure Appl. Chem. **85**, 1611 (2013)
8. T.T. Bui, G. Giovanoulis, A.P. Cousins, J. Magnér, I.T. Cousins, C.A. deWit, Sci. Total Environ. 541, 451 (2016)
9. ECHA, 2017. Information on chemicals. http://echa.europa.eu/information-on-chemicals
10. W.J. Tic, Przem. Chem. **84**, 32 (2005)
11. W.J. Tic, J. Guziałowska-Tic J., Przem. Chem. **95**, 1529 (2016)
12. Regulation (EC) No. 1907/2006 of the European Parliament and of the Council (2006)
13. OECD Guideline for testing of chemical substances No. 404 (2002)
14. J. Guziałowska-Tic, Chemik, **68**, 834 (2014)
15. Regulation of the Minister of Health. Journal of Laws 2012, item 1018 (2012)
16. OECD Guideline for testing of chemical substances No. 405 (2012)
17. J. Guziałowska-Tic, W.J. Tic, Ecol. Chem. Eng. A **22**, 2 (2015)
18. OECD Guideline for testing of chemical substances No. 471 (1975)
19. K. Mortelmans, E. Zeiger, Mut. Res. **455**, 61 (2000)
20. B.N. Maron, B.N. Ames, Mut. Res. **113**, 172 (1983)
21. OECD Guideline for testing of chemical substances No. 476 (1997)

 DOI: 10.1051/e3sconf/20171902030

Additives for reducing the toxicity of respirable crystalline silica. SILIFE project

Eliseo Monfort[1,*], *Ana* López-Lilao[1], *Alberto* Escrig[1], *Maria Jesus* Ibáñez[1], *Guliana* Bonvicini[2], *Otto* Creutzenberg[3], and *Christina* Ziemann[3]

[1] Instituto de Tecnología Cerámica, Universitat Jaume I, Castellón, Spain
[2] Centro Ceramico di Bologna (CCB), Bologna, Italy
[3] Fraunhofer Institute for Toxicology and Experimental Medicine (ITEM), Hannover, German

Abstract. Prolonged inhalation of crystalline silica particles has long been known to cause lung inflammation and development of the granulomatous and a fibrogenic lung disease known as silicosis. The International Agency for Research on Cancer (IARC) has classified Respirable Crystalline Silica (RCS) in the form of quartz and cristobalite from occupational sources as carcinogenic for humans (category 1). In this regard, numerous studies suggest that the toxicity of quartz is conditioned by the surface chemistry of the quartz particles and by the density and abundance of silanol groups. Blocking these groups to avoid their interaction with cellular membranes would theoretically be possible in order to reduce or even to eliminate the toxic effect. In this regard, the main contribution of the presented research is the development of detoxifying processes based on coating technologies at industrial scale, since the previous studies reported on literature were carried out at lab scale. The results obtained in two European projects showed that the wet method to obtain quartz surface coatings (SILICOAT project) allows a good efficiency in inhibiting the silica toxicity, and the preliminary results obtained in an ongoing project (SILIFE) suggest that the developed dry method to coat quartz surface is also very promising. The development of both coating technologies (wet and a dry) should allow these coating technologies to be applied to a high variety of industrial activities in which quartz is processed. For this reason, a lot of end-users of quartz powders will be potentially benefited from a reduced risk associated to the exposure to RCS.

1 Introduction

Total European usage of crystalline silica (i.e. quartz and cristobalite) is measured in thousands of millions of tonnes per annum. It is used in many manufacturing industries [1-3] such as quarrying, mining, mineral processing (e.g. drying, grinding, bagging, and handling [4]); slate working [5], stone crushing and dressing; foundry work; brick and ceramic tile making; some ceramic refractory processes; construction work [6-7], including work with stone [8], concrete, brick and some insulation boards; tunnelling; building restoration; and the pottery and ceramic industries [9-12]. Hence, a vast number of European workers (4 million [13]) are potentially exposed to crystalline silica dust at their workplace.

In this regard, it is long established that prolonged inhalation of crystalline silica particles causes lung inflammation and development of the granulomatous and fibrogenic lung disease known as silicosis. Furthermore, the International Agency for Research on Cancer (IARC) classified Respirable Crystalline Silica (RCS) in the form of quartz and cristobalite from occupational sources as carcinogenic for humans (category 1, [14]). In this regard, numerous studies suggest that the toxicity of quartz is conditioned by the surface chemistry of the quartz particles and, in particular, by the density and abundance of silanol groups [15].

Blocking these groups to avoid the interaction with cellular membranes would theoretically be possible in order to reduce or even to eliminate the toxic effect. This postulate has been repeatedly verified by various researchers [15-23] who managed to reduce its toxicity at lab scale, on treating quartz with different substances: polyvinylpyridine-N-oxide (PVPNO) [15], organosilanes [16-17], aluminium lactate [18- 22] and strong Lewis acids, such as $AlCl_3$ or $FeCl_3$ [23].

The mechanism by which these substances act consists of the formation of a surface coating on the quartz so that its silanol groups are no longer accessible for interaction in biological media. Fig. 1 shows an example of this process using propyltrimetoxysilane (PTMO) as coating agent, in which two consecutive chemical reactions take place: 1) hydrolysis of one methoxy group of PTMO releasing methanol as co-reaction product; and 2) silanization of quartz, in this stage the hydrolysed PTMO reacts with the silanol groups of quartz to form a very stable covalent bond.

Bearing in mind this background, this study was therefore undertaken to develop coating application

* Corresponding author: eliseo.monfort@itc.uji.es

DOI: 10.1051/e3sconf/20171902030

processes at industrial scale to reduce the RCS toxicity. Regarding this fact, by the integration of these coating processes into the quartz producer manufacturing process, a great number of end user of quartz powders may potentially be benefit from a reduced risk associated to the exposure to RCS.

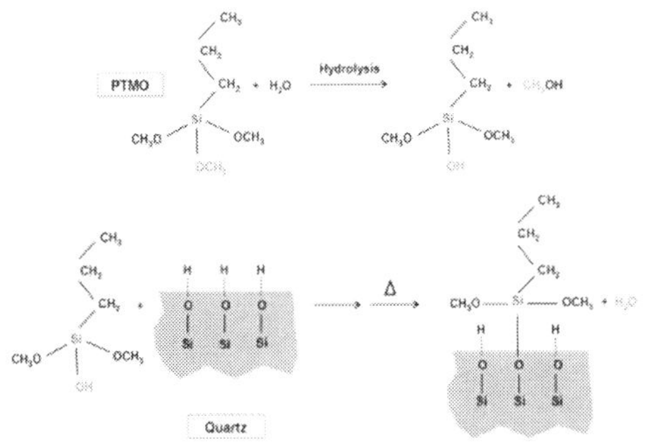

Fig. 1. Hydr.olysis and silanization reactions.

2 Materials and methodology

2.1 Materials

Although a wide range of compounds were studied, for the sake of brevity only the results obtained with SIVO160, the additive that exhibited the most promising behaviour, are presented. **SIVO160** is a commercial aminosilane polymer used for coating siliceous as well as metallic surfaces.
An industrial **quartz, used in a sanitary ware composition**, was selected. Its specific surface area was 0.87 m^2 g^{-1} and its particle size distribution (PSD) is shown in Fig. 2.

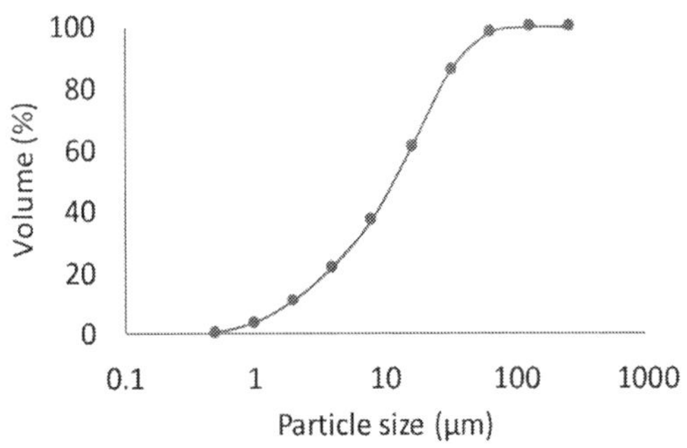

Fig. 2. PSD of the selected quartz.

2.2 Methodology

The followed methodology can be divided into two blocks: quartz coating and coating effectiveness assessment.

2.2.1 Quartzes coating

2.2.1.1 Wet method (SILICOAT)

1. **Dispersion of the additive in water**. The additive was incorporated into the quantity of water required in the formula. Certain organosilanes required preliminary **hydrolysis,** as hydrolysis took place more quickly in acid medium, this preliminary step was performed in the relevant cases, adding a certain quantity of HCl 0.01 M to the additive and stirring until a homogeneous liquid was obtained (Fig. 3, left).

3. **Addition of the quartz**. The resulting suspension was stirred for the corresponding reaction time (Fig. 3, right).

4. **Filtering the suspension and washing with ethanol**. Once the reaction time had ended, the suspension was filtered with a Büchner flask and the resulting powder was repeatedly washed with ethanol to remove any surplus additive.

Fig. 3. Hydrolysis with HCl (left). Stirring suspension with quartz (right).

2.2.1.2 Dry method (SILIFE)

The coating has been performed using a high shear-rate plow mixer LÖDIGE M5R with a capacity up to 4 l. The silane feeding has been carried out by an injector feeder (Fig. 4).

Fig. 4. Silane feeding (left) and mixer (right).

As in the wet method, after this process, the obtained quartz is rinsed with ethanol to remove any excess of silane and the obtained samples is dried with infrared lamps.

 DOI: 10.1051/e3sconf/20171902030

2.2.2 Coating effectiveness assessment

Zeta potential (ς potential) is a measure of the electrical charge in the surface of the suspended particles. It is thus related with the density of silanol groups in the quartz particles surface. As a consequence, the addition of the coating agent to quartz suspensions should increase or reduce the ς potential of the quartz (which is negative at pH > 2) and coating effectiveness can be measured in this manner. In this sense, it should be highlighted that some authors pointed out that the maximum inhibition of quartz toxicity is produced by rendering it iso-electric [24]. Moreover, the measurement of ζ potential was found to be extremely sensitive (modifications in surface charge due to the addition of less than 0.05% w/w of aminosilanes can be perfectly detected) and ζ potential is well correlated with toxicological results.

3 Results and discussion

The obtained results are divided into three blocks: wet method application results and validation of dry method.

3.1 Coating effectiveness

Thee variation of zeta potential with the proportioned quantity is represented in Fig. 5. From this figure, it can be observed that iso-electric point was associated with approximately 0.2% of SIVO160.

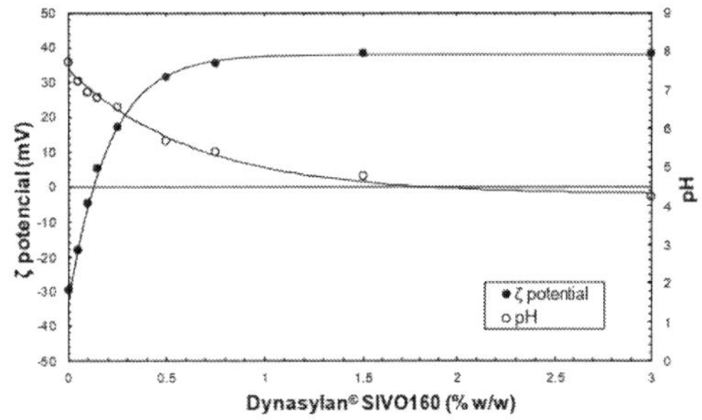

Fig. 5. Variation of zeta potential.

The most promising coating treatments at different concentrations were tested for toxicity. Fig. 6 shows the in vitro determination results of coating effectiveness. In these tests Al_2O_3 was used as particulate negative (PNC) and DQ12 as quartz positive control, this quartz is widely used in the literature as positive control, therefore the results may be compared with other studies on RCS toxicity. The toxicological tests presented in Fig. 6 are LDH release (Fig. 6, top) and the tail intensity in the alkaline comet assay (Fig. 6, bottom), which were used to detect membrane and direct DNA damage, respectively.

The in vitro toxicological results evidenced that the oligomeric, amino-modified siloxane SIVO160 seemed to possess considerable quartz toxicity-quenching potential. The toxicological study of the treated quartzes was completed in the project with in vivo tests, a more comprehensive information can be found in a specific paper [24].

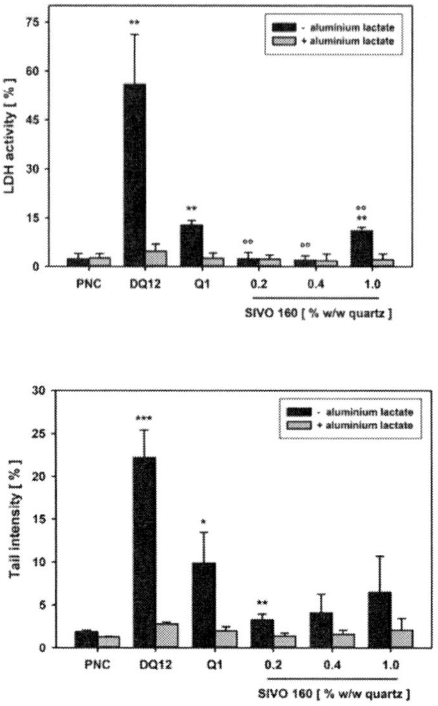

Fig. 6. Toxicological tests. Data represent means ± SD of three independent experiments.
*/**/***Significantly different from PNC: P ≤ 0.05/0.01/0.001, respectively, Student's t-test for unpaired values, two tailed.
°/°°/°°°Significantly different from uncoated Q1: P ≤ 0.05/0.01/0.001, respectively, Student's t-test for paired values, two tailed.

3.2 Dry method versus wet method

By the wet method, the selected treatments provided very satisfactory results in terms of toxicity, but unfortunately a lot of industrial processes using quartz do not include a wet stage, therefore in order to allow this coating technology to be implemented of in a wide variety of industrial applications, a new European project was proposed. Hence, the aim of the SILIFE project is to develop a dry method, based in the process proposed in SILICOAT, for reducing RCS toxicity at industrial scale. According to the surveyed literature, the use of a dry application method to this aim has not been previously reported.

Fig. 7 shows the preliminary results obtained so far applying this methodology and the comparison with those obtained by the wet application method (previous study).

The lab outcomes reported in Fig. 7 suggest that, surprisingly, from the physicochemical viewpoint, the dry method may be developed using similar dosages to those employed in the wet method. If these results are verified at industrial scale (nowadays under development) and by the toxicological assays, it would be a very promising technology.

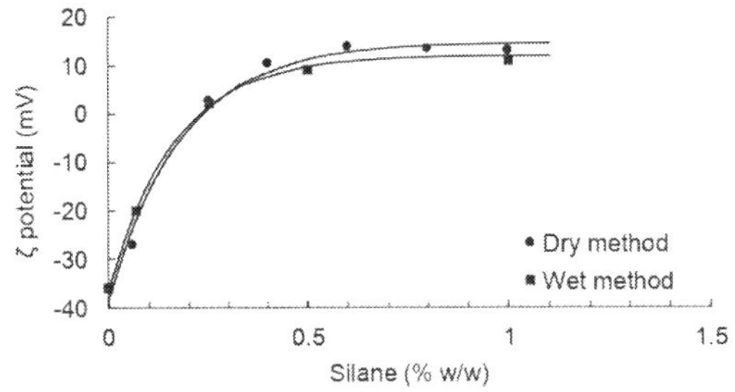

Fig. 7. Dry method versus wet method.

4 Conclusions

The results of a previous project (SILICOAT) showed that the wet method to industrially obtain quartz surface coatings allows a good efficiency in inhibiting the toxicological effect of silanol groups (verified by means of "in vitro" and "in vivo" tests) at relatively low doses. The developed process may be readily implemented at industrial scale, provided that they include a wet stage.

The lab results obtained in the SILIFE ongoing project suggest that the developed dry method to coat quartz surface is also very promising. Nevertheless, these preliminary physicochemical results must be scaled up and tested from the toxicological point of view.

The development of both coating technologies (wet and a dry) should allow these coating technologies to be applied at industrial scale in a high variety of industrial activities in which quartz is processed. For this reason, a lot of end-users of quartz powders will be potentially benefited from a reduced risk associated to the exposure to RCS.

Acknowledgements

This study was part of the SILICOAT project (FP7/2007-2013) and SILIFE project (LIFE14 ENV/ES/00238).

References

1. European Network for Silica (NEPSI): Good Practice Guide on Workers Health Protection through the Good Handling and Use of Crystalline Silica and Products Containing It, (2011). Available from: www.nepsi.eu (accessed April, 2015)
2. European Network for Silica (NEPSI): Agreement on Workers' Health Protection Through the Good Handling and Use of Crystalline Silicas and Products Containing It, (2006). Available from: www.nepsi.eu (accessed April, 2015)
3. D.L. Radnoff, M.K. Kutz, Ann. Occup. Hyg. **58**, 19–27 (2014)
4. M.H. Fulekar, Ann. Occup. Hyg. **43**, 269–73 (1999)
5. BE, Bang, H. Suhr, Ann. Occup. Hyg. **42**, 557–63 (1998)
6. M.E Lumens, T. Spee, Ann. Occup. Hyg. **45**, 585–95 (2001)
7. E.Van Deurssen, A. Pronk, S. Spaan, et al., Ann. Occup. Hyg. **58**, 724–38 (2014)
8. C.B. Healy, M.A. Coggins, M. Van Tongeren et al., Ann. Occup. Hyg. **58**, 6–18 (2014)
9. E. Monfort, M.J. Ibáñez, A. Escrig, et al., cfi/Ber. DKG **85**, 36–42 (2008)
10. E. Monfort, M.J. Ibáñez, A. Escrig, In Safety, reliability and risk analysis: Theory, methods and applications. London: Taylor and Francis, pp. 2743–46 (2009)
11. E. Monfort, A. Mezquita, E. Vaquer et al., In Comprehensive Materials Processing. Bassim: Elsevier, **8**, pp. 71–102, (2014)
12. C. Ziemann, O. Creutzenberg, M.J. Ibañez, et al., cfi/Ber. DKG **91**, 63–9 (2015)
13. European Trade Union Confederation (2007) Response to the 2nd phase consultation on Directive 2004/37/EC
14. International Agency for Research on Cancer (2012): IARC Monographs on the evaluation of the carcinogenic risk of chemicals to humans: A review of human carcinogens. Part C: arsenic, metals, fibres, and dusts. IARC Monographs, pp. 196-211, (2012)
15. H.W. Schlipköter, A. Brockhaus, J. Molec. Medicine, **39**, 22, 1182–1189 (1961)
16. Castranova, V. (Eds.) Silica and Silica-Induced Lung Diseases. Boca Raton: CRC Press, pp. 293–298 (1996)
17. Castranova, V.; van Dyke, K.; Wu, L.; Dalai, N.S.; Vallyathan, V. (Eds.) Silica and Silica-Induced Lung Diseases. Boca Raton: CRC Press, pp. 283–291 (1996)
18. R. Duffin, P.S. Gilmour, R.P.F. Schins, A. Clouter, K. Guy, D.M. Brown, W. MacNee, P.J. Borm, K. Donaldson, V. Stone, Toxicol. App. Pharm. 176(1),10–17 (2001)
19. R. Bégin, S. Massé, M. Rola-Pleszczynski, M. Martel, Y. Desmarais, M. Geoffroy, L. Le Bouffant, H. Daniel, J. Martin, Exp. Lung Res. **10**, 385–399 (1986)
20. G.M. Brown, K. Donaldson, D.M. Brown, Toxicol. App. Pharm. **101**, 1, 95–105 (1989)
21. C. Albrecht, A.M. Knaapen, A. Becker, D. Höhr, P. Haberzettl, F.J. van Schooten, P.J.A. Borm, R.P.F. Schins, Resp. Res., **6**, 129, (2005)
22. C. Ziemann, S.M. Reamon-Buettner, T. Tillmann, H.T. Hansen, M.J. Ibáñez, E. Monfort, G. Bonvicini, A. Escrig, O. Creutzenberg, In: Deutsche Gesellschaft für experimentelle und klinische Pharmakologie und Toxikologie (2014)

DOI: 10.1051/e3sconf/20171902030

23. R.P. Nolan, A.M. Langer, J.S. Harington, G. Oster, I.J. Selikoff, Environ. Res. **26, 2**, 503–520, (1981)
24. C. Ziemann, A. Escrig, G. Bonvicini, M.J. Ibáñez, E. Monfort, A. Salomoni, O. Creutzenberg, Annals of Work Exposures and Health, **61, 4**, 468-480 (2017)

Environmentally friendly paint and varnish additives based on isobutyrate aldehyde condensation products

Joanna Guziałowska-Tic[1,*] and *Wilhelm Jan* Tic[1]

[1]Opole University of Technology, Department of Environmental Engineering, Mikołajczyka 5 Street, 45-271 Opole, Poland

Abstract: The demand for exploitation of new plasticizers and coalescents to be used for production of water dispersion adhesives based on poli(vinyl acetate), turned to be recently of particular significance because the use of all kind of toxic phthalates was banned and certain restrictions were made to use glycol derivatives classified as volatile organic compounds. An alternative for toxic plasticizers used for production of adhesives is hydroxyester (HE-1) obtained from isobutanal during the process of aldol condensation under subsequent Cannizaro and Tischenko reactions.
The paper presents selected results of ecotoxicological tests on the environmental impact of hydroxyester HE-1. It was found that the substance is biodegradable and has no negative impact on algae growth rate, however it may inhibit algae biomass growth. The results of Daphnia immobilization test shows that the substance is safe for the aquatic environment. Whereas, at higher concentrations, hydroxyester HE-1 may exhibit acute toxicity to rainbow trout.

1 Introduction

The solvents are mainly applied for production of protective coatings. Their large chemical variety includes liquid hydrocarbons and organochlorines, which due to their physicochemical properties are a significant burden to the natural environment [1]. Therefore, they are being consequently replaced with more environmentally friendly new-generation products.

An important legislation in this field is the European Directive 2004/42/EC, so-called "Decopaint Directive", which entered into force on 1 January 2007 [2]. It concerns emission reduction of solvents from decopaints and automotive lacquers classified as volatile organic compounds (VOC), which were previously not covered by the restrictions of the Directive no. 1999/13/EC on reduction of solvent emission [3].

According to the 2004/42/EC, a volatile organic compounds is every substance of onset boiling point at temperature of 250 °C or lower measured under pressure of 101.3 kPa.

Environmental hazards related to the VOC result from the fact that in the stratosphere they interact with stratospheric ozone leading to the depletion of ozone layer, which serves as a natural UV filter. While in the troposphere, at increased temperature and exposed to solar radiation, VOCs react with nitrogen oxides (NO_x) producing tropospheric ozone, which is classified as a greenhouse gas and is one of smog components. Reaction rate, wind and possibility of delayed reaction due to the insufficient amount of sun light, means that ozone and generated smog can form with a few days delay, hundreds of kilometres from the source of VOC emission [4]. According to the recommendation of the World Health Organisation, the average eight-hour standard for ozone concentration is 120 μg/m³, while in Europe there are constantly reported concentrations exceeding this value. In order to reduce the quantity of generated tropospheric ozone, regulations were prepared aiming to reduce VOC emissions from various sources, including solvent-containing products.

The recipes for paints and varnishes commonly include plasticizers that are harmful to aquatic environment, persistent, bioaccumulative, toxic and classified as VOC [5,6]. The literature describes many examples of environmentally friendly plasticizers that can serve as an alternative for toxic ones [7-11]. The available literature focuses on the plasticization properties of these additive, however, there is not much data on their physicochemical, toxicological and ecotoxicological properties [10]. Table 1 presents selected alternative plasticizer with boiling temperature above 250° C, i.e. not classified as VOC. These plasticizers are usually used as PVC additives, but also for other polymers. These materials can be also used as additives for lubricating oils and vinyl-acrylic dispersions [12].

* Corresponding author: j.guzialowska@po.opole.pl

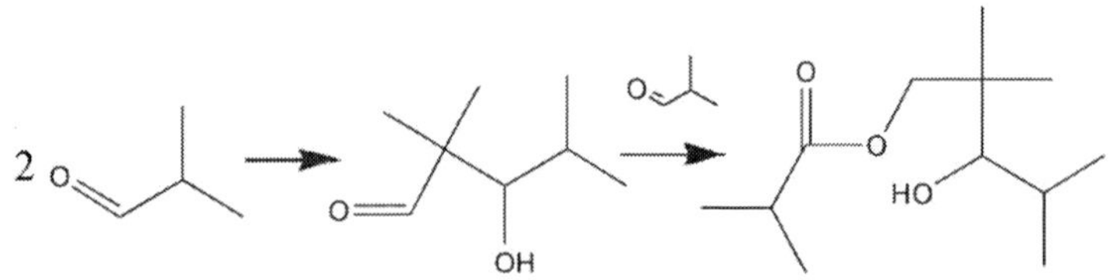

Fig. 1. General scheme of isubutanal condensation towards obtaining Hydroxyester HE1.

An alternative for withdrawn from use and limited plasticizers and coalescents for paints and varnishes is hydroxyester HE-1 obtained in a sequence of chemical reactions, where isobutyrate aldehyde is a main feed in the aldol condensation with subsequent Cannizzaro-Tishchenko reaction as shown in Fig. 1. Hydroxyester (HE-1) constitutes a mixture of two isomeric products, it is 3-hydroxy-2,2,4-trimethylpropyl isobutyrate and 3-hydroxy-2,2-dimethyl-1-(1-methylo)propyl isobutyrate that are in the equilibrium.

Hydroxyester HE-1 is a hydrophobic organic solvent, which is mainly used as a coalescent in formulations of water borne facade paints. The product has boiling point of 255 °C, which places it outside the VOC classification [13-15].

2 Material and methods

Due to the fact that it is planned to launch production of hydroxyester HE-1 in volume of 100 Mg per year, it is required to evaluate toxicological and ecotoxicological properties of the product [16]. The list of all ecotoxicological tests conducted for hydoxyester HE-1 is as fallows:

- *Pseudokirchneriella subcapitata,* growth inhibition test according to OECD guideline 201 [17],
- *Daphnia magna*, acute immobilization test according to OECD guideline 202 [18],
- Rainbow trout, acute toxicity test according to OECD guideline 203 [19],
- Earthworm, acute toxicity test according to OECD guideline 207 [20],
- Active sludge, respiration inhibition test according to OECD guideline 209 [21],
- Soil microorganisms: nitrogen transformation test according to OECD guideline 216 [22],
- Soil microorganisms: carbon transformation test according to OECD guideline 217 [23],
- Ready biodegradability according to OECD guideline 301 [24].

The test required to assess environmental risks were performed in accordance with good laboratory practice at the Institute of Industrial Organic Chemistry (IPO) in Pszczyna.

The paper presents results of four selected ecotoxicological tests including biodegradability, acute toxicity test for rainbow trout, algae growth inhibition test and acute immobilization test for Daphnia. The description of how the test was conducted can be found below.

2.1 *Pseudokirchnereilla subcapitata*, growth inhibition test according to the OECD Guideline No. 201

The test aimed to determine concentrations of the test material (hydroxyester HE-1) causing 50% inhibition of average specific growth rate and biomass growth for algae *Pseudokirchneriella subcapitata* E_rC_{50} and E_yC_{50}, respectively, after 72 h exposure), as well as determination of the lowest concentration (LOEC) of the test material causing growth inhibition compared to the control and the highest material concentration causing no effect compared to the control (NOEC).

At the beginning of the experiment, algae density was 1 x 10^4 cells/cm^3. The test used the following concentrations of the test material: 5, 10, 20, 40 and 80 mg/dm^3 and the control (0.0 mg/dm^3). Algae density was determined by means of spectrophotometry measuring absorbance of algae suspension at wavelength λ 670 nm after 24, 48 and 72 hours of exposure. The measured absorbance values were converted to cell density using the regression curve (nomogram). The nomogram was prepared by diluting 72 h algae culture and obtaining series of samples. For each sample, cell density was counted in Bűrker chamber by microscopy and the absorbance was measured. Cell density was calculated from the linear regression equation y=0.6003x-0.0003 with correlation coefficient R^2= 0.9953.

For calculation of algae growth rate and algae biomass growth, the following equations (1) and (2) were used:

$$\text{Growth rate } (0-72\text{h}) = \frac{[\ln(\text{cell density at 72 h})]-[\ln(\text{cell density at 0 h})]}{3\text{ days}} \quad (1)$$

$$Biomass\ growth\ [10^6\ cells\ in\ cm^3] = (cell\ density\ at\ 72h) - (cell\ density\ at\ 0h) \quad (2)$$

DOI: 10.1051/e3sconf/20171902031

Table 1. Chemical structures and boiling points of alternative plasticizers [10, 25].

Compound name	Abbreviation	CAS Number	Boiling point [°C]	Chemical structure
Bis(2-ethylhexyl) adipate	DEHA	103-23-1	377.88	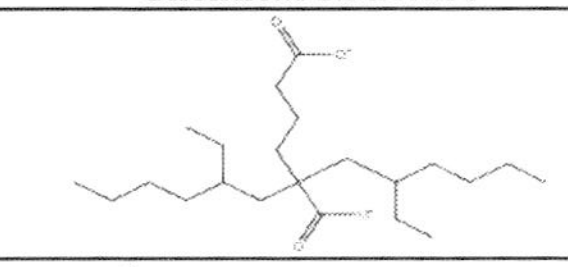
Dibutyl sebacates	DBS	109-43-3	344.5	
Bis(2-ethylhexyl terephthalate)	DEHT	6422-86-2	375	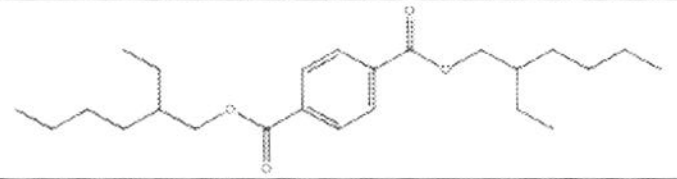
Alkylsulfonic phenyl ester	ASE	91082-17-6	300-400	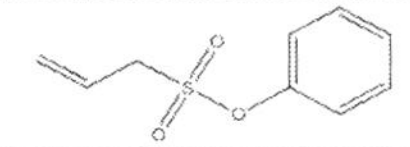
Bis(2-ethylhexyl)sebacates	DOS	122-62-3	256	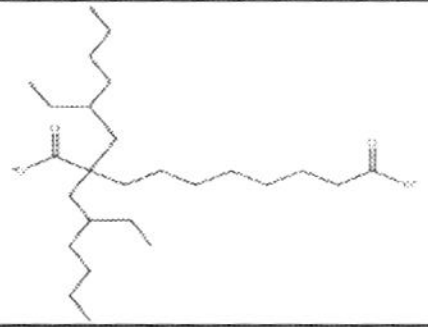
Tris-2-ethylhexyl trimellitate	TOTM	3319-31-1	355	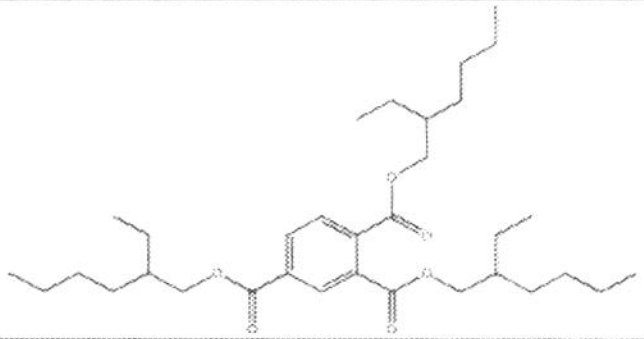
Glycerin triacetate	GTA	102-76-1	258	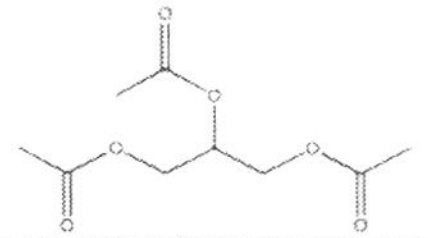
Trimethyl pentanyl diisobutyrate	TXIB	6846-50-0	281.5	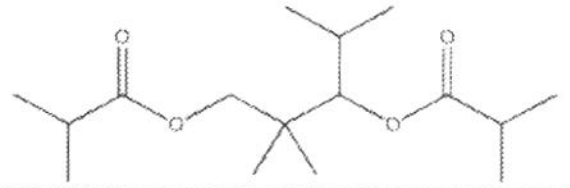

The calculations and statistical analysis were performed using commercial software ToxRat Professional.

According to the Regulation of the Minister of Health on criteria and classification methods concerning chemical substances and their mixtures, the chemical substance has no negative impact on algae growth, if values E_rC_{50} and E_yC_{50} after 72 h are higher than 100 mg/dm^3 [26].

2.2 Daphnia magna, acute immobilization test according to the OECD guideline 202

The aim of test was to determine toxic effect of hydroxyester HE-1 on Daphnia and determination of average concentrations causing 0, 50 and 100% Daphnia immobilization (determination of EC_{50}, EC_0 and EC_{100} in mg/dm^3) after 24 h and 48 h of exposure.
The main experiment used limiting concentration of 100 mg/dm^3 obtained by dissolving 50 mg of tested material in 500 cm^3 of water in measuring flask. The experiment included four repetitions with 5 tests organisms in each. Daphnia sensitivity was also checked in the reference test for reference substance, i.e. potassium dichromate ($K_2Cr_2O_7$).

According to the Regulation of the Minister of Health on criteria and classification methods concerning chemical substances and their mixtures, the chemical substance is safe to the aquatic environment (Daphnia), if value EC_{50} after 48 h is higher than 100 mg/dm^3 [26].

2.3 Rainbow trout, acute toxicity test according to the OECD guideline 203

The aim of the test was to determine acute toxicity for rainbow trout (*Oncorhynchus mykiss* Walb.) and value of LC_{50} (concentration causing 50% lethality in tested population) after 24, 48, 72 and 96 h, as well as determination of LC_0 (highest tested concentration causing no lethality by the end of the test) and LC_{100} (the lowest tested concentration causing 100% lethality after exposure time).

Rainbow trout, acute toxicity test for hydroxyester HE-1 was conducted in a static system. The fish was exposed to test material at concentrations: 5.6; 10.0; 18.0; 32.0 and 56.0 mg/dm^3 plus the control (0.0 mg/dm^3). The experiment was conducted in 35 dm^3 glass aquaria on four months old fish. Each aquarium contained 7 fish, while observations of lethality and intoxication symptoms were conducted after 3, 6, 24, 48, 72 and 96 hours of exposure.Trout sensitivity was also concerning chemical substances and their mixtures, the chemical substance is safe to the aquatic environment (trout), if value LC_{50} after 96 h is higher than 100 mg/dm^3 [26].

2.4 Ready biodegradability according to the OECD guideline 301

The aim of test was to determine by laboratory means, ready biodegradability of hydroxyester HE-1 in the aquatic environment after 28 days of experiment. As inoculum in the test, active sludge filtrate from biological wastewater treatment plant was used. The measuring system consisted of 0.3 dm^3 bottles for BOD determination filled with mixture composed of specific quantities of water, culture medium, active sludge and reference substance solutions and tested materials selected in the accordance with [24]. Sodium acetate was used as a reference substance at concentration of 5 mg/dm^3, while the test material had concentration of 2 mg/dm^3.

Oxygen concentration was measured using oxygen probe placed directly in the BOD bottles. Two samples were taken from each series for oxygen concentration analysis in one week intervals during the 28 days of incubation.
The obtained results were used to calculate biochemical oxygen demand (BOD), i.e. amount of oxygen consumed by microorganisms for metabolism of chemical.

While the %biodegradation of test material and reference material was calculated using the following equation:

$$\%\, biodegradation = \frac{BOD(mg\ O_2/mg\ of\ test\ material)}{TOD\ (mg\ O_2/mg\ of\ test\ material)} \times 100 \qquad (3)$$

Theoretical oxygen demand (TOD) was calculated based on the molecular formula and is a total amount of oxygen required to oxidize the chemical compound.
The theoretical molecular formula of hydroxyester HE-1($C_{12}H_{24}O_3$) was used to calculate TOD, which was 2.44 mgO_2/mg, while for sodium acetate ($C_2H_3NaO_2$) - 0.78 mgO_2/mg.

According to the Regulation of the Minister of Health on criteria and classification methods concerning chemical substances and their mixtures [26], the chemical substance is considered readily biodegradable, if after 28 day biodegradability test, a 60% biodegradation is achieved in tests based on oxygen loss [27].

3 Results and discussion

3.1 *Pseudokirchnereilla subcapitata*, growth inhibition test according to the OECD Guideline No. 201

Fig. 2 presents effect of the test material on *Pseudokirchneriella subcapitata* growth rate and biomass growth.

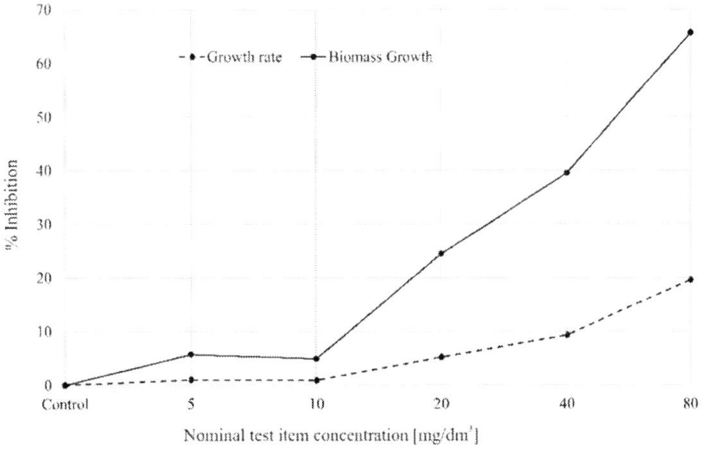

Fig. 2. Inhibition of biomass growth and growth for various concentrations of hydroxyester HE-1 in comparison to the control after 72 h.

Fig. 2 leads to the conclusion that hydroxyester HE-1 has greater effect on biomass growth than on algae growth rate. The greatest effect on biomass growth was observed for test material concentration above 10 mg/dm^3, while for the growth rate it was for concentration exceeding 40 mg/dm^3.

ECx values were determined based on nominal concentrations of the test material for biomass growth and growth rate (Table 2).

The concentration causing 50% growth rate inhibition in *Pseudokirchneriella subcapitata* (E_rC_{50}/72 h) was above 80 mg/dm^3, i.e. for the highest used in the experiment. Whereas E_rC2_0/72 h was also > 80 mg/dm^3, and E_rC_{10}/72 h was 36.46 mg/dm^3. The obtained E_rC_{50} was above 80 mg/dm^3, which means that hydroxyester HE-1 has no negative impact on the aquatic environment. Whereas average concentration causing 50% algae biomass growth E_yC_{50}/72 h was 48.10 mg/dm^3, E_yC_{20}/72 h – 16.47 mg/dm^3, while E_yC_{10}/72 h was 9.37 mg/dm^3. The obtained value of E_yC_{50} is greater than 10 mg/dm^3, but lower than 100 mg/dm^3. According to the Regulation of the Minister of Health of on criteria and classification methods concerning chemical substances and their mixtures, this means that hydroxyester HE-1 at higher concentrations may affect algae biomass growth [26].

LOEC and NOEC after 24, 48, 72 h were calculated next for nominal concentration of the tested preparation. LOEC for biomass growth and average specific growth rate was 20 mg/dm^3, while NOEC for biomass growth and average specific growth rate was 10 mg/dm^3.

3.2 Daphnia magna, acute immobilization test according to the OECD guideline 202

Total lethality for the tested material (hydroxyester HE-1) at concentration of 100 mg/dm^3 after 24 and 48 hours of exposure was 0%. For comparison, total lethality for the reference material ($K_2Cr_2O_7$) at concentration of 3.2 mg/dm^3 after 24 hours of exposure was 60%, and after 48 h – 100%.

The results of Daphnia acute immobilization test for the tested material are presented in table 3. Average concentration causing 50% Daphnia immobilization (EC_{50}) after 24 h is above 100 mg/dm^3. EC_0 after 24 h is equal to 100 mg/dm^3 or greater, while EC_{100} after 24 h is above 100 mg/dm^3.

Average concentration causing 50% Daphnia immobilization (EC_{50}) after 48 h is above 100 mg/dm^3. EC_0 after 48 h is equal to 100 mg/dm^3 or greater, while EC_{100} after 48 h is above 100 mg/dm^3 (Table 3).

Table 2. E_yC_x and E_rC_x values calculated according to the method based on the biomass growth analysis and the average specific growth rate for hydroxyester HE-1 concentrations, where N/A – not calculated due to the mathematical resons.

E_rC_x [mg/dm^3]	Exposure time [h]		
	24	**48**	**72**
E_rC_{10}	N/A	25.30	36.46
E_rC_{20}	< 5.00	74.84	> 80.00
E_rC_{50}	< 5.00	> 80.00	> 80.00
E_yC_x	**Exposure time [h]**		
	24	**48**	**72**
E_yC_{10}	< 5.00	6.33	9.37
E_yC_{20}	< 5.00	13.67	16.47
E_yC_{50}	10.46	59.09	48.10

Table 3. EC_x values for hydroxyester HE-1 and for the reference substance – $K_2Cr_2O_7$

Exposure time	EC_x [mg/dm^3]					
	EC_{50}		EC_0		EC_{100}	
	HE-1	$K_2Cr_2O_7$	HE-1	$K_2Cr_2O_7$	HE-1	$K_2Cr_2O_7$
24 h	> 100	2.58	≥ 100	1.0	> 100	> 3.2
48 h	> 100	1.26	≥ 100	0.56	> 100	3.2

As the value of EC_{50} after 48 hours of exposure is greater than 100 mg/dm^3, according to [26], hydroxyester HE-1 has no negative environmental impact (Daphnia). In turn, for the reference material, EC_{50}/48h was 1.26 mg/dm^3, EC_0 – 0.56 mg/dm^3, and EC_{100} – 3.2 mg/dm^3 (Table 3).

3.3 Rainbow trout, acute toxicity test according to the OECD guideline 203

Hydroxyester HE-1 toxicity to trout was assessed based on the observation of lethality and intoxication symptoms. After 3 h from the start of the experiment, intoxication symptoms were observed. For concentration of 18 mg/dm^3 1 fish showed loss of balance, 2 fish – unusual swimming, while the increased pigmentation was observed in all the fish. While at concentration of 32 mg/dm^3, all the fish exhibited loss of balance, unusual swimming and increased pigmentation.

At the highest used concentration (56 mg/dm^3), 100% lethality was observed after 30 minutes from the start of the experiment. At other concentrations, no lethality was observed during the whole experiment. For comparison, 100% lethality after 48 hours was observed the reference substance (potassium dichromate) at concentration of 32 mg/dm^3. Then, hydroxyester HE-1 toxicity to trout was assessed based on the observation of lethality and intoxication symptoms.

It was shown that LC_0 and LC_{100} after 96 h of exposure were 32 mg/dm^3 and 56 mg/dm^3, respectively. Whereas, the average concentration causing 50% fish lethality (LC_{50}) after 96 h was above 41 mg/dm^3. This means that according to [26], hydroxyester HE-1 at high concentrations may have harmful impact on aquatic organisms and may cause long-term adverse effects in the aquatic environment.

3.4 Ready biodegradability according to the OECD guideline 301

Average oxygen consumption and biochemical oxygen demand (BOD) on 0, 7, 14, 21 and 28 day of analysis for the test material and for the reference material are presented below.
For the tested material, BOD varied from 0.19 mg O_2/dm^3 on the seventh day of the experiment, up to 1.51 mg O_2/dm^3 on the last day of the experiment. Whereas, for the reference material (sodium acetate), BOD varied from 0.57 mg O_2/dm^3 on the first day of the experiment, up to 0.72 mg O_2/dm^3 on the last day of the experiment.
The calculated values of TOD and BOD were used in the equation (3) to determine biodegradability for hydroxyester HE-1 and for the reference material, i.e. sodium acetate (Fig. 3).

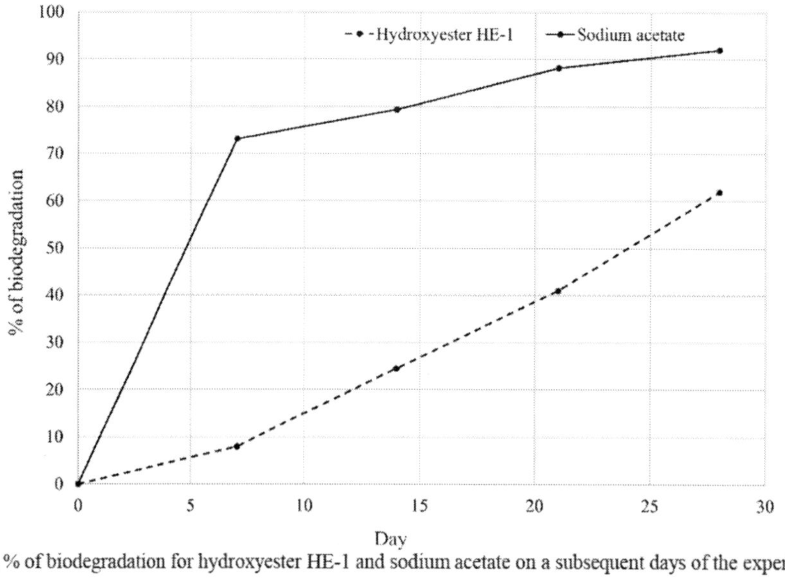

Fig. 3. % of biodegradation for hydroxyester HE-1 and sodium acetate on a subsequent days of the experiment.

Table 4. Comparison of selected ecotoxicological properties for hydroxyester HE-1 and other plasticizers [25], where: * – instead of toxicity to algae test acc. to the OECD guideline no. 211, algae growth inhibition test was performed acc. to the OECD guideline no. 201.

Chemical compound	Toxicity to fish LC50/96h	Toxicity to *Daphnia magna*; EC50/48h	Toxicity to aquatic algae; EC50/72h	Biodegradability BOD5
DEHA	>100 mg/dm^3	0.66 mg/dm^3	>500 mg/dm^3	94%
DBS	> 10,000 mg/dm^3	>1000 mg/dm^3	>1000 mg/dm^3	89.6%
DOS	>1000 mg/dm^3	>1000 mg/dm^3	>1000 mg/dm^3	65%
DEHT	>984 mg/dm^3	>984 mg/dm^3	>0.86 mg/dm^3	73.05%
TOTM	>100 mg/dm^3	>180 mg/dm^3	>100 mg/dm^3	biodegradable
ASE	no data	no data	no data	24%
GTA	no data	770 mg/dm^3	> 940 mg/dm^3	59.4%
TXIB	>1.55 mg/dm^3	>1.55 mg/dm^3	>7.49 mg/dm^3	32%
HE-1	41 mg/dm^3	> 100 mg/dm^3	E_rC_{50} > 80 mg/dm^3 * E_yC_{50} = 48.10 mg/dm^3 *	61.9%

3.5 Comparison of ecotoxicological properties for hydroxyester HE-1 and alternative plasticizers

Table 4 presents comparison of ecotoxicological properties determined for hydroxyester HE-1 and 8 state-of-the-art plasticizers, which names and structures are presented in Table 1. The data on ecotoxicological properties of known plasticizers were taken from ECHA website [25]. The following properties were compared: toxicity to fish (LC_{50}/96h), toxicity to Daphnia (EC_{50}/48h), toxicity to algae (EC_{50}/72h) and biodegradability after 28 days of the experiment.

Compounds: ASE, GTA and TXIB are not biodegradable, as their biodegradation after 28 days of the experiment was 24%, 59.4%, 32%, respectively, i.e. below the required 60%. DEHA plasticizer is biodegradable and shows no negative impact on rainbow trout and algae.
However, Daphnia acute toxicity test gave EC_{50}/48h equal to 0.66 mg/dm^3, which means that DEHA is highly toxic to aquatic organisms and may cause long-term adverse effects in the aquatic environment.

While, DEHT plasticizer is biodegradable and shows no negative impact on rainbow trout and Daphnia. However, algae acute toxicity test gave EC_{50}/72h equal to 0.86 mg/dm^3, which means that DEHT is highly toxic to aquatic organisms and may cause long-term adverse effects in the aquatic environment.

DBD, DOS and TOTM plasticizers are biodegradable and safe to the aquatic environment, as all the results of ecotoxicological tests presented in Table 4 are below the minimum allowable values.

Whereas, hydroxyester HE-1 is biodegradable, has no negative effect on Daphnia and algae growth rate. However, for algae biomass growth and toxicity to rainbow trout, hydroxyester HE-1 may have harmful effect to aquatic organisms, but only at high concentrations.

4 Conclusions

The paper presents results of ecotoxicological tests, including four selected examples: acute toxicity test for rainbow trout, algae growth inhibition test, acute immobilization test for Daphnia, as well as ready biodegradability test.

In case of testing the effect of hydroxyester HE-1 on algae growth rate inhibition, E_rC_{50}/72 h was above 80 mg/dm^3, i.e. the highest used in the experiment, which means that the tested material does not affect algae growth rate. On the other hand, it was found that hydroxyester HE-1 may have negative effect on algae biomass growth, as E_yC_{50}/72 h was 48.10 mg/dm^3. The obtained test results led to the conclusion that hydroxyester HE-1 can affect the aquatic environment, but only at its highest concentrations.

In case of Daphnia acute immobilization test, EC_{50}/48h was above 100 mg/dm^3, which, according to [26], means that hydroxyester HE-1 has no negative impact on Daphnia.

In case of acute toxicity test for rainbow trout, LC_{50} after 96 h of exposure was 41 mg/dm^3, which, according to [26], means that hydroxyester HE-1 can affect the aquatic environment, but only at its highest concentrations.

In turn, in ready biodegradability test, % of biodegradation of hydroxyester HE-1 after 28 days of experiment was 61.9%, which means it is biodegradable, thus safe to the environment.

Comparison or ecotoxicological results for hydroxyester HE-1 and other commercially available plasticizers leads to the conclusion that hydroxyester HE-1 is an environmentally friendly alternative to these plasticizers that are volatile organic compounds or are harmful to the environment.

Based on the presented results and according to the Appendix to the Regulation of the Minister of Health on criteria and classification methods concerning chemical substances and their mixtures

 DOI: 10.1051/e3sconf/20171902031

[26], it may be concluded that the tested substance – hydroxyester HE-1 is a safe substance and poses no threat to the environment.

This research work was financed as a R&D project of Inicjatywa Technologiczna I No. 13856, support was granted in the years 2007-2010.

References

1. P. Ventrice, D. Ventrice, E. Russoa, G. De Sarro, Environ. Toxicol. Pharmacol. **36**, 88 (2013)
2. Directive 2004/42/CE of the European Parliament and of the Council (2004).
3. Council Directive 1999/13/EC (1999)
4. T.E. Burghardt, A. Pashkevich, L. Żakowska, Transport. Res Procedia. **14**, 714 (2016)
5. Toxicity review for diisobutyl phthalate. U.S. Consumer Product Safety Commission (2011)
6. Ch.A. Staples, D.R. Peterson, T.F. Parkeron, W.J. Adams, Chemosphere **35**, 667 (1997)
7. J.V. Barbosa, F. Oliveira, J. Moniz, F.D. Magalhães, M.M.S.M. Bastos, J. Am. Oil. Chem. Soc. **89**, 2215 (2012)
8. J.V. Barbosa, E. Veludob, J. Monizc, A. Mendesa, F.D. Magalhăesa, M.M.S.M. Bastosa, Prog. Org. Coat. **76**, 1692 (2013)
9. E.S. Beach, B.R. Weeks, R. Stern, P.T. Anastas, Pure Appl. Chem. **85**, 1611 (2013)
10. T.T. Bui, G. Giovanoulis, A.P. Cousins, J. Magnér, I.T. Cousins, C.A. deWit, Sci. Total Environ. **541**, 451 (2016)
11. A. Schwieck, M.-C. Bock, Build. Environ. **85**, 243 (2015)
12. Wypych, G. Handbook of Plasticizers, 2nd ed. (ChemTec Publishing, Toronto 2012)
13. W.J. Tic, W.J., Przem. Chem. **84**, 32 (2005)
14. W.J. Tic, J. Guziałowska-Tic, J., Proceedings of the 2nd South East European Conference on Sustainable Development of Energy, Water and Environment Systems, SDEWES.SEE2016. 0038, 1-12 (2016)
15. W.J. Tic, J. Guziałowska-Tic, E. Sutor, W. Butor, Patent EP 2 436 667 A1 (2012)
16. Regulation (EC) No. 1907/2006 of the European Parliament and of the Council (2006)
17. OECD Guideline for Testing of Chemicals No 201 (2011)
18. OECD Guideline for the Testing of Chemicals No. 202 (2004)
19. OECD Guideline for the Testing of Chemicals No. 203 (1992)
20. OECD Guideline for the Testing of Chemicals No. 207 (1984)
21. OECD Guideline for the Testing of Chemicals No. 209 (2010)
22. OECD Guideline for the Testing of Chemicals No. 216 (2000)
23. OECD Guideline for the Testing of Chemicals No. 217 (2000)
24. OECD Guideline for the Testing of Chemicals No. 301D (1992)
25. ECHA, 2017. http://echa.europa.eu/information-on-chemicals
26. Regulation of the Minister of Health. Journal of Laws 2012, item 1018 (2012)
27. J. Guziałowska-Tic, W.J. Tic, Pol. J. Environ. Stud. **25**, 1515 (2016)

DOI: 10.1051/e3sconf/20171902032

Technical evaluation of photobioreactors for microalgae cultivation

Małgorzata Płaczek[1,*], *Agnieszka* Patyna[1], and *Stanisław* Witczak[1]

[1]Opole University of Technology, Faculty of Mechanical Engineering, Department of Process Engineering, 5 Mikołajczyka Street, 45-271 Opole, Poland

Abstract. This paper undertakes the description and assessment of various solutions applied for the design of photobioreactors as the type of apparatus, which can provide high output of green algae biomass. The design of such apparatus plays an important role in the context of the concurrent fulfillment of ecological and economic requirements, which are necessary to conduct an efficient and effective technology using cheap and easily accessible resources to produce different goods. Nowadays, algae is seen as one of the most promising sustainable way to produce energy in the future (biofuels, electricity, thermal energy) but technologies of biomass production and processing are still under development particularly to increase biomass and energy output. The cultivation costs in closed systems are still high, limiting their commercial applications to high-valued compounds but they can be reduced by efficient bioreactor designs, which are able to achieve high areal biomass productivities. This paper focuses on the advantages and drawbacks associated with the application of the particular types of bioreactors in algae production, description of their operation parameters and area for practical application, pointing of the constructions (tubular, flat panel, bubble column) that can contribute to improvement the profitability of large-scale production.

1 Introduction

Uncontrolled anthropogenic activities and technological revolution increase fossil fuel consumption at higher rate that prompted research related to the search for alternative energy sources to cover the current need and future demand of energy. In last few decades biomass of terrestrial crops has been studied as a promising source of renewable energy but large scale production is still debatable issue because of indirect use of fossil fuel, lower productivity, lack of availability of land which are usually intended for food production. These limitations of land based systems fetch opportunity to look into untapped potential energy source such as algae.

Nowadays algae are considered not only as a valuable organic resource with a potential application in the production of a wide range of products, starting with foodstuffs and animal feed, through beauty products, pigments and pharmaceuticals to fertilizers for agriculture and isotopically labeled compounds but as a source of biodiesel, biogas and potentially important component of the wastewater treatment system [1-8]. The current technologies applied for biofuel production on the basis of algae are not sufficiently effective and economically viable so as to be able to prove competitive in comparison to the existing solutions in petrochemical industry based on the use of renewable resources. However, in the consideration of the growing problems of the transport sector and the requirements associated with effective gaining resources for various industry branches, new ideas with regard to the use of integrated biomass cultivation and processing systems are bound to develop continuously. The effective production and processing systems are going to be determined by the development of advanced apparatus, which can provide reduction of the production cost and increase of its competitiveness in relation to the conventional systems.

Due to their small size, measured in micrometers, microalgae are cultivated in dedicated land installations [9-11] and ones that float on a water surface [12]. The open systems, including natural reservoirs (lagoons, lakes) and ponds (raceway ponds) are listed among the oldest configurations used in industry processes designed for production of microalgae [1,8-9,13]. At present, around 90% of the global algae production is conducted in the open systems [14-15]. Estimates indicate that the global annual production of microalgal biomass is equal to 15,000 tonnes of dry mass with the market value in the range from 30 to 300 euro per kilo [16].

The typical open systems of algae cultivation require large areas with considerable light exposition; hence, they are common on desert places, where there is no competition of agricultural production designed for food production purposes. Concrete and PVC tanks are relatively cheap to build and service, however, they production capabilities are limited, e.g. the production capabilities of *Chaetocereos* microalgae is equal to only 300,000 cells·ml^{-1} [17]. Apart from this, open systems in algae farms are susceptible to water loss from

* Corresponding author: m.placzek@po.opole.pl

evaporation and in desert places another problem is associated with the sand carried by sandstorms and volcanic ash transported with the wind over long distances (estimated at 1–10 $ton \cdot km^{-2} \cdot month^{-1}$). Parasites (including mites and competing algae strains) can easily enter the cultivation and cause a serious threat to the biosecurity of the farm due to the lack of safety measures in the vessel installations. In addition, in the areas with high level of insolation, the incident light (>100,000 lux) can led to the thermal shock and algal cells bleaching, which causes the decrease of production efficiency [7, 17]. In this respect, microalgae cultivation conducted in the closed system (photobioreactors) can prove to be more effective. This is due to the possibility of better control over the production process (temperature, light, pH, CO_2 concentration) and the potential increase of the productivity by several and even a dozen times compared to the open systems [18].

The efficiency ratio of biomass in relation to the substrate and its productivity (i.e. parameters representing the economic efficiency of production) are higher in the closed systems compared to the farms consisting of open tanks [19]. Wijffels et al. [20] report that the productivity of closed systems can be as much as 40–80 tonnes of dry mass per hectare per year accompanied by a comparable cost of production in both cultivation systems [19]. One can note, however, that in large photobioreactor systems forming outdoor installations, the productivity of algal biomass can vary depending on the scale of production [2]. For example, the productivity of *Haematococcus pluvialis* algae at a level of 0.05 $g \cdot l^{-1} \cdot d^{-1}$ can be achieved both in the cultivation conducted in a tube bioreactor with the capacity of 25,000 l as well as in a 55 liters bubble column [21-22]. The productivity of algae cultivation is relative to the applied technology and farm location, design of the photobioreactor, strain as well as sensitivity of the algae to the external environment [23].

Bubble column and airlift photobioreactor designs, as well as others, such as fermenter, flat panel and spiral tube designs are more complex than the open equivalents, and the cultivations realized in them are considered as much more cost-intensive [9,16]. Such a cost is associated with the necessity of gaining sufficient space inside the room for installations, higher energy consumption, need of medium sterilization and the necessary service cost conducted by staff with adequate skills and qualifications. Some of the photobioreactor types cannot be applied for algae cultivation on an industrial scale due to the scale and exploitation problems (clogging of pipelines and limited light penetration) [24-25]. The majority of this apparatus also requires the application of artificial indoor lighting, which leads to an increase in energy consumption followed by an increase of the operation cost of the installation [15,24,26].

The up-to-date, sustainable and environmentally-friendly designs of photobioreactors in terms of the use of technical solutions and technology include a number of combined technologies of algae processing. This, in turn provides their application both in environmental protection (for wastewater treatment, bioremediation, CO_2 sequestration), as well as a source of valuable substances with a potential use in various industrial branches and agriculture [12,27]. The currently built biorefineries with algae processing capabilities are able to concurrently produce biodiesel, animal feed, biogas and electricity [28]. The intensive development of algae biotechnology is aimed at the improvement of large-scale algae production in cost-intensive closed cultivation systems. In connection with this, the studies conducted nowadays focus primarily on the search for more and more effective microorganism strains, their genetic modifications, optimization of the design and the operation of various reactor types in order to ensure the high productivity of algal biomass [16,23]. Apart from this, activities are undertaken with the aim of reducing the high cost associated with the investment and fluid pumping, light demand, cooling, draining and drying of biomass in such a manner that the resulting system could be competitive in relation to the open systems, which have a predominant share in the market.

2 General characteristic of biological material

Algae form the group of photosynthetic microorganisms, as they possess the ability to convert solar energy into chemical one [29-30]. They form a source of valuable substances, i.e. proteins, lipids, hydrocarbons, vitamins and microelements. Water forms the principal component of the biomass, as it forms 75–90% of the harvested mass before processing. The remaining ratio of the content is made by mineral salts and hydrocarbons (30–50%) and lipids (7–15%) of the dry mass [4, 9]. The specific composition of microalgae makes them applicable in various branches of the food, as well as beauty and pharmaceutical industries. The industrial processes can apply both microalgae (single cell algae – Fig. 1a) as well as multiple cell ones (Fig. 1b). The long list with the wide range of useful products based on algae, include: agar, alginate, pigments, biosorbents, anti-inflammatory, antifungal agents, beauty products and creams, as listed in the literature of the subject [9,4]. As a consequence of anaerobic digestion of biomass, it is possible to derive biomethane, and biodiesel on the basis of oil extracted from algae – as well as bioethanol – after saccharification and fermentation [2,7,24,28-29,31-32]. The capability to absorb volatile organic compounds, remove CO_2, treat wastewater has led to the use of algae in the controlled ecological life support system (CELSS) in conjunction with photosynthesizing bacteria and higher plants, as this can solve the problem of oxygen supply to astronauts and CO_2 removal during long-term space missions.

The growth of algae is determined by a number of factors, which are defined as abiotic, biotic, and process related. Abiotic factors include the light, as the growth of biomass requires both light exposition over a given time, its adequate intensity, temperature, concentration of nutrients, O_2 and CO_2 content, pH value, salinity and presence of toxic chemicals [28]. Biotic factors include the presence of pathogens (bacteria, fungi, viruses) and the mutual effect of other algae strains [34]. An important group of parameters affecting the growth of algae is imposed by such process parameters as: intensity of mixing, biomass concentration and the related frequency of algae harvesting.

a)

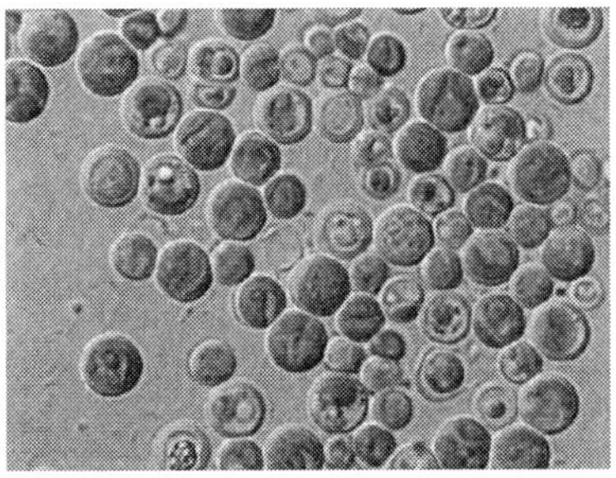

b)

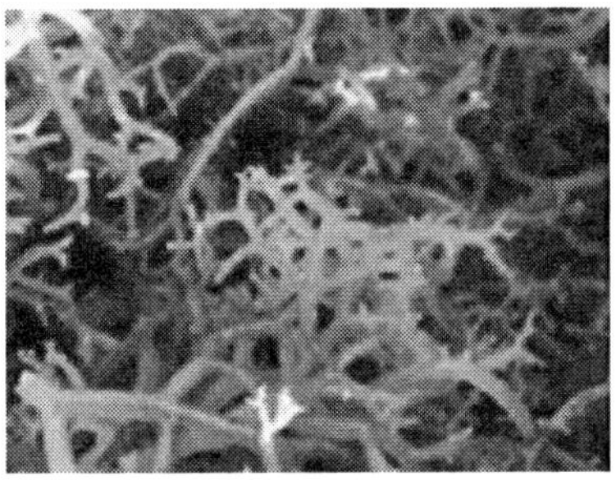

Fig. 1. Algae with potential application in industry:
a) *Chlorella vulgaris* – single-cell alga; b) *Laminaria japonica* – multi-cellular algae [33].

An indispensable component of each biomass growth system is associated with the medium in which the organisms develop and reproduce. Water forms the basic component of the algal cultivation system. The medium used for algae cultivation needs to include not only a source of carbon but also a variety of other necessary mineral compounds, which are parts of the algal cells. On the basis of an approximate molecular formula of algae biomass, Chisti [28] listed the basic elements forming this biomass: nitrogen (N), phosphorus (P), iron (Fe) and in some cases also silicon (Si). Apart from these elements, the medium contained in the water should include manganese (Mg), calcium (Ca) and sulfur (S) [8,28]. In general, the cost of cultivation is small when compared with the post-production procedures, such as draining and drying of biomass. In the closed system, supplementation of the medium is used, which means that nutrients are added during the course of the cultivation. Carbon forms the prevalent component in the photosynthetic cells of microalgae, as it forms around 50% of the dry mass in the content. The effective algal biomass cultivation systems include a constant supply of CO_2 as the principal source of carbon [28]. Experimental data indicates the increase of carbon assimilation coefficient along with the rise in the CO_2 concentration in the atmospheric air [35]. The generation of 100 tonnes of algae uses 183 tonnes of CO_2 supply [36].

An important aspect of the cultivation is associated with agitation of microalgal suspension, which prevents the sedimentation of cells, positively affects the distribution of CO_2, O_2 and nutrients contained in the liquid. In addition, this process can provide similar access to light by transferring cells from dark areas to light ones and promoting heat exchange [2,32,37-38]. An important exploitation aspect of microalgae cultivation is related to the ensuring and keeping the necessary temperature. Beside light intensity, temperature forms one of the parameters which can promote or limited the growth of the algae both in the closed and open cultivation system. Both the deficiency as well as surplus of heat can slow down the growth of microalgae, and in extreme case, the growth can be stopped. There are several ways in which energy can be fed into the system – by pumping air or exhaust gases rich in CO_2, agitation of algae as well as use of external heat source. Some cultivation systems require cooling during the day, e.g. by application of heat exchangers and cooling coils submerged in the deaeration chamber, as it is the case in the tubular photobioreactor design. In the dry climate, it is also possible to apply evaporation cooling involving water atomization along the surface of photobioreactor tubes. Water is used to naturally extract heat from the tube surface and results in the temperature decrease of the medium. Large photobioreactors can be located in greenhouses, thus securing the maintenance of a controlled temperature [28].

We can stress at this point that all of the above parameters, i.e. photobioreactor size, its design, light intensity and source, depth of light penetration, technique of feeding CO_2 and agitation mechanism can be adjusted to ensure the maximum growth of algae [39].

3 Design and operation parameters of photobioreactors

One of the basic parameters which considerably affects the design of a photobioreactor is associated with need to maintain adequate light penetration into the algal culture inside the reactor. This can be achieved by ensuring the considerable ratio of its surface area to its volume [9,10]. Smaller light intensity does not need to mean lower algae production efficiency because the majority of algae species can adapt to the low light penetration and its surplus (i.e. amount above the needs of photosynthesis) is emitted by microorganisms in the form of fluorescent light and dissipated in the form of heat [40]. This fact should be accounted for during the design stage and operation of photobioreactors to the biological processes occurring inside them. These, in turn, affect the efficiency of the photosynthesis and, consequently, the productivity of the biomass and its quality [39,41]. In particular, two papers explored the subject of effectiveness of light exposition [42-43]. These studies report that the maximum efficiency of solar light conversion into biomass on the way of oxygenic photosynthesis can be equal to 8–9%; however, with a note that the authors suggest the adoption of a lower value, i.e. 4.5% as a more realistic figure. Other aspects, which need to be taken into account during the design of

 DOI: 10.1051/e3sconf/20171902032

photobioreactors include the potential of cultivating various algal species, possibility of maintaining biological purity of the species, keeping fast CO_2 and O_2 transfer, ability to clean and keep the environment sterile, keeping and controlling the important exploitation parameters from the biotechnology point (pH, temperature, medium composition, gas concentration) and adequate operation without the formation of foam [8,10].

A number of various photobioreactor designs were developed with a purpose of cultivating different algae meeting the requirements listed above. In general, such apparatus can be divided into tubular, flat, fermenter and hybrid reactors. The first of the designs was designed particularly with the purpose of effective absorption of solar radiation, whereas the two further designs require the use of artificial light (from an external or internal source) [9]. Fig. 2 contains a diagram with an original classification of reactors designed for algal cultivation with the shape as the criterion for classification. The further part of this work contains a detailed characteristic of photobioreactors along with the list of their advantages and drawbacks as well as the range for their practical applications.

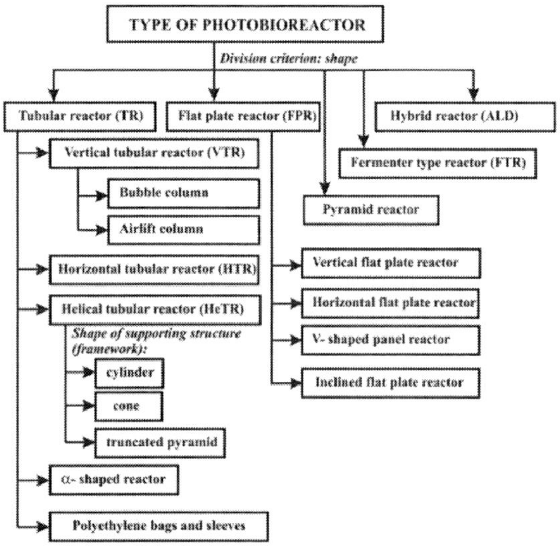

Fig. 2. Designs of bioreactors adapted to algal cultivation.

3.1 Tubular photobioreactors

Tubular reactors include three basic configurations, i.e. simple vertical column bioreactors with bubble and airlift columns, vertical tubular reactors and spiral reactors. This category of bioreactors also includes photobioractors with inclined pipes under a given angle in relation to the ground and the reactors in the shape of the letter α [44] as well as bag and sleeve photobioreactors made of polyethylene installed on arrangements based on supporting structures.

The first type of reactors, i.e. airlift and bubble columns are built in the form of vertical, transparent tubes, through which light penetrates the inside of the reactors thus ensuring an adequate level of light penetration into the cultivation.

3.1.1 Bubble columns

Bubble columns are cylinder-shaped vessels with the internal diameter often exceeding 20 cm, in which the geometrical condition $H>2D$ is fulfilled (Fig. 3a). Usually this shape factor, i.e. the ratio of the height to the diameter is in the range from 4–8, as such a design can ensure an optimum relation of the reactor's surface to its volume. As they are produced from common materials, such designs are cheap and are suitable for both outdoor and indoor algae cultivation. In addition, there are no moving mechanical parts in the design, which could potentially cause shear stresses and promote cell degradation. Besides, satisfactory heat and mass exchange is maintained and uniform environment for cultivation is promoted accompanied with effective O_2 release. In such reactors, the agitation of the suspension is realized by the presence of the gas phase, which is introduced into the culture in the form of small bubbles formed by a distributor located in the lower part of the column, which guarantees a sufficiently long duration of the contact between the algae and the gas phase.

For the case of tall columns, perforated plates are applied to gain a greater dispersion of the gas phase [10] or rubber membrane diffusers [45]. The mixing process achieved in this manner leads to a uniform distribution of algal cells in the reactor volume and the uniform access of algae to CO_2 as well as enables the successful removal of oxygen production by the algae during photosynthesis [46]. The effectiveness of photosynthesis can be increased by a greater rate of gas feed; thus the duration of the light and dark cycles is reduced by keeping the algae in the sunny and shady areas of the photobioreactor for a shorter period.

The list of the drawbacks of this solutions include the susceptibility to formation of biofilm on the column walls, which limits light penetration inside the apparatus and reduces the effectiveness of photosynthesis as well as the need to set aside large plots of land for leading high-efficiency installation. The remaining limitations primarily concern outdoor cultivations. These include difficulties in the control of the temperature of the culture and the occurrence of photoinhibition phenomenon, which involves stopping of algae growth as a result of too large light exposition that exceeds the light saturation constant [28, 47]. The regulation of cultivation temperature can apply thermostats, but this leads to a considerable cost increase.

3.1.2 Airlift photobioreactors

Airlift photobioreactors are formed by mutually interconnected cylinder-shaped vessels with two separate zones, i.e. riser and downcomer zones (Fig. 3b). There are three design solutions of such apparatus, i.e. one with internal circulation (with an overflow pipe inside), cylinder with a partition to separate the riser and downcomer zone and a configuration with external circulation. In this type of apparatus, liquid circulation is obtained as a result of differences in the fluid density in the particular parts of the apparatus. In the risen zone, where gas feed is provided, the density of the mixture (i.e. suspension with dissipated gas phase) is lower than the one in the downcomer zone. As a consequence, the suspension rises in the aeration zone and it flows down in the downcomer zone. There are also lesser known designs, including rectangular airlift photobioreactors, which demonstrate better agitation characteristics and better efficiency of photosynthesis. However, they are unsuitable for large-scale cultivation [10].

The principal factor responsible for cost generated in bubble and airlift columns is associated with the cost of gas that needs to be supplied and this considerably affects the profitability of algae production. A comparison of the cultivation cost of *Porphyridium* algae in bubble and airlift columns is reported by Merchuk et al. [48]. As a result of adding a promoter that causes spiral gas flow, the cost of gas used to produce 1 kg of biomass dropped by 50% in relation to the cost in the bubble column on condition that the same specific algae growth rate is maintained. This study also demonstrated that in bubble and airlift photobioreactors with the dimension of up to 0.19 m, it is possible to gain a final biomass concentration and an adequate growth rate comparable with tubular photobioreactors with a narrow design.

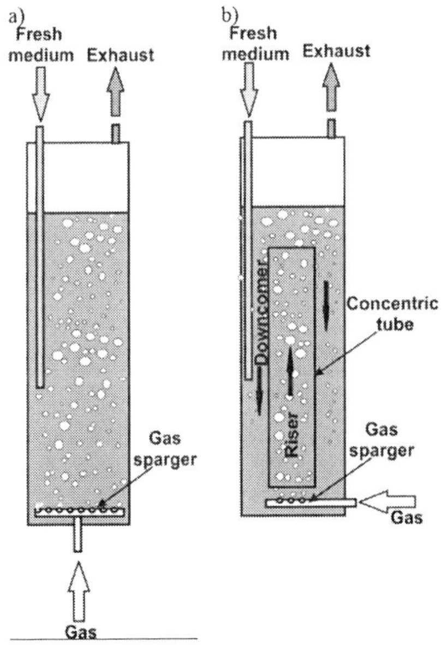

Fig. 3. Column type photobioreactor: a) bubble column, b) internal loop airlift.

3.1.3 Polyethylene bags and sleeves

Due to their low cost, transparency and sterility at the time of starting cultivation (high temperature of film extrusion), photobioreactors in the form of polyethylene bags are often applied (Fig. 4). These photobioreactors take the form of vertical bags, hanging individually or in the form of parallel series of bags mounted on a support (rack). A circulation pump is used to drive the flow of liquid through the particular bags, feed air inside the bag and provide medium. Such systems are characterized by good surface area to volume ratio. Cohen et al. [49] reported results with regard to bags with the capacity of 25 l (32x250 cm) to cultivate *Porphyridium* algae strain in which three times greater concentration of cells was obtained in relation to the typical value recorded in cultivation ponds. For the case of the use of *Spirulina*, the maximum productivity of such systems can be equal to 0.6 $g \cdot l^{-1} \cdot t^{-1}$ under the assumption of a reactor capacity of 100 ton (with the necessary surface for the cultivation of 1200 m^2). Tredici and Rodolfi [50] developed the idea of a photobioreactor in the form of a transparent elastic bag (foil) contained in a stiff, metal frame. In turn, authors in [31] applied polyethylene sleeves with a bubble column for cultivation of *Scenedesmus obliquus* microalgae. They discussed the advantages resulting from the outdoor cultivation and indicated the applicability of this bioreactor design for commercial production of algae in the context of gaining biomass designated for use in biofuels. Although the cultivation in such systems is easy and therefore quite common (also for multiplication of strains, there are considerable limitations resulting from the brittleness and low strength of the material exploited over prolonged periods of time. In addition, an increase in the volume of the algae cultivation systems is often associated with a drop in the efficiency of biomass production [9].

Fig. 4. Polyethylene sleeves.

3.1.4 Tubular photobioreactors

Tubular photobioreactors are formed by series of parallel tube sections, which can be joined in loops or systems inclined at a given angle against the ground (Fig. 5a-b). They are particularly suitable for the case of continuous outdoor cultivation, where the possibility of arranging reactors under an angle to the sun (in the north-south or east-west arrangement) provides adequate capture of solar radiation and ensures a high coefficient of light conversion. Such systems are designed to contain large volumes of cultivation media (with the tube diameter 10–60 mm, and a length of several hundred meters) since

 DOI: 10.1051/e3sconf/20171902032

they are not susceptible to microbiological contamination [40]. In general, we can distinguish two types of tube arrangement in vertical tubular photobioreactors, i.e. solar tubes installed in parallel to one another low above the ground and tube installed on top of one another supported on a horizontal ladder-shaped rack. The purpose of this design is to increase the number of tubes per surface area of the available ground. The tubes are coupled with a gas exchange system (degasifying column) and a heat exchange system. The role of degasifying column is concerned with removal of O_2 produced during photosynthesis as the factor responsible for photooxidization [51]. The internal diameter of the pipe of the solar collector is limited and should be equal to up to 0.1 m due to the difficulties in the sun penetration into the deeper layers of the high-density algae suspension, which is necessary to gain a considerable biomass growth [28].

a)

b)

Fig. 5. Design of tubular reactor [9]: a) horizontal b) stacked tubular algae reactor consisting of several connected loops (Algae PARC Wageningen University).

The algal culture is circulated from a vessel (degasifying column) into the solar panel (which receives sunlight) and returned into it, which prevents the photoinhibition phenomenon. On the basis of data by Chisti [28], the maximum rate of oxygen generated in photosynthesis in a tubular photobioreactor can even exceed 10 $g{\cdot}m^{-3}{\cdot}min^{-1}$ in the conditions of high insolation. The dissolved oxygen cannot be removed directly from the tube only in the degasification column, which means that the length of a straight pipe section in such installations cannot exceed 80 m. The ground under the sun collector is painted white or covered with a sheet of white material in order to increase light reflection and that is how the total light flux received by the tubes can be increased [7]. A solution involving artificial lighting (fluorescent lamps, light emitted by LED diodes) of the tubular photobioreactors is technically viable but this is an expensive option compared to the natural light, hence, thus it is mainly used for biomass production on a large scale with regard to top-of-the-range products [52]. The adverse phenomenon of algae sedimentation is prevented by maintaining turbulent flow of the algal suspension induced by a mechanical pump or by application of airlift devices. It is also possible to realize the mixing by the application of static agitators, which however reduce the light penetration into reactor. The relatively high linear velocities of the fluid in the range of 0.20–0.50 $m{\cdot}s^{-1}$ require large expenditure associated with its pumping and this value can be in the range 2000 $W{\cdot}m^{-3}$ (for flat panel reactors and bubble column ~ 50 $W{\cdot}m^{-3}$). The turbulent flow of the thick algae suspension from the dark to the light zones effectively reduces the duration of its stay in the particular zones, and reduces the occurrence of photoinhibition, which is adverse for algal growth [10].

In the systems without temperature control, large volumes of heat can be generated (with the fluctuations in the temperature in the range of up to 20°C during the day). For the purpose of the reduction of this exploitation problems, reactors are often located indoors (in greenhouses) made of metal, glass or plexiglas. There are also available options of providing shadow to tubes by using dark plastic covering, tube overlapping, spraying tube surfaces with water and an option of tube submersion in a water tank. The latter solutions are effective though expensive, since the maintenance of a low temperature of the medium requires the use of considerable amounts of water [9]. A simple way of increasing the scale of the bioreactor processes is associated with installing a new set of tubes to the already existing installation while maintaining an adequate linear velocity of the medium and gas exchange at a given level. The only problem can be related with the size of plot needed for this installation. Industrial production with the capacity of 5000–10000 liters in systems with high light concentration could only be realized by installing large tube numbers arranged in a vertical system, which is questionable from the cost-effectiveness angle. The greatest bioreactor in the world (serpentine reactor) with vertical arrangement of horizontally running tubes consists of 20 independent modules with the total length of 500 km and a capacity of 700 m^3 is designed for cultivation of *Chlorella* microalgae strain on an industrial scale (with the annual output of 130–150 t of dry mass per year) and is situated in Klötze (Saxon-Anhalt, Germany). The required cultivation temperature is secured by enclosed space in a greenhouse with the surface area of 1.2 ha and adequate heating and cooling systems [7,40]. The highest productivity equal to 25 $g{\cdot}m^{-2}{\cdot}d^{-1}$ was, however, obtained in a reactor with a relatively small capacity of around 10 m^3 [8]. For the case of *Spirulina* production in a tubular reactor with the capacity of 100 ton (with the required plot of land for cultivation of 1200 m^2), the maximum productivity is 0.8 $g \cdot l^{-1} \cdot t^{-1}$. Due to the problems with light penetration, these bioreactors are designed for tube diameters not exceeding 30 cm.

Another stage in the development of algae production systems was associated with the design of tubular bioreactors with a set of parallel tubes arranged under the angle of 5–7° in relation to the horizontal plane [9]. These tubes are made from flexible PVC (with a length of 6.4 m, diameter of 43 mm, and a thickness of 0.15 mm) and they are coupled by collectors made of PVC. The top collector plays the role of a degasifying column, whereas the lower one includes a perforated tube and is applied as a gas distributor to feed the particular tubes in the system. The temperature of the cultivation is regulated automatically by opening a valve and launching the spray. The largest experimentally examined bioreactors had a capacity of 4000 l and consisted of a set of 8 parallel tubes with the length of 44 meters. The mean productivity during the cultivation of Nannochloropsis strain was at a level of 0.7 $g \cdot l^{-1} \cdot d^{-1}$. The advantages of such photobioreactors are basically the same as other tubular designs, yet, the drawbacks are associated with insufficient gas transfer and no possibility of a satisfactory temperature control.

3.1.5 Helical tubular photobioreactors

Other solutions of bioreactor design are based on the potential application of flexible, transparent tubes with the internal diameters from 2.4 to 5 cm made of polyethylene, which are spirally wound on a cylinder-shaped (e.g. Biocoil), truncated pyramid or truncated cone shaped support structure. Several systems involving parallel running tubes are coupled by collectors with a system for liquid pumping (i.e. a pump or airlift layout). The use of centrifugal pump for microorganism pumping could form an element which excludes the possibility of cultivating selected algae species in a spiral photobioreactor, in particular the ones that are susceptible to shear forces, as they can lead to destruction of algal cells.

A bioreactor could also be equipped with a gas exchange system [34]. This type of photobioreactor is considered as one of the most effective designs, dedicated to the cultivation of small volume of microalgal suspension, i.e. for multiplication of strains for further use in larger tubular reactors [28]. The principal disadvantage of this design is related to the possibility of the occurrence of the fouling phenomenon, which involves tubes blocking as a result of the excessive growth of algal biofilm. The ways of reducing algal biofilm thickness include purposeful induction of air plugs in the tube, circulation of balls matching the internal diameter of the pipe in size, inducing flow of very turbulent fluid, as well as flow of sand and coarse particles suspended in the liquid with abrasive characteristics. In turn, the advantages of this design include the good relations between the energy consumption and the effectiveness of photosynthesis, uniform mixing, effective CO_2 from the gas to the liquid phase, smaller degree of algae exposition to mechanical stress and the relatively small requirements with regard to surface needed for the installation in relation to the relatively large volume of the cultivation medium [9]. In addition, there is a possibility of applying artificial light, whose source is located inside the apparatus, as it enables the effective control of light absorption by the culture to be performed. This measure can compensate for the not necessarily good adjustment of the apparatus in relation to the sun. The scale-up of the system is easy and involved an increase of the number of the parallel tube layer in a spiral.

In a spiral bioreactor in the truncated cone shape, it is possible to utilize light and solar radiation falling on the tube surface in a more effective manner. This shape provides a better spatial distribution of light and its absorption by microorganisms. The main disadvantage is associated with difficulties in rescaling, as the geometrical parameters (angle, height) are strictly defined. The only way of achieving a high pace of photosynthesis is related to the increase in the number of coupled apparatus, which contributes to the energy losses in the complex of network with liquid flow. The productivity in such complex system is also considerably lower calculated per size of the necessary plot of land.

3.2 Plate (panel) photobioreactors

Bioreactors in the shape of flat panels form some of the designs with the best parameters in terms of strength (Fig. 6).

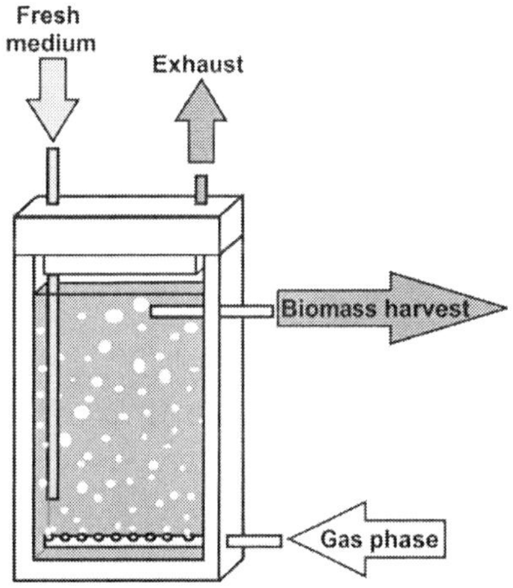

Fig. 6. Plate photobioreactor for algae cultivation.

They are formed by two adhesive sheets of transparent material (glass, plexiglas, polycarbonate) with the length of the route of light penetrating through it (d) in the range of a few to 70 mm. The large surface of the exposition offers high efficiency of photosynthesis [2]. The geometry of a single flat panel provides a high ratio of the reactor surface to its volume, which results in a higher efficiency of biomass production [35]. In accordance with the current state-of-the-art in the area, it is beneficial to maintain this ratio around 400 $m^2 \cdot m^{-3}$. In the cultivations, in which this condition is fulfilled, the

specific efficiency of biomass production is gained in a small volume of the cultivation medium and in the conditions of low energy expenditure for mixture agitation. In this bioreactor design, the parameter defining the ratio of their surface to the area of the plot which is occupied by the cultivation, should be equal to at least 10. The optimum value of this parameter is obviously relative to the type of the microalgae strain and the region, in which the apparatus is located [35].
The agitation is realized by use of gas phase (air) circulating along a perforated tube located in the bottom part of the apparatus (1 liter of air per 1 liter of reactor volume per 1 minute) or by application of a rotating mechanical module [19]. Some algae species in this type of reactor can be exposed to hydro-mechanical stress and form biofilm on the apparatus walls. There are suggestions that the plates should be arranged in a stack, perpendicular to the angle of the light falling on the reactor. This procedure is intended to identify the areas with various intensity of light penetration in the apparatus, which is desirable for the case of microalgae cultivation with various levels of photoacclimatization or intention of synthesizing specific products that can only be formed in the conditions of considerable insolation.
This apparatus are relatively cheap and easy to clean but difficulties can arise in connection with temperature maintenance. Large-scale cultivation can require the use of numerous modules and installation of support structures.

3.3 Fermenter type photobioreactors

Tank photobioreactors (fermenters), including commonly designs with a circular cross-section (Fig. 7) (sometimes with a rectangular as well), are made of steel, glass and organic glass applied predominantly in cultivations on a laboratory scale.

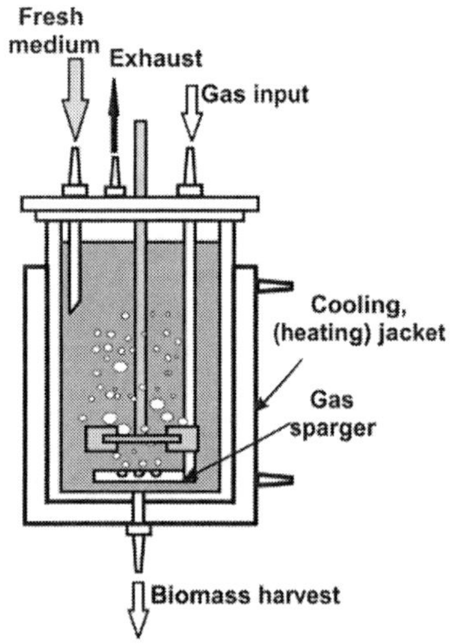

Fig. 7. Fermenter type photobioreactor.

They are included in the conventional system family, in which the medium is agitated mechanically by use of stirrers with a variable shape and size. Air feed together with CO_2 is supplied through the agitation system or directly under it. Beside natural light, it is possible to apply artificial lighting (fluorescent lamps, optic fibers), as this can ensure a uniform distribution of the rays of light. As a consequence of the insignificant ratio of the reactor surface to its volume (which is characteristic for this design), the effectiveness of light absorption is also small. The cultivation can be performed periodically, as well as in a semi-continuous and continuous manner. The complete control of the process parameters coupled with the many years' experience in biotechnology processes and running them on a various scale in the food and pharmaceutical industries also can provide the procedure realized in an axenix culture (i.e. free of undesired organisms). These microorganisms can be maintained in a reactor for a longer period, which plays a key role in the production of selected high-value metabolites [9]. The productivity of the biomass in such systems is very small and typically equal to 30–50 $mg \cdot l^{-1} \cdot d^{-1}$; therefore, the design of this reactor is continuously enhanced with the purpose of improving the efficiency of algal biomass growth.

3.4 Hybrid photobioreactors

Hybrid systems of algae cultivation involve the integration of various phases of microalgae growth in a single two-stage system. The initial part of algae growth takes place in a closed system (i.e. photobioreactor), whereas the subsequent ones in the open system. Hence, the hybrid systems include components and design solutions that are characteristic of the various cultivation systems, i.e. open and closed ones. The first phase of the cultivation is realized in a photobioreactor, which is meant is minimize the hazard associated which the contamination of the cultivation by foreign organisms while the beneficial conditions for cell multiplication are maintained. The second stage of production is designed to expose the cells to the stress of nutrients, which should initiate the synthesis of desired metabolites and lipid products and it is realized in cultivation ponds. Such system can include the production of oil and astaxanthin by *Haematococcus pluvialis*. Huntley and Redalje [53] reported that the mean amount of oil that can be gained from algae in this manner was equal to above 10 $tonnes \cdot ha^{-1} \cdot year^{-1}$ and the maximum was even 24 $tonnes \cdot ha^{-1} \cdot year^{-1}$.

3.5. Pyramid photobioreactors

Pyramid photobioreactors are some of the newest systems of algal cultivation, as they are completely computerized and automated. The fluid mixing is realized by an airlift system. The maximum production of *Spirulina* is equal to 1.45 $g \cdot l^{-1} \cdot t^{-1}$ and this value is over 4 greater in relation to the productivity in the open systems and it well exceeds the value gained in the tubular photobioreactors. The advantages of this system include the small area of plot required, i.e. around 60 m^2 for the cultivation medium with the mass of 100 tonnes. However, such designs are still mainly at an experimental stage. For the purpose of indicating the potential applicability of selected types of photobioreactors, Table 1 contains a brief characteristic of each of the designs, which includes the list of their advantages and drawbacks. The most effective designs of closed systems of photobioreactors utilized nowadays include bioreactors in the form of flat panels, tubular reactors and bubble columns [9, 52-54]. The popularity of bubble columns stems from their low maintenance cost, beneficial ratio between the surface of the reactor and its volume, as well as lack of moving parts and effective O_2 removal. In outdoor cultivations, the most effective systems involve the use of horizontal tubular reactors, as this design promotes CO_2 diffusion and the use of solar radiation is maximized, which can provide better productivity in relation to flat reactors. Finally, adequately modified flat photobioreactors are capable of offering maximum productivity even in the conditions of smaller light intensity.

Table 1. Comparison of advantages and drawbacks corresponding to selected types of photobioreactors.

Column photobioreactors (bubble and airlift column)
Advantages: • good biomass growth, high efficiency of photosynthesis, high potential of scalability, limited photoinhibition and photooxidization • cheap, compact and easy to maintain • potential of exposition to alternating dark and light cycles • small prerequisites of space demanded for cultivation • suitable for outdoor cultivation • low energy use • suitability for algae immobilization
Drawbacks: • small area of light exposition that is additionally reduced with the increase of column diameter • low ratio of reactor surface to its volume • possibility of biofilm formation on reactor walls
Applicability: Not suitable for algae cultivation with a low specific weight, e.g. strains containing large volume of lipids, since the cells of these algae are highly susceptible to flotation.
Flat (panel) photobioreactors
Advantages: • large surface of exposition to light • high ratio of surface to volume • suitable for outdoor cultivation • high productivity of biomass • ability to maintain uniform access to light across entire volume of cultivation • relatively cheap • easy to build, maintain, clean and operation • high efficiency of photosynthesis • suitability for algae immobilization • small concentration of dissolved oxygen
Drawbacks: • any increase of production scale requires the use of numerous modules and support structure • difficulties in control of cultivation temperature • risk of fouling • potential for occurrence of hydrodynamic stress in some algae species
Applicability: Production of algal strains with a high lipid content at limited access to nutrients.

Table 1 (cont.)

Tubular photobioreactors
Advantages: • large area of light exposition • suitability for outdoor cultivation • good productivity of biomass • relatively low cost of build
Drawbacks: • considerable requirements in terms of plot area for installation • possible fluctuations in pH, concentration of dissolved O_2 and CO_2 along the tube length (risk of photoinhibition) • risk of biofilm accumulation on tube walls • poor mass transfer
Applicability: Industrial production of valuable dyes, astaxanthin and for production of *Haematococcus*, *Nannochloropsis*, *Chlorella* strains.
Fermenter type photobioreactors
Advntages: • possibility of running axenic cultivation • control of all process parameters
Drawbacks: • low ratio of surface to volume • low efficiency of light absorption • low productivity
Applicability: Cultivation of marine algal species, i.e. *Chondrus crispus*, *Gracilari*, processes of biological wastewater treatment, production of high-value metabolites.

4 Conclusions

Concerns about global warming, increasing CO_2 emission and environmental pollution caused by fossil fuels, combined with their depletion and energy security are important factors in the development of alternative energy sources that is renewable, clean and environment friendly. In this context, algae biomass is becoming important potential resource for renewable energy production. Despite the progress that has been made in the recent years in the field of bioengineering and biotechnology, which promotes the improvement of the efficiency of algal growth in photobioreactors, open systems are dominant in algae cultivation on the industrial scale. This is principally due to a number of unresolved technical problems, high cost of investment and production in photobioreactors. The basic advantages listed in this paper associated with the use of photobioreactors on an industrial scale include the possibility of gaining high quality biomass, in which the volume of production often exceeds the productivity of the open system by a few times, reduction of water use by several orders of magnitude, direct exposition of algae to sunlight and control of CO_2 concentration. The designs of photobioreactors (e.g. tubular, panel) and optimization of operation parameters are continuously improved with the purpose of gaining maximum efficiency of biomass production while high efficiency of the use of supplied resources is concurrently preserved (e.g. water, energy, nutrients). These processes are optimized continuously in terms of sustainability and environmental friendliness.

References

1. P. Spolaore, C. Joannis-Cassan, E. Duran, A. Isambert, J. Biosci. Bioeng. **101**, 2 (2006)
2. C.U. Ugwu, H. Aoyagi, H. Uchiyama; Bioresour. Technol. **99** (2008)
3. C. Posten, C. Walter, *Microalgal Biotechnology: Potential and Production* (Walter de Gruyter & Co, 2012)
4. D. Kępska, Ł. Olejnik, Chemik **68** (2014)
5. S.P. Slocombe, J.R. Benemann, *Microalgal Production for Biomass and High-Value Products*, (CRC Press Taylor&Francis, 2016)
6. E. Eroglu, A. Melis, Int. J. Hydrogen Energy **41** (2016)
7. A.E-F. Abomohra, W. Jin, W. Tu, S-F. Han, M. Eid, H. Eladel, Renew. Sust. Energ. Rev. **64** (2016)
8. R. Voloshin, M.V. Rodionova, S.K. Zharmukhamedov, T.N. Veziroglu, S.I. Allakhverdiev, Int. J. Hydrogen Energy **41** (2016)
9. A.P. Carvalho, L.A. Meireles, F.X. Malcata, Biotechnol. Prog. **22**, 6 (2006)
10. R.N. Singh, S. Sharma, Renew. Sust. Energ. Rev. **16** (2012)
11. D.A.M.P. da Ponte, M.M. Werneck, D.A. Aranda, Nat. Sci. **8** (2016)
12. P. Wiley, L. Harris, S. Reinsch et. al., J. Sust. Bioenergy Syst. **3**, 1 (2013)
13. O. Pulz, W. Gross, Appl. Microbiol. Biotechnol. **65** (2004)
14. M.A. Borowitzka, N.R. Moheimani, *Developments in Applied Phycology,*vol **5** (2013)
15. J. Pruvost, F. Borgne, A. Artu, J-F. Cornet, J. Legrand, Chapter Five – *Industrial Photobioreactors and Scale-Up Concepts*, Adv. Chem. Eng., Photobioreaction Engineering **48** (2016)
16. G.A. Andrade, M. Berenguel, J.L. Guzman, D.J. Pagano, F.G. Acien, J. Proc. Control **37** (2016)
17. M.R. Kitto, Global Aquaculture, Feed Sustainability, 2016
18. A.G. Silva, R. Carter, F.L.M. Merss, D.O. Correa, J.V.C. Vargas, A.B. Mariano, J.C. Ordonez, M.D. Scherer, GCB Bioenergy **7** (2015)
19. N.H. Norsker, M.J. Barbosa, M.H. Vermuë, R.H. Wijffels, Biotechnol. Adv. **29** (2011)
20. R.H. Wijffels, M.J. Barbosa, M.H.M. Eppink, Biofuel. Bioprod. Bior. **4** (2010)
21. M.C. Lopez, E.R. Sanchez, J.L. Lopez, F.G. Fernandez, et al., J. Biotechnol. **123**, 3 (2006)
22. M. Olaizola, J. Appl. Phycol. **12** (2000)
23. P.M. Slegers, P.J.M. Beveren, R.H. Wijffels, G. Straten, A.J.B. Boxtel, Appl. Energy **105** (2013)
24. Ch.Y. Chen, K.L. Yeh, R. Aisyah, D.J. Lee, J.S. Chang, Bioresour. Technol. **102**, 1 (2011)
25. F. Qiu, Algae Architecture, Integration of algae in architecture (2013) www.repository.tudelft.nl.
26. M. Krzemieniewski, M. Dębowski, M. Zieliński, Czysta Energia **9** (2009)
27. A. Lewicki, D. Janczak, W. Czekała Technika Rolnicza Ogrodnicza Leśna (2013) (in polish)
28. Y. Chisti, Biotechnol. Adv. **25** (2007)
29. T. Holtermann, R. Madlener, Appl. Energ. **88** (2011)
30. S.F. Mohsenpour, N. Willoughby, Bioresour. Technol. **142** (2013)
31. A.E-F. Abomohra, M. El-Sheekh, D. Hanelt, Biomass Bioenerg. **64** (2014)
32. T.M. Mata, A.A. Martins, N.S. Caetano, Renew. Sust. Energ. Rev. **14** (2010)
33. http://caplexhealth.blogspot.com/2015/08/medicinal-plants-used-for-treatment-of.html
34. M. Dębowski, M. Zieliński, M. Krzemieniewski, M. Dudek, A. Grala, Pol. J. Natur. Sc. **27**, 2 (2012)
35. C. Yoo, S.Y. Jun., J.Y. Lee, C.Y. Ahn, H.M. Oh, Bioresour. Technol. **101**, 1 (2010)
36. J. Doucha, F. Straka, K. Livansky, J Appl. Phycol. **17** (2005)
37. K. Zhang, N. Kurano, S. Miyachi, Bioproc. Biosyst. Eng. **25**, 2 (2002)
38. X. Chai, X. Zhao, Bioresour. Technol. **116** (2012)
39. I. Seo, I. Lee, H. Hwang, S. Hong et al., Biosyst. Eng. **113** (2012)
40. C. Posten, Eng. Life Sci. **9**, 3 (2009)
41. J.W.F. Zijffers, M. Janssen, J. Tramper, R.H. Wijffels, Mar. Biotechnol. **10** (2008)
42. R. Razeghifard, Photosynth. Res. **17** (2013)
43. A. Melis, Plant Sci. **17** (2009)
44. Y.K. Lee, S.Y. Ding, C.S. Low, Y.C. Chang, J. Appl. Phycol. **7** (1995)
45. L. Xu, P.J. Weathers, X-R. Xiong, Ch-Z Liu, Eng. Life. Sci. **9**, 3 (2009)
46. K. Kumar, C.N. Dasgupta, B. Nayak, P. Lindblad, D. Das, Bioresour. Technol. **102** (2011)
47. A. Richmond, *Handbook of Microalgal Culture: Biotechnology and Applied Phycology* (Blackwell Sciences Ltd, Oxford, UK, 2004)
48. J.C. Merchuk, M. Ronen, S. Giris, S. Arad, Biotechnol. Bioeng. **59** (1998)
49. E. Cohen, S. Arad (Malis), Biomass **18** (1989)
50. M.R. Tredici, L. Rodolfi, (*PCT WO 2004/074423 A2, 2004*)
51. A.S. Miron, A.C Gomez, F.G. Camacho, E.M. Grima, Y. Chisti, J. Biotechnol. **70** (1999)
52. O. Pulz, Appl. Microbiol. Biotechnol. **57** (2001)
53. M. Huntley, D. Redalje, Mitig. Adapt. Strat. Gl. **12**, 4 (2007)
54. S. Hindersin, M. Leupold, M. Kerner, D. Hanelt, Bioprocess Biosyst. Eng. **36** (2013)

 DOI: 10.1051/e3sconf/20171902033

Energy and emergy analysis of mixed crop-livestock farming

Anna Kuczuk[1,*], *Janusz* Pospolita[1], and *Stefan* Wacław[2]

[1] Opole University of Technology, ul. Prószkowska 76, 45-758, Opole, Poland
[2] Opole Agricultural Advisory Centre, ul. Główna 1, 49-330, Łosiów, Poland

Abstract. This paper contains substance and energy balances of mixed crop-livestock farming. The analysis involves the period between 2012 and 2015. The structure of the presentation in the paper includes: crops and their structure, details of the use of plants with a beneficial effect on soil and stocking density per 1ha of agricultural land. Cumulative energy intensity of agricultural animal and plant production was determined, which is coupled the discussion of the energy input in the production of a grain unit obtained from plant and animal production. This data was compared with the data from the literature containing examples derived from intensive and organic production systems. The environmental impact of a farm was performed on the basis of emergy analysis. Emergy fluxes were determined on the basis of renewable and non-renewable sources. As a consequence, several performance indicators were established: Emergy Yield Ratio EYR, Environmental Loading Ratio ELR and ratio of emergy from renewable sources P_R. Their values were compared with the parameters characterizing other production patterns followed in agricultural production. As a consequence, conclusions were derived, in particular the ones concerning environmental sustainability of production systems in the analyzed farm.

1 Introduction

Farming, along with other types of production, is associated with the need of application of a variety of means of production. In addition, farming is associated with the need of considerable energy input in it. This input takes the form of both the means of production (machines, equipment, fuel, fertilizers, and protection agents) as well as the ones that are extracted from the natural environment. In the case of these, we talk about energy derived from renewable sources (sun, wind, water and soil). Due to the fact that the discussion of energy aspects is often accompanied by the analysis of cost indices of production, whereas the use of the energy derived from free sources is often disregarded in the overall balance. This results in the fact that the calculations performed in such a manner do not account for the energy and cost associated with the decrease of the content of soil organic matter. In addition, such account does not involve the assessment of the sustainability of a given type of production.

The sun forms the fundamental source of energy for agricultural production. The mean energy flux absorbed by the earth is equal to 3.6E+4 $GJ \cdot ha^{-1} \cdot year^{-1}$ for the conditions in Poland [1]. The major portion of this flux participates in the heat exchange occurring between the soil, atmosphere and the background processes, whereas only a small fraction is used in the photosynthesis.

The process of agricultural production involves an input of a certain amount of energy of anthropogenic origin. For example, its cumulative values in the production of wheat is equal to 23.93 $GJ \cdot ha^{-1}$ in the conventional farms and 11.25 $GJ \cdot ha^{-1}$ in the organic ones [2]. A portion of this energy is directly delivered to the ground in the form of seeds, natural and mineral fertilizers as well as plant protection agents. Such an assessment of the cumulative energy intensity of production accounts solely for the use of anthropogenic components. The emergy calculation proposed by Odum [3] provides grounds for a more extensive analysis, which involves the use of renewable energy sources. This analysis involves the reference of all energy fluxes to solar energy, which forms a primary source of all resources and their products.

This paper undertakes both the energy as well as emergy analysis of mixed crop-livestock farming. The farm that was taken for the analysis forms a representative one for the conditions in the Opole province. The application of the emergy analysis in the calculations provides an assessment with regard

* Corresponding author: a.kuczuk@po.opole.pl

to the sustainability of the production and use of natural resources on the farm.

2 Materials and methods

The results and analysis presented in this paper are based on data obtained from a farm located in the Opole province. The analysis of the data partly applied the data presented in detail in [4], after they were corrected and updated with details with regard to a further year. The selection of the farm was justified by the possibility of gaining comprehensive data for the period of four years (2012-2015), which provided grounds for a series of calculations with regard to the plant and animal production in it. The analysis was conducted with the purpose of determining the cumulative energy intensity and the use of the environmental resources. Such assessment can be applied from the practical perspective as it could serve in the decision making process regarding the directions applied in the production. The calculations of the cumulative energy intensity of crop production applies equation (1) [5-7].

$$E_c = E_m + E_f + E_{mat} + E_l \text{ [GJ]} \quad (1)$$

where:
E_c –cumulative energy intensity of crop production [GJ],
E_m – cumulative energy intensity in tractors, machinery, means of transport [GJ],
E_f – cumulative energy intensity of the fuel [GJ],
E_{mat} – cumulative energy intensity of materials (fertilizers, sowing materials, protection agents) [GJ],
E_l – cumulative energy intensity of human labor [GJ].

The following energy fluxes were applied for the calculation of the energy intensity of animal production (2):

$$E_a = E_m + E_f + E_{mat} + E_l + E_{en} \text{ [GJ]} \quad (2)$$

where:
E_a – cumulative energy intensity of animal production [GJ],
E_m – cumulative energy intensity in tractors, machinery, means of transport [GJ],
E_f – cumulative energy intensity of the fuel [GJ],
E_{mat} – cumulative energy intensity of materials (fodder) [GJ],
E_l – cumulative energy intensity of human labor [GJ],
E_{en} – cumulative energy intensity of electricity [GJ].

The value of the two types of production was expressed in grain units (GU), and their cumulative energy intensity in $GJ \cdot GU^{-1}$. In addition, the cumulative energy intensity index was calculated and expressed in $GU \cdot GJ^{-1}$.

The emergy analysis forms a more objective measure. It is also used to determine the degree of loading and utilization imposed on the natural environment during the process of production [3,8]. The application of emergy analysis involves the conversion of each production process in relation to the use of solar energy, and emergy is defined by the product of exergy Ex of a given substance and its solar transformity τ (3) [3, 9, 10]:

$$Em = Ex \cdot \tau \text{ [seJ]} \quad (3)$$

The unit of emergy is seJ, in turn exergy is defined as the minimum input of work required to derive a given substance in a specific time on the basis of common components available in the surrounding environment [11]. For the case of more complex products, e.g. machines, the study applies the notion of thermo-ecological cost, which is associated with the impact resulting from all stages of the manufacturing process. It expressed the cumulative use of exergy of non-renewable components of the environment [12]. The calculations of the emergy of the agricultural production applied the specific values of production and the renewable sources are presented in Table 1.

Table 1. Emergy of means of production and renewable sources.

Type of energy	Value	Unit	Source
Solar radiation	0.989E+11	$seJ \cdot ha^{-1} \cdot day^{-1}$	[13]
Evaporated water	1.28E+8	$seJ \cdot kg^{-1}$	[3,9,10]
Wind	2.79E+10	$seJ \cdot ha^{-1} \cdot day^{-1}$	[3, 10, 14]
Sowing material	6.50E+12	$seJ \cdot kg^{-1}$	own calcul.
Nitrogen fertilizers	4.04E+13	$seJ \cdot kg^{-1}$	[9]
Phosphorus fertilizers	3.69E+13	$seJ \cdot kg^{-1}$	[9]
Potassium fertilizers	0.187E+13	$seJ \cdot kg^{-1}$	[9]
CaO fertilizers	1.68E+12	$seJ \cdot kg^{-1}$	[9]
Plant protection agents	1.48E+13	$seJ \cdot kg^{-1}$	[9]
Human labor	4.26E+13	$seJ \cdot h^{-1}$	[9]
Fuel	5.13E+12	$seJ \cdot kg^{-1}$	[9,15]
Machines and equipment	7.29E+14	$seJ \cdot h^{-1}$	[3,12]
Organic matter – decay	10.02E+14	$seJ \cdot ha^{-1}$	[9,10,16]
Electricity	1.6E+5	$seJ \cdot J^{-1}$	[9]

 DOI: 10.1051/e3sconf/20171902033

3 Results and discussion

3.1 Overview of data on analyzed farm

The farm subjected to the analysis supports both animal and crop production. The average surface of the agricultural area (AL) of the farm was equal to 29.54 ha throughout 2012-2015, including: the area of arable land: 26.10 ha and permanent pasture 3.44 ha. The crop production in the farm is both oriented for commercial purposes and partly used for production of animal feed. The animal production in the analyzed years involved beef cattle and pork production in a closed cycle.

The crop production is dominated by the cultivation of plant which promote the degradation of soil organic matter: cereals, rape, maize, potatoes (Table 2), whose total ratio in the overall production was equal to 81.61% in the analyzed period.

Table 2. Crops obtained from production and their structure – mean results for period between 2012-2015.

Type of crop	Sown area [ha]	Structure [%]	LSU·ha^{-1}
Total cereals	13.18	51.24	0.51
Grain maize	3.80	14.54	
Mixed cereals and grain legumes	4.44	16.98	
Rape	3.70	14.15	
Potatoes	0.40	1.53	
Grass in arable land	0.37	1.42	
Silage maize	0.04	0.14	

Concurrently, the production of plants with a positive impact on the soil environment was low. These include mixed cereals and grain legumes and grass cultivated on arable land (18.39%). This data suggests clearly that the crop rotation leads to excessive environment loading and the content of soil organic matter decreases as a consequence. Natural fertilizers (manure and slurry) are applied on the farm with the purpose of reducing the effect of soil degradation, which is accompanied by ploughing down straw and green mass grown as an aftercrop.

The stocking density per 1 ha of AL was equal to 0.51 expressed in livestock unit (LSU) in the analyzed period. In relation to the surface area of AL, this parameter is low and does not guarantee the balance between the animal and plant production.

An increase of the stocking density could contribute to the use of the greater proportion of natural fertilizers in the place of chemical ones.

3.2 Cumulative energy intensity of farming production

The assessment of cumulative energy of farming production is applied as a tool to demonstrate how much energy is consumed by the processes and particular phases of a production process. The analysis involves a comprehensive insight into the energy intensity of the production. Table 3 contains the averaged results of production and cumulative energy intensity of the production in the analyzed farm between 2012-2015.

Table 3. Cumulative energy intensity and results of farming production – mean for 2012-2015.

Specification	Crop production	Animal production
Cumulative energy intensity of the components of the production process [GJ]:		
Human labor	37.61	41.35
Machines and equipment	68.88	2.46
Fuel	111.17	5.07
Mineral fertilizers: NPK and CaO	276.04	
Sowing material and planting stock	74.71	
Protection agents	23.6	
Purchased fodder		97.25
Electricity		7.92
Cumulative energy intensity – total	627.38	154.05
Cumulative energy intensity per 1 ha – total	21.24	5.21
Remaining indicators and quantities:		
Generated GU – total	1512.12	394.0
Generated GU·ha^{-1}	51.19	13.34
Cumulative energy intensity [GJ·GU^{-1}]	0.41	0.39
Energy efficiency of production [GU·GJ^{-1}]	2.41	2.56

The farming production yielded on average 51.19 GU·ha^{-1} from the plant production and 13.34 GU·ha^{-1} from the animal production. This means that plant production plays a principal role on the farm (in particular production of cereals, rape and maize), which provided nearly four times more production expressed in grain units. Besides, the cumulative energy intensity of plant production is also four times greater from animal production. The greater energy intensity of labor in the animal production results from the everyday clean-up activities in animal housings (manure handling). We can note a high ratio of the used fertilizers and protection agents (responsible for total of 48% ratio in the cumulative energy intensity of production). Besides, the use of fuel accounts for 15% of the overall cumulative energy intensity of production on the farm (Fig. 1), coupled with 12% of fodder.

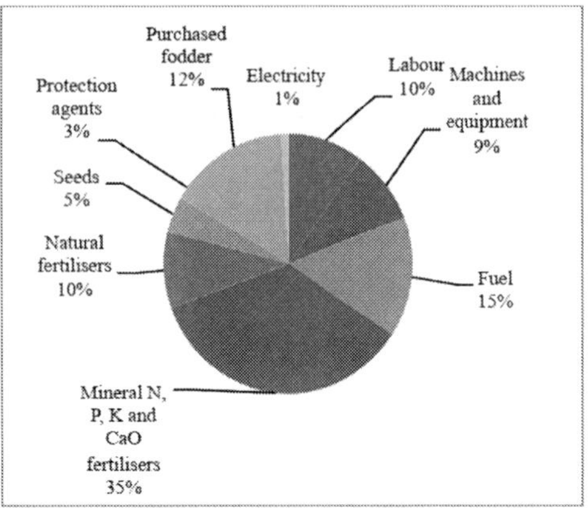

Fig. 1. Ratio of components of energy input of agricultural production.

However, similar results are demonstrated by other reports in this field [17-19].

Conventional system of production is followed in the analyzed farm and the use of chemical means of production significantly increases the energy intensity of the production in it.

Conventional system of production is followed in the analyzed farm and the use of chemical means of production significantly increases the energy intensity of the production in it. The results of various comparative studies performed with regard to selected types of conventional (intensive and extensive) farming as well as organic farming indicates the existence of considerable difference in the ratios of energy intensity of the components of the two systems. For the case of organic production, most commonly the cumulative energy intensity of the overall production is lower (Table 4). This is due to the avoidance of the considerable energy input contained in the chemical means of agricultural production.

Table 4. Examples of energy intensity of conventional and organic production systems.

Production	Energy input [$MJ \cdot ha^{-1} \cdot y^{-1}$]
Winter wheat (USA)[4]	1.77E+04
Wheat (India)[3]	1.18E+04
Corn –conventional (USA)[4]	3.26E+04
Corn – organic (USA)[4]	2.25E+04
Soyabean –organic (USA)[4]	1.05E+04
Soyabean –conventional (USA)[4]	1.26E+04
Winter wheat - conventional (Poland)[1]	2.39E+04
Winter wheat –organic (Poland)[1]	1.12E+04
Buckwheat – conventional (Poland)[2]	7.97E+03
Buckwheat –organic (Poland)[2]	1.12E+04

[1][2], [2][20], [3][21], [4][22]

3.3 Emergy analysis

The diversity of production in the analyzed farm makes it difficult to obtain a detailed calculation regarding the use of emergy. This calculation needs to include a few simplifying assumptions. First, the mean duration of the plant stay in the fields was adopted to be equal to 168 days. On this basis, and using data from Table 1., we determined the emergy of the solar radiation and that of the wind. For the case of water, its use was determined on the basis of total precipitation in the area of the farm and the duration of plant vegetation.

The thermoecological cost of machines and equipment was derived to be equal to 11.76E+6 $J \cdot \$^{-1}$ on the basis of [12]. The mean value of new machinery was taken to be equal to USD 30,000 and its depreciation period to be 12,000 hours. Hence, the calculation of the cumulative use of exergy gave 2.94E+7 $J \cdot h^{-1}$ related to the machine operation. The solar transformation was taken to be equal to τ = 6.2E+7, accounting for the earth sedimentation cycle. Due to the fact that the proportion of iron ores in steel production in Poland is equal to 0.4, the value of τ was multiplied by this value. The resulting use of emergy associated with machine operation was calculated to be Em = 7.29E+14 $seJ \cdot h^{-1}$. The emergy of the fuel, mineral fertilizers and protection agents was derived by analogy to the data from the literature, as given in Table 1. The emergy of the purchased fodder and sowing material was taken to represent the emergy derived from renewable sources. Its specific value was adopted at the same level as for cereals. The emergy of electricity for the domestic conditions was regarded as non-renewable quantity. Table 5 below contains a summary of emergy representing the use of renewable and non-renewable sources during the production process, whereas Fig. 2 contains a summary of their per cent ratios.

The emergy calculations lead to the statement of the impact of the production on the environment. This purpose applied the following indicators [23-25]: P_R – to determine the ratio of emergy use from renewable sources (4), ELR (Environmental Loading Ratio) – to express the environmental load of the activities (5), EYR (Emergy Field Ratio) – efficiency parameter (6) and use of emergy (Y) per GU of the obtained production. The value of Y (7) expresses the total use of emergy.

Below is a presentation of the formulae used to express the particular parameters:

$$P_R = \frac{Em_R}{Y} \tag{4}$$

$$ELR = \frac{Em_{MAT}+Em_{NM}+Em_{NS}+Em_{NF}}{Em_R} \tag{5}$$

$$EYR = \frac{Y}{Em_{MAT}+Em_{NF}+Em_{NM}} \tag{6}$$

Table 5. Use of particular components of production and the use of emergy respective to them [seJ·ha-1·y-1].

Specification	Mean values: 2012-2015	
Operating times of machines and equipment	13.71 [h]	9.99E+15
Fuel	102.49 [l]	4.36E+14
Fertilizers (N)	102.68 [kg]	4.15E+15
Fertilizers (P)	13.52 [kg]	4.99E+14
Fertilizers (K)	38.62 [kg]	7.22E+13
Fertilizers (CaO)	841.31 [kg]	1.42E+15
Fodder	1007.11 [kg]	2.51E+15
Plant protection agents	2.88 [kg]	4.26E+13
Electricity	86.4E+6 [J]	1.38E+13
Human labor	26.73 [h]	1.14E+15
Time of solar radiation	168 [day]	1.66E+12
Time of using wind	168 [day]	4.69E+12
Mass of evaporated water	3.05E+5 [kg]	3.90E+14
Mass of decayed organic matter	503 [kg]	1.10E+15
Seeds	157.68 [kg]	1.02E+15
Total (Y)		2.28E+16

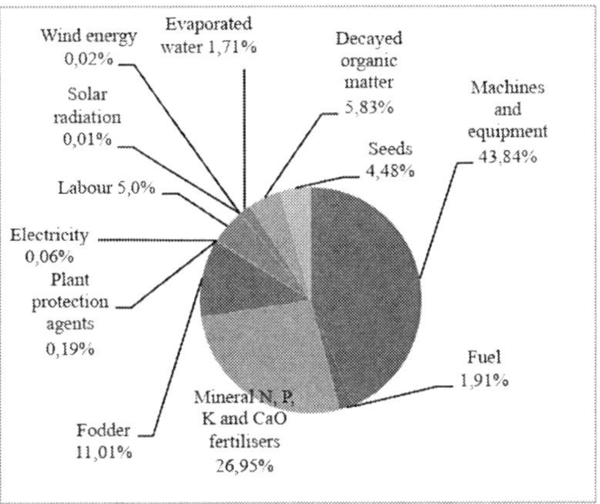

Fig. 2. Percent ratios of components of emergy in agricultural production on a farm.

$$Y = Em_R + Em_{MAT} + Em_{NM} + Em_{NS} + Em_{NF} \quad (7)$$

where:
Em_R – total emergy derived from renewable sources: sun, wind, water, human labor, sowing material, fodder,
Y – total use of emergy,
Em_{MAT} – emergy of protection agents and fertilizers,
Em_{NM} – emergy of machines and equipment,
Em_{NS} – emergy of decayed organic matter,
Em_{NF} – emergy of fuel and electricity.

The values of the estimated indicators P_R, ELR, EYR and $Y \cdot JZ^{-1}$ are presented in Table 6.

Table 6. Values of the emergy parameters of the production in the examined farm.

Parameter	Value
P_R	0.22
ELR	3.50
EYR	1.37
$Y \cdot GU^{-1}$	1.20E+13

The emergy analysis demonstrates that the greatest degree of the emergy loading results from the operation of machines and equipment. This, in turn, is due to the considerable thermoecological cost (and high exergy demand) of the machines, equipment as well as high solar transformation for the case of the applied means of production. Machines and equipment apply a considerable load on the environment during each phase of their production and this load is accounted for in the emergy analysis at the stage of the application of these means in the agricultural production.

Similar as for the case of cumulative energy intensity, mineral fertilizers form a second important item in the account of the emergy use. We can note that in the examined example, emergies of the fuel and electricity are represented by relative low values. The ratio of emergy derived from renewable sources (P_R) is responsible for 20% of the total emergy use and this value is considerable for the conditions in Poland. This is attributable to the relatively low value of the ELR indicator (3.50), which describes the environmental load. We can bear in mind that each manner of cultivation characterized by a value of $ELR > 1$ is considered to be responsible for environmental load [10]. The relatively low values of the P_R and ELR indicators in the analyzed farm result to a large degree from the substitution of natural fertilizers by mineral ones as well as the considerable ratio of fodder originating from renewable source in relation to the total emergy input. We can note at this point that the mean results of ELR in the domestic production of cereals are in the range of 12, although this value can be even a few times bigger.

4 Conclusions

The conducted study offers the statement of the following conclusions.

Mineral fertilizers represent the greatest proportion in the cumulative energy intensity. Other, comparable items on this list include: fuel, purchased fodder, human labor, natural fertilizers and machines and equipment. On average, the total energy intensity of plant production is equal to 21.24 GJ·ha-1 and this value corresponds to the standard energy intensity of the conventional production of cereals.

The energy intensity of both animal and plant

production is equal to around 2.5 GU·GJ^{-1}. This is a typical value for conventional farms and a lower value in comparison to an example of organic winter wheat production (3.98 [2]).

The emergy analysis offers a more comprehensive insight of the impact of a given type of production on the environment. When it is applied for this purpose, solar transformation demonstrates the environmental value of the resources used in the generation of means of production. This type of analysis also accounts for the value of the renewable sources used in the agricultural production. From this analysis, it is clear that the highest ratio is represented by the use of emergy associated with machines and equipment (around 44%). We can also note the considerable proportion associated with the use of mineral fertilizers (around 27%). By comparison, the use of the emergy of the fuel is low, as it accounts for around 2%.

The *ELR* indicator determined for the farm is low and equal to 3.50. This is affected by: limitation of the use of mineral fertilizers and their partial replacement by natural ones and the use of fodder which is regarded as originating from renewable sources. As a consequence, the ratio of renewable sources is equal to around 20%.

References

1. http://www.enis-pv.com/naslonecznienie-w-polsce.html (access on-line 15.05.2017)
2. A. Kuczuk, J. Agri. Sci., **8**, 4, 140-155 (2016)
3. H.T. Odum, *Environmental accounting – Emergy and Environmental Decision Making*. (John Wiley&Sons, Inc., 1996)
4. A. Kuczuk, S. Wacław, J. Res. Appl. Agri. Eng. **59**, 4, 31-36 (2014)
5. K. Sławiński, J.Res. Appl. Agri. Eng. **55**, 4, 99-101 (2010)
6. K. Sławiński, Inż. Rol. **4**,129, 43-249 (2011)
7. Z. Wójcicki, *Poszanowanie energii i środowiska w rolnictwie i na obszarach wiejskich*, (IBMER, Warszawa,2 007)
8. S. Ulgiati, H.T. Odum., S. Bastianoni. Ecol. Model. **73**, 215-268 (1994)
9. S.L. Brandt-Williams, *Handbook of Emergy Evaluation. A Compendium Data for Emergy Computation Issued in Series of Folios* (Centre for Environmental Policy, Environmental Engineering Sciences, University of Florida, 2002)
10. J. Jankowiak, E. Miedziejko, J. Agri. Rur. Devel. **2**, 12, 75-84 (2009)
11. J. Szargut, *Termodynamika techniczna*. (Wydawnictwo Politechnika Śląska Gliwice, 2000)
12. W. Stanek, *Metodyka oceny skutków ekologicznych w procesach cieplnych za pomocą analizy egzergetycznej* (Wyd. Polit. Śl., Gliwice (2009)
13. E. Klugmann-Radziemska, *Practical application of solar energy. Renewable Energy Sources in Opole Province*. 1/POKL/8.2.1/2008:1-14 (2008)
14. Ochrona Środowiska, GUS, Warszawa (2014)
15. R. Burges, *The use of emergy analysis for measuring the environmental costs and benefits of agriculture practices in Scotland* (in): Emergy synthesis **6**: *Theory and Applications of the Emergy Methodology*, CEP:265-271 (2011)
16. J. Kuś, *Systemy gospodarowania w rolnictwie. Rolnictwo integrowane*. Materiały szkoleniowe, **49**, IUNG, Puławy (1995)
17. J. Bojarszczuk, Stow. Ekon. Rol. i Agrob. Rocz. Nauk. **XVI**, 5, IUNG-PIB:15-21 (2014)
18. T. Dworakowski, *Plonowanie w różnych systemach produkcji – plonowanie roślin w różnych systemach produkcji*, www.odr.pl (2011), (access on-line 15.05.2017)
19. S. Parafiniuk, Prob. Inż. Rol. **1**, 79, 23-30 (2013)
20. A. Kuczuk, J. Res. Appl. in Agri. Eng. **61**, 4 6-14 (2016)
21. D. Pimentel, H.M. Pimentel, *Food, Energy and Society* (Third Edition, Taylor and Francis Group, Boca Raton, London, New York, 2008)
22. D. Pimentel, *Impacts of Organic Farming on the Efficiency of Energy Use in Agriculture* - An Organic Center State of Science Review (2006)
23. D. Chen, Z. Luo, Ecol. Indic. **34**, 623-626 (2013)
24. A.C. Haden, *Emergy evaluations of Denmark and Danish agriculture – Assessing the Limits of Agricultural Systems to Power Society*, Ekologisk Lantbruk **37**, SLU (2003)
25. R.J. Singh, B.N. Ghosh, N.K. Shrama, S. Patra, K.S. Dadhwal, P.K. Mishra, Ecol. Indic. **61**, 753–765 (2016)

The use of QSAR methods for determination of n-octanol/water partition coefficient using the example of hydroxyester HE-1

Joanna Guziałowska-Tic[1,*]

[1]Opole University of Technology, Department of Environmental Engineering, Mikołajczyka 5 Street, 45-271 Opole, Poland

Abstract. According to the Directive of the European Parliament and of the Council concerning the protection of animals used for scientific purposes, the number of experiments involving the use of animals needs to be reduced. The methods which can replace animal testing include computational prediction methods, for instance, the quantitative structure-activity relationships (QSAR). These methods are designed to find a cohesive relationship between differences in the values of the properties of molecules and the biological activity of a series of test compounds. This paper compares the results of the author's own results of examination on the n-octanol/water coefficient for the hydroxyester HE-1 with those generated by means of three models: Kowwin, MlogP, AlogP. The test results indicate that, in the case of molecular similarity, the highest determination coefficient was obtained for the model MlogP and the lowest root-mean square error was obtained for the Kowwin method. When comparing the mean logP value obtained using the QSAR models with the value resulting from the author's own experiments, it was observed that the best conformity was that recorded for the model AlogP, where relative error was 15.2%.

1 Introduction

The chemical safety assessment of products for the purpose of their registration requires tests to determine their physico-chemical, toxicological and ecotoxicological properties [1]. Unfortunately, some of those tests involve the use of animals. After the coming into force of the Directive of the European Parliament and of the Council No. 2010/63/EU of 22 September 2010 concerning the protection of animals used for scientific purposes, the number of experiments involving the use of animals has to be reduced so as to ease their pain and suffering in connection with the tests [2].

According to the provisions of the Directive, the target is to eliminate the unnecessary tests on animals. REACH requires that the registrants should restrict new tests with the use of vertebrates for the purpose of registration and use them only as a last resort [3]. First, it is necessary to collect and assess all existing data concerning the test substance and then, to detect any data gaps and consider the possibility of filling them with the results of *in vitro/ex vivo* tests or other, alternative concepts (Fig. 1).

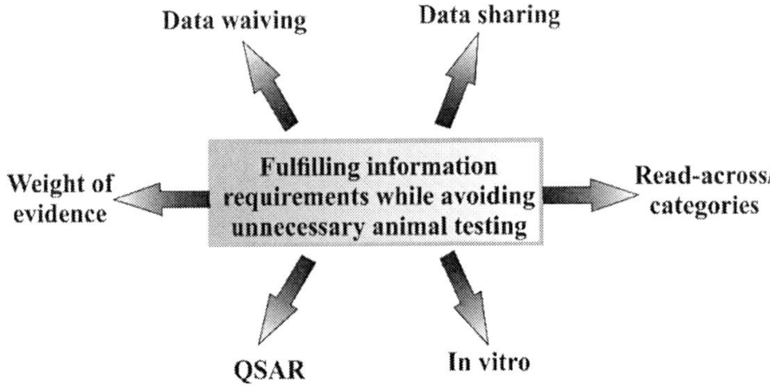

Fig. 1. Strategies to avoid unnecessary animal testing [3].

* Corresponding author: j.guzialowska@po.opole.pl

This means that all available data obtained from various sources, including tests made *in vivo* (on live animals), *ex vivo* (on animal tissue) and *in vitro* (with the use of cultured cells or bacteria) need to be collected. Another option is to use information based on tests involving human exposure, predictions based on available information about substances with similar structures (e.g., "cross-sectional approach" or "chemical categories"), as well as expectations based on reliable quantitative prediction methods, such as quantitative structure-activity relationships (QSAR) [5].

The QSAR methods are used, first of all, in the development and designing of new medicines. However, it is becoming an ever more frequent practice to use them for determination of the toxicological and ecotoxicological properties of chemical compounds within the REACH system [6]. The staff of the European Chemicals Agency (ECHA) have presented the use of various – not only experimental – methods for the assessment of the ecotoxicity of substances. Based on the results, it was found that only in just over 1% of the substances were the QSAR methods used for the assessment of ecotoxicological properties [7-11].

The biological activity of a chemical compound is directly related to its physical properties. Lipophilicity is the most useful parameter in the assessment of biological activity of substances and in the predicting their toxic activity. It is essential in the distribution of chemical compounds taking place in the body. Specifically, focus is on how a given substance behaves when penetrating the lipophilic membranes which separate aqueous media. Lipophilicity is described by partition processes taking place between two phases: non-polar (organic) and polar (typically water).

The polar/non-polar phase system in biological structures and in the natural environment is best illustrated by n-octanol and water as solvents. After mixing, the two solvents form two separate phases although, due to their mutual partial solubility, the resulting system comprises octanol saturated with water and water saturated with octanol.

The value of the n-octanol/water partition coefficient (logP) depends not only on the chemical structure of dissolved molecule of the compound (number and type of various functional groups, contribution of unsaturated sites, dipole moment value) but also – and to a high degree – on its size. The n-octanol/water partition coefficient, as measured for typical pollutants, is in a very wide range from 0.01 for high-polarity compounds to 10^{10} for highly hydrophobic substances. The span of values being so wide, the partition coefficient is expressed in the logarithmic form:

$$\log P = logC_{okt} - logC_w \quad (1)$$

where:
C_{okt} – molar concentration of substance in octanol (mol/L),
C_w – molar concentration of substance in water (mol/L).

Log Kow (log P) is an important parameter in the assessment of substances in for chemical safety, CLP classification (classification, labeling and packaging of substances and preparations) and PBT assessment (permanent/bioaccumulating/toxic) – Fig. 2.

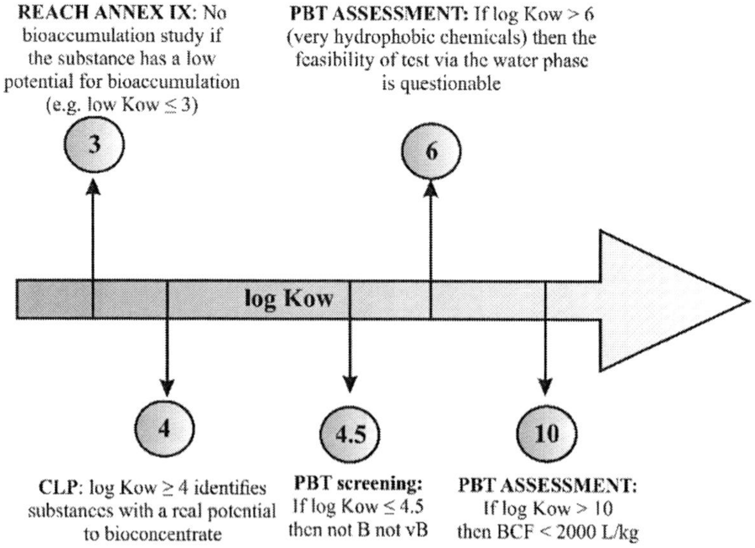

Fig. 2. Regulatory requirements for log Kow [12].

According to REACH, log Kow tests are required for all substances with production volumes of more than 1 Mg/year [1]. If log Kow ≤ 3, then (according to REACH, Annex IX), bioaccumulation tests are not required for substances with low bioaccumulative potential. For the purpose of CLP classification, log Kow ≥ 4 is the limit. The parameter is intended to identify substances with bioconcentration potential (BCF). The ability of substances to bioconcentrate is linked with their lipophilicity, in turn, is related to log Kow [12].

The use of QSAR methods for the assessment of the n-octanol/water partition coefficient using different models was reported earlier for polychlorinated biphenyls, thiophenols, benzene, ethanol, and other substances [12-17].

To assess how reliable the QSAR methods are in determination of the physico-chemical, toxicological and ecotoxicological properties of chemical substances, the results of tests on animals are compared with those performed using *in silico* methods. Models, enabling the assessment of the effect of chemical substances on the human health and natural environment are commonly available. Based on data from the ANTARES project, which promoted the use of numerical methods for determination of the properties of chemical substances, it was established that the following tools were applicable in the log Kow assessment: ToxPredict, VEGA, Epi Suite, SPARC, VCCLAB [18].

In this report, the results of own tests are compared with those obtained using different QSAR methods in log Kow determination for the hydroxyester HE-1. Three different models, provided by the VEGA tool *in silico*: Kowwin, MlogP, AlogP, were used.

2 Methods and materials

2.1 Materials

The hydroxyester HE-1 (3-hydroxy-2,2,4-trimethylpentyl isobutyrate) is used as an additive to paints and varnishes and is obtained in a sequence of chemical reactions in which isobutyric aldehyde is the basic starting material in the aldol condensation process with the subsequent Cannizaro and Tishchenko reaction [19] (Fig. 3).

The hydroxyester HE-1 (Fig. 3) is a hydrophobic organic solvent of which the main use is as coalescent in water-based architectural paint formulations [20].

HE-1 has a boiling point of 255°C which means, it is not a volatile organic compound (VOC). Its density is 0.9477 g/cm^3, flash point is 128°C, autoignition temperature is 410°C.

2.2. Method to determine logP

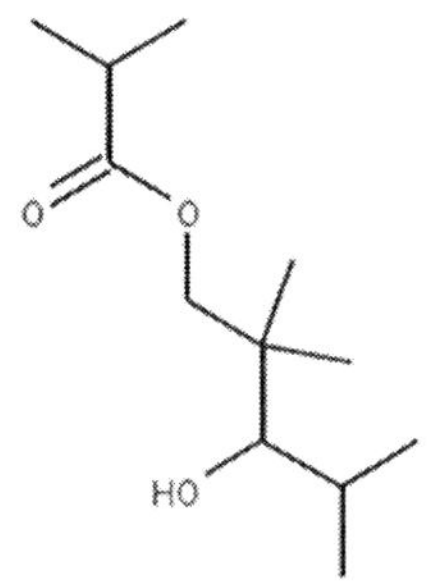

Fig. 3 Semi-structural formula for hydroxyester HE-1.

The n-octanol/water partition coefficient for the hydroxyester HE-1 was found to be 2.1 in own tests, performed in accordance with the OECD Test Guideline No. 117 [21].

LogP was found using the high-performance liquid chromatography apparatus HP 1090 with injector with loop capacity 20μl, UV detector with diode array, Purospher RP-18 column for HPLC, analytical balance, and measuring glass. The partition coefficient was determined using the following control substances: bromobenzene, thymol, diphenylamine, isopropylbenzene, benzyl benzoate, 1,2,4-trichlorobenzene, n-butylbenzene, dibenzyl and triphenylamine.

2.3 VEGA *in silico*

VEGA *in silico* renders accessible a number of QSAR models for the assessment of hazards, posed by chemical substances or for the development of individual models for research purposes. The platform is based on the CAESAR or T.E.S.T. data bases but it also uses other ones, for instance: EUTOXRISK, Toxbank, Caleidos, Orchestra, Antares-life, Life-prosil, LIFE-edesia.

The VEGA platform uses algorithms which are independent of the QSAR models and which enable the search for similar compounds. Such algorithms are used both for the identification of similar compounds and for the analysis of the role of descriptors (numerical value for the physicochemical parameters which characterize the specific ligand) and fragments for the chemicals searched for as well as for similar compounds.

The VEGA tool uses an algorithm which takes into account the molecular similarity of elements and molecules of chemical compounds in the aspect of relationship between their structure and properties [22]. In the VEGA tool, two compounds or elements are considered to be similar if their chemical structures and reactivities are similar in comparable conditions.

In this paper, the author used the latest version of VEGA *in silico* platform 1.1.3, available from www.vega-qsar.eu [23].

The input was the unambiguous notation of the structure of chemical compounds, based on the following character series (smiles) for HE-1: CC(C)C(C(C)(C)COC(=O)C(C)C)O.

The models Kowwin, MlogP and AlogP are discussed in this paper as ones used for the calculation of the n-octanol/water partition coefficient.

In the VEGA platform, the coefficients AlogP and MlogP were used as part of the Meylan model, implemented in the VEGA tool. Methods for the calculation of the n-octanol/water partition coefficient are based on similarities in the structure or properties of chemical compounds.

2.3.1 LogP model (Meylan/Kowwin) version 1.1.4

The model is based on the method of average fragment consistency (AFC), also known as Kowwin [24]. Generally, according to the method, an atom other than hydrogen is the core for a molecule fragment. Each fragment of a compound depends on the type of atoms which are connected with the core. The AFC method uses 250 correction coefficients for steric effects, H bonds, and influence of polar structural fragments. The correction coefficient is calculated from multiple linear regression (MLR). Regression analysis takes into consideration the size of the molecule fragment and its frequency in the structure as well as the correction coefficient and its frequency in the structure.

2.3.2 MlogP (Moriguchi logP) version 1.1.2

The MlogP model is based on the regression equation with 13 structural parameters [25,26]. The regression coefficients were assessed on the basis of 1230 organic compounds, including aliphatic, aromatic and heterocyclic compounds comprising the following elements: C, H, O, N, S, P, F, Cl, Br, I. The model variables include: the sum of lipophilic and hydrophilic atoms, N/O atom proximity effect, number of unsaturated bonds, number of polar aromatic substitutes, presence of ring-type structures, number of nitrogen groups, presence of intramolecular hydrogen bonds, amphoteric properties.

2.3.3 AlogP (Ghose-Crippen-Viswanadhan logP) version 1.1.2

The AlogP model is based on the regression equation, describing the hydrophobic properties of 120 atom types [27]. Compounds comprising such atoms as: C, H, O, N, S, Se, P, B, Si and halogens are considered. For each compound, logP is calculated as the sum of the number of all atoms, multiplied by their corresponding hydrophobicity constants.

2.4 Evaluation of models

The results were evaluated using the determination coefficient (R^2) and root-mean square error (RMSE), which are calculated as follows:

$$R^2 = 1 - \frac{\sum_{i=1}^{n}(y_i - \hat{y}_i)^2}{\sum_{i=1}^{n}(y_i - y_{avg})^2} \quad (2)$$

$$RMSE = \sqrt{\frac{\sum_{i=1}^{n}(y_i - \hat{y}_i)^2}{n}} \quad (3)$$

where:
y_i – experimental value,
$\hat{y}_i$ – value, obtained using QSAR method,
y_{avg} – average value of all the values obtained using the QSAR methods,
n – number of substances tested.

The QSAR model is regarded as satisfactory when R^2 is equal to or as close to 1 as possible and RMSE is as close to 0 as possible.

3 Results and discussion

Using the VEGA *in silico* tool, the values of logP were calculated by means of the three QSAR models: Kowwin, MlogP, AlogP. Table 1 shows the compounds with –molecular similarity to HE-1 which were used in the logP calculations for the three different models. The data in Table 1 indicate that chemical similarity of the compounds to HE-1 varied between 0.805 and 0.878.

Then, using three models, the results of experiments (originating from different data bases) were compared with those obtained by means of the VEGA tool and substances with chemical similarity to HE-1 (Fig. 4 - 6). Determination coefficients (R^2) and root-mean square errors (RMSE) were then calculated for each of the three models. The R^2 was high for each of the three models: 0.9878 for MlogP, 0.9596 for AlogP, and 0.9718 for Kowwin.

The RMSE was obtained from formula (3) for all the three QSAR models. For MlogP, RMSE was the highest: 1.03. For AlogP and Kowwin, RMSE was 0.72 and 0.39, respectively.

Taking into consideration molecular similarity, the results for R^2 and RMSE indicate that the Kowwin method is the optimum solution for determination of the n-octanol/water partition coefficient.

The average value of logP was then found for each of the three models (average of 6 values, obtained for compounds with molecular similarity) and compared with the value obtained in own experiments (logP = 2.1), as well as relative error was calculated (Table 2).

The results shown in Table 2 indicate that the best conformity between the results of own experiments and those originating from the QSAR models was obtained for the AlogP model.

 DOI: 10.1051/e3sconf/20171902034

Table 1. List of substances with chemical similarity to HE-1

CAS number of chemical compound	Smiles	Similarity	Semi-structural formula
50405-44-2	O=C1OC(C)C(C)C(O)C1C	0.878	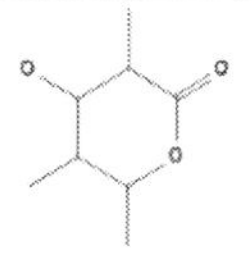
16409-45-3	O=C(OC1CC(C)CCC1C(C)C)C	0.829	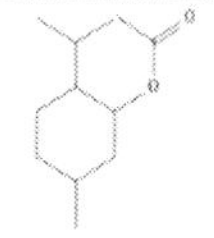
688-84-6	O=C(OCC(CC)CCCC)C(=C)C	0.825	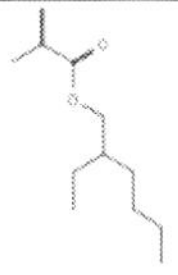
557-25-5	O=C(OCC(O)CO)CCC	0.82	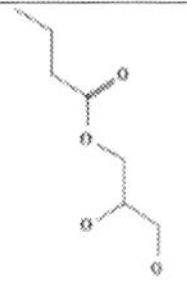
96358-98-4	O=C(OCC(=C)C)C1C(C=C(C)C)C1(C)(C)	0.816	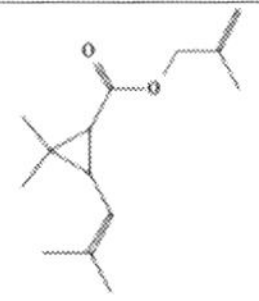
999-10-0	O=C(OCC)CCCO	0.805	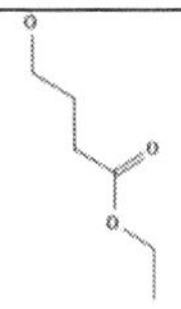

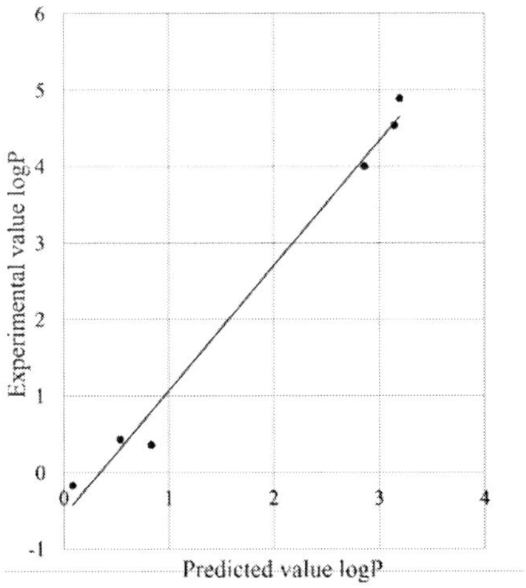

Fig. 4 Comparison between experimental results with those obtained using the MlogP model for substances with molecular structures similar to that HE-1

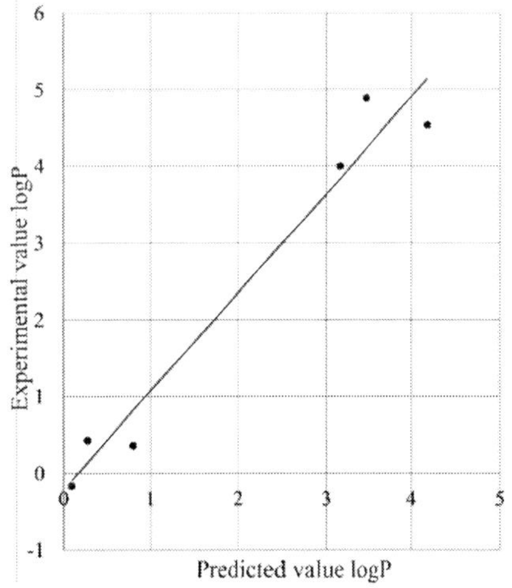

Fig. 5 Comparison between experimental results with those obtained using the AlogP model for substances with molecular structures similar to that HE-1

DOI: 10.1051/e3sconf/20171902034

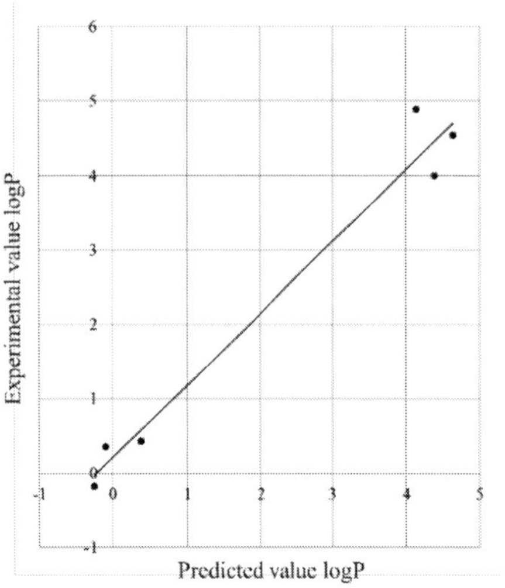

Fig. 6 Comparison between experimental results with those obtained using the Kowwin model for substances with molecular structures similar to that HE-1

Table 2 Comparison of average logP values provided by different QSAR models with the value obtained in own experiments for HE-1

Method	logP	Relative error, %
Own experiments	2.1	-
MlogP	2.75	30.9
AlogP	2.42	15.2
Kowwin	3	42.8

4 Conclusions

A comparison between the results of experiments on the calculation of the n-octanol/water partition coefficient for the additive to paints and varnishes – hydroxyester HE-1, and those obtained using the VEGA *in silico* tool was presented. Three models: logP: MlogP, AlogP and Kowwin were compared.

The results indicate that, for the substance with molecular similarity to HE-1, the highest determination coefficient was obtained for the model MlogP. When comparing the root-mean square error, the lowest value was obtained for the model Kowwin.

Taking into consideration the mean logP value, obtained using the QSAR models with those obtained in own experiments, it was observed that the optimum conformity (lowest relative error) was recorded for the model AlogP.

References

1. Regulation (EC) No. 1907/2006 of the European Parliament and of the Council (2006)
2. Directive of the European Parliament and of the Council No. 2010/63/UE (2010)
3. European Chemical Agency. How to use alternatives to animal testing to fulfil your information requirements for REACH registration (2016)
4. W. Lilienblum, W. Dekant, H. Foth, T. Gebel, J.G. Hengstler, R. Kahl, P.-J. Kramer, H. Schweinfurth, K.-M. Wollin, Arch. Toxicol. **82**, 211 (2008)
5. H.M. Patel, M.N. Noolvi, P. Sharma, V. Jaiswal, S. Bansal, S. Lohan, S.S. Kumar, S. Dhiman, V. Bhardwaj, Med. Chem. Res. **23**, 12 (2014)
6. A. Cherkasov, E.N. Muratov, D. Fourches, A. Varnek, I.I. Baskin, M. Cronin, J. Dearden, P. Gramatica, Y.C. Martin, R. Todeschini, V. Consonni, V.E. Kuzmin, R. Cramer, R. Benigni, Ch. Yang, J. Rathman, L. Terfloth, J. Gasteiger, A. Richard, A. Tropsha, J. Med. Chem. **57**, 12, (2014)
7. M.A. Sobanska, R. Cesnaitis, T. Sobanski, B. Versonnen, V. Bonnomet, J.V. Tarazona, W. De Coen, W., Sci. Total Environ.. **470-471**, 1225 (2014)
8. J.V. Tarazona, M.A. Sobanska, R. Cesnaitis, T. Sobanski, V. Bonnomet, B. Versonnen, W. De Coen, W., Sci. Total Environ. **472**, 137 (2014)
9. R. Cesnaitis, M.A. Sobanska, B. Versonnen, T. Sobanski, V. Bonnomet, J.V. Tarazona, W. De Coen, W. Sci. Total Environ. **475**, 116 (2014)
10. B. Versonnen, J.V. Tarazona, R. Cesnaitis, M.A. Sobanska, T. Sobanski, V. Bonnomet, W. De Coen, W., Sci. Total Environ., **475**, 123 (2014)
11. J. Guziałowska-Tic, W.J. Tic, Przem. Chem. **95**, 8 (2016)
12. C.I. Cappelli, E. Benfenati, J. Cester, Environ. Res. **143**, 26 (2015)
13. M. Rucki, M. Tichý, QSAR Comb. Sci. **25**, 4 (2006)
14. A.A. Toropov, A.P. Torpova, C.I. Cappelli, E. Benfenati, Fluid Phase Eqiulibr. **397**, 44 (2015)
15. Shi J.-Q., J. Cheng, F.-Y. Wang, A. Flamm, Z.-Y. Wang, X. Wang, S.-X. Gao, Ecotoxicol. Environ. Saf. **78**, 134 (2012)
16. F.-L. Zhang, X.-J. Yang, X.-L. Xue, X.-Q. Tao, G.-N. Lu, Z. Dang, J. Chem. Article ID 740548 (2013)
17. R.Z. Chen, W.Z. Tang, M. Sillanpää, Environ. Process. **3**, 73 (2016)
18. www.antares-life.eu (22.03.2017)
19. W. J.Tic, Advances in Chemical Engineering and Science. **6**, 4, (2016)
20. Research work, funded from the resources for science in the years 2007-2010 as the development project Technological Initiative I No. 13856.
21. Test No. 117: Partition Coefficient (n-octanol/water), HPLC Method (2004)
22. M. Floris, A. Manganaro, O. Nicolotti, R. Medda, G. F. Mangiatordi, E. Benfenati, J. Cheminform. **6**, 39 (2014)
23. www.vega-qsar.eu (25.03.2017)
24. W.M. Meylan, P.H. Howard, J. Pharm. Sci. **84**, 83 (1995)
25. I. Moriguchi, S. Hirono, Q. Liu, I. Nagakome, Y. Matushita, Chem. Pharm. Bull. **42**, 976 (1994)
26. I. Moriguchi, S. Hirono, I. Nagakome, H. Hirano, Chem. Pharm. Bull. **40**, 127 (1992)
27. A.K. Ghose, V.N. Viswanadhan, J.J. Wendolski, J. Phys. Chem. A **102**, 3762 (1998)

 DOI: 10.1051/e3sconf/20171902035

Energy value of soil organic matter and costs of its restoration

Anna Kuczuk*

Opole University of Technology, ul. Prószkowska 76, 45-758, Opole, Poland, a.kuczuk@po.opole.pl

Abstract. From the point of view of the sustainable soil management, the most important characteristic of soil organic matter (SOM) is associated with the energy content in it. This paper reports the results of an estimation of SOM resources and its energy value for the arable land in a selected farm. For this purpose, soil samples were taken in two fields from a soil depth profile of 30 cm. The testing regarding humus content were conducted at District Chemical and Agricultural Station in Opole. The study involved the assessment of organic matter content at a depth of 30 cm converted per 1 ha, energy value of the SOM resources and the theoretical energy potential was determined. In addition, an example of crop rotation was provided for the analyzed soils, which could be applicable in the process of restoring SOM resources. The cost of restoring the SOM resource was estimated and this value was compared with the energy value of fuel. The total cost of SOM restoration over the period of five years was equal to 3122.26-7845.86 PLN·ha^{-1} depending on the value of the lost revenue of commercial production, and simultaneously equal to the value of 6.2-16 Mg thermal coal.

1 Introduction – soil organic matter

There is a number of various interpretations of the notion of soil organic matter (SOM). Some sources state that SOM is formed by all animals and plant remains both dead and in various stages of decomposition as well as almost completely decomposed humus [1-4]. Other sources also account for the presence of soil organisms in it. In such an approach, SOM is formed by all organic matter in the soil, regardless of its origin and state of decay or alternatively, collection of all carbon compounds in the soil and concurrently coupled with the remaining soil components [5-9]. Sometimes soil organic matter is perceived to be equal to compounds formed by the final decay or even as the organic soil fraction excluding undecomposed organic remains [10].

Since the dead organic matter (organic residue and humus) can form as much as 85-90% of the total organic matter, with the humus content of up to 70-80%, of it, very commonly the notions of organic matter and humus are taken as synonyms. For the purposes of this elaboration, an equivalence between the meanings of SOM and humus was adopted.

The functions of the soil organic matter are invaluable both with regard to the soil environment as well as the environment in general. In addition, soil organic matter plays a role in term of the productivity of agriculture. These are two most important functions, which result from the SOM impact on a variety of chemical, physical and biological soil properties [7, 11, 12]. In particular, the last of these of underestimated in the context of the more numerous and stable populations of microorganisms inhabiting soil that are richer in humus. In turn, their activity promotes the processes of the decomposition of organic matter, which is associated with the supply of valuable nutrients to plants. The colloidal products of the organic matter decay and consequently, the reversed humus contribute to the preservation of soil aggregates [7, 13]. It is postulated that the soil organic matter content determined the quality and heath of soils. In particular, the light fraction of POM (*particulate organic matter*) forming the active SOM and subjected to rapid transformations is believed to play the most important role as it provides energy for the biological processes occurring in soils. As given by Fließbach et al. [14], as a consequence of the high content of soil microorganisms, the ratio of the light fraction remains low in the soil. This means that biomass is capable of regulating the organic residue in soil and the amount of this fraction could indicate the level of decay of the organic remains. The environmental role of the soil organic matter reaches far beyond the soil ecosystem. It was demonstrated that there is a considerable relation between the SOM content and heavy metals. Its high content in soils offers a considerable capacity to give an origin to establish bonded forms with metals, thus protecting the soil, water and organic world from the toxic impact of some of the metals [15-17].

The soil organic matter content in soils is given as a result of determination of soil organic carbon (C_{org}), On average it forms 58% of its composition. Soil forms a considerable reservoir and therefore becomes the potential source of CO_2, which is released during decomposition in a similar manner to energy, water and nutrients available for plants. Some soils can lose as much as from 20 to 80 tons of carbon per 1 ha, which is principally emitted into the atmosphere [18].

* Corresponding author: a.kuczuk@po.opole.pl

DOI: 10.1051/e3sconf/20171902035

The content of soil organic matter varies according to the type of the soil. For example, for the agricultural land, the lowest $C_{org.}$ content is noted for the case of chernozem (1.5-2.3%), whereas the lowest values are noted in luvisols (0.5-1.3%) [7]. Even a precautionary approach to the study of organic carbon $C_{org.}$ in soils suggests that its worldwide resource are in the excess of 1.5E+12 Mg [18, 19]. In the steady state, the volume of the CO_2 released into the atmosphere can be equal to 7.50E+10 Mg annually, of which human activity contributes to the emission of around 8.0E+08 Mg annually. We can note at this point that the soils subjected to agricultural production most commonly lose 20-50% of the carbon content in them [19].

The purpose of this article is to evaluate the content of SOM in soil samples taken from selected arable land. Besides, its energetic value was estimated, and cost of SOM restoration. It was noted that the soil can be a reservoir for significant amounts of energy, which is essential to sustaining the metabolic processes. By determining the value of the costs for SOM restoring, its value was referred to the value of thermal coal.

1.1 Energy accumulated by soil organic matter

From the point of view of the biological function performed by soil organic matter, beside the supply of nutrients, it forms a source of considerable amounts of potential energy. This energy can be perceived as:

a) energy of the accumulated organic material forming the soil organic matter,

b) metabolic energy, which drives the biological processes and originating from the organic material decomposed in the soil.

In fact, the two types of the energy accumulated in the soil organic matter are equivalent. They both originate in the solar energy and the process of photosynthesis, and the energy accumulated in the organic matter forms a resource comprising nutrients and energy needed for the functions of the soil organisms.

The metabolic energy is directly linked to the carbon cycle in the environment: use of CO_2 from the atmosphere by autotrophs, primary production, secondary production and finally decomposition of dead organic matter and reversal of certain amounts of energy (mainly in the form of heat) and return of CO_2 again into the atmosphere (respiration) and accumulation of carbon in the soil in the form of stabilized compounds. The decomposition process involving organic matter is estimated to produce heat, which is expressed as an equivalent of carbon with the value of 1-12 tons per 1 ha, depending on the type of agricultural production [20].

The content of energy in SOM can vary depending on the type of the organic material delivered into the soil. Graham and Sanger [21] identified clear differences in the caloric value of SOM depending on the type of the habitat. In the humus layer of the forested areas, it is equal to 5.04 $kcal{\cdot}g^{-1}$ (21.17 $MJ{\cdot}kg^{-1}$), whereas in marshy soils – 4.87 $kcal{\cdot}g^{-1}$ (20.41 $MJ{\cdot}kg^{-1}$), and in mud soils in ponds and lakes – 5.24 $kcal{\cdot}g^{-1}$ (21.96 20.41 $MJ{\cdot}kg^{-1}$). Concurrently, the energy value of the organic matter of the soil in forested areas corresponds to the mean value for the majority of soils. Similar results are offered by insight provided in other reports in the area [13, 22, 23].

Under the assumption of the above, the energy value of the soil organic matter can be expressed in terms of fuel equivalent (thermal or anthracite coal). Apparently, such a comparison is strictly theoretical. The energy value of the SOM only plays a role in the metabolic processes occurring in soils, and it does not directly serve economic purposes in terms of energy production. Even a comparison of the energy density of fuels (thermal coal: 24 $MJ{\cdot}kg^{-1}$, anthracite coal: 32.5 $MJ{\cdot}kg^{-1}$) and soil (e.g. for the $C_{org.}$ content= 2%: 0.66 $MJ{\cdot}kg^{-1}$ of soil) demonstrates the existence of considerable differences in the energy potential. Such a comparison also reveals the existence of a pool of the energy gathered in the soil and the value of the energy potential of the organic matter. It can be different depending on the type of soil. In the chernozem in the arable land, for the content of $C_{org.}$ equal to 1.5-2.3%, the resources of the SOM are equal to 200-270 $tons{\cdot}ha^{-1}$. For luvisols, for the content of $C_{org.}$ equal to 0.5-1.3%, it is 30-90 $ton{\cdot}ha^{-1}$ [7]. Therefore, such soils have different energy values corresponded to the heat, that can be derived from the combustion of 140 and 35.6 Mg of anthracite coal, respectively.

The quality and health of soils are relative to the high organic matter content in them. From the point of view of the maintenance of its high and constant level, it is necessary to ensure that the rate of the respiration process does not exceed the accumulation of carbon in soils [12]. Such conditions guarantee the maintenance of the nutritional (energy) demand of the soil organisms, while concurrently determines the effective course of the biological processes in them as well as decides on the health and quality of the soil.

Under the assumption of adequate soil management, the high content of the soil organic matter contributes to the accumulation of considerable amounts of stable forms of carbon (as humus) as well as improves physical and chemical properties of soils. As a consequence, energy in preserved in the stable form of SOM, which is subjected to continuous decomposition leading to the release of the nutrients available for plants. Hence, from the perspective of sustainable agricultural production, the maintenance of a steady supply of the organic matter to soil and the accumulation of SOM (energy) can be considered as an investment in the soil capital [24, 25]. Cong et al. [26] emphasize that the natural soil capital, which is represented by the soil organic carbon resource, can mitigate a number of threats facing the agriculture in the future. Crop diversification and primarily, the correct management of the soil organic matter can contribute to the increase of farm revenues and limit the future agricultural hazards, resulting from e.g. price fluctuations and variable weather conditions. All this is due to the direct relations between the biological soil diversity and the constant supply of organic matter.

2 Materials and methods

Soil samples were extracted from two arable lands with the purpose of estimating the energy value of SOM. The fields applied for the collection of soil samples are situated in the Opole province, Krapkowice commune. The surface area of the fields is as follows:

a) field A: 1.40 ha;
b) field B: 0.62 ha.

On the areas subjected to the analysis, the dominant crops include grains, maize and rape, which are listed as plants with a potential to degrade the soil organic matter. The crop rotation includes the use of cereal-legumes mixtures and ploughing down catch crops.

For each of the soil, laboratory testing (by accredited District Chemical and Agricultural Station in Opole) was performed with the purpose of determining granulometric composition (the Prószyński method), as well as content of humus (the Tiurin method) and total nitrogen (N_{tot}, the Kjeldahl method). The measurements of the bulk density [27] of the soils was also performed with the purpose of determining the humus content. The soil samples were taken over the period comprising 2014 and 2015. The samples in each field were extracted two times every year (before introduction of fertilizers and after the main crop was harvested). The first extraction was performed in March 2014, whereas the final – in September 2015. Due to the short duration of the tests and the small number of samples, the laboratory tests were aimed at determination of the mean SOM content and N_{tot} for the period involving two years.

The assessment of the organic matter resource was performed for the soil depth profile of 0-30 cm. In addition, information was gathered for each farm regarding the sowings and amount of organic matter that is supplemented into the soil.

For the purpose of determination of the energy value of SOM, it was assumed that the content of the organic matter is equal to 4.7 $kcal \cdot g^{-1}$ (19.7 $MJ \cdot kg^{-1}$) [23]. This was coupled with determination of its energy potential. The resource of the organic matter was established for the soil depth profile of 0-30 cm, together with the energy value expressed in $MJ \cdot ha^{-1}$, and the theoretical energy potential expressed as an equivalent of thermal fuel. The study focused attention on the relation between the crop structure and type of organic matter introduced into the soil and the existence of the variable amounts of the organic matter in the soil itself.

The results of the calculation encouraged the author to draw up a detailed scenario regarding the activities required to restore the organic matter content in soils. This was coupled with the estimation of an approximate cost that it needed to restore the soil organic matter by comparing it with the value of the thermal coal.

For this scenario the balance of SOM was calculated [28] and the amount of the required organic material for SOM restoring during five years. The estimated costs of SOM restoring include the chosen expenditures spent on soil-enriching plants and the value of added manure. In addition, the lost revenues and the avoiding costs of discarding the crop (wheat) were calculated [29-31].

3 Results and discussion

3.1 Energy accumulated in soil organic matter and its potential

The basic information derived from laboratory analysis involving soil samples and crops in the analyzed fields are summarized in Tables 1 and 2.

The analysis of the granulometric composition provided the information needed to determine the granulometric groups: for fields A and B: sandy loam with considerable fraction of sand by volume (57%-60%). The bulk densities of the analyzed soils are characterized by low values, which can demonstrate that the air passage through the soil is adequate. Concurrently, these values result from the greater proportion of the sand fraction than from the content of organic matter itself. The values of the densities are however, characteristic for mineral soils.

From the analysis of this data and on the basis of the data from European Soil Bureau [32] we can state that the humus content in the analyzed soils is very low. Such conditions can be hazardous for the quality and health of the soil ecosystem, since the minimum level of C_{org} that is required to sustain the stability of the soil is equal to 2% [33]. The many years of the analysis conducted by the team at IUNG-PIB [34], concerning the variability of the humus content in farmlands, demonstrate the boundary humus levels of 1.84% (1.07% C_{org}) for which the volume can remain constant. The soils with the higher humus content tend to lose it, whereas for the case of a lower level – it tends to increase on condition that the cultivation is performed so that the application of sustainable techniques of cultivation. Concurrently, the analysis provided insight into the annual rate of the organic matter loss from the soil at a level of 2.2%.

For the case of the examined soils, there is concurrently a very narrow C:N ratio (around 8:1), which is associated with the low SOM content and attests to the greater amount in nitrogen in the soil. A considerable higher value of this ratio for the case of field B can be attributed to the introduction of natural fertilizers to it. In both fields, aftercrop (mustard seed) was ploughed down, while in the following year, aftercrop was ploughed down only in the field B. In addition, manure was brought in twice in the amount equal to 35 and 20 $Mg \cdot ha^{-1}$, respectively (Table 2).

Table 1. Results of laboratory analysis of soil samples, humus content, carbon-to-nitrogen ratio (C:N).

Field	Soil type	$N_{tot.}$	Humus	$C_{org.}$	C:N	Assessment of $C_{org.}$ content		Mean bulk density $[g \cdot cm^{-3}]$	Humus resources $[Mg \cdot ha^{-1}]$
		[%]	[%]	[%]		based on ESB data*	based on IUNG data**		
A	sandy loam	0.12	1.568	0.91	7.6:1	v. low	low	1.05	49.46
B		0.11	1.573	0.91	8.2:1	v. low	low	1.09	51.34

*ESB – European Soil Bureau; **IUNG-PIB – Institute of Soil Science and Plant Cultivation – State Research Institute

Table 2. Crop rotation, use of natural fertilizers, organic mass ploughed down, 2014-2015.

Year	A	B
2013	forecrop – winter wheat – harvested straw	forecrop - potatoes
2014	aftercrop– field ploughed in spring	aftercrop – field ploughed in spring
	main crop - mixed cereals and grain legumes, harvested straw	manure – 35 $Mg \cdot ha^{-1}$
		main crop - mixed cereals and grain legumes, harvested straw
2015	main crop – rape: harvested straw	aftercrop – field ploughed in spring
		manure – 20 $Mg \cdot ha^{-1}$
		mixed cereal seeds – harvested straw

The application of the simplified assessment of the balance of SOM conducted by application of the reproduction and degradation coefficients by Eich and Kundler [28] demonstrate the loss of organic matter in field A (-0.01 $Mg \cdot ha^{-1}$) and its increase in field B (+8.45 $Mg \cdot ha^{-1}$) (Table 3). The results of laboratory analysis also demonstrate the changes in such a direction. Due to the fact that it does not account for the depth of the cultivation profile [35] and other parameter with a potential effect on the SOM content in soil (e.g. temperature, mineralization rate, water, water storage capabilities, amounts and diversity of soil organisms), such simplified assessment cannot give precise results. The results can only be applied with the purpose of the correct crop rotation in terms of the maintenance of a positive balance of the organic matter. More precise results are provided by the results of measurements with regard to humus content by application of laboratory test carried over a long-term span.

The resources of humus accumulated in the analyzed soil are characteristic for arable lands e.g. luvisols and brown soils [7, 36]. In turn, the energy accumulated in SOM results both from its total volume in the soils as well as from the type of the organic matter introduced into the soil. For the analyzed fields, the value of energy in SOM and expressed in $MJ \cdot ha^{-1}$ is in the range from 9.74E+05 to 1.01E+06. These values correspond to the amount of heat that can be gained from combustion of 40.58-42.08 Mg of thermal coal, with a caloric value of 24 $MJ \cdot kg^{-1}$ (Table 4). The energy accumulated in SOM seems to be considerable, however, it has a potential for application only in the sense of biogeochemical transformations due to its considerable dissipation. For the case of soils with the greater and concurrently more desirable fraction of organic matter, e.g. 4%, its energy value is relatively greeter and equal to 3.94E+06 $MJ \cdot ha^{-1}$ [37].

Table 3. Organic matter balance for the examined soil for the period 2014-2015.

Field	Reproduction (+) or degradation (-) of organic matter [Mg]			Loss/gain of organic matter [Mg]	Loss/gain of organic matter $[Mg \cdot ha^{-1}]$
	aftercrop ploughed	manure	main crop		
A	+0.98	0.00	-0.74 +0.49 -0.74	-0.01	-0.01
B	+0.43 +0.43	+3.06 +1.75	-0.33 +0.22 -0.33	+2.52	+8.45

Table 4. SOM resources, its energy and energy value expressed as an equivalent of thermal coal, mean results for 2014-2015.

Field	Resources of organic matter $[Mg \cdot ha^{-1}]$	Energy resources in soil organic matter $[MJ \cdot ha^{-1}]$	Energy value of organic matter as equivalent to thermal coal [Mg]
A	49.46	9.74E+05	40.58
B	51.34	1.01E+06	42.08

The results concurrently indicate the theoretical energy potential accumulated in the organic matter when the results of the calculations are compared with the cumulative energy intensity of the selected types of farm production (Table 5). Depending on the intensity of the production as well as involved means of production and types of crops, the value of the energy intensity can be different.

From the data in the table, we can see that – theoretically - energy accumulated in the organic matter of the analyzed soils could constitute an equivalent of the energy used for the production of various crops over the period of long years to come. For example, the SOM energy resources (for analyzed soils) are theoretically equivalent to the energy consumption of organic buckwheat production during 91-95 years. In the case of conventional wheat cultivation, depending on the intensity of production, the SOM energy is theoretically equivalent to the energy that could be used for a period of 41-86 years in wheat cultivation.

Table 5. Energy input in production and theoretical potential of soil organic matter.

Production	Energy input [$MJ \cdot ha^{-1} \cdot y^{-1}$]	Theoretical energy potential of SOM equivalent to energy intensity during the years of production	
		A	B
Winter wheat (USA)**	1.77E+04	55	57
Wheat (India)*	1.18E+04	83	86
Corn – conventional (USA)**	3.26E+04	30	31
Corn – organic (USA)**	2.25E+04	43	45
Soyabean – organic (USA)**	1.05E+04	93	96
Soyabean – conventional (USA)**	1.26E+04	77	80
Winter wheat - conventional (Poland)***	2.39E+04	41	42
Winter wheat - organic (Poland)***	1.12E+04	87	90
Buckwheat - conventional (Poland)****	7.97E+03	122	127
Buckwheat - organic (Poland)****	1.07E+04	91	95

[***38, ****39, *40, **41]

3.2 Cost of restoring SOM

Since the content of organic matter (and C_{org}) in the analyzed fields is very low, it is necessary to ensure that measures are taken so as to protect the soil resources. For the case of arable land, it needs to be based on an adequately selected crop rotation including long plant alternation, including leguminous fodder plants, grass, introduction of mineral fertilizers, ploughing down green masses and reduced tillage. In extreme case, the restoration of soil capital needs to involve excluding land from production of crops for many years, which consequently results in the loss of revenues. In reality, in particular in large commercial farms, whose aim is to gain profits within a short time span, such activities are rarely employed. Although the crop rotations involving extended periods of plant alternation can contribute to the greater benefits than the shorter periods, such as two years [42], the majority of farms apply short plant alternation cycles, in which the dominant role is played by cereal plants.

Under the assumption of the annual rate of organic matter loss equal to 2.2% [34], we need to primarily attempt to stop its degradation. For the analyzed fields this means the necessity of supplementing additional amounts of organic matter equal to around 1 Mg annually. This should not intervene in the normal operation of a commercial farm. Clearly, when we consider restoring the organic matter content in soils to the level determined by ESB, which corresponds to the level that is safe for the soil ecosystem (SOM $\geq$ 3.5%; $C_{org} \geq 2\%$), or at least to the level given by IUNG-PIB (SOM $\geq$ 1.84%; $C_{org} = 1.07\%$), we have to realize that the process will take a long time, and this involves the need to set aside the arable land from commercial production for a certain period in time.

An estimate regarding the cost of restoring soil organic matter in the analyzed soils should account for the value of the necessary procedures that have to be taken and the use of adequate means of production along with the lost revenues resulting from the land set-aside from production. Table 6 contains an example of crop rotation determined for the period including five years. From the calculated and simplified soil organic matter balance, we know that within 5 years the soil can gain around 6.5 Mg of organic matter (i.e. 1.3 Mg per year) on average; however, this calculation does not account for the losses resulting from mineralization. Such an amount corresponds to the amount that is indispensable for the removal of the impact of its annual losses. Hence, the restoration of SOM to its higher levels requires longer periods of time accounting for the time when the land is set aside from commercial production coupled with the introduction of greater volumes of organic and natural fertilizers. Concurrently, we need to bear in mind the admissible levels that do not pose a hazard to the environment.

On the basis of selected information derived from calculations performed by Warmia–Mazury Agricultural Advisory Centre, an estimation of the approximated cost which is indispensable to introduce the necessary amount of the organic matter into the soil. This calculation accounted for the value of manure, clover seeds, accompanying work during the introduction of manure and clover cultivation.

The cost that was incurred is equal to 3852.66 PLN·ha^{-1} over the period of five years. By accounting for the loss revenue from the set aside land (capable of producing wheat in III and IV year), on the basis of the 2016 wheat price (612.60 PLN·t^{-1}), we can conclude that the cost of restoration is equal to 3122.26 PLN·ha^{-1} (excluding agricultural payments). The price of the wheat seed was at a relatively low level, which resulted in losses from commercial plant production (-730.40 PLN·ha^{-1}]. The production of clover with the purpose of restoring organic matter proves profitable in such circumstances. This is due to the fact that it leads to minimization of the agricultural risk, as stated by Cong et al. [26].

Table 6. Example of crop rotation coupled with the cost of restoring soil organic matter.

<table>
<tr><td colspan="5">Years</td></tr>
<tr><td>I</td><td>II</td><td>III</td><td>IV</td><td>V</td></tr>
<tr><td>roots +35 tons of manure</td><td>spring barley + undersown clover, straw collected for litter</td><td>clover</td><td>clover</td><td>winter wheat, straw is harvested for litter</td></tr>
<tr><td colspan="5">Balance of organic matter, Mg·ha^{-1}</td></tr>
<tr><td>+1.66</td><td>+1.43</td><td>+1.96</td><td>+1.96</td><td>-0.53</td></tr>
<tr><td colspan="5">Total: 6.48 Mg·ha^{-1} = mean 1.3 Mg·ha^{-1} · year^{-1}</td></tr>
<tr><td colspan="5">Restoration cost involving prices as of 2016* [PLN·ha^{-1}] calculated for the period of 5 years</td></tr>
<tr><td colspan="2">Cost incurred</td><td colspan="3">Lost revenues from stopping wheat production</td></tr>
<tr><td colspan="2" rowspan="3">3852.66</td><td colspan="3">-730.40 (for wheat price of 612.60 PLN·t^{-1})</td></tr>
<tr><td colspan="3">2593.20 (for wheat price of 850.0 PLN·t^{-1})</td></tr>
<tr><td colspan="3">3993.20 (for wheat price of 950 PLN·t^{-1})</td></tr>
<tr><td colspan="5">Total restoration cost [PLN·ha^{-1}] calculated for the period of five years</td></tr>
<tr><td colspan="5">3122.26</td></tr>
<tr><td colspan="5">6445.86</td></tr>
<tr><td colspan="5">7845.86</td></tr>
</table>

*Estimate on the basis of selected agricultural calculations performed by: Warmia–Mazury Agricultural Advisory Centre [29-31]; the mean price of 1 ton of manure on the basis of information gained from agricultural portals.

Since farmers are interested in gaining both high efficiency of production, as well as high revenues, the calculations were additionally conducted for two price alternatives: 850.0 and 950.0 PLN·t^{-1}. The adoption of higher prices results in the increase of the restoration cost more than twice.

The values of the cost incurred in connection of restoring the organic matter can be compared with prices of thermal coal. Under the assumption that the cost of one Mg of thermal coal is 500 PLN·Mg^{-1}, the value of restoring the degraded resource of SOM (6.5 Mg) is equal to the value of 6.2-16 Mg of thermal coal depending on the adopted scenario of lost revenues.

4 Conclusions

The results of the research offer the statement of the following conclusions.

Soil organic matter forms a resource whose role determines the capability of agricultural production. It forms a reservoir containing considerable amounts of carbon and this is the principal role that it plays in the environment. It also forms a dispersed yet relevant energy resource (from the point of view of agricultural production and environment).

In general, soils in Poland are poor in terms of the organic matter content. This is also the case for the analyzed fields, where its resources are equal to 49.46 and 51.34 Mg·ha^{-1}, respectively, and $C_{org.}$ was determined at a level of 0.91%.

Despite this, the potential accumulated in the analyzed soils is in the range of 9.74E+05-1.01E+06 MJ·ha^{-1}, which corresponds to the potential of around 40 Mg of thermal coal.

On the basis of a comparison of this potential with the energy efficiency of various types of agricultural production, we can state that it could contribute the required energy input over the period of tens of years to come.

The restoration of an adequate and safe environmental level of SOM is associated with a long and cost generating enterprise. As demonstrated by the adopted scenario of activity, the attempt at introducing 6.5 Mg of organic matter over five years results in the generation of a considerable cost.

This, in turn, results from the application of adequate means of production and operations as well as lost revenues from production that was not obtained from set aside land.

For the developed scenario, the total cost of SOM restoration over the period of five years was equal to 3122.26-7845.86 PLN·ha^{-1} depending on the value of the lost revenue from wheat production.

The total value of the lost SOM restoring is simultaneously equal to the value of 6.2-16 Mg thermal coal, depending on price of a wheat commercial production.

Under conditions of predicting of low wheat prices (or other agricultural risks), the farmers may consider temporary exclusion of arable land from commercial production for the purpose of renewing the soil organic matter.

The costs incurred to enrich the soil in organic matter (3122.26 PLN ha^{-1}) appear to be relatively higher than the potential loss in revenue from commercial production (-730.40 PLN·ha^{-1}). However, the expenditures on organic matter contribute to improving the quality and health of the soil over an extended period of time.

References

1. G.D. Buchan, J. Soil and Water Cons. **65**, 2, 48A-54A (2010)
2. H. Dziadowiec, *Glebowa materia organicza – pojęcia podstawowe* (in): *Badania ekologiczno-gleboznawcze* [red. Bednarek R i in.]. PWN, Warszawa (2005)
3. M.M. Kononova, *Soil organic matter. Its nature, its role in soil. Formation and soil fertility*. The Acad. of Soil Sci. USSR. The V.V. Dokuchaev Soil Institute. Izd. Akad. Nauk. SSSR. Moscow (1966)
4. R.J. Manlay, Ch. Feller, M.J. Swift, Agric. Ecosyst. Environ. **119,** 217-233 (2007)
5. J.A. Baldock, J.O. Skjemstad, *Soil organic carbon/soil organic matter*, (in): *Soils analysis: an Interpretation Manual*, [Eds K. I. Peverill, L.A. Sparrow, D.J. Reuter.], CSIRO Publ.: Collingwood (1999)
6. A. Bot, J. Benites, *The importance of soil organic matter – Key to drought-resistant soil and sustained food production*, FAO Soils Bull. **80** (2005)
7. S. Gonet, H. Smal, J. Chojnicki, *Materia organiczna gleb* (in): *Gleboznawstwo* (red. Mocek A.), PWN (2015)
8. W. Horwath, *Carbon cycling: the dynamics and formation of organic matter*. Chapter 12, (in): *Soil microbiology, Ecology and Biochemistry* (red. Paul E.A.). 4th Edition. Nat. Res. Ecol. Lab. and Dep. Soil and Crop Sci., Colorado State University:339-382 (2015)
9. R.E. Pettit, *Organic matter, humus, humate, humid acids, fulvic acids and humin. Their importance in soil fertility and plant health.* www.humates.com/pdf/ORGANICMATTERPettit.pdf, (access on-line: 01.12.2015)
10. Soil Sci. Soc. of Am. (SSSA), 2008. Glossary of Soil Science Terms (2008) by Soil Science Glossary Terms Committee, (access on-line: 24.11.2014: https://www.soils.org/publications/soils-glossary#)
11. K. Eleki, *Soil management, crop rotations, and biomass removal effects on soil organic matter content*, IOWA State Univ. – retrospective Theses and Dissertations, paper 15526 (2007)
12. E.S. Krull, J.O. Skjemstad, J.A. Baldock, *Functions of Soil organic matter and the effect on soil properties*. CSIRO Land and Water, PMB2, Glen Osmond SA 5064, (2004)
13. H.C. Buckman, N.C. Brady, *Gleba i jej właściwości*, PWRiL, Warszawa (1971)
14. A. Fließbach, F. Eyhorn, P. Mäder, D.I. Rentsch, R. Hany, *„DOK" Long-term Farming, Systems Trial: Microbial Biomass, Activity and Diversity Affect the Decomposition of Plant Residues* (in): *Sustainable management of soil organic matter*, CABI Publishing, CAB International: 363-369 (2001)
15. R.J. Haynes, M.S. Mokolobate, Nutr. Cyc. Agroec. **59,** 47-63 (2001)
16. K. Quenea, I. Lamy, P. Winterton, A. Bermond, C. Dumat, Geoderma, **149**, 3–4, 217–223 (2008)
17. G.W. Thomas, Soil Sci. Soc. Am. Proc. **39**, 591, (1975)
18. R. Lal, The Final Frontier – View Point, Special Section, Science **34**, 1623-1627 (2014)
19. W.H. Schlesinger, *Soil organic matter: a source of atmospheric CO_2, The role of Terrestrial Vegetation in the global carbon cycle* (Chapter 4); Measurement by Remote Sensing (ed. Woodwell G.M.), John Willey&Sons Ltd, (1984)
20. E.J. Russel, E.W. Russel, 1950. *Soil conditions and plant growth*, Longmans London, (1950), (in): E.S. Krull, J.O. Skjemstad, J.A. Baldock, *Functions of Soil organic matter and the effect on soil properties*, CSIRO Land and Water, PMB2, Glen Osmond SA 5064. GRDC, (2004)
21. E. Gorham, J. Sanger, Ecol. Society America Ecol. **48**, 3, 492-494 (1967)
22. M.S. Darling, Oecol. **23**, 2, 127-139 (1976)
23. C.F. Jordan, *A World Pattern in Plant Energetics*, Am. Sci. **59**:425-433 (1971)
24. J.J. Hoorman, R. Islam, Agr. Nat. Res. Fact Sheet. The Ohio State University, (2010)
25. L. Zimny, R. Wacławowicz, A. Zych, *Materia organiczna gleby-polisa na przyszłość*. Nasza rola – gazeta rolnicza (2015) (access on line: 04.04.2017)
26. R-G. Cong, K. Hedlund, H. Andersson, M. Brady, Agr. Sys. **129**, 30-39 (2014)
27. Łachacz A. (red.), Właściwości gleb - Zeszyt edukacyjny, Wyd. UWM Olsztyn (2007).
28. *Kodeks Dobrej Praktyki Rolniczej*. MRiRW, IUNG-PIB, FAPA, Warszawa (2004)
29. *Kalkulacja dochodu bezpośredniego jęczmienia jarego*, www.w-modr.pl (access on-line: 10.05.2017),
30. *Kalkulacja dochodu bezpośredniego koniczyny czerwonej*, www.w-modr.pl (access on-line: 10.05.2017), (wrzesień 2016)
31. *Kalkulacja dochodu bezpośredniego pszenicy ozimej*, www.w-modr.pl (access on-line: 10.05.2017), (wrzesień 2016)
32. European Soil Data Base Maps – Primary properties (in:) European Soil Database Centre: www.esdac.jcr.europa.eu/resource-type/european-soil-database-maps (access on-line: 20.01.2017)
33. B.D. Kay, D.A. Angers, *Soil structure*, (w): *Soil physics companion*, Chapter 7 (ed. Warrick A.W.), CRC Press:249-295 (2001)
34. T. Stuczyński, J. Kozyra, A. Łopatka, G. Siebielec, J. Jadczyszyn, P. Koza, A. Doroszewski, R. Wawer, E. Nowocień, Studia i Rap. IUNGPiB, **7**, 77-115 (2007)
35. L. Zimny, R. Wacławowicz, A. Zych, Zeszyty Probl. Post. Nauk Roln. **582,** 137-146 (2015)
36. H. Dziadowiec, *Zasoby materii organicznej i morfogenetyczne typy próchnic leśnych* (in):

Badania ekologiczno-gleboznawcze [red. Bednarek R. i in.], PWN, Warszawa (2005)
37. C.F. Jordan, *An Ecosystem Approach to Sustainable Developmnet. Energy Use Efficiency in the American South*. Environmental Challenges and Sollutions 1, Springer (2013)
38. A. Kuczuk, J. Agric. Sci. **8**, 4, 140-155 (2016)
39. A. Kuczuk, J. Res. Appl. Agric. Eng. **61**, 4:6-14 (2016)
40. D. Pimentel, H.M. Pimentel, *Food, Energy and Society*, Third Edition, Taylor and Francis Group, (2008)
41. D. Pimentel, *Impacts of Organic Farming on the Efficiency of Energy Use in Agriculture* - An Org. Cent. State of Sci. Review, (2006)
42. A.M. Johanns, C. Chace, M. Liebmann, *Energy and Economic Returns by Crop Rotation,* Ag. Dec. Maker, File **A1-90**. Iowa State University, (2012) www.extension.iastate.edu/agdm (access on-line 09.05.2017)

Selected problems of gas-liquid flow through the channels filled with metal foams

Roman Dyga[1], and *Małgorzata* Płaczek[1,*]

[1]Opole University of Technology, Faculty of Mechanical Engineering, Department of Process Engineering, 5 Mikolajczyka Street, 45-271 Opole, Poland

Abstract. Open-cell metal foams are relatively unknown type of cellular material, which is increasingly being used as structural packing in the industrial equipment. The paper presents an analysis of experimental results on heat transfer and hydrodynamics of gas-liquid two-phase flow through the channels filled with metal foams. The research included the registration of temperature and pressure changes on fluid flow path. Furthermore the phase void fraction was measured and flow patterns in present in the channel were observed. It was found that in two-phase flow, both from heat transfer and hydrodynamics of flow point of view, the liquid plays the dominant role. It was also found a significant influence of flow patterns on frictional pressure loss value and phase void fraction. Whereas the flow patterns and geometry of the foam in much lesser extent influence on the intensity of heat transfer. Type of gas-liquid flow patterns primarily depends on flow conditions, including the velocity and properties of fluids. On the other hand, it wasn't stated any significant effect of geometrical parameters of foam on the type of flow and the value of phase void fraction. Among many flow patterns observed during the study, there have been identified four basic structures: plug, semi-slug, slug, and stratified.

1 Introduction

Various processes in technology require a maximum surface applied for heat or mass transfer. For this purpose, process apparatus often contains a variety of materials with a considerable specific surface, such as open-cell metal foams. Metal foam materials are the ones in which metal occurs in the form of thin fibers forming a cellular skeleton. Cells are the large empty areas between the fibers. The adjacent cells are interconnected by a number of "windows". The size (diameter) of a cell is usually greater by an order of magnitude from the cross-sectional size of the fibers in the skeleton. Such a design provides a very high porosity of foams, as it often exceeds 90% and it enables relatively free transport of fluids through the materials. The high thermal conductivity of metals results in the relatively high value of the effective thermal conductivity coefficient. Due to the continuous structure of the skeleton fibers, there is no thermal resistance on the contact surface between them, which normally occurs in structural packings. With the current technology, foams can be manufactured from a wide range of metal alloys. The available range of the foams includes ones from the materials, which enable their application in extremely adverse environmental conditions, such as high temperature and chemically corrosive environment.

Due to the above characteristics of metal foams, these materials are most commonly applied to play the role of elements increasing the active surface of the compact heat exchangers [1,2], heat regenerators, [3,4] and heat storage systems [5,6]. Besides, foams are applied as packings in chemical reactors [7–9]. The existing technology of catalyst deposition on the surface of the skeleton is also at a satisfactory level [10–12]. This, in turn, provides a potential to apply foams in catalytic reactors. The examples of the applications of foams in the process apparatus are given in the literature [e.g. 13,14,15].

1.1 Gas-liquid flow through open-cell metal foams

Two-phase gas-liquid mixture flow can be applied in industrial apparatus in which the structural packing comprises metal foams. This occurs e.g. in evaporator channels [16–18] and in column apparatus [19,20].

The results of various works contain a conclusion that the presence of foam on the boiling surface reduces the superheating temperature of this surface and shifts the boiling crisis in the direction of the greater densities of the heat flux [21,22]. Hu et al. [16] state that the value of the heat transfer coefficient registered during the flow of a boiling coolant through a flat channel with a metal foam packing is between 1.5-2.6 times greater in comparison to a channel that does not contain such a packing. Even a greater difference, i.e. by 2–4 times for the advantage of the foam occurs for the case of boiling in round, tubular channels. The authors in the work [16]

* Corresponding author: m.placzek@po.opole.pl

also note that the presence of metal foam results in a considerable increase of the pressure drop of the fluid, which in turn contributes to the decrease of the efficiency of the cooling system. The assessment by Hu et al. [23] provides an insight that the pressure drop of a boiling mixture tend to increase along with its greater flow rate and in the conditions when the mass flow rate of the steam is greater. Similar conclusions were derived by Ji and Xu [17]. These authors registered a linear change in the pressure drop in the function of the vapor mass fraction and greater pressure drop values in a case of foams with a greater pore packing density.

A report by Wang et al. [18] contains a conclusion that the increase of the intensity of the heat transfer during boiling in an area occupied by a metal foam results from disturbances of the flow within near-wall region, more intensive mixing of the steam and the liquid as well as an increase of the effective heat transfer coefficient.

A comparison of the heat transfer in the column with various types of packing, conducted by Pangarkar et al. [19] demonstrates that metal foams can be put on the list of the category of packings in which the heat transfer occurs with a high intensity. Another advantage resulting from the application of foams in the packed columns (used in the place of other packings) is associated with the decrease of the pressure drop. According to a study by Weijiang Zhang et al. [24], pressure drop resulting from the flow accounts for only around 20 % of the total pressures drop in a wire mesh packing. The relatively small pressure drop was also registered by other authors [25]. The reports based on research conducted with regard to the columns with metal foam packings generally contain a conclusion that the flow is generally much more stable in this case in comparison to flows through columns with other types of packing, e.g. in the form of a fixed bed. Due to the high porosity, it is possible to apply a greater load of the gas and liquid, whereas the volume fraction of the liquid can reach as much as the porosity of the foam, i.e. above 90% [25].

The above information is derived from the scarce sources containing reports into two-phase flow through metal foams. Due to the scarcity of such studies, the subject of gas liquid flow through metal foams can be considered as insufficiently studied and researched. The literature does not contain information regarding the flow patterns resulting from the gas-liquid flow through metal foams. The current state-of-the-art in the area faces considerable difficulties in the design, optimization and development of the apparatus containing metal foam packings. All of this formed an incentive to perform a study with regard to the identification of the phenomena that are specific with regard to the gas-liquid flow through channels with metal foam packings. A particular emphasis was placed on the determination of the impact of the flow conditions and geometrical parameters of the foams on the characteristics of the heat exchange, pressure drop and volume fraction of the phases. Another important issue taken into consideration was associated with the identification of the flow patterns during gas-liquid flow and the conditions.

2 Scope of the experiment

The scope of the experiment involved an experiment with regard to flow of two types of gas-liquid mixtures through horizontal channels with various types of metal foam packing. The study applied three foams made of aluminum alloys – two made of AlSi7Mg alloy and one made of Al 6101 alloy. The classification of the foams applied the density of pore packing, which according to the manufacturer's specifications was equal to 20 and 30 PPI (pore per inch) for the case of AlSi7Mg foams and 40 PPI for Al 6101 foam. All foams were characterize by similar values of the porosity ϕ and similar geometrical cellular skeleton, with a note that for the case of 40 PPI foam there were local deformations of the skeleton, as it is visible in Fig. 1d. The foams were distinct in terms of the cell diameter d_c as well as pore diameter d_p (the term pore used here applies to the window between the adjacent cells – Fig. 2c). The characteristic foam parameters are summarized in Table 1.

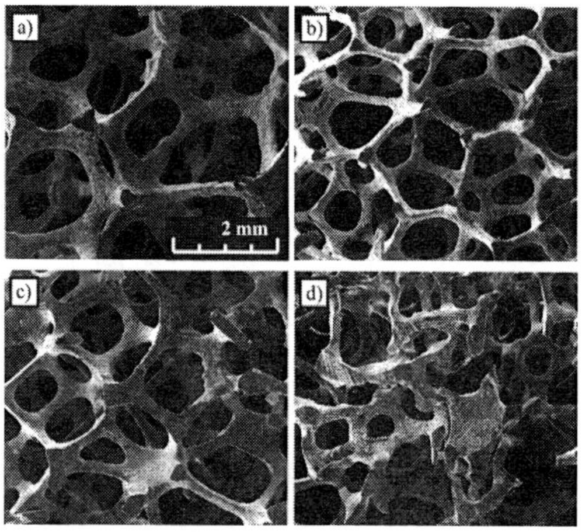

Fig. 1. Structure of the foams: a) 20 PPI, b) 30 PPI, c) 40 PPI, d) local deformations of the foam skeleton in 40 PPI foam.

Table 1. Characteristics of the foams.

Foam (alloy)	ϕ [-]	$d_c \cdot 10^{-3}$ [m]	$d_p \cdot 10^{-3}$ [m]	k_s [W/(m·K)]
20 PPI (AlSi7Mg)	0.9336	3.452	1.094	150.4
30 PPI (AlSi7Mg)	0.9435	2.255	0.712	150.4
40 PPI (Al 6101)	0.9292	2.386	0.824	189.4

The two-phase mixtures were formed by the air and water and air and Velol-9Q machine oil. From the point of view of the focus of this study, the most meaningful aspects are associated with the differences in the density, viscosity, specific heat and thermal conductivity of the components of the two-phase mixture. For oil at 20°C, these properties are as follows: $\mu_{ol} = 0.0086$ Pa·s, $\rho_{ol} = 859.4$ kg/m^3, $c_{ol} = 1848.8$ J/(kg·K), $k_{ol} = 0.128$ W/(m·K).

 DOI: 10.1051/e3sconf/20171903001

The study was conducted by application of an experimental installation, whose basic component was a horizontal channel with the internal diameter of 0.02 m completely filled with metal foam (individual channel was used for each type of the foam). In the central part of the channel, there was a measurement section *K2* with a length of 1.27 m. This section was preceded by a section used for flow development, i.e. *K1*. Following the measurement section, there was a section comprising an outlet, *K3* made from plexiglas. The transparent section made it possible to conduct observations of the patterns formed during the two-phase flow. The measurement section was heated from outside along the length of 1.18 m by means of resistance heater wound around the channel. In the heated channel section, the metal foam is fixed to the wall of the channel by means of an epoxy-aluminum thermo-adhesive. The measurement section was insulated by mineral wool with a thickness of 0.07 m. The diagram of the installation including the location of the measurement apparatus is presented in Fig. 2, whereas the actual view on the measurement section is found in Fig. 3.

Prior to entering the channel, the volume flow rates of the fluids forming the two-phase mixture V_f was measured by means of various types of flowmeters with a high level of accuracy. The pressure drop $(\Delta P/\Delta L)_{2f}$ was measured as the difference of the measured pressure in two points in the channel located at a distance of 1 m from each other. The measurements of the pressure drop were realized by the application of five piezoresistance sensors with the total measurement range of 0÷150000 Pa.

For the purposes of determining the air density, super pressure was measured in the channel as well.

The volume fraction of the phases ε_f were measured by the so-called trap method. By application of the pneumatically controlled membrane valves *VP*, the entry and outlet region of the measurement section were closed at the same instant in time. Throughout such measurements, the measured volume of the liquid in relation to the volume of the measurement section corresponded to the volume fraction of the liquid in the gas-liquid mixture. The flow patterns were identified on the basis of visual observations. The presence of foam in the channel resulted in a considerable deterioration of the visibility, hence, the attempt to visualize the results by use of the photographic and motion picture techniques did not offer satisfactory results.

The scope of the research into heat transport involved the measurement of liquid temperature t_f and the channel wall t_b by means of type K thermocouples with the diameter of 1 mm. The temperatures of the liquid and the wall were measured at four points along the channel length (sections I, II, III and IV in Fig. 2). Each of the sections comprised eight thermocouples. The tips of five thermocouples were located inside the channel at various distances from its axis (in the vertical plane). In turn the tips of three thermocouples were located inside the channel wall at a distance of 0.5 mm from its internal surface.

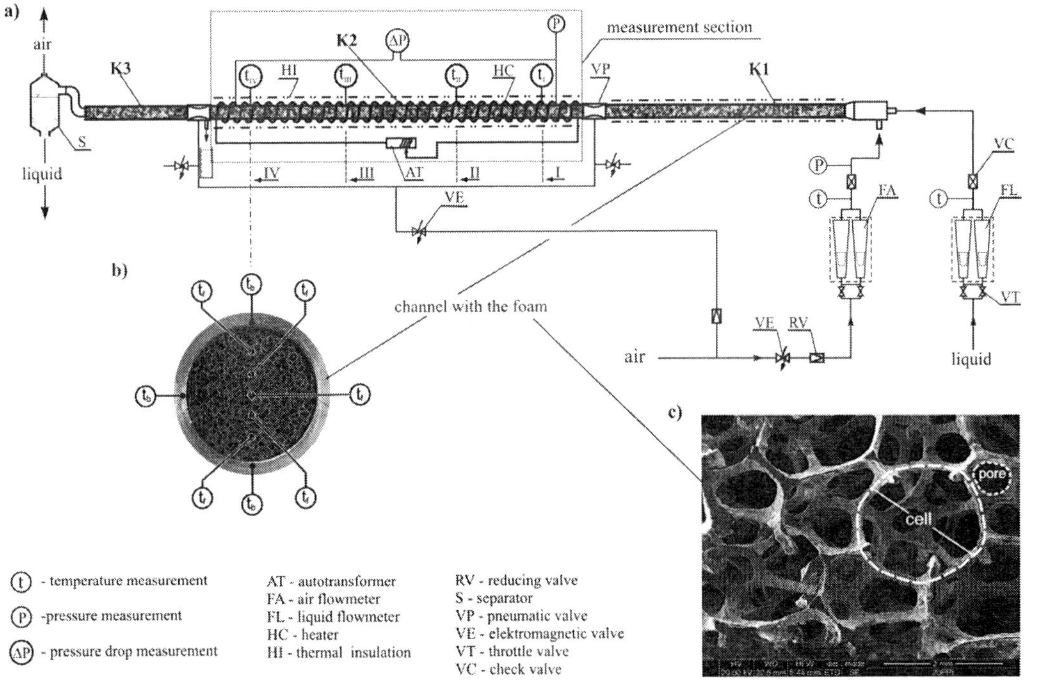

Fig. 2. Experimental setup, a) diagram of measurement section, b) location of thermocouples in the channel cross-section, c) details of cell and pore diameters.

DOI: 10.1051/e3sconf/20171903001

Fig. 3. Measurement section of the experimental setup.

The measurements were conducted within a considerable range of the fluid velocities (Table 2) so that laminar and turbulent flow conditions were recorded. The superficial phase velocity $w_{f,o}$ was applied to characterize the flow conditions. This notion denotes the mean velocity of the fluid flow through an empty measurement channel (i.e. when the impact of the presence of the other fluid and foam is disregarded),

$$w_f = \frac{4V_f}{\pi d^2}, \quad f \equiv a, ol, w \qquad (1)$$

where: V_f – volume flow rate of fluid [m^3/s], d – internal channel diameter [m].

Table 2. Conditions of conducted experiment.

fluid, f	w_f [m/s]	ζ_f [–]	t_f [°C]
air, a	0.028–2.39	0.313–0.998	21–95
water, w	0.006–0.061	0.002–0.988	24–88
oil, ol	0.006–0.061	0.002–0.988	19–93

ζ_f – inlet volume fraction of phases, i.e. ratio of the volume flow rate of a given phase in relations to the volume flow rate of gas-liquid mixture at the entry region to the channel.

3 Analysis of the results

3.1 Flow patterns

A number of various flow patterns were registered during the experiment. It was stated that the types of the forming patterns depend on the mutual relations of the flow rates and their properties. In the same flow conditions, i.e. for the same velocity and volume fraction of the phases, nearly identical flow patterns were observed for all foams (for given gas and liquid mixtures). Hence, the study did not demonstrate the impact of the geometrical parameters of foams on the gas-liquid flow patterns.

Four basic types of flow were identified on the basis of the formed patterns – stratified, plug, semislug and slug flows. In addition, other, transient forms of these flows were observed, such as stratified-semislug and plug-slug flows.

In the stratified flow, both phases flow in the form of continuous layers (Fig. 4a), with a thickness that is relative to the volume flow rate of the fluids at the inlet to the channel. Depending on the phase velocities, the interfacial surface can demonstrate wavy characteristics to a various degree. In extreme cases, the accumulation of the liquid can occupy the entire cross-section of the channel, forming semislug flow pattern. This type of flow was observed for the case when the superficial velocity of the liquid was in excess of 0.03 m/s, and the air velocity was greater than 1 m/s.

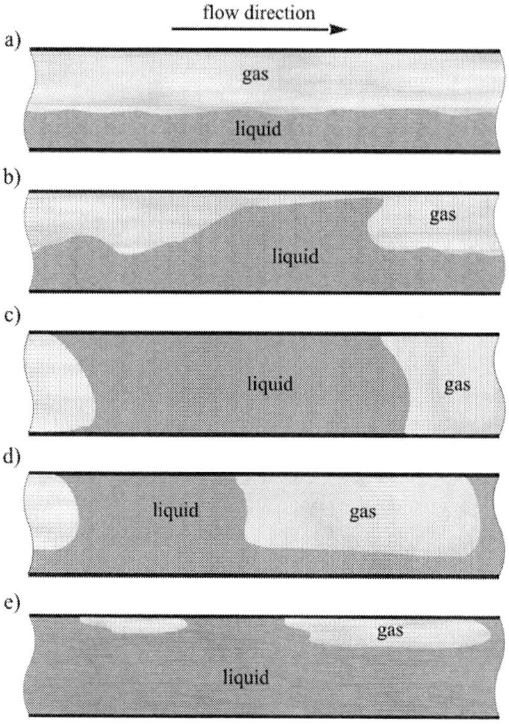

Fig. 4. Gas-liquid flow patterns in a channel with aluminum foam packing: a) stratified, b) semislug, c) slug, d) e) plug flow.

Very similar conditions can lead to the formation of the slug flow. Periodically, every few seconds, large liquid portions are observed in the channel, as they flow across the entire cross-section of the channel with a considerable velocity. Semislug and slug flows occur particularly frequently during the pumping of the air-oil mixture. For the case of a relatively large flow rate of oil, stratified flow was not observed at all, although it was the dominant type of flow for the air-water mixture. Stratified flow was only observed in the tests applying oil only for the case when only a small section of the channel was occupied by this liquid.

For the case of flows comprising large volume fraction of liquids and concurrently small air fraction, we have to do with the plug flow in the channel, as presented in Fig. 4d and 4e. In such a case, air occupies the top section of the channel in the form of extended bubbles, the so-called plugs. The dimensions of the plugs increase along with the increase in the flow rate of the air. Plug flow was registered for the case when the

DOI: 10.1051/e3sconf/20171903001

superficial air velocity did not exceed 0.2 m/s, and the ratio of the liquid to gas velocities was at least equal to 0.5.

On the basis of the sole values of the phase velocities, or else their volume fractions in the mixture, there is no possibility to unambiguously indicate the type of flow that is formed in the given flow conditions. The formation of the patterns is determined by a number of factors, including the physical properties of the fluids. During an attempt to identify a manner in which the particular flow patterns are formed in the channels with metal foam packing, it was established that this is possible to achieve it in a reference system proposed by Baker $(g_c/g_g)\gamma_B\Psi_B - g_g/\gamma_B$. In such a system, g_c and g_g denote the density of the mass flow rate of liquid and gas, respectively, and dimensionless parameters γ_B and ψ_B account for the relations between the physical properties of the gas and liquid to air and water,

$$\gamma_B = \left[\left(\frac{\rho_g}{\rho_a}\right)\left(\frac{\rho_l}{\rho_w}\right)\right]^{0.5}, \quad (2)$$

$$\psi_B = \left(\frac{\sigma_w}{\sigma_l}\right)\left\{\left[\left(\frac{\mu_l}{\mu_w}\right)\left(\frac{\rho_w}{\rho_l}\right)\right]^2\right\}^{1/3}, \quad (3)$$

where: ρ, μ, σ denote the density [kg/m^3], viscosity [Pa·s] and surface tension [N/m], respectively, while the subscripts refer to: l – liquid, g – gas, a – air, w – water.
As we can see in Fig. 5, the point denoting the type of the flow pattern form the compact regions (in a given reference system) determining the flow conditions in which the particular flow patterns can occur.

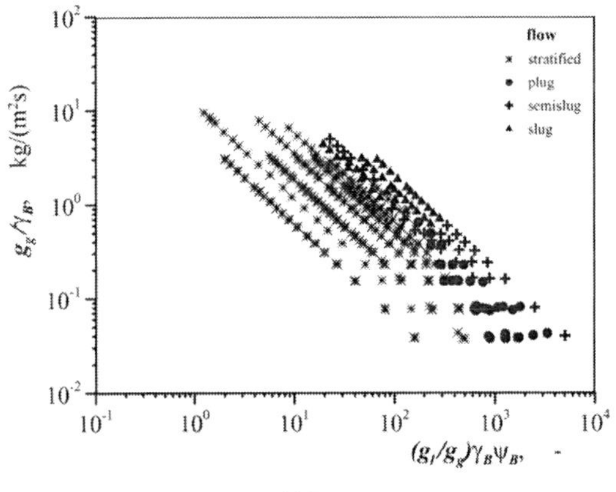

Fig. 5. Conditions of gas-liquid flow patterns occurrence during flow through metal foams.

3.2 Volume fractions of phases

The results of the measured volume fraction of the phases demonstrate that the volume fraction of gas ε_a increases along with the increase of the velocity of this phase. Concurrently, an increase of the liquid velocity contributes to decrease of the gas volume fraction. This observation is illustrated in Fig. 6, on the basis of an example involving air-water flow. The increase of the velocities of the phases results from the increase of the fluxes in the entry region of the channel; hence, the increased flow rate of the liquid in the channel leads to the reduction of the free space available for the gas (and the reverse).

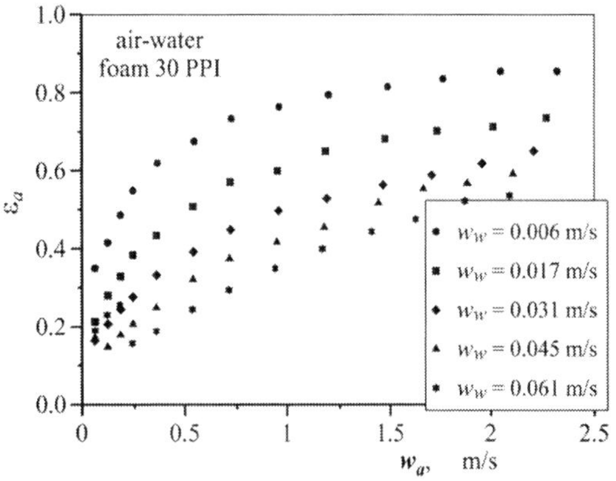

Fig. 6. Impact of phase velocities on the volume fraction of air in the mixture containing water – flow through 30 PPI foam.

For all three foams, in the same flow conditions (for a given gas-liquid mixture), the experiment offered very similar results with regard to the volume fraction of the phases (Fig. 7), which demonstrates the lack of the impact of the geometrical structure of the foam on the registered volume fractions.

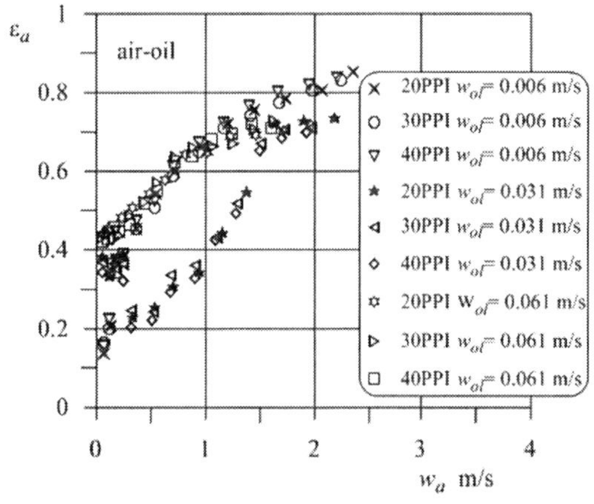

Fig. 7. Comparison of air volume fraction in the air-oil mixture during flow through 20, 30 and 40 PPI foams.

Foam forms the reason for the considerable interfacial slip, which is evidenced by the considerable differences between the inlet volume fraction in the entry region to the channel ζ_f and volume fraction of the phases in the region occupied by the foam ε_f. The air passes relatively freely through the space occupied by the foam due to its small viscosity. The foam skeleton tends to stop the more viscous liquid (in particular this regards oil). As a consequence, the volume fraction of the air in the space occupied by the foam could be over four times smaller from its inlet volume fraction (Fig. 8). In addition, as demonstrated in a work by [26], the volume fraction of the gas in the mixture pumped through the foams is smaller than for the case of the identical flow through empty channels.

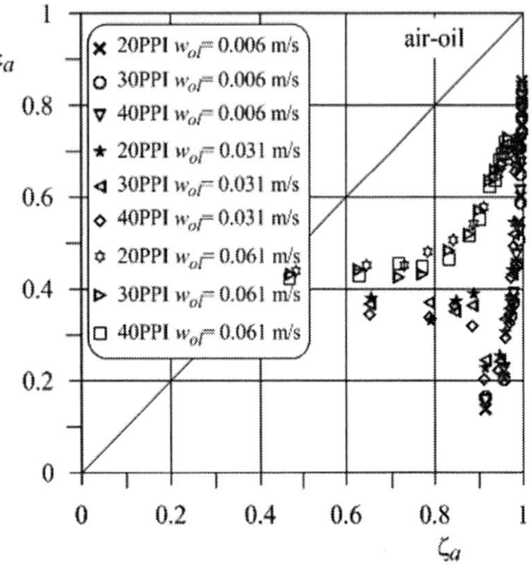

Fig. 8. Comparison of the measured volume fraction of air with its inlet volume fraction.

Despite the monotonous increase along with the increase of the velocities of the phases, the volume fraction of the phases tends to change abruptly when a change of the flow pattern occurs. This phenomenon is particularly evidenced for the change of the plug flow into the stratified one (Fig. 9); nevertheless, similar phenomena could also be registered for other flow patterns.

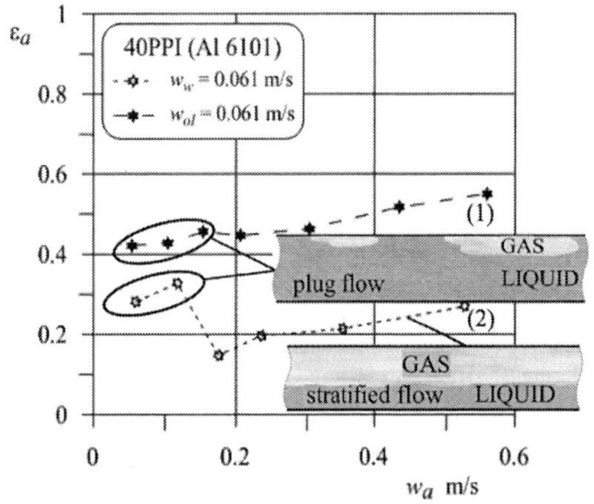

Fig. 9. Course of the changes in the air volume fraction in the conditions of flow patterns change during air-water and air-oil flow.

3.3 Pressure drop

Pressure drop during two-phase flow $(\Delta P/\Delta L)_{2f}$ increases significantly along with the increase of the gas and liquid velocities. For a small flow rate of the air, pressure drop tends to vary almost linearly and only when this velocity exceeds 1 m/s, the variations in the pressure drop take on exponential values. The deviations from the monotonous increase of the pressure drop occur (similarly as in case of the volume fraction) for the case when a change of the flow patterns occurs. An example of this could be associated with the apparently unjustified decrease of the value $(\Delta P/\Delta L)_{2f}$ despite an increase of the air velocity in the conditions of the transformation of plug flow into the stratified one (Fig. 10). The decrease of the pressure drop, which occurs in this case results from the decrease of the channel and foam surface that is wetted by water. In the plug flow, the liquid is in contact with a considerably greater surface than for the case of the stratified flow. Because water is a more viscous phase than air, it decides on the values of the pressure drop during two-phase flow to a greater degree than air does.

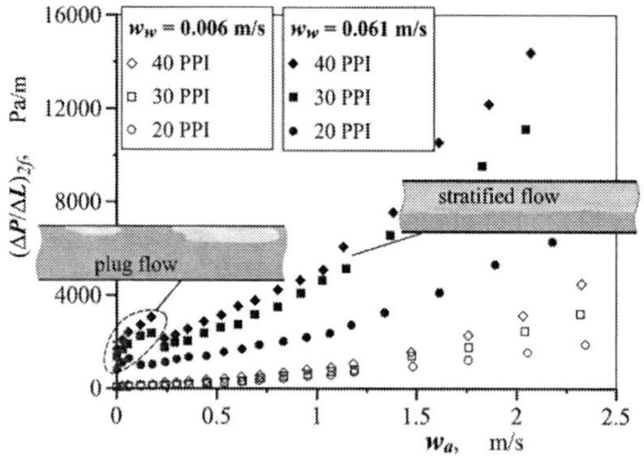

Fig. 10. Air-water pressure drop (accounting for the reduction effect) during flow through 20, 30 and 40 PPI foams.

This type of peculiarities are more frequently noted for the case of the air-oil mixture flow. This is due to the greater diversity of the flow patterns during this type of flow (as air-water flows mainly take the form of a stratified flow), which in turn leads to a more irregular course of pressure drop. The values of the pressure drop during air-oil mixture flow are also considerably greater than in the case of the air-water flow. This is due to the greater oil viscosity.

During flow of two-phase mixture through channels with various types of foams, different values of pressure drop were registered (in the same conditions). This means that the value of the pressure drop is affected by the geometrical parameters of foam. The smallest values of pressure drop were registered during flow through 20 PPI foam, whereas the greatest – for the flow through the 40 PPI foam (Fig. 10). The 20 PPI foam is characterized by the biggest cells diameter, which means that the flow occurs more freely than for the case of foams with the smaller cells. Although the cell diameter in 40 PPI foams is slightly larger than for the case of 30 PPI foam, the value of the pressure drop during flow through the first of them is considerably greater than for the case of the 30 PPI foams. The reason is likely to be associated with the specific geometrical structure of the 40 PPI foam skeleton. The large nodes at the contact points of the fibers in this foam, which are discernible in Fig. 1c, along with the numerous deformations of the skeleton (Fig. 1d) undoubtedly lead to an increase of the pressure drop.

3.4 Heat transfer

The measurement of temperature offered the determination of heat transfer coefficient value α_{2f} from the heated channel walls to the two-phase fluid,

$$\alpha_{2f} = \frac{q_{2f}}{t_b - t_f}, \tag{4}$$

where t_b and t_f denote the mean temperatures of the wall and the fluid calculated on the basis of data from all temperature measurement points located along the heated channel section (i.e. 32 in total).
The heat flux q_{2f} is given by the total of the densities of the heat fluxes absorbed by the liquids (q_w or q_{ol}) and that of the air q_a. This value was derived on the basis of the heat balances of the particular fluids

$$q_f = m_f(i_f^k - i_f^p), \quad f \equiv a, ol, w \tag{5}$$

The increase of the enthalpy of fluids $\left(i_f^k - i_f^p\right)$ was calculated on the basis of the temperature of the two-phase mixture at the entry and outlet regions of the measurement section. For the case of the air-water flow, the heat balance accounts for the latent heat of the evaporated water as well.

For all instances of flow that were investigated, the heat transfer coefficient increases along with the increase of the velocity of each of the phases. In addition, as we can observe in Fig. 11, the value of the heat transfer coefficient during two-phase flow is greater from the case when single-phase liquid flow is taken into account ($w_a = 0$).

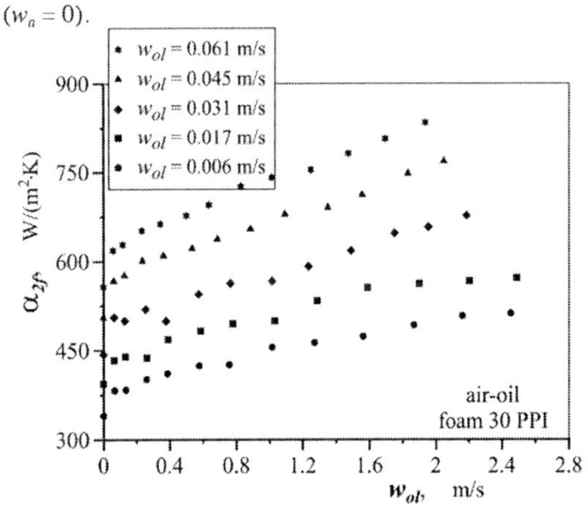

Fig. 11. Heat transfer coefficient during water-oil flow through a channel with 30 PPI foam packing

In contrast to the measurements with regard to the pressure drop and volume fraction, for the case of the studies involving heat transfer, the studies did not confirm the significant impact of the flow patterns on the intensity of heat transfer.
The value of heat transfer coefficient depends on the type of foam – its geometrical parameters and heat conductivity.
The greatest values of α_{2f} were recorded for the flow through 20 PPI foam, whereas the lowest- for the flow through 30 PPI foam (Fig. 12). In the consideration of the specific surface of the foam, which in combination with the internal area of channel form the surface of the heat exchange, the heat transfer coefficient should assume the greatest values for the case of the 30 PPI foam. This is due to the greater specific surface in this foam compared to 20 PPI foam.
The reverse result obtained in the study was explained in another paper [27]. In this paper, it was demonstrated that in the channels with metal foam packing, the heat exchange occurs primarily as a result of the heat transfer from the channel walls directly into the fluid, i.e. excluding the role of the foam. Foam itself participates in the heat transfer mainly through the conduction. During the flow through 40 PPI foam, the heat transfer coefficient assumes intermediate values, i.e. ones between the results gained for 20 and 30 PPI foams. In terms of its geometry, 40 PPI foam is similar to 30 PPI (i.e. they both have similar cell and pore size); however, it has a considerably greater value of the thermal conductivity k_s equal to 189 W/(m·K), compared to k_s = 150.4 W/(m·K) for 20 PPI and 30 PPI foams. A part of the heat is conducted through the foam skeleton and it is transferred to the fluid through it. Hence, the considerable thermal conductivity of 40 PPI foam contributes to the greater intensity of heat transfer compared to the case of flow through 30 PPI foam with a similar geometry.

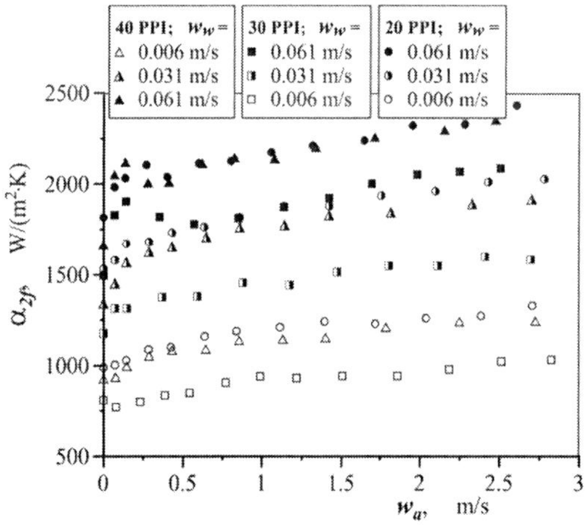

Fig. 12. Comparison of heat transfer coefficient for 20, 30 and 40 PPI foams in the conditions of air-water flow

4 Conclusions

The conducted experiments and analysis of the results demonstrate that the hydrodynamic and thermal phenomena related to the two-phase gas-liquid flow through channels packed with open-cell metal foams are determined principally by the flow conditions, including velocities of the fluids and their physical properties. The structure of foams affects the quantities characterizing the flow to a smaller degree – in some cases the impact of the foam structure can be even disregarded. For instance, the presence of foam forms a reason for the occurrence of interfacial slip which is the same for various foams (within the range of the performed experiment); hence, the structure of the foam does not affect the value of the volume fraction of the phases. The foam does not lead to the dispersion of the phases either, which is familiar from the flow through porous packings. During the flow through the foams, standard patterns are formed, such that are typical for flow through empty channels.

Four principal types of flow patterns were identified: plug, semislug, slug and stratified flow. The type of the flow patterns is determined only by the type of the flow conditions (i.e. velocities, properties and volume fraction of the phases). In addition, the formed patterns are independent of the geometrical parameters of the foams through which the flow occurs.

The parameters of the foams have a moderate impact on the characteristics of heat transfer and considerable effect on pressure drop during two-phase flow. From heat transfer point of view, the most important aspect is associated with thermal conductivity of foam material. The pressure drop during fluid flow through channel with metal foam packing is relative to the porosity of the foam, cell size and layout of the cell skeleton (apart from the well- recognized effect of the velocity and fluid properties). The pressure drop during two-phase flow and volume fraction of the phases change monotonously along with the changing velocities of the phases. Considerable deviations from this regularity can be observed in the conditions of the changes in the flow patterns.

References

1. E.J. Cookson, D.E. Floyd, A.J. Shih, Int. J. Mechanical Sci. **48** (2006)
2. B. Ozmat, B. Leyda, B. Benson, Mater. Manuf. Processes **19**, 5 (2004)
3. Z.Q. Chen, P. Cheng, T.S. Zhao, Int. Commun. Heat Mass **27**, 3 (2000)
4. K. Wang, Y.L. Ju, X.S. Lu, A.Z. Gu, Cryogenics, **47** (2007)
5. Y. Tian, C.Y. Zhao, Int. J. Heat Mass Transf. **58** (2013)
6. P.H. Vadwala, *Thermal Energy Storage in Copper Foams filled with Paraffin Wax* (Master of Applied Science, Mechanical & Industrial Engineering University of Toronto, 2011)
7. J. Lévêque, D. Rouzineau, M. Prévost, M. Meyer, Chem. Eng. Sci. **64** (2009)
8. R. Tschentscher, M. Schubert, A. Bieberle, T.A. Nijhuis, J. van der Schaaf, U. Hampel, J.C. Schouten, Chem. Eng. Sci. **66** (2011)
9. G. Incera Garrido, F.C. Patcas, S. Lang, B. Kraushaar-Czarnetzki CheM. EGg. Sci., **63** (2008)
10. M. Mitov, R. Rashkov, N. Atanassov, A. Zielonka, J Mater. Sci. **42** (2007)
11. A. Sirijaruphan, J.G. Goodwin jr., R.W. Rice, D. Wei, K.R. Butcher, G.W. Roberts, J.J. Spivey, Appl. Catal. A: Gen. **281** (2005)
12. H. Yu, H. Chen, M. Pan, Y. Tang, K. Zeng, F. Peng, H. Wang, Appl. Catal. A: Gen. **327** (2007)
13. J. Banhart, Progress Mater. Sci. **46** (2001)
14. X.-H. Han, Q. Wang, Y.-G. Park, C. T'joen, A. Sommers, A. Jacobi, Heat Transf. Eng. **33**, 12 (2012)
15. C.Y. Zhao, Int. J. Heat Mass Transf. **55** (2012)
16. H. Hu, Y. Zhu, H. Peng, G. Ding, S. Sun, Appl. Therm. Eng. **62** (2014)
17. X. Ji, J. Xu, Heat Mass Transf. **48** (2012)
18. P. Wang, D.Y. Liu, C. Xu, Appl. Energy **102** (2013)
19. K. Pangarkar, T.J. Schildhauer, J.R. van Ommen, J. Nijenhuis, J.A. Moulijn, F. Kapteijn, Chem. Eng. Sci. **65** (2010)
20. S7] C.P. Stemmet, M. Meeuwse, J. van der Schaaf, B.F.M. Kuster, J.C. Schouten, Chem. Eng. Sci. **62** (2007)
21. L. Tadrist, M. Miscevic, O. Rahli, F. Topin, Exp. Thermal Fluid Sci. **28** (2004)
22. Z.G. Xu, Z.G. Qu, C.Y. Zhao, W.Q. Tao, Int. J. Heat Mass Transf. **54** (2011)
23. H. Hu, Y. Zhu, G. Ding, S. Sun, Int. J. Refrigeration **36** (2013)
24. W. Zhang, B. Zhang, Z. Shi, Procedia Eng. **18** (2011)
25. S6] C.P. Stemmet, J.N. Jongmans, J. van der Schaaf, B.F.M. Kuster, J.C. Schouten, Chem. Eng. Sci. **60** (2005)
26. M. Płaczek, R. Dyga, Chem. Eng. Equipment **4** (2013)
27. R. Dyga, L. Troniewski, Arch. Thermodyn. **36**, 1 (2015)

The influence of the internal microbiome on the materials used for construction of the transmission natural gas pipelines in the Lodz Province

Agnieszka Staniszewska[1,*], *Magdalena* Jastrzębska[2], and *Krzysztof* Ziemiński[1]

[1]Lodz University of Technology, Faculty of Biotechnology and Food Sciences, Institute of Fermentation Technology and Microbiology, Division of Technical Microbiology, 171/173 Wólczańska Str., 90-924 Łódź, Poland
[2]Lodz University of Technology, Faculty of Mechanical Engineering, Department of Automation, Biomechanics and Mechatronics, 1/15 Stefanowskiego Str., 90-924 Łódź, Poland

Abstract. This paper presents investigation results of the influence of gas microbes on the biocorrosion rate of the materials used for gas pipelines construction in the Lodz Province. Samples of two types of carbon steel and cast iron were stored in the laboratory pipeline model reflecting the real conditions of working natural gas pipelines were. In the next step the influence of cathodic protection with parameters recommended for protection of underground structures was tested. Analyses of biological corrosion products generated on the test surface were carried out using a scanning electron microscope with an X-ray analyzer. The level of ATP was measured to confirm presence of the adsorbed microorganisms on the observed structures. Corrosion rates were determined by gravimetric methods. In the course of the study it was revealed that the rate of biocorrosion of steel is lower than that for cast iron. Our results also proved that the weight corrosion rate depends on the number of adhered microorganisms. In addition, it has been found that application of the carbon steel cathodic protection decreases its weight corrosion rate. The information obtained will help to increase the knowledge on the rate of biological corrosion causing losses/pits inside gas pipline.

1 Introduction

As a result of technological progress, rise in the living standards and broadly understood economic development, increasing usage of the electric energy is observed. Moreover, some of newer regulations, e.g. in the EU concerning the climate, require decrease of the usage of the fossil fuels for the electricity production. This promotes an increase in usage of the natural gas, which stands out among others conventional fuels due to its low greenhouse gas emissions (see Table 1) and high calorific value. In 2014, share of the natural gas in the electricity market was 2.06%, but it is expected to increase even up to 10% [1] by year 2030. Presently, the domestic natural gas reserves cover only 30% of total demand, so it is necessary to cover the deficit through import [2].

Gas pipelines are the most common way of natural gas transportation because it is relatively inexpensive, safe and environmentally friendly [3]. In Poland According to General Statistic Office data for 2013, the distribution and transmission network of natural gas pipelines length was over 140 000 km [4]. In the Lodz Province the length of gas pipeline system was only 4 158.9 km [5]. The existing gas connections allow to reach over 6.6 million individual, industrial and wholesale customers in total. Even with ongoing investments in the construction of new gas lines, they are not able to compensate the aging transmission grid. This results in the fact that about 60% of the operating gas pipelines are 30 years old [6].

Table 1. Comparison of greenhouse emissions generated by different types of fuel during production of energy [7].

Fuel	CO_2	CO	NO_x	SO_2	Dust
	kg/GJ				
Coal	120	1.096	0.177	0.840	1.003
Fuel oil	82	0.034	0.110	0.132	0.005
Natural gas	63	0.015	0.054	0.000	0.009

By now, steel remains the basic construction material of the pipes in the gas transmission lines [8]. Currently in use are both alloyed and unalloyed steel pipes. Their composition is strictly regulated by both Polish and European standards, which determine the contend of alloying elements (such as chromium, nickel, manganese, tungsten, copper, molybdenum, titanium). However, production processes result in presence of unwanted impurities, such as nitrogen, phosphorus,

* Corresponding author: 800075@edu.p.lodz.pl

 DOI: 10.1051/e3sconf/20171903002

sulfur and sulfur oxides.

Steel composition, together with aggressive factors of soil on the underground gas pipelines at large depths makes possible an accelerated external corrosion. Moreover, the humidity, constant temperature and composition of the flowing gas, create a convenient place for the development of microorganisms and, consequently, the biodeterioration of internal surface of the gas pipelines.

In practice, gas systems are protected from outside mainly by coatings (i.e. polyethylene and bitumen) or by cathodic protection in accordance with Regulation of the Minister of Economy of July 30, 2001 on technical conditions of the gas networks [9]. However, those methods do not eliminate the problem of corrosion which is induced by microorganisms on the internal surfaces of pipelines [10].

It is commonly known that microorganisms are able to colonize nearly all habitats. Into the gas networks they can be introduced during replacement or repairs of pipelines as well as during construction of new sections of the network. At that time microorganisms inhabiting ecosystems such as soil, water and air, environmentally non-specific natural gas such as *Bacillus, Clostridium, Pseudomonas* can colonize gas pipielines. These microorganisms may origin also from natural gas deposits or storage areas, where microbes are transported along with the tonnage of fuel.

Microbiological studies of natural gas wells indicate a wide variety of microbial communities, especially in unconventional gas resources [11-13]. Identification of the microorganisms found in the underground gas storage (UGS) indicates the biological origin of natural gas. In consequence they are also be present in the transported gas [14,15]. Some of microorganisms inhabiting the drilling mud or fracturing fluids, belonging to the phylum *Acidobacteria, Actinobacteria, Cyanobacteria, Proteobacteria, Verrucomicrobia* can grow up in the natural gas environment [16].

Lack of direct access to the natural gas transmission lines hinders their maintenance and revision. Changes resulting from the development of microorganisms inside gas pipelines cause disturbances in the correct operation of the gas system. In addition, they reduce the mechanical properties of the materials and even cause their biodeterioration [17].

Changes caused by the presence of microorganisms may lead in extreme to rupture of the pipeline and rapid gas leak, which generates both danger for gas users and the great losses in the industries using gas as the basic raw material.

The physico-chemical conditions, combined with the properties of the microorganisms themselves, their adaptive and defensive capabilities promotes the development of microbial corrosion in an anoxic environment [18-20]. Metabolic activity of microorganisms can start and/or increase speed of the chemical and electrochemical processes in material resulting in corrosion. This phenomenon is called Microorganisms Inducted Corrosion (MIC) or biodeterioration. Deposits loosely bonded with the surface of the metal, growths, pitting and loss of material are typical symptoms of the MIC occurring inside of the pipelines for natural gas transmission [21].

Depending on the type of microorganisms the corrosion deposits may be of different color and odor. For example presence of the iron bacteria results in reddish brown deposits, whereas "black powder", often accompanied by the smell of hydrogen sulphide, it is indicate the sulphate-reducing bacteria involvement [22, 23]. This deposit (sediment) primarily consists of oxides and hydroxides of iron, siderite ($FeCO_3$) and impurities such as elemental sulfur, hydrocarbons, metal debris, sand and dust. It yields the problems for gas pipeline operators in the dry and wet form, as it accelerates erosion of the gas pipelines materials. In addition, the accumulation of "black powder" leads also to clogging and reduction in pipeline throughput.

The losses generated in the gas supply sector due to the growth and properties of microorganisms are not easy to estimate. It has been estimated that about 40% of all corrosion in the industry is biological-origin corrosion [19,20]. Estimated damage caused by biological corrosion accounts for up to 20-30% of the total cost of corrosion of gas pipelines [24].

In addition, indirect costs associated with interruptions in installations and/or equipment operation increase the volume of corrosion damage.

The aim of the study was to provide the knowledge about the biological corrosion of the internal surface of natural gas pipelines. Investigations of the influence of commonly applied cathodic protection on the microbial growth in the natural gas pipelines can lead to new solutions for more effective protection against the corrosion. It should significantly decrease the costs generated by biocorrosion. This paper presents the influence of the microorganisms on materials constructing gas pipelines in the Lodz Province. The susceptibility of the selected materials to biocorrosion was investigated in the laboratory pipeline model reproducing conditions of the real gas pipeline. In addition, the influence of cathodic protection on the adhesion and growth of microorganisms in the environment of gas pipelines has been tested.

2 Materials and Methods

For this study it was possible to collect samples from the real pipelines during their dismantling (Z) or maintenance (S1 and S2) works carried out by the operator. Sampling sites were selected in agreement with an operator of the gas pipelines in the Lodz Province.

Samples were collected from two carbon steel pipelines (S1 and S2) in the Lodz province, what allowed to use in this study microbes naturally inhabiting a gas pipelines of this region.

The samples were cut out from two carbon steel (S1 and S2) and one cast iron (Z) pipelines. Characteristic of two type of carbon steel and one of type cast iron are presented in Table 2.

Preparation of the samples consisted of cutting material into rectangular plates with the active (adhesive) surface of about 1.5 cm^2. They were clean of existing corrosion.

 DOI: 10.1051/e3sconf/20171903002

Table 2. Chemical composition of the Z, S1 and S2 materials.

Material	C	Mn	Si	P	S	Cu	Fe
	Weight percentage of the elements [%]						
Z	3.59	0.48	3.15	0.92	0.082	-	91.778
S1	0.11	0.46	0.19	0.060	0.041	0.097	99.042
S2	0.10	0.48	0.10	0.012	0.013	-	99.295

They were clean of the existing corrosion. The entire surface of the cut pieces was purified in a digestion solution (PN-78 / H-04610), and additionally, the grinding surface was standardized by sanding on ASTM scale 180, 350, 600 grinding papers. Finally, the samples were degreased, sterilized, dried and weighed with accuracy of 10^{-4} g.

Corrosion resistance studies were carried out in a laboratory gas pipeline model for 4 months. Character of the corrosion process required long time of incubation of the samples, therefore the time interval between controls was 1-month long (28 days).

The laboratory set used in this experiment has been projected and constructed by authors. It was consisted of two parallel, cathodically protected and unprotected tubes of length 1 m and diameter 0.1 m each with an external power source (see Fig. 1).

a)

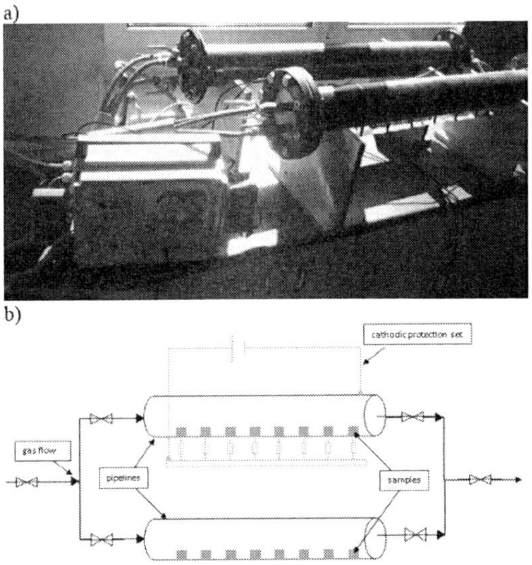

Fig. 1. Laboratory set used for experiment to conduct parallel research on corrosion with and without cathodic protection. (a) photo, (b) scheme.

The model was powered by natural gas without a microbiological filter, what allowed for free migration of the microbes in the system. Assumed parameters of the cathodic protection are in accordance with current cathodic protection criteria for underground construction without or with highly damaged surface protection. In this case the suggested maximal current density was 50 mA/m^2.

For observation and illustration of the changes in active surfaces of materials and bacterial cell adhesion to steel surfaces investigation the Scanning Electron Microscope S-3400N (Hitachi) was used.

In order to confirm the number of living cells in the sample the biochemical method using Portable luminometer type HY-LITE (Merck) was applied. The bioluminescent measurement of ATP (adenosine triphosphate) facilitated estimation the number of molecules of ATP produced during cellular metabolism. It should be noted that at the moment of the cells death, synthesis of ATP is inhibited and all already existing molecules are degraded by intracellular enzymes, ATPase and phosphatase.

Calculation of the corrosion rate of the sample was made by gravimetric method. The corrosion rate was determined directly from the loss of mass depending on the time of microbial activity per sample. Measurements were done with accuracy up to 0.1 mg using an analytical balance of AS 220/X (Radwag). The weight corrosion rate V [g/(m^2day)] was calculated according to formula (1).

$$V = \frac{m_0 - m_1}{S * t} \tag{1}$$

where: m_0 - initial mass of sample [g], m_1 - mass of sample after exposure [g], S-surface of sample [m^2], t-time of exposure [day].

3 Results and discussion

Samples used in this study differed in the content of alloying elements: C, Si, P and S, resulting in their the mechanical properties, including corrosion resistance. Nevertheless, macroscopic observation for both sets of samples (with and without applied cathodic protection) showed delicate black sediment on all active surfaces. After 4 months differences between corrosions occurring in steel (S1 and S2) and cast iron (Z) materials were observed. For S1 and S2 pits and deposits differing in their color from black to brown depending of time of exposure to corrosive conditions were visible, while for Z material corrosion products remained in the form of uniform black coating. However, the apparent differences between samples of the same material with and without presence the cathodic protection were impossible to observe in macroscopic scale.

Given that it was necessary to prove presence of the living microbes to confirm participation of the microorganisms in the occurring degradation processes. For this reason the ATP measurements were used and the obtained results are presented in Table 3. It should be noted all samples were sterile at the beginning of the experiments. It was assumed that all detected microbes were introduced with the gas. The increasing in time ATP production proved also that detected microbes were able not only to survive, but also to grow and develop in the natural gas environment.

DOI: 10.1051/e3sconf/20171903002

Table 3. Comparison of the ATP measurements results for samples incubated in systems with and without cathodic protection.

Material	Exposure time [day]	ATP [RLU]	
		with cathodic protection	without cathodic protection
Z	28	24	25
	56	21	23
	84	25	35
	112	32	37
S1	28	25	22
	56	23	21
	84	37	31
	112	46	32
S2	28	21	23
	56	23	25
	84	25	31
	112	35	34

Presence of the microorganism was also confirmed by SEM observations, where visible are structures that can be considered as adhering bacteria (Fig. 2).

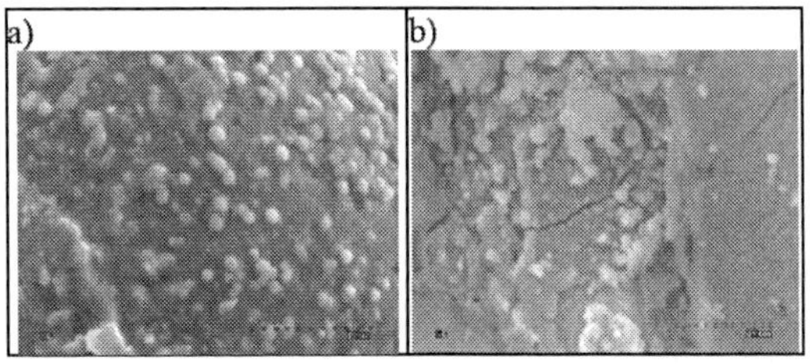

Fig. 2. Example of SEM image of adhered microorganisms on the surface of S1 material after 4 month cultivation in natural gas environment with cathodic protection (a) and without cathodic protection (b) (magn. 6000x).

Adhesion of microorganisms to metal surfaces stimulates microbial corrosion, which is a complex multi-stage metal degradation / degradation process. In the course of the ongoing destructive transformations, physical, chemical and biological agents have cumulative effect. Using the gravimetric method, the rate of biocorrosion of the materials tested was determined. Obtained results are presented in Tables 4-6.

According to the obtained results the highest rate of mass loss was observed after the first month of experiment and the lowest in the 4 month, what resulted in the time depending decreasing velocity of gravimetric. For all tested materials determined speed was lower for samples without cathodic protection.

All results obtained during 4 months of the experiment indicated that the highest rate of corrosion was for the cast iron. Calculated gravimetric corrosion velocity for cast iron was approximately 100 times higher than for other investigated materials.

Table 4. Mean values of corrosion rate for iron cast (Z) obtained during 4-month incubation of samples.

Material		*t* [day]	m_0 [g]	m_1 [g]	Δm [g]	V [g/(m²day)]
Z	With protection	28	6.3500	6.0034	0.3466	104.0116
		56	6.8359	6.3466	0.4892	67.7233
		84	6.9567	6.2316	0.7251	60.3674
		112	6.4787	5.7510	0.7276	51.9738
	Without protection	28	7.2636	6.9407	0.3228	80.0678
		56	6.5401	6.0696	0.4705	65.9999
		84	6.9552	6.2883	0.6669	54.0060
		112	7.1906	6.4274	0.7631	44.2447

Table 5. Mean values of corrosion rate for steel S1 obtained during 4-month incubation of samples.

Materia l		*t* [day]	m_0 [g]	m_1 [g]	m [g]	V [g/(m²day)]
S 1	With protection	28	5.7660	5.7602	0,0058	1,5033
		56	6.3861	6.3785	0,0076	0,8988
		84	6.7824	6.7705	0,0119	0,8720
		112	7.3371	7.3238	0,0134	0,6594
	Without protection	28	5.9769	5.9712	0,0057	1,3403
		56	6.7482	6.7394	0,0088	0,9524
		84	7.6246	7.6126	0,0120	0,7722
		112	6.6636	6.6497	0,0139	0,7568

Table 6. Mean values of corrosion rate for S2 material obtained during 4-month incubation of samples

Material		*t* [day]	m_0 [g]	m_1 [g]	m [g]	V [g/(m²day)]
S2	With protection	28	5.0348	5.0277	0.0071	1.4475
		56	5.7137	5.7073	0.0065	0.5774
		84	5.5244	5.5128	0.0116	0.7030
		112	5.9689	5.9562	0.0127	0.5466
	Without protection	28	4.4904	4.4851	0.0053	1.2134
		56	4.2578	4.2507	0.0071	0.8492
		84	4.5710	4.5611	0.0099	0.7230
		112	4.1325	4.1220	0.0105	0.6398

In the case of the iron, the highest mean gravimetric corrosion velocity V (104.0116 g/(m^{2}day)) was observed after first month of incubation, while the lowest value – 44.2447 g/(m^{2}day) was obtained for the fourth month of incubation for samples without cathodic protection.

The weight corrosion rates for both steel materials were similar with only slightly higher corrosion velocity for material S1. Similarly to the cast iron the highest corrosion velocities for both S1 and S2 materials were

 DOI: 10.1051/e3sconf/20171903002

observed in the first month of experiment for samples stored with cathodic protection (V=1.5033 g/(m^{2}day) and V=1.4475 g/(m^{2}day) respectively). After 4 months of experiment duration for S1 and S2 the lowest rate V=0.6594 g/(m^{2}day) and V=0.5466 g/(m^{2}day, respectively was obtained.

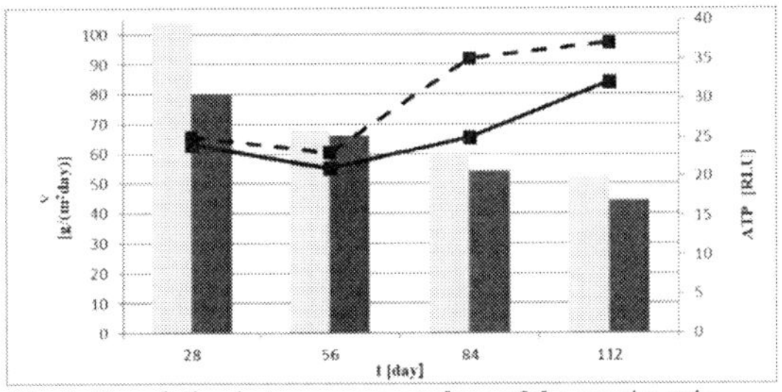

Fig. 3. Correlation between mean values of the gravimetric corrosion rate and ATP measurements for Z material with protection (light bar and continuous line, respectively) and without cathodic protection (dark bar and dashed line, respectively).

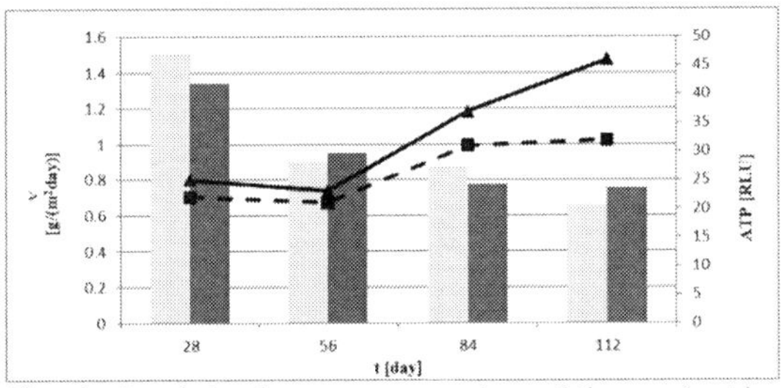

Fig. 4. Correlation between mean values of the gravimetric corrosion rate and ATP measurements for S1 material with protection (light bar and continuous line, respectively) and without cathodic protection (dark bar and dashed line, respectively).

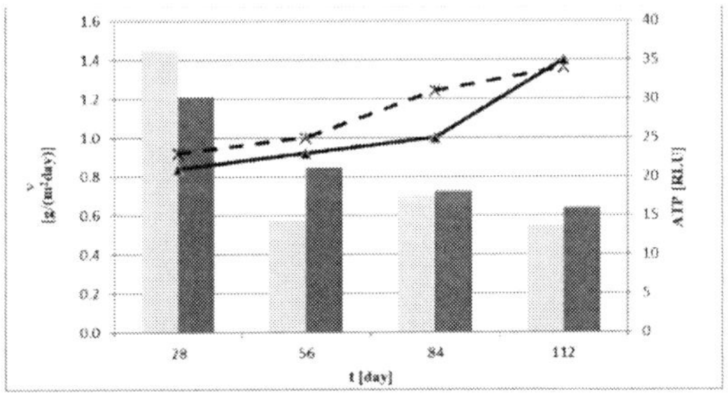

Fig. 5. Correlation between mean values of the gravimetric corrosion rate and ATP measurements for S2 material with protection (light bar and continuous line, respectively) and without cathodic protection (dark bar and dashed line, respectively).

In Figs. 3-5 correlation between corrosion velocity (V) and relative number of microbial cells (ATP) is presented. In all figures the following notation was used: samples with cathodic protection – bright bar for cathodic protection velocity and continuous line for ATP, and for samples without cathodic protection – dark bar and dashed line, respectively.

By comparing the values obtained so far it has been observed that with the development of microorganisms, the weight corrosion rate at which corroded materials are tested is reduced. At the same time, the growth of microorganisms on the studied surfaces was observed in the lag phase.

4 Conclusions

The research results allow for better understanding of the process of biocorrosion in materials used for the construction of gas pipelines in the Lodz Province. Experiments performed revealed that microorganisms involved in biocorrosion of the internal surfaces of the gas lines can migrate with gas flow. Applied ATP measurement and SEM observations for investigation of the bacterial growth rate indicated nearly constant rate of adhesion and growth of microorganisms. It was also noticed that the corrosion rate was decreasing in time. It could be due to the biofilm formation, which in turn could block access to the pipeline surfaces for other corroding agents. The highest gravimetric rate of biocorrosion was observed for cast iron. It is suspected that the observed nearly 100 times higher rate of biocorrosion was caused by much higher content of the carbon in the cast iron. In addition, the results suggest that unsuitable parameters applied for cathodic protection might stimulate the growth of microorganisms and catalyze the biocorrosion.

However, the conclusions based on present research results should not be treated as general concerning all gas pipelines. It is suggested that for better understanding of the biocorrosion processes in the gas lines there should be carried out studies taking into account also different materials and different cathodic protection parameters.

References

1. P. Bućko, Rynek Energii **3**, 17 (2015)
2. M. Kaliski, S. Nagy, S. Rychlicki, J. Siemek, A. Szurlej, Górnictwo i Geologia **5**, 27 (2010)
3. Sankara Papavinasam, *Corrosion Control in the Oil and Gas Industry* (Elsevier, Amsterdam, 2014)
4. General Statistic Office, http://www.stat.gov.pl
5. Statistical Office in Lodz, http://lodz.stat.gov.pl
6. M. Kaliski, S. Sikora, A. Szurlej, P. Janusz, Nafta-Gaz **2**, 125 (2011)
7. http://www.slawno.pl/files/0/20956/Dlaczego%20gaz%20ziemny.pdf (13.05.2017)
8. Marzena Więcek, *Wykonywanie i eksploatacja sieci gazowych,* 311[39].Z1.03. (Wyd. Instytut Technologii Eksploatacji – Państwowy Instytut Badawczy, Radom 2007)
9. Regulation of the Minister of Economy of July 30, 2001 on technical conditions of the gas networks [Dz. U. 2001 No. 97 pos. 1055]
10. J. Rządkowski, Rurociągi **34**, 12 (2004)
11. J.P. Davis, C.G. Struchtemeyer, M.S. Elshahed, Microb. Ecol. **64**, 942 (2012)

12. H. Mochimaru, H. Yoshioka, H. Tamaki, K. Nakamura, N. Kaneko, S. Sakata, H. Imachi, Y. Sekiguchi, H. Uchiyama, Y. Kamagata, Extremophiles **11**, 453 (2007)
13. H. Yoshioka, H. Mochimaru, S. Sakata, H. Takeda, S. Yoshida, Chemical Geology **419**, 92 (2015)
14. A.E. Ivanova, I.A. Borzenkov, A.L. Tarasov, E.I. Milekhina, S.S. Belyaev, Microbiology **76**, 461 (2007a)
15. A.E. Ivanova, I.A. Borzenkov, A.L. Tarasov, E.I. Milekhina, S.S. Belyaev, Microbiology **76**, 453 (2007b)
16. C.G. Struchtemeyer, P.D. James, M.S. Elshahed, Appl. Environ. Microbiol. **77**, 4744 (2011)
17. T.S. Khan, M.S. Al-Shehhi, J Nat Gas Sci Eng. **25,** 66 (2015)
18. J. Jan-Roblero, J.M. Romero, M. Amaya, S. Le Borgne, Appl Microbiol Biotechnol. **64**, 862 (2004)
19. J. Jan-Roblero, A. Posadas, J. Zavala Díaz de la Serna, R. García, C. Hernández-Rodríguez, World J. Microbiol. Biotechnol. **24**, 1775 (2008)
20. A. Rajasekar, B. Anadkumar, S. Maruthamuthu, Y.P. Ting, P.K.S.M. Rahman, Appl. Microbiol. Biotechnol. **85**, 1175 (2010)
21. Ch.I. Ossai: Advances In Asset Management Techniques: An overview of corrosion mechanisms and mitigation strategies for oil and gas pipelines. International Scholarly Research Network ISRN Corrosion, DOI: 10.5402/2012/570143 (2012)
22. Bronisław Zyska, Zofia Żakowska, *Mikrobiologia materiałów* (Wyd. Politechniki Łódzkiej, Łódź, 2005)
23. A. El-Sherik, J. Pipeline Eng. **1**, 19 (2016)
24. F.M. AlAbbas, J.R. Spear, A. Kakpovbia, N.M. Balhareth, D.L. Olson, B. Mishra, J. Pipeline Eng. **1**, 63 (2012)

E3S Web of Conferences **19**, 03003 (2017)
EEMS 2017

DOI: 10.1051/e3sconf/20171903003

Assessment of porous material anisotropy and its effect on gas permeability

Grzegorz Wałowski[1,*]

[1]Institute of Technology and Life Sciences, Poznan Branch, Department of Renewable Energy Sources, street Biskupińska 67, 60-463 Poznań, Poland,

Abstract. The results of experimental research upon the assessment of porous material anisotropy and its effect on gas permeability of porous materials with respect to the gas flow. The conducted research applied to natural materials with an anisotropic gap-porous structure and - for comparative purposes - to model materials such as coke, pumice and polyamide agglomerates. The research was conducted with the use of a special test stand that enables measuring the gas permeability with respect to three flow orientations compared with symmetric cubic-shaped samples. The research results show an explicit impact of the flow direction on the permeability of materials porous, which results from their anisotropic internal structures. The anisotropy coefficient and permeability effective coefficient of such materials was determined and an experimental evaluation of the value of this coefficient was conducted with respect to the gas stream and the total pressure drop across the porous deposit. The process of gas permeability was considered in the category of hydrodynamics of gas flow through porous deposits. It is important to broaden the knowledge of gas hydrodynamics assessment in porous media so far unrecognised for the development of a new generation of clean energy sources, especially in the context of biogas or raw gas production.

1 Introduction

The porous medium is a natural or artificial material characterised by a vast number of channels and empty spaces with relatively dimensions in comparison with the weight of the material itself. Empty spaces, irrespective of their shape, are called pores whose form and size are affected by a kind of granular deposit. The porous structure is characteristic for both natural and artificial materials, including soil, pumice (volcanic soil), ceramic construction materials, metal agglomerates and the so-called metal foams. Porous media are divided into two fundamental groups, viz.:

a) porous media constituting loose deposits comprising granular materials that take the space resulting from the geometrically-shaped volume - Fig. 1a;

b) frame-based media that are a solid and stiff porous structure in the form of the homogeneous material deposit - Fig. 1b.

In both cases, in a particular porous medium there usually occur pores with diversified and irregular shapes that most frequently create the gap-porous space as illustrated in Fig. 1b. For circular-shaped pores (channels) or similar ones the characteristic dimension of pores is the diameter, whereas for gap-shaped channels this is width. At the same time, the size of pores and flow gaps determines the structure of porous materials.

Pursuant to the guidelines of the International Union of Pure and Applied Chemistry (IUPAC) [3] there is applied the classification of gap-porous spaces set forth in Table 1.

a) b)

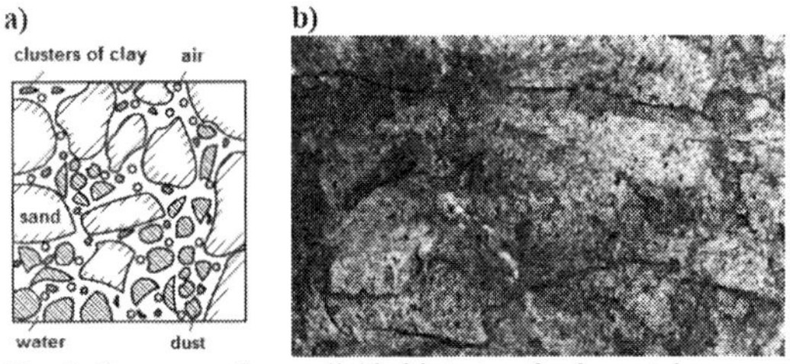

Fig. 1. Porous media are divided into two fundamental groups: a) scheme of natural soil structure [1], b) gap-porous structure of coal char [2].

As for the permeability of liquids porous media may be considered in two ways [4], viz.:

a) isotropic, if deposit filtration features have identical features in all directions of the flow;

b) anisotropic, if deposit filtration features are diversified compared to the selected direction of the flow.

The first instance proves the homogeneous structure of the porous deposit, whereas in the latter one in a specific volume there occurs a diversified directional structure of pores and flow channels. The example of isotropic and anisotropic structures is shown in Fig. 2. The fundamental characteristic of each porous material is its porosity as the dimension of part of deposit volume occupied by pores, i.e. free space [5, 6].

* Corresponding author: g.walowski@itp.edu.pl

Table 1. Classification of pores and gaps as per [3].

Pores		Gaps	
effective diameter *) d_{ef}, mm		width s, mm	
micro-pores	$d_{ef} \leq 2\ 10^{-3}$	sub-capillar	$s < 1\ 10^{-3}$
meso-pores	$2\ 10^{-3} < d_{ef} \leq 0.5$	capillar	$0.25 < s < 1\ 10^{-3}$
macro-pores	$d_{ef} > 0.5$	over-capillar	$s > 0.25$
*) pore dimension in the effective flow cross-section			

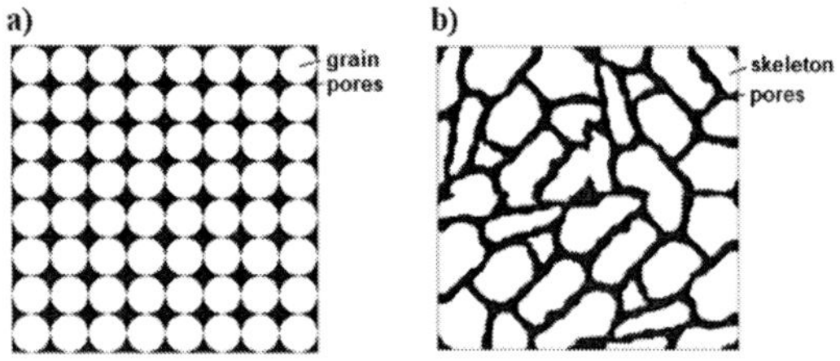

Fig. 2. Porous structure: (a) isotropic, (b) anisotropic (black: free space; white: frame-based structure).

Definition of porosity (1) is:

$$\varepsilon = \frac{V_o}{V} \quad (1)$$

where: ε - porosity, V_o – apparent volume (value calculated for the full section of the bed), m^3; V – volume, m^3;

The porosity is not generally affected by dimensions of particles of the porous material but it is considerably subject to the shape of grains (pores) of this material. The example is shown in Fig. 3 – for the isotropic deposit comprising circular elements of the identical size, the porosity changes from the maximum value (2):

$$\varepsilon_{max} = \frac{V_o}{V} = \frac{\left(d_p^3 - \frac{\pi d_p^3}{6}\right)}{d_p^3} = 0,4764 \quad (2)$$

where: ε_{max} – maximum porosity; V_o – apparent volume (value calculated for the full section of the bed), m^3; V – volume, m^3, dp – diameter of the spherical particle, m; for the regular cubic arrangement (Fig. 3a), up to the value of ε=0.2595 - for the rhombohedral structure "sphere on three spheres". Obviously, for real structures (natural materials) their porosity may exceed the maximum value and fall below the minimum value and the more the deposit structure differs from the isotropic one the greater the deviation is.

In can be assumed that the volume of free spaces directly proves the quantity (volume) of liquid to be stored in a particular deposit.

On the other hand, the pore arrangement structure and the shape of channels and of free space affect the deposit permeability.

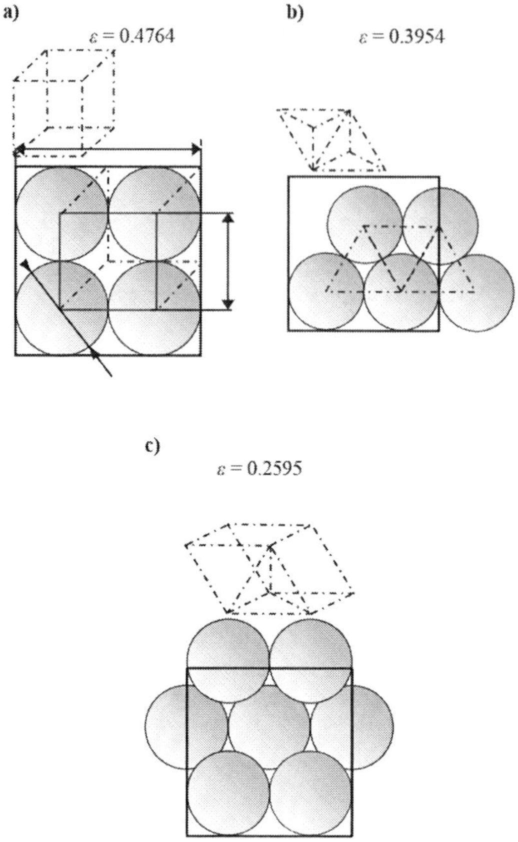

Fig. 3. Examples of isotropic structures of the porous materials: a) cubic, b) rhomboidal, c) rhombohedral [7].

In each case, this permeability which is the liquid stream going through the layer of the porous deposit is affected by the influence of pressure forces [6,8,9]. Of huge significance is whether the liquid flows because it is under the influence of gravitational forces or it is additionally forced by external forces due to the generated pressure.

2 Materials and method

The permeability research was conducted upon a number of diversified types of materials, the average porosity of which ranged from 22% to 56%. Most of them were coal chars (cokes) from the thermal processing of hard coal and there are also materials like partially melted waste rocks (including volcanic ones), natural and synthetic pumice and porous agglomerate. The research material comprised various types of solid frame structures thoroughly analysed in the study by Wałowski and Filipczak [10].

To obtain the research objective, the detailed experimental tests were conducted to assess of porous material anisotropy and its effect on gas permeability of porous materials with the diversified structure and the diversified process characteristics [2].

The research was conducted on a specially-designed stand [11], the fundamental component of which is the flow channel inside which the porous material sample is inserted, Fig. 4a.

a) **b)**

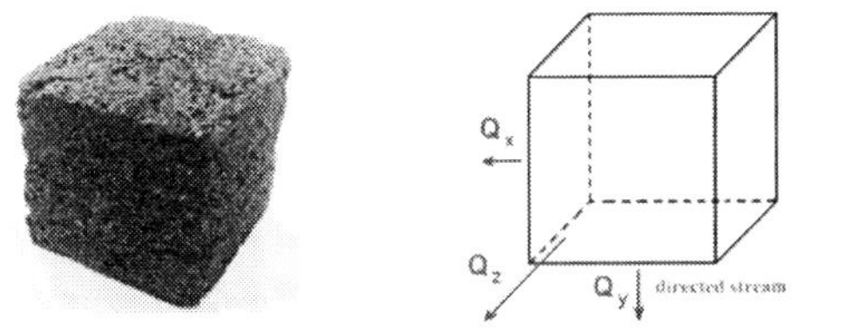

Fig. 4. Sample of research material [10]: a) 20 x 20 x 20 mm porous material, b) diagram of the gas flow through the sample.

The tested samples were cubic and the structure of the flow channel enabled measuring the gas permeability for each of the main flow directions (X, Y, Z) (Fig. 4b) through rotating the cubic sample in the selected plane of the measuring cell [10]. The gas permeability research was conducted by using air as a working medium.

The permeability measure was assumed to be a gas volume flow resulting from the allowable differential pressure forcing the gas flow on a given axis on the porous material sample.

3 Results and discussion

The significant feature of the porous deposit that directly results from its structure is anisotropy of the porous material and its influence on its permeability. As for its definition this feature is identified with the so-called coefficient of anisotropy [12,13,14], which is described as the root of the quotient of the permeability coefficient determined with respect to the mutually perpendicular flow planes (3):

$$\beta = \sqrt{\frac{K_h}{K_v}} \qquad (3)$$

where: β – anisotropy coefficient; K_h - coefficient of horizontal permeability, m^2; K_v - coefficient of vertical permeability, m^2;

In specific hydrodynamic conditions this coefficient deviates from the gas flow uniformity that results from a particular flow direction - an increase in the value of this coefficient leads to a decrease in deviation from the anisotropic structure of the porous deposit.

For the horizontal plane (Fig. 5) the permeability coefficient K_h (4):

$$K_h = \sqrt{K_X K_Z} \qquad (4)$$

where: K_h - coefficient of horizontal permeability, m^2; K_X - coefficient of permeability in the X direction, m^2; K_Z - coefficient of permeability in the Z direction, m^2;
is most frequently determined as the geometrical average from the value of the permeability coefficient in two horizontal directions, i.e. K_X and K_Z.

On the other hand, the permeability coefficient for the vertical plane K_v (5):

$$K_v = K_Y \qquad (5)$$

where: K_v - coefficient of vertical permeability, m^2; K_Y - coefficient of permeability in the Y direction, m^2;
is the value resulting from the flow of gas towards the Z axis

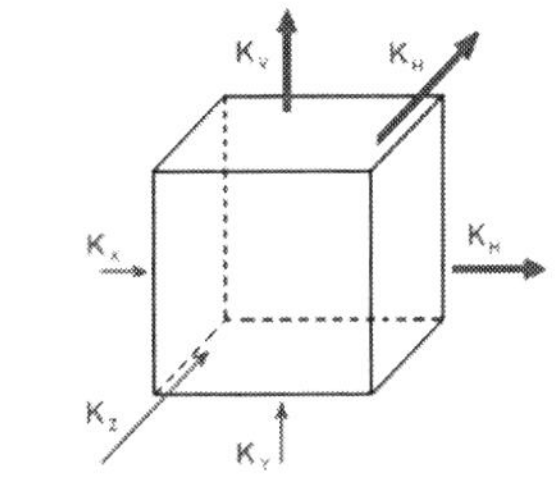

Fig. 5. Marking of flow directions and their referring calculation planes of permeability coefficient - horizontal K_h and vertical K_v.

The anisotropic structure characteristic for the porous material together with the graphical model of heterogeneity of free spaces are shown in Fig. 6., and illustrated by coal char in situ. Although this image may indirectly be used to evaluate the effective porosity but it is not sufficient to describe anisotropy on a quantitative basis as it only relates to the top layer of the material, whereas the real measurement of anisotropy is the potential permeability of the deposit.

a)

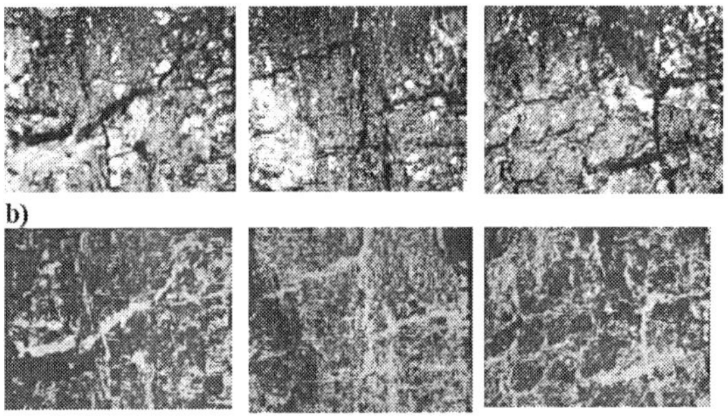

X direction Y direction Z direction

Fig. 6. Image of surface of coal char in situ with diversified spatial structure.

Therefore, it is necessary to determine directional permeability coefficients as shown in Fig. 7. In each case, however, anisotropy is considerably affected by structural features of the porous material.

The research results characterising the coefficient of anisotropy for various porous materials with respect to their permeability are shown in Fig. 8. This figure is a kind of map that identifies the material structure with respect to its permeability.

It may be noticed that for the porous polyamide agglomerate, anisotropy does not practically occur (6):

$$\beta = 1 \qquad (6)$$

where: β –anisotropy coefficient.

For the remaining materials (coke, pumice, coal char ex situ) relatively much smaller values of the coefficient of anisotropy prove the strong influence of the flow direction on their permeability but, at the same time, it does not affect the intensity of gas permeability - the coefficients of anisotropy have similar values throughout the entire range of the measurement.

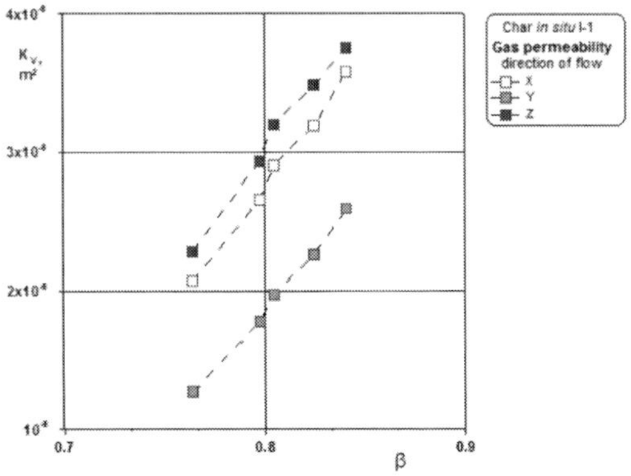

Fig. 7. Gas permeability of coal char in situ as function of material anisotropy.

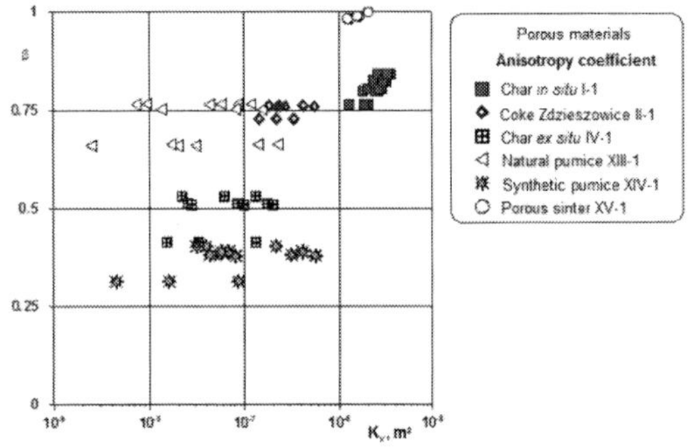

Fig. 8. Coefficient of anisotropy referred to the entire permeability of porous materials.

This may prove the repeatable structure of this group of materials but their hydrodynamic characteristics are closely related to this structure.

On the other hand, the measurement results shown in Fig. 9 (average values) prove a falling tendency for changes to the coefficient of anisotropy with respect to the effective porosity (7):

$$\varepsilon_{ef} = (0.5 \div 0.8)\varepsilon_b \tag{7}$$

where: ε_{ef} - effective porosity; ε_b - absolute porosity.

Three structural groups of materials may be classified: The first (independent) group consists of porous agglomerate (polyamide) with an effective porosity of around 25% and a high coefficient of anisotropy, undoubtedly due to the homogeneous structure of this material with the constant dimension and curvature of micro-channels in the whole volume.

Another group contains materials resulting from the thermal carbon processing - coal chars and coke. Their effective porosity is around 30-40% and the coefficient of anisotropy ranges from 0.5 to 0.8.

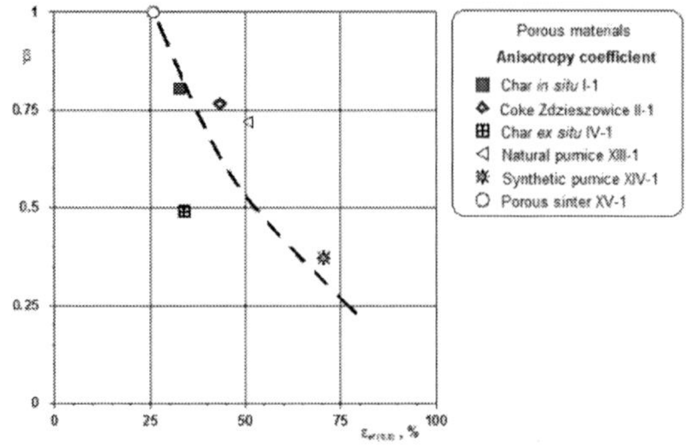

Fig. 9. Coefficient of anisotropy referred to effective porosity.

Coal char in situ has the much more gap-shaped structure whose dimensions of micro-channels are much greater compared to coal char ex situ which has a lot of small and meandering micro-channels with variable diameters, that are sometimes connected with micro-gaps. On the other hand, coke whose structure is considerably different is moderately anisotropic (at the level of coal char). This is, undoubtedly, affected by the meandering structure of coke channels with a greater dimension of pores compared to coal chars but, at the same time, those pores are frequently blind and closed for the flow. The third group is pumices with the effective porosity of 50-70%. Their structural features considerable differ from one another but the anisotropy of natural pumice is moderate (at the level of coke), whereas synthetic pumice has the smallest level of anisotropy. Undoubtedly, this is caused by the directional structure of this material obtained in the process of blowing pumice foam.

Similar conclusions are drawn by analysing the level of anisotropy of porous materials with respect to the so-called coefficient of effective permeability[15]. It is described by dependency (8):

$$K_{ef} = \sqrt{K_h K_v} \tag{8}$$

where: K_{ef} - coefficient of effective permeability, m^2; K_h - coefficient of horizontal permeability, m^2; K_v - coefficient of vertical permeability, m^2;
this coefficient characterises average conditions resulting from gas permeability in the horizontal and vertical flow configuration.

The results for the same materials as for the previous once sorted on the basis of this coefficient of anisotropy are shown in Fig. 10. Those results prove that the materials with the excessive fissuring such as coal chars in situ have a considerably anisotropic structure and a very good permeability. On the one hand, it proves that this group of coal chars is characterised by considerable fissuring but, on the other hand, is has a relatively large free area for the gas flow in comparison with the porous polyamide. On the other hand, the materials such as coke, natural pumice and coal char ex situ belonging to the group of materials with moderate anisotropy have relatively limited permeability. The very opposing anisotropy is characteristic for synthetic pumice whose structure restricts gas flow conditions with the aforesaid consequences.

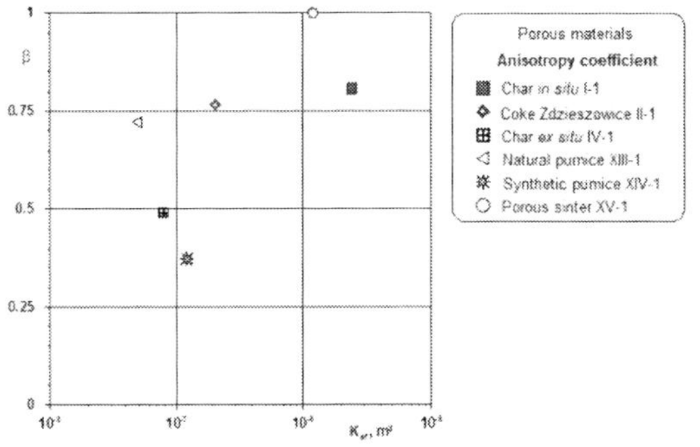

Fig. 10. Coefficient of anisotropy of porous materials - average values.

The distribution of experimental points show in Fig. 9 and Fig. 10 proves that the scale of anisotropy of the porous materials may be very large, which depends on the material structure and its effective porosity.

4 Conclusions

The conducted experimental research results in the following conclusions.

It has been proven that conditions of the quantitative assessment of hydrodynamics are affected by such characteristic parameters of the porous deposit as anisotropic properties of its structure. It considerably affects the gas permeability of the tested materials, which is considered in the theoretical assessment of this hydrodynamics;

The experimental evaluation of flow factors resulting from effective permeability of the porous deposit makes it possible to materialise the gas flow conditions, which is essential for the practical use of the measurement results.

The results of our own research quite well correlate those features in different process aspects resulting from the source models.

The research results on hydrodynamics of gas flow through frame-based porous deposits, and the proved usefulness of the process assessment of that research may, in many cases, be practically used to immobilise of methanogenic microorganisms in adhesive deposits

The work done in the framework of a project funded by the National Research and Development Centre implemented the program BIOSTRATEG, contract No. BIOSTRATEG2 / 298357/8 / NCBR / 2016 dated 05.18.2016 r Warsaw

References

1. Z. Wiłun Z, *Zarys geotechniki [Geotechnical outline]*, (WKiŁ, Warszawa, 1987)
2. G. Wałowski, *Hydrodynamika przepływu gazu przez złoże porowate: praca doktorska [Hydrodynamics of gas flow through porous deposits: doctoral thesis]*, Politechnika Opolska, Opole (2015)
3. J. Demichowacz-Pigoniowa, *Obliczenia fizykochemiczne, Termodynamika chemiczna i nauka o fazach, [Physicochemical calculations, Chemical thermodynamics and phases learning]*, Wrocław (1975)
4. T. Strzelecki, S. Kostecki, S. Żak, *Modelowanie przepływów przez ośrodki porowate [Modeling flows through porous media]*, (Dolnośląskie Wydawnictwo Edukacyjne, Wrocław, 2008)
5. G.A. Aksielrud, M.A. Altszuler, *Ruch masy w ciałach porowatych [Movement of mass in porous bodies]* (WNT, Warszawa, 1987)
6. A. P. Mozhaev, J. Eng. Phys. Thermophys. **75**, 2, 371-383 (2002)
7. J. Hapanowicz, Przepływ ciekłych układów dyspersyjnych. Studia i monografie, z. 204 [Flow of liquid dispersion systems. (Studies and monographs, no. 204, Politechnika Opolska, Opole, 2007)
8. K. Liszka, Podstawy eksploatacji złóż ropy. Skrypty uczelniane 869, [Basics of exploitation of oil fields. (University scripts 869, AGH, Kraków, 1982)
9. G. Wałowski, G. Filipczak, Inżynieria i Aparatura Chemiczna, **51**, 6, pp. 396-397 (2012)
10. G. Wałowski, G. Filipczak, J. Sustainable Min. (2016) http://dx.doi.org/10.1016/j.jsm.2017.03.003
11. G. Wałowski, G. Filipczak, E. Krause, *Hydrodynamika przepływu gazu w porowatych strukturach karbonizatu w aspekcie technologii podziemnego zgazowania węgla. Młodzi dla Techniki 2013. Young for Technic 2013*, Politechnika Warszawska, Płock, pp. 253 (2013)
12. J. Lu, D. Tiab, F.H. Escobar, J. Eng. Appl. Sci. **8**, 6, pp. 425-434, (2013)
13. E.E. Oaikena, J. Oloro, *Greener* J. Phys. Sci. **3**, 3, pp. 97-109, (2013)
14. A. Olufemi Bamidele, A.F. Olugbenga, O. Olusegun, Petroleum & Coal **51** 2, pp. 91-99, (2009)
15. A. Okwananke, S.O. Isehunwa, Mat. Conference: Analysis of water cresting in horizontal wells. Society of Petroleum Engineers Inc., 32nd Annual SPE International Technical Conference and Exhibition in Abuja, SPE 119733, pp. 1-10, Nigeria, August 4-6 (2008)

DOI: 10.1051/e3sconf/20171903004

Forming of film surface of very viscous liquid flowing with gas in pipes

Krystian Czernek[1,*] and *Stanisław* Witczak[2]

[1]Opole University of Technology, Faculty of Mechanical Engineering, Department of Process Engineering, 5 Mikołajczyka Street, 45-271 Opole, Poland
[2]Opole University of Technology, Faculty of Mechanical Engineering, Department of Process Engineering, 5 Mikołajczyka Street, 45-271 Opole, Poland

Abstract. The study presents the possible use of optoelectronic system for the measurement of the values, which are specific for hydrodynamics of two-phase gas liquid flow in vertical pipes, where a very-high-viscosity liquid forms a falling film in a pipe. The experimental method was provided, and the findings were presented and analysed for selected values, which characterize the two-phase flow. Attempt was also made to evaluate the effects of flow parameters and properties of the liquid on the gas-liquid interface value, which is decisive for the conditions of heat exchange and mass transfer in falling film equipment. The nature and form of created waves at various velocities were also described.

1 Introduction

Two-phase gas- very viscous liquid flow is quite common in technology applied in the up-to-date industry, both with regard to transport processes as well as thermal processing of liquid substances. The concurrent involvement of both gas and liquid phases with a very viscous in such systems results in a number of difficulties in ensuring adequate operating conditions for various types of apparatuses (heat exchangers, evaporators, reactors) in which the processes of exchange of momentum, heat and mass are realized. The basic reasons for such circumstances are associated with large stochastic characteristics of two-phase gas-liquid flow. Additionally, depending on the streams of the gas phase and liquid phases and the volume fractions of the liquid and solid phase various flow structures and forms of flow are encountered there. The process conditions corresponding to these structures constitute the fundamental obstacle in the process of ensuring maintenance adequate flow regimes and, hence, preservation of process parameters along the course of the flow.

In these circumstances important role is attributed to such solutions that ensure adequate operation of process apparatus within a maximum time frame. Concurrently, the solutions applied have to eliminate the disadvantageous phenomena associated with the course of the processes (instability of flow, origin of residue on the surfaces of the heating pipes in exchangers and evaporators, bubbly structures on liquid surface, instability of the interfacial surface, etc.). One of the ways in which solutions are sought is by indication of such solutions that enable the processes being realized in strictly defined flow regime, and, consequently, by indication of the possibility of conducting the processes of heat exchange within an most effective range of flow regimes.

The focus in this paper is on the issues associated with the possibilities of applying annular form of two-phase gas-liquid flow as one of the most effective ways in performing heat exchange processes, in particular with regard to heat and mass with a concurrent possibility of using substances whose physical properties are variable in the course of these processes. Phenomenological assessment of this type of flow is related to annular flow of gas and viscous liquid, whereas the latter term refers to such liquid in which the dynamic viscosity coefficient (at the temperature of 20°C) exceeds the corresponding value for water.

The thin layer of liquid, also known as film, may be formed in various manners. For instance, in terms of the distribution of the liquid over the heating surface of evaporative apparatus, basic three manners are distinguished by Czernek [4]: static, mechanical and hydraulic.

In the consideration of the extensive range and various application of two-phase gas-liquid annular flow and, consequently, extensive application of various forms of it in apparatuses that apply patterns encountered in this type of flow, this paper focuses on the analysis of applicability of annular gas-liquid flow with regard to the most comprehensive and verifiable range of liquid viscosities possible to realize in original studies.

In the literature in this subject Czernek et al. [3, 5], Czernek [2, 4], Xu et al. [15], Oriol et al. [10], Jassim et al. [9] the items devoted to two-phase gas-very viscous liquid flow are relatively scarce despite the wide range of existing applications of such systems. The results

*Corresponding author: k.czernek@po.opole.pl

presented up-to-date focus on examinations and descriptions of the phenomena occurring during flow of model mixtures, mostly involving air and aqueous solutions of sugar and polymers or oily liquids with relatively low viscosity. These papers involve identification and descriptions of flow structures, flow regime maps, pressure drops and determination of volume fractions of the particular components of the mixture, as well as thickness of liquid films and characteristics of their wavy motion. These issues are typical for the description of hydrodynamics of the two-phase gas-liquid streams. A fair share in this research has been conducted by Department of Process Engineering, Opole University of Technology, where the work mainly focuses on two-phase gas-liquid flow of liquids with the viscosity of over 100 mPa·s, presented in papers Czernek et al.[3], Troniewski et al. [12], Troniewski et al. [13], Witczak et al. [14].

By definition, such research is conducted on systems containing two media, i.e. for air-oil mixture flow, which in most cases comes down to assessment of hydrodynamics of such flow with emphasis of the various aspects of it. However, due to the characteristics of the liquid phase (i.e. for a phase with a high viscosity), in such flow we have to do with different characteristics of the flow when compared with two-phase single-component gas-liquid. Primarily, this is manifested by the occurrence of non-homogeneity of structures during flow of two-phase two-component mixture, which determine different volume fractions of these phases. Consequently, this leads to various hydrodynamic conditions along the flow route.

2 Experiments

The diagram of the experimental set-up applied for testing annular downward flow of gas-very viscous liquid is presented in Fig. 1. The media applied in this research involve air and machine oil with various viscosity coefficients. Flow channels 12 and 13 (with the diameter of 22mm) were fed with compressed air from a central pressure installation 4. The airstream passed through a reducing valve 5, and was regulated by means of a battery of rotameters 11. Constant gas pressure was maintained through a reducing valve, which secured correct measurement of its stream. The air stream was subsequently directed into the feeding system in the front section of each of the channels that formed a system with a central nozzle. Liquid was fed from tank 2, which could hold around 80 litres of oil in total. The tank was equipped with a agitator and heaters (with the heating power of 8 kW) with a thermostat that ensured maintenance of a constant temperature. Oil was pumped into the flow channel by means of a gear pump driven by a DC motor, which enabled the control system to adapt rotational speed of the pump continuously to ensure an adequate feed of oil stream. From the pump the oil was directed to a battery of rotameters 10, which were used to assess the stream of this component of the mixture. Subsequently, the oil was pumped into the feed tank (the one containing gas phase) where the two-phase mixture was formed. The gas and liquid streams were selected in such ratios that ensured that annular flow of two-phase gas-liquid mixture was obtained at its output.

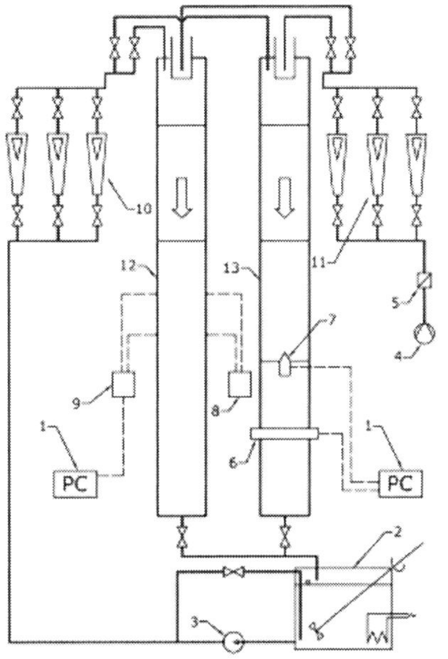

Fig. 1. Diagram of test stand: 1 – PC, 2 – oil tank with agitator and heater, 3 – oil pump, 4 – air compressor, 5 – reducing valve, 6 – laser knife,7 – camera, 8 – laser illumination, 9 – light detector, 10 – oil rotameter assembly, 11 – air rotameter assembly, 12 – measurement channel with laser illumination, 13 – measurement channel with laser knife.

In the same manner, other flow structures could be formed, which served for the purposes of comparison and search for the boundary between one structure and another. The measurement channel (12, 13) consisted of two parts – a non-transparent section and a transparent one. The non-transparent section (in the initial section of the pipe) ensured that stability of flow was obtained in the part of the channel, which served for the measurements. The latter was equipped with control and measurement instrumentation for optoelectronic evaluation of flow parameters 6÷9. By means of these optoelectronic sensor linked to as PC (with a card for data storage and processing), it was possible to measure the local thickness of the liquid films and characteristics of their waves. A linear measurement system was applied for these purposes (Fig. 2) consisting of four single-axis measurement probes distributed in various cross-sections of the channel.

These probes apply the principle of light spectrum dissipation when crossing an oil film. In turn, the optoelectronic assembly 6, 7 in this case formed a complementary part of the measurement system and applied the method of the so called light knife (Fig. 3), which enables the assessment of parameters that are specific for such flow in the cross-section of the channel. The mixture leaving the non-transparent section of the pipe was directed into the transparent section of the channel made of plexiglas. The observations regarding flow patterns were performed in this section and the

DOI: 10.1051/e3sconf/20171903004

structures are recorded by means of a digital video camera and a photo camera. Subsequently, the mixture passed on to the tank/separator 2, in which two phases are separated.

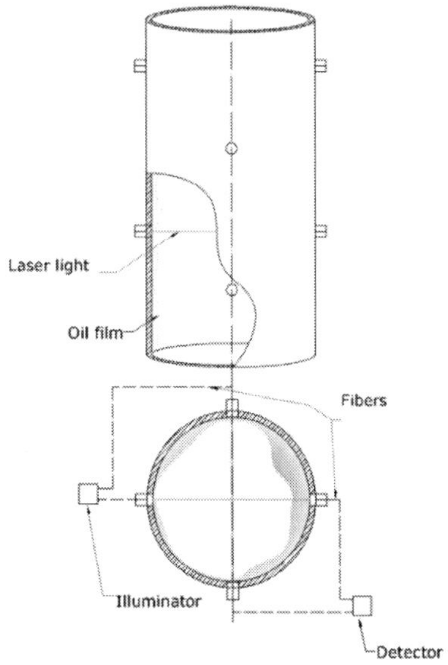

Fig. 2. Optoelectronic single-axis probe.

The applied measurement and control systems enable on to conduct research within a wide range of parameters of the flowing films. In the design phase of the research an attempt was made to select the range of the streams in a manner that ensure the occurrence of both laminar and turbulent flow in each of the measurement channels, for laminar liquid flow. Flow parameters were adapted throughout the research for air in the range$(1.8-384)\cdot 10^{-5}\,m^3/s$ and for oil $(2.92-965)\cdot 10^{-7}\,m^3/s$. The temperature of the mixture varied, accordingly, which was later accounted for during calculations of specific parameters.

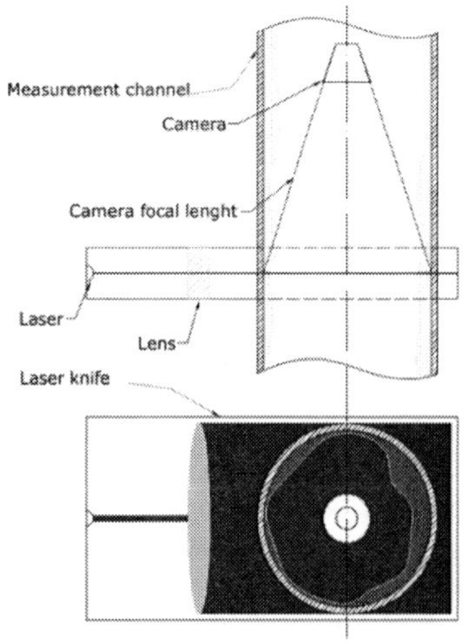

Fig. 3. Diagram illustrating the operating principle of laser knife.

The testing of the two-phase mixture flow were conducted in series for a dozen or so various air streams, which for each case was accompanied by several dozen measurements regarding oil streams for the pre-established flow parameters. During measurements conducted in this way simultaneous observations of the forming flow structures were made.

The circulating media applied in these hydrodynamic researches were air and oil. The rationale for this selection was associated with ensuring the possibility of gaining the highest possible value of dynamic viscosity coefficient. In order to obtain a wide range of oil viscosity, this research was performed for the range of temperatures (10-40)°C, for which the value of the dynamic viscosity coefficient varies in the range 90-3500 mPa·s. Low temperatures were obtained as a result of oil cooling until the liquidity state (around 10°C). The gas phase applied in the testing was air, which was extracted from an external compressed air installation. An adequate measurement of its pressure and temperature enable one to determine its physical properties indispensable for conducting further calculations.

Rotameters were applied to adjust air stream. Prior to testing they were calibrated to ensure that it would be possible to read the values of the air stream directly from rotameters. In turn, to determine the characteristics of oil streams it was necessary to calibrate rotameters to match the specific liquid viscosities depending on their temperatures (adiabatic flow), as this value has a considerable effect on the efficiency of rotameters.

A digital video camera or a photo camera to provide a more comprehensible documentation performed the registration of the structures.

3 Results and discussion

The results of tests were applied for classification of flow patterns and verification of the flow regime map.

A special characteristic of two-phase gas-very viscous liquid flow is associated with the fact that for the identified patterns of annular flow or patterns that are similar to annular flow in terms of spatial distribution of the liquid and gas phase (AL, AW, AH regimes, see Fig. 4), the observed thickness of liquid film is definitely smaller than the diameter of the gas root. The remaining regimes noted for this case essentially correspond qualitatively to forms described in the literature for liquids with much smaller viscosity of the liquid phase. This was indicated by Andritsos et al. [1], whose research was performed within the range of liquid viscosity variability up to 80 mPa·s. In particular, this pertains to the aspect of the smoothening effect of rising liquid viscosity on the geometric configuration of the resulting patterns.

These remarks are further confirmed by author's original experimental research Troniewski et al. [12], Czernek et al. [5] conducted within the range of liquid viscosities up to 3500 mPa·s. An example of registered flow structure is presented in Fig. 4.

Such a classification can be fully recommended as universal for the entire range of the occurrence of various flow patterns in a downward gas-very viscous liquid flow. In contrast to classification presented by Da Silva et al. [6], Geraci et al. [7], Hernandez et al. [8], Szalanski et al. [11] the novelty here involves the observation of a stalactite flow structure, in which the gas phase occurs in the liquid phase in the form of large bubbles with toroidal shape and a length that is comparable to the diameter of the channel and some of the liquid in the form of a thin stream is additionally transferred in the vicinity of the gas torus.

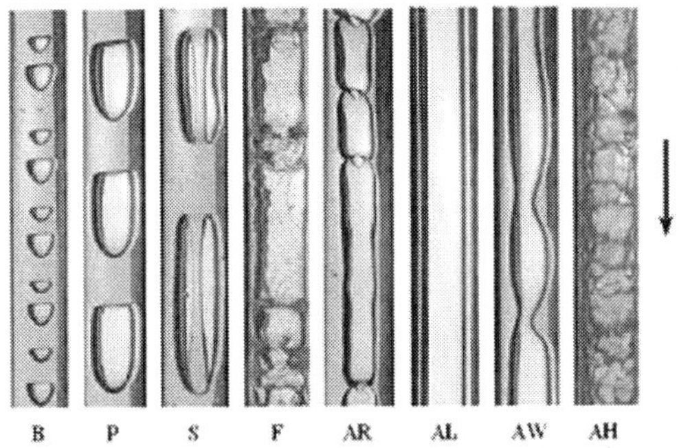

Fig. 4. Two phase gas-very viscous liquid downward flow structures: B – bubbly flow; P – plug flow; S – stalactite flow; F – foam flow;AR – annular core flow; AL – annular low wave flow; AW – annual wave flow; AH – annular hydraulic flow.

With regard to the structures of annular flow (AL-AH), for which the largest divergence was noted from the regimes occurring during flow of a liquid with varied viscosity, an independent identification of them was undertaken in order to specify the forms and range of their occurrence, as presented in Fig. 5

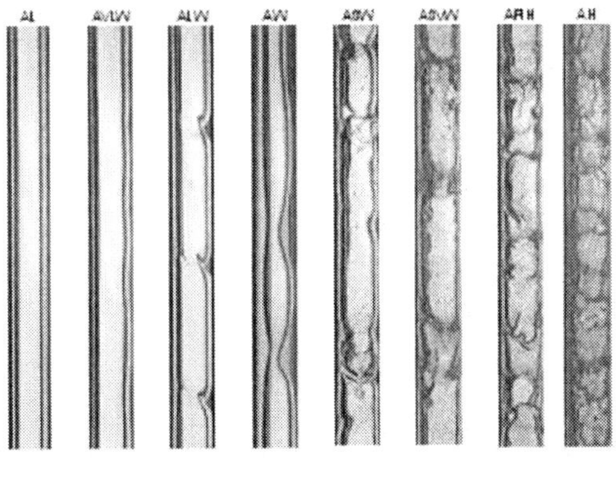

Fig. 5. Classification of two-phase annular gas-very viscous liquid downward flow patterns: AL-smooth film, AVLW-very low wavy annular flow, ALW-annular low wavy film, AW- wavy film, ASW- strongly wavy film, AVSW-very strongly wavy film, APIH-flow passing into hydraulic, AH-hydraulic film flow – for the range: (90÷3500) mPa·s.

The rationale for the selection of this classification is associated with the fact that the structures resulting from it in each case are characterized by possibility of unequivocal assessment of some special characteristics of the flowing liquid film. It is possible to distinguish three basic ranges of such flow structures. The first one (AL-AVLW-ALW) is specific for smooth or slightly wavy interfacial structure surface, second one (AW-ASW) is always associated with the occurrence of waves in this surface and the third one – specific for hydraulic forms (ASVW-APIH-AH), with highly dispersed configuration of wave crests and occurrence of interference associated with formation of capillary waves.

In addition, one can note that the first group of flow structures (smooth film and low waves) occurs for small liquid velocities and for large range of gas velocities (in particular for a liquid with high viscosity). In turn, wavy structures are encountered for medium range of the values of these velocities, while hydraulic ones (irregular) forms occur when the values of phase streams, an in particular, ones of the gas, are in the highest range. In the latter case these phenomena are accompanied by a considerable effect of the condition of liquid viscosity – the lower the value of it, the more limited range of reciprocal overlapping of the waves (and the smaller formation of capillary waves).

The ranges of the occurrence of particular annular flow structures registered throughout this research have been identified on the basis of flow maps, whose examples are presented in Figs. 6-7 with regard to a channel 22 mm in diameter.

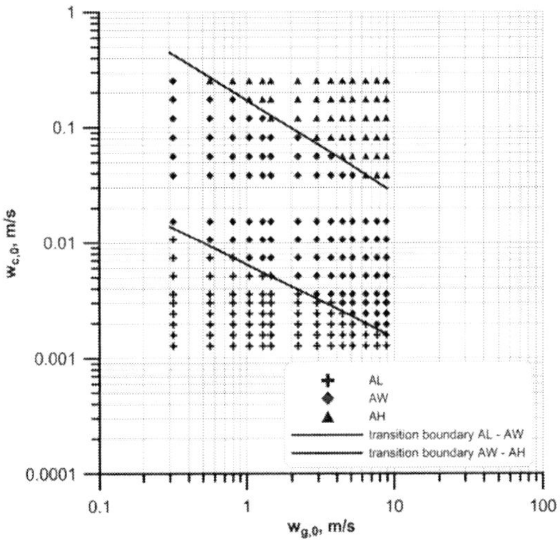

Fig. 6. Ranges of two-phase gas-high viscosity liquid downward flow in a pipe with the diameter of 22 mm, and oil viscosity of 90 mPa·s.

The analysis of the occurrence of a given flow regime involved, subsequently, determination of the effect of liquid viscosity and phase streams (by reference to the superficial velocities) on the range of the specific structures formed. These maps also contain boundaries of the transfer between areas that are specific for smooth films and wavy structures (AL–AW transition boundary) and the ones between the wavy structures and

 DOI: 10.1051/e3sconf/20171903004

completely hydraulic flow (AW–AH transition boundary).

The foundation for the assumption of the first boundary transition (AL–AW) was the appearance of waves on its surface with a geometry that could be described in terms of a specific wavelength and amplitude.

The passing of the other boundary (AW–AH) was taken for the case of occurrence of irregular waves, whose nature was difficult for unequivocal assessment.

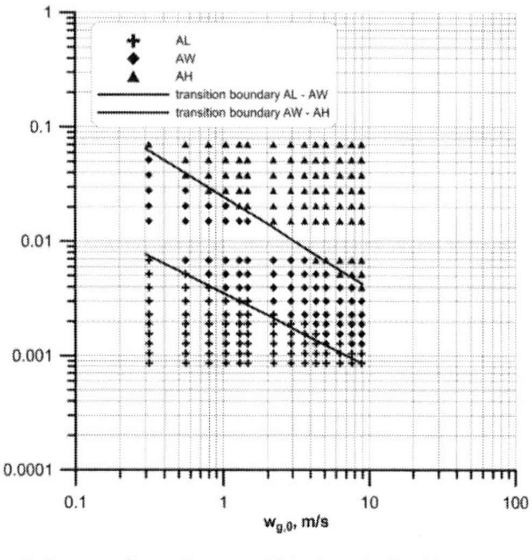

Fig. 7. Ranges of two-phase gas-high viscosity liquid downward flow in a pipe with the diameter of 22 mm, and oil viscosity of 1520 mPa·s.

Qualitative and quantitative analysis indicates that it is possible to determine the ranges of occurrence of various structures of annular flow of very viscous liquid. This has led to the development of a generalized map of two-phase gas-very viscous liquid flow, which is presented in Fig. 8 together with quantitative description of the transition boundaries.

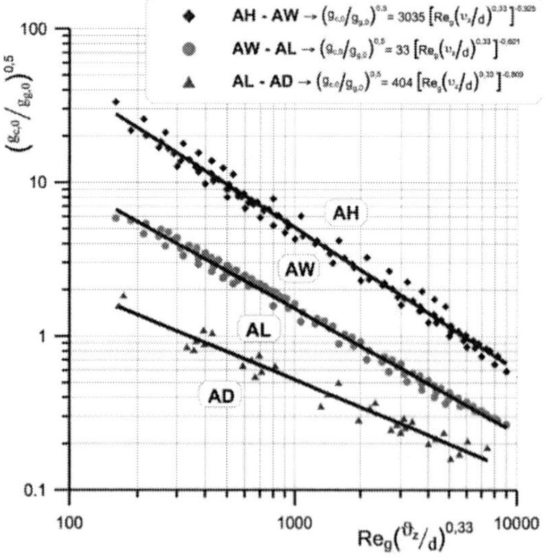

Fig. 8. Universal map of two-phase gas-very viscous liquid downward flow in a pipe.

From the point of view of hydrodynamics of downward flowing liquid, with regard to both theoretical research and conducted analysis, it was interesting to record parameters such as forming liquid films, state of the waves on its surface and value of the interfacial surface. In accordance with the adopted methodology, the measurement of liquid film thickness was undertaken using optoelectronic methods, both with regard to linear (point) and planar system of measurement.

Research involved determination of the effect of flow parameters on the variability of measured values with a particular emphasis on the effect of the variable liquid viscosity. The thickness identified during studies – for the reasons of the applied measurement systems – was determined for local states and averaged for specific flow conditions. This, in turn, enabled one to assess the nature of the downward flowing film. Independently, trap method was applied, whose results provided the results of mean liquid thickness over the entire length of the channel. In each of these cases the averaged local states were integrated means referred to time and length of the measurement channel.

From the analysis of the results one can conclude that the hydrodynamics of annular gas-liquid flow is mostly affected by liquid viscosity and gas stream. An increase in the velocity of the liquid always results in formation of film with greater thickness while an increase in the velocity of the gas phase for a constant liquid flow rate per unit periphery usually leads to a decrease in the oil film thickness. This tendency was observed with regard to all examined pipe diameters over the total range of changing viscosities. One can note, however, that this tendency is particularly discernible for turbulent flow of the gas phase. However, over the range of small velocities of the gas phase, the thickness of liquid film is close to the value specific for its gravitational flow.

This is illustrated by the results of research presented in Figs. 9-10, whose additional interpretation (in 3D) is presented in Figs. 11 to 12.

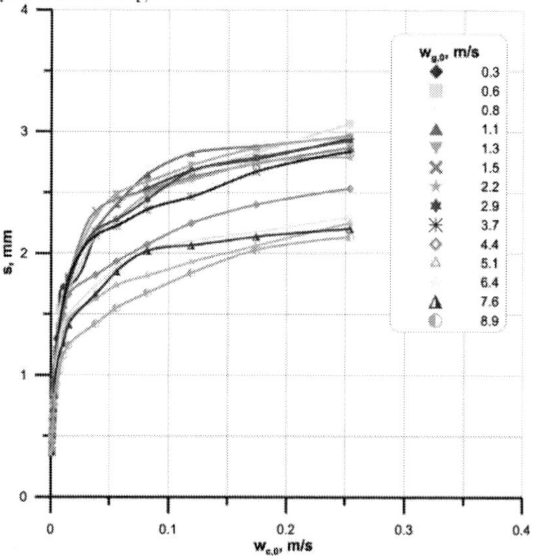

Fig. 9. Distribution of experimental points thickness of liquid film η_c = 90 mPa·s.

DOI: 10.1051/e3sconf/20171903004

These results describe the characteristics of film thickness variability observed during annular flow of oil for various flow conditions.

From the distribution of experimental points (Figs. 9-10) it stems that an increase in superficial velocity of gas does not always promote a decrease in oil film thickness and that the viscosity of the liquid and its superficial velocity affects it.

An increase of the viscosity of the liquid forms a reason for formation of its thicker film, which tends to increase faster following an increase in the liquid stream.

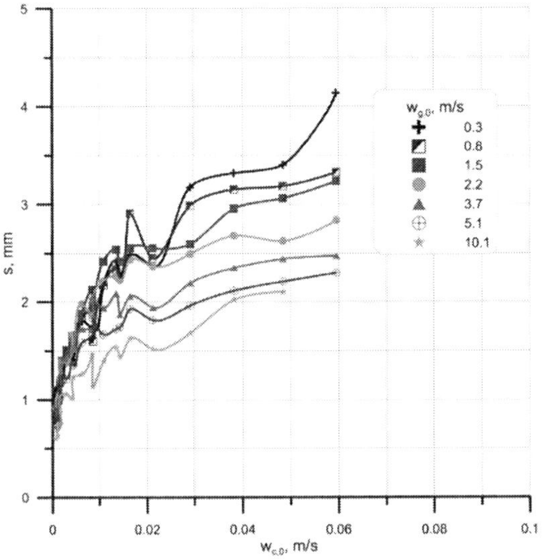

Fig. 10. Distribution of experimental points characterizing thickness of liquid film $\eta_c = 2200$ mPa·s.

A direct increase of the liquid's superficial velocity (for remaining parameters remaining constant) promotes greater wavy structure in the interfacial surface (Figs. 11-12), which consequently leads to formation of hydraulic flow with increasingly curvy interfacial surface.

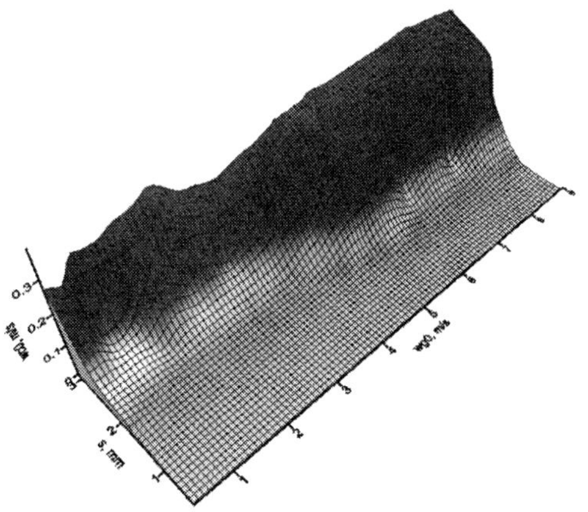

Fig. 11. Surface representation of variability of liquid film thickness $\eta_c = 90$ mPa·s (respective to Fig. 9).

An additional illustration of these types of phenomena is found in Figs. 13-14, which contain a set of exemplary profiles of changes in the thickness of the oil film with respect to Reynolds number for gas and liquid.

The distribution of the lines of these profiles indicates that for the highest gas streams (Fig. 14), regardless of liquid flow rate per unit periphery we will always have to do with hydraulic forms of viscous liquid flow, while in the remaining cases the transfer to such a form is considerably relative to the liquid stream.

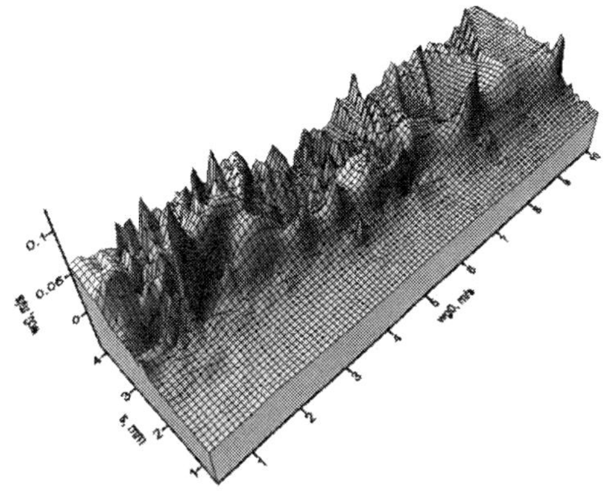

Fig. 12. Surface representation of variability of liquid film thickness $\eta_c = 1070$ mPa·s (respective to Fig. 10).

The relatively low values of it lead to the formation of wave structures in the flow (Figs. 13-14), which transform into hydraulic flow along with an increase of the gas stream.

It was concurrently concluded that the relative wave velocities (in relation to the mean velocity of the liquid film) tends to considerably drop following an increase of equivalent Reynolds number.

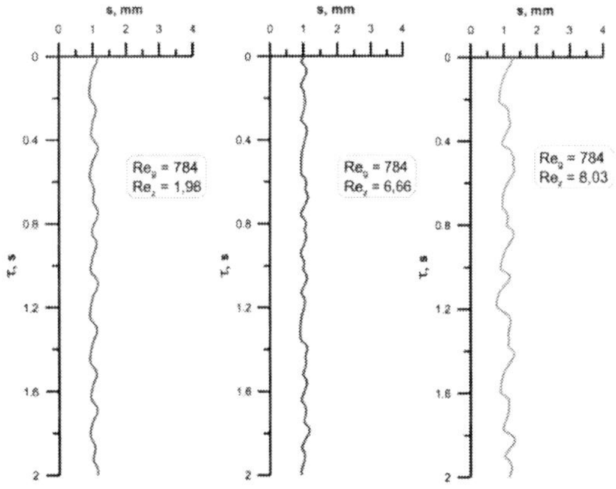

Fig. 13. Representation of variable liquid film thickness along the flow route: (D=22 mm, Re_g=785, $\eta_c = 1070$ mPa·s).

The interpretation of the research results indicates a considerable tendency of variation of the relation F_{2F}/F_p with respect to void fractions resulting from the criteria

DOI: 10.1051/e3sconf/20171903004

involving criteria based Reynolds number for liquid and gas, as presented in Fig. 15. From the distribution of the experimental points it stems that for a constant value of Re_g along with an increase of the value of equivalent Reynolds number we have to do with considerable decrease in the value of the actual contact surface on the boundary of the phases. This results from a concurrent rise of liquid film thickness (for increasingly higher Re_z) on one hand, and on the other, from the changes in the characteristics of waves of downward liquid film flow for higher liquid volumes, which is particularly discernible for very wavy and hydraulic annular flow structures.

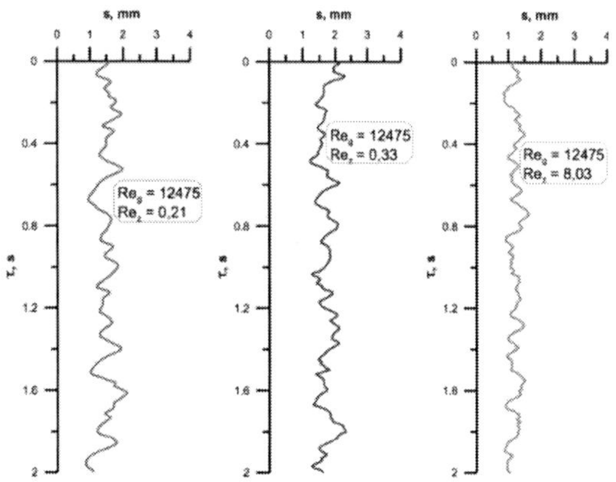

Fig.14. Representation of variable liquid film thickness along the flow route: (D=22 mm, Re_g=12474, η_c = 1070 mPa·s).

Additionally, as presented in Fig. 15 in these flow conditions an important role can be attributed to the characteristics of the gas phase.

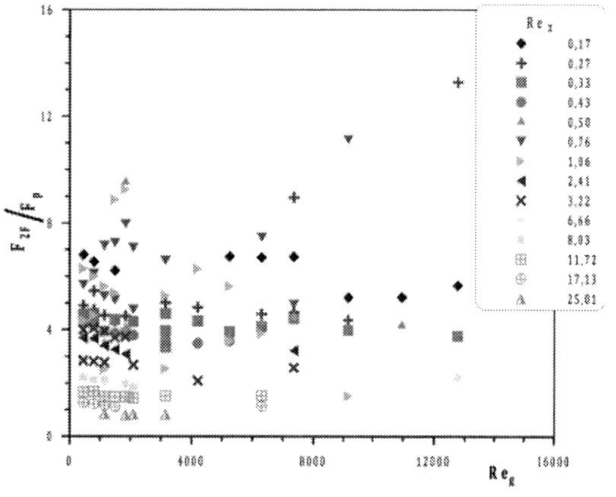

Fig. 15. Relative value of interfacial surface (F_{2F}/F_p) for various process conditions.

Whereas for the area of laminar flow we can observe relatively proportional increase of the relation F_{2F}/F_r for decreasing equivalent Reynolds number (Re_z), for the case of the turbulent flow of this phase (e.g. with regard to strongly wavy films and hydraulic flow) such an increase is not very unequivocal.

A closer investigation of the results has indicated that it is possible to determine the characteristics of these changes relative to any interpreted void fraction both in accordance with (1) and (2). Hence, just as referred to in the case of film thickness, a proposition is made to apply two alternative relations in order to determine the interfacial surface in two-phase gas-very viscous liquid flow. These are:

a) for adiabatic conditions

$$\frac{F_{2F}}{F_p} = 47.3\left(\frac{\varepsilon}{1-\varepsilon}\right)^{0.636} Re_g^{-0.683} Re_z^{-0.109} \qquad (1)$$

b) with respect to mass void fraction – one that is more beneficial in terms of conditions of heat exchange

$$\frac{F_{2F}}{F_p} = 2.85 \cdot 10^3 \left(\frac{x}{1-x}\right)^{0.627} Re_g^{-0.669} Re_c^{-0.111} \qquad (2)$$

where:

$Re_z = 4\Gamma/\eta_c$, $\Gamma = \dot{m}_c/\pi d$, $Re_g = g_T x D/\eta_g$,
$Re_c = g_T(1-x)D/\eta_c$, $\varepsilon = V_g/(V_g + V_c)$, $x = g_g/(g_g + g_c)$

A comparison between the results of measured interfacial surface and ones calculated in accordance with (1) and (2) indicates that over 85% of points are found to be within the boundaries of ±15% relative error (for the correlation coefficient of over 0.95 in both cases). For relation (2) it is illustrated in Fig. 16. Therefore, one can assume that these relations serve for a satisfactory description of interfacial surface, which is encountered for co-current downward flow of gas-very viscous liquid in vertical pipes.

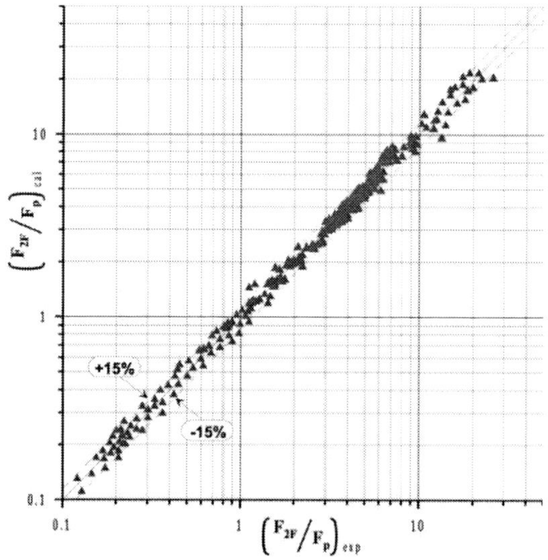

Fig. 16. Distribution of experimental points – in accordance with (2).

4 Summary

On the basis of analysis of the identified annular downward flow structures of viscous liquid flow determined during experiment a conclusion has been

made that the forms of annular structure can differ considerably depending on liquid viscosity and mutual relations of velocities of two phases – from smooth liquid film to severely dispersed hydraulic form. Finally, for this flow regime a classification of flow structures found in Fig. 4 is proposed. In general, this classification reflects the characteristics of annular gas-very viscous liquid two-phase flow and involves identification of structures in downward flow of oil-air mixture in a vertical pipe with a diameter of 22 mm within the range of viscosity coefficient of (90-3500) mPa·s.
Accounting for the similarity of structures formed, for the phenomenological assessment three boundaries at which different flow patterns emerging for annular gas-very viscous liquid two-phase flow are distinguished, which has enabled these authors to present a universal flow structure map found in Fig. 8.
On the basis of assessing the effect of specific process parameters on forming of qualitatively different forms of annular flow, a conclusion is made that liquid viscosity forms the dominant parameter affecting particular structure formation. In consideration of the fact that this map has been developed for a relatively small range of variations in process parameters, one can conclude that it is can fully applied with regard to calculations of tubular thin film apparatuses, in which the annular gas-very viscous liquid flow.
It was additionally indicated that an increase in the velocity of gas flow facilitates the origin of increasingly greater changes on the surface of liquid film. Except for films, which are smooth or slightly wavy, these changes are irregular and are demonstrated by various development of interfacial surface. The waves and turbulence found on the separation surface have amplitudes and wavelengths that are time variable and spatially varied. This has made the quantitative description of hydrodynamics of such flow even more complicated.
From the analysis of experimental data it stems that within the entire range of process parameters all of them have a considerable effect on the formation of liquid film. An increase in the viscosity of the liquid always promotes an increase of the mean thickness of liquid layer, while increase of the gas phase steam for a constant liquid flow rate per unit periphery generally results in a decrease of the film thickness. This, however, occurs only in the area of the flow with strongly developed interfacial surface, just as it takes place in hydraulic flow. These conditions are often accompanied by the phenomenon of wave interference, which promotes their accumulation or results in formation of additional capillary waves, which in turn, brings their greater local thickness (in comparison to mean values). This idiosyncrasy is especially characteristic for a liquid with a high viscosity.
In the relations forming the model of hydrodynamics of two-phase gas-very viscous liquid flow the wide range of variability of liquid viscosity coefficient (90÷3500) mPa·s is covered, which has not been fully considered in literature of this subject yet. This work involves a precise verification of experimental data, which can promote qualitative and quantitative assessment of flow phenomena that accompany such flow conditions. The scope of this research also enables one to recommend the developed model for engineering and applied design work in assessing the operation of thin film equipment with hydraulically generated liquid film.

Nomenclature

F – surface, m^2
Re – Reynolds number,
V – volume flow rate, m^3/s
d – diameter, m
g –mass flux, $kg/(m^2s)$
$\dot{m}$– mass flow, kg/s
s – liquid film thickness, m
Γ– wetting rate, kg/(ms)
η –viscosity, Pa·s
ρ – density, kg/m^3
ϑ_z – equivalent linear dimension, m

Subscripts and superscripts

0 – superficial (calculated for full cross-section area),
c – liquid,
cal – calculated,
exp – experimental,
g – gas,
p – pipe internal surface area,
z – equivalent values for free falling liquids
2F- two-phase flow.

References

1. N. Andritsos, T. J. Hanratty, Int. J. of Mult. Flow, **13**, 5 (1987)
2. K. Czernek, Chem. Eng. and Equ., **6** (2009)
3. K. Czernek, S. Witczak, Chem. and Proc. Eng., **34** (2013)
4. K. Czernek, Studies and Monographs **347**, Opole, (2013)
5. K. Czernek, G. Filipczak, S. Witczak, Przem. Chem., **87/2**, (2008)
6. M.J. Da Silva, S. Thiele, L. Abdulkareem, B.J. Azzopardi, U. Hampel, Flow Meas. and Instr., Volume **21**, 3 (2010)
7. G. Geraci, B.J. Azzopardi, H.R.E. van Maanen, Chem. Eng. Sc., **62**, 11 (2007)
8. V. Hernandez Perez, B.J. Azzopardi, R. Kaji, M.J. da Silva, M. Beyer, U. Hampel, Int. J. of Mult. Flow, **36**, 11-12 (2010)
9. E.W. Jassim, T.A. Newell, J.C. Chato, Exp. Therm. and Fluid Sc., **32**, 5 (2008)
10. J. Oriol, J.P. Leclerc, C. Jallut, P. Tochon, P. Clement, Chem. Eng. Sc., **63**, 1 (2008)
11. L. Szalinski, L.A. Abdulkareem, M.J. Da Silva, S. Thiele, M. Beyer, D. Lucas, V. Hernandez Perez, U. Hampel, B.J. Azzopardi, Chem. Eng. Sc., **65**, 12 (2010)
12. L. Troniewski, S. Witczak, K. Czernek, Chem. and Proc. Eng., **27** (2006)

DOI: 10.1051/e3sconf/20171903004

13. L. Troniewski, W. Spisak, Int. J. of Mult. Flow, **13**, 2 (1987)
14. S. Witczak, K. Czernek, Turbulence, **11** (2005)
15. J.Y Xu, Y.X Wu, Z.H Shi, L.Y Lao, D.H Li, Int. J. of Mult. Flow, **33**, 9 (2007)

 DOI: 10.1051/e3sconf/20171903005

Features of single tracks in coaxial laser cladding of a NI-based self-fluxing alloy

Eugene Feldshtein[1,*], *Oleg* Devojno[2], *Marharyta* Kardapolava[2] and *Nikolaj* Lutsko[2]

[1] Faculty of Mechanical Engineering, University of Zielona Góra, Prof. Z. Szafrana 4, 65-516 Zielona Góra, Poland
[2] Faculty of Mechanical Engineering, Belarusian National Technical University, Khmelnitsky str., 9, build. 6, Minsk 220013, Belarus

Abstract. In the present paper, the influence of coaxial laser cladding conditions on the dimensions, microstructure, phases and microhardness of Ni-based self-fluxing alloy single tracks is studied. The height and width of single tracks depend on the speed and distance of the laser cladding: increasing the nozzle distance from the deposited surface 1.4 times reduces the width of the track 1.2 – 1.3 times and increases its height 1.2 times. The increase of the laser spot speed 3 times reduces the track width 1.2 – 1.4 times and the height in 1.5 – 1.6 times. At the same time, the increase of the laser spot speed 3 times reduces the track width 1.2 – 1.4 times and the height 1.5 – 1.6 times. Regularities in the formation of single tracks microstructure with different cladding conditions are defined, as well as regularity of distribution of elements over the track depth and in the transient zone. The patterns of microhardness distribution over the track depth for different cladding conditions are found.

1 Introduction

Laser cladding (LC) is one of efficient methods to deposit on machine parts different coatings including functionally graded, multilayered, etc. It is possible to mix different powders while injecting them into a cladding zone and produce coatings with desired content, to create engineered internal structures [1–3].

Several groups can be identified among the factors affected laser cladding [4–10]. These groups include the following:

- a dependence on laser radiation conditions (laser power, pulse energy, pulse duration, pulse frequency, laser wave length, laser beam spot diameter, focal distance),
- a dependence on conditions of layer forming (laser scan speed, scan strategy rotation, number of spots, scanning space, stripe width, stripe overlap, hatch or track distance, layer thickness, gas content control),
- a dependence on characteristics of powders that was used (size of the powder particles, powder mass flow, temperature of the powder).

A number of studies have been carried out to investigate the properties of single tracks when laser cladding. In particular, in [11] the relationships between geometrical characteristics (width, height, area, and penetration depth) and the processing parameters (laser power, scanning speed and powder mass flow) for 316L stainless steel single tracks were discussed. In [12] the surface morphology and microstructure of single tracks and layer of this steel after introduction of a NiB additive were described. The microstructure and porosity evaluation in single tracks of the Ni60 self-fluxing alloy deposited by laser-cladding technique are discussed in [13]. The microstructure and nano-mechanical properties of single tracks and layer of AlSi10Mg alloy after SLM were researched in [14], and the density, roughness and microstructure of AlSi10Mg layer were described in [15]. The microstructural and mechanical properties and surface quality of CoCrMo alloy and their relationship with some of the cladding parameters were studied in [8,16] for single tracks and the surface.

In the latter case, different coating materials can be used, and self-fluxing alloys are widely applied among them. Three groups of these alloys are known now: Fe-based, Ni-based, and Co-based self-fluxing alloys [13,17–22]. The Ni- and Co-based self-fluxing alloys are characterized by wettability, deoxidization and by a fluxing effect as well as high physical and mechanical properties, and a good resistance to wear, oxidation and high temperature corrosion. The multiphase structure that is formed in the layer of Ni-based self-fluxing alloy during the reflow process ensures a high wear resistance of such coatings. It was reported in the above-mentioned research studies that carbides, chromium borides, a solid solution, and a eutectic phase are the constituents of the coating. It was also reported that chromium-molybdenum carbides and chromium-molybdenum borides, nickel borides, chromium carbides and borides were present in the nickel matrix of the coating.

The aim of the present investigation is to study the formation features, microstructure and chemical composition and microhardness of single laser tracks of a NiCrBSi self-fluxing coatings.

* Corresponding author: E.Feldsztein@ibem.uz.zgora.pl

DOI: 10.1051/e3sconf/20171903005

2 Experimental procedures

2.1 Material

The research was carried out with NiCrBSi self-fluxing hard alloy. The chemical composition of the powder is: 0.3–0.6 % C, 1.7–2.5 % B, 1.2–3.2 % Si, 8–14 % Cr, 1.2–1.3 % Fe and Ni is the rest. The hardness of the material is equal HRC 35–40.

2.2 Laser cladding technology

The single tracks were applied to a substrate of AISI 1045 steel. Cladding was performed using a continuous-wave CO_2 laser "Comet 2" with the power of 1 kW (Fig. 1). Powder was fed to the work area through a lens with a nozzle for spot coaxial laser cladding. The compressed air was used as a transporting gas. The focal length was of 200 mm, and a focused spot was of 1.0 mm.

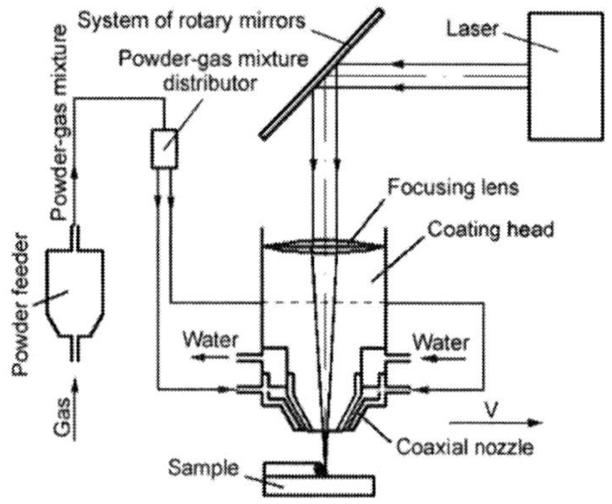

Fig. 1. The scheme of a single track deposition.

2.3 Preparation of samples

Cladding tracks of the Ni-based alloy were tested in the speed range of 40–120 mm/min when the distance between the nozzle and the surface of sample was of 10–14 mm. The diameter of the laser spot was of 1 mm, which corresponds to a power density of $1.27 \cdot 10^5$ W/cm^2. After cladding the samples were cut in the direction perpendicular to the tracks, and then their transverse metallographic sections were made and the geometric parameters of the single track, i.e., width *B* and height *H*, were determined.

2.4 Measuring equipment

The track sizes were measured using an optical microscope with accuracy of 0.001 mm. The microstructure, morphology and chemical composition were analyzed by a scanning electron microscope "Mira". The microhardness was measured using a MICROMET II Hardness Tester. Statistica 12 software was used for a graphical interpretation of some results.

3. Results and discussion

3.1 Influence of laser cladding conditions on the single track geometry

The effect of laser cladding conditions on the height *H* and width *B* of single tracks is shown in Fig. 2. It can be seen that increasing the nozzle distance from the deposited surface 1.4 times reduces the width of the track 1.2 – 1.3 times and increases its height 1.2 times. At the same time, the increase of the laser spot speed 3 times reduces the track width 1.2 – 1.4 times and the height 1.5 – 1.6 times.

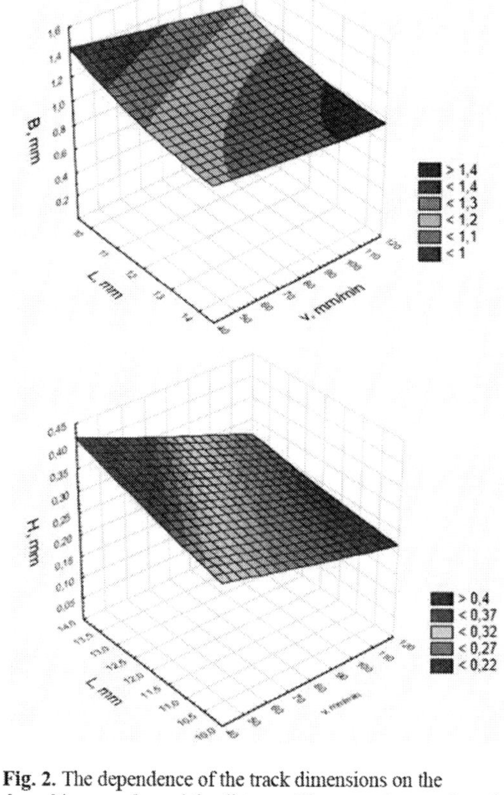

Fig. 2. The dependence of the track dimensions on the deposition speed *v* and the distance *L* between the nozzle and the sample surface.

3.2 Microstructure and phase composition of single tracks

Studies of the single track structure have shown that the deposited tracks have a dense composition with distinct structural components and consist of three regions (Fig. 3): the deposited material (1), heat affected zone (2) and the base material (3).

The structure of the deposited track consists of γ-Ni (Ni-based solid solution) as well as chromium borides and carbides.

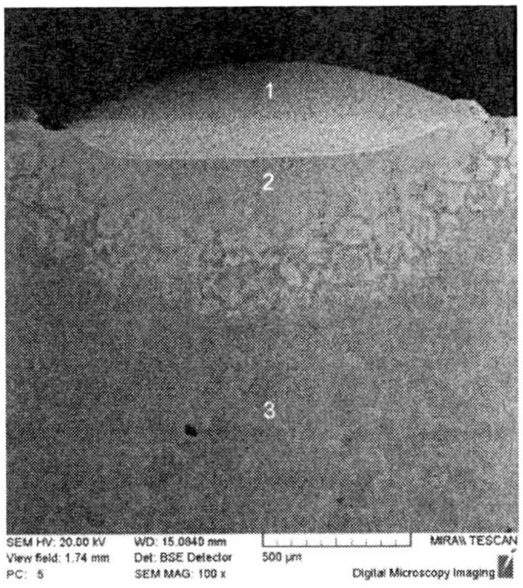

Fig. 3. The microstructure of a single track and the heat affected zone.

The structure is formed under the influence of high temperatures, high cooling speed and directional heat removal. Dendrites of the first and second orders are seen in the upper region of the deposited layer and they are arranged at an angle of 45° to the surface. The self-fluxing alloy powder is melted at high temperatures and crystallized in the pseudo-dendrites form and complex quasi-eutectic γ-Ni + Ni_3B is formed between dendrite branches at temperatures 950 – 1000 °C. First-order branches of dendrites are crystallized in the first place and they contain the maximum amount of the powder constituent elements. The branches of the second and third orders contain a reduced amount of reinforcing components. The lower the rate of deposition, the more slowly a eutectic mixture is crystallized. This mixture is characterized by the ability to cure the pores and moreover it reduces the possibility of cracking. The lower part of the deposited layer is at high temperatures a long time and the dendrite-globular structure with a eutectic mixture of γ-Ni + Ni_3B is formed there. The heat affected zone is formed between the deposited single track and the substrate material and the thickness of it is equal to 10 – 40 μm. Its presence indicates the chemical interaction between the coating material and the substrate. During contact between double surfaces the restoration of metal oxide films is observed due to the presence of strong oxidants: boron and silicon. The formed borosilicate glass floats to the surface of the coating or evaporates. The bimetal bond strength becomes equal to the strength of the substrate metal in this case but with increasing residence times in the liquid pond the interdiffusion of iron and nickel and a significant decrease in hardness can be observed. Structures of tempered martensite and troostite are formed in the upper region of the heat affected zone under high temperatures. Directly at the interface the temperature effect is insufficient for dissolution of the grain boundaries, but quenching and tempering processes occur inside grains.

3.3 The distribution of elements over the depth of single tracks

The analysis of the elements distribution over the depth of PG-12H-01 alloy single track showed that at a depth of 0.1 mm below the coating, i.e. in the substrate material, almost pure iron is present and the content of other elements is close to zero (Fig. 4). At the boundary of the coating and substrate, i.e. in the transient zone, the iron content is reduced to 65 %, the nickel content is increased to 27 % and chromium up to 6 %. Thus, an active mixing of the coating and substrate materials takes place in the molten pond. Further away from the substrate material surface, the content of elements practically does not change, but does not reach the values typical for the in-situ powder. The reason for this discrepancy is the strong mixing of the substrate and coating materials when a single track is formed.

Features of mixing and interaction of the coating and substrate materials in the transient zone of the single track are shown in Fig. 5. At a depth of more than 5 μm under the coating the material composition consists of practically pure iron, whereas at the boundary between the coating and the substrate material the composition is: 79 % iron, 17 % nickel, 4 % chromium and 1 % silicon, indicating that an active interaction of materials takes place. Further, the material composition of the deposited track practically does not change throughout the depth of the transient zone, and its thickness is negligible. The fact that the material composition in the transient zone of the single track is almost the same as throughout the coating depth is caused by strong mixing of the coating and substrate material.

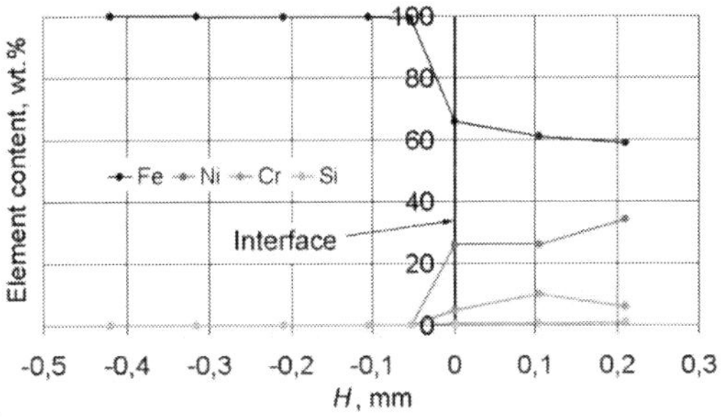

Fig. 4. The distribution of elements at a depth of a single track.

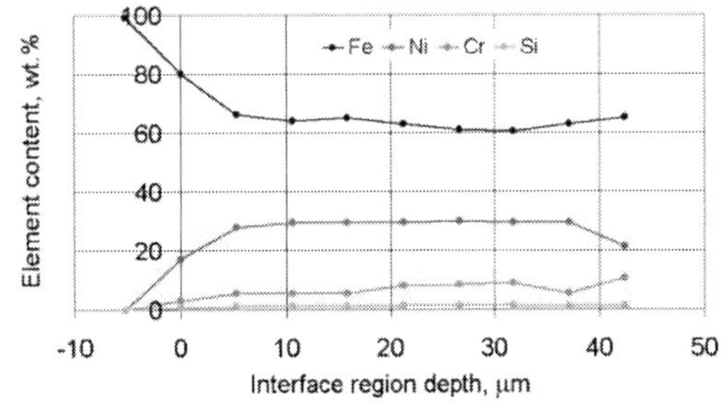

Fig. 5. The distribution of elements in the transient zone of a single track.

3.4 Microhardness of single tracks

Changes of the microhardness in the depth of single tracks for various cladding conditions are shown in Fig. 6. Based on the measurements made, the following can be noted:

- microhardness near the track surface is not high (150 – 500 HV100), then its growth is observed, and then its gradual reduction to the microhardness of the substrate;
- lowest microhardness (120 – 150 HV100) is observed at the cladding speed of 40 mm/min for all cladding distances;
- when the cladding speed increases up to 100 mm/min, an increase of microhardness up to 500 – 650 HV100 is observed, but the cladding speed of 120 mm/min decreases the microhardness again to a value, which was observed at the cladding speed of 80 mm/min (300 – 350 HV100);
- some increase in the microhardness is observed with decreasing the cladding distance;
- the highest microhardness (700 HV100) is observed at the cladding speed of 100 mm/min and a cladding distance of 10 mm;
- at low deposition rates, less than 60 mm/min, the overall level of microhardness is low and equals 150 – 350 HV100.

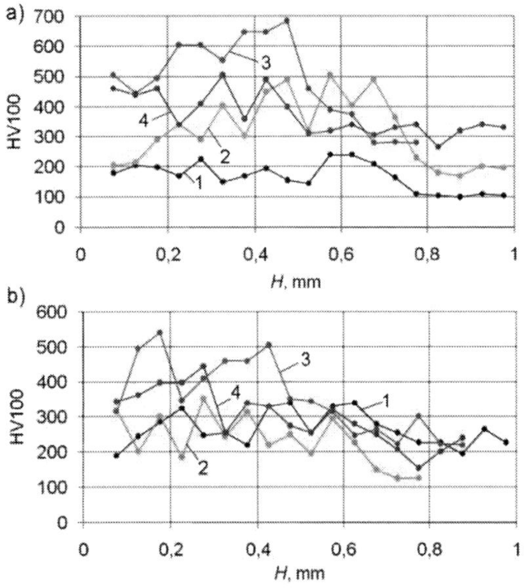

Fig. 6. Microhardness changes in the single track depth for the distance from the nozzle to the sample surface of 10 mm (a) and 14 mm (b): 1 – v = 40 mm/min, 2 – v = 80 mm/min, 3 – v = 100 mm / min, 4 – v = 120 mm/min.

4 Conclusions

The novelty of this investigation is the complex analysis of single tracks of an Ni-based self-fluxing alloy that have been formed by laser cladding using the CW CO_2 laser. The height and width of single tracks depend on the speed and distance of the laser cladding: increasing the nozzle distance from the deposited surface 1.4 times reduces the width of the track 1.2 – 1.3 times and increases its height 1.2 times. The increase of the laser spot speed 3 times reduces the track width 1.2 – 1.4 times and the height in 1.5 – 1.6 times. At the same time, the increase of the laser spot speed 3 times reduces the track width 1.2 – 1.4 times and the height 1.5 – 1.6 times. Regularities in the formation of single tracks microstructure with different cladding conditions are defined, as well as regularity of distribution of elements over the track depth and in the transient zone. The patterns of microhardness distribution over the track depth for different cladding conditions are found. On the basis of the investigated features of the single tracks formation, it is possible to provide a quality coating as a whole. On the basis of the investigated features of the single tracks formation, it is possible to provide a quality coating as a whole.

References

1. J.C. Ion, *Laser Processing of Engineering Materials. Principles, Procedure and Industrial Application* (Elsevier Ltd., Oxford, 2005)
2. W.M. Steen, *Laser Material Processing* (Springer, 2003)
3. E. Toyserkani, A. Khajepour, S. Corbin, *Laser Cladding* (CRS Press, 2005)
4. D. Dimitrov, K. Schreve, N.D. Beer, Rapid Prototyp. J. **12** (2006)
5. I. Yadroitsev, P. Bertrand, I. Smurov, Appl. Surf. Sci. **253** (2007)
6. J. Ciurana, L. Hernandez, J. Delgado, Int. J. Adv. Manuf. Technol. **68** (2013)
7. M. Averyanova, E. Cicala, P. Bertrand, D. Grevey, Rapid Prototyp. J. **18** (2012)
8. Y. Pupo, K.P. Monroy, J. Ciurana. Int. J. Adv. Manuf. Technol. **80** (2015)
9. L.E. Criales, Y.M. Arısoy, B. Lane, S. Moylan, A. Donmez, T. Özel, Add. Manuf. **13** (2017)
10. A.J. Dunbar, E.R. Denlinger, J. Heigel, P. Michaleris, P. Guerrier, R. Martukanitz, T.W. Simpson, Add. Manuf. **12** (2016)
11. H. El Cheikh, B. Courant, S. Branchu, J.-Y. Hascoët, R. Guillén, Opt. Lasers Eng. **50** (2012)
12. J. Stašić, D. Božić, Surf. Coat. Technol. **307** (2016)
13. C. Zeng, W. Tian, W.H. Liao, L. Hua, Surf. Coat. Technol. **294** (2016)
14. N.T. Aboulkhair, I. Maskery, C. Tuck, I. Ashcroft, N.M. Everitt, J. Mater. Proc. Technol. **230** (2016)
15. K. Kempen, L. Thijs, J. Van Humbeeck, J. Kruth, Mater. Sci. Technol. **31** (2014)
16. K.P. Monroy, J. Delgado, L. Sereno, J. Ciurana, N.J. Hendrichs, Met. Mater. Int. **20** (2014)
17. X. Tong, F.-h. Li, M. Liu, M.-j. Dai, H. Zhou, Opt. Laser Technol. **42** (2010)
18. M. Aghasibeig, H. Fredriksson, Surf. Coat. Technol. **209** (2012)
19. I. Hemmati, R.M. Huizenga, V. Ocelík, J.Th.M. De Hosson, Acta Mater. **61** (2013)

DOI: 10.1051/e3sconf/20171903005

20. E. Feldshtein, M. Kardapolava, O. Dyachenko, Surf. Coat. Technol. **307** (2016)
21. G.M. Krolczyk, P. Nieslony, J.B. Krolczyk, I. Samardzic, S. Legutko, S. Hloch, S. Barrans, R.W. Maruda, Measurement **70** (2015)
22. L. Janka, J. Norpoth, S. Eicher, M.R. Ripoll, P. Vuoristo, Mater. Des. **98** (2016)

 DOI: 10.1051/e3sconf/20171903006

Experimental studies of thin films deposition by magnetron sputtering method for CIGS solar cell fabrication

Sławomir Gułkowski[1], and *Ewelina* Krawczak[1,*]

[1]Faculty of Environmental Engineering, Lublin University of Technology, Nadbystrzycka 38, 20-618 Lublin, Poland

Abstract. Among a variety of the thin film solar cell technologies of second generation, copper-indium-gallium-diselenide device (CIGS) with the latest highest lab cell efficiency record of 22.4 % seems to be the most promising for the power generation. This is partly due to the advantages of using low cost films of few microns thick not only as a metallic contacts but also as a main structure of the solar cell consisted of high quality semiconductor layers. This paper reports the experimental studies of the CIGS absorber formation on Soda Lime Glass substrate covered by thin molybdenum film as a back contact layer. All structures were deposited with the use of magnetron sputtering method only. Technological parameters of the deposition process such as deposition power, pressure and deposition time were optimized for each layer of the structure. Mo back contact was examined in terms of resistivity. EDS measurements were carried out to verify stoichiometric composition of CIGS absorber. Thin film of Al was used as a top contact in order to examine the quality of *p-n* junction. The I-V electrical characteristic of the *p-n* junction was analysed in terms of solar cell application.

1 Introduction

$CuInGaSe_2$ solar cells based on copper-indium-gallium-diselenide (CIGS) absorber become one of the promising devices in the photovoltaic industry mainly due to attractive properties of this *p*-type material such as direct band gap, high absorption coefficient and inherent *p*-type conductivity [1,2]. $CuInGaSe_2$ absorber was also found to be high tolerant material to variations in compositions [3,4]. What is more, the recent highest noticed efficiencies of the lab solar cells equals 22.4% are close to the efficiencies of the silicon devices, which makes possible to replacing Si crystalline solar cells with the CIGS [5]. For a large scale CIGS module the efficiency of 15 % is comparable to crystalline silicon devices as well [6]. More than 80% of the current solar cells production requires cutting of large silicon crystal. The technology which could avoid the material losses can be a method for future developing cost-efficient solar cells [7,8]. Basic device structure of these two technologies is the same. Crystalline silicon solar cells are built on the basis of *p-n* junction of silicon material with different types of doping. CIGS solar cells based on the *p-n* junction of $CuInGaSe_2$/CdS structure. CIGS/CdS junction is deposited on soda lime glass substrate (SLG), covered by Mo layer of one micron thick [1]. Because of its high conductivity and relative stability at the high processing temperature molybdenum was found to be the best back contact material for CIGS solar cells. Mo layer is usually deposited using DC magnetron sputtering technique. Properties of the grown films play a crucial role in the performance of the CIGS device [9]. For this reason, parameters of the deposition process such as power, time and pressure should be carefully chosen. In terms of CIGS absorber, several deposition methods have been reported. Nowadays, the highest efficiency devices are produced using co-evaporation technique [1,10]. In the process the Cu, In, Ga and Se are delivered from open-boat sources under different temperature to the heated substrate. The main advantage of this method is flexibility to choose the process specific and control the film composition and bandgap [1]. However, the main disadvantage of this technique is to obtain the uniform deposition over large areas [11]. Another method of CIGS structure preparation is sputtering of metallic precursor consisted of Cu, In and Ga with required stoichiometric ratios of Cu/(In+Ga) and Ga/(In+Ga) [12,13]. In order to crystallize the absorber, the post-selenization process is followed by heating precursor layers at temperature up to 600°C under the Se atmosphere [14,15]. Promising alternative method is the sputtering deposition of CIGS absorber directly from the single quaternary target consisted of metals as well as Se under the high temperature without additional post-selenization step [11,16,17]. In order to form a *p-n* junction, CdS film has to be deposited on the CIGS absorber. The highest performance of CIGS solar cells were achieved with the use of chemical bath deposition (CBD) of the CdS layer [18,19]. The CBD process is simple and very attractive for devices of small, laboratory size cells. However, in commercial application CBD method is slow and produces liquid

* Corresponding author: e.krawczak@pollub.pl

wastes that need to be recycled. For this reason, interest of other deposition techniques as magnetron sputtering has grown over the last few years [20,21]. Formed *p-n* junction must have top conductivity contact. The obvious choice is transparent oxide. The main materials that can be used are as follows: SnO_2, In_2O_3:Sn (ITO) or ZnO. The most common is ZnO due to low material costs. The transparent conductive oxide can be deposited by RF magnetron sputtering from ceramic target [1]. Aluminum contact grid completes the device.

In this work fabrication of $CuInGaSe_2$/CdS junction on the Mo covered SLG substrate was shown. Aluminum and molybdenum contacts as well as *p*-type absorber and *n*-type CdS layer were deposited using DC/RF magnetron sputtering method. Single quaternary $CuInGaSe_2$ target was used for absorber deposition under temperature of 400°C. Technological parameters of the film deposition process were carefully chosen to achieve the thickness of the particular layer as well as to control the stoichiometric concentration of the absorber. I-V dark curve of the SLG/Mo/CIGS/CdS/Al structure was measured in order to examine the quality of the *p-n* junction.

2 Materials and methods

Thin films of the CIGS structure were formed using only sputtering system, RF or DC in dependence of conductive or nonconductive material of the target. Size of the target was 4'' in diameter and 6 mm thick. The structure was formed on soda lime glass (SLG) substrate of 1 cm x 1 cm size. Prior to deposition process the glass substrates were chemically cleaned in acetone, distilled water and ethanol for 10 min respectively, and dried in a N_2 gas. The vacuum chamber was evacuated to a background pressure of 1.0 x 10^{-2} Pa. Argon working gas of high purity (99.99%) was introduced into deposition chamber in presputtering and sputtering process. In order to find the deposition rates of particular materials, several experiments were carried out at constant deposition power of 80 W (1 W/cm^2) and pressure of about 2 Pa for different deposition times. Precise thickness measurements were carried out using Bruker DektakXT profiler. The four point probe RM 3000+ system of Jandel Engineering was used to find the best conductivity of the back and top contact film. Both $CuInGaSe_2$ and CdS structures were examined in terms of deposition rate as well as film composition using energy dispersive spectroscopy (EDS) of Hitachi SU3500 Scanning Electron Microscopy. Table 1 presents optimal set of parameters that was chosen for each material to form final device.

Table 1. Technological parameters of the magnetron sputtering process of particular films deposition.

	Power [W]	Pressure [Pa]	Deposition time [min]
Molybdenum	80	1.99 Pa	45
CIGS	80	1.99 Pa	90
CdS	80	2.09 Pa	1 min 22 sec

I-V dark characteristic of the obtained structure was measured in a solar cell simulator.

3 Results and discussion

Fig. 1 shows the thickness measurements results of the molybdenum thin films as a function of the deposition time. The technological parameters such as deposition power and pressure were kept constant equals 80 W and 1.98 Pa respectively.

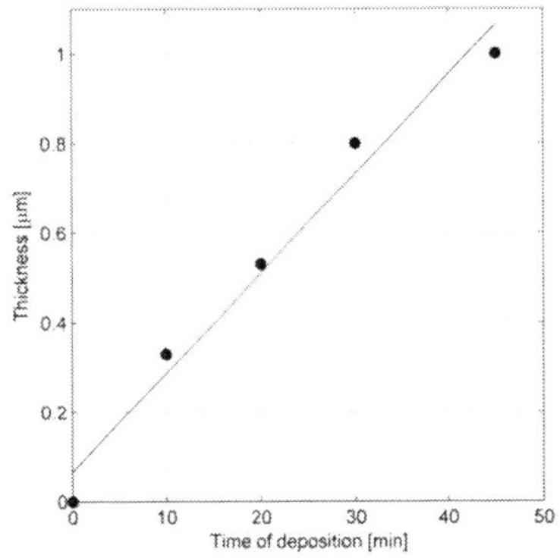

Fig 1. Thickness of the Mo layer as a function of deposition time at constant deposition power of 80 W and pressure about 2 Pa.

From the figure, linear correlation between deposition time and film thickness can be observed. On the basis of linear regression analysis Mo deposition rate was estimated to be 0.38 nm/s.

The four point probe resistivity measurements were carried out in order to examine the conductivity of the Mo thin film. Results of the examination were shown in Fig. 2.

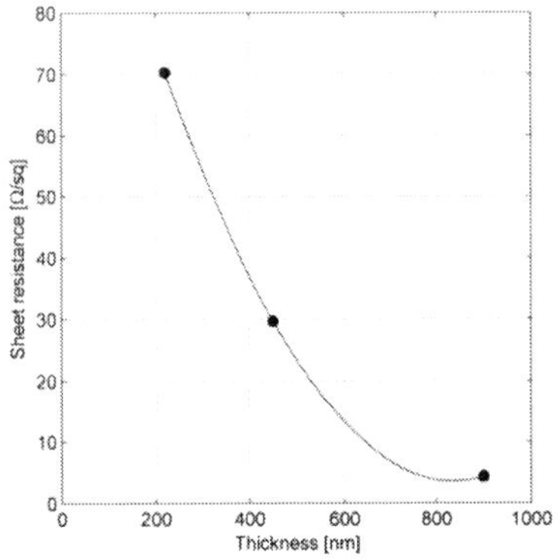

Fig. 2. Sheet resistance dependence on the layer thickness at constant deposition power of 80 W.

Fig. 2 depicts that sheet resistance of the films decreases with the increase of its thickness. One micron thick Mo layer prepared at deposition power of 80 W was characterized by sheet resistance equal to about 4 Ω/sq. The resistivity of the film was found to be 2 mΩ·cm.

DOI: 10.1051/e3sconf/20171903006

This value is within the range of resistivity values for various deposition power reported by [8].
Fig. 3 shows the dependence of CIGS layer thickness on deposition time at constant deposition power of 80 W and pressure of 1.99 Pa.

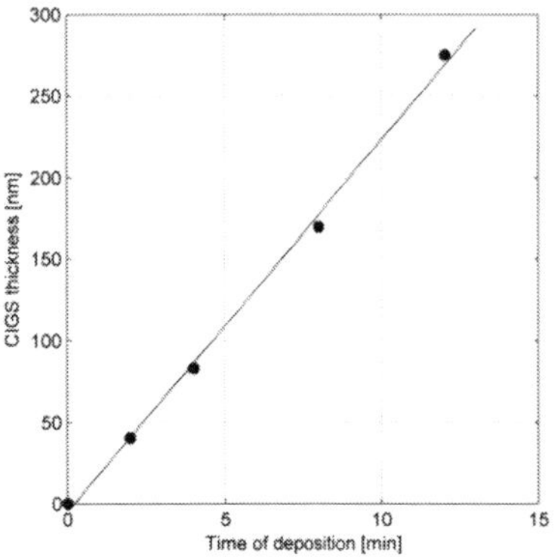

Fig. 3. $CuInGaSe_2$ film thickness versus deposition time at constant power of 80W. Temperature of glass substrate was equal to 400°C during the deposition process.

The temperature of the glass substrate was kept constant at 400°C throughout the deposition process. From the results of linear regression applied to the data the deposition rate of CIGS film was determined. Deposition time of at least 2 microns layer was found to be about 90 min. SEM image of the top surface of the CIGS film deposited at parameters described above was shown in Fig. 4.

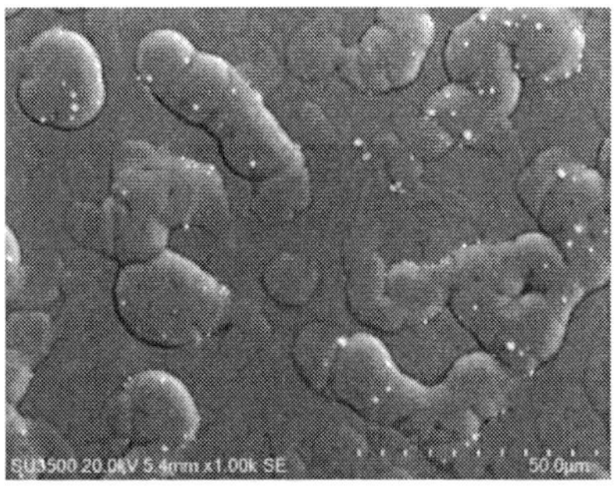

Fig. 4. SEM image of the top surface of two microns CIGS thin film deposited at power of 80 W and substrate temperature of 400°C.

In order to obtain the chemical composition of CIGS absorber, energy dispersive spectroscopy measurements (EDS) were carried out. Sample was observed under 10 kV mode. Fig. 5 shows EDS profile of the sample. Atomic composition of the film was as follows: Cu - 14.8 at %, In – 18.35 at %, Ga – 2.86 at %, Se – 25.95 at %. On the basis of measured values the relative ratios of Cu/(In+Ga) and Ga/(In+Ga) were found to be 0.7 and 0.13 respectively. Calculated values are within the ranges of CIGS absorber composition ratios [12].
In order to examine the quality of the *p-n* junction the high conductivity Al contact was deposited on the top of the structure and dark electrical I-V characteristic was measured.

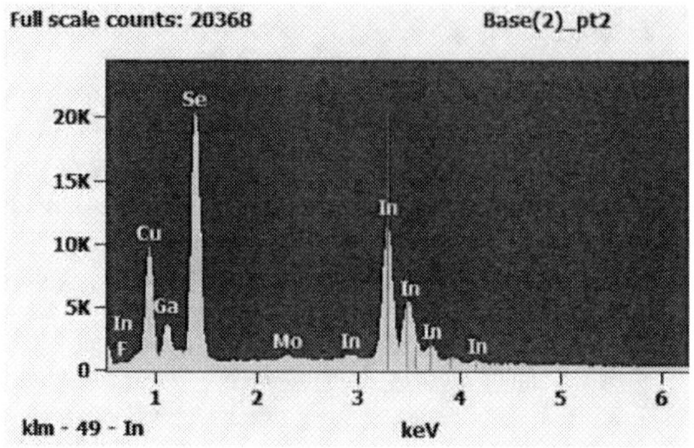

Fig. 5. EDS spectral analysis of the top surface of CIGS absorber.

Similar to previously described procedure of deposition rate determination measurements of thickness versus deposition time were carried out for CdS material. From the line of best fit the deposition rate was found to be 0.77 nm/s. CdS film of 60 nm thick needs about 78 s of the deposition at power of 80 W and pressure of 2.09 Pa.

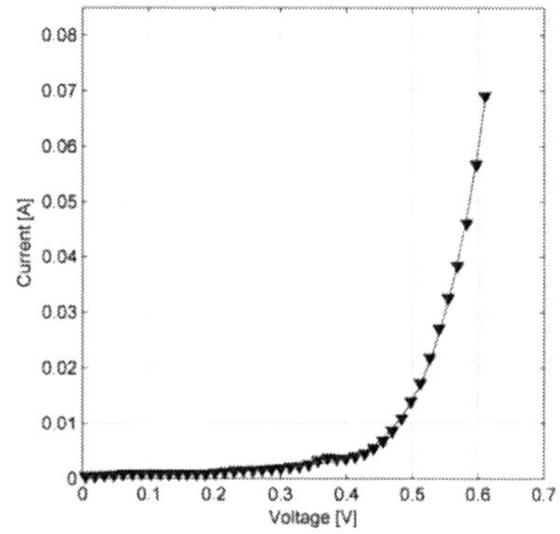

Fig. 6. Experimental dark I-V characteristics of $CuInGaSe_2$/CdS junction.

The exponential shape of the I-V curve (Fig. 6) was found to be typical characteristic of the high quality diode. The presence of *p-n* junction of the $CuInGaSe_2$/CdS structure inside the device was proved. Further analysis of the device demands a transparent conductive oxide deposition on the top of the structure to have transparent contacts that allow the light curve measurements and the main parameters of the solar cell extraction [22].

4 Conclusions

CIGS solar cells become a promising alternative solution for common Si-crystalline devices in terms of high efficiency as well as low production costs. However, $CuInGaSe_2$ *p*-type absorber, the main part of the structure needs to be improved in a large scale production. Magnetron sputtering deposition seems to be a good solution for this purpose because allows to obtain a complete solar device in one technological process. This study reported the process of CIGS device formation from Mo back contact to aluminum top contact with the use of sputtering deposition technique. Technological parameters of sputtering process such as deposition power, pressure and time were found to be optimal. Deposition rates of particular materials were established to obtain demanded thickness of each layer. Mo contact layers were examined in terms of resistivity. Sheet resistance of one micron Mo layer was found to be about 4 Ω/sq. EDS measurements were carried out to verify stoichiometric composition of the CIGS absorber. Relative ratios of Cu/(In+Ga) and Ga/(In+Ga) were found to be 0.7 and 0.13 respectively. I-V dark characteristic was measured. The shape of the curved was found to be typical diode exponential shape which proved the high quality *p-n* junction presence inside the structure.

Authors would like to thank Alain Fave from Institut National des Sciences Appliquées de Lyon for possibility to measure film profiles.

References

1. S. Hegedus and A. Luque, *in Handbook of Photovoltaic Science and Engineering*, 2nd ed., edited by A. Luque and S. Hegedus, (John Wiley & Sons, Chichester, 2011)
2. C. Men, Z. Tian, Q. Shao, H. Zhang, Z. An, Appl. Surf. Sci. **258** (2012)
3. R. Rocheleau, J. Meakin, R. Birkmire, Proc. 19th IEEE Photovoltaic Specialist Conf. (1987)
4. K. Mitchell, C. Eberspacher, J.H. Ermer, K. L. Pauls, D. N. Pier. IEEE Trans. Electron Devices **37**, 2 (1990)
5. P. Jackson, R. Wuerz. D. Hariskos, E. Lotter, W. Witte, M. Powalla, Phys. Status Solidi RRL **10** (2016)
6. K.A.W. Horowitz, R. Fu, M. Woodhouse, Sol. Energ. Mat. Sol. C **154** (2016)
7. S. Gulkowski, J.M. Olchowik. Annu. Set Environ. Prot. **15** (2013)
8. J.M. Olchowik, S. Gulkowski, K. Cieslak, I. Jozwik, A. Fave, A. Kaminski, J. Non-Cryst. Solids **354** (2008)
9. S. A. Pethe, E. Takahashi, A. Kaul, N. G. Dhere, Sol. Energ. Mat. Sol. C **100** (2012)
10. P. Jackson, D. Hariskos, E. Lotter E, S. Paetel, R. Wuerz, R. Menner, W. Wischmann, M. Powalla, Prog. Photovolt. Res. Appl. **19** (2011)
11. J.A. Frantz, R.Y. Bekele, V.Q. Nguyen, J.S. Sanghera, A. Bruce, S.V. Frolov, M. Cyrus, I.D. Aggarwal, Thin Solid Films **519** (2011)
12. S. Gulkowski, E. Krawczak, J.M. Olchowik. Proc. of the 31st EU PVSEC Conf. (2015)
13. S. Gulkowski, J. Civ. Eng. Environ. Archit., JCEEA, XXXIII **63** (2016)
14. C.Y. Hsu, P.C. Huang, Y.Y. Chen, D.C. Wen. Int. J. Photoenergy, **2013** (2013)
15. M. Acciarri, S. Binetti, A. Le Donne, B. Vodopivec, L. Miglio, S. Marchionna, M. Meschia, R. Moneta. Proc. of the 27th EU PVSEC Conf. (2012)
16. H. Kong, J. He, L. Huang, L. Zhu, L. Sun, P. Yang, J. Chu, Mater. Lett. **116** (2014)
17. G.-X. Liang, P. Fan, C. - M. Chen, Z.-H. Zheng, D.-P. Zhang, J. Alloy Compd. **610** (2014)
18. P. Lisco, P.M. Kaminski, A. Abbas, K. Bass, J.W. Bowers, G. Claudio, M. Losurdo, J. M. Walls, Thin Solid Films **582** (2015)
19. S. Lee, Eun Soo Lee, Tae Yeon Kim, Jun Sik Cho, Young Joo Eo, Jae Ho Yun, Ara Cho, Sol. Energy Mater. Sol. Cells **141** (2015)
20. A. Punoose, M. Marafi, G. Prabu, F. El Akkad, Phys. Stat. Sol. A **177** (2000)
21. D. Kim, Y. Park, M.Kim, Y. Choi, Y.S. Park, J. Lee. Mater. Res. Bull. **69** (2015)
22. A. Zdyb, E. Krawczak, P. Lichograj, Opt. Appl. **46** (2016)

Effects of hydrogen-charging on the properties of S235JR steel

Izabela Pietkun-Greber [1,*]

[1]Opole University of Opole, Independent Department of Process Engineering, ul. Dmowskiego 7-9, 45-365 Opole, Poland

Abstract. The paper presents the test results of the S235JR steel susceptibility to damage under the influence of hydrogen. The test of mechanical properties was performed on the basis of a static stretch test of non-hydrogenated samples and after cathodic polarization. Electrochemical measurements for the assessment of corrosion resistance of non-hydrogenated and hydrogenated steels were carried out using open circuit potential measurement and registering of potentiodynamic polarization curves in a three-electrode measuring system. Hydrogenation was carried out for between 3 and 24 hours in a solution of 0.1 N sulfuric acid (VI) with the addition of 2 mg/dm^3 of arsenic oxide (III) at an electric current density of 10 mA/cm^2. The hydrogen content in the steel before and after saturation with hydrogen was determined using the analyzer. Fracture samples after tensile test were observed using scanning electron microscope. The results of the research showed that as the hydrogen concentration in the examined steel increased (the lengthening of the saturation time), the deterioration of its mechanical and electrochemical properties occurred.

1 Introduction

Hydrogen penetration into the construction material occurs both at product's manufacturing stage and during its exploitation. In the welding zone, hydrogen can come from several sources, i.e. from moisture, water vapour, degradation of cellulose products from the electrode cover, protective gas and also from the grease present on the surface of the welded material or electrode [17]. In the marine environment, a biofilm containing anaerobes may be the source of hydrogen which reduce sulphates to sulphides and hydrogen sulphide.

The concentration of hydrogen in metals and alloys contributes not only to changes in their mechanical, physical and electrochemical properties, but also to the change in the performance properties of constructions or devices made of them. The level of property changes depends on factors that characterize the state of the alloy, its stress state, the level of deformation associated with it, and the time of being subjected to the surrounding environment's influence.

The effect of hydrogen in steels and alloys is a complex phenomenon. The level of mechanical properties degradation is the basic criterion for material destruction. Slow Strain Rate Test (SSRT) with obtained narrowing changes as its result is particularly informative, or, to a slightly lesser degree, the length changes [1,2]. Structural criteria are also used in assessing the hydrogen degradation level. The presence of trapped hydrogen in defects in the microstructure affects the change of materials' fracture mechanisms [3]. Differences in hydrogen concentration are not only reflected on the surface of the breakthroughs after the tensile test [4,5] but also in deterioration of corrosion properties [6-8]. In addition, recent research has shown that the method of quantitative assessment of breakthroughs can be used in assessing the hydrogen degradation level of alloys [9].

There are many research results of hydrogen degradation in various materials (steels, alloys) [10-13], which can be found in reference books. The authors of these works have evaluated the destruction of materials based on only one criterion, i.e. mechanical or structural properties. It should also be noted that there are not many works presenting the results of research on hydrogen influence on steel degradation which take the change in mechanical properties along with corrosion resistance into account.

The S235 steel belongs to unalloyed structural steels. It is widely used for welded, load-bearing and dynamically loaded constructions [14]. Due to operating conditions, this type of steel may undergo hydrogen degradation during the production of welded constructions (in the form of cold cracks) or in the process of utilizing the structure (in the form of hydrogen induced cracks HIC).

Since there is little information about the hydrogen degradation of unalloyed structural steels and about the assessment of their degradation according to mechanical properties criterion including corrosion resistance level, these issues have become the subject of the undertaken research. The results obtained may be useful to researchers interested in the hydrogen degradation of iron alloys. What is more, they can be used by engineers

* Corresponding author: ipietkun@uni.opole.pl

who are involved in the process of designing materials resistant to hydrogen degradation, and to help improve the efficiency of choosing the material during designing process.

2 Research material and methodology

2.1 Material

S235JR steel in the form of 2.0 mm thick metal sheet was used (Table 1).

Table 1. The chemical composition of the tested S235JR steel (Fe rest).

Steel	Chemical elements [wt.%]							
	C	Mn	Si	P	S	Cu	Cr	Ni
PN-EN 10025-2	0.17	1.40	-	0.025	0.025	0.55	-	-
Spectra analysis	0.13	0.98	0.01	0.013	0.015	0.02	0.03	0.02

The analysis of the chemical composition was conducted using an optical spectrometer. Table 1 lists the content requirements for S235JR steel according to PN-EN 10025-2 [14] and the average content of the elements in the check chemical analysis of the tested S235JR steel.

2.2 Hydrogenation of samples and determination of hydrogen content

Hydrogenation of rectangular specimens of 10 x 10 x 2 was carried out in a solution of 0.1 N H_2SO_4 with the addition of arsenic oxide (III) (2 mg/dm^3). An electric current density of 10 mA/cm^2 was used for between 3 and 24 hours. The cathodic polarization conditions were chosen in such a way as to ensure the same hydrogen concentration across the sample.

The amount of hydrogen in the samples of the tested steel before and after polarization was determined using the LECO ONH836 analyzer. The hydrogen content of the steel was determined in ppm. The hydrogen content in steel was measured on three samples at delivery state and after saturation with hydrogen. In order to limit the risk of hydrogen desorption in the alloy until measuring its concentration in the material, the samples were stored in liquid nitrogen.

2.3 Study of mechanical properties

A tensile test was used to evaluate the susceptibility of the examined steel by using flat samples with an initial measuring length of 50 mm and a cross section of 2 x 12.55 mm in accordance with PN-EN ISO 6892-1 [15]. The test specimens were wet ground with sandpaper, gradation of 240 to 1200, and then degreased with acetone. The tests were performed on the ZWICK/ROELL endurance machine with a deformation rate of 0.00003 s^{-1}. Due to the significant effect of the hydrogen desorption process from the alloy on the obtained results of mechanical properties (change of hydrogen concentration), the tensile tests were carried out immediately after the hydrogenation was completed. Measurements were made at 21÷23°C. Each study was carried out on 3 samples with characterizing the state of the material without hydrogen and after hydrogenation.

The assessment of the hydrogen damage of steel was made on the basis of the so-called hydrogen embrittlement index HEI. The HEI index is determined by the percentage change of the determined parameter after the sample is saturated with hydrogen in relation to this parameter before saturation [3,16].

The higher the value of the HEI index, the greater the susceptibility of steel to hydrogen degradation.

$$HEI = \left|\frac{X_H - X_{wH}}{X_{wH}}\right| \cdot 100\% \qquad (1)$$

where:

X_H - value of the parameter in the presence of hydrogen,
X_{wH} - value of the parameter without hydrogen.

2.4 Microscopic observations

The microstructure of the investigated S235JR steel was evaluated on metallographic polished sections digested with 3% Nital using the Hitachi S-3400N scanning electron microscope. After the tensile test, the assessment of hydrogen influence on the change in the nature of the fracture was conducted. Contamination from samples was removed in an ultrasonic scrubber using acetone. Fracture samples were observed on the Hitachi S-3400N scanning electron microscope.

2.5 Electrochemical research

Electrochemical measurements for assessing the corrosion resistance of non-hydrogenated and cathodic polarized steel were conducted using open circuit potential measurement (E_0) and registering the potentiodynamic polarization curves in a three-electrode measuring system. The measuring system consisted of a measuring vessel, an AMEL PSW01 System 5000 potentiometer and a computer with "CorrWare" software. The auxiliary electrode was made of platinum, while the calomel saturated electrode (NEK) was used as the reference electrode. A 3% aqueous NaCl solution at 23°C was used for the study. Depending on the type of the test, the work electrode (WE) was respectively non-hydrogenated and hydrogenated steel samples of approximately 50 mm^2 surface. The test specimens were ground with waterpaper of 1200 grit and then thoroughly rinsed with distilled water, avoiding greasing the ground surfaces. The ground samples were purified in ethyl

DOI: 10.1051/e3sconf/20171903007

alcohol using ultrasonic washer, rinsed with distilled water and dried with blotting paper. The samples prepared in such a way, depending on the type of the test, were placed directly in the electrochemical vessel or in the electrolyser to saturate with hydrogen. Before conducting polarization measurements of non-hydrogenated and hydrogenated samples, the measurements of the open circuit (E_{OCP}) were carried out to determine the value of the initial corrosion potential (E_{corr}). Once the open circuit potential has stabilized, they were polarized at a potential scanning rate dE/dt = 1 mV/s in the area of ±300 mV from the open circuit potential value (E_{OCP}). The corrosion current density (I_{corr}), corrosion potential (E_{corr}) and polarization resistance (R_p) were determined using the Stern method.

3 Results and discussion

3.1 Microstructure of steel

The Hitachi S-3400N electron microscope studies show that the microstructure of the S235JR steel is composed of ferrite and perlite grains (Fig. 1).

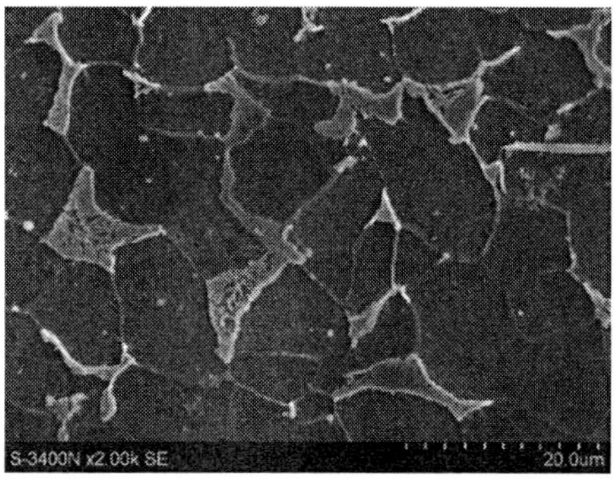

Fig. 1. S235JR steel structure.

3.2 The amount of hydrogen in steel under different cathodic polarization conditions

Fig. 2 shows the results of the measurements of hydrogen content in the steel samples tested before and after saturation with hydrogen.

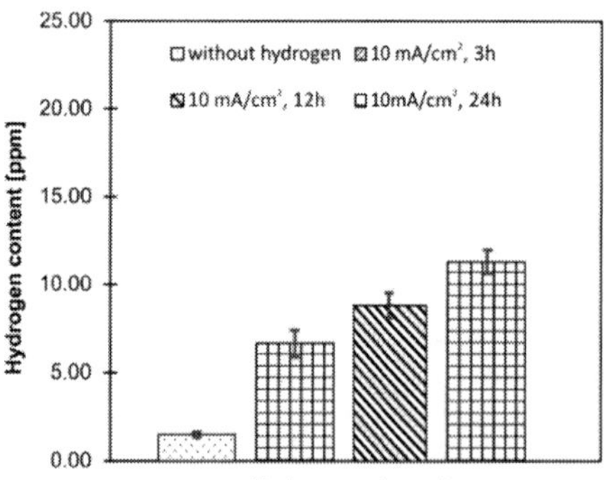

Fig. 2. Hydrogen content in the researched S235JR steel.

From the analysis of the data in Fig. 2, it was observed that with the increase of cathodic polarization time, the concentration of absorbed hydrogen in the steel increased. In the samples of non-polarized steel, the hydrogen content was 1.50 ppm.

3.3 The effect of hydrogenation conditions on mechanical and electrochemical properties

The following parameters were used to evaluate the effect of hydrogen on the mechanical properties of the tested S235JR steel: percentage elongation (A), percentage reduction (Z), tensile strength (R_m) and the yield strength ($R_{p0.2}$). The stress-elongation curves of the steel samples before and after the hydrogenation are shown in Fig. 3. Table 2 summarizes the results of the strength and plastic properties tested in the tensile test.

Based on the experimental results (Table 3) from relation (1), the degree of hydrogen damage was determined taking into account the percent reduction of area of non-hydrogenated and hydrogenated samples. The results are shown in Fig. 4.

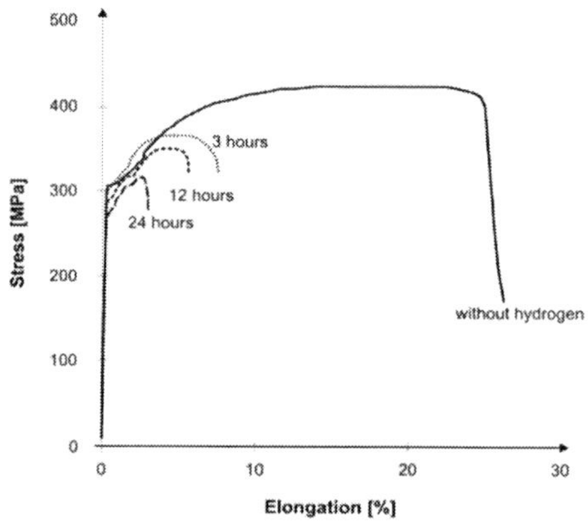

Fig. 3. S235JR steel stress-elongation curves before and after hydrogenation.

The analysis of the data in Table 3 shows that the change in the cathodic polarization conditions of S235JR steel has a different effect on the change of its mechanical properties, i.e. strength and plasticity. The analysis of the mechanical properties of the post-hydrogenated steel showed a decrease in the yield point from 324 MPa (the sample stretched in air without hydrogen interaction) to 281 MPa after cathodic polarization lasting 24 hours. The tensile strength of steel was reduced from 425 MPa (without hydrogen interaction) to 320 MPa with hydrogen content of 11.30 ppm. The analysis of the stress-elongation curves (Fig. 3) of steel showed that hydrogen saturation contributes to reducing its percentage elongation. After 3 hours of cathodic polarization, the percentage elongation of steel has decreased about 3 times (from 26.4% to 8.9%), and 4 times after the polarization period has been extended to 12 hours. The steel percent reduction of area decreased from 45% to 19.6% at the hydrogen content of 6.66 ppm and about 11 times at the hydrogen content of 13.30 ppm. HEI's index ranged from 56% to 86% and increased along with the polarization time (Fig. 4).

surface of S235JR steel after hydrogenation showed a change in the fracture character. The samples of steel prior to hydrogenation were macroscopically ductile (Fig. 5). The samples of steel subjected to 3 hours of

Table 2. Mechanical properties of S235JR tested before and after hydrogenation.

Sample	$R_{p0,2}$ [MPa]	R_m [MPa]	A_{50mm} [%]	Z [%]	HEI [%]
as-received	324	425	26.4	45.0	-
3 hours	300	383	8.9	19.6	56
12 hours	297	379	6.6	11.6	74
24 hours	281	320	5.2	6.4	86

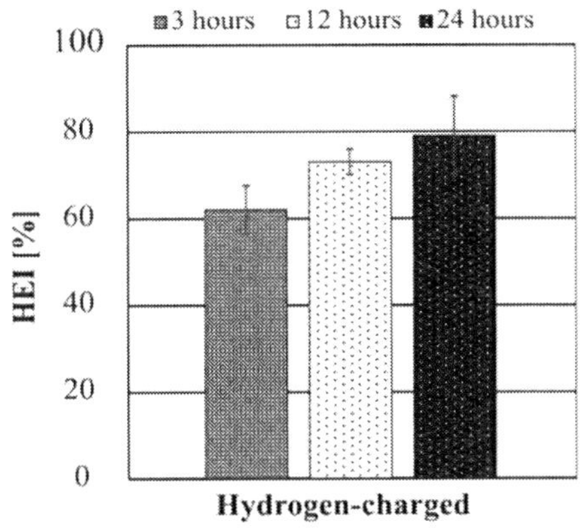

Fig. 4. HEI value S235JR steel.

The fractographic analysis revealed that the surface of the S235JR non-hydrogenated sample was characterized by the morphology typical for the breakthroughs of a plastic nature. Cracking occurred in a plane perpendicular to the direction of maximum tensile stresses (Fig. 5).

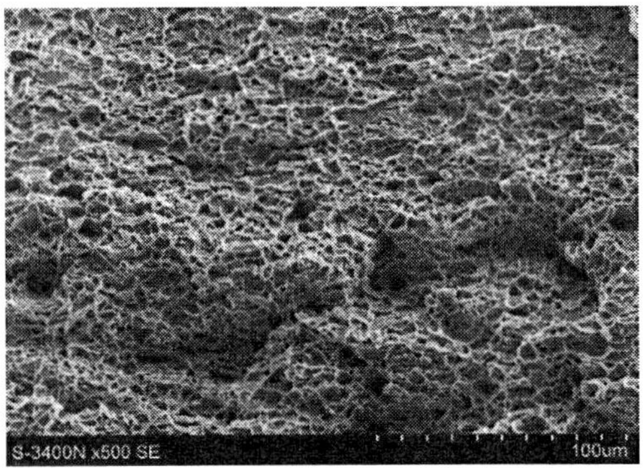

Fig. 5. View of the fracture surfaces of samples of steel unhydrogenated and stretched.

Surface fractographic studies of the surfaces of hydrogenated steel samples and then subjected to tensile tests revealed changes in the morphology of the fractures. The fractographic analysis of the fractal

polarization at cathodic current density showed a ductile and transcristalline mixed fracture (Fig. 6). At the surface of the fractures there were cracks initiated around non-metallic inclusions.

The chemical composition of the non-metallic inclusions was determined on the Hitachi S-3400N scanning microscope using energy dispersive X-ray spectroscopy (EDS). The analysis showed that the extracts consist mainly of sulfur and manganese (Fig. 7).

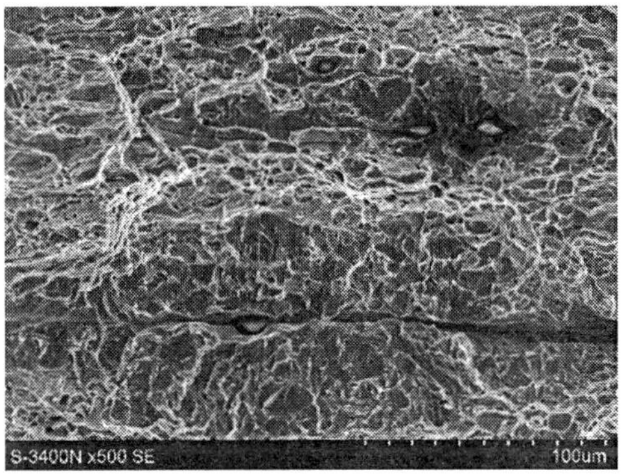

Fig. 6. View of the fracture surfaces of samples of steel hydrogenated for 3 hours and stretched. Cracking with sulfide inclusions.

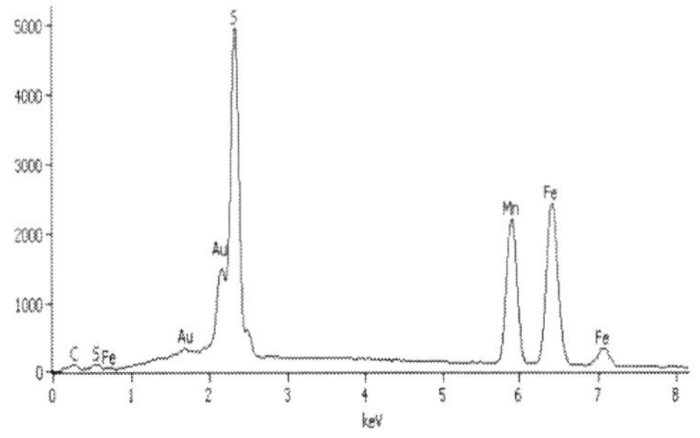

Fig. 7. X-ray spectrum of non-metallic inclusions.

For longer hydrogenation times of steel samples S235JR (i.e. 12 and 24 hours), numerous blisters and cracks were observed on the side surfaces of the samples. The crack mechanism was transcristalline (Figs. 8-9).

The course of the potentiodynamic curves of the non-hydrogenated and cathodic polarized S235JR steel is shown in Figure 10. The potentiodynamic results are summarized in Table 3. The S235JR value of the corrosion potential (E_{corr}) of hydrogen-free S235JR steel measured in 3% NaCl solution was -507 mV. On the potentiodynamic curve of the samples tested without hydrogen saturation no passive range was observed (Fig. 10). The analysis of the data in Table 3 shows that hydrogen absorbed by the steel causes a change in its value of the corrosion potential (E_{corr}).

Potentiodynamic polarization curves for steel samples subjected to hydrogenation in the period of 3 to 24 hours showed no passive range. The increase in the value of the potential caused a continuous increase in the current density. The potential value measured in 3% NaCl solution, depending on the amount of hydrogen absorbed during the cathodic polarization, ranged from -581 mV to -623 mV. The decrease in polarization resistance was accompanied by an increase in the corrosion current density from 13.90 µA/cm^2 to 19.09 µA/cm^2.

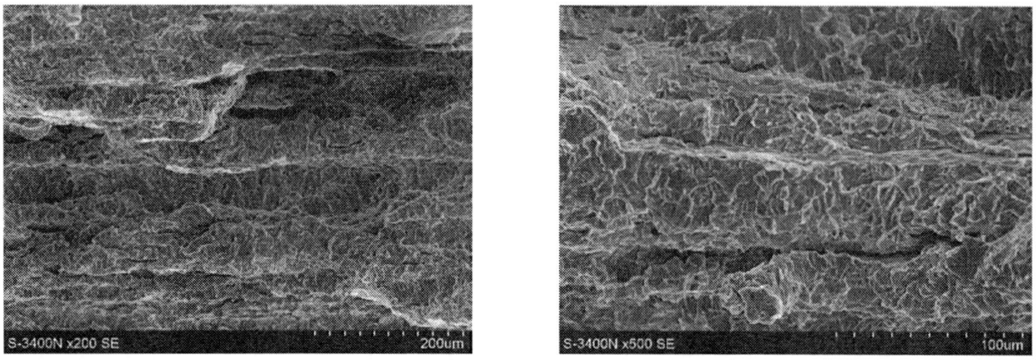

Fig. 8. View of the fracture surfaces of samples of steel hydrogenated for 12 hours and stretched.

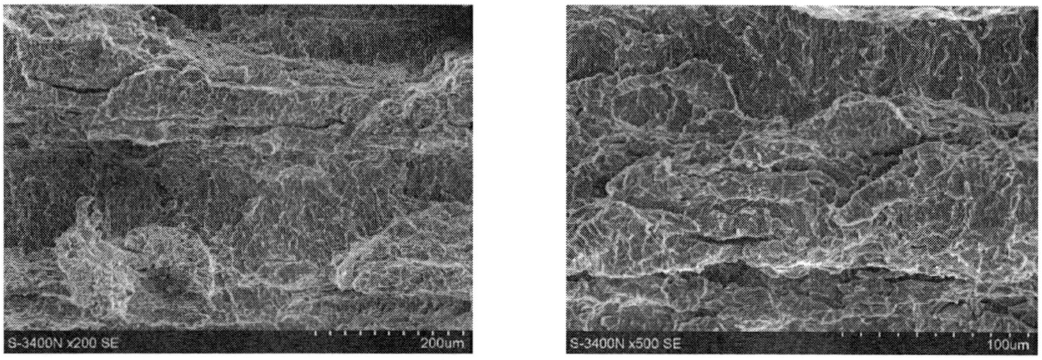

Fig. 9. View of the fracture surfaces of samples of steel hydrogenated for 24 hours and stretched.

Table 3. The results of potentiodynamic measurements S235JR.

Sample	E_{OCP} [mV]	E_{corr} [mV]	I_{corr} [µA/cm^2]	R_p [kΩ/cm^2]
as-received	-606	-507	13.90	1.88
3 hours	-619	-581	15.25	1.70
12 hours	-623	-600	17.37	1.44
24 hours	-662	-623	19.09	1.25

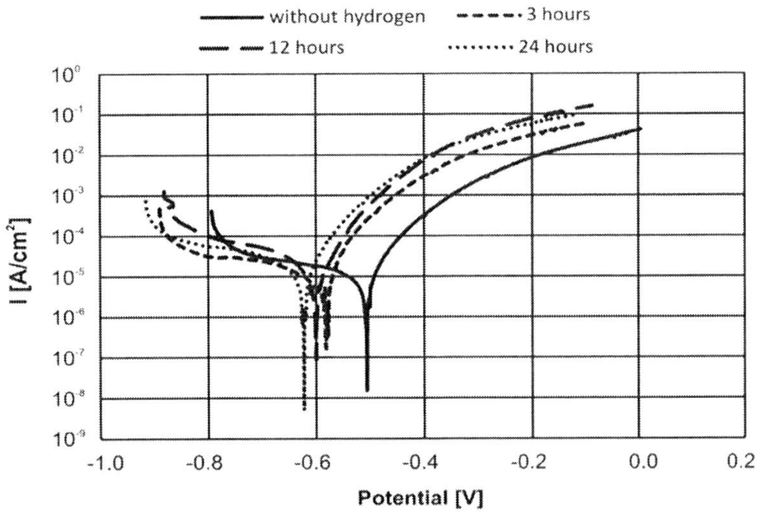

Fig. 10. Polarization curves of S235JR unhydrogenated and hydrogenated steel, studied in the 3% NaCl solution.

4 Summary

The study showed that the duration of cathodic polarization influences the amount of hydrogen absorbed by the tested S235JR steel. With the increase in hydrogenation time, the hydrogen concentration in the steel samples increased. The impact of hydrogen in S235JR steel leads to reduction of its strength and plastic properties, i.e. the elongation and constriction measured in the tensile test. The drop in plastic properties caused by hydrogen has changed the nature of the fracture. Samples stretched in the air had a ductile fracture, and after hydrogen interaction, a mixed and transcrystalline fracture. At the surface of the fractures there were cracks initiated around non-metallic inclusions. The effect of hydrogen in S235JR steel also leads to a deterioration of its corrosion resistance measured in 3% aqueous NaCl solution. It has been found that the hydrogen saturation time contributes to the change in corrosion resistance of the tested steel measured in 3% NaCl solution. It was shown that, as the hydrogen content increased, the value of the corrosion potential (E_{corr}) decreased in the tested samples. The decrease in polarization resistance was accompanied by an increase in the corrosion current density.

References

1. A.P. Moon, R. Balasubramaniam, B. Panda, Mater. Sci. Eng. A **527** (2010)
2. J. Ćwiek, J. Ach. Mater Manufa. Eng. **47**, 1 (2011)
3. M. Sozańska, The "fish-eye" – like hydrogen damage in selected group of steel used in power industry (Pol. Śląska, Gliwice, 2006)
4. H. Yashiro, P. Bound, N. Kumagai, and K. Tanno, Corros. Sci. **40** (1998)
5. Y. Liu, M. Wang, G. Liu, Mater. Sci. Eng. A **594** (2014)
6. I. Pietkun-Greber, Ecol. Chem. Eng. A **23**, 3 (2016)
7. J. Michalska, B. Chmiela, J. Łabanowski, W. Simka, J. Mater. Eng. Pernorm. **23**, 8 (2014)
8. I. Pietkun-Greber, A. Mościcki, M. Sozańska, Proc. ECOpole **10**, 1 (2016)
9. M. Sozańska, A. Mościcki, B. Chmiela, Arch. Metall. Mater. **62** (2017)
10. R.A. Siddiqui, H.A. Abdullah, J. Mater. Pro. Tech. **170** (2005)
11. C.F. Dong, Z.Y. Liu, X.G. Li, Y.F. Cheng, Int. J. Hyd. Energ. **34** (2009)
12. J. Sanchez, S.F. Lee, M.A. Martin-Rengel, J. Fullea, C. Andrade, J. Ruiz-Hervias, Eng. Failure Analysis **59** (2016)
13. J. Ćwiek, Hydrogen degradation of high strength weldable steels (Pol. Śląska, Gliwice, 2006)
14. PN-EN 10025-2:2007
15. PN-EN ISO 6892-1:2016
16. V. Tkachov, Mater. Sci. **36**, 4 (2000)

DOI: 10.1051/e3sconf/20171903008

Klinkenberg effect in hydrodynamics of gas flow through anisotropic porous materials

Grzegorz Wałowski[1,*] and *Gabriel* Filipczak[2]

[1]Institute of Technology and Life Sciences, Poznan Branch, Department of Renewable Energy Sources, street Biskupińska 67, 60-463 Poznań, Poland
[2]Faculty of Mechanical Engineering, Chemical Engineering Department, Opole University of Technology, street Mikołajczyka 5, 45-001 Opole, Poland

Abstract. This study discusses results of experiments on hydrodynamic assessment of gas flow through backbone (skeletal) porous materials with an anisotropic structure. The research was conducted upon materials of diversified petrographic characteristics, both natural origin (rocky, pumice) and process materials (char and coke). The study was conducted for a variety of hydrodynamic conditions, using air, as well as for nitrogen and carbon dioxide. The basis for assessing hydrodynamics of gas flow through porous material was a gas stream that results from the pressure forcing such flow. The results of measurements indicate a clear impact of the type of material on the gas permeability, and additionally – as a result of their anisotropic internal structure – to a significant effect of the flow direction on the value of gas stream.

1 Introduction

The gas flow through porous media, both in the aspect of their application in the industrial technology and their presence in the natural environment, is a very complex issue that is not still fully recognised. As for a phenomenological aspect this flow may be subject to various hydrodynamic criteria, which is affected by a medium structure, a kind of fluid (one- and multi-phase) and a flow forcing method (gravitational, pressure). A vast array of publications pertaining to this issue and thoroughly analysed, among others, in studies by Strzelecki [1], Piecuch [2], Orzechowski [3] or Błaszczyk [4] describing this issue on a research and analytical basis *de facto* refer to the phenomenon of filtration and they are generally identified with the phenomenon of laminar fluid flow through granular deposits according to Darcy's law [5]. Obviously, it does not exhaust many other examples of the flow of fluids through porous media. For the turbulent flow of fluids the Forchheimer model [6] and the Ergun model should be distinguished [7]. The more advanced description of the flow for the spatial layout of capillaries in the form of the meandering channels is also found in the Kozeny-Carman model [8].

The reference books frequently discuss other models of hydrodynamics of one- and multi-phase fluids flowing through porous media, considering the impact of fluid features and a kind of porous medium on the flow through granular deposits [2–4, 9]. However, the vast majority of those models pertain to granular deposits but only a few studies analyse gas permeability through backbone (skeletal) porous materials.

In this context, our own research assesses conditions of hydrodynamics of the gas flow through backbone (skeletal) porous materials with an anisotropic structure. The results of research upon the assessment of gas permeability of various solid porous materials have been presented and the assessment of process conditions concerning hydrodynamics of the gas flow through materials with a diversified internal structure has been conducted.

2 Scope and research methodology

To familiarise with hydrodynamic conditions of the gas flow through porous materials, detailed experimental tests were conducted to assess the gas permeability of porous materials with the diversified structure and, at the same time, the diversified process characteristics.
The research material comprised various types of solid skeletal constructions, including those natural and those deriving from the thermal carbon gasification technology. Most of them were coal char *in situ*, coke, melted waste rock *ex situ*, and polyamide agglomerate of symmetrical spatial structure.

All the types of materials applied in research underwent the assessment of selected parameters describing features characteristic for porous materials resulting from their porosity and physical structure as basic process quantities affecting the hydrodynamics of the gas flow through porous materials.
The quantity-based assessment applied to such parameters as the apparent density and porosity of a specific type (sample) of the porous material. In this regard, structural research upon the tested porous

* Corresponding author: g.walowski@itp.edu.pl

materials conducted on the basis of the SEM scanning image [9] was helpful.

The experiments pertained to two different measurement systems thoroughly analysed in other own works [10, 11]. The first system was used to assess the permeability of porous materials in the barbotage conditions. In this case, the shape of samples resulted from naturally obtained parts of the native material with an unspecified sample shape - Fig. 1a.

The latter one was applied to analyse the permeability on the basis of the samples configured to the shape of the cubic solid - Fig. 2a. In this system the gas flow might be directed with respect to the arbitrarily selected X, Y and Z axes. This required the development of a special measurement system that is currently being patented [12]. In their geometrical form those cubic-shaped samples were parts of volume samples and were compared with them with respect to their internal structure.

Fig. 3 shows some examples of materials used in the experiments. We can see that these materials are very different in their structure, which in addition is anisotropic in every case. Furthermore, for char (Fig. 3a) the slot-porous structure is dominant. Of course, each material has such a different structure in directional cross section (X, Y, Z), as for the coke is shown in Fig. 4.

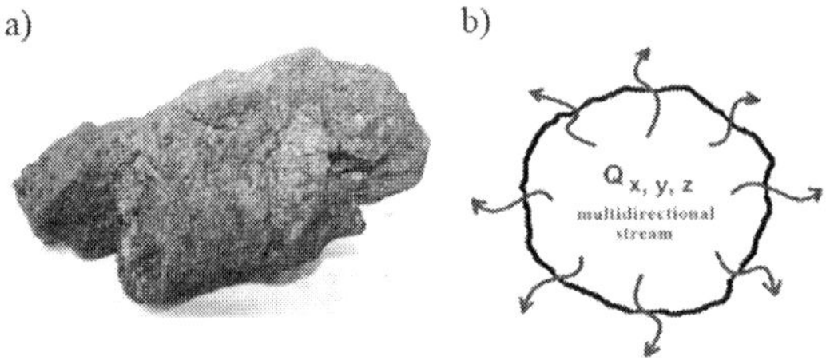

Fig. 1. Sample of unspecified shape: a) research material, b) flow chart (multi-directional - fractal flow).

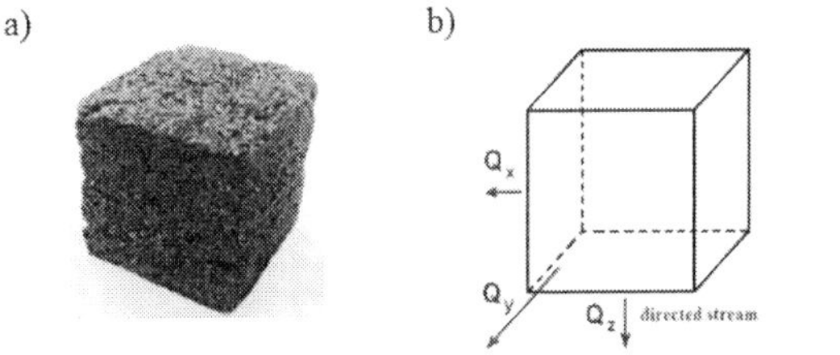

Fig. 2. Cubic-shaped sample: a) research material, b) flow char (X, Y, Z direction flow).

The investigation tests were conducted with reference to various gases (air, nitrogen, carbon dioxide) to the extent of the permeability stream resulting from the reference pressure. The permeability function of the pressure decline in the porous deposit was independently carried out, assuming the so-called multi-directional (fractal) system for the gas flow through samples with unspecified shapes (Fig. 1.b) and the directional flow XYZ characteristic for cubic-shaped samples (Fig. 2.b).

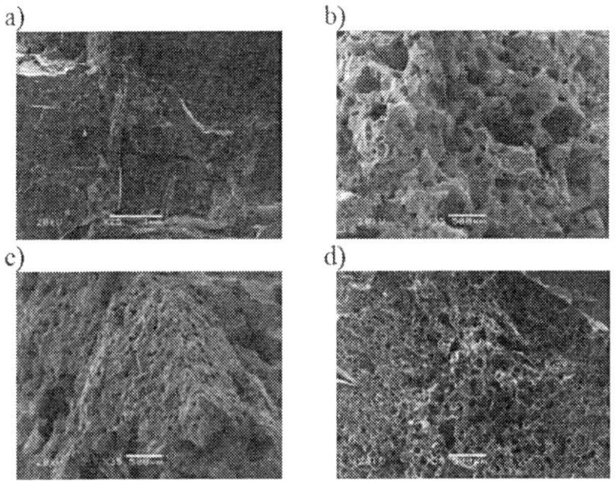

Fig. 3. Structure of tested materials: a) char *in situ*, b) coke, c) partially melted waste rock, d) char *ex situ*.

Fig. 4. Sample of coal-char structure in a different direction.

A detailed description of the experimental set-up is given in source work [9], where a description of the test conditions is also provided.

3 Research results and their analysis

The basis for assessing the hydrodynamics of the gas flow through deposits and porous materials is the characteristic of their permeability resulting from the pressure inducing this flow. In each case, this characteristic is determined by calculating the impact of the available overpressure on the obtained gas stream or vice versa - the impact of the gas stream on the value of this overpressure that corresponds to a decline in this stream pressure. In the latter case, this corresponds to the determination of complete resistances of the gas flow through such deposit.

The results shown in Fig. 5 prove that with respect to porous materials in the form of coal char the nature of changes to gas permeability functions is highly diversified. For the same coal char *in situ* (I-1, I-2, I-3) there are obtained highly different permeability characteristics and their common deviation is expressly affected by the structure of the porous material.

Moreover, those characteristics are parabolic, which proves their similarity to the hydrodynamics of the flow through the closed channels. On the other hand, the non-linear tendency of those characteristics proves the dominance of the turbulent flow, which is also associated with the deviation from Darcy's law [13,14].

The comparison to other materials as shown in Fig. 5 shows that coal char *in situ* (I-1, I-2, I-3) are porous structures which are much more permeable compared to coke (II-1, II-2, II-3) or melted waste rock (III-1, III-2, III-3).

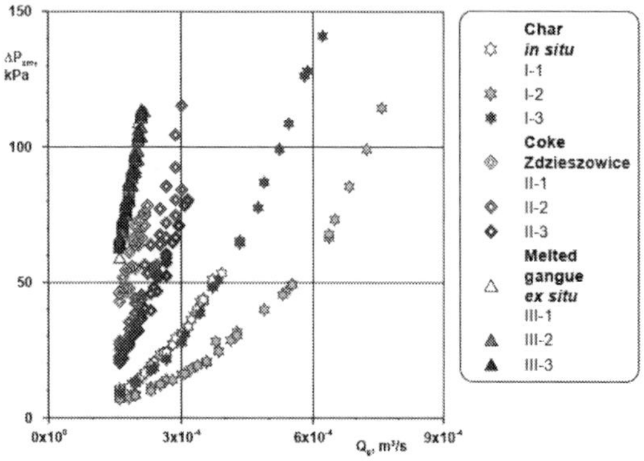

Fig. 5. Permeability of various kinds of porous materials (volume sample); I- coal char *in situ*, II- coke (Zdzieszowice), III- melted waste rock *ex situ*.

As for coke which has the highest porosity in this group this proves that a large part of its pores is closed for the gas flow and, concurrently, that the more complex structure of coal chars has characteristics of the gap medium, which at the same reference pressure ensures a much better gas permeability for this medium. On the other hand, in comparison with the aforesaid materials the permeability of the waste rock is undoubtedly justified by a relatively smaller porosity of this material and with the less participation of pores open for the flow [13].

The analogous characteristics of permeability were made for the cubic specimen (20×20×20 mm) by using the measurement system assessing permeability in the directional flow (Fig. 2). In this measurement, example results of measurements for coal char *in situ* and sintered polyamide agglomerate are illustrated in Fig. 6, which shows characteristics of the air permeability in three independent flow directions (X, Y, Z).

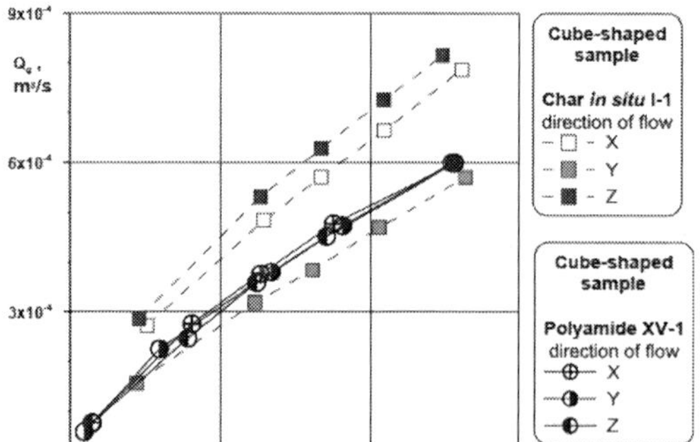

Fig. 6. Layout of experimental points characterizing the asymmetry of air flow with respect to three axes of cube specimen (XYZ) for coal char and sintered polyamide.

The layout of experimental points shows that the permeability of coal char is considerably affected by the direction of the gas flow. This proves the explicit effect of the flow asymmetry with respect to the selected flow direction (axis) and, consequently, the explicit anisotropic structure of this type of material (Fig. 4). On the other hand, for the porous polyamide that forms the agglomerate of spherical particles of identical dimensions (diameter of 0.1 mm), the permeability characteristics is not practically affected by the gas flow direction. At the same time, the determined characteristics prove that this porous polyamide, despite the fact its porosity is much smaller compared to coal char (30% [9]), has similar characteristics of the gas flow. This proves the observation that the greater effect of the permeability of various kinds of coal char is rather the result of their porous and gap structure rather than the result of their porosity. It is interesting that the permeability characteristic of the porous polyamide is also of non-linear nature, which - with respect to the measurements - proves the advantage of the turbulent gas flow.

At the same time, the research results shown in Fig. 7 prove that the permeability of the porous material is not affected by the sample shape but by its internal structure. The layout of the experimental points shows that in the same conditions of the reference pressure between the volume coal char sample and the cubic-shaped sample (the figure shows averaged air flow values), the permeability characteristics of this material are of a similar nature and within the same scope of values.

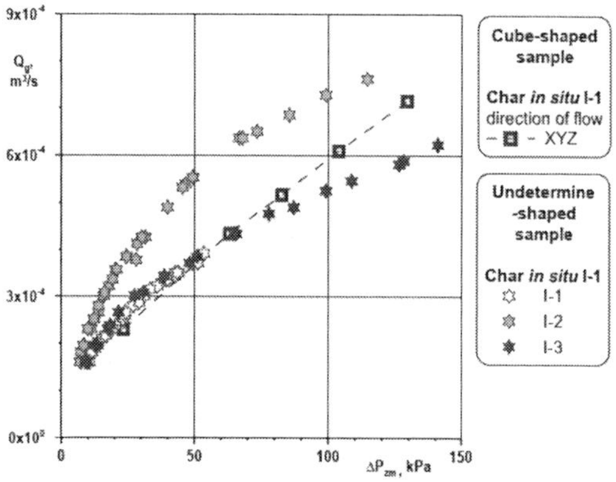

Fig. 7. Gas permeability of coal char for samples of various shapes.

The comparison of permeability characteristics of coal char *in situ* with respect to the flow of air, nitrogen and carbon dioxide for cubic-shaped specimens (with dimensions of 20×20×20 mm) is shown in Fig. 8.

Regardless of the detected anisotropy of this material, it may also be observed here that to the entire extent of the reference pressure, the permeability of this coal char *in situ* is less for air than for nitrogen but considerably higher than for carbon dioxide. For this latter gas the limited permeability characteristics was also observed. Undoubtedly, this refers to the choking phenomenon that limits the permeability growth and increases the reference pressure. This may be affected by the fact that the density is higher compared with the remaining ones but according to the fact proven in reference books [15] it is more reliable that the hyphenation choking effect expressed by a pressure decline in the porous material is greater, the greater resistance gas flowing the dense network of micro channels encounters (adiathermal conditions). On the other hand, other researchers [16] suggest that this may be affected by the so-called Klinkenberg's effect – as a phenomenon that limits the movement of gas molecules with sizes of flow (pores) channels. Furthermore, this may be due to the different scale of gas expansion when flowing through the microcapillaries. In the latter case, attention is drawn to the significance of this phenomenon, among others, in the aspect of the sequestration of carbon dioxide in geological porous deposits.

On the other hand, the observed differences in permeability for the different flow directions of the gas are undoubtedly the result of a differentiated flow pattern in each direction. This is revealed itself of each different flow direction, different for each direction, as well as the different geometry of the capillaries, the shape and size of which are for flowing gas very different and at the same time which are difficult to assess unequivocally.

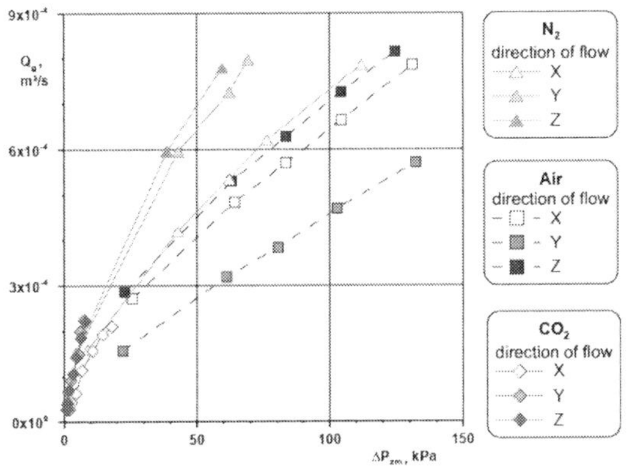

Fig. 8. Permeability of char *in situ* depending on a working medium (cubic specimen with dimensions of 20×20×20 mm).

4 Conclusions

The recognition of the issue of hydrodynamics of the gas flow through skeletal porous media shows that the reference books contain very little information. In this respect, relevant experiments on porous materials were conducted and hydrodynamic phenomena assessments resulting from gas flow resistances were carried out.

The hydrodynamic results of gas flow through porous beds with skeletal structure can in many cases be used in the process calculations of hydrodynamics of gas flow through porous beds, especially in the context of biofuel of shale ore [17], biofilm formation by methanogenic microorganisms [18]. Retention in the process of underground bioconversion when selecting strains of microorganisms for maximum biogas yield [19].

The study conducted as part of the project financed by the National Centre for Research and Development conducted in the BIOSTRATEG program, contract No BIOSTRATEG2/298357/8/NCBR/2016 dated 18 May 2016.

References

1. T. Strzelecki, S. Kostecki, S. Żak, *Modelling of flows through porous media* (Dolnośląskie Wydawnictwo Edukacyjne, Wrocław, 2008) [in Polish]
2. T. Piecuch, Rocznik Ochrona Środowiska **11** (2009)
3. Z. Orzechowski, J. Prywer, R. Zarzycki, *Mechanics of fluids in engineering and environmental protection* (WNT, Warszawa, 2009) [in Polish]
4. M. Błaszczyk, *Study of migration processes of petroleum substances and their emulsions in porous structures*, PhD dissertation, Dep. of Chemical Eng., Lodz University of Technology, Łódź, 2014 [in Polish]
5. B. Bębenek, *Energy losses in fluid flows*, (Wydawnictwo Politechniki Krakowskiej, Kraków, 1987) [in Polish]
6. A.M. Amao, *Mathematical model for Darcy and Forchheimer flow with applications to well performance analysis*, Master's thesis, Department of Petroleum Eng., Texas Tech. University, 2007
7. S. Ergun, Chemical Eng. Prog. **48**, 2, 89-94 (1952)
8. Z. Kembłowski *Theoretical fundamentals of chemical and process engineering* (WNT, Warszawa, 1985) [in Polish]
9. G. Wałowski, *Hydrodynamics of gas flow through porous materials*, PhD dissertation, Faculty of Mechanical Engineering, Opole University of Technology, Opole 2015 [in Polish]
10. G. Wałowski, G. Filipczak, Inż. Ap. Chem. **51**, 6, 396-397 (2012)
11. G. Wałowski, G. Filipczak, Inż. Ap. Chem. **52**, 6, 581-582 (2013)
12. G. Wałowski, G. Filipczak, E. Krause, *System for determination the coefficient of gas permeability through porous materials with anisotropic structure*, Patent RP 225980, Warszawa 2016.
13. G. Wałowski, G. Filipczak, J. Sustainable Min (2017), http://dx.doi.org/10.1016/j.jsm.2017.03.003
14. G. Wałowski, G. Filipczak,. Inż. Ap. Chem. **55**, 6, 245-250 (2016)
15. S. Michałowski, K. Wańkowicz, *Termodynamika procesowa* (WNT, Warszawa, 1993)
16. J. Blicharski, R. Smulski, AGH Drilling Oil Gas Vol. **29** 1, 89-96 (2012)
17. C. Ernst et al. J. Chem. Technol. Biotechnol. **88**, 2115-2132 (2013)
18. D. Chudy, S. Jabłoński, M. Łukaszewicz, Acta Sci. Pol. Biotechnologia, **10**, 4, 17-28 (2011)
19. A. Stachowiak i in., *Brown coal deposit in the Legnica-Ścinawy region and technologies of their development* (Mezozoik i Kenozoik Dolnego Śląska, WIND, Wrocław, 121-135, 2011) [in Polish]

Wear of carbide inserts during turning of C45 steel in dry cutting conditions and in presence of emulsion mist

Eugene E. Feldshtein[1,*], and *Szymon* Wojciechowski[2]

[1]University of Zielona Gora, 4 Prof. Z. Szafrana street, 65-516 Zielona Gora, Poland
[2]Poznan University of Technology, Faculty of Mechanical Engineering and Management, Institute of Mechanical Technology, Poznan, Poland

Abstract. This paper describes the wear features of indexable carbide P25 inserts when turning C45 steel with cutting speeds of 250–450 m/min in dry cutting conditions and in the presence of an emulsion mist (MQCL method). As the active component of the emulsion mist 5%, the aqueous emulsion of EMULGOL concentrate was used. While analyzing the size, shape and formation dynamics of tool wear, the larger dimensions were found for dry cutting. The statistical analysis of results allowed calculating the dependencies to determine the wear rates. The SEM analysis revealed significant oxygen contents at the contact areas that are due to high temperatures in the cutting zone. When cutting with the emulsion mist the presence of phosphorus and copper entering in EMULGOL concentrate was fixed directly on the wear areas. It suggests the possibility of using compounds included in the emulsion mist to form AW and EP films at the contact areas in the cutting zone.

1 Introduction

Under present conditions the characteristic factor for the production intensification is the requirement of the productivity growth and reducing the processing cost. This can be achieved, while ensuring efficient operation of the cutting tools and with the use of materials with good machinability. Improved machinability of the material can be achieved in various ways, in particular through effective lubricating and cooling technological mediums and methods for their introduction into the cutting area [1, 2].

In latter case, various technical solutions can be used. In particular, the technological fluid can be introduced into the cutting zone with an elevated pressure through the cutting tool or from the outside [2], or the machining is realized in dry or "almost dry" conditions (near dry cutting).

There are two possible ways in this case: the use of oil mist (minimum quantity lubrication, MQL) or emulsion mist (minimum quantity cooling, MQC). Since the particles of oil, soap, various additives are always contained in the emulsion, the method called MQCL (minimum quantity cooling and lubrication) can be discussed.

Cooling methods based on very low amounts of coolants and lubricants, in comparison to dry cutting, enable the reduction of surface roughness [3-5], tool wear [6, 7], cutting force [5, 8], cutting temperature [9] and advantageous chip shape modification, as well as the [10] beneficial stress distribution [11, 12]. Moreover, in order to improve the lubricating conditions of the MQL method, the nanoparticles made of silver, copper and molybdenum disulfide, as well as the carbon nanotubes are being provided [8, 13]. However, in case of MQCL method, the extreme pressure EP and anti-wear AW additives are being applied [10, 14]. In these techniques, the oil and emulsion mist formation parameters play an important role. The researches [15, 16] revealed that the lower droplets can penetrate the cutting zone in a better way (MQL method), and enable the faster heat dissipation from the cutting zone (MQCL method) [17, 18].

During the machining processes, the tool wear is usually manifested by the abrasion wear on the flank face and the crater wear formed onto the rake face [6, 19]. Therefore, the prediction of tool wear is necessary to avoid its catastrophic failure, and thus affecting the machining efficiency [20].

Chetan et al. [15] investigated the tool wear of inserts coated with TiN layer during dry and MQCL turning of Nimonic 90 and Ti6Al4V aircraft alloys. The biodegradable emulsion consisting of sunflower oil mixed with water in 10:1 proportions has been used as an active medium. The application of MQCL method caused the lower tool wear in comparison to that obtained during dry turning (especially for Ti6Al4V alloy). This can be attributed to the improved penetration and wettability of machined surface by the active medium's droplets due to increased wetting angle. The wetting angle obtained on the surface of NIMONIC 90 alloy was higher than that obtained on the surface of Ti6A14V alloy, which consequently leads to the

* Corresponding author: E.Feldsztein@ibem.uz.zgora.pl

appropriate wetting of surface and formation of lubricating layer.

The aim of this study is to evaluate the effect of the emulsion mist on the character and intensity of the carbide inserts wear when C45 steel finish turning.

2 Experimental details

The C45 EN 10083-2:1991 carbon structural steel was used to evaluate the cutting tool life [21]. Turning was performed using the lathe with CSDBM 2020-M12 toolholder and SNUN120408 insert, made of P25 ISO 513: 2012 cemented carbide.

Turning was performed with a cutting depth of 0.5 mm, feed rate of 0.1 mm/rev and cutting speeds in the range of 250–450 m/min. Insert's wear were analyzed in dry cutting conditions and in the presence of the emulsion mist. As the active component of the emulsion mist, the 5 % aqueous emulsion of EMULGOL concentrate was used.

To determine the chemical composition of the concentrate the MiniPal X-ray spectrophotometer was used which is based on X-ray fluorescence analysis method. The measurement results are shown in Table. 1.

To generate emulsion mist, the Mikronizer 1 LN Micro Unit Wlenox Nozzle device was used [17]. The compressed air pressure was of 70 PSI (0.48 MPa). Emulsion mist was fed to the cutting zone with the three nozzles of 1.2 mm diameter, wherein the side nozzles were located at an angle of 30° relative to the central nozzle. The distance from the nozzles plane to the cutting zone was 0.3 m. The compressed air flow was 5.8 l/min and the emulsion flow was 26.4 g/h. These conditions are optimal from the point of view of the mist droplet size and conditions of heat removal from the cutting zone [17]. analysis of dynamical properties has been conducted on the three ball end mills with various lengths (see Tab. 1, Fig. 1).

Table 1. The chemical composition of the EMULGOL concentrate.

Element	C	H	O	Mg	S	K	Ca	P	Cu	Fe	Zn
Wt. %	72.2	16.7	2.5	0.03	1.5	1.2	2.1	1.0	2.4	0.3	0.07

Wear of carbide inserts was estimated with a resolution of 0.01 mm, using an optical microscope with a digital camera after every 4 min, until the critical tool wear has been reached. Limit values of the wear were recorded with the Dino-Lite AM7013MZT universal digital microscope. The 3D images of wear areas were obtained using an IFM 6.1 optical microscope (Infinite Focus Measurement Machine). The SEM studies of worn surfaces were carried out using the JEOL JSM-5600LV microscope with X-ray microanalyzer.

3 Results and discussion

The analysis of worn surfaces of carbide inserts on the flank faces, and craters on the rake faces, reveals that their dimensions are greater during dry cutting (Fig. 1). Registered 3D images of worn surfaces confirmed the results of the measurements and revealed a more complex character of inserts wear when dry turning.

The dynamics of carbide inserts wear is illustrated in Fig. 2, which shows the changes in time for the maximum height of wear VBmax of the main flank face. It is easy to notice the advantages for turning in an emulsion mist environment. The Fig. 2 shows that independently on cutting speed value, the tool wear growth is lower during turning in MQCL conditions. In case of turning in the highest investigated cutting speed, the tool wear rate obtained for the MQCL method is 40 % lower than that reached during the dry turning.

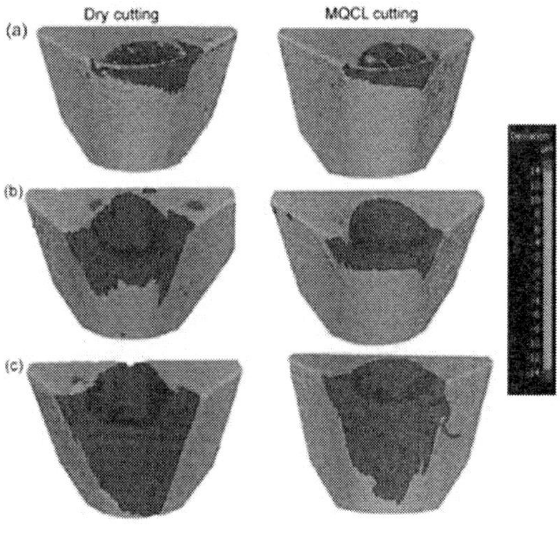

Fig. 1. 3D images of wear areas: a) v_c = 250 m/min; b) v_c = 350 m/min; c) v_c = 450 m/min.

To determine the causes of the emulsion mist effectiveness, the SEM analysis of the worn surfaces, which include the main flank (Fig. 3) and rake faces (Fig. 4), was carried out. Studies were realized for the cutting speed of 250 m/min, which is recommended as typical for finish turning of structural steels. The analysis fixed the presence of iron on worn surfaces which indicates the prevalence of the adhesive nature of the wear, as well as high, 16–20% oxygen content.

It can be assumed that the high temperatures in the cutting zone cause the formation of oxides on contact surfaces. During cutting with the emulsion mist, the presence of phosphorus and copper was recorded directly on the wear areas that are included in the EMULGOL concentrate. These elements are known for their ability to form compounds on the friction surfaces that reduce friction coefficient and wear rate.

 DOI: 10.1051/e3sconf/20171903009

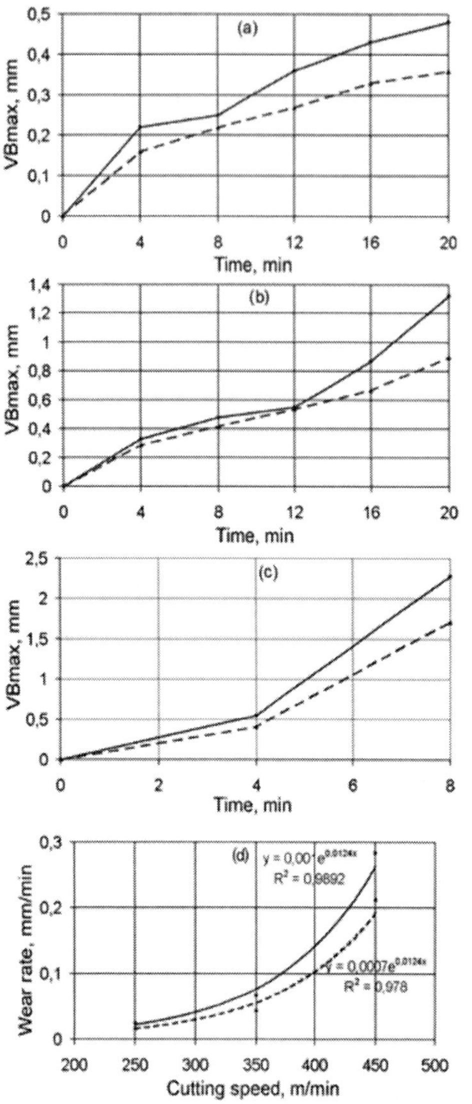

Fig. 2. Wear dynamics (a–c) and the wear rate (d) for carbide inserts: a) $v_c = 250$ m/min; b) $v_c = 350$ m/min; c) $v_c = 450$ m/min, continuous line – dry turning, dashed line - MQCL.

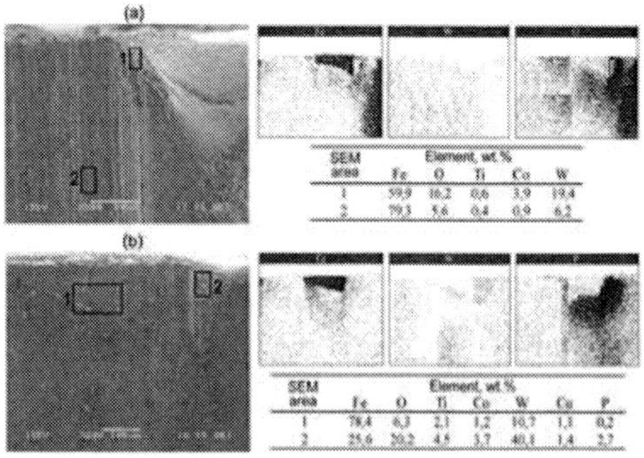

Fig. 3. SEM results for worn areas of the main flank face when cutting speed of 250 m/min: a) after dry cutting; b) after MQCL cutting.

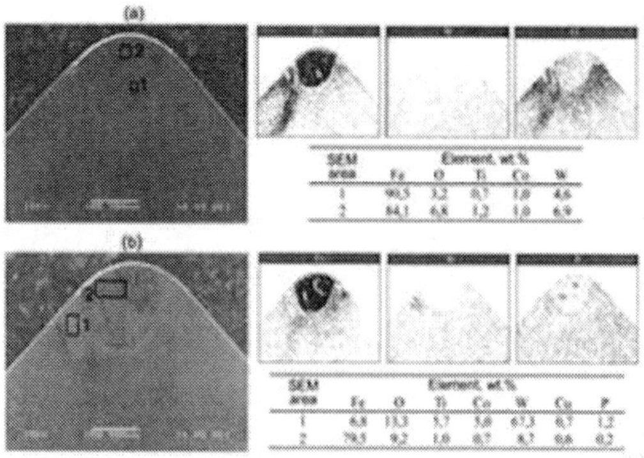

Fig. 4. SEM results for worn areas of the rake face when cutting speed of 250 m/min: a) after dry cutting; b) after MQCL cutting

This confirms the possibility of using the compounds included in the emulsion mist to form anti-wear and extreme pressure films on the contact areas in the cutting zone [14].

4 Conclusions

Machining of steels in emulsion mist environment has significant advantages over the dry cutting, in particular, in the case of the cutting tools wear rate. In conditions of emulsion mist the dimensions of the wear area are significantly smaller, and this difference is the more, the more there is the cutting speed. Wear rate is decreased by 40 % when cutting with the highest of cutting speeds investigated, while at lower speeds the difference is reduced to 30%. The EDS analysis showed, these positive changes are due to the appearance of active chemical compounds in the cutting area, which are included in the emulsion mist and form anti-wear and extreme pressure films on the contact surfaces. Based on the results obtained, the following general scientific and technological trends can be proposed:

- the maximum efficiency of emulsion mist using is ensured at high cutting speeds,
- the introduction of the active additives in the emulsion mist improves the operating conditions of this method.

References

1. K. Weinert, I. Inasaki, J.W. Sutherland, T. Wakabayashi, CIRP Ann. – Manuf. Techn. **53**, 2 (2004)
2. V.P. Astakhov, Editor. J.P. Davim. Springer-Verlag Limited, London, (2008)
3. R.W. Maruda, S. Legutko, G.M. Krolczyk, C. Lukianowicz, A. Stoic, Teh. Vjes. **22**, 5 (2015)
4. G.M. Krolczyk, R.W. Maruda, P. Nieslony, M. Wieczorowski, Measurement, **94**, (2016)

5. V.N. Gaitonde, S.R. Karnik, J.P. Davim, J. Mater. Process. Tech. **204**, 1–3 (2008)
6. R.W. Maruda, G.M. Krolczyk, E. Feldshtein, P. Nieslony, B. Tyliszczak, F. Pusavec, Wear **372–373**, (2017)
7. N.R. Dhar, M. Kamruzzaman, M. Ahmed, J. Mater. Process. Tech. **172**, 2 (2006)
8. Y. Zhang, C. Li, D. Jia, B. Li, Y. Wang, M. Yang, Y. Hou, X. Zhang, J. Mater. Process. Tech. **232**, (2016)
9. M. Hadad, B. Sadeghi, Int. J. Mach. Tools Manuf. **63**, (2012)
10. R.W. Maruda, G.M. Krolczyk, P. Nieslony, S. Wojciechowski, M. Michalski, S. Legutko, J. Manuf. Processes, **24**, (2016)
11. R.W. Maruda, G.M. Krolczyk, M. Michalski, P. Nieslony, S. Wojciechowski J. Mater. Eng. Perform. **26**, 1 (2017)
12. H. Hassanpour, M.H. Sadeghi, A. Rasti, S. Shajari, J. Clean. Prod. **120**, (2016)
13. P. Krajnik, A. Rashid, F. Pušavec, M. Remškar, A. Yui, N. Nikkam, M.S. Toprak, J. Clean. Prod. **112**, (2015)
14. R.W Maruda, S. Legutko, G.M. Krolczyk, S. Hloch, M. Michalski, Int. J. Surf. Sci. Eng. **9,** 5 (2015)
15. B. Chetan, C. Behera, S. Ghosh, and P.V. Rao, Ceram. Int. **42**, 13 (2016)
16. M.B.G. Jun, S.S. Joshi, R.E. DeVor, S.G. Kapoor, Trans. ASME, J. Manuf. Sci. Eng. **130**, 3 (2008)
17. R.W. Maruda, G.M. Krolczyk, E. Feldshtein, F. Pusavec, M. Szydlowski, S. Legutko, A. Sobczak-Kupiec, Int. J. Mach. Tools Manuf. **100**, 1 (2016)
18. R.W. Maruda, E. Feldshtein, S. Legutko, G.M. Krolczyk, Teh. Vjes. **22**, 5 (2015)
19. G. Poulachon, A. Albert, M. Schluraff, I.S. Jawahir, Int. J. Mach. Tools Manuf. **45**, 2 (2005)
20. G.K. Dosbaeva, M.A. El Hakim, M.A. Shalaby, J.E. Krzanowski, S.C. Veldhuis, Int. J. Refract. Met. Hard Mater. **50**, (2015)
21. ISO 3685:1993. Tool-life testing with single-point turning tools

Numerical simulation of micro-crack occurring in pipe made of stainless steel

Daria Wotzka[1,*]

[1]Opole University of Technology, 45-758 Opole, Poland

Abstract. Research works carried out regard to studies aiming at determination of the effect of cumulative duty operation on the development of micro-cracks in pipelines for transport of chemical substances. This paper presents results of computer simulations of a pipeline made of stainless steel. The model was investigated using the COMSOL Multiphysics environment. The object under study was divided into sub areas and then discretized according to the FEM method. The physico-chemical parameters of individual areas were defined based on measurement data. The main aim of research works was the modeling of acoustic emission wave, which is emitted in the vicinity of the tip of micro-crack as a result of its development. In order to solve the task, heterogeneity in the structure of the material, referred to damage/micro-crack, causing local stresses was assumed. The local stresses give rise to elastic waves, which propagate in the material in all directions. When the emission waves reach the boundaries of the pipe they are then transferred into acoustic waves and propagate in the surround air, until their natural attenuation. The numerical model takes into account the effect of high pressure (3.6 MPa) and negative temperature (-100°C) of the gas, transported inside the pipe. The influence of changes of these values in the range of ± 20% on the obtained results was investigated. The main contribution of the works is the multiphysical simulation model of transportation pipe made of steel, coupling structural mechanics, thermal conductivity and acoustic waves.

1 Introduction

The main aim of works, results of which are presented in this paper, was to develop a numerical model for enabling investigation of cracks generated acoustic emission (AE) signals occurring in pipelines based on computer simulations. The issue regard to non destructive testing of pipelines for transmitting gaseous chemical substances in temperature below minus 100°C. Micro-cracks, when present in the material structure, initiates development of discontinuities, which under operational load develop to dominant crack in the fatigue damage and plastic strain process and finally may lead to uncontrolled damage. This in turn causes technical and economic hazard for the production process and environmental protection.

The non-destructive testing method is commonly applied for detection and localization of damages occurring in technical various facilities [1,2,3]. Theoretical analyses and computer modeling of defect existing in the structure of pipelines made of steel, on their creation and propagation, were investigated e.g. in [4-8].

The main contribution of this paper lies in presentation of simulation results, performed by using a multiphysical model of a transportation pipe, coupling structural mechanics, thermal conductivity and AE waves. Chosen model parameters were changed and their impact on the achieved results was depicted and discussed.

2 The numerical model

2.1 Modelling assumptions

The modelling task was performed using the COMSOL Multiphysics environment. It was assumed that the AE wave was emitted at the surface of the pipe due to load created by micro-crack located within its structure. The tube is made of austenitic steel type 304 (A312TP304). For modelling purposes the crack was assumed as heterogeneity in the material structure, which leads to local stress. This stress induces elastic waves which propagate in all directions in the pipe. Reaching the boundary of the pipe, the elastic waves transform to acoustic emission waves, which propagate to the surrounding. It was further assumed that (1) the displacements at the crack tip are small, (2) due to small value of strain, linearization of equations, describing the relationship between the strain and displacement is possible, (3) constant coefficients of the constitutive equations describing the relations between the stress and strain, due to linear elastic type material. The numerical model calculates the distribution of stress and displacement caused by pressure acting on the inner pipe surface, negative gas temperature and the crack presence in the stationary state. The advantage of the applied software lies in the possibility of integration of partial linear differential equations describing the displacement

* Corresponding author: d.wotzka@po.opole.pl

 DOI: 10.1051/e3sconf/20171903010

field in solid bodies, heat flow and acoustic wave equations. The dependent variables are the displacement vector **u**=[u,v], temperature T and acoustic pressure p. Simulations were performed in the two-dimensional space with axial symmetry. The independent variables in the model include geometric coordinates x and y, time t and acoustic wave frequency f. The physical parameter values for steel and air are given in Table 1.

Tab. 1. Physical parameter values of the materials applied.

Parameter name	Value	Unit
Steel - within the solid domain		
Young's modulus	211e9	Pa
Poisson ratio	0.3	1
Density	80300	kg/m^3
Thermal expansion ratio	16e-6	1/K
Thermal conductivity	15	W/(m*K)
Heat capacity	500	J/(kg*K)
Material type	izotripic	-
Air- within the pressure acoustic domain		
Density	1.225	kg/m^3
Acoustic wave velocity	343	m/s

The governing equations used for numerical calculations using FEM are described in detail in an earlier publication [8], in which other analysis results using similar model are depicted.

2.2 The object under study - geometry

The object under study is a cross section of pipe of length equal to 50 mm, including axis of symmetry, in center of which the micro crack is located. The resulting model dimension is 6x50 mm. Its geometry is depicted in Fig. 1, where also the applied boundary conditions are presented, which are as follows. For the mechanical domain: free condition at the outer surface of the pipe (edges no 5&6), displacement condition at the crack tip (point no 4), pressure condition at the inner pipe wall (edge no 2), symmetry condition at the vertical axial symmetry. For the thermal conductivity domain: temperature at edge no 2, surface to ambient radiation at edge no 5, open boundary at edge no 6 and also symmetry condition at the vertical axial symmetry. The micro crack tip is marked as red dot in the point no 4.

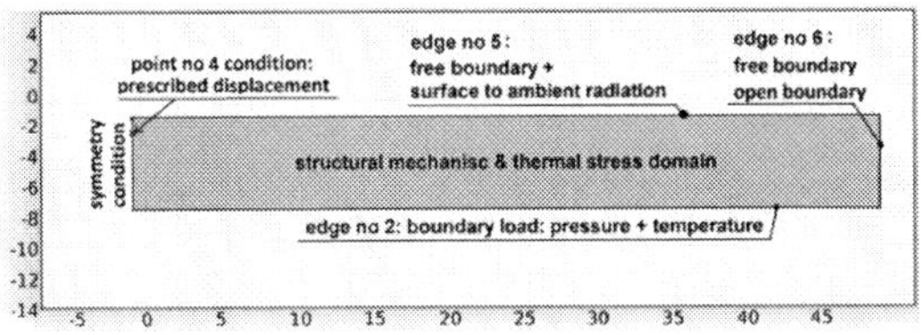

Fig. 1. Presentation of boundary conditions applied in model

Free condition denotes that there is no force and limitations at the boundary of the area. The boundary load means the effect of the pressure of gas pressing the inner pipe wall. Prescribed displacement means a growing crack. The model assumes displacement of 0 towards the pipe interior (y axis). Such a condition corresponds to the 'fixed'-type condition and stands for displacement u=0. Temperature at a boundary means in the considered case the effect of the negative temperature of gas pressing the inner pipe wall. Surface-to-ambient radiation means the infiltration of ambient heat inside the pipe.

Tab. 2. Range of the analyzed parameter values.

Parameter	Range of the analyzed values	Unit
Temperature	{88, 99, 110, 121, 132}	K
Pressure	{2.58, 2.91, 3.23, 3.55, 3.88}	MPa
Freuquency	{100 : 10 : 200}	kHz
Time	{0 : 0.001/29 : 0.0010}	s

In order to simulate acoustic phenomena the geometry was extended by 10 mm of surrounding air domain of room temperature and 10 mm of the so called perfectly matched layer (PML), an area that simulates unrestricted space, as depicted in Fig. 2. The model was applied for determination of the impact of the gas pressure and temperature variations on the dependent values within the analyzed solid domain and for calculation of acoustic pressure and sound power level in the immediate vicinity of the pipe, thus in the air domain. Additionally to the boundaries described earlier, acoustic pressure conditions was set at edge no 5 as source of AE wave, which is a derivative of the 'von Mises' stress at the outer pipe wall. The acoustic wave propagating in PML is not reflected and does not return to its bordering center. A continuity condition is set at the boundary between the PML and acoustic domain, which is fulfilled at all the inner boundaries.

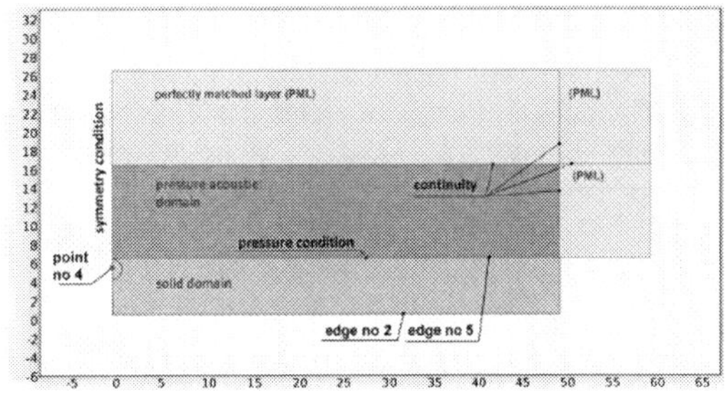

Fig. 2. Boundary conditions regarding the acoustic domain.

3 Example results

Fig. 3 present example calculations depicting local distribution of 'von Mises' stress, N/m^2, obtained at the vicinity of the crack tip for p=3.234e6 MPa, T=110 K. The greatest decrease in stress values can be observed up to 30 microns from the crack tip on the y axis and up to 10 microns on the x axis. The maximal calculated stress is at the tip and equals 4.338e10 N/m^2.

Fig. 4 present values of the 'von Mises' stress along edge no 5 (see markings in Fig. 1) for two temperature and five pressure values. In particular, results for two extreme gas temperature values, i.e. T = 88 K and T = 132 K, and all the considered gas pressure values are presented.

 DOI: 10.1051/e3sconf/20171903010

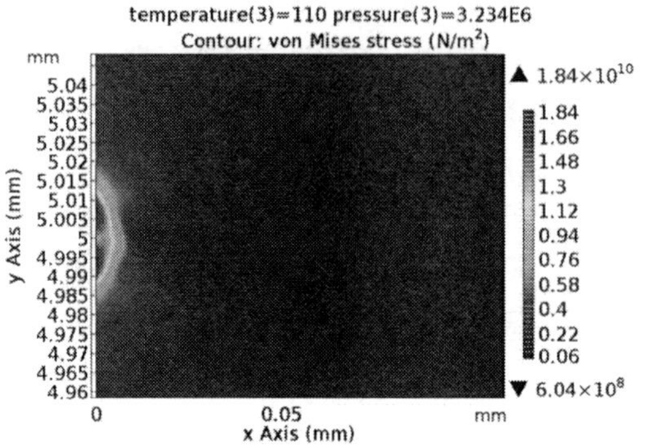

Fig. 3. Values of the 'von Mises' stress in the vicinity of the crack tip.

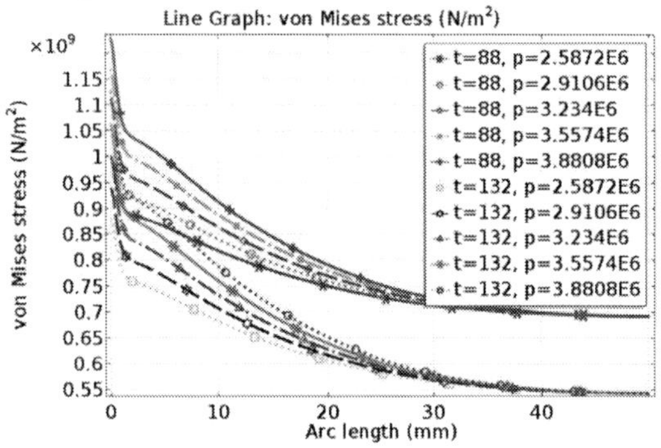

Fig. 4. Values of the 'von Mises' stress along edge no 5 for two temperature and five pressure values.

From the graph it can be recognized that on the outer surface of the pipe (edge no 5) the stress decreases exponentially with increasing distance from the crack tip. The impact of pressure is significant up the distance of about 30 mm. The calculated values do not decrease to zero and at the furthest distance from the crack tip (50 mm), the impact of gas pressure disappears. The temperature of the gas transported in the pipe has a significant impact on the calculated stress values, which are higher by lower temperatures.

In Fig. 5 total displacement values obtained along the inner pipe surface (see edge no 2 in Fig. 1) for the same parameters as above mentioned are presented.

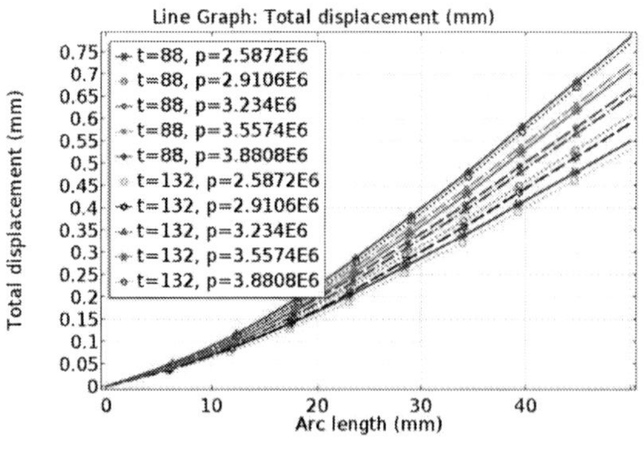

Fig. 5. Values of the displacement along edge no 2 for two temperature and five pressure values.

From Fig 5 it follows that the value of structural displacement increases together with the rising distance from the crack tip. Far from the gap, at a distance of 50 mm, both the impact of the gas pressure and the influence of the temperature are visible, while the latter showing a much lower impact.

In Fig. 6 total acoustic pressure distribution in the considered area is presented. The graph depicts example calculations for sinusoidal AE wave with a main frequency component equal to 100 Hz. In the PML domain the waveform is completely suppressed.

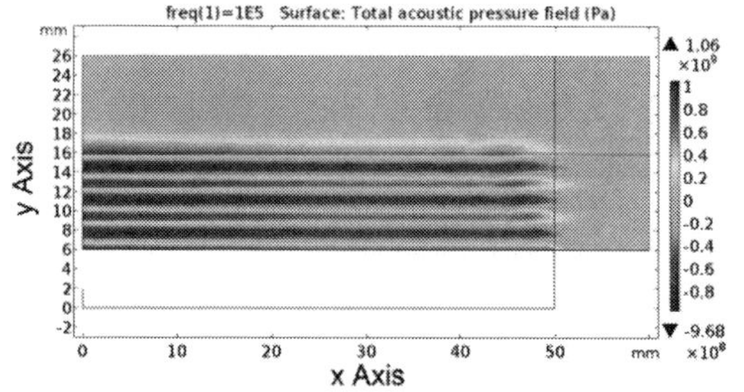

Fig. 6. Distribution of the total acoustic pressure field within the acoustic domain.

In Fig. 7 transient analysis results depicting dynamics of acoustic pressure values, calculated at points 8-12, which are located across the acoustic domain in its center, are presented. As AE source a Gaussian function, of parameters determined based on experimental measurements, was applied. The moving ridge of the EA wave can be clearly seen.

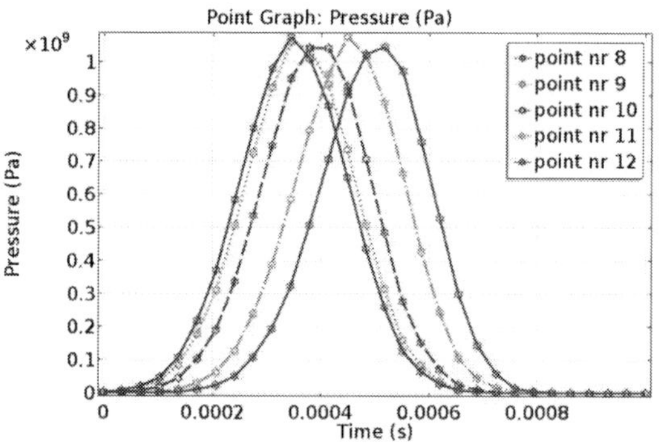

Fig. 7. Values of the acoustic pressure calculated in various points.

4 Results and discussion

Space distributions of the 'von Mises' stress, generated in the pipe for gas pressure changes in the range of 3.234 MPa ± 20% and temperature variations in the range of 110 K ± 20% and the sound pressure distributions in the air around the pipe were analyzed in the task. Based on the obtained values the following conclusions were driven:

- at the crack tip, the stress is constant and does not depend on the temperature of the gas inside the pipe. On the other hand, its value increases proportional to the pressure of the gas transported in the pipe. The changes in the gas pressure have smaller impact as compared to those of the temperature changes;

- on the outer surface of the pipe the stress decreases exponentially with increasing distance from the crack tip. This value does not decrease to zero but stabilizes at a value, which varies from the model parameters. At the points furthest from the crack tip, the difference in gas pressure difference that is visible in the immediate vicinity of the crack tip, disappears. The stress value depends on the temperature of the gas transported in the pipe, this dependence is however not linear relative to distance;
- differences in stress values for the same gas pressure values and higher temperatures are in the crack tip surrounding smaller than at the distance further than 50 mm from the crack tip;
- along the distance simulating the crack, at a distance of up to 20 μm from its tip, the stress value decreases rapidly. On the remaining length, up to the edge of the object, the stress is stabilized. There was also no significant influence of both temperature and gas pressure on the stresses over the entire length of the crack;
- along the edge furthest from edge no 6, the gas pressure does not affect the stress value. On the other hand, its value is determined by the temperature of the gas; the higher the temperature, the less the stress;
- there was no effect of gas pressure on the temperature distribution in the pipe. The temperature range was determined by the temperature of the gas. In addition, there was a slight diffusion of the ambient temperature to the inner structure of the pipe;
- the largest displacements occur at the locations furthest from the crack tip. At a distance of 50 mm from the tip, the displacement is strongly dependent on the gas pressure and to a much lesser extent on the temperature;
- the sound pressure level in the immediate vicinity of the pipe does not depend on the frequency of the acoustic wave emitted on the outer wall of the pipe and reaches a maximum of 272.46 dB. In contrast, the distribution of total acoustic pressure in the air domain depends on the frequency with which the acoustic wave emits from the source;
- within 10 mm of the pipe surface there is no noticeable amplitude-damping that exceeds 1 GPa. Throughout the length of the analyzed section, the ridges and valleys of the acoustic wave propagating in the domain are clearly visible;
- based on the amount of stress, it is possible to infer where local stresses may cause the development of the cracks and thus the generation of the EA wave. The value of the 'von Mises' parameter may be an indicator of the amplitude of the EA wave.

5 Conclusions

This paper is related to improvement of the AE method, applied for detection and localization of defects occurring in transportation pipes. During the theoretical studies COMSOL Multiphysics was used, which made it possible to perform numerical simulations with the Finite Element Method. The aim of the research task was to study the influence of changes in pressure and temperature values of the gas, transported inside a pipeline made of stainless steel, on the values of the stresses, generated in the structure of the pipe, due to a discontinuity, which was modelled in the form of a micro crack. The stress level results from the load acting on the micro crack, and directly affects the intensity of sound pressure, which is emitted during this phenomenon. The task was realized by integrating differential equations describing the field of displacement, conduction of heat in solids and the equations describing acoustic waves, propagating in the air.

The work was founded by The National Centre for Research and Development within the LIDER Project no LIDER/18/44/L-3/NCBR/2012.

References

1. A. Mostafapour, S. Davoudi, Appl. Acoust. **74** 2013)
2. C. Nayak, Int. J. Appl. Eng. Res. **9** (2014)
3. H. Majchrzak, A Cichoń, S Borucki, Arch. Acoust. **42** (2017)
4. J. Abolle Okoyeagu, J. Palacio Torralba, Y. Chen, R. Reuben, Proc. Of 31 Conf. on European Working Group on Acoustic Emission, Fr.3.B.2, Sept. 3-5, Germany, (2014)
5. E. Paffumi, K.-F. Nilsson, Z. Szaraz, Eng. Failure Anal. **47** (2015)
6. C. Ruggieri, F. Dotta, Eng. Struct. **33** (2011).
7. K.R. Jayadevan, E. Berg, C. Thaulow, E. Østby, B. Skallerud, Int. J. Solids Struct. **43** (2006)
8. D. Wotzka, T. Lusa, J. of Mech. Sci. Technol. **29** (2015)

Electrical properties of aluminum contacts deposited by DC sputtering method for photovoltaic applications

Ewelina Krawczak[1,*], and *Sławomir* Gułkowski[1]

[1]Faculty of Environmental Engineering, Lublin University of Technology, Nadbystrzycka 38, 20-618 Lublin, Poland

Abstract. The use of aluminum contacts is common in the process of silicon solar cells production because of low contact resistivity. It has also a great importance in thin film technology for photovoltaics, especially in copper-indium-gallium-diselenide (CIGS) devices. The final stage of CIGS cell production is the top contact deposition of high conductivity layer for lateral current collection. Such material has to be highly optically transparent as well. In order to make a contact, metal is deposited onto TCO layer with minimum shadowing to allow as much light as possible into device. The metal grid contact is being made by deposition of few microns of aluminum. The resistivity of the deposited material as well as resistance between the metal grid and TCO layer plays a great role in high quality solar cell production. This paper presents the results of four point probe conductivity analysis of Al thin films deposited by direct current (DC) magnetron sputtering method. Influence of technological parameters of the Al deposition process on sheet resistance of deposited layers has been showed. In order to obtain the lowest resistivity of the thin contact layer, optimal set of sputtering parameters, i.e. power applied, deposition time and deposition pressure was found. The resistivity of the contact between two adjacent Al metal fingers deposited onto transparent conductive Al-doped zinc oxide film has been also examined.

1 Introduction

Recently, materials for thin film solar cells are the subject of researcher interest. Their good technological parameters and application in low-cost, high-rate semiconductors with a thickness of few microns are main objectives to be achieved. Solar cells based on chalcopyrite materials, such as $Cu(InGa)Se_2$ (CIGS), achieved the highest conversion efficiency of up to 22.4% [1]. Direct band gap, high absorption coefficient for solar radiation and inherent *p-type* conductivity make CIGS structure one of the most promising absorber material forsolar cell application. Despite of this, the optimization process need to be implemented for improving homogeneity in large-scale manufactured cells, increasing efficiency and reducing cost of production process. The basic CIS/CIGS cell structure is deposited on soda lime glass, but can be also obtained on flexible substrates made of metal [2], ceramic [3], ultra-thin glass [4] or thin stainless steel [5]. The model is based on $Cu(InGa)Se_2/CdS$ junction and Mo back contact. The core of the CIGS solar cell is CIGS *p-type* absorber layer with the large variety of deposition possibilities, such as magnetron sputtering method [6], electrodeposition [7,8], co-evaporation [9], low-temperature pulsed electron deposition (LTPED) [10,11] and sputtering with post-selenization [12-13]). The top layers (front contact) are made of transparent conductive oxide (TCO) with high-conductivity. TCO materials are mainly based on tin oxide (SnO_2), indium tin oxide (In_2O_3), zinc oxide (ZnO) or theirs combinations, but the most widely applied is indium tin oxide (ITO). This material is characterized by good optical and electrical properties, but its quantity is limited, because of uniqueness and cost of indium element. Zinc oxide (ZnO) and ZnO structures doped with the third group element, such as B, Ga, In, Al can be appropriate replacement of ITO layers as a front electrode [14]. The transparent films can be made by use of variety techniques, such as magnetron sputtering deposition (RF or DC mode), chemical vapour deposition, sol-gel deposition, spray coating, pulsed laser deposition (PLD), atomic layer deposition (ALD) or metalorganic chemical vapor deposition technique (MOCVD) [15-19]. The above mentioned layers are characterized by the high transmittance, wide band gap and low resistivity [20-21]. Completion of the CIGS structure is current-collecting grid, which can be made of aluminum.

This paper presents the magnetron sputtering process optimization in terms of aluminum deposition on thin film transparent conductive layer as a top contact for photovoltaic applications. Influence of the process parameters such as power, pressure and deposition time was found. Sheet resistance of the films deposited under various technological parameters were measured. Dependency of the TCO layer conductivity on the resistance between adjacent fingers was also examined.

* Corresponding author: e.krawczak@pollub.pl

DOI: 10.1051/e3sconf/20171903011

2 Materials and methods

The transparent conductive Al doped ZnO films and aluminum contacts were deposited onto 1 mm thick soda lime glass slides by using Alliance Concept AC450 magnetron sputtering system. Firstly, the ZnO:Al layer was deposited and then Al metal grid was obtained by using glass mask on the top of Al doped Zinc Oxide film. The glass substrates were ultrasonically and chemically cleaned with acetone, ethanol and dried with nitrogen.

After the cleaning process, samples were immediately loaded inside the magnetron deposition chamber under the target directly. The 10.16 cm (4 inches) in diameter and 6 mm thick targets with purity of 99.999% were used. The RF power was applied to ZnO:Al ceramic cathode made by 98% wt.% ZnO and 2 wt. % Al and the DC power to aluminum target. The distance between the target and the substrate was fixed at 90 mm. The films were obtained at room temperature in Ar atmosphere. Overheating of the substrate holder was avoided due to water cooling system. After mounting the substrates, the sputtering chamber was evacuated to the base pressure equals $1.0x10^{-2}$ Pa in order to carry out the presputtering process and remove undesired impurities from the target surface.

The Al doped ZnO films were obtained with optimal parameters, such as power of 290 W, working pressure of 0.26 Pa and deposition time – 45 min. In order to find optimal technological parameters, several series of aluminum top contact layers were deposited. In each series, technological parameters as discharge power, deposition time and deposition pressure were varied during the deposition process.

Sputtering power density was ranged from 0.75 W/cm^2 to 1.25 W/cm^2.

Deposition time was fixed at 15, 30, 45 min and the working pressure varied from 0.2 to 1.57 Pa. The sheet resistance of each layer and the metal – TCO contact was determined at room temperature using RM3000+ four point probe equipment of Jandel Engineering Limited, shown in Fig. 1.

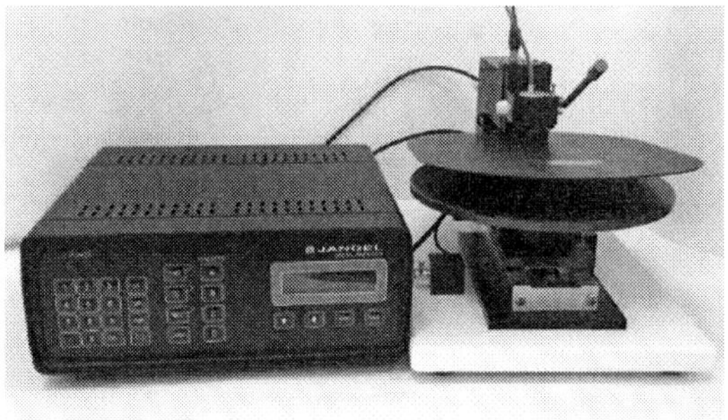

Fig. 1. RM3000+ four point probe equipment of Jandel Engineering Limited.

The optimal parameters of Al layers were chosen to deposition as a contact fingers on the top of ZnO:Al film. Then the resistances (R_1, R_2) between adjacent metallic contacts were measured by using multimeter, which is shown in Fig. 2.

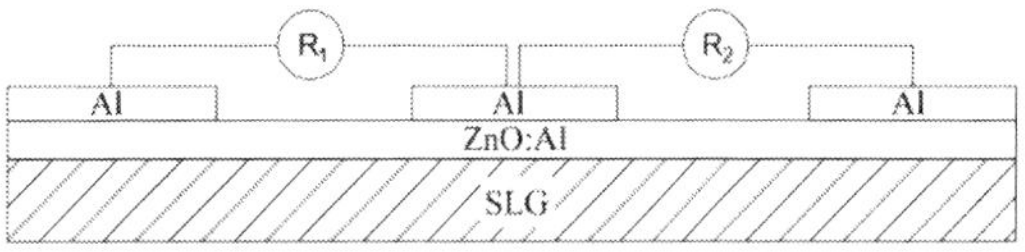

Fig. 2. A schematic representation of the prepared structure with measurement points.

3 Results and discussion

Four probe method were used to measure electrical parameters of the prepared samples. Fig. 3 shows the comparison of sheet resistance (Rs) dependency on deposition time at different powers of 60 W, 80 W and 100 W applied to the aluminum cathode. The working pressure was kept constant at 1.57 Pa.

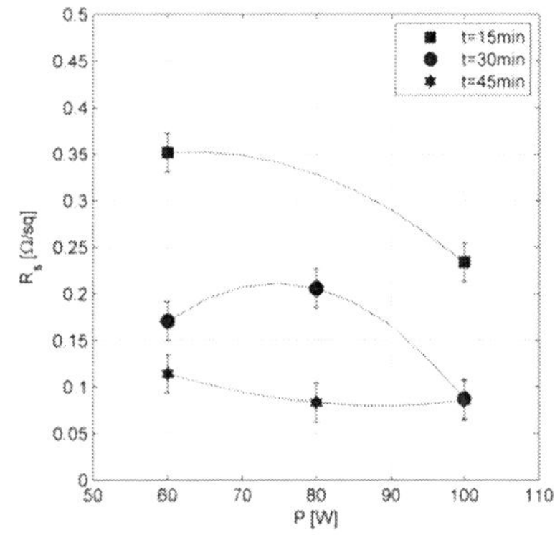

Fig. 3. Sheet resistance dependency on the deposition time at different deposition power applied to the Al cathode.

In each case the sheet resistance was found to decrease with increasing deposition time. This can be explained by the fact that resistance of the layer decreases with the increase of its thickness that depends on the time of deposition as it was reported in [22]. As it can be seen in Fig. 3 for power equals 60 W the sheet resistance decreases rapidly from 0.35 Ω/sq to about 0.1 Ω/sq in 30 minutes of the deposition process. In case of higher power this drop is even faster. For power equals 100 W the level of sheet resistance value of about 0.1 Ω/sq was achieved after less than 30 minutes of the deposition. Further increase of the deposition time does not lead to better conductivity of the layer.

Fig. 4 compares the sheet resistance of aluminum thin films prepared at two different deposition powers and deposition times. The power varied from 60 W to 100 W. Fig. 4 depicts that sheet resistance of the deposited layer is more sensitive to the power applied in the case of the values of the deposition times lower than 30 min. In case of higher values of the deposition time (i.e. 45 min) the influence of the deposition power on sheet resistance can be negligible. From the foregoing discussion the conclusion is that the lowest value of sheet resistance can be obtained for deposition time and power equal to 30 min and 100 W respectively. Lower value of power applied (80 W) demands longer duration

DOI: 10.1051/e3sconf/20171903011

of deposition process (45 min) to obtain optimal conductivity of the contact layer.

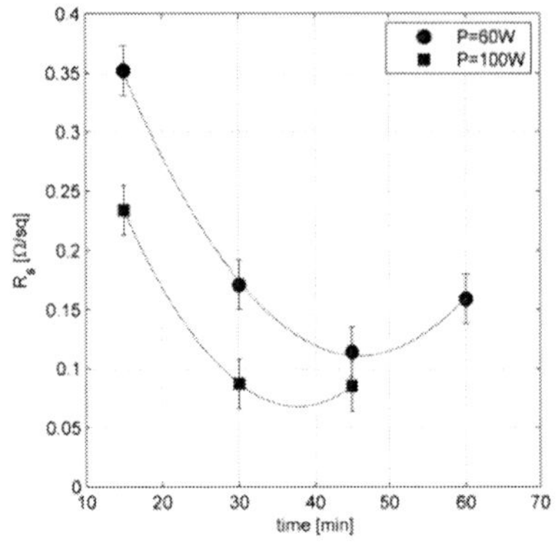

Fig. 4. Sheet resistance dependency on the deposition power at different deposition times.

For further optimization of the deposition process technological parameters dependence of the sheet resistance on the deposition pressure has been investigated. Four samples of the thin films deposited at power of 80 W and time equals 45 min have been prepared at working pressure which varied from 0.3 to 1.2 Pa. From Fig. 5, a correlation of the resistance on the deposition pressure was not observed.

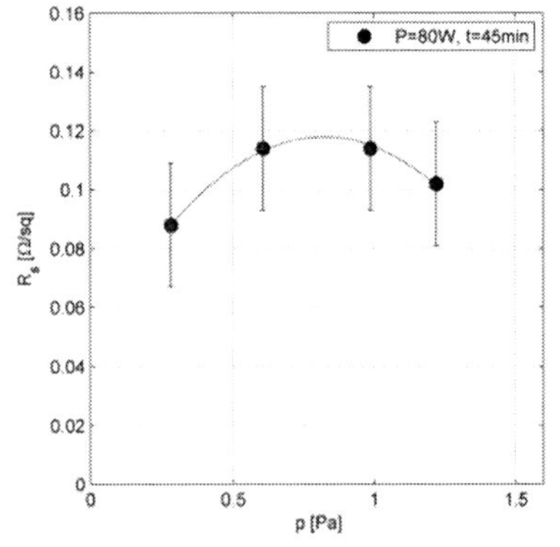

Fig. 5. Sheet resistance as a function of the deposition pressure. The deposition power and time were equal to 45 min and 80 W respectively.

Sheet resistance of the given films varied from 0.09 Ω/sq to 0.12 Ω/sq in this case. However, it can be assumed that sheet resistance decrease with decreasing of the pressure and better conductivity of the film should be obtained at the deposition pressure not greater than 0.4 Pa.

From Figs. 3 – 5, the optimal parameters of the deposition have been established in order to obtain Al film of the highest conductivity. Optimal parameters were chosen as follows: deposition power, 100 W; deposition time, 30 min and pressure, 0.4 Pa. However, the top contact of the CIGS solar cells consists of thin layer of transparent conductive oxide (i.e. zinc oxide) and metallic fingers on the top of the layer.

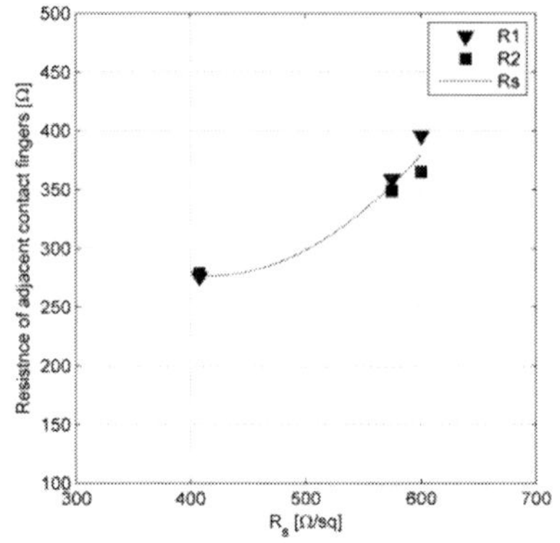

Fig. 6. Correlation between sheet resistance of the transparent layer and resistance of adjacent Al contact grids.

In order to examine the influence of the particular layers on the conductivity of the complete structure, the samples were prepared according to the scheme presented in Fig. 2. Figure 6 shows the results of resistance measurements between two adjacent aluminum fingers deposited on the top of the Al-doped zinc oxide film. It can be clearly seen that this resistance strongly depends on the quality of the transparent oxide and increases with the increase of the sheet resistance of the ZnO/Al film measured by four point probe method.

4 Conclusions

Influence of the Al deposition technological parameters on the sheet resistance of the Al films were studied. These parameters were as follows: deposition power, deposition time and pressure during the process. Metal layers were deposited on the soda lime glass (SLG). Sheet resistance was measured with the use of four point probe method. It was found that the sheet resistance decreases with the increase of the deposition time as well as power applied. After about 30 minutes of the deposition with use of power equal to 100 W, sheet resistance was found around 0.1 Ω/□ which is small enough value in solar cell applications What is more, after this time, significant changes of the sheet resistance with power applied was not noticed. In the second stage of work Al metal were deposited onto TCO in order to measure the influence of the TCO layer on the conductivity of the structure. It was found that conductivity of the contact strongly depend on the conductivity of the transparent oxide. In the conclusion, optimized parameters of deposition process were found to obtain the best quality aluminum front-contact for CIGS solar cells.

References

1. P. Jackson, R. Wuerz. D. Hariskos, E. Lotter, W. Witte, M. Powalla, Phys. Status Solidi RRL, **10** (2016)

2. T. Yagioka, T. Nakada, Appl. Phys. Express, **2**, 072201 (2009)
3. S. Ishizuka A. Yamada, K. Matsubara, P. Fons, K. Sakurai, S. Niki, Curr. Appl. Phys. **10** (2010)
4. A. Gerthoffera, C. Poulainb, F. Rouxa, F. Emieuxa, L. Greneta, S. Perrauda, Sol. Energ. Mat. Sol. C **166** (2017)
5. W.-S. Liu, H.-C. Hua, N.-W. Pu, S.-C. Liang, J. Alloy Compd. **631** (2015)
6. S. Gułkowski, E. Krawczak, J. Olchowik, Proc. of 31st Eur. Photovoltaic Solar Energy Conf. and Exhibition (2015)
7. C. Adel, B.M. Fethi, B. Brahim, Optik **127** (2016)
8. M. Ganchev, J. Kois, M. Kaelin, S. Bereznev, E. Tzvetkova, O. Volobujeva, N. Stratieva, A. Tiwari, Thin Solid Films **511 – 512** (2006)
9. Repins, M. A. Contreras, B. Egaas, C. DeHart, J. Scharf, C. L. Perkins, B. To, R. Noufi, Prog. Photovolt. **16** (2008)
10. S. Rampino, N. Armani, F. Bissoli, M. Bronzoni, D. Calestani, M. Calicchio, N. Delmonte, E. Gilioli, E. Gombia, R. Mosca, L. Nasi, F. Pattini, A. Zappettini, M. Mazzer, Appl. Phys. Lett. **101**, 132107 (2012)
11. N. Cavallari, F. Pattini, S. Rampino, F.Annoni, M. Barozzi, M. Bronzoni, E. Gilioli, E. Gombia, C. Maragliano, M. Mazzer, G. Pepponi, G. Spaggiari, R. Fornari, Appl. Surf. Sci. **412** (2017)
12. X. Peng, M. Zhao, D. Zhuang, L.Guo, L. Ouyang, R. Sun, L. Zhang,Y. Wei, S. Zhan, X. Lv, Y. Wu, G. Ren, J. Alloy Compd. **710** (2017)
13. J.-K. Sim, S.-K. Lee, J.-S. Kim, K.-U. Jeong, H.-K. Ahn, C.-R. Lee, Appl. Surf. Sci. **389** (2016)
14. J.-H. Cha, S.-M. Kwon, J.A. Bae, S.H. Yang, C.-W. Jeon, J. Alloy Compd. **708** (2017)
15. S. Kuprenaite, A. Abrutis, V. Kubilius, T. Murauskas, Z. Saltyte, V. Plausinaitiene, Thin Solid Films **559** (2016)
16. T. Schuler, T. Krajewski, I. Grobelsek, M.A. Aegerter, Thin Solid Films **502** (2006)
17. S. Kuprenaite, T. Murauskas, A. Abrutis, V. Kubilius, Z. Saltyte, V. Plausinaitiene, Surf. Coat. Tech. **599** (2016)
18. C.G. Granqvist, Sol. Energy Mat. Sol. C **91** (2007)
19. E. Guziewicz, I.A. Kowalik, M. Godlewski, K. Kopalko, V. Osinniy, A. Wójcik, S. Yatsunenko, E. Łusakowska, W. Paszkowicz, M. Guziewicz, Jpn. J. Appl. Phys. **103**, 033515 (2008)
20. J. Yoo, J. Lee, S. Kim, K. Yoon, I.J. Park, S.K. Dhungel, B. Karunagaran, D. Mangalaraj, J. Yi, Thin Solid Films **480–481** (2005)
21. B.G. Choi, I.H. Kim, D.H. Kim, K.S. Lee, T.S. Lee, B. Cheong, Y.-J. Baik, W.M. Kim, J. Eur. Ceram. Soc. **25** (2005)
22. A. Zdyb, E. Krawczak , P. Lichograj, Opt. Appl. **46** (2016)

Impact of Fe powder sintering and soldering in production of porous heating surface on flow boiling heat transfer in minichannels

Wojciech Depczyński [1], *Artur* Piasecki[2], *Magdalena* Piasecka[1,*] and *Kinga* Strąk[1]

[1] Kielce University of Technology, Faculty of Mechatronics and Mechanical Engineering, Al. 1000-lecia P.P. 7, 25-314 Kielce, Poland

[2] contractor of the project from the Polish National Science Centre (No. DEC-2013/09/B/ST8/02825)

Abstract. This paper focuses on identification of the impact of porous heated surface on flow boiling heat transfer in a rectangular minichannel. The heated element for Fluorinert FC-72 was a thin plate made of Haynes-230. Infrared thermography was used to determine changes in the temperature on its outer smooth side. The porous surface in contact with the fluid in the minichannel was produced in two processes: sintering or soldering of Fe powder to the plate. The results were presented as relationships between the heat transfer coefficient and the distance from the minichannel inlet and as boiling curves. Results obtained for using a smooth heated plate at the saturated boiling region were also presented to compare. In the subcooled boiling region, at a higher heat flux, the heat transfer coefficient was slightly higher for the surface prepared via soldering. In the saturated boiling region, the local heat transfer coefficients obtained for the smooth plate surface were slightly higher than those achieved from the sintered plate surface. The porous structures formed have low thermal conductivity. This may induce noticeable thermal resistance at the diffusion bridges of the sintered structures, in particular within the saturated boiling region.

1 Introduction

The increase in heat flow intensity as used in electronic systems and other modern technologies, with a focus on minimizing the size of the heat exchangers at maximum heat flux, has led to the need for heat transfer intensification. One of the solutions is the use of boiling phenomena, the other - using enhanced heated surfaces produced by different methods [1–3]. Studies have been conducted on pool boiling [4–6], flow boiling [7–11] heat transfer and confined spaces as face seals [12,13]. Primarily passive methods are being developed for use in porous coating applications and in metal surface mechanical processing [14,15].

Metal foams are a new type of material with a wide range of applications because of their excellent properties including light weight, impact energy absorption capacity, specific thermal acoustic properties and low thermal conductivity [16–18]. Open porous metallic foams have been used in heat and mass transfer applications, for instance in heat pipes, vapour chambers, and loop heat pipe elements. In the past few years, there has been a growing interest in metal foams. Both the foam structure and the pore morphology depend on the fabrication method. Ashby et al. [16] classify the metal foam production methods according to the process and the state of matter. Similar classification was presented in the paper by Davies et al. [17]. The superior thermal conductivity of copper foams is particularly important. Porous sintered structures can also be produced by sintering small diameter wires. In the article [19] the alternative low-cost technology for fabrication of foams from metals, alloys, intermetallics, was proposed. The metal foam was formed by a typical base metal like Al, Cu and other [20–23].

The subject of this paper is the application of porous metal surfaces produced by soldering and sintering iron powder to the heated surface of the minichannel in flow boiling heat transfer experiments with a refrigerant flow. One of the porous surface types was produced with metal foams formed by reduction of metal oxides during sintering. The mixture was sintered in a dissociated ammonia atmosphere. Fe foam was prepared according to the method described in the Polish patent [24]. However crucial influence on the porosity has the ratio between quantity of metal oxide powder and amount of matrix metal powder, which is a basic structure of produced sinter [15]. This allows for compose of irregular cellular structures with pores open or closed. The foam material can be stacked and co-sintered with top layers to sandwich structures. The porous layers were formed by sintering technique the powder mixture and the reduction of iron powder ASC 100.29, ASC 100.24, DISTALOY SE and iron oxide of different granulation. The second type of porous surface was produced by soldering the same iron powder to the tested surface (made of Haynes-230) according to the method

* Corresponding author: tmpmj@tu.kielce.pl

described in the Polish patent [25].

A review of literature relating to flow boiling heat transfer in minichannels with microstructured surfaces, smooth and enhanced, was presented in [7–9]. Focusing research on microstructured surface of heat exchangers with minichannels seems interesting because these surfaces provide a potential for further heat transfer enhancement.

2 Experimental stand and methodology

2.1 Experimental stand

The experimental stand with the main loops: flow loop, a data acquisition system and a supply and control system is represented schematically in Fig. 1.

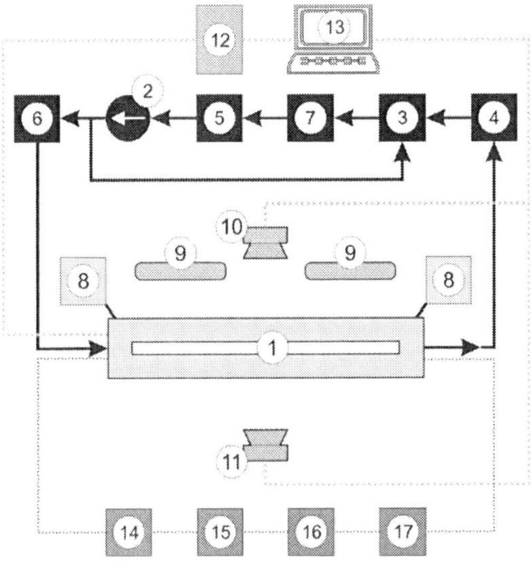

Fig. 1. The schematic diagram of the main loops and systems at the experimental setup, 1-measurement module; 2-gear pump; 3-compensating tank; 4-tube-type heat exchanger, 5-filter, 6-mass flow meter, 7-deaerator, 8-pressure transducer, 9-infrared camera; 10-fast camera, 11-high power LEDs, 12-data acquisition station, 13-computer, 14-inverter welder, 15-shunt, 16-ammeter, 17-voltmeter.

The flow loop comprises: a measurement module with a minichannel (1), a gear pump (2), a compensating tank (3), a heat exchanger (4), a filter (5), mass flow meter (6), a deaerator (7) and pressure transducers (8). The data and image acquisition system consists of: an infrared camera (11) and a fast camera (10), a lighting system (9), a data acquisition station (12) and a computer with specialized software (13). The supply and control system consists of a inverter welder (14), a shunt (15), an ammeter (16) and a voltmeter (17). The experimental setup was presented in detail in [9].

2.2 A measurement module with a minichannel

The study was conducted for FC-72 boiling in the test section with a rectangular, vertical and asymmetrically heated minichannel with a depth of 1.7 mm, a width of 18 mm and a length of 180 mm, Fig. 2. The heated element for fluid flowing in the minichannel was the plate made of Haynes-230 alloy with thickness of 0.45 mm which was enhanced on one side indirect contact with the fluid flowing in the channel. The temperature of the smooth side of the heated plate was measured by infrared camera in the central axially symmetric part of the minichannel. Thermal accuracy of the infrared camera E60 manufactured by FLIR was ± 1 °C or ± 1% within the temperature range of 0÷120°C [11]. It was possible to observe the plate surface through a glass pane using a fast camera. K-type thermocouples and pressure transducers were installed at the inlet and outlet of the minichannel.

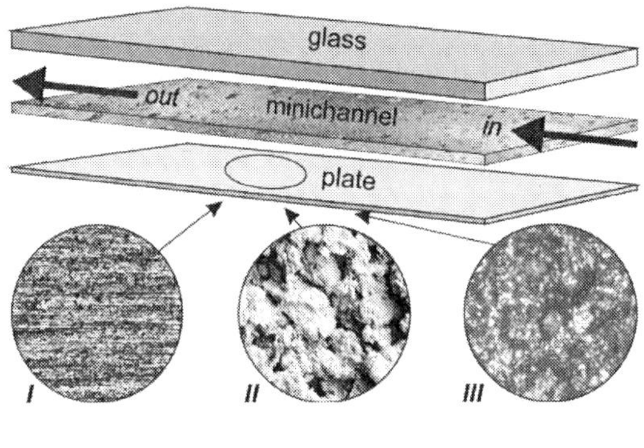

Fig. 2. The schematic diagram of a measurement module with a minichannel, I,II, II - fragment of the plate surface contacting fluid in a minichannel: I - smooth surface, II, III - the porous surfaces produced in two processes: sintering (II) and soldering (III) of Fe powder to the plate.

2.3 The porous heated wall of the minichannel

2.3.1 Properties of Haynes-230 alloy

Haynes-230 is a nickel-chromium alloy with tungsten and molybdenum additions. This alloy has excellent oxidation resistance (1150°C) and superior strength at high temperature, outstanding resistance to oxidizing environments up to 1150°C for prolonged exposures, resistance to nitriding environments, and excellent long-term thermal stability. Haynes-230 also provides excellent service in clean and moderately-dirty environment. It is readily fabricated and formed. Haynes-230 is also mark by lower thermal expansion characteristics than most alloys, and a pronounced resistance to grain coarsening with prolonged exposure to high-temperatures [26].

2.3.2 The tested porous surfaces

The tested porous surfaces were produced by sintering iron powder (in the dissociated ammonia atmosphere) and by soldering the powder to the 0.45 mm thick plate made of Haynes-230.

To produce the first surface type, the powder mixture was sintered in a dissociated ammonia atmosphere at 1180 °C for 45 minutes. The Fe foam was prepared from ASC 100.29 powder (by Höganäs company), Cu powder and iron oxide Fe_2O_3 (ASC100.29 +12%Fe_2O_3+6%Cu by weight percent). After sintering the samples situated in furnace were moved to an area where it was cooled in a protective atmosphere. Cooling took place in an average speed of about 25 degrees per minute. To determine pore size and shape from the foam specimens, the image analyser software was used. For further study there was selected the foam with an approximating 60% of porosity. The relative volume of particles of the foam was determined in accordance with the Cavalieri-Hacquert principle [14]. The result of the porous metal foam from ASC 100.29 showed porosity approximately 60% with bulk density. The shape and chemical composition of sintered material particles affect the creation of the diffusion bridges, between which pores arise. This significant value of porosity has an important impact on the heat transfer conditions.

Fig. 3 and Fig. 4 present 3D topographies and roughness profiles for porous surfaces produced by sintering (Fig. 3) and soldering (Fig. 4) iron powder to plate being a heated wall of a minichannel.

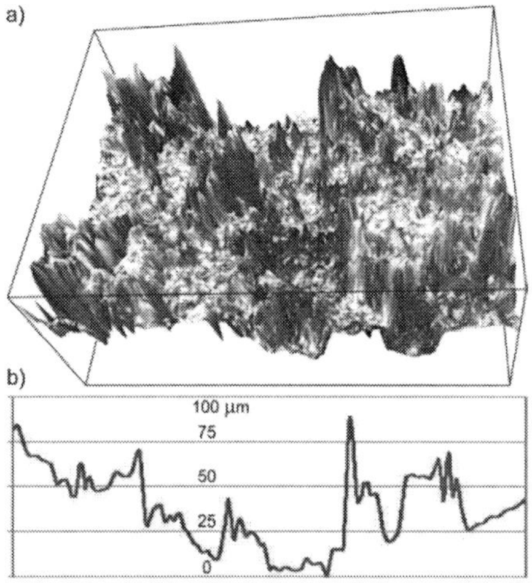

Fig. 3. Porous surface produced bysintering iron powder, a) 3D topography, b) roughness profile.

Soldering was carried out in according to the patent [25] application: once the surface was covered with a thin film of solder and flux, the fragmented structures of metallic material (powders) were joined by soldering to the plate. The main parameters of this porous surface are: the diameter of a solder granule - approx. 15÷80 µm (with an average being 40÷65 µm); the density of the solder paste – 2990 kg/m^3, the thickness of the soldered layer – approx. 100 µm and the maximum height of the soldered layer – 200 µm [27].

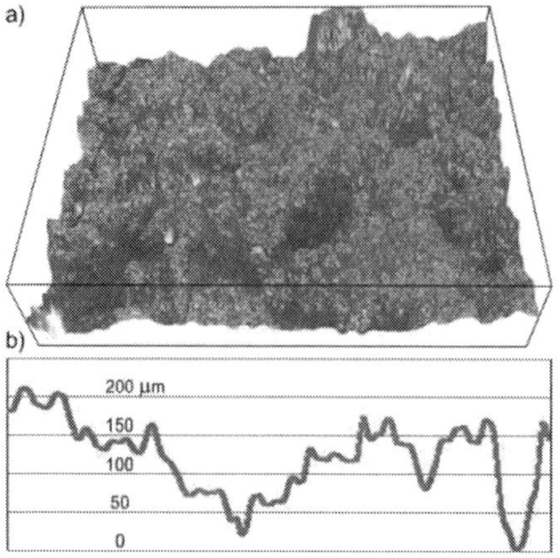

Fig. 4. Porous surface produced bysoldering iron powder, a) 3D topography, b) roughness profile.

2.4 Experimental methodology

After the desired values of the pressure and flow rate were reached, the electric power supplied to the heated plate was increased gradually to achieve an increase in the heat flux transferred to the fluid. It caused a change in the heat transfer between the plate and the fluid from single phase convection to nucleate boiling. The distribution of temperature on the heated plate was measured on the smooth side coated with a black paint layer of known emissivity [28]. Simultaneous observations of the two-phase flow structures were conducted on the opposite side of the minichannel.

Main thermal and flow parameters of the tested experimental series are listed in Table 1.

Table 1 Main thermal and flow parameters of the tested experimental series.

Parameter	Average or range*			
	surface with Fe powder/smooth			
	sintering/soldering		sintering/smooth	
Mass flux [kg/(m^{2}s)]	423	422	405	402
Absolute pressure, the inlet, [kPa]	107	111	156	170
Absolute pressure, the outlet [kPa]	104	102	144	173
Heat flux q_w [kW/m^2]	27	38	108	128
Inlet liquid subcooling [K]	41	42	45	45
Fluid temperature T_f* [K]	289÷300	291÷303	297÷334	297÷341
Heated plate temperature T_p* [K]	316÷323	319÷331	352÷359	359÷370

* range, at the inlet÷at the outlet

3 Heat transfer coefficient determination

The local values of the heat transfer coefficient for stationary state conditions; were calculated using simple one-dimensional method. The main assumptions in this method are as follows: (i) the heat is transferred between the plate and the flowing fluid through the plate-fluid contact surface, (ii) the temperature changes in plate and fluid along the minichannel width are negligible, as are the changes on the remaining side surfaces of the plate and on the external surface of the glass, (iii) only one direction of heat flow is taken into account i.e. perpendicular to the direction of flow related to the heated plate thickness.

The resulting local heat transfer coefficients were obtained similarly as in [9, 29] from the following equation:

$$\alpha(x) = q_w / (T_P(x,\delta) - T_f(x) - q_w \cdot \frac{\delta_P}{\lambda_P}) \quad (1)$$

where x – distance from the minichannel inlet, q_w – density of the heat flux transferred from the heated plate to the fluid, T_p – plate temperature measured by infrared thermography, δ_p – thickness of the plate (0.00045 m), λ_P – thermal conductivity of the plate (8.3 W/(m·K)), T_f – temperature of the fluid, assuming that $T_f = T_l$ in the subcooled boiling region and $T_f = T_{sat}$ in the saturated nucleate boiling region, T_l – fluid temperature calculated from the assumption of the linear distribution of the fluid temperature along the minichannel, T_{sat} – liquid saturation temperature determined from the assumption of the linear distribution of the fluid pressure along the channel. It was assumed that the saturated boiling region starts when the temperature of the heated surface is higher than the saturation temperature.

The heat transferred to the fluid in the minichannel was assumed to be equal to the difference between the heat generated by the heated plate and the heat loss to the surroundings according to the equation [9, 29]:

$$q_w = (I \cdot \Delta U)/A - q_{w,loss} \quad (2)$$

where I – current, ΔU – voltage drop, A – surface area of the heated plate, $q_{w,loss}$ – heat loss to the surroundings.

The central part of the heated plate where the surface temperature is measured with the infrared camera (4 mm x 180 mm) is not insulated. Therefore, the loss of heat to the surroundings is the greatest. It was estimated as 1.3% of the heat flux density [9].

4 Results

The results are represented graphically as a relationship between the heat transfer coefficient and the distance from the minichannel inlet. Separate plots were generated for the subcooled boiling region (Fig. 5) and the saturated boiling region (Fig. 6). In Fig. 6 results obtained for using a smooth heated plate in experiments was also presented to compare.

The values of the heat transfer coefficient calculated for the porous surface produced by soldering, sintering iron powder and for the smooth plate were marked with circles, diamonds, and squares, respectively.

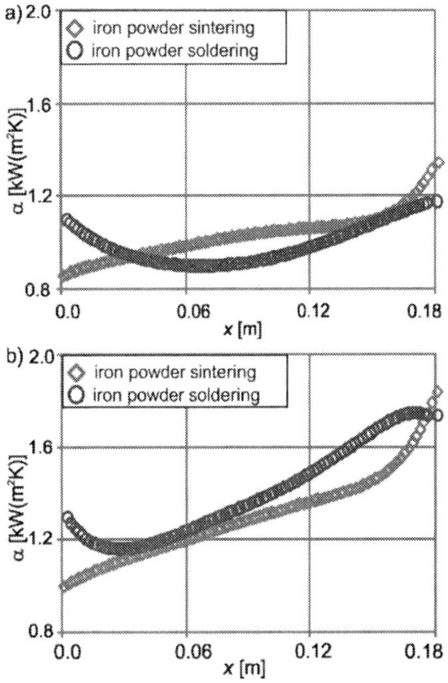

Fig. 5. Heat transfer coefficient vs. the minichannel inlet; data for the subcooled boiling region: a) q_w = 27 kW/m², b) q_w = 38 kW/m²

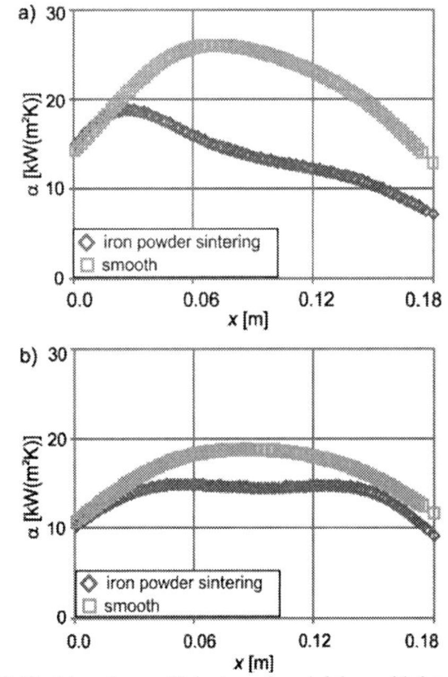

Fig. 6. Heat transfer coefficient vs. the minichannel inlet; data for the saturated boiling region, a) q_w = 108 kW/m², b) q_w = 128 kW/m².

In the subcooled boiling region, the heat transfer coefficient was low. The local values of the heat transfer coefficient increased slightly with the increase of the distance from the minichannel inlet (Fig. 5) for two heat fluxes. At lower heat fluxes (q_w = 27 kW/m^2), heat transfer coefficients at subcooled boiling region were slightly higher for surfaces prepared via sintering, except for the channel inlet area (Fig. 5a). At a higher heat flux (q_w = 38 kW/m^2), local heat transfer coefficients were higher for the surface prepared via soldering (Fig. 5b).

In the saturated boiling region the heat transfer coefficient was high, with values up to over a dozen times greater (Fig. 6) than those obtained for the subcooled boiling region (Fig. 5).

Due to lack of data (the liquid in the minichannel evaporates at the set heat fluxes), the heat transfer coefficient in the saturated boiling region for the soldered surface was not calculated. Fig. 6 shows also local heat transfer coefficients from the experiments with smooth surfaces. In the saturated boiling region, the heat transfer coefficient was increasing from the inlet up to the distance of about 1/3 to 1/2 of the distance from the channel inlet (depending on the surface type), then it began to decrease on the way to the channel outlet. The local heat transfer coefficients during saturated boiling from the smooth surface were slightly higher than those obtained from the sintered surface. The difference is negligible at a higher heat flux, q_w = 128 kW/m^2 (Fig. 6b).

It can be underlined that the porosity of the powder-based iron structure may produce high values of thermal resistance at diffusion bridges and a decrease in the surface wettability. Finally, the formation of vapour bubbles can be responsible for the reduction in heat transfer efficiency [27]. In the case of the surface formed by iron powder soldering, thermal resistance additionally occurs at the contact zones between the substrate and solder and between the solder and Fe powder grains. Low conductivity of the material in the diffusion bridges due to sintering may contribute to relatively low heat transfer efficiency in the saturated boiling region.

Analysis of the results suggests that small dimension of the pores in both surface types may exacerbate surface wettability during the refrigerant flow and generate vapours „plugs" thereby inhibiting heat transfer in the minichannel. This hypothesis is confirmed by the results of pool boiling heat transfer experiments, where capillary processes play a major role in the transport of refrigerants.

Boiling curves were drawn from the data collected during tests, which involved increasing the heat flux supplied to the heated plate. Boiling curves frequently represent the relationship between the heat flux density and the difference in temperature between the heated surface and the bulk fluid in the case of flow boiling in minichannel. The boiling curves in Fig. 7 were generated for three distances from the minichannel inlet: 0.06 m (Fig. 7a), 0.105 m (Fig. 7b) and 0.12 m (Fig. 7c). They were plotted for the two porous surfaces and the smooth plate.

The boiling curves shown in Fig. 7 were a basis for the heat processes analysis. Upon the inflow of the subcooled liquid (section A-BI), increased heat flux density leads to the onset of boiling (BI - nucleate boiling incipience). Boiling regions start with the subcooled boiling region. The liquid is superheated in the area adjacent to the plate and subcooled in the flow core. The spontaneous nucleation results in a temperature drop on the heated surface (section BI-C). The bubbles absorb a large amount of energy transferred to the liquid and act as internal heat sinks [8]. In the boiling region that follows, the saturated boiling region, the temperature of the fluid in the core of the flow is equal to the saturation temperature. Furthermore, in the fully developed nucleate boiling region, there were temperature differences T_P-T_f at high heat flux (section C-D). No significant differences in shape were observed between the boiling curves plotted for the smooth surface and for the surface enhanced by soldering iron powders.

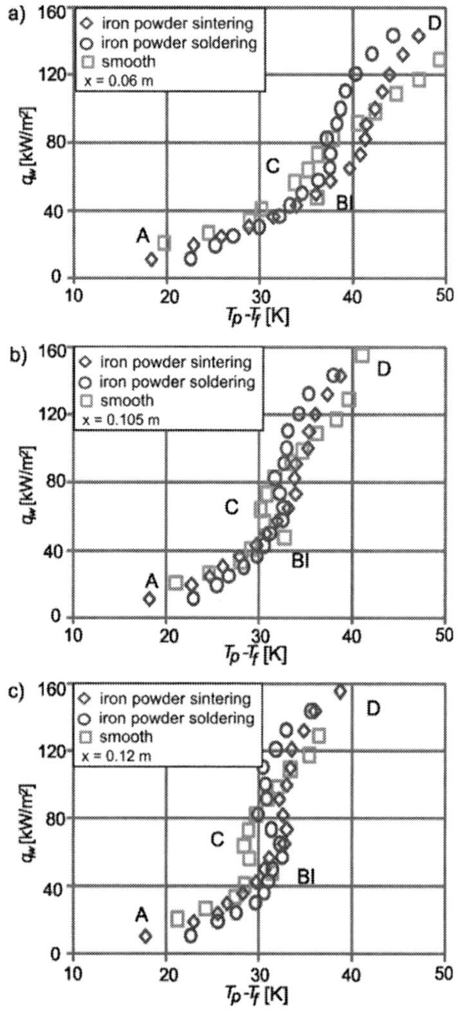

Fig. 7. Boiling curves for three distances from the minichannel inlet: a) 0.06 m, b) 0.105 m, c) 0.12 m.

5 Conclusions

This article focuses on identification of the impact of porous heated surface on flow boiling heat transfer in minichannel. The porous surface in contact with the fluid in the minichannel was produced in two processes: sintering or soldering of Fe powder to the plate made of Haynes-230. The results were presented as relationships between the heat transfer coefficient and the distance from the minichannel inlet and as boiling curves constructed for selected distances along the channel length. Results obtained for using a smooth heated plate in experiments was also presented to compare for the saturated boiling region.

In the subcooled boiling region, at a higher heat flux, the local heat transfer coefficients were slightly higher for the surface prepared via soldering. In the saturated boiling region, the local heat transfer coefficients obtained for the smooth plate surface were slightly higher than those achieved from the sintered plate surface.The porous structures formed from Fe powders bonded with the heated surface have low thermal conductivity. This may induce noticeable thermal resistance at the diffusion bridges of the sintered structures, in particular within the saturated boiling region.

The boiling curves were generated for three distances from the minichannel inlet. No significant differences in shape were observed between the boiling curves plotted for the smooth surface and for the surface enhanced by soldering iron powders.

Acknowledgements

The research reported herein was supported partially by a grant from the Polish National Science Centre (No. DEC-2013/09/B/ST8/02825).

References

1. B. Grabas, Exp. Therm. Fluid Sci. **74**, 100–109 (2016)
2. N. Radek, K. Bartkowiak, Phys. Procedia, **12** (1), 499–505 (2011)
3. N. Radek, J. Orman, Ł., L.A.D. J.I. Shalapko (Ed.), Sci. Basis Mod. Technol. Exp. Prospect., Khmelnytskyi National University, Jaremche, Ukraine, 236–245 (2011)
4. T. Orzechowski, A. Tyburczyk, K. Ziętala, Sci. Lett. Rzesz. Univ. Technol. - Mech **31**, (86(2/2014))
5. R. Pastuszko, M. Piasecka, J. Phys. Conf. Ser. **395**, 12137(2012)
6. R. Pastuszko, R. Kaniowski, J. Phys. Conf. Ser. 7th Eur. Therm. Conf. **25**, 2019 (2016)
7. M. Piasecka, Ann. Nucl. Energy. **73**, 282–293 (2014)
8. M. Piasecka, Int. J. Refrig. **56**, 198–212 (2015)
9. M. Piasecka, K. Strąk, B. Maciejewska, Heat Transf. Eng. **38** (3), 332–346 (2017)
10. M. Piasecka, K. Strąk, B. Maciejewska, B. Grabas, J. Phys. Conf. Ser. **745**, 3212 3(2016)
11. M. Piasecka, K. Strąk, B. Grabas, Arch. Metall. Mater. **62** (2017) (in print)
12. S. Blasiak, A. V. Zahorulko, Tribol. Int. **94**, 126–137 (2016)
13. S. Blasiak, Int. J. Heat Mass Transf. **81**, 90–102 (2015)
14. W. Depczynski, R. Kazala, K. Ludwinek, K. Jedynak, Materials **9** (7), 1–12 (2016)
15. W. Depczyński, *Proc. Met. 2014 23rd Int. Conf. Metall. Mater.*, 1219–1224 (2014)
16. M.F. Ashby, A.G. Evans, N.A. Fleck, L.J. Gibson, J.W. Hutchinson, H.N.G. Wadley, *Metal Foams: A Design Guide* (Buterworth-Heineman, 2000)
17. G.J. Davies, S. Zehn, J. Mater. Sci. **18**, 1899–1911 (1983)
18. M.A. Suarez, I.A. Figueroa, G. Gonzales, O. Lara-Rodriguez, G.A. Novelo-Perlata, I. Alfonso, I.J. Calvo, J. Alloy. Andg. Agent. or Spac. Compd. **585**, 318–324 (2014)
19. T.M. Wójcik, EPJ Web Conf. 2501100 (2012)
20. A.M. Parvanian, M. Sadatfar, M. Panjepour, A. Kingston, A.P. Sheppard, Mater. Des. **53**, 681–690 (2014)
21. A.M. Parvanian, P. M., Mater. Des. **49**, 834–841 (2013)
22. Y. Hangai, K. Zushida, H. Fuji, R. Ueji, O. Kuwazuru, N. Yosikawa, Mater. Sci. Eng. A. **585**, 468–474 (2013)
23. Y. Hangai, S. Koyama, M. Hasegawa, T. Utsonomiya, Metal. Mater. Trans. A. **41A**, 2184–2186 (2010)
24. W. Żórawski, R. Chatys, W. Depczyński, A method of producing porous structures, Patent of the Republic of Poland, No 199720 (2002) (in Polish)
25. M. Piasecka, A method for porous structures producing on the surface of metal products, Patent of the Republic of Poland, No 225512 (2015) (in Polish)
26. HAYNES® 230® alloy information http://www.haynesintl.com.
27. M. Piasecka, K. Strąk, *Proc. 12th Int. Conf. Heat Transf. Fluid Mech. Thermodyn. HEFAT 2016*, Torremolinos-Malaga, Spain, 34–41 (2016)
28. K. Strąk, M. Piasecka, Trans. Inst. Fluid-Flow Machinery**128**, 97–118 (2015)
29. M. Piasecka, B. Maciejewska, Exp. Therm. Fluid Sci. **68**, 459–467 (2015)

The use of modern measurement techniques for designing pro ecological constructions

Michał Wieczorowski[1,*], *Bartosz* Gapiński[1], *Maciej* Szymański[2], and *Artur* Rękas[3]

[1]Poznan University of Technology, Institute of Mechanical Technology, Piotrowo 3, 60-965 Poznan, Poland
[2]Ster Sp. z o.o., Czluchowska 12, 60-434 Poznan, Poland
[3]AGH University of Science and Technology, Faculty of Non-Ferrous Metals, Mickiewicza 30, 30-059 Krakow, Poland

Abstract. In the paper some possibilities of application modern length and angle metrology techniques to design constructions that support ecology were presented. The paper is based on a project where a lighter bus and train car seat was developed. Different options were presented including static and dynamic photogrammetry, computed tomography and thermography. Research related with dynamic behaviour of designed structures gave input to determine deformation of a seat and passengers sitting on it during communication accidents. Works connected to strength of construction elements made it possible to optimize its dimensions maintaining proper durability. Metrological actions taken in relation to production machines and equipment enabled to better recognize phenomena that take place during manufacturing process and to correct its parameters, what in turns also contributed to slim down the construction.

1 Introduction

Contemporary constructions are required not only to fulfil functional parameters and assure utility features on a certain level of aesthetics but also to support pro ecological activities [1]. One of the directions connected with such a thinking is designing light solutions, still maintaining all the necessary strength parameters. It is particularly visible when various parts of vehicle equipment is concerned, where every reduced kilogram saves fuel and reduces emission of polluting substances to the atmosphere.

In the paper these works were discussed on an example of a seat manufactured by company STER, who is a leading and renowned in the world supplier of that kind of products to many leading bus (for different purposes) and train car manufacturers. The aim of a project realized by a company and research institutions was then elaboration of a seat with a mass significantly smaller than the versions manufactured so far. As an effect reduction of nearly 10 kg was achieved, what makes about 30% of total mass of a seat and – as an example – gives a 400 kg less total weight of an average bus. That in turns reduces fuel consumption even up to as much as 2l per 100 km.

For research and inspection purposes advanced metrology solutions and special constructions were applied, helping to visualize on-going processes, verification of placed research hypotheses obtaining data for simulation purposes with very low measurement uncertainty. This topic is not widely discussed in references, the manufacturers of bus seats and structures have to comply with different formal regulations, regardless how this is achieved or measured.

Technological progress in XX century enabled development of coordinate measurement technique. Thus, it is possible to obtain information regarding shape and dimensions of an inspected workpiece basing on a set of measurement points described by means of coordinates in space. These data are transformed by measurement software into related elements that are a reference for designation of dimensions and deviations of a measured part in accordance with its geometrical product specification (GPS). A measurement of free form surfaces and their assessment in relation to a CAD model is also possible. Coordinate technique facilitates these tasks to a great extent or even makes them possible to happen or execute. Without it e.g. determination of many position deviations would not be possible, just because of difficulty in definition centre of a circle (Fig. 1).

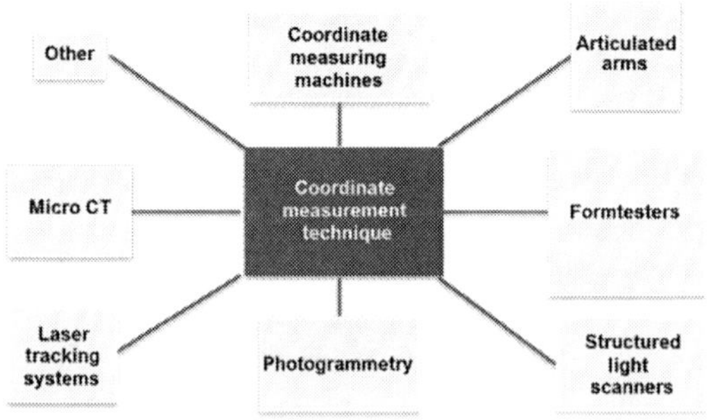

Fig. 1. Application of coordinate measurement technique in measurement devices [2].

* Corresponding author: michal.wieczorowski@put.poznan.pl

That is why we have a bunch of application possibilities of coordinate measurement technique in different devices, what was in details depicted on Fig. 1. This graph presents a general division of measurement instruments in which analysis of points coordinates on measurement surfaces was used to determine geometrical specification of an element. Such a differentiation of measurement instruments has its source in complexity of measurement tasks and tendency to obtain as versatile measurement information as possible, with reliable credibility and including economic aspect. These devices have then different measurement methods, number and way of collecting points and accuracy parameters and these features significantly contribute to decision which of them should be chosen in particular metrology task.

2 Static and dynamic photogrammetry

Seats for public transportation must ensure specific needs depending on their designation and function. Separate set of requirements features the ones that are manufactured for tourist buses equipped with seat belts, another one for city buses and still another one for train cars. Still, in each case, though to a different extent, it is necessary to fulfil strength requirements regarding static and dynamic deformation. When inspecting both: single elements and whole construction sets of seats it is necessary to use photogrammetry techniques. Accuracy parameters of these systems assure metrologically correct measurement results, while their flexibility makes it possible to map even very complicated free form surfaces. Photogrammetry itself can be defined as a method to obtain three dimensional data about shape and location of objects basing on one or many two dimensional images [3, 4, 5]. In length and angle metrology while measuring geometric features of workpieces, photogrammetry means measurement of clearly coded characteristic elements on objects. They are usually markers: points, lines or fringes projected onto the object by a projector. Photogrammetric systems used in measurement techniques allow for digitization of a scanned object. There are two basic groups of methods: static and dynamic ones. In relation to this project both of them were used. To inspect complex elements as seats for public transportation it is necessary to verify geometry of single pieces and the whole structure. This allows to assess manufacturing process and final assembly. Knowledge about real shape of a seat gives opportunity to confirm its dimensional compatibility with specification. It is also a necessary condition to ensure possibility of applying a certain seat to static and dynamic deformation tests.

To assess seat geometry as well as effects of research a scanning system with structural light was used. It was composed of a projecting light source and two CCD cameras registering layout of projected fringes on the object surface. Working principle of this device is based on triangulation [6], what was shown on Fig. 2. Basing on optical transformation equations a processing unit calculates 3D coordinates for each camera pixel. A result of such an action is a cloud including up to several million points for each single measurement shot.

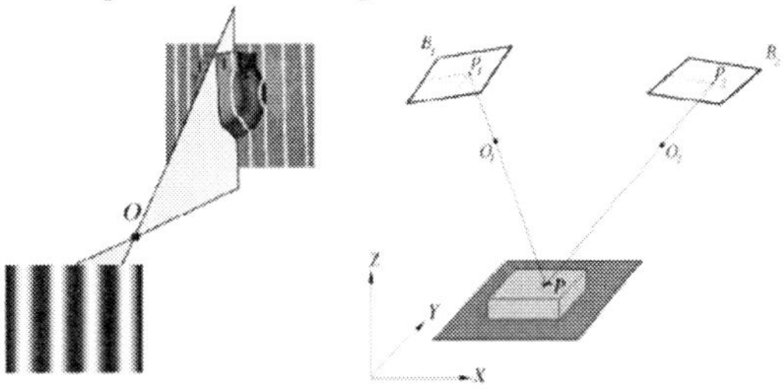

Fig. 2. Projection of fringes and triangulation.

A projecting system and cameras can be integrated into one sensor head. Thanks to triangulation it is possible to compare real data with nominal ones and workpieces before and after strength tests. This allows for estimation of plastic deformations of particular parts of seat structure and deformation of the whole geometry in relation to mounting places. An example of a seat with a list of measurement points and dimensional deviations was shown on Fig. 3.

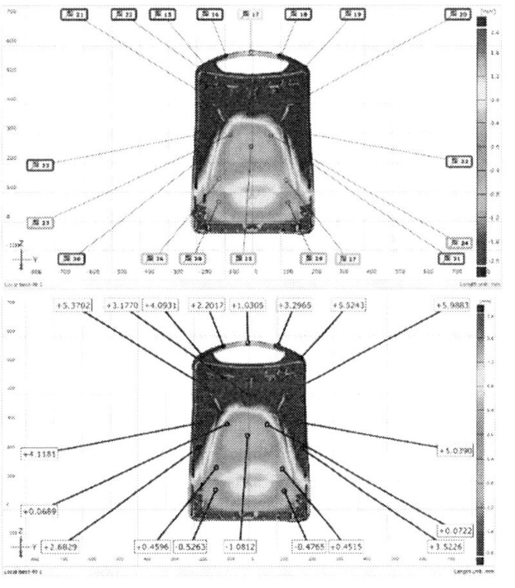

Fig. 3. A photogrammetric image of a bus seat with color map of deviations.

Thanks to these measurements seat sections were modified and reinforcement areas were defined, as well as a cooling method was elaborated that would not cause significant dimensional changes.

In dynamic measurements, fast cameras register displacements of point located on measured object. Its full dynamic behavior is obtained from many images recorded with great frequency. Because of a great number of measurement points a system enables to get a precise analysis of deformation. The first presented test shows analysis of closing process of a mould with different closing forces. An injection moulding machine had different possibilities from 500 to 1200 kN, so measurements were taken in 100 kN intervals. For all measurements reports were made with identically defined coordinate system. Points were analyzed in X axis, i.e. in mould closing direction. A measurement result for 1200 kN closing force is shown on Fig. 4.

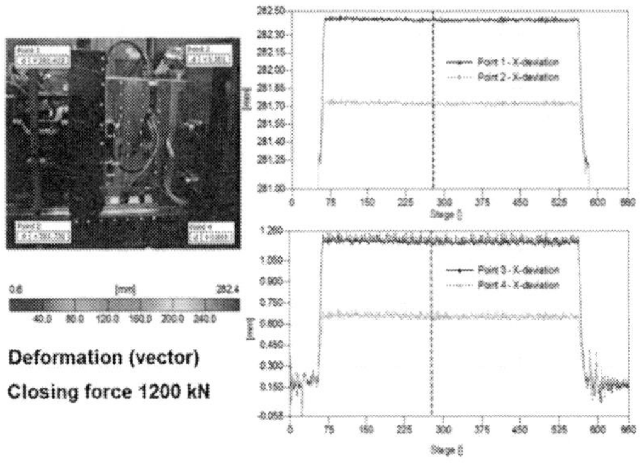

Fig. 4. Closing force measurement report for 1200kN, X axis.

Basing on this procedure (together with thermography research shown in chapter 3) it was possible to determine proper working parameters of injection moulding machine to obtain a workpiece with expected geometry.

Dynamic photogrammetry was also used to analyze seat behavior during static and dynamic tests connected with its strength. To ensure proper dynamics of a measurement displacement of special appropriately located markers is considered. It also allows to calculate velocity and acceleration. All this analysis was performed in relation to independent reference points as well as relatively e.g. for lower part of seat and support. In research a measurement space in a form of cube with 2m side was used [7]. Such a configuration enables for maintaining proper accuracy parameters (particular points were identified with uncertainty smaller than 0,05 mm) and including the whole area of interest where investigated phenomenon takes place. It is particularly important as tests with static and dynamic loads are single ones and their repetition for the same seat is not possible. In static tests a force is applied to a stationary seat while in dynamic ones a seat with mannequins is accelerated to a certain speed and than rapidly stopped to imitate an accident. In both cases a measurement takes place during the test, so – regardless the name – both, static and dynamic measurement really has a dynamic factor.

Dynamic measurements were performed on a specially constructed research track enabling for a full crash test. It allows for accelerating a carriage to 50 km/h, though a maximum speed of 32 km/h was used to fulfil directives and regulations. These tests are made to verify, whether a sitting person is properly kept by a seat belt and a seat before.

As an effect it is possible to determine what kind of injuries can happen to a passenger when accident takes place and if a seat construction with mounting system is strong enough.

On Fig. 5 mannequins during a dynamic test can be observed. Their displacement to the front was blocked on a knee level after a contact with the seat before. In such a situation knees work like hinges and the mannequin body rotates to the front in relation to an axis going through both knees. It effects in a dynamic contact of a head with a seat construction. Because of this, a seat before is plastically deformed.

Fig. 5. Heads of mannequins strike seats before.

Fig. 6 shows the first stage of deformation analysis of a seat on the level of bolster after being hit by a mannequin not fastened by a seat belt. Two point are analyzed – one at each bolster.

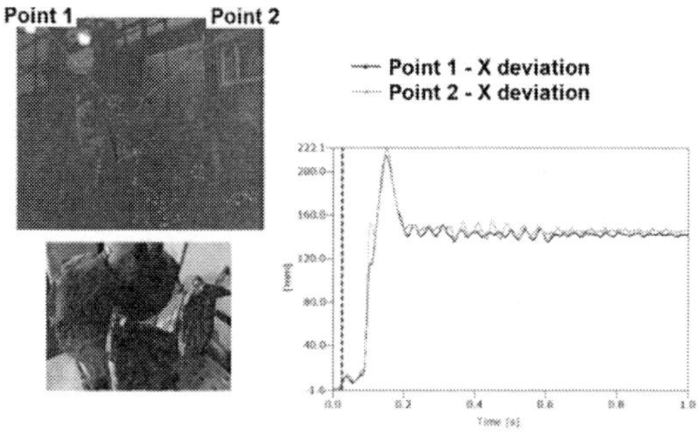

Fig. 6. Bolster and seat deformation after being hit by a mannequin not fastened by a seat belt.

As it was mentioned before, in static tests a force is applied to a stationary seat. This test is executed to determine whether a person is properly kept by a seat and seat belts and would not be seriously injured and whether a seat and its mounting are strong enough. The test has a positive result when a after applying a force a seat would not be elastically deformed more than a certain limit and if after releasing a force plastic deformation will appear as a crash zone. This zone has to take a part of collision energy from a passenger saving his life. Furthermore, seat elements can not be separated and the seat itself must remain stiffly mounted. After such a test there can be no cracks and sharp edges on any seat part. They could cause serious injuries and wounds.

On Fig. 7 a moment is depicted when a swingle on a specially designed research setup striking a seat causes its maximum deflection and deformation. Maximum

DOI: 10.1051/e3sconf/20171903013

deformation value in highest point of support reaches nearly 29 mm. Value of deviation goes successively down in the seat direction. It is a proper reaction concluding from a fact that the closer is the swingle from the seat the smaller is the arm force on which it operates.

Fig. 7. A research setup with a swingle designed for energy absorption test – a moment of maximum inclination of a seat support.

On Fig. 8 a plot showing change of location of a marker in upper part of support was shown. This place is the most exposed to deformation.

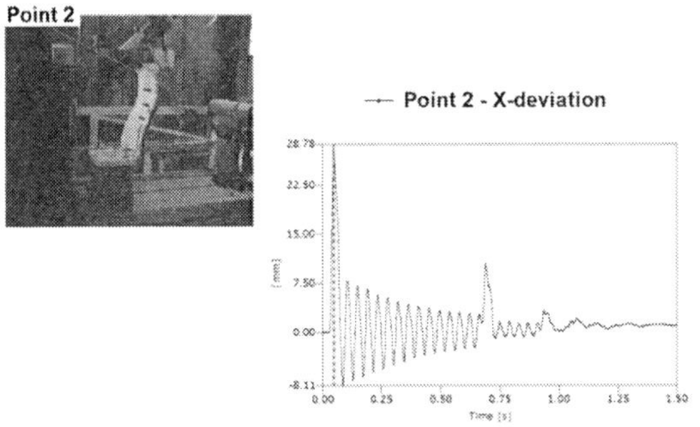

Fig. 8. A research setup with a swingle designed for energy absorption test – a plot showing displacement of a point located in upper part of support.

Analysis of that plot very precisely shows a course of energy absorption test. As it was shown in the moment of strike of a swingle its upper part moves at as much as 29 mm to the front. Elastic deformation causes push back of a swingle and a support reaction causing its deflection in the opposite direction of 8 mm from initial position. So the support covered a distance of 37 mm. From that moment it makes a wobbly movement with fading amplitude until a swingle strikes again. Then a deformation is much smaller i.e. about 8 mm. Again it causes a support reaction caused by elastic deformation. A support comes back to the initial position making a wobbly movement with a small amplitude. After 1.25 seconds from a first strike of a swingle a return of a support to initial position can be observed, and deformation value does not exceed 1mm. This proves a proper construction of a seat, because during test it had nearly only elastic deformation.

Another application of dynamic photogrammetry was testing strength of elements of seat structure. In the research load and deformation parameters from a tensile strength machine were recorder and simultaneously data from photogrammetric system were analyzed. It was necessary to paint the whole sample with white paint and successively put irregular black dots on it to make this kind of measurements. This kind of marking is generally similar to markers in static or dynamic photogrammetry described in chapter two e.g. for geometry inspection. However, for strength tests traditional markers would not allow to get such a versatile information as it is when white paint and black dots are considered. An image of a sample after bending (total deformation) and analysis of changes that took place inside it were shown on Fig. 9.

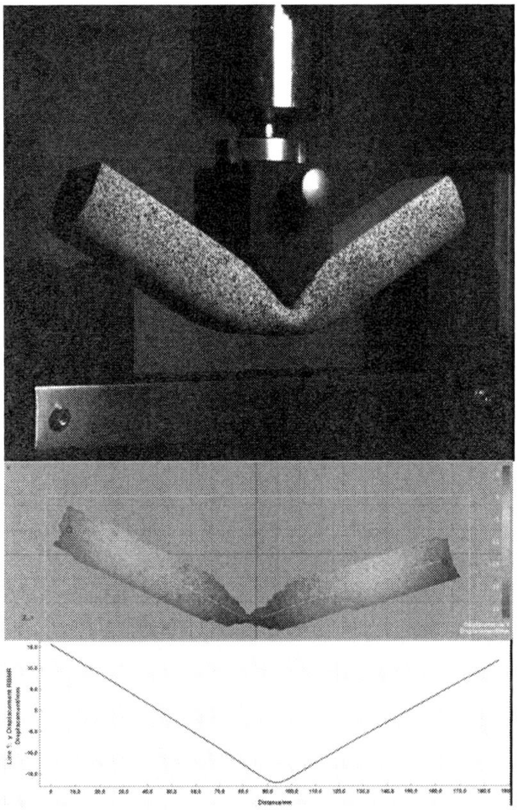

Fig. 9. A sample after total deformation and image of displacements.

Basing on this analysis possibilities of load transfer of a chosen profile was assessed.

3 Thermography

A basic feature of thermography systems is the one that they indirectly measure temperature of an inspected workpiece; what is measured directly is a power of infrared radiation emitted by an object. This radiation is transferred by a detecting structure of a measurement system into electric signal that carries information about temperature of a part. Temperature in turns is determined with the use of thermometric characteristics of a thermograph written in the memory of a specialized computer designated among others to control process of reproduction of temperature fields seen by a camera of a thermovision system. Apart from radiation emitted by the object (this radiation carries information about its temperature) another radiations gets to a detection structure of a thermograph: reflected from an object and own radiation of atmosphere generated on the segment of path between object and camera. Furthermore, radiation emitted by interior of the camera falls on the detector as well. Each of these above mentioned components of radiation has a different distribution of radial power spectrum density, that depends among others on conditions of atmosphere (humidity, pressure, carbon oxide contents, fog, pollination, emissive properties of an object, emissive properties of camera elements etc.). All this makes that reproduction of object temperature is realized as an effect of computation procedures, including influence that these factors have on useful measurement signal.

A research stand for seat thermal diagnosis was made basing on thermovision camera. What was investigated was the influence of different cooling systems on a ready plastic made seat. An example of a seat during cooling after molding injection is shown on Fig. 10.

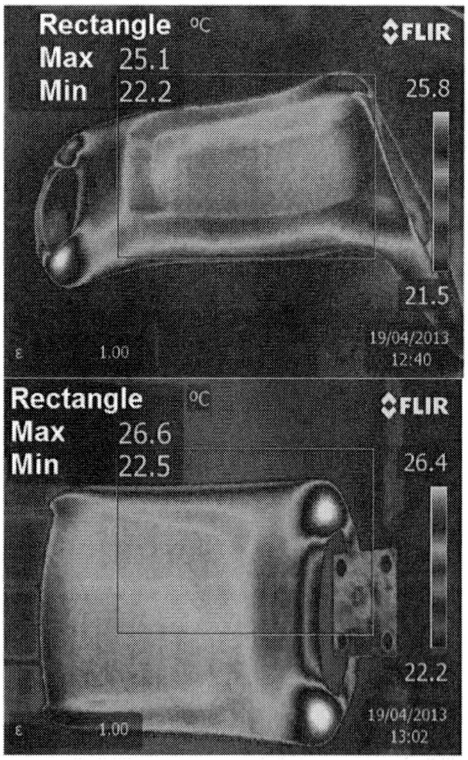

Fig. 10. Images of a seat during a cooling process.

Basing on that kind of thermovision research (including dynamic photogrammetry described in chapter 2) it was possible to optimize injection molding process and cooling method. This in turns dramatically improved repeatability of dimensions and geometrical deviations of a ready seat.

4 Computed tomography

Traditional optical or tactile measurement methods quite often suffer from lack of access to places where taking measurements or collecting points is necessary. This can be even more difficult for some of coordinate techniques. Measuring some of the features may require cutting a workpiece which from one side is not effective and from the other one this process itself causes modifications of both: an edge and layers located beneath the surface. This in turns significantly influences on measurement accuracy parameters. A good solution of this problem is using a computed tomograph (CT). This method of measurement has recently allowed to extend measuring possibilities [8]. It applies X-rays going through the object and giving an image on a detector. CT scanners are known in medical applications for a long time, but for technical imaging of 3D objects and extracting geometrical features, these devices are still a kind of novel solution. The image from CT scanner allows for estimation of the geometry of manufactured product as well as internal closed surfaces. It also allows for analysis of pores inside material or estimation of deformation [9, 10]. Voxel is a unit of 3D spatial image (volumetric element) – and it corresponds to pixel for 2D images. Voxel can be then treated as workpiece diameter divided by a number of pixels. A picture of a tomograph with its working principle was depicted on Fig. 11.

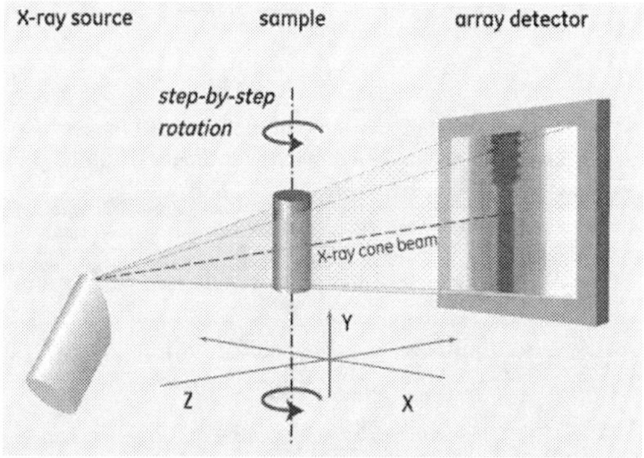

Fig. 11. A CT scanner and its functioning principle.

Preparing tomographic image consists of measurement of radiation absorption penetrating the object. Its volume is divided into voxels which have the same linear coefficient of radiation absorption. The reconstructed sectional image is a quantitative map of the linear coefficient of radiation absorption in voxels which are a part of the scanned slice [11,12,13]. The obtained cloud of points allows to conduct the analysis of dimensions and internal structure.

Computed tomography in the research concerning bus and train car seats was used to analysis of a bended profile [14]. A change in thickness of walls on inner and outer side of arc (being a result of bending) can be observed (Fig. 12). What is understood a wall on the outside surface gets thinner. As it is visible from the drawing a change of thickness is uniform on the whole length of the arc and there are no important local changes, which of course means a positive assessment of the process.

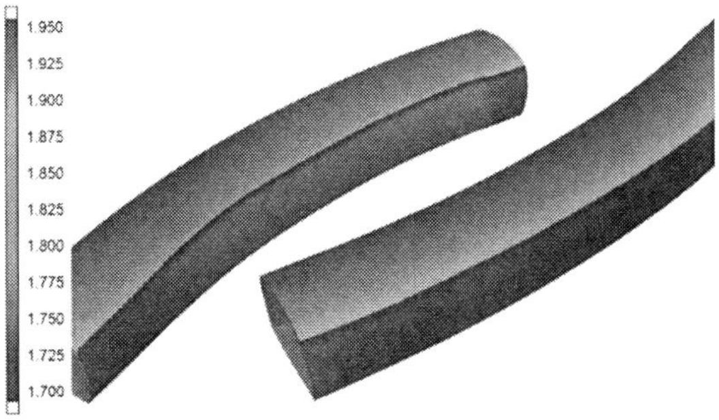

Fig. 12. Wall thickness measurement in the bending area – a view from inner and outer side.

Wall thickness measurement particularly in hardly accessible places is a very good task for computed tomographs. As a result of CT measurements of seat elements a 3D model as a result of inspection was obtained. Basing on that geometry assessment in relation to nominal CAD model is possible (Fig. 13).

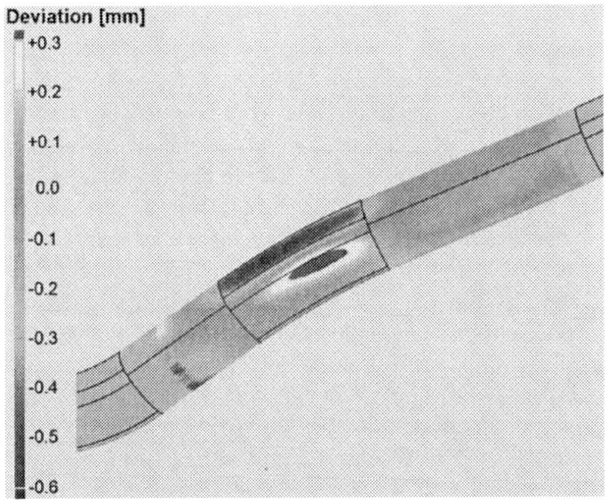

Fig. 13. A map of deviations of a bended profile from nominal CAD model obtained by means of a CT.

These measurements and research made it possible to obtain results that were a base to assess a degree of necking of a wall section. It was also a feedback to design of a profile that would fulfil all the dimensional and strength requirements.

5 Conclusions

Modern techniques of length and angle metrology used to aid construction of a new bus seat helped to create lighter solutions, but maintaining required strength parameters. Research related with dynamic behavior of designed structures gave input to determine deformation of a seat and passengers sitting on it during communication accidents. Works connected to strength of construction elements made it possible to optimize its dimensions maintaining proper durability. Metrological actions taken in relation to production machines and equipment enabled to better recognize phenomena that take place during manufacturing process and to correct its parameters, what in turns also contributed to slim down the construction. The paper shows that nowadays designing pro ecological construction requires a lot of metrological actions to prove their functionality.

Part of this work was supported by the Polish National Centre of Research and Development (project contract No. Innotech In-Tech K2/IN2/58/182896/NCBR/12; "Elaboration of manufacturing technology of new generation ultralight seats for public transportation fulfilling requirements of UE directives, UN regulations and American White Book").

References

1. G.M. Królczyk, Niesłony P. Maruda R.W., Wojciechowski S., J. Cleaner Prod., **142** (2017)
2. L. Marciniak - Podsadna, *Methodics of accuracy testing of optical coordinate scanners*, PhD dissertation, Poznań University of Technology, 2014
3. R.J. Campbell, P.J. Flynn, Comput Vision Image Understanding, **81**, 2 (2001)
4. D. Caspi, et al., IEEE T Pattern Anal, **20**, 5 (1998)
5. R. Majchrowski, Grzelka M., Wieczorowski M., Sadowski L., Gapiński B., Metrology & Meas Syst, **XXII**, 4 (2015)
6. M. Wieczorowski, M. Ruciński, R. Koteras, Arch Foundry Eng, **10** (2010)
7. M. Wieczorowski, Gapiński B., Szymański M., Grzelka M., Rękas A., Szymański S., Mechanik, **11** (2016)
8. B. Gapiński, Wieczorowski M., Marciniak-Podsadna L., Dybała B., Ziółkowski G., Procedia Eng, **69** (2014)
9. J.P. Kruth, Bartscher M., Carmignato S., Schmitt R., De Chiffre L., Weckenmann A., CIRP Annals, **60** (2011)
10. S. Carmignato, E. Savio, CIRP Annals, **60** (2011)
11. E. Ratajczyk, Mechanik, **2** (2011)
12. A.G. Ramm, *Inverse Problems, Tomography and Image Processing*, Springer, 1998
13. N. Sorokin, *High-Performance Reconstruction in Computer Tomography: Formal Specification and Implementation*, VDM Verlag, 2008
14. Gapiński B., Wieczorowski M., Szymański M., Szymański S., Grzelka M., Rękas A., Mechanik, **11** (2016)

DOI: 10.1051/e3sconf/20171903014

The study on dynamic properties of monolithic ball end mills with various slenderness

Szymon Wojciechowski[1,*], *Maciej* Tabaszewski[1], *Grzegorz M.* Krolczyk[2], and *Radosław W.* Maruda[3]

[1] Poznan University of Technology, Faculty of Mechanical Engineering and Management, Institute of Mechanical Technology, Poznan, Poland
[2] Opole University of Technology, 76 Prószkowska St., Opole 45-758, Poland
[3] University of Zielona Gora, 4 Prof. Z. Szafrana street, 65-516 Zielona Gora, Poland

Abstract. The reliable determination of modal mass, damping and stiffness coefficient (modal parameters) for the particular machine-toolholder-tool system is essential for the accurate estimation of vibrations, stability and thus the machined surface finish formed during the milling process. Therefore, this paper focuses on the analysis of ball end mill's dynamical properties. The tools investigated during this study are monolithic ball end mills with different slenderness values, made of coated cemented carbide. These kinds of tools are very often applied during the precise milling of curvilinear surfaces. The research program included the impulse test carried out for the investigated tools clamped in the hydraulic toolholder. The obtained modal parameters were further applied in the developed tool's instantaneous deflection model, in order to estimate the tool's working part vibrations during precise milling. The application of the proposed dynamics model involved also the determination of instantaneous cutting forces on the basis of the mechanistic approach. The research revealed that ball end mill's slenderness can be considered as an important milling dynamics and machined surface quality indicator.

1 Introduction

Ball end milling is usually applied in the production of dies, molds and turbine blades. These processes are very often carried out in finish conditions, thus the high quality of the machined surface is required. Nevertheless, many researches [1-3] show that fundamental problems occurring during ball end milling in finish conditions are the excessive surface roughness and surface location errors, as well as the intense tool wear [22]. The deterioration of surface quality during machining can be attributed to machine tool-cutting process interactions, as: milling kinematics [4, 5], the loss of process stability [6, 23], elastic-plastic deformations of workpiece [7, 8], as well as the tool's working part displacements (vibrations) [9, 10]. Nevertheless, the vibrations seem to be the most influential factor affecting the milling accuracy.

According to Jun et al. [11] and Wojciechowski et al. [12], tool's working part instantaneous displacements during precise milling are significantly affected by the tool deflections induced by cutting forces and machine-toolholder-tool system's geometrical errors (e.g. tool run out). The consequence of the excessive tool displacements during milling are the unacceptable surface errors. Research carried out by Lopez de Lacalle et al. [13] revealed that during finish curvilinear ball end milling of hardened steel, the cutting forces significantly affect the values of surface location errors (SLE). The occurrence of SLE is correlated directly with the tool deflections. Desai and Rao [14] observed that during end milling of curved surfaces, the variation of surface error induced by the tool deflection is not uniform along the tool path. This is mainly due to the change of workpiece curvature along tool path which affects radial engagement of cutting teeth and thereby changing surface error magnitude and the nature of error profile itself. Moreover, during precise milling with relatively low cutting forces or machining with the application of the rigid tools, the geometrical errors of the spindle-toolholder-milling tool system affect influentially the vibrations, and thus the surface quality. According to Sun and Guo [15], the static run out can significantly affect cutter's sweep surface during five-axis flank milling, and thus machined surface's quality. In addition, Buj-Corral et al. [16] developed surface topography model for the peripheral milling with the consideration of machining kinematics, eccentricity and tool geometry. It was noticed that the roughness profiles vary along the work material's height during milling with the occurrence of an eccentricity.

The values of tool deflections generated during machining are correlated directly with the selected cutting parameters (mainly depth of cut, feed per tooth and cutting speed), as well as the dynamical properties of machine-toolholder-tool system. The dynamical properties of this system are very often depended on the selected milling tool's overhang and thus its slenderness. It is well known, that the growth of tool's slenderness causes the growth of its flexibility which can result in an excessive vibrations during milling. However, in many applications, especially in the aerospace industry, the

* Corresponding author: szymon.wojciechowski@put.poznan.pl

DOI: 10.1051/e3sconf/20171903014

need to use long end mills with high length to diameter ratios is necessary. These tools are frequently required for the manufacturing of parts with deep pockets and relatively thin ribs. The researches carried out by Tlusty et al. [17] and Mendes de Aguiar et al. [18] show that the appropriate selection of milling parameters together with the tool slenderness can improve the material removal rate and surface finish. Nevertheless, the direct relations between the ball end mill's slenderness, its dynamical (modal) parameters and the instantaneous tool deflections generated during precise milling are still insufficiently recognized.

This work focuses on the evaluation of dynamical properties of the monolithic ball end mills with various overhangs. The carried out experiments involve the measurements of tool's modal mass, damping and stiffness coefficient, and subsequently the prediction of tool's instantaneous displacements generated during milling, with the application of dynamic model. The obtained results can be further applied to the optimal selection of tool's overhang allowing the minimization of vibrations and improvement in surface finish.

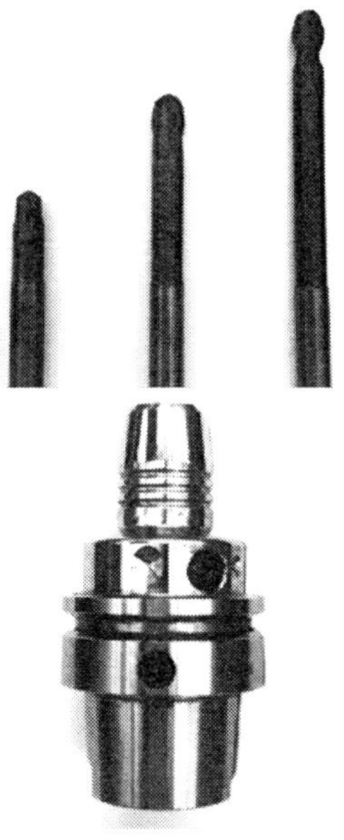

Fig. 1. Ball end mills and hydraulic toolholder.

2 Experimental details

2.1 Ball end mill – toolholder system

The analysis of dynamical properties has been conducted on the three ball end mills with various lengths and thus slenderness values (see Tab. 1, Fig. 1). According to López de Lacalle et al. [9], the slenderness of the two-toothed ball end mill can be calculated based on equation:

$$\lambda = \frac{l^3}{(0.8D)^4} \tag{1}$$

The investigated tools, made of cemented carbide with TiAlN coating are intended to finish milling of hardened alloy steels. The tools have been clamped in the hydraulic toolholder. Consequently, the three various overhangs have been obtained.

The selection of hydraulic toolholder was motivated by its relatively high stiffness and damping coefficients, which allow the selection of higher overhangs during milling.

Table 1. Dimensions of ball end mills applied in the research.

Variant	Overhang l [mm]	Diameter D [mm]	Tool length l_c [mm]	Slenderness λ [mm^{-1}]
1	32	10	72	8.8
2	63	10	105	70.2
3	95	10	135	216

According to work [19], the stiffness and damping coefficients of carbide tool-hydraulic toolholder system are respectively 240% and 35% higher than those measured in carbide tool-shrink fit toolholder system.

2.2 Identification of system's modal parameters

The conducted research involved the evaluation of milling tool-hydraulic toolholder system's dynamical properties, including modal mass m, stiffness coefficient k and damping coefficient c. These quantities have been estimated on the basis of modal test (Fig. 2) enabling the determination of frequency response function (FRF) for the 1 degree of freedom (1 DOF) system.

The FRF function for the 1 DOF system is defined in the following form [20]:

$$H(j\omega) = \frac{X(j\omega)}{F(j\omega)} = \frac{\frac{1}{m}}{-\omega^2 + j\omega\frac{c}{m} + \frac{k}{m}} = \frac{\frac{1}{m}}{-\omega^2 + 2j\omega h + \omega^2{}_o} \tag{2}$$

where: $H(j\omega)$ is the dynamic compliance, $X(j\omega)$ is system's response in the frequency domain, $F(j\omega)$ is a force in the frequency domain, j – is an imaginary unit, $\omega = 2\pi f$, $2h = c/m$, ω_o is the freqeuncy of free, undamped vibrations.

The applied identification method was based on experimental determination of H_1 characteristics, according to definition:

$$H(j\omega) = H_1(j\omega) = \frac{G_{FX}(j\omega)}{G_{FF}(j\omega)}, \tag{3}$$

where: $G_{FX}(j\omega)$ is a cross spectral density, $G_{FF}(j\omega)$ is auto spectral density.

The experimental stand of modal test (Fig. 2) consisted of impact hammer, 1-directional piezoelectric accelerometer (connected with amplifier), A/D transducer and a PC equipped with software dedicated to the analysis of dynamical properties.

DOI: 10.1051/e3sconf/20171903014

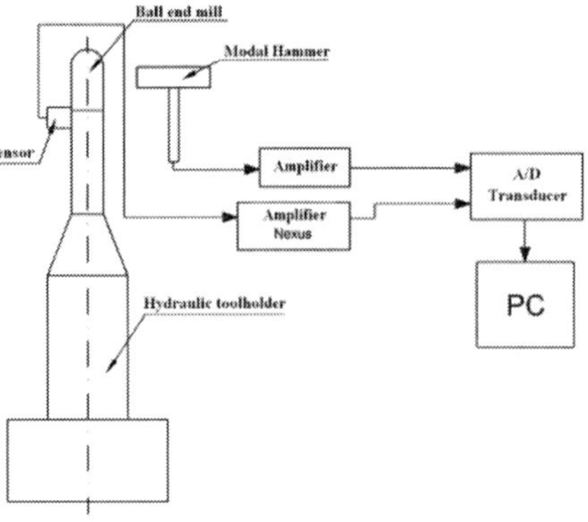

Fig. 2. The scheme of modal test experimental set-up.

The sampling frequency has been selected at the level of 20 480 Hz, and the vibration amplifier's sensitivity was equal to 1 mV/(m/s^2). During the carried out measurements, the maximal acquired system's natural frequency did not exceed the 6000 Hz. Thus, the Nyquist-Shannon criterion, towards the sampling frequency selection has been fulfilled. Moreover, the conducted measurements did not involve the use of filters.

In order to calculate the statistical measures of the acquired signals, the 10-repetitions of measurements for the each investigated variant have been conducted.

2.3 Ball end mill dynamic deflection model

The estimated modal parameters of the ball end mill – toolholder system were further applied in the dynamic deflection model. The instantaneous deflections $y(t)$ of the ball end mill (Fig. 3), induced by cutting forces can be calculated on the basis of the differential motion equation [12]:

$$m \cdot \frac{d^2 y}{dt^2} + c \cdot \frac{dy}{dt} + k \cdot y(t) = F_z(t) \cdot \sin\alpha + F_y(t) \cdot \cos\alpha \quad (4)$$

where: $y(t)$ – instant tool working part deflection, tangent to the plane defined by the feed motion vector and tool's rotational axis, α – surface inclination angle, $F_z(t)$ – instantaneous force component in the thrust direction, $F_y(t)$ – instantaneous force component in the feed direction.

The equation (4) has been solved numerically in **MATLAB Simulink** software with the application of a variable step Runge-Kutta method (**ode45** solver). The instantaneous values of F_y and F_z forces, can be calculated on the basis of the mechanistic approach, described by equations:

$$F_y(t) = \sum_{j=1}^{z_c} \begin{bmatrix} -\left(K_{re} l_j + K_{rc} A_{zj}\right) \cdot \sin\varphi_{rj} \sin\varphi_j + \\ -\left(K_{ae} l_j + K_{ac} A_{zj}\right) \cdot \cos\varphi_{rj} \cdot \sin\varphi_j + \\ -\left(K_{te} l_j + K_{tc} A_{zj}\right) \cdot \cos\varphi_j \end{bmatrix} \quad (5)$$

$$F_z(t) = \sum_{j=1}^{z_c} \begin{bmatrix} \left(K_{re} l_j + K_{rc} A_{zj}\right) \cdot \cos\varphi_{rj} + \\ -\left(K_{ae} l_j + K_{ac} A_{zj}\right) \cdot \sin\varphi_{rj} \end{bmatrix} \quad (6)$$

where: K_{te}, K_{re}, K_{ae} are the edge specific coefficients, K_{tc}, K_{rc}, K_{ac} are the shear specific coefficients, l_j is the length of cutting edge of the j-th tooth, A_{zj} is the cross sectional area of cut of the j-th tooth, φ_j, φ_{rj} are the instantaneous positioning angles of the j-th cutting edge, z_c is the active number of teeth.

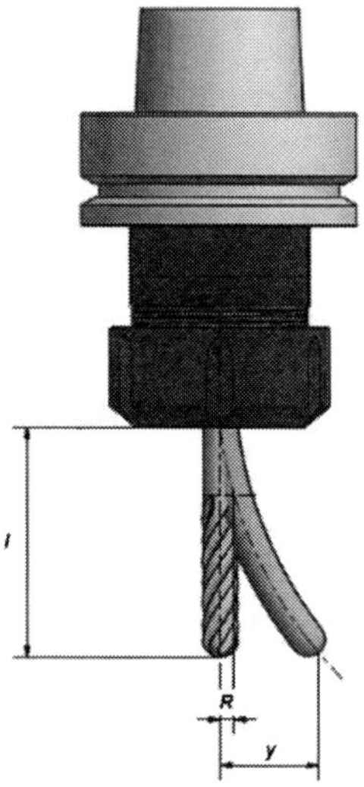

Fig. 3. Ball end mill deflection in the y direction.

In order to calculate the instant cutting forces (F_y, F_z) it is necessary to determine cross sectional area of cut and active length of cutting edge, as well as calibrate specific coefficients. These parameters' values can be estimated with the application of methods and data presented in work [21], which was focused on the cutting forces modeling during ball end milling of hardened alloy steel.

3 Results and discussion

Figures 4 – 7 depict the modal parameters' (k, c, m, f_o) values determined for the investigated tool slenderness values. It can be seen that the growth of tool overhang (and thus the slenderness) causes the monotonic decrease of system's damping coefficient c, stiffness coefficient k and natural frequency f_o. Moreover, the c, k, f_o = f(λ) dependencies are described by the power functions, which means that in a range of higher tool slenderness values ($\lambda \geq 70.2$ mm^{-1}), the differences in modal parameters' values are decreasing and stabilizing. It should be noted, that the decline in c, k, f_o values can be undesirable because it can lead to the growth of tool deflections during milling (induced by the decrease of stiffness coefficient), reduction of process dynamical stability (induced by the decrease of damping coefficient), as well as the higher risk of resonance phenomenon (induced by the decrease of natural frequency).

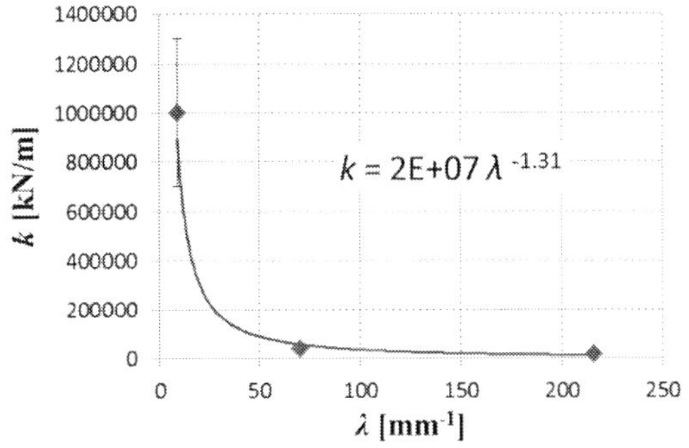

Fig. 4. The influence of the tool slenderness on the system's stiffness coefficient.

Nevertheless, in case of a modal mass m, a non-monotonic influence of tool slenderness value is seen. The lowest m values are reached for the moderate investigated slenderness: $\lambda = 70.2$ mm^{-1}. The modal mass is a parameter which describes the mass of the investigated system which is subjected to the deflections induced by the variable force. Thus, the lower modal mass, the lower ball end mill's volume is subjected to the deflections during milling.

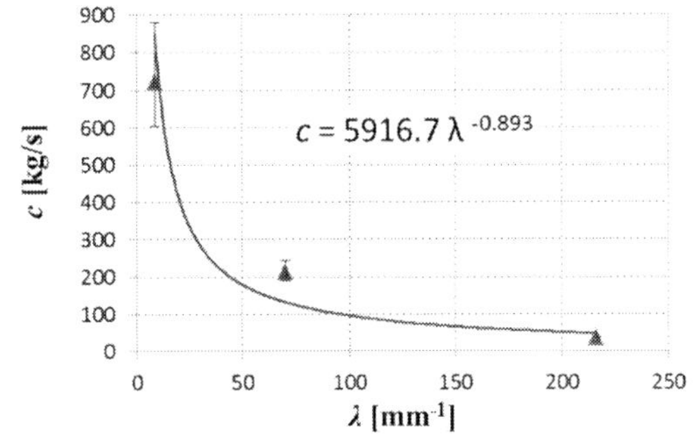

Fig. 5. The influence of the tool slenderness on the system's damping coefficient.

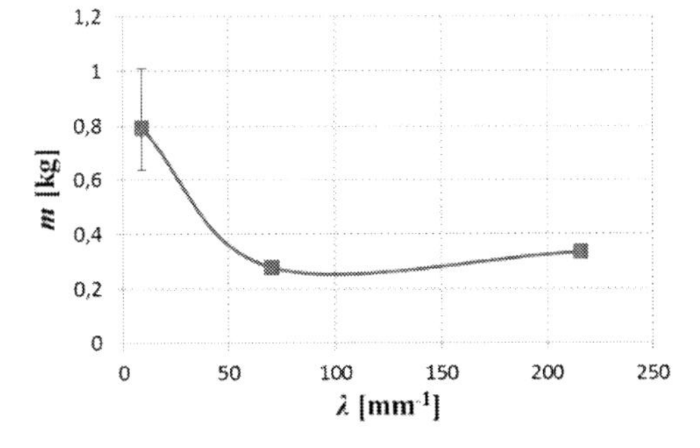

Fig. 6. The influence of the tool slenderness on the system's modal mass.

Figures 4 – 7 show also that the highest error bars' values of the determined m, c, k parameters are reached for the lowest investigated slenderness (which corresponds to highest measured system natural frequency). It can be caused by the relatively low sampling rate (20 480 Hz), which in case of a system with the average natural frequency equaled to 5693 Hz results in approx. 3 measurements per vibration period. Thus, it affects the appearance of some inaccuracies in the obtained digital signal's form.

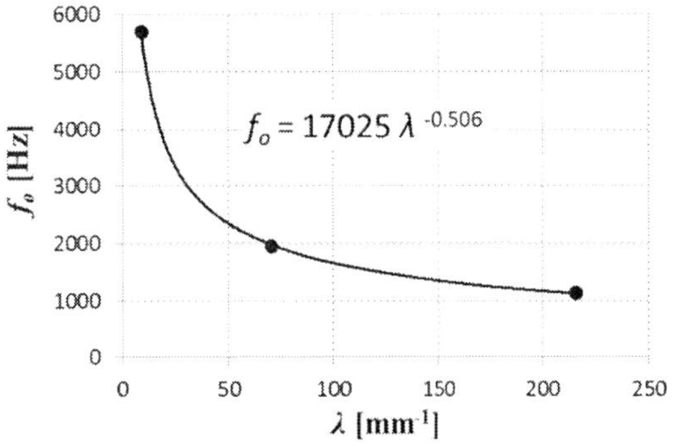

Fig. 7. The influence of the tool slenderness on the system's natural frequency.

The next step of analysis involved the simulations of ball end mill's working part instantaneous deflections $y(t)$. During this stage, the upper recommended cutting parameters' values for the precise milling of hardened steel have been selected as: a_p = 0.3 mm and $f_z = 0.1$ mm/tooth. Figures 8-10 show that independently on the investigated system, the instantaneous deflections reach alternately the positive and negative values, which indicates that the tool's working part is periodically indenting inside the workpiece, (exceeding the nominal depth of cut value), as well as moving outside the material which is being cut (reducing the nominal depth of cut value).

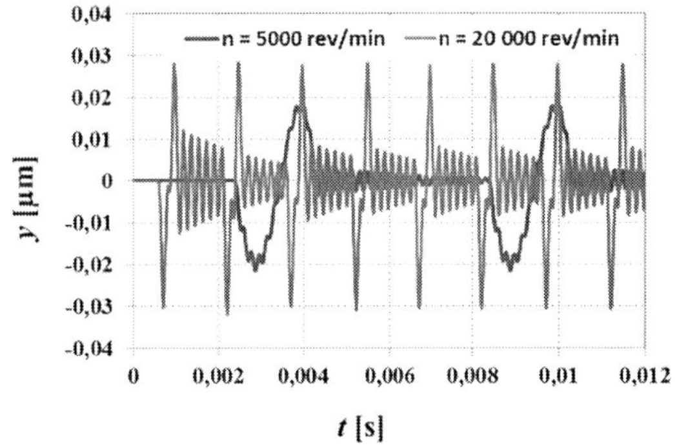

Fig. 8. The time course of instantaneous tool deflections for $\lambda = 8.8$ mm^{-1}.

In case of the system with the lowest slenderness (λ = 8.8 mm^{-1}), the deflection amplitudes are not exceeding the 0.04 μm, independently on the selected rotational speed, which means that they have insignificant contribution to the formation of surface finish, even during the precise milling. Nevertheless, in case of the moderate tool overhang ($\lambda = 70.2$ mm^{-1}), the

 DOI: 10.1051/e3sconf/20171903014

deflection amplitudes are significantly higher and reach almost 2 μm for the n = 20 000 rev/min and 0.5 μm when the n = 5 000 rev/min. It is worth indicating that further growth of tool overhang/slenderness is not affecting influentially the growth of tool deflection amplitudes. In case of the system with the highest investigated overhang/slenderness, the deflection amplitudes are lower than 3 μm for the n = 5 000 rev/min and 2 μm when the n = 20 000 rev/min. This finding reveals that in a range of a higher rotational speeds, the deflection amplitudes are similar for the systems with λ = 70.2 mm^{-1} and λ = 216 mm^{-1}. Nevertheless, when the system with λ = 216 mm^{-1} is applied, the forced vibrations reach the steady state after longer cutting time (~0.03 s), in comparison to milling conducted with less slender systems ($\lambda \leq$ 70.2 mm^{-1}).

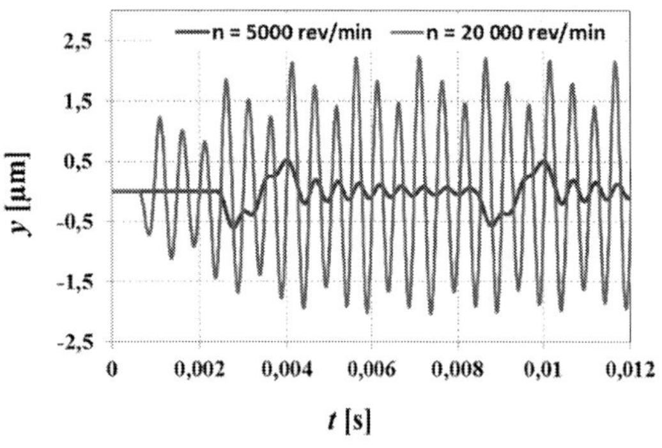

Fig. 9. The time course of instantaneous tool deflections for λ = 70.2 mm^{-1}.

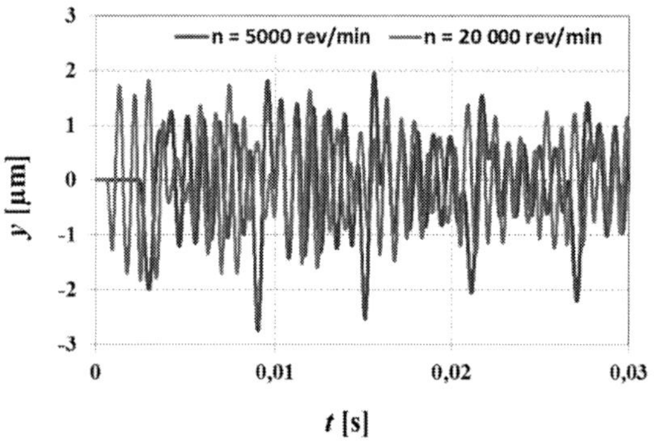

Fig. 10. The time course of instantaneous tool deflections for λ = 216 mm^{-1}.

It should be noted that instantaneous tool y deflections with the relatively high frequencies (higher than tooth passing frequency – $f_t = (zn)/60$), found for the systems with λ = 70.2 mm^{-1} and λ = 216 mm^{-1} can contribute to the formation of surface roughness profile in the direction perpendicular to the feed motion vector.

Figures 8-10 reveal also that system's slenderness value affect the relations between the maximal tool deflections and the selected rotational speed. In case of systems with $\lambda \leq$ 70.2 mm^{-1}, the growth of rotational speed induces the growth of deflection amplitudes, however the opposite trend is seen in case of a system with the highest investigated slenderness (λ = 216 mm^{-1}). These relations are strictly correlated with the superposition of forced vibrations dominant harmonic's frequency to the system's natural frequency. In case of systems with $\lambda \leq$ 70.2 mm^{-1}, the growth of rotational speed causes that the ratio of forced vibration dominant harmonic's frequency to system's natural frequency aims at 1. Thus it contributes to the growth of vibrations amplitudes – according to the resonance curve. Nevertheless, in case of a system with λ = 216 mm^{-1}, for the higher rotational speed value (n=20 000 rev/min), the ratio of forced vibration dominant harmonic's frequency to system's natural frequency is greater than 1. Consequently, it contributes to the decline of vibration amplitude during milling with the higher rotational speed.

4 Conclusions

On the basis of the carried out research the following conclusions have been formulated.

- The growth of tool overhang (slenderness) causes the monotonic decrease of system's damping coefficient c, stiffness coefficient k and natural frequency f_o, described by the power functions. However, in case of a modal mass m, a non-monotonic influence of tool overhang is seen.
- The application of the ball end mill with slenderness of λ = 216 mm^{-1} can cause almost 75-fold growth of vibration amplitude in relation to system with λ = 8.8 mm^{-1}. On the other side, the simulated maximal tool deflections are not exceeding 3 μm, independently on the investigated tool slenderness. This suggests that ball end mills with overhangs in a range from the 32 mm to 95 mm, clamped in the hydraulic toolholder can be applied during precise milling of hardened steel. However, the vibrations generated during the milling process can be affected also by the machine-toolholder-tool geometrical errors (e.g. run out).
- Simulations based on the solution of differential motion equation reveal also that system's slenderness value affect the relations between the maximal tool deflections and the selected rotational speed. In case of systems with $\lambda \leq$ 70.2 mm^{-1}, the growth of rotational speed induces the growth of deflection amplitudes, however the opposite trend is seen in case of system with the highest investigated slenderness (λ = 216 mm^{-1}). This finding suggests that tools with higher overhangs can be applied during precise milling in HSM conditions. Nevertheless, the forced vibrations dominant harmonic's frequency should be higher than the system's natural frequency.

References

1. Y. Pan, Y. Changfeng, X. Shaohua, Z. Dinghua, D. Xingtang, Procedia CIRP **56** (2016)

2. M.A. Salgado, L.N. Lopez de Lacalle, A. Lamikiz, J. Munoa, J.A. Sanchez, Int. J. Mach. Tools Manuf. **45**, 6 (2005)
3. G.M. Kim, B.H. Kim, C.N. Chu, Int. J. Mach. Tools Manuf. **43** (9) (2003)
4. S. Wojciechowski, P. Twardowski, M. Wieczorowski, Metrol. Meas. Syst. **21** (1) (2014)
5. S. Wojciechowski, P. Twardowski, M. Pelic, R.W. Maruda, S. Barrans, G. Krolczyk, Precis Eng. **46** (2016)
6. E. Ozturk, E. Budak, J. Manuf. Sci. Eng. **132** (2), (2010)
7. Schultheiss F., Hagglund S., Bushlya V., Zhou J., Stahl J-E, Procedia CIRP **13** (2014)
8. S. Wojciechowski, R.W. Maruda, P. Nieslony, G.M. Krolczyk, International Int. J. Mech. Sci. **119** (2016)
9. L.N. López de Lacalle, A. Lamikiz, J.A. Sanchez, M.A. Salgado, Int. J. Adv. Manuf. Technol. **24** (2004)
10. S. Wojciechowski, P. Twardowski, M. Pelic, Procedia CIRP **14** (2014)
11. M.B.G. Jun, R.E. DeVor, S.G. Kapoor, J. Manuf. Sci. Eng.-Trans. ASME **128** 2006
12. S. Wojciechowski, T. Chwalczuk, P. Twardowski, G.M. Krolczyk, Arch. Civ. Mech. Eng. **15**, 4, (2015)
13. L.N. López de Lacalle, A. Lamikiz, J.A. Sanchez, M.A. Salgado, Int. J. Mach. Tools Manuf. **47** (2007)
14. K.A. Desai, P.V.M. Rao, J. Mater. Process. Technol. **212** (2012)
15. Y. Sun, Q. Guo, J. Manuf. Sci. Eng.-Trans. ASME **134** (2012)
16. I. Buj-Corral, J. Vivancos-Calvet, H. Gonzalez-Rojas, Int. J. Mach. Tools Manuf. **51** (2011)
17. J. Tlusty, S. Smith, W.R. Winfough, CIRP Ann-Manuf. Technol. **45** (1996)
18. M. Mendes de Aguiar, A.E. Diniz, R. Pederiva, Int. J. Mach. Tools Manuf. **68** (2013)
19. Z. Nowakowski, S. Wojciechowski, Mechanik **08/09** (2015)
20. S. Gade, H. Herlusfen, Technical Review, **2**, Bruel & Kjear A/S (1994)
21. S. Wojciechowski, Int. J. Mach. Tools Manuf., **89** (2015)
22. J. Kuczmaszewski, P. Pieśko, Eksploat. Niezawodn. **16**(1) (2014)
23. R. Rusinek, M. Wiercigroch, P. Wahi, Procedia CIRP **31** (2015)

Effect of Heat Treatment on the Fatigue Life of Steel-Titanium Bimetal

Andrzej Kurek[1,*]

[1]Opole University of Technology, Mechanical Engineering Faculty, 5 Mikolajczyka Str., Opole, Poland

Abstract. The work presents results of explosively welded steel-titanium bimetal fatigue tests, as well as the native and clad-on materials. The tests were applied to specimens made of bimetal constituting fusion of S355J2 steel with SB 265 G1 titanium, where the latter was the material being clad-on during plating process. Both the material prior to plating (the titanium) and following plating was tested, as well as prior to and following the heat treatment process. The tests result in fatigue life of specimens made of native material, acquired from the plater, being lower than that of the steel-titanium bimetal specimens fatigue life, while the fatigue life of SB 265 G1 titanium does not change significantly in result of plating. Furthermore, the state of stress in the plater native layer following welding (compressive stress), positively influences its fatigue life, this effect being insignificantly reduced by heat treatment.

1 Introduction

Technology of explosive welding enables permanent fusion of materials that have very different properties, difficult to achieve with another welding techniques [1]. The claddings obtained with this method are materials exhibiting highly gradient properties and have complex nature of fusion zone. One of the clad materials key utilisation areas are process equipment constructions (power engineering and chemical industries). Wide use of titanium and its alloys in power engineering (condensers, steam condensers, heat exchangers and steam turbines in power plants and combined heat and power plants) makes the problem of the steel-titanium bimetals fatigue life more and more important. Both in these and other applications, the fatigue life of the clads is the key parameter. In a view of the problem specifics, both tests specified by adequate standards and non-standardized tests are performed. The strength tests presented in the work relate to metallic composite, called cladding, obtained in process of so-called explosive cladding. The problem of residual stress in explosively cladded materials is extensively shown in papers [2,3], The aim of the work is comparing results of fatigue tests on specimens made of native materials (steel and titanium) and the bimetal in two states, prior to, and after heat treatment, followed by determining its impact on fatigue life of resulting product.

2 Materials and methods

The work presents results of fatigue testing material, being a fusion of S455J2 steel with SB265 G1 titanium as the material being clad on, as well as steel being the native material and titanium SB265G1. Considering thickness of the layer being clad on – titanium (6 mm), specimens having non-standardized dimensions and shape have been used in fatigue testing. Prior to commencing the test, the shape of specimens should have been determined, which was accomplished with use of FEM calculations aimed at determining a size and shape of the specimen, ensuring sufficient stress level in the materials.

The specimens of elaborated shape (100x9x9 mm) were used for fatigue tests, that were performed on the MZGS-100 fatigue test stand, constituting an equipment of fatigue testing laboratory in Department of Mechanics and Machine Design of Opole University of Technology [4] (Fig. 1), the test were conducted under controlled bending momentum with constant frequency (30Hz).

Fig. 1. MZGS-100 fatigue test stand.

The specimens made of clad material have been produced such that bimetal joint would run precisely in

*Corresponding author: a.kurek@po.opole.pl

the middle of their cross-section, dividing the specimen into two equal halves – the steel one and the titanium one (Fig. 2).

In the first series of tests the specimen was fastened in the machine so that bending would be performed according to the scheme (Fig. 3a), in the plane perpendicular to the bimetal joint. For test of such type, initial fatigue tests were done that have been presented in the works [5,6]. The results of this research are shown in Table 1a and Figs. 4-6. They have been depicted as specimens loaded perpendicularly to the bimetal joint plane. The subsequent series was performed with another clad loading configuration, the specimen was fastened in the fatigue tester so that moment vector would act within bimetal joint plane (Fig. 3b). The results of tests that have been presented in the works [5,6], are compiled in Table 1b and Figs. 5, 6 and depicted as specimens loaded parallel to the bimetal joint plane.

The third series of tests has been done on specimens prepared from native material (the S355J2 steel) that were acquired from plates previously explosively welded. In order to create possibility of correlating obtained test results with the two previous test series, the specimens of the same shape and dimensions as used for the first and the second test series have been employed. Results of those test that were presented in the works [5,6], have been placed in Table 3b and in Fig. 6.

The subsequent series of tests was performed on specimens made of native material, SB265G1 titanium, in condition as prior to cladding. The results of tests performed with specimens, having the same shape and dimensions as in previous series of tests, are listed in Table 3a and in Figs. 4-6.

The last series of tests related to clad material, sampled from the same test plate, but prior to subjecting it to heat treatment. The specimens for testing were prepared similarly, as in case of the first and the second series of tests. The results are shown in Table 2 and Fig. 7.

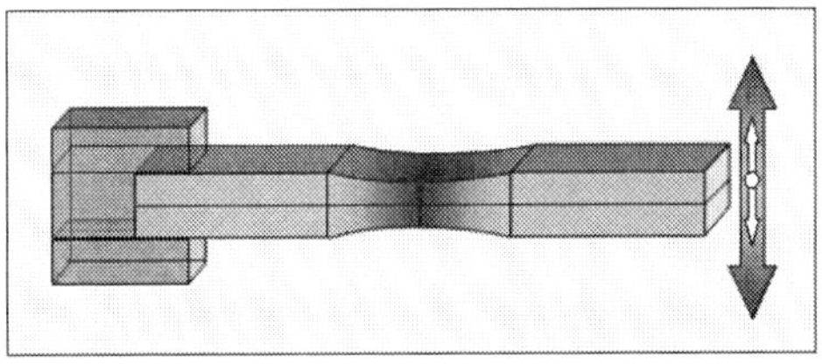

Fig. 2. Schematic of specimens fastening and loading [7].

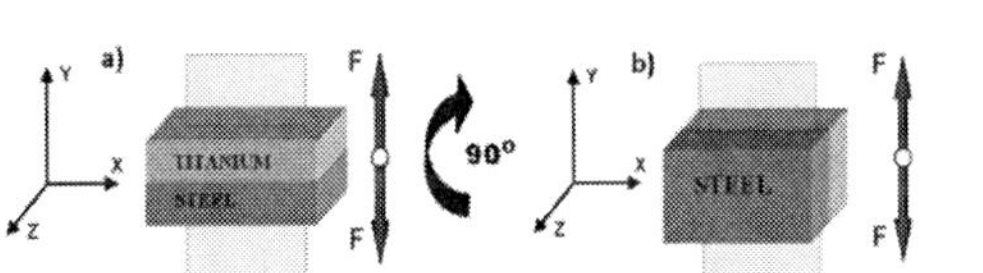

Fig. 3. Schematics of loading manner dependent on specimen fastening: "perpendicular" (a) and "parallel" (b) [7].

Considering that MZGS-100 fatigue test stand enables execution of tests at constant bending moment value, the stress magnitude values shown in the tables have been determined by means of FEM calculations, assuming the linear-elastic material model.

Table 1. Test parameters and fatigue life of the steel-titanium bimetal, for loading bimetal perpendicularly a) and parallel b) to the joint plane, the material after heat treatment (HT).

a) Perpendicularly

no.	σ_a, MPa steel	σ_a, MPa titanium	N_f
1	380	266	56000
2	367	257	77000
3	360	252	133000
4	347	243	95000
5	343	240	101000
6	328	230	156000
7	321	225	238000
8	294	206	345000
9	274	192	1760000
10	287	201	172000

b) Parallel

no.	σ_a, MPa steel	σ_a, MPa titanium	N_f
1	371	185	132000
2	361	180	79000
3	351	175	98000
4	339	169	235000
5	336	168	143000
6	316	157	241000
7	280	140	368000
8	278	139	453000
9	260	130	5039000
10	259	129	965000

The fatigue test results for steel-titanium bimetal, and titanium in condition preceding the joining action, depending on σ_a stress magnitude and N_f number of cycles, are shown in Fig. 4. In case of the steel-titanium bimetal, the diagram presents results obtained for the specimen loading case in the plane perpendicular to the bimetal joint (Fig. 3a), with consideration given to stress amplitude, applied in the test conditions, present in SB265 G1 titanium. On the grounds of tests, distinguish differences between fatigue lives obtained from material tests prior to and following the fusion action, has not been found, therefore a common fatigue characteristics is applied. The Basquin stress-life curve was used and the equation is presented on figures.

DOI: 10.1051/e3sconf/20171903015

Table 2. Test parameters and fatigue life of the steel-titanium bimetal, for loading bimetal perpendicularly to the joint plane, the material prior to heat treatment (HT).

no.	σ_a, MPa steel	σ_a, MPa titanium	N_f
1	385	270	355000
2	407	285	129900
3	355	249	680980
4	351	246	423776
5	332	233	793401
6	321	225	1281800
7	396	278	184800
8	409	287	80095
9	317	222	4000000
10	385	270	400006
11	364	255	270000
12	372	261	276674
13	430	302	95250
14	387	272	289900
15	428	300	119000
16	377	264	284500
17	332	233	2500000

Table 3. Test parameters and fatigue life of SB265 G1 titanium (a) and native material, S355J2 steel (b).

a) SB265G1 titanium

no.	σ_a, MPa	N_f
I	311	8900
II	306	14900
III	283	38100
IV	262	27800
V	245	52100
VI	235	119400
VII	230	491300
VIII	228	241800
IX	214	1227800
X	201	327800
XI	196	2262400
XII	195	3708800

b) S355J2 steel

no.	σ_a, MPa steel	N_f
I	329	78000
II	314	84000
III	303	156000
IV	291	175000
V	272	165000
VI	271	195000
VII	253	612000

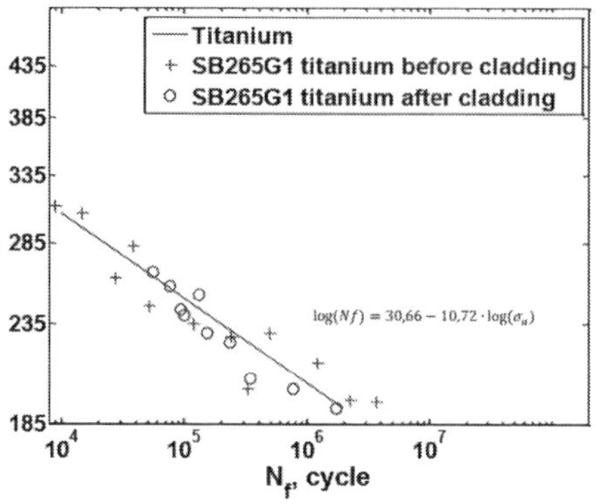

Fig. 4. Diagram of stress amplitude dependence σ_a=f(N_f), present in titanium.

In Fig. 5 the fatigue test results are shown for steel-titanium bimetal, presenting σ_a stress amplitude (in steel and titanium) dependence on N_f fatigue life for two manners of clad material loading and for native material, SB265 G1 titanium, together with characteristics considering stress amplitude applied in test conditions, present both in S355J2 steel and SB265 G1 titanium. On the grounds of tests, distinguish differences between fatigue lives obtained from the steel-titanium clad tests depending on loading type, has not been found, therefore a common fatigue characteristics is applied.

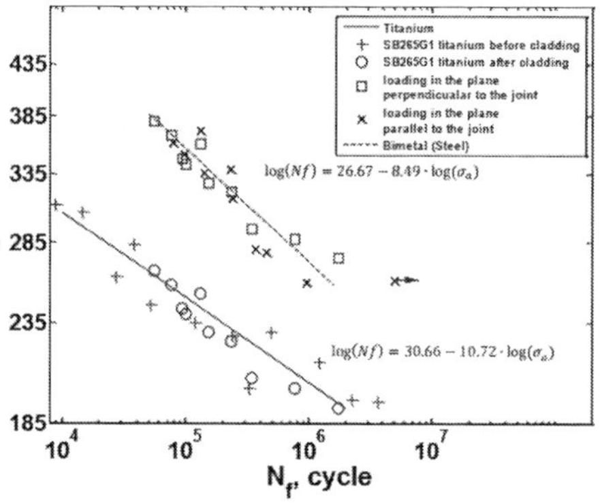

Fig. 5. Diagram of stress amplitude dependence σ_a=f(N_f), present in steel and titanium.

The dual-log diagram (Fig. 6) presents results of fatigue tests showing stress amplitude σ_a (in steel) dependence on N_f fatigue life, for two types of clad material loading (presented by means of common characteristics), and for native material (S355J2 steel) and native material (SB265 G1 titanium).

The dual-log diagram (Fig. 7) presents collectively results of all fatigue tests as a stress amplitude σ_a (in steel) dependence on N_f fatigue life, those results are contained in Tables 1-3.

DOI: 10.1051/e3sconf/20171903015

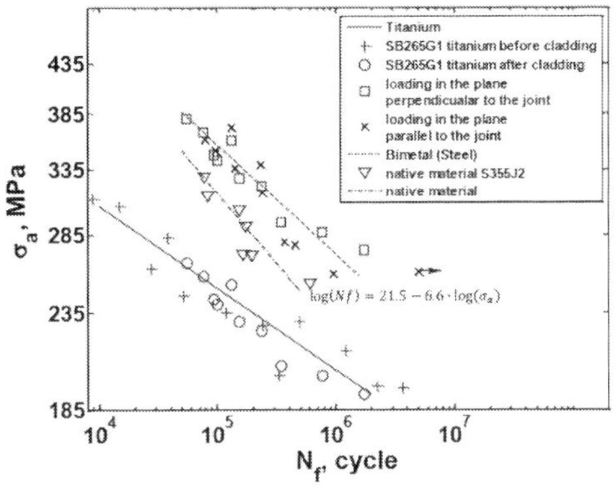

Fig. 6. Diagram of stress amplitude dependence $\sigma_a=f(N_f)$, present in steel, titanium and clad material.

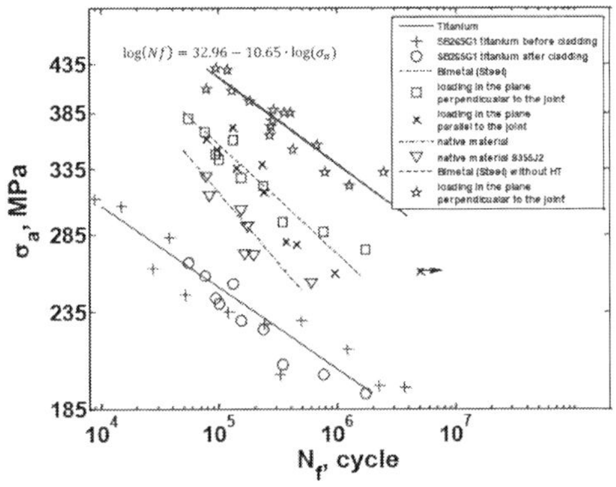

Fig. 7. Collective diagram of stress amplitude dependence $\sigma_a=f(N_f)$.

It should be noted, however, that indispensable part of fatigue test are scatters of experimental results, which are also present for uniform materials, and in case of bimetallic materials, clad with explosive method, we are facing non-uniform material structure, which may strengthen differences in test results for similar loads.

3 Conclusions

For the majority of tested, heat treated bimetal specimens, loaded perpendicularly to the joint plane, it was the SB265 G1 titanium material in which fracturing was initiated, which despite high difference of joined materials Young moduli, determining lower stress value, was characterised with lower fatigue life. The obtained in such manner values of the clad fatigue life, with consideration given to stress amplitudes within titanium layer, are comparable to SB265 G1 titanium fatigue lives in basic condition, preceding the explosive cladding stage. This enables to claim, with high likelihood, that it is possible to use fatigue characteristics determined for the material in basic condition, when evaluating fatigue life for elements made of clads containing this material. In case of specimens being bent in plane parallel to the plane of joint, fracture initiation occurs in the steel layer of the bimetal. This results from the higher difference of stress amplitudes present in steel and in titanium in such case, which entails fracture initiation process occurring earlier in the steel part of the bimetal. However, the fatigue life of such bimetal is markedly higher than fatigue life of native material.

It comes from the tests, that fatigue life of specimens made of native material, acquired from the clad, is lower than that of titanium-steel specimens.

Furthermore, the test results enable to claim that heat treatment used by manufacturer negatively influences fatigue life of the bimetal. Such condition may be explained by removing internal stresses from material or by introduction of new, lower compressive stresses, considering difference in thermal expansion parameters of joined materials. This hypothesis however requires additional research.

References

1. M. Gloc, M. Wachowski, T. Płocinski, K. J. Kurzydlowski, J. Alloys Compd. **671**, 446-451, (2016)
2. A. Karolczuk, M. Kowalski, Key Eng. Mater. **592-593**, 594-597 (2014)
3. A. Karolczuk, M. Kowalski, G. Robak, Solid State Phenom. **199**, 460-465 (2013)
4. A. Kowalski, W. Ozgowicz, M. Kurek, A. Kurek, T. Łagoda, Fratt. Integr. Strutt. **35**, 449-455 (2016)
5. A. Kurek, M. Wachowski, A. Niesłony, T. Płocinski, K.J. Kurzydłowski, Arch. Metall. Mater. **59**, 4, 1565-1570, (2014)
6. A. Kurek, A. Niesłony, Mater. Sci. Forum. **726**, 106-109, (2012)
7. A. Niesłony, A. Kurek, R. Bański, L. Cížek, Sci. Fasc. Opole Uni. Tech., Mech. **97**, 337, 41-42, (2010)
8. A. Karolczuk, K. Kluger, M. Kowalski, F. Zok, G. Robak, Mater. Sci. Forum, **726**, 125-132 (2012)

DOI: 10.1051/e3sconf/20171903016

Designing ad hoc temperature fixed point cells

Orest Kochan[1,*], *Jun* Su[2], and *Roman* Kochan[1,3]

[1]Lviv Polytechnic National University, Department of measuring information technologies, 79013 Karpinskoho Str. 1, Lviv, Ukraine
[2]School of Computer Science, Hubei University of Technology, Wuhan, China
[3]University of Bielsko-Biala, Bielsko-Biala, Poland

Abstract. The design of the ad hoc (made of hand-on materials) temperature fixed-point cell (TFPC) is considered in this paper. There was developed a TFPC filled with lead and an apparatus for its investigation. The feature of the considered TFPC is minimization of the stray thermal fluxes distorting the temperature plateau during a phase change. Conclusions about the influence of the design of a TFPC on the temperature plateau of phase change are described.

1 Introduction

According to [1], the share of temperature measurements among all measurements in industry is 40%, but in energetics this percentage is about 70%. The accuracy of temperature sensors in most cases is decisive for high-quality production [2-4]. Also, a proper temperature mode is crucial for durable operation and safety of equipment [5-8]. Thus, the accuracy of measurements has a considerable effect on the world economy [8]. The estimation of the market of metrological services is approximately 4% of GDP in developed countries [9]. The progress in metrology means accuracy increase and development of techniques that increase metrological reliability of measurements as well as implementation into praxis new metrological procedures and instruments that improve metrological service [6,7]. The progress in temperature measurements is important too [10]. Its accomplishments can give a considerable economic gain. For instance, the increase of temperature measurement accuracy in power plants will gain $400 million worldwide [8] with no additional expenses of fuel. The analysis of literature [1,11,12] shows that accuracy of measuring instruments is constantly growing up [13,14], however, the error of temperature measurements has decreased very little because the error of sensors dominates in temperature measuring chains [1,11,12]. Despite all efforts in recent decades, the error of sensors has decreased very little.

Thermocouples (TCs) are the most widely used sensors for temperature measurements above 650°C, despite the fact that their error is much bigger than that of the rest of the measuring chain [1,11,12].

At first glance, it is reasonable to make the operation time as long as possible, because it saves the sensor itself and cuts down the expenses connected with a sensor replacement. However, this is connected with losses due to big measurement error. On the other hand, it is not always possible to ensure metrological service in-situ to decrease the error of a sensor.

During long-term operation of a TC, its legs become inhomogeneous [15-17] under the influence of time and operation temperature. The rate of inhomogeneity strongly depends on operating conditions, thermal shocks, vibration, contaminations, etc. There is no direct method for determining the degree of TC inhomogeneity. This inhomogeneity is the main source of error when measuring a temperature by thermocouples [15,16]. It changes the Seebeck coefficient of a TC sections. Thus, the Seebeck emf of the TC will drift with time in accordance with the change of inhomogeneity in the TC legs, therefore the error appears due to the deviation of the actual conversion characteristic of a TC from the nominal one.

The traditional method for determining the error of TCs is a periodic calibration in a calibration laboratory [17]. However, under laboratory conditions the error of TCs is very likely to be different than in-situ [17,18]. In addition, this procedure usually requires the complex multistep procedure: stop of operation, removal of measuring equipment from the operating place, transfer it to the calibration laboratory, calibration, return the measuring equipment back to the operating place, installation of equipment again and launch the operation again. The methods for calibration of TC in situ seem to be the best ones. Therefore, recently a lot of attention has been paid to developing the temperature fixed-point cells (TFPCs) [10, 19] for in situ calibration.

The TFPC is a capsule containing a reference material [20] and is put on the measuring junction of the TC to determine the thermocouple error during melting or freezing. Reference materials are pure metals or eutectic alloys whose temperatures of phase change are accurately known and remain constant during the phase change [21]. This constant temperature of the phase change is called "a temperature plateau" (Fig. 1): a- the

* Corresponding author: orestvk@gmail.com

plateau while melting a reference material; b – the plateau while freezing; c – the plateau while freezing with supercooling. The temperature of a plateau is known with high accuracy and can be converted into emf [1]. Comparison of this emf with the one measured by a TC allows determining the TC error. It is reasonable to carry out this procedure in situ. TFPCs are not yet widely used in industry because of the technological problems.

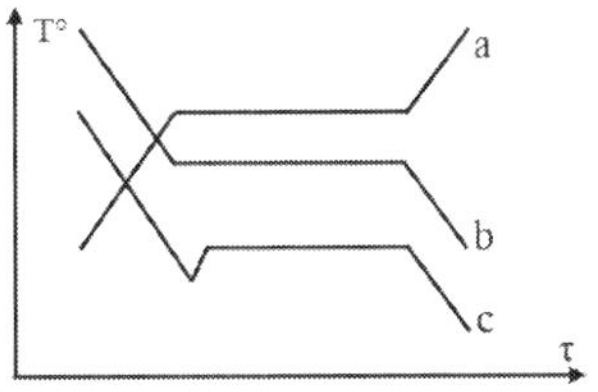

Fig. 1. Different types of a temperature plateaus.

They are usually used in laboratories. However, in the last decade there appeared many publications in leading journals and reports on conferences in this area. Such interest of researchers to TFPCs gave experts the reason to predict a significant increase of TFPCs usage in the range of medium and high temperatures [10, 19].

2 State of the art

The typical implementation of a TFPC is described in [20]. Fig. 2 shows the change of Seebeck emf in time in modes of sustainable heating and cooling as given in [20]. Fig. 2 demonstrates that the temperature plateau at the phase change is difficult to identify. Furthermore, it is clear that the temperature of melting is not equal to that of freezing. It should be noted that this type of a temperature plateau (Fig. 2) directly contradicts physics [21], indicating that the temperatures of melting and freezing are, stable and equal. The graphs [20] can be explained by the presence of a significant stray thermal flux (Fig. 3) in the studied TFPC, which heats the measuring junction of the TC while melting (when the environment temperature is above than that of the phase change) and cools while freezing (when the temperature is below than that of the phase change).

The stray thermal flux distorting the temperature plateau is shown in Fig. 3 with dashed lines. It does not come across the phase change boundary of the reference material [20]. Therefore, the temperature of the measuring junction of the TC does not equal to that of the phase change. Thus, the increase in the accuracy of the temperature measurements using the TFPC described in [20] is doubtful. Thus, as seen from the analysis, the main cause of the poor temperature plateau in a TFPC may be its ineffective design. Designs of the known TFPCs are different [22-24], but many of them have the similar drawbacks.

The theory of designing of TFPCs, including calculation of thermal processes in a TFPC, is proposed in [24], where the basic features of their designs are explored. However, as seen in [20,22,23], structural factors are often not taken into account when designing a TFPC.

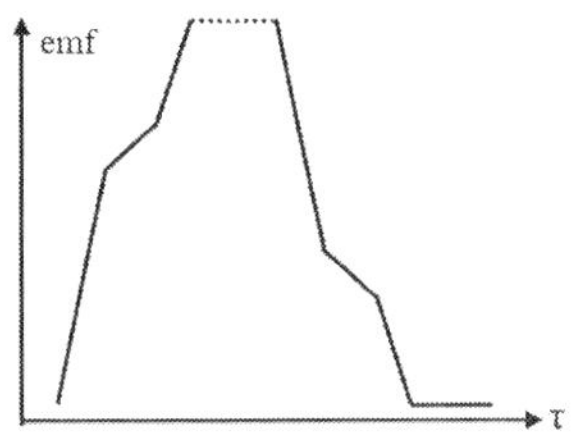

Fig. 2. The actual temperature plateau, [26]

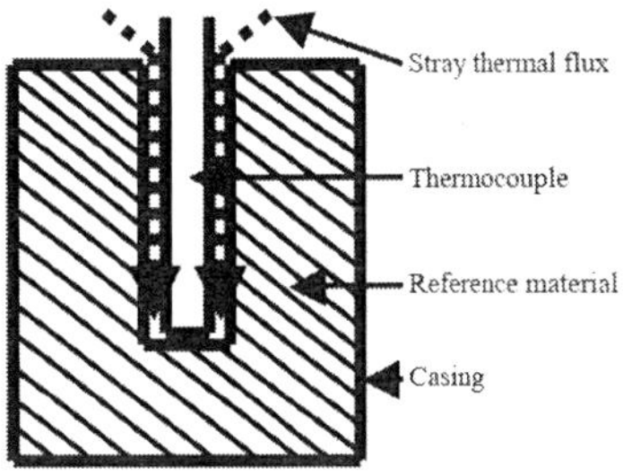

Fig. 3. The cross section design of the TFPC [25].

The aim of this paper is to design an ad hoc (made of improvised materials) TFPC to demonstrate that reduction of the stray temperature flux allows obtaining high quality temperature plateaus.

3 Designing the TFPC

In order to make the temperature plateau, that forms the Seebeck emf of the TC, accurately consistent with the phase change temperature of the TFPC, there should be reduced (preferably completely eliminated) the stray thermal flux that flows toward/outward the measuring junction of a TC from the outer environment (see. Fig. 3). The TFPC should ensure the adiabatic conditions for the measuring junction of a TC; heat exchange must take place between the TC measuring junction and a reference material, and between the reference material and the outer environment. Heat transfer between the TC measuring junction and the outer environment has to approach zero.

This can be achieved, by maximizing the coverage area of the TC measuring junction with the liquid reference material (at the temperature of phase change), and by maximizing the distance for the stray heat flux to flow. How to ensure these conditions is shown in Fig. 4. The TC measuring junction is deeply immersed into the reference material that covers it from all sides. When heating is sustainable, the medium (the furnace) temperature increases as the arrows show in Fig. 4. The TC with the TFPC have much higher heat capacity than the environment, thus the temperature of the TFPC is lower and the stray thermal flux flows in the way to heat the TFPC and the TC measuring junction. When the reference material begins to melt, the temperature of phase change remains constant. Then the thermal flux heating the TC is split into two ones. The first flux comes from the furnace and ensures the continuation of melting. The second flux goes from the surface of the phase transition to the TFPC center and the TC measuring junction. The latter thermal flux provides equalization of the TC temperature (see Fig. 4, the temperature change direction is indicated by the arrows within the surface of the phase change). The direct thermal flux from the furnace toward the TC measuring junction is only possible through the TC legs, but they should be long enough and heat exchange between them and the reference material should be good. Therefore, the stray thermal flux, which distorts the temperature plateau (Fig. 2 and 3), has a minimum value.

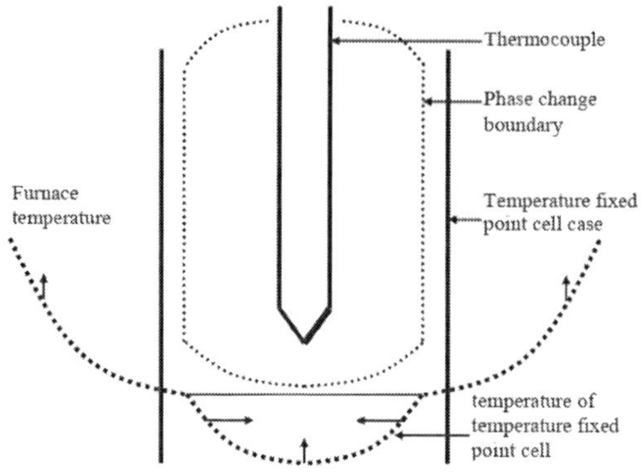

Fig. 4. The proposed TFPC and its thermal fluxes.

4 The design of the proposed ad hoc TFPC

The simple design and technique of designing the TFPC are offered. There were taken precautions to minimize the effect of the stray thermal flux on temperature plateau toward or outward the reference material.

Firstly, as it is shown in Fig. 5 there was prepared a TC. There was used type K TC made of chromel-alumel with 1.2 mm diameter of legs. It was covered with a long (20 cm) ceramic insulating bead with the outer diameter of 8 mm. The legs between the bead and the measuring junction of the TC were bare for approximately 35 mm. This area of the TC was covered with two thin layers of a mixture of sodium silicate (liquid glass) with gypsum. The total coating thickness was approximately 0.5 mm.

After this, the TFPC was prepared. Its body was made of the copper tube with the inner diameter of 8 mm and its length was 100 mm; it was folded from one side. A 35mm-long screw of 4mm in diameter was twisted in with washers and nuts to cork the folded end. The tube was filled with finely chopped lead, which was then melted on a gas burner. The amount of lead was chosen in the way to fill the tube on 75...80%. Then the lead in the case was melted again and the TC was immersed into it with tension at the maximum possible depth (at least 20 mm) (see. Fig. 5). The prepared TFPC with the immersed TC is shown in Fig. 6. The proposed design of the TC with TFPC for study corresponds to Fig. 4. The TC measuring junction is almost completely covered with the reference material. The TC legs are immersed into the reference material in about 20 their diameters and coated with a thin layer of insulation, i.e. they have good thermal contact with the reference material. The ratio of the case length of the TFPC to its diameter is about 10, so according to [26], the case can be considered infinitely long. The stray thermal flux must have a minimum effect on the temperature plateau. The TC measuring junction is like in a "thermos" that stores inside the TFPC the phase change temperature regardless of changes in temperature in the furnace [25]. It should be noted that the above considerations about the form of the temperature plateau in the proposed design of the TFPC, are mainly qualitative, therefore, they require experimental confirmation.

5 The apparatus to study of the proposed TFPC

There was constructed an apparatus for experimental studies of the proposed design of the TFPC. The main element of the apparatus is a shaft furnace, made of a tubular wire resistor of type PEV with the nominal power rating of 50 watts. It is located in the center of a glass jar which then is filled with gypsum. Power circuit consists of an autotransformer of type LATR-1, the output voltage is measured with the analog multimeter of type C20. The temperature of the furnace was measured with the additional type K TC. The temperature measurement of the reference junctions of both TCs was carried out using a thermistor of type KMT4 with the nominal resistance of 30kΩ. It was previously individually calibrated using the mercury thermometer of type TL18 (its error does not exceed 0.1°C). The tested TFPC with the TC was placed in the center of the furnace. The distance to the bottom of the furnace was set using the 35mm-long screw of 4mm in diameter and a centering plate was fixed at its end. To reduce the temperature gradient in the furnace and temperature changes of the reference junctions the following steps were taken: 1) the furnace is closed with a plaster lid with holes for both TCs; 2) the lid is sealed additionally with plaster solution and covered with cotton wool; 3) the reference junctions of both TCs and the thermistor are placed in a copper box; 4) the copper box is covered with a layer of cotton wool. The measuring system includes a TC equipped with the TFPC, a TC measures the temperature in the furnace, the thermistor Rt to measure the temperature of the two reference junctions of the TCs, the switch S1 for the TCs, the switch S2 between the TCs and the thermistor as well as the precision digital multimeter (DMM) of type UT70D. When measuring the Seebeck emf, the UT70D operates in the mode of voltage measurement under improved resolution (the resolution of 1 μV corresponding to approximately 0.025°C for a type K TC). When measuring the resistance of the thermistor with the nominal resistance of 30 kΩ, the UT70D operates in the resistance measurement mode with the resolution of 1Ω, which corresponds to approximately 0.0012°C. It should be noted that the instability of the temperature during the temperature plateau while melting or freezing of the reference material is calculated as the difference between the current temperature and the initial temperature (the temperature at the point, which is assigned to be the beginning of the phase change).

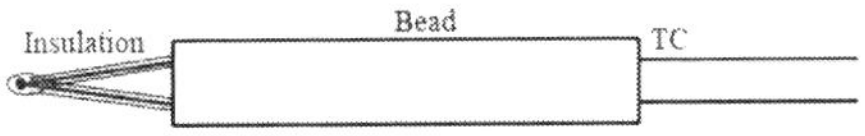

Fig. 5. The TCprepared for immersion into a TFPC.

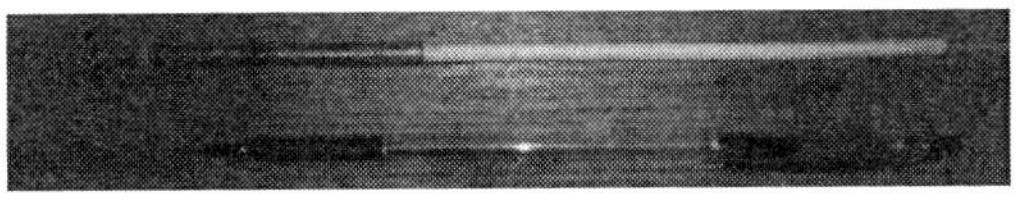

Fig. 6. TC with a TFPC.

In this case, the systematic errors of the measuring system, including the TCs and the DMM will not distort temperature plateau. Thus, only the random errors of the measurement circuit affect the measurement error of the Seebeck emf instability of the temperature plateau. The study before the experiments showed that this error does not exceed ± 1 μV. The systematic measurement error of the reference junction temperature is also compensated by the fact that the purpose of the experiment is to measure the temperature change during the phase change of the reference material. The random error of the correction of the reference junction temperatures, when the resolution of the thermistor resistance measurement is of 0.0012°C, can be neglected.

6 Conclusions

When designing a TFPC stray thermal flows have to be taken into account to keep the slope of a temperature plateau of the order of a few tenth of a Celsius degree to fulfil the European roadmap for thermometry [10]. As has been shown in this paper, the proper design of the TFPC, which prevents the heat exchange between the reference material and environment, makes it possible to achieve high quality temperature plateau even for a relatively simple ad hoc TFPC made of on hand materials. Such ad hoc TFPCs can be used for calibration of TCs in situ.

The study of metrological parameters of a TFPC (in particular, the stability of the phase change temperature) makes sense only when a TFPC ensures a high quality temperature plateau.

This work was supported by Foundation of Wuhan Science and technology Bureau (2015030809020370), Doctoral Scientific Research Fund from Hubei University of Technology (No. BSQD14037), National Natural Science Foundation of China (Nos. 61602162, 61502155), Green Industry Technology Leading Project of Hubei University of Technology (No. ZZTS2016004).

References

1. Y. Lutsyk, O. Hook, O. Lakh, B. Stadnyk, *Temperature Measurements: Theory and Praxis* (Lviv, 2006) (In Ukrainian)
2. A. Glowacz, Z. Glowacz, Appl. Acoust. **117 (Part A)**, 20-27 (2017)
3. P. Hreha, A. Radvanska, L. Knapcikova, G.M. Krolczyk, S. Legutko, J.B. Krolczyk, Hloch S., Monka P., Metrol. Meas. Syst. **22,** 2, 315-326 (2015)
4. V. Perzel, M. Flimel, J. Krolczyk, A. Sedmak, A. Ruggiero, D. Kozak, A. Stoic, G. Krolczyk, S. Hloch, Thermal Sci. **00**, 46 (2016) DOI: 10.2298/TSCI150212046P
5. H. R. Ashrafi, K. Shahbazian, S. Bidmeshki, S. Yaghooti, P. Beiranvand. Adv. Sci. Technol. Res. J. **10**, 29, 1-12 (2016)
6. A. Glowacz, Z. Glowacz, Measurement, **93**, 86-93 (2016)
7. J. Su, O. Kochan, R. Kochan, Int. J. Thermop. **37**, 4, 1 (2016)
8. T. Wang et all, Stabilized metal sheathed type K and E thermocouples improve turbine efficiency, ISA tech/expo technology update, **1**, 5, 439-448 (1997)
9. J. Birch, *Benefit of legal metrology for the economy and society*, (A study for the International Committee of Legal Metrology, 2003)
10. G. Machin, et all. Int. J. Thermop. **35,** 3-4, 21-30 (2014)
11. J. Webster, *Measurement, Instrumentation, and Sensors Handbook* (CRCnetBase, 1999)
12. P.R.N. Childs, *Practical Temperature Measurement* (Elsevier, 2001)
13. A. Aly, A. El-Lail, I.J. Inf. Technol. and Comput. Sci. **03**, 56-60 (2013)
14. http://www.thermo-electra.com/en/Products/Instruments
15. D.J. Southworth, *Temperature Calibration with Isotech Block Baths* (Southport, Isotermal technology ltd, 1999)
16. K. C. Sloneker, Ceram. Ind. **159**, 4, 13-18 (2009).
17. J. Su, O. Kochan, Meas. Tech+ **57**, 10, 1160-1166 (2015)
18. S. Chang, O. Kochan, Sensors & Transducers **160,** 12, 55-61, (2013)
19. G. Machin, Twelve years of high temperature fixed point research: a review, Temperature: Its Measurement and Control in Science and Industry, **8(1)**, 305-316 (2013)
20. O. Ongrai, J.V. Pearce, G. Machin, S. Sweeney, Meas. Sci. Technol. **22**, 015104 (2011)
21. D. Halliday, *Fundamentals of Physics* (Wiley, 2011)
22. S. Rudtsch, A. Aulich, C. Monte, Stray thermal influences in zinc fixed-point cells. Temperature: Its Measurement and Control in Science and Industry, **8**, 1, 265-270 (2013)
23. M. Kalemci, et al., Int. J Thermophys. **36**, 1968-1979 (2015)
24. A. Sachenko, V. Milchenko, V. Kochan, *Measuring Temperature Using Sensors with Built-In Calibrators* (Moscow, 1986)
25. S. Chang, O. Kochan, et all, Demands to the design of temperature fixed point cells, Proc. of the workshop TEMPMEKO, 244 (2016)
26. J.IV Lienhard, and J.V. Lienhard, *A Heat Transfer Textbook* (Cambridge: Phlogiston Press, 2008)

DOI: 10.1051/e3sconf/20171903017

Verification of new model for determining the critical plane orientation angle

Marta Kurek[1,*]

[1]Opole University of Technology, Mechanical Engineering Faculty, 5 Mikolajczyka Str., Opole, Poland

Abstract. The principal objective of this study is to establish a method which can enable the assessment of fatigue life at the stage of the design and building of machine parts and equipment. The paper contains a description of a manner in which the angle of a critical plane can be determined, which is coupled with its adequate verification. The study in the paper is based on an assumption of an idea of a critical plane which adopts that the occurrence of a fatigue crack of a material is caused by the occurrence of strain in the critical plane of a material. The conclusion from this study takes the form of a comparison between the experimental and calculated results performed by adoption of the developed model. The results demonstrate satisfactory level of conformity between the two.

1 Introduction

For over two hundred years the fatigue of structural materials has been the subject of interest to scholars and engineers. The subject of material fatigue is extensively studied and research due to the hazard of enormous material losses and safety risks that can be caused by a fatigue failure.

Many tragic catastrophes have been caused by fatigue cracks in construction elements, whose occurrence made it impossible for a load to be transferred. This, in turn, resulted in a complete failure of a structure. Another equally important aspect of material fatigue is the pursuit of reducing the mass and production costs as well as operation of machines and structures.

The material fatigue and construction phenomenon is a significant problem of our time. It appears in many branches of industry e.g. aerospace, machinery, mining and transport, and is especially visible in the case when an emergency or failure people results in deaths and injuries.

The main objective of the majority of research works [1-6] concerning the projected fatigue life is to find a method which can be applied for the estimation of the fatigue life already at the stage of the design and throughout the course of constructing elements of machines and equipment.

The aim of the work is to present a new model applicable for assessment of fatigue life on the basis of a concept of the critical plane accompanied by the analysis of the scatter of the results. Every model requires verification, best on the basis of experimental data.

2 Multiaxial critical plane-based criterion

Stress criteria for a significant group of the fatigue criteria can be derived from the classic damage theory. During their evolution a collection of criteria arose out of the concept of a critical plane. This group of criteria will be the main subject of this study. The critial plane concept assumes, that material fatigue cracking is caused by stress acting along the materials critical plane. The potential use of the critical plane in the research of multiaxial fatigue was first proposed by Stanfield in 1935 [7]. Today, there is a steadilly growing interest in this concept. In a paper datigng back to 2001 [8], the authors proposed a new multiaxial fatigue criterion, for which the angle of critical plane orientation is expressed using the sum of two angles. The first is the maximum angle defined by normal stresses, whereas the second one is angle β, whose value is expressed in relation to the direction given by the maximum in the normal direction and is expressed as

$$\beta = \frac{3}{2}\left[1-\left(\frac{\tau_{af}}{\sigma_{af}}\right)^2\right]45^\circ \quad (1)$$

Several years later in literature [9] another proposition appeared as to how to express the given angle, i.e.:

$$\beta_1 = \frac{9}{8}\left[1-\left(\frac{\tau_{af}}{\sigma_{af}}\right)^4\right]45^\circ \quad (2)$$

---

* Corresponding author: ma.kurek@po.opole.pl

$$\beta_2 = \frac{3\sqrt{3}}{3\sqrt{3}-1}\left[1-\left(\frac{\tau_{af}}{\sigma_{af}}\right)^3\right]45^\circ \quad (3)$$

$$\beta_3 = \frac{\sqrt{3}}{\sqrt{3}-1}\left[1-\left(\frac{\tau_{af}}{\sigma_{af}}\right)\right]45^\circ, \quad (4)$$

$$\beta_4 = \frac{3}{\left(\sqrt{3}-1\right)^2}\left[1-\left(\frac{\tau_{af}}{\sigma_{af}}\right)\right]^2 45^\circ \quad (5)$$

where: σ_{af} – fatigue limit for bending, τ_{af}- fatigue limit for torsion

A graphical interpretation of the equations (1-5) is presented in Fig. 1. Depending on the adopted model the estimated fatigue life may vary considerably.

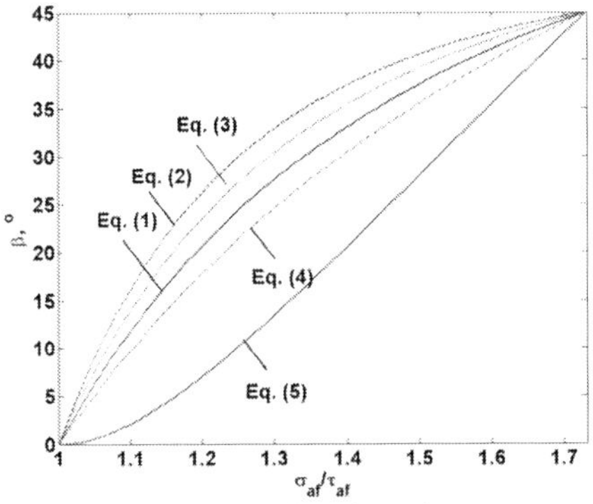

Fig. 1. Graphical representation of equations 1-5.

Each of the presented propositions refers in its expression to the fatigue life. It's worth noting that the equations (1-5) have a limited range. The adopted range is from 1 to √3. There is a number of various materials for which the ratio of the normal stress to the shear stress exceeds the range that is presented above.

Following the analysis of the variability of the calculated fatigue life that was performed in relation to the value of angle β [10], a new expression was derived in the form:

$$ctg(4\beta) = 22.5\left[\left(\frac{1+\sqrt{3}}{2}\right) - \frac{\sigma_a(N_f)}{\tau_a(N_f)}\right] = 22.5\left[\frac{1+\sqrt{3}}{2} - B_2(N_f)\right], \quad (6)$$

where:

$$B_2(N_{fi}) = \frac{\sigma_a(N_{fi})}{\tau_a(N_{fi})}. \quad (7)$$

A new relation presented below was developed on the basis of fatigue testing into several construction materials.

A graphical interpretation of formula (6) is given in Fig. 2, whereas the broken line marks the proposition developed by Carpinteri (1).

In its current form, the newly proposed relation (formula 6) accounts for the ratio of normal to shear stresses. This forms a significant novelty in relation to the earlier statements, which involve the ratios of the fatigue boundaries.

For a variety of materials, in particular the ones with non-parallel characteristics, the above statement does not prove to be justified.

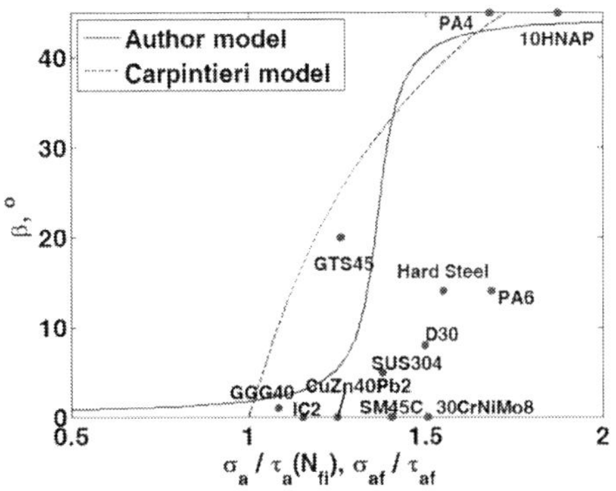

Fig. 2. The relationship between the angle β and the parameter B_2 (ratio of normal to shear stresses) according to different models

3 Scatter analysis of fatigue life

Every model needs to be verified. The analysis of the scatter bands of the estimated fatigue life can be applied for this purpose, both in relation to the proposed angle as well as depending on the ratio of the normal to shear stresses.

The analysis applied the values of the scatter bands for the particular materials just as it was calculated in [10] for angle β in the range <0°, 45°>. Fig. 3 presents the methodology followed during the analysis of the proposed expression (6) with regard to the relation of angle β and scatter T.

Assumption (1) : The values of angles β were calculated so that the scatter T is in the range $<T_{min}, 1.1T_{min}>$. On this basis, β_{opt} □ $<\beta_{min}, \beta_{max}>$.

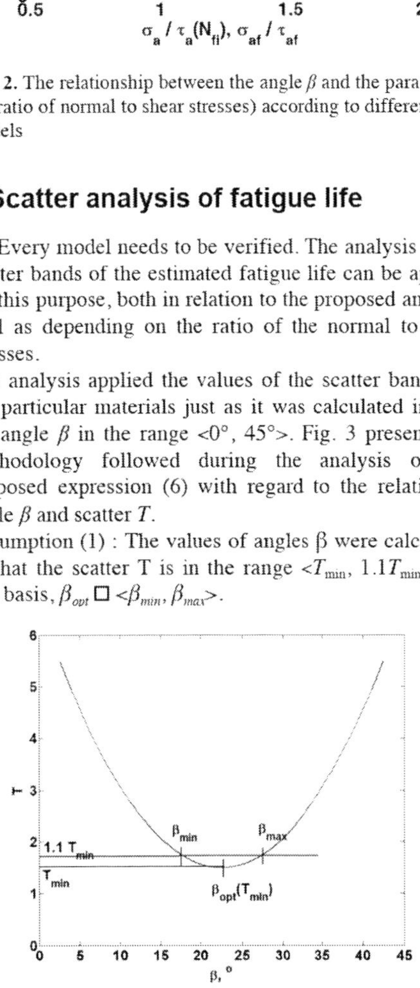

Fig. 3. Methodology of presented calculations – assumption 1

Fig. 4 illustrates the relation between scatter values T and the angle β for aluminum alloy 2017A (known under the name PA6 [11]).

DOI: 10.1051/e3sconf/20171903017

Assumption (2): Ratio of normal to shear stresses is based on scatter for pure bending and pure torsion for selected materials in accordance with Fig. 5.

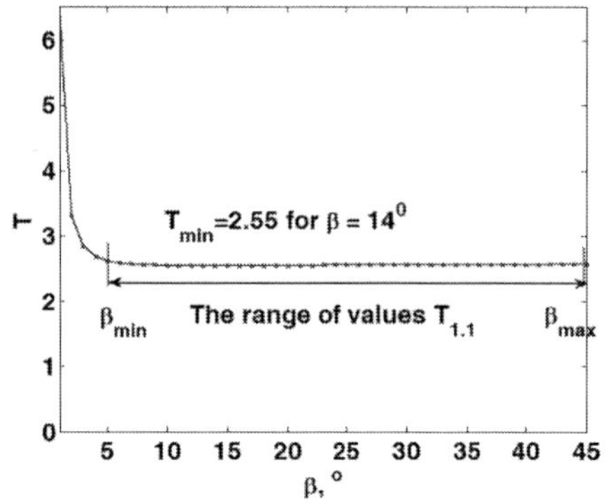

Fig. 4. Relationship between scatter values T and the angle β for PA6.

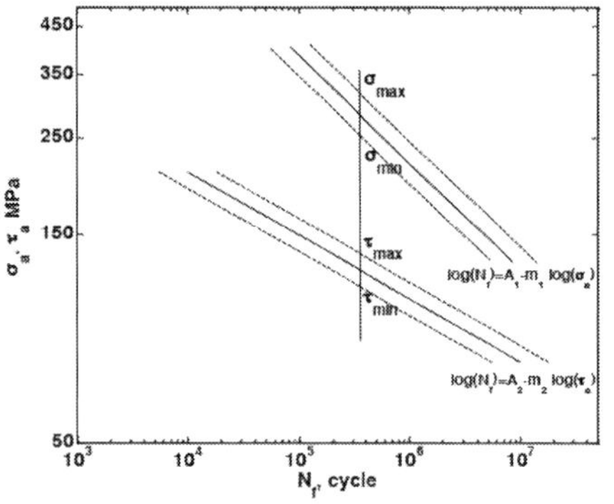

Fig. 5. Methodology of presented assumption 2.

The logarithmic dependence of the ratios of experimental and calculation strength should be used with the purpose of performing an adequate analysis of the fatigue life scatter.

A new method of determination the fatigue life scatter has been proposed by [9], where it was defined as the root mean square error

$$E = \sqrt{\frac{\sum_{i=1}^{n} \log^2 \frac{N_{exp}}{N_{cal}}}{n}} \quad (8)$$

Therefore, the scatter can be determined as

$$T = 10^E \quad (9)$$

where:

$$B_{opt} = \frac{\sigma_a(N_{fi})}{\tau_a(N_{fi})}, \quad (10)$$

$$B_{max} = \frac{\sigma_{max}(N_{fi})}{\tau_{min}(N_{fi})}, \quad (11)$$

$$B_{min} = \frac{\sigma_{min}(N_{fi})}{\tau_{max}(N_{fi})}. \quad (12)$$

Fig. 6 contains the results of an experimental study for oscillatory bending and bilateral bending for the analyzed aluminum alloy. This chart also contains scatter bands calculated on the basis of relations 8 and 9, which were applied for the calculations of B_{opt}, B_{max}, B_{min} parameters in accordance with relations (10-12). Table 1 contains a summary of the analyzed parameters for several materials selected from various material groups.

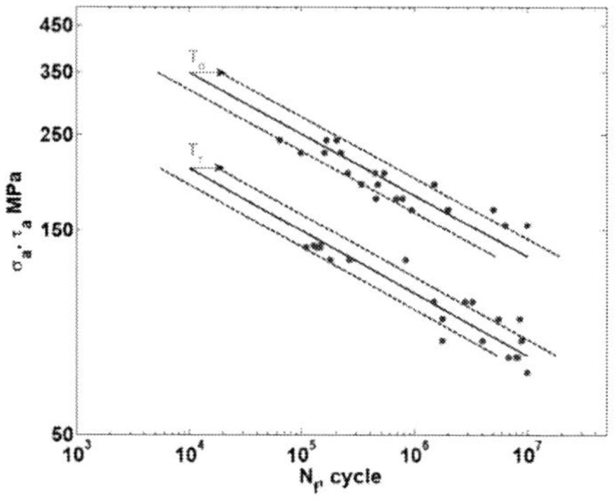

Fig. 6. Wöhler (Basquin) curves with scatter for pendulum bending and torsion for the analyzed alloy aluminum PA6.

Table 1. Analyzed parameters for selected materials.

Material	T_σ	T_τ	B_{max}	B_{min}	B_{opt}
10HNAP	2.2	1.62	2.21	1.61	1.87
CuZn40Pb2	1.49	1.44	1.37	1.15	1.25
PA6	1.91	1.82	2.02	1.41	1.69
GGG40	1.6	1.76	1.18	0.98	1.02
SUS304	1.62	1.2	1.5	1.22	1.38

The results that were gained, were subsequently plotted onto the diagram representing the relation of angle β depending on the ratio of normal to shear stresses (Fig. 7).

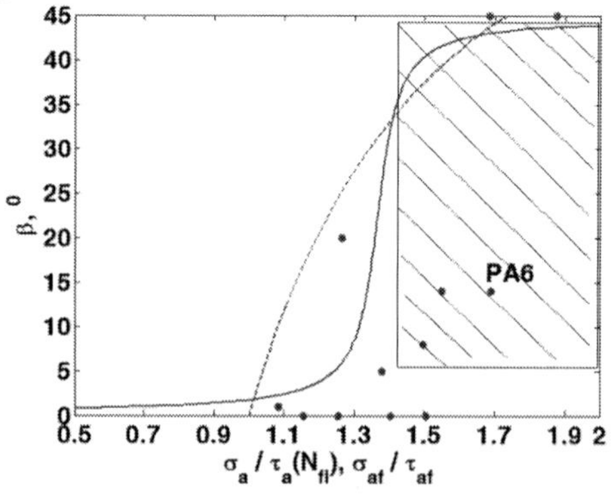

Fig. 7. Exemplary analysis of material PA6.

Such an analysis was conducted with regard to all test materials. Fig. 8 presents the results of the analysis for selected materials: GTS45, GGG40, 10HNAP, SM45C. The hatched field is the area for which the results of the fatigue life are the same or differ insignificantly.

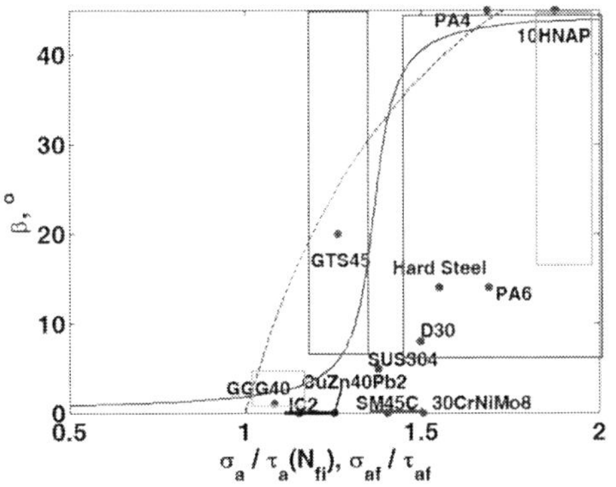

Fig. 8. Analysis results of selected materials.

On the basis of the analysis of Fig. 8, we can note that for some of the materials (e.g. GTS45, PA6), both models provide accurate results. The hatched fields involve both proposals for determination of the orientation angle of the critical plane. However, in some cases (e.g. SM45C, 10HNAP), the applicability fields only apply or are closer for the case of the proposition postulated by this author.

4 Verification of the proposed criterion

Fig. 9 presents comparison of the experimental fatigue life with the calculated fatigue life for Carpinteri's and Author's model. In this case the obtained results are satisfactory. A majority of the results are included into the scatter band of the coefficient 3.0. Such a system of points was easy to predict, as it was demonstrated by the analysis presented in Fig. 8.

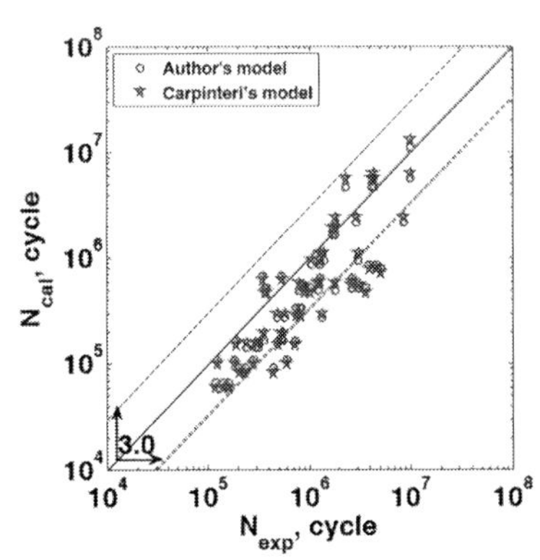

Fig. 9. Comparison of the calculated fatigue life N_{cal} with the experimental fatigue life N_{exp} for aluminium alloy 2017A.

5 Conclusions

This paper was concerned with the verification of a new model applicable for the verification of the orientation angle of the critical plane. Under the assumption of two conditions, the study determined the applicability ranges of the proposed model. After calculations and analysis, the following conclusions can be stated:

1. The proposed verification confirmed the assumptions adopted with regard to the analyzed model.
2. The new expression for β angle has a wider range than $<1, \sqrt{3}>$.
3. The analysis performed in the paper involved the ratio of normal to shear stresses in the place of the traditional ratio of the fatigue limits. This is particularly important with regard to the materials with non-parallel characteristics.
4. Further analysis is needed with regard to successive materials in the future.

Project financed by the National Science Centre. Decision number: 2016/21/D/ST8/02007.

References

1. Ł. Pejkowski., D. Skibicki, Acta Mech. Sinica **32,** 4, 696-709 (2016)
2. A. Carpinteri, C. Ronchei, A. Spagnoli, S. Vantadori, Theor. Appl. Fract. Mec. **73**, 120-127 (2014)
3. M. Ciesla, G. Junak, A. Marek, Arch. Metal. Mater. **61**, 1, 271-274 (2016)
4. G. Szala, B. Ligaj, Key Eng. Mater. **598**, 231-236 (2014)
5. D. Skibicki, Ł. Pejkowski, Int. J. Fatigue **102**, 18-36 (2017)
6. B. Zhong, W. Yanrong, W. Dasheng, W. Jialiang, Int. J. Fatigue **102**, 241-251 (2017)
7. G. Stanfield, *Discussion on "The strength of metals under combined alternating stresses"*, by H. Gough and H. Pollard. *Proc. Institution of Mechanical Engineers*, 131 (1935)
8. A. Carpinteri, A. Spagnoli, Int. J. Fatigue **23**, 135-145 (2001)
9. K. Walat, T. Łagoda, Int. J. Fatigue **67**, 73-77 (2014)
10. A. Carpinteri M. Kurek, T. Łagoda, S. Vantadori Int. J. Fatigue **100**, 512-520 (2017),
11. A. Niesłony, T. Łagoda, K. Walat, M. Kurek, Mat.-wiss. U. Werkstofftech, **45 (10)**, 947-952 (2014)

DOI: 10.1051/e3sconf/20171903018

Determination of mechanical properties of carbon/epoxy plates by tensile stress test

Paul Bere[1], and *Jolanta B.* Krolczyk[2,*]
[1] Technical University of Cluj-Napoca, Faculty of Machine Building, B-dul Muncii 103 – 105, RO-400641 Cluj-Napoca, Romania
[2] Opole University of Technology, Faculty of Mechanical Engineeing, Proszkowska 76, 45-758 Opole, Poland

Abstract. The polymeric composite materials used in aerospace, military, medical or racing cars manufacturing end up being used in our daily life Whether we refer to the performing vehicles, subassemblies or parts for aircrafts, wind, telegraph poles, or medical prostheses they all are present in our lives and they are made of composite materials (CM). This paper presents research regarding three different composite materials, plates by carbon fiber, in epoxy matrix. Starting with materials presentation, manufacturing methodology and determination of mechanical properties at carbon fiber/epoxy were done. Vacuum bag technology to obtain the composite structure offer opportunity to get a very compact and homogeny composite structure. For the moment this technology are adequate for high performances pieces. The mechanical characteristics of plates made of composite materials reinforced presented indicates closed value like metal materials. Based on the results, a comparative study between the reinforced materials typically used to manufacture the plates of CM is carried out.

1 Introduction

The design of new applications requires investigating in area of new materials, including composites or bimetals [1-3]. Vast studies on current issues facing the use of CM [4-6] have been concentrated on the extensive investigations in which relate to aspects of composite materials in vehicle construction [7-9]. The CM since 1975 has a very impressive development. These materials entered our everyday life after a time made in the top areas of global technology. Further development of composite materials, particularly in the last 40 years has led to some investigations that reveal both their properties and shortcomings.

The migration of CM of high-performance areas to daily areas is in a continuous and sustained rate. Used in a military, aviation, motorsport, or medical domain the CM arrives in the domestic areas like agriculture, furniture's, arts, or building constructions. The physical-mechanical properties of these materials require the use more frequently. Polymeric reinforced Composite materials successfully replace the traditional metallic materials.

The use and manufacture of polymer composite plates has seen a steady and also impressive development. Replacing the plates made of metal, with plates made of reinforced composite materials has led to the reduction of the weight while obtaining high mechanical characteristics. The usually technologies used for manufacturing reinforced polymeric plates are hand lay-up process, hand lay-up and pressing and vacuum bag technology. The physical and mechanical properties of carbon/epoxy composite presented in open literature [10-16]. In the same time the presented results indicate a low density of materials in condition where mechanical properties are similar to metals.

This paper obtain the performant and compact plates by CM and use in the vacuum bag technology. For obtaining the mechanical properties, the samples were subjected the tensile testing. Three different reinforced materials to obtain the composites plates were tested by carbon fiber/epoxy matrix.

The goal of this paper is to determinate how the type of materials architecture has influence the tensile strength of carbon fibre. The paper presented the materials and the manufacturing methodology to obtain the carbon fibre/epoxy plates, and the mechanical properties evaluation in tensile strength case. The morphology of the damaged area of the carbon samples are investigates. At the final, the authors presented a comparative study regarding the results obtained for carbon /epoxy composite plates.

2 Manufacturing method

2.1. Materials

Experimental research aimed at determining the tensile stress characteristics of composite plates. Epoxy resin, type Epiphen RE 4020/DE 4020 was used for the matrix. The mixing ratio parts by weight were 100:30, according to manufacturer recommendations.

Epoxy resin Epiphen RE 4020 / DE 4020 characteristics:

- Aspect: transparent,

* Corresponding author: j.krolczyk@po.opole.pl

- Pot life at 20°C - 45 min,
- Brookfield Viscosity at 25°C – 300 MPa.s,
- Gelling time (thin layer in hours at 20°C), 8 - 9 hours,
- Density la 25°C - 1150 kg/m3,
- Curing at 20°C, 24 hours.

The reinforcement material used was a biaxial carbon fiber fabric by 0-90° plain, twill and unidirectional weave fabric 3K, 200 g/m^2). The samples codes used in the paper for types of fabric are: plain P, twill T and unidirectional UD. The technology to obtain the plates keeps initial architecture of reinforced material.

2.2 Samples parts manufacturing

For obtaining the composite samples were used the vacuum bag technology (Fig. 1). In initial stage the wet technology and hand lay-up process was used. This procedure realize a good impregnation of the reinforced material.

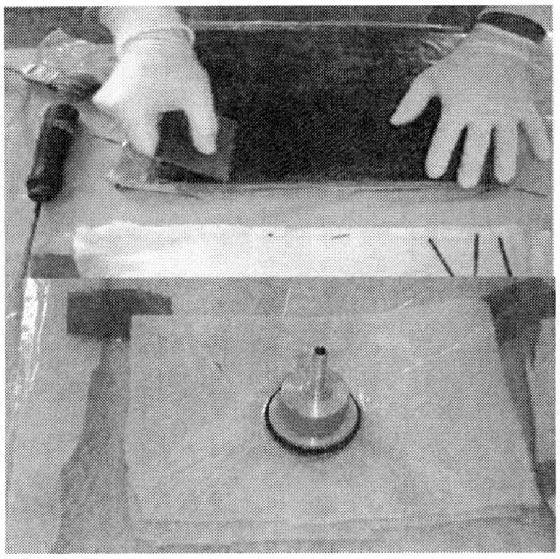

Fig. 1. Composite plates manufacturing process.

Different plates from carbon fibre/epoxy fabrics were manufactured. The impregnation of the carbon fibre material was done on the metal plates mould. Initially the surface of the mould was traded using the mould sealer tip Frekote B15 Aero by Loctite. Several times of mould sealer were applied. This agent prevents the bonding of the carbon fibre/epoxy on the mould surface.

The hand lay-up technology was used to have a good impregnation of the carbon fibre. All the layers were applied a mould and covered with a perforated plastic folia and breather fabric. All the system was covered by a vacuum bag foil. At the border the bag was closed by vacuum sealing type.

On the bag has been applied a vacuum pressure of - 0.9 Bar. This pressure eliminates all the air bubbles from composite materials and the excess of the resin. The atmospherics pressure creates a compact CM with a very thin and uniformed structure.

During the polymerization time of composite material applied vacuum was maintained. The polymerization process was made in an oven at 80° time to 8 hours.

After polymerization time the auxiliary materials are rejected and the plates are release. To determinate the psychical and mechanical properties of the plates the borders are cute until homogeny structure.

From all CM plates P, T, UD, five samples sets for tensile stress tests were manufacturing. The dimension of samples (Fig. 2) are:
L = 250 mm (including TABs), l = 200 mm, b = 20 mm, h = 0.7 - 2.3 mm.

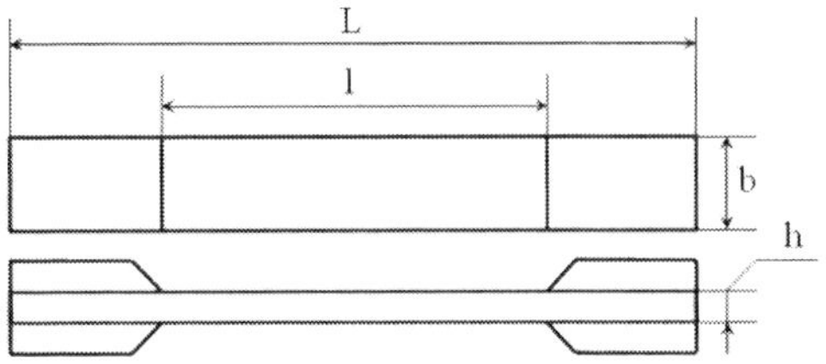

Fig. 2. Sample dimension according to the EN ISO 527-4 and ISO 527-5 standard.

The samples were cut by water jet technology. The water jet machine type OMAX 2626 was used. To prevent the samples delamination during the water jet penetration time of the carbon composite plates, the penetration water jet area was positioned at 40 mm distance from the samples border [17-19].

The samples dimensions were compliant according standards EN ISO 527-4 and ISO 527-5 [20, 21].

The ends of the specimens were reinforced with a 2 mm thick, 20-35 mm glass fabric TABs in order to avoid the slippage and breakage of specimens at clamping. The TABs were fixed on the specimens with a structural adhesive (3M Scotch-Weld 9323 B/A). The resistive force recorded during the tensile test was divided by the cross-sectional area of the specimen in order to determine the tensile stress.

For fracture, maximum resistance force recorded during the tensile test was related to cross-sectional area of the specimen:

$$\sigma_r = F_{max} / A \qquad (1)$$

The samples were loaded to failure with constant crosshead speed (5 mm/min) using an Instron 8801 (100 KN) servohydraulic testing system All measurements were carried out at room temperature (20°C) and humidity in the range of 45-50%.

The samples are coded from 1 to 5 from the border to interior of the plates.

3 Experimental results

The obtained surfaces of composite plates copied the surface of the mould. It is not observing the air bubble on the surface. A good pressed structured are obtained.
The obtained characteristics of the plates are:
- weight fraction ratio 72%,

- density of composite material 1087.97 kg/m^3 for carbon fiber plates.

The weight fraction ratio was calculated using equation:

$$W_f = \frac{w_f}{w_c} \times 100[\%] \tag{2}$$

Where was note by:
W_f–weight fiber fraction,
w_f– weight of reinforced material,
w_c – weight of composite materials.
Obtained values of tensile tests it was processed. The sets of samples were marked.
For carbon-epoxy samples the main values of tensile stress are:

UD -1050 MPa; Plain -570 MPa; Twill 525 MPa

The breaking area of the carbon/epoxy samples is different. In the carbon/epoxy case the maximum stress produce like on explosion at the samples. The sample was broken on many pieces. In all the cases the stress of the samples is accompanied by acoustic emission and this predicting catastrophic breakage.

A force/displacement diagram is presented in Fig. 3 carbon fiber (plain weave fabric 3K by 200g/m^2) /epoxy.

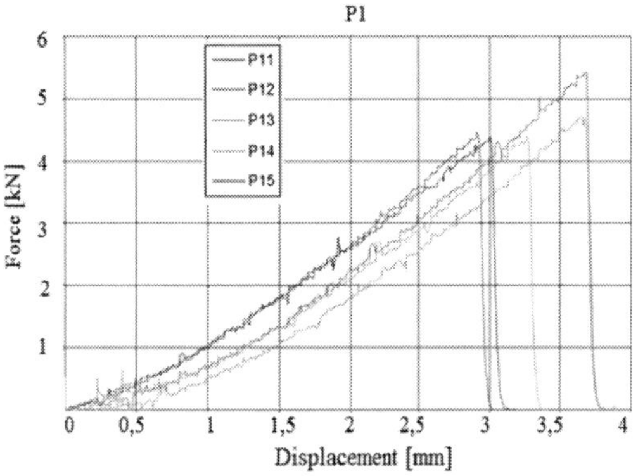

Fig. 3. Force-displacement diagrams experimentally determined for samples of carbon fiber (plain weave fabric 3K by 200g/m) /epoxy.

In Fig. 3 are presented the comparative diagrams for all five P samples subjected by tensile stress test (P11 represent the first sample cut form carbon plate, and the P 15 the last sample). The samples are noted with different colours. On X axe are presented variation of displacement (elongation) of the samples and on Y axe the variation force during the traction request. The force/displacement diagram shows a linear elastic behaviour at the CM until fracture zone when they are depreciating suddenly. The force/displacement diagram shows a linear elastic behaviour at the CM until fracture zone when they are depreciating suddenly.

This behaviour is typical for fiber reinforced composite materials in which the direction of the fibers is the direction of the request. In this case longitudinally arranged filaments have contributed to the high value of fracture strength. It should be noted that for the balanced used fabric only half of the filaments used opposed traction application. The other halves of the filaments were arranged transversely to the direction of application. A normal behaviour for CM is indicating in the diagram where we can see an elastic area. The elongation of the samples is 0.2% for a maximum stress.
The analysis of microstructure fracture area from fibers carbon/epoxy sample relieve the carbon fiber monofilament simultaneously grouped and broken (Fig. 4).

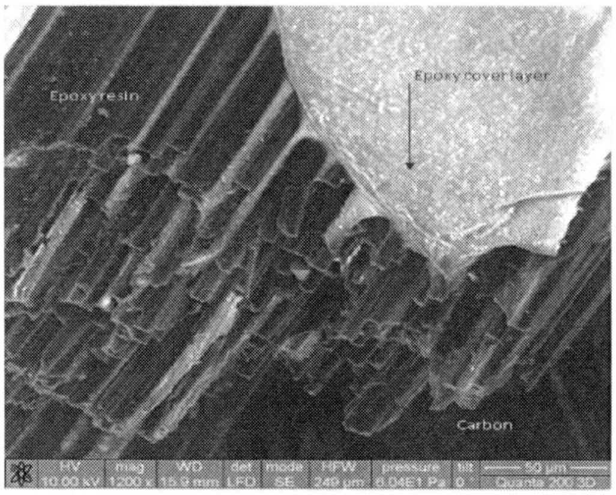

Fig. 4. Microstructure fracture area of CM sample subjected by tensile test.

No individual monofilaments are observed, tensile request being taken over by groups of filaments that have acted together. It appears a good impregnating matrix and a good compatibility between filaments and matrix. The epoxy matrix parts remain bonded on carbon fibre monofilaments after tensile test. In the same time in the figure presented we can see the matrix impregnate the monofilaments and cover it.

The morphology of the structure was investigated using scanning electron microscopy SEM type Quanta 200 3D DUAL BEAM.

If compare this three types of carbon/epoxy the maximum tensile stress (1050 MPa) were obtained in UD Carbon/epoxy case. The results presented in Fig. 5 classify these CM at tensile stress from this point of view.

In the fabrics type of materials cases it can be observed a normal behaviour where the tensile strengths are in the fabrics of carbon fibre structure only half of monofilaments are on force direction applied approximatvely half of the UD carbon fibre.

Half of these are applied perpendicular on force direction. In the same time we can observe that in the carbon fibre twill tissue the tensile strength increase. This type of material have very good facilitates in the complex three dimensional surfaces. The twill fabrics fit very well on the complex surfaces and respect the curve of surface (the creases not appear).
The tensile strength obtained are very high comparable with metals. However we must to thinking that the density of material is 7 times lower.

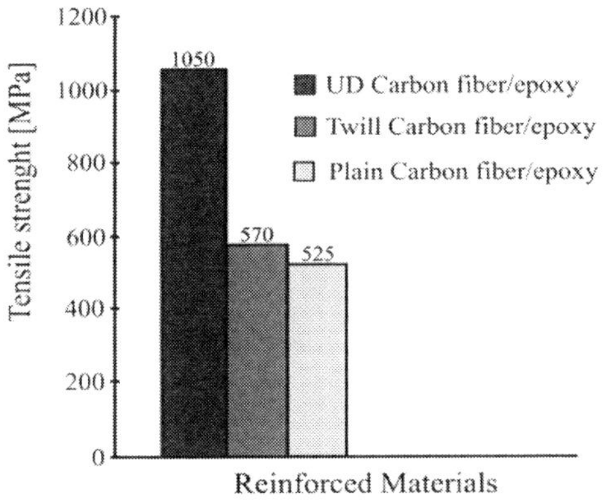

Fig. 5. Comparative diagram between tensile strength and materials type of CM

4 Conclusions

Three different architecture types of carbon fibre/epoxy matrix plates were investigated. The manufacturing methods by vacuum bag technology indicate a height performance obtained by CM. The density of carbon/epoxy obtained plates is seven times lower than steel and twice as aluminium alloys. Microstructure of the material indicates a very good impregnation of the carbon fibre and the good compatibility between fibres and the matrix. The technology used, materials, the polymerisation and thermal treatments indicate a special material class obtained.

According to the mechanical properties in tensile strength case, the authors relive the different between types of fibre architectures for the same materials.

The novelty of the problem is to emphasize for the composite designers the different results between the same types of the materials but using different architecture type. In the same time showing which technology can be applied to obtain these results.

Following the results of tensile stresses in composite plates we see that the highest value is obtained by UD carbon/epoxy CM. The values of these materials are ultimate tensile stress at 1050 MPa at UD type case. In this case all the monofilaments are arranged by request direction. The tensile strength of the composite material obtained, has a high value compared to high quality steel in terms of the density decrease of approx. 7 times in the case of M.C.

Compare the biaxial fabric, plain and twill type is observed closer values of tensile strength. For carbon twill tissue the results are 54% lower value like UD case. In this case only half on monofilaments are positioning on request direction. Another half of these are in the transversal position. For carbon/epoxy plain tissue, obtained results regarding the tensile stress test is much closed to twill tissue. The results indicate 8% lower in balanced plain tissue case.

This work was supported by a grant of the Romanian National Authority for Scientific Research and Innovation, CNCS/CCCDI–UEFISCDI, project number PN-III-P2-2.1-BG-2016-0210, within PNCDI III.

References

1. C.V. Opelt, J.M.F. Paiva, G.M. Cândido, M.C. Rezende, Eng. Fail Anal. **79**, 342-350 (2017)
2. P. Nieslony, P. Cichosz, G.M. Krolczyk, S. Legutko, D. Smyczek, M. Kolodziej, Measurement **78**, 129 – 137 (2016)
3. P. Nieslony, G.M. Krolczyk, K. Zak, R.W. Maruda, S. Legutko, Precis Eng. **47**, 104 – 110 (2017)
4. E.E. Feldshtein, L.N. Dyachkova, Compos. Part B-Eng. **58**, 16-24 (2014)
5. P. Valášek, A. Ruggiero, M. Müller, Compos Part B- Eng. **122**, 79-88 (2017)
6. A. Ruggiero, P. Valášek, M. Müller, Compos Part B- Eng. **104**, 9-16 (2016)
7. R.M. Jones, *Mechanics of Composite Materials. Taylor and Francis* (Philadelphia, 1999)
8. L.P. Kollar, G.S. Springer, *Mechanics of Composite Structures* (Cambridge University Press, Cambridge, 2003)
9. V.V. Vasiliev, E.V. Morozov, *Mechanics and Analysis of Composite Materials* (Elsevier Science, London, 2001)
10. T. Sathishkumar, S. Satheeshkumar, J. Naveen, J Reinf, Plast. Compos. **33,** 1258-1275 (2014)
11. J. M. Stickel, M. Nagarajan, Int J Appl Glass Sci **3,** 122-136 (2012)
12. S. C. Amico, C. C. Angrizani, M. L. Drummond, J Reinf. Plast. Compos. **29,** 179-189 (2010)
13. M. S. M. Jusoh, C. Santulli, M. Y. M. Yahya, N. S. Hussein, H. A. I. Ahmad, Mate. Sci. Eng. Adv. Res. **1,** 19-25 (2016)
14. E. H. Portella, D. Romanzini, C. C. Angrizani, S. C. Amico, A. J. Zattera, Mater. Res. **19,** 542-547 (2016)
15. P. Bere, Theoretical and experimental research on the manufacturing and mechanical behaviour of the tubes made from polymeric composite materials, PhD thesis, Technical University of Cluj-Napoca, Romania (2009)
16. G. L. Enachescu, M. F. Stefanescu, J. Plastic Mater. **53,** 198-201 (2016)
17. V. Miron, N. Balc, A. Popan, C. S Borzan, P. Bere, AJME **11**, 2, 87-92 (2013)
18. Popan, N. Balc, A. Carean, A. Luca, AJME **10**, 3, 97-102 (2012)
19. I.A. Popan, N. Balc, A. Carean, A. Luca, V. Ceclan, AJME 11, 3, 17-22 (2011)
20. ISO 527-4 2: 1997, Plastics – Determination of tensile properties. Test conditions for multidirectional fiber – reinforced plastic composites.
21. ISO 527-5 A, B: 1997. Plastics – Determination of tensile properties. Test conditions for unidirectional fiber – reinforced plastic composites.

DOI: 10.1051/e3sconf/20171903019

Materials, used in historical buildings, analysis methods and solutions puroposals

M.Sami Döndüren[1], and *Ozlem* Sişik[1]

[1]Department of Civil Engineering, Selcuk University, 42060 Konya, Turkey

Abstract. Most of historical buildings are built with pressure principle and have the characteristics of masonry structures. Therefore, the structure components of buildings are constituted bearing walls, columns, buttresses, vaults and domes. Natural stone, cut stone, rubble stone brick or alternate materials were used in the bearing elements. Brick-dust and mortar with more binding feature were used as combination elements. In time, some problems were occurred in used materials and in structure as a result of various effects. Therefore, it is necessary to apply various applications in framework of repair and strengthening of buildings. In this study, restoration of historic buildings and the control of the adequacy of the bearing systems as one most important part of structure were examined. For this purpose, static analysis of Edirne-Merkez Demirtaş (Timurtaş) mosque located in Edirne was tested. Testes could give suggestions and be applied if buildings needed be revealed. The structure was modelled with finite element model of sap2000 package program and the forces generated under various loads and stresses, the occurred deformation due to that, overflow of allowable stress of this deformation and stresses were investigated. As the results of this study can be note that the maximum compressive stress at the construction is calculated as 1.1 MPa.

1 Introduction

Historical buildings are the most important ruins to describe history of a society. Buildings exposed to harsh effects of years and to natural disasters are generally under effect of very serious problems and they have risk of collapse and extinction. Historic buildings, which are the most important parts of our cultural heritage, must be best protected and repaired [1].

Most of historical buildings are built with pressure stress and have the characteristics of masonry structures. Therefore, the structure components of buildings are constituted bearing walls, columns, buttresses, vaults and domes [2]. Natural stone, cut stone, rubble stone brick or alternate materials were used in the bearing elements. Brick-dust and mortar with more binding feature were used as combination elements. The tensile forces of historical buildings have been met by using with wood elements or steel clamp, steel tensioner elements [3].

2 Materials and Characteristics Used in Historical Buildings

The building materials play a decisive role in the behaviors of the structures they use. Historical agglomerations use different materials such as natural stone, brick, wood, mortar. Understanding the physical and mechanical properties of these materials is an indispensable part of the work to be done to evaluate historical structures [4-6,7].

3 Static Estimation of Historical Constructions

3.1 The Parameters Used In The Models For Structural Analysis

Interpretation of the results of structural analyzes of historic buildings using the finite element method differences from the interpretation of the calculations of structures produced by today's engineering technologies. Because of it is not always possible to carry out experiments that will determine material properties by taking samples from historical structures, it is sometimes very difficult to determine the bearing capacities of structural members according to the calculation results.

The parameters used in the calculation of the structural earthquake forces are given below:

- A0 (Earthquake Region Coefficient) = 0.1 (Region 4)

- If the ground class is not foreseen, is taken according to DBYBHY S (T) = 2.5
- I (Structure Importance Coefficient) = 1.0
- R (Carrier System Behavior Coefficient) = 2

In describing the masonry wall material, the Elasticity Module of the masonry units used for wall construction according to the directive is 200 times the character pressure resistance of the material,so;

$$Ed = 200 * fd. \quad (1)$$

According to TS 2510 (Turkish Standarts 2510), the pressure resistance of natural stones used in construction of bearing walls should not be smaller than 350 kgf/cm^2.

$$f_d = \text{free pressure resistance} *0.5 \quad (2)$$

$$f_d = 0.5 * 350 = 175 \ kg/cm^2$$

$$E = f_d * 200$$

$$E = 175*200 = 35000 \ kg/cm^2$$

The pressure safety tension for the stone masonry walls, f_{em} = 0.3 MPa is recommended.
Tension safety tensions can be accepted as 15% of the value determined as pressure safety tension. In this case, the tensile safety strain for the stone wall is calculated as f_m = 0.3× 0.15 = 0.045 MPa.

The earthquake force coming from the wall will be divided into the horizontal cross-sectional area of the wall to calculate the shear stress in the wall and to be compared with the wall slip safety stress τ_{em}, which will be denoted as Equation (3) [5].

$$\tau_{em} = \tau_o + \mu\sigma \quad (3)$$

$$\tau_{em} = 0.10+0.5(0.3/2) = 0.175 \text{ MPa}$$

3.2 Timurtaş (Demirtas) Mosque

It is one of the earliest examples with a square-shaped, single-volume type of mosque. It is a monumental view with a monumental harim which is covered with Turkish triangles and sitting on a dodecagon pulley, and the architectural structure located on the northwest corner of it.
The Turkish triangles, which provide passage to the domes covering the Harimi, have been widely used in various forms in Turkish architecture. The building material was made of rubble stone,coarse stone and brick.
As an alternative wall technique was applied on the ground form walls with an alternate covering of a row of stones and two rows of bricks.
Brick material was used completely on bridges, belts, mihraptas and domes [3].

3.2.1 Pre-restoration architectural featuresPre-restoration architectural features

In this section informations given was taken from the restoration,pre- restoration and restoration reports prepared for the Timurtaş (Demirtaş) Mosque of Edirne foundation general in 2007 prior to restoration of the building. Figs. 1-3 show the pictures of the tomb before and after repair.

Fig. 1. Photographs before and after restoration of the Timurtaş(Demirtaş) Mosque (2007-2008).

Fig. 2. Photographs before and after restoration of the Timurtaş(Demirtaş) Mosque (2007).

Fig. 3. Photographs before and after restoration of the Timurtaş(Demirtaş) Mosque (2007).

The material decrement on the groun form walls have been completed according to the original material and mesh system. Filling the capillary cracks in the structure with epoxy plaster, 4 cm wide cracks are stitched with iron plates and they are filled with epoxy again. For cracks larger than 4 cm, it is recommended that the perimeter be disinfected and filled in accordance with the original material and mesh system. Except base of the minaret, the mosque was destroyed. Since the old photo about the minaret could not be found, it was completed with the guidance of periodical analyzes. As a suggested at the restitutions projects, the rostrum with alternated had been built of bricks. Transition from the ground form to the base was made with triangles. As a result of the periodical analyzes, this decision was made (Saruca Pasha

Mosque, Bird's Dogan Mosque, Kadi Bedrettin Mosque). The height of the minaret was determined in proportion to the height of the dome as a result of the analysis. Minare Base height is determined from existing stone profiling. Minarets lug height is determined from the tracks in front again.

3.2.2 Structural model and analysis

SAP 2000 V18 finite element program is used for modeling. The building walls consist mainly of at least three main materials, stone, brick and mortar (plaster). However, since the general behavior of the structure in the model is concerned, it is assumed that the carrier elements are formed from a single material and the related unit volume weight, modulus of elasticity and Poisson's rate are used Sketches of the plan of the structure is given in Fig. 4.

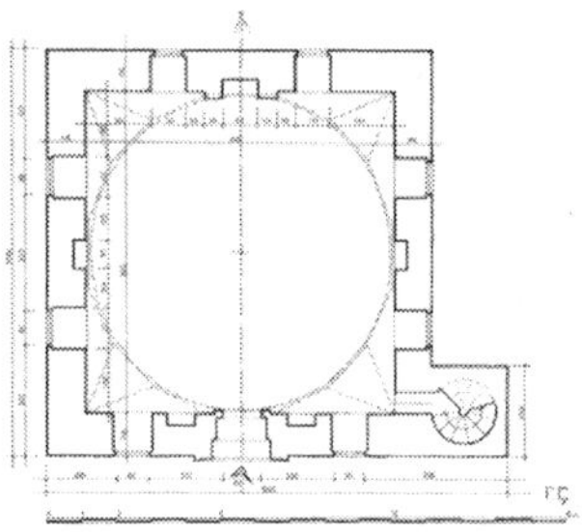

Fig. 4. Sketches of the plan of the Timurtaş (Demirtaş) Mosque.

Based on the build-up relay project, in the numerical model, the wall thickness is defined as 1.1 m, and the thickness of the cove is defined as 0.25 m. Material properties are defined; The material of the wall and minaret is stone, the material of the dome is brick. The carrier elements of the structure are modeled as shells. 559 shells (area) were created using 558 knot points in the prepared structure model. 51 fixed supports are defined in the points that are transferred to the ground. The structural modeling is given in Fig. 5.

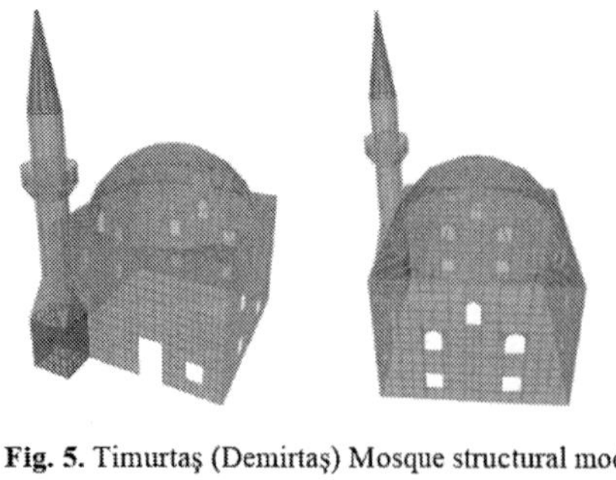

Fig. 5. Timurtaş (Demirtaş) Mosque structural modeling.

Since the results of the analysis obtained are very difficult to show for each knot and each element, the results obtained are presented in terms of colored stress distribution and graphs.

3.2.2.1 The analysis of the structure under the dead load

The weight effect of the structure is taken into account by using the values calculated in Unit volume weight (γ), elasticity modulus (E) and Poisson ratio (ν) of building materials.

The weight of the structure is G = 10351.5 kN.

As a result of the static analysis of the 3D final element model of Timurtaş (Demirtaş) Mosque under its own weight, the possible stress distributions in the structure, the strain values were reached and critical locations where cracks could occur were detected.

Under G force, the greatest displacement in the structure is about 0.979 mm in the vertical direction at the top of the dome. R = 2 is used in this analysis, the elastic displacement should be calculated as 0.979 mm × 2 = 1.958 mm (Fig. 6). It is about 1.53 mm in the vertical direction at the top of Minaret. R = 2 is used in this analysis, the elastic displacement should be calculated as 1.53 mm × 2 = 3.06 mm.

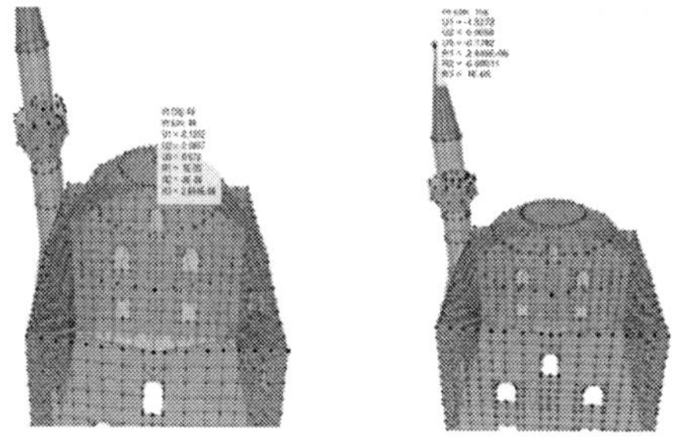

Fig. 6. Under G force, the greatest displacement in the Timurtaş (Demirtaş) Mosque (mm).

The behaviour of the structure according to the G force acting, prepared separately S22 and S11 (tensile and compressive) stress value graph (Fig. 7).

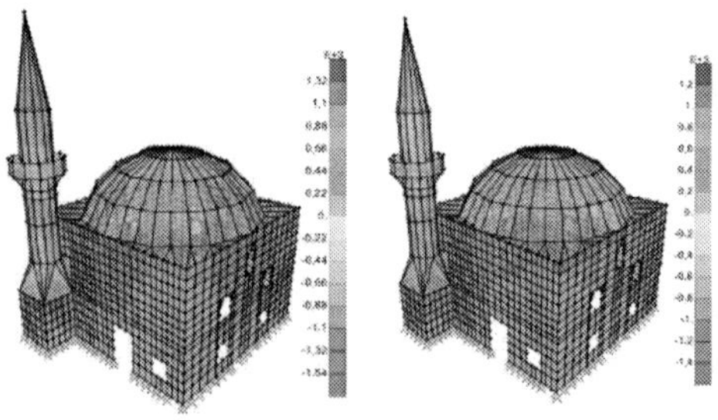

Fig. 7. Under G force, S11 and S22 Stresses in the Timurtaş (Demirtaş) Mosque (KN/m^2).

 DOI: 10.1051/e3sconf/20171903019

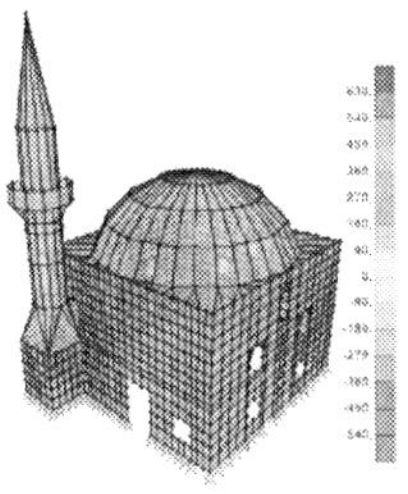

Fig. 8. Under G force, S12 Stresses in the Timurtaş (Demirtaş) Mosque (KN/m^2).

The most unfavourable compressive stresses calculated were 1.54 MPa from S11 Graph and the most unfavorable tensile stresses were also found to be 1.32 MPa on S11 graph (Fig. 8). However, it should be kept in mind that the values of the strain distributions in the graphs S11 and S22 are very close to one another. Both values exceed the safety stress values of the stone walls given in Table. The maximum shear strain determined from the S12 graph is 0.630 MPa. This value is also above the safe dislocation stress value.

As a results under G force when the effects of structural analysis are examined, on the carrier walls of the structure. In the Turkish Earthquake Regulation, it has been observed that the compression tension and dislocation stress values proposed for masonry structures are exceeded. As a matter of fact, the size of the dome of the structure and the cracks formed is an indication that the structure is not stable (Fig. 9).

Fig. 9. Timurtaş (Demirtaş) Mosque prior to the restoration of damaged photos.

3.2.2.2 Modal analysis

The fixed spectral coefficient S(T) is assumed to be 2.5 and the effective earthquake coefficient Ao = 0.1 is assumed in the specular seismic solution. In defining the earthquake effects, the method of joining the modal effects is adopted and using the exact quadratic joining method. under the vertical structure and earthquake is intended to obtain the elastic behavior. The earthquake load reduction factor R = 2 is considered in all periods. Using modal analysis and mass and rigid matrices of the structure system. Mode shapes and periods are obtained.The mass participation rate of the fourth mode showing the lateral displacement movement of the main mass in the X direction is calculated as 54% and the mass participation rate of the third mode showing the lateral displacement movement in the Y direction is calculated 56%. Since the structure is symmetrical, the mass participation rates give almost the same value. The highest natural vibration period already obtained is 0.24 and the building is very rigid.

In Fig. 10, the first 6 mode graphs of the structure are given.

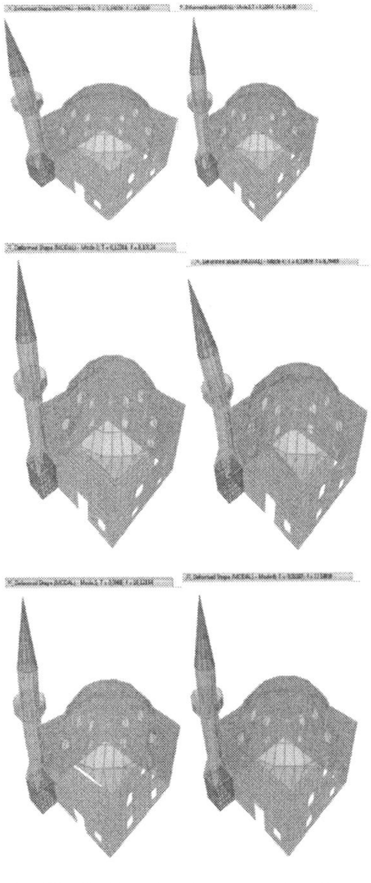

Fig. 10. The first 6 mode graphs of the Timurtaş (Demirtaş) Mosque.

3.2.2.3 G+Ex ve G+Ey earthquake loads

Shape changes occurring in the modal spectral analysis resultant structure under earthquake load reduction (R = 2) and earthquake loads acting in the direction of X and Y with dead load are given in Figs. 11 and 12.

G+E$_x$: The earthquake loading brings a lateral displacement of 12.75 mm in the X direction, 1.19 mm in the Y direction, 0.98 mm in the Z direction,

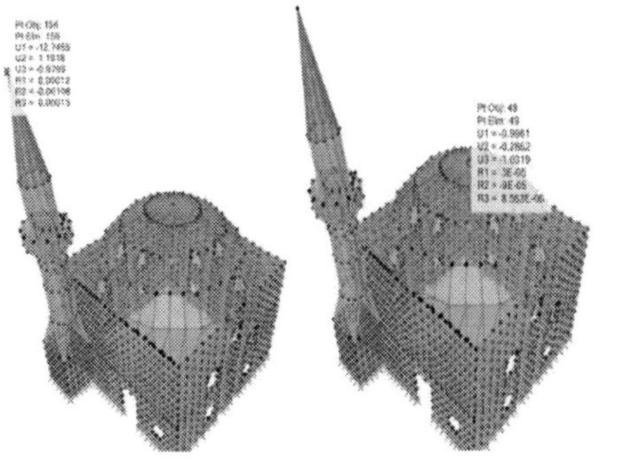

Fig. 11. Under G+E_x force, the greatest displacement in the Timurtaş (Demirtaş) Mosque (mm).

and 0.999 mm in the X direction and 0.29 in the Y direction and 1.03 mm in the Z direction.

G+E_y: The minaret in the X direction is 3.51 mm in the direction of the earthquake, 11.11 mm in the Y direction, 0.85 mm in the Z direction and 0.36 mm in the X direction, 1.19 in the Y direction and 1.08 mm in the Z direction.

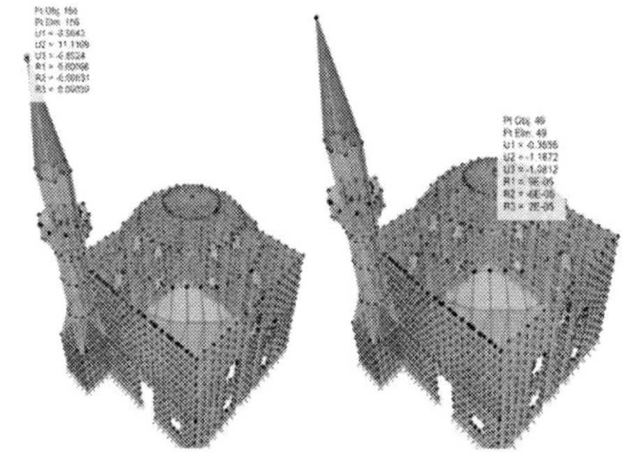

Fig. 12. Under G+E_y force, the greatest displacement in the Timurtaş (Demirtaş) Mosque (mm).

As a result of the analyzes made, it has been determined that the most difficult parts of the structure in the static state of the structure are the edges of the upper window cavities, the ground form that the dome and the dome 8,5 are sitting on.

G+Ex ve G+E_y: It is seen that in the case of earthquake loading, the regions, where the stresses of S11 (Fig. 13) and S22 (Fig. 14) in the structure are unfavorable there are at the corner points of the dome and wall joints and windows and door gaps. The greatest compressive stress in the structure is calculated as 1.1 MPa, which exceeds the pressure safety stress of 0.3 MPa. The tensile stresses determined at 2.16 MPa in the construction exceed the tensile safety stress and reinforcement is require.

The calculated maximum displaced stresses were G+E_x ve G+E_y (Fig.15.) At 0.99 MPa. This value has exceeded the value of safe shear stress calculated as 0.175 MPa.

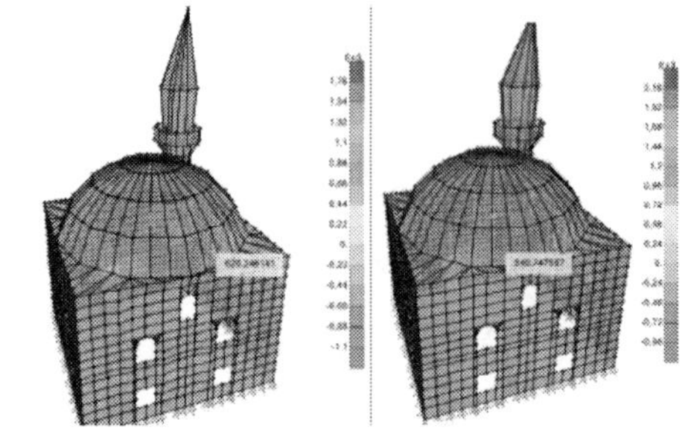

Fig. 13. Under G+E_x and G+E_y forces, S11 Stresses in the Timurtaş (Demirtaş) Mosque (KN/m^2).

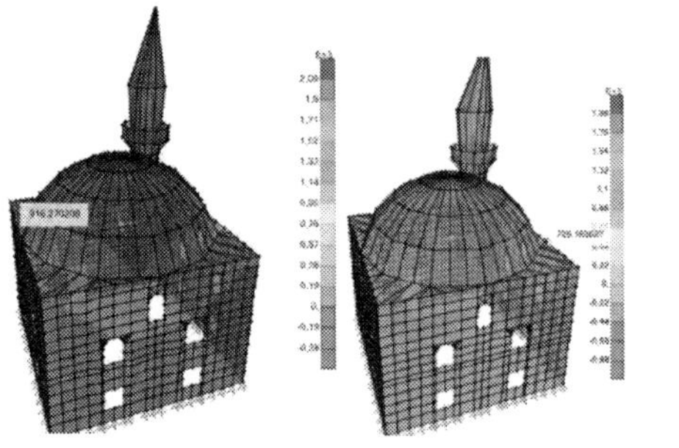

Fig. 14. Under G+E_x and G+E_y forces, S22 Stresses in the Timurtaş (Demirtaş) Mosque (KN/m^2).

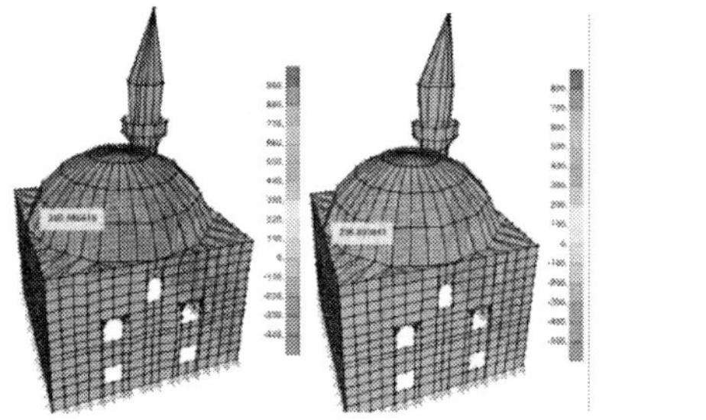

Fig. 15. Under G+E_x and G+E_y forces, S12 Stresses in the Timurtaş (Demirtaş) Mosque (KN/m^2).

4 Conclusions

From the early Ottoman architectural examples, they are historical structures covered by a square-shaped, single-volume dome. As a result of the analyzes, the following results were obtained.

• Timurtaş (Demirtas) Mosque: The maximum compressive stress at the construction is calculated as 1.1 MPa, which exceeds the pressure safety tensile strength of 0.3 MPa. The tensile stresses determined at 2.16 MPa exceed the tensile safety stress and reinforcement is required. The calculated maximum displace stress was 0.99 MPa. This value exceeds the safe displace stress value is calculated as 0.175 MPa. As of January 2017, the current situation has been examined and it has been determined that there are spills and dissociations in the stones forming the body walls in particular. Despite the high stresses in the finished element model, there are no large and risky cracks in the areas where the cowbag merges with the body walls and in the window and door openings. This can be interpreted to mean that the structure still retains its stability. However, maintenance of the damaged stones and filling of the gap between the joints will be favorable for the continuation of the stable condition of the building. When the analysis results are evaluated, it is observed that there are great stresses in the walls forming the main building, in the dome, in the doors, in the window spaces and in the minaret. Especially for historic buildings built with materials that are inadequate for tensile strength, these high values of stress have a considerable risk. It is a necessity to improve these stresses if a child of the works studied in the study is thought to have collapsed the upper body walls of the dome without the dome before the restoration. Static problems that need to be solved with minimal intervention, while remaining faithful to historical structures, require the development of new methods. Otherwise, the historical buildings can not be kept up for a long time just by cleaning and repairing works.

References

1. A. H. Aköz, Deprem Etkisi Altindaki Tarihi Yiğma Yapilarin Onarim Ve Güçlendirilmesi, (2008)
2. N. Bayülke, Yiğma Yapilarin Deprem Davranişi Ve Güvenliği, 1.Türkiye Deprem Mühendisliği ve Sismoloji Konferansı, (2011)
3. Ö. Şişik, Edirne'de Bulunan 15.Yy Ve 16.Yy'da İnşaa Edilmiş Tarihi Cami Ve Türbelerin Taşıyıcı Sistem Analizi Ve Çözüm Önerileri, Konya, (2017)
4. M. Şahin, Eski yapıların korunmasında 'ICOMOS' ilkeleri, Bursa, (2015)
5. Türkiye Bina Deprem Yönetmeliği. DBYBHY2007, Türkiye, (2016)
6. S. Uğuz, Tarihi Yiğma Bir Binanin Deprem Güvenlik Analizi: Tarihi Konya-Gazi Lisesi (Darü'l Muallim) Örneği, Konya (2016)
7. U.C.Yavuz, Tarihi Yapilarda Statik Güçlendirme Teknikleri, Ankara, Kültür ve Turizm Bakanlığı Kültür Varlıkları ve Müzeler Genel Müdürlüğü (2012)

Surface phenomena accompanying plasma gouging of S355 steel

Anatol Kałasznikow[1,*], and *Mariusz* Michalski[2]

[1]AK ANATOL, 4 Handlowa Street, 68200 Żary, Poland
[2]Faculty of Mechanical Engineering, University of Zielona Gora, 4 Prof. Z. Szafrana Street, 65516 Zielona Gora, Poland

Abstract. The paper presents the results of investigation of surface changes taking place after the process of gouging with air plasma. The investigation included the assessment of the plasma current influence on the structure and hardness of metal sheets with the thickness of 6, 12 and 16 mm. It has been found that, in the process of gouging, both the value of the plasma stream intensity and the thickness of the material significantly influence the properties of the top layer and the size of the formed grooves. The existence of two surface zones has been proved: the zone of melted metal and the heat affected zone. It has been determined that increase of the plasma stream intensity, with constant sheet thickness, results in an increase of the gouging depth. . It has been proven that the increase the sheet thickness for constant plasma currents influence for decrease in the depth of the gouging grooves.

1 Introduction

Plasma gouging is a process of making a groove on a metal surface with the use of plasma arc. The heat of the plasma arc arising between the electrode and the object being processed rapidly melts the metal and the gas from the nozzle blows off the melted semi-automatic or automatic (CNC) device for plasma cutting. Plasma cutters can be provided with the gouging mode but the process can be affected in the cutting mode, too (with adequate selection of parameters).

In the process cutting, as well as in that of plasma gouging, the material under the process is subjected to both thermal action (plasma arc) and mechanical one (gas pressure). Such action, with additionally occurring high cooling speeds, results in modification of the structure and properties of the top layer of the material being processed.

Literature survey shows that plasma cutting of unalloyed and low-alloy steels results in formation of three zones with different properties. The first one is the zone of melted metal (MM), the second one is the heat affected zone (HAZ) and the third one is the zone with the structure of the basic material. The melted metal zone – white layer [1] has hard-to-etch structure of high hardness and brittleness [2]. In that zone (about 20 μm thick), changes of the chemical composition (as compared to the initial material) can be observed, as well as saturation of metal with gases (mainly nitrogen when cutting with the use of air plasma) [2, 3]. In works [4, 5], an effect of strong hardening of the layer situated near the surface of cutting performed with air plasma has been observed. In microscopic inspections, microanalysis and x-ray examinations, the presence of phases ε (Fe_2N) and γ '(Fe_4N) has been found. The heat affected zone is also characterized with high hardness which gradually decreases towards the core to obtain the hardness of the basic material. In the case of cutting unalloyed and low-alloy steels (with various thicknesses), the use of air plasma results in formation of heat affected zone with an average thickness of about 0.6 – 0.7 mm [1, 6, 7]. However, increase of the sheet thickness results in structure changes and increase of hardness in the heat affected zone. This effect can be explained by the increased cooling speed resulting from the heat conduction by the material core. In work [8], plasma cutting of S355 steel has been used as an example to prove that the thickness of the heat affected zone depends directly on the cutting speed and the distance from the cutting edge. Incrise of cutting speed result in the SWC decrise. It has also been determined that the thickness of the heat affected zone increases with the increase of the distance from the top cutting edge.

Both in the process of machining, cutting and plasma gouging, there is a hazard of micro cracks at the processed surfaces due to hard and brittle structure of the heat affected zone [2, 9]. In such case, subsequent processes or dynamic loads can initiate development of cracks and, consequently, lead to failure of elements. In order to prevent it, the layer susceptible to cracking should be removed [2, 10]. This makes it very important to know the influence of the plasma gouging or cutting parameters on the properties of the material being processed.

The theme of the plasma cutting on the properties of materials being processed is widely described in the

* Corresponding author: anatol@zsanatol.com.pl

professional literature, however, there is shortage of data concerning plasma gouging.

Literature data imply that, in the case of plasma gouging, the authors concentrate on the problems related to the influence of the processing parameters on the shape and size of the formed grooves. Work [11], shows the possibility of using air plasma for gouging carbon steel as a cheaper alternative to the use of nitrogen.

Furthermore, the relationship between the groove shape and size and the gripping angle and scanning speed has been determined. The change of the groove shape from "U" to "V" takes place with reduction of the grip inclination angle and increase of the scanning speed. It has also been found that the presence of nitrides occurring on the surface of the gouged metal does not influence formation of gas porosity in the process of welding.

Work [10] uses alloy steel and argon – helium plasma with cooling gas in the form of argon to examine the influence of scanning speed and method (linear and oblique) and of the plasma current intensity on the depth of the gouged groove. It has been found that increase of the plasma current intensity results in an increase of the groove depth with an insignificant increase of its width.

The purpose of the present work was to determine the influence of the stream intensity in the process of plasma gouging on the structure and properties of S355JR steel with three different thicknesses.

2 Materials and methods

The test material used was unalloyed constructional steel grade S355JR in the form of a standard sheet with the thicknesses of 12 and 16 mm.

The chemical composition of the steel and its designation have been shown in Table 1; an example of microstructure can be seen kin Fig. 1.

Table 1. Designation and chemical composition of the steel (max), % wt. (acc. to PN-EN 10025).

Steel	C	Mn	Si	P, S	N
PN S355JR EN 1.0045	0.24	1.6	0.5	0.045	0.009

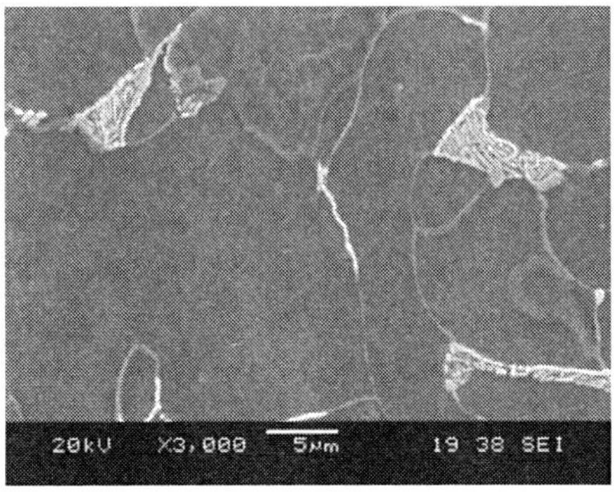

Fig. 1. Ferritic - pearlitic microstructure of the S355JR steel (scanning microscope).

The process of gouging has been performed with the use of a plasma cutting device with the working table size of 6000x 2000 mm with the plasma source, Formica ForCut133 WDM. The processing parameters have been selected to be constant for all the sheet thicknesses with only the plasma current intensity variable (15, 20, 25 A). The parameters of the plasma gouging process can be seen in Table 2.

Table 2. The parameters of gouging.

The parameters of gouging	Sheet 6, 12, 16mm
Amperage, A	15, 20, 25
Arc voltage, V	130
Speed of gouging, mm/s	66
Plasma gas	air
Plasma gas pressure, MPa	6,4
Cooling gas	air
Cooling gas pressure, MPa	2,2
Electrode	Hafnium (HF)
Nozzle diameter, mm	1
Torch distance, mm	3

The process of gouging the sheets consisted inn making parallel 100 mm long grooves on the sheet surfaces, at intervals of 5 mm. For each of the three sheet thicknesses, three sets of samples have been made (for plasma current intensity of 15, 20, 25 A; numbering in table3). The assessment of the obtained effects of gouging has been based on the samples taken in the form of cross sections of the gouged sheets. In order to perform metallographic examination, microsections have been prepared. Etching has been effected with the use of the chemical reagent, Nital (3% water solution of HNO_3). Microscopic examination has been performed with the use of an optical microscope, AxioObserver A1.m and a scanning microscope, Jeol 5600LV. Chemical composition has been assessed with the use of an x-ray microanalyzer, EDS, coupled to a scanning microscope. The HV0.1 microhardness has been measured with the use of a ZVICK ZHV-10 hardness tester, with the load of 100 g.; indenter operation time 15 s.

3 Results and discussion

The next subsections show the results of microscopic studies with the use of X-ray microanalysis and hardness measurements of samples.

3.1 Microscopic examination

The purpose of microscopic examination was to find structural changes of the material as result of plasma

DOI: 10.1051/e3sconf/20171903020

gouging and to determine the shape and size of the formed grooves.

The variable values of plasma current intensity applied in the subsequent experiments, as well various sheet thicknesses of the sheet subjected to plasma gouging have resulted in the acquisition of grooves with various shapes and dimensions. Microscopic examinations have allowed us to determine that the cross sections of the grooves consist of three zones: 1) melted metal, 2) heat affected zone. Fig. 2 shows examples of microstructures of the gougings under examination obtained for a 12 mm thick sheet, with variable plasma current intensity of 15, 20 and 25 A, respectively.

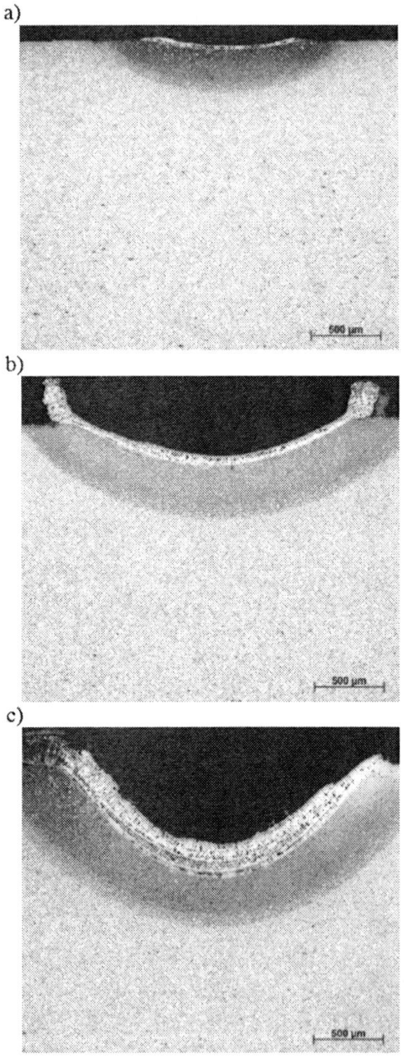

Fig. 2. Microstructure of the gouged steel sample with the thickness of 12 mm, magnification 50 x. Plasma current intensity: a) 15 A; b) 20 A; c) 25 A.

The groove size has been described in three dimensions: a) groove width – s(mm), b) groove depth – g (mm), c) melted layer thickness – g_1 (mm). The value of the groove width and depth is stated with the incorporation of the melted layer thickness. The groove size measurement results shown in table 2 are average values obtained from three grooves

Table 3. Assembled results of the gouged groove size measurements.

Nr	b, mm	Amperage, A	Groove width, mm	Melted g_1, mm	Groove depth g, mm
1	6	15	1.0	0,05	0,09
2		20	2,2	0,18	0,52
3		25	2,5	0,60	1
4	12	15	1,0	0,04	0,06
5		20	2,1	0,07	0,35
6		25	2,3	0,2	0,83
7	16	15	1	0,02	0,05
8		20	1,8	0,09	0,30
9		25	2,2	0,29	0,80

Increase of the plasma current intensity (with the same sheet thickness) results in an increase of the dimensions of the melted metal zone and the heat affected zone, as well as the dimensions of the grooves themselves (Fig. 2). In accordance with Table 2, one can state that, with the same sheet thickness (and maximum plasma current intensity) the groove width grows over two times and its total depth increases grows even ten times. With the increase of the groove depth, thicker melted metal zone is obtained. In industrial conditions, this effect can be minimized by increasing the pressure of the auxiliary (cooling) gas or by increasing the nozzle inclination angle. Furthermore, the growth of the melted metal zone is accompanied by formation of characteristic flashes (Fig. 2). The investigation performed shows that the effects of gouging depend also on the thickness of the sheet under processing. Increase of the sheet cross section (in plasma gouging) causes an increase of the cooling speed (heat conduction through the material core), which results in obtaining grooves with smaller width and depth. The tests have also shown that that, gouging sheets of various thicknesses (with the same plasma current intensity) results in different sizes of the melted metal zone and heat affected zone. Increase of the sheet thickness (6 mm in Fig. 3 and 12 mm in Fig. 2c) causes reduction of the melted metal zone and the heat affected zone.

Microscopic examinations have revealed significant changes in the structure of the steels gouged with air plasma. In the samples in which the melted metal zone is in tye form of a thin layer (up to about 0.05 mm), it constitutes a white layer.

Fig. 3. Microstructure of the gouged 6 mm thick steel with the plasma current intensity of 25 A, magnification 50x.

The structure of that hard-to-etch layer consists of low carbon martensite and iron nitrides [4, 5]. It is formed as result of high temperature influence of the air plasma on the top layer of the melted material and significant cooling of the liquid solution due to the intensive conduction of heat through the material core. In the analysis of the linear element distribution of a sample with revealed white layer (Fig. 4), it has been determined that formation of that layer is accompanied by an effect of its decarbonizing and nitriding.

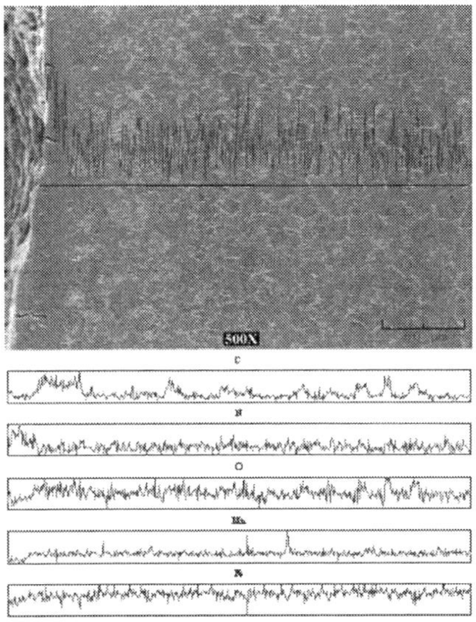

Fig. 4. Linear distribution of elements, C, N, O, Mn. Fe, on a cross section of a groove gouged in 6 mm thick steel with the current intensity of 15 A.

With small thickness of the melted layer (low value of the plasma current intensity), short time of the gouged material heating, as well as quick cooling, significantly influence the degree of homogenization of austenite in the heat affected zone. In this case, homogenous martensitic structure appears only in the zone located directly under the zone of melted material (Fig. 5). The thickness of that layer depends on the temperature which has to be high enough for diffusion homogenization of carbon in austenite. In the deeper layers of the heat affected zone, the time of austenitizing and carbon diffusion rate allow for austenite homogenization only in the grains of pearlite. At the same time, low carbon austenite is formed from ferrite. As result of cooling, martensitic structure arises, with visible boundaries between the phases of former ferrite and pearlite. With larger depth and decreasing cooling rate in the heat affected zone, bainite and ferrite have been found in addition to martensite. In the subsequent layer of the heat affected zone, just before the core structure is reached, local presence of cooled pearlite can be found.

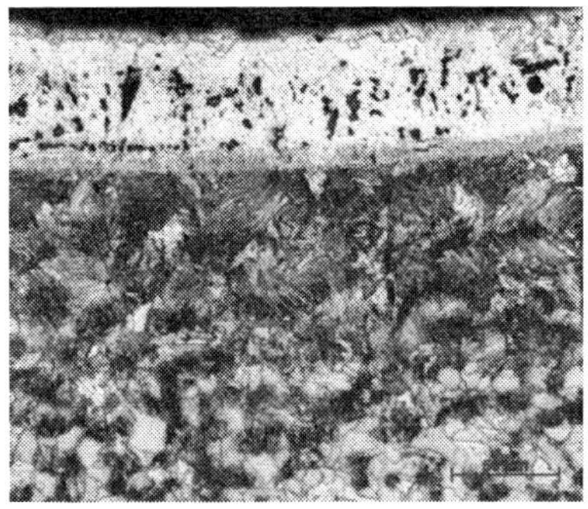

Fig. 5. Microstructure of the gouged 6 mm thick steel with the plasma current intensity of 15A, magnification 500x.

When the melted metal zone is thicker (Fig. 3), its structure differs from that of a thin layer. Fig. 6 shows the results of an analysis of the linear distribution of elements in a sample with revealed thick zone of melted metal. Linear analysis (Fig. 6) shows that the thick zone of melted metal consists of two layers: outer one, with elevated nitrogen concentration and inner one, with elevated concentration of carbon. Probably, the elevated carbon concentration is a result of high processing temperature with the initial crystallizing outer layer of metal make good conditions of carbon diffusion from the deeper layer of the material to the inner layer of the melting zone. Analysis of the structures of that zone shows that its outer layer is low carbon strip martensite with nitrides (Fig 7a – top part); the inner layer of that zone is bainite with high dispersion (Fig. 7b).

It has also been determined that the melted zone shows significant porosity resulting from high nitrogen saturation. Due to the high cooling rate, nitrogen (N2) is trapped in the solidified metal during solidification [11]. It has been determined that the thicker melted zone the stronger is the phenomenon. The structure of the heat affected zone of samples with thick melted zone is similar to that of samples with thin melted zone. However, the thinner melted zone (higher plasma current intensity) the higher austenitizing temperature and longer its time, that is why austenite grows. Due to that, the beginning of the heat affected zone of the samples with thick melted zone has more coarse-grained structure than that of the samples with less thickness of the e. melted zone.

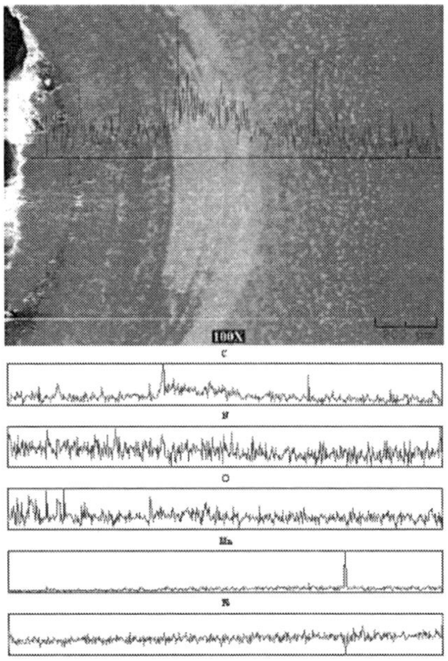

Fig. 6. Linear distribution of elements, C, N, O, Mn. Fe, on a cross section of a groove gouged in 6 mm thick steel with the current intensity of 25 A.

a)
b)

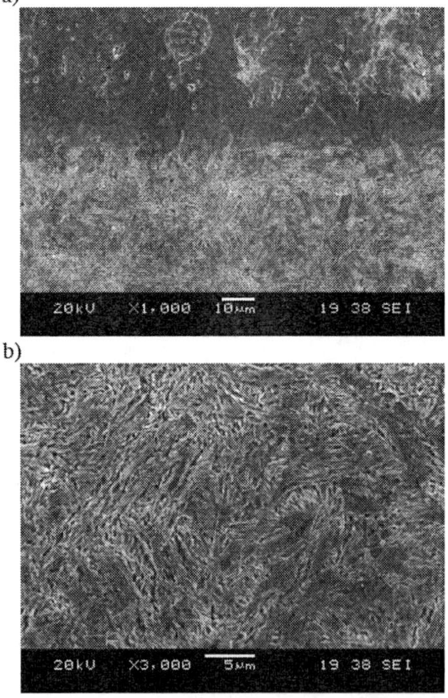

Fig. 7. Microstructure of the melted zone of 6 mm thick steel gouged with the plasma current intensity of 25A: a) boundary of the outer and inner layer, magnification 1000 x; b) bainite of the inner layer of the melted zone, magnification 3000x.

3.2 Measurement of hardness

The purpose of hardness measurement was to determine the influence of the plasma current intensity on the changes of hardness in the heat affected zones of the gouged sheets with various thicknesses. On the groove cross section, from its bottom into the material, two rows of indentations have been made. The indentations have been made alternately at intervals of 50 μm. The values of hardness changes of all the nine samples are shown in Table 4. Hardness measurements are only those of the heat affected zone because no repeatable results could be obtained in the melted zones (porosity and brittleness).

Table 4. HV0.1 hardness distribution in the cross section of the gouged samples.

SWC, µm	No. samples								
	1	2	3	4	5	6	7	8	9
	HV0.1								
0*	371	525	609	450	483	490	519	558	714
50	334	313	411	460	434	474	513	474	507
100	281	310	344	468	430	434	513	496	513
150	265	244	291	430	414	400	411	485	279
200	236	229	276	331	344	291	470	294	334
250	198	213	227	276	294	256	222	325	387
300	197	207	203	212	248	224	197	313	302
350	181	197	203	193	209	198	-	244	291
400	177	193	196	179	196	189	-	227	250
450	-	179	196	-	187	190	-	211	248
500	-	-	194	-	181	182	-	198	212
550	-	-	191	-	-	-	-	-	209
600	-	-	189	-	-	-	-	-	200
650	-	-	189	-	-	-	-	-	-
700	-	-	182	-	-	-	-	-	-

*- first measurement located at the boundary of the melting zone and the heat affected zone (SWC).

Considering the hardness distribution results (Table 4), it can be stated that, for the individual sheet thicknesses, the highest values of hardness have been obtained with the plasma current intensity of 25A. The highest value, 714HV, has been recorded in sample (9) with the thickness of 16 mm. In this case, high temperature of processing facilitates the processes of carbon diffusion in austenite while large cross section of the sheet ensures quick cooling. However, it has to be noted that those maximum values go rapidly down with larger depth. The maximum size of the heat affected zone, 700 μm, has been found in sample (3) with the least thickness and with the highest plasma current intensity. This is a result of high temperature of processing with relatively slow cooling. Similarly, the lowest size of the heat affected zone has been found in sample 7 – the largest sheet thickness and the lowest value of the plasma current intensity. The influence of plasma current intensity on hardness changes in the heat affected zone has been shown in Fig. 8. The shape of the hardness distribution curves shows that increase of the plasma current intensity (with the same sheet thickness) causes extension of the heat affected zone and increase of the hardness values.

The thickness of plasma gouged sheet has also a significant influence on the hardness and size of the heat affected zone (Fig. 9). The recorded changes of hardness in the samples under examination, gouged with identical plasma current intensity, but with different thicknesses (Table 4) result from various rates of heat conduction into the material.

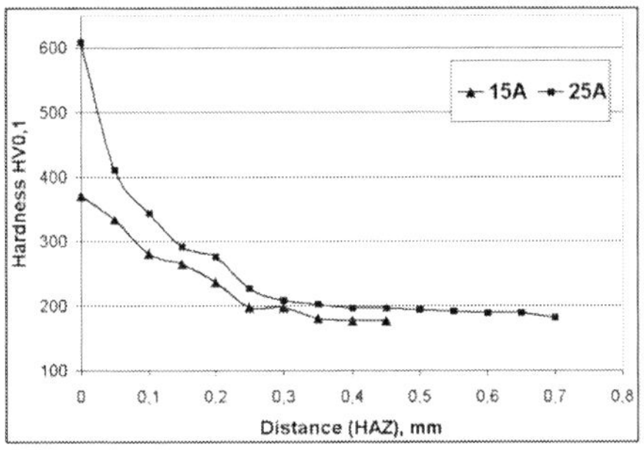

Fig. 8. The influence of the plasma current intensity on the hardness of heat affected zone in a 6 mm thick sheet.

The thickness of plasma gouged sheet has also a significant influence on the hardness and size of the heat affected zone (Fig. 9). The recorded changes of hardness in the samples under examination, gouged with identical plasma current intensity, but with different thicknesses (Table 4) result from various rates of heat conduction into the material.

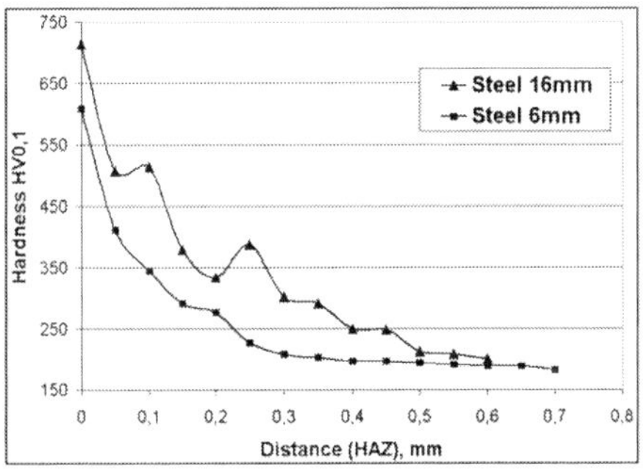

Fig. 9. The influence of the thickness of a sheet gouged with the plasma current intensity of 25A on the hardness of the heat affected zone.

With a bigger cross section of the gouged sheet (Fig. 9, curve – Steel 16 mm), heat conduction is faster. Faster heat conduction results raising of the hardness value and reduction of the heat affected zone.

4 Conclusions

Basing on the performed tests, it has been stated that, in the process of gouging with the use of air plasma, the value of the plasma current intensity and the thickness of the material being processed significantly influence the properties of the top layer and the size of the executed grooves.

Analysis of the test results presented in the paper has made it possible to draw the following conclusions:

- with the increase of the sheet thickness, with the given plasma current intensity, the depth of the gouged grooves decreases. This results in reduction of the heat affected zone and growth of its hardness;
- increase of the plasma current intensity results in increasing the depth of gouging. This is a result of higher processing temperature which also facilitates the carbon diffusion processes and, consequently, causes increase of the heat affected zone hardness and size;
- increase of the plasma current intensity and gouging depth influences the growth of the melted zone and its structure. Thin melted zones (up to about 0.05 mm) consist of low carbon martensite with iron nitrides. This structure is formed as result of quick cooling of decarburized but nitrogen saturated layer of metal. Thicker melted zones are characterized by laminar structure. In the outer layer, the structure of low carbon martensite with nitrides appears. In the inner layer, one can find bainite with higher carbon content arising due to slower cooling of that layer;
- the heat affected zones show structure gradually changing towards the material depth. Starting from the boundary of the melted zone, the following structures occur: martensite, martensite with bainite, and cooled pearlite and with secretions of ferrite, as well as ferrite with cooled pearlite. The depth of the heat affected zone and the width of the individual structural layers depends on the sheet thickness and the plasma current intensity.

References

1. S. Zaborski, T. Stechnij, Inżynieria Maszyn, E **16**, 4, (2011)
2. M.S. Węglowski, T. Pfeifer, Biuletyn Instytutu Spawalnictwa, **1**, (2014)
3. E. Tasak, Przegląd Spawalnictwa, **2–3**, (1982)
4. T. Pfeifer, Budowa strukturalna i własności SWC elementów wycinanych plazmowo ze stali o wysokiej wytrzymałości - Part I. Stal, 7–8, (2013)
5. T. Pfeifer, Budowa strukturalna i własności SWC elementów wycinanych plazmowo ze stali o wysokiej wytrzymałości - Part II. Stal, 9–10, (2013)
6. J. Górka, R. Skiba, Przegląd spawalnictwa, **2**, (2013)
7. R. Pakos, Przegląd spawalnictwa, 8, (2012)
8. D.J. Thomas, Int. J. Steel Struct. **11**, 4, (2011)
9. R.W. Maruda, G.M. Krolczyk, M. Michalski, P. Nieslony, S. Wojciechowski, J. Mater. Eng. Perform. **26**, 1, (2017)
10. T. Hamada, T. Kato, K. Aoyama, T. Aikawa, S. Asai, T. Kurokawa, H. Takebayashi, Weld. World **58**, (2014)
11. H. Nohara, T. Jpn. Soc. Mech. Eng. **78**, (2012)

Computational compliance criteria in water hammer modelling

Kamil Urbanowicz[1,*]

[1]West Pomeranian University of Technology, Szczecin, Faculty of Mechanical Engineering and Mechatronics, Al. Piastów 17, 70-310 Szczecin, Poland

Abstract. Among many numerical methods (finite: difference, element, volume etc.) used to solve the system of partial differential equations describing unsteady pipe flow, the method of characteristics (MOC) is most appreciated. With its help, it is possible to examine the effect of numerical discretisation carried over the pipe length. It was noticed, based on the tests performed in this study, that convergence of the calculation results occurred on a rectangular grid with the division of each pipe of the analysed system into at least 10 elements. Therefore, it is advisable to introduce computational compliance criteria (CCC), which will be responsible for optimal discretisation of the examined system. The results of this study, based on the assumption of various values of the Courant-Friedrichs-Levy (CFL) number, indicate also that the CFL number should be equal to one for optimum computational results. Application of the CCC criterion to own written and commercial computer programmes based on the method of characteristics will guarantee fast simulations and the necessary computational coherence.

1 Introduction

When water flow is started or stopped faster than the system can respond acoustically, a pressure surge commonly referred to as a water hammer event will occur. The same principle is also applicable when the flowing medium is another fluid (oil, steam etc., in the case of steam, the pressure surge is known as a steam hammer event). In the mentioned transient state, water hammer waves – also called pressure waves – are transmitted through the system at the speed of sound of the water-pipe combination. In typical steel pipes, the pressure wave speed is greater than 1200 m/s. This pressure waves can cause major problems, from noise and vibration to pipe collapse and total system failure.

This is the reason why the protection of water systems or nuclear power plants cooling systems against water hammer is so important. Designers of such a pressurised systems normally use expensive additional equipment to protect them against water hammer damage: hydro-pneumatic or surge tanks, special valves, air chambers, rapture disks [1, 2]. But the simplest, effective and nearly cost free protection is to control the valve-closing time and the pump speed. For optimal control procedures at the stage of designing such systems, a numerical simulation analysis is needed [3, 4].

It is assumed that the flow of this type is being described by the hyperbolic system of partial differential equations (sequentially by the continuity equation and the motion equation) [1-6]:

$$\frac{\partial p}{\partial t} + \rho c^2 \frac{\partial v}{\partial x} = 0, \tag{1}$$

$$\frac{\partial p}{\partial x} + \rho \frac{\partial v}{\partial t} + \rho g sin\gamma + \frac{2}{R}\tau_w = 0, \tag{2}$$

where:

p – pressure [Pa], t – time [s], ρ – density [kg/m^3], c - pressure wave speed [m/s], v - flow velocity [m/s], x - axis coordinate [m], g - acceleration due to gravity [m/s^2], γ - pipe tilt [°], R - inner radius [m], τ_w - wall shear stress [Pa].

Direct solutions of this equations are not known at present, therefore the method of characteristics (MOC) is used most often which, when initial and boundary conditions are properly selected, allows numerical solution of these equations. The MOC technique is now widely known but yet is not entirely clarified, which will be shown by the tests performed in this study.

Like any numerical method, it must meet the necessary requirements that will ensure the stability and adequate accuracy (compliance) of the results being obtained. The literature [1, 2, 5, 6] presents the stability criterion in the form of:

$$CFL = \frac{c\Delta t}{\Delta x} \le 1, \tag{3}$$

where:

Δt – time step [s], Δx – single reach length (horizontal distance between neighbouring grid nodes) [m].

The exemplary simulations performed in this study indicate that this criterion should be rather written in the form of equality. As for the compliance criterion, it is difficult to find in the current literature on this subject any condition which would have to be satisfied to maintain necessary simulation accuracy. As shown by the results of the comparative tests performed and

Corresponding author: kamil.urbanowicz@zut.edu.pl

DOI: 10.1051/e3sconf/20171903021

discussed in this paper, an additional criterion should be associated with compaction on the length of pipe of the numerical grid of the characteristics.

The objective of this study is therefore to answer the question on what division should be considered appropriate that the simulation results being observed are characterised by an acceptable compliance with those obtained by experimental tests.

2 Computational compliance criteria

2.1 Difference between stability and compliance of results

Stability is one of the most important properties of dynamic systems. In the case of unsteady flow, especially during the water hammer event examined in this study, the pipe flow which after brining it from the steady state tends over time to another steady state, being characterised by another flow velocity, is considered the stable one. The concept of stability of the unsteady flow caused by rapid valve closure can also be defined by setting relevant requirements for the courses of pressure changes (as well as for the courses of average velocity value) in the examined cross-sections. Nevertheless, the pressure changes that take place in the systems where during water hammer events no cavitation caverns are formed should be always accompanied with time by a gradual decrease in maximum pressures at successive amplitudes. However, in the pressure runs with cavitation, maximum pressures at the second or the third amplitude (that occur after the first column separation) can not be higher than twice the value being determined using the Joukowsky equation. The said oversized increase in pressure is being explained by the fact of superposition of a greater number of pressure waves [7], because any closure of the cavitation area is a source of the next wave. Superposition of waves results in the fact that the stability of exponential maximum pressure envelope, typical for the pressure runs without cavitation, is not observed in the cavitation courses. Also in the case of pipe flows with cavitation, there is a typical pressure drop at the time of flow deceleration after the occurrence of maximum amplitude.

Within the problem being analysed in this study, the stability of numerical solutions of the differential equations that describe the water hammer event should be mentioned as well. And here, the stability of solution should be understood as continuous dependence of the solution on set initial conditions.

In this connection, the stability criterion is not responsible for the accuracy of modelling but only for its correctness in the mathematical sense. Following the analysis of the method of characteristics (which is used to numerically solve the partial equations that describe the water hammer), as well as the results being obtained themselves, it is reasonable therefore to introduce another criterion, the objective of which would be to guarantee compliance between simulated results and those being observed experimentally.

2.2 Purpose and scope of the study

Unfortunately, the studies that used the method of characteristics to model the unsteady pipe flows have not indicated at what minimum division the results being obtained will be characterised by an acceptable error. Simpson and Bergant [8] modelled the pipe flows with cavitation assuming different computational elements (Table 1). They observed, when applying the division into 32 elements and more, that there are significant deviations and a significant variation in the results being obtained. This fact is explained by them that the greater the number of divisions into grid reach lengths, the more cavitation areas is modelled and, at the same time, the greater is the likelihood of a superposition of waves. They suggested that the number of reaches (grid points) over the pipe length (N) is to be chosen so that the ratio of the maximum cavitation volume being modelled to the volume of fluid contained in a single numerical element is to reach the value below 10%. Interestingly, they did not notice any modelling problems when applying division into 4 or 8 elements. In 2001, in another study – this time on the pipe flows without cavitation-related column separation, Bergant et al. [9] applied the division only into N = 8, 16 and 32 elements and showed that different number N on the diamond (staggered) grid, using the friction model according to Brunone, does not affect the quality of simulated pressure runs (Fig. 1).

Table 1. Discretisation and cavitation and friction models applied in other studies.

Year [Paper]	Author's initials	Grid	Friction model	Number of reaches N	Cavitation model
1994 [8]	S.A.R. – B.A.	R; S	Q-S	4,8,16,32, 64,128,256	SM; DVCM, DGCM
1996 [10]	B.M. – A.A.*	R	BM	3,5,8,10,12 14,16,20	DVCM
2001 [9]	B.A. et al.	S	CT (Z); BM	8,16,32	No cavitation
2006 [11]	A.A. – L.M.	S	Q-S; CT (Z; V-B); BM	12,24,36	No cavitation
2007 [13]	H-T.L. et al.	FD L-W	Q-S	10,20,40	DBCM
2009 [14]	H-T.L.- H-T.E.	FD L-W	Q-S	6,12,24,40	DBCM
2012 [12]	A.A. – L.M.	S	Q-S CT (Z; V-B)	17,33,65	DACM
2014 [15]	K.U. et al.	S	CT (Z; V-B; V)	54,108,216 432, 864	DGCM
2015 [16]	B.A. – K.U.	S	CT (Z; V-B; V)	32, 64, 96	DGCM

Legend:
Initials: S.A.R.-Simpson Angus R.; B.A.-Bergant Anton; A.A.-Adamkowski Adam; L.M.-Lewandowski Mariusz; H-T.L-Hadj-Taieb Lamljed; H-T.E-Hadj-Taieb Ezzeddine; K.U.-Karadzic Uros; B.M.-Bughazem Mohamed; A.A.*-Anderson Alexander
Grid: R-rectangular; S-staggered (diamond); FD L-W – Finite difference Lax- Wendroff scheme
Friction models: Q-S – quasi-steady; BM – Brunone's model; CT – Convolution type; Z – Zielke model; V-B – Vardy-Brown model; V – Vítkovský model
Cavity models: SM – Streeter model; DVCM – Discrete vapour cavitation model; DBCM – Discrete bubbly cavitation model; DGCM – Discrete gas cavitation model; DACM – Discrete Adamkowski cavitation model

DOI: 10.1051/e3sconf/20171903021

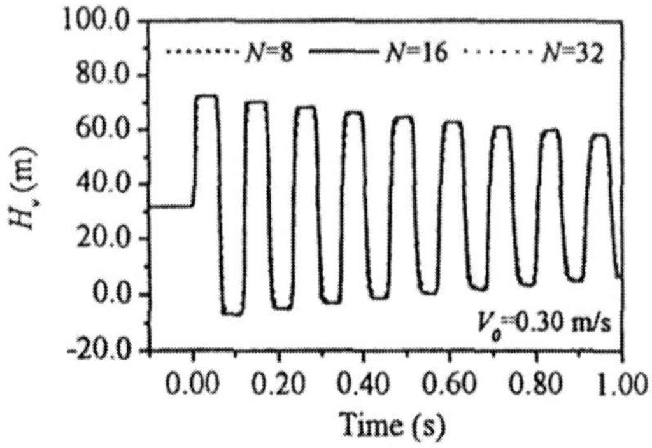

Fig. 1. Bergant et al.'s results for different N [9].

Bughazem and Anderson [10] pointed to a significant effect of N on simulation results based on a series of tests performed on the rectangular grid of the characteristics (they used the friction model according to Brunone). Adamkowski and Lewandowski, when examining the effect of friction on simulation results of the pressure runs without cavitation [11], observed no effect of the grid density on the simulation compliance and the stability of numerical computations when applying the division into N = 12, 24 and 36. In their next comparative study on the pipe flows with cavitation [12], the same authors showed that the division of N, however, affects the time of modelling the transition from the course with cavitation into the course without it (Fig. 2).

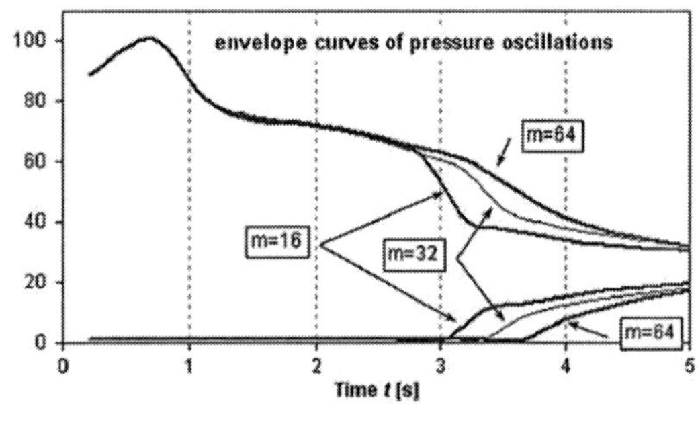

Fig. 2. Envelope of the analysed pressure curve obtained for different N [12].

The significant effect of N on simulation results (Fig. 3, Fig. 4) has been also observed by Hadj-Taieb et al. [13, 14], but these authors did not use the method of characteristics and solved the basic equations with the finite-difference method (Lax-Wendroff method). However, while analysing the results obtained by Karadžić et al. [15], it is clear that also for the relatively dense division N = 54, 108 and 216 significant difference between the simulated results can be seen (Fig. 5).

In other studies, their authors often forget to specify which numerical discretisation was used to perform comparative tests [17-20] or they only show the applied number of divisions without commenting its effect on the results being obtained [21-23].

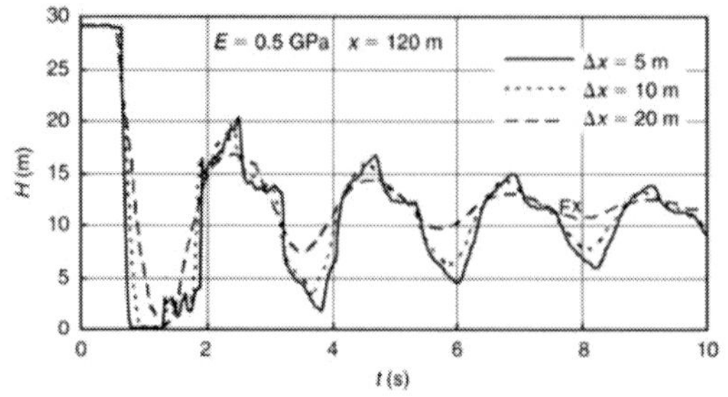

Fig. 3. Hadj-Taieb's et al. 2007 results [13].

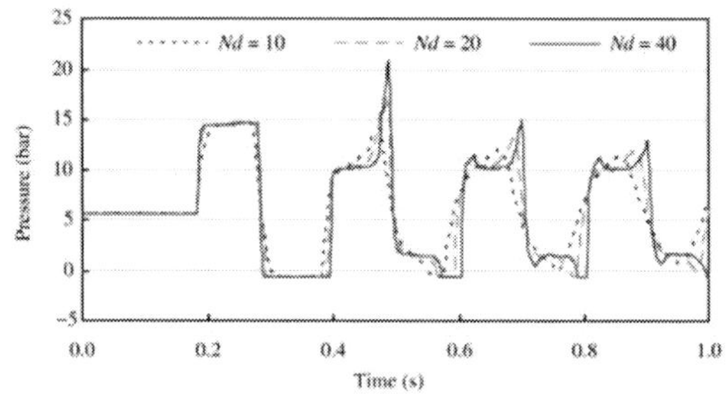

Fig. 4. Hadj-Taieb's 2009 results [14].

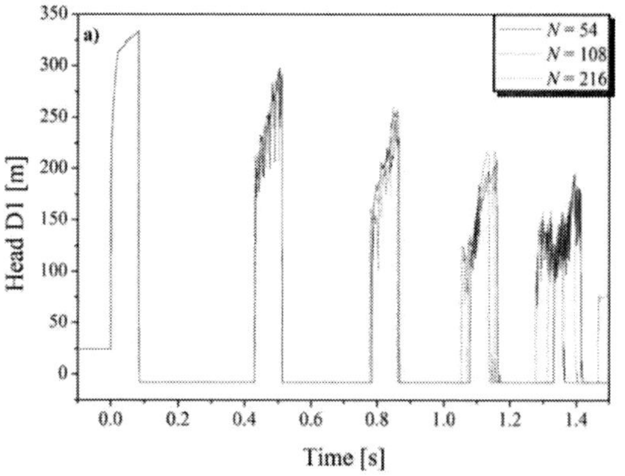

Fig. 5. Variation of results for large N [15].

Summing up, the following three important conclusions can be drawn from the studies being analysed:

– no significant effect of N on simulation results is being noticed in the pressure runs without cavitation [9, 11];

– the effect of N seems to be much larger in the pressure runs with cavitation, although this does not refer to the modelling of the first amplitudes being characterised by maximum pressures but rather to the time of transition of the courses with cavitation-related column separation into the courses without cavitation [8, 12, 15];

– the method of characteristics is less sensitive to the change of N factor than other methods (for example, the finite-difference method – the studies by Hadj-Tajeb [13, 14]).

On the basis of the literature being reviewed above, the objective and scope of the study could be defined.

The purpose of tests performed and described in this study was to determine the optimum value (number) of the numerical division of hydraulic pipe (of the sample length L) into computational elements (marked in this study as N). This division should be as small as possible so that the computer calculations that use the method of characteristics are to be made as fast as possible.

Owing to the fact that the rectangular grid of the characteristics is commonly used and the effect of N on them was analysed according to Table 1 only in two studies, the tests being described in the next section will be carried out just on it.

Some authors, including Vítkovský [24-26], have suggested that the rectangular grid is the result of the overlapping of two diamond grids (staggered-diamond grid) and, according to them, this affects simulation results. Own experiments presented in this study as well as in previous ones [17, 19, 22] do not show, however, any problem associated with the rectangular grid. Commonly, a sharp toothed increase in pressure is being observed on the peak of the first pressure amplitude on the rectangular grid. This drawback can be minimised, however, by showing on the graph illustrating the variation of pressure in a given pipe cross-section being analysed every second simulated result. Then, a smoothed simulation waveform will be obtained, very similar to the results being obtained on the diamond grid.

The scope of the study will refer to the comparative analysis of a number of curses being modelled (with and without cavitation; laminar, transient and turbulent: $1100<Re_i<40350$) with the experimental ones being obtained in a typical hydraulic system.

2.3 Effect of number of reaches on simulation results

In the tests presented below, the hydraulic pipe being examined was divided into N computational elements, starting from N = 2, every 1, to N = 64. Carrying out computer simulation in such a wide range guaranteed an accurate picture of the variation of two basic quantitative parameters: Et and Ep, being determined for each simulated pressure run. Parameters Ep (maximum pressure compliance) and Et (time compliance) characterise compatibility of the pressure runs being modelled [19]. They are calculated using the experimental and numerical results (Fig. 6) from the following formulas:

$$Ep = \frac{\sum_{i=1}^{k}\left|\frac{p_{is}-p_{ie}}{p_{ie}}\right|\cdot 100\%}{k}, \qquad (4)$$

$$Et = \frac{\sum_{i=1}^{k-1}\left|\frac{t_{is}-t_{ie}}{t_{ie}}\right|\cdot 100\%}{k-1}, \qquad (5)$$

where:

k - number of analysed pressure amplitudes; p_{is} and p_{ie} - maximum pressure on the i-th analysed amplitude from the simulated and experimental graph, respectively; t_{is} and t_{ie} - time of occurrence of the maximum pressure on the i-th analysed amplitude from the simulated and experimental graph, respectively.

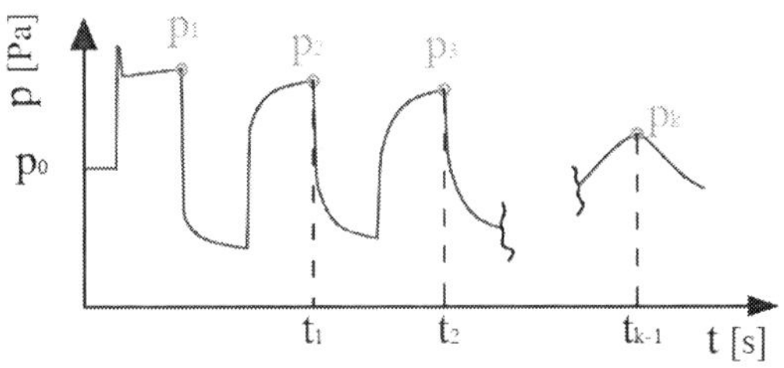

Fig. 6. Analysed pressure and times values in the quantitative analysis.

In his study, when determining parameter Et, the time compliance on the first amplitudes was omitted because the chosen experimental results used for comparisons were characterised by a noticeable effect of the vibrations of the valve being closed on the initial phase of pressure curve (fluid structure interaction (FSI) effect).

Experimental tests necessary for comparative analysis have been performed by Adamkowski and Lewandowski at the Institute of Fluid-Flow Machinery of the Polish Academy of Sciences in Gdańsk. In their paper of 2006 [11], these authors showed courses, in which cavitation-related column separation did not occur. From that study, nine courses were selected for the comparisons being made in this study (Table 2). However, in the study of 2012 [12], the same authors showed different results obtained on a slightly modified test bench, in which cavitation-related column separation occurred. In the comparative tests presented below, six courses obtained then were used (Table 3). The details necessary for numerical modelling are compiled in the presented tables (Table 2 and 3). The values of pressure wave propagation speed c, which were used in computations, collected in the Tables 2 and 3 were not calculated from theoretical equations but determined by the analysis of experimental pressure runs.

Table 2. Experimental details for cases without cavitation.

NO CAVITATION (L = 98.11 m; D = 0.016 m; ρ = 997.65 kg/m³; ν = 9.493·10⁻⁷ m²/s; k = 16)		
v_0 [m/s]	c [m/s]	p_T [Pa]
0.066	1305	$1.2647\cdot10^6$
0.162	1304	$1.2642\cdot10^6$
0.340	1304	$1.265\cdot10^6$
0.467	1308	$1.253\cdot10^6$
0.559	1306	$1.264\cdot10^6$
0.631	1307	$1.2641\cdot10^6$
0.705	1305	$1.263\cdot10^6$
0.806	1306	$1.263\cdot10^6$
0.940	1304	$1.264\cdot10^6$

DOI: 10.1051/e3sconf/20171903021

Owing to the fact that this study examined the effect of N on the pressure runs with cavitation, it was necessary to implement additional conditions for determination of the volumes of emerging cavitation areas in the computer programme being used to solve the basic equations describing the unsteady flow of that type. As a model of cavitation-related damage of the flow continuity, the model according to Adamkowski-Lewandowski was used, which has been described in detail in the paper of 2009 [27].

Table 3. Experimental details for cases with cavitation.

WITH CAVITATION (L = 98.56 m; D = 0.016 m; ρ = 1000 kg/m³; ν = 1.09·10⁻⁶ m²/s)			
v_0 [m/s]	c [m/s]	p_T [Pa]	k
0.47	1260	$2.44\cdot10^5$	13
0.82	1250	$2.42\cdot10^5$	12
1.03	1230	$2.43\cdot10^5$	11
1.7	1240	$4.23\cdot10^5$	12
1.9	1240	$4.23\cdot10^5$	11
2.74	1260	$7.03\cdot10^5$	13

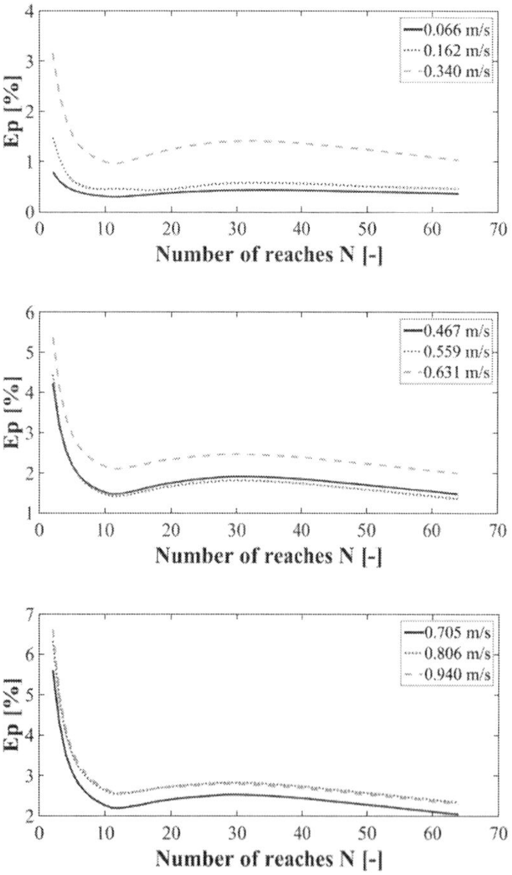

Fig. 7. Results for Ep (cases without cavitation).

The Table 2 shows that the average value of pressure near the reservoir p_T was $1.263\cdot10^6$ [Pa] and that the differences observed in the speed of pressure wave propagation (1304-1308 m/s) were slight (from the average value of 1305.44 m/s, the error was only ± 0.2%). For the tests with cavitation-related column separation (Table 3), it can be concluded that the differences being observed experimentally in the speed of pressure wave propagation were significant in relation to those without cavitation (from the average value of 1246.667 m/s, the error was ± 1.36%). In the tests being performed, dispersion of the values of pressure wave propagation speed c was as much as 30 m/s, suggesting that there are bubbles of air in the flowing fluid in the chosen pressure runs, in which the propagation speed was the lowest because this is mainly them that decrease the speed of pressure wave propagation.

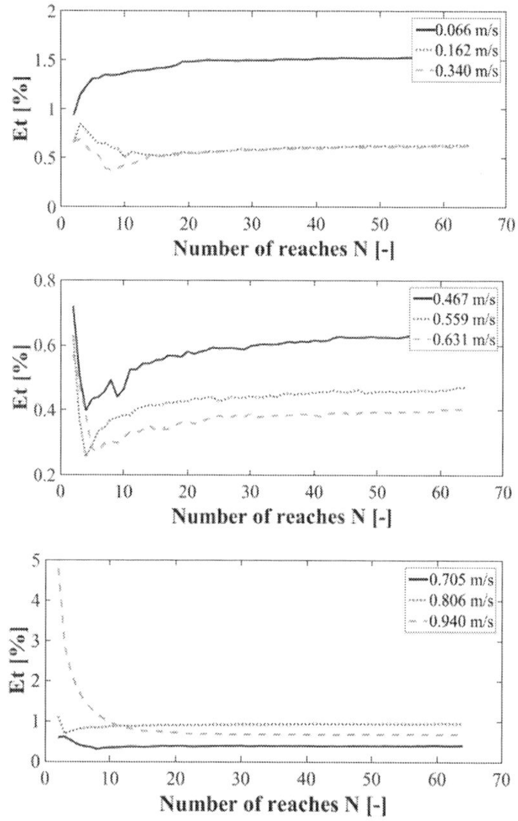

Fig. 8. Results for Et (cases without cavitation).

The detailed results of the variation of parameters Ep and Et in the function of changes in the grid density N of the characteristics are shown by two presented graphs (Fig. 7 and Fig. 8). In the comparative tests, the time and the amplitude compliances of these pressure amplitudes which occurred in first five seconds from the valve closure were analysed. In the courses without cavitation, the number of the analysed amplitudes was the same for all carried out runs, i.e. k=16, whereas in the courses with cavitation their number varied depending on the boundary and initial conditions (for k values see Table 3).

On the other hand, the next two graphs show the variation of the same parameters but for the courses in which column separation occurred, i.e. there were cavitation areas appearing (Fig. 9 and Fig. 10).

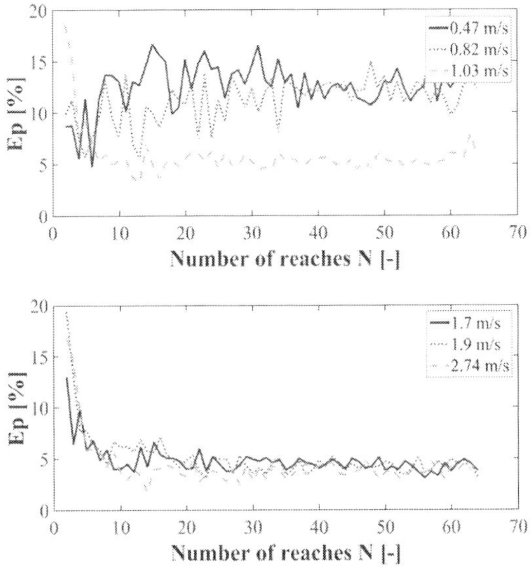

Fig. 9. Results for Ep (cases with cavitation).

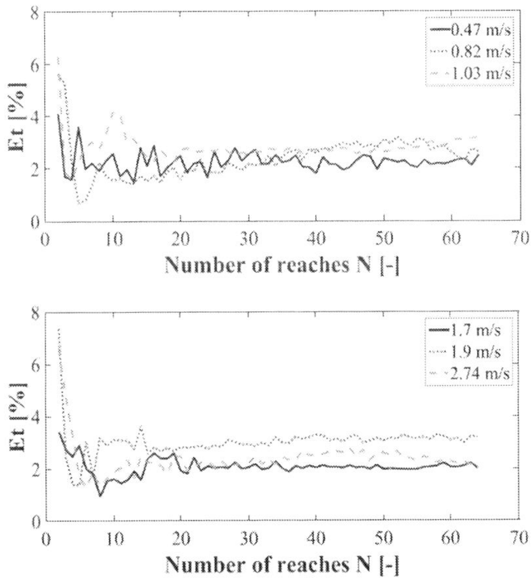

Fig. 10. Results for Et (cases with cavitation).

From the results of quantitative tests shown in a graphical form (Fig. 7 to Fig. 10), the following conclusions can be drawn:

- in the pressure runs without cavitation, the acceptable compliance of simulations with the experimental results was obtained for the division into N = 10 (Fig. 7 and Fig. 8). Analysis of the variation of parameter Ep (Fig. 7) shows that this error will be smaller for N > 60. On the other hand, analysis of the variation of parameter Et (Fig. 8) showed that with the increasing N in five runs (v_0 = 0.162; 0.34; 0.467; 0.559; 0.631) there was a slight but steady increase in the error. However, it can be considered that the Et results being obtained for N = 10 were characterised by acceptable values;
- analysis of the results obtained for courses with cavitation (Fig. 9 and Fig. 10) showed that in these runs the results can be accepted if division of the pipe is assumed at 10 elements (N = 10) although smaller dispersion of the results in these analysed runs is visible for N > 20;
- pressure runs with cavitation in the transitional range (v_0 = 0.47; 0.82) are characterised by unacceptable modelling errors Ep within the ranges of 8% to 16% and a high sensitivity to changes in parameter N (Fig. 9). This may suggest the need of further work on improving the applied cavitation model for the range of transitional pipe flows.

3 Effect of the Courant-Friedrichs-Lewy (CFL) number

The Courant-Friedrichs-Lewy (CFL) number expresses the ratio of the distance traveled by a disturbance in one time step to the length of a computational distance step. As pointed in many textbooks, the CFL number in the method of characteristics should be less than or equal to unity [1, 2, 5, 6] so as to ensure that the solution remains within the computational domain. However, the exemplary numerical simulations performed in this study, consisting in the application of the CFL number from the following range 0.5<CFL≤1 (the following values were adopted for the tests: CFL = 1, 0.95, 0.9, 0.85, 0.8, 0.7, 0.6 and 0.5) for the same simple system, showed that the application of lower value of the CFL number results only in deterioration of the modelling errors – higher values of parameters Ep and Et. The input values for the simulation being performed were assumes as follows: c = 1245 [m/s], ρ = 1000 [kg/m^3], ν = $1.09\cdot10^{-6}$ [m^2/s], N = 32 [-], L = 98.56 [m], R = 0.08 [m], v_0 = 1.9 [m/s], and $p_T = 4.23\cdot10^5$ [Pa]. The selected simulation results are presented in Figure 11.

It is clear from this graph that with the decreasing CFL number the next simulated pressure amplitudes appear faster. The transformed CFL condition L (Eq. 3), while assuming appropriate discretisation of the grid over the pipe length Δx and the pressure wave propagation speed c, has the following form:

$$\Delta t \leq \frac{\Delta x}{c} \Rightarrow \Delta t \leq \frac{L}{N\cdot c}, \text{ because: } \Delta x = L/N, \qquad (6)$$

where:
L – pipe length [m], N – number of reaches (grid nodes) [-].

It is clear from this formula that each reduction of time step (equivalent to adoption of lower CFL number) results in numerical calculations such as adoption of another length of the hydraulic pipe being examined. In order to avoid errors of the above type, the CFL condition in the case of using the rectangular grids of the characteristics with fixed computational mesh must be associated with the adoption CFL number equal to unity:

$$\frac{\Delta t \cdot c}{\Delta x} = 1. \qquad (7)$$

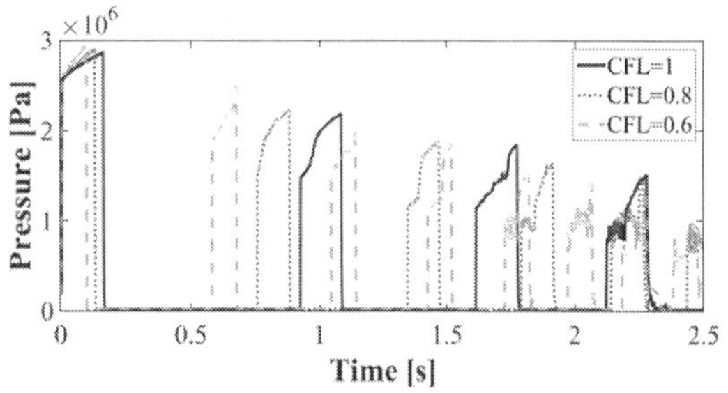

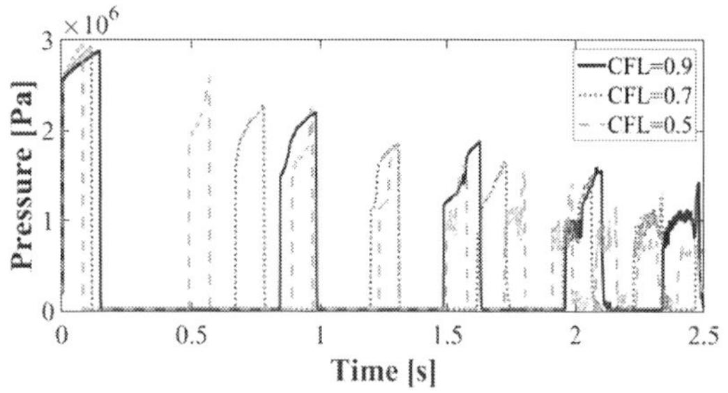

Fig. 11. CFL number effect.

4 Conclusions

A number of simulations performed in this study helped to determine a numerical criterion of the compliance of simulated results. The computation compliance criterion CCC enforces at the beginning of numerical calculations the necessary division of pipe into computational elements. According to the test results being observed in the method of characteristics for rectangular grids, the following condition must be satisfied:

$$N \geq 10.$$

An additional condition, which was demonstrated in section 3, is to maintain the stability criterion in the form of the Courant-Friedrichs-Lewy number equal to unity (CFL = 1).

In order to confirm the condition defined above in a wider range, further tests for pipes with other lengths, as well as other internal diameters, are necessary.

In addition, the performed tests clearly show that the unsteady pressure runs without cavitation are simulated with a greater accuracy; this may suggest that there is a need for modification of the cavitation model according to Adamkowski-Lewandowski so that the error for the Reynolds number, especially from the range of transitional flow (multiple transition from the laminar flow in the turbulent one during the pressure pulsation dumping is a phenomenon that accompanies the water hammer event) is also below the acceptable range of 5%.

References

1. E.B. Wylie, V.L. Streeter, L. Suo, *Fluid Transients in Systems* (Prentice Hall, 1993)
2. M.H. Chaudhry, *Applied Hydraulic Transients* (Springer, 3rd ed., 2014)
3. Z. Zarzycki, S. Kudźma, K. Urbanowicz, J. Theor. App. Mech-Pol. **49**, 135 (2011)
4. K. Urbanowicz, Z. Zarzycki, J. Theor. App. Mech-Pol. **50**, 487 (2012)
5. S. Mambretti, *Water Hammer Simulations* (WIT Press, 2014)
6. R.H. Pletcher, J.C. Tannehill, D. Anderson, *Computational Fluid Mechanics and Heat Transfer* (Taylor and Francis, 3rd ed., 2013)
7. A.R. Simpson, E.B. Wylie, J. Hydraul. Eng. **117**, 1310 (1991)
8. A.R. Simpson, A. Bergant, J. Hydraul. Eng. **120**, 361 (1994)
9. A. Bergant, A.R. Simpson, J.P. Vítkovský, J. Hydraul. Res. **39**, 249 (2001)
10. M.B. Bughazem, A. Anderson, *Proc. 7th Int. Conf. Pressure Surges, BHR Group, Harrogate, England*, 537 (1996)
11. A. Adamkowski, M. Lewandowski, J. Fluids Eng. **128**, 1351 (2006)
12. A. Adamkowski, M. Lewandowski, J. Hydraul. Eng. **138**, 935 (2012)
13. L. Hadj-Taieb, E. Hadj-Taieb, C. Thirriot, Int. J. Simul. Model. **27**, 347 (2007)
14. L. Hadj-Taieb, E. Hadj-Taieb, Int. J. Simul. Model. **29**, 263 (2009)
15. U. Karadžić, V. Bulatović, A. Bergant, Stroj. Vestn. J. Mech. E. **60**, 742 (2014)
16. A. Bergant, U. Karadžić, *Proc. 12th Int. Conf. Pressure Surges, BHR Group, Dublin, Ireland*, 639 (2015)
17. K. Urbanowicz, Z. Zarzycki, S. Kudźma, J. Theor. App. Mech-Pol. **50**, 889 (2012)
18. A. Keramat, A.S. Tijsseling, Q. Hou, A. Ahmadi, J. Fluid Struct. **28**, 434 (2012)
19. K. Urbanowicz, Z. Zarzycki, J. Theor. App. Mech-Pol. **53**, 295 (2015)
20. S. Henclik, J. Theor. App. Mech-Pol. **53**, 543 (2015)
21. K. Weinerowska-Bords, J. Fluids Eng. **137**, 054501 (2015)
22. K. Urbanowicz, M. Firkowski, Z. Zarzycki, J. Phys.: Conf. Ser. **760**, 012037 (2016)

DOI: 10.1051/e3sconf/20171903021

23. H. Wang, L. Zhou, D. Liu, B. Karney, P. Wang, L. Xia, J. Ma, C. Xu, J. Hydraul. Eng. **142**, 04016036 (2016)
24. J.P. Vítkovský, M. Lambert, A.R. Simpson, A. Bergant, *Proc. 8th Int. Conf. Pressure Surges, BHR Group, The Hague, Netherlands*, 471 (2000)
25. J.P. Vítkovský, M. Stephens, A. Bergant, M. Lambert, A.R. Simpson, *Proc. 9th Int. Conf. Pressure Surges, BHR Group, Chester, UK*, 405 (2004)
26. J.P. Vítkovský, M. Stephens, A. Bergant, A.R. Simson, M. Lambert, J. Hydraul. Eng. **132**, 709 (2006)
27. A. Adamkowski, M. Lewandowski, J. Fluids Eng. **131**, 071302 (2009)

Numerical analysis of the states of stress in the steel-titanium bimetal during cyclic bending

Grzegorz Robak[1,*] and *Artur* Cichański[2]

[1]Opole University of Technology, ul. Mikolajczyka 5, 45-270 Opole, Poland
[2]University of Science and Technology in Bydgoszcz, Al. prof. S. Kaliskiego 7, 85-796 Bydgoszcz, Poland

Abstract. This paper presents the results of experimental study into steel-titanium bimetal obtained by means of explosive welding, which involved the development of fatigue cracks. A particular emphasis in the report was placed on the analysis of development of the crack in the area in which the two metals are joined. The study was accompanied by a numerical analysis of the distribution of stresses in the weld area in terms of the elastic-plastic characteristics. The numerical analysis performed in terms of the elastic-plastic characteristics demonstrated that the titanium layer is completely plasticized as a result of the acting loading, when the crack is still located in the steel part of the examined plater.

1 Introduction

Composite materials are a common group of construction materials that allow to combine the properties of materials with significantly different strength properties and chemical com-position. There are many types of composite materials as well as manufacturing processes.

One of the ways to obtain a composite resulting from merging e.g. steel with titanium is explosion welding [1]. This process enables to combine materials, such as, e.g., steel-titanium, steel-zirconium or steel-aluminium, which cannot be combined using other methods or are very difficult to combine. During the detonation of the explosive placed on the cladding (e.g., titanium, zirconium), a detonation wave is created whose front moves continuously across the cladding. Plates collide at the front of the detonation wave, resulting in a lasting bond. Due to the high dynamics of detonation, the cladding is deformed to form a permanent bond with the substrate, usually in a corrugated shape. The shape of the bond, i.e., height and density of the wave, depends largely on detonation parameters and the thickness of the combined layers. Bimetals made in this way are used in process and electric power apparatus, such as heat ex-changers, reactors, columns, pipes, and others [2-6]. Due to the high speed of creating a bond between the metals during explosion bonding, the bonding zone has a heterogeneous structure [3, 7, 8] which can affect mechanical properties, in particular resistance to cracking.

It is necessary to conduct testing in order to determine the behaviour of these materials under certain loading conditions. It is also important to know the mechanisms associated with the growth of cracks, particularly in the bonding site.

This paper presents the results of experimental studies regarding the analysis of the growth of fatigue cracks in a bimetal made of two layers of steel (S355J2+N) and titanium (Grade 1), with particular attention to the behaviour of the growth of cracks within the bond site in tension – compression and cyclic bending tests.

2 Experimental research

Experimental studies included tension cyclic bending with a controlled displacement. The growth of cracks was recorded using a stereomicroscope with a maximum magnification of x45 connected to the camera. Specimen made of steel-titanium bimetal was used in the study, unilaterally notched in the steel part by spark erosion, shown in Fig. 1. Specimen was made of plated sheet obtained from the corner of the bimetallic plate. After the welding process, the sheet has been subjected to heat treatment by annealing at 600°C for 90 min, then cooled in the furnace to the temperature of 300°C at a rate of 100°C/h. Cooling in the open air was the final stage. Mechanical properties and chemical composition of bimetal components are shown in Tables 1 and 2.

Table 1. Mechanical properties of the bimetal.

Material	R_e (steel), $R_{0,2}$(titanium), MPa	R_m, MPa	E, GPa	ν
Steel S355J2+N	382-395	598-605	206-220	027-0.30
Titanium Grade 1	189-215	308-324	100-105	0.37-0.39

* Corresponding author: g.robak@po.opole.pl

 DOI: 10.1051/e3sconf/20171903022

Table 2. Chemical composition of the bimetal.

Material	Steel S355J2+N					
Chemical element	C	Si	Mn	P	S	Cu
Max. contents, % weight	0.22	0.55	1.60	0.025	0.025	0.45
Material	Titanium Grade 1					
Chemical element	C	Fe	H	N	O	Ti
Max. contents, % weight	0.10	0.20	0.015	0.03	0.18	99.50

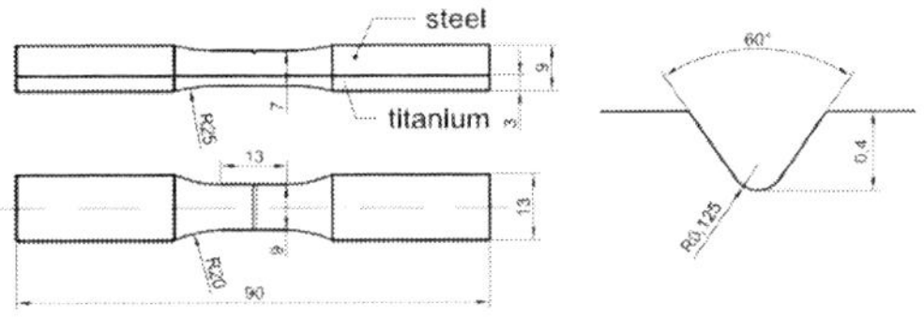

Fig. 1. Shape and dimensions of the notched specimen for the cyclic bending test.

As a result of explosion bonding, the bond between steel and titanium in the studied bi-metal takes an undulating shape with variable parameters [7]. Fig. 2 shows a diagram of the steel–titanium bond with indicated wave height in the specimen subjected to cyclic bending.

a)

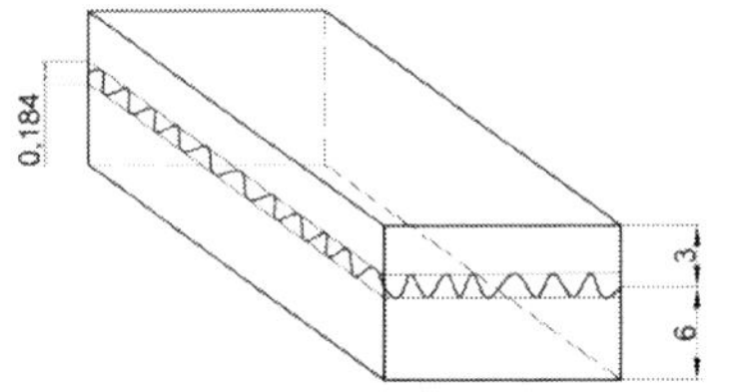

b)

Fig. 2. a) Diagram of the steel–titanium bond on a specimen subjected to cyclic bending (the height of the titanium layer: 3 mm, the height of the steel layer: 6 mm, the height of wave in the bond zone for the studied specimen 0.184 mm); b) Image (x30) of the fracture of the test specimen with visible bonding in the form of a wave.

Experimental studies were carried out for one types of load. Specimens were subjected to cyclic bending at a controlled displacement. The study was carried out at a fixed frequency f equal to 7 Hz. Displacement amplitudes Δl were taken at levels of Δl = 2.4 mm, Δl = 2.6 mm, Δl = 2.7 mm and Δl = 2.9 mm, with corresponding initial values of bending moments M_{b0} = 13.0 Nm, M_{b0} = 14.4 Nm, M_{b0} = 15.2 Nm and M_{b0} = 16.0 Nm. Since the force took place at a constant displacement Δl, the values of bending moments changed during crack propagation as a result of decreasing sectional area of specimens.

Fig. 3 shows the change in the values of bending moments as a function of the crack length a. Testing was interrupted when the crack after exceeding the bond zone reached the length of 1.5 mm. As the result of twice lower value of Young's modulus for titanium, the remaining section of the specimen showed much lower stiffness, which prevented further study. The dashed line in Fig. 3 is the boundary between the substrate (steel) and cladding (titanium).

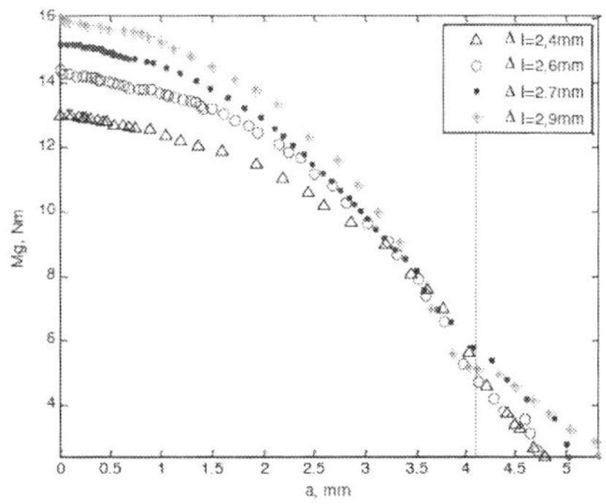

Fig. 3. The change of the values of bending moments in the specimens subjected to cyclic bending as a function of the crack length a.

3 Analysis of the experimental results

Fig. 4 shows the resulting graphs of the change of fatigue crack length a in function of the number of cycles N for specimens subjected to cyclic bending. The difference in the initiation and further propagation of cracks on both sides of specimens did not exceed 6%, so it was decided to measure the growth of cracks on only one side of specimens. Note that the growth of cracks in the bond zone of two materials (dashed line, Fig. 4) followed without much difference despite changes in material properties and an increase in hardness of the material (titanium) in the bond zone [7].

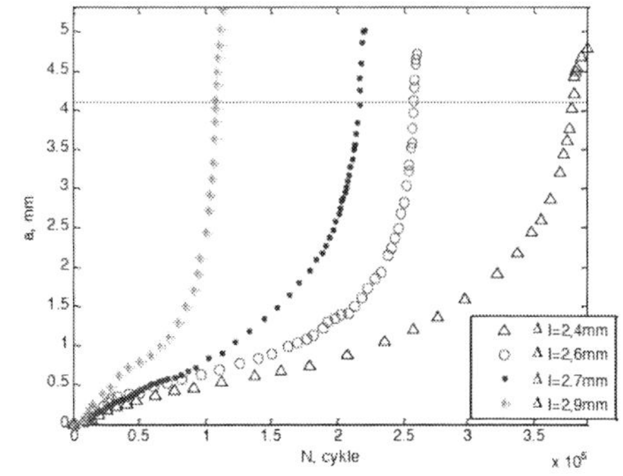

Fig. 4. Fatigue crack length a as a function of the number of cycles N for four displacement amplitudes Δl.

DOI: 10.1051/e3sconf/20171903022

A reduction in the propagation rate was observed in the last phase of the growth of the crack (a = 4.5-5.0 mm), which can be explained by formation of large zones of plastic deformations visible on the surface of the specimen and a significant decrease in the amplitude of the bending moment.

The growth of cracks took place in the direction perpendicular to the axis of the specimen (Fig. 5a-b). However, at a distance of 0.1 mm from the connection line, the crack changed the direction by about 30°, assuming the direction perpendicular to the surface of the wave forming the bond (Fig. 5c-d).

This observation is consistent with the well-known knowledge of the directions of propagation of fatigue cracks [9].

After crossing the bond within 0.15 mm, the crack again changed its direction to perpendicular to the axis of the specimen (Fig. 5e-f). Changing the direction of crack propagation was observed for each specimen irrespective of the load level, wherein the angle varied depending on the arrangement of the bond wave in a given specimen. Deflection in the direction of crack propagation may be due to changes in material properties within the bond zone.

a)

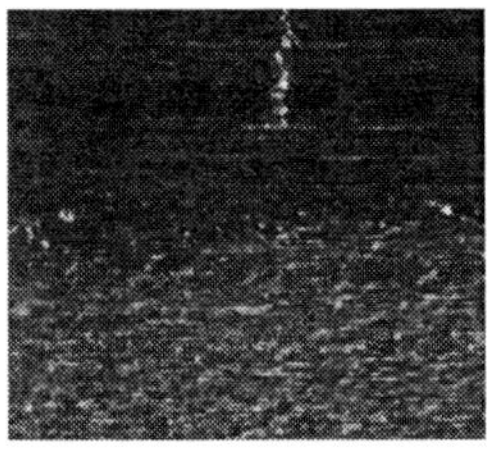

b)

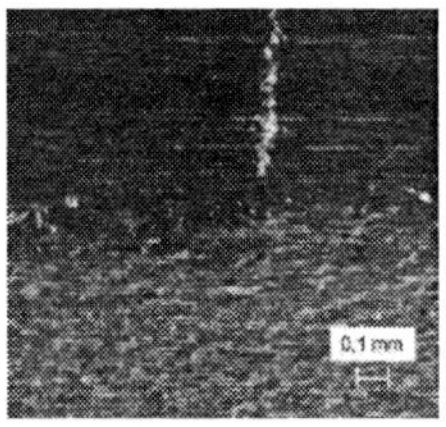

c)

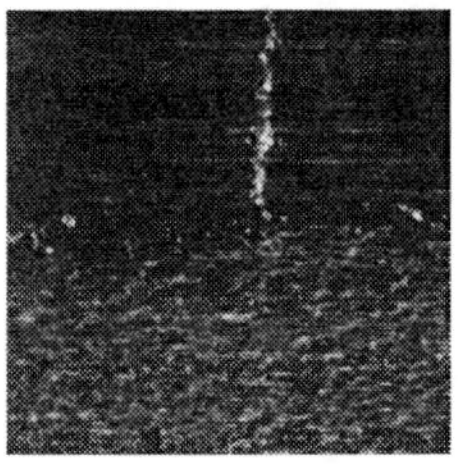

d)

e)

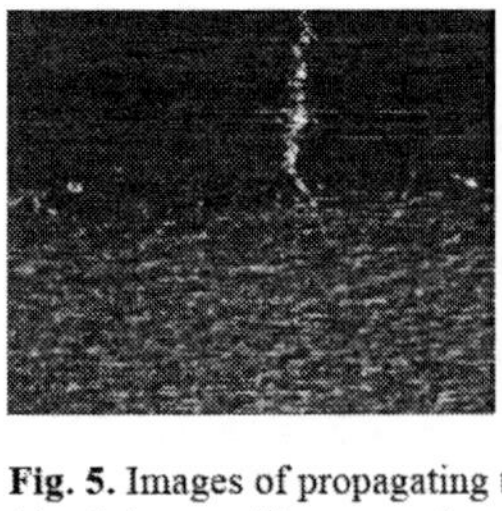

f)

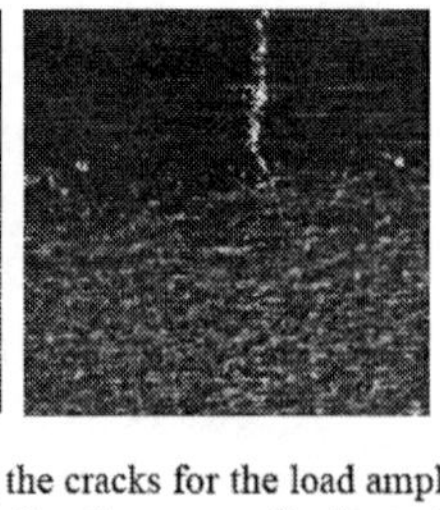

Fig. 5. Images of propagating the cracks for the load amplitude Δl = 2.4 mm a-b) propagation direction perpendicular to the axis of the specimen, c-d) a change in the direction of propagation, e-f) propagation of the crack in titanium.

Fig. 6 shows graphs of propagation velocity as a function of crack length for the four amplitudes of displacement Δl.

It can be noted in the graphs shown that the propagation rate increases with the increase of the displacement amplitude.

a)

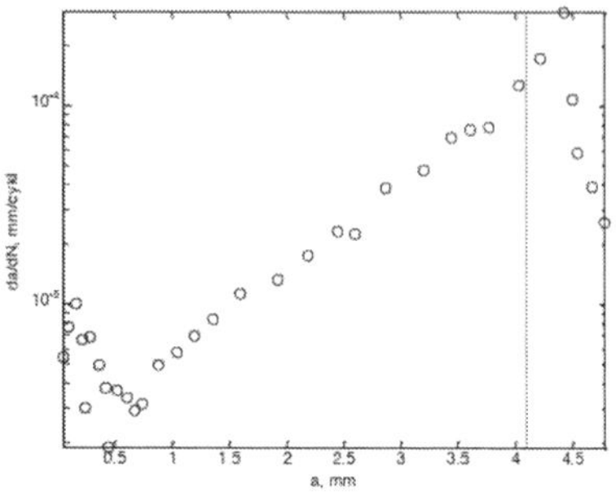

b)

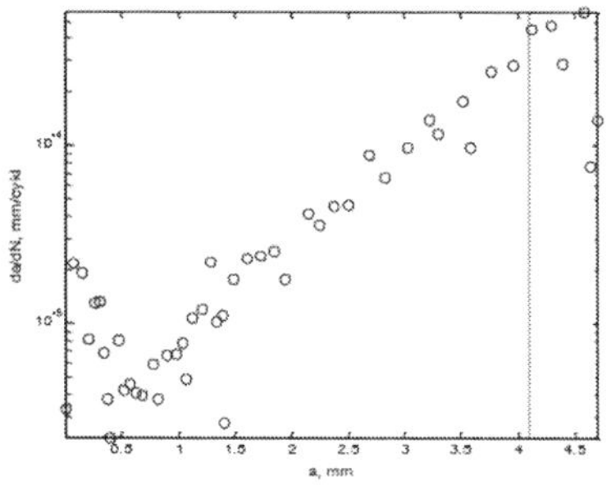

c)

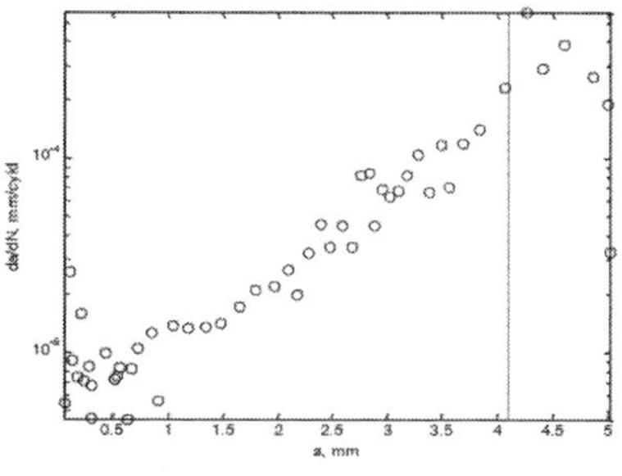

d)

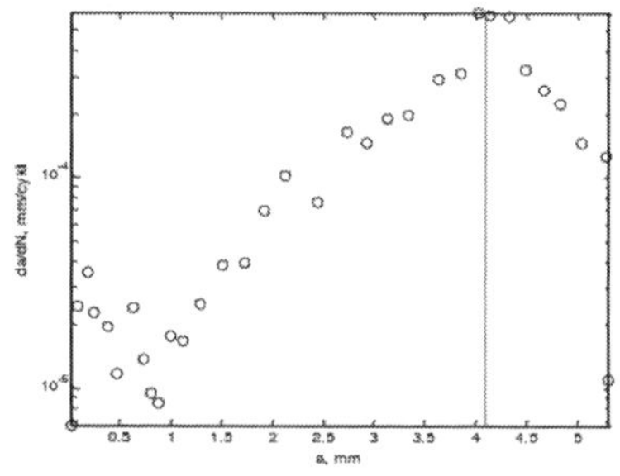

Fig. 6. Graphs of the crack propagation rate as a function of crack length for specimens subjected to cyclic bending and four displacement amplitudes: a) Δl = 2.4 mm, b) Δl = 2.6 mm, c) Δl = 2.7 mm, and d) Δl = 2.9 mm.

However, the crack growth rate varies during the whole period of propagation. In the initial stage of crack growth, i.e., from zero to the length of the crack of about 0.2 mm, the propagation rate (Fig. 6d) reaches the Δl = 2.9 mm equal to about da/dN = $0.3 \cdot 10^{-4}$ mm/cycle, and then the speed is reduced to a value of da/dN = $0.19 \cdot 10^{-4}$ mm/cycle in the range of 0.2 mm to 0.5 mm. Starting from the length of the crack of 0.5 mm, the propagation rate increases uniformly, reaching the maximum value in the material bond zone equal to da/dN = $0.29 \cdot 10^{-3}$ mm/cycle. After it is exceeded, the propagation rate decreases again, which is caused by a sharp reduction in bending moment and a decrease in stiffness of the specimen. The same variation of propagation rate was observed for other levels of the displacement amplitude Δl. The phenomenon of the decrease in the growth rate of crack propagation after its initial rapid growth may be due to the occurrence of compressive residual stresses [10,11] occurring in the titanium layer as the result of heat treatment.

4 Numerical analysis

Nonlinear FEM analyses were performed in ANSYS 18.0 environment. Cyclic properties of both metals were used to describe plastic behaviour of materials Table 2. Computations were performed as two-dimensional problem in plane strain conditions.

Table 2. Cyclic properties of the bimetal.

Material	K', MPa	n'
Steel S355J2+N	720.94	0.1258
Titanium Grade 1	190	0.0641

Numerical simulations were performed for 6 different crack length a = 4.218÷4.786 mm observed in last phase of crack development during test made with displacement amplitude Δl = 2.4 mm. The force values recorded during test for each crack length was converted to bending moment and shear force taken as boundary conditions for FEM analysis. The crack tip was modeled as a half of an arc of a radius 10 μm. In the computations there were used second order 8 nodes two-dimmensional finite elements PLANE183 [12]. In order to ensure the regularity of the dividing grid in the specimen model in the vicinity of the crack tip, a circular subarea divided into 4 quarters of appropriately selected lengths of edges was introduced (Fig. 7). The subarea thus prepared was mapped meshed [13]. In order to transition from the rounded edge of the crack tip subarea to the straight edges across of rest of the specimen model, a transition subarea was introduced (Fig. 7). This transition subarea was free meshed [14]. Along the α = 0° direction FE size increases proportionally to the radius increase so as to maintain a homogeneous aspect ratio (Fig. 9a). Map of the opening stress σ_{22} near the crack tip for a = 4.217 mm is shown in the Fig. 9a). Map of the opening stress σ_{22} in the crack tip area for a = 4.217 mm is shown in the Fig. 9b).

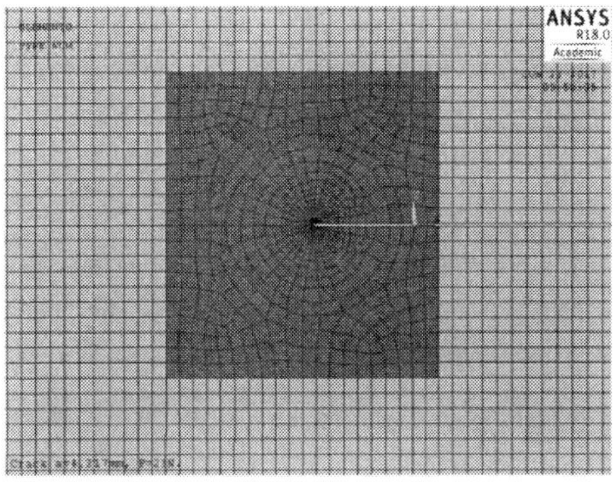

Fig. 7. Mesh morphology in crack tip area.

For the FEM analysis there is influence of the number of FE between the crack tip and the maximum location of the opening stress σ_{22} along the α = 0° direction on the stress distribution [15]. In the model there is 7 FE between the crack tip and the stress maximum location (Fig. 9a). In contrast to the radial direction, in the angular direction the FE size does not affect the stress distribution in front of the crack at α = 0° [15].
On the map of the opening stress σ_{22} in the crack tip area Fig. 9b) the titanium layer is strongly emphasized by relative high stress values.
The stresses in the titanium layer are close to the plastic limit for this material (Fig. 8). This results in an increase in the opening stress σ_{22} gradient in the steel part of specimen.

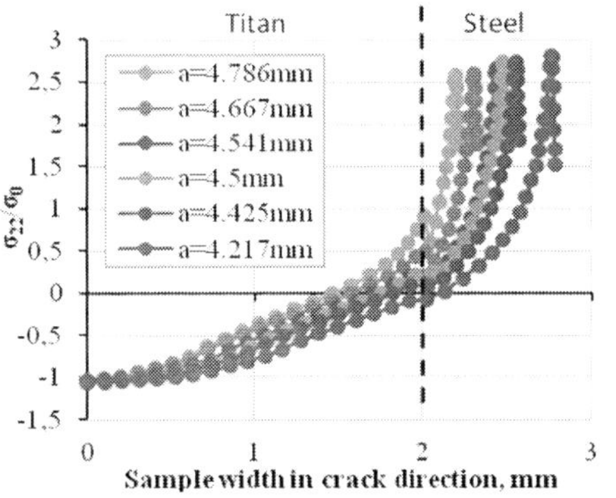

Fig. 8. Normalized opening stress distribution across of specimen width in crack direction.

The value of the maximum opening stresses σ_{22} reaches the highest level at the distance about 0.018 mm from crack tip (Fig. 10). The values of maximum opening stresses σ_{22} changes in range from 2.58÷2.8 σ_0 with the change of fatigue crack length from a = 4.786 mm to a=4.218 mm. The maximum opening stresses decrease with decrease of distance to a free surface of the specimens. Intensity of this decrease is higher for deeper cracks (e.g. a = 4.786 mm) than for shallower cracks (e.g. a = 4.217 mm).

DOI: 10.1051/e3sconf/20171903022

a)

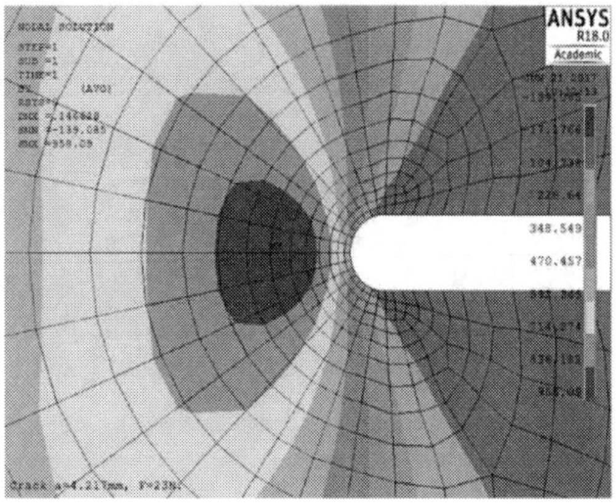

b)

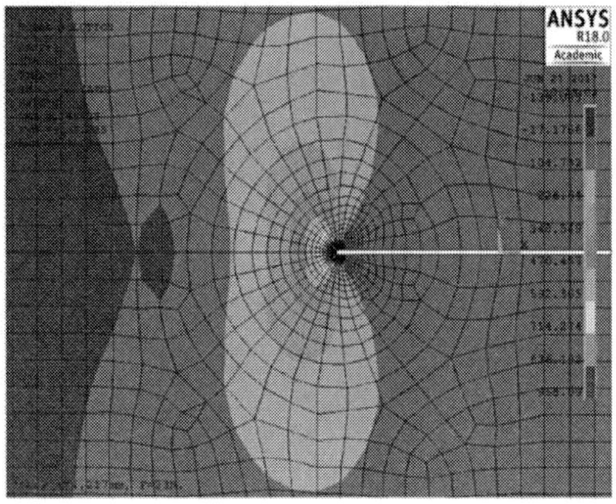

Fig. 9. Map of the opening stress σ22 for a=4.217mm: a) near the crack tip; b) in the crack tip area.

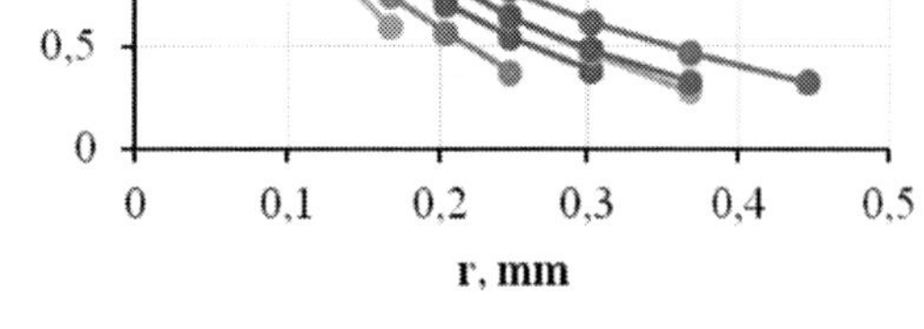

Fig. 10. Normalized opening stress distribution near the crack tip.

5 Conclusions

As a result of the research into steel-titanium bimetal, it was concluded that the occurrence of the plastic deformation fields in the steel before the cracking front did not result in a considerable stratification of the joint under the effect of a load comprising cyclically varied bending with a controlled value of the displacement. However, we noted a two-times change in the direction of the crack propagation in the area of the joint. The crack changed its direction in the steel component to perpendicular to the line of the joint and after it was trespassed, in the titanium section the direction was again perpendicular in relation to the axis of the sample. The numerical analysis performed in terms of the elastic-plastic characteristics demonstrated that the titanium layer is completely plasticized as a result of the acting loading, when the crack is still located in the steel part of the examined plater. We also noted an irregular change in the velocity of the crack propagation during its development, which can be attributed to the occurrence of residual stresses resulting from the thermal machining of bimetals.

References

1. S. Akbari Mousavi, S. Al-Hassani, A. Atkins, Mater. Des. **29**, 1334-1352 (2008)
2. H. Dyja, A. Maranda, R. Trebliński, *Wydawnictwo Wydziału Metalurgii i Inżynierii Materiałowej Politechniki Częstochowskiej* (2001)
3. S. Król, R. Bański, Z. Szulc, A. Gałka, Adv. Mater. Sci. 7, 50-56 (2007)
4. L. Bower, J. Banker, *Proceedings of Corrosion Solutions Conference*, (2003)
5. G. Young, Hydrocarbon Eng. (2005)
6. K. Bujis, Super Yacht Ind. **3/5**, 70-73 (2008)
7. A. Karolczuk, M. Kowalski, R. Bański, F. Żok, Int. J. Fatigue **48**, 101-108 (2013)
8. S. Akbari Mousavi, P. Farhadi Sartangi, Mater. Sci. Eng. A-Struct. Mater. Prop. Microstruct. Process, **494**, 329-336 (2008)
9. Ł. Pejkowski, D. Skibicki, J. Sempruch, Stroj Vestn-J. Mech. E, **60/9**, 549-560 (2014)
10. A. Karolczuk, K. Kluger, M. Kowalski, F. Żok, G. Robak, *Materials Science Forum* **726**, 125-132 (2012)
11. A. Karolczuk, M. Kowalski, K. Kluger, F. Żok, Arch. Metall. Mater. **59**, 1129-1133 (2014)
12. A. Cichański, *Proceedings of 17th International Conference on Engineering Mechanics*, 79-82 (2011)
13. A. Cichański, *Proceedings of 22nd International Conference on Engineering Mechanics*, 110-113 (2016)
14. A. Cichański, AIP *Conference Proceedings* **1822**, (2017)
15. M. Graba, J. Gałkiewicz, J. Theor. Appl. Mech. **45**, 225-237 (2007)

Designing an accurate system for temperature measurements

Orest Kochan[1,*], *Jun* Su[2], and *Roman* Kochan[1,3]

[1]Lviv Polytechnic National University, Department of measuring information technologies, 79013 Karpinskoho Str. 1, Lviv, Ukraine
[2]School of Computer Science, Hubei University of Technology, Wuhan, China
[3]University of Bielsko-Biala, Bielsko-Biala, Poland

Abstract. The method of compensation of changes in temperature field along the legs of inhomogeneous thermocouple, which measures a temperature of an object, is considered in this paper. This compensation is achieved by stabilization of the temperature field along the thermocouple. Such stabilization does not allow the error due to acquired thermoelectric inhomogeneity to manifest itself. There is also proposed the design of the furnace to stabilize temperature field along the legs of the thermocouple which measures the temperature of an object. This furnace is not integrated with the thermocouple mentioned above, therefore it is possible to replace this thermocouple with a new one when it get its legs considerably inhomogeneous.. There is designed the two loop measuring system with the ability of error correction which can use simultaneously a usual thermocouple as well as a thermocouple with controlled profile of temperature field. The latter can be used as a reference sensor for the former.

1 Introduction

Technical progress is challenging for measurement technologies and instruments due to the constant improvement and complication of technological processes in industry. Measurement chains are necessary to component in production quality management, therefore the market of metrological services is important branch of economy in developed countries. It is estimated to be about 4 % of GDP [1,2].

The concept of temperature as well as phenomena and techniques on which temperature measurements are based, evolved together with the development of science. The measuring instrument by which the temperature of bodies is measured is called a thermometer. The branch of science connected with construction and usage of thermometers is called thermometry [3,4]. The history of thermometry is an important part of history of science [5]. The first ideas about heat and temperature appeared long time ago, but Empedocles and Aristotle are the first scientists whose works are now known. Therefore they are considered as early founders of thermal science. Almost two thousand years after these two Galilei and most prominent scientists of the XVII century including Hooke, Von Guericke, Boyle, Huygens, Newton and many others contributed to the concept of temperature and its measurement [6]. Their works became the theoretical base for practical temperature scales invented in the XVIII century by Fahrenheit, Celsius, Romer, Reaumur and others. Some of these scales survived till now and are widely used in modern praxis. These ideas of thermometers and scales were based on common sense and practical needs but not on deep theoretical notions. Therefore constructions of thermometers as well as temperature units varied considerably.

Explosive development of industry during the XVIII-XIX centuries is now known as the Industrial Revolution [7] required accurate and reliable techniques of temperature measurements. Development of thermodynamics allowed physics to develop deep theoretical notions of temperature and heat. There was possible to adjust existing temperature scales using explanations of thermal behavior of matter. The newly developed theory of thermodynamics led to the concept of absolute thermodynamic temperature introduced by William Thomson (Lord Kelvin) [8]. This concept is based on the Carnot cycle and fundamental thermodynamic laws. According to modern understanding, temperature is a quantitative expression of the average kinetic energy contained in a piece of matter [8,9]. This is the most familiar definition. In general, for any substance, the higher the temperature, the faster its atoms or molecules move.

The unit of thermodynamic temperature, (also often referred to as Kelvin temperature or absolute temperature), is kelvin (K) and is defined in terms of the interval between the absolute zero of temperature and the triple point of pure water (273.16 K). One kelvin is the fraction 1/273.16 of that temperature interval. In addition to the thermodynamic temperature, the Celsius (°C) temperature scale is

* Corresponding author: orestvk@gmail.com

 DOI: 10.1051/e3sconf/20171903023

used in everyday life. The Celsius temperature is defined as equal to the thermodynamic temperature minus 273.15 whereas the Celsius degree (1°C) is numerically equal to that of Kelvin.

Intensification of international trade and collaboration led to the need for accurate measurements using unified instruments and practical realization of temperature scale which could be internationally recognized. Therefore, after long negotiations and discussions, there was established the International Bureau of Weights and Measures (IBWM) in 1875. Thermometry was in focus of interest of the IBWM since its establishment but as an auxiliary measure required for measurement of the temperature of accurate standards of length. Difficulties with measurement reproducibility led to the need of unification of temperature scales for the sake of accurate measurements of temperatures and subsequently to the promotion of temperature to a main physical quantity along with length, mass and other quantities. The studies of gas thermometers carried out by P. Chappuis [10] are now considered as the origin of modern thermometry. Based on these studies, the normal hydrogen scale was established in 1887. The main drawback of this thermometer was the approximation of the ideal gas with hydrogen therefore the scale deviated from the thermodynamic temperature. A competing sensor – the resistance temperature detector appeared in this time invented and researched by Callendar [11]. Later this invention played a very important role in the thermometry because this invention became a base of later temperature scales due to its convenience in practical use. The period between 1887 and 1927 was dedicated to the study of a more convenient temperature scale [12]. The new scale was adopted in 1927 under the title "the International Temperature Scale of 1927" [12]. This scale was replaced by the one adopted in 1948 under the title "the International Practical Temperature Scale 1948" [12]. It was amended and extended toward low temperatures in 1968. The new version got the title "the International Practical Temperature Scale 1968" [12]. The next step was made in 1990 when a new temperature scale came into action. Its title is "the International Practical Temperature Scale 1990" [12, 13]. This scale together with the Provisional Low Temperature Scale of 2000 are valid now [12]. They are based on a set of fixed points and are split into a number of overlapping ranges and sub-ranges; [12, 13]. Nowadays temperature is one of the most frequently measured physical quantities. The accuracy of temperature measurements often directly determines the quality, effectiveness and safety of industrial manufacturing processes [14-18]. According to [4]: "Statistical data show that the percentage of temperature measurements is 40 % of all measurements in industry. In some branches of industry this percentage may be even considerably higher, in particular in power industry it equals about 70 %." That is why so many methods, techniques and sensors for temperature measurements exist [3, 4].

The expansion of the sphere of temperature measurements [3, 4, 18-20], the increase in the requirements for their quality [17, 21] and the need for non-stationary measurements [15, 18, 19] imposes additional requirements on sensors [1, 4, 22], equipment [23, 24] and methods of signal processing [20, 21].

2. Thermocouples

However, despite the variety of existing sensors, thermocouples (TC) dominate in measurement praxis [3,4]. Base metal TCs are the most popular TCs. Noble metal TCs dominate among the standard instruments whereas refractory metal TCs are used for high temperatures (above 1300°C) [3,4].

The phenomenon of thermoelectricity, that is what the thermocouple is based on, was discovered in 1821 by Seebeck [25]. The first attempt to measure temperature using thermocouples was made in 1826 by A. C. Becquerel. He also discovered that the Seebeck's coefficient does not depend on the wire diameter and length [25]. However, due to lack of accurate and sensitive instruments, the thermocouple was not popular because its electromotive force (emf) vs temperature dependence was not accurately determined. In addition, it was found that either small contamination or permanent strain in thermocouple (TC) legs considerably alter the emf vs temperature dependence [25]. The practical needs of the cement manufacture industry led to studies in the field of high temperature measurements using thermocouples carried out by Le Chatelier [26]. Due to his efforts the thermocouple became an accurate and reliable temperature sensor. It was spread throughout the World very fast and adopted in other branches of industry, e.g. metallurgy [25]. By this time, all TCs were made of noble metals and their alloys, however in the beginning of the XX century appeared much cheaper and more suitable for practical use base metal TCs [27]. One of them, the type K TC, is now the most widely used temperature sensor.

The merits of TCs are: (i) wide range of measurements [3, 4], (ii) simple design, [3, 4], (iii) low price [28]. However, there is one considerable drawback such as degradation of TC legs [22,29,30]. Degradation processes are prone to all types of used TCs [31-37]. They appear due to long term use at high temperatures [29, 36]. According to [22], the degradation of TC legs leads to two interrelated errors: (i) the error due to the change in the developed emf with time in a constant temperature field along TC legs. It is referred to as the error due to drift of a conversion characteristic (CC) of a TC [34,38,39]; (ii) the error due to the change in the developed thermal emf in a changeable temperature field along TC legs. It is referred to as the error due to thermoelectric inhomogeneity of TC legs acquired during operation [22,30,37,40-42].

For the first time, the problem of thermoelectric inhomogeneity was considered in 1906 [43], but according to [29,30,32,44] it still remains one of the main sources of error in temperature measurements using TC. Despite the studies in this field [40,42,44-47] the problem has still not been solved yet. Thermoelectric inhomogeneity is considered to be the main source of TC error [30, 43]

The ways to increase the accuracy of temperature measurements by TCs with significant acquired during long-term operation thermoelectric inhomogeneity of legs were proposed in the paper [48]. The method of correction of this error was considered and studied in the paper [49]. However, the efficiency of the method [49] decreases with the change in the nature of drift of CC of TC sections as a consequence of a change in the temperature field. This occurs for long-term changes of a temperature field, when temperatures of TC sections after the change in the temperature field, will be perceived by these TC sections as new operating temperatures. In this case the method of compensation of changes in the temperature field along the legs of the TC which measures the temperature of an object [48,50] is more effective. The method is based on one of the fundamental properties of the inhomogeneity [22] - if there is no change in the profile of a temperature field along TC legs, then their acquired inhomogeneity, no matter how large it is, can neither manifest itself nor cause error of temperature measurements. In this case, the degradation processes running in TC legs will manifest themselves exclusively as the drift of the TC CC with time. Thus, all the methods for drift correction mentioned in [48] are fully suitable for effective improvement of the measurement accuracy. Generally speaking, the error due to acquired inhomogeneity of TC legs is the error of a random nature with respect to the operation time (it corresponds to random changes in the temperature field along TC legs). This error in a constant temperature field converts into a systematic error of the drift of the TC CC with time, which in turn can be corrected. The correction methods listed in point 2 of paper [48] are considered in [51-52]. They can use results of a calibration either in a calibration laboratory [51], or in situ [53,54].

The goal of the paper is to develop the method which will counteract the manifestation of the error due to acquired thermoelectric inhomogeneity of TC legs in a changeable temperature field of operation.

3 The method of compensation of the error due to acquired thermoelectric inhomogeneity of TC legs

It is necessary to stabilize the profile of a temperature field along the legs of the TC, which measures the temperature of an object, to implement the method of compensation of the effect of acquired inhomogeneity of TC legs proposed in [48,50]. The stabilization can be achieved by creating an independent temperature field for the TC, measuring the temperature of the object (the main thermocouple), by means of auxiliary temperature control systems. The heaters and sensors of these systems should be located coaxially with the main thermocouple (MTC) and must be shifted along the axis of the MTC legs. The structure of the proposed TC with controlled profile of temperature field (TCPTF) is shown in Fig. 1 [50].

The TCPTF consists of: (i) the main thermocouple (MTC), whose goal is to measure the temperature of an object; (ii) the subsystems to control the temperature field are along the legs of the MTC. Each subsystem consists of a heater (H1...Hn) and a corresponding thermocouple (TC1...TCn). All TCs have their reference junctions wired to a measuring and controlling system (MCS). In the bottom part of Fig. 1 there is shown the temperature field ABC. This field is established and maintained by the heaters H1...Hn. Therefore, even if the object's temperature field changes within the fields E...F, the field ABC remains stable and constant. This prevents the manifestation of the error due to acquired inhomogeneity of the MTC. Therefore the methods of correction for the error due to acquired inhomogeneity can be used on the objects with changeable temperature field, because the proposed TCPTF maintains its own stable temperature field.

4 The furnace with controlled profile of temperature field

The design of TCPTF proposed in [50], presented in Fig. 1.

If the TCPTF is made as an integrated sensor, it has significant drawbacks: (i) lack of unification with existing TCs due to complexity and uniqueness of the design; (ii) impossibility of using already produced TCs; (iii) difficulties with replacement of the MTC when it get its legs significantly degraded (inhomogeneous); (iv) the need for metrological certification as a new type of sensor. These drawbacks are significant as they hamper implementation of the TCPTF. However, these drawbacks are not prone to the method of compensation of the effect of changes in the temperature field of the object on the temperature field of the MTC. They are only prone to the design of the TCPTF as an integrated sensor.

To eliminate them, it is reasonable to split the TCPTF [55,56] into: (i) the MTC itself. As a MTC it is reasonable to use a usual TC. (ii) the mean for stabilizing the temperature field (the special furnace) along the legs of the usual TC mentioned in point 1.

One of the problems in creating such a special furnace is the need to bring a large number of lead-outs to the butt of the furnace. For instance, for 10-section furnace it is necessary to place 40 lead-outs,

 DOI: 10.1051/e3sconf/20171903023

insulated by ceramic beads due to high temperature, on the butt of the furnace. The main idea is to combine a heater and a TC that measures the temperature of a section by making a heater from one thermocouple material, and its lead-out from the complementary one. This allows reducing the number of lead-outs,

A structure of the special furnace proposed in [55,56] is shown in Fig. 2. It consists of the outer 1 and inner 2 tubular shells (the diameter of the inner

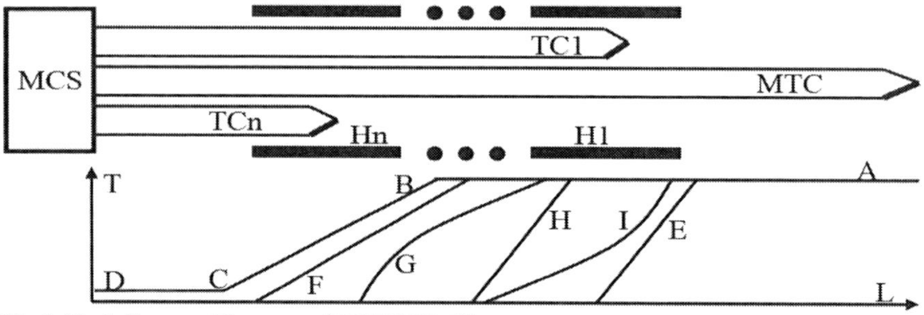

Fig. 1. Block diagram of the proposed TCPTF [47, 49].

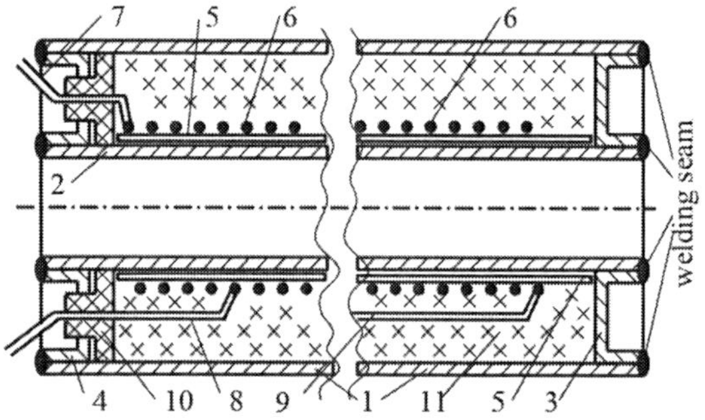

Fig. 2. Structure of a special furnace.

case corresponds to the outer diameter of a usual TC thermowell). Shells 1 and 2 are welded together by walls 3 and 4. There is an insulating (ceramic) cylinder 5 inside of the furnace. Heater 6 is either wounded or sprayed on cylinder 5. Terminals 7 ... 9 of heater 6 are led out through insulating sleeves 10. The space between shells 1 and 2 is filled with thermal insulation 11.

The heaters and TCs are combined by making heater 6 and one of the outermost lead out 7, for instance from chromel, and the other lead outs (8 and 9) from the complementary TC material alumel. In this case, the heater together with the lead outs form the chromel-alumel TCs, or in other words, type K TCs. With 10 sections of heaters, the number of lead outs (sleeves 10) will be 11 [55,56].

5 The structure of the system using the TCPTF

The direct use of the TCPTF for temperature measurement is hampered by its considerable thermal inertia. This happens because of the need to wait until the end of the transient processes, not only for the measuring junction of the MTC, but also to stabilize the entire present temperature field along the legs of the MTC to get measurement results without the influence of the error due to the acquired thermoelectric inhomogeneity of the MTC.

The time of such stabilization, taking into account the thermal connection between sections of the TCPTF, can, in the worst case, exceed 2 ... 3 hours. Such a time can cause additional dynamic errors when measuring temperature of large objects. Such a time constant is inappropriate with the use of TCPTF in control systems. Therefore, it is proposed to add the error correction circuit equipped with a usual TC of the usual inertial properties to a measuring chain, based on a TCPTF.

The structure of such a two-loop control system is shown in Fig. 3. It consists of a TCPTF and its measuring chain MC1, a usual TC and its measuring chain MC2, the correction unit CORR, the adder AD, the set unit establishing the temperature to maintain SET, the regulator REG, and the heater of an object HEAT. The loop TC - MC2 - AD - REG – HEAT works as a normal control system and implements any of the known laws of control (P, PI, PID). The measuring junctions of the TC and the TCPTF are located closely to each other so that there is virtually no temperature difference between them on a particular object.

The TCPTF, its measuring chain MC1 and the correction unit CORR form a correction channel that compares the results of measurements carried out by the TC with that of carried out by the TCPTF. It quite rarely (for example, when the change of a temperature field of the object exceeds the permissible value, but not less than once a day) corrects the TC error by submitting the appropriate correction signal to the adder AD. In this case, the correction signal can be obtained practically without a time limit for its formation.

Thus, the dynamic properties of the control system will not differ from that of traditional systems that use usual TCs.

 DOI: 10.1051/e3sconf/20171903023

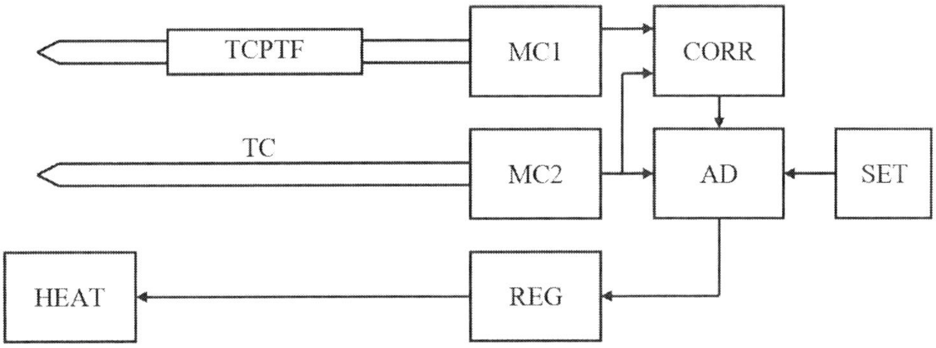

Fig. 3. The structure of the two-loop control system using a TCPTF.

Conclusions

The proposed method for compensating changes in a temperature field along the legs of a TC measuring the temperature of an object has a rather high efficiency. In contrast to the method of correction of the error due to acquired thermoelectric inhomogeneity considered in [48], this method does not correct the error due to drift a TC CC. However, it makes it possible to apply the known methods of correction of the error due to drift of a TC CC listed in [47], generalized in [50] and developed in [51]. In this case, in contrast to the correction method [48], the considered method of compensation creates almost ideal conditions for the use of the methods for correcting the error due to drift of TC C, because all sections of a TC under any changes in the temperature field of the object, remain at the temperature of constant operation. There is also no danger of changing the character of the error due to drift of a TC CC (or its sections) and the associated danger of an increasing the residual error of correction. These advantages determine the prospects of the proposed method.

The proposed is designed for objects where high accuracy of measurement is important either from the safety point of view or from the economical point of view. For instance, improved accuracy of temperature measurements in thermal power plants can save 400 million US dollars worldwide a year [57]. The direction of further studies lies in modelling using the metrological software test [58] to study errors of the whole proposed system or its main components.

This work was supported by Foundation of Wuhan Science and technology Bureau (2015030809020370), Doctoral Scientific Research Fund from Hubei University of Technology (No. BSQD14037), National Natural Science Foundation of China (Nos. 61602162, 61502155), Green Industry Technology Leading Project of Hubei University of Technology (No. ZZTS2016004).

References

1. J. Birch, Benefit of Legal Metrology for the Economy and Society. (A study for the International Committee of Legal Metrology, 2003) https://www.oiml.org/en/files/pdf_e/e002-e03.pdf
2. J. Birch, Role and impact of legal metrology for consumer protection. (ISO/COPOLCO Workshop, 2012) http://www.iso.org/iso/livelinkgetfile?llNodeId=400882&llVolId=-2000
3. J. Webster, *Measurement, Instrumentation, and Sensors Handbook* (CRCnetBase, 1999)
4. Y. Lutsyk, O. Hook, O. Lakh, B. Stadnyk, *Temperature Measurements: Theory and Praxis* (Lviv, 2006) (In Ukrainian)
5. F. Sherwood Taylor, **5(2)**, 129-156 (1942)
6. H. Chang. *Inventing Temperature: Measurement and Scientific Progress* (Oxford, 2004)
7. R Allen. *The British Industrial Revolution in Global Perspective* (Cambridge and New York, 2009)
8. R. Wolfson. *Essential University Physics*, 2nd Edition (Middlebury College, 2012)
9. S. Gibilisco. *Physics demystified*. (New York, 2002)
10. P. Chappuis. Trav. Mem. Bur. Int. Poids Mes. **VI**, 1-19 (1888)
11. H. L. Callendar. Philos. Mag. **34**, 104-111 (1891)
12. http://www.bipm.org/en/measurement-units/history-si/temperature-scales/
13. H. Preston-Thomas. Metrologia **27**, 3-10 (1990)
14. H. R. Ashrafi, K. Shahbazian, S. Bidmeshki, S. Yaghooti, P. Beiranvand. Adv. Sci. Technol. Res. J. **10**, 29, 1-12 (2016)
15. V. Perzel, M. Flimel, J. Krolczyk, A. Sedmak, A. Ruggiero, D. Kozak, A. Stoic, G. Krolczyk, S. Hloch, Thermal Sci. **00**, 46 (2016) DOI: 10.2298/TSCI150212046P
16. A. Glowacz, A. Glowacz, Z. Glowacz, Meas. Sci. Rev. **15**, 3, 119-126 (2015)

17. R. W. Maruda, G. M. Krolczyk, E. Feldshtein, F. Pusavec, M. Szydlowski, S. Legutko, A. Sobczak-Kupiec, Int. J. Mach. Tool. Manu. **100**, 81-92 (2016)
18. T. Habisreuther, T. Elsmann, Z. Pan, A. Graf, R. Willsch, M.A. Schmidt. Appl. Therm. Eng. **91**, 860-865 (2015).
19. P. Hreha, A. Radvanska, L. Knapcikova, G.M. Krolczyk, S. Legutko, J.B. Krolczyk, S. Hloch, P. Monka. Metrol. Meas. Syst. **22**, 315-326 (2015)
20. A. Glowacz, A. Glowacz, P. Korohoda, Arch. Metall. **59(1)**, 31-34 (2014)
21. A. Glowacz, Z. Glowacz, Measurement **93**, 86-93(2016)
22. J. Su, O. Kochan, Meas. Tech+ **57**,10, 1160-1166 (2015)
23. A. Aly, A. El-Lail, I.J. Information Technology and Computer Science, **03**, 56-60 (2013)
24. R. Kochan, O. Berezky, A. Karachka, O. Bojko, I. Maruschak, IEEE T. Instrum. Meas. **51(1)**, 96-101 (2002)
25. L. B. Hunt, Plat. Metals Rev. **8**, 23-28 (1964)
26. H. Le Chatelier, Compt. rend. **102**, 819-826, (1886)
27. I. A. Rogelberg, V. M. Beylin, *Alloys for thermocouples: Reference book* (Moscow, 1983)
28. P. R. N. Childs, J. R. Greenwood, C. A. Long, Rev. of Sci. Instrum. **71**, 2959- 2978 (2000)
29. K. C. Sloneker, Ceram. Ind. **159(4)**, 13-18 (2009).
30. D.J. Southworth, *Temperature Calibration with Isotech Block Baths* (Southport, Isotermal technology ltd, 1999)
31. R. M. Park, *Manual on the use of thermocouples in temperature measurement* (ASTM, New York, 1993)
32. Y. Abdelaziz, F. Edler, Meas. Sci. Technol., **20(5)**, 055102, (2009) doi:10.1088/0957-0233/20/5/055102
33. J. V. Pearce, Meas. Sci. Technol. **18**, 11 3489-3495 (2007)
34. R. Strnad, M. Jelínek, G. Failleau, T. Deuzé, M. Sadli, N. Arifovic, A. Diril, P. Pavlasek, M. Langley, J. V. Pearce, Automa. **6**, 28-31 (2014)
35. T. J. Quinn, *Temperature* (London, 1983)
36. V. P. Pavlov, *Thermoelectric inhomogeneity of thermocouple legs* (Publishing House of Standards, Moscow, 1979)
37. M. Holmsten, J. Ivarsson, R. Falk, M. Lidbeck, L.-E. Josefson, Int. J Thermophys. **29(3)**, 915-925 (2008)
38. L. Kortvelyessy, *Thermoelement praxis* (Vulkan-Verlag, Essen, 1981)
39. N. Rogelberg, A. Nuzhnov, G. Pokrovskaya et all, *Investigation of alloys for thermocouples* (Metallurgiya, Moscow) **29**, 33-41 (1969)
40. J. Tamba, K. Yamazawa, S. Masuyama, H. Ogura, M. Izuchi, Int. J. Thermophys. **32**, 2436-2451 (2011)
41. E. S. Webster, D. R. White, H. Edgar, Int. J. Thermophys. **36**, 444-466, (2015)
42. M. Hiti, J. Bojkovski, V. Batagelj, J. Drnovšek, Elektrotehniški vestnik/Electrotechnical Review. **72**, 4, 189-194 (2005)
43. W. P. White, Phys. Rev. **23**, 4, 449-474 (1906)
44. M. Holmsten, S. Ljungblad, L. E. Josefson, IProc. of the int. Symposium TEMPMEKO. Portorož (Slovenia), **A**, 87 (2010)
45. Y.G. Kim, K.S. Gam, J.H. Lee, Meas. Sci. Technol. **8**, 317-321 (1997)
46. K. D. Hill, D. J. Gee, Proc. of the 9-th int. temperature symposium temperature: its measurement and control in science and industry, Los Angeles, (USA), **8(1552)**, 520-525 (2012)
47. G. Zaid, Instrumentasi **27**, 1, 18-21 (2003)
48. J. Su, O. V. Kochan, V. S. Jotsov, Meas. Tech+. **58**, 3, 327-330 (2015)
49. Jun Su, O. Kochan, N. Vasylkiv, R. Kochan, Meas. Tech+, **58**, 8, 904-910 (2015)
50. O. Kochan, R. Kochan, O. Bojko, M. Chyrka, Proc. of the 4th IEEE International Workshop IDAACS'2007, Dortmund, Germany. 47-51 (2007)
51. A. Sachenko, *Accuracy increasing methods development and precision measurement system design for industry. Dissertation abstract fo the doctor of engineering degree* (Leningrad, 1988)
52. A. Sachenko, V. Kochan, V. Turchenko, V. Holovko, Y. Savitski, T. Laopoulos, Patent of Ukraine №50830.
53. G. Failleau, C. J. Elliott, T. Deuzé,· J. V. Pearce, G. Machin, M. Sadli, Int. J. Thermophys. **35**, 1223-1238 (2014)
54. D. Zvizdic, D. Sestan, Int. J. Thermophys. **36**, 336-346 (2015)
55. O. Kochan, J. Su, V. Kochan, Proc. of the 13th IMEKO TC10 Workshop on technical diagnostics advanced measurement tools in technical diagnostics for systems' reliability and safety Warsaw (Poland), **105-110** (2014)
56. J. Su, O. Kochan, V. Kochan, Chunzhi Wang, Int. J Thermophys. **37**, 10 (2016) DOI 10.1007/s10765-015-2025-x
57. T. Wang, D. Bediones, P. Swirla, H. Henrikson, E. Janhunen, K. Bachalo, Stabilized metal sheathed type K and E thermocouples improve turbine efficiency. ISA, 439-448, (1997)
58. Kochan R, Proc. of the 2nd IEEE international workshop of IDAACS'2003, **168 – 173** (2003).

DOI: 10.1051/e3sconf/20171903024

The use of 3d scanner for testing changes in shape of human limbs under the influence of external mechanical load

Kamila Kasperska[1], *Michał* Wieczorowski [1,*], and *Jolanta B.* Krolczyk[2]

[1]Poznan University of Technology, Faculty of Mechanical Engineering and Management, 3 Piotrowo Street, 60-965 Poznan, Poland
[2]Opole University of Technology, Faculty of Mechanical Engineering, 76 Proszkowska Street, 45-758 Opole, Poland

Abstract. Three-dimensional scanning is used in many fields: medicine, architecture, industry, reverse engineering. The aim of the article was to analyze the changes in the shape of the limbs under the influence of a mechanical external load using the method of three-dimensional scanner uses white light technology. The paper presents a system of human movement, passive part - skeleton and active part - the muscles, and principles of their interaction, which results in a change of the position of the body. Furthermore, by using the 3D scan, the differences in appearance of the arm and leg depending on the size of the external load in different positions have been presented. The paper shows that with increasing load, which muscles must prevent, increases the volume of certain parts of the legs, while another parts of them will be reduced. Results of the research using three-dimensional scanner allow determining what impact on changing the legs shape has an external mechanical load.

1 Introduction

3D scanning is becoming increasingly popular as a measuring tool [1]. It makes it possible to check the quality of the produced elements by comparing the 3D mesh of the product with the Computer Aided Design (CAD) model. It provides information on the areas where the differences are bigger which leads to the improvement in the technological process and elimination of the causes of the resulting errors. It may be applied to many areas: medicine, architecture, industry, reverse engineering. 3D metrology has been known also for many years as an inspection method [2-5]. 3D scanning has become also a tool used in almost every area of life. People are constantly striving to improve the imaging tools in order to obtain the results that are almost identical to the real objects and the preparation of the test surface was unnecessary. The 3D Go!Scan scanner used in the research is one of the latest solutions and the image is collected and set based on the colour and texture.

The optical method, lighting with structural light has a number of advantages, i.e.: precision of measurement by an order better than in the case of tracing, non-contact measurement, possibility of automating the measurement [6-7].

Automation progress is related to computer aided manufacturing. CAD is one of the kinds of computer aided production. It is often combined with Computer Aided Manufacturing in order to obtain fluent transition from the stage of design to the stage of production [8]. Computer Integrated Manufacturing (CIM) is a synthesis of the two previous kinds [9-11].

The aim of the article was to analyze the changes in the shape of the limbs under the influence of a mechanical external load using the method of three-dimensional scanner uses white light technology.

2 Materials and methods

The research part included the determination of the differences in the appearance of the arm and lower leg depending on the size of the external load which they were subjected to in the given position. It is assumed that when the load which the muscles need to counteract is increased, the volume of some parts of limbs will increase and other parts will be reduced.

The manual 3D Creaform scanner using the technology of white light was used for the research. The scanner takes a quick measurement of spatial objects with complex geometry. The advantage is that the tested object must not be mounted rigidly, its small movements are allowable. The device is up to 0.100 mm accurate.

The objective of the research part was to examine how the load affects the external shape of the limb muscles. Assumptions for the research in the field of weight training:

- exercise, which will cause the change of shape of the given part of the limb and it will be possible to illustrate its entire circumference,
- time to stand still in the given position should range between 45 and 60 s,
- test surface should be smooth; removal of body hair is necessary,
- use of available sports equipment, namely assembled

* Corresponding author: michal.wieczorowski@put.poznan.pl

DOI: 10.1051/e3sconf/20171903024

dumb-bells.

Two parts of the body were used in the research: a calf and an arm.

- for a calf: calf raises with dumb-bells - the three-headed muscle of the calf becomes visible,
- for an arm: dumbbell arm curls or pull-downs, standing position - mainly the biceps brachii works.

Table 1. Load plan for individual persons.

	Arm load - biceps brachii
Subject No. 1	no load
	3 kg
	5 kg
	13 kg
	Arm load - biceps brachii
Subject No. 2	no load
	3 kg
	5 kg
	13 kg
	Lower leg - calf
Subject No. 2	natural position, standing
	muscle flexed, no load
	muscle flexed, 11 kg

Two people took part in the tests: subject No. 1: a man who attends the gym for pleasure and subject No. 2: a woman who professionally trains powerlifting. In theory, there was no need for special preparation of the test surface but in order to shorten the time of model creation it was decided to tape the part of the limbs with markers which help to find the correct location of the newly obtained data. In the case of the upper limb it turned out to be sufficient but there was a problem with the lower limb. It was a big surprise, because the geometry of lower limb is far less complex than the geometry of arm and it seemed to be an easier object to scan with a 3D scanner. Additional, colourful patterns were drawn, apart from the points stuck on the limb, in order to facilitate the operation of the device. Two models were scanned for different loads, which are presented in Tabela 1, and adjusted to the capabilities of the research participants. The largest weight constituted about 60% of the first maximum.

The results of each measurement were acquired in the form of a cloud of points representing the part of the limb when doing a specific exercise flexing the working parts of the muscles. Most of the programs that have the function of editing the point clouds represent it omitting the actual colours. The scanner that was used creates a three-dimensional model based on colour and texture of the object, therefore, the ViewShape program was used to show the results as per Fig. 1 and 2.

Editing the measurement data was carried out in the GOM Inspect program. It is used primarily for analysis of information obtained during the 3D imaging during the quality control, product preparation and production.

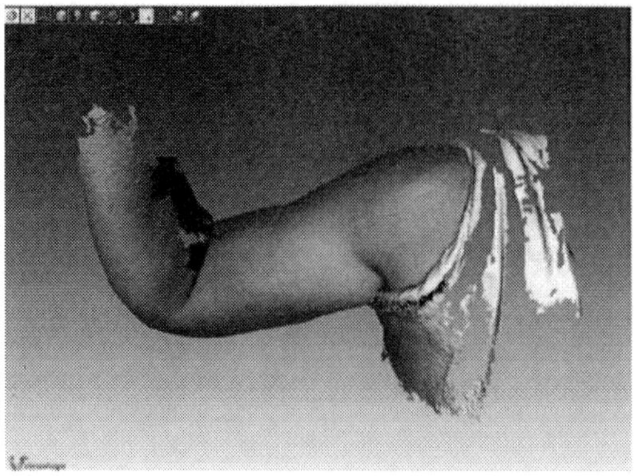

Fig. 1. Obtained point clouds presenting the examined parts of the body in the ViewShape program: arm.

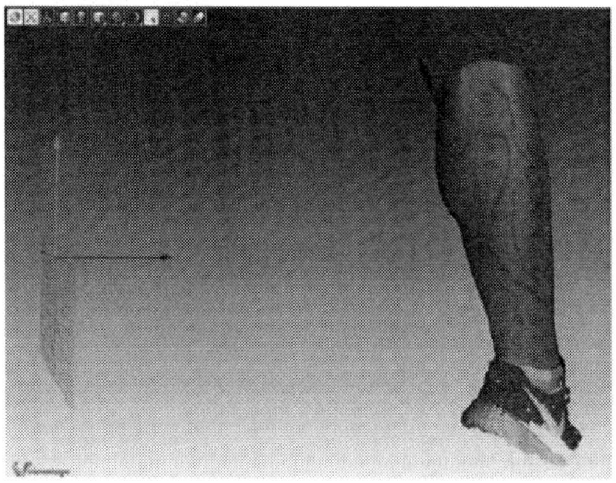

Fig. 2. Obtained point clouds presenting the examined parts of the body in the ViewShape program: lower leg.

3 Results

The main task to be carried out in the research part was the determination of the influence of the lifted weight on the appearance of the limbs. In order to do this they were scanned with the use of 3D method, which made it possible to determine the length of the cross sections of the examined parts in the significant areas, that is in the areas where, in theory, the biggest changes should occur. We use the GOM Inspect program to examine the circumferences, especially its function of determining the cross section in a specific area and then automatic generating of its length. Determining the circumference of the given part of the body using the GOM Inspect V8 SR2 program is shown in Fig. 3-4.

In Fig. 5 the changing shape of the arm (biceps) is shown - a 3D image - when doing exercises without load and a 3D image of the arm when doing exercises with the load of 3 kg (Fig. 6).

The exercise performed is characterised by the most intensive work of the biceps brachii, therefore, mainly the fact that this muscle is convex will have an influence on changing the shape of the upper limb.

Table 2 presents the results of the research.

DOI: 10.1051/e3sconf/20171903024

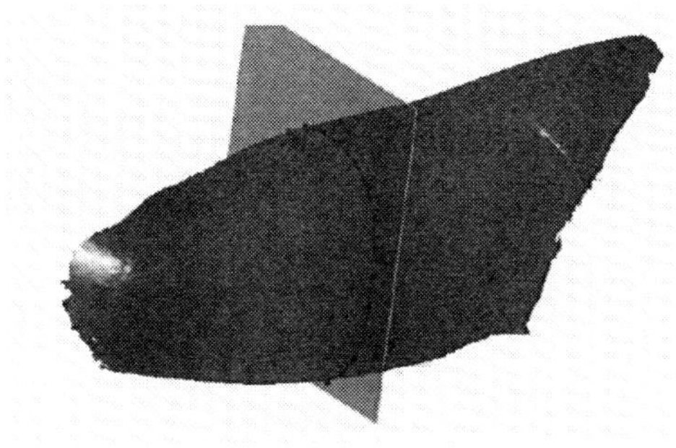

Fig. 3. Determining of the area of the cross section.

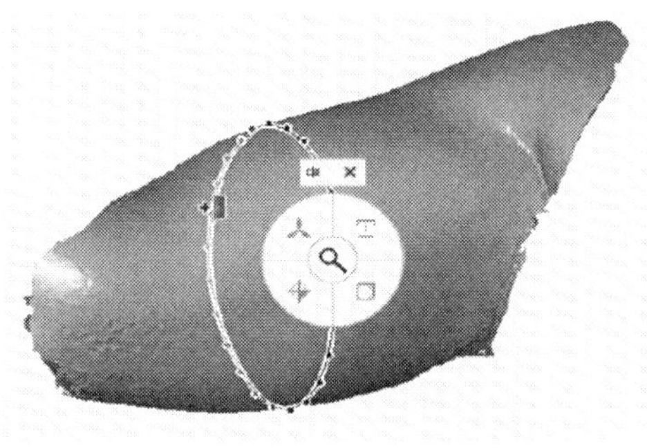

Fig. 4. Measuring the length of the cross section circumference.

Table 2. The arm circumference when exercising with different loads.

Subject No. 1	
Load [kg]	Circumference [mm]
no load	386.2
3	389.1
5	392.5
13	unsuccessful
Subject No. 2	
Load [kg]	Circumference [mm]
no load	386.7
3	388.1
5	389.9
13	393.4

A similar operation was carried out to observe changes of the calf shape when standing and performing the calf raise exercise with a load of 11 kg. Calf raises engage the muscles of the calf, namely the soleus and inner and outer head of the gastrocnemius muscle. A significant change of the shape during the exercise is noticed, even without load. Tab. 3 presents the results of the research.

Table 3. The length of the calf circumference in different positions for subject no 2.

Load [kg]	Circumference [mm]
Standing position	431.4
Calf rises	440.5
Calf rises + load of 11 kg	441.2

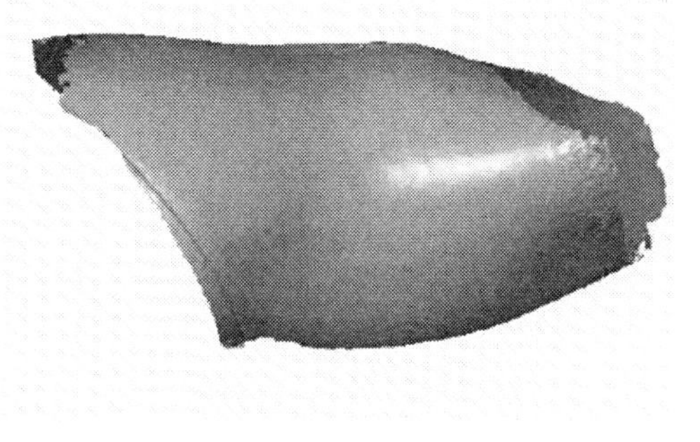

Fig. 5. Change of the arm shape when exercising without load.

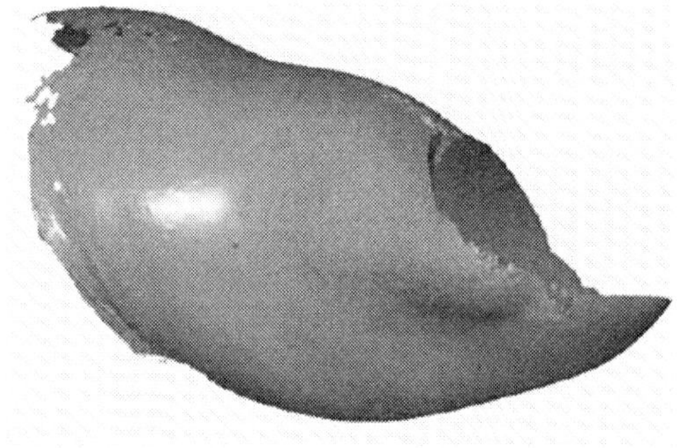

Fig. 6. Change of the arm shape when exercising with a load of 3 kg.

Apart from the examination of the changes in limb circumference during the movement without load or with a load, also the specification of their transformations was considered. The option of the Geomagic Control was used here. It makes it possible to fit two different 3D meshes in the STL format with the use of "Best Fit Alignment" method and then to compare them by means of 3D method and to create colourful maps of deviations. It shows the parts of the limb in which the modifications were the largest as well as the level on which they were located. Due to the fact that each of the muscles works differently during the particular movement, there are surpluses in some areas and losses in other areas.

In Fig. 7 and 8 the selected research results are presented. It was observed for the map shown in Fig. 7 that the biceps brachii muscle became much more visible in the front view and the triceps brachii muscle became slightly more visible in the back view. The shape change in the area of the larger muscles caused cracks in the lower and upper sections of the limb.

Results in the form of colourful map of deviations for the lower leg when standing and during calf raises shown in Fig. 8.

Analysing the obtained results it may be concluded that the flexion of the three-headed calf muscle caused the cracks on the outer side along the tested muscle and on the inner side under the muscle as well as a slight increase in its volume.

DOI: 10.1051/e3sconf/20171903024

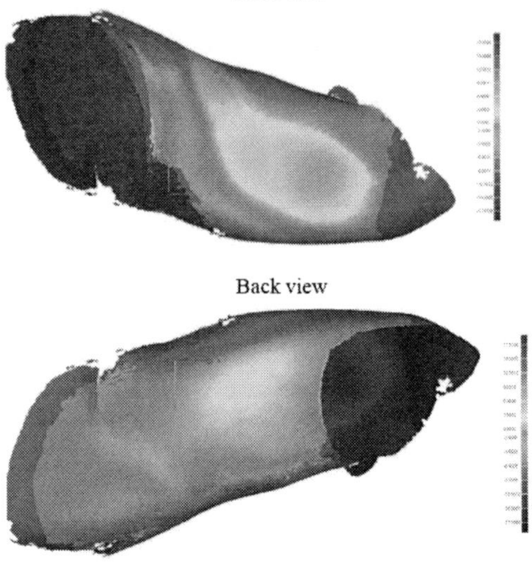

Fig. 7. Map of deviations for subject No. 1 comparing the data arising during the performance of the arm exercises without load and with the load of 3 kg.

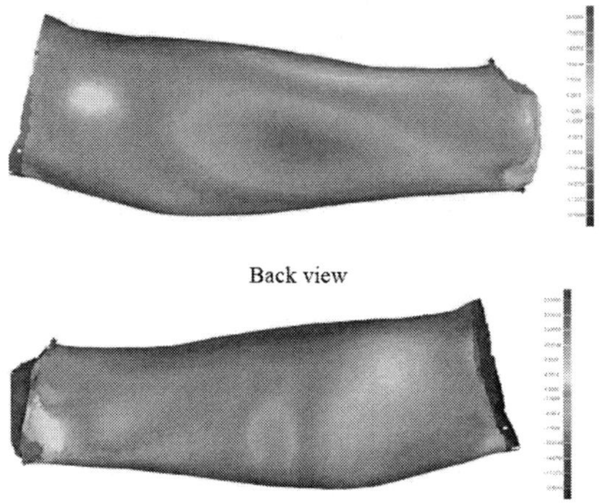

Fig. 8. Map of deviations for subject No. 2 for the lower leg when standing and during the calf raise.

4 Conclusions

1. The results obtained thanks to the research performed make it possible to determine the influence of the external mechanical load on the shape change of the limb. It may be noted that in no case two identical 3D meshes were obtained.
2. The first step was the determination of the length of the limb circumference in the relevant area, that is in the area where the volume change was the most significant. It made it possible to advance a thesis assuming that when the external load influencing the arm and lower leg is increased, the tested length also increases. This process is influenced by many factors which will be described later.
3. Determination of the areas where the changes were the most significant was possible thanks to the creation of the colourful maps of deviations, which identified the areas of modifications. It enabled us also to realize that the muscle flexion may lead to the increased volume in one area but also to the losses of volume in another area.
4. In the case of upper limb the biceps branchii muscle is slightly more convex for both subjects, with a small increase in the volume of the triceps branchii muscle at the same time. The flexion of these muscle areas during the exercise performance made the cracks in the upper and lower area of the arm more visible.
5. This was not the case for the lower leg. The first measurement was taken when the subject was in the standing position with both feet on the ground and the other during the calf raise without weights. It made it possible to show how the lack of muscle flexion influences the shape of the calf. A slight increase in its volume was also noted, with significant losses on the edges of the parts which conduct the greatest work during the performed activity. In this case, the additional load only made the changes that appeared before more visible.

References

1. T.G. Mathia, P. Pawlus, M. Wieczorowski, Wear **271** (2011)
2. P. Nieslony, G.M. Krolczyk, K. Zak, R.W. Maruda, S. Legutko, Precis Eng. **47** (2017)
3. G.M. Krolczyk, J.B. Krolczyk, R.W. Maruda, S. Legutko, M. Tomaszewski, Measurement **88** (2016)
4. C. Zhang, Z. Li, C. Hu, S. Chen, J. Wang, X. Zhang, Meas. Sci .Technol.. **28** (2017)
5. J.B. Krolczyk, B. Gapinski, G.M. Krolczyk, I. Samardzic, R.W. Maruda, K. Soucek, Y. Javadi, S. Legutko, P. Nieslony, L. Stas, The. Vjesn. **23** (1), (2016)
6. B. Curless, *Overview of active vision techniques* (SIGGRAPH 99 Course on 3D Photography, 1999)
7. J. Geng, *Structured-light 3D surface imaging: a tutorial* (Optical Society of America, 2011)
8. S. Wojciechowski, P. Twardowski, M. Pelic, Procedia CIRP **14**, 113-118 (2014)
9. S. Wojciechowski, R.W. Maruda, P. Nieslony, G.M. Krolczyk, Int. J. Mech. Sci. **119**, 360-369 (2016)
10. S. Wojciechowski, P. Twardowski, M. Pelic, P.W. Maruda, S. Barrans, S., G.M. Krolczyk, Precis Eng. **46**, 158-165 (2016)
11. S. Jun, O. Kochan, Meas. Sci. Rev. 14 (1), 29-34 (2014)

Identification of internal defects of hardfacing coatings in regeneration of machine parts

Jerzy Józwik [1,*], *Krzysztof* Dziedzic[2], *Mykhalo* Pashechko[3], and *Andrzej* Łukasiewicz[4]

[1]Lublin University of Technology, Mechanical Engineering Faculty, Department of Production Engineering, 36 Nadbystrzycka Street, 20-816 Lublin, Poland
[2] Lublin University of Technology, Electrical Engineering and Computer Science Faculty, Institute of Computer Science, 20-618 Lublin, Nadbystrzycka 36B, Poland
[3] Fundamentals of Technology Faculty, Lublin University of Technology, 38 Nadbystrzycka Street, 20-618 Lublin, Poland
[4] POLFER Podzespoły Indukcyjne S.A., 25 Woźniki, 08-200 Łosice, Quality Department, Poland

Abstract. The quality control of hardfacing is one of the areas where non-destructive testing is applied. To detect defects and inconsistencies in the industrial practice one uses the same methods as in the testing of welded joints. Computed Tomography is a type of X-ray spectroscopy. It is used as a diagnostic method that allows to obtain layered images of examined hardfacing. The paper presents the use of Computed Tomography for the evaluation of defects of hardfacing parts and errors. Padding welds were produced using GMA consumable electrode welding with CO_2 active gas. The padding material used were cored wires FILTUB DUR 16, and ones produced from a Fe-Mn-C-Si-Cr-Mo-Ti-W alloy. The layers were padded on to different surfaces: C45, 165CrV12, 42CrMo4, S235JR steel. Typical defects occurring in the pads and the influence of the type of wire on the concentration of defects were characterized. The resulting pads were characterized by occurring inconsistencies taking the form of pores, intrusions and fractures.

1 Introduction

The main reason of machine part defects which cause, already high, costs incurred by companies to rise, is wearing of work surfaces. Currently, one of the most used methods of regenerating machine parts is using hardfacing, i.e. the layering of liquid metal onto the object being regenerated [1-2]. In comparison to welding, hardfacing is not a process of joining elements, but one aiming to create a surface layer on the object, of a material supplied usually using an electrode or a wire. Often, hardfacing is used for regeneration, repairs or modernization elements of tools and machines, the replacement of which would usually be too work or cost intensive. At Polfer Podzespoły Indukcyjne S.A., injection molds are being repaired using this method. The wear caused by the form being burned out or by cracked edges on the elements forming the object, lead to flashes being produced, which can cause the product to be rejected at later stages of processing. Scrapping an injection form is economically unreasonable. Other methods of repairing a worn tool which include cutting out the offending fragment, manufacturing and mounting a compensating one, are often costly and very difficult, if not impossible. In those cases, hardfacing and processing the worn element is quite reasonable, economically and technically. However, situations do occur, in which a defect in the hardface can cause the surface of the form, after the aforementioned repair work has been carried out, to become worn quickly, local irregularities or pits to appear. Those are visible on the surface of the product after injection, as localized micro-flashes. Therefore, detecting and dealing with defects in hardfacing is quite important in industrial practice.

Positive technical and economic effects of regenerating and modifying surfaces using the hardfacing method, in particular regard to elements, which require high resistance to wear, and corrosion, caused the dynamic development of technologies, tools and materials used for hardfacing. The welding techniques being used currently allow for repairs to be carried on practically every damaged or worn part, by way of welding or hardfacing. Some of the most modern and most often used hardfacing materials are electrodes, or solid and cored wires. The materials used currently are produced from Fe, Ni, Co, Ti with additions of different elements; among others: Si, Cr, B, Mn, V, Mo, W, Nb, which influence the mechanical, tribological and corrosive properties of these materials [2-7].

In industrial practice, detecting defects and inconsistencies in faced layers, is done using the same methods as the ones used in studying welded joints.

Modern computer tomography, with its advantages, is steadily rising in availability and usage in industrial applications, including welding. Computer Tomography (CT) is a type of x-ray spectroscopy. It is used a diagnostic method, allowing the user to obtain images of the layers of the faces being studied [8-11]. It utilizes

* Corresponding author: e-mail: j.jozwik@pollub.pl

a composite projection of the object, produced from images taken from different directions to create section (2D) and spatial (3D) images [12]. In relation to the above, computer tomography has been used in welding for non-invasive studies of welded joints and hardfaces. The aim of these studies is to determine the qualitative and quantitative occurrence of welding defects, their spacing and sizes [13]. CT is more accurate method when compared to the Roentgen methods used previously. It is however limited by high equipment cost, mobility and the size of the measurement surface [14-18].

2 Materials and methods

2.1 Hardfacing coatings

The subject of the study was hardfaced coatings. These coats ere made using GMA arc hardfacing with CO_2, with the wire being fed manually. The hardfacing was conducted using the ESAB POWER COMPACT 250 welding device. To ensure identical facing conditions, all processes were conducted using one type of shielding gas (CO_2); the other facing parameters were also unchanged for all other samples (Fig. 1).
The following parameters were used for facing:

- amperage: 230 A
- welding power: DC +
- wire feed speed : 7 m/min

In industrial practice, a common occurrence is the need to repair or regenerates parts of machines produced from different materials. Quite often, these materials are classified as difficult or impossible to weld, such as: 165CRV12 and 42CrMo4 steel. To check if such elements can be fixed or regenerated, with the use of cored wires being studies in this work, samples have been obtained by hardfacing different types of steel, starting with typical weldable steels, f.ex. C45 or S235JR, and ending with those marked as impossible to weld, 165CRV12 and 42CrMo4. For comparison purposes, the hardfacing was done using two different cored wires, intended for that process. The cored wires used, were FILTUB DUR 16 (trade name) and a Ø1,2 mm diameter wire marked as N-4, produced from the Fe-Mn-C-Si-Cr-Mo-Ti-W alloy. The chemical composition of the cored wires used for hardfacing can be found in Table 1.

Table 1. Chemical composition (w %) of the cored wires being used (FILTUB DUR 16 and N-4).

	Element (wt. %)						
	C	Si	Mn	Cr	Mo	W	Ti
FILTUB DUR 16	0.45	0.60	1.6	5.5	0.8	-	-
N-4	0.98	0.89	2.05	5.57	1.92	2.29	0.97

The FILTUB DUR 16 wire is an alkaline, intermediate alloy cored wire used for hardfacing coats resistant to grinding wear on elements such as: grabs of a digger, teeth of scoop, jaws, hammers, mixer blades, etc.). The cored wire produced from the Fe-Mn-C-Si-Cr-Mo-Ti-W alloy, is used for hardfacing coats resistant to grinding wear.

The coats were faced onto different surfaces, such as:

- C45 carbon steel - used in the production of machine bodies, form for processing plastics. It also used in the production of support elements of tools, such as: plates, spacers, etc. It is characterized by high strength and ductility.
- 165CrV12 steel (NC10) - a ledeburite chrome steel, intended, among others, for threading tools, high efficiency cutting tools, forming rollers, wire drawing dies, steel cutting knives. A steel characterized by a low tendency to crack.
- 42CrMo4 steel - heat enhancement steel with high hardenability. It is characterized by a high resistance to grinding wear. Used in the production of machine elements which require high strength and ductility, such as: turbine parts, cogs, rotors, shafts, pipe joints.
- S235JR steel - used in welded constructions. Utilized in support structures, as well as those carrying dynamic loads, such as: crane pillars, bodies of ore processing machines, heavily loaded bridges, pulleys.

Fig. 1. Samples faced using cored wired marked as N4 (left side) and FILTUB DUR 16 (right side), on: a) C45, b) 165CrV12, c) 42CrMo4, d) S235JR steel.

Welding was conducted down hand. Faces were layered spirally, in two layers, clockwise, starting from the center of the sample. Manually feeding the wire is characterized characterizes itself by relatively low thickness of the whole area being faced. The thickness of the faces being layered on in a single passes was, on average, 3 mm. The second layer was applied after a period of time required to remove the carbon deposits left, after applying the first layer.

The faced samples can be seen in Fig. 1. The samples were processed mechanically by grinding and electric discharge cutting, to appropriately correct and align their shape and size (diameter: 33.5 mm, height: 10 mm). The grinding process has been carried on a KENT surface grinder.

2.2 Microtomographic identification of defects in hardfaced layers

The inconsistencies in structure and internal defects has been carried out using an X25 (North Star Imaging) computer tomography machine, shown in Fig. 2. Source of x-ray radiation: Hamamatsu L12161-07 (voltage: 50 kV, current: 200 μA, focal spot size: 20 microns), detector – Varian (pixel pitch: 127 x 127 microns, mode: 01_1x10.5pFVG1 framerate: 10 fps. flip: none, rotation: 90 degree), Distances (tube to detector: 1014.228 mm, tube to part: 348.689 mm, calculated Ug: 0.03817 mm, zoom factor: ×2.91. The study consisted of directing an x-ray at the object, and registering the radiation intensity on the other side of the detector panel. All hardfaced samples have been exposed in identical conditions.

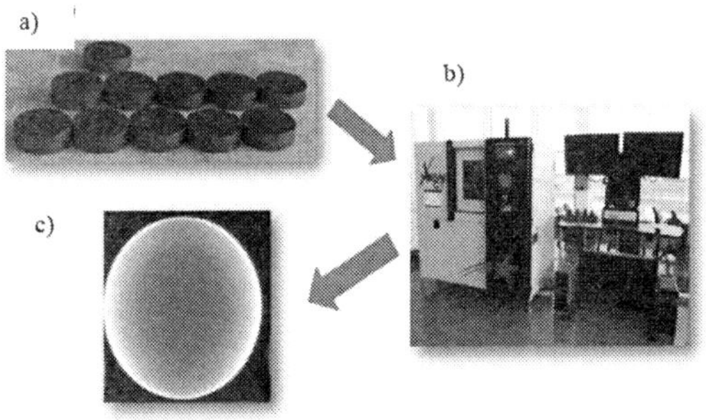

Fig. 2. Images from the study: a) samples after mechanical processing, b) X25 (North Star Imaging) computer tomography system, c) X-Ray image of the face.

3 Results and discussion

Samples, the base material of which was C145 steel, were faced using FILTUB DUR 16 cored wire. These did not show any defects, in the form of "white grains", which are areas of reduced radiation permeability. No scratches or cracks were observed either. The only defects were a few areas of slightly reduced density (Fig. 3). The size of these areas did not exceed 2 mm.

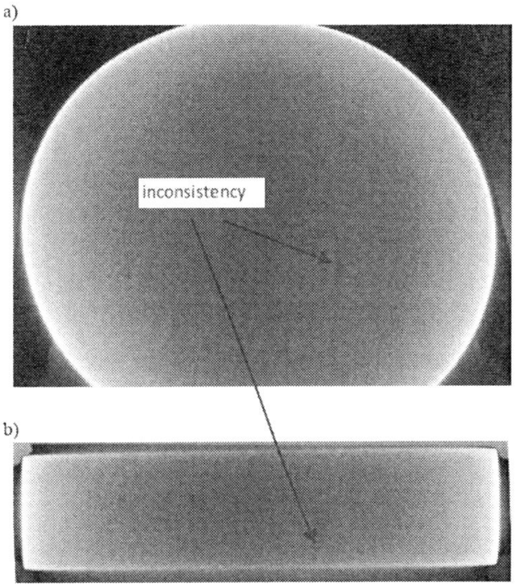

Fig. 3. Sample faced with FILTUB DUR 16 cored wire on C45 steel: a) horizontal cross-section (front view), b) vertical cross-section.

In the case of samples, the base material of which was C45 steel, faced with N4 cored wire, very slight intrusions can been seen on the whole surface of the face (Fig. 4).

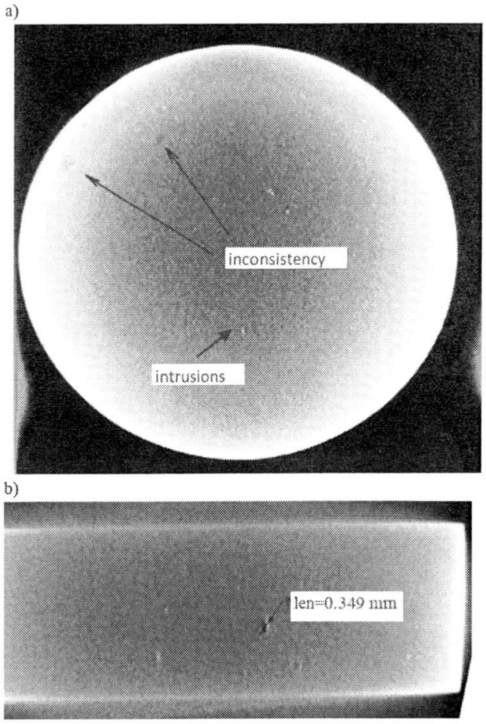

Fig. 4. Sample faced with N-4 cored wire on C45 steel: a) horizontal cross-section (front view), b) vertical cross-section.

These intrusions show up on the image as significantly brighter than the surrounding material. This is evidence of their higher density and significantly higher x-ray radiation absorption. The largest observed intrusion had a diameter of less than 0.4 mm. No signs of a layered structure were observed, which indicates proper melting of the base material during facing. However, individual small, darker areas have been observed, which indicates the presence of inconsistencies, such as air bubbles.

Similarly to samples on a C45 steel base, 165CRV12 (NC10) based samples faced with FILTUB DUR 16 cored wire did not exhibit any of the following defects: scratches, cracks or grains of a significant size and radiation permeability (Fig. 5). Defects consisting of areas of reduced density were rare and small. The largest face structure inconsistencies observed were smaller than 1.5 mm.

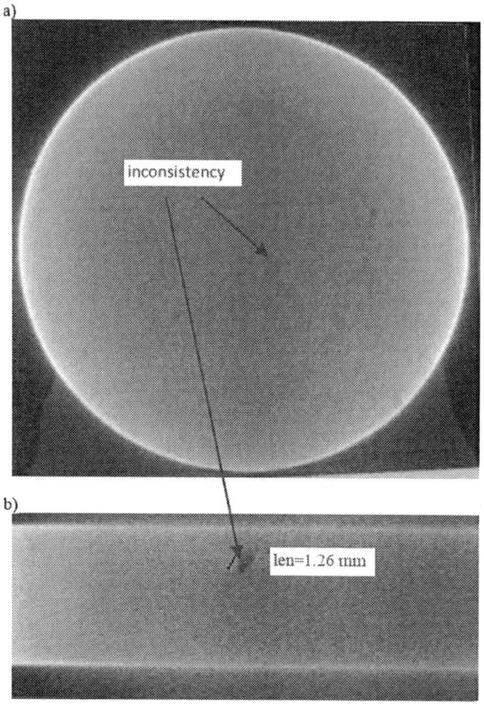

Fig. 5. Sample faced with FILTUB DUR 16 cored wire on 165rV12 steel: a) horizontal cross-section (front view), b) vertical cross-section.

In samples faced with N-4 cored wire, on a 165CrV12 steel, one can observe (similarly to C45 steel) very slightly, significantly large intrusion, absorbing X-rays, as well as intrusions with inconsistent distribution (Fig. 6). Layered structures have not been found. A smaller amount of lower density areas has been observed, however they were larger. The largest inconsistency area measured: 0.981 x 1.72 mm.
Also visible are scratches and cracks. The largest crack measure 8.97 mm in length and 0.243 mm in width. The largest crack was 0.563 mm deep. The base material remained untouched.

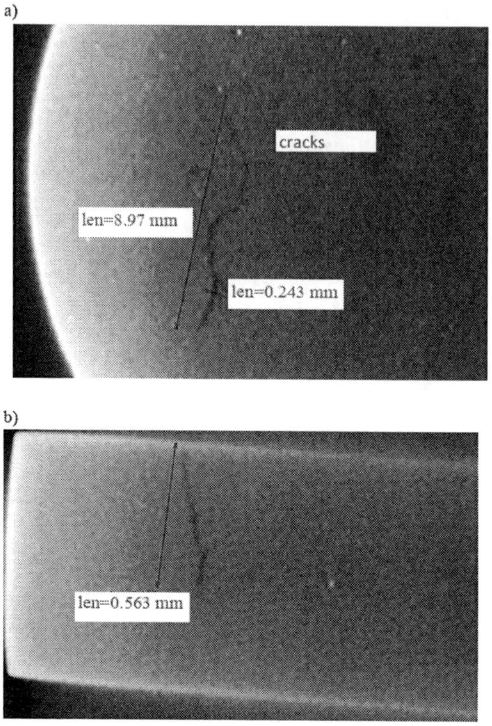

Fig. 6. Sample faced with N-4 cored wire on 165CrV12 steel: a) horizontal cross-section (front view), b) vertical cross-section.

Samples faced using FILTUB DUR 16 cored wired (base material 142CrMo4), do not exhibit any visible defects (Fig. 7).

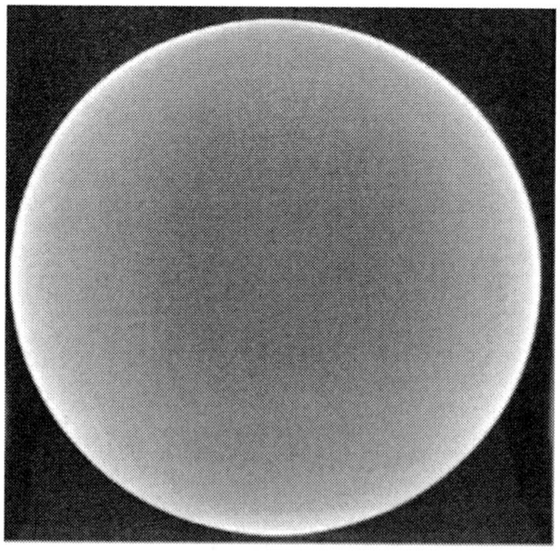

Fig. 7. Sample faced with FILTUB DUR 16 cored wire on 42CrMo4 steel- horizontal cross-section (front view).

The samples faced using N-4 cored wire on 42CrMo4 base are characterized by small, irregularly spaced intrusions, characterized by high density and low X-ray permeability appearing in the face, as well as smaller areas of much lower density, identified as air bubbles.

 DOI: 10.1051/e3sconf/20171903025

The sizes and charcteristics of these defects is similar to those on 165CrV12 steel. In the case of 42CrMo4 steel, these cracks appear in larger numbers, and are of similar size when compare to those on 165CrV12 steel. These cracks can be as deep as the whole face, in both layers (Fig. 8). However, they are less visible and narrower (about 0.1 mm wide).

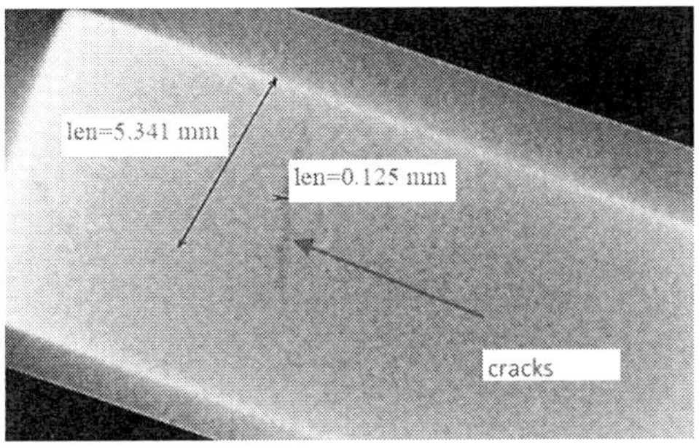

Fig. 8. Sample faced with M-4 cored wire on 42CrMo4 steel- vertical cross-section.

Similar to all previous samples faced with FILTUB DUR 16 cored wire, the sample with a S235JR steel base does not exhibit any defects manifesting as small areas with very high density. No cracks or scratches are visible either.

However, the area of reduced density can be observed in all areas of the sample, located at a depth corresponding to the first layer of the hardface (Fig. 9).

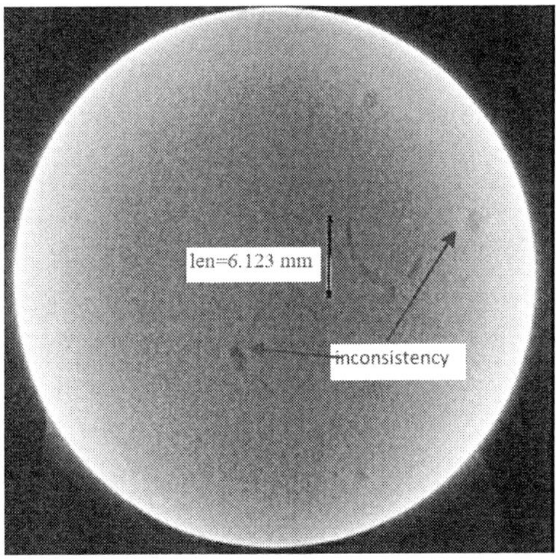

Fig. 9. Sample faced with FILTUB DUR 16 cored wire on S 235 JR steel.

As with all samples faced with N4 wire, the samples with S235JR steel exhibit very small intrusion spaced irregularly along the whole mass of the face, characterized by small X-ray permeability, visible in photographs as very small, white grains.

Reduced density defects were also observed. They occur at a depth corresponding to the first hardface layer. Largest horizontal cross-sections: 1.465 x 7.369 mm (Fig. 10)

Fig. 10.. Sample faced with N-4 cored wire on S 235 JR steel.

A compilation of analysis results for the studied samples can be found in Table 2. The analysis showed, that less defects were found in the faced layers when using FILTUB DUR 16 cored wired. In regards to the N-4 wire: modifying the hardfacing process was recommended in most cases.

Table 2. Compilation of analysis results for the studied samples.

Face/ Base	Types of exhibited defects	Notes
FILTUB DUR 16/ C45	Individual bubbles in the first hardface layer	Decision to be made dependent on intended use
N-4/ C45	Grains of unidentified intrusions.	Decision to be made dependent on intended use
FILTUB DUR 16/ 165CrV12	Individual bubbles in the first hardface layer	Decision to be made dependent on intended use
N-4/ 165CrV12	Grains of unidentified intrusions. Cracks	Process modification recommended
FILTUB DUR 16/ 42CrMo4	No faults	No notes
N-4/ 42CrMo4	Grains of unidentified intrusions. Cracks	Process modification recommended
FILTUB DUR 16/ S 235JR	Groups of bubbles in the first hardface layer	Process modification recommended
N-4/ S 235JR	Grains of unidentified intrusions. Groups of bubbles in the first hardface layer	Process modification recommended

4 Conclusions

On the basis of the carried out research the following conclusions have been formulated:

- The computer tomography method described in this paper offers a significantly better process of diagnostics and controlling of hardfaced layers and welded joints when compared to rational radiography.
- CT allows for better accuracy when detecting

DOI: 10.1051/e3sconf/20171903025

welding defects, which allows the user to evaluate the hardfaces and determine their operative usability.

• All hardfaces produced on samples using FILTBU DUR 16 cored wires exhibit inconsistencies in the form of pores. All of the faces did not exhibit any cracks.

• All hardfaces produced on samples using N4 cored wires exhibit small, high density intrusions of an unidentified material. The intrusions are visible as white grains on the grey background on the face.

• Hardfaces produced on samples using N4 cored wires on 165CrV12 and 42CrMo4 wires, identified as difficult to weld exhibit multiple cracks. A solution to this problem can be a modification of the process, i.e. changing hardfacing parameters, using a buffer layer or heating the base material.

References

1. R. Kolbusz, A. Tabor, W. Wojciechowski, AFE **6**, 21 (2006)
2. M. Pashechko, K. Dziedzic, M. Barszcz, Powder Metall. Met. Ceram. **52**, 7-8 (2013)
3. J. Józwik, K. Dziedzic, I. Usydus, P. Raos, G. Krolczyk, Teh. Vjesn. **5**, 23 (2016)
4. N. Yüksel, S. Sahin, Mater. Des. **58** (2014)
5. E. Badisch, C. Katsich, H. Winkelmann, F. Franek, R. Manish, Tribol. Int. **43** (2010)
6. R. Kumar, S. Chattopadhyaya, S. Hloch, G. Krolczyk, S. Legutko, Eksploatacja i Niezawodnosc – Maintenance and Reliability **18**, 1 (2016)
7. M. Tabatabaeipour, J. Hettler, S. Delrue, K. Abeele, NDT&E International **80** (2016)
8. E. Ratajczyk, Pomiary Automatyka Robotyka **5** (2012)
9. S. K. Dinda, J. M. Warnett, M. A. Williams, G. Roya, P. Srirangam, Mater. Des. **96** (2016)
10. J.B. Krolczyk, B. Gapinski, G.M. Krolczyk, I. Samardzic, R.W. Maruda, K. Soucek, Y. Javadi, S. Legutko, P. Nieslony, L. Stas, Teh. Vjesn. **23** (1) (2016)
11. J. Zapata, R. Vilar, R. Ruiz, Expert Syst. Appl. **38** (2011)
12. M. Yang, S.M. Xiong, Z. Guo, Acta Materialia **92** (2015)
13. U. Hampel, A volume in Woodhead Publishing Series in Electronic and Optical Materials (2015)
14. A. Kułaszka, M. Chalimoniuk, M. Wieczorowski, D. Brzozowski, Przegląd Spawalnictwa **12** (2015)
15. J. Józwik, K. Dziedzic, W. Wierzba, Przegląd spawalnictwa **88**, 7 (2016)
16. J. Jozwik, M. Czwarnowski, ASTRJ **9**, 28 (2015)
17. J. Jozwik, L. Semotiuk, I. Kuric, ASTRJ **9**, 28 (2015)
18. G. M. Krolczyk, J. B. Krolczyk, R. W. Maruda, S. Legutko, M. Tomaszewski, MEAS **88** (2016)

E3S Web of Conferences **19**, 03026 (2017) DOI: 10.1051/e3sconf/20171903026
EEMS 2017

AlSi21CuNi silumin modification with phosphor and strontium micro additions

Mariusz Michalski[1,*], and *Ferdynand* Romankiewicz[1]

[1] Faculty of Mechanical Engineering, University of Zielona Gora, 4 Prof. Z. Szafrana Street, 65516 Zielona Gora, Poland

Abstract. The paper presents the results of investigation of the influence of AlFeP and AlSr modifiers on the structure of the hypereutectoid silumin, AlSi21CuNi. The silumin has been modified in two stages. The first stage was melt with the addition of phosphor in the quantity of 120ppm. The second stage was combined application of phosphor and strontium. The strontium addition amounted 150, 250 and 350 ppm. At each melt, the metal has been cast after 10, 30, and 60 minutes from the introduction of the modifiers. The effectiveness of the modification has been assessed basing on qualitative and quantitative analysis of the microstructure of castings. It has been determined that the applied addition of phosphor causes significant comminution of the primary silicon secretions. Maximum comminution of primary silicon has been obtained after the longest time of the modified alloy soaking. The most advantageous effects of combined modification have been obtained with the application of 120ppm addition of P and 350ppm addition of Sr after 60 minutes from the introduction of the modifiers to the metal bath.

1 Introduction

The AlSi21CuNi alloy belongs to the group of piston hypereutectoid silumins. The structure of the alloy consists of hard primary secretions of phase â (Si) and relatively soft eutectic mixture (á +â) and intermetallic phases. In addition to many advantages, the hypereutectoid silumins have a negative feature related to their tendency to form coarse grained structure directly influencing deterioration of mechanical properties. Phase â in the form of very big brittle crystals creating clusters appears on the background of coarse-grained lamellar eutectic mixture. The proved features of such material prevent meeting the requirements concerning surface smoothness and dimensional accuracy of the processed castings, mostly pistons, and significantly deteriorate their tribological properties [1, 2].

In order to eliminate the indicated drawbacks of the structure, hypereutectoid silumins atrre subjected (at the stage of casting) to the process of modification with theμ use of phosphor addition. After modification, the structure of the obtained castings consists of significantly comminuted (down to the sizes below 50 μm) primary silicon secretions with compact wall structure, evenly distributed in the matrix of the irregular lamellar eutectic mixture ($\alpha + \beta$) [3].

The up-to-date development of the production technology forces engineers to improve the quality of the materials used constantly [4,5]. This concerns, among others, pistons which are subjected to more and more severe thermal and mechanical loads and longer time of exploitation in the automotive industry. A survey of literature [6-8] shows that, in the case of hypereutectoid silumins, there is a possibility of further structure improvement and, consequently, obtaining better mechanical properties of the castings. This is to be achieved due to the change of the morphology of the irregular lamellar eutectic mixture into one with fibrous structure while maintaining the adequate degree of comminution and even distribution of the primary silicon secretions. Unlike lamellar eutectic mixture, fibrous one has minimum interfacial distance, ë, and rounding of the eutectic silicon crystal contours, which is of much importance for the improvement of impact strength and wear resistance [7]. That is why more and more researchers working on improvement of the hypereutectoid silumin properties think that it is purposeful to elaborate complex modifiers containing, in addition to phosphor, other additives aiming at modification of the eutectic mixture. The change of irregular lamellar eutectic mixture into fibrous one is achieved due to the modification of silumin with sodium or strontium. An important feature of strontium as a modifier is its resistance to oxidation. The mechanism of phosphor acting in modification of hypereutectoid silumins is different from that of strontium acting in the process of eutectic mixture modification in hypoeutectic alloys. That is why combined modification of hypereutectoid silumins with the addition of phosphor and strontium is problematic and its effectiveness depends mainly on the application of adequately selected technological parameters of the modification. In the foundry practice, manufacturing a casting with the desired structure and, consequently, desired properties, depends on many factors, such as: the temperature and

* Corresponding author: m.michalski@ibem.uz.zgora.pl

time of the liquid alloy heating, kind and quantity of the applied modifier, modification temperature, time of maintaining liquid alloy after modification, casting temperature, solidification rate.

Although the technology of combined (phosphor and silicon) modification of hypereutectoid silumins is widely described in literature, the results of those investigations concern mostly alloys with less content of silicon. As far as high silicon alloys are concerned, there is shortage of information about the influence of specific parameters; this is particularly related to the influence of: the quality of the modifying agents, quantity of the applied modifiers, time of the liquid alloy maintenance after modification on the effectiveness of the performed modification. Complex treatment of the process of modification of hypereutectoid silumins is, therefore to be understood as creation of such conditions of the alloy crystallization in which the above partial elements are summed and make it possible to obtain advantageous structure and, consequently, advantageous properties of castings.

In recent years, investigations aiming at comminution of the structure of castings by the application of magnetic field or advanced casting techniques are also performed [9, 10]. The use of those techniques, however, is complicated and, therefore, limited.

2 Investigation methodology

The tests have been performed with the use of the popular hypereutectoid piston silumin, AlSi21CuNi with the chemical composition to be found in Table 1.

Table 1. Chemical composition of the tested AlSi21CuNi alloy

Percentage [%]					
Si	**Cu**	**Ni**	**Mg**	**Mn**	**Fe**
21.170	1.200	0.870	0.680	0.170	0.450
Ti	**Zn**	**Pb**	**Sn**	**Al**	
0.020	0.200	0.020	0.007	residual	

The melts have been performed at the temperature of the metal bath soaking, 850°C for a period of 30 minutes. The operation of single modification has been performed with the use the initial alloy, AlFeP, with the chemical composition shown in Table 2. The quantity of the modifying agent has been selected so as to ensure phosphor addition at the level of 120ppm.

Table 2. Chemical composition of the modifying agent, AlFeP.

AlFeP- percentage [%]						
Fe	**P**	**Si**	**Mn**	**Ti**	**C**	**Al**
14.36	4.69	0.49	0.30	0.15	0.05	residual

Complex modification of silumin has been performed by combining agent AlFeP with the modifying agent AlSr10 (10% of Sr, Al – the rest). Maintaining the quantity of the first agent (120ppm of P), three quantities of strontium addition have been applied: 150, 250, 350ppm referred to the mass of the alloy being tested. In order to determine the influence of the time of modified alloy soaking on the effectiveness and durability of the modification results, the alloy has been cast after 10, 30 and 60 minutes from the moment of the modifier introduction to the metal bath. In order to obtain samples of unmodified alloy, a trial of melting has been performed applying the procedure as in the case of modified alloy with the omission of the operation of modification.

The silumin has been melted in a resistance furnace with type use of a chamotte – graphite crucible. Maintenance of the desired metal temperature has been ensured by a microprocessor temperature control device type TC6 made by METROL company coupled to a thermocouple installed in the cylindrical chamber of the furnace. The liquid metal temperature has been monitored by dipping in it the measurement end of the thermocouple type K (NiCr-NiAl), entering it through a hole made in the insulation cover of the furnace. The mass of the metal charge was 1.2 kg for each melt. After the predetermined temperature of (850°C) has been reached, the liquid metal was soaked for 30 minutes. Next, at 750°C, a soaked graphite sinker was used to introduce the modifier to the metal bath. The tests have been performed with the use of a cylindrical split gravity die (with the inner diameter of 27 mm, wall thickness of 17 mm and the height of 45 mm), located on a steel base. In order to obtain samples for metallographic examination, when the casting cooled down a part of it was cut off at the level one third of the height from the mould bottom (in the direction perpendicular to the casting axis). Microsections have been made on the sample cutting surface by means of an automatic grinding and polishing machine made by "STRUERS" company. Polished samples have been subjected to etching for 10 seconds with reagent 0.5% HF with the composition of (0.5 ml HF + 99.5 ml H_2O).

Metallographic examination has been performed with the use of a scanning microscope, JSM-5600LV made by JEOL company, coupled to an x-ray microanalyzer, EDS. Qualitative analysis of the samples has been performed basing on the image obtained with the use of a detector of backscattered electrons (BEI). For a more full description of the structure of castings, a quantitative analysis of the primary silicon secretions and eutectic silicon secretions has been performed. Those tests have been performed with the use of quantitative metallographic methods basing on computer analysis of image. The analyses have been performed with the use of the "Particle Analysis" module of the EDS 2000 microanalyzer software. Metallographic examinations (qualitative and quantitative) of all samples have been performed in micro-areas located at the middle of the microsection radius. In the quantitative description of the microstructure, the following parameters have been used: N_A – relative amount of secretions (number of primary or eutectic silicon crystals in a unit of the microsection surface, mm^{-2}), D_2 - average diameter of primary silicon secretions (calculated as the diameter of a circle with the same area as the particle, μm) and ë-average interfacial distance (formula 1).

$$\lambda = 1/N_A^{1/2} \quad (1)$$

The results of quantitative metallography, in the form of the values of stereological parameters of the primary silicon crystals, have been statistically elaborated. For each sample, the values of N_A, D_2, being arithmetical average ones of those parameters from N analysed micro areas of the microsection and then, for each stereological parameter, standard deviation (S) has been calculated, as well as the width of the confidence interval (δ) [10]. Since the number of the analysed micro images has always met the condition of $N < 30$, t-Student distribution has been adopted for the calculations; the $t_{\alpha,n-1}$ value for ut has been taken from tables, for the adopted level of confidence, P= 0.95.

3 Test results

Analysis of the structure of unmodified alloy has shown that the crystals of primary silicon have big dimensions and irregular shapers. The crystals often form clusters (Fig. 1a, b), which indicates uneven distribution of those secretions in the volume of the casting. another negative feature of those secretions is their not compact structure as evidenced by "voids" visible in the crystal cross sections. In an analysis of the quality of the eutectic silicon structure, occurrence of micro areas with secretions in the small lamellar and fibrous form has been found (Fig. 1a, b). This is due to the quick cooling of the casting (application of the gravity die) and the cluster character of the primary silicon secretions. Those observations prove the results of quantitative metallography shown in Table 3. Basing on those results, one can also state that, as the time of the bath soaking flows, more comminution of the primary silicon secretions is obtained. The average diameter of the silicon crystals after the first casting time (10 minutes) was about 65.7 μm; in a casting obtained after 60 minutes it was only about 50.5 μm. When comparing the N_A parameter, i.e. relative amount of primary silicon crystals, small value of it has been found for all castings, however, for the subsequent casting times, the value increases from N_A = 33.6 mm^{-2} to 52.5 mm^{-2} in a casting obtained after 60 minutes. In the case of eutectic silicon, it has been found that the degree of its comminution decreases with the extension of the metal bath soaking time (Table 3) while the interfacial distance increases.

Table 3. The results of measurements of stereological parameters of primary and eutectic silicon of silumin in the unmodified state, after modification with the AlFeP agent and after complex modification with AlFeP and AlSr.

Melt designation	Casting time, min.	PRIMARY SILICON									EUTECTIC SILICON	
		Stereological parameters										
		D_2	S	δ	N_A	S	δ	λ	S	δ	N_A	λ
		µm			mm^{-2}			µm			mm^{-2}	µm
Unmodified alloy	10	65.7	5.5	± 3.1	33.6	9.9	± 5.7	178.4	28.9	± 16.5	235873.2	2.1
	30	63.8	6.4	± 4.0	34.8	7.6	± 4.7	172.5	20.2	± 12.6	220126.7	2.1
	60	50.5	4.2	± 2.8	52.5	10.1	± 6.0	140.0	15.1	± 9.0	145260.9	2.6
AlFeP 120ppm P	10	28.4	2.1	± 1.6	184.8	43.0	± 31.9	75.1	9.5	± 7.1	60885.2	4.1
	30	21.0	1.0	± 0.7	317.3	58.7	± 43.6	56.8	5.1	± 3.8	17961.7	7.5
	60	17.7	0.8	± 0.5	411.9	19.1	± 55.0	49.9	5.0	± 3.5	21323.8	6.8
AlFeP 120ppmP **AlSr** 150ppm Sr	10	70.6	9.0	± 5.1	28.8	9.2	± 5.3	193.3	31.7	± 18.0	167662.9	2.4
	30	43.4	1.9	± 1.2	70.2	10.6	± 7.0	120.3	8.7	± 5.7	98347.5	3.2
	60	25.0	1.8	± 1.3	227.5	23.2	± 17.2	66.5	3.4	± 2.5	99954.8	3.2
AlFeP 120ppmP **AlSr** 250ppm Sr	10	65.1	6.1	± 4.0	27.9	5.5	± 3.6	192.7	23.6	± 15.5	367606.5	1.6
	30	47.3	2.7	± 1.9	59.8	12.2	± 8.5	131.4	15.2	±10.6	136119.3	2.7
	60	32.0	2.4	± 1.8	135.4	25.8	± 19.2	87.0	8.5	± 6.3	119644.4	2.9
AlFeP 120ppmP **AlSr** 350ppm Sr	10	42.5	4.5	± 3.6	87.6	12.6	± 10.1	107.6	7.8	± 6.2	540217.8	1.4
	30	31.4	4.1	± 3.0	153.2	39.3	± 29.2	82.6	10.5	± 7.8	100457.1	3.2
	60	28.2	2.5	± 1.9	160.9	35.3	± 26.2	80.1	8.4	± 6.2	107589.5	3.0

 DOI: 10.1051/e3sconf/20171903026

a) unmodified

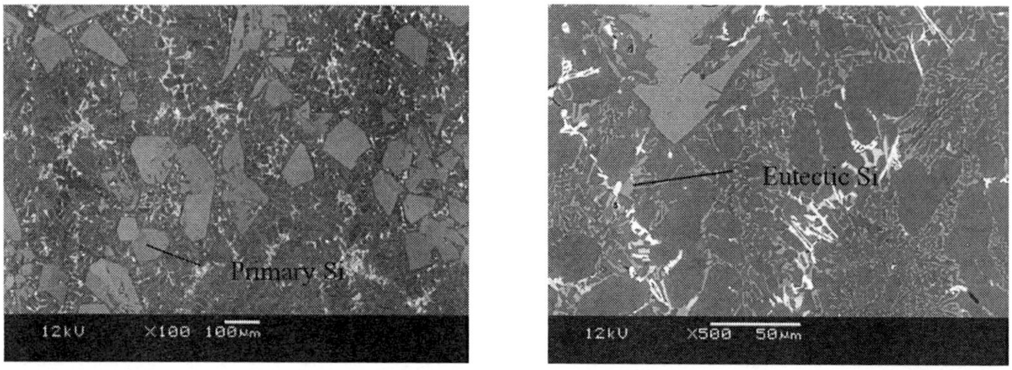

b)

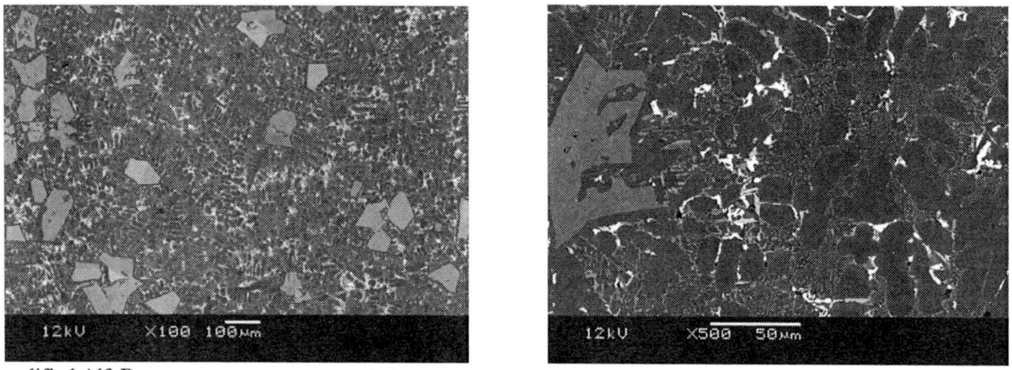

c) modified AlfeP

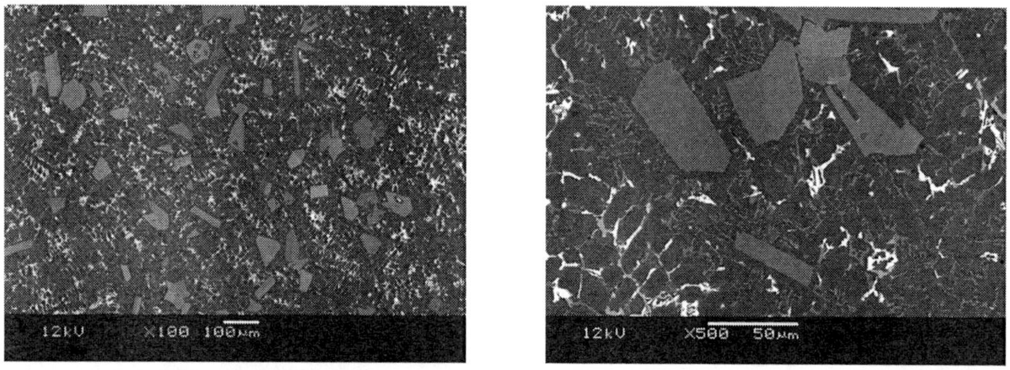

d)

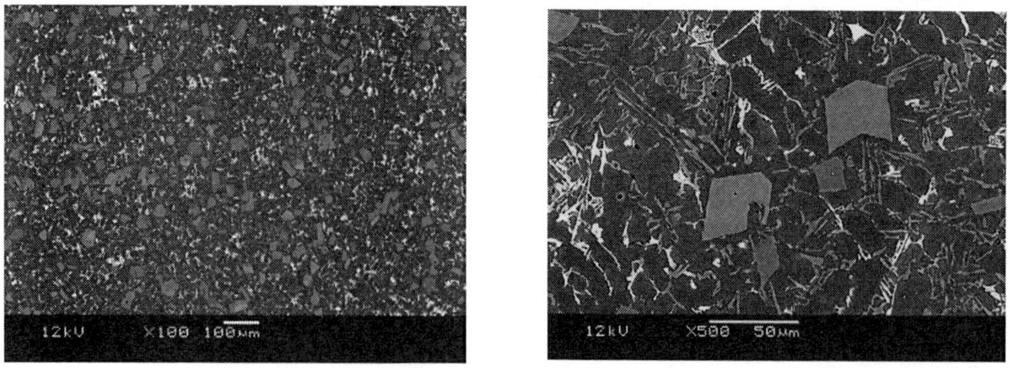

Fig. 1. Microstructure of AlSi21CuNi silumin: unmodified, time to casting: a) 10 minutes, b) 60 minutes; modified with phosphor, time to casting: c) 10 minutes, d) 60 minutes. Soaking temperature 850°C, magnification 100x and 500 x.

 DOI: 10.1051/e3sconf/20171903026

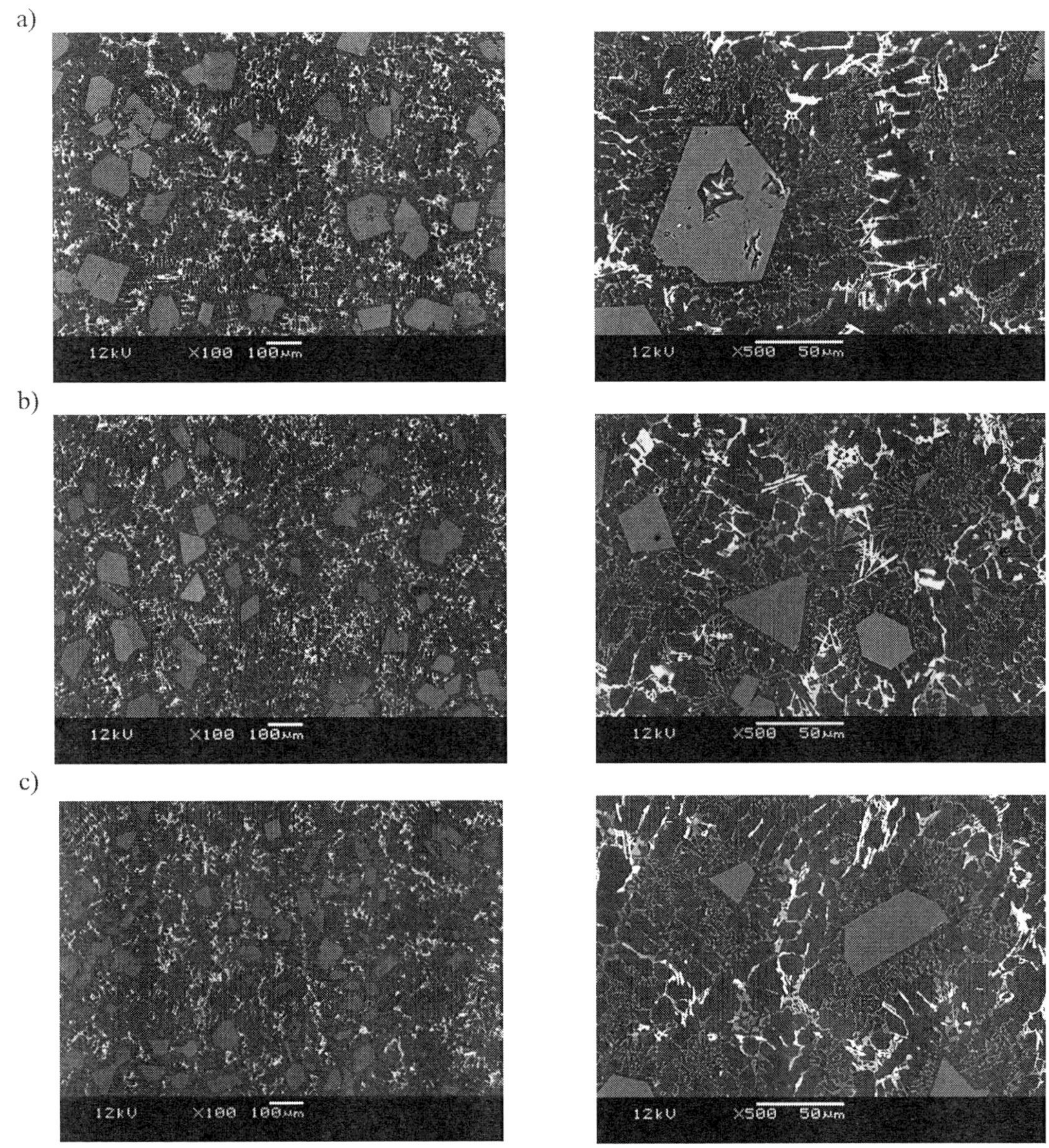

Fig. 2. Microstructure of AK20 silumin after combined modification with AlFeP agent in the phosphor quantity of 120 ppm and AlSr agent in the strontium quantity of 350 ppm., time to casting: a) 10 minutes, b) 30 minutes, c) 60 minutes. Magnification 100x and 500x.

Modification with the use of phosphor has resulted in advantageous changes in the silumin structure (fig.1c, d) as compared to unmodified alloy. In the sample cast after 10 minutes (fig.1c) from the moment of modifier introduction, primary silicon had fine and compact structure. The structure analysis shows that the effect of comminution of the primary secretions of silicon increases with the extension of the time of the modifier contact with the metal bath. As regards comminution of the primary silicon, the most advantageous structure has been obtained in the sample cast after 60 minutes (fig.1d). However, longer time of the modified metal soaking deteriorates the structure of eutectic silicon. From fine crystal structure (magnification 500x, fig. 1c), eutectic silicon changes to lamellar (magnification 500x, fig. 1d). These observations are proved by the results of quantitative metallography (table 3). With the comminution of the primary secretions of silicon, their amount grows significantly. The N_A value in the sample cast after 60 minutes from the introduction of phosphor to the bath is 411.9 mm^{-2} and is more than twice as high as that recorded in a sample cast after 10 minutes. opposite effect has been observed for eutectic silicon. In the sample cast after 10 minutes, the N_A value was 60885.2 mm^{-2}; in a sample cast after 60 minutes, the relative amount of eutectic silicon crystals has decreased by about three times.

The examination results of complex modification

(with phosphor and strontium) of AlSi21CuNi silumin have been shown in the form of examples of microstructure photographs (fig. 2) and table 3 in which the values stereological parameters of all melts of the investigation cycle under discussion are presented. The structure analysis shows that combined modification with the addition of phosphor and strontium, regardless of the addition quantity applied (150, 250, 350 ppm) has brought the most advantageous effects of the primary silicon comminution after casting time of 60 minutes (fig. 2c). It has also been proved by the results of quantitative metallography.

The smallest average diameter of the primary silicon crystals, $D_2 = 25.0\ \mu m$, has been found after casting time of 60 minutes with the presence of the smallest applied quantities of strontium (together with phosphor). Increase of the addition of strontium has resulted in an increase of that value, i.e. deterioration of the effectiveness of the primary silicon modification. In an analysis of the results of the melts under discussion in relation to the modification of eutectic mixture, the effect of strong refinement of it has been observed. T his has been proved not only by microstructure photographs but also by the obtained values of N_A and λ. In this aspect, the most advantageous effects have been obtained in the casting modified with the additions of 120 ppm of P and 350 ppm of Sr, cast after10 minutes where $N_A = 540217.8\ mm^{-2}$ and $\lambda= 1.4\ \mu m$. The analyses presented above show that, with extension of the modified alloy soaking time, the degree of comminution of the primary silicon secretions increases, but the effect of eutectic mixture refinement gets weaker. In connection with that, obtaining of complex modified structure of the AlSi21CuNi silumin structure can depend on reaching a "compromise" between satisfactory comminution of the primary silicon and adequate quality of the eutectic mixture. Such assumption indicates that the most advantageous effects of complex modification have been obtained with the addition of 120 ppm of phosphor and each of the applied quantities of strontium, in castings obtained after 60 minutes from the introduction modifiers to the metal bath.

4 Conclusions

Basing on the investigation of AlSi21CuNi silumin modification, it has been stated that the obtained structure of the castings has been influenced not only by the kind and quantity of the added modifiers but also by the time of the modified alloy soaking.

The performed tests are a basis to formulate the following conclusions:

- the structure of unmodified alloy castings consists of big secretions of primary silicon and fine crystal eutectic silicon forming clusters. As the soaking time of the liquid alloy is extended, the primary secretions get slightly smaller with the simultaneous deterioration of the effect of the eutectic mixture comminution;
- maximum comminution of primary silicon has been found in a sample cast after 60 minutes from the introduction of the AlFeP agent (120 ppm of P) to the metal bath. It has been determined that the longer time of soaking the alloy modified with phosphor, the more disadvantageous (thick lamellar) structure has the eutectic silicon;
- complex modification of AlSi21CuINi silumin (with phosphor and strontium) is a "compromise" between a satisfactory effect of primary silicon comminution and adequate quality of the eutectic mixture. From such point of view, the most advantageous effects of modification have been obtained in the sample cast after 60 minutes from the introduction of 120 ppm of P and 350 ppm of Sr.

References

1. Z. Poniewierski, WNT, Warszawa, (1989)
2. P. Wasilewski, PAN, Katowice, (1993)
3. M. Michalski, F. Romankiewicz, Editor. K. Świątkowski, Wydaw. Naukowe "Akapit", Kraków, (2014)
4. P. Niesłony, G.M. Krolczyk, K. Zak, R.W. Maruda, S. Legutko, Precis. Eng. **47** (2017)
5. R.W. Maruda, S. Legutko, G.M. Krolczyk, C. Lukianowicz, A. Stoić, Teh. Vjesn. – Tech. Gaz. **22**, 5 (2015)
6. F. Binczyk, E. Krzemień, J. Piątkowski, II Ogólnopolska Konferencja Naukowa „Quality '99", Odlewnictwo, Kraków, (1999)
7. K. Müller, W. Reif, B. Pol. Acad. Sci-Tech. **28** (1996)
8. F. Romankiewicz, R. Romankiewicz, M. Michalski, Archiwum Odlewnictwa, **4**, 11 (2004)
9. K. Al-Helal, I.C. Stone, Z. Fan, Trans. Indian Inst. Met. **65**, 6 (2012)
10. X. Chen, Y. Zhong, T. Zheng, Z. Shen, J. Wang, L. Fan, Y. Zhai, M. Peng, B. Zhou, W. Ren, Z. Lei, Z. Ren, Q. He, J. Alloy, Compd. **714** (2017)

Evaluation of tribological properties of selected engine oils during operation of the friction pairs of steel-on-steel

Marcin Barszcz[1,*], *Jerzy* Józwik[2], *Krzysztof* Dziedzic[3], and *Kamil* Stec[1]

[1]Lublin University of Technology, Fundamentals of Technology Faculty, Department of Fundamental of Technology, 38 Nadbystrzycka Street, 20-618 Lublin, Poland
[2]Lublin University of Technology, Mechanical Engineering Faculty, Department of Production Engineering, 36 Nadbystrzycka Street, 20-618 Lublin, Poland
[3]Lublin University of Technology, Electrical Engineering and Computer Science Faculty, Institute of Computer Science, 36B Nadbystrzycka Street, 20-618 Lublin, Poland

Abstract. The paper includes an assessment of the tribological properties of mineral and synthetic Lotos oil marked SAE 15W/40 and SAE 5W/40 at ambient temperature and 100 °C. The evaluation was based on the analysis of the tribological properties of friction couple consumables. Tribological tests were performed using the Anton Paar THT 1000 high temperature tribotester according to ASTM G133. Tribological properties were investigated using the "ball on disc" method. The change of friction coefficient, friction couple temperature, volume wear of samples and counter-samples and Hertz stresses were evaluated. In addition, hardness tests of the friction couple materials as well as surface roughness before and after friction were performed. On the basis of tribological studies, it was noted that Lotos Synthetic 5W/40 oil has better cooling properties compared. For both oils the coefficient of friction was lower at ambient temperature than at 100 °C. The highest value of volume wear of the sample was noted for the combination lubricated with Mineral Oil 15W/40 at 100 °C (0.0143 mm^3) while for counter-sample lubricated with synthetic oil at ambient temperature (0.0039 mm^3). The highest sample wear coefficient was recorded for the mineral oil lubricated at temperature of 100 °C ($3.585*10^{-7}$ $mm^3/N/m$) while for counter-sample lubricated with synthetic oil at ambient temperature ($9.8768*10^{-8}$ $mm^3/N/m$). The Hertz stress for each test couple had a value of 1.787 GPa.

1 Introduction

Wear of machine and equipment parts resulting from friction has always been a problem for engineers involved in the design and selection of materials for friction couples. It determines the durability of the whole machine or device, and the period of failure-free operation, which ultimately affects the quality of their work [1].

In many cases, the elements of machines and devices parts cooperating in terms of friction work in very heavy and variable conditions. This involves the use of materials of appropriate utility properties, the use of specialized surface treatment methods, lubricants and lubrication techniques. This certainly has a significant impact on the work of tribological systems and their suitability for specific working conditions. Therefore, the current work and research is aimed at finding new construction and material solutions that reduce the wear and production and operation costs of machines and equipment [2-6].

One of the methods to reduce the friction and wear of the cooperating elements is to use suitable lubricants. In recent years, intense works have been carried out to improve their properties. Many researchers are attempting to introduce various additives to engine oils. Authors of many works have confirmed that adding nanoparticles to the oil leads to improved tribological properties. Nanoparticles such as carbon nanoparticles and, in particular, fullerene nanoparticles [7] diamond nanoparticles [8], nanotubes [9] and graphene nanoparticles [10,11] were introduced. These exhibited promising properties thanks to the ease of shear of individual layers. Nanofulerenes and nanotubes showed a reduction in friction and wear from macro- to nanoscale, due to the formation of tribofilm. Research with addition of oxides nanoparticles were carried out by the following authors: copper oxides CuO [12,13,14], titanium dioxide TiO_2 [13], silicon dioxide SiO_2 [15] and metal nanoparticles, i.e. copper [16,17], iron, cobalt and their combinations [17]. According to [17], the use of Cu nanoparticles provides the most effective friction and wear reduction in each of the nanoparticles studied. Surface analysis indicates that the nanoparticle components have precipitated on the contact surface during the application of oils with nano-additives. In turn, the authors [18,19] investigated the effect of modifying the Lotos Dynamic 0W/40 oil with the chemical treatment of Motor-Life Professional. The authors of the paper [20] have noted that a suitable lubricant has an influence on the self-organization of the friction surface. In the paper [20], based

* Corresponding author: jjozwik@pwsz.chelm.pl

on the results of the various lubricants, the authors have indicated that the wear intensity is low. According to [21], this effect of the lubricant and its properties is related to the self-organization of the surface in the friction process.

The authors of this paper and the team are engaged in the development and testing of wear resistant materials [22-25]. In recent years, they have attempted to investigate the phenomenon of surface self-organization in the process of friction. As a result, basic tribological studies were carried out to determine tribological properties of mineral oil and synthetic Lotos oil marked SAE 15W/40 and SAE 5W/40 at ambient temperature and 100 °C. The authors of this paper will use these oils for further research to modify and study self-organization of the friction surfaces.

2 Experimental details

2.1 Research object

Two engine oils were analyzed: Lotos Synthetic 5W/40 and Lotos Mineral 15W/40, which were subjected to tribological evaluation at ambient temperature and temperature of 100 °C. The temperature of 100 °C was selected because it is similar to the operating temperature of the engine oil. The specifications for engine oils are shown in Tables 1 and 2.

Table 1. Specification of synthetic engine oil Lotos Synthetic 5W/40.

Parameter name	Parameter
API quality class	SN/CF
SAE viscosity grade	5W/40
Kinematic viscosity at 100 °C	13.7
Viscosity index, min	165
Flow temp., max (°C)	-36
HTHS viscosity at 150 °C	3.6
Basic number (mgKOX)	10
Evaporation, NOACK test, max	11.5

Table 2. Specification of mineral engine oil Lotos 15W/40.

Parameter name	Parameter
API quality class	SJ/CF
SAE viscosity grade	15W/40
Kinematic viscosity at 100 °C	15
Viscosity index, min	130
Flow temp., max (°C)	-30
HTHS viscosity at 150 °C	-
Basic number (mgKOX)	6.9
Evaporation, NOACK test, max	-

In order to carry out the study and evaluation of the tribological properties, elements of the friction couple in the form of samples and counter-samples were prepared. The samples were made of C45 steel, the chemical composition of which is shown in Table 3. They were prepared in the form of a disc with dimensions of φ30x5 mm. The samples thus prepared were subjected to heat treatment in the form of hardening and high tempering. Tempering was performed at temperatures between 820 °C and 860 °C. Water was used as a cooling center. In order to remove the stresses caused by hardening, the samples were subjected to tempering at temperatures of 550 °C to 660 °C. Prior to tribological studies, the sample surfaces were ground.

Table 3. C45 steel chemical composition.

C	Mn	Si	P	S
0.42-0.5	0.5-0.8	0.1-0.4	max 0.04	max 0.04
Cr	**Mo**	**Ni**	**Cu**	**Al**
max 0.3	max 0.1	max 0.3	max 0.3	max 0.5

The counter-samples used in the tests were made of 100Cr6 bearing steel. The chemical composition of the steel is shown in Table 4. The diameter of the beads was 6 mm.

Table 4. 100Cr6 steel chemical composition.

C	Mn	Si	P
0.95-1.1	0.25-0.45	0.15-0.35	max 0.025
S	**Cr**	**Ni**	**Cu**
max 0.025	1.3-1.65	max 0.3	max 0.3

2.2 Tribological studies

Tribological studies were carried out using the Anton Paar THT 1000 high temperature tribotester according to ASTM G133, Fig. 1. It allows to determine the coefficient of friction of the co-operating materials, their working temperature at the contact point, and the amount of wear depending on the applied surface pressure, slip velocity of the shield and sliding distance. It is equipped with heating elements so that one can conduct research at temperatures up to 1000 °C. The "ball on disc" association was used to study tribological properties, Fig. 2.

Fig. 1. Tribotester THT 1000.

Tribological studies were conducted at a steady slip velocity of the shield (sample) of 0.4 m/s and an F_n=20 N load at a radius of 10 mm. The study was carried out at ambient temperature and temperature of

100 °C for two engine oils, Lotos Synthetic 5W/40 and Mineral Oil 15W/40. In each test, the friction couple was placed in a container and immersed in equal volume of oil (50 ml). The friction coefficient, frictional force and friction couple temperature were recorded during the tests. The temperature was measured with a thermocouple of 1 mm from the friction surface. The sampling frequency was 10 Hz.

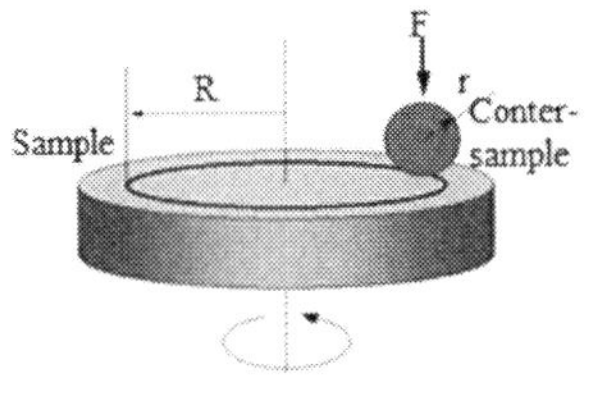

Fig. 2. Friction couple; R – radius of friction, r – ball radius, F – normal force.

After the tribological studies, volume wear of samples and counter-samples was determined. Volume wear of the samples was calculated as the product of the mean value of the sample wear area and the circumference of the friction wheel formed in the test as described by the formula (1):

$$V_f = A \cdot L \tag{1}$$

where: V_f – volume wear, A – mean value of the sample wear, L – friction circumference.

The volume wear of counter-samples was calculated using the following formulas (2 and 3):

$$V = \frac{\pi h^2}{3}(3r - h) \tag{2}$$

$$h = r - \sqrt{r^2 - \left(\frac{d}{2}\right)^2} \tag{3}$$

where: r – ball radius, d – ball wear diameter, h – ball wear height.

2.3 Hardness tests

Hardness tests were performed using a Rockwell hardness tester. The C45 steel samples were subjected to heat treatment using a C scale. A 120° diamond tipped indentation probe was used which was loaded with a 98 N pre-force and a total of 1471 N. For each sample, 5 measurements were performed and then the arithmetic mean of the results was calculated.

2.4 Surface roughness tests

Surface roughness testing of the samples was carried out before and after tribological studies. They were made using Surtronic 3+ needle gauge. The tests were perpendicular to the direction of the machining and friction traces. They were made at an elementary length of 4 mm with an accuracy of 0.02 µm. Surface roughness was determined on the basis of the average arithmetic deviation of the profile from the Ra average line. The test was repeated 5 times for each sample and then the arithmetic mean of the results was calculated.

3 Results and discussion

The hardness of the samples after heat treatment (hardening and high tempering) varied between 49-53 HRC and 62-63 HRC. In turn, the surface roughness parameter Ra of samples before the tribological studies varied within the range of 0.32-0.46 µm.

After the hardness and surface roughness tests, tribological studies were performed. They were conducted in order to evaluate the tribological properties of the two mineral oils Lotos Mineral 15W/40 and Lotos Synthetic 5W/40 at ambient temperature and 100 °C. One of the values that were analyzed during the tribological tests was the temperature of the test compound. Fig. 3 shows a graph showing the temperature distribution for friction couples lubricated with mineral oil Lotos Mineral 15W/40 and Synthetic 5W/40 oil at ambient temperature. By analyzing the graph, it can be seen that the temperature of the friction couple (under accepted operating conditions) when lubricated with mineral oil Lotos Mineral 15W/40 is higher than in case of lubrication with synthetic oil Lotos Synthetic 5W/40. The difference is about 4 °C. In the case of synthetic oil Lotos Synthetic 5W/40, as the sliding distance increases, the temperature of the contact is increasing.

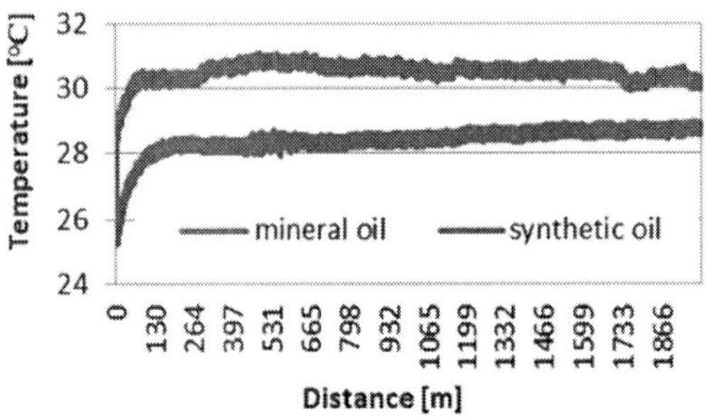

Fig. 3. Temperature distribution for couple lubricated with Lotos mineral oil 15W/40 and synthetic 5W/40 Lotos oil at ambient temperature.

By analyzing the graph shown in Figure 4, some similarities can be noted as with ambient temperature lubrication. Therefore, the temperature of the combination of lubricants with mineral oil Lotos Mineral 15W/40 is also higher than in case of lubrication with synthetic oil Lotos Synthetic 5W/40. The difference is around 11 °C. In both cases, the temperature at the beginning of the work increases to 101 °C and then decreases, which is an evidence of the lapping process that occurs in the initial phase of the test. Then the temperature rises again and after about 1/3 of the road it stabilizes. However, in the case of synthetic oil, the increase in temperature after stabilization was 1-2 °C higher than the initial temperature, i.e. the couple over the entire working period oscillated within the input temperature. This gives the opportunity to say that this oil well protects the molten surfaces and prevents the heat

DOI: 10.1051/e3sconf/20171903027

from precipitating during friction. On the other hand, when the mineral oil was immersed in the mixture, a temperature increase of about 10 □ above its initial value was observed, i.e. during the stabilization the friction couple worked at a temperature of about 110 °C. The temperature difference between the two oils during stabilization was about 10 °C.

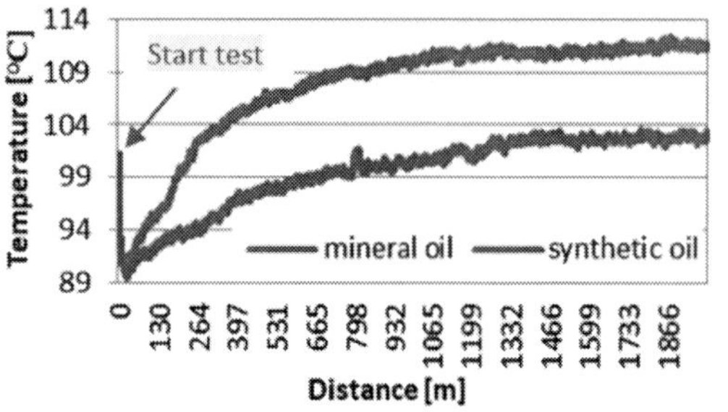

Fig. 4. Temperature distribution for the tested couple lubricated with mineral oil Lotos 15W/40 and Lotos Synthetic 5W/40 at temperature of 100°C.

The results shown in Fig. 5 describe the mean coefficient of friction obtained from laboratory tests for the test compounds at ambient temperature and 100 °C. They oscillate within 0.125-0.146. The highest coefficient of friction was achieved for the compound lubricated with Lotos Synthetic 5W/40 oil at 100 °C (0.146). Referring to the temperature tests for the test compound, despite the highest friction coefficient, the temperatüre rose only by 2 °C from the input temperature (that is, 102 °C). In turn, for the same input parameters, i.e. operating temperature of 100 °C, for the combination lubricated with Mineral Oil 15W/40, despite the lower coefficient of friction, which was 0.132, its temperature reached 111 °C. This allows us to say that synthetic oil is much better at cooling the surface at high temperatures. The opposite is true for ambient operating conditions. The lower coefficient of friction was reached by the couple lubricated with synthetic oil, while the higher with mineral oil lubrication. The coefficient of friction for synthetic oil lubrication had the lowest value compared to the other three cases.

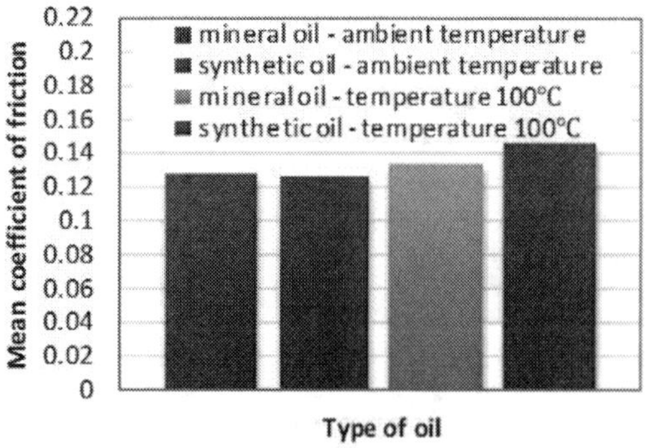

Fig. 5. Mean coefficient of friction for the testes couples.

Referring to the results of the tests shown in Fig. 6a and Fig. 6b, which show the roughness of the surface of the sample, the differences before and after friction can be clearly observed. In the highlighted section, where there are large differences (Fig. 6b – between two vertical lines) compared to the rest of the histogram, wear of the material in the form of a friction trace can be observed. This demonstrates the microscopic effect that occurs between the sample and the counter-sample and the formation of a furrow on the surface of the sample. For the sample before friction, this cannot be observed. Therefore, it is possible to observe traces of wear during the operation of friction couple.

a)

b)

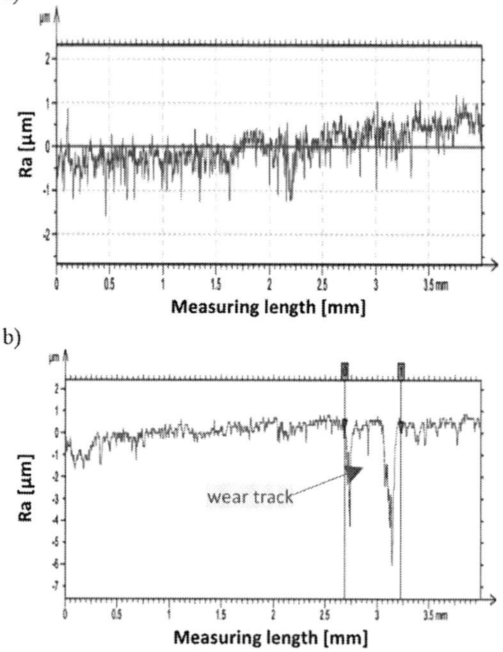

Fig. 6. Surface roughness test results: a) before friction, b) after friction.

Referring to the test results of the volume wear of the samples and the counter-samples shown in Figs. 7a and b, there is some analogy between them for the three samples. The values are close to each other, which may indicate a steady steam consumption. Only in the case of compound lubricated with synthetic oil operating at ambient temperature, one can see a significant difference. Volume wear of this sample is the lowest in comparison to the rest of the samples tested, whereas in the case of the counter-sample it is 9 times higher than the others. The volume wear of this counter was 0.0039 mm^3, while the others ranged from 0.00042 to 0.00076 mm^3.

By analyzing the results of the sample wear coefficient and the counter-samples, which were presented in the graphs (Fig. 8a and Fig. 8b), the highest value in the counter-samples was obtained by a ball lubricated with synthetic oil at ambient temperature. It has reached a value $9.8768*10^{-8}$ $mm^3/N/m$ which was nine times higher than for other counter-samples. Other values range from 1.049 to $1.68*10^{-8}$ $mm^3/N/m$. In the case of samples, the highest value was obtained for the

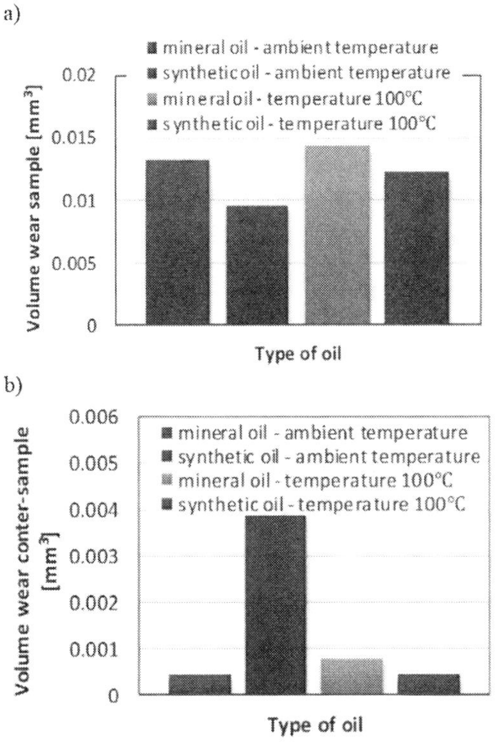

Fig. 7. Values of volume wear of a) samples, b) counter-samples.

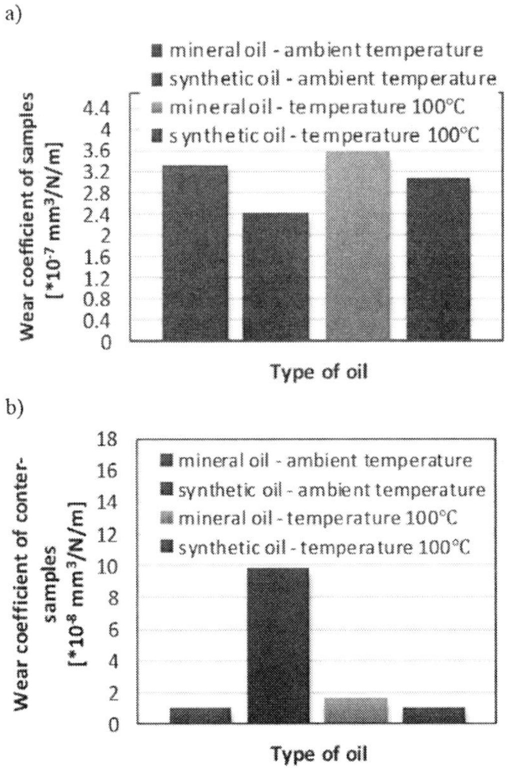

Fig. 8. Wear coefficient of a) samples, b) counter-samples.

couple lubricated with mineral oil at temperature 100 °C. It had a value of 3.585*10^{-7} mm^3/N/m. The lowest value (2.408*10^{-7} mm^3/N/m) was obtained by a sample lubricated with synthetic oil at ambient temperature. The maximum Hertz stresses for each test case were 1.787 GPa.

4 Conclusions

The following conclusions can be formulated on the basis of the research carried out (under the experimental conditions assumed) and the results obtained:

• Lotos Synthetic 5W/40 oil has better cooling properties compared to Lotos mineral oil 15W/40 tested.

• The average coefficient of friction for the test compounds at ambient temperature and mineral oil lubrication 15W/40 was 0.182 and when lubricated with Lotos Synthetic 5W/40 – 0.125. On the other hand, in temperature of 100 °C, it was 0.133 and 0.147 respectively.

• In the tested friction couples there was a friction wear, as proved by the traces of micro-machining and chasing on the samples surfaces.

• The highest value of volume wear of the sample was noted for the combination lubricated with Mineral Oil 15W/40 at 100 °C, which was 0.0143 mm^3. On the other hand, the lowest value amounting to 0.0096 mm^3 was achieved by a sample lubricated with Lotos Synthetic 5W/40 at ambient temperature.

• The values of volume wear for three counter-samples varied from 0.00042 to 0.00076 mm^3. The highest value was achieved by a ball lubricated with synthetic oil at ambient temperature (0.039 mm^3).

• The highest sample wear was recorded for the mineral oil lubricated at temperature of 100 °C. It had a value of 3.585*10^{-7} mm^3/N/m. The lowest value amounting to 2.408*10^{-7} mm^3/N/m was obtained a sample lubricated with synthetic oil at ambient temperature.

• The highest counter-sample wear was recorded when lubricated with synthetic oil at ambient temperature. It was 9.8768*10^{-8} mm^3/N/m which is 9 times higher than the other counter-samples. Other values ranged from 1.049 to 1.68*10^{-8} mm^3/N/m.

• The Hertz stress for each test couple had a value of 1.787 GPa.

References

1. P. Adamiec, J. Dziubiński, Wydawnictwo Politechniki Śląskiej, Gliwice 2005
2. T. Burakowski, T. Wierzchoń, Wydawnictwo Naukowo-Techniczne, Warszawa 1995
3. J. Jozwik; M. Czwarnowski, ASTRJ, **9/28**, (2015)
4. J. Jozwik, L. Semotiuk, I. Kuric, ASTRJ, **9/28**, (2015)
5. E. Feldshtein, J. Józwik, S. Legutko, ASTRJ, **10/30**, 2016
6. A. Glowacz, Z. Glowacz, Infrared Phys. Technol. **81**, 2017

7. B.C. Ku, Y.C. Han, J.E. Lee, J.K. Lee, S.H. Park, Y.J. Hwang, Int. J. Precis. Eng. Manuf. **11**, (2010)
8. B.S. Shenoy, K.G. Binu, R. Pai, D.S. Rao, R.S. Pai, Tribol. Int. **45**, (2012)
9. Y. Hwang, C. Lee, Y. Choi, S. Cheong, D. Kim, K. Lee, J. Lee, S. Kim, J. Mech. Sci. Technol. **25**, (2011)
10. V. Zin, F. Agresti, S. Barison, L. Colla, E. Mercadelli, M. Fabrizio, C. Pagura, Tribol. Lett. **55**, 1, (2014)
11. G. Yan-Bao, Z. Si-Wei, Lubricants, **4**, (2016)
12. Y. Sayed, Akl Ahmed, A. Abdel-Rehim, A. Esraa, A. Khafagy, SAE Tech. Paper **01-0487**, (2016)
13. Q. Xiaowen, J. Zhining, Y. Yulin, F. Bingli, Tribol. Int. **44**, (2011)
14. M. Asrul, N.W.M. Zulkifli, H.H. Masjuki, M.A. Kalam, Procedia Eng. **68**, (2013)
15. K. Prabu, B. Nalini, B. Suresh, M. Sekar, J. Achiev. Mater. Manuf. Eng. **2**, (2016)
16. A. Kotnarowski, Logistyka **2**, (2010)
17. J. Padgurskas, R. Rukuiza, I. Prosycevas, R. Kreivaitis, Tribol. Int. **60**, (2013)
18. S. Laber, A. Laber, Tribologia **3**, (2015)
19. A. Laber, Wyd. Uniwersytet Zielonogórski, (2012)
20. M. Barszcz, M. Paszeczko, K. Lenik, Arch. Metall. Mater. **60** (4), (2015)
21. I. Gershman, E.I. Gershman, A.E. Mironov, G.S. Fox-Rabinovich, S.C. Veldhuis, Entropy **18**, 365, (2016)
22. M. Paszeczko, K. Dziedzic, M. Barszcz, ASTRJ **31**, 10, (2016)
23. M. Barszcz, M. Paszeczko, K. Lenik, Prace Szkoły Inżynierii Materiałowej **5**, (2014)
24. M. Pashechko, J. Józwik, K. Dziedzic, M. Karolus, I. Usydus, Mater. Sci, **6**, (2017)
25. J. Józwik, K. Dziedzic, I. Usydus, P. Raos, G.M. Krolczyk, Teh. Vjesn. **5**, (2016)

DOI: 10.1051/e3sconf/20171903028

Analytical and artificial neural network models to estimate the discharge coefficient for ogee spillway

Alpaslan Yarar[1,*]

[1] Selcuk University, Engineering Faculty, Civil Engineering Department, Hydraulics Division, 42031 Konya, Turkey

Abstract. In this study, analytical and Artificial Neural Network (ANN) model were used for determine the discharge coefficient of Ogee Spillways. For this aim, discharge coefficients of 11 different heads were calculated by using a test flume having 7.5 cm width, 15 cm depth and 5 m length, in the laboratory. Discharge coefficients were also computed by the formula for the same heads measured in the laboratory to investigate the accuracy of experimental setup. An ANN model was set by using the experimental results in order to estimate the discharge coefficient. Then, the performance of the ANN model was investigated. As the result, the coefficient of determination between ANN model and experimental setup is found R^2= 0.98. ANN model is show a good consistency with experimental results.

1 Introduction

Spillways are used to hold water at a certain level in reservoir besides to store water for different aims, such as energy, irrigation, flood control, etc. Ogee-shaped spillways let excessive water in reservoir pass to downstream. The ideal form of the spillway recommended by The United States Bureau of Reclamation (USBR) [1] is seen in Fig. 1.

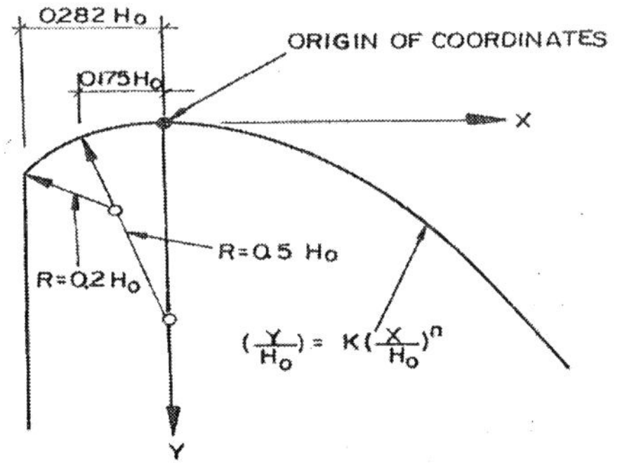

Fig. 1. Standard crest profile of an overflow spillway [1].

The downstream of the spillway can be defined by Eq.1.

$$\frac{Y}{H_0} = K\left(\frac{X}{H_0}\right)^n \tag{1}$$

K and n are constant, depend on the upstream inclination and the velocity and H_0 is the design head.

The discharge over the spillway can be calculated by Eq. 2.

$$Q = CLH^{3/2} \tag{2}$$

Q (m^3/s) is the discharge, C is the discharge coefficient, L (m) is the length of the crest and H (m) is the head over the crest.

Spillways are rarely operated with their design head since the design head corresponds to very large return periods having relatively small risks. Therefore, the discharge coefficient for an existing total operated head should be determined by the Fig. 2 [2].

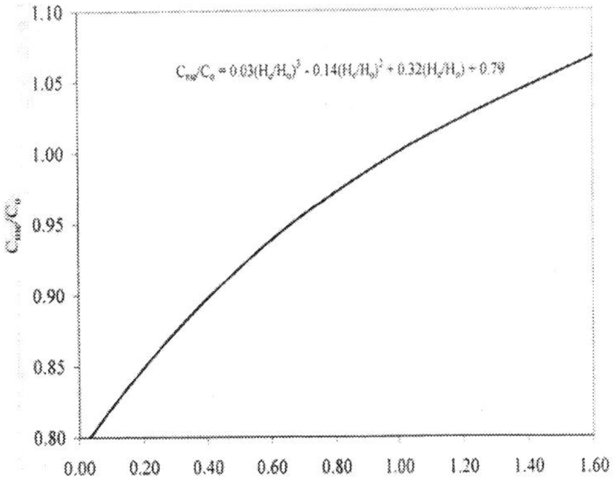

Fig. 2. Discharge Coefficients for varying heads.

Many researchers have studied for numerical and experimental studies on spillways [3-7]. And ANN

* Corresponding author: ayarar@selcuk.edu.tr

model have also used for studies on spillways. Azamathulla et. al. [8], studied for scour depth prediction for ski-jump type of spillways. They also used some traditional methods. Their study showed that the traditional equation-based methods of predicting design scour downstream of a ski-jump bucked could better be replaced by one of the soft computing schemes [8]. Salmasi and Özger [9], investigated the applicability and accuracy of the neuro-fuzzy approach in estimating the proper values of energy dissipation of skimming flow regime over stepped spillways. They were also developed multi regression equations. They found that neuro-fuzzy model is more successful than regression equations [9].

In this study, it was aimed that to investigate the capability of ANN model to estimate the discharge coefficient for Ogee spillway.

2 Materials and Methods

In this study, an experimental setup containing a simple flume which was made plexiglass and having 7.5 cm width, 15 cm depth and 5 m length, was used in order to calculate discharge passing over ogee-shaped spillway for different heads (Fig. 3). The flume had closed loop water system and the flow to the flume was supplied from constant head water tank and the flow discharge was regulated by a valve. That the transparent flume also enables to see the flow conditions. Flow depth measured with a measurement device holded along the flume to an accuracy of ± 1 mm.

Fig. 3. General view of the flume used in the study.

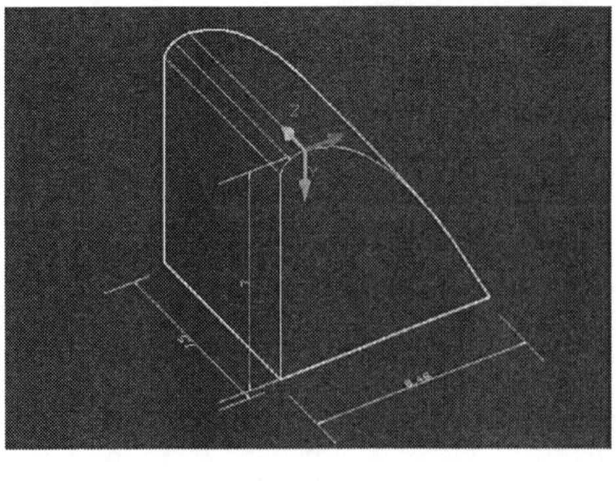

Fig. 4. Ogee spillway model.

A tank, made by smooth plastic which has 63 x 67 x 72 cm dimensions, was used to measure the discharge. An ogee spillway model was modeled by Eq. 1 according to USBR type and manufactured with steel material to use in the experiments (Fig. 4).

3 Artificial Neural Networks

The composition of ANN is inspired from biological neural networks. A neuron is one of the basic components of neural networks. It can vary in terms of size and shape, according to its function and mission in neural systems [10].

The system consisting of three layers, an input layer, an hidden layer and an output layer, named Multi Layer Perceptron (MLP) (Fig. 5).

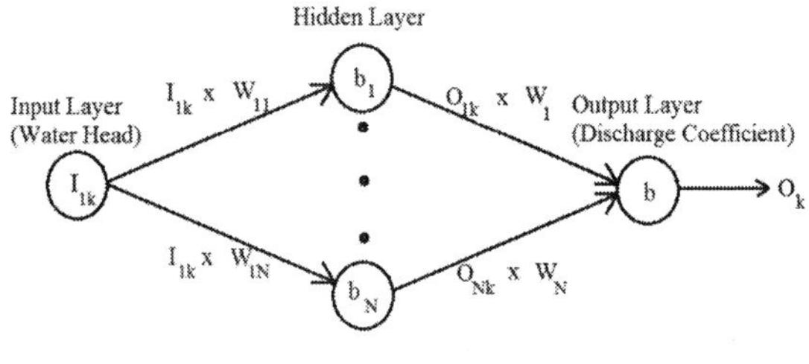

Fig. 5. A structure of Multilayer Perceptron.

The mathematical expression of a three-layer feed forward ANN for prediction is

$$O_{jk} = f_1\left(b_j + \sum_i W_{ij} I_{ik}\right) \quad (3)$$

$$O_k = b + \sum_i W_j O_{jk} \quad (4)$$

where:
f_1 is the activation function for the hidden layer,
I_{ik} is the i_{th} input for the k_{th} sample point,
O_{jk} is the output of j_{th} node of the hidden layer,
W_{ij} and W_j are the weights that control the strength of connections between layers,
b_j and b are the biases that are used to adjust the mean value for the hidden layer and the output layer, respectively [11]

4 Results and discussion

The flow depths over the spillway were measured for different discharge. The discharges were determined by the tank for each water heads to calculate the discharge coefficients. This process was done with holding time until filling certain volume (Fig. 6).
Discharge coefficients were calculated by Eq. 2 using the water heads.
Then the discharge coefficients obtained from both experimental and analytical results were compared with each other.

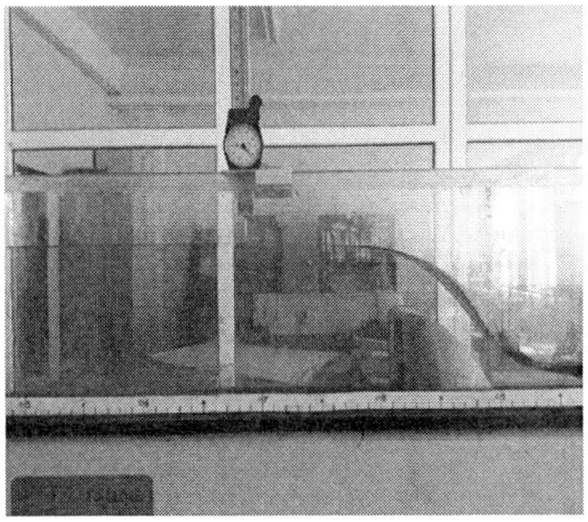

Fig. 6. Water depth and discharge measurement points on the experimental set up.

Discharge coefficients were calculated by Eq. 2 using the water heads. Then the discharge coefficients obtained from both experimental and analytical results were compared with each other. Measured water heads and corresponding discharge coefficients for both experimental and analytical studies were given in Fig. 7.

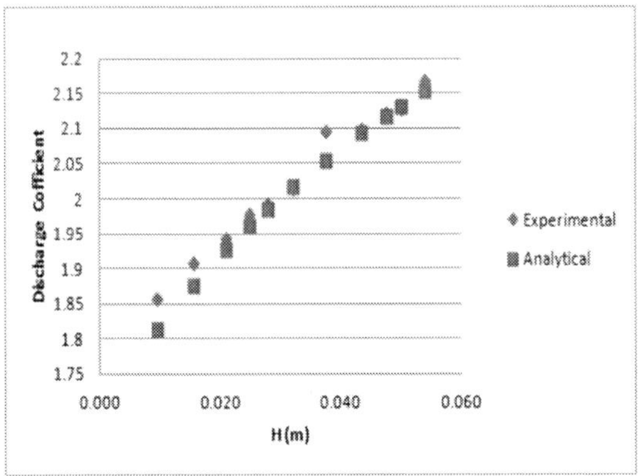

Fig. 7. Comparison of the experimental and analytical results.

Experimental results were used for ANN modeling. 6 data set of the 11 data set were used for training and remaining 5 data set were used for testing. Feed forward back propagation ANN model and Scaled Conjugate Gradiant (SCG) algorithm were used for modeling. Three layered ANN formed with 1 input layer 1 hidden layer and 1 output layer was used in ANN model. The relationship between ANN results and experimental results is shown in Fig 8.

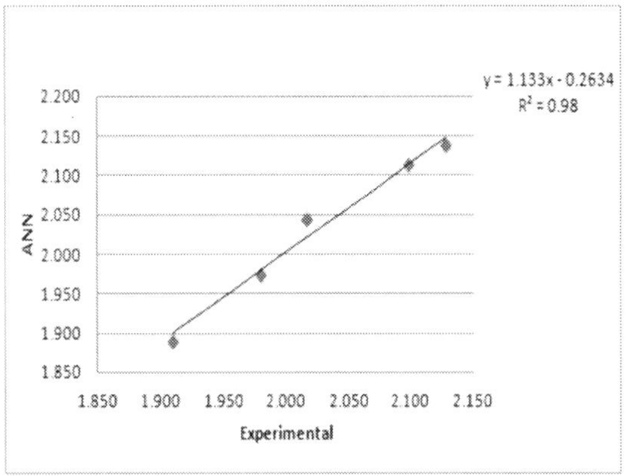

Fig. 8. Relationship between the results of experimental and ANN test.

5 Conclusions

Hydraulic design of a spillway is one of the most studied subjects in hydraulic engineering. Properly designed spillways will be able to pass flood flows efficiently and safely to downstream of dams. So, determining the discharges correctly becomes important for the water structures. To determine the discharges depends on to determine the discharge coefficients. In recent years, artificial intelligence methods based on learning from examples have been widely applied in hydraulic studies. In this study, it was aimed to investigate the ANN performance on determining the discharge coefficients. For this purpose both experimental and analytical studies were done. It was seen that ANN model produced very successful results. Overall, the studies presented in this paper showed us that ANN model can be an alternative method to determine the discharge coefficients. Using ANN model, we can determine the proper coefficients quickly, without using the graphs and the equations.

References

1. USBR, 1987, Design of small dams, Third Edition, Washington: Water Resources Technical Publication
2. A. M. Yanmaz, *Applied water resources engineering* (METU Press, 2001).
3. B.M. Savage, M.C. Johnson, J. Hydraul. Eng. ASCE; **127**, 8, 641–9 (2001)
4. C. Song, F. Zhou. J Hydraul Eng, ASCE, **125**, 9, 959–67 (1999)
5. J.M. Sanchez, J. Pomares, J. Dolz, Proceedings, International Workshop on Hydraulics of Stepped Spillways, Zurich, Switzerland, 2000.
6. B. Dargahi, J. Hydraul. Eng, ASCE, **132**, 98, 99–907 (2006).
7. B. M. Savage, M.C. Johnson, Flow over ogee spillway, J. Hydraul. Eng. **127**, 8640–649 (2001)

DOI: 10.1051/e3sconf/20171903028

8. H. Azamathulla, M.C. Deo, P.B. Deolalikar, Advances in Engineering Software **39**, 689-698 (2008)
9. F. Salmasi, M. Özger. Arab. J. Sci. Eng. **39**, 6099–6108 (2014)
10. A. Dorum, A. Yarar, M.F. Sevimli, M. Onüçyildiz, Expert Syst .Appl. **37**, 6587–6593 (2010)
11. N. Chitsazan, A. A. Nadiri, F. T. C. Tsai. J. Hydrol. **528**, 52–62 (2015)

Testing of crimp connections made on a prototype stand

Nikodem Wróbel[1], *Michał* Rejek[1], and *Grzegorz* Krolczyk[2,*]

[1] „PRO-ZAP" Grupa Introl, Mechanical Design Department, 63-400 Ostrów Wlkp., Grabowska 47a, Poland,
[2] Opole University of Technology, Faculty of Mechanical Engineering, 45-001 Opole, St. Mikołajczyka 5, Poland

Abstract. The paper presents a project of a prototype stand for making inseparable crimped joints through crimping operation, which is commonly used in the construction of heat exchangers. The joint is formed by tools in form of crimping jaws, which are considered as a punch stamp for pressing two materials into each other. In effect these two materials are mutually "crimped". This elaboration contains a case study of joints being made by the workstand. Furthermore, the paper presents an analysis of features of the joints by examining them by destructive testing: Micrography, Tensile strength test, and non-destructive testing: Leakage test. The elements to be joined have been made of aluminium 6060, which is well matched to indirectly or directly make joints for all kinds of coolers or condensers.

1 Introduction

Making inseparable joints for the automotive or aerospace industry, determines particular approach to design, implementation and testing of them. By approach to economy of this question, the final product with made joints shall beleakproof and of appropriate durability, at the lowest possible costs of the entire product and technological process. Economical operation of various devices strongly depends on many factors, such as process parameters or type of material [1, 2]. Aluminum alloy is a material difficult to weld with the resistance spot welding. It is still a challenge to join the aluminum sheet and the steel sheet together [3-5]. There are many other alternatives of aluminum alloy joining such as friction stir welding [6], adhesive joining or mechanical clinching [7-9]. The method of mechanical clinching has the advantage of no damage on the sheet surface, no heat, no pretreatment procedure, etc. [10].

Selecting the method of forming joints, it is necessary to consider whether the tightness of joints shall be obtained by forming processes. the other question is whether the crimping operations are considered only as indirect step prior to the final processes as bonding or soldering. Considering the mentioned aspects, the requirements of the forces which joints must hold out should be clearly stated.

A significant difficulty in joining two materials is usually a limited space for mechanism used to crimp them, in relation to the forces necessary to be provided in order to ensure requested quality of the joint.

2 Materials and methods

The manufactured device is intended for forming two components, the first one in shape of pipe and second one in the shape of connecting block. The operation of crimping is performed by jaws adapted to the inner diameter of a hole of connecting block. The Figs. 1 and 2 shows the tested case of joint before the "joining" operation.

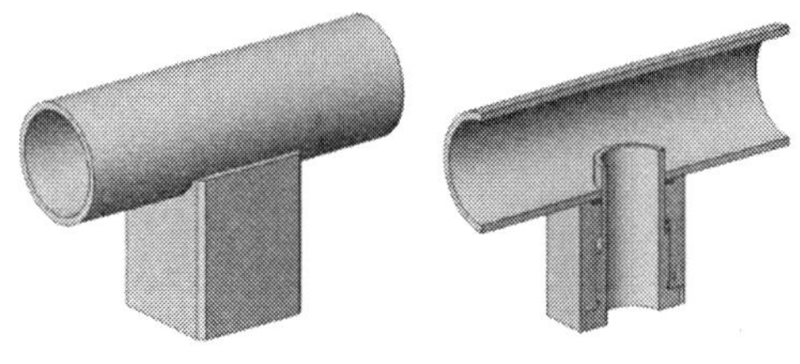

Fig. 1. View of the connecting block and the pipe.

The joint is designed to increase its durability and tensile strength force, by counteracting the torque through the locking shape of joining elements.

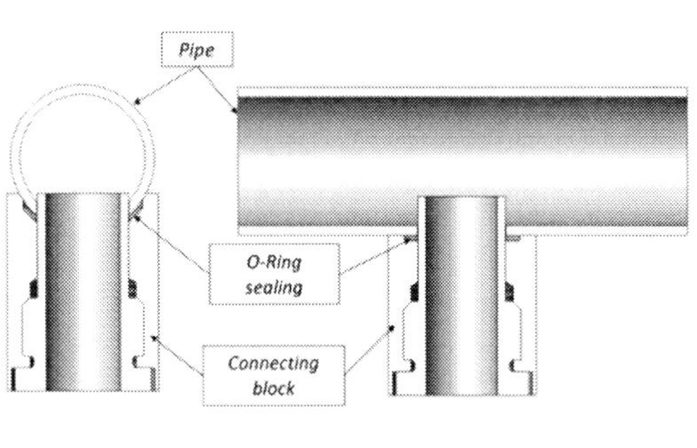

Fig. 2. View of the pipe and connecting block in two planes.

For designing process there should be assumed that between the tube and the connecting block, the jaws will spread in four directions, and their shape will be

* Corresponding author: g.krolczyk@po.opole.pl

DOI: 10.1051/e3sconf/20171903029

different for perpendicular cross-sections due to differences in the height of wall of hole in pipe element. Joining elements trough the forming process of the joint must be pressed against each other to compensate possible gap between them. These gaps are as a result of differences in tolerances of used components. The joint scheme is presented in three steps (see Fig. 3).

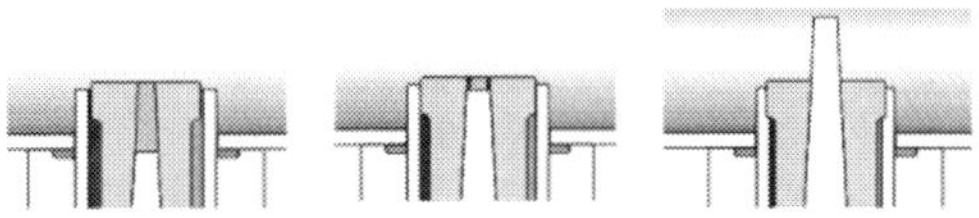

Fig. 3. Theorethical scheme of joint making proces: 1st phase – prior to jaws (no contact with components (a); 2nd phase – contact with comopnents (b), 3rd phase - interaction of jaws into material (c)

Forming jaws in described test were made of third-generation powder steel (K490) with high load carrying capacity and high impact resistance. The jaws through the applied technological processes have been hardened to a hardness of 56 HRC.

3 Stand concept

Designing of prototype stands involves a series of inconveniences related with requirement to fulfill defined functions of stand while maintaining the quality of a product, assure the ergonomy of the operator's work, and repeatability of process, all of above at meeting pre-established cost level..

In order to form the joint, a stand powered by pneumatic actuator has been designed, developed by a mechanical transmission, which allows to increase the forces transmitted to the system. As a consequence of used mechanism, despite of limited access to joining elements, the strength of the joint is being increased. However, application of pneumatic actuator not allows to full control of the connection making process between components, therefore it would be more accurate in further research to develop the system with electric servomotor. Nevertheless, the designed concept allows transmitting requested values of forces and executing the joint. Repeatability of forces provided to joint can be controlled by force or torque sensors and displacement sensors located on actuators or jaws.

Limited access has forced to equip the stand with mechanism of mechanical transmission in the form of wedges, manufactured on spindle spread 90° apart. On each fabricated surface the jaws were guided. Applied system converts vertical motion generated by the actuator onto the horizontal motion of the punch jaws, with simultaneous transmission ratio of 1:19. The Fig. 4 shows a 3D view of the stand with the elements of acting and sensing.

The actor generating force is a pneumatic servomotor, which at the value of pressure medium equal to 6 bars, provides 3kN of force to the system makes it possible to generate a force of 18.80 kN.

The sensing part of the system consists of three sensors:

• Force sensor HBM RSCC3/2T with an RM4220 transducer, the range of 2 tons and accuracy of ± 0.25%,
• Turck displacement sensor WIM45-UNTL-LIU5X2-0,3-PSG4M,
• SMC pressure sensor supervising its magnitude during the tests..

In order to collect data of the carried out tests a PC with a PC LAB transducer has been used, which allows to collect signals during the test, in the form of current values (4 - 20 mA – displacement sensor) and voltage values (0 – 10 V – force sensor).

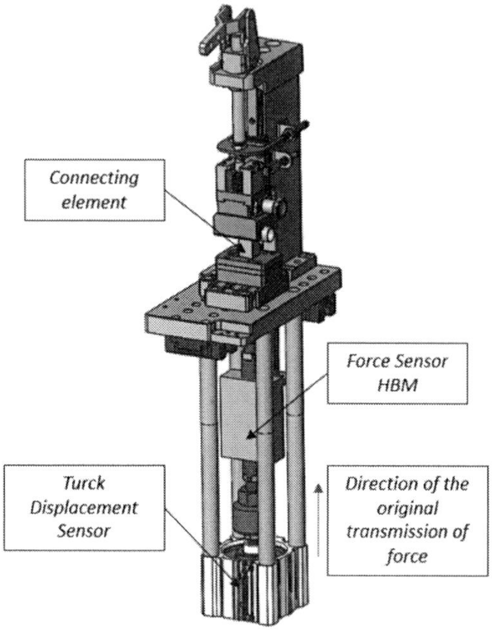

Fig. 4. View onto the main assembly of the driving machine

4 Methodology of the experiment

The tests have been made on the basis of six sample sets, consisting of a pipe and a connecting block each. Process of making crimped joints has been performed for two pipe diameters. Each of them has had also two values of the flange diameter, the samples have been provided equally, three with larger diameter and three with smaller one. The samples have been selected for application of sealings of o-ring form and for soldering operations. A view of the joint with the dimensions has been shown in Fig. 5 and Table 1.

In order to make the joint, 2 kinds of matched jaws to each inner diameter of pipe (larger and smaller) have been used (Tables 2 and 3).

The processing of each joint has been recorded by a data collection system. Logged values of forces required to form joint give possibility to match properly a size of used actors and designed transmission ratios.

Obtained joints can be verified by using destructive and non-destructive tests. The most common validation tests are tightness tests, endurance tests, micrographs and industrial tomography.

 DOI: 10.1051/e3sconf/20171903029

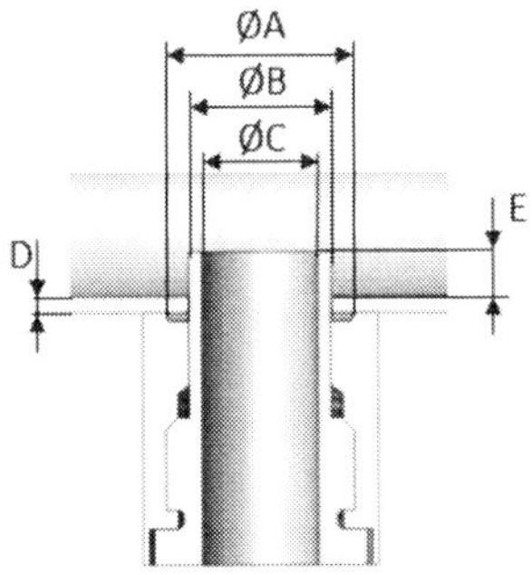

Fig. 5. Connection view with key dimensions marked on.

Table 1. Design of experiment.

No	Case	ØA	ØB	ØC	D	E	O-ring
1	A	15.8 h11	12 H7/h6	9.9	1.3	3.7	Ø14x1.0
2	B	13.8 h11	10 H7/h6	7.9	1.3	3.7	Ø12x1.0

Table 2. Design of experiment.

No	Case	Tool	Test 1 leaktest	Test 2 Tensile strength	Micro-graph	Other
1	A	S1 S2	X		X	no O-ring
2	A	S1 S2	X			no O-ring
3	A	S1 S2	X	X		O-ring
4	B	S3 S4	X		X	no O-ring
5	B	S3 S4	X			no O-ring
6	B	S3 S4	X	X		O-ring

Tested samples were a subject of air leakage test and a tensile strength test after crimping operation, but prior to start the soldering process. The performed joints were additionally tested by micrography studies to determine the dependence of crimp action to breaking strength achieved.

Table 3. The pinching stamps used in the case study

	S1	S2
Case „A"		
	S3	S4
Case „B"		

The ATEQ F520 device has been used to perform a leakage test. Two types of sample sets have been tested with O-ring and after carrying out the operations of crimping and soldering. The tensile strength test was performed on a machine based on a ball screw, along with a force and displacement sensor. To perform tensile tests in connecting block inner threads have been made and the pipes were gripped with prisms.

5 Results and discussion

Each sample has been a subject to an analysis of tightness. The medium pressure of leakage test was preset at 0.6 MPa, the admissible tipping point for a leakage was set at the level of 4 [Pa/s], which is the standard level of leakage for oil or water coolers in the automotive industry. Testing with air testers is affected by the temperature of the test agent as well as the sample itself, the effect on the result can also be observed against the number of consecutive leak tests for the same set of sample. The obtained results for each sample sets of the joints have met requested range of test parameters, regardless of the type of seal (O-ring, Solder). The results have been listed in Table 4.

Table 4. Results of the connection leaktighteness.

Test no	1	2	3	4	5	6
Pressure (MPa)	0.6	0.6	0.6	0.6	0.6	0.6
Leakage ((Pa x l)/s)	2	2	3	3	2	3

On the basis of selected tested sample sets micrography test has been performed (Fig. 6), which allows to determine the interaction of jaws into material. The average interaction was 0.6mm (dimension „I") per side, of which the crimping of wall of connecting block in the wall of pipe was 0.25 mm (dimension G). The tests and analyzes show that the alpha-angle and L-distance have a significant impact on the strength of the joint. Higher contact point of the jaw with connecting block allows to reduce the required force, however the quality of joints is unsatisfactory.

When the punching jaws were affecting on material in bottom part of samples, the force provided by stand to the joint was not sufficient to make the joint. The thickness of wall of connecting block had also significant impact on the quality of the joint. The inner edge of the pipe either should be considered, if it is sharp, that is, with a small radius, it is easier to form the material.

Fixed joints made with native material and using cold forming processes, in the field of manufacture industry of heat exchangers are particularly exposed to pressure of the refrigerant, tensile forces and vibrations. Joining two elements by crimping operations, may be considered as the final execution of the joint, or as an intermediate operation in advance of soldering or bonding.

When crimping process is treated as an intermediate operation, joint must also be resistant to the forces and any other impact of extrinsic factor associated with the handling process. Executing joint was recorded for each of the two diameter groups, via the force sensor

DOI: 10.1051/e3sconf/20171903029

and displacement sensor.

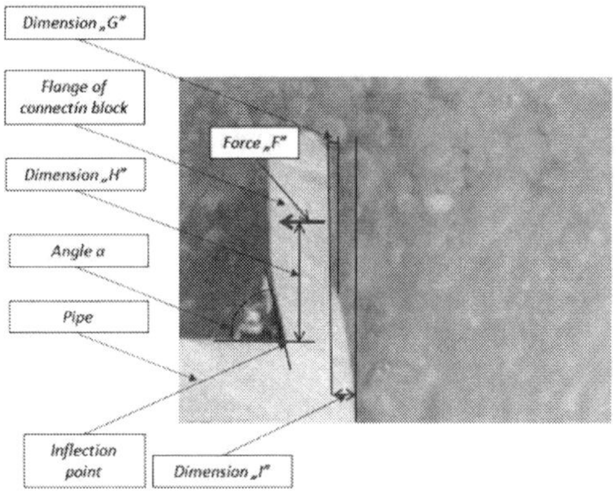

Fig. 6. Microsection of the connection

Analysis of the obtained test results is presented at the Fig. 7 and the Fig. 8.

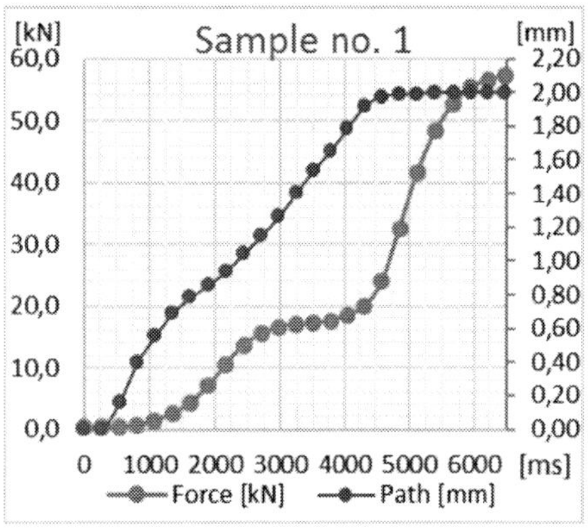

Fig. 7. Diagram of the force-at-time and force-at-path ratio for sample no. 1.

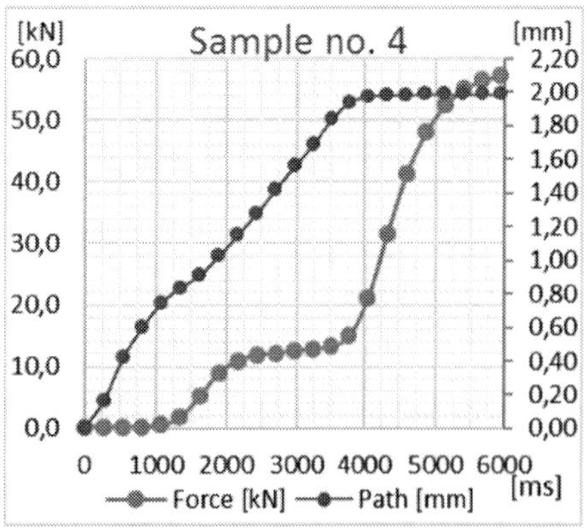

Fig. 8. Diagram of the force-at-time and force-at-path ratio for sample no. 4

The conclusion is that there is dependence between the inner diameter of the flange of connection block and the reaction force generated during the joint forming process.

The presented diagrams allow to study stages of formation the joint, as well as learning phenomenon associated with the forming process. The process is a slightly different than assumed as a theoretical one. Executed micrographs as well as analysis of data shown at charts allowed to determine first step of process, which will be defined as elimination of clearances between the jaws and inner flange of connecting block. For the first sample it is noticeable within the range of the value of force 0 to less than 1.4kN, for the "B" test cases it is noticeable within the range of force equal 0 to less than 1.8kN. Second step of process is defined as start of interference between jaws and inner flange of connecting block, which is connected with smoothing of the roughness and the bending of the inner flange of connection block (point of inflection occurs on the inner edge of the tube). The second step presented at diagram for test case "A" is noticeable within the range of force value 1.4kN to 15.6kN and for second sample "B" is noticeable within the range of force value 1.8kN to 11kN. The last step of the process is penetration of jaws into material with less bend of flange of connection block. Results for described step can be noticed in case of the first sample set within the range of force 15.6kN to 18.6kN, for second test case within the range of force of 11kN to 13.3kN. Obtained values of forces for both test cases points out that for smaller inner diameter of connecting block was less energy consuming that for larger diameter of connecting block.

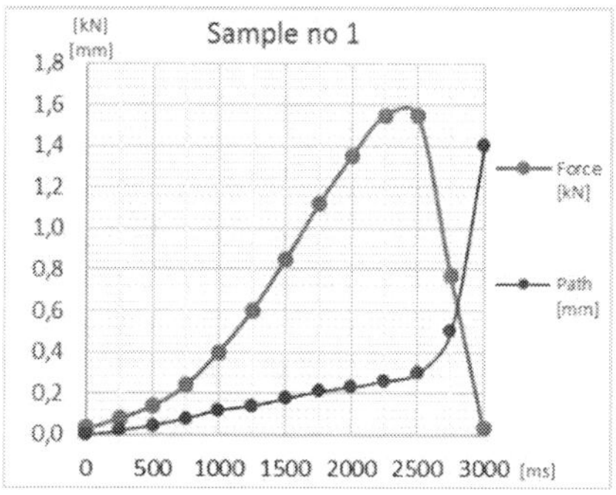

Fig. 9. Joint tensile chart depending on force applied within the requested time and the path for the sample 1.

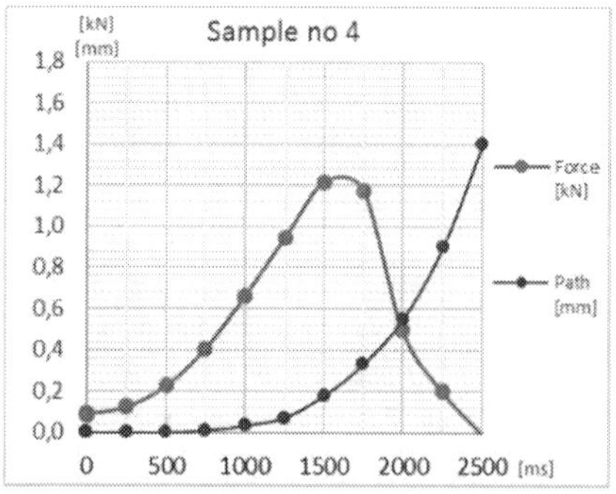

Fig. 10. Joint tensile diagram depending on force applied within the requested time and the path for the sample 4.

Analysis of the obtained test results for stress tests shown in Fig. 9 and Fig. 10 points out that the joint is beginning to loss his endurance and stiffness in case of test "A" at 0.75kN and in the case of test "B" at 0.25kN. Test samples were completely destroyed at forces of values 1.6kN and 1.2kN.

6 Conclusions

Executed tests and studies have allowed to formulate the following conclusions:

I. The difference in diameter between cases "A" and "B" was 2mm, which means that the ratio of the smaller diameter to the larger one was 0.79%. Transmission of this shape dependence onto the forces obtained in the various phases of joint forming was as follows:

- The first phase turned out to be less energy-consuming for the larger diameter of the connector and the difference was 0.4kN. That means that the play ratio and movement resistance between the jaw and inner diameter in cases "A" and "B" were slightly larger in favor of the smaller diameter.
- The second phase of the joint forming, by bending and partial pressing into the material of the collar, absorbed 4.6 kN in the extreme of this compartment for the "A" cases, what was observed as an increase by 30%. The requested result for the second phase was more difficult to be achieved in case of larger diameters.
- The final phase of the joint creating of the second development was also more energy-consuming in the "A" cases. The percentage ratio of this dependence was 28%, with a value of 5kN.

Analysis of the difference in absorbed forces at different stages of joint formation shows that increasing the circumference between "A" and "B" by 6.28 mm while maintaining the same height of the flange has increased the needed force by 30%.

II. The analysis of the joint strength showed that loosening in sample "A" occurred at a force three times lower than in the connector "B", the figure difference was 0.5 kN. Total destruction of the sample was however only at 25% greater for sample "A" than in the case "B".

III. In order to increase the strength of the joint with similar energy consumption, it is necessary to consider changing the shape of the jaws.

IV. The resulting leakage for the samples was in the correct range and did not exceed 3 [(Pa * l) / s]. Showing no difference between different type of the seals.

V. In order to use the crimp connector as an O-ring target, with no necessity of making plans for subsequent operations, increase the applied force to produce the joint about 30% - 40%.

VI. Full control of the crimping process would be possible with a servo drive instead of a pneumatic actuator, which would allow, among other things, to control the speed of the joint creation. Unfortunately, it increased the cost of the device by 30%.

References

1. R.W. Maruda, G.M. Krolczyk, P. Nieslony, S. Wojciechowski, M. Michalski, S. Legutko, J. Manuf. Process, **24**, (2016)
2. G. Krolczyk, J. Królczyk, S. Legutko, A. Hunjet, Tehnički Vjesnik - Technical Gazette **21**, 2 (2014)
3. C. Chen, X. Han, S. Zhao, F. Xu, X. Zhao, T. Ishida, Int. J. Adv. Manuf. Technol. (2017) (in press)
4. F. Lambiase, Int. J. Adv. Manuf. Technol. **80**, 9 (2015).
5. X.C. He, Int. J. Adv. Manuf. Technol. 48, 5 (2010)
6. R. Kumar, S. Chattopadhyaya, S. Hloch, G. Krolczyk, S. Legutko, Eksploatacja i Niezawodnosc – Maintenance and Reliability **18**, 1(2016).
7. J. Mucha, L. Kaščák, E. Spišák, Arch. Civ. Mech. Eng. **11**, 1 (2011)
8. J. Mucha, L. Kaščák, E. Spišák, Adv. Mech. Eng. **5** (2013)
9. N. Wrobel, M. Rejek, G. Krolczyk, S. Hloch, Lecture Notes in Mechanical Engineering (2017) (in press)
10. K. Mori, T. Maeno, S. Fuzisaka, J. Mater. Process Technol. **212**, (2012)

DOI: 10.1051/e3sconf/20171903030

Behaviour of Strengthened RC Frames with Eccentric Steel Braced Frames

Mehmet Kamanli[1,*], and *Alptug* Unal[1]

[1]Selcuk University, Engineering Faculty, Civil Engineering Department, Konya, Turkey

Abstract. After devastating earthquakes in recent years, strengthening of reinforced concrete buildings became an important research topic. Reinforced concrete buildings can be strengthened by steel braced frames. These steel braced frames may be made of concentrically or eccentrically indicated in Turkish Earthquake Code 2007. In this study pushover analysis of the 1/3 scaled 1 reinforced concrete frame and 1/3 scaled 4 strengthened reinforced concrete frames with internal eccentric steel braced frames were conducted by SAP2000 program. According to the results of the analyses conducted, load-displacement curves of the specimens were compared and evaluated. Adding eccentric steel braces to the bare frame decreased the story drift, and significantly increased strength, stiffness and energy dissipation capacity. In this strengthening method lateral load carrying capacity, stiffness and dissipated energy of the structure can be increased.

1 Introduction

There are a lot of studies about strengthening of Reinforced Concrete (RC) buildings, but no adequate detailed studies conducted for this subject. Reinforced concrete buildings can be strengthened by steel braced frames. These steel braced frames may be made of concentrically or eccentrically indicated in Turkish Earthquake Code 2007 (Fig. 1).

Such systems are generally used in order to supply stiffness and strength against lateral loads in low and medium height buildings. In addition to saving of material in these systems, it is possible to restrain storey drift effectively by providing high lateral stiffness. The braced frames provide energy consumption under the effect of big lateral loadings with changing direction by pressure-wrenching and by flowing under tensile loads [1,2]. Also there are some analytical and experimental studies related to concentrically and eccentrically steel braced frames in the literature [3-14].

In this study pushover analysis of the 1/3 scaled 1 reinforced concrete frame and 1/3 scaled 4 strengthened reinforced concrete frames with internal eccentrically steel braced frames were conducted by SAP2000 program. In the study, Eccentrically Steel Braced Frames (ESBF) described in Turkish Earthquake Code 2007 were designed and tested with pushover analyses, which are 4 types in total being V- braced, Λ- braced, \- braced and /- braced. For this purpose, 4 pieces of 1/3 scaled strengthened RC frames with ESBF models made up of 60x60x4mm cross-section box profile were formed. In the analytic study, pushover analyses of the RC frames strengthened with these ESBFs were conducted by the SAP2000 program, and according to the results of the analyses conducted, load-displacement curves of the specimens were given and compared.

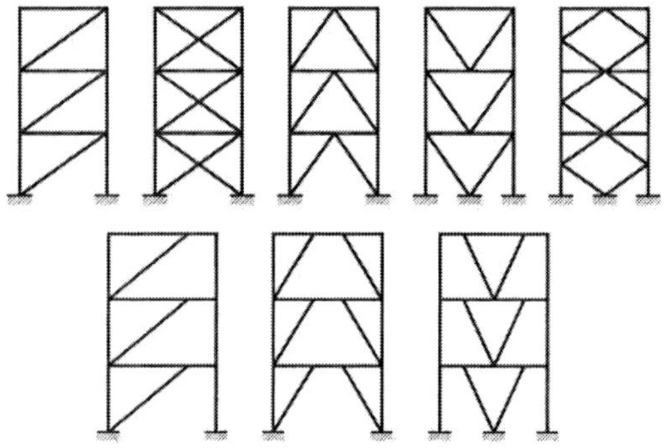

Fig. 1. Steel Braced Frames.

2 Material and Methods

In building systems formed in order to represent the buildings in literature, the buildings were modelled by generally forming 1/3 scaled systems. The cross-sections in 1/3 scaled ESBF group were determined as 60x60x4 mm. ESBFs can be used to meet lateral loads in current building systems, to limit storey drifts and to improve stiffness of the building as well as they can be added to the system as reinforcement elements afterwards. For this purpose, it was considered to strengthen 1/3 scaled reinforced concrete frame with designed 1/3 scaled ESBFs by means of transverse beam. In Fig. 2, dimensions of empty reinforced concrete frame were given. In the bare frame the concrete compressive strength is taken as 20 MPa, reinforcing yield strength is

* Corresponding author: mkamanli@selcuk.edu.tr

taken as 420 MPa and reinforcing tensile strength is taken as 550 MPa. The columns in experiment members were dimensioned as 100x150 mm and the beams were dimensioned as 150x150 mm. The dimensions of groundwork were selected as quite big to prevent any damage occurrence during experiment.

In the foundations, 3ϕ14 mm (3 rebar and rebar diameter =14mm) reinforcement in the top, 3ϕ14 mm reinforcement in the bottom and 2ϕ14 mm web reinforcement were used. The stirrups were arranged as ϕ8/100 mm in the foundation. Column longitudinal reinforcement was prepared as 4ϕ10 mm and ϕ6/100 mm stirrup was used. Confinement zone was not formed in columns and the stirrups were not continued at the connection zones of foundation and beams. In the beams, 3ϕ10 mm reinforcement at the top, 3ϕ10 mm longitudinal and ϕ6/100 mm lateral reinforcements at the bottom were used. The stirrups were not continued in the zones where beams enter to the columns and stirrup compaction was not performed at the places close to connection zones.

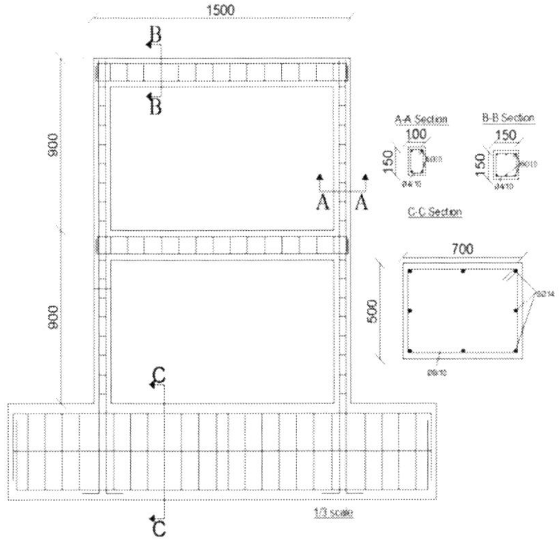

Fig. 2. 1/3 scaled base reinforced concrete frame.

In order to investigate the strength properties of concentric steel braced frame specimen, steel extrusion experiment was carried out according to TS EN ISO 6892-1st of January 2010 standard and its results were investigated [15].

According to test results, it was determined that maximum load in the cross section was 20.5 kN, tensile stress was 400.2 N/mm^2, yield strength was 337.9 N/mm^2 and elongation at rupture was 29.1%.

In this study, pushover analysis of 1/3 scaled reinforced concrete frame with inadequate earthquake resistance and 5 reinforced concrete frames reinforced with CSBF were carried out with SAP2000 program (Fig. 3) [16,17]. In pushover analysis performed with SAP2000 program, FEMA 356 American Regulations were applied. It is necessary to make different definitions for these columns, beams and braced.

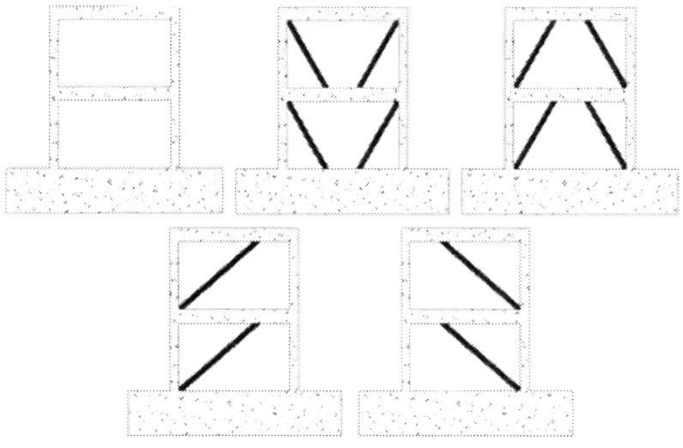

Fig. 3. Specimens.

The length of plastic hinge was taken as Lp=0.5 h in FEMA 356 and TEC-2007. In SAP2000 program, the hinge choices for column, beam and braced were presented. P-M3, M3 and P hinge definitions were given for steel columns, steel beams and steel braced, respectively. Moreover, moment curvature value was calculated for reinforced concrete frame and it was defined in SAP2000 [16-19]. Moment curvature graphs for reinforced concrete frame are given in Fig. 5.

SAP2000 implements the plastic hinge properties described in FEMA-356. As shown in Fig. 4, five points labeled as A, B, C, D, and E define the force–deformation behavior of a plastic hinge.

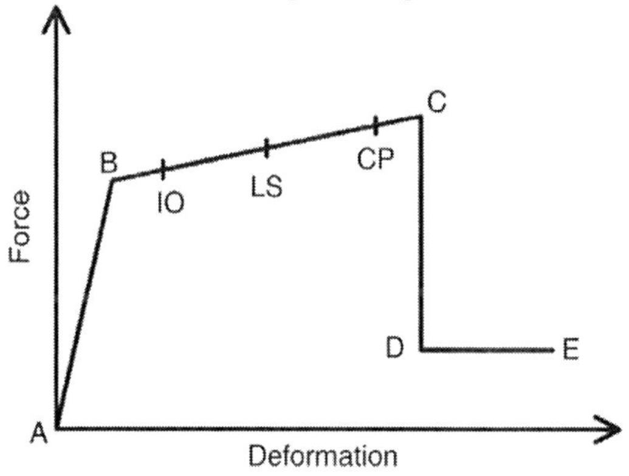

Fig. 4. Force–deformation relationship of a typical plastic hinge.

The values assigned to each of these points vary depending on the type of element, material properties, longitudinal and transverse steel content, and the axial load level on the element [18].

Immediate Occupancy (IO) – yielding of steel, significant cracking of concrete and nonstructural damage will arises.

Life Safety (LS) - damage of structural and nonstructural components will starts. We have to make essential circulation routes accessible to minimize risk of injury and causality for this stage.

Collapse Prevention (CP) – This point ensure a small risk of partial or complete building collapse by limiting

DOI: 10.1051/e3sconf/20171903030

structural deformations and forces to the onset of significant strength and stiffness degradation.

Point C is the indication of ultimate capacity of the structure and Point D indicate residual strength for the structure. Complete failure will occur at point E.

In TEC-2007, Immediate Occupancy (IO) limit is given as $(\varepsilon cu)MN=0.0035$, Life Safety (LS) limit is given as

$(\varepsilon cg)GV=0.0135$ and Collapse Prevention (CP) limit is indicated as $(\varepsilon cg)GC=0.018$ and pushover analysis was carried out with SAP2000 program based on these data.

Since equivalent seismic load was considered in pushover analysis and the weights of upper and ground storey were equal, 2 units of lateral loads were defined for upper storey and 1 unit of lateral load was defined for ground storey.

3 Results and Discussion

In this part, the results of analytical studies are presented and compared.

Moment-curvature and PM Interaction relationship for reinforced concrete frame are presented in Fig. 5.

A meaning of the specimens were given below:

RS: Reference Specimen (Bare Reinforced Concrete Frame)

VBS: V-Braced Specimen (Strengthened by Eccentrically Steel V-Braced Specimen)

TVBS: Transverse V-Braced Specimen (Strengthened by Eccentrically Steel Transverse V-Braced Specimen)

DBS: /-Braced Specimen (Strengthened by Eccentrically Steel Diagonal-Braced Specimen)

TDBS:\-Braced Specimen (Strengthened by Eccentrically Steel Transverse Diagonal-Braced Specimen)

Base shear-top displacement graphs obtained after pushover analysis of RS, VBS, TVBS, DBS and TDBS types are given in Fig. 6. In Table 1, on the other hand, maximum load and top displacement values for each specimen were given. According to these data and graphs, it was observed that TVBS type had more load-carrying capacity than the other types.

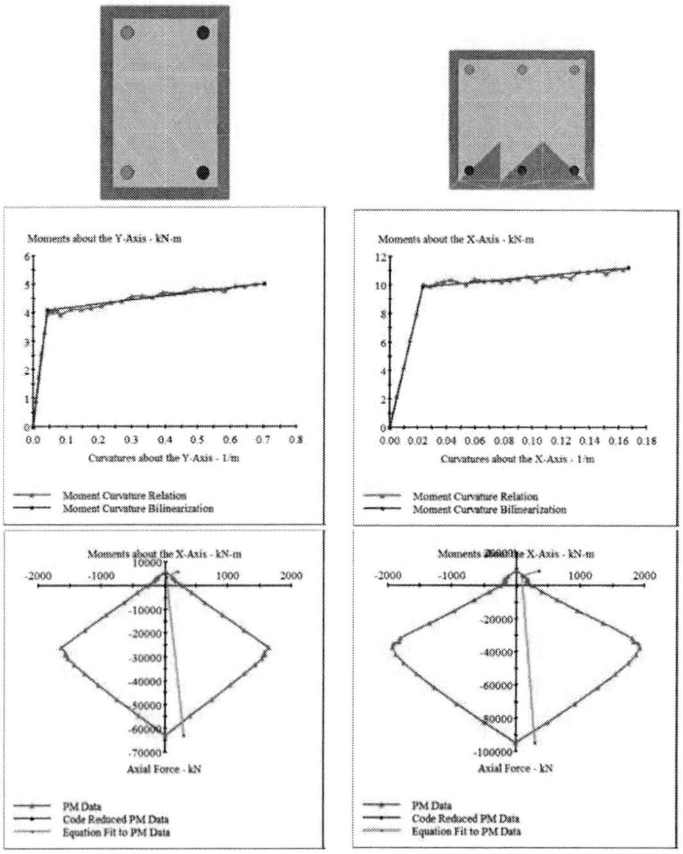

(a) (b)

Fig. 5. Moment-curvature and PM Interaction relationship for reinforced concrete frame (a) column (b) beam.

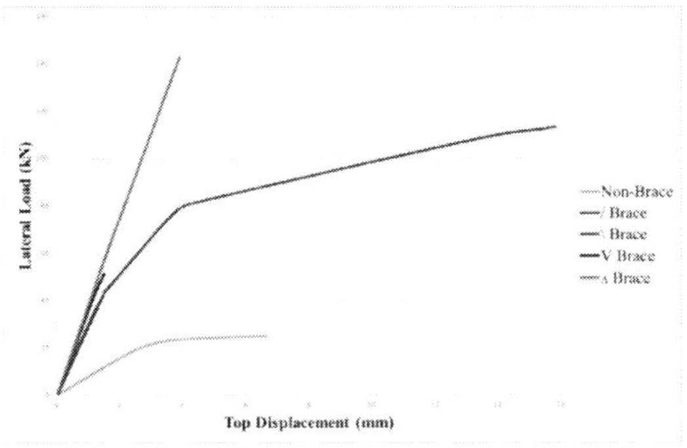

Fig. 6. Lateral load-carrying capacities of specimens

Table 1. Maximum load and top displacement values for each specimen.

	RS	VBS	TVBS	DBS	TDBS
Max. Load (kN)	24.87	51.19	142.51	113.03	38.26
Max. Displacement (mm)	0.66	0.15	0.39	1.59	0.14

4 Conclusions

Lateral load-carrying capacities are ordered in descending order as TVBS, DBS, VBS, TDBS and RS.
Lateral load carrying capacity of strengthened frames to the lateral load carrying capacity of the bare frame varied between 1.54 to 5.73. The largest strength increase was obtained in TVBS. Lateral strength of VBS was 2.06 times, TVBS was 5.73 times, DBS was 4.54 times and TDBS was 1.54 times greater than those of bare frame. The results show that the strengthened frame by eccentric steel braces applicable in buildings.
Adding concentrically steel braces to the bare frame decreased the story drift, and significantly increased strength, stiffness and energy dissipation capacity.
In this strengthening method lateral load carrying capacity, stiffness and dissipated energy of the structure can be increased.

References

1. A.Dogan, Research about concentrically steel braced frames according to design specifications for buildings in seismic regions 2007, Master Dissertation; Istanbul Technical University, Istanbul, Turkey (2007)
2. TEC-2007, Turkish Code for the Seismic Design of Structures, (2007)
3. C.D.Annan, M.A. Youssef, M.H. El Naggar, Eng. Struct. **31**, 7, (2009)
4. G. Brandonisio, M. Toreno, E Grande, E. Mele, A. De Luca, J. Construct. Steel Res. **78** (2012)
5. E. Grande, A. Rasulo, Eng. Struct. **49** (2013)
6. I. Hajirasouliha, A. Doostan, Adv. Eng. Software **41**, 3 (2010)
7. R.A. Jazany, I. Hajirasouliha, H. Farshchi, J. Construct. Steel Res. **88** (2013),
8. G. Metelli, Eng. Struct. 46 (2013)
9. A.E. Ozel, E.M. Guneyisi, Struct. Safety, **33**, 1 (2011)
10. C.W. Roeder, E.J. Lumpkin, D.E. Lehman J. Construct. Steel Res., **67**, 11 (2011)
11. Maheri, M.R., Kousari, R. and Razazan, M. (2003), "Pushover tests on steel X-braced and knee-braced RC frames", Eng. Struct., 25(13), 1697-1705.
12. Z. Qu, S. Kishiki, Y. Maida, H. Sakata, A. Wada Eng. Struct. **105** (2015)
13. A. Khampanit, S. Leelataviwat, J. Kochanin, P. Warnitchai, Eng. Struct. **81** (2014)
14. T. Görgülü, Y.S. Tama, S. Yilmaz, H. Kaplan, Z. Ay, J. Construct. Steel Res. **70** (2012)
15. TS EN ISO 6892-1st, Metallic materials- Tensile Testing- Part-1: Method of test at room temperature, Turkish Standards Institution, Ankara (2010)
16. SAP2000, version 15, CSI Company, Berkeley, California, USA.
17. SAP2000 User Manual, CSI Company, Berkeley, California (2000)
18. FEMA 273/356, Federal Emergency Management Agency, 1996 a, Guidelines for the Seismic Rehabilitation of Buildings, Redwood City, California, USA (1996)
19. M., Inel, H. B., Ozmen, Struct. Eng. Mech. **30**, 5 (2008)

Influence of long-lasting heat treatments on the structure and properties of the zirconium-steel bond

Mariusz Praźmowski[1,*], *Henryk* Paul[2], *Michał* Najwer[1], and *Anna* Pocica[1]

[1]Opole University of Technology, Faculty of Mechanical Engineering, 5 Mikolajczyka St., Opole, Poland
[2]Polish Academy of Sciences, Institute of Metallurgy and Materials Science, 25 Reymonta St., Krakow, Poland

Abstract. The explosive welding technology is related to the dynamic impact of the clad onto base material. These results in significant microstructural changes of the bonded materials, the increase of internal stresses and the strain hardening of layers near the interface. In order to eliminate these unfavorable effects, it is advisable to carry out proper heat treatment. The paper presents the results of structural examinations and mechanical properties of bimetal zirconium - carbon steel, after long-lasting heat treatment. Bimetallic samples were annealed at 600°C for 1, 10, 100, 500, 1000 hours. Microscopic examination were used to the analysis of microstructural changes, whereas systematic microhardness measurements and static tensile test to evaluation of the mechanical properties. It was found that the temperature has a beneficial effect on the changes in the bond zone and for longer annealing times a systematic decrease of the strength properties of the tested bimetal was observed. However, essential for the clad mechanical properties changes were observed in carbon steel plate. After short annealing (times of 1h and 10h) only small microstructural changes due to processes of recovery and recrystallization were observed in layers near the interface. The increase of annealing time was conducive to grain growth and carbon diffusion in the interfacial layer from steel towards zirconium.

1 Introduction

The construction of modern and corrosion-resistant chemical and processing apparatus with a planned long service life and relatively low production costs requires reactive metals, such as titanium, zirconium, tantalum and their alloys, to be used. These materials in the form of rolled flat products, i.e. plates or tapes are relatively expensive. The application of layered composites may be a solution to the reduction of their use costs. In such systems the base plate (usually steel of a proper grade) is the material that complies with construction requirements, then on top of it a thin layer of material with special properties (more expensive) is cladded. So far the only technology that enables materials with very different physical and chemical properties to be join together is explosion welding. It enables to achieve adequately high strength parameters of the bond that ensure the wide industrial processing and use of clad elements in all sorts of constructions.

Technological aspects of explosion bonding were demonstrated in many authors' research [1-3]. A distinguishing feature of the technology is joining together materials as a result of the oblique collision of plates at high velocity. Consequently, there is considerable plastic deformation and strengthening of joined materials in layers near the interface. The structural changes that occur in layers near the interface (the weld zone) were described in detail in many earlier publications [1-9]. The strain hardening of these layers and increase of internal stresses are unfavorable from the point of view of further processing of clad, i.e. straightening, drawing and stamping, machining, etc. [10-13]. In this case annealing is an important issue in the clads production. The purposefulness, or even necessity, of carrying out this type of treatment was documented in an earlier works, e.g. [14-17]. In the case of layered materials, the selection of proper annealing parameters constitutes a big technological problem. On the one hand we are dealing with a basic material that requires high temperature and long heat treatment time. On the other hand, the use of reactive material as the overlay layer largely limits these parameters as it is generally sensitive to high temperature oxidation [18-19]. What is more, a long-lasting impact of a high temperature under operating conditions does have an effect on the structural changes in the weld zone which most often decreases the strength properties of the clad.

This study presents the results that form the basis for the quality evaluation of zirconium-carbon steel bond achieved with the use of explosion welding. The impact of a long-lasting effect of a high temperature on the structural changes and mechanical properties of zirconium (the flyer plate) and carbon steel (the base plate) plates were analysed. Simulation heat treatments were performed at a temperature of 600°C for 1, 10, 100, 500 and 1000 hours in vacuum. Based on macro- and microstructural observations, as well as on mechanical properties measurements the effect of heat treatments on

* Corresponding author: m.prazmowski@po.opole.pl

the properties of the zirconium-carbon steel clads, was shown.

2 Research material and techniques

The material for this study was produced by means of explosion welding in the form of testing plates with dimensions 300 x 500 mm. The sheets of steel of 22 mm thickness (type P355NL2) as the base plate and sheets of Zr700 alloy of 3.15 mm thickness as a flyer plate were used for the clad preparation. Table 1 present the chemical composition whereas Table 2 the basic mechanical properties of the joined materials in the initial state.

Table 1. Chemical composition (in wt.%) of carbon steel P355L2N and zirconium Zr 700

	Material	Zr 700	P 355L2N
Chemical composition [%]	**C**	< 0.002	0.170
	Si	-	1.130
	Mn	-	0.345
	P	-	0.008
	S	-	0.001
	Ni	-	0.285
	Cr	0.050	0.150
	H	< 0.0003	-
	N	< 0.002	0.004
	Al	-	0.045
	Mo	-	0.035
	O	0.050	-
	Nb	-	0.019
	Fe	0.050	rest

Table 2. Mechanical properties of carbon steel and zirconium

Material	R_e, [MPa]	R_m, [MPa]	E, [GPa]	A_5, [%]
Zr 700	143	300	100	31
P 355L2N	402	511	210	27

After cladding, the entire surfaces of all materials underwent ultrasonic tests in order to determine the continuity of the bond. The material for further microstructural analyses was cut-off from regions without visible traces of discontinuity; the section perpendicular to sheet surface and parallel to the detonation front motion was analyzed. A detailed description of the procedure of zirconium-carbon steel clads preparation with the use of the explosive bonding process was presented in the works of [4-5].

Five samples were 'closed' in a vacuum quartz capsule (Fig. 1a) and then they were heat treated. The heat treatment was carried out at a temperature of 600 °C by annealing the samples for 1, 10, 100, 500 and 1000 hours, followed by air-cooling. An electric muffle furnace was used for the annealing, as presented in Fig. 1b. Heat treatment conditions and sample marking are shown in Table 3.

Table 3. Annealing conditions and sample marking

Sample designation	Condition	Treatment
P 355	steel in the state before bonding	cold rolled sheet and fully recrystallized
Zr 700	zirconium in the state before bonding	cold rolled sheet and fully recrystallized
W	condition after bonding	not heat-treated
1 h	condition after bonding and long-lasting heating at 600° C	heating time 1 hour
10 h		heating time 10 hours
100 h		heating time 100 hours
500 h		heating time 500 hour
1000 h		heating time 1 000 hour

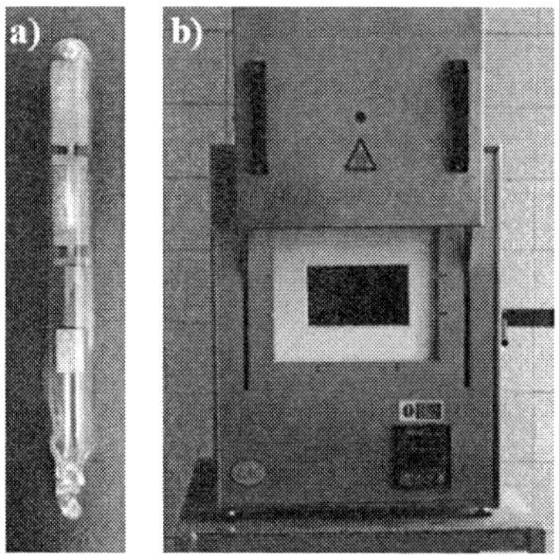

Fig. 1. Bimetal heat treating: a) the sample 'closed' in quartz capsule, b) the LMH 04/12 electric muffle furnace produced by LAC.

2.1 Metallographic examinations

Before the welding, the samples of 22 x 20 x 10 mm (for steel) and 3,15 x 20 x 10 (for zirconium) in size were cut-off from the starting plates. These samples were marked as P 355 and Zr 700, respectively. For the purposes of metallographic examinations, the

DOI: 10.1051/e3sconf/20171903031

metallographic specimens were prepared from these materials. The samples were polished manually with the use of abrasive papers, with a grit size up to 600. Then the samples were polished with a „Struers DPU2" mechanical polisher using the Al_2O_3 suspension. The final 8 h polishing was performed with a VibroMet vibratory polisher made by Buehler with the use of a colloidal suspension based on silicon monoxide. The final vibratory polishing stage allowed us to remove the strengthening caused by the grinding and manual polishing process. The last stage involved chemical etching with a 3% HNO_3 solution in C_2H_5OH in order to reveal the carbon steel structure and the interface between the welded plates. (There was no chemical etching of the zirconium plate since the application of very intense etching agent it is necessary to use. This damage the carbon steel and interfacial structures).

The metallographic examinations were performed with the use of an OLYMPUS IX 70 inverted microscope that was equipped with OPTI-TECH image analysis software. The detailed characterization of the interface between plates was made by measuring their basic parameters, i.e. the length 'n' and height 'h' of the wave and the surface area of the melted areas 'P_i'. In all cases, the measurements were obtained based on the microstructures observed in longitudinal sections. The measurements were made according to the scheme presented in Figs. 2a-d and the quantitative share of melted areas ('RGP' coefficient) was determined on the basis of equation 1.

$$RGP = \frac{\sum_{i=1}^{n} P_i}{L}, [\mu m] \qquad (1)$$

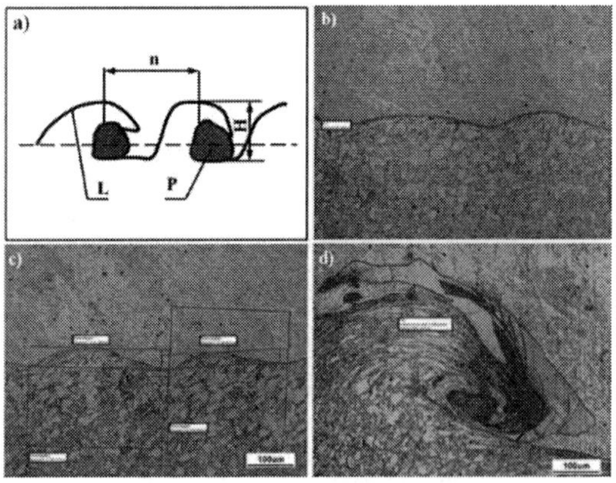

Fig. 2. a) Schematic presentation of the interface showing basic parameters of the wave:
H - wave height, L- length of the joining line, n- wave length, P- surface area of the melted areas.
(b-d) Examples of the microstructures showing the interfaces of different morphology:
b) flat interface, c) semi-flat interface, c) interface with melted zones, where: P_i - is the summary surface area of melted zones.

2.2 Microhardness tests

Microhardness changes in layers near the interface were carried out with the use of the Vickers microhardness tester - LECO MHT Series 200. The measurements were carried out in the longitudinal section along 3 'line scans', perpendicular to the interface. Indentations were carried out for each of the joined material in the 'direction' from the interface, i.e. outside the melting zone. The first measurement was carried out 0.02 mm from the bottom of the wave. The tests were conducted in accordance with the guidelines provided by ISO 6507-3:1996 using 50G and 100G loads. Moreover, the microhardness measurements were done at the selected points in layers near the interface for samples 'after welding' and 'after annealing'.

2.3 Tensile strength test

The tensile strength tests were performed on both groups of clads (in the state after welding and annealing) on the basis of non-standard samples; their shape and dimensions are shown in Fig. 3a, whereas, schematic presentation of the sample fixing, in Fig. 3b.

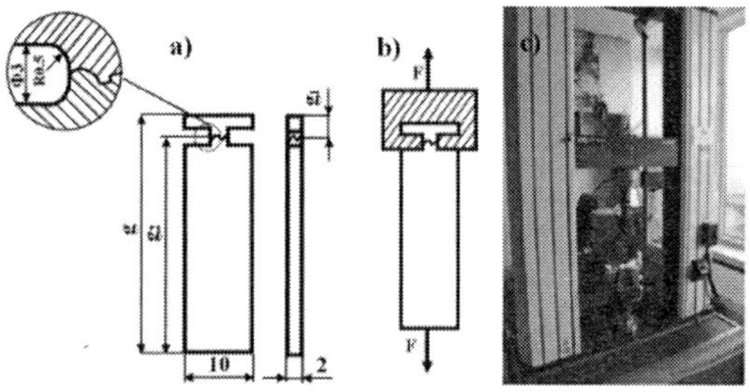

Fig. 3. a) Sample shape and dimensions for tensile test, b) sample fixing during test, c) tensile testing machine.

The measurements of the tensile strength (and the determination of R_m value) were carried out using the tensile testing machine - Instron 3382, shown in Fig. 3c. The value of tensile strength for each of the clads was determined based on the average of three tests.

3 Results and discussion

3.1. Optical microscopy observations

The microstructure observations were carried out using optical microscopy. Fig. 4a shows the structure of Zr 700 plate in the initial state, i.e. before the welding process. This material is characterized by an approximately equiaxial structure of oriented clusters of alpha-phase grains of a size ranged between 70 µm and 170 µm. Fig. 4b shows the structure of the base material, i.e. carbon steel P355 before the welding process. This material was characterized by the medium-sized grains: 4 µm - 11 µm for pearlite and 10 µm - 20 µm for ferrite. In the case of pearlite, a band of refined grains, typical of materials after hot deformation was observed. The microstructures of both materials before welding, show fully recrystallized states.

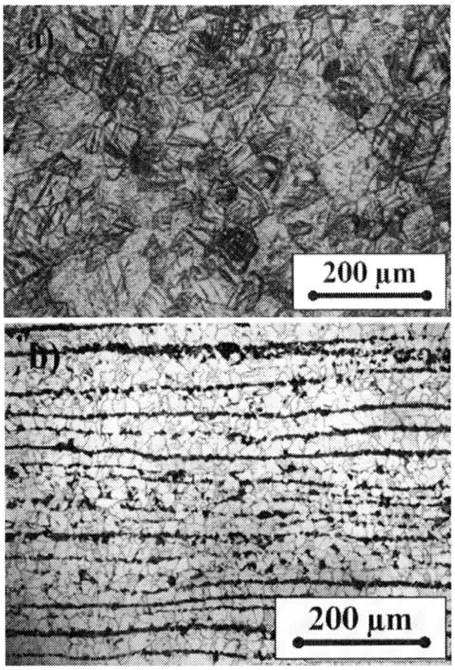

Fig. 4. Structure of materials in the initial state: a) Zr 700, imaging by using the polarized light, b) steel P355 N2L.

The microscopic observations of clad in state 'after welding' made it possible to determine the parameters describing the interface (wave) between the welded materials. As a result of the observations and measurements, the nature of the joint was determined as 'wavy' (Fig. 5a) with wavelength - n = 500 µm and height - H = 51µm. The detailed analysis showed the occurrence of layers of flat grains and small number of melted areas of non-continuous form; they were mostly found in the vortex and on the crest of the waves, Figs. 5b, c. The average values of parameters describing the wave are presented in Table 4.

Table 4. Parameters describing the shape of the wave and the quantity of melted areas for clad in 'after welding' state.

Designation sample	Length of the bond line L [µm]	Wave height H [µm]	Wave length n [µ m]	Melt surface area P [µ m^2]	Melt depth equivalent RGP [µ m]
W	10 632	51	500	3 366	0.35

In layers near the interface on the steel side, a distinct plastic deformation of the ferrite and perlite grains was observed; they were situated along the direction of the detonation front movement. The strongest deformation of grains was observed on the crests of waves, where the structure of very flat grains forms a layer of 25 µm -100 µm in width. As the distance from the interface increases the effects connected with strong deformation progressively disappeared. It was observed that at a distance of 450 µm - 850 µm from the interface only the structure of nearly equiaxed grains were observed, as also typically observed in the carbon steel plate before welding.

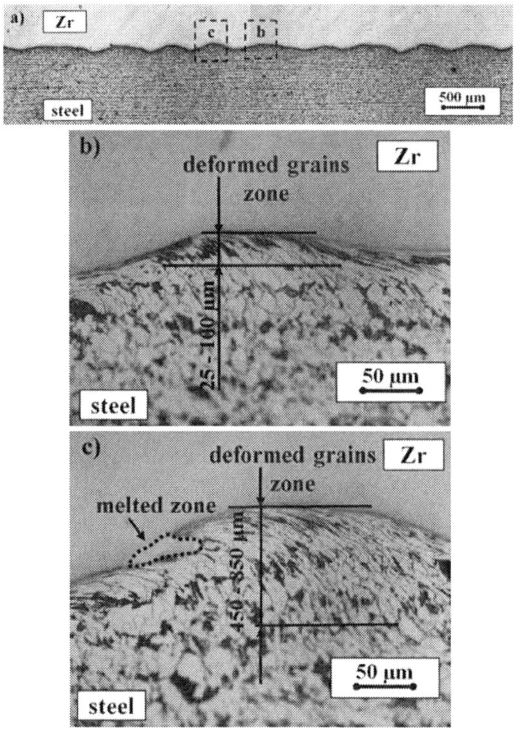

Fig. 5. Bond zone of zirconium- carbon steel bimetal: a) interface between the welded materials, b) zone of deformed grains, c) melted area

Since the heat treatment does not lead to changes of the wave parameters the analysis related to the characteristics of the weld zone was not carried out in the annealed samples. Structural changes due to the influence of temperature shows that in the case of the '1h' sample, i.e. the sample annealed for 1 hour, no significant microstructural changes were observed. Essentially, the microstructure was similar to that of observed in the clad is state 'after welding' (W). Figure 6a shows ferrite and pearlite grains of strongly elongated shape (in the direction of the detonation front movement). They were identified in carbon steel plate in areas near the crest of the waves. Only small quantity of melted areas were identified; they were the privileged places of the macro-/micro- cracks nucleation, as observed in Fig. 6a.

Recrystallized ferrite grains (the evidence of which is their equiaxial shape) and a few pearlite grains were observed in the steel plate after 10h annealing (Fig. 6b). The layer of the ferrite grains (free of cementite), adjacent to the interface, can be the evidence of decarburization of this zone as a result of a long-term

DOI: 10.1051/e3sconf/20171903031

annealing at high temperature. This phenomenon intensifies for longer annealing times. In the case of

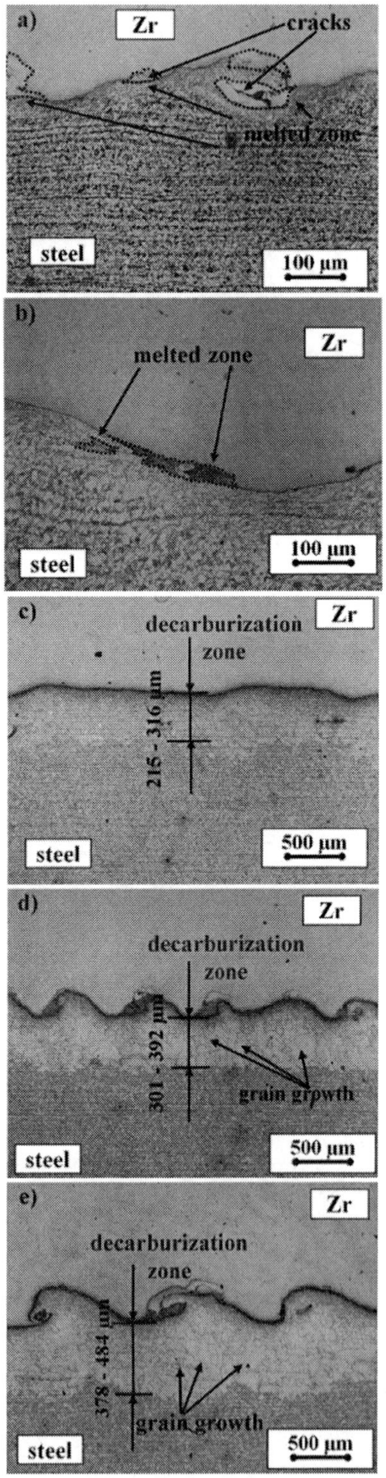

Fig. 6. Structure in the weld zone after annealing at temperature of 600 °C for: a) 1 h, b) 10 h, c) 100 h, d) 500 h, e) 1000 h.

annealing for the period of 100 h (100 h sample), the width of the decarburized zone is 215 - 316 µm (Fig. 6c). The extension of the annealing time up to 500 hours (500 h sample) resulted in the increase of this zone width up to 301-392 µm (Fig. 6d). As expected, the largest width of the decarburized zone was obtained for the annealing time of 1000 hours (1000h sample); in this case the width of the zone of low carbon content was ranged between 387 µm and 484 µm (Fig. 6e). Moreover, a significant growth of ferrite grains was observed for two longest annealing times in relation to other cases; from the point of view of strength properties this is strongly undesirable phenomenon.

3.2 Strain hardening of the layers near the interface

During explosive welding the plates are brought together under very high pressure and at high velocity. The deflection of the flyer plate leads to oblique collision and 'provokes' shear strains near-the-collision point/line. These phenomena contribute to the strain hardening of both materials in areas near the interface, as presented in a number of earlier works [4-8]. However, it is widely also accepted that the significant part of the kinetic energy of the jet, the impact energy and the work of plastic deformation are dissipated to heat in a very short time [6-8]. This leads to local melting and then to the zones of solidified melt formation.

The microhardness measurements made it possible to determine the changes of strain hardening, in the both near the interface (up to 0.5 mm) and in areas above 3 mm from the interface. The results were compared with the microhardness values carried out in plates before the explosion welding. They are presented in Fig. 7.

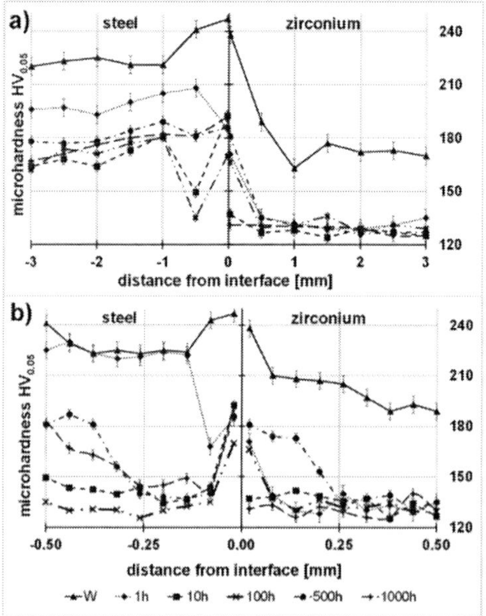

Fig. 7. Average results of the microhardness measurements in the clads after welding and after annealing at a distance of: a) 3 mm, b) 0.5 mm from the interface.

DOI: 10.1051/e3sconf/20171903031

The microhardness measurements carried out in samples subjected to long-term heat treatments were compared to the measurements of bimetal hardness after welding. This opens the possibility the analyzis of the influence of heat treatment on strain hardening and strength properties of the tested samples. Fig. 7a shows the microhardness of the flyer and base plates in areas distanced ~ 3 mm from the interface.

The largest microhardness (strain hardening) in both welded plates, was measured 0.02 mm from the interface. The microhardness decreases systematically as the distance from the interface increases and the microhardness of steel reaches its stable level ~ 1mm from the interface, as presented in Fig. 7b). The analysis of microhardness changes in the zirconium plate showed that the annealing time (at this temperature) does had any significant impact on microhardness in areas near the interface.

In the case of 1h, 100h and 500h samples the microhardness of zirconium plate decreases of ~30%-40%, whereas in the case of 10h and 1000h samples the microhardness decreses of~55% in relation to the highest microhardness (value of the first measured point) in the zirconium, in state 'after bonding'. A similar level of decrease in the strengthening was observed throughout the analyzed area (up to 5 mm from the boundary). It is important to note that for annealed samples, the microhardness of zirconium plate was always below the microhardness measured in state 'before the welding', in the entire analysed area (175 HV).

Similar changes were observed in the carbon steel; its microhardness decreased as the annealing time increased. In the case of the shortest annealing time of 1h, the microhardness of layers near the interface decreased by 20% with respect to the state 'after welding'. The increase of the annealing time up to 10 hours led to the 25% decrease of microhardness, whereas in the case of sample annealed for 100 hours - up to 30% (in relation to the microhardness at this point of the material after welding). The impact of the annealing time was more evident at a distance of 0.5 mm from the interface; after annealing for 1h the decrease of microhardness was ~10% in relation to the microhardness (at this point) of the material in state 'after welding'. The increase of the annealing times up to 10 hours and 100 h led to the 40% and 45% decrease of microhardness, respectively, as compared to the microhardness in state 'after welding'.

The microhardness measurements in the melted areas for the sample in state 'after welding' (Fig. 8a) and after 1 hour annealing (Fig. 8b) showed that their values changes significantly. In the case of samples without annealing, the minimum microhardness was 477 HV, whereas the maximum - 896 HV. The maximum value of microhardness in zirconium plate is nearly 4 times greater than the most strain hardened layers of steel plate (~247 HV). Annealing for shorter annealing times (1h and 10h) did not cause significant changes in the microhardness of the melted areas with respect to the microhardness of sample 'after welding'. As an example, the microhardness of the sample annealed for 1h ranged between 354 and 1053 HV (Fig. 8b).

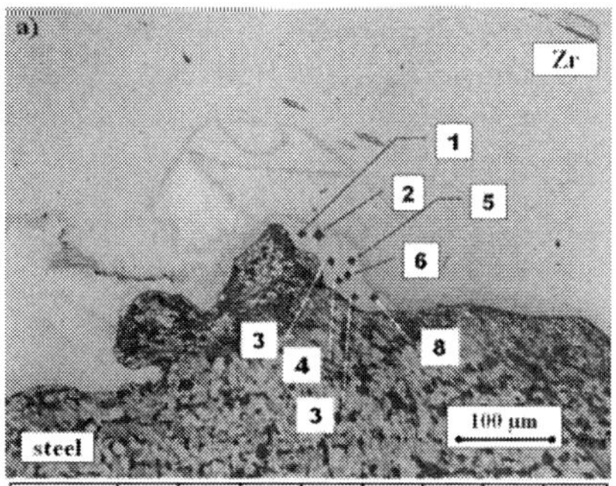

Point	1	2	3	4	5	6	7	8
$HV_{0.05}$	688	477	756	832	831	833	896	792

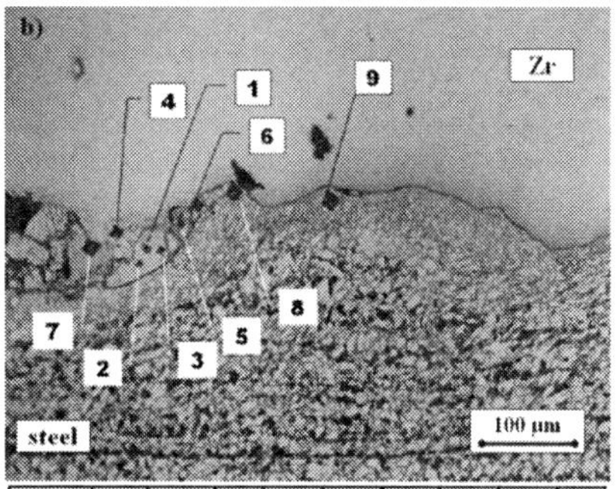

Point	1	2	3	4	5	6	7	8	9
$HV_{0.05}$	874	1053	791	354	833	536	224	205	239

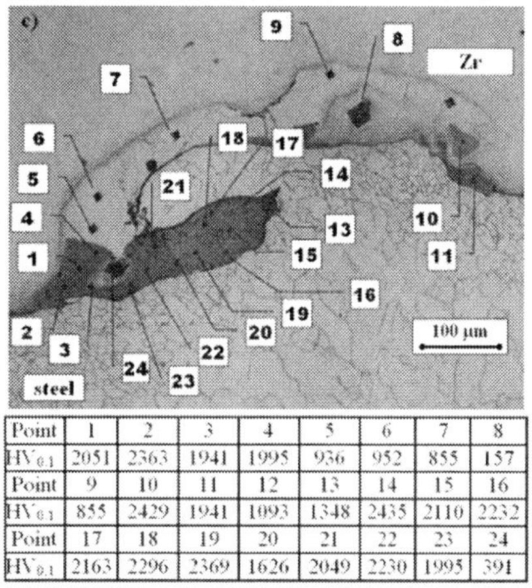

Point	1	2	3	4	5	6	7	8
$HV_{0.1}$	2051	2363	1941	1995	936	952	855	157
Point	9	10	11	12	13	14	15	16
$HV_{0.1}$	855	2429	1941	1093	1348	2435	2110	2232
Point	17	18	19	20	21	22	23	24
$HV_{0.1}$	2163	2296	2369	1626	2049	2230	1995	391

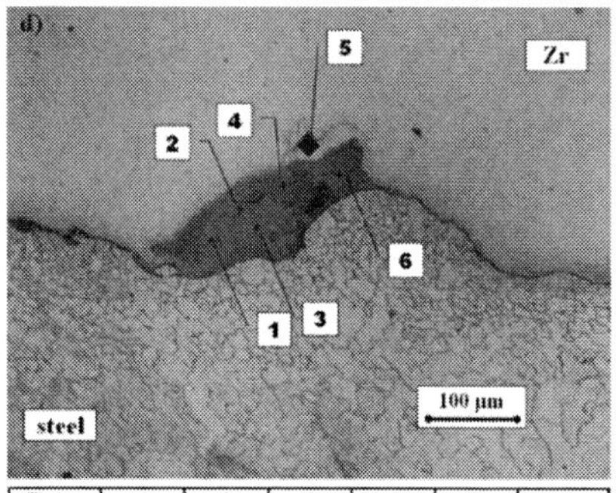

Point	1	2	3	4	5	6
$HV_{0.1}$	2363	2360	2357	2508	270	2357

Fig. 8. Microhardness in the melted areas of zirconium-carbon steel clad: a) state after welding and state after annealing at 600°C for: b) 1h, c) 100h, d) 500h.

Long annealing times (i.e. 100h and 500h) resulted in over 1.5 times larger values of microhardness in the melted areas relative to the state after welding. The maximal value of microhardness in the melted areas for the 100h sample was 2 436 HV (Fig. 8c, point 14) whereas for the 500h sample reached 2508 HV (Fig. 8d, point 4).

3.3 Strength properties

The tensile test was carried out on samples in states 'after welding' and 'after annealing' for determination of the tensile strength (R_m) od the clad; it makes possible to compare changes in the mechanical properties. Fig. 9 shows the average values of three tests for each case.

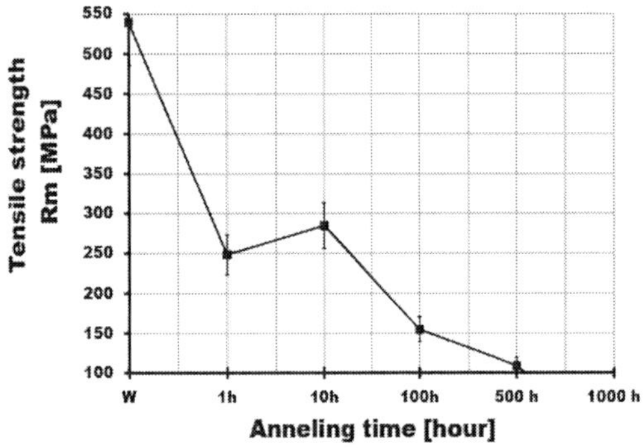

Fig. 9. Impact of the annealing time on the tensile strength of zirconium - carbon steel clad.

The tests show that the annealing time significantly affects the value of the tensile strength (R_m) of the clad. The average strength of the bimetal after welding (sample W) was 537 MPa. This value confirms high mechanical properties of the joint, which are similar to the properties of steel before welding (Table 4). The annealing of the clads at a temperature of 600°C for 1h and 10h (1h and 10h samples) leads to the strength decrease of approximately 50% up to 248 MPa and 284 MPa, respectively (Fig. 9). This results from the microhardness decrease due to microstructural changes caused by the recrystallization of severely deformed layers near the interface. The strength decrease of ~70% was observed in the case of the 100h sample annealed for the 100 hours, and ~85% decrease in the case of annealing time of 500 hours (500h sample), in relation to the microhardness of the sample after the welding process. The strength test was not carried out for the sample annealed for the 1000 h, since the tested material undergo delamination during heating. Such low strength properties for the longest annealing times are associated with a significant growth of grains in the layers near the interface and almost complete decarburization of this zone on the side of the carbon steel plate.

4 Conclusions

The obtained results make it possible to draw the following conclusions:

- The correct selection of explosion welding process parameters, i.e. detonation velocity and initial distance between the plates being welded, makes it possible to obtain a corrugated interface with a high amplitude and length of the wave.
- Brittle and hard areas of solidified melt were identified in layers near the interface, but their participation in the joint did not exceed the limit values of the 'correct' RPG coefficient, which is assumed to be 10μm.
- The microhardness of the melted areas, which occur near the interface, was always several times higher that the one of the most strain hardened layers of both sheets. Brittle nature of the melted zones contributes to the occurrence of cracks, which are visible during the microscopic examination. For longer annealing times (100h and 500h) more than 1.5 times increase of the microhardness of the melted zone was observed.
- The heat treatment significantly reduces the strain hardening of layers near the interface, and the largest changes were observed in the carbon steel plate.
- The annealing induces significant microstructural changes in layers near the interface. Shorter annealing times, i.e. 1h and 10h, have a positive impact on the mechanical properties due to the formation of small (equiaxed) recrystallized grains. In the case of longer annealing times (100h and 500h), mechanical properties decreases; there is observed a significant growth some of the grains and the decarburization of steel plate in layers near the interface.
- Microstructural changes occurring during the annealing affect the mechanical properties of the tested clad. An extended annealing time causes a systematic decrease of the tensile strength in relation to the values measured in the state 'after welding'.

The presented results have allowed to assess the effect of time and temperature on the heat treatment on the tensile strength, microhardness and microstructure of zirconium / steel bimetal.

Acknowledgment:
This work was partially supported by the Polish Ministry of Science and Higher Education, project no: NN507 457839.

References

1. B. Crossland, *Explosive Welding of Metals and its Application*, (Clarendon Press, UK, 1982)
2. H. Dyja, A. Maranda, R. Trębiński, *Technologie wybuchowe w inżynierii materiałowej*, (Wyd. Wydz. MiIM Pol. Częstoch. 2001)
3. W. Walczak, *Zgrzewanie wybuchowe metali i jego zastosowanie*, (WNT, Warszawa 1989)
4. M. Prażmowski, Arch. Metall. Mater. **59**, 1137-1142 (2014)
5. M. Prażmowski, H. Paul, Arch. Metall. Mater. **57**, 1201-1210 (2013)

6. H. Paul, T. Baudin, F. Brisset, M. Prażmowski, Mat. Sci. Engin. **82** (2015)
7. H. Paul, J. Morgiel, T. Baudin, F. Brisset, M. Prażmowski, M. Miszczyk, Arch. Metall. Mater. **59**, 1129-1136 (2014)
8. H. Paul, J. Morgiel, M. Faryna, M. Prażmowski, M. Miszczyk, Int. J. Mat. Res. **106**, 782-792, (2015)
9. M. Prażmowski, D. Rozumek, H. Paul, Engn. Failure Anal. **75**, 71-81 (2017)
10. M. Acarer, B. Gulenc, F. Findyk, Mater. Des. **24**, 659-664 (2003)
11. P. Nieslony, P. Cichosz, G.M. Krolczyk, S. Legutko, D. Smyczek, M. Kolodziej, Measurement **78**, 129-137 (2016)
12. P. Nieslony, G.M. Krolczyk, K. Zak, R.W. Maruda, S. Legutko, Prec. Eng. – J. Int. Soc. Prec. Eng. Nanotechnol. **47**, 104-110 (2017)
13. T.Z. Blazynski (ed.), *Explosive welding, forming and compaction*, (Applied Science Publishers, London and New York, 1983)
14. M. Prażmowski, H. Paul, F. Żok, Arch. Metall. Mat. **59**, 1143-1149 (2014)
15. F. Findik, R. Yilmaz, T. Somyurek, Sci. Res. Essays **6**, 9, 4141-4151 (2011)
16. C. Prothe, S. Pauly, C. Toth, *Corrosion Solutions Conference 2005*, 53-59.
17. M. Raudbari, A. Mehdipoor, R. Azarafza, Int. Res. J. Appl. Basic Sci. **7**, 10, 687-692 (2013)
18. T. Tetsui, Y. Miura, Tech. Rev. **1**, 1-9 (2002)
19. J. Małecka: Int. J. Mat. Res. **9**, 104, 885-891 (2013)

E3S Web of Conferences 19, 03032 (2017) DOI: 10.1051/e3sconf/20171903032
EEMS 2017

Investigation of the influence of coolant-lubricant modification on selected effects of pull broaching

Krzysztof Adamczuk[1], *Stanisław* Legutko[2,*], *Alicja* Laber[1], and *Wojciech* Serwa[1]

[1]Faculty of Mechanical Engineering, University of Zielona Gora, 4 Prof. Z. Szafrana Street, 65-516 Zielona Gora, Poland
[2]Faculty of Mechanical Engineering and Management, Poznan University of Technology 3 Piotrowo Street, 60-965 Poznan, Poland

Abstract: The paper presents the results of testing the wear of the tool (pull broach) and a gear wheel splineway surface roughness after the friction node of pull broach/gear wheel (CuSn12Ni2) had been lubricated with metal machining oil and the same oil modified with chemically active exploitation additive. To designate the influence of modifying metal machining oil by the exploitation additive on the lubricating properties, anti-wear and antiseizure indicators have been appointed. Exploitation tests have proved purposefulness of modifying metal machining oil. Modification of the lubricant has contributed to reduction of the wear of the tools – pull broaches and to reduction of roughness of the splineway surfaces.

1 Introduction

Strong worldwide competition in production industry forces producers to constantly improve and seek solutions reducing production costs while maintaining or raising the product quality [1-4]. Reduction of production costs allows for reduction of the price of ready-made products and, consequently to become an attractive supplier in the market. However, high quality of products must be maintained because its lowering may result in drop of sales of products. Properly designed and generated surface structure can, for example, minimize the friction coefficient. Modifications aiming at improvement of the surface quality are a result of thermal and mechanical loads of the process of machining, e.g. tool wear OUPN system stability or machining parameters. The machined surface is generated on the basis of the tool profile and the actual surface profiles depends on various disturbances in the process of machining [5, 6].

In order to improve the working conditions of highly loaded gears, thin and hard anti-wear coatings were applied [7-10]. Identification of lubrication conditions in concentrated contact have also been dealt with [11, 12].

Fontanari and others [13] have examined the influence of the friction combination of CuSn12 bronze / 42CrMo4V steel on the tribological properties of a worm gear. The tests have been performed with various loads of the friction node on a friction path of about 1800 km, interrupting it periodically in order to measure the sample mass. Basing on the tests, curves of wear as a function of the friction path have been made. Surface damages have been analysed with the use of scanning microscopy. Various pinholes and splinters have been found on the friction surface; their size depended on the load applied to the friction pair.

The influence of load, speed and kind of lubricating agent on the tribological properties of CuSn10 bronze has also been tested [14]. Basing on the tests, the influence of those parameters on the friction coefficient, friction area temperature and the wear of the friction node has been determined. The tribological properties are also influenced by the environment. T. Kucukomeroglu and others [15] have found that the tribological properties of the CuSn10 alloy vary depending on the conditions of testing. They have compared the results obtained in vacuum to those obtained in atmospheric conditions. The test results have shown that the friction coefficient increases in vacuum and wear resistance grows. A. Laber and others [16] have also performed tests using casting bronze and sintered bronze CuSn10Ni2, for the friction node. The friction node has been greased with various lubricants modified with chemically active exploitation additive. Modification of the oils has variously influenced the lubricating properties (standardized tests, tester T-02) and tribological properties (not standardized tests, tester T-05) of the bronzes under investigation.

The wear of wedges of pull broaches for forming internal surfaces has also been investigated by K. Wieczorowski and others [17]. The authors have found excessive wear on the side edges of the pull broaches, which has been the reason of damages of the splineway side surfaces.

The purpose of the present investigation was to solve the problem of wear of pull broaches in the process of pull broaching of the splineways of gear wheels made of CuSn12Ni2 bronze and to reduce the value of roughness parameters of the splineway. Modification of the lubricant (metal machining oil) with chemically active

* Corresponding author: stanislaw.legutko@put.poznan.pl

exploitation additive was aimed at improvement of the working conditions of the pull broaches.

2 Experiment

2.1 Test materials

The material taken to the tests was CuSn12Ni2 bronze with the following mechanical properties: Brinell hardness 110 HB, tensile strength Rm = 550 MPa. This grade of bronze is very resistant to static loads, corrosion, abrasion and elevated temperatures. It has very good cast ability. In Table 1, one can find a view of the structure and the average chemical composition of the CuSn12Ni2 bronze.

Table 1. Chemical composition and metallographic structure of CuSn12Ni2 bronze.

Chemical composition				
Cu %	Sn %	Zn_{max} %	Pb_{max} %	Ni_{max} %
84.5-87.5	10-12	0.25	0.16	2.5

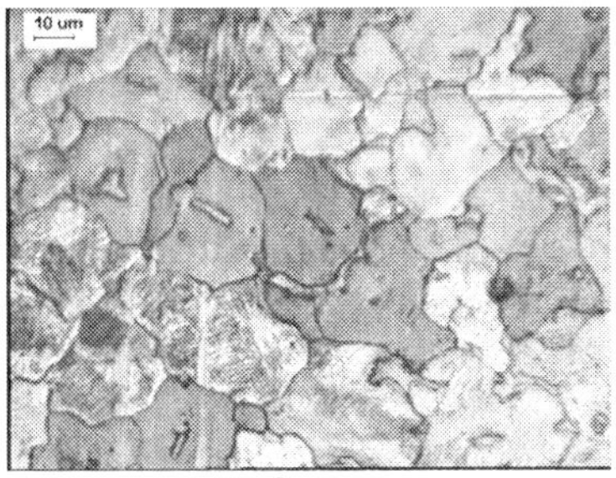

Magnification x500

For comparative analysis of the pull broaches, two tools with the same dimensions and parameters in the initial condition have been selected (Fig. 1). Using those tools, 6000 splineways in gear wheels have been made.

Fig. 1. Pull broaches used in the tests.

The lubricating properties were tested in the presence of the lubricating agent for metal machining with the following physical and chemical properties: density at 20°C < 865 kg/m^3, kinematic viscosity at 40°C = 22 mm^2/s, ignition temperature 190°C, flowing temperature - 15°C.

The lubricating agent has been modified with an exploitation additive with the following properties: kinematic viscosity at 40°C = 117 mm^2/s, kinematic viscosity at 100°C = 5.63 mm^2/s.

Table 2 shows the designations of the tested pull broaches according to the producer.

L.p.	
1	2
Our own designation	
P1	P2
Producer's designation	
6P9x3.7x2.8 RN1060 Id.-Nr:5980640; SC 6-5-2 RL16-40 250.3095 11-0221 6; SWZ 08/11	6P9x3.7x2.8 RN1060 Id.-Nr:5980640; SC 6-5-2 RL16-40 250.3095 11-0221 3; SWZ 08/11
The cooling – lubricating agent applied	
Industrial lubricating oil	Modified lubricating oil

The additive is conveyed to the friction zone by means of the lubricating agent forming a boundary layer on the friction surfaces; the layer is created as result of chemisorption; the layer has high resistance to high dynamic and temperature loads and resistance to breaking.

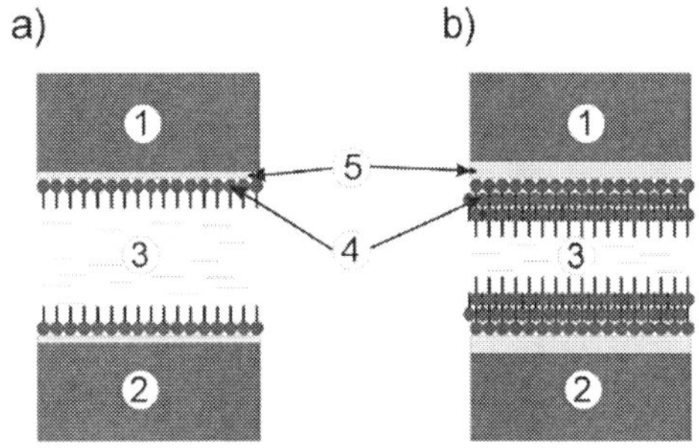

Fig. 2. A model of friction node lubricated with: a) market available lubricating agent, b) market available lubricating agent with exploitation additive; 1, 2 - friction elements, 3 – lubricating agent, 4 – boundary layer formed due to physical sorption, 5 – boundary layer formed due to chemisorption [18].

2.2. Methodology of investigation

Tests of the lubricating properties of the lubricant for metal machining have been performed according to PN-76/C-04147 by means of a four ball apparatus. The following indices of the lubrication properties: weld point P_z; index of wear under load I_h; not seizing load P_n; seizing load P_t; limit wear load G_{oz}, and the average diameter of defects. The determination of the P_z, P_n and I_h indices has been effected with stepwise increasing load to the moment of welding of the friction node. The P_t index has been determined with the friction node load increasing from 0 to 800 daN at the rate of 408.8 N/s the determination of the limit wear load with the friction node load of 150 daN. A view of the four ball apparatus and a diagram of the friction node can be seen in Fig. 3.

The assessment of the effect of the lubricating agent modification in order to improve the working conditions of the pull broaches required tests under exploitation conditions. Exploitation tests have been performed with the use of two CNC broaching machines (Fig. 4). In one of them, CNC1, metal machining oil has been applied; in the other, CNC2, the same oil modified with the exploitation additive. The friction node for the CNC1 machine consisted of a CuSn12Ni2 gear wheel/P1 pull broach; the node for the CNC2 machine included a gear wheel of CuSn12Ni2 bronze/ pull broach P2 (Fig. 5). After the exploitation tests, the wear of the pull broaches has been analysed with the use of a workshop microscope, MWD, with the magnification of 50 times.

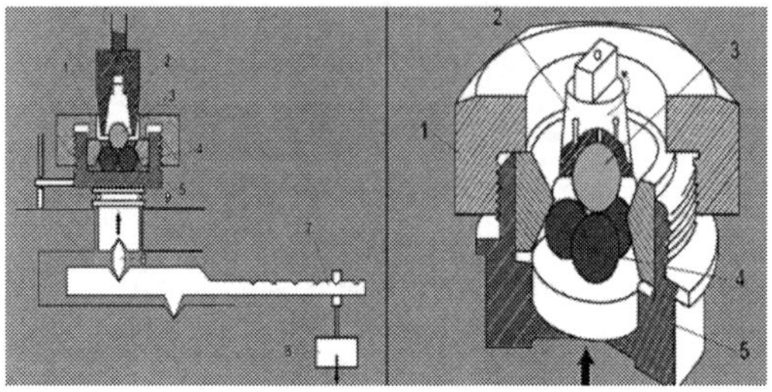

Fig. 3. The four ball apparatus: kinematic diagram of the friction node: 1 – cover fixing the bottom balls, 2 – upper ball holder, 3 – upper ball (turned one), 4 – bottom balls (immobile), 5 – vessel with the tested oil, 6 – prism, 7 – lever, 8 - weight, 9 –tested oil; b) a diagram of the friction node.

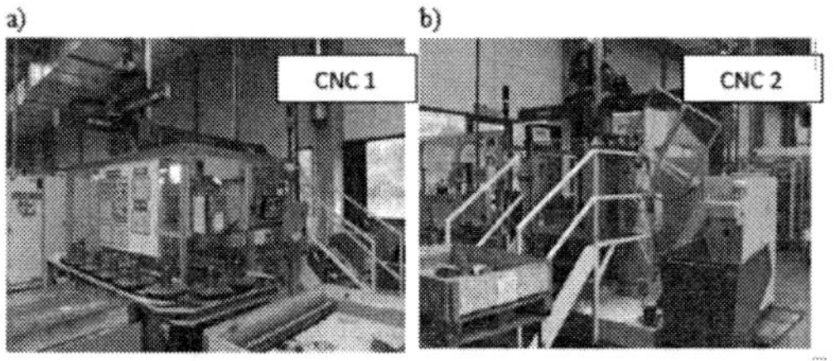

Fig. 4. Broaching machines used in the exploitation tests: a) with metal machining oil, b) with metal machining oil modified with exploitation additive.

The geometry of the cutting wedges of the broaches has been determined by means of a contour device made by Mahr, model MahrSurf XC 20 PCV. The collected results of dimension values for the individual tools have allowed us to perform a comparative analysis in respect of changes in the geometry of the tested tools.

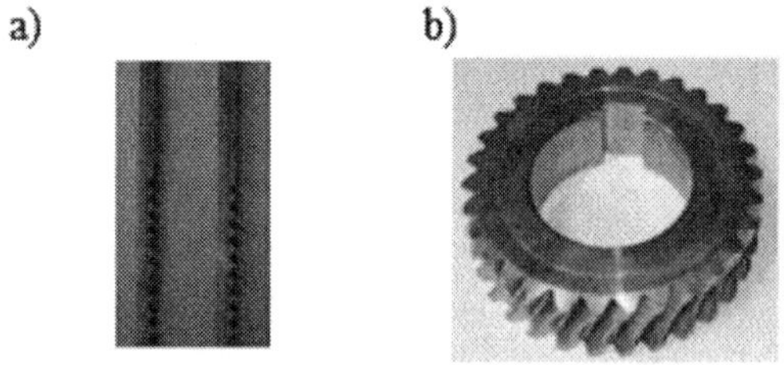

Fig. 5. Elements of the friction node: a) pull broach, b) gear wheel.

3 Investigation results

3.1 Results of the tests of lubricating properties

In Tables 3 and 4, one can find the results of lubricating property tests of the lubricant under investigation and of the same lubricant modified with exploitation additive.

Basing on the test results shown in Tables 3 and 4, it has been determined that modifying the metal machining oil with chemically active additive has not influenced improvement of the weld point *Pz,* and not seizing load *Pn*. On the other hand, an increase of the seizing load index P_t, has been obtained (from the value of 201.99 to the value of 270.23 daN). As a result of the modification the lubricating layer durability has been improved by 34%. The limit wear load G_{oz}, has also been found to grow from the value of 96.30 to the value of 138.67 daN/mm^2, which has contributed to the improvement of anti-wear properties of the lubricant.

Table 3. Lubricating properties of the lubricants under investigation.

Lubricant	P_t **[daN]**	P_z **[daN]**	P_n **[daN]**	I_h **[daN]**
Metal machining oil	201.99 ±10	500 ±25	100	62.509 ±3.1
Metal machining oil + exploitation additive	270.23 ±13.5	315 ±15.7	80	62.291 ±3

Table 4. Values of the limit wear load, G_{oz}, of the agents under investigation.

Lubricant	G_{oz} **[daN/mm^2]**
Metal machining oil	96.30 ±4.8
Metal machining oil + exploitation additive	138.67 ±6.9

3.2 Verification tests

The lubricant influence on the tool wear has been determined by measurement of the changes of the geometry of the tool cutting edges. For that purpose, measurements have been performed and then the following dimensional magnitudes have been compared to those of a new tool (Fig. 6):

- clearance angle (α);
- rake angle (γ);
- the angle of the linear section of the chip groove (β);
- cutting edge rounding radius (r);
- chip groove rounding radius (R);
- height increase of the individual edges (a).

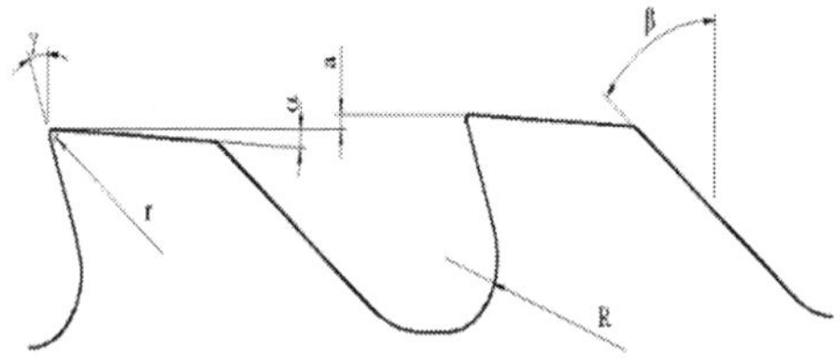

Fig. 6. Designation of the compared dimensions of the pull broach cutting edges: a – height increase of the individual edges; α – clearance angle; β - angle of the linear section of the chip groove; γ – rake angle; r – radius of the cutting edge roundoing; R – radius of the chip groove rounding.

The measurements of the dimensional values of the geometry have been performed on two sections: D1 and D2, at the beginning of the tool (8 first wedges – the first measurement section) and in its end part (wedges 26 to 33 – second measurement section). The test results can be found in Fig. 7.

The performed calculations show that the average change of the cutting edge height increase *a*, in the tested section D1, as compared to a new edge, is smallest in the case of tool P1 working in the presence of industrial oil. The same result has been recorded for the pull broach P1 in the case 3 of comparative analysis of the changes of the rake angle γ. Analysing the other dimensional magnitudes (i.e. clearance angle α, and the angle of the linear section of the chip groove β) in the same way as before, the smallest deviations from the model have been observed for pull broach P2 which had been working in the presence of lubricating oil with the modifying additive. The presented analysis neither excludes nor unambiguously proves positive or negative influence of the applied lubricants on the maintenance of the tool geometry or extension of its life. The analysis result is also influenced by the dimensional tolerance with which the tested pull broaches had been made. With the measurements performed with so high accuracy, the analysis result could be made false due to dimensional deviations adopted for the pull broaches or inaccurate execution of them.

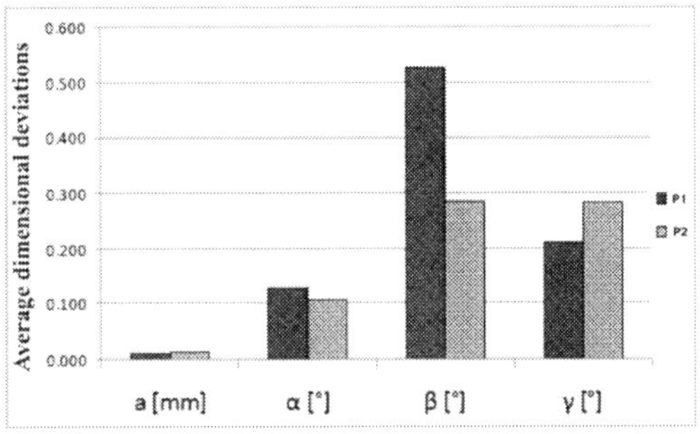

Fig. 7. Arithmetical average of dimensional deviations of the geometry of tools P1 and P2 on section D1.

The analysis of the tool wear by assessment of the changes of the cutting wedge geometry has been performed with use of microscopic examination of the rake face which is exposed to various kinds of changes (e.g. blunting, damages) resulting from high load during operation. Test results can be seen in Fig. 8.

An analysis of the cutting edge rounding outline and the condition of the side surface of the broach wedges has proved positive influence of the exploitation additive on the improvement of P2 tool life by:

- more precise maintenance of the cutting edge geometry;
- burns not occurring in vast areas;
- lower number of\accretions;
- lack of significant traces of machining in accordance with the direction of it;
- lack of damages of the protective coating;
- lack of material chipping from the corner.

Microscopic examination of the rake face and the flank face has also been performed. The results can be found in Fig. 9.

The cutting edge wear of tool P1 is larger than that of the cutting edge of tool P2.

The roughness of the splineway surfaces made in the presence of the tested lubricant has also been measured. The measurements of the splineway surface roughness have been performed on randomly selected gear wheels. The average value of the roughness parameter *Ra* of a splineway lubricated in the manufacturing process has reduced from 0.71 μm to 0.38 μm after the application of a modified lubricant. Table 5 presents the value of average surface roughness measurements, according to *Ra* parameter, of splineways made in the presence of the lubricants under investigation.

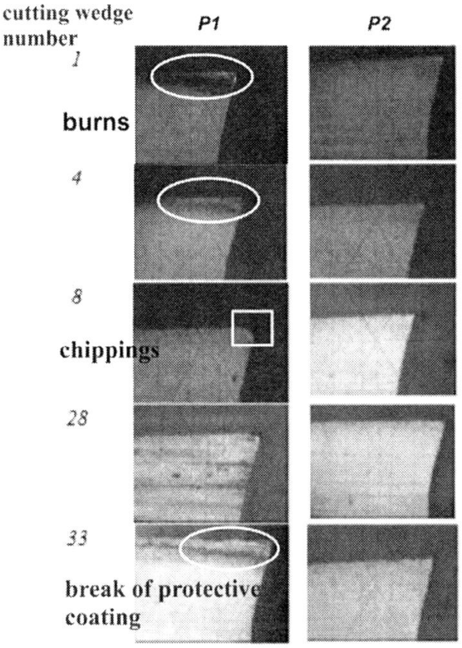

Fig. 8. Photographs of used pull broach cutting edges made during microscopic examination.

a tool	Image of the rake edge of the individual edges, combined of three photographs
P1	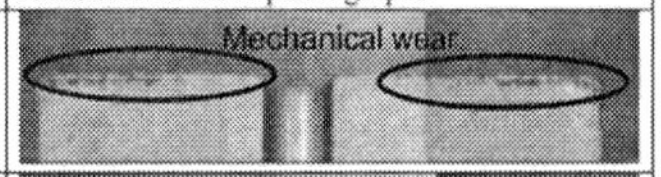
P2	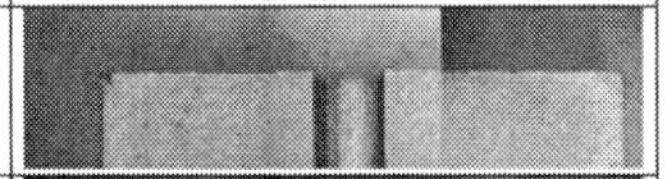
P1	
P2	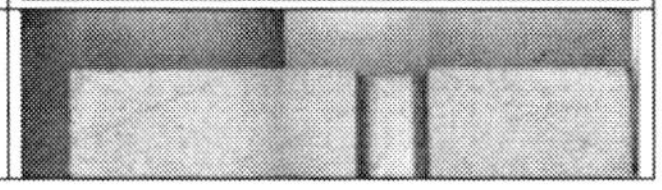

Fig. 9. A view of the rake face of the cutting edges.

The obtained results prove positive influence of the exploitation additive on improvement of the surface quality of machined splineways by reduction of the average value of the roughness parameter *Ra*. The value of the *Ra* parameter has reduced by 47% with the use of modified lubricant.

Table 5. *Ra* roughness parameter of splineways executed with the application of various lubricants.

Measurement number	Industrial lubricating oil *Ra*	Modified lubricating oil *Ra*
R_a average	**0.71 μm±0.04**	**0.38 μm±0.01**

4 Conclusions

The results of the tests of lubricating properties of the lubricants under examination do not unambiguously prove a positive influence of the exploitation additive on the improvement of all the indices of the lubricating properties. Some of those indices, i.e. the weld point Pz, the highest not seizing load Pn, and the index of wear under load I_h, do not show improvement as result of modification. On the other hand, an increase of the seizing load P_t, and the limit wear load index G_{oz}, has been obtained. The increase of the limit wear load index has significantly improved the anti-wear properties of the modified oil.

The application of the modifying additive has influenced reduction of the tool wear. This effect has been obtained by:

- reduction of the value of the coefficient of friction between the flank face and the material being machined;
- reduction of the value of the coefficient of friction between the rake face and the formed chip;
- reduction of temperature in the cutting zone;
- occurrence of lower pressures and stresses during machining.

The performed exploitation tests prove positive influence of modification of metal machining oil on improvement of the surface quality of the splineway of a gear wheel made of casting bronze CuSn12Ni2, by reduction of the average value of the roughness parameter *Ra*. The *Ra* value has dropped by 47%. As result of the modification, the durability of the cutting tool has also increased by 40%.

References

1. R.W. Maruda, S. Legutko, G.M. Krolczyk, C. Lukianowicz, A. Stoic, Teh. Vjes. **22**, 1219 (2015)
2. G.M. Krolczyk, J.B. Krolczyk, R.W. Maruda, S.Legutko, Measurement **88**, 176 (2016)
3. S. Wojciechowski, P. Twardowski, M. Pelic, R.W. Maruda, S. Barrans, G. Krolczyk, Precis. Eng. **46**, 158 (2016)
4. S. Wojciechowski, R.W. Maruda, P. Nieslony, G.M. Krolczyk, Int. J. Mech. Sci. **119**, 360 (2016)
5. G.M. Krolczyk, R.W. Maruda, P. Nieslony, M. Wieczorowski, Measurement **94**, 464 (2016)
6. R.W. Maruda, G.M. Krolczyk, P. Nieslony, S. Wojciechowski, M. Michalski, S. Legutko, J. Manuf. Process **24**, 107 (2016)
7. R. Michalczewski, W. Piekoszewski, M. Szczerek, W. Tuszyński, Problems of Exploitation **1**, 247 (2003) (in Polish)
8. R. Michalczewski, M. Szczerek, W. Tuszyński, Problems of Exploitation **4**, 247 (2004) (in Polish)
9. R. Michalczewski, W. Piekoszewski,W. Szczerek, M. Tuszyński, Tribology **5**, 81 (2007) (in Polish)
10. R. Michalczewski, W. Piekoszewski, M. Szczerek, W. Tuszyński, Tribology **6**, 133 (2007) (in Polish)
11. R.W. Maruda, E. Feldsthein, S. Legutko, G.M. Krolczyk, J. Frict. Wear **36**, 548 (2015)
12. M. Paslieclико, K. Dziedzic. M. Barszcz, Powder Metall. Met. C **52**, 469 (2013)
13. V. Fontanari, M. Benedetti, G. Straffelini, Ch. Girardi, L. Giordanino, Wear **302**, 1520 (2013)
14. B.S. Unlu, E. Atik, J. Alloy. Compd. **489**, 262 (2010)
15. T. Kucukomeroglu, G. Purcek, O. Saray, L. Kara, Jamme **30**, 172 (2008)
16. S. Laber, K. Adamczuk, Tribology; Theory and Practice **6**, 49 (2010) (in Polish)
17. K. Wieczorowski, S. Legutko, T. Kędzierski, ATMiA **2**, 31 (2002) (in Polish)
18. A. Laber, *Study of the use of exploitation additives to lubricating oils in tribological systems* (Publishing House of Zielona Góra University, Zielona Góra, 2012) (in Polish)

 DOI: 10.1051/e3sconf/20171903033

The measurements of temperature and deformations of car radiators

Katarzyna Peta[1,*], and *Karol* Grochalski[1]

[1]Poznan University of Technology, Institute of Mechanical Technology, Piotrowo 3, 60-965 Poznan, Poland

Abstract. In the study main factors influencing the exploitative durability of the aluminum radiators used in motorization were classified. Special attention was put to thermal strains occurring during the usage of a car. The causes of theirs formation were identified, including disproportionate distribution of temperature in the construction of radiator, which changes cyclically depending on the characteristics of cooling fluid flow. In order to evaluate the influence of temperature and deformations of radiators on their durability, resistance tensometry method was used supplemented with temperature measurements with the use of thermoelements. Three linear resistive tensometers and three thermoelements were placed in the key areas of radiator (the inlet and outlet of cooling liquid to the heat exchanger and separator of the areas). Measurements were carried out during the examination of the durability of radiators on thermal shocks, which is one of the most basic examinations that imitates conditions of their work and verifies mechanical durability of products. Critical areas in the radiator were located, which are the most vulnerable to damages, including cracks. After the conducted research measurements of tightness were carried out, which verification is one of the most important requirements set for products in contact with intermediary medium in heat exchange. The study was supplemented with the observation of metallographic structures of the areas of fatigue cracks.

1 Introduction

Radiators used in the engine cooling system have to meet a number of exploitative requirements resulting from the conditions of their work, of which the most important are tightness, thermal efficiency, mechanical durability (endurance) and corrosion resistance. They are exposed mainly to activity of the medium fluid intermediary in heat exchange with rapidly changing temperature of the range from -20^0C to 100^0C and the pressure of about 1 bar. Therefore, resistance to thermal shocks is the basic durability criterion of the radiators, which determines the construction of choices, production technology, the precision of making parts and the quality of their assembly [1-4].

With the increase of the time of exploitation of heat exchangers sectional plastic deformations develop, mainly brazed joints which consequently lead to crack initiations in these areas. The primary cause of their occurrence is thermal overloads, which intensity and caused results directly relate to working conditions of radiators in the engine cooling system. The main consequence of receiving and giving heat by radiator is increase or decrease of its temperature which leads to changes in linear measurements of the product according to the phenomenon of thermal expansion of materials. The limited possibility of deformation of construction results in the appearance of sectional heat strains usually caused by uneven distribution of temperature in radiator, inhomogeneity of mechanical properties, chemical composition and the metallographic structure of materials, especially brazed joints. The continuous activity of heat loads results in the thermal fatigue of radiator characterized by the net of cracks on its surface [5-11].

The determination of the degree of construction deformation, especially using nondestructive measurements methods is important in the assessments of its exploitative fatigue, in particular in the conditions of thermal loads. This group includes mainly tensometry methods and the methods of temperature measurements with the use of thermoelements. On this basis it is possible not only to determine the deformations of radiator but also to predict the location of possible cracks, which is important in industrial practice [12-17].

2 Research methodology

The measurements of temperature and distortions were carried out on car radiators made of clad aluminum tapes based on alloy of aluminum-manganese AA3003 (tape core material) and aluminum-silicon alloy (AA4343) (brazed material). The latter begins to melt at a temperature of 582^0C and after the crystallization it forms inseparable joints in the heat exchanger. The chemical composition of the materials used in the productions of radiators are shown in Table 1, their basic mechanical properties in Table 2.

* Corresponding author: katarzyna.peta@put.poznan.pl

 DOI: 10.1051/e3sconf/20171903033

Table 1. The chemical composition limits of aluminum alloys used in the study.

Alloy	Fe	Si	Cu	Mg	Mn	Zn	Al
AA3003	0.7	0.6	0.05-0.2	0.05	1.0-1.5	0.1	rem.
AA4343	0.8	6.8-8.2	0.25	0.05	0.1	0.2	rem.

Table 2. Basic mechanical properties of aluminum alloys used in the study.

Material	Tensile strength (Rm) [N/mm^2]	Yield point ($Rp_{0.2}$) [N/mm^2]	Elongation (A_{50}) [%]
AA4343 4% (clad) AA3003	115.00	58.00	37.00

Brazing is the main operations of the manufacturing process of radiators influencing the formation of insepa-rable joints and their durability. They were carried out in the protective atmosphere of nitrogen in a radiant-convective furnace, which main areas included:

- thermal degreasing (the removal of processing oils),
- application of non-corrosive flux,
- heating in a convective chamber,
- brazing in a radiant chamber,
- cooling in the protective atmosphere of nitro-gen,
- cooling in atmosphere air.

The flow of cooling fluid in the examined radiator from its inlet to outfall proceeded in the shape of "U" letter, according to the diagram in Fig. 1. The direction of the circulation of cooling fluid has a significant effect on the distribution of temperature in the heat exchanger, consequently distribution of deformations.

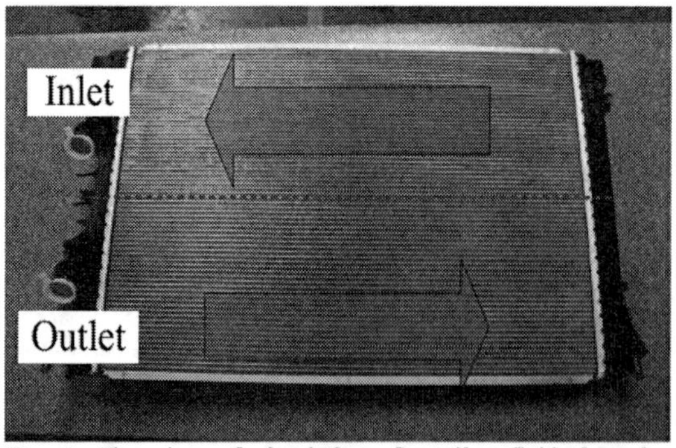

Fig. 1. The view of circulation of cooling fluid through radiator.

The examination of endurance of radiators consisted of temperature measurements and deformations in envi-ronment imitating the conditions of their work. For this purpose, endurance test of radiators on activity of thermal shocks was performed, based on the circulation of cooling liquid (aqueous solution of ethylene glycol) through the heat exchanger which the temperature-time cycle included (Fig. 2 and Fig. 3):

- the increase of temperature of cooling liquid from -20^{0}C to 100^{0}C within 30 s,
- the stabilization of cooling liquid of 100^{0}C within 30 s,
- the decrease of temperature of cooling liquid from 100^{0}C to -20^{0}C within 30 s,
- the stabilization of cooling liquid of -20^{0}C within 30 s.

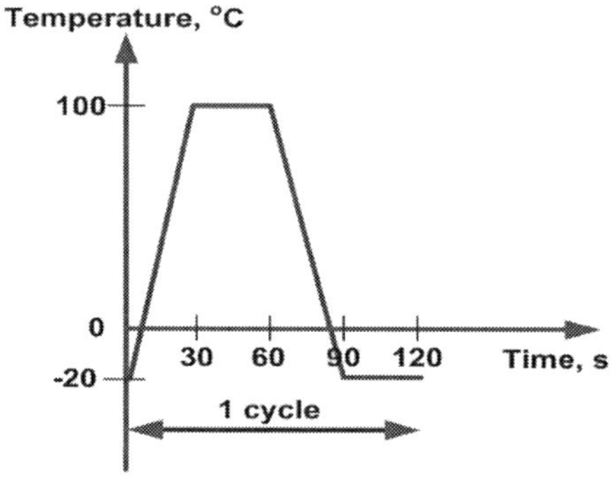

Fig. 2. The temperature-temporal profile of cooling liquid during the endurance examination of the heat exchanger on thermal shocks.

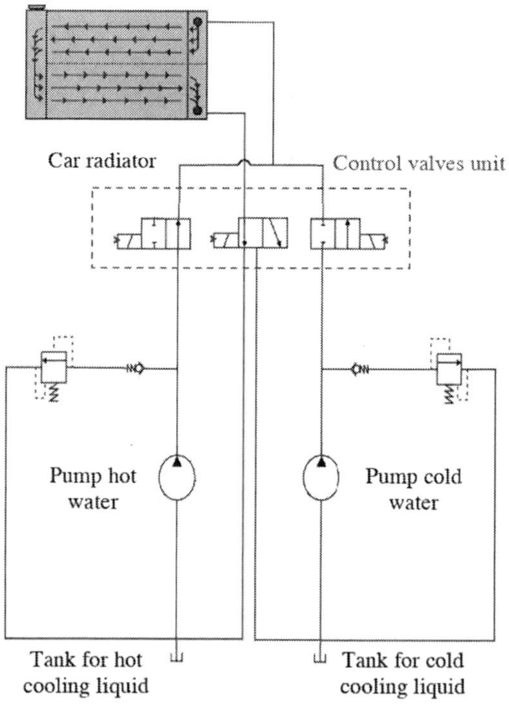

Fig. 3. The temperature-temporal profile of cooling liquid during the endurance examination of the heat exchanger on thermal shocks.

 DOI: 10.1051/e3sconf/20171903033

The study was conducted in the temperature of environment of 23^{0}C and the strength of the flow of cooling liquid was 4000kg/h.

The measurements of temperatures and deformations of radiators were carried out during the endurance examination on activity of thermal shocks, which reflects the actual working conditions. For this purpose, tensometry method was used, which is based on the determination of tensometer deformations caused by acting on its external forces and temperature measurement method with the use of thermoelements. Three linear resistive tensometers were used with self-compensating properties of the nominal resistance of 120Ω±0.35%, gauge factor 2.09±1.0%, transverse sensitivity 0.2% and temperature compensation modulus for aluminum of $\alpha = 23 \times 10^{-6}$/K. They were placed in the key areas of radiator, i.e. the inlet and outlet of cooling liquid to the heat exchanger and separator of the areas. In the same areas thermoelements were installed so as to determine dependence of the deformations of the heat exchanger from the temperature on its surface (Fig. 4). Thermoelements from NiCr-NiAl alloy of 0.5 mm diameter and a length of 200 mm were used.

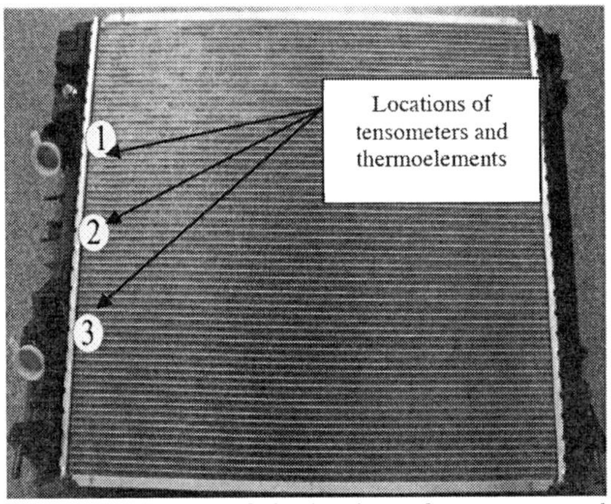

Fig. 4. The diagram of the locations of tensometers and thermoelements in radiator.

3 Discussion of obtained results

During the test of endurance of radiator on thermal shocks, the temperature and deformations measurements of the construction were carried out. The obtained results allowed to determine the effects of temperature changes of the radiator on the formation of deformations of its construction. It has been determined that with the increase of radiator temperature what follows is the extending of its deformations, thereby the change of linear measurements of the product. It is directly connected with the phenomenon of thermal expansion of materials.

The diagrams shown in Fig. 5, Fig. 6 and Fig. 7 present the profile of temperature dependence and deformations from the time for three considered areas in the radiator, which deformation changes largely in a linear degree with the change of its temperature which is related to the thermal conductance of aluminum. The effect of sectional concentration of strains and consequently the biggest changes of the linear measurements of construction occurs in temperature of cooling fluid of 100^{0}C, which is noticeable in the form of characteristic peak in the inserted diagrams. What was obtained is the repeatability of the results of temperature measurements and radiator deformations in the following cycles of measurements their endurance for thermal shocks. The greatest value of the deformations was recorded in the area of inlet of cooling fluid to radiator, which is associated with the occurrence of the biggest changes of temperature in this area, ranging from -20^{0}C to 100^{0}C. In the area of its outfall the intensity of the influence of thermal shocks on the radiator decreases because the heat of cooling liquid is gradually channeled by the heat exchanger. It determines a smaller thermal impact in this area of the radiator, resulting in smaller deformations of construction.

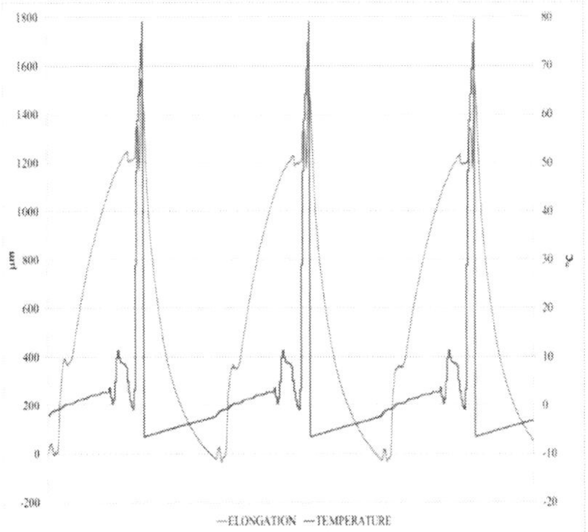

Fig. 5. The diagram of temperature and deformations in the function of time for the area of inlet of cooling liquid to the heat exchanger.

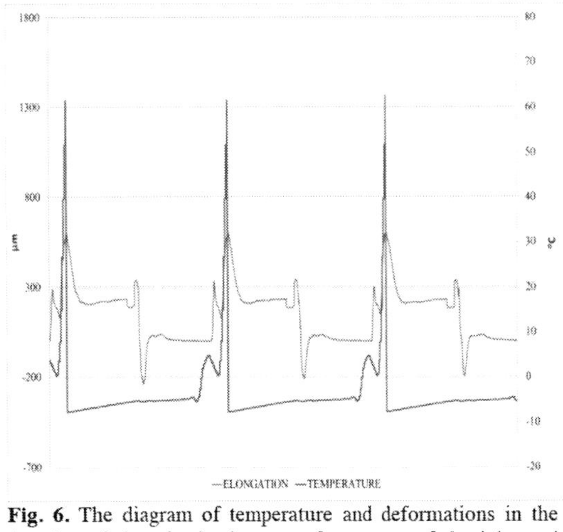

Fig. 6. The diagram of temperature and deformations in the function of time for in the area of separator of the inlet and outfall of cooling liquid areas.

After the completion of 1500 measurements cycles in the endurance examination of the radiator on thermal shocks according to manufacturing requirements, the measurements of its tightness were carried out, which did not show the leak of cooling liquid.

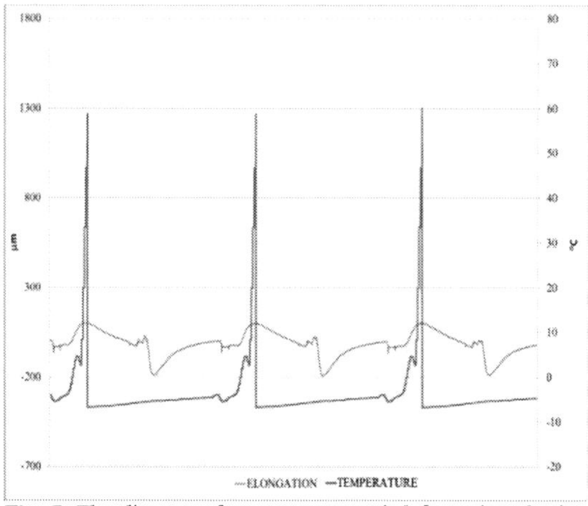

Fig. 7. The diagram of temperature and deformations in the function of time for outlet area of cooling liquid from the heat exchanger.

The heat exchanger fulfilled the established quality requirements. In order to identify the mechanism of forming of the permanent distortion of the contraction further 1500 cycles of endurance examination on thermal shocks were performed. The result of the prolonged exposures to cyclical thermal loads was weakening of the material, consequently the formation on fatigue cracks (Fig. 8). They appeared in the tube-tank joints area, mainly in zone of inlet of cooling liquid to radiator.

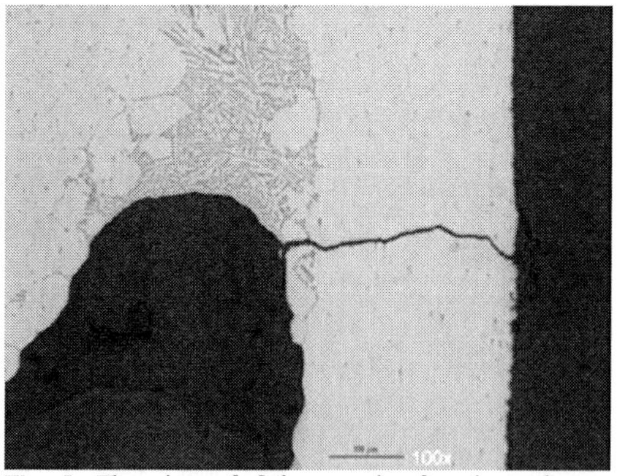

Fig. 8. The view of fatigue crack after the endurance examination of radiator of fluids on thermal shocks.

4 Conclusions

The most important final conclusions concerned the achieved results are:

- the effect of cooling liquid temperature was determined based on the distortions of radiators,
- uneven distribution of cooling liquid temperature values in different areas of the heat exchanger was observed,
- temperature measurements with the use of thermoelements were supplemented by tensometry measurements of distortions allowing to locate the areas of the radiator that are the most exposed to the effects of heat activity, which consequently leads to fatigue cracks,
- increased closing control of the heat exchangers should in particular focus on the area of inlet of the cooling fluid to the radiator, where there are the highest values of deformation during their exploitation,
- the obtained results are useful for industrial practice by indicating the method to identifying radiators construction deformations.

References

1. A. Mutai, Appl. Therm. Eng. **1** (2015)
2. R. Prakash., M. Selvam, A. Pandian, S. Palani, K. Harish, Indian J. Sci. Technol. **9** (2016)
3. K. Thulukkanam, Heat Exchanger Design Handbook. Second Edition (CRC Press, Boca Raton, 2013)
4. G. Prudhvi, G. Vinay, G. Babu, Int. J. Eng. Adv. Technol. **2** (2013)
5. J. Nowacki, M. Kawiak, Przegląd Spawalnictwa, **81** (2009)
6. H. Kim, S. Lee, J. Mech. Sci. Technol. **26** (2012)
7. A. Kłyszewski, J. Żelechowski, A. Frontczak, P. Rutecki, W. Szymański, Z. Zakotowicz, M. Nowak, Arch. Metall. Mater. **59** (2014)
8. K. Czapczyk, P. Siwak, P. Jabłoński, Ł. Furmański, P. Grobelny, S. Legutko, Adv. Sci. Technol. Res. J. **10** (2016)
9. A. Saigal, E. Fuller, Comp. Mater. Sci. **21** (2001).
10. S. Kahl, H. Ekström, J. Mendoza, Metall. Mater. Trans. A **45A** (2014)
11. T. Miller, S. Adamczak, J. Świderski, M. Wieczorowski, A. Łętocha, B. Gapiński, B. Pol. Acad. Sci-Tech. **65** (2017)
12. Y. Chen, A.H. Clausen, O.S. Hopperstad, M. Langseth, Int. J. Solids Struct. **46** (2009)
13. G. Owolabi, D. Odoh, A. Peterson, A. Odesh, H. Whitworth, World J. Mech. **3** (2013)
14. T.S.B. Naser, G. Krallics, Acta Polytech. Hung. **11** (2014)
15. J. Maszybrocka, A. Stwora, B. Gapiński, G. Skrabalak, M. Karolus, B. Pol. Acad. Sci-Tech. **65** (2017)
16. N. Rossini, M. Dassisti, B. Khaled, O. Abdul, Mater. Design **35** (2012)
17. P. Jabłoński, P. Czajka, Proceedings of 5th International Conference on Advanced Manufacturing Engineering and Technologies. Belgrade-Serbia (2017)

AUTHOR INDEX

AUTHOR INDEX